Sea Shells of Tropical West America

A. MYRA KEEN

With the Assistance of James H. McLean

Sea Shells of Tropical West America

Marine Mollusks from Baja California to Peru

SECOND EDITION

Stanford University Press, Stanford, California

1971

Stanford University Press, Stanford, California
© 1958, 1971 by the Board of Trustees of the Leland Stanford Junior University
Printed in the United States of America
ISBN 0-8047-0736-7   LC 70-143786
Original edition 1958   Second edition 1971

To the many colleagues who have aided in the preparation of this book and especially to Dwight W. Taylor, who underwrote a substantial part of the publication costs

Preface to the Second Edition

We can be quietly amused at George Eliot's appraisal of our field: "Why, you might take up some light study—conchology now; I always thought that must be a light study." Whether the study of mollusks is "light" or not, one is gratified to mention that it generates enough interest in enough people that not once but twice a donor has felt justified in offering a publication subsidy for a book on Panamic sea shells; the underwriting of part of the costs of the first edition by Mr. Bauer has been matched for this enlarged version by Dr. Dwight Taylor.

Three of my students have been especially helpful in the assembling of this revision: Dr. Eugene Coan has read the entire text critically, besides contributing in ways that are acknowledged at appropriate places therein; Dr. James McLean has assisted at several points, also as acknowledged, and has taken full responsibility for two sizable sections of text; and Dr. Judith Terry Smith turned a pile of accumulated notes into a coherent typescript.

Those segments of text for which collaborators have taken entire responsibility are: Turridae and Archaeogastropoda (up to the Neritacea), Dr. McLean; Terebridae, Twila (Mrs. Ford) Bratcher and Mr. Robert Burch; Polyplacophora (chitons), Mr. Spencer Thorpe; and Marginellidae, Dr. Coan and Mr. Barry Roth.

The new color plates have been subsidized by donations from Dr. and Mrs. Thomas Burch; Mrs. John Q. Burch; the Conchological Club of Southern California; Mrs. Faye Howard; the Santa Barbara Malacological Society; the Southwestern Malacological Society; Mr. Lawrence Thomas; the Yucaipa Shell Club; Miss Margaret Cunningham; and Dr. Taylor.

The acknowledgments of the first edition still apply, with some additions: Authors who were publishing descriptions of new Panamic species during the past two years have generously sent me carbon copies of their typescripts and prints of illustrations in advance, especially Dr. William Emerson, Dr. McLean, Mr. Richard Petit, Dr. George Radwin, Dr. Donald Shasky, and Dr. Emily Vokes. A paper by Dr. A. A. Olsson (Olsson, 1971, in the Bibliography) could not, because of timing problems, be taken fully into account here nor all the species cited for comparison. Two papers that were to have been published (by Mr. Thorpe and by the team of Coan & Roth) have been delayed for reasons beyond the authors' control; the species they expect to describe are figured and characterized here, but the specific names are omitted, being cited only as in manuscript. A paper by Knudsen (1970) was received too late for pertinent data on the Pelecypoda (range extensions, name changes, and eleven new species) to be properly accommodated in the text; his new data are given on p. 303.

Institutions and individuals lending material for photographing or otherwise supplying figures are in the main acknowledged under "Sources of Illustrations." The Paleontological Research Institution, on request of Dr. Olsson, loaned the plates from his work on Panamic-Pacific Pelecypoda, and his excellent figures have been freely reprinted here as supplements to the other available illustrations. Thanks are due also to the museum officials of the British Museum (Natural History) and the University of Copenhagen for having made available to me facilities for photographing type specimens in their collections.

Most of the line drawings and a considerable amount of the photographic work have been done by Mr. Perfecto Mary, staff artist at Stanford. The maps were drafted by Mr. Roth, and the photograph on the inside covers is from a color slide provided by Dr. Coan.

In the earlier edition I commented that the next task should be a canvass of European collections for a study of types. This has since been made possible for me by a fellowship from the John Simon Guggenheim Foundation, and I was thus able in 1964 and 1965 to search out and photograph the Mörch types at the University of Copenhagen and most of the eastern Pacific types of Hinds, Broderip, Sowerby, Reeve, Orbigny, and Carpenter at the British Museum (Natural History). Still to be studied are the types in the French and German museums, and one may hope that some young malacologist will undertake this project. The Panamic types in American museums have already been documented, and Dr. McLean has photographed most of them. Many of his pictures are used herein as supplements to the previously published illustrations, but unfortunately it is not feasible to indicate the full extent to which he has thus contributed.

For all of this collaboration I am very grateful. As any author knows, it is too much to hope that all errors of fact and interpretation will have been detected, but I am sure that they are far fewer here because of the vigilant help of my many co-workers.

What was called in the 1958 Preface the "other and longer task"—the study of types and the compilation of a complete list—is, in the present edition, largely accomplished. The lists of microscopic forms, which had been available in card-file form at Stanford University since the 1930's when they were prepared under the auspices of a former graduate student, Dr. Don L. Frizzell, but which had been vetoed by the committee as not of sufficient popular interest, are incorporated in the new text, and the offshore and island forms have been included in the canvass of recognized species. Much revisionary work has been done by authors during the past fourteen years, because there has been an active interest in molluscan systematics. Presumably, then, the groundwork for a proper study of our fauna has been fairly well laid, and the rough outlines of the picture sketched in. What remains now is the challenging task of observing the animals themselves—studying and interpreting their ecology, anatomy, physiological adaptations, behavior, and means of dispersal, and all the other aspects of their relationships to each other and to other organisms in their physical environment.

M. K.

Stanford, California
March 1971

Preface to the First Edition

Science has been called the only true democracy, in that knowledge acquired by one person is always freely shared. Such sharing has been manifest in the preparation of this book, for when I was elected to the task, those individuals who had most cause for resenting my intrusion upon their especial fields of work were the most generous in giving me the benefit of their discoveries. In particular, Dr. S. Stillman Berry, who is making an intensive study of the northern part of the Gulf of California, and Dr. Leo G. Hertlein, who has in manuscript a checklist on the entire Panamic province molluscan fauna, took time to give me advice, to read what I had written, and to lend material for illustration. I have tried to acknowledge this help in the text, wherever practicable, but as a complete accounting would only bore the reader, I have had to let much of it fall into the pattern of impersonal sharing. Parts of the manuscript have been read by Messrs. Curt Dietz, George Harrington, and Allyn Smith and by Professor Hubert G. Schenck, the final draft having profited from their criticisms. My former student, Mr. Robert Robertson, now at Harvard University, in addition to reading the Introduction, has supplied needed references, procured illustrations, and offered useful suggestions. To all of these persons I offer thanks, as also to the many correspondents who answered inquiries and the colleagues who assisted in less obvious but no less real ways. It is inevitable that, in spite of care, errors of fact and judgment will have crept into the book. For these I must accept responsibility, but I am glad that their total number has been reduced by the vigilant help of my co-workers.

For illustrative material, my debt to the California Academy of Sciences is especially great. Not only was I given freedom to use any of the large file of negatives and prints, but also was loaned specimens, and the staff photographer made several needed pictures. Dr. E. Fischer of the Muséum d'Histoire naturelle de Paris had the extant type specimens of Mabille and of Rochebrune searched out and photographed for my use. Mr. E. P. Chace and the San Diego Society of Natural History loaned a number of rare specimens for photographing; Dr. Norman Mattox of the Allan Hancock Foundation, University of Southern California, supplied a needed paratype; and Dr. J. P. E. Morrison of the U.S. National Museum loaned a plate for making new line drawings. The staff of the Paleontology-Stratigraphy section, U.S. Geological Survey, at Menlo Park, made photographic darkroom facilities available. Author and reader alike are indebted to these people. Photographs of ten previously unfigured type specimens in the British Museum (Natural History) were made for my use through the kind offices of Mr. I. C. J. Galbraith, curator, and Dr. Harald A. Rehder of the U.S. National Museum (Smithsonian Institution) likewise made available photographs of some fifty unfigured type specimens. A set of prints of the C. B. Adams type specimens, recently published by the Museum of Comparative Zoölogy of Harvard University, was supplied by Dr. W. J. Clench.

Except as noted above, the original photographs here published were made by Messrs. Alexander Tihonravov, technician, and Morris Fiksdal, graduate student, of Stanford University, and by Mr. David H. Massie, of Oakland. Line drawings are, in the main, the work of Mr. Perfecto Mary, staff artist at Stanford.

As first planned, this book was to be prepared by a committee under my supervision, the principal member to have been Mr. David Vernon of Los Altos. His death interrupted that program. He had, however, written out usable drafts on several families (Pectinidae, Cardiidae, Strombidae, Cypraeidae, and Conidae) and had drawn text figures for the last two. With the permission of Mrs. Vernon, I have incorporated these summaries and drawings in the present book, having revised and rearranged the material to harmonize with the style finally adopted. Because Mr. Vernon's sections fall at widely separated places in the book, I acknowledge their authorship here rather than in the text. Having lost its key members by death, by graduation of some of the students, and by employment of others in far places, the committee eventually disbanded, in spite of the optimistic notice published by one member (Grace P. Johnson, "Plan for a collector's guide . . ." [Abstract], Amer. Malacological Union, *Ann. Rept. for 1956*, p. 22).

A book of this size and one with so many illustrations could not have been produced at a price within the budget of the average collector had not its publication been generously subsidized by Mr. Harry J. Bauer of Los Angeles. His insistence that a work of this sort is needed and his willingness to underwrite the initial costs finally persuaded me to attempt its preparation.

As it stands, the book forms a sort of progress report, a catalogue of the pertinent literature of the past two centuries on tropical West American mollusks and a census of most of the larger forms. Much revisionary work remains to be done, for there are doubtless many misidentifications and errors of allocation that remain undetected. A few years hence, when the curators of some of the collections in Europe have finished the task upon which they are now engaged—reviewing and searching out type specimens—it may be possible to illustrate many unfigured types, notably those of Mörch and Carpenter. Diligent search may uncover other material that is at present lost to science, such as the type specimens of Valenciennes, Menke, and Philippi. What we already have, however, bulks large enough that the other and longer task may well be left for the future.

M.K.

Stanford, California
May 1958

Contents

Color Plates

XV. 1. *Cypraea annettae.* 2. *Titiscania limacina.* 3. *Ficus ventricosa.*
4. *Carinaria japonica.* 5. *Cypraea albuginosa.* 6. *Cypraea arabicula.*
Sponsored by the Santa Barbara Malacological Society, Santa Barbara,
California.

XVI. 1. *Cassis centiquadrata.* 2. *Lamellaria sharonae.* 3. *Trivia pacifica.*
4. *Cassis coarctata.* 5 & 6. *Casmaria vibexmexicana.* Sponsored by the
Conchological Club of Southern California, Los Angeles, California.

XVII. 1. *Oliva polpasta.* 2. *Oliva julieta.* 3. *Olivella dama.* 4. *Olivella zanoeta.*
5 & 6. *Oliva incrassata.* 7. *Agaronia testacea.* 8. *Oliva porphyria.*
Sponsored by Rose Burch (Mrs. John Q. Burch), Seal Beach, California.

XVIII. 1. *Jenneria pustulata.* 2. *Cyphoma emarginatum.* 3. *Nassarius tiarula.*
4. *Cancellaria tessellata.* 5. *Bulla punctulata.* 6. *Simnia aequalis.*
Sponsored by Faye B. Howard, Santa Barbara, California, as a
memorial to her son, Patrick John Howard, 1937–1968.

XIX. 1. *Oxynoe panamensis.* 2. *Tylodina fungina.* 3. *Aplysia parvula.*
4. *Aplysia californica.* 5. *Berthellina quadridens.* 6. *Pleurobranchus
areolatus.* 7. *Lobiger souverbii.* 8. *Conualevia alba.* Sponsored by
Dwight W. Taylor.

XX. 1. *Hypselodoris californiensis.* 2. *Cadlina evelinae.* 3. *Nembrotha eliora.*
4. *Chromodoris norrisi.* 5. *Chromodoris sedna.* 6. *Tridachiella diomedea.*
Sponsored by Dwight W. Taylor.

XXI. 1. *Aegires albopunctatus.* 2. *Okenia angelensis.* c. *Navanax inermis* and
Laila cockerelli. 4. *Antiopella barbarensis.* 5. *Dendrodoris krebsii.*
6. *Melibe leonina.* Sponsored by Dwight W. Taylor.

XXII. 1. *Flabellinopsis iodinea.* 2. *Spurilla chromosoma.* 3. *Armina californica.*
4. *Dirona picta.* 5. *Berghia amakusana.* 6. *Hermissenda crassicornis.*
Sponsored by Dwight W. Taylor.

Sea Shells of Tropical West America

Introduction

Among the coastlines of the world the west coast of the Americas, sweeping boldly and diagonally across the globe, is unique in the evenness of its contour, almost unbroken by embayments. It is in general a precipitous coast, with rugged land close by the sea, a narrow continental shelf, and few offshore islands. Political subdivisions may be sharply drawn, but from the Alaskan Gulf southward no obvious geographic barriers, such as long seaward-extending promontories, intersect it. One might expect, therefore, that the marine fauna would be more or less uniform or at least gradational along the entire coast, except, perhaps, at the cold-water extremes of the Arctic and Antarctic. The sea shells, however, do not bear out so easy an assumption. True, a few groups, like *Haliotis*, range from their metropolis in California southward, usually in dwindling numbers, and others, like the large keyhole limpets, range from an optimal area in South America northward, but the great bulk of the Central American molluscan fauna bears a distinctive tropical or exotic stamp. Students of distribution have in fact divided the western American coast into several faunal provinces, with a cool-temperate Aleutian or Oregonian province in the north, a temperate Californian province (cooled in part by the upwelling of deep and cold water), and in the central part, a Panamic or tropical province, bordered to the south by another cool-temperate region, the Peruvian province (again, cooled by upwelling).

Tracing the history of the Panamic fauna is an interesting exercise in earth history. The ocean expanse to the west, though dotted in places with tropical islands, is too broad to be crossed by any but very long-lived larvae; most floating larvae from the East Indies would metamorphose and sink to the bottom long before reaching American shores. The depths are too great to be inhabited by any but specially adapted deepwater forms, and migration across the sea floor is thus beyond reasonable possibility. The harsh conditions of the northern and southern ends of the American continents would preclude immigration by any but the most hardy of mollusks, and the steep and narrow continental shelf along most of the coast does not afford much living space. Thus, paradoxically, the source area of the ancestral stock for most of the present-day Panamic fauna lies to the east, walled off now by a narrow land bridge, the Isthmus of Panama. Geologically speaking, this is only a temporary barrier, for from time to time in the past there has been free interchange between the western Atlantic and the eastern Pacific—and it is this interchange that accounts for the large number of similar species occurring in the two areas. But since late Miocene time the Isthmus has been a barrier, and the faunas of the two sides have had a few million years to become somewhat—though not yet completely—differentiated.

One who knows the modern Panama shells has only to leaf through the illustrations of, say, Brown and Pilsbry (1913) on the Miocene of eastern Panama or the reports by Woodring (1925, 1928) on the Miocene of Jamaica to feel a sense of familiarity. Here are the Strombinas, the Terebras, the Cancellarias, the Chiones, and many others that today are well developed in terms of numbers of species on the west coast but that are represented by only a few in the Caribbean. Woodring (1966) has coined the term "paciphiles" for these, and he has given an illuminating discussion of the part the intermittent land barrier has played in faunal development.

A precise comparison of the two faunas should prove a rewarding exercise for biogeographers and the like, as well as for malacologists. Now that a full list of the tropical Pacific molluscan fauna is available, in the pages that follow, one may hope that a similar work on the Atlantic will appear, for it would then be possible to pursue such a comparison. At the moment, preliminary lists suggest that the Pacific side, in spite of its narrow continental shelf, has more species. Further documentation, obviously, is needed. A few species, however, do seem to be common to the two coasts, probably unchanged from the ancestral Miocene stock. But the pairs of similar or twin or geminate or cognate species (the term "analogous" of earlier authors now is frowned on as not in harmony with usage elsewhere in biology) need a more substantial review. It would also be interesting to speculate on why some groups, such as the arks, should have speciated more rapidly in the Atlantic, and others, like the columbellids, should have fared better in the Pacific.

The ancestral tropical American fauna itself (Atlantic cum Pacific) seems to have descended from the still more ancient Tethyan fauna that developed in the great east-west sea lapping the southern shores of what is now Eurasia during early Tertiary time, some 60 million years ago. The eastward spread of the Tethyan fauna gave rise to the modern Indo-Pacific and Japanese faunas, while the westward spread reached across the Atlantic to the southern shore of North America. Modern geologic evidence suggests, of course, that the Atlantic was not always so deep or so wide as it is at present; continental drift and sea-floor spreading have altered the shorelines and changed the shape of the land masses. Thus with the continued encroachment of the Americas into the Pacific, the offshore islands to the west became, in fairly recent geologic time, the final outposts of the faunal waves—a meeting place predominantly populated from the American shores, but with a fringe of stragglers from the far-off Indo-Pacific area. Such sea-floor changes and faunal movements have been imperceptibly slow by human standards, of course, as year by year the currents shift a little and larval mollusks drift and settle a little beyond the range of the parent stock.

In the main, shells from the eastern Pacific were great rarities in the seventeenth- and eighteenth-century natural history cabinets of Europe. The early period of collecting has been well reviewed by Olsson (1961) and by Dance (1966). According to C. B. Adams (1852), the first serious collector of marine shells in the Panamic province was Joseph Dombey, a French botanist who arrived in Peru in 1778 and returned to Europe with some shells that were later described by Lamarck. As late as 1825, the London dealer John Mawe, who published a small book entitled *The Voyager's Companion, or shell collector's pilot; with instructions and directions where to find the finest shells . . .* , had this to say about the western American coast (pp. 22–24):

From Peru, however, many beautiful and rare varieties of shells have been brought, more especially from the shores of Callao and farther north. Hence, until we reach the bay of Panama, we are equally unacquainted with the shells that may be produced on the coasts and rivers.

In the cluster of islands in the bay of Panama, there is a fishery for pearls, which are large and well formed, but of bad color. Many fine murices and other rare shells have been found there.

The Gallipagos [*sic*] islands are rich in shells;—it is astonishing that more varieties have not been brought from thence. I have received, through the favor of some of the captains and officers of whale ships, some choice specimens.

Farther north we meet with no shells, until we arrive at Ceres [Cedros] Island, off the coast of California. From this island the finest earshells (Haliotis Splendens) [*sic*] have been brought; for those which fell into the Author's hands, in less than a year, he paid above £100; they were generally brought to him by sailors belonging to ships that went there for seal skins. It may be remarked that the [shells] of these seas . . . differ from the same species in the Atlantic, particularly the limpets, muscles [mussels], and clams.

Thus, Mawe was one of the first, if not indeed the first, to notice the similarity but not identity of eastern Pacific and western Atlantic forms.

With the return to England in 1831 of Hugh Cuming—that voyager who has been nominated by Dance (1966) as "the prince of shell collectors"—the picture changed abruptly. Cuming, after several years of residence on the west coast of South America, took back with him to London an unrivaled suite of material. When the Cuming collection was finally acquired by the British Museum, in 1865, his associates had been eagerly examining and describing the material for some 33 years, and in the process demonstrating the real richness of the Panamic fauna.

Other British workers were busy, also. Sir Edward Belcher collected assiduously during his surveying expeditions, and R. B. Hinds, a naturalist who had accompanied him on one such occasion, published in his report (Hinds, 1844–45) what may be a first discussion on the whole topic of molluscan distribution.

Stimulated by the potentialities of careful fieldwork in the American tropics, Professor C. B. Adams, from Amherst, Massachusetts, undertook an extended study of the Panamic area, with a more detailed faunal analysis than had yet been attempted. By boat and on horseback he traveled to the west coast of Panama and spent six weeks there, in November and December, 1850. His report, which appeared in 1852, is the first major work on the molluscan fauna of any local section of the Panamic province. But though it was sound in generalities, it was faulty in some of its identifications, owing to Adams's lack of comparative material and literature, as was soon pointed out by Philip P. Carpenter (1856–57), who had access to such material in Britain.

Carpenter was a British clergyman with a taste for natural history. In the early 1850's he had acquired a large part of the Reigen collection, from Mazatlán, Mexico. While working up this collection, Carpenter compiled a monumental analysis of the literature on the entire west coast of America and published a series of books and papers that still are basic to a study of the fauna. Although he prepared illustrations of his new species—especially camera lucida drawings of the small forms—few were published during his lifetime. He had lamented that the type specimens of C. B. Adams were unfigured (a situation that was not remedied for a century [see Turner, 1956]), but the same fate befell his own material. Only within the past few years have most of these types been figured (Palmer, 1958 and 1963; Brann, 1966; Keen, 1968).

The century since Carpenter's time has been a busy one; hardly a year has passed, and certainly no decade, that has not seen some record of significant discoveries. From 1925 on, however, interest seems to have been intensified, on the part both of enthusiastic amateurs (H. N. Lowe, for example, made a number of one-man expeditions along the entire Central American coast in the 1930's) and of institutions, notably the California Academy of Sciences, the Los Angeles County Museum, and the Allan Hancock Foundation, which have accumulated much dredged as well as intertidal material.

This book, which of course draws heavily on these past studies, is intended to be something more than a popular handbook, for it deals not only with the large, the common, and the showier shells, but also with the small, the rare, and the ordinary—in fact, with all of the species of mollusks (except some cephalopods) that have been recorded from the Panamic province. The treatment is not, however, in monographic style, for although many synonyms are cited, there are no formal synonymies, and citations are pared down to author and date. The reader who needs more information must consult the original references given only by implication (titles to these are in the Bibliography) or later available monographs, such as Olsson's (1961) on the pelecypods, and others that are mentioned at appropriate places in the text. And for studies in such areas as anatomy, physiology, biochemistry, behavior, and ecology—all of them important fields but beyond the scope of the present systematic review—the reader must refer to the standard literature-recording serials, such as the *Zoological Record* and *Biological Abstracts*. The Bibliography given here, though extensive, by no means exhausts all aspects of malacology.

Further to the task of keeping the book's length within reasonable bounds, some other delimitations have been necessary:

Geographic boundaries. To stress the Panamic fauna and to exclude from the list most of the stragglers from adjacent provinces, the boundaries of complete coverage have been set at what are essentially the natural limits of the Panamic province—Magdalena Bay (Bahía Magdalena), Baja Cailfornia, and Punta Aguja, Peru, with a seaward extension including the Galápagos archipelago and the other, nearer, offshore islands. An occasional species from north or south of these boundaries may be mentioned, but few of these are illustrated, especially if they are well documented in other works.

Size. No size minimum has been set for inclusion of species in the text, but illustrations are in the main only of those forms larger than 5 mm in length; sample representatives are shown for the genera and subgenera of what are predominantly microscopic groups among the gastropods. Citation of these species is generally a simple listing of name, author, date, and type locality, with no statement of range and with original generic allocation in square brackets. Readers interested in these groups will find that good illustrations usually accompanied the original descriptions, but later or better figures are mentioned in this list. Much work remains to be done on these small snails.

Depth. No limitations have been set for coverage of the offshore fauna, either pelagic or benthic, but only a scattering of the deepwater forms is here illustrated, and listing usually does not go beyond type locality and depth. Though it is unlikely that most collectors will obtain material from depths greater than 200 m, modern photographs of the type specimens have been utilized, when available.

Fauna. Coverage is primarily of mollusks from the marine environment, but a few groups that have become adapted to brackish-water conditions are included, particularly if the habitat is such that the shells might become intermixed with those of strictly marine forms. No distinctions have been drawn between shelled and soft-bodied or shell-less mollusks such as the nudibranchs, which often are omitted in molluscan faunal lists (and were, in fact, not included in the first edition). The illustrations of the latter are not as comprehensive as for the shelled groups, and no new line drawings of them have been attempted, but some 18 species are figured in the color plates. The cephalopods, as noted above, have not been given full treatment, for work on these is already under way by specialists. With the other taxa, the emphasis here is of necessity on the shell, leaving for future workers the challenging task of correlating with shell details the information (little of it yet published) about the soft-bodied animals (mollusks) that formed the shells. But in a great many cases, especially following the descriptions of the better represented families and genera, the text does comment briefly on the animals themselves.

The book is divided into five principal sections: the Introduction; the main or systematic text, which discusses and illustrates the Panamic members of the seven molluscan classes; four Appendixes (brief mentions of rejected and doubtful species, a glossary, maps and other geographic aids, and sources of illustrations); a fairly comprehensive Bibliography that attempts to list every paper in which recognized species were described (a few titles of works in which synonymous names were proposed may have been omitted); and a full Index to the taxonomic names given in the text, both recognized and synonymous.

Systematics. The systematic part is arranged in a more or less conventional order, beginning with the presumably simplest forms in each class and proceeding toward the most highly developed. Because molluscan classification is still a matter of some debate, the exact order of arrangement, down to the sequence of families, must be a compromise among the principal systems in current use. Within this larger systematic framework, an alphabetical sequence has in general been used. Thus, the order of species under any subgenus or genus is always alphabetical. The order of subgenera within genera (when there is more than one) and of genera within families is also generally alphabetical, except that the type subgenus or genus is usually listed first. In the very few cases where a systematic sequence has been adopted, the rationale for the distinction is given. The order of families within superfamilies follows the same pattern. Here alphabetization ends.

Most of the Panamic species were named during the nineteenth century, a period of rapid exploration. The naturalists who did the describing often had little access to the publications of others, and some of them had scant respect for what few rules of nomenclature there were, so that names multiplied in the literature. During the present century, with the development of a standard procedural code—the International Code of Zoological Nomenclature—a period of what to the outsider seems capricious name-changing has ensued. New species will be described, and the name-changing will continue, but a greater stability and uniformity of interpretation would seem to be in the offing. Explanations of the necessary changes are given in the text of this book, and definitions of the special terminology of nomenclature form one part of the Glossary. However, since scientific nomenclature (taxonomy) is a study in itself, the reader who would like a better understanding of the subject should consult either or both of the following works and the Code itself:

R. E. Blackwelder. *Taxonomy: A Text and Reference Book.* Wiley, New York, 1967. 698 pp. $19.95.

Ernst Mayr. *Principles of Systematic Zoology.* McGraw-Hill, New York, 1969. 428 pp. $12.50.

International Code of Zoological Nomenclature. N. R. Stoll, ed. International Trust for Zoological Nomenclature, London, 1961. (2d ed., with minor changes, 1964.) 176 pp. $3.00.

This being a handbook on a faunal province, it need not also be an introduction to the study of Mollusca. There are already books on that subject available from the booksellers or at public libraries. A glossary of morphological and other terms is included, however, in the Appendixes. For the beginner to the field, the following introductory works should prove useful:

R. T. Abbott and H. S. Zim. *Sea Shells of the World.* Golden Press (a Golden Nature Guide), New York, 1962. 160 pp., 790 figs. in color. $1.25.

R. T. Abbott. *How to Know the American Marine Shells.* New American Library (a Signet Key Book), New York, 1961 (reissued, 1970). 222 pp., 159 figs., 12 pls. in color. $1.25.

Kathleen Y. Johnstone. *Collecting Seashells.* Grosset & Dunlap, New York, 1970. 198 pp., 8 pls. in color, numerous figs. $5.95.

J. E. Morton. *Mollusca: An Introduction to Their Form and Functions.* Harper Torchbooks (the Science Library), New York, 1960. 232 pp., 23 text figs. $1.40 (more technical and advanced than the other three books).

The more extensive discussions of physiology, life histories, ecology, etc., must be sought elsewhere. A bibliographic summary of what has been written on aspects of malacology other than systematics would be an interesting project but is beyond the scope of this book. A start, however, is made in a Finding List or Topical Summary at the beginning of the Bibliography, which calls attention especially to papers published during the past two decades that may not otherwise have been cited in the systematic part of the book.

Descriptions. The descriptions of species are arranged in a fairly standard pattern, with the following parts:

1. A species number, which is used also to identify the illustration (on color plates XI–XXII, as well as on the black-and-white plates).

2. The scientific name, the name of the author (the person who proposed the species), and the date of publication of his description. Complete references to page, plate, and figure are not cited, but titles of publications are given by author and date in the Bibliography. The use of parentheses around the author's names for some species is a convention required under the International Code to show that the species was originally assigned by the author to some genus other than the one now in use and that it has since been transferred—a new combination. In the lists of microscopic gastropods and others not illustrated herein, the original generic allocation is indicated—in square brackets—for the convenience of the reader who wishes to consult the original figure, but otherwise original allocations and the names of the authors who have initiated the reallocation are not given.

3. Synonyms—other scientific names that have been applied or that may apply to a given species. Many of these may prove to be useful when further study of types and of populations shows that consistent differences are recognizable.

4. A characterization, giving notes on color and clues for distinguishing closely similar forms, with little or no mention of shell features that can be seen in the illustrations. It would be useful if characterizations of the soft parts could be included here also, but these are so rarely available that no attempt was made to comb the literature in search of them. The color plates, however, do show the soft parts of a number of the forms.

5. Dimensions, in metric units, much simpler than fractional inches. Length is defined as the greatest dimension parallel to the hinge in bivalves or to the axis of coiling in gastropods. In bivalves, height is the greatest dimension perpendicular to length; in gastropods it is the same as length. Diameter is a measure of convexity ("thickness" of some authors) for bivalves, but for the gastropods it is the greatest dimension perpendicular to length. The term semidiameter or semithickness may be used to describe the convexity of a single valve of a bivalve shell. The reader must remember that mollusks, like all other organisms, are variable; thus the dimensions given herein are merely samples to indicate the general proportions and size of presumably mature specimens, and may not always be average sizes.

6. Range or distribution. The northernmost occurrence is cited first, the southernmost last. For purposes of range description, the Gulf of California is treated generally as a shoreline running from around the tip of Baja California north to the head of the Gulf, then east and south along the Sonoran coast or Mexican mainland. Depths are cited in meters, and a conversion chart is given in the Geographic Aids in the Appendixes for the convenience of the reader who still likes to think in terms of feet or fathoms. A geographic index of place names and a series of location maps are also given in the Geographic Aids. The ranges are compiled from published literature and from such unpublished records as are available. It is to be expected that further extensions may come to light with additional study of collections.

7. Relative abundance. Local conditions may show wide variations not only seasonally but from year to year. Thus, trustworthy assessments of relative abundance can be made only after study of large suites of material taken from numerous areas. This rarely being feasible, the attempt, previously made, of furnishing abundance ratings for all of the cited species is abandoned in the present edition; however, notes by collaborators who have been able to make field studies are included.

Characterizations at higher taxonomic levels. Although these are rarely verbalized elsewhere, except in formal treatises, the bits and pieces of information included here for the taxa higher than families are a part of the mental furnishings of any professional malacologist, who is apt to consider them so self-evident as not to require statement. They are supplied here as a matter of editorial policy and should be useful to readers unfamiliar with the implicit structure of zoological classification.

Keys. Dichotomous keys are included for most groups (families, subfamilies, genera) in which there are more than a few genera or subgenera. For the most part, keys are attempted only if there are four or more taxa to be distinguished. In a few cases the genera from two or more families are keyed out together, always with a covering explanation in the preceding paragraph.

Common names. Few mollusks, especially in the Panamic province, are well enough known to have received genuine common or popular names. True, the books as early as a century ago (for example, Reeve's *Conchologia Iconica*) have dutifully presented them, mostly as English translations of the Latin specific names, but such names rarely have become part of our vernacular. There would be some advantage to continuing them, as a memory device, if this could be consistently done, but there are complications. The Latin names may be meaningless —as in *Chione gnidia.* The word *gnidia* literally means "a dweller in Gnidos." A descriptive English word such as "frilled" might be used for this particular shell, but unless one took some trouble to explain, would not the implication be that "gnidia" means "frilled"? Again, several Latin words of the same meaning may be used for species in one genus, and there are not always sufficient English synonyms. A few names, indeed, are downright misleading if translated literally. Sooner or later some authors arrive at the dubious device of merely reversing generic and specific names, as: *"Macoma affinis* . . . The Affinis Macoma." We have therefore decided not to attempt common names, except incidentally in some of the discussions, and we refer the reader who really wants to know the meanings of the Latin and Greek words to the many classical dictionaries in public libraries or to the following helpful books:

R. W. Brown. *Composition of Scientific Words: A Manual of Methods.* Smithsonian, Washington, D.C., 1954. 882 pp. $14.00.

E. C. Jaeger. *A Source-Book of Biological Names and Terms.* Thomas, Springfield, Ill., 1944. (2d printing, 1947.) 256 pp. $6.75.

Figures. The format used here makes separate plate explanations unnecessary, but brief notes on the sources of the figures are given in the Appendixes. Included in these notes are the citation of the original reference for copied figures or the locality and repository (place where the specimen is housed) for photographic illustrations. Magnifications are not consistently shown in the explanations, for approximate sizes of average or of type specimens are a part of the text. Many photographs that were not earlier available, of type specimens at the British Museum or the United States National Museum, are given in this edition; these replace or supplement the lithographic and other previous illustrations.

Production of figures at exactly natural size is rarely practicable. Hence, the student must learn to think to scale, to recognize proportions and forms whether they are magnified or reduced. In comparing specimens with figures larger than natural size, it is convenient to close one eye and move the specimen up and down over the illustration, the differences in shape thus becoming readily evident. A reading glass can be used over a figure that is smaller than the specimen with somewhat the same effect, except that the image must be compared with the specimen side by side. Some of the figures, especially the line drawings of radulas and soft parts, carry a separate size scale, a line of a stated length. This can be used as a ruler to determine the length of the original specimen. For example, a line (no matter what its actual length) that is labeled "5 mm" standing beside a specimen roughly four times its length indicates that the actual length of the shell is about 20 mm.

The trend in malacology today is a shifting from the purely descriptive to the more interpretive types of study, now that most of the species new to science have been discovered (a few new ones, of course, will continue to be recognized, especially among the microscopic and deepwater forms). The question of ecological relationships among mollusks, and among marine animals and plants generally, is drawing particular attention today, as man becomes more aware of his dependence upon his environment. Perceptive descriptions of terrain, such as those of Berry (1956a) and C. B. Adams (1852), quoted in the first edition of this book, are few indeed. The amateur can contribute much useful information by learning to observe carefully and to go into the field with sharpened vision, having some foreknowledge of what to seek. Most of the older faunal lists, if they mention habitat at all, monotonously repeat such phrases as "under [or on] rocks at low tide." This information may be useful for the collector, but it tells nothing of what the mollusks were doing there (whether feeding or being fed upon, moving about or withdrawn, spawning, and so on), nor does it give a clue to the interrelationships with other kinds of organisms. Even the spawning dates are known for only a few forms. The collector, therefore, can return with much more than some choice specimens if he goes out prepared to observe and to interpret. There are no general studies of the Panamic ecosystem; there is, however, a masterly study of the intertidal fauna and ecosystem of the Californian province, in which many comparable ecological considerations are discussed at length:

E. F. Ricketts and Jack Calvin. *Between Pacific Tides* (4th ed., revised by J. W. Hedgpeth). Stanford University Press, Stanford, Calif., 1968. 614 pp. $10.95.

Whether or not the publication of a guide to the Panamic mollusks is in the interests of conservation remains to be seen. The last introductory paragraph of the first edition sounded a note of caution:

This book will have been of scant service to the science of malacology if it stimulates unrestrained collecting. The apparent wealth of a tropical coast suggests a limitless reserve, but such an area can be ruined by overcollecting just as surely as can one anywhere else. Already the more accessible collecting spots in the northern part of the Gulf are being denuded by overeager amateurs who carry away shells by the sackful; by commercial collectors; and by thoughtless persons who needlessly destroy the habitat, whether collecting or not. The careful and conscientious collector may take a representative sample, but not an entire colony or population, and he tries to leave the site as undisturbed as possible, replacing every overturned rock in its original position as quickly as he can, so as not to disrupt the living conditions for the host of plants and animals nestled there. It is the delicate balance between these organisms—some large enough to see, others microscopic in size—that keeps an area continuously productive.

A book reviewer in 1959 commended that paragraph as a "strong plea for conservation." Now, in retrospect, one realizes painfully that it was not strong enough. Whether the book itself was the stimulus or whether the trend of the times would in any case have increased the activity of collectors, the malacological traffic has burgeoned, on the part of both amateurs and dealers, and not only automobile loads but truckloads of material continue to cross the southern border of the United States, Mexico being especially vulnerable to collecting forays on the part of the Norteamericano.

On the other hand, there is also some developing awareness on the part of many collectors. A "Shell Collector's Code of Ethics" was formulated in 1963 by the Pacific Northwest Shell Club, which has been subscribed to by other clubs and by individuals. It needs still wider publicity and adherence and is here repeated, in slightly amended and abridged form:

Because I appreciate our heritage of wildlife and natural resources,
I WILL make every effort to protect and preserve them, not only for my own future enjoyment, but also for the benefit of generations to come.
I WILL make sure that I leave things as I found them.
I WILL return rocks, boulders, kelp, and seaweeds to their original positions after looking beneath them.
I WILL refill the holes I dig.
I WILL take only those specimens that I know I can clean and use.
I WILL avoid taking juvenile or flawed specimens, leaving them to grow to maturity and to propagate.
I WILL never knowingly deplete an area of an entire species-population.
I WILL respect the property rights of others; treat public land as I would the property of my friends and collect on private beaches only with the owner's permission.
I WILL leave behind no trash or litter and discard no burning material.

Self-restraint on the part of conservation-minded collectors may not be enough to stem the tide of increased collecting pressure, especially if our population continues to expand and prosper. It may well be that licensing and stiff fees for permits will be necessary. There is some evidence that a few of the Latin American governments are awakening to the problems. Legal controls, such as closed areas to protect against overfishing, and rigorous border inspection, may be the only way to preserve in usable form the great natural resource of the Mexican and Central American shores until such time as the techniques of aquaculture and mariculture can be developed to replace our present irresponsible exploitation.

Deplorable as the heedless fisherman's and collector's acts may be, they pale beside the greatest threat of all to the marine fauna—pollution. This unanticipated repercussion of our expanding technology could wipe out marine life altogether, as it already has in some limited areas, such as certain estuaries. Pollution may come from dumping of sewage and chemical and heavy-metal effluents into the sea, from the spilling of oil, from the runoff of agricultural pesticides, or from contamination of the sea's surface by settling of particles from a polluted atmosphere. Radioactive fallout and slow recirculation of dangerously toxic gases and materials jettisoned far offshore add further hazards to the whole marine ecosystem. Chilling documentation of this is available in the following reference work:

P. R. Ehrlich and Anne H. Ehrlich. *Population, Resources, Environment: Issues in Human Ecology*. Freeman, San Francisco, 1970. 383 pp. $8.95.

To counter the menace of pollution, the first step seems to be the education of those who are not yet aware of the brief time left to fend off the disastrous consequences of our technology; the next step, stiff legislation to restrain those who would blindly continue our headlong course. International agreements and regional planning programs by the separate governments can be used with effect. Control of human population increase and a new technology aimed at ameliorating the effects, redressing the errors of the past and present, and restoring some sort of ecologic balance are not only needed and advisable; they are vital to our continued enjoyment of the earth's resources, both sea and land.

PHYLUM MOLLUSCA

Phylum Mollusca

Until recent years, the phylum Mollusca had been divided into five classes: Gastropoda (snails), Pelecypoda (or Bivalvia or Lamellibranchiata: clams), Cephalopoda (squids, etc.), Amphineura (chitons, etc.), and Scaphopoda (tusk shells). Additional names have been proposed, however, for classes that are known only in the fossil record, and paleontologists predict eventual recognition of as many as ten classes. Modern zoologists are accepting seven, splitting the Amphineura into their two components, the Aplacophora (solenogastres) and the Polyplacophora (chitons), and recognizing the Monoplacophora as distinct from the Gastropoda, with which there had been confusion. The Monoplacophora were considered to be extinct until, in 1957, the sensational discovery of *Neopilina* off the west Central American coast proved that an otherwise early Paleozoic stock is still living.

The word mollusk comes from the Latin *mollis*, soft-bodied. Not all mollusks have shells, but all do have certain features of the soft parts in common: The mantle, as the outer layer of tissue, has a number of functions, such as water circulation, and it is responsible for the secretion of the shell. All mollusks have a foot, which serves as a means of locomotion in at least some stage of each individual's life. All have a fairly simple nervous system, with nerve fibers and ganglia; and circulatory, digestive, and reproductive systems, variously modified. Most have a gill or ctenidium for respiration, although in some the gill takes on different functions and in others respiration is carried on by different organs. All except the pelecypods or bivalves have a radula, a rasping organ in the mouth. These basic features of the mollusks have been used in determining formal subdivisions, and as such are reflected in the names that workers have chosen for major categories—names ending in *-neura* (nerves), *-poda* (foot), *-branchia* (gills), *-glossa* (radula—literally, "tongue"), and the like.

All of the molluscan classes are represented in the marine environment. Two, the Gastropoda and the Pelecypoda, have also moved out of the sea and into freshwater conditions, but only some of the gastropods have managed the further achievement of crawling out on land and becoming air breathers.

The Mollusca have a long geologic history, for they were present almost from the time the first shelled organisms left fossil impressions in the rocks and the curtain really went up on the drama of life. The Gastropoda seem to have some claim to the title of earliest class, though a case can also be made for the Monoplacophora. Both appeared in the early part of Cambrian time, about 450 million years ago. This suggests that both have descended from a common ancestor that had no shell and that thus left no trace in the fossil record. The Pelecypoda, Cephalopoda, Polyplacophora, and probably Scaphopoda all appeared within the next 100 million years of recorded fossil history, in the early part of the Ordovician. The Aplacophora, lacking shells, have left no fossil record, but they seem not to be similar in form to the ancestral mollusk; they appear rather to be highly specialized. In numbers of species and individuals the Gastropoda, Pelecypoda, and Cephalopoda have been the most flourishing, leaving a continuous fossil record. Insects may surpass them in numbers of species and invasion of varied habitats, especially on land, but for the length of their history and their ability to make progressive change, the Mollusca, and especially the Gastropoda, are unrivaled in the animal kingdom.

OUTLINE OF CLASSIFICATION

[Class Gastropoda: Order Nudibranchia]
    Suborder Aeolidiida (= Cladohepatica ["divided liver"], in part), 835
        Infraorder Pleuroprocta ["marginal anus"], 835
            Family Flabellinidae, 836
        Infraorder Acleioprocta ["unconcealed anus"], 837
            Families Eubranchidae, 837; Fionidae, 837
        Infraorder Cleioprocta ["concealed anus"], 837
            Superfamily Aeolidiacea, 837
            Families Aeolidiidae, 839; Facelinidae, 840; Glaucidae, 840
    Order Gymnophila ["loving exposure"] (=Soleolifera ["sandal-bearing"]),
      841
        Superfamily Onchidiacea, 841
            Family Onchidiidae, 841
  Subclass Pulmonata ["lunged"], 843
    Order Basommatophora ["carrying eyes basally"], 843
        Superfamily Melampacea (= Ellobiacea), 843
            Family Melampidae, 843
        Superfamily Siphonariacea, 850
            Families Siphonariidae, 850; Trimusculidae, 852

Class Monoplacophora ["bearing one plate"], 857
    Order Tryblidioidea, 857
        Superfamily Tryblidiacea, 857
            Family Tryblidiidae, 857

Class Aplacophora ["bearing no plates"] (= Amphineura ["nerves on both
    sides"], in part), 859
    Order Neomeniida, 859
        Family Neomeniidae, 859

Class Polyplacophora ["bearing many plates"] (= Loricata ["clad in mail"];
    Amphineura ["nerves on both sides"], in part), 861
    Order Lepidopleurida, 863
        Family Lepidopleuridae, 863
    Order Chitonida, 863
        Families Chitonidae, 863; Acanthochitonidae, 866; Ischnochi-
           tonidae, 868; Lepidochitonidae, 879; Mopaliidae, 880

Class Scaphopoda ["boat-footed"], 883
        Families Dentaliidae, 883; Siphonodentaliidae, 888

Class Cephalopoda ["head-footed"], 893
  Subclass Coleoidea (Dibranchia), 893
    Order Octopoda (= Polypoidea), 893
        Superfamily Octopodacea, 893
            Family Octopodidae, 894
        Superfamily Argonautacea, 894
            Family Argonautidae, 894

CLASS PELECYPODA

Class Pelecypoda

(BIVALVIA, LAMELLIBRANCHIA, and LAMELLIBRANCHIATA of authors)

The Pelecypoda (or "hatchet-footed ones") are bivalved mollusks, with the two parts of the calcareous shell joined together by a hinge. Most of the terms used in classification of the group refer in some way to this linkage, whether to the cartilaginous part (*-desma*), to the teeth along the hinge margin (*-donta*), or to the muscles that pull the valves shut (*-myaria*). Classification in terms of gill structure has also been proposed. The gills, or ctenidia, are not so much breathing devices as they are water-circulatory organs, the outer surface being composed of cilia, microscopic filaments that beat in unison and set up water currents to bring food to the mouth and to help excrete wastes. Shape and complexity of the gills differ in different groups of the bivalves.

The earliest pelecypods seem to have had nacreous shells. Hence, we conclude that nacreous material in the shell today indicates relationship to the primitive stocks. Pearly inner layers occur among part of the nuculids, most of the mussels, all of the relatives of the pearl oysters, and some others. The shells that have differentiated cardinal and lateral hinge teeth are porcelaneous, from which we might surmise—and the fossil history confirms—that the highly developed or heterodont hinge marked a major advance in pelecypod construction. Ribbing was also something of an advance, for the earliest shells were smooth.

Pelecypods have become adapted to many environments, but lacking a head or a highly efficient nervous system, they have been unable to leave the water. They inhabit marine, brackish, and fresh waters, and can endure a surprising degree of cold, as in the Arctic or in great ocean depths. Most of them live below the surface of the sea floor (or lake or river floor), some constructing more or less permanent burrows, some moving about; a few, like the mussels, can survive periods of exposure during low tides.

Subclass CRYPTODONTA
(PROTOBRANCHIA)

Edentulous or nearly so; shell equivalve.

Order SOLEMYOIDA
(LIPODONTA)

Shell material aragonitic, homogeneous; shell gaping, anterior end longer than posterior.

Superfamily SOLEMYACEA

Shell cylindrical; foot adapted for burrowing; siphons well developed.

Family SOLEMYIDAE

The most conspicuous characteristic of the family is the varnish-like periostracum that projects in long fringes beyond the margins of the shell. The hinge is edentulous, but the soft parts show a relationship to the taxodont families Nuculidae and Nuculanidae.

Genus SOLEMYA LAMARCK, 1818
(SOLENOMYA, SOLENIMYA of authors)

This is mainly a deepwater group, records of 3,290 m being not uncommon, and the species may have wide geographic ranges.

Subgenus ACHARAX DALL, 1908

Ligament external; shell relatively large.

1. Solemya (Acharax) johnsoni Dall, 1891 (Synonym: **S. agassizii** Dall, 1908). The two species described by Dall are both from very deep water and have the same geographic range, the only cited differences being of features that are variable. Length, 100 to 150 mm. Puget Sound, Washington, to Peru, in depths of 110 m in the north to 3,270 m in Panama Bay.

Subgenus PETRASMA DALL, 1908

Ligament internal, behind the beaks.

2. Solemya (Petrasma) panamensis Dall, 1908. As in all Solemyas, there are no hinge teeth, but there is a small posterior prop supporting the internal ligament. Length (not including periostracum), 39 mm; height, 15 mm; diameter, 8 mm. Lowe found this form on the mud flats at La Paz at extreme low tide. Santa Barbara, California, through the Gulf of California and south to Panama, in depths to 1,650 m.

3. Solemya (Petrasma) valvulus Carpenter, 1864. The chondrophore or support for the internal ligament is without any props. The shell is small (about 20 mm long), thin, transparent, and ornamented by slender, brownish radial lines. San Pedro, California, to Sonora, Mexico.

Subclass PALAEOTAXODONTA

Shell equivalve, with closed margins; shell material nacreous to crossed lamellar; hinge taxodont.

Order NUCULOIDA

Ligament generally amphidetic, gills protobranchiate; foot grooved, not byssiferous in adult.

Superfamily NUCULACEA

Hinge with few to many chevron-shaped teeth or denticles; ligament central, partly or entirely internal.

Family NUCULIDAE

The inside of the shell is brilliantly nacreous. Hinge teeth are in two series, anterior and posterior, separated by a resilifer that lies just under the opisthogyrate (back-

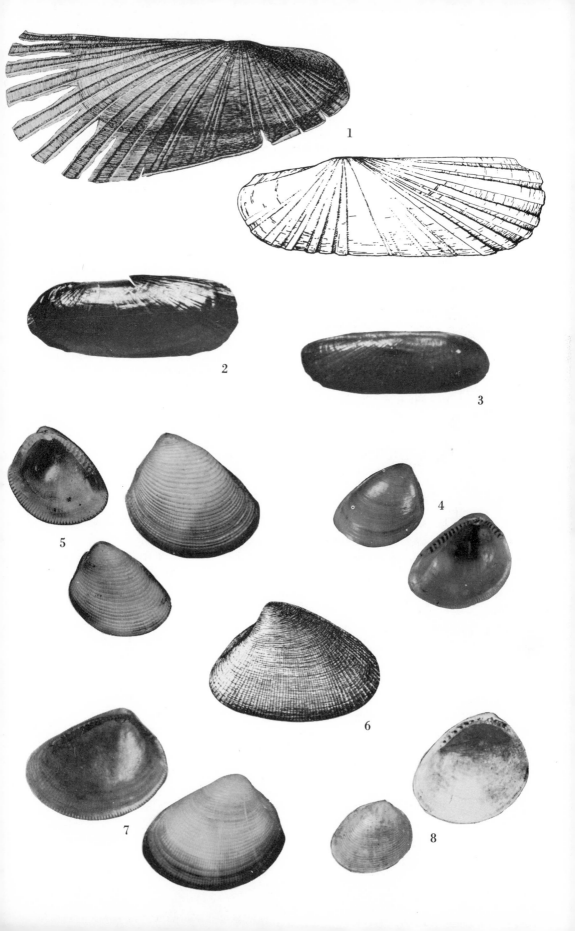

ward-pointing) beaks. The pallial line is entire. Nuculids are in general cold-water forms, mostly offshore; thus, tropical species are apt to occur in very deep water. Only a few have been reported from the shallower waters of the Panamic province.

Genus NUCULA LAMARCK, 1799

Beaks opisthogyrate; internal ligament in a resilifer; shell material nacreous. Two Panamic subgenera.

Subgenus NUCULA, s. s.

Inner margin of shell crenulate, reflecting the presence of radial ribs in the middle or outer layers.

4. Nucula (Nucula) declivis Hinds, 1843. The oblique shell is ornamented by very fine radial riblets. Length, 5 mm; height, 4 mm; diameter, 3 mm. Puerto Peñasco, Sonora, Mexico, to Panama, in 7 to 9 m. 𝒈 - 𝒔𝒔ᵐ

5. Nucula (Nucula) exigua Sowerby, 1833. Ornamentation is both concentric and radial, concentric predominating. Length, 6 mm; height, 5 mm; diameter, 4 mm. San Bartolomé (Turtle) Bay, Baja California, through the Gulf of California and south to Peru, in depths of 11 to 1,900 m.

6. Nucula (Nucula) iphigenia Dall, 1908. Among the largest of all Nuculas, this is known only from deep water. Length, 35 mm; height, 22 mm; diameter, 16 mm. Panama Bay, 474 m.

7. Nucula (Nucula) paytensis (A. Adams, 1856). Like *N. exigua* but with less pointed beaks and a more quadrate outline. Length, 6 mm. Northern Peru.

8. Nucula (Nucula) schencki Hertlein & Strong, 1940. Minute, rounded, the shell not as inflated as in the other species. Sculpture is concentric, with faint radial crenulations of the inner ventral margins. Length, 2 mm; height, 1.7 mm; diameter, 1 mm. Bahía de los Angeles, Gulf of California, to Port Guatulco, Mexico, in 13 to 45 m.

Subgenus ENNUCULA IREDALE, 1931

Inner margin of shell smooth.

9. Nucula (Ennucula) colombiana Dall, 1908. Small, smooth, with 14 teeth in the anterior series, 7 in the posterior. Length, 4.5 mm; height, 3 mm; diameter, 2.2 mm. Panama to Chile, 45 to 730 m.

10. Nucula (Ennucula) linki Dall, 1916. An inflated little shell, dark olive in color; hinge with 11 anterior and 6 posterior teeth. Length, 6 mm; height, 5 mm; diameter, 3.6 mm. Queen Charlotte Sound, British Columbia, to Guaymas, Sonora, Mexico, in depths to 44 m.

Genus NUCULA, s. l.

11. Nucula agujana Dall, 1908. Peru, 1,900 m.

12. N. chrysocome Dall, 1908. West Mexico to Peru, 733 to 4,060 m.

13. N. panamina Dall, 1908. Gulf of Panama, 3,050 m.

14. N. savatieri Mabille & Rochebrune, 1889. Gulf of Panama to Peru, 589 to 1,900 m.

15. N. taeniolata Dall, 1908. Acapulco, Mexico, 900 m.

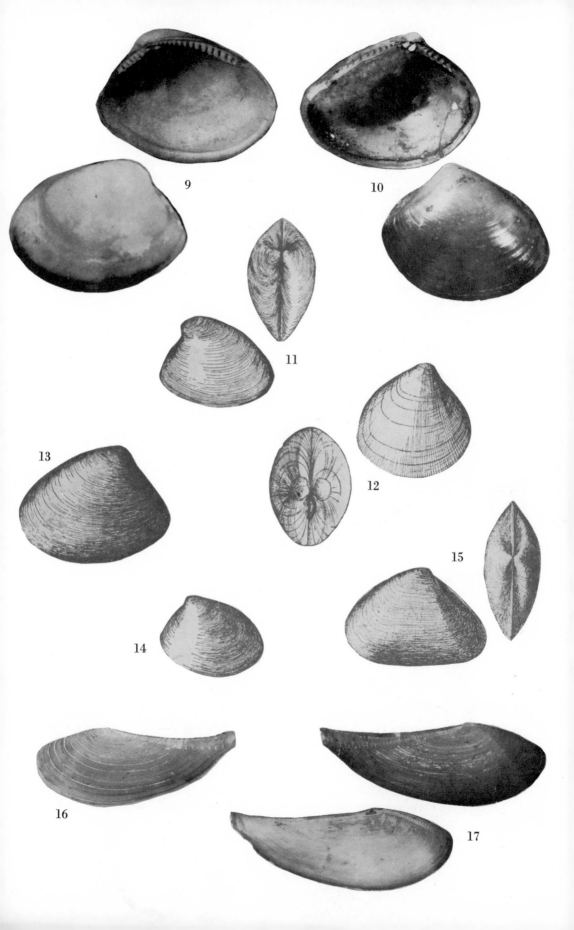

Superfamily NUCULANACEA

Shell elongate posteriorly, many forms beaked or rostrate; a resilifer present in most, ligament entirely external in a few; pallial sinus usually present.

Family NUCULANIDAE
(LEDIDAE)

More elongate than the Nuculidae and more inequilateral, the posterior end usually rostrate. The shells are less nacreous than the Nuculidae, lacking brilliant luster in the inner layer. Pallial sinus small to well developed. Seven genera are reported in the Panamic province, several being confined in their distribution to very deep water. Most forms occur well offshore. The key includes the subgenera of *Nuculana*.

1. Dorsal margin nearly straight.................................*Adrana*
   Dorsal margin sloping downward or concave posteriorly................ 2
2. Posterior end conspicuously rostrate............................. 3
   Posterior end rounded to pointed but not rostrate..................... 4
3. Pallial line entire...*Spinula*
   With a pallial sinus.......................................*Thestyleda*
4. Posterior end gaping...*Yoldia*
   Posterior end tightly closed....................................... 5
5. Posterior teeth much more numerous than anterior series...........*Malletia*
   Posterior teeth not conspicuously numerous......................... 6
6. Hinge teeth few...*Sarepta*
   Hinge teeth about 10 to 20 in number............................. 7
7. Ligament external ....*Tindaria*
   Ligament internal 8
8. Surface smooth or with only faint ribs near margins.............*Politoleda*
   Surface with regular concentric ribs............................... 9
9. Shell elongate, height about one-third the length...............*Costelloleda*
   Shell short, height about one-half the length.....................*Saccella*

Genus NUCULANA LINK, 1807
(LEDA SCHUMACHER, 1817)

Posterior end pointed, posterior dorsal margin curved; pallial sinus small and rounded. Several of the subgenera are easy to recognize, but among the deepwater species many are difficult to assign and are given as *Nuculana, s. l.* Like the Nuculas, the Nuculanas are more characteristic of the cold, northern waters. There are no species in the Panamic province that can be cited as *Nuculana, s. s.*, the type species of which is an Arctic form. Four other subgenera are recognized (see the key to the Nuculanidae).

Subgenus COSTELLOLEDA HERTLEIN & STRONG, 1940

Long and slender; the surface delicately sculptured.

16. Nuculana (Costelloleda) costellata (Sowerby, 1833). The sharp-edged ribs and the long, slender outline are distinctive characteristics. Length, 22 mm; height, 8 mm; diameter, 4.4 mm. Santa Inez Bay, Baja California, to Colombia, in depths of 18 to 42 m.

17. Nuculana (Costelloleda) marella Hertlein, Hanna & Strong, 1940 (Synonym: **Leda cestrota** Dall, 1890, of authors, not of Dall—an Atlantic form).

This is larger, more rostrate, and with more even sculpture than *N. costellata*. Length, 32 mm; height, 11 mm; diameter, 5 mm. The Gulf of California to Panama, in 65 to 75 m depth.

Subgenus **POLITOLEDA** HERTLEIN & STRONG, 1940

The smooth shell shows only some fine wavy incised lines but no ribs.

18. Nuculana (Politoleda) polita (Sowerby, 1833). The wavy incised lines run obliquely across the lower part of the shell and the posterior end. Length, 29 mm; height, 14 mm; diameter, 9.6 mm. Champerico, Guatemala, to Panama, in depths of 13 to 73 m.

Subgenus **SACCELLA** WOODRING, 1925

The shell, which is shaped something like the classic Grecian lamp, has strong, regular concentric ribs throughout.

19. Nuculana (Saccella) acrita (Dall, 1908) (Synonym: **Leda laeviradius** Pilsbry & Lowe, 1932). This has a smooth area in front of the posterior ridge, like *N. eburnea*, but it is a smaller shell, with a strong hinge. Length, 6 mm; height, 3.3 mm. Punta Peñasco, Sonora, Mexico, to Panama, in depths to 53 m.

20. Nuculana (Saccella) bicostata (Sowerby, 1871) (Synonym: **N. (S.) dranga** Olsson, 1961). Perhaps only a variant of *N. elenensis* in which the posterior radial rays appear doubled; the sculpture seems to be stronger and sharper. Length, about 14 mm; height, 7 mm. Panama to Ecuador.

21. Nuculana (Saccella) callimene (Dall, 1908). A rare deepwater form, less pointed posteriorly than most others. Length, 14 mm; height, 9 mm. The Gulf of Nicoya, Costa Rica, to Tomé, Chile, in depths of 180 to 470 m.

22. Nuculana (Saccella) eburnea (Sowerby, 1833) (Synonym: **Nucula lyrata** Hinds, 1843). The shell has a smooth area in front of the posterior keel. Length, about 12 mm; height, 6.4 mm. El Salvador to Ecuador, in depths of 13 to 112 m.

23. Nuculana (Saccella) elenensis (Sowerby, 1833) (Synonyms: **? Leda excavata** Hinds, 1843 [not Goldfuss, 1837]; **L. crispa** Hinds, 1843; **L. acapulcensis** Pilsbry & Lowe, 1932). Most abundant of the Panamic Nuculanas, this is a variable form, with wide differences in outline and sculpture at any station where it has been collected. Olsson (1961) regards *N. acapulcensis* as distinct because of its slightly larger size, but the differences are minor. He considers *N. crispa* (Hinds) a morphologic subspecies of *N. elenensis* that has the anterior ray strongly indented. Length, 14 mm; height, 7 mm; diameter, 5 mm. Bahía de los Angeles, Gulf of California, to Bahía Sechura, Peru, in 4 to 82 m.

24. Nuculana (Saccella) fastigata Keen, 1958 (Synonym: **N. gibbosa** Sowerby, 1833 [not Fleming, 1828]). A long, thick, posteriorly pointed shell with prominent umbones, the concentric sculpture of somewhat upturned ribs separated by smooth-bottomed grooves. Length, 37 mm; height, 19 mm; diameter, 16 mm. Guaymas, Mexico, to Tumbez, Peru, in depths of 9 to 82 m.

25. Nuculana (Saccella) hindsii (Hanley, 1860) (Synonym: **N. redondoensis** Burch, 1944). The original figure is reproduced here, showing a small, pointed, ovate-oblong shell with a conspicuous anterior indented ray and with concentric sculpture. The type was described as being probably from the Gulf of

Nicoya, Costa Rica. As Abbott (1954) has suggested, Hanley's name is probably the correct one for the small and fairly common West American form that has been identified by authors under the name of an Atlantic species (*N. acuta* Conrad, 1832). It is more pointed posteriorly than its Atlantic twin. Length, 6 mm; height, 4 mm; diameter, 3 mm. Alaska to Panama.

26. Nuculana (Saccella) impar (Pilsbry & Lowe, 1932). The sculpture is of fine and close concentric ridges on the beaks that become more widely spaced in the middle of the valve and closer again near the ventral margin. Length, 15 mm; height, 8 mm; diameter, 6 mm. Punta Peñasco, Sonora, Mexico, to Port Parker, Costa Rica, in depths of 4 to 37 m.
   See Color Plate XII.

27. Nuculana (Saccella) ornata (Orbigny, 1845). A Peruvian species that ranges north to Ecuador. Olsson (1961) has discussed and illustrated it.

28. Nuculana (Saccella) oxia (Dall, 1916). The small shell is very acute posteriorly, with subcentral beaks; the scupture is of regular concentric ridges, and a depressed ray runs from beak to base anteriorly. The hinge has about 8 teeth on either side of the resilifer. Length, 4.5 mm; height, 3 mm; diameter, 1 mm. Off Santa Rosa Island, California, to the Gulf of California, in depths to 88 m.

29. Nuculana (Saccella) taphria (Dall, 1897) (Synonym: **Nucula caelata** Hinds, 1843 [not Conrad, 1833]). Fairly common in California, this form is characterized by the strong and uniform concentric sculpture, the nearly central beaks, and the bluntly pointed rostrum. Length, 17 mm; height, 11 mm; diameter, 8 mm. Bodega Bay, California, to Arena Bank, Gulf of California, in depths of 11 to 82 m.

Subgenus THESTYLEDA IREDALE, 1929
Posterior end drawn out into a spoutlike rostrum.

30. Nuculana (Thestyleda) hamata (Carpenter, 1864). Not uncommon in the Californian fauna, this is a species of wide distribution. The long posterior rostrum is squarely cut off at the end, and the shell is strongly concentrically sculptured. Length, 12 mm; height, 7 mm. Puget Sound, Washington, to southern Baja California, in depths to 82 m; earlier reports south to Panama are probably in error.

Genus NUCULANA, *s. l.*
31. Nuculana agapea (Dall, 1908). Gulf of Panama to Ecuador, 2,870 to 3,050 m.

32. N. cordyla (Dall, 1908). Panama to Ecuador, 590 to 733 m.

33. N. lobula (Dall, 1908). Off Acapulco, Mexico, 260 m.

34. N. loshka (Dall, 1908). Gulf of Panama, 2,320 m.

35. Nuculana lucasana Strong & Hertlein, 1937. Small, olive-colored, the sculpture of regular rounded concentric threads, with an anterior radial depression. Length, 12 mm; height, 8 mm; diameter, 7 mm. Cape San Lucas, Baja California, depth 37 to 400 m.

36. N. pontonia (Dall, 1890). California, Gulf of Panama, and the Galápagos Islands, 1,485 to 3,050 m.

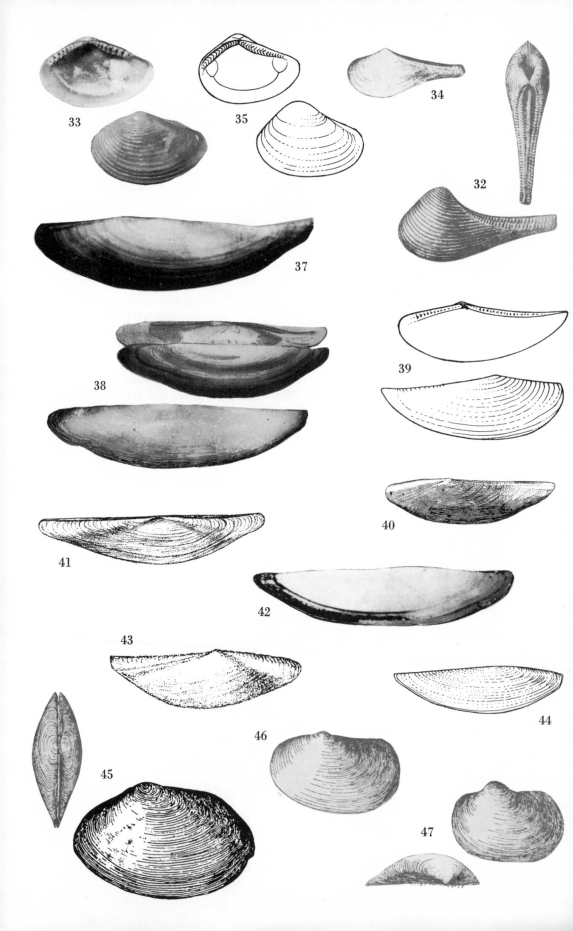

Genus ADRANA H. & A. ADAMS, 1858

The long, nearly straight dorsal margin with low umbones and the smoothly arched ventral margin are the distinctive features of this genus, which seems to be confined to the tropical Americas. Sculpture is mainly concentric or slightly oblique.

37. Adrana crenifera (Sowerby, 1833). This is a slender and posteriorly pointed form said by Olsson to be locally fairly common. Length, 35 mm; height, 9 mm. Ecuador to Peru.

38. Adrana cultrata Keen, 1958 (Synonym: **Nucula elongata** Sowerby, 1833 [not Bosc, 1801]). Intermediate in size among the West American Adranas, this has the umbones a little in front of the center and only the slightest sinuosity of the ventral margin toward the ends. Length, 50 mm; height, 12 mm; diameter, 5.5 mm. Acapulco, Mexico, to Ecuador, in 22 to 26 m.

39. Adrana exoptata (Pilsbry & Lowe, 1932). An Adrana that resembles the Nuculanas and might be confused with them. The form is more slender, and the ventral margin more smoothly arched. The posterior dorsal area is smooth and the concentric sculpture is of coarse riblets. Length, 20 mm; height, 6 mm; diameter, 3 mm. Guaymas, Sonora, Mexico, to Oaxaca, Mexico, in 7 to 48 m depth.

40. Adrana penascoensis (Lowe, 1935). White with a straw-colored glossy periostracum; this species resembles *A. exoptata* but has finer sculpture. Length, 37.5 mm; height, 9.4 mm. The species has been reported only in the northern end of the Gulf of California, in depths of 18 m.

See Color Plate XII.

41. Adrana sowerbyana (Orbigny, 1845) (Synonym: **Nucula lanceolata** Lamarck, of authors, preoccupied). Closest to *A. cultrata*, this has sinuations of the ventral margin near the ends. Length, 67 mm; height, 12 mm. Panama to Ecuador.

42. Adrana suprema (Pilsbry & Olsson, 1935). As its specific name implies, this is one of the largest of the Adranas. The beaks are more nearly central than in *A. cultrata*. Specimens have been found in beach drift in Nayarit, Mexico (Mrs. E. E. Wahrenbrock), and in Los Santos province, Panama (type locality). Length, 107 mm; height, 24 mm; diameter, 9 mm.

43. Adrana taylori (Hanley, 1860). According to Sowerby (1871), who had seen the type specimen, there is a peculiar narrowness and almost a constriction in the ventral margin at the anterior end. It has not been reported since first described from Guatemala. Length is about 28 mm.

44. Adrana tonosiana (Pilsbry & Olsson, 1935). Fine denticles bead the edge of the posterior dorsal margin. Length, 33 mm; height, 9.4 mm; diameter, 5.5 mm. Off Mazatlán, Mexico, to Panama, in depths of 26 to 37 m.

Genus MALLETIA DES MOULINS, 1832

Hinge with posterior teeth smaller and more numerous than anterior teeth; ligament external. Three Panamic subgenera.

Subgenus MALLETIA, *s. s.*

Compressed, smooth or with weak concentric sculpture; ends of shell blunt.

45. Malletia (Malletia) peruviana Dall, 1908. Off Peru, 1,900 m.

46. M. (M.) truncata Dall, 1908. Gulf of Panama, 2,690 to 3,330 m.

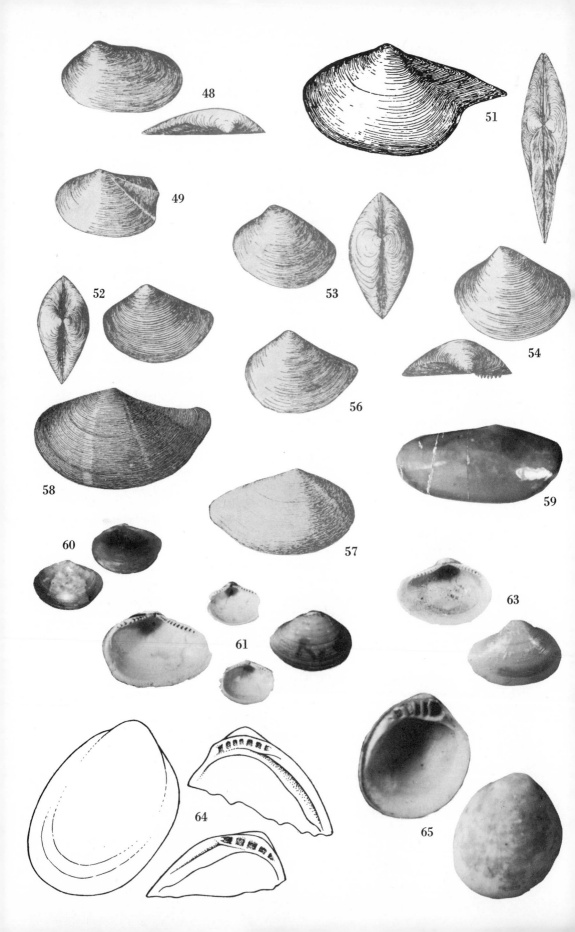

Subgenus MINORMALLETIA DALL, 1908

Small, smooth, inflated; ligament amphidetic; pallial sinus large.

47. Malletia (Minormalletia) arciformis Dall, 1908. Acapulco, Mexico, 900 m.

48. M. (M.) benthima Dall, 1908. Acapulco, Mexico, 900 m.

Subgenus NEILO A. ADAMS, 1854

Surface with well-developed concentric sculpture; posterior end somewhat pointed.

49. Malletia (Neilo) goniura Dall, 1890. Panama to Ecuador, 1,355 to 3,050 m.

Genus SAREPTA A. ADAMS, 1860

Ovate, ligament in a fossette; beaks central; hinge teeth few.

50. Sarepta abyssicola E. A. Smith, 1885. North and South Pacific, 3,750 to 4,350 m.

Genus SPINULA DALL, 1908

Shell posteriorly rostrate; ligament amphidetic; no pallial sinus.

51. Spinula calcar (Dall, 1908). North Pacific to Peru, to 4,330 m.

Genus TINDARIA BELLARDI, 1875

Ligament external; anterior and posterior sections of hinge equal.

52. Tindaria atossa Dall, 1908. Gulf of Panama, 1,870 m.

53. T. compressa Dall, 1908. Gulf of Tehuantepec, 4,080 m.

54. T. mexicana Dall, 1908. West Mexico, 1,210 m.

55. T. panamensis Dall, 1908. Gulf of Panama, 2,320 m.

56. T. smirna Dall, 1908. Gulf of Panama, 3,050 m.

Genus YOLDIA MÖLLER, 1842

With a posterior gape; ligament external in most, internal in a few. Four Panamic subgenera. No species of *Yoldia, s. s.,* occur in this area.

Subgenus KATADESMIA DALL, 1908

Shell pointed posteriorly; ligament amphidetic.

57. Yoldia (Katadesmia) vincula Dall, 1908. Gulf of Panama, 589 to 3,050 m.

Subgenus MEGAYOLDIA VERRILL & BUSH, 1897

Pallial sinus large and well marked.

58. Yoldia (Megayoldia) martyria Dall, 1897. Shell olive green in color, smooth; hinge with 21 anterior and 17 posterior teeth. Length, 26 mm. Kasaan Bay, Alaska, to the Gulf of California in depths of 65 to 250 m.

Subgenus ORTHOYOLDIA VERRILL & BUSH, 1897

Shell long, ends rounded.

59. Yoldia (Orthoyoldia) panamensis Dall, 1908. Panama Bay, 315 to 589 m.

Subgenus **YOLDIELLA** VERRILL & BUSH, 1897

Ligament internal; posterior end of shell rounded; shell small, quadrangular, external ligament weak; pallial sinus small and indistinct.

60. Yoldia (Yoldiella) cecinella Dall, 1916. Shell minute, polished, with no lunule or escutcheon. Hinge with 6 to 7 minute teeth on each side of the small resilifer. Length, 5 mm; height, 2.6 mm; diameter, 1.5 mm. The Aleutian Islands to the Gulf of California, in depths of 48 m or more.

61. Y. (Y.) dicella Dall, 1908. Acapulco, Mexico, 1,207 m.

62. Y. (Y.) leonilda Dall, 1908. Gulf of Panama, 3,050 m.

63. Y. (Y.) mantana Dall, 1908. Ecuador, 733 m.

Family **NUCINELLIDAE**

Shell minute, non-nacreous. Hinge with one or two lamellar lateral teeth and several cardinal teeth. There is no resilifer. Instead, the ligament is in a pit or fossette. The pallial line is entire. Two Panamic genera.

Genus **NUCINELLA** WOOD, 1851
(**PLEURODON** WOOD, 1840 [not HARLAN, 1831];
NEOPLEURODON HERTLEIN & STRONG, 1950)

Ovate, inequilateral, with an angulate expansion of the hinge margins and a small ligamental fossette.

64. Nucinella subdola (Strong & Hertlein, 1937). Minute, regularly oval, with a persistent dark periostracum. Length, 1.9 mm; height, 2.5 mm. Bahía de los Angeles, Gulf of California, to Mazatlán, Mexico, in depths to 22 m.

Genus **HUXLEYIA** A. ADAMS, 1860
(**CYRILLA** A. ADAMS, 1860)

Ligamental fossette large and round, impinging on the cardinal area.

65. Huxleyia munita (Dall, 1898). A southern California species; Gulf of California records of it have probably been misidentifications of *Nucinella subdola*.

Subclass **PTERIOMORPHIA**

Shells variously shaped; mainly sedentary forms with free mantle margins; generally with byssal fixation or cementation.

Order **ARCOIDA**
(**PRIONODONTA**; **EUTAXODONTA**)

Gills filibranchiate; hinge with numerous teeth; shell material of crossed-lamellar structure.

Superfamily **ARCACEA**

Trapezoidal in outline, mostly with radial sculpture and a velvety periostracum; ligament elongate, not confined in a resilifer. Hinge with taxodont dentition. Differing from the Nuculacea and Nuculanacea by the porcelaneous texture of the shell; by the straight rather than chevron-shaped hinge teeth; and by the gill structure, which also shows no close relationship. A key may be useful in distinguishing the families of the arks.

1. Beaks opisthogyrate (backward-pointing) . . . .*Noetiidae*
   Beaks not opisthogyrate. 2
2. Shell markedly inequilateral in outline. . . . .*Arcidae*
   Shell nearly equilateral or symmetrical. 3
3. Small; hinge edentulous. . . . .*Philobryidae*
   Medium-sized to large; hinge with a row of teeth. 4
4. Ligament in chevron-shaped grooves. . . . .*Glycymerididae*
   Ligament mainly in a central triangular resilifer. . . . .*Limopsidae*

Family ARCIDAE

Hinge long, the ligament distributed in grooves or pits above it. Muscle scars equal in size; pallial line entire. Shells often with a fibrous or velvety periostracum. Some may be anchored to the substrate by a byssus that protrudes through a gape on the ventral margin. Many arks occur intertidally on mud flats or under rocks.

Until the type species of *Arca* was fixed by an Opinion of the International Commission on Zoological Nomenclature, there was a diversity in the classifications used by authors. Studies by Reinhart (1935, 1943), Rost (1955), Olsson (1961), and Newell in the "Treatise on Invertebrate Paleontology" (1969) have been utilized here. The key is to the genera in the three subfamilies that occur in the Panamic province.

1. Inner margins of valves smooth. 2
   Inner margins of valves crenulate. 5
2. Hinge with an edentulous pit or gap in the middle. . . . .*Arcopsis*
   Hinge with a continuous series of teeth. 3
3. Ligamental area extremely wide and almost flat. . . . .*Arca*
   Ligamental area narrow and V-shaped. 4
4. Umbones dividing ligamental area unequally. . . . .*Barbatia*
   Umbones at posterior end of ligament. . . . .*Litharca*
5. Shell thin and fragile in texture. . . . .*Bathyarca*
   Shell sturdy and solid in texture. 6
6. Ligamental area extending equally on either side of beaks. . . . .*Anadara*
   Ligamental area mainly posterior to beaks. . . . .*Lunarca*

Subfamily ARCINAE

With a ventral byssal gape; sculpture of fine and mostly irregular-sized ribs; inner margin of shell not crenulate. Three Panamic genera.

Genus ARCA LINNAEUS, 1758
(NAVICULA BLAINVILLE, 1825; BYSSOARCA SWAINSON, 1833)

The type species of *Arca* has been fixed by the International Commission on Zoological Nomenclature as *Arca noae* Linnaeus, 1758—Noah's Ark, a boat-shaped shell from the Mediterranean, with a wide ventral gape and irregular radial sculpture.

Subgenus ARCA, *s. s.*

Hinge plate as long as the entire shell, straight, with evenly spaced denticles.

66. Arca (Arca) mutabilis (Sowerby, 1833) (Synonym: **Arca crossei** Dunker, 1870). As the specific name suggests, there are wide variations in the shape of this shell, some specimens having a very wide ligamental area, others having only a moderately wide one. The four to six ribs on the posterior slope are coarser than the rest and may be dark in color. The shell is brownish white under a yellowish-brown fringelike or almost scaly periostracum, which is especially developed along the sharp ridge that divides the posterior area from the central slope of the shell. Length, 38 mm; height, 23 mm; diameter, 21 mm. Specimens may commonly be found under rocks at low tide from Bahía Magdalena, Baja California, through the Gulf of California and south to Ecuador and may be dredged to depths of 82 m.

67. Arca (Arca) pacifica (Sowerby, 1833). The shell is irregularly ribbed and is variable in shape, mostly with an expanded posterior portion, notched near the posterior-dorsal margin. Narrow, rounded V-shaped brown bands ornament the whitish shell. A very similar species, *A. zebra* (Swainson, 1833), lives in the Caribbean area. A large specimen of *A. pacifica* measures: length, 128 mm; height, 72 mm; diameter, 80 mm. Specimens may be found adhering to the undersides of rocks in the intertidal zones or dredged in depths to 137 m from Scammon's Lagoon, Baja California, to Paita, Peru.

68. Arca (Arca) truncata (Sowerby, 1833). This species may be only subspecifically distinct from *A. pacifica*. It is proportionately longer at the posterior end, with a sharply truncate posterior margin. Length, about 80 mm. Galápagos Islands, Ecuador. A specimen from the syntype lot has been illustrated as "holotype" (thus becoming a lectotype) by Olsson (1961, pl. 4, fig. 1).

Genus BARBATIA GRAY, 1842

Barbatia differs from *Arca* in the much narrower ligamentary area, the cancellate sculpture, and the more evenly distributed hairy periostracum. The key is to subgenera.

1. Byssal gape wide and conspicuous.........................*Cucullaearca*
   Byssal gape narrow... 2
2. Ribs of posterior slope wide, prominent.......................*Calloarca*
   Ribs of posterior slope similar to central slope ribs..................... 3
3. Posterior slope set off by a low ridge.............................*Acar*
   Posterior slope not set off by a ridge............................... 4
4. Sculpture somewhat irregular to decussate......................*Barbatia*
   Sculpture of uniform fine riblets.............................*Fugleria*

Subgenus BARBATIA, *s. s.*

Ventral gape narrow; cardinal area low; ligamental grooves closely spaced.

69. Barbatia (Barbatia) lurida (Sowerby, 1833) (Synonym: **Byssoarca vespertilio** Carpenter, 1856). A brownish shell with a cancellate surface. Fresh specimens have a bearded periostracum with spaced radial rows of longer hairs. Length, 52 mm; height, 28 mm; diameter, 22 mm. Occasional specimens may be found intertidally, attached to rocks, or dredged from depths to 22 m. Bahía San Luis Gonzaga, Gulf of California, to Zorritos, Peru.

Subgenus ACAR GRAY, 1857

Small quadrate shells, with strongly cancellate sculpture and prominent muscle scars. Periostracum weak or wanting.

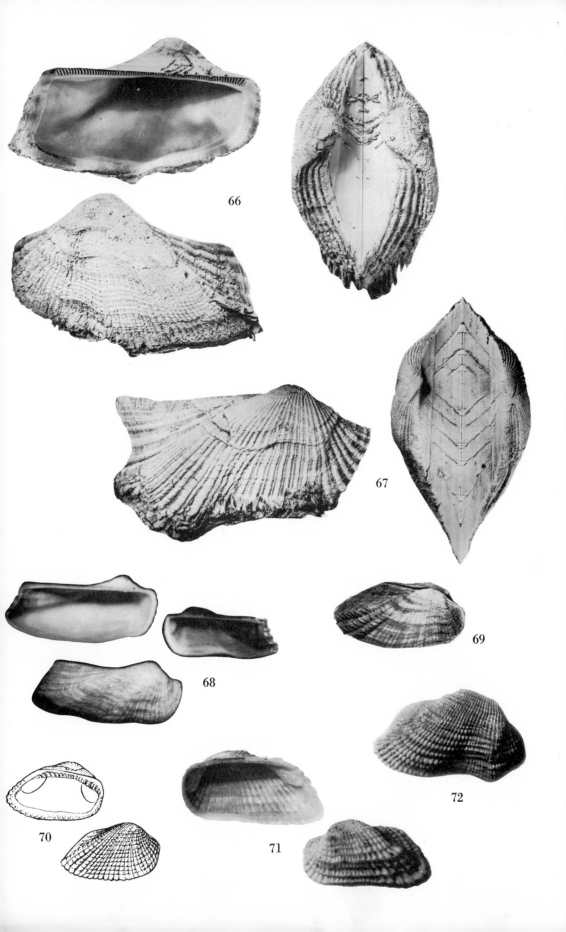

66

67

68

69

70

71

72

70. Barbatia (Acar) bailyi (Bartsch, 1931). This is a small form: length, 8 mm; height, 5 mm; diameter, 5 mm. The shell is brownish, without evident periostracum. Southern California to the Gulf of California, intertidally and to depths of a few meters.

71. Barbatia (Acar) gradata (Broderip & Sowerby, 1829) (Synonym: **Arca panamensis** Bartsch, 1931). The most coarsely sculptured of the West American Acars, this is also the largest. Length, 21.4 mm; height, 11 mm; diameter, 12 mm. Scammon's Lagoon, Baja California, to Negritos, Peru, intertidally on rocky shores or dredged in depths to 37 m.

72. Barbatia (Acar) rostae Berry, 1954 (Synonym: **Arca pholadiformis** C. B. Adams, 1852 [not Orbigny, 1844]). Long considered merely a finely sculptured variant of *B. gradata*, this is now separated as a distinct species, with the following differences: the sculpture is finer and more even than that of *B. gradata*; the concentric ridges do not turn up at their edges; the posterior end is more pointed; the hinge area is narrower. It differs also in the deeper, more arched hinge plate, which has fewer, longer, and more oblique teeth. Length, 22 mm; height, 13 mm; diameter, 11 mm. The distribution is much the same as that of *B. gradata*, from Scammon's Lagoon, Baja California, to Ecuador, intertidally or in shallow-water dredgings.

Subgenus CALLOARCA GRAY, 1857

Quadrate, with heavier ribs on the posterior umbonal ridge and an edentulous gap in the middle of the hinge.

73. Barbatia (Calloarca) alternata (Sowerby, 1833). The shell is thin and rather long, with the posterior area set off by a ridge almost sharp enough to be called a keel. The ribs of the middle portion of the shell are less coarse, are close set and divided down the center by a fine incised line. Length, 33 mm; height, 16 mm; diameter, 15 mm. This is not a common species, but specimens have been taken, mostly beyond the intertidal zone, in depths of 7 to 27 m, from Punta Peñasco, Sonora, Mexico, to Ecuador.

Subgenus CUCULLAEARCA CONRAD, 1865

The byssal gape is well developed.

74. Barbatia (Cucullaearca) reeveana (Orbigny, 1846) (Synonyms: **B. nova** Mabille, 1895; **B. r. lasperlensis** and **B. r. velataformis** Sheldon & Maury, 1922. This is a fairly large, coarsely reticulate form, with a habit of attaching between rocks, so that the shell may be much distorted in shape or even worn smooth in spots by friction with its surroundings. The shell itself is white, with a dark grayish-brown periostracum. Length, 82 mm; height, 45 mm; diameter, 29 mm. One of the commonest of the arks, this may be found from Manuela Lagoon, Baja California, and throughout the Gulf of California, south to Zorritos, Peru, intertidally and to depths of 120 m.

Subgenus FUGLERIA REINHART, 1937

With fine radial ribbing.

75. Barbatia (Fugleria) illota (Sowerby, 1833) (Synonym: **Arca tabogensis** C. B. Adams, 1852). The threadlike radial ribbing is distinctive. The shell itself is a grayish or ivory white, with a dark periostracum that consists of fine hairs

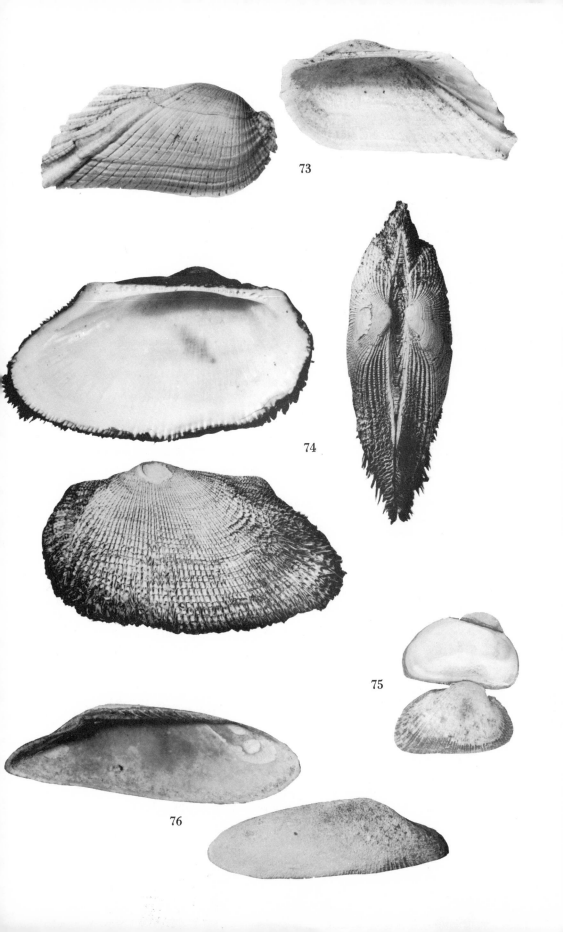

73

74

75

76

connected by a foliaceous membrane. Length, 35 mm; height, 24 mm; diameter, 21 mm. Specimens are not uncommon in the intertidal zone, attached to rocks, and dredging records show an offshore occurrence to depths of 70 m, from Angel de la Guarda Island, Gulf of California, to Lobitos, Peru.

Genus LITHARCA GRAY, 1842

Hinge teeth numerous, in two series, the anterior series longer, posterior series sloping downward; sculpture of fine radial riblets, with a periostracum in the young, mostly eroded away in adults; ventral margin with an elongate medial gape for attachment of a byssal plug; ligament in oblique grooves, entirely in front of beaks.

76. Litharca lithodomus (Sowerby, 1833). One of the rarest of arks in collections, this is mostly restricted to the northwest part of South America. The long, slender, posteriorly pointed shell is unique among arks both for its construction of the ligamentary area and for its evident boring habit. Specimens are apt to be so worn by contact with the narrow holes in which the shells are lodged that sculpture is obliterated except on the posterior end. Length, 70 mm; height, 23 mm; diameter, 21 mm. Nicaragua to Peru; living specimens have been reported mainly in Ecuador.

Subfamily ANADARINAE

Without byssal gape; beaks prosogyrate; hinge line somewhat arched. Three Panamic genera.

Genus ANADARA GRAY, 1847

Radial sculpture regular, valves tightly closing, without a byssal gape. The classification adopted here is based in part upon the work of Olsson (1961) but departs from his arrangement in several details. The key is to subgenera.

1. Shells equivalve, valves exactly meeting on margins................... 2
   Shells inequivalve, left valve overlapping right........................ 5
2. Shells massive, with relatively few ribs.......................*Grandiarca*
   Shells not markedly heavy, ribs numerous............................ 3
3. Ligament area trigonal, wider than long......................*Esmerarca*
   Ligament area narrow, longer than wide.............................. 4
4. Posterior margin not sinuate.................................*Anadara*
   Posterior margin sinuate, especially in young......................*Rasia*
5. Sculpture discrepant ....*Cunearca*
   Sculpture the same on both valves.................................. 6
6. Hinge teeth converging at ends, diverging in center..............*Larkinia*
   Hinge teeth roughly parallel throughout......................*Scapharca*

Subgenus ANADARA, s. s.

Hinge teeth in a continuous series of uniform size; ligament amphidetic.

77. Anadara (Anadara) adamsi Olsson, 1961. The relatively small shell has about 30 ribs, with sculpture varying from strongly noded in some to almost smooth; its outline is more rectangular than *A. similis*, which it resembles. Length, 23 mm; height, 16 mm; diameter, 14.6 mm. Panama.

78. Anadara (Anadara) concinna (Sowerby, 1833) (Synonym: **Arca cumingiana** Nyst, 1848 [*A. concinna* is not preoccupied by *Cucullaea concinna*

Phillips, 1829]). The shell is rather long, with 30 fine ribs, of which the most anterior ones have a groove down the middle. The periostracum is bristly in the interspaces of the anterior and posterior ribs. Length (large specimen), 52 mm; height, 29 mm. Dredged in depths of 9 to 90 m on sand or mud bottoms from San Luis Gonzaga Bay, Gulf of California, to Ecuador.

79. Anadara (Anadara) mazatlanica (Hertlein & Strong, 1943). This differs from *A. formosa* in having a shorter anterior end, more sharply angled at the junction with the hinge, and ribs showing no indication of a groove. The ventral margin is rounded, not straight as in *A. formosa*. Length, 62 mm; height, 36 mm; diameter, 31 mm. Santa Inez Bay, Gulf of California, to off Mazatlán, Mexico, in 33 to 102 m depth, and south to Peru. A large specimen 75 mm long, 42 mm high, was dredged in 1969 by Dr. Enrique del Solar, off Puerto Pizarro, Peru. The periostracum is dark brown, moderately heavy, with longer hairs down the crests of the ribs.

80. Anadara (Anadara) obesa (Sowerby, 1833). The numerous ribs (39 to 44) distinguish this from all other radially ribbed arks. Fresh specimens show a bristly periostracum in the interspaces of the ribs. Length, 26 mm; height, 18 mm; diameter, 16 mm. It is also an uncommon shell, dredged in fine sand or mud at depths of 22 to 112 m from off Cape San Lucas, Baja California, to Negritos, Peru.

81. Anadara (Anadara) similis (C. B. Adams, 1852). From *A. tuberculosa*, this species is distinguished by the greater proportionate length, the rounded ridges running down from the umbones, and the dorsal margin that is rounded at each end. The whitish shell is covered by a brownish-olive periostracum, and there are about 40 ribs. Length, 50 mm; height, 32 mm; diameter, 30 mm. Corinto, Nicaragua, to Guayaquil, Ecuador, mostly offshore in depths to 24 m.

82. Anadara (Anadara) tuberculosa (Sowerby, 1833). Shell large, ovate, rather thick, with 33 to 37 ribs, the dorsal margin somewhat angulate at each end. Nodes on the ribs, especially on the anterior end, are the reason for the specific name. Length, 56 mm; height, 42 mm; diameter, 40 mm. One of the important food mollusks of the Central American coast, it is eagerly sought by fishermen even as far south as Peru. Judging from archaeologic remains, it was a highly regarded food source in the past also. Ballenas Lagoon, Baja California, through the Gulf of California and south to Tumbes, Peru; abundant in mangrove swamps.

Subgenus CUNEARCA DALL, 1898

Sculpture on one valve markedly different from that on the other.

83. Anadara (Cunearca) aequatorialis (Orbigny, 1846) (Synonyms: **Arca ovata** Reeve, 1844 [not Gmelin, 1791]; **Arca subelongata** Nyst, 1848). The shell is thin, the posterior end pointed. Ribs are 30 to 32 in number, those on the left valve broader than those on the right. The ribs on the anterior area are coarsely nodose, on the posterior area flat and smooth. Length, 34 mm; height, 30 mm; diameter, 24 mm. This seems to be an offshore form, ranging from Mazatlán, Mexico, to Zorritos, Peru, in 11 to 73 m depth.

84. Anadara (Cunearca) bifrons (Carpenter, 1857) (Synonyms: **Arca cardiiformis** Sowerby, 1833 [not Basterot, 1825]; **Arca cordata** Deshayes, 1857 [not Benett, 1831]; **Arca corculum** Mörch, 1861). Like *A. aequatorialis*, this has about 30 ribs, but the shell is thicker and is only slightly extended at the

posterior end. The anterior ribs are wrinkled but not noded; the posterior ribs are flat, hardly visible, and separated by very narrow interspaces. Length, 44 mm; height, 41 mm; diameter, 35 mm. The species is not uncommon, recorded from the Gulf of California to Paita, Peru.

85. Anadara (Cunearca) esmeralda (Pilsbry & Olsson, 1941). The cardinal area of this form has chevron-shaped grooves, a distinctive feature. The ribs number 26 to 28, those on the left valve strongly nodose; on the right valve the anterior 8 to 10 ribs are strongly nodose, the remainder weakly nodose to smooth. Length, 57 mm; height, 53 mm; diameter, 51 mm. Isabel Island, Mexico, to Panama, mostly offshore, in sandy mud, in depths of 55 to 73 m.

86. Anadara (Cunearca) nux (Sowerby, 1833). The small size and fewer ribs (22 to 23) characterize this species. The ribs are noded and the beaks and umbones are toward the anterior end of the shell. In fresh specimens the periostracum is retained; brown bristles grow in the interspaces of the ribs. Length, 20 mm; height, 18 mm; diameter, 16 mm. Concepcion Bay, Gulf of California, to Zorritos, Peru, in shallow water, sandy mud bottom, in depths of 4 to 73 m.

87. Anadara (Cunearca) perlabiata (Grant & Gale, 1931) (Synonym: **Arca labiata** Sowerby, 1833 [not Lightfoot, 1786]). The heart-shaped shell has 28 slightly nodulous ribs and is more symmetrical than in *A. nux*, the beaks and umbones being nearly central and the ligament area diamond-shaped. Length, 33 mm; height, 30 mm; diameter, 29 mm. This is not an uncommon form, and living specimens may be found on sandbars at extreme low tide or dredged in shallow water (to depths of 82 m), from Magdalena Bay, Baja California, and the Gulf of California to Tumbes, Peru.

Subgenus ESMERARCA OLSSON, 1961

With slightly discrepant sculpture, somewhat as in *Cunearca*, ligament area high, trigonal, flattened.

88. Anadara (Esmerarca) reinharti (Lowe, 1935). This form, though rather recently described, is more common than some of the others, probably not having been recognized as a good species because it was confused with *A. (Larkinia) multicostata*. The number of ribs is 26 to 29, the ribs nodulose on the anterior and central slopes of the shells, especially in the left valve. On mature specimens grooves develop down the ribs. A tendency toward discrepant sculpture (as in *Cunearca*) may be detected, so that this species is rather an anomalous one. Compared with *A. multicostata* it is longer and has more chevron-shaped grooves on the ligament. Also, the posterior and ventral margins are slightly sinuous. Olsson has made this species the type of his new subgenus. Length, 45 mm; height, 40 mm; diameter, 38 mm (the original specimen was a small one). There seem to be no intertidal records, but it has been dredged in depths of 2 to 91 m from Punta Peñasco, Gulf of California, to Ecuador.

Subgenus GRANDIARCA OLSSON, 1961

More oblique and massive than *Larkinia*, with fewer ribs.

89. Anadara (Grandiarca) grandis (Broderip & Sowerby, 1829) (Synonym: **Arca quadrilatera** Sowerby, 1833). By all counts this is the largest of the Panamic province arks. The white shell, with a smooth, dark periostracum and about 26 ribs, thickens with age until it becomes impressively massive. A single

88

89

90

91

valve 145 mm (6 inches) long, in the Stanford collection, weighs 1½ pounds. Length, 96 mm; height, 79 mm; diameter, 75 mm. The size and relative abundance of the species make it a staple food clam, especially in the southern part of its range. Specimens may be taken at extreme low tide on sandbars from Magdalena Bay, Baja California, through the Gulf of California and south to Tumbes, Peru.

Subgenus LARKINIA REINHART, 1935

With the hinge teeth sloping differently at the ends than in the center, diverging in the center, converging at the ends.

90. Anadara (Larkinia) multicostata (Sowerby, 1833). This has more ribs than *A. grandis* (31 to 36 rather than 25 to 27). The shell is thinner and more quadrate. Unlike *A. grandis*, it is a little inequivalve, the left valve overlapping the right, as in *Cunearca* and *Scapharca*. The shell is white under a dark-brown, hairy periostracum. Length, 59 mm; height, 52 mm; diameter, 50 mm. Although specimens have been taken in tropical waters on sandbars at very low tide, they are more commonly gotten by dredging in shallow water, where the mollusk apparently lives free upon the bottom, in depths to 128 m. Newport Bay, California (rare); outer coast of Baja California through the Gulf of California and south to Panama and the Galápagos Islands.

Subgenus RASIA GRAY, 1857
(CARA GRAY, 1857)

Shells elongate, with the posterior margin sinuate, especially in the younger specimens.

91. Anadara (Rasia) emarginata (Sowerby, 1833). The posterior margin is sinuate, with a somewhat eared effect, though not as markedly as in the juvenile stages of *A. formosa*. The shell is long and thin, with a short anterior end, white, with a black ray down the beaks of most specimens. There are 28 to 30 flat and close-set ribs. Length, 50 mm; height, 32 mm; diameter, 14 mm. This is also an uncommon shell, dredged from depths of 5 to 24 m from Magdalena Bay, Baja California, and the Gulf of California to Paita, Peru.

92. Anadara (Rasia) formosa (Sowerby, 1833) (Synonyms: **Arca auriculata** Sowerby, 1833 [not Lamarck, 1819]; **Arca aviculoides** Reeve, 1844 [not deKoninck, 1842]; **Arca aviculaeformis** Nyst, 1848). In the young stages, the shell has a heavy and bristly periostracum and has been considered a separate species, but Olsson (1961) has shown that *A. aviculaeformis* is really only the young of *A. formosa*. Gray had made this juvenile form the type of his synonymous subgenus *Cara*; even the subgenus *Rasia* is not very markedly separated from *Anadara, s. s.* The adult of *A. formosa* is elongate, the posterior dorsal area rather broad. There are about 35 to 38 flat-topped ribs, the posterior ones wider, the anterior ones finely nodulous and grooved down the middle. A large specimen measures in length, 121 mm; height, 58 mm; diameter, 64 mm. This is not a common species, though it is known to occur from Cedros Island, Baja California, to Paita, Peru. It has been dredged in depths of 11 to 82 m.

Subgenus SCAPHARCA GRAY, 1847

Inequivalve (like *Cunearca*), but with the sculpture of the two valves similar in development throughout.

PLATE I · *Glycymeris gigantea* (Reeve)

92

93

94

95

93. Anadara (Scapharca) biangulata (Sowerby, 1833) (Synonyms: **Arca sowerbyi** Orbigny, 1846 [unneeded new name for **Arca biangulata,** not pre-occupied by *Arca biangula* Lamarck, 1805]; **Arca gordita** Lowe, 1935). The acute angle of the junction between anterior and dorsal or hinge margin is distinctive. Rost (1955) suggests that, on the basis of soft parts, the species might be placed in the subgenus *Rasia* Gray, of which *Anadara formosa* is the type, but that species has an equivalve shell, whereas *A. biangulata* is definitely inequivalve. Hence, we may conclude that much more study of these forms is needed. Olsson (1961) assigns this to the subgenus *Caloosarca,* the type of which is from the Florida Pliocene. Length, 30 mm; height, 20 mm; diameter, 15 mm. North end of the Gulf of California to Peru. *Anadara (Caloosarca) notabilis* (Röding, 1798) is a similar species in the western Atlantic.

94. Anadara (Scapharca) cepoides (Reeve, 1844). The shell is fairly large, inflated, with about 32 flat-topped smooth ribs, white under a brownish periostracum. Length, 59 mm; height, 53 mm; diameter, 45 mm. This is a rather rare form, known to occur from Bahía San Luis Gonzaga, Gulf of California, to San Miguel, Panama, offshore in depths to 84 m.

95. Anadara (Scapharca) hyphalopilema Campbell, 1962. One of the largest of west-coast Scapharcas, it is distinguished also by a dark-colored feltlike periostracum. There are 41 squarish ribs. Length, 60 mm; height, 60 mm; diameter, 55 mm. Off Guaymas, Mexico, depth about 90 m.

96. Anadara (Scapharca) labiosa (Sowerby, 1833). The shell has 34 to 39 flat-topped ribs and a thin periostracum. Length, 48 mm; height, 34 mm; diameter, 28 mm. Concepcion Bay, Gulf of California to Tumbes, Peru (type locality), in depths of 25 to 40 m.

Genus BATHYARCA KOBELT, 1891

The shells are thin, smooth or with fine radial ribs, the hinge long and straight, studded with small teeth. Species of this genus are confined to deep water.

97. Bathyarca nucleator Dall, 1908. Off southern California to the Gulf of Panama in depths to 2,320 m. [See further note, p. 303.]

Genus LUNARCA GRAY, 1842
(ARGINA GRAY, 1842, preoccupied; ARGINARCA McLEAN, 1951)

The ligamental area is narrow and entirely back of the beaks. The hinge teeth are in two groups, with a short and irregular anterior series in front.

98. Lunarca brevifrons (Sowerby, 1833) (Synonyms: **Arca vespertina** Mörch, 1861; **Argina bucaruana** Sheldon & Maury, 1922; **Arca melanoderma** Pilsbry & Lowe, 1932). Long misunderstood because of a typographical error by Reeve, who figured the species, this has acquired several synonyms. The type specimen in the British Museum has 36 ribs in each valve. It measures 32 mm in length, 19 mm in height, and 15 mm in diameter, and is evidently the same form Mörch later described, which has 36 to 38 ribs and a periostracum, in fresh specimens, tufted between the ribs. The beaks in most specimens have a black stain or streak. The species has been recorded from the west coast of Baja California in depths to 44 m and southward to northern Peru. Lowe reported specimens intertidally on a reef at Panama.

Subfamily STRIARCINAE

Cardinal area ligament-free; beaks nearly orthogyrate.

Genus ARCOPSIS VON KOENEN, 1885

Hinge teeth interrupted in the middle by a gap; ligament in a small triangular area below the beak, adjacent to this gap; beaks only slightly opisthogyrate to orthogyrate.

99. Arcopsis solida (Sowerby, 1833) (Synonym: **Barbatia digueti** Mabille, 1895). Resembling *Barbatia gradata* in shape and size and sculpture but readily distinguishable by the ligament, which forms a small, dark, diamond-shaped mass under the beaks, whereas the ligament of *B. gradata* is spread over most of the cardinal area. Length, 15 mm; height, 10 mm; diameter, 9 mm. Not uncommon in the intertidal zone, attached to rocks, from the west coast of Baja California and the Gulf of California to Paita, Peru. A similar species in the western Atlantic is *A. adamsi* (Dall, 1886). The group has traditionally been placed near *Barbatia*, but the striate structure, according to Newell (1969), warrants transfer to the Noetiidae.

Family NOETIIDAE

Shell without a byssal gape; hinge as in Arcidae; ligament vertically striate, with multiple strips of elastic layer embedded in a fibrous layer in contact with shell; adductor muscles somewhat bordered by a ridge.

Subfamily NOETIINAE

Strongly opisthogyrate (with backward-pointing beaks); ligament elongate and broad, in transverse grooves anterior to the beaks. Useful reference: MacNeil (1938).

Genus NOETIA GRAY, 1857

Ovate to trigonal, mostly sturdy shells; hinge teeth radial, anterior series somewhat longer than posterior series. Three Panamic subgenera.

Subgenus NOETIA, s. s.

Shell trigonal, with a fairly strong radial umbonal ridge; ribs even, broad.

100. Noetia (Noetia) magna MacNeil, 1938. Larger than the much more common *N. reversa*, with finer ribs and a less pointed posteroventral angle, umbones somewhat broader. Length, 60 to 80 mm. Described as a fossil; reported as living by Roth (in press). Nicaragua to Ecuador (living); Ecuador to Peru (fossil).

101. Noetia (Noetia) reversa (Sowerby, 1833) (Synonyms: **Arca hemicardium** Philippi, 1843; **N. triangularis** Gray, 1857). The shell is of medium size, inflated, equivalve, with about 36 ribs. The beaks are large, prominent, posterior, and curved backward. Length, 37 mm; height, 34 mm; diameter, 34 mm. Although specimens have been reported on intertidal mud flats, the most recent records are from dredgings, in depths of 22 to 73 m, Bahía San Luis Gonzaga, Gulf of California, to Peru.

Subgenus EONTIA MACNEIL, 1938

Posterior end not expanded; outline elongate.

102. Noetia (Eontia) olssoni Sheldon & Maury, 1922. The narrow, elongate shell has a medial depression on the posterior end. A large specimen may measure 37 mm in length, 22 mm in height, although the type specimen was smaller, less than 15 mm in length. Some specimens show an orange-yellow stain within. Mazatlán, Mexico, to Negritos, Peru, mostly offshore in depths to 24 m.

Subgenus SHELDONELLA MAURY, 1917

Shell with the posterior end strongly expanded or widened.

103. Noetia (Sheldonella) delgada (Lowe, 1935). There are about 30 ribs. The shell is rounded posteriorly, obliquely drawn out, with an outline unlike that of any other West American ark. The umbones are pink in fresh shells. Length, 12 mm; height, 9 mm. Off Guaymas and Manzanillo, Mexico, to Panama, dredged in depths of 37 to 55 m.

Superfamily LIMOPSACEA

Orbicular to obliquely oval, posterior slope not usually set off by an umbonal ridge; surface smooth or radially ribbed, mostly with a fibrous periostracum.

Family LIMOPSIDAE

Small; ligament short, mainly in a central triangular resilifer; hinge with several teeth (taxodont dentition), usually in two series; outline of shell oval to oblique; periostracum tending to be tufted and thick.

Genus LIMOPSIS SASSI, 1827

Mostly deepwater and cold-water forms. Oblique-ovate, compressed, with the periostracal tufts arranged in radial rows.

104. Limopsis compressus Dall, 1896. West Mexico to the Gulf of Panama, 1,960 to 4,080 m. [See further note, p. 303.]

105. L. diazi Dall, 1908. Acapulco, Mexico, 1,207 m.

106. L. juarezi Dall, 1908. Acapulco, Mexico, to the Gulf of Panama, 1,207 to 1,960 m.

107. L. panamensis Dall, 1902. Gulf of Panama, 1,870 m.

108. L. stimpsoni Dall, 1908. Gulf of Panama, 1,870 to 2,320 m.

109. L. zonalis Dall, 1908. Gulf of Panama, 1,015 to 1,430 m.

Family GLYCYMERIDIDAE

Subcircular, equivalve, more or less equilateral; valve margins not gaping; hinge with a row of teeth, somewhat chevron-shaped; ligament in grooves on the flattened cardinal area. Periostracum present, thin to velvety.

Genus GLYCYMERIS DA COSTA, 1778
(PECTUNCULUS LAMARCK, 1799)

Shell sturdy, solid, porcelaneous in texture; hinge plate arched, with part of the teeth somewhat chevron-shaped. The shell may be so symmetrical that one must examine the soft parts to be sure of orientation, the foot protruding anteriorly, the siphons posteriorly. In most shells the posterior muscle scar has a slight flange or buttress around it that also aids in the orientation. The ligament is distributed in chevron-shaped grooves below the beaks on the cardinal area and may be as much in front of the beaks as behind. In most other bivalves the major part of the ligament, if not all of it, is behind the beak. These are not deepwater forms but most occur offshore and must be taken by dredging or diving. Single valves are fairly common in beach drift.

A number of generic and subgeneric names have been proposed for glycymeridid groups, as Nicol (1945) has shown. The conservative classification adopted here uses three.

Subgenus GLYCYMERIS, s. s.

Radial sculpture of fine threads and very weakly developed low ribs. Periostracum heavy.

110. Glycymeris (Glycymeris) gigantea (Reeve, 1843). One of the largest forms in the whole family. A large specimen may attain a height of 100 mm. The nearly smooth surface shows radial striae and the white ground color is mottled with reddish brown in zigzag patterns. There are about 30 hinge teeth. The species seems to be confined to the Gulf of California area, from Bahía Magdalena, Baja California, to Acapulco, Mexico, dredged in depths of 7 to 13 m.
See Color Plate I.

111. Glycymeris (Glycymeris) lintea Olsson, 1961. Somewhat resembling *G. maculata*, this is longer for the height, lighter-colored, with only a few spots of brown color; the sculpture is of fine low riblets with a linen-like surface texture. Length, 58 mm; height, 48 mm; diameter, 33 mm. Panama to northern Peru.

112. Glycymeris (Glycymeris) maculata (Broderip, 1832). The color pattern, which is laid on in small spots of chestnut brown, especially on the earlier part of the shell, and the smaller size distinguish this form from *G. gigantea*. Length, 35 mm; height, 31 mm; diameter, 22 mm. The northern part of the Gulf of California and Bahía Magdalena, Baja California, to Zorritos, Peru.

Subgenus AXINACTIS MÖRCH, 1861

Ribs few, heavy, strongly striate, with wide interspaces.

113. Glycymeris (Axinactis) delessertii (Reeve, 1843). The ribs number 9 to 11, ornamented by smaller riblets. The shell is grayish white, spotted with brown and cream in irregular bands. Length, 32 mm; height, 31 mm; diameter, 22 mm. Mazatlán, Mexico, to Panama.

114. Glycymeris (Axinactis) inaequalis (Sowerby, 1833) (Synonym: **Pectunculus assimilis** Sowerby, 1833). This is distinguished from *G. delessertii* by having still fewer ribs (about 6). The ribs are broad and rounded, wider than the interspaces, and ornamented by incised smaller riblets and wide zigzag stripes of tan-brown color. Length, 35 mm; height, 38 mm; diameter, 24 mm. San Marcos Island, Gulf of California, to Peru, in depths of 4 to 24 m.

Subgenus TUCETONA IREDALE, 1931

Radial ribs well developed, interspaces narrow to moderately wide, ribs sometimes dividing terminally; periostracum thin or wanting.

115. ? Glycymeris (Tucetona) canoa Pilsbry & Olsson, 1941. This was described from the Pliocene of Ecuador and later recognized by Hertlein and Strong as living in the southern end of the Gulf of California. Possibly the Gulf form may prove to be a separate species, with about 22 broad ribs that are nearly smooth, interspaces wide, the periostracum heavier there than on the ribs. Length, 38 mm;

height, 39 mm. Off La Paz and Arena Bank, Gulf of California, to Manzanillo, Mexico, living; fossil, Ecuador.

116. Glycymeris (Tucetona) multicostata (Sowerby, 1833) (Synonyms: **Pectunculus bicolor** Reeve, 1843; **P. minor** Orbigny, 1846 [not Lea, 1833]; **G. chemnitzii** Dall, 1909). There are 35 to 40 flat-topped ribs on the grayish shell, and in occasional individuals these may divide, especially at the ends, into numerous riblets. Irregular blotches of dark brown color develop here and there, and the interior of the shell, especially the posterior part, is brown. In the southern part of the range the shells seem to have fewer ribs, and some have been confused in collections with *G. canoa*, which lacks the strong color. Length, 35 mm; height, 36 mm; diameter, 20 mm. Punta Peñasco, Sonora, Mexico, to Guayaquil, Ecuador, in depths to 90 m.

117. Glycymeris (Tucetona) strigilata (Sowerby, 1833) (Synonyms: **Pectunculus tessellata** Sowerby, 1833; **P. pectinoides** Deshayes, 1843). Authors have found it difficult to separate the two forms *G. strigilata* and *G. tessellata*. Acting as "first reviser," Olsson has synonymized them, selecting the name *G. strigilata* as the preferred one, although the name *tessellata* has had greater usage. Under the nomenclatural Code, his action is to be accepted unless the two are again separated. Compared with *G. multicostata* this has fewer and wider ribs, about 25 in number. The color is grayish white, checkered with rich purple spots. A large specimen measures in length, 33 mm; height, 33 mm. The range is Guaymas, Gulf of California, to Ecuador, in depths of 13 to 110 m.

? Family PHILOBRYIDAE

Small, equivalve, umbones projecting; shell material not nacreous. Ligament internal or only partly external; a short thin byssus along anterior margin; hinge edentulous but with strong vertical crenulations; anterior adductor muscle scar tending to be obsolete.

The placement of this family is open to question. Although the type species of the genus *Philobrya* is West American, little work has been done here on the group. Elsewhere the genera and species similar to it in form that have been studied show hinge structures and soft parts closest to those of the Limopsidae and Glycymerididae. However, the shape of the shell, the toothless hinge, the single muscle scar, and the byssus suggest an affinity to the Pterioida.

Genus PHILOBRYA CARPENTER, 1872
(BRYOPHILA CARPENTER, 1864 [not TREITSCHKE, 1825])

Small, pear-shaped, with a flaky periostracum that may be radially marked; hinge area nearly smooth; interior shiny, minutely punctate.

118. Philobrya setosa (Carpenter, 1864). The shell looks like a tiny mussel but with a frilled or spinous tan-colored and heavy periostracum. The hinge is straight and flat, with no denticles. A bundle of byssal threads emerges from the valves below the beaks for attachment of the shell to the substrate. Length, 5 mm; width, 3.5 mm. Alaska to the Gulf of California, on rocks.

Order MYTILOIDA
(DYSODONTA)

Equivalve, very inequilateral; generally with a byssus; adductor muscle scars of unequal size; ligament extended along posterior dorsal margin; siphons not developed, mantle margins free.

Superfamily MYTILACEA

Elongate, beaks near anterior end; shell material dense in the outer layer, somewhat pearly within; anterior adductor scar small; a periostracum commonly present, bristly to hairy; shell mostly anchored by a byssus.

Family MYTILIDAE

Like the Nuculas, the mussels have a long history, extending back in time to the Paleozoic era, and even that early had developed some of the shell features we use today in differentiating genera. For a long while after scientific naming had begun with Linnaeus—who introduced the name *Mytilus*—workers were content to place all pear-shaped bivalves either into this or into *Modiolus*, depending upon whether the beaks were at the tip of the shell or a little distance back. This distinction does not always prove easy to make, however, and gradually other generic and subgeneric names have crept into the literature. Few workers took the trouble to examine the inside of the shell or the soft parts of the animal that built the shell, so that much of the assignment of species was arbitrary. Even the most stable groups of organisms tend, with time, to show progressive change. This family has been in existence for millions of years. May it not be, therefore, that careful study would show differentiation to have taken place, here as elsewhere, though perhaps in small ways rather than in the larger and more conspicuous ways of, say, the Veneridae? Such a possibility seems all the more plausible when one studies the report of the Norwegian zoologist Dr. Toon Soot-Ryen, who spent a year in California critically examining the internal as well as the external features of West American Mytilidae. Although the number of divisions he would recognize in the family must disturb the conservative—who would like to pigeonhole all the forms under a few group names—we should make at least some attempt to understand the complexity of the family as this specialist has worked it out for us. Hence, in the present summary, only minor modifications of his scheme are suggested.

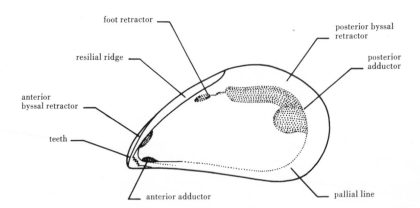

Although characteristically the mussels cling by the byssus to the exposed faces of surf-beaten rocks, where the churning water brings a maximum amount of food and oxygen, members of the family have become adapted to a number of other habitats, even to the extreme of a sheltered life in a burrow they have excavated for themselves in rocks or in other shells. The sedentary habits of the mussels and their tendency to cluster in colonies have made them a target for several predators, not the least of which is man. Carnivorous gastropods such as *Thais* and many kinds of starfish may be expected near these colonies, and mussels form a staple food item in some countries second only to oysters. Mussel poisoning, from which there have been many fatalities, is caused by the concentration of toxins from certain of the microscopic organisms caled dinoflagellates. Mussels (and other filter-feeders as well) somehow avoid being killed by the toxin but segregate it in their tissues, where it may stay for some time. Thus mussels might remain poisonous to a predator (such as man) even after a dinoflagellate bloom—a "red tide"—has ended.

The classification adopted here is based on the revision by Dr. Soot-Ryen in the *Treatise on Invertebrate Paleontology* (Moore, 1969). It recognizes four subfamilies. The key, adapted from Soot-Ryen (1955), is to genera in the Panamic province.

1. Resilial ridge pitted. . . . .*Mytella*
   Resilial ridge compact, dense. 2
2. Lunule with radial sculpture; anterior margin with teeth or crenulations. . 3
   Lunule smooth; anterior margin without teeth or crenulations; shell
      smooth or with irregular sculpture. 9
3. Shell either smooth or with concentric or irregular sculpture. 4
   Shell with more or less well-marked radial sculpture. 5
4. Anterior retractor muscle in front of beaks, scars of siphonal muscles
      visible; shell cylindrical, smooth or with irregular sculpture. . . . . . .*Adula*
   Anterior retractor muscle wanting, shell pear-shaped. . . . .*Choromytilus*
5. Dorsal margin with strong toothlike crenulations. 6
   Dorsal margin behind ligament smooth or no more crenulate than re-
      mainder of margin. 7
   Anterior adductor not on a septum. . . . .*Brachidontes*
   Anterior adductor placed on a septum in each valve. . . . .*Septifer*
7. Radial sculpture weaker or absent on median part of shell; periostracum
      strong, with hairlike protuberances. . . . .*Gregariella*
   Radial sculpture of equal strength over entire surface; periostracum
      mostly thin 8
8. Radial ribs relatively strong, tending to bifurcate anteriorly and posteri-
      orly; ligament deep-set. . . . .*Crenella*
   Radial ribs fine and numerous, not markedly bifurcating, weaker anteri-
      orly and posteriorly. . . . .*Megacrenella*
9. Shell elongate, cylindrical; periostracum strong; irregularly sculptured
      or with chalky incrustation; anterior retractor scar in front of umbo. . . .10
   Shell somewhat flaring; periostracum smooth or with hairlike fringes;
      anterior retractor scar normally behind beak. 11
10. Without chalky incrustation; margins slightly crenulate. . . . .*Adula*
    With chalky incrustation posteriorly; margins smooth. . . . .*Lithophaga*
11. Anterior retractor in front of beak. . . . .*Lioberus*
    Anterior retractor behind beak or in umbo. 12
12. Periostracum shiny, grayish to yellowish white. . . . .*Amygdalum*
    Periostracum brownish, not shiny. 13
13. Periostracum not hairy; beaks terminal, somewhat coiled. . . . .*Botula*
    Periostracum hairy; beaks distinctly behind anterior end. . . . .*Modiolus*

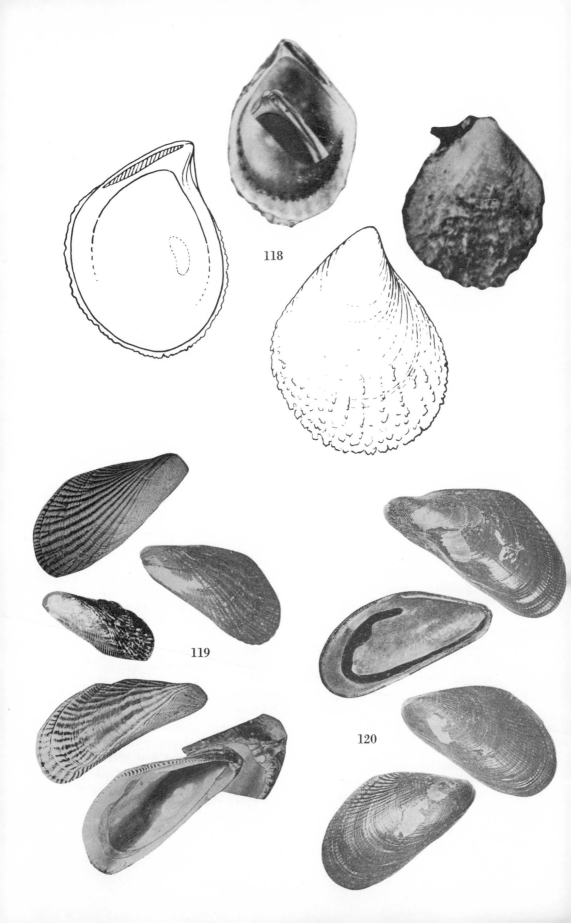

118

119

120

Subfamily MYTILINAE

Pear-shaped, with beaks at or near the anterior end; mostly with denticles along the anterior margin and radial sculpture on the lunular area; shells smooth to ribbed. Three, possibly four, genera in the Panamic province.

Genus MYTILUS LINNAEUS, 1758

Anterior retractor scar elongate; posterior retractors continuous; hinge with several small denticles; shell nearly smooth. No species of the typical genus have been cited in Panamic lists. However, a specimen of the Californian blue mussel, *M. californianus* Conrad, 1837, in the collection of the Cholla Bay Oceanographic Station, was taken in a rocky tide pool in Bahía de Adair, Sonora, Mexico. Apparently this represents a sporadic introduction, but collectors should watch to see whether the species becomes established.

Genus BRACHIDONTES SWAINSON, 1840
(HORMOMYA MÖRCH, 1853; SCOLIMYTILUS, AEDIMYTILUS OLSSON, 1961)

The small mytilids are so variable in form and sculpture that a subdivision into two genera, *Brachidontes* and *Hormomya*, frequently attempted, is difficult to make consistently. Inasmuch as Dr. Vida Kenk (*in litt.*) advises that she has not been able to find differences in the soft parts, a conservative classification is adopted here, recognizing only the one genus.

119. Brachidontes adamsianus (Dunker, 1857) (Synonym: **Mytilus stearnsi** Pilsbry & Raymond, 1898). The ribs of this dark purplish shell are elegantly granulated. About 15 mm in length, 7 mm in width. The northern end of the range is uncertain, perhaps California; thence southward through the Gulf to Ecuador and the Galápagos Islands, occurring intertidally on rocky shores.

120. Brachidontes playasensis (Pilsbry & Olsson, 1935) (Synonym: **Scolimytilus esmeraldensis** Olsson, 1961). The riblets are lower than in other species and nearly uniformly developed over the whole shell, which is wide and wedge-shaped, the ventral margin straight or bent slightly inward. In color the periostracum is a rich brown, the shell itself a purplish brown. This is a mud-flat dweller. Length, 21 mm; height, 9 mm; diameter, 9 mm. Ecuador and Peru.

121. Brachidontes puntarenensis (Pilsbry & Lowe, 1932) (Synonym: **B. multiformis houstonius** Bartsch & Rehder, 1939). On this shell the sculpture is more uniform, present even on the ventral side. The ribs are fine and close-set. Specimens may range in size to 17 mm long. Costa Rica to Ecuador and Galápagos Islands.

122. Brachidontes semilaevis (Menke, 1849) (Synonyms: **Mytilus multiformis** Carpenter, 1857; **Scolimytilus aequatorialis** Olsson, 1961). As Carpenter himself suspected, his *Mytilus multiformis* conforms precisely to Menke's description of a small mussel from Mazatlán; thus, Carpenter's name should have been synonymized long ago. As compared with *B. puntarenensis*, this is more irregularly sculptured. On the ventral side the shell is smooth, and there is another smooth area behind the lunule; now and then one finds a specimen that is entirely smooth. The color varies from dark purple to brown above, to yellowish, white, or greenish olive below. The hinge has one to three (usually two) dark purple teeth larger than the other denticles. Length, about 12 mm. The northern part of the Gulf of California to northern Peru, in the intertidal zone, among algae, and offshore in depths to 31 m.

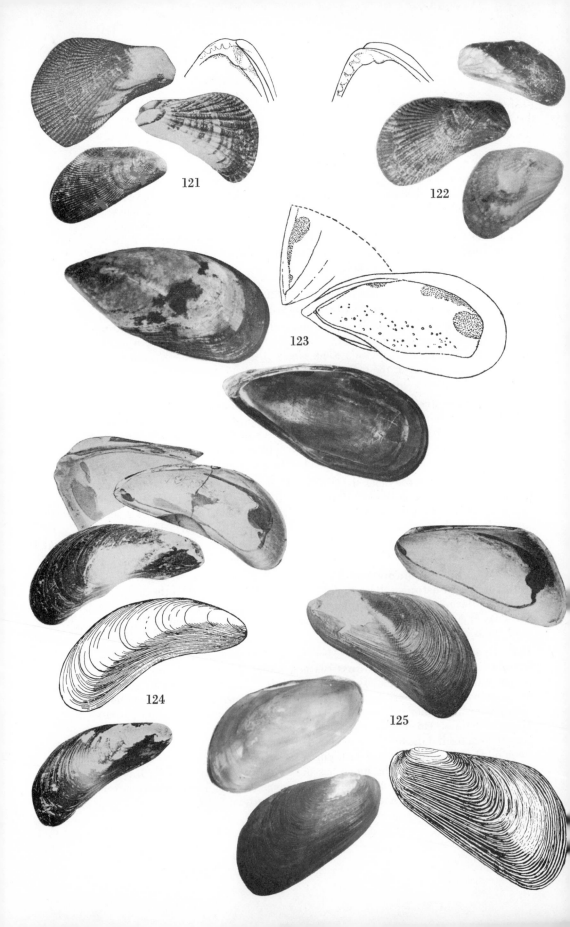

Genus CHOROMYTILUS SOOT-RYEN, 1952

Resilial ridge compact, not pitted; lunule bent inward to form one central tooth in the right valve and a groove in the left.

123. Choromytilus palliopunctatus (Carpenter, 1857). However worn the outside of the shell may be, the lustrous dark purple lining with a pattern of tiny scars like pinpricks along the ventral third makes identification of the species easy. It is one of the largest mussels of the province, measuring 68 mm in length, 31 mm in width, 24 mm in diameter. Occurrence seems to be confined to exposed-coast intertidal areas, where the mussels live fastened to the rocks by a very strong byssus. Magdalena Bay, Baja California, to Panama.

Genus MYTELLA SOOT-RYEN, 1955

Resilial ridge pitted, anterior retractor scar rounded, with another distinct scar below; color green.

124. Mytella arciformis (Dall, 1909). Although synonymized with *M. strigata* by Soot-Ryen (1955), the species is distinct according to Olsson (1961). The shell is slender and strongly arched, its ventral side flattened. The color is greenish brown under a thin periostracum. A mud-flat species more common in the southern part of its range. Length, 41 mm; width, 16.5 mm. El Salvador to Peru.

125. Mytella guyanensis (Lamarck, 1819) (Synonyms: **Modiola sinuosa** King, 1831; **Modiola mutabilis** Carpenter, 1857). In contrast to *M. strigata* this species has umbones not at the end of the shell. The color variation is similar, however, with the interior of the shell whitish, tinged with violet on the muscle scars. The anterior retractor scar is behind the umbo, the anterior adductor high up along the anterior margin. Fine concentric ridges develop on the posterior part of the older specimens. Length, 58 mm; width, 28 mm; diameter, 25 mm. This mussel is common nearly buried in muddy sand, attached to stones, from Puerto Peñasco, Mexico, to northern Peru. It has been reported from the outer coast of Baja California, but these records need confirmation. Like *M. strigata*, it occurs also in the Atlantic, from Venezuela to Brazil.

126. Mytella speciosa (Reeve, 1857). Rarest of the Mytellas, this is a thin-shelled form, more elongate than the others, with the anterior retractor scar in front of the umbo. Length, about 50 mm; width, 20 mm. It is intertidal, from Bahía Magdalena, Baja California, to northern Peru.

127. Mytella strigata (Hanley, 1843) (Synonyms: **Mytilus falcatus** Orbigny, 1846 [not Goldfuss, 1837]; **Mytilus charruanus** Orbigny, 1846). Except for color, this species resembles the common edible mussel of more northern waters. The color may vary from light green to nearly black; it may be uniform or shaded to yellowish brown at the anterior end and below, or it may even be banded in crisscross fashion; normally, the interior is dark purple. In most specimens there are distinct teeth on the hinge and a radiately sculptured lunule. This species occurs on mud flats or in shallow lagoons, like the more southern *Mytilus arciformis* Dall, which Soot-Ryen (1955) synonymized with it but which, according to Olsson (1961), is distinct. Length, 44 mm; width, 22 mm. The northernmost occurrence seems to be Guaymas, Sonora, Mexico; thence south to El Salvador and the Galápagos Islands. The type locality is Brazil, and the range along the eastern side of South America is from Venezuela to Argentina.

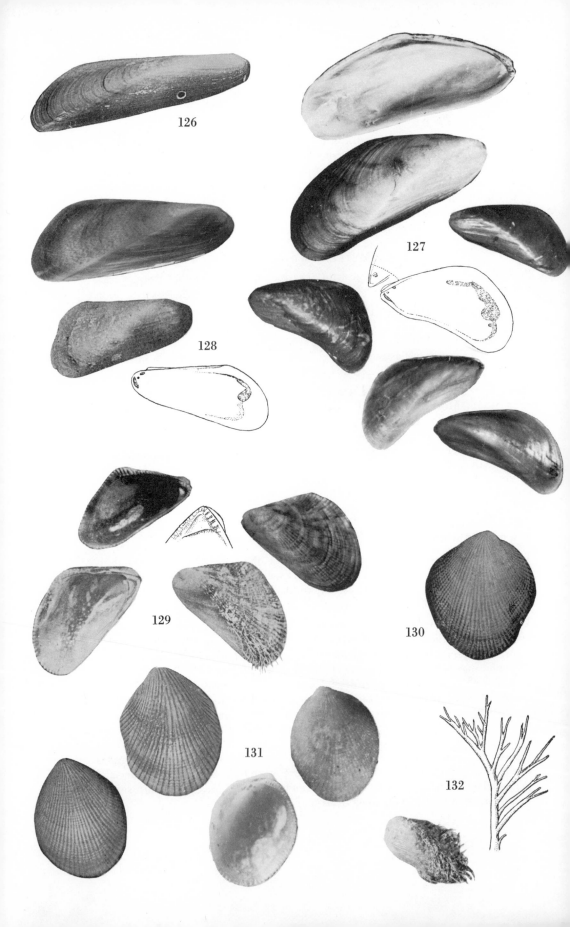

126

127

128

129

130

131

132

128. Mytella tumbezensis (Pilsbry & Olsson, 1935). Although synonymized with *M. speciosa* by Soot-Ryen (1955), this is considered by Olsson (1961) a distinct form that is shorter and wider for its length and is smaller. Length, 44 mm; width, 21 mm. It has the same geographic distribution but is more common southward, occurring on mud flats. Bahía Magdalena, Baja California, to Tumbes, Peru.

Genus SEPTIFER RÉCLUZ, 1848

The shelly deck across the inside of the tip of the valve is an unmistakable feature. Exteriorly the shell resembles *Brachidontes*.

129. Septifer zeteki Hertlein & Strong, 1946 (Synonym: **S. cumingii** of authors, probably not of Récluz, 1849, a South Pacific species). The shell is small (less than 12 mm long), greenish in color, with fine radial ribs. Living specimens are found on rocky or stony bottoms from the shore down to 90 m. Baja California to Peru.

Subfamily CRENELLINAE

Small shells, usually inflated, with radial sculpture that is commonly absent in the middle part. Four genera that occur in the Panamic province are recognized in the revised classification of Soot-Ryen (*in* Moore, 1969).

Genus CRENELLA BROWN, 1827

Small, nearly orbicular, with radiating sculpture, sunken ligament; dorsal margin finely striate.

130. Crenella caudiva Olsson, 1961. One can recognize this species by the rectangular and somewhat irregular shape of the shell, whereas the other small west-coast Crenellas are evenly ovate. Length, 2.3 mm; width, 2.2 mm. Reported only in Ecuador.

131. Crenella divaricata (Orbigny, 1846) (Synonyms: **? C. inflata** Carpenter, 1864; **C. ecuadoriana** Pilsbry & Olsson, 1941). The sculpture is fine, the interspaces narrower than the ribs, bifurcating; the shell is white. Length, 3 mm; width, 2.5 mm. Range on the west coast is from southern California through the Gulf of California and south to Peru, mainly offshore in depths of 4 to 450 m. The type locality of Orbigny's species is Cuba. Although most West American authors have not felt that Pacific specimens were distinct enough to warrant specific separation, there are some slight differences that may prove indeed to be significant. Olsson (1961) regarded *C. ecuadoriana* as being distinct, differing from the Caribbean species by being more symmetrical, with stronger sculpture. Whether this form is the same as the one named by Carpenter remains to be determined. Carpenter's name has priority, but unfortunately his type specimen, from Cape San Lucas, is lost, and a neotype would have to be found at or near that locality and formally selected as replacement; there was no illustration of Carpenter's type, and his description is not sufficient to make adequate comparison with the other nominal forms.

Genus GREGARIELLA MONTEROSATO, 1884

The periostracum is strong and hairlike, the middle part of the shell nearly smooth; beaks incurved; umbonal angle almost carinate.

132. Gregariella chenui (Récluz, 1842) (Synonym: **Modiola opifex** of authors, not Say, 1825). The small, white, cancellately sculptured shell has a yellow-

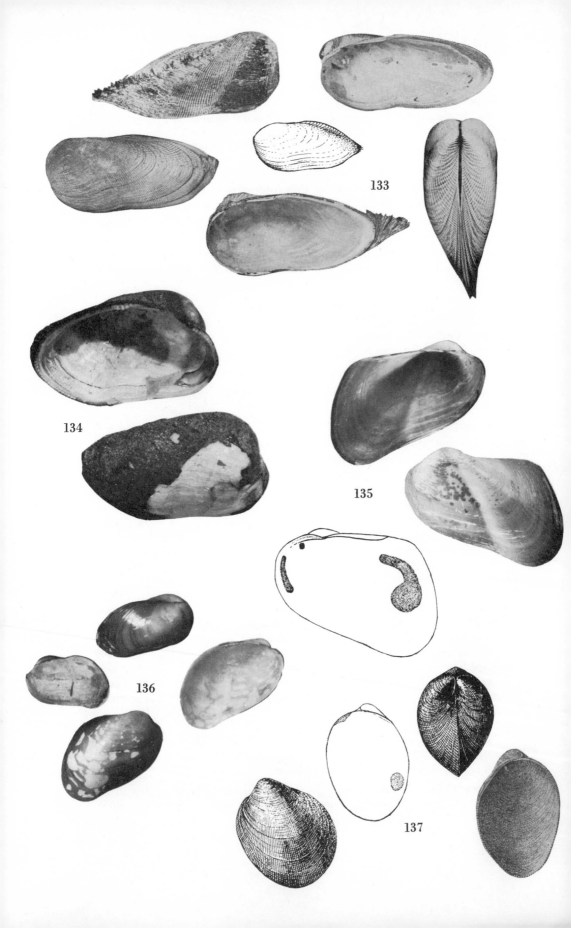

133

134

135

136

137

ish periostracum, ornamented posteriorly with long, branching hairs. Length, 6.5 mm; width, 3.5 mm. Single valves may be found in beach drift from Monterey, California, to Peru; dredging records range from 30 to 90 m. The species occurs also in the Atlantic, the type locality being Brazil.

133. Gregariella coarctata (Carpenter, 1857). Distinguished from *G. chenui* by the nearly smooth shell, strong keel, and smooth hairs on the periostracum. Length, 17 mm; width, 10 mm. Reported as burrowing into large shells. Scammon's Lagoon, Baja California, to Ecuador and the Galápagos Islands, in 4 to 16 m depth.

134. Gregariella denticulata (Dall, 1871). The bearded periostracum is brown to black, over a pearly-white shell that is tinged with purple on the umbones. A purple spot shows just behind the posterior adductor muscle scar. Length, 17.5 mm; height and diameter, 10 mm. Acapulco, Mexico, south probably to southern Peru.

Genus LIOBERUS DALL, 1898

Though superficially resembling *Modiolus*, this group has, according to Dr. Soot-Ryen, long siphons that suggest relationship to *Musculus*, a genus generally of more northern distribution.

135. Lioberus salvadoricus (Hertlein & Strong, 1946). This is a small offshore form, sculptured with raised concentric lines, the posterior part of the shell covered with a mass of mud particles matted with fine threads. Length, 24 mm; width, 14 mm; diameter, 12 mm. It was dredged from mud, sand, and shell bottoms, in 4 to 30 m of water. North of San Felipe, Baja California, and Cholla Cove, Sonora, Mexico, to Costa Rica.

136. Lioberus splendidus (Dunker, 1857). Though similar in size and texture to *L. salvadoricus*, this has a more rounded posterior dorsal slope and broader anterior end; the periostracum is smooth and shiny, a dark greenish brown in color, the posterior end free of incrustation. There are three specimens from the type lot in the British Museum, not previously illustrated. The type locality is indefinite—"California"—and the species has remained unrecognized by later workers. Length of the largest specimen, 21 mm; height, 12 mm.

Genus MEGACRENELLA HABE, 1965
(SOLAMEN IREDALE, 1924, of authors)

Distinguished by larger size, this differs also from *Crenella* by its finer sculpture, the lack of a toothlike process under the beaks, and the more marginal placement of the ligament.

137. Megacrenella columbiana (Dall, 1897) (Synonyms: **? Crenella megas** Dall, 1902; **C. rotundata** Dall, 1916). Like other Crenellas, this is an offshore form, small (maximum length, 21 mm), whitish, with mud and sand particles adhering to the surface. It has been dredged in 29 to 530 m depth, from the Aleutian Islands to southern Mexico—or to Panama, if *C. megas* is, as suspected, a synonym.

Subfamily LITHOPHAGINAE

Slender, cylindrical shells, smooth or with fine oblique striations; periostracum well developed. Mostly boring forms, eroding holes in rock or other shells. Two genera in the Panamic province.

Genus LITHOPHAGA Röding, 1798

These are cylindrical shells, tapering posteriorly, which bore into soft rock or other shells, eroding a cavity that exactly fits the shell. Several subgenera or "sections" have been proposed by authors for groups of species within the genus, chiefly on the basis of differences in the sculpturing on the posterior part of the shell and on chalky coverings or incrustations, these thicker on the posterodorsal triangle and usually projecting beyond the posterior margin of the valve. The key is to subgenera; no species of subgenus *Lithophaga, s. s.,* occur in the Panamic province.

1. Incrustation of valves not prolonged beyond posterior end. . . . . . .*Leiosolenus*
   Incrustation prolonged beyond posterior end. 2
2. Incrustation on one valve only. . . . .*Rupiphaga*
   Incrustation on both valves. 3
3. Ends of incrustation bladelike, twisted, overlapping. . . . .*Myoforceps*
   Ends of incrustation not twisted. 4
4. Incrustation roughened into a feathery pattern. . . . .*Diberus*
   Incrustation smooth to slightly granular. 5
5. Incrustation tapering to a rounded point. . . . .*Labis*
   Incrustation two-pronged, hollowed out within. . . . .*Stumpiella*

Subgenus DIBERUS Dall, 1898

With plumelike incrustations projecting beyond the valves.

138. Lithophaga (Diberus) plumula (Hanley, 1844). The incrustation is distinctly like a feather with elevated ribs, and the periostracum shows transverse wrinkles on the dorsal side. Length, 45 mm; width, 15 mm. Specimens may be found boring into *Spondylus* or other large shells or into masses of coral in shallow water down to depths of 37 m, from the Gulf of California to Peru.

Subgenus LABIS Dall, 1916

Incrustations smooth, long, ending in a median spine.

Lithophaga (Labis) attenuata (Deshayes, 1836) (Synonym: **Lithodomus inca** Orbigny, 1846). The incrustation on the posterior end is prolonged into a tube. A large specimen may measure as much as 100 mm in length. Two geographic subspecies may be distinguished:

139. Lithophaga (Labis) attenuata attenuata (Deshayes, 1836). Relatively the more slender of the two subspecies, an average specimen measuring 61 mm in length, 16 mm in height (not including the periostracal tube). Costa Rica to Chile.

140. Lithophaga (Labis) attenuata rogersi Berry, 1957. More robust than *L. a. attenuata*, this has a shorter incrustation, the valves more swollen around the flattened hinge area, with a decided angle near the middle of the dorsal margin. Length, 63 mm; height, 23 mm. Southern California through the Gulf to southern Mexico; boring into rocks or corals in shallow water, to a depth of 80 m.

Subgenus LEIOSOLENUS Carpenter, 1857

Incrustations thin, diffused, without projecting parts.

141. Lithophaga (Leiosolenus) hancocki Soot-Ryen, 1955. Here the incrustation is loose, thick, but not projecting beyond the edges of the shell, with a divaricate pattern on the posterodorsal triangle. The shell itself has a light yellowish-

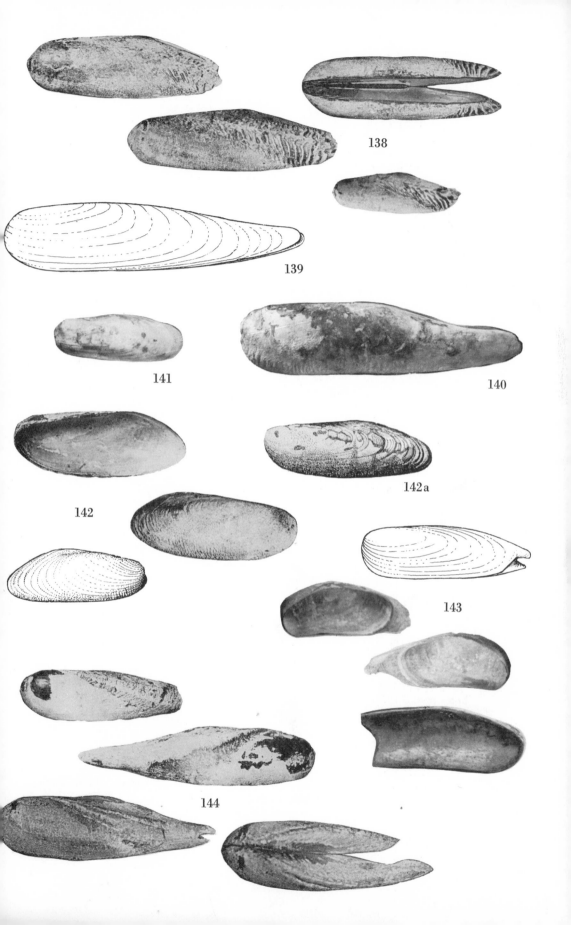

138

139

141

140

142

142a

143

144

brown periostracum. Length, about 24 mm. Panama to the Galápagos Islands, in depths to 24 m; boring in masses of coral.

142. Lithophaga (Leiosolenus) spatiosa (Carpenter, 1857) (Synonyms: **? Lithophagus rugiferus** Carpenter, 1857, *ex* Dunker, MS [invalid name, proposed only in synonymy]; **? L. abbotti** Lowe, 1935). This is the largest of the Panamic Lithophagas if *L. abbotti* Lowe, 1935 (fig. 142a), is a synonym, with lengths to 62 mm. The incrustation is thin, equal over all the surface, and tends to be arranged in rows of pustules. Boring in valves of *Pinctada* and *Ostrea* from shore down to 27 m. San Felipe, Gulf of California, to Ecuador.

Subgenus MYOFORCEPS FISCHER, 1886

Projecting posterior parts of incrustation crossed.

143. Lithophaga (Myoforceps) aristata (Dillwyn, 1817) (Synonyms: **Modiola caudigera** Lamarck, 1819; **L. caudatus** Gray, 1827; **Lithophagus aristatus gracilior** and **L. a. tumidior** Carpenter, 1857; **Dactylus carpenteri** Mörch, 1861). The ends of the incrustation are twisted around each other and look like bent scissors blades. Length, about 25 mm; width, 10 mm. Most specimens are to be found boring into other mollusks rather than into rock, in the intertidal area and down to depths of as much as 300 m. The species is widely distributed, from its type locality in west Africa to the West Indies, the Mediterranean, Red Sea, Australia, Japan, and from southern California to Peru.

Subgenus RUPIPHAGA OLSSON, 1961

Incrustation on only one valve.

144. Lithophaga (Rupiphaga) hastasia Olsson, 1961. The incrustation forms a thickened rib that may be prolonged as a one- or two-pronged blade beyond the end of the shell. Length, 38 mm; width, 9 mm. Panama to Ecuador.

Subgenus STUMPIELLA SOOT-RYEN, 1955

Incrustation closed behind but with dorsal and ventral openings.

145. Lithophaga (Stumpiella) calyculata (Carpenter, 1857). The incrustation has two ridges and protrudes posteriorly, where it is closed except for two narrow openings connected within. Length, about 16 mm; width, 5 mm. Guaymas, Gulf of California, to the Galápagos Islands; boring in old shells.

Genus ADULA H. & A. ADAMS, 1857

Although formerly considered a subgenus of *Botula*, *Adula* is actually not closely related, for the shape of the umbones and the placement of the anterior retractor muscles on the thickened margin show it to be closer to *Lithophaga*. With the exception of the first species discussed here, the Adulas are borers, excavating holes in soft rock.

146. Adula diegensis (Dall, 1911). Authors have had difficulty assigning this species to a genus. It was first thought to be a *Modiolus* on account of its free-living habit, on mud flats or piling, but the soft parts demonstrate its true relationship. Length, 42 mm; width, 18 mm. Oregon to Sonora, Mexico.

Adula soleniformis (Orbigny, 1842). This is the type species of the genus, divisible into two geographic subspecies:

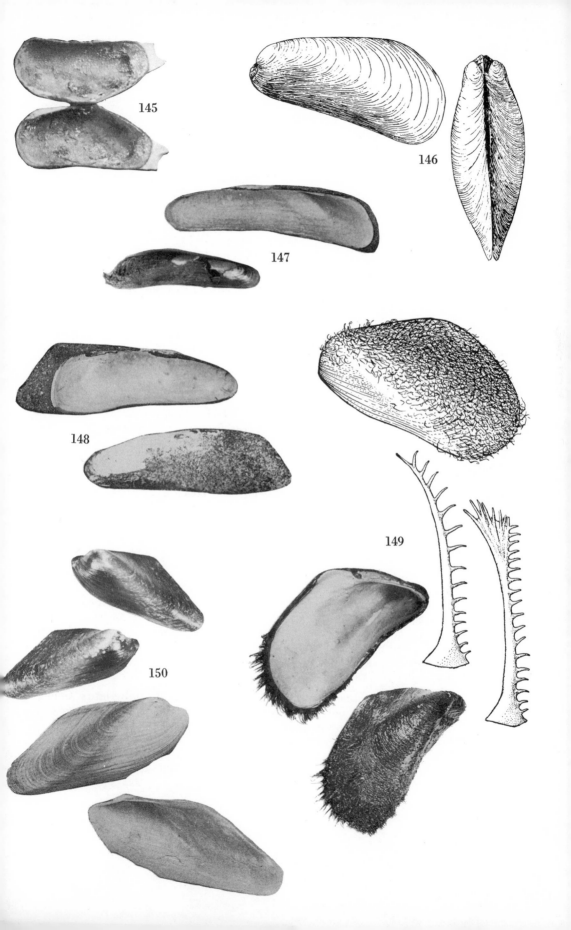

147. Adula soleniformis soleniformis. Specimens of the typical subspecies, from Ecuador and northern Peru, attain a length of 39 mm; height, 10 mm.

148. Adula soleniformis panamensis Olsson, 1961. Compared with the typical southern form, this is smaller, with a maximum length of 28 mm; height, 9.5 mm, the valves wider posteriorly. It bores into soft rock in Panama.

Subfamily MODIOLINAE

Lunular area smooth; anterior margin without teeth or crenulations; shell smooth or with irregular sculpture. Three genera in the Panamic province.

Genus MODIOLUS LAMARCK, 1799
(VOLSELLA SCOPOLI, 1777)

Lamarck's name *Modiolus* was conserved by the International Commission on Zoological Nomenclature (1955), mainly because of its wide usage in nineteenth-century literature, although *Volsella* has priority.

The classical distinction between the true mussels, *Mytilus*, and the horse mussels, *Modiolus*, is in the placement of the beaks—whether at the anterior tip of the shell or some little distance back. This is still a useful distinction but not completely reliable. However, all the species here grouped in *Modiolus* have the beaks well removed from the end of the shell.

149. Modiolus capax (Conrad, 1837) (Synonyms: **Modiola spatula** Menke, 1849; **Modiola subfuscata** Clessin, 1889). The bright orange-brown color of the shell, its tumid shape, and the serrate periostracal hairs all serve to mark this species. Length, 81 mm; width, 40 mm; diameter, 36 mm. Specimens may be found intertidally on rocks or boulders or dredged in mud to 46 m. Santa Cruz, California, to Paita, Peru.

150. Modiolus eiseni Strong & Hertlein, 1937. This is a deepwater form, rather small, clearly triangular in shape, with a sharp angle on the posterodorsal margin. Length, 29 mm; width, 13 mm; diameter, 12 mm. Outer Gorda Bank and Guaymas, Gulf of California, to Ecuador, on mud, sand, or shell bottoms, in depths of 4 to 360 m.

151. Modiolus pseudotulipus Olsson, 1961 (Synonym: **Modiolus americanus** Leach of authors, not Leach, 1815). Although Soot-Ryen (1955) had considered that the eastern Pacific and western Atlantic shells were indistinguishable, Olsson recommends separation on the basis of the greater posterior length of the Pacific form and its more subdued coloring. The shell is thin, with a light brown periostracum to which sand grains adhere in a mat, the periostracal hairs being broad and unbranched. The byssus is yellow and silklike. Length, 65 mm; width, 38 mm; diameter, 14 mm. Bahía Magdalena, Baja California, to Peru.

152. Modiolus rectus (Conrad, 1837) (Synonym: **M. flabellatus** Gould, 1850). The longest and most slender of the west-coast horse mussels, this form can be recognized not only by its shape but by its bluish-white shell covered by a yellowish-brown periostracum. Length, 118 mm; width, 40 mm; diameter, 33 mm. The animals live burrowing in mud, with the posterior margin above the surface, from the intertidal zone down to 45 m, from Vancouver Island, British Columbia, to Concepcion Bay, Gulf of California.

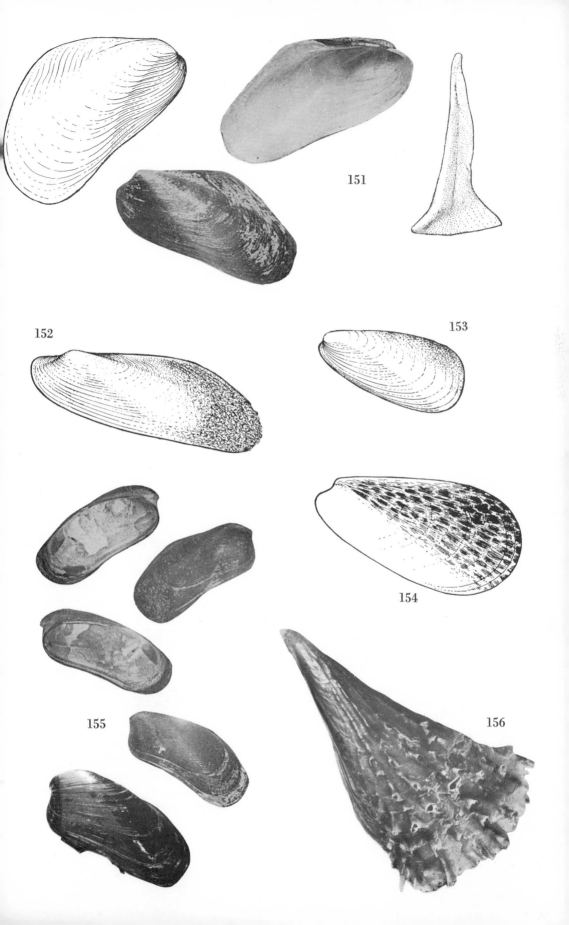

151

152

153

154

155

156

Genus AMYGDALUM MEGERLE VON MÜHLFELD, 1811

These are thin, flat shells, somewhat resembling *Modiolus*, but with a shiny periostracum. Instead of attaching itself, the animal builds a "nest" of mud and sand particles held together by fine byssal threads.

153. Amygdalum americanum Soot-Ryen, 1955. A small, offshore form, the yellowish-white shell marked with grayish-brown triangular spots on its posterior dorsal part. Length, 21 mm; width, 10 mm; diameter, 6 mm. Guaymas, Mexico, to Paita, Peru, in 4 to 37 m.

154. Amygdalum pallidulum (Dall, 1916). In this shell the anterior and central part is an opaque grayish white, with yellow stains; the dorsal and posterior areas are translucent, marked with crisscross stripes of white. Length, 21 mm; width, 15 mm; diameter, 10 mm. Specimens have been taken in shore collecting, though mostly they must be dredged, to 380 m depth, from northern California to Colombia. A similar species in the Atlantic is *A. sagittatum* (Rehder, 1943).

Genus BOTULA MÖRCH, 1853

The cylindrical shells have incurved or subspiral umbones, and the dorsal margin above the ligament is finely striate. The anterior retractor muscle scar is marginal, just below the umbones.

155. Botula cylista Berry, 1959. A dark brown to black shell with a smooth periostracum. This has been identified by authors as *B. fusca* (Gmelin, 1791) because of its similarity to that West Indian form, from which it differs by being higher for its length, with a somewhat more arched dorsal margin. Length, 26 mm; width, 12 mm; diameter, 14 mm. Mazatlán, Mexico, to Manta, Ecuador (rare).

Superfamily PINNACEA

Elongate, partially nacreous; muscle scars two, well separated.

Family PINNIDAE

The Pinnas or pen shells are slender, fragile, and so brittle that in life they must be protected. Accordingly, the animal that forms this shell buries itself in the mud of a quiet bay, with the pointed end of the shell down, anchored there by a tuft of fine fibers secreted by glands of the foot. The gaping posterior end of the shell protrudes just above the surface, where, at a very low tide, the observant collector may discover specimens. The fibrous anchor is called a byssus, a term that has an odd history. Aristotle, the first great Greek naturalist, noticed Pinnas in the Mediterranean and described their habit of life—how they were anchored below, in the depth, as it were, of the bay floor. The word he used—byssus—comes to us in another derivation: abyssal, meaning "without bottom" or "very deep." A Dutch researcher, P. van der Feen, has traced the sequel to Aristotle's observation. When, in the Middle Ages, a translation was made of Aristotle's work, the translator, being no naturalist, missed the point and decided that Aristotle had said the shell was anchored by a "byssus." Thus the Greek word for depth came to refer to a molluscan structure! The fine golden-brown byssal fibers of the Mediterranean Pinna were used to weave the legendary cloth-of-gold, a fabric that must have rivaled our finest synthetics, for a woman's scarf of this material was said to be so flexible it could be rolled into a ball the size of a walnut. So much work was required, however, to harvest enough of the Pinnas to make any significant amount of this fiber, and to process the fiber for weaving, that the cloth could be afforded only by royalty. Two Panamic genera.

Genus PINNA LINNAEUS, 1758

The shell is triangular in shape, with a groove down the middle of the inside that divides the nacreous portion into two lobes.

156. Pinna rugosa Sowerby, 1835. Spinose on the outside, the shells have about 8 rows of somewhat tubular spines, though in old specimens these may be worn down or almost obsolete. An unusually large specimen in the Stanford collection measures 59 cm (23 inches) in length. The more usual size is about 190 mm in length, 90 mm maximum width. Southern Baja California through the southern end of the Gulf of California to Guaymas, Mexico, and south to Panama, on mud bars.

Genus ATRINA GRAY, 1847

Shaped rather like a ham, these shells have no groove down the inside, and the nacreous area is undivided.

157. Atrina maura (Sowerby, 1835) (Synonym: **Pinna lanceolata** G. B. Sowerby, 1835 [not J. Sowerby, 1821]). The shell is somewhat flattened, with thin or slender spines in about 18 rows. Like all other Atrinas, the shell has a blackish or dark brown color. The flesh is used as food by Mexicans, and Lowe, who reported great heaps of the shells at Mazatlán in 1930, says that the large white muscle tastes like that of the Atlantic coast's giant scallop. Length, about 225 mm; width, 125 mm; diameter, 50 mm. Baja California to Peru, on offshore mud flats.

158. Atrina oldroydii Dall, 1901. This large grayish-brown Californian form has been reported from the outer coast of Baja California as far south as Bahía Magdalena. It is more trigonal, with finer sculpture than the Panamic species. Length, 237 mm; height, 135 mm; diameter, 68 mm.

159. Atrina texta Hertlein, Hanna & Strong, 1943. The shell is trigonal, more symmetrical than *A. oldroydii*, which it most resembles. The posterior end is wider and squarer than in *A. maura*. There are 26 radial rows of spines. The color is pale brown, darker on the beaks and around the muscle scar. Length, 140 mm; width, 80 mm; diameter, 35 mm. Gorda Bank off the tip of Baja California, depth 125 m, to the Galápagos Islands.

160. Atrina tuberculosa (Sowerby, 1835). More strongly convex than *A. maura*, the shell of this species has short and thick spines. Length, 200 mm; width, 125 mm; diameter, 37 mm. Gulf of California to Panama.

Order PTERIOIDA
(PTEROCONCHIDA)

Generally inequivalve and somewhat inequilateral; ligament variously situated in one or more sockets, usually extended by a secondary fusion layer; shell material prismatic-nacreous to foliate; pallial line entire.

Superfamily PTERIACEA

Right valve less convex than left valve; ligament external; shells anchored throughout life by a byssus; right valve usually with a byssal notch; pallial line anteriorly discontinuous.

Family PTERIIDAE

Shell usually with a triangular winglike projection at each end of straight hinge line, the anterior wing or auricle smaller, with a byssal notch below it in right valve; juvenile stages with two muscle scars, anterior adductor reduced in adult, posterior impression large, nearly central; interior of valves pearly; sculpture variable.

Pearls, the only gem of animal origin, are formed by members of this bivalve family, known as pearl oysters. They are only distantly related to the true oysters but, like them, are of great commercial importance—the true oysters as food, the pearl oysters as producers of pearls, iridescent ornaments, and buttons. True pearls form (as the Japanese culture-pearl industry has demonstrated) when a bit of the living outer tissue or mantle, which secretes the nacreous lining of the shell, is torn loose and lodged inside the oyster, where the mantle cells can continue to grow and to coat with nacre any irritant, such as a grain of sand, that may have entered at the time of injury. A pearl so formed is round and symmetrical—and rare. Nacre deposited over an irritant remaining next to the shell forms an unsymmetrical or blister pearl, commercially usable in making jewelry. Since mother-of-pearl buttons can be cut from the thick shells of many pearl oysters, the diver who recovers no pearls may yet have some return for his effort.

There was a thriving pearl-fishing industry in the Gulf of California, near La Paz, in the last century, but overintensive collecting and the development of richer fields elsewhere eventually made these operations unprofitable.

Two genera in the Panamic province.

Genus PTERIA SCOPOLI, 1777
(AVICULA BRUGUIÈRE, 1792)

The word "pteria" means wing in Greek. These shells have a winglike extension of the hinge margin, set off from the disk of the shell by a sinus or bend.

161. Pteria sterna (Gould, 1851) (Synonyms: **Avicula fimbriata** Dunker, 1852; **A. peruviana** Reeve, 1857; **A. vivesi** Rochebrune, 1895; **P. viridizona** Dall, 1916; **P. beiliana** Olsson, 1961). More than one species may be represented within the complex usually labeled *P. sterna*, and it is possible that one or more of the cited synonyms will prove valid. Dunker's type, supposed to be from western Central America, has a large anterior ear and a very short posterior wing; Reeve's species was described as having a large ear and a moderately long wing; Rochebrune's type has not been illustrated and may be lost; the shell described by Olsson from Panama, with an abnormally large anterior ear, is similar in outline to one figured by Carpenter from Mazatlán, Mexico, that he regarded as an aberrant young *P. sterna*. According to Dr. S. S. Berry, a distinct form that is somewhat short-winged occurs in clusters on mud flats. Obviously, more work is needed to interpret this species. The thin, brittle shell is dark brown outside and bluish nacreous within, under a shaggy periostracum. Length, 100 mm; height, 85 mm. In the Gulf of California, where the species was formerly abundant, the numbers have been greatly reduced by pearl fishing. Baja California through the Gulf of California and south to Peru, in shallow water offshore.

Genus PINCTADA RÖDING, 1798
(MARGARITIPHORA MEGERLE VON MÜHLFELD, 1811;
MELEAGRINA LAMARCK, 1819)

No winglike extension on hinge, the anterior margin smoothly rounded.

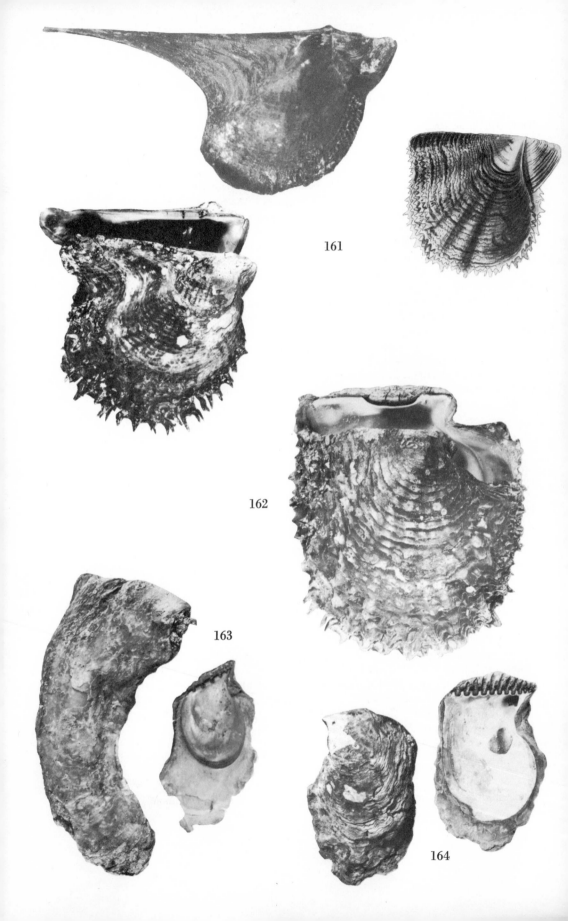

161

162

163

164

162. Pinctada mazatlanica (Hanley, 1856) (Synonym: **Avicula barbata** Reeve, 1857). The shell is heavier than that of *Pteria*, also dark-colored, with a very brittle periostracum that must be kept well oiled to prevent it from shredding away. Length, 100 to 125 mm. Like *Pteria*, this species has been depleted in the Gulf of California area by pearl fishing. The outer coast of Baja California, through the Gulf of California and south to Peru, in shallow water offshore.

Family ISOGNOMONIDAE
(ISOGNOMONTIDAE of authors)

Hinge with multiple ligamental grooves but without hinge teeth; left valve more convex than right in inequivalve forms; adult shell with a single central muscle scar; pallial line discontinuous, broken up into small pits; surface of shell smooth to irregularly sculptured; interior layer nacreous.

Genus ISOGNOMON [LIGHTFOOT], 1786
(PERNA Bruguière, 1792 [not PHILIPSSON, 1788]; MELINA PHILIPSSON, 1788; PEDALION of authors)

Smaller and more inequilateral than *Pteria* or *Pinctada*, which it resembles, this has the ligament divided and lodged in a series of pits or grooves.

The type species of *Isognomon, Ostrea perna* Linnaeus, 1758, presents problems to systematists, who cannot be sure of its identity; it seems in any case to be close to if not identical with the type of *Melina*, and modern practice seems to be to accept the type of *Melina* as representing *Isognomon*. The name *Melina* often is attributed to Retzius, 1788, but Retzius was merely Philipsson's supervisor or examiner. It was Philipsson who wrote and published the dissertation in which the name was proposed. The name *Pedalion*, used by many authors for this group, is invalid because it was proposed as a vernacular or common name and was not latinized until after other generic names had become available.

163. Isognomon janus Carpenter, 1857 (Synonym: **? Perna anomioides** Reeve, 1858). The principal distinction that Carpenter noticed in this shell was that there are fewer pits along the hinge margin—as many as 10 to 12 in *I. recognitus*, only 7 to at most 9 in *I. janus*. The color is lighter and more uniform, a tan or buffy brown, without the gray or purple of the other species. The shell tends to be higher for the length or more tongue-shaped. Radial sculpture may be more pronounced in *I. janus*, and concentric lamellae are lacking. *Perna anomioides* Reeve was described as from "California," but the present label on the type specimen in the British Museum says "Torres Straits" (i.e. northern Australia). Reeve's figure does resemble *I. janus* in outline. Length, 30 mm; height, 80 mm; diameter, 7 mm. San Ignacio Lagoon, Baja California, to Oaxaca, Mexico.

164. Isognomon recognitus (Mabille, 1895) (Synonym: **Perna chemnitziana** Orbigny of authors, not of Orbigny, 1846). A name change is unavoidable for the so-called Western Tree Oyster. It has long been known as *I. chemnitzianus* and considered identical with a Caribbean species. That species now proves to be a synonym of a prior Caribbean form, *I. bicolor* (C. B. Adams, 1845). As Dr. J. P. E. Morrison of the U.S. National Museum confirms (*in litt.*), there are small but consistent differences between the eastern Pacific and the western Atlantic species; therefore it seems better to use for the Pacific form a name based on Pacific material. The shell varies in shape from squarish to trapezoidal or rectangular; in color from grayish yellow to brown or purplish; and in sculpture from smooth to scaly.

The nacreous area within is set off by a slight ridge. The number of ligamental pits varies from 6 to 12. Length, 25 mm; height, 46 mm; diameter, 14 mm. Attached to rocks or other shells, intertidally or in shallow water from northern Baja California to Chile. The Pacific species averages slightly larger than its Atlantic twin, *I. bicolor*, and the ridge bounding the pallial area is less marked. The name *Perna quadrangularis* Reeve, 1848, sometimes cited as a synonym, seems to apply rather to a closely related Japanese form.

Family MALLEIDAE

Shell nearly equivalve; valve margins often gaping; ligamental area triangular, ligament in a central pit; adults monomyarian (one muscle scar); interior of shell nacreous. Two genera in the Panamic province.

Genus MALLEUS LAMARCK, 1799

Hinge line long, as in the Pteriidae, but ligament compressed in a single central pit. This is the hammer oyster of the western Pacific. No species of *Malleus, s. s.,* occur in the Panamic province.

Subgenus MALVUFUNDUS DE GREGORIO, 1885
(FUNDELLA DE GREGORIO, 1885 [not ZELLER, 1848]; PARIMALLEUS IREDALE, 1931; BREVIMALLEUS McLEAN, 1947)

Hinge line short, posterior wing blunt, anterior wing short or wanting, byssal notch a broad sinus; earlier growth stages with irregular concentric lamellae. The name *Parimalleus*, by which the group has been known in recent years, falls as a synonym of the earlier *Malvufundus*, owing to a type designation by Cox (in Moore, 1969).

165. Malleus (Malvufundus) rufipunctatus (Reeve, 1858) (Synonyms: **M. aquatilis** and **vesiculatus** Reeve, 1858; **M. panamensis** Mörch, 1861; **M. obvolutus** Folin, 1867). The waxy-white to brownish shell is irregular in outline, tending to be somewhat quadrate; the edges are thin. Maximum length, about 25 mm; maximum height, 35 to 50 mm; maximum diameter, 18 mm. Boss and Moore (1967) regard this as a synonym of the Caribbean *M. candeanus* (Orbigny, 1842) rather than a twin species. Mazatlán, Mexico, to Panama, attached to other shells, such as *Ostrea*, or commensal with sponges.

Genus VULSELLA RÖDING, 1798

Small shells with a ligament in a central resilifer and a short hinge line.

166. Vulsella pacifica Dall, 1916. This small, dark purple shell has a white margin and a few fine radial striations crossing some faint concentric ridges. Length, 9 mm; height, 11 mm; diameter, about 6 mm. It was collected in Nicaragua, but, since no further specimens have been reported, possibly it is not truly a West American form. The genus seems otherwise confined to the western Pacific.

Superfamily OSTREACEA

Sturdy shells, with porcelaneous texture, more or less equilateral, variously sculptured; hinge having a central ligament in a broad pit; adults with a single central muscle scar; interior lustrous in some, not nacreous; valves attached to substrate by cementation.

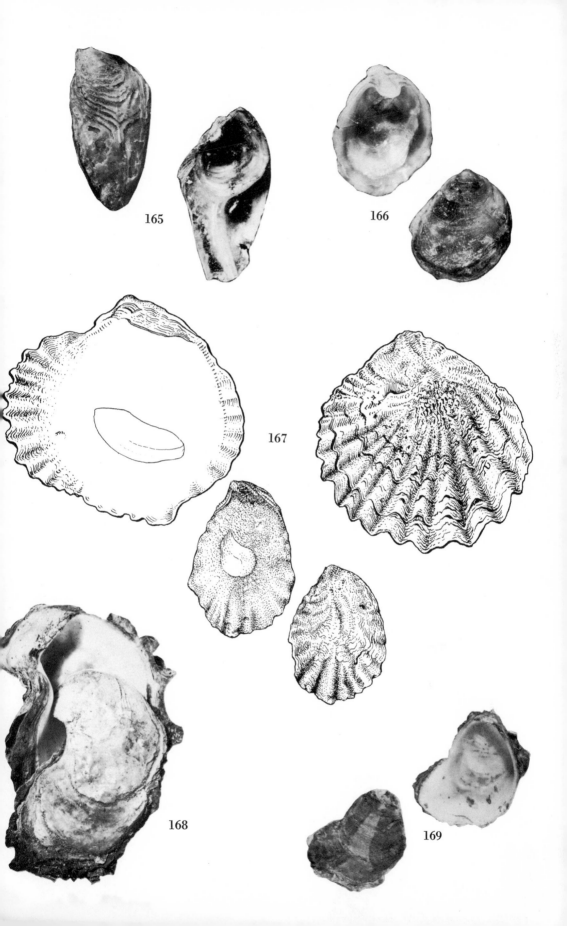

Family OSTREIDAE

Left (lower) valve generally larger and deeper; right or upper valve often nearly flat; hinge line without ridges or crura (singular, *crus*); shell margin with or without fine denticles. Radial ribbing irregular, tending to divide with growth.

Genus OSTREA LINNAEUS, 1758

With the characters of the family. Color and shape of the ligamentary area, size, color, and placement of the adductor muscle scar, and presence or absence of marginal denticles are useful features in recognition of species.

Oysters have been relished as food since prehistoric times. Although in popular lore they are associated with the formation of pearls, this is a myth, for the only pearl an *Ostrea* could produce would be porcelaneous like the interior of the shell. Oysters do not make the most attractive of cabinet specimens, for the color patterns are subdued, the surface texture rather coarse, and the sessile habit results in crowding and distortion. Identification of species is not always easy because of the variations of form induced by their manner of living. No attempt at subdivision of the genus *Ostrea* is made here, although some subgeneric names have been proposed; one such name is pointed out for an obvious candidate species, but for the most part proper subdivision awaits study by a specialist.

167. Ostrea angelica Rochebrune, 1895 (Synonyms: "**O. cumingiana** Dunker" of authors, not of Dunker, 1846; "**O. veatchi** Gabb" of authors, probably not of Gabb, 1866). The white shell has a number of strong radial plications extending to the margin, with a row of denticles near the hinge. The interior is tinged with green. The size is from 50 to 100 mm in length. Gulf of California to Ecuador.

168. Ostrea columbiensis Hanley, 1846 (Synonyms: **O. aequatorialis** Orbigny, 1846; **O. ochracea** Sowerby in Reeve, 1871). The white shell is about 75 mm long, with a wavy purple margin that is smooth inside. The adductor scar is kidney-shaped and purple in color. The upper valve may be rayed with purple or yellow on the outside. These oysters, which adhere to rocks or mangroves, are used for food, especially in Peru. San Bartolomé (Turtle) Bay, Baja California, to Chile; common in mangrove swamps.

169. Ostrea conchaphila Carpenter, 1857 (Probably the "**O. multistriata** Hanley" of authors, a preoccupied name). Typically, the shell is thin, nearly circular, and flat. The outside is radially striped with dark brown, purple, or even orange; the inside is white or greenish, with denticles along the margin for about a third of the distance from the hinge to the ventral edge. The ligament area is narrow. Length, 25 to 50 mm. Baja California to Panama, intertidally to a depth of at least 25 m.

170. Ostrea corteziensis Hertlein, 1951. Elongate, somewhat triangular, with a wide ligamentary area, rather flat. The surface shows faint radial furrows. Within, the shell is white, the margin smooth. This species, which forms one of the food oysters of the Gulf, was long known as *O. chilensis* Philippi, 1844, a misidentification; for Chilean specimens show a rounded shape and denticles along the hinge margin. An average shell is 150 mm high by 100 mm broad, but a size of 250 mm or more is not unusual. Head of the Gulf of California to Panama.

171. Ostrea fisheri Dall, 1914 (Synonyms: **O. turbinata** of authors, not of Lamarck, 1819; **O. jacobaea** Rochebrune, 1895 [not Linnaeus, 1758]). A nearly circular, flattish shell; the upper valve is roundly arched and the ligament small

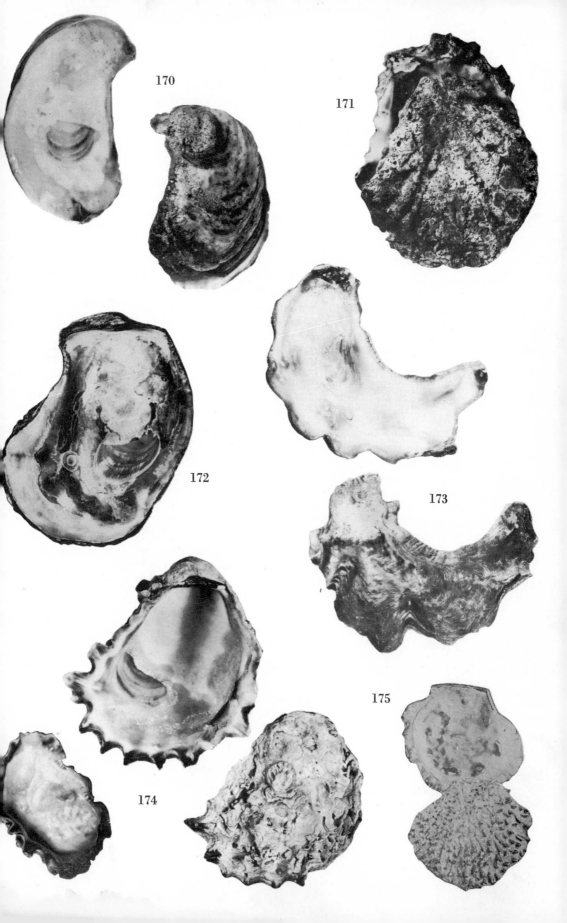

for the size of the shell, which may be 175 mm across. Within, it may be white or light in color but mostly is brownish to blackish purple. The smooth margin may have a few small folds. The southern part of the Gulf of California to Ecuador and the Galápagos Islands; common only in the Gulf area.

Ostrea hyotis Linnaeus, 1758. An Indo-Pacific species, reported living offshore on Clipperton Island.

172. Ostrea iridescens Hanley, 1854 (Synonyms: **O. spathulata** of authors, not of Lamarck, 1819; **O. lucasiana** and **turturina** Rochebrune, 1895). The rectangular form and the wide hinge are enough to distinguish this form; near the hinge it also has a row of large denticles that fit into sockets in the other valve, and, most striking of all, it has a brownish metallic luster in the interior. An average specimen is 75 to 100 mm, a large one 125 to 150 mm in length. On rocks exposed between tides from La Paz, Gulf of California, to northern Peru.

173. Ostrea (Lopha) megodon Hanley, 1846. Oysters in which the margin of the shell is folded into a few large projections that fit against the similar large saw-toothed projections of the other valve are placed in the subgenus *Lopha* Röding, 1798. On *O. megodon* there are four or five of these rounded plications, and the shell itself is curved in a wide arc. The exterior is dark-colored, the interior white with margins greenish and denticulate along about half their length. Maximum dimensions, 75 mm. Shells may occur in shallow water or be dredged offshore to 110 m, from Scammon's Lagoon, Baja California, to Paita, Peru.

174. Ostrea palmula Carpenter, 1857 (Synonyms: **O. amara** Carpenter, 1863; **O. mexicana** Sowerby in Reeve, 1871; **O. serra** Dall, 1914 [not Lamarck, 1819]; **O. dalli** Lamy, 1930). One of the most variable of the Panamic province oysters, this may be recognized by the flat or even concave upper valve that fits down into the plicate margins of the cup-shaped lower valve and by a dark-colored border. The shell color varies from green to purplish blue and the size from 50 to 75 mm in maximum dimensions. San Ignacio Lagoon, Baja California, and through the Gulf of California to Ecuador and the Galápagos Islands; attached to mangrove roots or to rocks, especially on reefs exposed to surf, in depths to 7 m.

175. Ostrea tubulifera Dall, 1914. A nearly circular, somewhat flat shell, white within, darker outside, with the color showing through in irregular blotches, the exterior surface ornamented with tubular spines of dark reddish purple color. Diameter, 45 mm. Olsson (1961) has figured Dall's holotype, which was from the "Gulf of California," and he records a specimen from Panama collected by Morrison. The species bears a remarkable resemblance to one in the western Pacific, *O. echinata* Quoy & Gaimard, 1835. Gulf of California to Panama.

Superfamily PECTINACEA

Adult shells subequilateral, nearly circular, with winglike extensions of the hinge line (auricles); monomyarian, with a single central adductor scar; with a byssal notch below the right ear, right valve underneath when at rest; sculpture mostly radial; interior porcelaneous; ligamental area with a central pit bounded by calcareous ridges or crura.

Family PECTINIDAE

Valves usually brightly colored; mantle margin with short filaments and light-sensitive organs called eyespots or *ocelli*.

Commonly known as scallops, the numerous members of this family have a wide distribution in both warm and cold waters. Some occur only in deep water. Pectens may attach temporarily to the bottom by a byssus, but most are free-swimming and can move about by clapping the valves together, propelling themselves by a jet of water forced out near the auricles. They have long been favorites of collectors, and their simple beauty of form has inspired many a work of art and architecture. The group is more abundant in the Atlantic than in the eastern Pacific. Only about five species are at all common in the Panamic province, and none is of commercial importance here.

The key is adapted from Hertlein & Strong (1946); it is to genera and subgenera, omitting *Propeamussium*, which in the Panamic province is restricted to very deep water.

1. Right valve arched; left valve flat or nearly so........................ 2
   Both valves convex... 3
2. Right valve only moderately arched........................*Flabellipecten*
   Right valve arched, strongly convex....................*Oppenheimopecten*
3. Ribs strong, corrugating the shell; shell sturdy......................... 4
   Ribs delicate, not corrugating the shell; shell very thin................. 9
4. Ribs and interspaces strongly and radially striate...................... 5
   Ribs and interspaces without strong radial striae...................... 6
5. Shell large (more than 50 mm in length); ribs 9 to 12...........*Lyropecten*
   Shell less than 50 mm in length; ribs 5 to 7................*Pseudamussium*
6. Ears unequal in length, the right ear markedly longer.............*Chlamys*
   Ears about equal in length... 7
7. Shell relatively thin, valves flattened....................*Leptopecten, s.s.*
   Shell relatively thick, valves convex................................. 8
8. Ribs rounded in section, more than 20 in number..............*Argopecten*
   Ribs triangular in section, fewer than 15.....................*Pacipecten*
9. Right valve with concentric, left valve with radial sculpture......*Cyclopecten*
   Both valves with radial or reticulate sculpture...............*Delectopecten*

Genus PECTEN MÜLLER, 1776

Right valve slightly to strongly arched, left valve flat to concave. Ears subequal in size and shape. Two Panamic subgenera; no species of *Pecten, s. s.*, occur in the Panamic province.

Subgenus FLABELLIPECTEN SACCO, 1897

Right valve slightly to moderately arched, ribs numerous, with interspaces narrower than ribs.

176. Pecten (Flabellipecten) lunaris Berry, 1963. Similar to *P. sericeus* but more solid, with heavier ribbing in right valve and slightly convex left valve, and brighter coloring. Length, 70 mm; height, 64 mm; diameter, 16 mm. Off Sonora coast of Mexico, depths 55 to 82 m.

177. Pecten (Flabellipecten) sericeus Hinds, 1845. Ribs number 22 to 23, are rather low and broad, triangular in section. The right valve is a little less brightly colored than the left, in shades of brown to pinkish. A large specimen measures 60 mm in height (beak to base). Angel de la Guarda Island, Gulf of California, to Caleta la Cruz, Peru (03° 38′ S), also the Galápagos Islands; offshore in depths of 13 to 155 m.

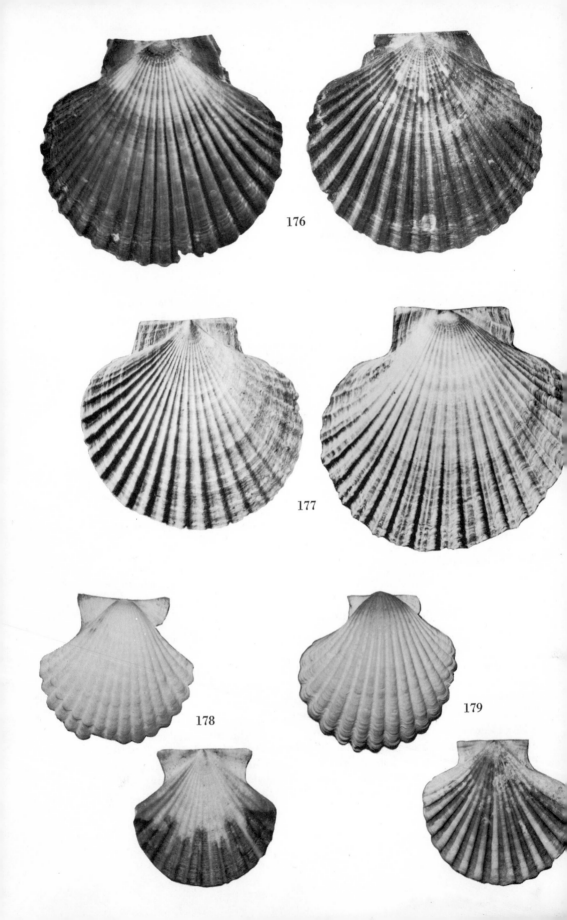

176

177

178

179

Subgenus **OPPENHEIMOPECTEN** VON TEPPNER, 1922

Right valve strongly arched, left valve flat to slightly concave.

178. Pecten (Oppenheimopecten) galapagensis Grau, 1959. Right valve with 15 ribs, left with 14; right valve white, interior pink. Length, 49 mm; height, 44 mm; diameter, 15 mm. Galápagos Islands, in 18 to 274 m.

179. Pecten (Oppenheimopecten) hancocki Grau, 1959. Right valve with 16 to 17 ribs, left with 15 to 16, shell white throughout. Length, 49 mm; height, 47 mm; diameter, 17 mm. Cocos Island, off Costa Rica, in 86 m.

180. Pecten (Oppenheimopecten) perulus Olsson, 1961. Right valve with 22 ribs, left with 17 to 18. Shell wine-red, lighter or even white on the umbones. Length, 35 mm; height, 32 mm; diameter, 11 mm. Panama to northern Peru.

181. Pecten (Oppenheimopecten) vogdesi Arnold, 1906 (Synonyms: **P. dentatus** Sowerby, 1835 [not Sowerby, 1829]; **P. excavatus** of authors, not of Anton, 1839; **P. cataractes** Dall, 1914). Right valve with about 21 low, rounded ribs, left with about 20. Color varying from buff to bright reddish brown, brighter on left valve; interior white to pinkish brown. Length and height about 100 mm. Punta Eugenia, Baja California, through the Gulf and south at least to Panama; more southern records may be confused with *P. perulus*. Mostly dredged from shallow water to depths of 155 m.

Genus **ARGOPECTEN** MONTEROSATO, 1889
(**PLAGIOCTENIUM** DALL, 1898

Sturdy shells, ears nearly equal in size, valves convex and approximately equal; ribs numerous, concentric sculpture reduced to fine striae in interspaces. Resembling the European genus *Aequipecten* Fischer, 1886, but with less convex valves and weaker concentric striae. Waller (1969) has shown that the type species is the same as that of *Plagioctenium*, which thus falls as a synonym.

182. Argopecten circularis (Sowerby, 1835) (Synonyms: **Pecten tumidus** Sowerby, 1835 [not Turton, 1819]; **P. ventricosus** Sowerby, 1842; **P. inca** Orbigny, 1846; **P. solidulus** Reeve, 1853; **P. filitextus** Li, 1930). A wide variety of color and markings is exhibited on the shells, from almost pure white through blotched and streaked patterns to solid dark orange and purple. The inflated form, with both valves convex and with about 21 ribs, is unmistakable. It is the commonest species of the family in the Panamic province. Length, about 50 mm. Cedros Island, Baja California, through the Gulf of California and south to Paita, Peru, in depths of 1 to 135 m. South of Paita it is replaced by the larger, flatter *A. purpuratus* (Lamarck, 1819). A similar species in the Atlantic is *A. gibbus* (Linnaeus, 1758). The species of this group have been allocated to *Aequipecten*, but Waller (1969) concludes that *Argopecten* is more appropriate. He demonstrates that the type species, *P. solidulus* Reeve, which was described without locality, is based on a juvenile specimen of the eastern Pacific species earlier named *circularis* by Sowerby. Thus *Argopecten* displaces *Plagioctenium*, which has the same type species.

The Californian species *A. aequisulcatus* (Carpenter, 1864) has been considered a subspecies of *A. circularis* by some authors, but the shell is larger and duller-colored, and the form tends to live in bays; it has been reported not only along the outer coast of Baja California but also into the southern end of the Gulf of California as far as La Paz (182*a*).

180

181

182

182a

183

184

185

Genus CHLAMYS RÖDING, 1798

With the valves equal in size but the ears markedly unequal.

183. Chlamys lowei (Hertlein, 1935). The species is distinguished by the very small posterior ears, anterior ears being well developed. It has 20 to 22 rounded triangular ribs, and the coloring varies from grayish brown, flecked with brown spots, to an orange-brown. Length, 12 mm; height, 14 mm. Catalina Island, California, to Ecuador and the Galápagos Islands, in depths to 146 m.

Genus CYCLOPECTEN VERRILL, 1897

Small, symmetrical, thin right valve somewhat flattened, with concentric sculpture; left valve radially sculptured.

184. Cyclopecten acutus Grau, 1959. Small, with transparent margins, rounder than *C. pernomus*, without radial ridges on the left valve and finer concentric ridges on the right valve; auricles sculptured. Length and height, 4 mm; diameter, 1.5 mm. Off western Colombia, 59 to 137 m.

185. Cyclopecten catalinensis (Willett, 1931). Mostly a Californian form, this has been dredged in the Gulf of California at Tiburon Island in 29 to 37 m. The shell surface is smooth and the left valve lacks folds or plications. Height, 7 mm; width, 7.5 mm; diameter, 1.6 mm.

186. Cyclopecten cocosensis (Dall, 1908). The color consists of white, red, and brown lines and zigzags or clouded patches. The posterior area of the shell is set off by a furrow. Height, 8.7 mm; length, 9 mm. Cocos Island, Costa Rica, and Gulf of Panama, 95 to 113 m.

187. Cyclopecten exquisitus Grau, 1959. Shell opaque, right valve concentrically ridged, left valve with both concentric and radial ridges, hinge nearly as long as disk, anterior auricle with 10 to 12 radial ridges. Height and length, 3.5 mm; diameter, 1 mm. Angel de la Guarda Island, Gulf of California, to Galápagos Islands and Callao, Peru, in depths of 22 to 274 m.

188. Cyclopecten incongruus (Dall, 1916). This is the largest of West American Cyclopectens. Height, 14 mm; width, 15 mm; diameter, 3 mm. Off Cedros Island, Baja California, 1,250 m.

189. Cyclopecten liriope (Dall, 1908). Panama Bay to Galápagos Islands, 1,463 to 2,320 m.

190. Cyclopecten pernomus (Hertlein, 1935) (Synonym: **Pecten rotundus** Dall, 1908 [not von Hagenow, 1842]). Smallest of the Panamic pectens, this little white shell is variously dotted with brown. The left valve, which is larger than the right, is radially striate, the striae bifurcating toward the margin. The left ear is slightly the larger, and both ears are larger than in *C. cocosensis*. Also, the valves are evenly curved, with no posterior furrow. Height, 3 mm. Cedros Island, Baja California, through the Gulf of California and south to Ecuador, in depths to 355 m.

Genus DELECTOPECTEN STEWART, 1930

Sculpture of fine, somewhat beaded radial ribs; shell thin and delicate.

191. Delectopecten polyleptus (Dall, 1908). Galápagos Islands, 550 m.

186

187

189

190

194

195

196

197

192. Delectopecten randolphi (Dall, 1897). This essentially Californian species was reported at Guaymas, Gulf of California, by Grau. The range is from Bering Sea to the Gulf of California, in 18 to 1,940 m.

193. Delectopecten vitreus (Gmelin, 1791). A deepwater species that occurs nearly worldwide, this has been identified from off Clipperton Island in 185 to 365 m by Grau; it had been confused with *D. zacae*.

194. Delectopecten zacae (Hertlein, 1935) (Synonym: **Pecten panamensis** Dall, 1908 [not Dall, 1898]). The surface is radially striate, with 40 to 65 minute scaly threads. Height, about 12 mm; length, 10 mm. Gulf of California to Panama; offshore islands, including Clipperton and the Galápagos Islands, in 10 to 1,840 m.

Genus LEPTOPECTEN VERRILL, 1897

Shell small, thin to medium in weight, valves subequal and only slightly convex, ribs narrow, with wider interspaces; concentric sculpture, if present, microscopically fine. Two subgenera in the Panamic province.

Subgenus LEPTOPECTEN, s. s.

Thin, the interior reflecting the exterior ribbing, ribs somewhat rounded.

195. Leptopecten (Leptopecten) biolleyi (Hertlein & Strong, 1946). Resembling *L. (L.) velero*, this has fewer ribs—only 12 to 13—that are of more uniform height. Length, about 7 mm. Punta Abreojos, Baja California, through the Gulf of California, south to Ecuador, in depths of 18 to 220 m. Type locality, Puerto Parker, Costa Rica.

196. Leptopecten (Leptopecten) camerella (Berry, 1968). One might take this for a southern race of the Californian *L. (L.) latiauratus* (Conrad, 1837— reported by Grau (1959) as living in the southern end of the Gulf of California. However, *L. camerella* lacks the microscopic concentric sculpture between the ribs that is well developed in *L. latiauratus*. Length and height, about 20 mm. Off southern Baja California.

197. Leptopecten (Leptopecten) euterpes (Berry, 1957). The shell is bright rose-red or apricot-yellow; ribs are 17 to 19 in number, some of them beaded toward the margin. For the size of the valves the ears are large. Length, 7.4 mm; height, 7.1 mm; diameter, 2 mm. Off Guaymas to Acapulco, Mexico, at depths of 11 to 183 m.

198. Leptopecten (Leptopecten) palmeri (Dall, 1897). The left valve of the brownish-pink shell is lighter-colored than the right, and both are blotched with white. There are about 14 to 15 ribs, rather narrow, with wider interspaces. Length, 37 mm. The species seems to be confined to the Gulf of California, intertidally and to depths of 90 m.

199. Leptopecten (Leptopecten) velero (Hertlein, 1935). The shell is small, with 16 roughened ribs, every third one of which is larger than the others. The type specimen measures only 6.4 mm in length. Las Animas Bay, Gulf of California, to Peru, in depths of 5 to 55 m.

Subgenus PACIPECTEN OLSSON, 1961

Ribs more angulate in outline and shell heavier than in *Leptopecten, s. s.*

200. Leptopecten (Pacipecten) tumbezensis (Orbigny, 1846) (Synonyms: **Pecten aspersus** Sowerby, 1835 [not Lamarck, 1918]; **P. sowerbyi** Reeve, 1852; **P. paucicostatus** Carpenter, 1864; **P. indentus** and **splendens** Li, 1930). The left valve varies from a light reddish brown to almost black, sprinkled with tiny light-colored dots, and many specimens have also larger white blotches or rays. The right valve is white to yellowish, with a tendency toward dark rays along the posterior margin. The 14 to 15 ribs are separated by interspaces of almost equal width. Diameter, about 32 mm. Gulf of California to Paita, Peru, just below low tide line to depths of 128 m.

Genus LYROPECTEN CONRAD, 1862

Large shells, the well-developed radial ribs secondarily sculptured with fine riblets. No species of subgenus *Lyropecten, s. s.,* occurs in the Panamic province.

Subgenus NODIPECTEN DALL, 1898

With irregular nodes on ribs, giving a somewhat humped appearance.

201. Lyropecten (Nodipecten) magnificus (Sowerby, 1835). There are 13 to 15 ribs, and the nodes may be small and widely scattered; the ears have a few radial riblets. Length, 175 mm; height, 165 mm; diameter, 65 mm. A southern species confined to the coast of Ecuador, more common on the Galápagos Islands.

202. Lyropecten (Nodipecten) subnodosus (Sowerby, 1835) (Synonym: **L. intermedius** Conrad, 1867). Most spectacular of the tropical West American scallops, this has colors ranging from dull purple or white with purple lines to brilliant shades of orange and magenta. The 10 to 11 ribs are very conspicuous, some of them with large knobs or nodules, which makes the species easy to recognize. Single valves may be found on beaches, but good specimens must be obtained by divers or trawlers, from deeper water. Length, 175 mm; diameter, 75 mm. Scammon's Lagoon, Baja California, to Peru. A similar species is *L. (N.) nodosus* (Linnaeus, 1758), from the Atlantic, more nodose and with one less rib.

Genus PROPEAMUSSIUM DE GREGORIO, 1884

Shell nearly flat, with internal ribs; ears equal; lacking a byssal sinus. Mostly in deep water, especially in the tropics.

203. Propeamussium malpelonium (Dall, 1908). Gulf of Panama and off Colombia, in 2,690 to 4,500 m.

Genus PSEUDAMUSSIUM MÖRCH, 1853 (ICZN Opinion 714)

Small, right valve more convex than left; hinge short, ears unequal; sculpture of variable radial riblets folded into several broad radial corrugations. The type of this genus has been fixed by the International Commission on Zoological Nomenclature as *Pecten septemradiatus* Müller, 1776, a species from the northeastern Atlantic with a few low radial ribs, brown in color. No species of the type subgenus occur in the Panamic province.

Subgenus PEPLUM BUCQUOY, DAUTZENBERG & DOLLFUS, 1889

Ribs are fewer and radial grooving of the ribs more marked in this subgenus, which has as type *Pecten clavatus* Poli, 1795, a Mediterranean species.

204. Pseudamussium (Peplum) fasciculatum (Hinds, 1845) (Synonyms: **? P. panamensis** Dall, 1898; **Pecten miser** Dall, 1908). The brownish-pink shell has about half a dozen broad ribs subdivided by fine radial grooves. In the left valve the margin may be turned down to meet the right valve. Length, about 30 mm. The southern part of the Gulf of California to Panama, offshore in depths of 31 to 333 m.

Family DIMYIDAE
(DIMYACIDAE of authors)

Small, thin, suborbicular, lacking ears, compressed, closed, attached by the right valve, which is less convex; externally lamellose; interior porcelaneous; with two subequal muscle scars.

Genus DIMYA ROUALT, 1850

The hinge has a row of small denticles and two feeble crura or ridges. The pallial line is faintly visible as a smoothly curved row of dots. An anomalous group with the central ligament of the Pectinidae and Ostreidae but with two muscle scars.

205. Dimya californiana Berry, 1936. The shell is small, nearly circular in outline, flattened, its outer surface laminated, looking like thin plates of mica. The hinge ridges join above a small round socket that holds the ligament. Around the inner margin of the shell, just back from the edge, is a line of fine denticles. Length of a specimen dredged in the Gulf of California, 11 mm; height, 9 mm; diameter, approximately 3.5 mm. Southern California to Angel de la Guarda Island, Gulf of California, in 89 to 1,227 m.

Family PLICATULIDAE

Small shells with a few radial folds, with the hinge much as in the Pectinidae, two crural ridges in each valve; auricles small or absent; valve of irregular shape, owing to attachment to the substrate near the umbo.

Genus PLICATULA LAMARCK, 1801

Right valve usually the more convex; the single adductor scar posterior to the center of the shell; sculpture coarsely radial.

206. Plicatula anomioides Keen, 1958. The species is not uncommon at its type locality but has been overlooked by collectors because from the outside it looks so much like the common *Anomia adamas*. It is a thin white shell with spots of green on the inside and with very fine radiating riblets, especially near the beaks. Diameter, about 35 mm. Guaymas to Mazatlán, Mexico; attached to flat surfaces of rocks.

207. Plicatula inezana Durham, 1950. Largest of the Panamic Plicatulas, this reddish-brown form tends to be triangular in shape, with few and rather heavy divaricating ribs. It has been confused by authors with *P. spondylopsis*. Length, about 50 mm. Southern part of the Gulf of California to southern Mexico. Described as a Pleistocene fossil.

208. Plicatula penicillata Carpenter, 1857. The small size and brown streaks or lines on a white background distinguish this species. Fine radial spines decorate many specimens. Length, 15 to 20 mm. Southern Mexico to Eucador, in rock crevices or the insides of dead shells.

209. Plicatula spondylopsis Rochebrune, 1895 (Synonyms: **P. dubia** Hanley of authors, not of Hanley; **P. ostreivaga** Rochebrune, 1895). The shells are brownish on the outside, white within. Photographs of Rochebrune's type speci-

205

206

207

208

209

mens, furnished by the Paris Museum, show both as having fine radial ribbing and a rounded-quadrate outline. Plicatulas being variable, a large series would be needed to separate the two; *P. ostreivaga* is here regarded as a synonym. Diameter, about 40 mm. Gulf of California to Ecuador.

Family SPONDYLIDAE

Medium-sized to large shells, strongly sculptured with spinose radial ribs; auricles small; without byssal notch; shells attached to substrate near the umbo of the right valve, which is convex.

Genus SPONDYLUS LINNAEUS, 1758

Muscle scar large, posterior to center of shell; cardinal area of hinge larger in right valve, ligament deeply sunken in a triangular pit. Adult with two crural ridges adjacent to ligament.

210. Spondylus calcifer Carpenter, 1857 (Synonyms: "**S. limbatus** Sowerby" of Reeve, 1856, not Sowerby, 1847; **S. radula** Reeve, 1856 [not Lamarck, 1806]; **S. smithi** Fulton, 1915). Largest of the American *Spondylus*, this is in its adult stage coarsely ribbed, often riddled by the burrows of sponges, worms, and small boring clams. A wide band of purplish red marks the inside margin of most specimens. Young shells may be hard to distinguish from those of *S. princeps*; generally, the spines are more numerous and evenly distributed. Large specimens may be 150 mm across and weigh 3 pounds or more. The name *"calcifer"* (lime-bearer) refers to the extensive use that was made by the Spanish settlers of Central America, who used the lime of these shells as a source of cement. Earlier Americans evidently valued the shells, too, for specimens have been found in Amerind grave sites in eastern Mexico. The Gulf of California to Ecuador.

Spondylus gloriosus Dall, Bartsch & Rehder, 1938. A species of the Hawaiian Islands, reported living offshore at Clipperton Island.

Spondylus princeps Broderip, 1833 (Synonyms: **S. dubius** and **leucacantha** Broderip, 1833). The thorny oyster is one of the most showy of West American bivalves. The shell may be large—as much as 100 to 150 mm in diameter—and the hinge ridges so interlocked in many of them that the valves cannot be separated without breaking the teeth. This shell is often labeled *S. crassisquama* Lamarck, 1819, or *S. pictorum* Chemnitz, 1784, in collections, but as Hertlein and Strong have shown (1946), both these names apply to species not West American. Broderip's name *S. princeps* is therefore the first to be applied to the Panamic form. The type specimen, however, was from off Ecuador and represents a southern subspecies not familiar to most collectors. Olsson (1961) has figured the type and has suggested recognition of geographic subspecies.

211. Spondylus princeps princeps Broderip, 1833. The color is a uniform coral red, and the spines are nearly uniform, crowded and spikelike, with no open spaces between. A color form that Olsson regards as a subspecies—*S. p. leucacantha* Broderip, 1883—has slightly less crowded spines, some rows of which are white, and is restricted to Ecuador. There is a small attachment scar on the lower valve. Length, 130 mm; height, 135 mm; diameter, 79 mm. Panama to northwestern Peru.

210

211

212

213

212. Spondylus princeps unicolor Sowerby, 1847. This is, according to Olsson, the common thorny oyster of the Gulf of California, although it was described without locality. The rows of spines are more widely spaced than in *S. princeps princeps*, with three smaller rows of spines between the primary rows. Attachment scars are small or absent, and the valves may be nearly alike in shape. The color may vary from white through orange or pink to red, and the colored inner band may be narrow or occasionally absent. Cedros Island, Baja California; Concepcion Bay, Gulf of California, to Jalisco, Mexico, in 7 to 30 m.

See Color Plate II.

213. Spondylus ursipes Berry, 1959. Similar in general texture to *S. calcifer* but smaller, more ovate, with an orange line or band within the inner margin. Length, 126 mm; width, 100 mm. Off Angel de la Guarda Island, Gulf of California, in 18 m.

Superfamily LIMACEA

Equivalve, ovate; hinge edentulous; central muscle scar obscure.

Family LIMIDAE

The Limas or file shells resemble the Pectens in having ears, but the shell is obliquely asymmetrical, with the anterior end the longer and the valves gaping widely. Although able to swim, the animals are less active than most pectens, tending to bury themselves on the sea floor in a nest constructed of bits of rubble agglutinated by mucus. The mantle edge is fringed, and the living animal is a beautiful object to watch in an aquarium when it is fully expanded or moving about.

Genus LIMA BRUGUIÈRE, 1797

Shell eared, gaping, oblique, with radial ribs, white in color. Six Panamic subgenera.

Subgenus LIMA, s. s.

Ribs heavy and coarse, with scaly ornamentation.

214. Lima (Lima) tetrica Gould, 1851. Largest and coarsest-ribbed of the Panamic Limas, the shining white shell is about 50 mm in height. The Gulf of California to Ecuador, offshore in depths of 9 to 110 m.

Subgenus ACESTA H. & A. ADAMS, 1858

Large, thin shells with numerous riblets, confined to deep water.

215. Lima (Acesta) agassizii Dall, 1902. Panama Bay, 589 m.

216. L. (A.) diomedae Dall, 1908. Galápagos Islands, 704 m.

Subgenus LIMARIA LINK, 1807

Ribs fine, not scaly; gape narrow; ears subequal in size; shell medium-sized.

217. Lima (Limaria) hemphilli Hertlein & Strong, 1946 (Synonym: **L. dehiscens** of authors, not of Conrad, 1837). The shell gapes a little more on the anterior than on the posterior side, and the longest part (parallel to the hinge) is above the middle of the anterior margin. Length, 17 mm; height, 27 mm; diameter, 12 mm. Monterey, California, to Acapulco, Mexico, in depths of 18 to 91 m.

See Color Plate XII.

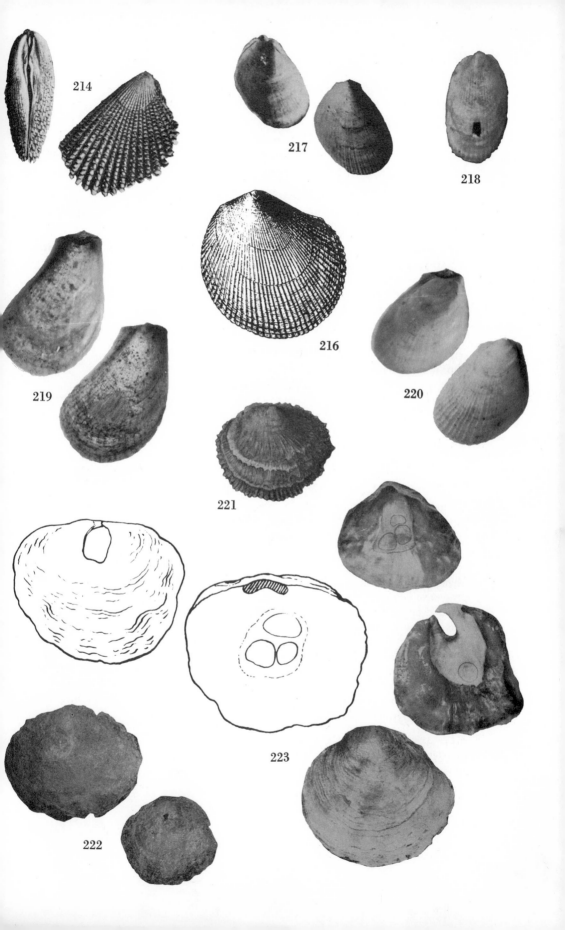

Subgenus **LIMATULA** WOOD, 1839

Small shells (length less than 10 mm), with rather glassy texture.

218. Lima (Limatula) similaris Dall, 1908. This minute form, only 4.5 mm in length, has been dredged offshore in depths of 55 to 106 m from Baja California to Panama. It is similar to the more northern *L. subauriculata* (Montagu, 1808) but has no furrow down the middle.

Subgenus **PROMANTELLUM** IREDALE, 1939

Widely gaping, flattened shells, the valves touching only at the hinge and the lower margin, with the ears unequal, anterior ear the larger.

219. Lima (Promantellum) pacifica Orbigny, 1846 (Synonyms: **L. arcuata** Sowerby, 1843 [not Geinitz, 1840]; **L. galapagensis** Pilsbry & Vanatta, 1902). The shell is long and narrow, of a dull texture and ivory-white color. Length, 19 mm; height, 30 mm; diameter, 9 mm. The northern end of the Gulf of California to Peru; under rocks at low tide.

See Color Plate XII.

Subgenus **SUBMANTELLUM** OLSSON & HARBISON, 1953

Gape narrowed to a mere chink; sculpture of fine radial threads.

220. Lima (Submantellum) orbignyi Lamy, 1930. (Synonym: **L. angulata** Sowerby, 1843 [not Muenster, 1841]). The longest part of this shell (parallel to the hinge) is at about the middle of the anterior margin. Length, 18 mm; height, 24 mm; diameter, 12 mm. The northern end of the Gulf of California to Chile, in depths of 7 to 22 m.

See Color Plate XII.

Superfamily **ANOMIACEA**

Monomyarian, with the anterior adductor muscle obsolete; gills filibranchiate; hinge lacking true teeth but with central ligament supported by crura or ridges in some forms; inner layer of shell lustrous or subnacreous.

Family **ANOMIIDAE**

Shells irregular in outline, mostly sessile, slightly to markedly inequivalve. Byssus present in young stages, later modified in most to become pluglike, horny, passing through an embayment or foramen in the lower (normally the right) valve. Adductor muscle scar subcentral, with one or more pedal and byssal retractor muscle scars above it, best seen on the left valve opposite the foramen.

The Anomias are often called jingle shells. Like the oysters they may be firmly attached to the substrate, so that the lower valve breaks if pried loose, but the attachment is not by cement. Rather, it is by the byssus, around which the lower valve has grown in a half-moon shape. On a smooth surface, such as the inside of a dead bivalve shell, the jingle may have a glassy texture, but on a rough surface the texture is much coarser, and the irregularities of the substrate may be perfectly mirrored in both valves. Such markings may be oblique to the normal direction of sculpture of the jingle, depending upon the position in which it has become attached; if it is on a ribbed shell, the ribbing may run counter to its own, with a complex resultant sculpture. Added to the variability of shape is considerable variation in color. Three Panamic genera.

Genus ANOMIA LINNAEUS, 1758

The muscle scar pattern in the unattached valve is a good index for recognition of *Anomia*: opposite the opening are one large muscle scar above and two smaller ones below. No hinge teeth or ridges reinforce the ligament.

221. Anomia adamas Gray, 1850 (Synonym: **A. simplex** Mabille, 1895). Possibly only a color form of *A. peruviana*, with which some authors synonymize it. However, a distinctive radial ribbing pattern shows up in some specimens that is independent of any on the base to which they were attached. Color of the shell varies from lustrous white to bright orange. About 40 mm across. The Gulf of California to Central America.

222. Anomia fidenas Gray, 1850 (Synonyms: **A. tenuis** C. B. Adams, 1852; **Placunanomia claviculata** Carpenter, 1857). Thin, small, round, flat, white or nearly glassy transparent, this occurs on the undersides of rocks or on other shells. Length and width, 30 mm. Panama.

223. Anomia peruviana Orbigny, 1846 (Synonyms: **A. alectus, hamillus, lampe, larbas,** and **pacilus** Gray, 1850; **Calyptraea aberrans** C. B. Adams, 1852). This protean shell, the Peruvian jingle, has no basic sculpture pattern of its own. Growing inside a smooth shell, it may be thin and nearly transparent. On a rock it may have a very irregular surface. A fully developed shell tends to be bluish green, especially on the lower valve, with a nacreous luster inside. Length, 46 mm; width, 34 mm; diameter, 9 mm. Monterey, California, to Paita, Peru; intertidally on rocks or other shells and offshore in depths to 110 m.

Genus PLACUNANOMIA BRODERIP, 1832

The presence of elevated ridges at the hinge margin of one valve, fitting into a socket in the other, separates this genus from the other two. Also, the shell is strongly plicate in three or four places around the margin.

224. Placunanomia cumingii Broderip, 1832. The shell is smooth in texture, somewhat pearly, whitish, especially within. It is the largest jingle of the family and is a showy shell, not common in collections. Length, 85 mm; diameter, 25 mm. Specimens are to be found in fairly shallow water intertidally or offshore in depths to 46 m from Carmen Island, Gulf of California, to Ecuador.

225. Placunanomia panamensis Olsson, 1942. Smaller than *P. cumingii*, flat, with the radial plications nearly obsolete, shell rounded. Length and width, 60 mm. First named as a Pleistocene fossil, later found living; Panama.

Genus PODODESMUS PHILIPPI, 1837

The muscle scar pattern in the unattached valve shows only two major scars, one above and the other diagonally below it. Otherwise, the shell is much like that of *Anomia*, although averaging larger. Three Panamic subgenera.

Subgenus PODODESMUS, *s. s.*

Byssal perforation or foramen small, partially or entirely covered by shell in adult; ligamental support not markedly elevated.

226. Pododesmus (Pododesmus) foliatus (Broderip, 1834) (Synonym: **Placunanomia pernoides** Carpenter, 1857). The shell was described as having a brown center in the unattached valve, but the color seems to vary from a uniform

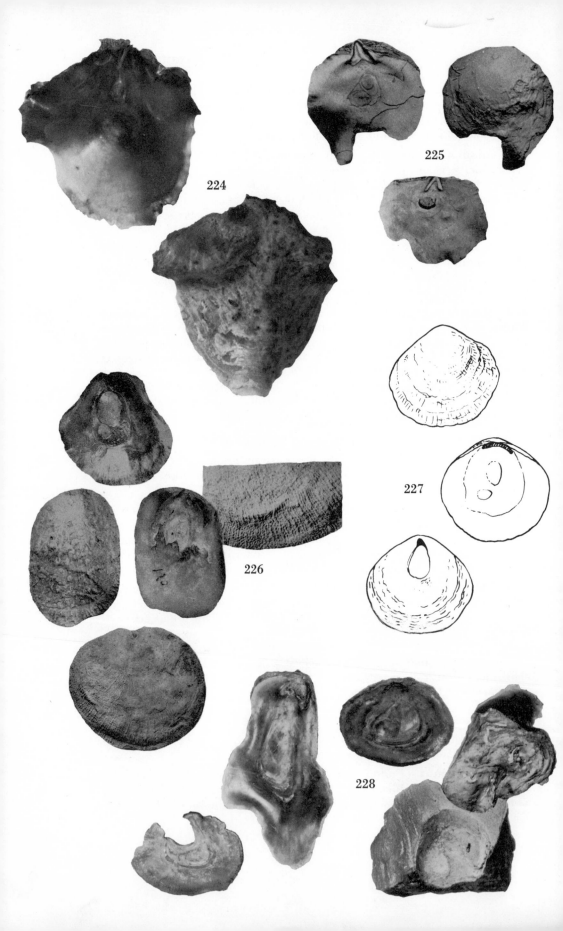

white to dark brown; the outer surface may have fine radial ribbing. The relationship of this species to *P. (T.) pernoides* (Gray, 1853) has not been clearly defined, nor has the geographic range. Length, 45 mm; width, 40 mm. Probably from Mazatlán, Mexico, or the southern end of the Gulf of California southward to Lobitos, Peru.

Subgenus **MONIA** GRAY, 1850

Byssal foramen large, rarely covered by shell in adult.

227. Pododesmus (Monia) cepio (Gray, 1850). This has been confused with *Anomia macrochisma* Deshayes, 1839 (specific name emended by Carpenter to *macroschisma*, which has become the most frequent spelling). Probably two subspecies should be recognized, *P. (M.) macrochisma macrochisma* from the north and western Pacific, a coarsely ribbed form, and *P. (M.) m. cepio*, the finer-ribbed southern form. In California this jingle is common on the red abalone, but where abalones are not available it will attach to other shells or to rocks. The shape, as in *Anomia*, is variable, but the color is fairly constant, a dirty-white exterior, shading from a dark green to greenish-white or even pure white interior. The whitish specimens have a texture resembling potato chips. Length and height, 67 mm. Southern Alaska to Baja California, questionably into the Gulf of California.

Subgenus **TEDINIA** GRAY, 1853

Byssal foramen closed in adult; ligament elevated on a pedestal.

228. Pododesmus (Tedinia) pernoides (Gray, 1853). A light to dark brown shell, often distorted by growth in crevices. Surface smooth to somewhat coarsely ribbed, not finely ribbed as in *P. foliatus*, with which its range may overlap. Southern California southward as far as Mazatlán, Mexico.

Subclass **HETERODONTA**

Shell with a hinge plate differentiated into cardinal and lateral areas, having true hinge teeth, at least in juvenile stages; shell material porcelaneous; gills eulamellibranch; mantle lobes joined, forming siphons.

Order **VENEROIDA**
(**TELEODONTA**)

Equivalve, with two subequal adductor muscle scars; posterior lateral hinge teeth, when present, behind ligament; hinge with one to three cardinal teeth (a few groups edentulous in adult stage); mostly active to nestling, rarely burrowing.

Superfamily **CRASSATELLACEA**

Cardinal teeth two in either valve; lateral teeth thin, elongate.

Family **CRASSATELLIDAE**

With wavy concentric sculpture over at least the upper part of the valve; ligament internal, like that of some more primitive bivalve groups, but with true hinge teeth (not simple crural ridges) on either side. Pallial line entire. Two Panamic genera.

Genus **EUCRASSATELLA** IREDALE, 1924

Inequilateral, medium-sized to large thickened shells with a smooth inner margin, concentric ribs obsolete toward margin.

This group has been known as *Crassatella* Lamarck, 1799, which paleontologists consider to be restricted to the lower Tertiary; its shell has a crenulate inner margin. No species of *Eucrassatella, s. s.*, occur in the Panamic province, the distribution being mainly Australian; it is a group with prosogyrate beaks.

Subgenus HYBOLOPHUS STEWART, 1930

Beaks slightly opisthogyrate; ventral margin sinuous; anterior lateral teeth long; inner margin of shell smooth.

229. Eucrassatella (Hybolophus) digueti Lamy, 1917 (Synonyms: **Crassatella undulata** Sowerby, 1832 [not Lamarck, 1805]; **Crassatellites laronus** Jordan, 1932). One of the large forms in the genus; shells may be as much as 92 mm long. A thick, dark brown periostracum covers the buff-colored shell; within, the buff shades to brown toward the truncate posterior end. Because the truncation grows more marked with age, shells of intermediate size resemble *C. (H.) gibbosa*, and sometimes it is not easy to make an identification. The two species may even prove to be synonymous. A similar Atlantic species is *E. antillarum* (Reeve, 1842). Rarely, beach shells may be found, but mostly specimens must be dredged in depths from 13 to 64 m, Gulf of California to Colombia.

230. Eucrassatella (Hybolophus) gibbosa Sowerby, 1832 (Synonym: **Crassatellites rudis** Li, 1930). The smaller size and pointed posterior end are distinguishing features of this cinnamon-brown shell. Length, 54 mm; height, 39 mm; diameter, 30 mm. Offshore in depths of 22 to 37 m, Gulf of California to Peru.

Genus CRASSINELLA GUPPY, 1874

The Crassinellas contrast sharply with the Crassatellas in size, for few specimens even approach a diameter of 10 mm. They are triangular, concentrically sculptured with undulating ribs of varying strength, the shell being white, sparsely dotted with brown or pink. The genus is one of the few in which the beaks are opisthogyrate; that is, the tip of the shell points posteriorly. It is exclusively American, mainly confined to tropical waters of the Caribbean and Panamic areas. Although these are variable shells and one sometimes feels that within any large suite all extremes could be found, there are some distinctive patterns that are here considered to constitute separate species.

231. Crassinella adamsi Olsson, 1961. Least trigonal of all the species, this could be mistaken for a small venerid but for its opisthogyrate beaks; sculpture varies from faint to strong concentric ribs. White with a few brown stains. Length, 3.5 mm; height, 2.7 mm. La Paz and Guaymas, Mexico, to Ecuador.

232. Crassinella ecuadoriana Olsson, 1961. The nearly trigonal shell has its posterior margin slightly the longer, indented below and ending in a snoutlike point. The shell is white to lilac, rayed with brown or purplish pink. Length, 2.9 mm; height, 2.5 mm; diameter, 1.4 mm. Banderas Bay, Nayarit, Mexico, in 9 m; to western Colombia and Ecuador.

233. Crassinella mexicana Pilsbry & Lowe, 1932. This resembles *C. pacifica* but is smaller, with somewhat finer concentric ribs and a sharper angle between anterior and posterior dorsal slopes, the outline nearly symmetrical. Color, white with flecks and stains of brown. Length, 3.4 mm; height, 3.3 mm; diameter, 1.8 mm. La Paz to Banderas Bay, Mexico.

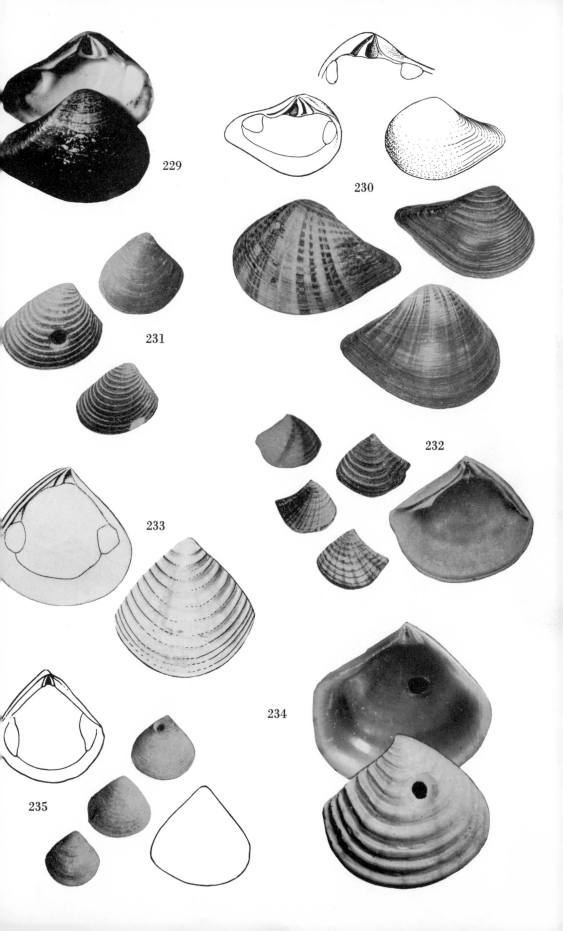

229

230

231

232

233

234

235

234. Crassinella pacifica (C. B. Adams, 1852). Most abundant of the Pacific Crassinellas, this averages somewhat larger. It is a little inequilateral, the posterior side slightly longer. The unworn surface has a fine radial pattern overlying the concentric ribs. Length, 4.1 mm; height, 3.7 mm. Cape San Lucas and through the Gulf of California, south to Peru.

235. Crassinella varians (Carpenter, 1857). The small shell is smooth or nearly so, almost equilateral, with a conspicuous brown blotch on the posterior side of most specimens. La Paz and Guaymas, Gulf of California, to Ecuador.

? Family CARDINIIDAE

Ovate to somewhat trigonal, inequilateral, thick-shelled forms with one or two cardinal teeth, strong lateral teeth in the right valve, weak laterals or none in the left; smooth to concentrically sculptured.

Genus TELLIDORELLA BERRY, 1963
(LIROTARTE OLSSON, 1964)

Resembling *Crassinella* in outline but with strong to recurved concentric lamellae and a hinge with one heavy cardinal tooth and two lateral sockets in the right valve, two cardinals and no laterals, only marginal ridges in the left. Chavan (in Moore, 1969) considers this a living fossil. There is one species in the Miocene of Ecuador and one living form in the eastern Pacific. The family Cardiniidae is known otherwise only in Paleozoic and Mesozoic strata.

236. Tellidorella cristulata Berry, 1963. The shell is nearly trigonal, with concentric ridges that end in pointed crests along the dorsal margin. Although resembling *Crassinella* in size and shape, *T. cristulata* lacks any color markings, the shell being pure white. Length, 6 mm; height, 5 mm; diameter, 1 mm. Puerto Libertad, Sonora (type locality) to Espíritu Santo Island, Baja California, and Banderas Bay, Jalisco, Mexico, in depths of 27 to 90 m.

Superfamily CARDITACEA

Inequilateral, heart-shaped shells, with radial ribbing in most; beaks prosogyrate, somewhat incurved; hinge with two cardinal teeth in either valve and more or less remote laterals; animal byssiferous.

Family CARDITIDAE

Medium-sized fairly sturdy shells, radial ribbing well developed, inner margin strongly crenulate; ligament external; anterior cardinal of right valve weak, the posterior cardinal long and thin; lateral teeth short; anterior adductor muscle scar somewhat raised on a platform.

Genus CARDITA BRUGUIÈRE, 1792

The Carditas or little heart shells are stout, mostly with strong radial ribbing and dull or spotted coloring. The hinge has one or two cardinal teeth and may or may not have lateral teeth at the ends of the hinge plate. A related genus, *Venericardia*, was widespread during Eocene time. Modern workers do not consider that it survived beyond that time and think that all the living forms so called can be classed as *Cardita*. The carditids are shallow-water mollusks, many attaching themselves under rocks by a byssus.

There has been some shifting of opinion with respect to the type species of *Cardita*. Under the Code of Zoological Nomenclature, it should be accepted as *C. variegata* Bruguière, 1792, member of a group that in the strict sense does not occur on American coasts. Several subgenera, based on outline and details of the hinge, are recognized, five in the Panamic province; *Cardita, s. s.*, does not occur on the Panamic west coast.

1. Shell elongate, at least twice as long as high........................ 2
   Shell ovate, less than twice as long as high.......................... 3
2. Lunule not evident; lateral teeth weak....................... *Byssomera*
   Lunule present; lateral teeth strong....................... *Carditamera*
3. Adults small, less than 10 mm high....................... *Pleuromeris*
   Adults medium-sized to large, more than 10 mm high.................. 4
4. Umbones high, pointed; hinge wide..................... *Strophocardia*
   Umbones not conspicuous; hinge moderate......................... 5
5. Periostracum not evident................................... *Cardites*
   Periostracum well developed, somewhat furry................. *Cyclocardia*

Subgenus BYSSOMERA OLSSON, 1961

Lateral teeth and lunule small to absent; posterior margin angulate.

237. Cardita (Byssomera) affinis Sowerby, 1833 (Synonyms: "**Cardita modulosa** Lam." Valenciennes, 1846 [not *C. nodulosa* Lamarck, 1819]; **C. californica** Deshayes, 1854). Shell elongate, length about twice width; ribs 15 or more in number, smooth to scaly, especially posteriorly; brownish white to brown. Length, about 50 mm. Living under stones or in crevices, intertidally and offshore to depths of 27 m. Authors have cited *C. californica* as a smoother northern subspecies of *C. affinis*. The type lots in the British Museum lend no support to such classification, for both show the same variations; both tend to have spinose posterior ribs. The large smooth-ribbed form that occurs in quieter bays along the Mexican coast (Keen, 1958, fig. 162) may, when more material is available to indicate whether it is a situs form or a genuine geographical isolate, prove worthy of a name (fig. 237*a*).

Subgenus CARDITAMERA CONRAD, 1838
(LAZARIA GRAY, 1854)

Lateral teeth strong; lunule in both valves; posterior slope rounded.

238. Cardita (Carditamera) radiata Sowerby, 1833 (Synonym: **Lazaria observa** Mörch, 1861). This species has about 17 ribs, of equal distinctness over the whole shell; the color pattern is strongly spotted rather than mottled. Length, 48 mm. Although it has been reported as far north as Baja California, it is more common to the south, from Nicaragua to Ecuador, on mud flats, to 24 m depth.

Subgenus CARDITES LINK, 1807

Lunule heart-shaped; ribs heavy, scaly; posterior slope truncate.

239. Cardita (Cardites) crassicostata (Sowerby, 1825) [not preoccupied by *C. crassicosta* Lamarck, 1819] (Synonyms: **C. cuvieri** Broderip, 1832; **C. michelini** Valenciennes, 1846; **C. sulcosa** Dall, 1908). The squared ribs and offset dorsal area are characteristic of this species. It is one of the most colorful of the genus; many worn specimens show a rich orange pink spotted with brown. Length,

236

237

237a

238

239

240

241

49 mm; height, 43 mm; diameter, 42 mm. The Gulf of California to Peru, intertidally and offshore to 55 m.

240. Cardita (Cardites) grayi Dall, 1903 (Synonym: **C. crassa** Sowerby, 1839 [not Lamarck, 1819]). The ovate inflated outline and the rounded ribs of this species distinguish it. Length is about 32 mm. Beach specimens have been found from the Gulf of California to Ecuador. Lowe reported it living at extreme low tide in sand under rocks.

241. Cardita (Cardites) laticostata Sowerby, 1833 [not preoccupied by *Venericardia laticosta* Eichwald, 1830] (Synonym: **C. tricolor** Sowerby, 1833 [first reviser, Dall, 1903]; **C. arcella** Valenciennes, 1846). The rectangular shape of the shell, the 22 to 23 high, square-edged ribs, and the spotted or mottled color pattern serve to distinguish it. Length, 40 mm. It is fairly common from the Gulf of California to Peru, living near the extreme low-tide line in sand under rocks, or offshore to depths of 27 m.

Subgenus CYCLOCARDIA CONRAD, 1867

Short, compressed; ribs triangular, beaded; beaks low.

242. Cardita (Cyclocardia) spurca beebei Hertlein, 1958. The shell differs from that of the typical *C. spurca* Sowerby, 1833, a Chilean form, by being smaller, thinner, and having finer ribs. The shell is white under an olive periostracum and may show a tinge of pink just under the beaks. Length, 18 mm; height, 15.5 mm; diameter, 11 mm. Carmen Island, Gulf of California, to Panama Bay, in 45 to 65 m depth.

Subgenus PLEUROMERIS CONRAD, 1867

Small, inflated; ribs granular; lunule elongate.

243. Cardita (Pleuromeris) guanica Olsson, 1961. The small white shell is nearly circular, slightly oblique, with about 17 low granular ribs and a flattened lunule in the right valve. The genus is known in the Tertiary and living fauna of the east coast, but Dr. Olsson's is the first report from the west coast. Length of the probably immature holotype, 3.2 mm; height, 3.3 mm; diameter (one valve), 1.1 mm. Known only from Panama.

Subgenus STROPHOCARDIA OLSSON, 1961

Ribs wide and low, interspaces linear; periostracum coarse.

244. Cardita (Strophocardia) megastropha (Gray, 1825) (Synonyms: **Venericardia flammea** Michelin, 1831; **Cardita tumida** and **varia** Broderip, 1832). Odd valves of this colorful form are much more likely to be found than live specimens. The high, pointed umbones, the low, rounded ribs, and the brownish-red color flecked with white or yellowish spots make it easily recognizable. Average length is about 48 mm. The Gulf of California to Ecuador and the Galápagos Islands, offshore to 100 m.

Family CONDYLOCARDIIDAE

These are minute shells with an internal ligament and a peculiar rectangular rim or collar that sets off the nepionic part of the shell from the main disk. The shell is inequilateral, with the anterior end longer than the posterior. Most of the species occur in the far south Pacific and the south Atlantic oceans, only a few in tropical American waters.

242

243

245

244

246

247

248

Genus **CONDYLOCARDIA** BERNARD, 1896
(**HIPPELLA** MÖRCH, 1861 [suppressed, ICZN, 1969]])

Ribs radial, few; shell subtrigonal; prodissoconch saucer-shaped, capping the umbones. The shells somewhat resemble those of *Verticordia*, with which the unillustrated genus long was supposed to be synonymous. In the mid-1960's Mörch's original material came to light. Because this generic name, *Hippella*, had 35 years' priority over *Condylocardia*, a petition to the International Commission on Zoological Nomenclature was necessary to have the *nomen oblitum* (forgotten name) suppressed.

245. Condylocardia digueti Lamy, 1916. Resembling a juvenile *Cardita* but with an internal ligament and with only about ten ribs. A French collector, Diguet, first found material on Espíritu Santo Island near La Paz in 1897, but it was not described until after he had again visited the locality and collected more specimens in 1914. Length, 1.5 mm; height, 1.3 mm; diameter, 0.65 mm. La Paz, Baja California, to Bahía de los Angeles and Guaymas, Gulf of California, and south to Cuastecomate, Jalisco, Mexico.

246. Condylocardia hippopus (Mörch, 1861) (Synonym: **C. panamensis** Olsson, 1942). More equilateral than *C. digueti*, posterior end pointed; glassy white, with about eight prominent ribs. Length, 2 mm. Puntarenas, Costa Rica, to Panama. Because Mörch's type material was not figured and was not studied by later authors, Olsson described his Panamic (Pleistocene) material as new. Later he found living specimens also.

Superfamily **CORBICULACEA**

Hinge with up to three cardinal teeth; pallial line entire or with a small sinus; mainly nonmarine in habitat.

Family **CORBICULIDAE**
(**CYRENIDAE** of authors)

Members of this family inhabit brackish to fresh water. Because the shells may be floated by streams down to the coast and intermixed with marine material, the larger Panamic forms are discussed briefly here. Much work remains to be done on them before the species are properly understood. The last formal monograph was by Prime (1865). Dr. J. P. E. Morrison of the United States National Museum, long a specialist on the nonmarine faunas, has made available for the present review his unpublished notes on the American Corbiculidae. His studies of type material have enabled him to take a fresh look at the old lists and to arrive at new synonymies. Several supposed West American species now prove to be Caribbean, and a number of others must be reduced to synonymy. His generous help is here gratefully acknowledged. Dr. Eugene Coan also assisted in preliminary organization of data.

Typically, the Corbiculas (*Cyrena* or *Cyclas* of earlier workers) have three cardinal teeth in either valve, two lateral teeth, and two equal adductor muscle scars. The type genus, *Corbicula*, which has long, serrate lateral teeth, is Oriental in distribution, except that one species—*C. manilensis* (Philippi, 1844), often erroneously listed as *C. fluminea* (Müller, 1774)—has been introduced into the Pacific coast states and has spread, helped by aquarists and by fishermen who may discard live bait clams into streams, eastward to Tennessee and southward to southern California and Arizona, but it has not spread into the streams draining into the Gulf of California, nor is it apt to become established there because of the salinity. Johnson (1959) lists Prime's corbiculid type specimens.

Genus POLYMESODA RAFINESQUE, 1828

Periostracum present; hinge with three cardinal teeth in either valve; lateral teeth two, smooth, not serrate, only moderately long. Three Panamic subgenera.

Subgenus POLYMESODA, s. s.

Shell nearly smooth; periostracum velvety, somewhat folded; pallial sinus narrow, ascending, fairly deep.

247. Polymesoda (Polymesoda) mexicana (Broderip & Sowerby, 1829) (Synonyms: **Cyrena nitidula** Deshayes, 1855; **C. fragilis** Sowerby, 1878, *ex* Deshayes, MS). Although the type specimen of this has been lost, authors seem to be in agreement on the form—a rather thin, quadrate shell, with a soft brown periostracum. Length, about 32 mm; height, 28 mm; diameter, 29 mm. Estuaries in the southern end of the Gulf of California to Puerto Vallarta, Mexico.

248. Polymesoda (Polymesoda) notabilis (Deshayes, 1855) (Synonyms: **P. pullastra** Mörch, 1861; **P. zeteki** Pilsbry, 1931). The whitish to flesh-pink shell has a purplish posterior margin, and the beaks are well forward. The dark periostracum tends to shred off except near the margin. Length, 43 mm; height, 34 mm; diameter, 24 mm. Costa Rica to northern Peru, in mud.

Subgenus EGETA H. & A. ADAMS, 1858
(ANOMALA DESHAYES, 1855 [not VON BLOCK, 1799])

Smooth, with velvety, slightly wrinkled periostracum; pallial sinus small to obsolete.

249. Polymesoda (Egeta) altilis (Gould, 1853) (Synonyms: **Cyrena fontainei** Philippi, 1851 [not Orbigny, 1844]; **C. olivacea** Carpenter, 1857; **C. triangula** of authors, not of Philippi, 1849). The periostracum is olive green and covers the shell, rising into concentric folds, especially on the anterior part; the shell is pink to purplish within. Length, about 63 mm; height, 52 mm; diameter, 32 mm. Estuaries of the Mazatlán area, Mexico.

250. Polymesoda (Egeta) anomala (Deshayes, 1855) (Synonyms: **Cyrena isocardioides** and **peruviana** Deshayes, 1855). A trigonal form with a ridge down the posterior slope, the periostracum light grayish green, the inside gray with violet markings. Length, 50 mm; height, 40 mm; diameter, 34 mm. El Salvador to northern Peru. Type of the subgenus *Egeta*.

251. Polymesoda (Egeta) inflata (Philippi, 1851) (Synonyms: **Cyrena maritima** C. B. Adams [not Orbigny, 1842]; **C. cordiformis** Récluz, 1853 [not Deshayes, 1824] = **C. recluzii** Prime, 1865; **C. angulata** Deshayes, 1855 [not Römer, 1835]; **? C. dura** Deshayes, 1855; **C. inflata** Deshayes, 1855 [not Philippi, 1851] = **C. panamensis** Prime, 1860; **C. tumida** Prime, 1860; **C. cardiformis** Sowerby in Reeve, 1878, *ex* Deshayes, MS; **P. joseana** Morrison, 1946). The hinge is relatively large and heavy, and the shell itself is sturdy. The anterior half is dingy white, the posterior part purplish, covered by an olive-colored periostracum that may be wrinkled. Length, 56 mm; height, 48 mm; diameter, 36 mm. Oaxaca, Mexico, to northern Peru.

PLATE II · *Spondylus princeps* Broderip (Thorny Oyster)

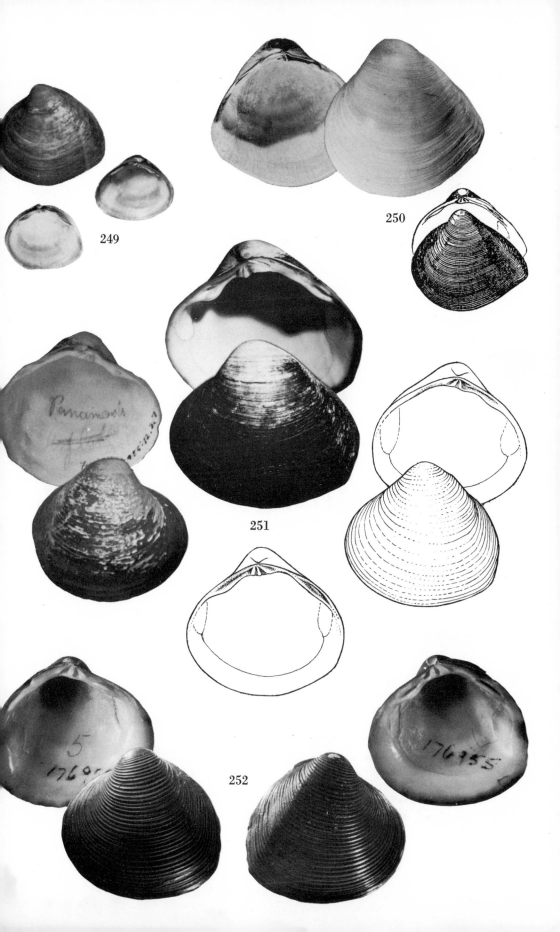

249

250

251

252

Subgenus **NEOCYRENA** CROSSE & FISCHER, 1894

Surface somewhat corrugated; periostracum shiny, not velvety. The species of this group seem to be more localized than in the other subgenera. Careful collecting in the estuaries of the Central American coast is needed before a really satisfactory list can be given.

252. Polymesoda (Neocyrena) boliviana (Philippi, 1851) (Synonyms: **Cyrena exquisita** Prime, 1867; **C. tribunalis** Prime, 1870). Rounded to angulate in outline, with well-marked corrugations. Length, 24 mm; height, 21.5 mm. Panama to Ecuador.

253. Polymesoda (Neocyrena) fontainei (Orbigny, 1844) (Synonyms: **?** **Cyrena fortis** Prime, 1861; **C. chilina** Prime, 1867). A somewhat trigonal shell, nearly smooth except for concentric threads, with a brown or green periostracum; the interior violet. Length, 51 mm; height, 42 mm; diameter, 26 mm. Ecuador.

254. Polymesoda (Neocyrena) meridionalis (Prime, 1865). A small, nearly globular shell with a dark greenish periostracum. Length, 35 mm; height, 35 mm; diameter, 26 mm. Tumbes to Paita, Peru.

255. Polymesoda (Neocyrena) nicaraguana (Prime, 1869) (Synonyms: **Cyrena solida** Philippi, 1847 [not Dunker, 1843]; **C. radiata** of authors, not of Hanley, 1844, an Atlantic species). This is the type species of the subgenus. The valves are convex and well rounded, with coarse but even concentric sculpture and a dark olivaceous periostracum. Length, 40 mm; height, 36 mm; diameter, 28 mm. Nicaragua; perhaps south to Panama and Ecuador.

256. Polymesoda (Neocyrena) ordinaria (Prime, 1865) (Synonym: **? Cyrena germana** Prime, 1867). The status of this species remains unclear, for the type localities of both *P. ordinaria* and *P. germana* are open to question. Dr. Morrison reports material that compares well with the type of *P. germana* from Nayarit and Puerto Vallarta, Mexico.

Superfamily **DREISSENACEA**

Mytiliform to quadrate, the beaks terminal; interior of shell not nacreous; ligament sunken; hinge edentulous; beak cavity bridged by a septum or myophore; posterior adductor muscle long; periostracum well developed. Mainly nonmarine in habitat.

Family **DREISSENIDAE**

Animal byssiferous; siphons two; gills reticulate.

Genus **MYTILOPSIS** CONRAD, 1858

As the generic name implies, the shells of *Mytilopsis* look like *Mytilus*, but they have a porcelaneous internal layer and a triangular cup-shaped myophore that buttresses the septum across the end of the shell, projecting into the cavity. The species are adapted for life in brackish to fresh water.

257. Mytilopsis adamsi Morrison, 1946. The shell is small and whitish or glassy under a thin horny periostracum that is pinched up into fringing folds along the lines of growth. Length, 12 mm; height, 6 mm; diameter, 5.5 mm. The type lot was found attached by the byssus to rocks in a freshwater lagoon and in a stream on San José Island, Panama Bay.

253

254

255

256

257

258

259

260

260a

258. Mytilopsis trautwineana (Tryon, 1866). A freshwater form reported by Olsson (1961) in the rivers of Ecuador.

259. Mytilopsis zeteki Hertlein & Hanna, 1949. Resembling *M. adamsi*, differing mainly in outline. Length, 25 mm; height, 13 mm; diameter, 14 mm. Although a dark-colored byssus is developed, the mussel apparently nestles in holes. The type specimens were in a block of wood taken from Miraflores Locks, Canal Zone, Panama. Olsson (1961) suggests possible synonymy with *M. adamsi* and also a close affinity to the Atlantic *M. leucophaetus* (Conrad, 1831).

Superfamily GLOSSACEA

Inequilateral, beaks well forward and somewhat incurved or spiral; shell surface smooth or nearly so; hinge with two to three cardinal teeth that tend to be parallel to hinge margin and with well-developed laterals; pallial line mostly entire.

Family VESICOMYIDAE

Ovate to elongate, usually with an incised lunule; hinge with teeth not clearly differentiated into cardinals and laterals, parallel to hinge margin.

Genus VESICOMYA DALL, 1886

Thin, ovate, known mostly as deepwater forms, especially in the Panamic province. Three Panamic sugenera.

Subgenus VESICOMYA, *s. s.*

Periostracum polished; lunule bounded by a groove.

260. Vesicomya (Vesicomya) donacia Dall, 1908. Gulf of Panama, 2,320 m.

260a. V. (V.) lepta (Dall, 1896). Length, 58 mm. Tillamook, Oregon, to Bahía Concepción, Gulf of California, in 1,440 to 1,570 m.

261. V. (V.) ovalis (Dall, 1896). Gulf of Panama, 3,050 m.

Subgenus ARCHIVESICA DALL, 1908

Elongate, inflated, lunule not set off; pallial line with a small sinus descending almost vertically from posterior adductor scar.

262. Vesicomya (Archivesica) gigas (Dall, 1896). Large, relatively heavy-shelled; length, 110 mm; height, 63 mm. Type of the subgenus. Gulf of California, 1,565 m.

Subgenus CALLOGONIA DALL, 1889

Lunule without a border; pallial sinus present, shallow but acute.

263. Vesicomya (Callogonia) angulata (Dall, 1896). Gulf of Panama, 2,320 to 3,050 m.

Superfamily ARCTICACEA

Inequilateral; mostly smooth; pallial line entire or nearly so; hinge with up to three cardinal teeth.

Family BERNARDINIDAE

Small to minute shells with the ligament partly or entirely internal; hinge with two to three cardinal teeth and two or more laterals; pallial line entire. Two Panamic genera.

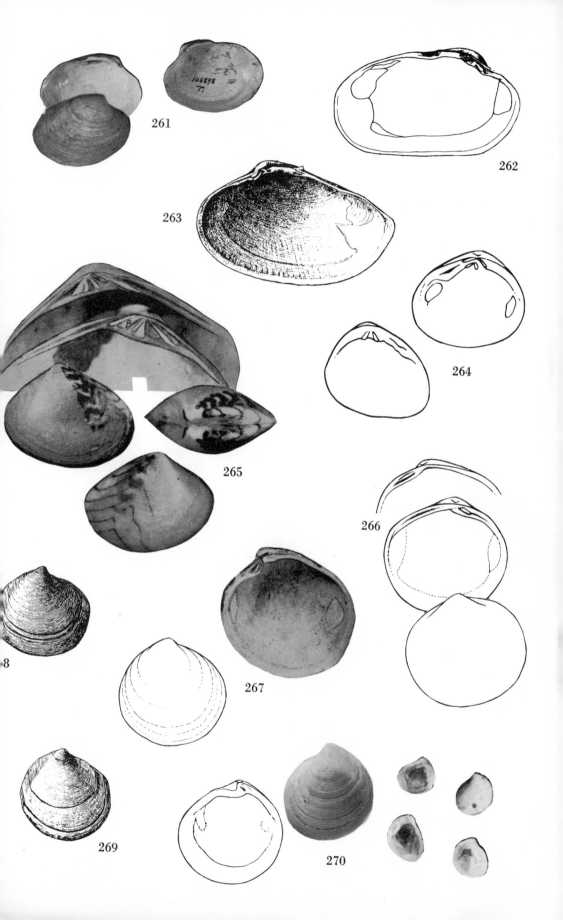

261

262

263

264

265

266

267

268

269

270

Genus BERNARDINA DALL, 1910

Sculpture concentric; prodissoconch set off by a concentric ridge.

264. Bernardina margarita (Carpenter, 1857). Minute rather than merely small, the shell is concentrically sculptured, with the nepionic part set off by a ridge in most specimens; hinge with two cardinal teeth in the right valve, three in the left, the ligament in a pit behind the cardinals. Color varies from all white or all brown to radially banded or white with pink umbones. Like many other small bivalves, this form is ovoviviparous, brooding the young within the mantle cavity. Length, about 1.5 mm. The genus was based on a Californian species, *B. bakeri* Dall, 1910. Mazatlán to Banderas Bay, Mexico, and the Tres Marias Islands, in bottom sediment brought up by divers.

Genus HALODAKRA OLSSON, 1961

Ovate, inequilateral, smooth or with concentric striae; ligament in a shallow resilifer, somewhat sunken; hinge with three or more teeth in cardinal area and a posterior lateral behind the resilifer, posterior margins of shell grooved to receive opposite valve.

265. Halodakra subtrigona (Carpenter, 1857). The conspicuous feature of this small shell is a series of chevron-shaped brown and white markings on the posterior slope. Both this species and the *Bernardina* were described as *"Circe,"* in Veneridae, by Carpenter, and it was not until recent years that their true affinities were worked out. An as yet unnamed species of *Halodakra* occurs along the Pacific coast of Baja California, and *H. brunnea* (Dall, 1916), of southern California (named as *Psephidia*), is the northernmost representative of the genus. Length, 4 mm; height, 3.3 mm. Intertidally, in fine sand or gravel, Cape San Lucas to Puerto Peñasco, and Mazatlán, Mexico, south to northern Peru.

Superfamily CYRENOIDACEA

Lenticular, inequilateral, somewhat compressed thin shells with prosogyrate beaks and external ligament, no lunule. Hinge of a modified lucinoid type, lacking posterior lateral teeth. Muscle scars large, elongate; pallial line entire. The relationships of this group are not clear; some authors place it near the Corbiculacea, and others assign it to the Lucinacea. A middle course of accepting a superfamily status, suggested by Olsson (1961), is taken here.

Family CYRENOIDIDAE

The hinge has two cardinal teeth in each valve, the anteriormost one in the left and both in the right fused to the upper ends of what were originally anterior lateral teeth, the resulting teeth being shaped like a figure 7.

Genus CYRENOIDA JOANNIS, 1835
(CYRENELLA DESHAYES, 1835; CYRENODONTA H. & A. ADAMS, 1857)

Thin shells resembling a compressed *Diplodonta* but with angulate laminar not bifid teeth. Sculpture of fine concentric striae under a thin horny periostracum. The pallial line is indistinct but entire. The genus has only a few species in the warmer waters of the eastern and western Atlantic and the eastern Pacific. Living specimens occur in the mud of brackish-water mangrove swamps.

266. Cyrenoida insula Morrison, 1946. The shell is blue-white under a horny periostracum. Length, 6.7 mm; height, 6.2 mm; diameter, 3.8 mm. The type locality is in mangrove swamps of Pearl Islands, Panama.

267. Cyrenoida panamensis Pilsbry & Zetek, 1931. Grayish white, inflated, with a chamois-skin periostracum, this form has conspicuous umbones, and somewhat prosogyrous beaks. Length, 18 mm; height, 17.5 mm; diameter, 10.3 mm. The type locality is Panama, in a brackish-water mangrove swamp.

Superfamily LUCINACEA

Equivalve, subcircular, beaks low; anterior and posterior areas somewhat set off by low radial folds.

Family LUCINIDAE

The hallmark of the family is the elongate anterior adductor muscle scar. The shells are mostly lenticular in shape, with low and nearly central beaks. The ligament may be entirely external or partly internal, and the hinge may vary from two cardinal teeth in each valve and one or more lateral teeth to none at all in the adult (juvenile shells may show traces). The pallial line is entire, with no sinus. The key is to the seven genera (in four subfamilies) in the Panamic province.

1. Ligament partially internal, sunken below dorsal margin. 2
   Ligament entirely external. 4
2. Shell surface smooth. . . . . *Miltha*
   Shell strongly sculptured 3
3. Sculpture of evenly reticulate concentric and radial ribs. . . . . *Codakia*
   Sculpture of divaricating radial ribs. . . . . *Ctena*
4. Hinge teeth wanting in adult shells. . . . . *Anodontia*
   Hinge teeth present in adult shells. 5
5. Sculpture of oblique radiating lines only. *Divalinga*
   Sculpture various but not of oblique radiating lines. 6
6. Posterior and anterior dorsal areas set off by a change in sculpture. . . . . *Lucina*
   Posterior and anterior dorsal areas not set off. . . . . *Lucinoma*

Subfamily LUCININAE

Lenticular, convex; sculpture mainly concentric, radial ribs weak. Three Panamic genera.

Genus LUCINA BRUGUIÈRE, 1797
(PHACOIDES of authors)

Anterior and posterior areas set off by a change of sculpture; predominant concentric sculpture overlain by weak radial riblets; both cardinal and lateral teeth present; inner margin crenulate.

Thorny nomenclatural problems afflict this group, owing to differences of opinion among authors on the interpretation of the type species. For a number of years during the early part of this century the name *Phacoides* was in vogue as a combining term for several widespread subgroups. However, as Chavan (1937–38) has shown, this name was proposed in the vernacular (as a French common name) by Blainville in 1825 but not adopted as a scientific name until Dall revived it in 1901, by which time substitutes were available. Chavan, who is a specialist on the group, recommends abandonment of *Phacoides* and recognition of several subgenera under *Lucina* as well as several genera under the family Lucinidae, the course adopted here, although not always in complete accordance with his classification. The matter of the correct type species for *Lucina* can only be resolved by a petition to the International Commission on Zoological Nomenclature. There are three species that have been interpreted by one author or another as type—*Venus edentula* Linnaeus, 1758, *V. pensylvanica* Linnaeus, 1758, and *L. jamaicensis* Lamarck, 1801 (= *V. pectinata* Gmelin, 1791). The first of these is type species by subsequent monotypy under a strict interpretation of the Rules of Zoological Nomenclature, but because accepting it as type would mean a major change of concept, few authors have seriously recommended this course. Either of the other two species could be adopted as type without affecting West American terminology, for the eastern Pacific has no species close enough to these two Caribbean forms to fall within the same subgenus. Thus, so far as we are concerned, *Lucina* in the broad sense can be used without formal solution of the problem.

The key is to the six subgenera here recognized as occurring in the Panamic province.

1. Surface with concentric sculpture only............................. 2
   Surface with both concentric and radial sculpture.................... 4
2. Sculpture of fine and regularly spaced coarser concentric ribs..........*Here*
   Sculpture of uniform-sized concentric ribs.......................... 3
3. Posterior area set off by a change in direction of concentric sculpture;
      lunule present but not deep...............................*Callucina*
   Posterior area not set off; lunule deep.........................*Cavilinga*
4. Radial sculpture of fewer than 7 very broad ribs...............*Pleurolucina*
   Radial sculpture of 10 or more ribs................................ 5
5. Radial and concentric sculpture about equal, strong.............*Lucinisca*
   Radial and concentric sculpture unequal, feeble................*Parvilucina*

Subgenus CALLUCINA DALL, 1901

Dorsal areas obsolete; lunule slightly asymmetrical; inner ventral margin finely crenulate.

268. Lucina (Callucina) lampra (Dall, 1901). Most other West American lucines are white, so that this species and the next may be recognized by the color, which in *L. lampra* varies from white to yellow or coral pink. The only sculpture is of fine concentric threads. The valves are slightly longer than high—about 17.5 mm long and 16 mm high. The head of the Gulf of California to Santa Cruz Bay, Mexico, intertidally and to depths of 55 m.

269. Lucina (Callucina) lingualis Carpenter, 1864. The height and length of this species are almost the same, about 18 mm, so that it is higher for its length than *L. lampra*. Shells may be white or suffused with a rich apricot color within. The sculpture is like that of *L. lampra*. Magdalena Bay throughout the Gulf of California to Acapulco, Mexico, intertidally and to depths of 24 m.

Subgenus CAVILINGA CHAVAN, 1937

Ligament a little sunken; lunule deep.

270. Lucina (Cavilinga) prolongata Carpenter, 1857. This shell is definitely higher than long, somewhat oblique in shape, but too small to be appreciated without a microscope. Length, 3 mm; height, 4 mm. The lunule is deep, almost like that of *L. excavata*. Scammon's Lagoon, Baja California, to Mazatlán, Mexico.

Subgenus HERE GABB, 1866

Lunule deep, distorting hinge; inner shell margin with denticles.

271. Lucina (Here) excavata Carpenter, 1857. The shell is white, with some well-spaced concentric riblets. A deeply sunken lunule crowds the anterior lateral teeth. A Californian species, *L. (H.) richthofeni* Gabb, 1866, has been considered synonymous. Now that a photograph of Carpenter's type is available, one can see differences, for *L. excavata* has a longer dorsal margin, with less rounding of the ends, the lunule is shallower, and the shell is thinner. Carpenter's type specimen, probably immature, is 10 mm in length; *L. richthofeni* may be as much as 25 mm long. Gulf of California to Mazatlán, Mexico, in depths to 110 m.

Subgenus LUCINISCA DALL, 1901

Sculpture reticulate; posterior area less distinct than anterior; inner margin denticulate.

272. Lucina (Lucinisca) centrifuga (Dall, 1901) (Synonym: **Phacoides liana** Pilsbry, 1931). Although this was named as a subspecies of the Californian *L. nuttalli* Conrad, 1837, which has regularly cancellate sculpture, it is apparently a distinct species. The holotype was a young specimen in which concentric lamellae are widely and irregularly spaced, with small scales at intersections of the radial ribs. In larger specimens the radial ribs are of uneven size, tending to alternate large and small as in the type of *L. liana*. Shells are pure white. Length of an average specimen, 15 mm. Gulf of California to Panama, intertidally and offshore to depths of 82 m.

273. Lucina (Lucinisca) fenestrata Hinds, 1845. One of the showiest and finest species of the genus in West American waters, with rasplike spines over the entire surface, a large specimen measuring 44 mm in height. Cedros Island, Baja California, to Peru, in depths of 13 to 73 m; mostly uncommon to rare.

Subgenus PARVILUCINA DALL, 1901

Small, rounded, with fine concentric sculpture; radial striae weaker in the middle.

274. Lucina (Parvilucina) approximata (Dall, 1901). Almost any handful of shell rubble from the beaches of southern California southward will contain valves of this abundant little lucine. Few specimens are more than 6 mm in length and height. Under a microscope they show fine concentric and radial ribbing. Southern California to Panama, intertidally and to depths of 1,024 m.

275. Lucina (Parvilucina) mazatlanica Carpenter, 1857. Resembling *L. (P.) approximata* but smaller, the anterior end longer, the lunule long and deep. Length, 4 to 4.5 mm. Gulf of California to Panama, in depths of 4 to 1,024 m.

Subgenus PLEUROLUCINA DALL, 1901

Radial sculpture of low folds, stronger at ends of shell; inner margin crenate.

276. Lucina (Pleurolucina) cancellaris Philippi, 1846. A small shell but beautiful under the microscope, with its ribs intersected by overriding concentric threads. Large specimens are as much as 6 mm high. Cedros Island, Baja California, to Panama in 7 to 70 m. Dall unfortunately assigned this species and its Caribbean twin, *L. amianta* (Dall, 1901), to the subgenus *Bellucina*, the type of which is an Indian Ocean form only superficially similar, and American authors have been slow to recognize the error.

277. Lucina (Pleurolucina) leucocymoides (Lowe, 1935). The center of the disk in this shell is a single wide fold, set off from the anterior and posterior areas. The concentric sculpture is of reflected ribs. Length 10.7 mm; height, 11 mm. Angel de la Guarda Island, Gulf of California, to Manzanillo and Tres Marias Islands, Mexico, in 37 to 110 m.

278. Lucina (Pleurolucina) undatoides Hertlein & Strong, 1945 (Synonym: **L. undata** Carpenter, 1865 [not *Venus undata* Pennant, 1777]). The three to five wide ribs of this species are crossed by concentric threads that appear to be undulating. The shell is longer and less convex than *L. leucocymoides* (length, about 12 mm). The species seems confined to the Gulf of California area. Lowe reported it as rare, on mud flats, at La Paz.

Genus CODAKIA SCOPOLI, 1777

These are large, flattened shells with strong radial and concentric sculpture. The ligament is in a deep groove channeled between the overlapping dorsal margins. The anterior lateral teeth are crowded close against the cardinals. Inner margin smooth.

279. Codakia distinguenda (Tryon, 1872) (Synonyms: **C. colpoica** Dall, 1901; **C. pinchoti** Pilsbry & Lowe, 1932). The elegantly sculptured saucer-shaped white valves of this shell, with their wide rose-red inner margin and creamy-yellow center, make an attractive souvenir for any collector, and they are not uncommon on the beaches of the whole province. This is the largest member of the family in West American waters, a large specimen measuring as much as 140 mm across (about 5½ inches). The East American *C. orbicularis* (Linnaeus) is very similar in form. *C. pinchoti* Pilsbry & Lowe, 1932, was proposed for Panama shells that were more tumid and had stronger sculpture, but the differences seem within the range of variation of the species. However, Olsson (1961) considers the Panamic form to be distinct. Magdalena Bay, Baja California, to Panama, in shallow water, on tide flats at extreme tide.

Genus CTENA MÖRCH, 1861
(JAGONIA RÉCLUZ, 1869)

Resembling *Codakia* but smaller, radial ribs increasing by bifurcation; shell margin finely denticulate within.

280. Ctena chiquita (Dall, 1901). The small flat yellowish-white shells have finely bifurcating threadlike ribs that are absent on dorsal areas. Length, about 14 mm. West coast of Baja California to La Libertad, El Salvador, in 10 to 120 m.

281. Ctena clarionensis Hertlein & Strong, 1946. The outline is more oblique in this white shell than in the other species. Length, about 14 mm; height, 12.5 mm; diameter, 8 mm. Clarion Island, off the west coast of Mexico.

282. Ctena clippertonensis Bartsch & Rehder, 1939. Slightly more nearly circular than the last, and thinner, but also having the ribbing continued onto the dorsal areas. Length, 15 mm; height, 14 mm; diameter, 7.5 mm. Clipperton Island, off Central America, to Panama, on rocks and offshore to 75 m.

283. Ctena galapagana (Dall, 1901). Larger size and coarse ribbing distinguish this form, which is ribbed throughout. Length, 20 mm; height, 9 mm; diameter, 9 mm. Off Nicaragua to the Galápagos Islands, Ecuador, intertidally and to 37 m.

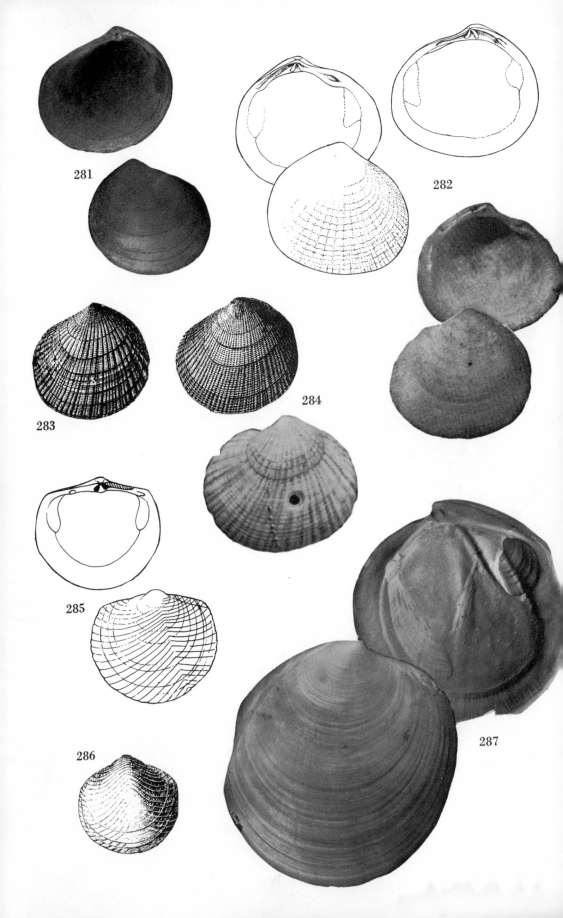

281

282

283

284

285

286

287

284. Ctena mexicana (Dall, 1901) (Synonym: **Lucina pectinata** Carpenter, 1857 [not C. B. Adams, 1852]). The most widely distributed of the Panamic Ctenas, recognizable by its fine and numerous radial ribs, which are evenly distributed over the entire white shell. Length, 22 mm; height, 19 mm. The type of the genus. *C. galapagana* may prove to be a geographical subspecies of *C. mexicana*. The Gulf of California to Ecuador, intertidally and to depths of 80 m.

Subfamily DIVARICELLINAE

Shell convex, rounded, with divaricate or undulating sculpture.

Genus DIVALINGA CHAVAN, 1951
(DIVARICELLA of authors, not of VON MARTENS, 1880)

The shells are white, nearly orbicular, marked with obliquely radiating lines that meet to one side of the center of the disk. Lateral hinge teeth well developed; inner margin of shell denticulate.

Long known as *Divaricella*, this was shown by Chavan (1951) not to be close to the type of that group, which lacks lateral teeth and is smooth within. Two Panamic subgenera.

Subgenus DIVALINGA, s. s.

Divaricate sculpture continuous; anterior adductor scar rounded at end.

285. Divalinga (Divalinga) eburnea (Reeve, 1850) (Synonyms: **Divaricella lucasana** Dall & Ochsner, 1928; **Divaricella columbiensis** Lamy, 1934; [not preoccupied by *Lucina eburnea* Deshayes, 1835, a *nomen nudum*, or by *Venus eburnea* Gmelin, 1791, a *Ctena*]). Nearly spherical, white, with well incised oblique sculpture. Magdalena Bay, Baja California, to Peru, intertidally and to depths of 55 m. A closely related species, *D. quadrisulcata* (Orbigny, 1846) occurs in the Caribbean.

Subgenus VIADERELLA CHAVAN, 1951

With a smooth area at junction of diverging oblique lines, radially from umbones down the disk; anterior adductor scar pointed at end.

286. Divalinga (Viaderella) perparvula (Dall, 1901) (Synonym: **Lucina pisum** Philippi, 1850 [not Sowerby, 1836]). Smaller than *D. (D.) eburnea*, with an incomplete pattern. There may also be well-developed radial striae. Length of adults about 20 mm. Southern part of the Gulf of California to Ecuador. The type specimen was a juvenile shell only 7 mm long.

Subfamily MILTHINAE

Shells relatively solid; sculpture obsolete to weakly concentric; anterior adductor muscle scars long. Inner margin of shell smooth. Two Panamic genera.

Genus MILTHA H. & A. ADAMS, 1857

Shells flat and discoidal, smooth or with irregular concentric striae, a little inequilateral, with an asymmetrical lunule; ligament long, somewhat sunken below the dorsal border, on an enlarged nymph; posterior and anterior areas set off by radial ridges; hinge without lateral teeth.

287. Miltha xantusi (Dall, 1905). Lenticular, solid, ivory white. Length, about 70 mm; height, 71 mm. One of the largest and rarest of the West American lucines, it is known by only a few specimens in the southern part of the Gulf of California, mostly off Cape San Lucas, in depths of 55 or more meters.

Genus PEGOPHYSEMA STEWART, 1930
(LISSOSPHAIRA OLSSON, 1961)

Inflated shells, lacking sculpture but with anterior area set off; lunule narrow, depressed; hinge plate triangular, without teeth, posterior end of ligament supported by a shelly ridge or nymph. Anterior adductor muscle scar relatively long and narrow.

288. Pegophysema edentuloides (Verrill, 1870). The hinge is without teeth except in very young specimens; mature adults are inflated and nearly smooth, but juvenile shells are flatter, with weak furrows setting off dorsal areas, especially in anterior area. Length, to 65 mm (average about 45 mm); average height, 40 mm. Cedros Island, Baja California, and through the Gulf of California to Tenacatita Bay, Mexico, in 33 to 165 m. The type of *Lissosphaira, P. spherica* (Dall & Ochsner, 1928), occurs as a fossil in the Galápagos Islands and Ecuador; reports of living specimens probably are based on the relatively flat young of *P. edentuloides*. A similar species in the western Atlantic is *P. schrammi* (Crosse, 1876), type of the genus. The name *Anodontia* Link, 1807, has been used for these species, but the type species of that group, which is western Pacific in distribution, has a number of significant points of difference—the ligament sinks down posteriorly into the shell cavity, without any supporting nymph; the hinge plate has a faint pustular cardinal tooth; and the anterior adductor muscle scar is relatively short. The type species of *Anodontia* is *Venus edentula* Linnaeus, 1758, an Oriental form, according to Chavan in Moore (1969).

Subfamily MYRTAEINAE

Shells thin, somewhat quadrangular, compressed, with concentric sculpture; anterior adductor muscle scars medium-sized.

Genus LUCINOMA DALL, 1901

Lenticular, lacking dorsal areas; lunule long, not sunken; hinge plate without lateral teeth; inner margins of shell smooth.

289. Lucinoma annulata (Reeve, 1850). Essentially a northern species, this form reaches its southern limit in the Gulf of California. The sculpture is somewhat like that of *Lucina excavata* in that between the widely spaced raised concentric ridges are fine growth lines, but this shell has a thin, dark olive-green periostracum and is much larger—as much as 58 mm in length. It is always an offshore species, dredging records being at depths of 55 to 90 m.

290. Lucinoma heroica (Dall, 1901). Described as a *Phacoides*, from deep water. Length, 71 mm. Southern part of the Gulf of California, in 1,836 m.

Family UNGULINIDAE
(DIPLODONTIDAE)

Moderately small shells, more or less globular, smooth—or at least not ribbed—with large muscle scars and an entire pallial line. The hinge has no lateral teeth but has two cardinals in each valve, one or both of which may be bifid. Though nestlers by habit, these little clams have undistorted shells because they form nests around themselves of mud particles agglutinated with mucus. Three Panamic genera.

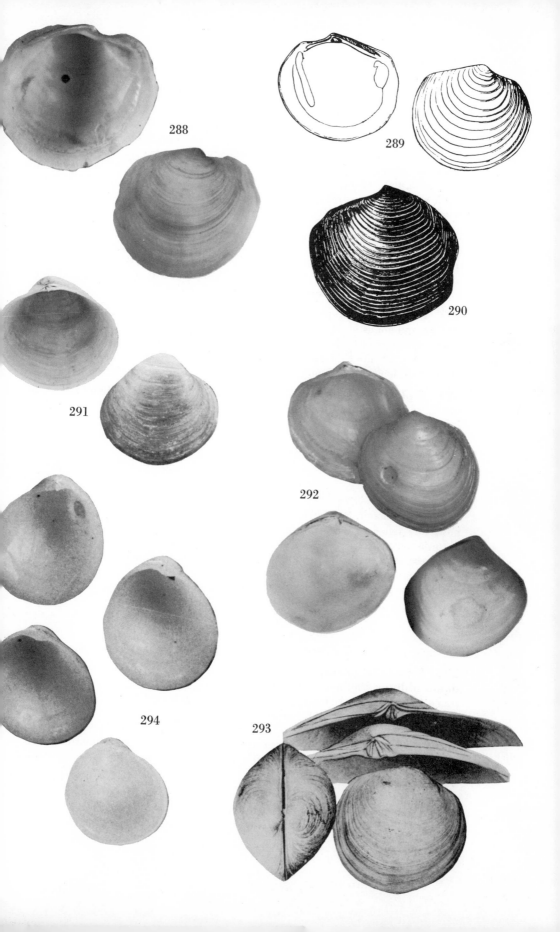

288

289

290

291

292

293

294

Genus DIPLODONTA BRONN, 1831

In recent years there has been an attempt to displace Bronn's generic name by the supposedly prior *Taras* Risso, 1826. Chavan (1952) has demonstrated, however, that Risso's type species is unidentifiable and may well belong to another family. Hence, we return to the use of *Diplodonta*, a word that means "double-teeth," an appropriate reference to the bifid cardinals. The ligament is external, on a small platform behind the beaks. It is of even width and may extend underneath and slightly in front of the beaks, an arrangement that is not common in bivalves. The shell is inflated, nearly equilateral.

291. Diplodonta inezensis (Hertlein & Strong, 1947). The white shell is thin, the posterior cardinal deeply bifid, the beaks and umbones narrow. Length, 18 mm; height, 17 mm; diameter, 16 mm—a nearly spherical shell. Only offshore localities of collection have been reported, from Santa Inez Bay, Gulf of California, to Pearl Islands, Panama, in depths of 11 to 64 m.

292. Diplodonta subquadrata (Carpenter, 1856). This is a thinner shell than the last, and not as nearly spherical. The outline is more quadrate. Dimensions of a large specimen: length, 29 mm; height, 25 mm. San Ignacio Lagoon, Baja California, through the Gulf and south to the Galápagos Islands, on mud flats and offshore to depths of 137 m.

293. Diplodonta suprema Olsson, 1961. Unusual among the living species of the genus on account of its size, this may reach a length of 37 mm. The shell is inflated, thin, and white or ivory-colored, with fine growth striae. It much resembles the Californian *D. orbella* (Gould), 1851, even to the short ligament bounded posteriorly by a small escutcheon. Length, 37 mm; height, 31 mm; diameter, 28 mm. Panama.

Genus FELANIELLA DALL, 1899

Lenticular, smooth; periostracum polished; anterior muscle scar narrower than posterior; ligament in a marginal groove.

In the type species of *Felaniella*, *F. usta* (Gould, 1861), from Japan, the ligament resembles that in *Diplodonta*, but the West American species that have the compressed shape and the shiny periostracum have the ligament as in the New Zealand group of *Zemysia*, broader just back of the beaks. Thus, the subgenus *Felaniella*, *s. s.*, is not represented in the Panamic province.

Subgenus ZEMYSIA FINLAY, 1927

Ligament sunken and widened anteriorly; muscle scars relatively broad.

294. Felaniella (Zemysia) calculus (Reeve, 1850) (Synonym: **? Diplodonta obliqua** Philippi, 1846). Philippi's type specimen has never been illustrated, and authors have interpreted *D. obliqua* from his brief description. Reeve's figure of *D. calculus* probably has provided the concept of the form. A photograph of Reeve's type material is given here; it was from Nicaragua. The exact range of the species has yet to be worked out, for many specimens so identified turn out to be other Diplodonts. Length, 6 mm; height, 7 mm.

295. Felaniella (Zemysia) sericata (Reeve, 1850) (Synonyms: **Lucina cornea, nitens,** and **tellinoides** Reeve, 1850; **Diplodonta artemidis** Dall, 1909; **D. serricata** of authors). Some workers would argue that, on the basis of strict priority, the name *L. cornea* (Reeve) should be used, because it appears first on the page, ahead of the other three, and a few later authors have adopted it. However, the International Code on Zoological Nomenclature establishes the "first reviser" rule—the first person to recognize that two or more names are synonyms selects one, which shall stand. The first worker to recognize that *Lucina cornea*, *nitens*, and *sericata* are identical forms was Dall (1901), and he selected *sericata* as the best-known name—a somewhat unfortunate choice, for that species was described without known locality, whereas the others were definitely West American. He did not realize that *Lucina tellinoides*, which he considered a *Pseudomiltha*, might fit into this complex. Had he chosen *F. cornea* as the valid name, a case might be made for recognizing *F. tellinoides* as subspecifically different, but *F. sericata* stands midway between the two forms in outline. The shell of this species is thin, white, semitranslucent, covered by a pale-olive periostracum. A large specimen measures 22 mm in height. There seems to be a tendency for larger and older specimens to be more inequilateral, with lower and less conspicuous umbones; concentric sculpture may also be more strongly developed. Olsson (1961) separated these as a species that he cited as *F. tellinoides*. Although *F. sericata* has been reported as far north as Monterey Bay, California, it is not common north of San Ignacio Lagoon, Baja California; from there it ranges south to northern Peru. It may be found intertidally, but more commonly is dredged in depths to 75 m on sand or mud bottoms.

Genus PHLYCTIDERMA DALL, 1899

The shell is shaped like that of *Diplodonta*, but the surface is beaded, pustulose, punctate, or even almost reticulate.

Subgenus PHLYCTIDERMA, s. s.

Ligament in a groove on dorsal margin but not sinking below; dorsal margin of even width.

296. Phlyctiderma (Phlyctiderma) discrepans (Carpenter, 1857) (Synonym: **Diplodonta semirugosa** Dall, 1899). Dall's specific name was based upon Carpenter's description of material he had taken from a cavity in a *Chama* valve, but for some reason Dall regarded Carpenter's species (proposed as a variety) as unidentifiable. The photograph of the type specimen in the British Museum, given here, confirms its similarity to other West American material identified as *P. semirugosum*, although—this specimen being young—its pustular sculpture is not well developed. The surface sculpture in this species is of fine to coarse concentric pustules or punctations. A large specimen measures 15 mm in height. The Gulf of California to Panama, mainly offshore, in depths to 18 m. A similar species in the Atlantic is *P. semiasperum* (Philippi, 1836).

297. Phlyctiderma (Phlyctiderma) elenense Olsson, 1961. Small, plump, with broad, high umbones projecting prominently above the hinge line. Sculpture of elongate concentric nodes. Length, 9 mm; height, 8 mm; diameter (one valve), 3.6 mm. Specimens of the California Academy of Sciences from Espíritu Santo Island, Baja California, and in the Stanford University collection from Tres Marias Islands, Mexico, extend the range northward from the type locality, Santa Elena, Ecuador.

298. Phlyctiderma (Phlyctiderma) insula Olsson, 1961. Resembling *P. caelatum* but much smaller, with the pustules rounded, more aligned concentrically. Length, 8.5 mm; height, 7.5 mm; diameter (one valve), 3 mm. Panama and western Colombia.

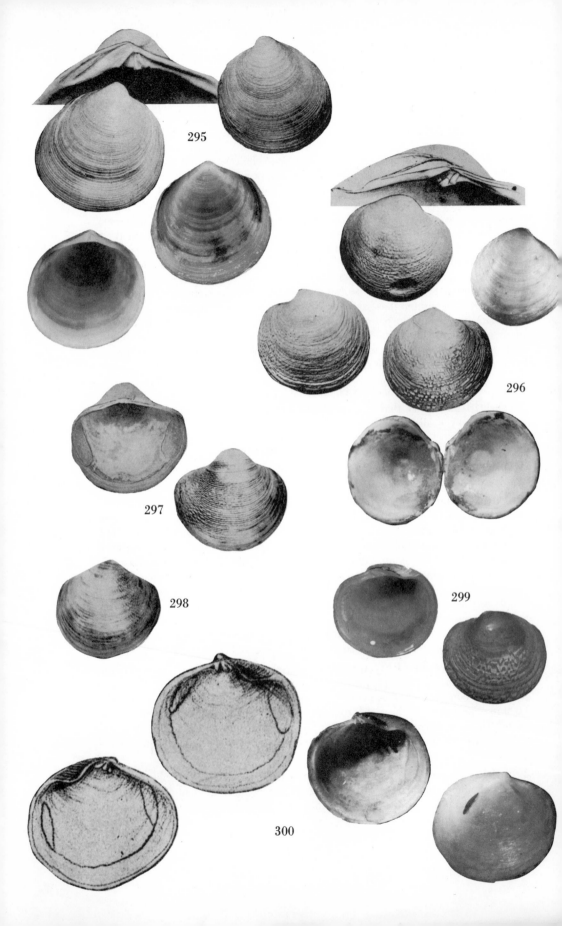

295

296

297

298

299

300

Subgenus **PEGMAPEX** BERRY, 1960

Ligament progressively lower posteriorly and dorsal margin widened.

299. Phlyctiderma (Pegmapex) caelatum (Reeve, 1850). Although the species has hitherto been considered a typical *Phlyctiderma*, the deeply sunken ligament is more like that of the type species of *Pegmapex*. The surface sculpture is rather coarse, of concentric ridges broken up into a zigzag pattern. Diameter, about 25 mm. The range is from Sihuatanejo, Mexico, to Guayaquil, Ecuador, intertidally and to 20 m.

300. Phlyctiderma (Pegmapex) phoebe (Berry, 1960). This is the type of the subgenus. The shell is thin, small, white, with low beaks. In front the hinge plate is narrow, but it flares downward posteriorly to accommodate the sunken ligament in a shelflike arrangement. As long ago as 1857 Carpenter found some of these shells among other Diplodontas in a cavity within a large *Chama* valve, but because of the variability of form of the lot, he did not name this as new, and its distinctness was not recognized until recent years. Length of holotype, 6 mm; height, 5 mm. Mazatlán, Mexico.

Family **THYASIRIDAE**

Thin, trigonal, edentulous, posterior area set off; muscle scars long, superficial.

Genus **THYASIRA** LAMARCK, 1818

Inequilateral, with one or more deep furrows or folds setting off the posterior part of the shell. Hinge without teeth.

301. Thyasira barbarensis (Dall, 1890). Lenticular, white, this form has one fairly deep radial furrow on the anterior slope and a weak depression toward the posterior dorsal margin. Length and height, 17 mm; diameter, 10 mm. Mostly in deep water from Puget Sound, Washington, to southern California; dredged in 90 to 110 m off Guaymas, Mexico.

302. Thyasira excavata Dall, 1901. The shell is thin and white under a pale yellow periostracum. The type is from very deep water and has three sharp and two or three obscure radial ridges. Specimens that may belong in this species have been dredged in recent years from shallower water, but they have less deep folds and weaker or obsolete radial ridges; also, the anterior margin is less convex. Perhaps these will later be deemed a separate species when there is more material for comparison. Length, 20 mm; height, 17.5 mm; diameter, 15 mm. Gulf of California between San Marcos Island and Guaymas in 1,830 m (type locality); also in deep water off Tillamook, Oregon, according to Dall; off La Paz, Guaymas, and Gulf of Tehuantepec, 18 to 90 m.

Superfamily **GALEOMMATACEA**
(**ERYCINACEA** and **LEPTONACEA** of authors)

Small bivalves, mostly symbiotic with other invertebrates, these clams have adapted to life in the burrows of crabs, worms, etc., where they find a food source in the material the host winnows out of the water. They are not parasitic, for they do not feed directly on the tissues of the associate. They may, however, attach themselves by a byssus on some special crevice of the host, and the valves may be distorted by pressure—adding to problems of recognition for the systematist. The group has been relatively little studied, and it is probable that many more species occur than are listed here.

The superfamily names Erycinacea or Leptonacea are commonly used for this complex of genera; technically, the earliest and therefore the correct name must be Galeommatacea, because the family-group name Galeommatidae Gray, 1840, has priority over Leptonidae Gray, 1847, and Erycinidae Deshayes, 1850.

The review of this group by Chavan in Moore (1969) recognizes several families. Because the distinctions between families (five, in the Panamic province) are mainly on details of the hinge, it seems preferable here to give a key—admittedly an artificial one—to all the 13 genera currently accepted as members of the superfamily in the province (two of uncertain status are omitted), basing it on shell features that are readily observable. The grouping of genera into families follows, in the main, the recommendations of Chavan in Moore (1969).

1. With well-marked radial sculpture in some part of the valve (may be
   shown only as marginal crenulations) 2
   Without radial sculpture; shells virtually smooth..................... 6
2. Radial sculpture strong, over entire valve.......................... 3
   Radial sculpture weak, crenulating ventral margin................... 4
3. Hinge teeth delicate, hinge plate narrow....................... *Cymatioa*
   Hinge teeth well developed, on a widened hinge plate........... *Solecardia*
4. Outline trigonal, somewhat inequilateral..................... *Pythinella*
   Outline quadrate, nearly equilateral............................... 5
5. Radial sculpture finer at ends of shell, somewhat divaricate..... *Galeommella*
   Radial sculpture even, not divaricate at ends of shell.......... *Tryphomyax*
6. Anterior lateral teeth strong, well removed from cardinals............. 7
   Anterior lateral teeth absent or only weakly developed, close to cardinals.. 9
7. Shell surface with wrinkled sculpture, outline ovate-quadrate........ *Lasaea*
   Shell smooth, outline oblique..................................... 8
8. Lateral teeth well removed from cardinals.................... *Amerycina*
   Lateral teeth close to cardinals................................ *Lepton*
9. Adductor muscle scars high inside valves....................... *Bornia*
   Adductor muscle scars in normal position........................10
10. Outline nearly equilateral...................................... *Kellia*
    Outline inequilateral 11
11. Right valve without cardinal teeth........................... *Mysella*
    Both valves with cardinal teeth.................................12
12. Cardinals thin, small, hinge plate narrow..................... *Aligena*
    Cardinals heavier, buttressed by widened hinge margin.......... *Orobitella*

Family GALEOMMATIDAE
(GALEOMMIDAE of authors)

Quadrate; hinge with small teeth; resilium not sharply defined; shell surface sculptured, usually with radial riblets. Three genera in the Panamic province.

Genus GALEOMMA TURTON, 1825

Sculpture of radial riblets that are somewhat differently spaced anteriorly and posteriorly; ventral margin gaping. The type species is *Galeomma turtoni* Sowerby *et al.* in Turton, 1825 (often cited more simply as of Sowerby, 1825). It is from the northeastern Atlantic. No closely similar species occurs in the Panamic province, for none shows a ventral gape, which is fairly wide in *G. turtoni*. Rather, the Panamic species seem to belong in *Tryphomyax*.

Genus CYMATIOA BERRY, 1964
(CRENIMARGO BERRY, 1963, not COSSMANN, 1902)

Rounded quadrate, somewhat inflated, with a fine surface pattern of minute punctations and a few faint and widely spaced radial ribs that show mainly as denticles on the shell margin, giving it a wavy appearance. Resembling the leptonid genus *Solecardia* but with a weaker and narrower hinge.

303. Cymatioa dubia (Deshayes, 1856). The shell is small, white, with the beaks at the anterior third instead of central. Radial markings are faint but show as minute crenulations of the ventral margin. Length, 7 mm; height, 4.5 mm. Guayaquil, Ecuador. The figure is of the type specimen, in the British Museum.

304. Cymatioa electilis (Berry, 1963). Shell milky-translucent, the adductor muscle scars, which are high up on the valves, showing through to the outside. Beaks are nearly central. The hinge has small and delicate teeth, one cardinal and one posterior lateral in the right valve, two cardinals and a posterior lateral in the left. Length, 16 mm; height, 12 mm; diameter, 7 mm. Guaymas, Mexico, and southern end of the Gulf of California in depths to 45 m. Manzanillo (type locality) and south to the Galápagos Islands (Stanford University specimens).

305. Cymatioa pulchra (Philippi, 1849) (Synonym: **? Scintilla cumingii** Deshayes, 1856). The shell, which Philippi described as a *Kellia* from the shores of West America, has never been illustrated. He said it was ovate, nearly equilateral, white, a little inflated, with the ventral margin slightly dentate. His description of the hinge as well as that of the shell seems close to the figure of Deshayes's shell from Panama. Philippi's shell was apparently smaller and proportionally higher, the length being 18 mm, height, 13 mm, diameter, 5.8 mm, whereas Deshayes's type, which is in the British Museum, measures 25 mm in length, 15 mm in height. Deshayes's species has been considered a synonym of *Solecardia eburnea* Conrad on the basis of a comparison of specimens by Carpenter a century ago, but since the illustration of the hinge seems more like *Cymatioa* than *Solecardia*, a revised synonymy is suggested here.

Genus GALEOMMELLA HABE, 1958

Quadrate, small, with radial and concentric sculpture, radial predominating and tending to become divaricate and finer at ends of shell. Valves not gaping. The type species, *G. utinomii* Habe, 1958, is from southern Japan.

306. Galeommella peruviana (Olsson, 1961). The ribbing is almost reticulate, for well-developed concentric riblets intersect the radials, but they fade out at the ends, where only fine divaricate radials fan out toward the posterior margin. Length, 4.8 mm; height, 2.7 mm. Puerto Peñasco, Sonora, Mexico, in beach drift (Mrs. G. Riley Hettick collection) to Salinas, Ecuador (Stanford University collection), and Zorritos, Peru (type locality). The species was assigned to *Solecardia* by Olsson, but the hinge is closer to that of *Cymatioa*; sculpture suggests allocation to the genus *Galeommella*, which hitherto was recorded only in Japan.

Genus TRYPHOMYAX OLSSON, 1961

Quadrate, with even radial sculpture not divaricate at ends of shell; ends slightly gaping in some specimens, and most show a radial sulcus and ventral notch.

307. Tryphomyax lepidoformis Olsson, 1961 (Synonym: **T. l. laevis** Olsson, 1961). Type species of the genus. Shell small, thin, flat, scalelike, nearly rec-

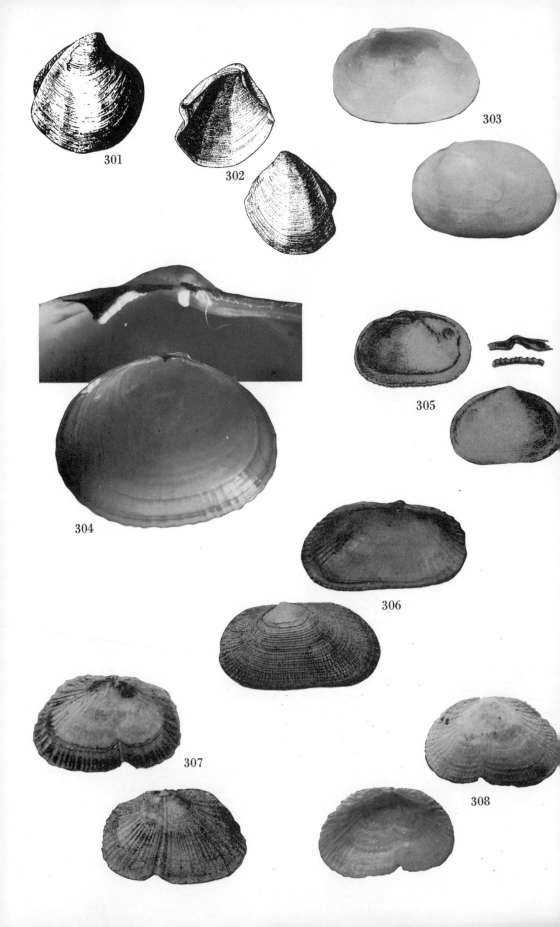

301

302

303

304

305

306

307

308

tangular in outline, with a radial channel dividing it into two nearly equal parts, probably resulting from pressing of the shell margin against some structure on the host organism; a smooth form lacking the sulcus was suggested as a morphological subspecies by Olsson. The shells are white, with a small brown cap on the tip of the beak (the prodissoconch). Hinge with one cardinal tooth in the right valve, two in the left. Length, 4 mm; height, 2.5 mm. Panama.

308. Tryphomyax mexicanus (Berry, 1959). Except for slightly larger size and finer ribbing, this is very similar to *T. lepidoformis,* and if further collecting fills in the geographic gap, this name will take precedence. Because of the wide geographic separation, the two are listed here as separate species. Length, 6 mm; height, 3.7 mm. Off Bahía San Luis Gonzaga, Baja California, near the head of the Gulf of California, in 5 to 7 m (type locality) to Puerto Peñasco, Sonora, Mexico, beach drift (Mrs. G. Riley Hettick collection). The species was allocated to *Galeomma* by Berry, the name *Tryphomyax* not at that time having been proposed.

Family ERYCINIDAE
(LASAEIDAE)

Small, somewhat compressed shells, the hinge plate with cardinal and lateral teeth in both valves, a triangular but ill-defined socket indenting the center of the plate, for the reception of the internal ligament. Two Panamic genera.

Genus AMERYCINA CHAVAN, 1959

Although resembling closely the European genus *Erycina* Lamarck, 1805, to which a number of West American species have in the past been assigned, this genus differs by having opisthogyrate beaks, the anterior end being longer than the posterior. The pivotal cardinal tooth, in the left valve, is bifid.

309. Amerycina colpoica (Dall, 1913). The shell is strongly inequilateral, with the beaks nearly at the posterior end. White, with a pale yellow periostracum. Length, 10 mm. Head of the Gulf of California to Corinto, Nicaragua, intertidally and to depths of 24 m.

310. Amerycina cultrata Keen, new species. The shell is somewhat inequilateral but not markedly so, shaped a little like a knife blade, with the anterior end longer and more broadly rounded than the posterior, the posterior end being almost truncate. Hinge of right valve with one small cardinal tooth; laterals, one anterior and one posterior, somewhat distant from the beaks; hinge of left valve with a large cardinal and narrow anterior and posterior lateral teeth. Anterior adductor muscle scar circular, high in the shell, posterior scar longer, paralleling posterior margin. Three valves are in the Stanford University collection, all dredged off Isla Partida, Espíritu Santo Island, near La Paz, Baja California, in depths of 5 to 33 m. Length (largest valve), 12 mm; height, 7 mm. The shell differs from *A. colpoica* in being only slightly inequilateral.

Genus LASAEA BROWN, 1827

The broad and somewhat inflated umbones, quadrate shape, and tendency toward reddish or brownish color are characteristics of these small shells; the hinge is heavy for the size of the shell, with large laterals and a slender thornlike cardinal in each valve.

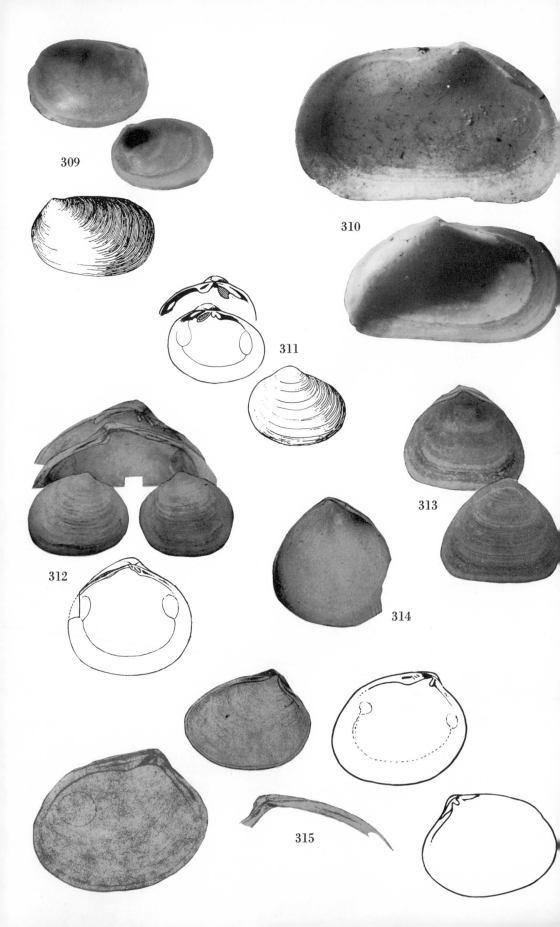

311. Lasaea subviridis Dall, 1899. The species is characterized by its light greenish or yellowish-gray shell color, suffused with pink at the beaks, and by the wavy concentric sculpture. Length, 3.1 mm; height, 2.5 mm; diameter, 1.4 mm. Shelter Cove, California, to the Gulf of California; intertidally, especially among *Mytilus*.

Lasaea sp. There are a number of records of *Lasaea* in the Gulf of California that may be of undescribed species. Much more work is needed on these and related small bivalves. The two Lasaeas described by Carpenter prove to belong in the Montacutidae.

Family KELLIIDAE

Hinge with two cardinal teeth and strong posterior laterals; ligament wholly internal, in a socket that forms a gap behind the cardinal teeth on the hinge plate. Two (possibly three) Panamic genera.

Genus KELLIA TURTON, 1822

Subquadrate, inflated, smooth, usually with a varnish-like periostracum; hinge plate with a wide notch below the beaks.

312. Kellia suborbicularis (Montagu, 1803) (Synonym: **? Montacuta chalcedonica** Carpenter, 1857). The inflated little shell is thin, nearly translucent, white, its surface smooth or polished, and, as other authors have noted, there seems no way to distinguish Pacific specimens from those in Atlantic waters, where the species is widely distributed. Carpenter's species, the type of which is in the British Museum, is based on a juvenile shell only a millimeter long, which was somewhat broken anteriorly, thus perhaps appearing shorter than it should; its hinge suggests *Kellia*. Length of an average specimen, 5 mm, height, 4 mm. Nestling in crevices of rocks and other shells, from British Columbia to Peru in the eastern Pacific; also Atlantic.

Genus BORNIA PHILIPPI, 1836

Subtrigonal, white, smooth or with minute striae; hinge plate narrow, with strong lateral teeth and sockets; adductor muscle scars unequal, placed high in the shell; interior punctate in some.

313. Bornia chiclaya Olsson, 1961. Rounded-trigonal, high; beaks slightly prosogyrate; surface concentrically striate. Length, 7 mm; height, 6 mm; diameter, 1 mm (one valve). Northwestern Peru.

314. Bornia egretta Olsson, 1961. High-ovate, with narrow beaks; surface nearly smooth. Length, 10 mm; height, 9.7 mm; diameter, 2 mm (one valve). Northwestern Peru.

315. ? Bornia obtusa (Carpenter, 1857). The type specimen in the British Museum is an immature shell only 1.7 mm long. Carpenter considered the species to be a *Montacuta* because of the inequilateral outline, but the hinge resembles that of *Bornia*. Mazatlán, Mexico.

316. ? Bornia papyracea (Deshayes, 1856). The thin shell is so transparent that the muscle scars may been from the outside. The type specimen, a photograph of which is given here, is in the British Museum; the hinge was not exposed for study. Length, 15 mm; height, 11 mm. Off La Paz and Loreto, Baja California (Stanford collection) to Ecuador, in depths to 46 mm. Type locality, Colombia.

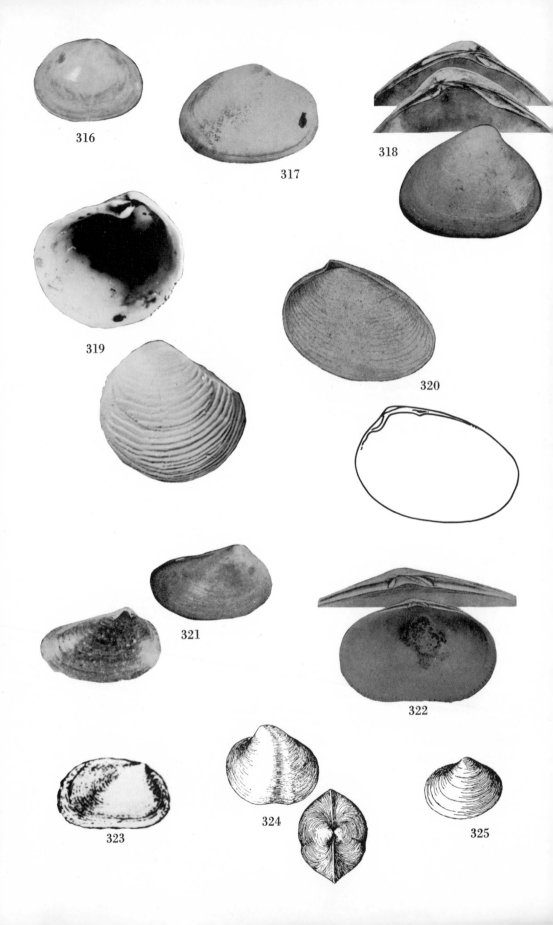

317. Bornia venada Olsson, 1961. Similar to *B. papyracea* in outline but smaller and slightly longer in proportion. Length, 9 mm; height, 6.5 mm; diameter, 2.4 mm. Panama.

318. Bornia zorritensis Olsson, 1961. Less trigonal than the other species in outline, with wider umbones; surface smooth. Length, 10.5 mm; height, 8 mm; diameter (one valve), 2.1 mm. Northwestern Peru.

Genus **ODONTOGENA** COWAN, 1964

Subcircular, with a strong cardinal tooth on the hinge of the right valve, a smaller one above it, and two cardinals in the left valve, the anterior merged with an anterior lateral tooth; posterior laterals weak to well developed.

The type species, *O. borealis* (Cowan, 1964), is from British Columbia, Canada, offshore. It approaches in form the Miocene fossil figured by Chavan in Moore (1969) as representing *Aligena*, a genus considered to belong in the Montacutidae by most authors. Until type material can be critically studied, the expedient of utilizing a West American name is here adopted, for at least one and perhaps two undescribed tropical West American species are involved.

319. ? Odontogena sp. A single valve of an unnamed species that probably represents an unnamed subgenus or even genus is in the Stanford University collection, from Ecuador. It comes closest to *Odontogena* among named generic taxa, but it has not the strong posterior laterals, at least in the one right valve seen, which is here figured. Description of the species should await further material and an opposite valve. The hinge has a strong central cardinal tooth and a smaller hook-shaped anterior cardinal above it, merging with an anterior lateral. There is a large socket behind the cardinal for the resilium. No posterior lateral is evident in this valve. The most remarkable feature of the solid little white shell is the sculpture, which is concentric, with irregular breaks in the ribs that make them appear dichotomous. Length, 6 mm; height, 5.9 mm. Salinas (Santa Elena), Ecuador, in beach drift.

Family **LEPTONIDAE**

Compressed, beaks small; hinge plate little indented by ligament or resilium; pallial line irregular and well removed from the ventral margin. Two Panamic genera.

Genus **LEPTON** TURTON, 1822

Small, smooth or with radial sculpture, ligament mostly internal; hinge normally with one or two cardinal teeth and two laterals.

320. ? Lepton ellipticum (Carpenter, 1857). The ovate little shell is so thin the faint concentric sculpture can be seen through to the inside on the holotype, which is in the British Museum. Length, about 3.5 mm. Mazatlán, Mexico.

321. Lepton lediforme Olsson, 1961. As the specific name implies, this small shell has the outline of a *Nuculana* (a genus long called *Leda*). The beaks are near the anterior end, and there are on the hinge two prominent teeth with a gap between, where the resilifer is lodged. Length, 2.5 mm; height, 1.6 mm. Panama.

Genus **SOLECARDIA** CONRAD, 1849

Elongate-quadrate, with a fine surface pattern of minute punctations. Hinge plate wide, hinge teeth well developed.

322. Solecardia eburnea Conrad, 1849. The largest member of the superfamily in west Central American waters. The shell is thin, oblong, and white, with fine surface punctations. The pallial line bypasses the adductor muscle scars. Length, 21 mm; height, 14 mm. Puerto Peñasco, Gulf of California, to Panama, intertidally and in shallow water to 5 m.

323. ? Solecardia obliqua (Sowerby, 1862). The outline of the shell is inequilateral, and the color was said to be brown. According to Olsson (1961), the species may be an *Orobitella*; a decision about its placement must await study of the hinge. It was described as a *Scintilla*. Length, 9 mm; height, 5.5 mm. Type locality, Ecuador.

Family MONTACUTIDAE

Shells slightly to moderately inequilateral, with a socket below the beak for the resilium; hinge with one somewhat hook-shaped tooth near the center (not a true cardinal tooth but the bent-up end of an anterior lateral) and one or more posterior lateral teeth. Four Panamic genera.

The type genus, *Montacuta* Turton, 1822, is not yet reported in the Panamic province; it occurs in the North Pacific and the Atlantic.

Genus ALIGENA LEA, 1846

Valves somewhat trigonal, the anterior end longer, beaks slightly prosogyrate; hinge plate rather narrow, with one small "cardinal" tooth in either valve and a long oblique socket for the resilium, ending in a widening of the hinge plate that simulates a posterior lateral tooth. Harry (1969) has reviewed the American species.

The genus has been interpreted by Chavan in Moore (1969) as having two cardinal teeth and belonging in the Kelliidae, on the basis of topotype material from the Miocene of Virginia. However, the conventional interpretation of *Aligena* as a montacutid seems justified because Lea in the original description emphasized the presence of only a single cardinal tooth in either valve.

324. Aligena cokeri Dall, 1909. A strong radial constriction runs down the middle of the shell, probably caused by pressure of the margin against some structural feature of the host, most of these small clams being symbiotic with such other invertebrates as segmented worms, crustaceans, and other burrowers. The hinge teeth in this species are weak to wanting. Length, 9.5 mm; height, 8 mm; convexity, 8 mm. Puerto Peñasco, Sonora, Mexico, to Peru, intertidally and to depths of 25 m.

325. Aligena nucea Dall, 1913. Small, white, less inflated than *A. cokeri*. Length, 4 mm; height, 3 mm; convexity, 2.2 mm. Puerto Peñasco, Sonora, Mexico, to Corinto, Nicaragua, intertidally and to depths of 25 m.

Genus MYSELLA ANGAS, 1877
(ROCHEFORTIA VÉLAIN, 1878)

Subquadrate to rounded-trigonal, with prominent umbones, anterior end longer; hinge plate stout, with a short central resilifer, bordered by two teeth in the right valve.

326. ? Mysella clementina (Carpenter, 1857). The minute and probably juvenile shell can be recognized only from the manuscript drawing of Carpenter, for

326

327

328

329

330

331

332

333

the holotype—which was only about a millimeter in length—has been lost. Mazatlán, Mexico.

327. Mysella compressa (Dall, 1913). The umbones are unusually high and pointed. Length, 7 mm; height, 6 mm. From the head of the Gulf of California (type locality) southward through the Gulf to Banderas Bay, Mexico (Stanford collection). (Alaska to Peru, *fide* Olsson, 1961.)

328. ? Mysella dionaea (Carpenter, 1857). The holotype (a single valve) is obviously a juvenile shell, only 1.8 mm in length. It is in the British Museum. Mazatlán, Mexico.

329. Mysella negritensis Olsson, 1961. More trigonal and larger than the other species. Length, 10 mm. Negritos, Peru.

330. ? Mysella umbonata (Carpenter, 1857). The type specimen in the British Museum is in poor condition, but Carpenter's manuscript drawing may help in recognizing the species. It seems to be a more quadrate form than the others. Length, about 2.8 mm. Mazatlán, Mexico.

Genus OROBITELLA DALL, 1900

Resembling *Aligena* in the arrangement of the hinge but with stronger teeth and a wider hinge plate; outline of shell tending to be more quadrate. A shaggy periostracum is characteristic.

331. Orobitella chacei (Dall, 1916). This is the most quadrate of the West American species and approaches in outline the ancestral group *Neaeromya* Gabb, 1873, from the Miocene of the Caribbean area, but the hinge has the left cardinal that is lacking in *Neaeromya*. Length, 5.3 mm; height, 3.5 mm; diameter, 1.8 mm. From southern California to Santa Inez Bay, Gulf of California, intertidally and to depths of 24 m.

332. Orobitella jipijapa Olsson, 1961. The outline is nearly equilateral in this relatively large shell, which has a somewhat chalky surface texture. Length, 12 mm; height, 9 mm; diameter, 3 mm. Ecuador.

333. Orobitella margarita Olsson, 1961. The shell is inequilateral, with the posterior side shorter and sloping down at a rather sharp angle. The ventral margin may be indented or sinuous in some specimens. Length of a large specimen, 11.5 mm; height, 7 mm; diameter, 2.2 mm. Panama.

334. Orobitella obliqua (Harry, 1969). This was described as an *Aligena*, but it has the quadrate outline and the hinge of *Orobitella*. Length and height, 6.2 mm; diameter, 2.4 mm. Guaymas area, Sonora, Mexico. If this is properly an *Orobitella* and if *Solecardia obliqua* (Sowerby, 1862) proves to be an *Orobitella*, Harry's species will require a new name.

335. ? Orobitella oblonga (Carpenter, 1857). Outline similar to that of *Orobitella*, but the hinge is delicate enough for *Aligena*. The holotype in the British Museum is figured. It still shows some shreds of periostracum. Length, 5.5 mm; height, 4 mm. Mazatlán, Mexico.

336. Orobitella peruviana Olsson, 1961. Ovate-quadrate, anterior end longer; relatively heavy and large; surface with concentric striae. Length, 13.6 mm; height, 10.8 mm; diameter (one valve), 2.5 mm. Ecuador (fossil) to Boca Pan, Peru.

334

335

336

337

338

339

340

341

337. Orobitella sechura Olsson, 1961. Higher and more trigonal than *O. peruviana*; surface nearly smooth except for growth lines, easily weathering to a chalky texture; periostracum pale brown, sometimes faintly rayed radially. Length, 10 mm; height, 9 mm; diameter (one valve), 2.5 mm. Sechura Bay, northwestern Peru.

338. Orobitella trigonalis (Carpenter, 1857). A solid, trigonal shell, described as having a well-developed periostracum. The outline is somewhat like that of *O. (Isorobitella) singularis* Keen, 1962, from Bahía San Quintin, outer coast of Baja California, but the hinge is less massive and the periostracum thinner. It is probable that the specimens reported as *Aligena cerritensis* Arnold, 1903, by Harry (1969) from Baja California and the Gulf may be *O. trigonalis* instead. *A. cerritensis* was described from the Pleistocene of southern California and has been recorded as living in that area and southward by Dall. The specimens reported by Harry show the periostracum well. Additional material in intermediate localities may eventually prove that there is only a single species involved, in which case Carpenter's name would have priority. Length of the lectotype in the British Museum, 7.5 mm; height, 6.6 mm. Mazatlán, Mexico (type locality), ranging, perhaps, from outer Baja California northward to southern California.

339. Orobitella zorrita Olsson, 1961. Quadrate, inequilateral, the anterior end slightly longer; surface nearly smooth, with microscopic roughening at the sides; periostracum thin, light brown. Length, 8.4 mm; height, 6 mm. Zorritos, northwestern Peru.

Genus PYTHINELLA DALL, 1899

Inequilateral, trigonal, with prominent umbones, anterior end longer and produced, the ventral margin straight or somewhat indented. Hinge as in *Mysella*.

340. Pythinella sublaevis (Carpenter, 1857). The hook-shaped triangular shell has an unmistakable outline. It occurs mostly offshore. Length, 2.2 mm; height, 1.4 mm. Mazatlán, Mexico, to Panama. Specimens were dredged at Banderas Bay, Mexico, in 11 m.

Superfamily CYAMIACEA

Equivalve, somewhat stout small shells with the ligament only slightly embedded in the hinge plate; hinge teeth in two series, appearing to consist of upper and lower elements. The typical family, Cyamiidae, occurs mainly in the southern hemisphere and is not represented in the Panamic fauna.

Family SPORTELLIDAE

Ligament external, on a small nymph or platform, in some with an additional small pit for the resilium on the hinge plate. The key is to genera.

1. Elongate, length about twice the height.........................*Ensitellops*
   Ovate to quadrate, length less than twice height.......................2
2. Outline oblique, inequilateral...............................*Basterotia*
   Outline quadrate, nearly equilateral...........................*Sportella*

Genus SPORTELLA DESHAYES, 1858

Ovate-quadrate, umbones medial or near anterior end; hinge with one or two large teeth; ligament mostly external, on a small nymph.

341. Sportella stearnsii Dall, 1899. The shell resembles *Orobitella* in size and shape, but the external ligament distinguishes it. Length, about 14 mm; height, 9 mm. Gulf of California to the Galápagos Islands, intertidally and to depths of 18 m.

Genus BASTEROTIA Hörnes, 1859

The shell has one thornlike cardinal tooth in each valve. Two Panamic subgenera.

Subgenus BASTEROTIA, s. s.

In shape the shell is trapezoidal, vaulted, with a sharp ridge delimiting the posterior slope. The shell surface is granular.

342. Basterotia (Basterotia) peninsularis (Jordan, 1936). A white shell with a knobby cardinal tooth on the hinge. Length, 13 mm; height, 10.5 mm. Described as a Pleistocene fossil from Magdalena Bay, Baja California. Living, Port Guatulco, Mexico, to Galápagos Islands, intertidally and to depths of 13 m.

Subgenus BASTEROTELLA Olsson & Harbison, 1953

More compressed than *Basterotia, s. s.*, posterior ridge rounded, not carinate, cardinal tooth smaller and set at its anterior end; surface of shell without granulations.

343. Basterotia (Basterotella) hertleini Durham, 1950 (Synonyms: **B. californica** Durham, 1950; **B. ecuadoriana** Olsson, 1961). The smooth shell is rounded anteriorly and posteriorly in the adult, with a weak ridge setting off the posterior slope; young shells are proportionally longer, more truncate posteriorly, with a sharper ridge. A growth series in the Stanford collection, dredged off the Mazatlán coast, bridges the difference in outline between *B. californica* and *B. hertleini*, both of which were described as fossils. At the southern end of the range the relative proportions change slightly, so that the shell seems shorter and higher; this is the *B. ecuadoriana*, which may prove to be a geographic subspecies. Intertidally, nestling in crevices and offshore to 46 m, Cape Tepoca and La Paz, Mexico, to Ecuador.

Genus ENSITELLOPS Olsson & Harbison, 1953

Elongate, the anterior end much the shorter, dorsal and ventral margins nearly parallel; prodissoconch distinct; surface of shell smooth or with scattered spines or pustules; hinge with a long slender posterior tooth in left valve.

344. Ensitellops hertleini Emerson & Puffer, 1957. Shell elongate, dull chalky white, and sculptured with irregular prickles or pustules. Length, 9.5 mm; height, 3 mm. San Felipe, near the head of the Gulf of California, to Panama and Ecuador, intertidally and to depths of 110 m.

345. Ensitellops pacifica Olsson, 1961. The white shell has the valves unequal in convexity; it is smaller and proportionately shorter than *E. hertleini*. Length, 5 mm; height, 2.8 mm. Panama to Ecuador.

Superfamily CHAMACEA

Sculpture well developed, concentric or radial or both; shell cemented to substrate by one valve, at least temporarily; beaks prosogyrate.

342

343

344

345

346

346a

Family CHAMIDAE

Like other attached mollusks, these are variable in shape and sculpture and there-fore difficult to identify. Calcareous algae may coat the spines, and the shells may be invaded by minute borers that seek shelter within the layers of the shell. The Latin word *chama* (pronounced cahm'-ma or kay'-ma) referred to a gaping cockle of some kind, not necessarily of this family. Three Panamic genera.

Genus CHAMA LINNAEUS, 1758

Of the three genera in the Panamic province, this has the most species. Odd valves, especially the unattached valve, may be found in beach drift, but live specimens must be searched for on rocks and other shells in the intertidal zone or in dredg-ings. The Chamas are attached by the left valve. Therefore, as one looks at the attached shell, the beak seems to be twisted to the right. The hinge in the left valve consists of two cardinal teeth, one heavy and roughened; in the right, of two widely separated small cardinal teeth. In both valves there is one small posterior lateral. The muscle scars are large, joined by a simple or entire pallial line.

346. Chama buddiana C. B. Adams, 1852 (Synonym: **C. rubropicta** Bartsch & Rehder, 1939). The ground color of the outside is dark red, with large and small white spines in irregular rows. Inside, the shell is white, with a pink line or band near the margin, which is finely crenate and shows at its edge the red of the outer layer. Average length is about 50 mm, though large specimens may attain twice that. The type locality is Panama, and although workers have reported it from north of there, most records may be confused with *C. mexicana*. From Clipperton Island and Panama south to Ecuador and the Galápagos Islands.

346a. Chama corallina Olsson, 1971. Resembling *C. sordida* but with finer radial riblets and fewer large spines, which develop on only an occasional speci-men. Length, 20 mm; height, 22.6 mm; diameter, 18.6 mm. Gulf of Panama, dredged in 55 m.

347. Chama echinata Broderip, 1835 (Synonym: **C. coralloides** Reeve, 1846). Not the spines but the color is the best diagnostic feature here, for even a fragment of the coral-pink hinge is unmistakable. The remainder of the interior is a strangely contrasting color: dark purple. Exteriorly the grayish-white shell is covered with short, irregular spines. Length, about 35 mm; height, 45 mm. Speci-mens may be found attached to rocks at low water from the Gulf of California to Panama.

348. Chama frondosa Broderip, 1835 (Synonym: **C. parasitica** Rochebrune, 1895). In its typical form this is a rather heavy grayish-pink shell with long, broad spines like fan-shaped leaves. Within, there is a pinkish-purple margin; the rest is white. Full development of the spines seems more common southward. Length and diameter, 75 mm; height, 90 mm. La Paz, Gulf of California (rare) to Ecuador and the Galápagos Islands, intertidally on rocks and offshore to depths of a few meters. Carpenter's naming of a distinct and fairly common Mexican species, which he called "*C. ? frondosa* var. *mexicana*," as a subspecies has re-sulted in some confusion.

349. Chama mexicana Carpenter, 1857 (Synonym: **C. ? frondosa fornicata** Carpenter, 1857). In its young and unworn condition this resembles *C. buddiana* much more than it does *C. frondosa*, with two rows of large white spines and several rows of smaller ones on a dark-red shell. These spines break off in older

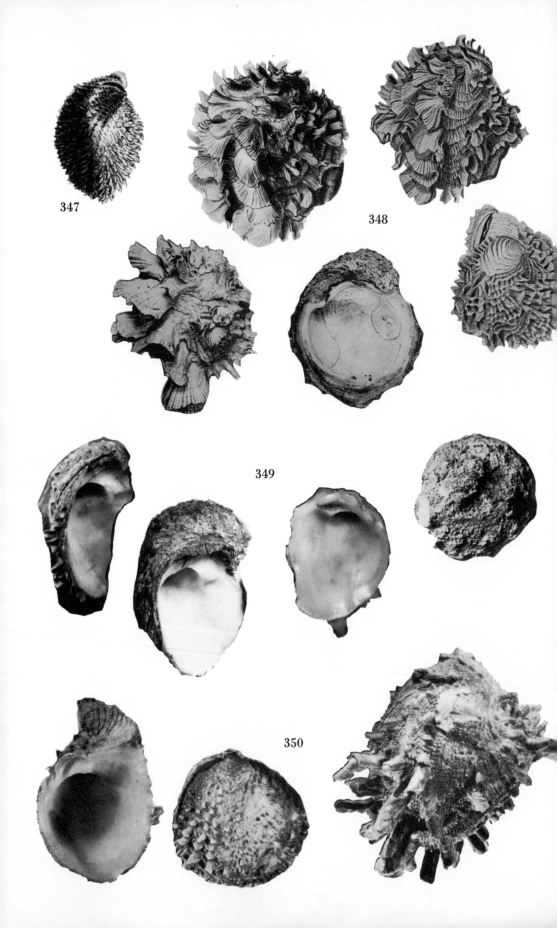

specimens, and thickened old shells become riddled by borers and incrusted with marine growths. The pink color band near the margin of *C. buddiana*, mentioned by C. B. Adams in describing that species, seems to be a good character for separating it from the similar-appearing *C. mexicana*, for the latter has instead a flush or stain of pink, especially around the anterior muscle scar, but no sharp line or band near the finely crenulate margin, except the red edge of the outer layer. The species seems to be restricted to the Mexican coast, at least judging from specimens in the Stanford collection. The size range is much as in *C. buddiana*, with large individuals reaching a length of 100 mm. Intertidally and offshore to 53 m, Puertecitos, near the head of the Gulf of California, to southern Mexico.

350. Chama sordida Broderip, 1835 (Synonym: **C. digueti** Rochebrune, 1895). A very attractive little shell, the outside of which is a rich coral red, with irregular stubby spines and a fine surface pattern of radial riblets. Within, the shell is translucent white, the exterior color showing through; the margins are finely crenulate. Length is 37 mm. *C. sordida* is not a common species, but it has been taken from the Gulf of California to Colombia, intertidally and to 82 m.

351. Chama squamuligera Pilsbry & Lowe, 1932 (Synonym: **C. spinosa** Broderip of authors, not of Broderip, 1833). This is the smallest West American form. The shells are rounded, both valves arched, and the sculpture consists of small scales that tend to join into concentric frills. The shell margin is finely crenulate and may even be granulose on fully adult shells, none of which are more than an inch in diameter. About 20 mm in height. Baja California to Galápagos Islands, intertidally and to depths of 13 m.

352. Chama venosa Reeve, 1847. A nearly smooth shell, whitish outside and inside, with a fine surface pattern of short, oblique, brown lines that look like the work of a skilled draftsman, so regularly are they drawn in. Length, about 25 mm; height, 45 mm. The species seems to be restricted to the Gulf of California, where, in recent years, it has been taken a few times on old wharf piling.

Genus ARCINELLA SCHUMACHER, 1817
(ECHINOCHAMA FISCHER, 1887)

The generic name *Arcinella* having been considered a homonym of *Arcinella* Oken, 1815, *Echinochama* has long been used instead. However, a ruling of the International Commission on Zoological Nomenclature (Op. 417, 1956) has suppressed Oken's work as nonbinomial; thus, *Arcinella* Schumacher—the type of which is *A. arcinella* (Linnaeus, 1767)—must be reinstated.

Arcinella represents forms that are lightly if at all attached, and the shells are much more symmetrical than those of the other two genera of the family. In the Panamic province there is only one rare species, which has a close resemblance to *A. arcinella* (Linnaeus, 1767), a fairly common Caribbean species.

353. Arcinella californica (Dall, 1903). The regular form and the symmetrical rows of long spines make this easy to recognize. Height, about 40 mm. Only a few specimens have been taken, those reported having been dredged in 22 to 46 m of water. Cedros Island, Baja California, to Panama.

Genus PSEUDOCHAMA ODHNER, 1917

The right valve in *Pseudochama* is attached to the substrate, firmly cemented to rock or another shell. As one looks at the specimen in place, the beaks seem to

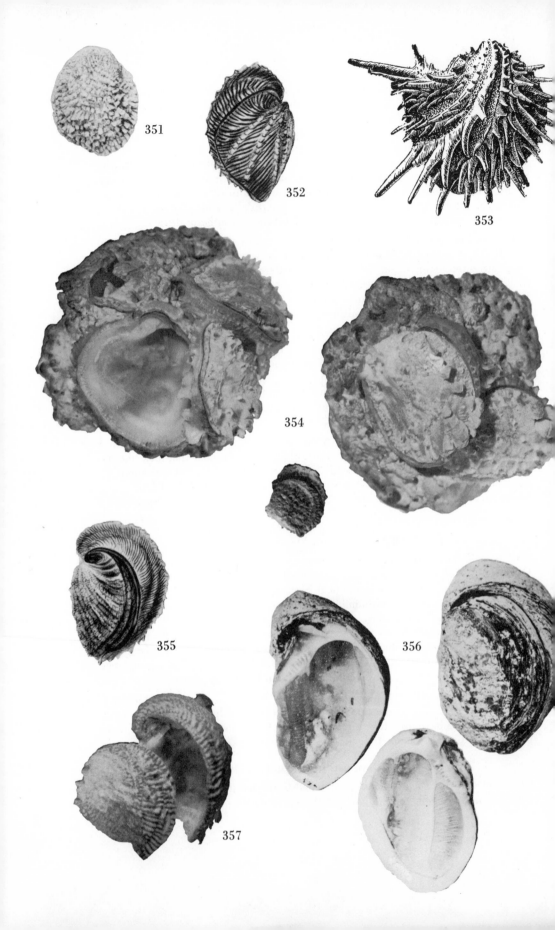

twist to the left, just the opposite of *Chama*. It was long thought that Chamas might attach by either valve, indifferently. Odhner (1919) concluded that there are significant differences of the hinge and that the two groups are not the simple reverse of each other. Not entirely agreeing with Odhner, Yonge (1967) regards *Pseudochama* more as a convenient group name for the chamid species that attach by the right valve than as a genus in the usual sense of the term. In the adult of *Pseudochama* the right or attached valve has one large granulose cardinal tooth on a prominent ridge that fits below a ridgelike cardinal tooth of the left valve. The dentition of young shells is much more regular, with two cardinals in either valve. The ecology of *Pseudochama* is similar to that of *Chama*. Often the shells may be effectively concealed by an overgrowth of other organisms attached to the outside.

354. Pseudochama clarionensis Willett, 1938. A bright red shell with the spines mostly white, some red, the inside white clouded with rose. The spines or scales are of nearly uniform size and the inner margin of the shell is crenate. Length about 23 mm. Off Clarion Island, 55 m depth.

355. Pseudochama corrugata (Broderip, 1835). The purple interior of the shell is a good guiding feature for identification. Length is about 60 mm. It is the most common of the species reported in the Panamic province and ranges from La Paz, Baja California, to Peru.

356. Pseudochama inermis (Dall, 1871). Only a few specimens have been reported in the literature. It is possible that this is only a variant of *P. panamensis*. It is white inside and out except for a stripe of purple or brown on the unattached valve. The attached valve is smooth and covered with a tough yellowish periostracum. Intertidally with other species of *Chama*. Length, about 60 mm. Gulf of California to Oaxaca, Mexico.

357. Pseudochama janus (Reeve, 1847). Similar in form to *P. panamensis*, this has an elegant pattern of brown lines on the whitish to yellowish shell surface. As in *P. panamensis*, there is some variation in the development of the spines or scales on the shell. Length, about 50 mm. Punta Peñasco, Sonora, Mexico, to the Galápagos Islands (type locality).

358. Pseudochama panamensis (Reeve, 1847). The shell is whitish, some specimens stained with brown within, others lightly striped with brown on the outside. The internal margin is smooth. Maximum length, about 40 mm. La Paz, Gulf of California, to Panama.

359. Pseudochama saavedrai Hertlein & Strong, 1946. This is a spinose form, the scalelike spines being laid on in two principal radial rows. The inner margin is denticulate. The color is light yellowish brown exteriorly and white within. Length, 41 mm; height, 46 mm; diameter, 30 mm. The head of the Gulf of California to La Paz, Baja California, and Manzanillo, Mexico (type locality), intertidally and offshore to depths of 46 m.

Superfamily **CARDIACEA**
(**CYCLODONTA**)

With radial sculpture showing a change of pattern on the posterior slope; hinge with two conical cardinal teeth in either valve, those in right valve somewhat fused, lateral teeth distant from cardinals; pallial line entire.

Family CARDIIDAE

Ligament external, short; adductor scars subequal; hinge teeth cruciform in arrangement; lateral teeth one anterior, one posterior in left valve, two anterior, one posterior in right valve (rarely wanting).

The members of the Cardiidae are commonly known as heart (Greek, *kardia*) cockles or simply as cockles. The color patterns and sculpture, being more constant than in most bivalves, offer a ready means of species separation. The shell is equivalve, but rarely equilateral, with porcelaneous texture. The internal margin is serrate to crenulate in all but the smoothest shells. The two adductor muscle impressions are of about the same size and shape, joined by an indistinct, smoothly curved pallial line. The siphons are short and the foot well developed and long, capable of moving the animal about in short leaps as it obtains leverage in the sand and contracts. The periostracum is conspicuous in some species, thin and transparent in others. The hinge shows a considerable amount of differentiation, and this, with the wide differences of sculpture and form, has justified the division of the old Linnaean genus *Cardium* into a number of less inclusive genera, which again are subdivided into subgenera.

Great adaptability is shown in this family, its members occurring in warm and cold seas, from mid-tide line to moderate depths, and, in Europe, even into brackish or almost fresh water. The common habitat is on sand or mud flats at or just under the surface.

The family is especially large in the fossil record. About 200 living species are known, of which 20 are recorded from the Panamic province. Four of the five currently recognized subfamilies are represented in this area, listed here in systematic rather than alphabetical sequence. There are no members of the subfamily Cardiinae in the Panamic province; the key is to genera and subgenera within the other four subfamilies.

1. Anterior lateral teeth lacking............................*Lophocardium*
   Anterior lateral teeth always present............................... 2
2. Shell longer than high..*Papyridea*
   Shell higher than long or subcircular............................... 3
3. Ribs beaded; shell small, subcircular, nearly equilateral......*Microcardium*
   Ribs not beaded; shell inequilateral............................... 4
4. Posterior slope smoother than remainder of shell; posterior margin without crenulations ....*Laevicardium*
   Posterior slope with ribs; crenulate within........................... 5
5. Hinge set at angle to the ventral margin........................... 6
   Hinge parallel to ventral margin................................... 8
6. Shell medium-sized; ribs 28–30, square-sided...............*Americardia*
   Shell small; ribs relatively few (18–22), rounded, nearly smooth........ 7
7. Shell outline markedly oblique...............................*Apiocardia*
   Shell nearly equilateral.........................*Trigoniocardia, s. s.*
8. Ribs with well-developed spines on posterior and anterior slopes........ 9
   Ribs nearly smooth on most of shell or not spinose....................11
9. Ribs with a frill along crest but without separated spines......*Phlogocardia*
   Ribs smooth or only slightly roughened along crest....................10
10. Hinge plate narrow, long, bent in middle...................*Acrosterigma*
    Hinge plate wide, short, and straight.......................*Mexicardia*
11. Ribs with thornlike spines along sides......................*Dallocardia*
    Ribs with imbricating scaly spines across crests........*Trachycardium, s. s.*

Subfamily TRACHYCARDIINAE

Posterior slope with spinose ribs, its margin dentate; hinge angulate to straight, short, wide. Two Panamic genera.

Genus TRACHYCARDIUM MÖRCH, 1853

With strong ribs, those of posterior slope ending in digitations of the margin; all ribs tending to develop spines; hinge relatively short, broad, teeth strong. Five Panamic subgenera (see key, above).

Subgenus TRACHYCARDIUM, s. s.

With spinose sculpture on all ribs; hinge straight, short.

360. Trachycardium (Trachycardium) consors (Sowerby, 1833) (Synonym: **Cardium laxum** Dall, 1901). The 30 to 34 strong ribs are covered by spines that are imbricate, fitting over one another like tiles on a roof; anteriorly these spines are more flattened and posteriorly more pointed. The shell is light buff to pink or yellowish, with brown stains, darker toward the posterior margin. The inside is reddish purple to flesh-colored, yellow or white around the margins. Length, 60 mm; height, 54 mm; diameter, 50 mm. The species is fairly common on tide flats from the northern part of the Gulf of California south to Ecuador and the Galápagos Islands, and has been dredged from depths of 45 m. It is the Pacific Coast relative of *T. (T.) isocardia* (Linnaeus, 1758), from the Caribbean. See Color Plate III.

Subgenus ACROSTERIGMA DALL, 1900

Ribs flattened, scaly only at edges; hinge angulate.

361. Trachycardium (Acrosterigma) pristipleura (Dall, 1901) (Synonyms: **Cardium maculosum** Sowerby, 1833 [not Wood, 1815]; **C. maculatum** Sowerby, 1840 [not Gmelin, 1791]; **C. hornelli** Tomlin, 1928). The ribs, numbering from 34 to 39, are flattened, with very narrow interspaces, and the anterior ribs show a filelike roughening along each edge. The color of the exterior is yellowish white, with reddish-brown markings and often with pinkish-purple concentric bands or blotches. The interior is white. There is a thin, light brown periostracum that persists, mainly toward the ventral margin. Height, about 65 mm. This is not a common species, although it may be found by shore collectors anywhere between the southern portion of the Gulf of California and Guayaquil, Ecuador.

Subgenus DALLOCARDIA STEWART, 1930

Ribs with thornlike spines, stronger on the posterior sides.

362. Trachycardium (Dallocardia) senticosum (Sowerby, 1833) (Synonym: **Cardium rastrum** Reeve, 1845). Compared with *T. consors*, this is smaller, lighter in color, with more ribs (35 to 40), which have only short, thornlike scales, most prominent on the anterior part of the shell. The coloring is basically white, inside and out, with reddish-brown or purple bands and spots exteriorly. Height, about 40 mm. The recorded range includes the whole of the Gulf of California, southward to Paita, Peru. Really good specimens can ordinarily be obtained only by dredging on shallow muddy bottoms. A very similar Atlantic species is *T. (D.) muricatum* (Linnaeus, 1758).

358

359

360

361

362

363

Subgenus MEXICARDIA STEWART, 1930

Scaly sculpture on ribs apparent only on posterior slope.

363. Trachycardium (Mexicardia) panamense (Sowerby, 1833) (Synonyms: **Cardium rotundatum** Carpenter, 1857 [not Dujardin, 1837]; **C. procerum** of authors; **Trigoniocardia eudoxia** Dall, 1916). There seem so many points of difference between this and the generally more southern form, *T. procerum*, that the two are here considered as separate species. The shell is much more massive and may attain a height of 90 mm, compared to 55 mm for the southern one. The ribs are triangular in outline, with fine nodes along the sharp crest, especially on the umbones, and the rather wide spaces between ribs are flat. The shell is white under a greenish, persistent periostracum, except for spots of brown on the umbones. The species is a common one in mud, from low-tide line to a depth of several meters, ranging from San Ignacio Lagoon, Baja California, throughout the Gulf of California but—in spite of its name—apparently no farther south than Costa Rica.

See Color Plate XII.

364. Trachycardium (Mexicardia) procerum (Sowerby, 1833) (Synonyms: **Cardium laticostatum** Sowerby, 1833; **C. dulcinea** Dall, 1916; **C. parvulum** Li, 1930). Some authors have placed this species in the subgenus *Ringicardium*, but that group is essentially European and African in distribution. *T. procerum* is of moderate size, with smooth, rounded ribs that are separated by linear interspaces. The shell is only moderately thick and the color is white, extensively mottled and spotted with brown. Length, about 50 mm. Although a few specimens have been found as far north as the Gulf of California, the geographic range seems mostly to be southward from Oaxaca, Mexico, to Chile.

Subgenus PHLOGOCARDIA STEWART, 1930

Rib scales coalesced as a vertical frill.

365. Trachycardium (Phlogocardia) belcheri (Broderip & Sowerby, 1829). Sharp saw-toothed ridges adorn the posterior side of each of the 23 to 25 ribs. The color of the shell, both inside and out, is a blend of white with pastel shades of yellow, brown, and rose. Maximum height is about 55 mm. Specimens are obtained mostly from fairly deep water (to 110 m), from Cedros Island and Guaymas, Mexico, south to Panama.

Genus PAPYRIDEA SWAINSON, 1840

Elongate, ribs numerous, coarser on posterior slope; hinge relatively narrow, with teeth weaker than in *Trachycardium*; posterior end of shell gaping.

366. Papyridea aspersa (Sowerby, 1833) (Synonym: **P. bullata californica** Verrill, 1870). The elongate form and the large posterior gape easily distinguish this species from any other west coast cockle. The shell is almost paper-thin, with numerous prickly ribs. The color is white, mottled with reddish brown, the pattern showing through to the inside. Length, about 45 to 50 mm. This species is found on sandbars or beaches from Manuela Lagoon, Baja California, south to Peru, including the whole of the Gulf of California. It is similar to *P. soleniformis* (Bruguière, 1789) from the Caribbean.

367. Papyridea crockeri Strong & Hertlein, 1937. The shell is shorter and more inflated than *P. aspersa*, with wider umbones and somewhat smoother ribs,

364

365

366

367

368

369

370

371

372

about 48 in number. The color is yellowish spotted with brown. Length, about 50 mm. First described from off the tip of Baja California, it has since been dredged as far north as off Guaymas in the Gulf of California, in depths to 175 m.

368. Papyridea mantaensis Olsson, 1961. A white shell with fewer ribs (about 36 in number) than *P. aspersa* or *P. crockeri,* the beaks more anterior in position, the ribs triangular in cross section rather than rounded. Mostly white, with some mottling of pale red or violet on the low umbones. Average length about 40 mm, large specimens 60 mm. Acapulco, Mexico, to northern Peru.

Subfamily FRAGINAE

Posterior slope set off by a ridge; rib sculpture of concentric threads or nodes.

Genus TRIGONIOCARDIA DALL, 1900

Relatively small, ribs few, with fine concentric striae in interspaces, beads or nodes on crests but no spines; hinge short, angulate. Three subgenera.

Subgenus TRIGONIOCARDIA, s. s.

Ovate, with beaded sculpture on ribs.

369. Trigoniocardia (Trigoniocardia) granifera (Broderip & Sowerby, 1829) (Synonym: **Cardium alabastrum** Carpenter, 1857). There are about 18 to 20 ribs on the small shell, those of the anterior part with prominent grainlike nodules. Wide interspaces are bridged by regularly spaced cross-threads. Most specimens are shiny white; some have a yellowish blotch on the inside. Length, 15 mm or less. Magdalena Bay, through the Gulf of California and south to Zorritos, Peru, on mud flats, below the low-tide line; offshore in depths to 25 m.

Subgenus AMERICARDIA STEWART, 1930

Quadrate, medium-sized; ribs and intercostal spaces smooth.

370. Trigoniocardia (Americardia) biangulata (Broderip & Sowerby, 1829) (Synonym: **Cardium magnificum** Carpenter, 1857). The solid and inflated shell has 27 or 28 broad, flattened ribs. The exterior is yellow or white, marked with large brown spots; the interior is partly white, partly reddish brown, lighter along the margins. Length, about 35 mm. From southern California the species ranges south to Guayaquil, Ecuador, intertidally and to depths of 155 m.

371. Trigoniocardia (Americardia) guanacastensis (Hertlein & Strong, 1947) (Synonyms: **Cardium planicostatum** Sowerby, 1833 [not Sedgwick & Murchison, 1829]; **C. magnificum** of authors, not of Carpenter, 1857, *ex* Deshayes, MS). Similar to the last in coloring, except for being white within, this has two or three more ribs and a more oblique outline. Length, 36 mm; height, 46 mm. The species has long been mistakenly labeled *C. magnificum,* and not until recent years has a valid name been proposed. Cape San Lucas, Gulf of California, to Paita, Peru. A similar Atlantic species is *T. (A.) media* (Linnaeus, 1758).

Subgenus APIOCARDIA OLSSON, 1961

Small, ribs nearly smooth; shell rounded-ovate in outline.

372. Trigoniocardia (Apiocardia) obovalis (Sowerby, 1833) (Synonym: **Cardium ovuloides** Reeve, 1845). A small white shell, distinguished by the oblique outline, with the beaks well forward and the anterior section of the hinge so short the laterals are crowded against the cardinals. The 20 to 22 ribs are roughened by numerous minute nodes. Maximum size attained is about 14 mm length, 20 mm height. Magdalena Bay and the Gulf of California to Salinas, Ecuador. Sometimes the shells are found on shore but mostly they are taken by dredging.

Subfamily PROTOCARDIINAE

Rounded-quadrate, beaks nearly central; posterior slope well defined by an umbonal ridge, its radial ribs tending to be spinose and coarser than on the anterior part of shell; hinge long, slightly arched. Two Panamic genera.

Genus LOPHOCARDIUM FISCHER, 1887

Shell thin and fragile, posteriorly gaping; umbones inflated. Anterior lateral teeth are wanting, unique among Cardiidae, for all others have anterior lateral teeth at the front margin of the hinge to aid in firm closure of the shell. These are offshore forms in the main.

373. Lophocardium annettae (Dall, 1889). The color is light salmon pink, deepening to rosy red at the posterior margin. The numerous weak radial ribs fade out anteriorly. The posterior end shows several concentric ridges and is set off by a raised rib fringed with periostracum that is light brown and persistent. Length, about 47 mm; height, 39 mm. Although specimens taken intertidally are reported at San Felipe, this species is mostly found by dredging, from the head of the Gulf of California south to Costa Rica.

374. Lophocardium cumingii (Broderip, 1833). The shells are even thinner than those of *L. annettae*, a lighter color, longer for the height, and without the concentric ridges on the posterior slope. Length, 28 mm. Acapulco, Mexico, to Colombia, always offshore, in 22 to 26 m of water. Rare.

Genus NEMOCARDIUM MEEK, 1876

Small, with numerous ribs, those of posterior slope set off by a difference in width or emphasis and ending in serrations at posterior margin. Hinge arched, relatively long. No species of *Nemocardium, s. s.,* occur in the Panamic province.

Subgenus MICROCARDIUM THIELE, 1934

Small, with secondary concentric sculpture on central and anterior slopes, both in intercostal spaces and as beads along ribs.

375. Nemocardium (Microcardium) panamense (Dall, 1908). White, with only 33 ribs. Panama, 333 m.

376. Nemocardium (Microcardium) pazianum (Dall, 1916). This little clam is fairly abundant in deep water but has never been found near shore. It is sculptured with some 60 fine ribs, the posterior ribs with minute scales. The color is yellowish white to red or orange, uniformly or in bands. Length and height, about 15 mm. Cedros Island, Baja California, to Panama, in depths of 25 to 90 m.

Subfamily LAEVICARDIINAE

Elliptic-oblique in outline; posterior slope relatively smooth; ribs never spinose.

Genus LAEVICARDIUM SWAINSON, 1840

Posterior slope smoother than remainder of shell; ribbing subdued throughout, showing mainly as crenulations within; hinge arched, moderately long.

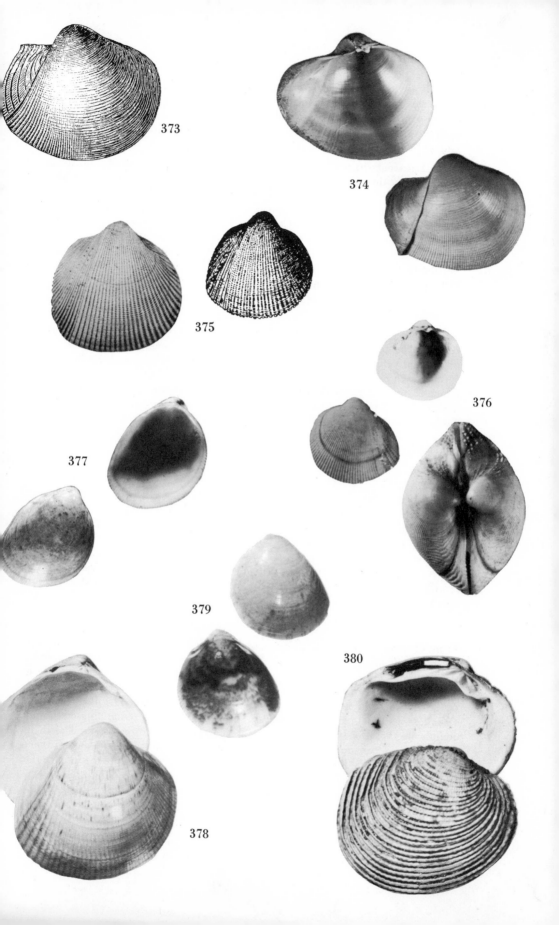

377. Laevicardium clarionense (Hertlein & Strong, 1947). A rare offshore form, this resembles the much larger *L. elatum* and might be mistaken for the young of that species except for the oblique outline. The shell is yellow, spotted with brown. Length, 24 mm; height, 31 mm; diameter, 16.5 mm. Santa Inez Bay, Gulf of California, to Clarion Island, Revillagigedo group, in depths of 64 to 155 m.

378. Laevicardium elatum (Sowerby, 1833). Under a smooth, horn-colored periostracum the shell is yellow, some specimens with a few brown streaks, the interior white. It is inflated and shows only faint traces of ribs. Length, to 150 mm—the largest although perhaps not the heaviest-shelled among the living species of the family. Southern California south as far as Panama, most common (though far from abundant) on mud flats in the Gulf of California and along the outer coast of Baja California. Young specimens are noticeably different in proportion from adult shells, being broader for their height, and are buff rather than yellow, with faint brown markings. The contrast in size between adults of this species and the other two is remarkable.

See Color Plate XII.

379. Laevicardium elenense (Sowerby, 1840) (Synonym: **Cardium apicinum** Carpenter, 1864). Shells are small, thin, and nearly smooth; externally brown and yellow with brown or purple dots, internally with reddish brown uniformly or in bands. Fine radial ribs can be made out in worn specimens. Length, about 25 mm. Magdalena Bay, Baja California, through the Gulf of California and south to Zorritos, Peru, rare intertidally, mainly offshore in depths to 90 m. Authors have sometimes recognized *L. elenense apicinum* (Carpenter) as a northern subspecies, said to be broader, with more spotted coloring, but the differences are within the limits of variation.

Superfamily VENERACEA

Ovate shells, with ornamentation predominantly concentric but also radial in some, with lamellae or spines, especially near the posterior slope; beaks anterior, prosogyrate; ligament external, entirely behind the beaks. Cardinal hinge teeth generally three in both valves (two in one valve in some). Pallial line usually with a sinus.

The arrangement of families within the superfamily and the sequence of subfamilies within families for this group is systematic rather than alphabetical, following the order that has been adopted in Moore (1969).

Family VENERIDAE

From the standpoint of color, beauty of form, and diversity, the Venus clams are the most advanced of the bivalves. The shell is porcelaneous, whether large or small, and likely to have a polished, highly finished appearance. The shape is mostly ovate or cordate (heart-shaped), inequilateral, with the beaks well in front of the midline. Sculpture may be radial or concentric or both. The ligament is external, on a special platform or nymph. Additional nymphs resembling teeth may complicate the already complex hinge, which consists of three cardinal teeth in each valve, one or more of which may be grooved or bifid. In some genera there is an anterior lateral hinge tooth in the left valve, fitting into a lateral socket in the right valve. The pallial line has a more or less deep sinus, reflecting the habit of burrowing below the surface of the sea floor. The key to the genera (in eight subfamilies) is adapted from Hertlein and Strong (1948).

1. Left valve with anterior lateral tooth or denticle.................... 2
   Left valve with no anterior lateral tooth........................... 9
2. Inner margin strongly crenulate.................................. 3
   Inner margin smooth... 4
3. Length greater than height; sculpture cancellate...............*Periglypta*
   Length and height nearly equal; sculpture concentric.........*Ventricolaria*
4. Shell orbicular and compressed, or lenticular...................*Dosinia*
   Shell trigonal to elongate in outline, not lenticular.................. 5
5. Sculpture reticulate; shell small; pallial sinus slight..............*Gouldia*
   Sculpture concentric only.. 6
6. Shell trigonal; beaks subcentral.............................*Tivela*
   Shell longer than high; beaks anterior to midline.................... 7
7. Inner ventral margin with oblique grooving................*Transennella*
   Inner ventral margin without oblique grooving..................... 8
8. Impression of pedal retractor muscle deep; shell large, thick.....*Megapitaria*
   Impression of pedal retractor muscle shallow; shell medium-sized.....*Pitar*
9. Inner margin crenulate, at least anteriorly..........................10
   Inner margin smooth...13
10. Shell elongate, anterior end narrowed, small; concentric sculpture con-
    spicuous ....*Irus*
    Shell roundly trigonal or ovately quadrate, medium-sized to large.......11
11. Ligamental area roughened, irregularly pitted...............*Mercenaria*
    Ligamental area not conspicuously roughened or pitted...............12
12. Hinge plate short, triangular; left posterior cardinal long, middle cardi-
    nal thick; pallial sinus short....................................*Chione*
    Hinge plate long and narrow; left posterior cardinal short, middle cardi-
    nal grooved; pallial sinus long.............................*Prototbaca*
13. Umbones with concentric undulations.....................*Clementia*
    Umbones smooth 14
14. Shell suborbicular or roundly quadrate.....................*Cyclinella*
    Shell not suborbicular, but rectangular to trigonal...................15
15. Adults medium-sized to large...........................*Compsomyax*
    Adults small, less than 10 mm.............................*Psephidia*

Subfamily VENERINAE

With inner margin crenulate, at least in the young, and with an anterior lateral hinge tooth. Two Panamic genera.

Genus PERIGLYPTA JUKES-BROWNE, 1914

Authors have confused this with *Antigona* Schumacher, 1817, a western Pacific group that has sculpture and a small pallial sinus like *Chione*, whereas the Periglyptas have a large pallial sinus and cancellate, not lamellar, ribbing.

380. Periglypta multicostata (Sowerby, 1835) (Synonym: **Venus thouarsi** Valenciennes, 1846). The heaviest if not the largest of the Panamic members of the family, this has a rounded-quadrate shell heavily sculptured with both concentric and radial ribs. The pallial sinus is wide and fairly long. The shell itself is an ivory white, the interior in some specimens pinkish or violet under the hinge. A large specimen measures: length, 118 mm; height, 115 mm; diameter, 78 mm. Gulf of California to Punta Verde, Peru, living in sand among rocks at extreme low tide. Beach shells are not uncommon.

Genus VENTRICOLARIA KEEN, 1954
(VENTRICOLA of authors, not of RÖMER, 1867)

The long-used name *Ventricola* must fall as a synonym of *Venus, s. s.*, a group that is confined to the Mediterranean. The inflated form and entirely concentric sculpture separate *Ventricolaria* from *Antigona*, to which many authors have considered it related.

381. Ventricolaria isocardia (Verrill, 1870). A well-grown specimen looks like a tennis ball with a dent on one side, this dent being the impressed lunule. On the outside, the shell is brown, with darker blotches in radial or zigzag bands. The interior is white, flushed with pink near the umbones. Length, 87 mm; height, 81 mm; diameter, 64 mm. The Caribbean species *V. rigida* (Dillwyn, 1817) is very similar in appearance. The Gulf of California to Gorgona Island, Colombia, in depths to 110 m.

382. Ventricolaria magdalenae (Dall, 1902). This has been regarded as a subspecies of *V. isocardia*. It differs in being smaller (maximum length, 48 mm), proportionately longer for the height, with more delicate and closely-spaced concentric sculpture and a color pattern of brown flecks on a pale yellowish ground. Outer coast of Baja California at Magdalena Bay and into the southern end of the Gulf of California in depths of 65 to 80 m.

Subfamily CIRCINAE

With some of the radial ribs subdivided or dichotomous.

Genus GOULDIA C. B. ADAMS, 1847

Small, sculpture fine; lunule long, escutcheon wanting; pallial sinus small.

383. Gouldia californica Dall, 1917 (Synonym: **Gafrarium stephensae** Jordan, 1936). The latticed sculpture of this minute form is distinctive. The shell is white with small brown spots near the hinge. Length, 6 mm; height, 5.5 mm; diameter, 3 mm. Gulf of California to Panama, mostly offshore in depths to 160 m.

Subfamily MERETRICINAE

With subdued sculpture; the hinge with cardinal teeth of similar size tending to radiate symmetrically. Two Panamic genera.

Genus TIVELA LINK, 1807

The shells of *Tivela* (pronounced Tiv′-ela, with a short *i*) are smooth, variously colored in tones of buff to brown, and triangular in outline. These clams are characteristic inhabitants of sandy beaches and bars. Two Panamic subgenera.

Subgenus TIVELA, *s. s.*

Convex, with full prominent umbones; lunule mostly strongly impressed.

384. Tivela (Tivela) argentina (Sowerby, 1835) (Synonym: **Cytherea aequilatera** Deshayes, 1839). Exteriorly the shell is yellowish white under a thin, light brown periostracum, with no color markings. Inside, the shell is white. A large specimen measures: length, 61 mm; height, 58 mm; diameter, 31 mm. Puerto Peñasco, Sonora, Mexico, to Panama, on sandbars.

385. Tivela (Tivela) byronensis (Gray, 1838) (Synonyms: **Cytherea radiata** Sowerby, 1835 [not Mühlfeld, 1811]; **Venus solangensis** Orbigny, 1845;

C. semifulva Menke, 1847; **C. gracilior** and **intermedia** Sowerby, 1851; **C. pulla** Philippi, 1851). This is an abundant and variable Tivela. One useful feature for recognition is a spot of dark blue on the beaks of most specimens. The shell is rather more inflated than the other Tivelas of the coast. Under an olive-brown velvety periostracum, the reddish-brown color pattern may vary from light to dark, mostly as radial stripes. The name *T. gracilior* (Sowerby) was given to a white variant. Length, 58 mm; height, 55 mm; diameter, 37 mm. A somewhat similar species in the Caribbean is *T. mactroides* (Born, 1778). The range is from Lagoon Head, Baja California, through the Gulf of California and south to Guayaquil, Ecuador, on sand beaches and offshore to depths of 73 m.

386. Tivela (Tivela) delessertii (Sowerby, 1854, *ex* Deshayes, MS) (Synonym: **Cytherea (Tivela) arguta** Römer, 1865). With valves relatively long, compressed, smaller than the other forms, and thick-shelled for its size. Coloring is a cream to light chestnut-brown ground with dark purplish-brown stripes. Length, 31 mm; height, 24 mm; diameter, 18 mm. Santa Inez Bay, Gulf of California, to Panama, on sand beaches.

387. Tivela (Tivela) hindsii (Hanley, 1844). Although often regarded as a synonym of *T. byronensis* by authors, this has a number of differences: it is smaller, has a more compactly trigonal and inflated shell, and has a color pattern of zigzag concentric bands rather than predominantly radial markings. The velvety periostracum is lacking. Length, 32 mm; height, 26 mm; diameter, 20 mm. West Mexico to Ecuador.

388. ? Tivela (Tivela) lineata (Sowerby, 1851). Olsson (1961) has suggested that this species, described from an unknown locality, may have come from Panama, for he has found specimens there that match the figure well. Length, 31 mm; height, 28 mm; diameter, 11 mm. The shell is white with feather-like markings at the ends and is shaped like *T. delessertii*.

Subgenus **PLANITIVELA** OLSSON, 1961

Umbones narrow, inconspicuous; lunule weakly impressed.

389. Tivela (Planitivela) hians (Philippi, 1851). As the name—*hians*, gaping—suggests, this may be recognized by the incomplete closing of the posterior part of the valves. The shell is relatively longer for its height and more compressed than any other species except *T. planulata*. The coloring runs to a reddish or purplish brown. The size is about the same as in *T. planulata*, but the anterior end is distinctly longer than the posterior, and the gape of the posterior margin gives the shell a truncate appearance. It is a southern species not ranging north of northern Peru.

390. Tivela (Planitivela) planulata (Broderip & Sowerby, 1830) (Synonyms: **Cytherea planulata suffusa** Sowerby, 1835; **C. undulata** Sowerby, 1851). The tightly closed trigonal, flattened shell with radiating color bands in shades of light to dark brown makes up a distinctive form that is only slightly less common than *T. byronensis*. Length, 55 mm; height, 43 mm; diameter, 23 mm. The Gulf of California to Ecuador.

Genus **TRANSENNELLA** DALL, 1883

Small shells, mostly less than 35 mm in length, the inner margin with more or less well-developed oblique grooves. Confined to the Americas, the type species being *T. conradina* (Dall, 1883) from the Caribbean.

387

388

389

390

391

392

393

391. Transennella caryonautes Berry, 1963. Largest among the Transennellas, this has the look of a *Pitar* but with the grooving of the margin that is the hallmark of *Transennella*. The color is brownish, rayed with somewhat rippled stripes of darker brown. Length, 40 mm; height, 35 mm; diameter, 24 mm. Southern end of the Gulf of California, from the tip of Baja California to Mazatlán, taken offshore by shrimp-dredging boats.

392. Transennella galapagana Hertlein & Strong, 1939. A small form with variable color pattern, ranging from pure white to purplish-brown, with tent-shaped markings; the valves are relatively thick. Length, 6 mm; height, 4 mm; diameter, 3 mm. Taken intertidally, Santa Cruz Island, Galápagos.

393. Transennella humilis (Carpenter, 1857). Study of the type in the British Museum has shown (Keen, 1968) that this is similar to the more northern *T. tantilla* (Gould, 1853) and might be considered a geographic subspecies—*T. tantilla humilis*. It differs from typical *T. tantilla* in being smaller, longer, smoother, and more brightly colored. It was illustrated by Römer in 1865, but no one could have recognized the form from his figure, which was natural-sized. Length, 4 mm; height, 3 mm. Common in shallow bays of the Gulf of California from La Paz to Mazatlán.

394. Transennella modesta (Sowerby, 1835) (Synonym: **Venus cumingii** Orbigny, 1845 [unnecessary new name for **Cytherea modesta** Sowerby, 1835, not *V. modesta* Dubois, 1831]; **T. sororcula** Pilsbry & Lowe, 1932). The shell is strongly sculptured concentrically, whereas the other species have only faintly incised concentric ribbing. The lunule is broadly heart-shaped. The color is cinnamon brown or cream, with radial markings that may form a network of chocolate-brown lines. Because some specimens are as long as 35 mm, this form has been confused with *T. caryonautes*, which is smooth exteriorly. Length of an average specimen, 16 mm; height, 13 mm; diameter, 10 mm. Santa Inez Bay, Gulf of California, to Ecuador, mostly offshore in depths to 90 m.

395. Transennella omissa (Pilsbry & Lowe, 1932). The glossy white shell is marked with buff to brown in zigzag patterns. A shorter and more triangular shell than *T. puella*, with which the authors compared it, concentrically grooved near the beaks and at the posterior end. The marginal grooving so characteristic of other species in the genus was not mentioned in the original description, nor does it show on the type specimens. Possibly this is not a valid *Transennella*, although in all other characteristics it seems to conform. Length, 8 mm; height, 6.4 mm; diameter, 4 mm. San Juan del Sur, Nicaragua.

396. Transennella puella (Carpenter, 1864). Coloration is of brownish zigzag markings on a cream or white ground, the interior white, flushed with purplish in some specimens. Length, 16 mm; height, 12 mm; diameter, 8 mm. Guadalupe Island, off Baja California, through the Gulf of California and south to Nicaragua, mostly offshore in depths to 80 m. Records of the South American *T. pannosa* (Sowerby, 1835) from Baja California probably are misidentifications of this species. The true *T. pannosa* is thicker shelled, longer posteriorly, with lower umbones; it reaches a length of 26 mm and occurs south of Callao, Peru.

Subfamily **PITARINAE**

With the cardinal teeth not tending to radiate symmetrically, the left anterior cardinal tooth smaller than the broadly wedge-shaped middle cardinal, joined to it at the upper end. Anterior laterals well developed. Two Panamic genera.

Genus **PITAR** RÖMER, 1857

Oval to subtrigonal, smooth or finely concentrically lamellate; lunule superficial, escutcheon not defined.

Because there are so many species of *Pitar* in the Panamic province, the use of subgenera will aid both in distinguishing and in grouping them. The key is adapted from one by Hertlein and Strong (1948).

1. With spines or scales along posterior umbonal angulation......*Hysteroconcha*
   Without spines or scales along posterior umbonal angulation........... 2
2. With fine zigzag sculpture on at least part of shell..........*Hyphantosoma*
   With no zigzag sculpture...................................... 3
3. Shell suborbicular to globose; inner margin irregularly crenate.....*Tinctora*
   Shell ovate to subtrigonal; inner margin smooth..................... 4
4. Shell with strong concentric ribs or lamellae...............*Lamelliconcha*
   Shell smooth or with fine concentric threads......................... 5
5. Left middle cardinal only slightly longer than anterior cardinal.....*Pitar, s. s.*
   Left middle cardinal decidedly longer and thicker than anterior cardinal ....*Pitarella*

Subgenus **PITAR**, *s. s.*

Smooth or finely striate; pallial sinus deep and pointed; left posterior cardinal confluent with nymph.

397. Pitar (Pitar) berryi Keen, new species. Shell medium-sized, sturdy, plump, ovate, smooth-surfaced, tinged with yellowish brown in darker and lighter concentric bands or uniformly, the color showing through to the upper part of the inner cavity as a flush of pinkish yellow. Lunule somewhat flattened but not depressed, bounded by an incised line. Posterior slope convex, with an ill-defined radial depression near the dorsal margin. Hinge teeth strong, the central cardinal of the left valve wide, the anterior cardinal narrow and slightly the longer. Pallial sinus small to medium-sized, rounded at the end. Inner margin of shell smooth, thin. This form was first recognized as an undescribed species by Dr. S. S. Berry on the basis of shrimp-boat material obtained by a number of collectors from boats operating out of Guaymas and Mazatlán, Mexico. Locality of the type lot (Stanford University collection): dredged off La Cruz, Banderas Bay, Jalisco, Mexico, depth 18 to 37 m, on a sand-mud bottom. Length of holotype, 47 mm; height, 42 mm; diameter (two valves), 31.5 mm. Length of a shrimp-boat specimen in the Stanford collection, 44 mm; height, 37 mm; diameter (both valves), 31 mm. The precise geographic range is not yet known; presumably it includes the southern end of the Gulf of California to southern Mexico, offshore. This form has larger, sturdier, and more uniformly colored shells than others in the subgenus.

398. Pitar (Pitar) consanguineus (C. B. Adams, 1852). The shell is white, with brown radial markings—especially on the upper half—and pinkish umbones. On the hinge, the anterior left lateral tooth stands up higher than the other teeth. Length, 32 mm; height, 26 mm; diameter, 17 mm. It is rather a rare species, ranging from Port Guatulco, Mexico, to Panama, mostly offshore in depths of 4.5 to 44 m.

399. Pitar (Pitar) elenensis (Olsson, 1961). Similar to *P. consanguineus* but larger, more trigonal, with the anterior end longer and less arched, the lunule larger and not impressed in the middle. Length, 46 mm; height, 39 mm; diameter, 30 mm. Panama to northern Peru.

400. Pitar (Pitar) fluctuatus (Sowerby, 1851). The specific name probably refers to the angular brown markings, for the concentric ribs are even, fine, and rounded. Length, 13 mm; height, 11 mm; diameter, 8 mm. Panama Bay to Santa Elena, Ecuador.

401. Pitar (Pitar) helenae Olsson, 1961. This species proves to have a wider range than Olsson realized. It has been misidentified as *P. newcombianus* (Gabb, 1865) in the Gulf of California area. Gabb's species is Californian and seems not to range south of the outer coast of Baja California, being replaced in the Gulf and southward by two species, *P. berryi* and *P. helenae*, which has a cream-colored shell heavily rayed with brown and with brown angular markings, the beaks stained with purple; there is a lighter median ray. The interior is white with a flush of purple in the umbonal cavity. Pallial sinus large but not long, broadly rounded. Length, 24 mm; height, 20 mm. Gulf of California to Panama, mostly offshore in depths to 45 m.
See Color Plate XIII.

402. Pitar (Pitar) hoffstetteri Fischer-Piette, 1969. White, with small irregular orange blotches along the growth lines, no escutcheon, lunule depressed. Differing from *P. helenae* principally in the size of the pallial sinus, which is broad and long. Length of shell, 23 mm; height, 19 mm; diameter (one valve), 6 mm. Isla Baltra, Galápagos.

403. Pitar (Pitar) perfragilis Pilsbry & Lowe, 1932. A small, thin, white shell, ovate-trigonal, with a large lunule and a broad pallial sinus rounded at the end. The small size suggests that the type specimens are immature. The species has not been recognized in later collections. Length, 10 mm; height, 9 mm; diameter, 7.5 mm. San Juan del Sur, Nicaragua.

Subgenus HYPHANTOSOMA DALL, 1902

Sculptured with zigzag concentric grooves.

404. Pitar (Hyphantosoma) aletes Hertlein & Strong, 1948. A handsome white shell with surface sculpture of fine zigzag lines, and a broader, heart-shaped lunule. The pallial sinus projects forward but for less than half the length of the shell. Length, 54 mm; height, 46 mm; diameter, 43 mm. Only a few specimens have been taken as yet, from Arena Bank and Guaymas, Gulf of California, to Judas Point, Costa Rica, in depths of 77 to 110 m.

405. Pitar (Hyphantosoma) hertleini Olsson, 1961. Although confused with *P. pollicaris* by authors, this has a thinner shell with more convex valves, and it is smaller. The color markings are much more evident, not the same on the two valves, with radial rays of brown on a cream-white ground and fine irregular zigzag markings of darker brown. The interior is white and the pallial sinus long. Length, 36 mm; height, 29 mm; diameter, 19.5 mm. Panama to Peru.

406. Pitar (Hyphantosoma) pollicaris (Carpenter, 1864). Several features distinguish this from *P. aletes*: the shell is larger, the pallial sinus is nearly half as long as the shell, and immature individuals have brown zigzag markings in addition to the zigzag sculpture. A large specimen measures: length, 80 mm; height, 60 mm; diameter, 39 mm. The Gulf of California, probably just beyond low-tide line; few specimens have been taken, even by dredging. One dredging record is at 13 m, sand bottom.

406

407

409

408

411

410

Subgenus **HYSTEROCONCHA** DALL, 1902
(**DIONE** GRAY 1847 [not HUBNER, 1817])

With concentric lamellae; posterior area bordered by spines.

407. Pitar (Hysteroconcha) brevispinosus (Sowerby, 1851). As figured by Olsson (1961), this is proportionately higher and more trigonal than *P. lupanaria*, with more uniform ribbing and short scales instead of spines at the posterior ridge. Length, 45 mm; height, 38 mm. Gulf of California to Ecuador.

408. Pitar (Hysteroconcha) lupanaria (Lesson, 1830) (Synonyms: **Cytherea semilamellosa** Delessert, 1841; **Dione exspinata** Reeve, 1863). Like the thorny oyster, *Spondylus*, this has spines so long and slender one wonders how the clam moves without breaking them. The shell is white, tinted with violet and with violet spots at the bases of the spines. Length (not including spines), 53 mm; height, 43 mm; diameter, 29 mm. Ballenas Bay, west coast of Baja California, through the Gulf of California and south to Negritos, Peru; common on sandy beaches, especially after storms and offshore to depths of 24 m. The name *Dione exspinata* Reeve, 1863, seems to be based on an individual in which the spines were undeveloped. *P. (H.) dione* (Linnaeus, 1758), type of the subgenus, is a Caribbean form resembling *P. lupanaria*.

409. Pitar (Hysteroconcha) multispinosus (Sowerby, 1851) (Synonym: **Callista longispina** Mörch, 1861). Previously considered a subspecies, this now is regarded as a distinct species. The spines are consistently more numerous than in *P. lupanaria* and more slender. The concentric ribs are sharp-edged and narrow, the color brown rather than purple. The shell itself is smaller and proportionately longer. Length, 40 mm; height, 32 mm; diameter, 25 mm. Gulf of California to northern Peru.

410. Pitar (Hysteroconcha) roseus (Broderip & Sowerby, 1829) (Synonym: **Cytherea lepida** Chenu, 1847). Except for a spinous white stripe on the posterior umbonal angle, the shell is a dull rose purple, compressed, and concentrically close-ridged. Length, 44 mm; height, 34 mm; diameter, 20 mm. The Gulf of California to Panama; to 73 m.

Subgenus **LAMELLICONCHA** DALL, 1902

With thin concentric lamellae; ligamental nymph striate.

411. Pitar (Lamelliconcha) alternatus (Broderip, 1835). There has been some confusion on the interpretation of Broderip's species in that it was based upon two distinct forms; the description came from one, and the name was coined from the other. The type specimen in the British Museum settles the problem. It has sharp, raised concentric ribs that alternate in height. The interior is white stained with violet to chestnut on the upper part. The lunule and posterior dorsal margins may be brownish red or there may be some narrow rays or lines of color. Length, 45 mm; height, 40 mm; diameter, 27 mm. Gulf of California to northern Peru, on sand beaches and dredged to depths of 55 m. *P. (L.) circinatus* (Born, 1778) is a closely related species in the Caribbean area. Olsson (1961) regards *P. alternatus* as a geographic subspecies of it.

412. Pitar (Lamelliconcha) callicomatus (Dall, 1902). Similar to *P. alternatus* in shape but more oval, this has one to three secondary lamellae between each two primary ribs, and the shell is dull white. Length, 47 mm; height, 36 mm; diameter, 22 mm. Acapulco, Mexico, to Ecuador, in depths of 26 to 73 m; only a few specimens have been found.

413. Pitar (Lamelliconcha) concinnus (Sowerby, 1835) (Synonym: **? Cytherea suppositrix** Menke, 1849). The posterior end of the shell is somewhat pointed and the concentric ribs are of similar size throughout. Specimens vary in color from white to brown, with radial stripes of purplish brown on many. Length, 35 mm; height, 28 mm; diameter, 18 mm. Magdalena Bay, Baja California, through the Gulf of California and south to Paita, Peru; fairly common on sand beaches and offshore to depths of 73 m.

414. Pitar (Lamelliconcha) frizzelli Hertlein & Strong, 1948. The rich color marking and heavy concentric ribbing are unique among West American Pitars. The whole surface of the shell is marked with light and dark brown blotches and tent-shaped lines. Inside, the shell is yellow in the center, shading to purple under the beaks and over the muscle scars and to white near the basal margin. Length, 46 mm; height, 33.5 mm; diameter, 24 mm. As yet only a few specimens have been collected, all dredged off the southern end of the Gulf of California, in depths of 82 to 110 m.

415. Pitar (Lamelliconcha) hesperius Berry, 1960. The shell is more rounded than *P. alternatus* or *P. concinnus*, with strong and crisp concentric ribs. In color it is white with a brown lunule, brown ribs, and rays of deeper brown curving downward from the umbones. Length, 42 mm; height, 37 mm; diameter, 25 mm. Mazatlán area, Mexico, from shrimp-dredging boats.

416. Pitar (Lamelliconcha) paytensis Orbigny, 1845 (Synonym: **Cytherea affinis** Broderip, 1835 [not *Venus affinis* Gmelin, 1791, also now considered a *Pitar*]). Of three somewhat similar species, *P. concinnus*, *tortuosus*, and *paytensis*, this has the most pointed shell. The concentric ribs are even and regularly spaced but varying in coarseness from shell to shell. Color mostly white, never strongly rayed or marked with reddish brown. Length, 37 mm; height, 25 mm; diameter, 15 mm. Gulf of California to Peru.

417. Pitar (Lamelliconcha) tortuosus (Broderip, 1835). Some authors consider this identical with *P. concinnus*, merely a white form with the ribs more irregular than usual; Lowe, however, who collected it on mud flats and sandbars, was convinced that it is distinct. Length, 40 mm; height, 32 mm; diameter, 23 mm. Guaymas, Mexico, to Panama, according to Lowe.

418. Pitar (Lamelliconcha) unicolor (Sowerby, 1835) (Synonyms: **Chione badia** Gray, 1838; **Cytherea ligula** Anton, 1839). The fine concentric sculpture is absent on the middle part of the shell. Another distinctive feature is the long pallial sinus that extends forward, not upward. In color the shell is mostly uniform—white or brown, but an occasional specimen is variegated. Length, 46 mm; height, 41 mm; diameter, 20 mm. Acapulco, Mexico, to Jipijapa, Ecuador, intertidally on sand beaches, though not common, and offshore to 11 mm depth.

419. Pitar (Lamelliconcha) vinaceus (Olsson, 1961). Resembling *P. alternatus* but smaller, more trigonal, beaks more anterior, the umbones fuller and more prominent; concentric ridges thin, often recurved; color brown or light purplish chestnut, rarely white, interior white or stained with violet. Length, 36 mm; height, 30 mm; diameter, 20 mm. Nayarit, Mexico, to Ecuador. This was considered a subspecies of the Caribbean *P. (L.) circinatus* (Born, 1778) by Olsson.

Subgenus PITARELLA PALMER, 1927

The relationships of this subgenus to *Callocardia* Adams, 1864, the type species of which is Indo-Pacific, and *Agriopoma* Dall, 1902, the type of which is western

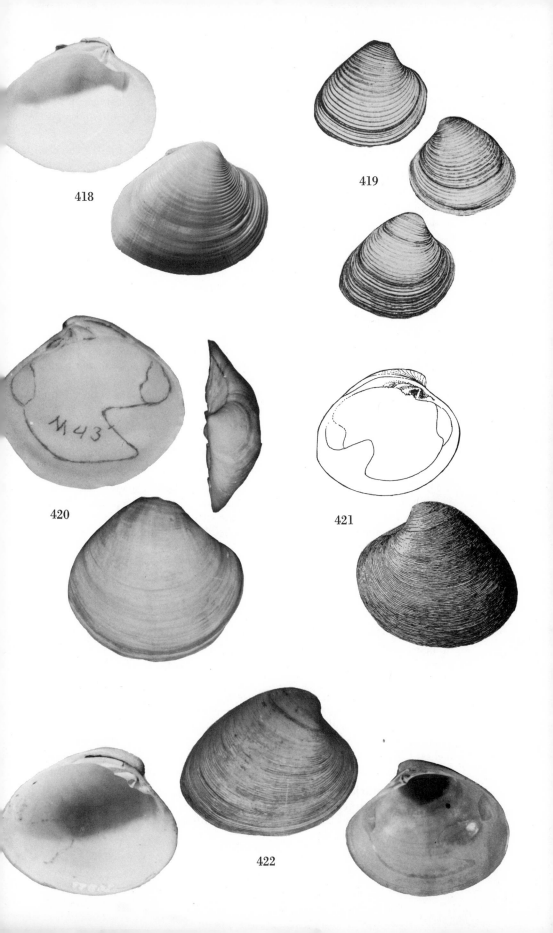

418

419

420

421

422

Atlantic, are unclear; the hinge suggests affinities, but other shell characters suggest closer ties with *Pitar*. Olsson (1961) regards *Pitarella* as a subgenus of *Agriopoma*. The cardinal teeth in the right valve are fused at the top and separated from the edge of the shell by a narrow gap.

420. ? Pitar (Pitarella) aequinoctialis (Fischer-Piette, 1969). The shell is white, with a rosy-orange flush at the umbones and a few orange radial markings, the inside pale pink above, lighter toward the edge. The shell is relatively shorter than in the other two species of *Pitarella*, with a more angulate pallial sinus. Length, 37 mm; height, 33 mm. Santa Elena, Ecuador. Following Olsson's classification, Fischer-Piette allocated this to *Agriopoma*.

421. Pitar (Pitarella) catharius (Dall, 1902) (Synonym: **P. tomeanus** of authors, not of Dall, 1902). The texture is described as earthy, that is, not smooth and glossy; solid and white, with a shallow radial depression back of the umbones, extending to the ventral margin. Length, 58 mm; height, 50 mm; diameter, 41 mm. An offshore form; Ballenas Bay, west coast of Baja California to the Gulf of California and south to Caleta la Cruz, northern Peru, in depths of 13 to 80 m.

422. Pitar (Pitarella) mexicanus Hertlein & Strong, 1948 (Synonym: **Pitar lenis** Pilsbry & Lowe, 1932 [not Conrad, 1848]). This shell differs from that of *P. catharius* by being longer, less globose, the posterior end less broadly rounded, the radial depression less evident, the left middle cardinal and the right middle and posterior cardinal teeth thinner, and the pallial sinus more bluntly rounded. Length, 42 mm; height, 33 mm; diameter, 25 mm. Near the head of the Gulf of California to Chitiqui, Panama, in depths of 4 to 80 m.

Subgenus **TINCTORA** JUKES-BROWNE, 1914

Thick-shelled, round, glossy; valve margins crenulate.

423. Pitar (Tinctora) vulneratus (Broderip, 1835). The creamy-yellow shell is marked with irregularly spaced violet rings or zones of color, especially toward the ventral margin. The inner margin of the shell has faint crenulations. A fairly large specimen measures: length, 45 mm; height, 40 mm; diameter, 27 mm. Magdalena Bay, Baja California, through the Gulf of California and south to Panama, on sand beaches and offshore to depths of 15 m. Beach valves are more common than live specimens.

Genus **MEGAPITARIA** GRANT & GALE, 1931

Like the subgenus *Tinctora* of *Pitar*, this group is restricted to the west coast of Central America. The species share with *Periglypta multicostata* the distinction of being among the largest shelled Veneridae. The Megapitarias are smooth and brightly colored, resembling the Pitars except for the larger size and thickness of the valves.

424. Megapitaria aurantiaca (Sowerby, 1831) (Synonym: **Cytherea aurantia** "Hanley," Sowerby, 1851). The periostracum is a dull, orange brown, beneath which the shell is pink to pinkish brown; the inside is white, the hinge plate tinted with purple. A large specimen measures: length, 112 mm (over 4 inches); height, 85 mm; diameter, 59 mm. The head of the Gulf of California to Salinas, Ecuador, living at or below extreme low water and offshore to 10 m.

425. Megapitaria squalida (Sowerby, 1835) (Synonyms: **Chione biradiata** Gray, 1838; **Cytherea chionaea** Menke, 1847). The periostracum is a shiny grayish brown, somewhat mottled or striped in many specimens. Beneath it the

PLATE III · *Trachycardium consors* (Sowerby)

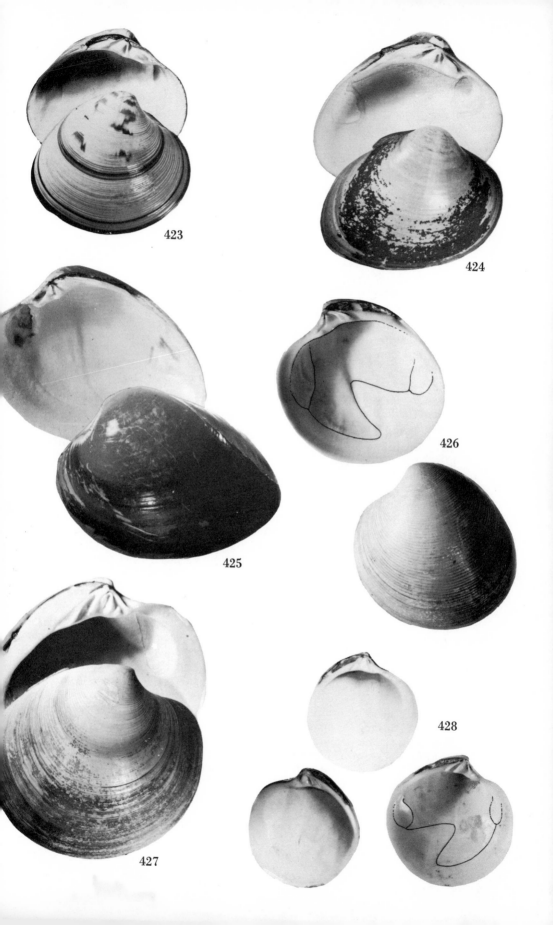

shell is a drab light brown, white within. A large specimen measures: length, 120 mm; height, 97 mm; diameter, 68 mm. Scammon's Lagoon, Baja California, through the Gulf of California and south to Mancora, Peru; common on sandy mud flats but occurring also offshore to depths of 160 m. The names *biradiata* and *chionaea* have been applied to striped and mottled color forms, respectively, but these seem not to be genuine subspecies.

See Color Plate XII.

Subfamily DOSINIINAE

Equivalve, lenticular, concentrically striate; hinge strong, with anterior lateral teeth.

Genus DOSINIA SCOPOLI, 1777
(ARTEMIS of authors)

Shell lenticular, with heart-shaped lunule and regular concentric ribbing. Hinge strong, with a pointed or pustular left anterior lateral tooth; ligamental area long and wide, somewhat beveled; pallial sinus deep, tongue-shaped.

426. Dosinia dunkeri (Philippi, 1844) (Synonym: **Artemis simplex** Hanley, 1845). Surface with fine and very regular concentric ribs. The pallial sinus is angular and points toward the middle or upper part of the anterior adductor muscle scar. The sculpture is finer than that of *D. ponderosa*. Length, 56 mm; height, 55 mm; diameter, 31 mm. Magdalena Bay, Baja California, through the Gulf of California and south to Zorritos, Peru; common on mud flats and offshore to depths of 55 m.

427. Dosinia ponderosa (Gray, 1838) (Synonyms: **Venus cycloides** Orbigny, 1845; **Cytherea (Artemis) gigantea** Philippi, 1847, *ex* Sowerby, MS). A well-developed specimen of this large, white, rounded shell, with its coarse concentric sculpture and heavy hinge, is unmistakable. Small specimens can be distinguished from adult *D. dunkeri* by the relatively thicker shell and coarser sculpture as well as by the rounder outline. A large specimen measures: length, 145 mm (nearly 6 inches); height, 139 mm; diameter, 75 mm. Scammon's Lagoon, Baja California, through the Gulf of California and south to Paita, Peru; in deeper water than *D. dunkeri*, offshore to 60 m. Though not rare, it is not as abundant a form as *D. dunkeri*.

428. Dosinia semiobliterata Deshayes, 1853. (Synonym: **D. annae** Carpenter, 1857). Fischer-Piette and Delmas (1967), who have compared the type specimens, have shown that Deshayes's species with an erroneous locality label of "Australia" is actually the West American form later described by Carpenter. Although Deshayes's name would qualify as a *nomen oblitum*, the facts that it has twice been cited as the senior synonym and adopted (Fischer-Piette & Delmas, 1967; Fischer-Piette, 1969); that Carpenter's type was not illustrated until recent years (Keen, 1968); and that *nomen oblitum* procedures are to be revised, make acceptance of strict priority seem in the long run to be the preferable course. Rarest of Panamic Dosinias, *D. semiobliterata* is slightly flatter and smoother than *D. dunkeri* and has a more horizontal pallial sinus, which points toward the lower part of the anterior adductor scar. Length, 50 mm; height, 48 mm; diameter, 29 mm. The species occurs mainly in the southern end of the Gulf of California (type locality, Mazatlán), but two valves in the Stanford collection extend the range to Panama.

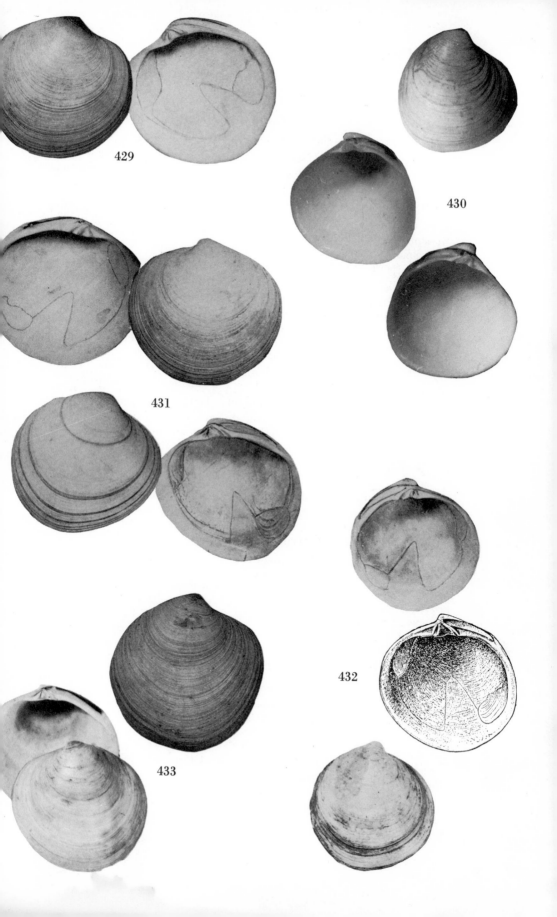

429

430

431

432

433

Subfamily CYCLININAE

Like Dosiniinae in shape but without anterior lateral teeth or incised lunule.

Genus CYCLINELLA DALL, 1902

Internal margins smooth, lunular area faintly indicated.

429. Cyclinella jadisi Olsson, 1961. A lenticular shell, white or with a reddish or brownish stain on the umbones, the surface of fine growth lines having a silky sheen. Posterior muscle scar nearer hinge than in other west coast species. Length, 62 mm; height, 60 mm; diameter, 26 mm. Panama to Ecuador.

430. Cyclinella producta (Carpenter, 1856). The type specimen in the British Museum was first illustrated by Palmer in 1963. The contracted dorsal margin gives a more trigonal outline to the shell than in most species. Now that figures are available, perhaps collectors will be able to recognize the species, which has a type locality of Panama. Length, about 40 mm; height, 40 mm.

431. Cyclinella saccata (Gould, 1851). This has been considered a synonym of the more southern C. *subquadrata* (Hanley, 1845) by authors, but Olsson (1961) has shown that it is distinct. It is smaller than C. *subquadrata*, the posterior-dorsal margin is shorter, and the whole posterior margin is more evenly rounded. The posterior adductor scar is low and the pallial sinus is directed upward toward the hinge. The shell is white, with a thin, pale yellowish periostracum. Length, about 40 mm; height, 35 mm. Gulf of California to Panama, in depths to 46 m.

432. Cyclinella singleyi Dall, 1902. The shell is inflated and white, the lunule broadly heart-shaped. Within, the pallial sinus is sharply angular at the end and projects nearly to the center of the shell. Length, 36 mm; height, 32 mm; diameter, 22 mm. Scammon's Lagoon, Baja California, to the head of the Gulf of California and south to Panama, mostly beyond the low-tide level especially in estuaries.

433. Cyclinella subquadrata (Hanley, 1845) (Synonym: **? Artemis maci-lenta** Reeve, 1850). The shell is large, lenticular, the posterior margin with an obtuse angle about halfway down. Pallial sinus low, relatively short. Length, to 95 mm; height, to 86 mm. Ecuador and northern Peru.

434. Cyclinella ulloana Hertlein & Strong, 1948. Olsson (1961) regards this as a distinct species, although it was proposed as a subspecies of C. "*kröyeri*" (a name that will, under the Code of Zoological Nomenclature, have to be written *kroeyeri*). The C. *kroeyeri* (Philippi, 1847), which it resembles in outline, is much more southern in distribution—from Peru southward to Chile. The subcircular C. *ulloana* is thin-shelled, with a brownish periostracum. Length, up to 75 mm; height, 77 mm; diameter, 38 mm. Known from dredgings, in depths to 46 m in the Gulf of California, and from beach drift off Nayarit, Mexico.

Subfamily CLEMENTIINAE

Inequilateral, thin, without escutcheon; sculpture subdued or wanting; inner ventral margin smooth, hinge without lateral teeth. Two Panamic genera.

Genus CLEMENTIA GRAY, 1842

Thin, ovate, with undulating concentric sculpture, especially on umbones; no lunule; pallial sinus deep.

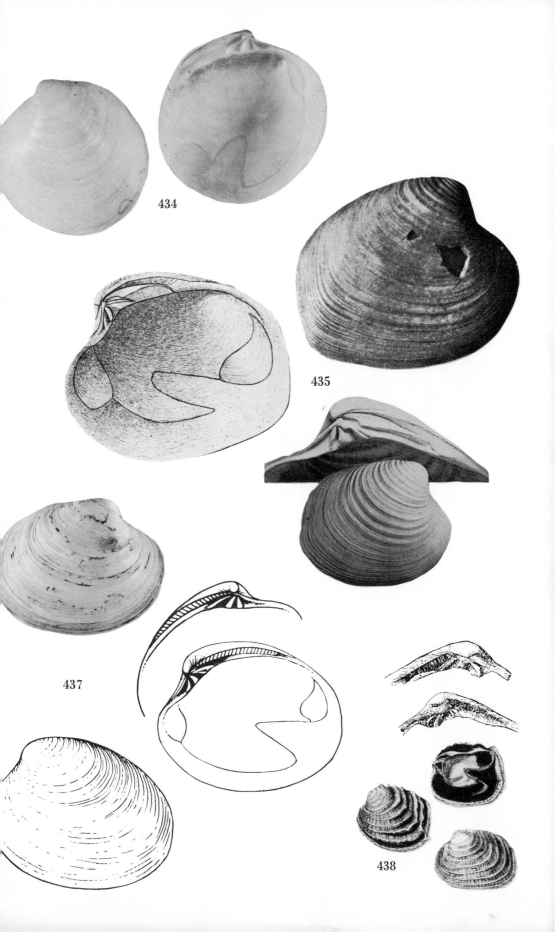

434

435

437

438

435. Clementia solida Dall, 1902. For many years this was known only by the single broken specimen that is the type, from Mexico. More recently additional specimens have been found; one perfect specimen in the Stanford collection, with both valves still joined by ligament, washed in on a beach near Mazatlán. Others have been found near Guaymas. The shell is ovate-quadrate in outline, thin, white, with concentric undulating sculpture, especially on the umbones. Length, 79 mm; height, 63 mm; diameter, 34 mm. Guaymas, Mexico, to Panama and Ecuador.

436. Clementia gracillima Carpenter, 1857. The name *C. gracillima* Carpenter, 1857, must be regarded as a *nomen dubium*. The type specimen was a minute shell only a little over 3 mm long. It was probably a juvenile and may not even have been a member of the family Veneridae. Carpenter did not prepare a drawing of it as he did of most of his other minute shells, and the type has now recrystallized and become unrecognizable. The description is too brief to be of use in interpreting the species.

Genus COMPSOMYAX STEWART, 1930

Shell smooth, thin, oblique-ovate.

437. Compsomyax subdiaphana (Carpenter, 1864). This species is commonly dredged in Puget Sound and occurs in fairly deep water off southern California. A few valves have been taken in the northern head of the Gulf of California by Mrs. Keith Abbott. Thus one more name may be added to the list of California species, such as *Macoma indentata*, that have become dispersed to the northern Gulf area. Length (average), 50 mm; height, 35 mm; diameter, 20 mm. Southern Alaska to the northern end of the Gulf of California.

Subfamily TAPETINAE

Ovate to elongate, the inner margins smooth, at least posteriorly, the hinge plate narrow, lacking lateral teeth. Two Panamic genera.

Genus IRUS SCHMIDT, 1818

Subquadrate, with raised concentric lamellae intersected by fine radial threads; hinge weak. The generic name *Irus* long was dated from Oken, 1815, but Oken's work recently has been officially rejected by the International Commission on Zoological Nomenclature as nonbinomial. Fortunately, *Irus* was validated in the same sense by Schmidt, as two British zoologists, Drs. J. Bowden and D. Heppell, have shown. No substitutes, therefore, need to be sought, although *Notirus* Finlay, 1928, was adopted by a few workers for a short time. No species of *Irus*, *s. s.*, occur in the Panamic province.

Subgenus PAPHONOTIA HERTLEIN & STRONG, 1948

Lunule bounded by an incised line, the inner margin crenulate on the anterior part of the shell.

438. Irus (Paphonotia) ellipticus (Sowerby, 1834) (Synonyms: **Petricola solida** Sowerby, 1834; **P. solidula** Sowerby, 1854; **Venerupis foliacea, P. oblonga**, and **P. paupercula** Deshayes, 1853). As with all shells that develop in rock cavities, these are greatly distorted and variable in shape—hence, the number of names that have been applied to what are probably purely environmental variations. A nondistorted specimen is ovate-oblong, rather compressed, inequilateral, with spaced concentric lamellae and closely set radial ribs, especially on the anterior part of the shell. The exterior is whitish, but the inside of the shell is brown and white or almost entirely brown. Specimens that have been in tight-fitting holes may become frilled and squarely truncate. These have been known as *I. foliaceus* (Deshayes). Large specimens may attain a length of 26 mm or more. Mazatlán, Mexico, to Arica, Chile, in hard mud or rock cavities and offshore to depths of 24 m.

Genus PSEPHIDIA DALL, 1902

Small, polished shells, trigonal, with pointed umbones.

439. Psephidia cymata Dall, 1913. A small white shell ornamented by fine concentric threadlike sculpture. Length, 6 mm; height, 5.5 mm. Santa Barbara Islands, California, to the Gulf of California, in shallow water and offshore to depths of 82 m.

Subfamily CHIONINAE

Ovate-trigonal, inequilateral, with cancellate sculpture; inner margins crenulate in most forms; hinge plate with no anterior lateral teeth. Three Panamic genera.

Genus CHIONE MEGERLE VON MÜHLFELD, 1811

Concentric sculpture strong; lunule and escutcheon well defined in most forms. The genus *Chione*, abundant in species, is almost entirely tropical American in distribution. The name is pronounced as three syllables—ki-oh'-nee (not ki-own'). The key is to subgenera.

1. Sculpture subdued, fading out toward ventral margin.............*Iliochione*
   Sculpture well developed, relatively even over entire shell...............2
2. With radial sculpture dominating the concentric..................*Timoclea*
   With concentric sculpture overriding or dominating the radial...........3
3. Radial sculpture weak to wanting............................*Lirophora*
   Radial sculpture evident between concentric ribs......................4
4. Lunule and escutcheon wanting...............................*Chionista*
   Lunule incised; escutcheon beveled................................5
5. Pallial sinus short to wanting; posterior margin rounded........*Chione, s. s.*
   Pallial sinus rounded at end; posterior area a little rostrate or pointed below ....*Chionopsis*

Subgenus CHIONE, s. s.

Concentric sculpture often frilled; pallial sinus small; cardinal teeth smooth.

440. Chione (Chione) californiensis (Broderip, 1835) (Synonyms: **Venus succincta** Valenciennes, 1827 [not Linnaeus, 1767]; **V. leucodon** Sowerby, 1835; **V. nuttalli** Conrad, 1837; **C. gealeyi** and **durhami** Parker, 1949). This species may be distinguished from *C. undatella*, with which it has been confused, by the coarser and more widely spaced concentric ribs and—on mature specimens—the flattening and thickening of the concentric lamellae toward the edge of the shell, so that they nearly cover the interspaces. The shell is whitish, irregularly freckled with brown on some specimens, tinged with blue inside, especially around the posterior part. Dimensions of a large specimen: length, 68 mm; height, 65 mm; diameter, 38 mm. This is a common clam from Point Mugu, Cali-

439

440

441

442

443

444

445

445a

fornia, to Panama, intertidally on mud flats at low tide and offshore to depths of 69 m, mud bottom. Parker (1949) recognized three subspecies that seem to have no geographic significance.

441. Chione (Chione) compta (Broderip, 1835) (Synonym: **C. meridionalis** I. Oldroyd, 1921). The few flattened and slightly upturned concentric lamellae, the flattened shell, the fine divaricating ribs on the anterior and posterior slopes, and the pallial line distant from the ventral margin are all diagnostic features of the species. Parker (1949) considered that *C. meridionalis* might be separable subspecifically in that the concentric ridges are less well developed on the umbones of the type specimens. Length, 33 mm; height, 31 mm; diameter, 10 mm. The Gulf of California to Bayovar, Peru, mostly offshore in depths of 22 to 27 m.

442. Chione (Chione) guatulcoensis Hertlein & Strong, 1948. A small form, with spaced concentric ribs that are elevated into sharp lamellae posteriorly. The shell is white, blotched and banded with brown lines and dots; the interior is white and rose. Length, 11 mm; height, 9 mm; diameter, 6 mm. This seems to be the West American representative of the Caribbean *C. mazyckii* Dall, 1902. Off Port Guatulco, Mexico, to Panama Bay, in depths to 13 m.

443. Chione (? Chione) subimbricata (Sowerby, 1835). A somewhat trigonal shell, dark-colored, with brown radial bands or small irregular zigzag markings on a light ground. The concentric ridges are regular and about 12 to 20 in number. Radial sculpture is weak, the ribs dividing or bifurcating as the shell increases in size. Length, 34 mm; height, 30 mm; diameter, 24 mm. La Paz and Guaymas, Gulf of California, to Paita, Peru, intertidally and offshore to depths of 9 m. Olsson (1961) has shown that this and the next (often considered a subspecies), both long assigned to *Anomalocardia* (a western Atlantic and Indo-Pacific genus), more properly belong in *Chione*. The ribbing is, however, not as lamellar as in many Chiones.

444. Chione (? Chione) tumens (Verrill, 1870). This form is distinct from *C. subimbricata* in having only about 6 to 8 rounded concentric ridges (one could hardly call them ribs) on the entire shell. Restricted to the southern part of the outer coast of Baja California and to the Gulf of California.

445. Chione (Chione) undatella (Sowerby, 1835) (Synonyms: **Venus neglecta** Sowerby, 1835; **V. entobapta** Jonas, 1845; **Cytherea sugillata** Jonas, 1846; **V. perdrix** Valenciennes, 1846; **V. simillima** Sowerby, 1853; **V. excavata** Carpenter, 1857; **V. bilineata** Reeve, 1863; **Chione undatella taberi** Parker, 1949). This is easy to confuse with *C. californiensis*, and the many names cited above testify to its variability. The best means of separating the two species is in the ribbing, which consists of more closely spaced, thinner, and sharper concentric lamellae in *C. undatella*—about 4 to 6 ribs per centimeter on the middle of the shell as compared with about 3 in *C. californiensis*. Forms in which the concentric ribs are marked with brown and are somewhat frilled have been called a subspecies, *C. (C.) undatella neglecta*, but until the hinges of this and other supposed subspecies or synonyms are studied, subdivisions seem unjustified. One form illustrated here (fig. 445a) has a narrower hinge, a longer shell, and a pattern of larger brown blotches. It may constitute a distinct species, for it is not geographically separated; whether any of the several available names will actually apply remains for future research. Length, 60 mm; height, 52 mm; diameter, 34 mm. Southern California to Paita, Peru, on protected sandy beaches and offshore to depths of 90 m.

Subgenus CHIONISTA KEEN, 1958

The following two species seem to constitute a subgenus intermediate in some ways between *Chione* and *Protothaca*. They lack the escutcheon of *Chione* and have much more subdued and polished concentric sculpture, but the hinge and pallial sinus are of the characteristic *Chione* pattern. Some workers have thought that these species might belong in the New Zealand group *Austrovenus*, but the shells of that genus have a lunule. In *Chionista* both lunule and escutcheon are wanting.

446. Chione (Chionista) cortezi (Carpenter, 1864, *ex* Sloat, MS) (Synonym: **C. gibbosula** of authors, not of Reeve, 1863). As compared with the more widely distributed *C. fluctifraga*, this is a larger and more trigonal form. The concentric ribs are more closely and evenly spaced, the radial ribs reduced to striae and to a few incised lines anteriorly and posteriorly. The interior has less of the bluish color, mainly at the ends. Length, 62 mm; height, 55 mm; diameter, 34 mm. The outer coast of Baja California (near Magdalena Bay) to San Felipe in the upper part of the Gulf of California and to Guaymas, Mexico (type locality), intertidally.

447. Chione (Chionista) fluctifraga (Sowerby, 1853) (Synonyms: "**Venus callosa** Conrad" of Sowerby, 1853, not of Conrad, 1837; **V. gibbosula** Reeve, 1863, *ex* Deshayes, MS). The concentric ribs are smooth and polished, irregularly broken up by radial furrows that are not of even width. The shell is a creamy to grayish white, stained dark blue within, especially at the posterior end. Length, 51 mm; height, 49 mm; diameter, 34 mm. San Pedro, California, through the Gulf of California to Guaymas, Sonora, Mexico, mainly intertidally.

Subgenus CHIONOPSIS OLSSON, 1932
(GNIDIELLA PARKER, 1949)

Shell somewhat inflated, with a tendency toward elongation of the posterior dorsal margin; pallial sinus well developed; hinge with at least the right posterior and middle left cardinal teeth grooved or bifid.

448. Chione (Chionopsis) amathusia (Philippi, 1844) (Synonym: **? Venus darwinii** Dunker in Römer, 1857). The sculpture is fine, with interribs between the primary radial ribs and with the concentric lamellae raised into prickly scales. The shells are mostly a light brown in color. Length, 40 mm; height, 36 mm; diameter, 26 mm. The Gulf of California to Mancora, Peru, mostly offshore in depths to 73 m.

449. Chione (Chionopsis) crenifera (Sowerby, 1835). The lightly frilled concentric lamellae cover the entire rather small shell but are not elevated. As in most of the Chiones, the brown markings of the buff- to tan-colored shell may vary considerably, from radial stripes to zigzag lines or solid blotches. The identity of this species remains somewhat in doubt. The type specimen, from Santa Elena, Ecuador, seems to have been lost. Sowerby in 1851 figured two specimens, one white with a brown lunule and escutcheon (figure reproduced here), the other brown with zigzag radial markings, presumably from Paita, Peru. Later workers have labeled a number of different forms under this name. Dall even considered that the species is identical with the Caribbean *C. subrostrata* (Lamarck, 1818).

446

447

448

449

450

451

The original description gives little help in precise determination—ovate, decussate, length, 37 mm, height, 33 mm. Specimens in the Stanford collection from Paita, Peru, purchased from a British dealer who had presumably compared them with authentic material, match Sowerby's figures well. These resemble *C. montezuma* but are less finely sculptured. Possibly *C. montezuma* is a northern representative or geographic subspecies of *C. crenifera*. Unfortunately, Olsson did not discuss the latter. Ecuador to Peru (?).

450. Chione (Chionopsis gnidia (Broderip & Sowerby, 1829). One of the most ornate of the Chiones, this is also one of the largest. Well-developed specimens have the concentric lamellae scalloped and rising into prickly scales. The shell itself is a matte white or beige, shiny white within. A large specimen measures: length, 101 mm; height, 95 mm; diameter, 68 mm. Cedros Island, Baja California, through the Gulf of California and south to Paita, Peru, in bays and offshore to depths of 33 m.

451. ? Chione (Chionopsis) jamaniana (Pilsbry & Olsson, 1941). In the southern part of the range of the species *C. gnidia* this form appears, first described as a Pliocene fossil. Sculpture is stronger than in *C. gnidia*, the concentric ridges are more solid and appressed, and the interspaces tend to become smooth in older shells. Off Punta Pasado, Ecuador, in depths to 18 m.

452. Chione (Chionopsis) montezuma Pilsbry & Lowe, 1932. Resembling the more common and widely distributed *C. pulicaria*, this is smaller, thinner, more elongate posteriorly, and the color pattern of zigzag brown lines is stronger in most specimens. Length, about 40 mm; height, 31 mm; diameter, 24 mm. Costa Rica to Panama.

453. Chione (Chionopsis) olssoni Fischer-Piette, 1969. Resembling *C. purpurissata*, the shell is higher for the length and lighter-colored within. Length, 48 mm; height, 46 mm. Ecuador.

454. Chione (Chionopsis) ornatissima (Broderip, 1835). (Synonym: **Chione traftoni** Pilsbry & Olsson, 1941). The shell somewhat resembles *C. gnidia*, but the concentric lamellae are thin, rather widely spaced, and lack the scales of that species. It is one of the most highly ornamented species in the genus, the raised concentric frills in a well-developed shell being delicately fluted by the radial sculpture. A large specimen measures: length, 53 mm; height, 45 mm; diameter, 33 mm. Panama to Ecuador, in depths of 13 to 27 m; also fossil in Ecuador.

455. Chione (Chionopsis) pulicaria (Broderip, 1835). Why the author of the specific name should have associated this attractive shell with fleas is anybody's guess, but that is what the name literally means—"of the fleas." The shell is comparatively large and inflated, with zigzag color markings in shades of brown, somewhat blurred and irregular. The white interior is variously stained with a bluish purple, either as a streak along the posterior part or as a diffuse area along the pallial line. A large specimen measures: length, 47 mm; height, 39 mm; diameter, 30 mm. The Gulf of California to Tumaco, Colombia, on sandbars and offshore to depths of 18 m.

456. Chione (Chionopsis) purpurissata Dall, 1902. (Synonym: **Venus lilacina** Carpenter, 1864 [not Gray, 1838]). Larger than *C. pulicaria*, the light-brown shell is mottled with darker brown but with no definite patterns or dots.

452

453

454

455

456

457

The distinguishing features are the less pointed posterior-ventral end and the bright rose-purple interior of the shell. A large specimen measures: length, 64 mm; height, 57 mm; diameter, 43 mm. This species seems to be most similar to the Caribbean *C. pubera* (Bory de Saint-Vincent, 1827, *ex* Valenciennes, MS). The Gulf of California to Guatemala, offshore in depths to 30 m.

Subgenus ILIOCHIONE OLSSON, 1961

Sculpture subdued, concentric ribs fading out toward ventral margin.

457. Chione (Iliochione) subrugosa (Wood, 1828). Formerly allocated to *Anomalocardia*, a Caribbean genus, this was made type of a new subgenus under *Chione* by Olsson. Except for the sculpture the shell resembles *Chione, s. s.* Coloring is variable, from off-white to cream or gray, usually with three or four wide radial bands of dark brown or with mottling of lines or dots. There is a thin yellow periostracum that is easily eroded, after which the shell takes on a polished look. Length, about 45 mm; height, 35 mm; diameter, 25 mm. Magdalena Bay, Baja California, through the Gulf of California and south to Peru, in lagoons or on mud flats; used as food in many places.

Subgenus LIROPHORA CONRAD, 1863

With wide or recurved concentric ribs, radial ribs weak or wanting. Pallial sinus short and angular.

458. Chione (Lirophora) discrepans (Sowerby, 1835). A white shell with closely spaced, even, and slightly upturned lamellae, lacking radial sculpture except for crenulations of the inner margin that reflect radial riblets in the middle layer of the shell. Length, 36 mm; height, 31 mm. The type locality is Islay, Peru; specimens found by Mrs. E. E. Wahrenbrock on the beach at Playa Novillero, Nayarit, Mexico, extend the range northward and add another species to the list of Panamic mollusks.

459. Chione (Lirophora) kellettii (Hinds, 1845). The shell is buff to yellowish brown, with ribs that are smooth and fused together over most of the shell but developed into prominent white lamellae at the ends. Length, 34 mm; height, 26 mm; diameter, 9 mm. The Gulf of California to northern Peru, offshore in depths of 46 to 73 m.

460. Chione (Lirophora) mariae (Orbigny, 1846) (Synonym: **Venus cypria** Sowerby, 1835 [not Brocchi, 1814]). The strong raised lamellae cross the shell without joining together at any point. The shell is light brown with a few stripes or spots of darker brown. Length, 23 mm; height, 17 mm; diameter, 12 mm. Cedros Island, Baja California, through the Gulf of California and south to Guayaquil, Ecuador, mostly offshore in depths to 110 m. This seems to be the West American twin to the Caribbean *C. (L.) paphia* (Linnaeus, 1767).

461. Chione (Lirophora) obliterata Dall, 1902. A heavy little shell, with faint purplish radial markings on a pale yellow-brown base. The concentric sculpture is irregular, especially on mature shells, the ribs coalescing here and there. The inside is yellow with a flush of purple near the hinge. Length, 24 mm; height, 18 mm; diameter, 14 mm. Described as from Panama Bay, this was considered by Olsson (1961) to be a Caribbean species; however, Dr. S. S. Berry (*in litt.*) has specimens from west Mexican shrimp boats that fit the description well.

458

459

460

461

462

463

464

Subgenus TIMOCLEA BROWN, 1827

Small shells with sculpture predominantly radial, crossed by concentric lamellae of varying strength; escutcheon varying from weak (in the European type species, *Venus ovata* Pennant, 1777) to beveled and bounded by a scaly ridge; pallial sinus small and angular.

462. Chione (Timoclea) effeminata (Stearns, 1890). The sculpture comes close to being evenly cancellate but with a slight dominance of the radial riblets. White, with pink beaks, it is rosy purple within. About 14 mm in length; 10 mm in height. Perhaps the small size has caused it to be overlooked by collectors, or perhaps it is a rare form. At any rate, the species seems to be known only by the type material, from Panama Bay.

463. Chione (Timoclea) squamosa (Carpenter, 1857) (Synonym: **Venus troglodytes** Mörch, 1861). The radial ribs are somewhat flattened, and the concentric lamellae are stronger on the posterior area, especially forming a scaly escutcheonal margin in young specimens; color ranges from pure white with purple or brown spots to light brown exteriorly; the inside is white, with a large brown blotch posteriorly. Length, 6 mm; height, 4 mm. Gulf of California southward to Peru. A similar but slightly larger form, *C. (T.) picta* Willett, 1944, ranges northward from Magdalena Bay, Baja California; it is fairly common as a Pleistocene fossil in southern California and averages more than 6 mm in length.

Genus MERCENARIA SCHUMACHER, 1817
(VENUS of authors)

Hinge with a roughened area under the ligament; sculpture predominantly concentric, with fine radial ribs.

464. Mercenaria apodema (Dall, 1902). This resembles the Atlantic *M. campechiensis* (Gmelin, 1791) but is a creamy white, has rounded concentric ribs that are not lamellar, and has a narrower roughened area on the hinge. Length, 47 mm; height, 43 mm; diameter, 28 mm. A single valve was collected nearly a century ago at Panama, and the species has not since been reported in the literature. A complete specimen in the Stanford University collection that matches well the original figure and description was found in the 1940's by Mr. A. Sorensen at Guaymas, Mexico. Unfortunately, this specimen was not taken alive, so that positive proof has yet to be furnished for the occurrence of *Mercenaria* in the Panamic province.

Genus PROTOTHACA DALL, 1902
(PAPHIA, TAPES, and VENERUPIS of authors)

With reticulate sculpture, a lunule, and with the escutcheon either wanting or in the left valve only; hinge with the middle cardinal teeth bifid, the hinge plate widened near the anterior end so that its lower edge makes a nearly right-angled bend just back of the junction with the margin of the shell; pallial sinus pointed, mostly moderate to long.

The type species of the genus, *P. thaca* (Molina, 1782), is from Chile; there are no members of *Protothaca, s. s.*, that range northward into the Panamic province. The key is to subgenera.

1. Escutcheon wanting 2
   Escutcheon present in left valve only................................ 3
2. Sculpture divided into three areas of different emphasis...........*Colonche*
   Radial sculpture fairly even over entire shell..................*Tropithaca*
3. With strong concentric lamellae............................*Antinioche*
   Sculpture mainly radial (concentric lamellae weak if present)........... 4
4. Radial ribs fine, especially in middle of shell...................*Leukoma*
   Radial ribs heavy and coarse throughout.....................*Notochione*

Subgenus ANTINIOCHE OLSSON, 1961

Superficially this resembles *Periglypta* in shape but with sculpture more as in *Chione*. The hinge reveals the relationship to *Protothaca*.

465. Protothaca (Antinioche) beili (Olsson, 1961) (Synonym: **Chione antiqua** of authors, not *Venus antiqua* King & Broderip, 1832). The rounded-ovate shell has wavy concentric lamellae well spaced throughout. The color is creamy white to light brown with rays and blotches of darker brown; lunule brown, escutcheonal area with alternating spots of brown and white. Inner margins crenulate except posteriorly. Length, 45 mm; height, 39 mm; diameter, 29 mm. Panama to Ecuador, deeply buried among rocks.

Subgenus COLONCHE OLSSON, 1961

Like *P.* (*Antinioche*) but lunule and escutcheon wanting.

466. Protothaca (Colonche) ecuadoriana (Olsson, 1961). Ovate, white, with fine radial and concentric sculpture; lunule and escutcheon wanting. Surface usually chalky in texture. A specimen of this was erroneously figured by Keen (1958, fig. 343) as *P. tumida* (Sowerby, 1853). Colombia to Ecuador, probably on mud flats of bays.

Subgenus LEUKOMA RÖMER, 1857
(NIOCHE HERTLEIN & STRONG, 1948)

Radial ribs fine; lunule incised, ribbed; escutcheon in left valve. The name *Nioche* was proposed when it was thought that *Leukoma* was invalidated by a prior *Leucoma*, but under the present Code of Zoological Nomenclature, the difference in spelling permits *Leukoma* to stand. The type species was selected in 1881 by Kobelt as *Venus granulata* Gmelin, 1791, a Caribbean form very similar to *P. asperrima*, which is type of *Nioche*.

467. Protothaca (Leukoma) asperrima (Sowerby, 1835) (Synonyms: **Venus histrionica** and **intersecta** Sowerby, 1835; **Tapes tumida** Sowerby, 1853). The intersections of the fine ribs of this shell give a rough surface that has suggested the specific name, which means "rasplike." Color and sculpture vary in this form, and Olsson (1961) would recognize two subspecies, which, however, are not clearly separable geographically. The shell tends to be more coarsely sculptured and duller-colored in the southern part of its range, and the most colorful specimens occur in Panama, where shells with fine ribs and bright brown markings on a creamy-white ground predominate; rays and blotches of dark brown and a brown lunule are characteristic. A similar species in the Caribbean is *P.* (*L.*) *pectorina* (Lamarck, 1818). Length of an average specimen, 45 mm; height, 37 mm; diameter, 25 mm. Gulf of California to Peru.

465

466

467

468

469

468. Protothaca (Leukoma) mcgintyi (Olsson, 1961). Similar to *P. metodon* but smaller, more pointed posteriorly, more compressed. Length, 31 mm; height, 27 mm; diameter, 20 mm. Panama.

469. Protothaca (Leukoma) metodon (Pilsbry & Lowe, 1932). The small, brightly marked shell resembles a *Chione*, but the hinge has the characteristic shape of *Protothaca*. It is plump and solid, light buff in color with darker spots, the lunule and escutcheonal area dark brown or, on the posterior part, with brown bars. The pallial sinus is relatively small. The ribs are low and not conspicuously cancellate; a few anterior ribs are coarser than the rest. The shell is rounder and more inflated than that of *P. asperrima*. Length, 40 mm; height, 38 mm; diameter, 32 mm. Guaymas, Mexico, to Panama.

470. Protothaca (Leukoma) subaequilateralis (Fischer-Piette, 1969). Externally the shell resembles *P. asperrima* in sculpture and coloring. The hinge, however, differs by having the anterior cardinal in the left valve remarkably long and lamellar and the corresponding socket in the right valve long and slitlike. Length, 30 mm; height, 25 mm. Ecuador.

471. Protothaca (Leukoma) zorritensis (Olsson, 1961). The rounded-ovate shell is creamy white to brown, either uniformly or with a zigzag pattern of darker color, the interior white with a flush of violet posteriorly. Pallial sinus small but distinct. Length, about 25 mm; height, 20 mm; diameter (one valve), 12 mm. Zorritos to Paita, Peru.

Subgenus NOTOCHIONE HERTLEIN & STRONG, 1948

Ribs heavy, predominantly radial; shell rounded-trigonal in outline.

472. Protothaca (Notochione) columbiensis (Sowerby, 1835). The shape and the beveled escutcheon in the left valve are reminiscent of *Chione*, but the radial ribbing and the hinge are those of *Protothaca*. The yellowish-gray shell is mottled with brown, the interior white tinged with a streak of purple. Length, 55 mm; height, 48 mm; diameter, 32 mm. Mazatlán, Mexico, to Pacasmayo, Peru, intertidally; more common in the southern part of its range.

Subgenus TROPITHACA OLSSON, 1961

Escutcheon reduced in size, radial sculpture even, not divided into three zones; shell smaller and more brightly colored than in other subgenera.

473. Protothaca (Tropithaca) grata (Say, 1831) (Synonyms: **Venus discors, fuscolineata,** and **tricolor** Sowerby, 1835; **V. pectunculoides** Valenciennes, 1846; **V. muscaria** Reeve, 1863). Description of the color patterns of this variable form would be a task, for no two specimens are quite alike. The ground color is an off-white on which are laid intricate designs in brown and purplish black, or wide areas may be a solid color. Within, the shell is white to wholly purple, the margin weakly crenulate. Length, 40 mm; height, 34 mm; diameter, 23 mm. Cape Colnett, Baja California, through the Gulf of California and south to Chile, on mud flats and offshore in depths to 390 m.

474. Protothaca (Tropithaca) pertincta (Dall, 1902). Sculpture coarser than in *P. grata*. Interior white to purple. A large specimen measures 38 mm in length, 33 mm in height. Galápagos Islands.

470

471

472

473

474

475

Family PETRICOLIDAE

This is a family of nestling clams. The name literally means "dweller in rocks," which is a good description. Some forms may actively excavate dwelling places in hard clay, but others appropriate any convenient crevice and let the shell grow to fit the shape of the hole, resulting in a variety of shapes.

The Petricolidae differ from the Veneridae by lacking a lunule and escutcheon, by lacking lateral teeth on the hinge, and by having only two cardinal teeth in the right valve. The pallial sinus is well developed.

Genus PETRICOLA LAMARCK, 1801

Ovate, mostly solid shells; sculpture, if present, mainly radial. Shells usually distorted by nestling habit in rock cavities. Three Panamic subgenera.

Subgenus PETRICOLA, s. s.

Radial sculpture fine, uniform, with sporadic zigzag patterns, especially on the umbones or on immature specimens.

475. Petricola (Petricola) charapota Olsson, 1961. A medium-sized shell with fine radial riblets rather evenly distributed except toward the ends, where they divide or are bent or become hook-shaped. The color is a faded brown. Length, 30 mm; height, 20 mm; diameter, 20 mm. Ecuador.

476. Petricola (Petricola) exarata (Carpenter, 1857) (Synonyms: **Rupellaria linguafelis** Carpenter, 1857; **Naranio scobina** Carpenter, 1857; **Cypricardia noerri** De Folin, 1867; ? **P. botula** Olsson, 1961). The type material of Carpenter's three species consist of small, perhaps juvenile, shells, although material collected and reported by Coan (1962) is not much larger. Ribbing tends to be radial, with a few zigzags, finer in the type specimens of *P. linguafelis*. Shells are white, the type of *P. exarata* showing a few brown spots. With growth the shells tend to become more elongate. Length of Carpenter's specimens, 4.5 to 5 mm. The shells described by Olsson as *P. botula* are larger (length about 15 mm) and, like those described by De Folin, come from Panama. Altata, north of Mazatlán, Mexico (the latter is the type locality), to Panama.

477. Petricola (Petricola) lucasana Hertlein & Strong, 1948. The shell is somewhat rounded in shape, white, with bluish-tinged areas and some reddish brown near the beaks; within, it is dark brown (orange-brown in worn specimens). The pallial sinus is short, broad, and rounded. Length, 25 mm; height, 25 mm; diameter, 17 mm. Gulf of California, from Cape San Lucas to Punta Peñasco, Sonora, and southward to Oaxaca, Mexico.

Subgenus PETRICOLARIA STOLICZKA, 1870

Subcyindrical, with radial ribs much coarser on anterior part of shell.

478. Petricola (Petricolaria) cognata C. B. Adams, 1852. The white shell has a flattened lunule-like area in front of the beaks. It is less cylindrical than some of the other members of the subgenus, the height being nearly one-half the length. The shell is relatively thick, and the hinge teeth of the right valve are remarkably heavy. The adductor scars and the pallial sinus are strongly impressed. Length, 23 mm; height, 10 mm; diameter, 11 mm. The species has not been authentically reported outside the type area, near Panama.

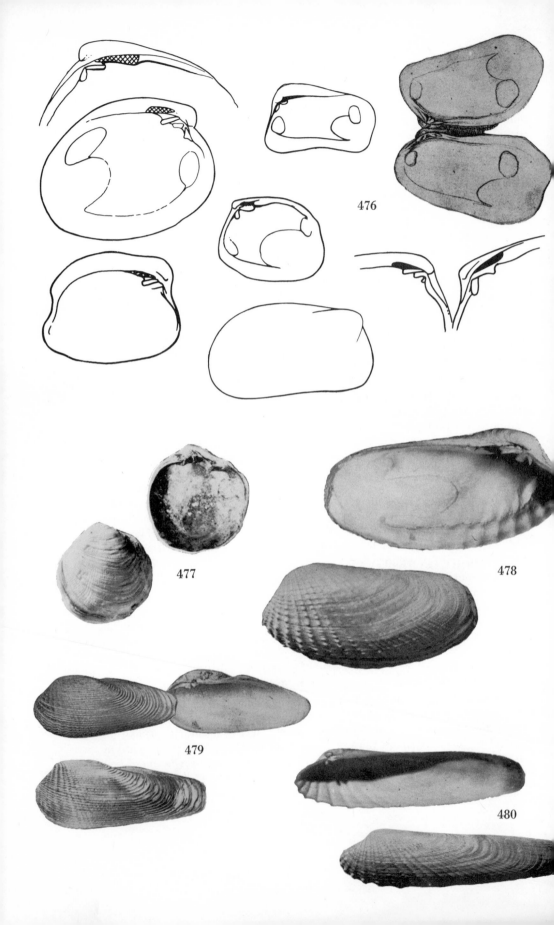

476

477

478

479

480

479. Petricola (Petricolaria) concinna Sowerby, 1834. The white shell has a few threadlike radial ribs and fine concentric ribs that tend to become lamellar on the posterior slope. The anterior end of the shell is rounded, the posterior end elongate and pointed. Length about 25 mm; height, 15 mm. Ecuador to Arica, Chile.

480. Petricola (Petricolaria) parallela Pilsbry & Lowe, 1932. The most slender of the Panamic Petricolas, this is white, sculptured with a few coarse nodulous ribs anteriorly. The pallial sinus is long and of equal width throughout, rounded at the end. Length, 60 mm; height, 15 mm; diameter, 10 mm. Scammon's Lagoon, Baja Californià, to the Gulf of California and south to Corinto, Nicaragua, intertidally and to depths of 15 m.

Subgenus **RUPELLARIA** FLEURIAU DE BELLEVUE, 1802

Radial ribs uniform or coarser posteriorly, not developing zigzag patterns.

481. Petricola (Rupellaria) denticulata Sowerby, 1834 (Synonyms: **Venerupis peruviana** Jay, 1839; **P. ventricosa** Deshayes, 1853 [not Krauss, 1848]). The anterior end of the thick white shell tapers to a narrow point. Inside, the shell is a dark purplish brown, with the pallial sinus angular and pointed. Length, 28 mm; height, 15 mm; diameter, 14 mm. La Paz, and through the Gulf of California, south to Paita, Peru, boring in hard clay.

482. Petricola (Rupellaria) peruviana Olsson, 1961. The ribbing on this white to cream-colored shell is fine enough for *P. (Petricola)*, which it much resembles, but there is no evidence of zigzag growth. Length, 30 mm; height, 21 mm; diameter, 14 mm. Ecuador to northern Peru.

483. Petricola (Rupellaria) robusta Sowerby, 1834 (Synonyms: **P. sinuosa** Conrad, 1849; **P. bulbosa** Gould, 1851). The radial ribs on the posterior part of the shell are distinctive. The shell itself is white, becoming orange or even brownish black on the margins of some specimens. Length, 23 mm; height, 20 mm; diameter, 16 mm. Puerto Peñasco and the Gulf of California south to Guayaquil, Ecuador, boring in hard clay.

Family COOPERELLIDAE

Without lateral teeth; ligament depressed, on a laminar nymph.

Genus **COOPERELLA** CARPENTER, 1864
(**OEDALIA** CARPENTER, 1864; **OEDALINA** CARPENTER, 1865)

These are thin, quadrate, rather small shells. The hinge of the right valve has two thin cardinal teeth, that of the left has three. The pallial sinus is large and wide.

484. Cooperella panamensis Olsson, 1961. The shell is similar in form to the southern Californian *C. subdiaphana*, but it has a more rounded outline, with the two ends almost alike in shape. Extremely thin and fragile, smooth, polished, with a few concentric undulations, especially on the anterior slope. Length, 10 mm; height, 8 mm; diameter, 5 mm. Guerrero, Mexico, to Panama.

485. Cooperella subdiaphana (Carpenter, 1864) (Synonym: **Oedalia scintillaeformis** Carpenter, 1864). The shell is white and smooth, with bifid cardinal teeth on the hinge. Though best known in southern California, the species ranges south to the Gulf of California, and a similar species, *C. atlantica* Rehder, 1943,

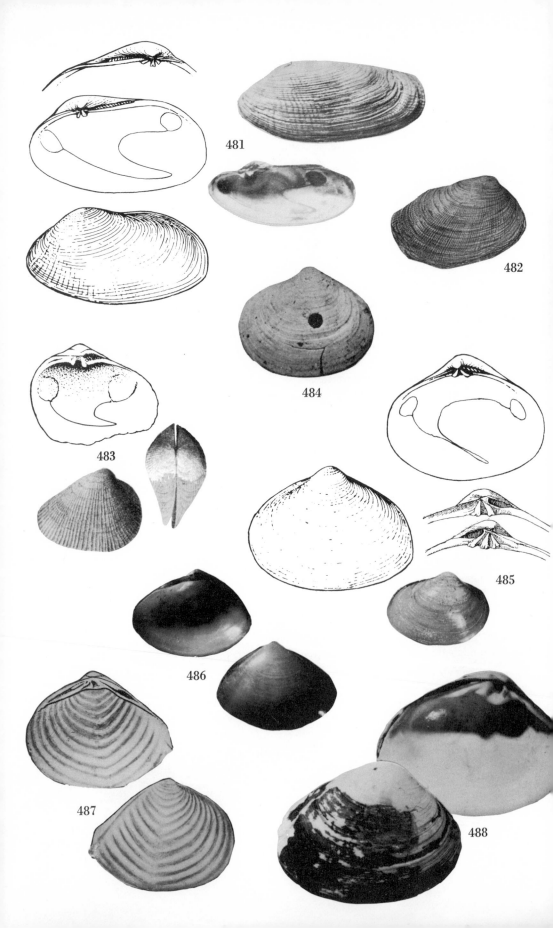

481

482

483

484

485

486

487

488

has been discovered in the Caribbean. Length, 15 mm; height, 12 mm; diameter, 7 mm. Queen Charlotte Islands, British Columbia, to the head of the Gulf of California at San Felipe. Specimens may be found intertidally in California constructing "nests" of agglutinated mud or sand in sheltered crevices, and dredging records show a range offshore to depths of 24 m.

Superfamily MACTRACEA

The characteristic that sets the Mactracea apart is the large socket-like chondrophore or resilifer that encloses the internal portion of the ligament, the resilium. A smaller external ligament may or may not be present, on a small nymph. The chondrophore is just back of the ∧-shaped cardinal teeth. Lateral teeth are also developed along the dorsal margin of the hinge. The pallial sinus is mostly deep and rounded.

Family MACTRIDAE

Shells are equivalve, thin, inflated, trigonal to ovate in shape, of medium to large size, porcelaneous, the outer layer of periostracum being either shiny or matte-surfaced, tinged with some shade of olive or buff.

A number of subdivisions of the family and of the principal genus, *Mactra*, have been made, but as most of these are based on details of the hinge that may be difficult to observe, only a few divisions will be cited here, and the key characteristics will be to features of West American species, not necessarily to species of these groups that occur elsewhere. The key is to genera and (in *Mactra*) subgenera.

1. Pallial sinus shallow or nearly obsolete..........................*Rangia*
   Pallial sinus moderate to deep and rounded.......................... 2
2. Shell with undulating concentric sculpture........................... 3
   Shell smooth or nearly so, concentric sculpture when present not undulating 7
3. Undulations on umbones only (wanting in some specimens)....*Micromactra*
   Undulations over entire shell..................................... 4
4. Posterior end elongate or pointed.............................*Raeta*
   Posterior end short, not pointed................................. 5
5. Posterior slope not set off by a keel......................*Tumbeziconcha*
   Posterior slope set off by a keel................................. 6
6. Anterior lateral teeth very short.............................*Harvella*
   Anterior lateral teeth well developed.......................*Mactrinula*
7. Ligament entirely internal...................................*Mulinia*
   Ligament partly external... 8
8. Posterior and dorsal margins meeting at nearly a right angle.......*Anatina*
   Ovate-trigonal or elongate but posterior end not truncate.............. 9
9. Periostracum coarse; shell large.........................*Mactroderma*
   Periostracum not conspicuously coarse; shell medium-sized to small......10
10. Ovately trigonal, posterior slope set off by a keel or angulation...*Mactrellona*
    Ovately elongate, posterior area not sharply set off...................11
11. Anterior and posterior slopes concentrically sculptured.........*Mactra, s. s.*
    Anterior and posterior slopes not concentrically sculptured..............12
12. Anterior end somewhat pointed, longer than posterior.........*Mactrotoma*
    Anterior end evenly rounded, not longer than posterior............*Spisula*

Genus MACTRA LINNAEUS, 1767

The ligament is divided into two parts, the major part housed in the resilifer. Between the resilifer and the outer ligament there is a narrow shelly plate, a lamina. The shell itself is of a rather brittle porcelaneous texture, variously sculptured concentrically but without radial sculpture.

Subgenus MACTRA, *s. s.*

Lunule and escutcheon not set off by a groove; pallial sinus round, not deep.

486. Mactra (Mactra) williamsi Berry, 1960. The small shell is thin and white under a yellowish-gray periostracum that is a little fringed on the posterior slope. Concentric ridges like well-spaced heavy growth lines develop anteriorly and posteriorly, as in many Indo-Pacific and Tertiary species of *Mactra* but not in any other eastern Pacific species. Length, 15 mm; height, 10 mm; diameter, 3 mm. Off La Libertad, Ecuador, in 18 m.

Subgenus MACTRINULA GRAY, 1853

Undulating concentric sculpture covers the entire shell.

487. Mactra (Mactrinula) goniocyma (Pilsbry & Lowe, 1932). A small shell, with regular concentric undulations that sag in a broad V near the posterior part of the central slope. Only a few specimens have been found. Length, 16 mm; height, 12 mm; diameter, 6 mm. Acapulco, Mexico, to Ardita Bay, Colombia, dredging records being to depths of 40 m, mud bottom.

Subgenus MACTRODERMA DALL, 1894

Large, inequilateral, with coarse periostracum.

488. Mactra (Mactroderma) velata Philippi, 1849. The shell is large and thick, the posterior end evenly, almost acutely rounded, but not truncate, and the posterior dorsal margin slopes down rather abruptly. Length, 78 mm; height, 56 mm; diameter, 27 mm. The Gulf of California to Panama, according to Dall, but the species seems to be rare north of southern Mexico. Lowe reported it on mud flats, C. B. Adams on the reef at Panama.

Subgenus MACTROTOMA DALL, 1894

Posterior dorsal area with an impressed radial band and darker periostracum.

489. Mactra (Mactrotoma) dolabriformis (Conrad, 1867). The shell is somewhat flattened, of a polished white under a dull brown periostracum. Length, 85 mm; height, 58 mm; diameter, 24 mm. Southern California to Panama, on mud flats.

490. Mactra (Mactrotoma) nasuta Gould, 1851 (Synonyms: **M. californica** Reeve, 1854 [not Conrad, 1837] = **M. deshayesi** Conrad, 1868; **M. hiantina** Deshayes, 1855; **Mactrotoma revellei** Durham, 1950). The similarity between this and the Atlantic *M. brasiliana* Lamarck, 1818, has been mentioned by several authors. The beaks are a little back of the middle, the anterior end of the shell tapering, the posterior end truncate. The shell is white under a straw-colored periostracum. Dimensions of a large specimen: length, 118 mm; height, 82 mm; diameter, 41 mm. San Pedro, California, to Colombia, but the shell is rare everywhere. Lowe reported odd valves on mud flats, and the Beebe-Crocker Expedition dredged two valves in 55 to 80 m.

Subgenus MICROMACTRA DALL, 1894

The undulation of the beaks is a sure clue when it is developed, but unfortunately a specimen now and then fails to show the concentric waves. The small size and details of shape and hinge must then be used in identification of the species.

491. Mactra (Micromactra) angusta Reeve, 1854 (Synonym: **M. atacama** Pilsbry & Olsson, 1941). The anterior dorsal margin of this thin white shell is concave, and the pallial sinus is relatively short, not extending to a line drawn vertically through the beaks. A large specimen measures: length, 42 mm; height, 27 mm; diameter, 13 mm. Guatemala to Zorritos, Peru, intertidally (rare) and to depths of 26 m, in mud.

492. Mactra (Micromactra) californica Conrad, 1837. Compared to *M. angusta*, this is shorter, the anterior dorsal margin nearly straight. It is not a common shell, but not so rare as *M. angusta*, with a range from Puget Sound, Washington, to Costa Rica, on mud flats. Length, 46 mm; height, 29 mm; diameter, 15 mm.

493. Mactra (Micromactra) fonsecana Hertlein & Strong, 1950 (Synonym: **"M. angusta** Deshayes" of Pilsbry & Lowe, 1932). In contrast to the two species above, this form has a long pallial sinus, extending to or past a line drawn vertically through the beaks. A brownish-gray periostracum covers the white shell. The anterior dorsal margin is straight, as in *M. californica*, but the shape is more that of *M. angusta*. The shell is longer for the height than *M. isthmica*. Length, 53 mm; height, 34 mm; diameter, 17.5 mm. Nicaragua to Ecuador.

494. Mactra (Micromactra) isthmica Pilsbry & Lowe, 1932. The largest of the West American Micromactras, this has a long pallial sinus, extending even with the beaks. The shell is white under a buff to olivaceous periostracum, with a raised cord from the beaks to the posterior ventral margin. Length, 55 mm; height, 38 mm; diameter, 20 mm. In describing this shell, Pilsbry and Lowe state, facetiously: "The corrugation at the beaks varies, being well developed both back and front [as in *M. fonsecana*], or only posteriorly, or sometimes practically obsolete, though faint traces of corrugation are visible when looked for with determination." Gulf of Fonseca, Nicaragua, to Panama.

495. Mactra (Micromactra) vanattae Pilsbry & Lowe, 1932 (Synonym: **M. v. acymata** Pilsbry & Lowe, 1932). The pallial sinus is even shorter than that of *M. californica*, and the wavy concentric sculpture is developed only on the posterior part. The periostracum is thin and grayish drab in color. Pilsbry and Lowe described a variety from Panama, *M. vanattae acymata*, in which the beak undulations are obsolete. As they pointed out, such shells technically should be placed in *Mactrotoma*, but all other features indicate merely an extreme variation from the *M. vanattae* pattern. Length, 47.5 mm; height, 30 mm; diameter, 17.5 mm. Gulf of Fonseca, Nicaragua, to Panama City, on the beach.

Subgenus TUMBEZICONCHA PILSBRY & OLSSON, 1935

The lack of a keel or ridge at the junction of posterior and central slopes sets this subgenus apart from *Mactrinula*.

496. Mactra (Tumbeziconcha) thracioides (Adams & Reeve, 1850). Exteriorly this very closely simulates the thracioid group *Cyathodonta*, but a glance at the hinge, with its sturdy chondrophore, reveals the true relationship. Length, 34 mm; height, 28 mm; diameter, 18 mm. El Salvador to Tumbes, Peru, mostly offshore.

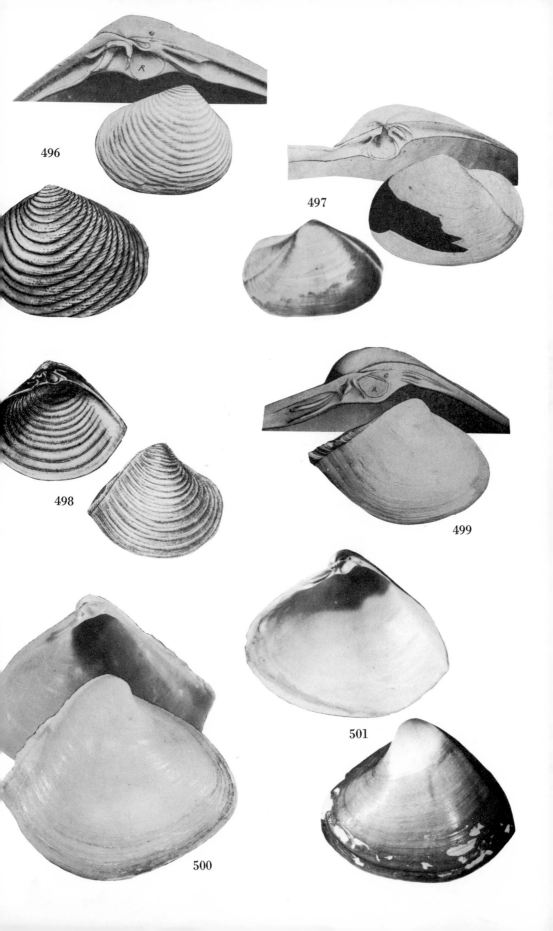

496

497

498

499

500

501

Genus ANATINA SCHUMACHER, 1817
(LABIOSA MÖLLER, 1832, of authors)

The shells are fairly large, thin and brittle, trigonal to elongate-quadrate. The beaks are opisthogyrate; that is, they point toward the posterior end of the shell, and the umbones are low and inconspicuous, with weak concentric sculpture. The pallial sinus is short, broad, and rounded.

497. Anatina cyprinus (Wood, 1828) (Synonym: "**Anatina anatina**" of authors [not *Mactra anatina* Spengler, 1802, an Atlantic species]). First collected before the year 1800 and described in 1828 as probably from Peru, this species was not firmly established as West American until the early 1960's, when it was dredged in the Gulf of California (Keen, 1961) and reported from Ecuador (Olsson, 1961). The shell is thin and translucent white and can be separated from the Atlantic relative, *A. anatina* (Spengler, 1802), by its more rectangular outline. Length of an average specimen, 42 mm; height, 28 mm; length of a large specimen, 70 mm. Off Bahía San Luis Gonzaga, Guaymas, and Carmen Island, Gulf of California, depths 25 to 45 m; south to Ecuador and possibly Peru.

Genus HARVELLA GRAY, 1853

As in *Raeta*, the surface is plicate or folded, but the valves are much more inflated, the beaks prosogyrate, and the posterior margin has a lamellar keel.

498. Harvella elegans (Sowerby, 1825) (Synonyms: **H. pacifica** Conrad, 1867; **Raeta maxima** Li, 1930). Between the keel and the dorsal margin the shell is smooth, and there is also a smooth area in front of the beaks. Although beach valves are not uncommon, entire specimens must be dredged. Length, 57 mm; height, 48 mm; diameter, 32 mm. The Gulf of California to Zorritos, Peru, offshore in depths of 26 to 70 m, mud.

Genus MACTRELLONA MARKS, 1951
(MACTRELLA of authors, not of GRAY, 1853)

Unfortunately, the familiar name *Mactrella* for this group must be abandoned, for it was originally applied without a careful study of the type species, which turns out to be a *Mactrinula*.

These rather large, thin shells are markedly inequilateral, with high umbones and a posterior slope set off by an angulation or by a raised shelly ridge. The surface of the shell is otherwise smooth.

499. Mactrellona alata (Spengler, 1802). This is a Caribbean species that is reported as occurring locally in the tropical eastern Pacific by Olsson, who figures a specimen from Panama (1961, p. 327, pl. 56, fig. 4). It has the posterior area bordered by a keel, the ventral margin evenly arched, and has no posterior gape. Length, 40 mm. Nicaragua to Ecuador, according to Olsson, who regards *M. subalata* (Mörch, 1860) as a synonym.

500. Mactrellona clisia (Dall, 1915). The distinctive feature of this rather uncommon shell is a strong elevated keel along the margin of the flattened posterior slope. The valves gape slightly on the posterior margin. Length, 77 mm; of a large specimen, 95 mm; height, 60 mm; diameter, 32 mm. The Gulf of California to Ecuador.

501. Mactrellona exoleta (Gray, 1837) (Synonym: **Lutraria ventricosa** Gould, 1851). The posterior slope is bounded by a sharp angulation, not a keel.

A large specimen measures: length, 121 mm; height, 90 mm; diameter, 67 mm. The Gulf of California to Peru, mostly offshore in depths to 24 m.

502. Mactrellona subalata (Mörch, 1860). The holotype of this was not figured until recently (Keen, 1966). Olsson (1961) had suggested that the species might be synonymous with *M. alata*, but the holotype does not match either his figure or Atlantic material in the Stanford collection. *M. subalata* has a well-developed second ridge between the keel and the dorsal margin, and the shell is more broadly rounded anteriorly. Further study is needed before a decision can be reached about the synonymy of the two forms. There is no posterior gape in either one. Length, 55 mm; height, 40 mm. Nayarit, Mexico (Wahrenbrock collection) to Nicaragua (type locality).

Genus MULINIA GRAY, 1837

Both parts of the ligament being sunken below the dorsal margin of the hinge, the two valves meet smoothly along their entire dorsal edges, with only a tiny hole or crevice just back of the beaks.

503. Mulinia coloradoensis Dall, 1894 (Synonyms: "**M. byronensis** Gray" of authors; **M. c. acuta** and **M. modesta** Dall, 1894). Compared with the better known *M. pallida*, this is a heavier, thicker shell, the umbones wider and lower, the umbonal ridge less marked. There is a tendency toward radial ribbing on the ends of the shell. The name *M. c. acuta* was given by Dall to a long variant. Length, 49 mm; height, 37 mm; diameter, 32 mm. Restricted to the Gulf of California and west Mexican area.

504. Mulinia pallida (Broderip & Sowerby, 1829) (Synonyms: **M. donaciformis** Gray, 1837; **Mactra angulata** Reeve, 1854; **Mactra goniata** Deshayes, 1855; **Mactra laciniata** Carpenter, 1856; **Mactra bistrigata** Mörch, 1860; **Mulinia bradleyi** Dall, 1894; **Corbula altirostris** Li, 1930). A roundly trigonal shell with high, narrow umbones, this is white under a smooth, yellowish-olive periostracum, fringed in the young and in an occasional adult. The posterior slope is a little flattened and set off from the central slope by angulation. The short and narrow pallial sinus is about two-fifths as long as the shell. Length, 56 mm; height, 44 mm; diameter, 34 mm. Intertidally and to depths of 25 m. Southern part of the Gulf of California to northern Peru.

Genus RAETA GRAY, 1853

The outline is trigonal, the umbones narrow and fairly high; the surface is not simply concentrically sculptured, but the shell itself is thrown into low concentric ridges that tend to be weaker near the margin. There is no periostracum.

505. Raeta undulata (Gould, 1851). Similar to the Atlantic *R. plicatella* (Lamarck, 1818)—better known as *R. canaliculata* (Say, 1822)—but differing from that species by having the beaks farther forward. A large specimen measures: length, 120 mm; height, 95 mm; diameter, 70 mm. Beach valves are fairly common; entire specimens are hard to find, even by dredging. San Pedro, California, to Peru.

Genus RANGIA DESMOULINS, 1832
(GNATHODON SOWERBY, 1831 [not OKEN, 1816])

The brackish-water habitat of these shells led to a long dispute about their true relationships. Dall, in 1894, finally established their affinity to *Mactra*. The several

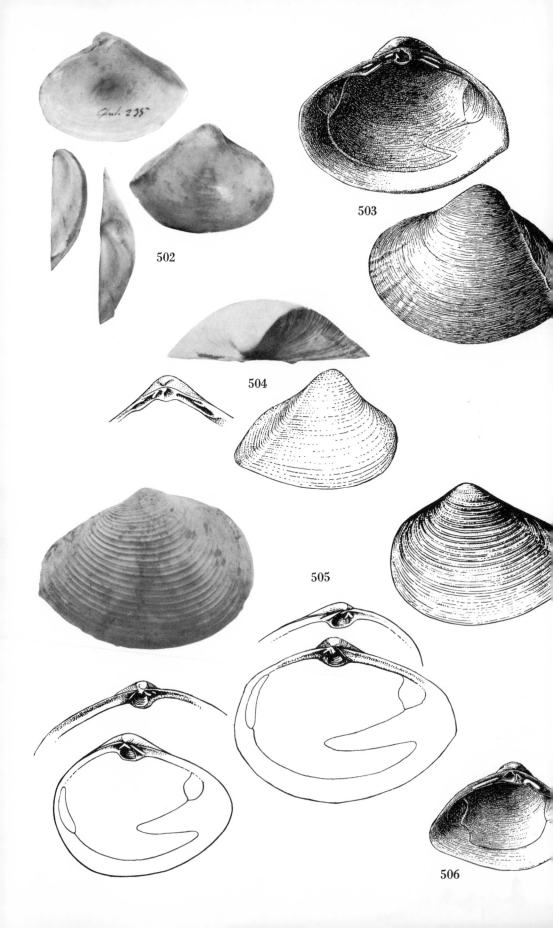

502

503

504

505

506

species in the Americas are now ranked in two subgenera—*Rangia, s. s.*, not represented in the Panamic province, with long posterior lateral teeth and a short pallial sinus, and *Rangianella*, with short posterior lateral teeth and an almost obsolete pallial sinus.

The generic name's having been dedicated to the nineteenth-century malacologist Sander Rang, it should be pronounced with a hard *g*.

Subgenus RANGIANELLA CONRAD, 1863

506. Rangia (Rangianella) mendica (Gould, 1851) (Synonyms: **Gnathodon trigonum** and **truncatum** Petit, 1853). A straw-colored periostracum covers the white shell, darker and more fibrous on the posterior slope. Length, 25 mm; height, 18 mm; diameter, 12.5 mm. The shells are common in mangrove swamps or brackish water in the Gulf of California area at least as far south as Mazatlán, Mexico.

Genus SPISULA GRAY, 1837

Small to large shells, mostly ovate-trigonal in shape. Hinge with the external ligament separated from the internal resilium by a flat space, not a ridge or plate as in *Mactra*. Lateral teeth tending to be striate.

507. Spisula adamsi Olsson, 1961. Small, nearly symmetrical, white under a thin shiny yellowish-gray periostracum. Pallial sinus small, quadrate at its end. Length, 36 mm; height, 27 mm; diameter, 18 mm. Panama to northern Peru. This is the first species to be assigned to *Spisula* in the eastern tropical Pacific area; the genus is more common in the northern Pacific and northwestern Atlantic, but it also occurs in the western Pacific.

Superfamily TELLINACEA

Shells mostly inequilateral, with external ligament, never with a chondrophore; cardinal teeth two in either valve, laterals more or less well developed; pallial line widely sinuate.

Family TELLINIDAE

The tellens are small to medium-sized marine forms that burrow into the sea floor and feed by means of two slender siphons that circulate seawater and surface sediment—with the contained microscopic nutrients—through the mantle cavity and expel it again after the food has been sorted out. Because the shells lie partly on one side, the posterior end may be twisted a little to the right, but the valves do not gape. The hinge has two rather small cardinal teeth in each valve, and one or two lateral teeth in most genera. The ligament is external in all but one group. The pallial sinus is, as one would expect, deep and wide. It may be discrepant— that is, the shape of one valve may not always be the same as the shape of the other valve.

Because there are such wide differences in form and outline in this family and so many species, it will be necessary to use keys to separate the major groups. The classification adopted here follows in part one worked out by Hertlein and Strong (1949), modified by more recent studies of Olsson (1961), Boss (1966–69), Keen (1969), and Coan (1971). It is conservative, the only wise course until someone makes a thoroughgoing study of the family as a whole, taking into account worldwide distribution and geologic history. Afshar (1969) has discussed the systematics of the family. The key is to genera.

1. Lateral teeth present, at least in the young............................ 2
   Lateral teeth wanting.. 4
2. Outline trigonal; dorsal margins serrate........................*Tellidora*
   Outline rounded to elongate; dorsal margins smooth, at least anteriorly.... 3
3. Shell rounded, with fine divaricating sculpture...................*Strigilla*
   Shell elongate, sculpture various but not divaricate................*Tellina*
4. Right valve with a posterior flexure; valves not compressed........*Florimetis*
   Right valve not more flexed than left; both valves somewhat flattened..... 5
5. Sculpture of fine oblique riblets...........................*Temnoconcha*
   Sculpture various or wanting, not oblique.......................... 6
6. Ligament simple, external, on a small nymph....................*Macoma*
   Ligament partially immersed, with an inner portion set off by a
   groove ....*Psammotreta*

Genus TELLINA LINNAEUS, 1758

The tellens are colorful shells, displaying many shades of pink, red, and white. They are shiny shells, and the slightly twisted posterior area, marked by some differences in sculpture patterns, and the characteristically shaped pallial sinus make them fairly easy to recognize. Identification of species—especially of the numerous small kinds—is not so simple. The separation of a number of well-marked subgenera is a first step. The beginner may be content to stop at this point and not to attempt to discriminate between the many very similar but slightly and constantly different species that comprise each subgenus. The type species of *Tellina* is a smooth form restricted to the West Indian area; there are no species in the Panamic province that can be cited as *Tellina, s. s.*

1. Posterior slope with strong radial sculpture.................*Elliptotellina*
   Posterior slope smooth or with concentric or decussate sculpture........ 2
2. Shell surface obliquely grooved................................. 3
   Shell surface not obliquely grooved............................. 4
3. Outline ovate, ends rounded, nearly equilateral.................*Hertellina*
   Outline trigonal to trapezoidal, not rounded at ends..............*Scissula*
4. Right anterior lateral well removed from cardinals.................... 5
   Right anterior lateral close against cardinals....................... 8
5. Ligament somewhat sunken or immersed......................*Laciolina*
   Ligament entirely external....................................... 6
6. Posterior lateral strong in right valve; interior of shell with a reinforcing
   swelling or rib ....*Eurytellina*
   Posterior lateral weak in right valve; no internal rib................. 7
7. Shell small to medium-sized, outline mostly trigonal to quadrate.....*Angulus*
   Shell moderately large (45 mm), inequilateral, posterior end pointed,
   anterior rounded ....*Elpidollina*
8. Pallial sinus not connected to anterior adductor by linear scar........... 9
   Pallial sinus connected to anterior adductor by scar...................10
9. Beaks prosogyrate ....*Lyratellina*
   Beaks orthogyrate ....*Merisca*
10. Flexure in right valve well developed.............................11
    Flexure in right valve not evident...............................12
11. Outline inequilateral, posterior end somewhat pointed...........*Tellinella*
    Outline equilateral, posterior end rounded...................*Tellinidella*
12. Posterior slope not set off by scales at its margin...............*Phyllodella*
    Posterior slope set off by raised scales on concentric ribs.........*Phyllodina*

Subgenus ANGULUS MEGERLE VON MÜHLFELD, 1811

Mostly small trigonal to quadrate shells, with the posterior lateral tooth weaker than in other subgenera, showing only as a small denticle below the ligamental nymph in some species.

The name *Angulus* was commonly applied to the nondistinctive small tellens of the west coast for many years, and then authors shifted to *Moerella* Fischer, 1887. Actually, the tropical American species, both Atlantic and Pacific, do not fit well into either of these genera. The type species of *Angulus* is a Philippine form, that of *Moerella* Mediterranean. Were there not already so many subgeneric categories in the family it would seem simplest to provide a special one for these American species; perhaps someday someone will. At present the specialists such as Boss (1966–69) recommend return to the use of *Angulus*.

508. Tellina (Angulus) amianta Dall, 1900. Small, white to yellowish shells, with fine, close concentric ribbing that becomes sharper and more irregular on the posterior slope. The rounded anterior end is much longer than the flexed and pointed posterior end. A large specimen measures: length, 14 mm; height, 6.8 mm; diameter, 4.4 mm. The Gulf of California to Ecuador, offshore in depths to 28 m. *T. (A.) sybaritica* Dall, 1881, is a similar species in the Atlantic.

509. Tellina (Angulus) carpenteri Dall, 1900 (Synonym: **T. arenica** Hertlein & Strong, 1949). The posterior end of the shell is broad, abruptly truncate, the color rose-pink with irregular zones of white and salmon. Length, to 25 mm; height, to 15 mm; diameter, 5 mm. Southern Alaska to Panama, offshore in depths to 390 m. Attempts to divide this species on the basis of range have not been successful, for the same variations occur northward and southward, and there is no clear line of separation, according to Coan (1971).

510. Tellina (Angulus) cerrosiana Dall, 1900. A small, white, sharply concentrically striate form, nearly trigonal in outline. Length of the type specimen, 5.2 mm; height, 3.2 mm; diameter, 1.5 mm; but it may not be mature. Off Cedros Island, Baja California, to the Gulf of California, in 15 to 48 m.

511. Tellina (? Angulus) chrysogona Dall, 1908. A bright yellow shell, shading to white near the margins, from deep water. It was described in the subgenus *Moerella*. Length, 13 mm; height, 10 mm; diameter, 5 mm. Off the Galápagos Islands, depth 550 m.

512. Tellina (Angulus) coani Keen, new species. A small, thin, inequilateral shell, with the beaks at about the posterior one-third, the posterior end with the margin sloping abruptly downward, then slightly truncate as it joins the ventral margin; nearly smooth except for faint concentric lirae near the center of the valves. The shell is mostly translucent white, a few specimens pink, with two or more interrupted pink rays from the beaks to the ventral margin; hinge with two cardinal teeth in each valve, laterals evident in right valve, obsolete in left; pallial sinus long, reaching nearly to the anterior adductor scar. Length, 6.5 mm; height, 3.7 mm. Candelero Bay, near La Paz, Baja California (type locality) northward through the Gulf of California to Cholla Cove, Bahía de Adair, Sonora, Mexico. A not uncommon form in beach drift and shallow water offshore, this has probably been mistaken for the young of a larger species. Specimens in all the lots studied attain about the same size, the largest shell measuring only 10 mm in length; this suggests maturity. The small size and the rays of color are the distinctive features.

507

508

509

510

511

512

513

513. Tellina (Angulus) felix Hanley, 1844. A small, flattened, smooth, brilliantly rose-red form, this is one of the prettiest of the smaller tellens. Length, 17 mm; height, 9.4 mm; diameter, 4.3 mm. Mazatlán, Mexico, to Zorritos, Peru; mostly in shallow water offshore in depths to 24 m.

514. Tellina (Angulus) guaymasensis Pilsbry & Lowe, 1932. This resembles *T. macneilii*, a little lower and not so brightly colored, tending to be white with streaks of pink. Length, 15 mm; height, 9 mm; diameter, 5 mm. The species is as yet known only from the type locality, Guaymas, Sonora, Mexico.

515. Tellina (Angulus) hiberna Hanley, 1844 (Synonyms: **?** **T. donacilla** Carpenter, 1857; **T. panamensis** Dall, 1900 [not Philippi, 1849]; **T. tabogensis** Salisbury, 1934). Most noticeable features of this species are the ivory-white color, an iridescent surface sheen in fresh shells, and a flexuous posterior dorsal margin. Some specimens are faintly tinged with pink on the umbones. Length, 13 mm; height, 10 mm; diameter, 5 mm. The Gulf of California to northern Peru, mostly offshore in depths of 4 to 55 m. *T. (A.) gibber* von Ihering, 1907, is a similar species in the Atlantic.

516. Tellina (Angulus) macneilii Dall, 1900. Resembling T. (A.) *felix*, this is more inflated, shorter, the posterior end more abruptly truncate. The color is deep rose, slightly zoned, paler toward the basal margin. Length, 11.5 mm; height, 7.3 mm; diameter, 4.2 mm. Guaymas, Mexico, to the Gulf of Nicoya, Costa Rica; on mud flats and offshore in depths to 73 m.

517. Tellina (Angulus) meropsis Dall, 1900 (Synonym: **T. (Moerella) paziana** Dall, 1900). A thin white shell with the anterior end longer than the posterior and well rounded. The pallial sinus ascends to a high, rounded point just back of the beaks, then descends obliquely, curving in an arc well separated from the anterior adductor scar. Length, 13 mm; height, 11 mm; diameter, 5 mm. Southern California to Costa Rica; offshore in depths to more than 180 m. *T. (A.) mera* Say, 1834, is a similar species in the Atlantic.

518. Tellina (Angulus) recurvata Hertlein & Strong, 1949 (Synonym: **T. recurva** Dall, 1900 [not Deshayes, 1855]). An elongate shell, with low beaks, the posterior dorsal margin dipping a little below the ligament, the end obliquely truncate. Length, 12 mm; height, 7.5 mm; diameter, 2.9 mm. Point Firmin, Baja California (near the head of the Gulf), to Octavia Bay, Colombia; offshore in depths of 27 to 48 m.

519. Tellina (Angulus) straminea Deshayes, 1855. A species overlooked by West American authors, it was omitted from the first edition of this book because the only available illustration seemed unrecognizable. Photographs of the type lot in the British Museum (reproduced here) now can give collectors an opportunity to search for specimens. The shell is small, thin, and white, the size and outline being remarkably similar to that of *T. suffusa* Dall (which may therefore eventually fall as a synonym), the only point of difference being that Dall's species has a lunule or flattened area in front of the beaks. This may be present in *T. straminea* but does not show clearly in the photographs. Length, 12 mm; height, 8.5 mm. Type locality, Gulf of California.

520. Tellina (Angulus) subtrigona Sowerby in Reeve, 1866. (Synonyms: **T. puella** C. B. Adams, 1852 [not Hanley, 1845]; **T. erythronotus** Pilsbry & Lowe, 1932; **T. puellula** Salisbury, 1934). The creamy white, wedge-shaped

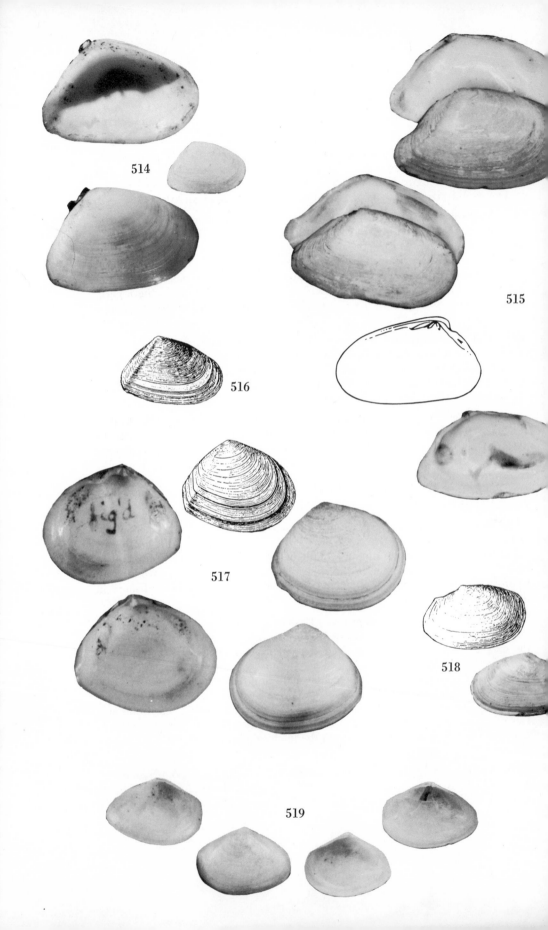

shell is flushed with pink around the margins, especially anteriorly and posteriorly. A large specimen measures: length, 25 mm; height, 14.5 mm; diameter, 7 mm. Magdalena Bay, Baja California, to the Bay of Panama, intertidally and to a depth of 13 m. *T. (A.) exerythra* Boss, 1964, is a similar Atlantic species.

521. Tellina (Angulus) suffusa Dall, 1900. Characteristic features are an unusually large lunular area, a short, blunt anterior end, and a pointed posterior end. The color varies from white to pinkish or yellowish. Length, 13.5 mm; height, 9 mm; diameter, 4.7 mm. San Ignacio Lagoon, Baja California, to Corinto, Nicaragua; no habitat records available. *T. (A.) tampaensis* Conrad, 1866, is a similar Atlantic species.

522. Tellina (Angulus) tumbezensis (Olsson, 1961). The white to pink shell somewhat resembles *T. suffusa* but with the beaks more posterior and the posterior slope more arched. There is no lunule. The shell is nearly smooth except for fine radial striae posteriorly. Length, 28 mm; height, 18 mm; diameter, 9 mm. Ecuador to Peru, on mud flats.

Subgenus ELLIPTOTELLINA COSSMANN, 1886

Small, oval shells, both ends rounded, with two lateral teeth in the right valve, the laterals of the left weak to wanting.

523. Tellina (Elliptotellina) pacifica Dall, 1900. The oval, rounded outline and the radial grooves on the central slope are good markers of this small and not common tellen. The color is yellowish white, with a rosy spot near each end of the hinge margin. Length, 8 mm; height, 4.5 mm; diameter, 2.5 mm. Santa Inez Bay, Gulf of California, to Panama, in depths of from 7 to 33 m. *T. (E.) americana* Dall, 1900, is a similar Atlantic species.

Subgenus ELPIDOLLINA OLSSON, 1961

Trigonal, thin-shelled; posterior end shorter than anterior; pallial sinus large, confluent below with pallial line, extending across above to the anterior adductor scar.

524. Tellina (Elpidollina) decumbens Carpenter, 1865 (Synonym: "**T. peasii** Carpenter" Sowerby in Reeve, 1868). The shell is creamy white, rounded-oblique, some specimens flushed with rose on the umbonal area, especially within; pallial sinus touching the anterior adductor scar. Length, 45 mm; height, 32 mm; diameter, 17 mm. The species seems to be confined to the Panama area.

Subgenus EURYTELLINA FISCHER, 1887

These are the largest and sturdiest members of the genus *Tellina* in the Panamic province. Most are elongate and concentrically sculptured. As with *Elliptotellina*, the lateral teeth are weak or wanting in the left valve.

Tellina (Eurytellina) eburnea Hanley, 1844 (Synonyms: **T. panamanensis** Li, 1930; **T. liana** Hertlein & Strong, 1945). The shell is heavy, white, with coarse, even surface sculpture. The dorsal areas are wide and have coarse ridges and wrinkles parallel to the margin; the left valve has a large lunular area, the right none. Interior white, the end of the pallial sinus not reaching the anterior adductor scar. Two geographic subspecies may be recognized:

525. Tellina (Eurytellina) eburnea eburnea. Length, 49 mm; height, 32 mm; diameter, 15 mm. Panama to northern Peru. *T. (E.) angulosa* Gmelin, 1791, is a similar Atlantic form.

526. Tellina (Eurytellina) eburnea askoyana Hertlein & Strong, 1955. A smaller northern form, with the pallial sinus extending almost to the anterior adductor scar. Length, 27 mm; height, 18 mm. Champerico, Guatemala, to Pinas Bay, Panama, offshore in 13 to 25 m.

527. Tellina (Eurytellina) ecuadoriana Pilsbry & Olsson, 1941. The shell is rose red, with whitish zones, rather large, and trigonal. The pallial sinus does not quite touch the anterior adductor scar, which, together with color, separates it from *T. hertleini*, its closest relative. Length, 50 mm; height, 29 mm; diameter, 8 mm. Corinto, Nicaragua, to Santa Elena, Ecuador.

528. Tellina (Eurytellina) hertleini (Olsson, 1961) (Synonym: "**T. planulata** Sowerby, 1867", of authors [probably not of Sowerby and not from Panama]). The white shell is large, trigonal, with a depressed area in the middle of the ventral margin. The pallial sinus touches the anterior adductor scar, as in *T. laceridens*. Except for color and slight differences of outline, it closely resembles *T. ecuadoriana*. Length, 59 mm; height, 33 mm; diameter, 12 mm. El Salvador to the Gulf of Dulce, Costa Rica.

529. Tellina (Eurytellina) inaequistriata Donovan, 1802 (Synonyms: **T. gemma** Gould, 1853; **T. leucogonia** Dall, 1900). A red to orange-red shell, this is distinguished by its pattern of concentric sculpture: the posterior third of the shell has coarse concentric lamellae that turn into fine striae anteriorly. Within, there are fine crenulations near the anterior-ventral margin. Length, 23 mm; height, 12.5 mm. The Gulf of California to Guayaquil, Ecuador, mainly offshore at depths of 18 to 33 m. *T. (E.) nitens* C. B. Adams, 1845, is a similar Atlantic species.

See Color Plate XIII.

530. Tellina (Eurytellina) laceridens Hanley, 1844. One of the larger members of the subgenus, this white shell may have a flush of pink on the umbones and be yellow-spotted within. The cardinal teeth are grooved, almost jagged-looking. The pallial sinus is so deep that it touches the anterior adductor muscle scar. Length, 53 mm; height, 35 mm; diameter, 11.5 mm. Corinto, Nicaragua, to Tumbes, Peru, on mud flats at lowest tide. *T. (E.) alternata* Say, 1822, is a similar Atlantic species.

531. Tellina (Eurytellina) mantaensis Pilsbry & Olsson, 1943. This rather rare southern form resembles the much more common *T. simulans* but is longer, and the dorsal margins slope downward at a lower angle. Length, 21 mm; height, 15 mm. Gulf of Chiriqui, Panama, to Caleta la Cruz, northern Peru, mud bottom.

532. Tellina (Eurytellina) prora Hanley, 1844. A polished, ovately trigonal shell, the posterior area set off by an angulation, with fine incised concentric striae. The color is rose pink, with whitish concentric bands. The pallial sinus does not touch the anterior adductor scar. Length, 46 mm; height, 17 mm; diameter, 12 mm. The fine sculpture separates this from the more abundant *T. rubescens* and *T. simulans*. La Paz, Gulf of California, to Guayaquil, Ecuador, offshore in depths of 11 to 42 m.

533. Tellina (Eurytellina) regia Hanley, 1844. A rare form that has concentric sculpture somewhat irregular, as in *T. inaequistriata*, but here the posterior third of the shell is smooth or nearly so, in one or both valves, the anterior end of the shell sculptured with spaced concentric grooves. Length, 17 mm; height, 10 mm; diameter, 3.5 mm. Corinto, Nicaragua, to Panama, to depths of at least 13 m.

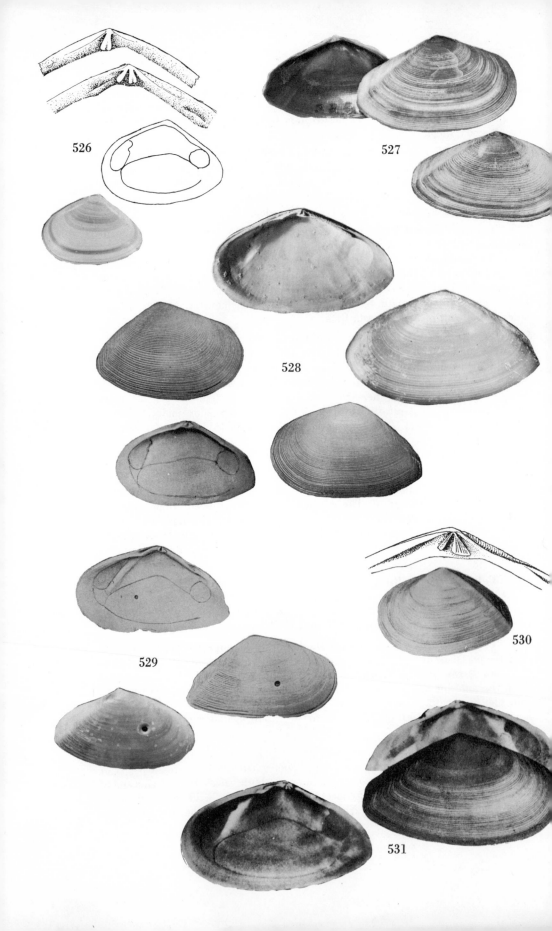

526

527

528

529

530

531

534. Tellina (Eurytellina) rubescens Hanley, 1844. The rose-red shell is commonly confused with the much more abundant *T. simulans*. It is more quadrate and is distinguished from the other red tellens by the pallial sinus, which touches the anterior adductor muscle scar. Length, 40 mm; height, 35 mm; diameter, 9 mm. Tenacatita Bay, Mexico, to Peru, intertidally and to depths of 10 m.

535. Tellina (Eurytellina) simulans C. B. Adams, 1852. This is likely to be confused with *T. rubescens*, as the name suggests, but it is a much more abundant form. Differences are in the pallial sinus, which does not touch the anterior adductor scar, in the greater proportionate length, more trigonal outline, and in the more widely spaced concentric sculpture. Length, 45 mm; height, 27 mm; diameter, 11 mm. Scammon's Lagoon, Baja California, through the Gulf of California and south to Peru, intertidally and to depths of 24 m. *T. (E.) punicea* Born, 1778, is a similar Atlantic species.
See Color Plate XIII.

Subgenus HERTELLINA Olsson, 1961

Shell shaped like *Sanguinolaria* but with a hinge much as in *Eurytellina*. Surface of shell finely sculptured with oblique striae.

536. Tellina (Hertellina) nicoyana Hertlein & Strong, 1949. The elongate-ovate outline and the pale rose color are distinctive, to say nothing of the oblique sculpture. Length, 34 mm; height, 19 mm; diameter, 7.8 mm. As yet it is known only from the type locality, Ballena Bay, Gulf of Nicoya, Costa Rica, dredged in 64 m.

Subgenus LACIOLINA Iredale, 1937

Somewhat resembling *Macoma* in shape and in the partially sunken ligament, but hinge with well-developed lateral teeth.

537. Tellina (Laciolina) ochracea Carpenter, 1864. The large thin shell is white or light sulphur yellow, the color stronger near the beaks. Sculpture is of fine concentric striae. Length, 38 mm; height, 28 mm; diameter, 9 mm. Cape San Lucas through the Gulf of California; on tide flats and offshore at least to 80 mm. This has been confused with *T. viridotincta*, which proves to be a *Psammotreta*. A similar species in the Atlantic is T. (*L.*) *laevigata* Linnaeus, 1758.

Subgenus LYRATELLINA Olsson, 1961

With prosogyrate beaks and well-spaced somewhat elevated concentric riblets.

538. Tellina (Lyratellina) lyra Hanley, 1844. The anteriorly directed beaks and the widely spaced concentric ribs of this dull white shell are characteristic. A large specimen measures: length, 50 mm; height, 35 mm; diameter, 12 mm. Baja California to Tumbes, Peru, mostly offshore in depths to 26 m. *T. (L.) juttingae* Altena, 1965, is a similar Atlantic species.

539. Tellina (Lyratellina) lyrica Pilsbry & Lowe, 1932. The beaks are not so strongly prosogyrate as they are in *T. lyra*, and the concentric sculpture is not so widely spaced. Both species are uncommon in collections. A large specimen measures: length, 40 mm; height, 30 mm; diameter, 13 mm. The Gulf of California to Caleta la Cruz, northern Peru, offshore in 26 to 70 m.

Subgenus MERISCA Dall, 1900

Small, trigonal, with a strong posterior flexure and spaced concentric ribs.

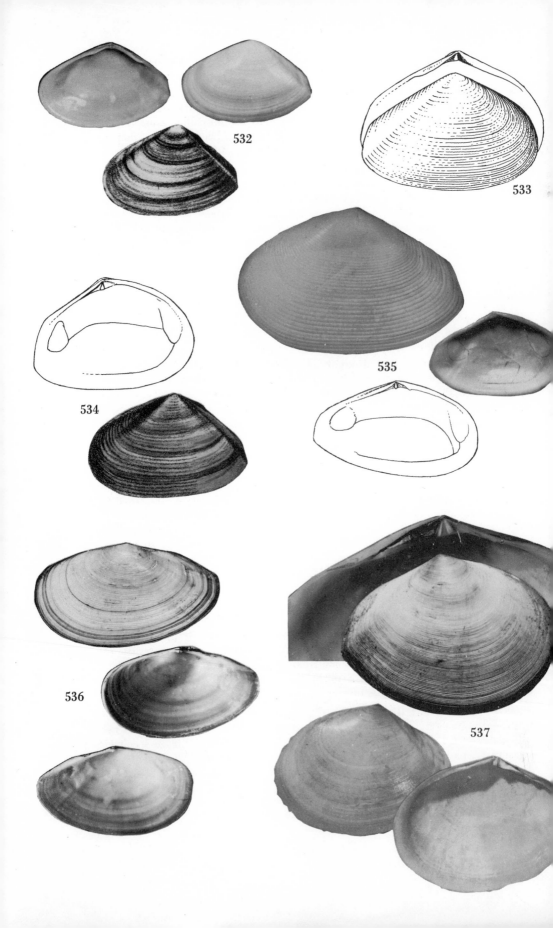

532

533

534

535

536

537

540. Tellina (Merisca) brevirostris Deshayes, 1855 (not preoccupied by the invalid *T. brevirostris* Oken, 1815) (Synonyms: **T. brevicornuta** Salisbury, 1934; **Merisca margarita** Olsson, 1961). Sculpture fine, shell more elongate than *T. rhynchoscuta*, both valves nearly equally convex. Distinguished from *T. reclusa* by the short snout into which the posterior end is drawn; in *T. reclusa* the posterior margin is blunt beyond the flexure. Length, 20 mm; height, 15 mm. El Salvador to Panama.

541. Tellina (Merisca) reclusa Dall, 1900. Although the posterior flexure is pronounced, it is not drawn out into a rostrum, as in *T. rhynchoscuta*, and the concentric sculpture is much finer. The shell is proportionately longer than that of *T. ulloana*. Length, 20 mm; height, 15 mm; diameter, 7 mm. San Ignacio Lagoon, Baja California, to Panama, mostly offshore in depths of 5 to 70 m. A similar Atlantic species is *T. (M.) aequistriata* Say, 1824.

542. Tellina (Merisca) rhynchoscuta (Olsson, 1961) (Synonym: "**T. cristallina** Spengler, 1798" of authors, often erroneously emended to **T. crystallina**). The posterior margin is pulled out into a rostrum, the concentric lamellae well separated, with minute radial striae between. The shell is pure white. Length, 25 mm; height, 19.5 mm; diameter, 6 mm. Scammon's Lagoon, Baja California, throughout the Gulf of California and south to Ecuador, intertidally and to depths of 24 m. The Atlantic species *T. (M.) cristallina* has a straight posterior dorsal margin, whereas in the eastern Pacific species the margin is incurved and the posterior end drawn upward into a decided snout.

543. Tellina (Merisca) ulloana Hertlein, 1968 (Synonyms: "**Tellina proclivis** Hertlein & Strong, 1949" of authors and **T. declivis** Sowerby, 1868, of authors). The spaces between the concentric ribs do not have the minute radial striae of *T. rhynchoscuta*. The shell is small, white, trigonal, and inflated. Length, 9 mm; height, 7.8 mm; diameter, 4.8 mm. Magdalena Bay, Baja California, to Panama, mostly offshore in depths of 22 to 48 m. The name *T. proclivis* was proposed as a replacement for the preoccupied name *T. declivis* Sowerby, which was presumed to be West American, but it has since been proved to be Atlantic; thus, a second name was needed to apply to the West American species. A similar Alantic species is *T. (M.) martinicensis* Orbigny, 1842.

Subgenus PHYLLODELLA HERTLEIN & STRONG, 1949

With hinge, pallial sinus, and concentric sculpture of anterior part like *Eurytellina* but with the concentric lamellations on the posterior part of *Phyllodina*.

544. Tellina (Phyllodella) insculpta Hanley, 1844. The posterior area with small scales, delicate lamellae, or even somewhat cancellate sculpture and the elongate compressed outline are characteristic of this white form. Length, 34 mm; height, 18 mm; diameter, 5.8 mm. Champerico, Guatemala, to Santa Elena Bay, Ecuador, offshore in depths of 6 to 26 m.

Subgenus PHYLLODINA DALL, 1900

The free pallial line distinguishes this subgenus from *Phyllodella*.

545. Tellina (Phyllodina) fluctigera Dall, 1908. The elongate shell is concentrically ridged, with the sculpture subdued in the center but standing up as lamellae at the ends. Length, 32 mm; height, 20 mm; diameter, 9 mm. As yet known only from the type locality, 333 m in Panama Bay. *T. (P.) persica* Dall & Simpson, 1901, is a similar species in the Atlantic.

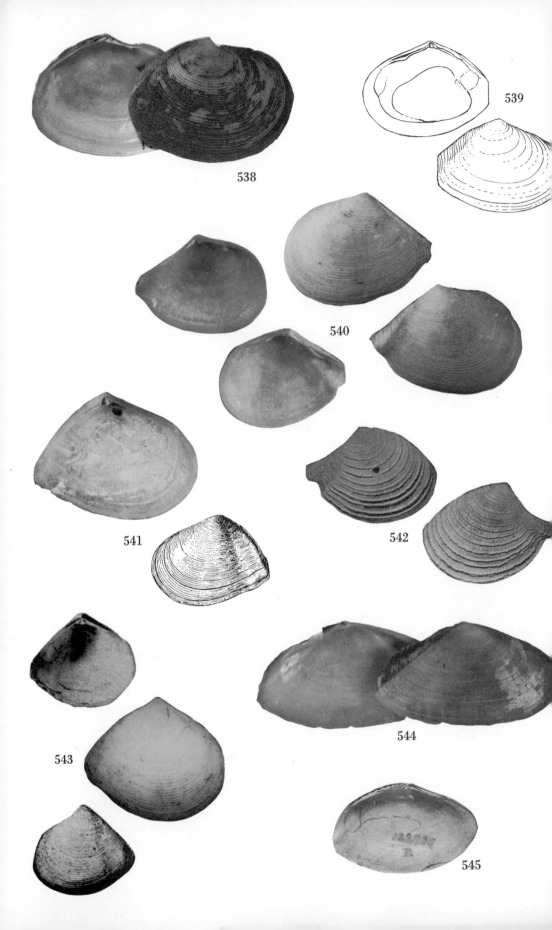

538

539

540

541

542

543

544

545

546. Tellina (Phyllodina) pristiphora Dall, 1900. The medium-sized, rather chalky shell is yellowish white in color, shading to pale salmon in the anterior dorsal area of some specimens. A large specimen measures: length, 36 mm; height, 23 mm; diameter, 10 mm. Santa Inez Bay, Gulf of California, to Punta-renas, Costa Rica, offshore in 22 to 155 m. *T.* (*P.*) *squamifera* Deshayes, 1855, is a similar species in the Atlantic.

Subgenus SCISSULA DALL, 1900

Hinge with a single strong right anterior lateral tooth, near the cardinals, other lateral teeth wanting. Surface of shell with oblique grooves.

547. Tellina (? Scissula) delicatula Deshayes, 1855. Strongly inequilateral, the white shell is flushed with rose and has dark irregular lines crossing the oblique striae. The type specimen at the British Museum is too badly broken for photographing. Length, 17 mm; height, 10 mm. The type locality is Mazatlán, Mexico.

548. Tellina (Scissula) esmeralda (Olsson, 1961). The shell is thin and white to translucent, and the microscopically fine oblique sculpture shows through to the inside; the pallial line shows to the outside. The posterior slope is smooth and the posterior end shorter than the anterior. The pallial sinus touches the anterior adductor scar. Length, 20 mm; height, 10 mm; diameter, 3 mm. Ecuador. A similar species in the Atlantic is *T.* (*S.*) *sandix* Boss, 1968 (the *T. exilis* Lamarck of authors).

549. Tellina (Scissula) varilineata Pilsbry & Olsson, 1943. The lines are finer and cross the shell at a lesser inclination than those of the next species. Length, 17 mm; height, 9.8 mm; diameter, 4 mm. Los Santos Province, Panama, to Peru.

550. Tellina (Scissula) virgo Hanley, 1844 (Synonym: **? T. deshayesii** Carpenter, 1856 [not Hanley, 1844]). The small shells are thin, glassy, pink or white in color, finely striate obliquely except on the posterior area. A large specimen measures: length, 20 mm; height, 12 mm; diameter, 4 mm. Magdalena Bay, Baja California, to Peru, intertidally and to depths of 15 m. A similar species in the Atlantic is *T.* (*S.*) *iris* Say, 1822.

Subgenus TELLINELLA MÖRCH, 1853

Posterior area with a strong flexuous rib. Color patterns laid on in bold radial stripes.

551. Tellina (Tellinella) cumingii Hanley, 1844. It is fitting that so showy a form should be named for the man who first systematically collected in the Panamic province. The yellowish-white shell is marked with radiating brown or purplish streaks or spots and sculptured with fine, close concentric lamellae. Large specimens reach a length of 55 mm or more and a height of 23 mm. Magdalena Bay, Baja California, through the Gulf of California and south to Colombia, mostly offshore in depths of 9 to 73 m. A similar Atlantic species is *T.* (*T.*) *listeri* Röding, 1798.

551a. Tellina (Tellinella) cumingii argis Olsson, 1971. This was named as *T.* (*Tellina*) *argis*. It seems to be closely related to *T. cumingii* and may prove to be an offshore subspecies, with coarser sculpture and less color marking. Length, 32 mm; height, 18 mm; diameter of one valve, 3 mm. Gulf of Panama, dredged.

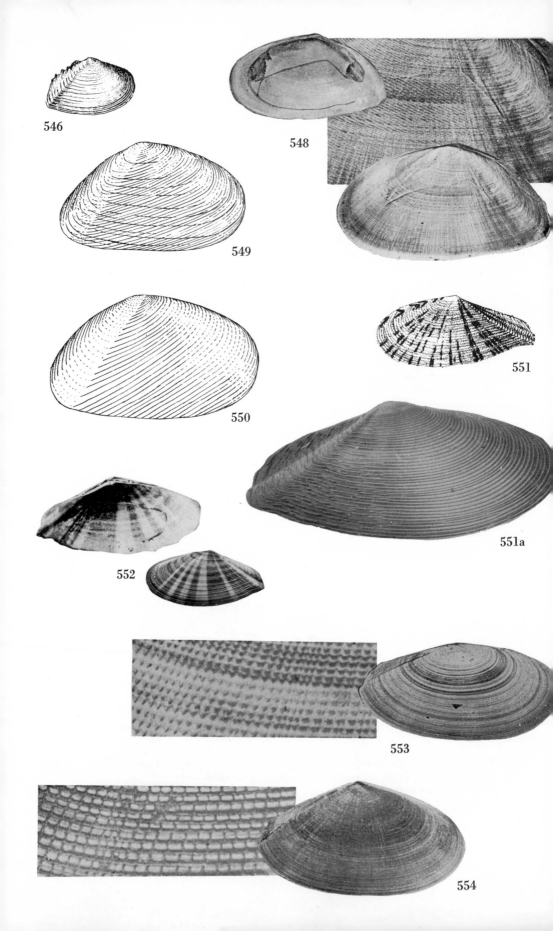

546

548

549

551

551a

550

552

551

553

554

552. Tellina (Tellinella) zacae Hertlein & Strong, 1949. Smaller than *T. cumingii*, this is white with golden-orange radiating color bands of varying widths. Length, 33 mm; height, 15 mm; diameter, about 8 mm. The species is as yet known only from a few dredging stations in the southwestern part of the Gulf of California in 64 to 165 m.

Subgenus **TELLINIDELLA** HERTLEIN & STRONG, 1949

With a strong posterior flexure; concentric sculpture decussated by radial grooves; hinge with a bifid posterior cardinal in right valve and the posterior lateral distant from the cardinal teeth.

553. Tellina (Tellinidella) mompichensis (Olsson, 1961). Narrowly elliptical, the shell is more than twice as long as high, white or rose-colored, sometimes a little banded; surface reticulation fine. Length, 57 mm; height, 27 mm; diameter, about 10 mm. Ecuador and Peru.

544. Tellina (Tellinidella) princeps Hanley, 1844. The color of the relatively large shell is dark rose-pink; compared with the more abundant *T. purpurea*, it has a heavier hinge, is larger, and has the ventral margin nearly straight. Length, 78 mm; height, 40 mm. Panama to northern Peru.

555. Tellina (Tellinidella) purpurea Broderip & Sowerby, 1829 (Synonym: *T. broderipii* Carpenter, 1857). No other Panamic province tellen is so brilliantly purple-red as this, and the ovate-elongate outline is also distinctive. The thin shell is faintly ornamented with a fine network of concentric and radial markings, and the dorsal margins are white. Length, 50 mm; height, 26 mm; diameter, 8 mm. The Gulf of California to Colombia; off sand beaches.

Genus **FLORIMETIS** OLSSON & HARBISON, 1953
(**APOLYMETIS** and **METIS** of authors)

Lateral teeth wanting; posterior flexure strong, especially evident in right valve; umbones inflated; shell inequilateral; ligament somewhat sunken but not completely internal. The type species of *Apolymetis* proves to be only superficially similar to American forms. A western Pacific group, *Leporimetis* Iredale, 1930, is close, and possibly authors may eventually submerge *Florimetis* in it either as a subgenus or as a synonym. The type of *Florimetis* is *Tellina intastriata* Say, 1826, from the Caribbean; that of *Leporimetis* is *T. spectabilis* Hanley, 1844.

556. Florimetis asthenodon (Pilsbry & Lowe, 1932). A thin white shell, with a buff periostracum, the anterior end longer than the posterior and widely rounded. The hinge is weaker than that of other species in the genus, the pallial sinus high and large, the anterior adductor muscle scar long and narrow. Length, 52 mm; height, 39 mm; diameter, 22 mm. El Salvador to Panama.

557. Florimetis cognata (Pilsbry & Vanatta, 1902) (Synonym: **Apolymetis cognata clarki** Durham, 1950). Of the Panamic species, this is nearest to the better known Californian form, *F. obesa* (Deshayes, 1855) (synonyms *Tellina biangulata* Carpenter, 1856, and *T. alta* Conrad, 1837, preoccupied), which occurs as far south as Magdalena Bay. The posterior end is broadly rounded, but the shell is thin, with a narrower hinge, and it is a gray white in color. Fairly strong radial and a few concentric threads show, especially on the anterior part of worn specimens. Length of a large specimen, 80 mm; height, 70 mm; diameter, 20 mm. Gulf of California to northern Peru and the Galápagos Islands (type locality), intertidally and offshore to 24 m. The form *F. cognata clarki* (Durham) was

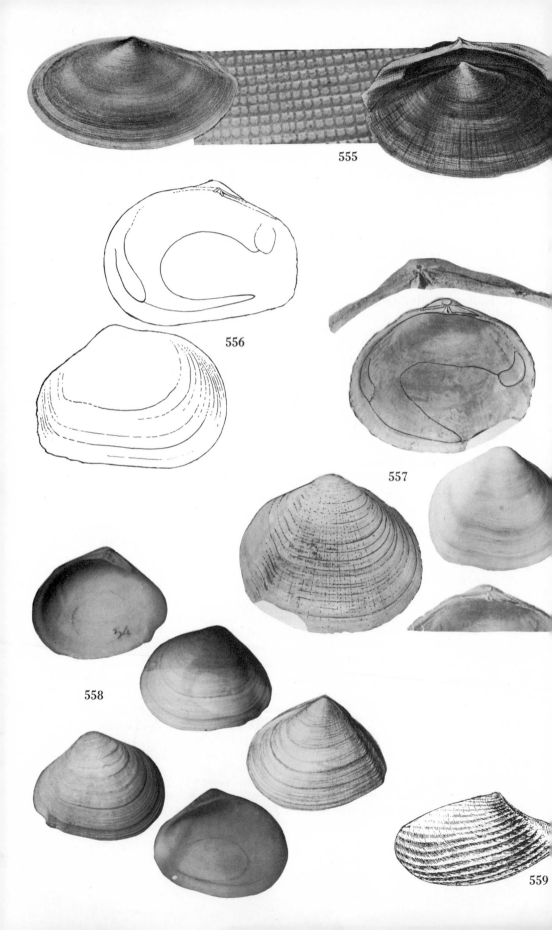

555

556

557

558

559

named as a Pleistocene fossil from Santa Inez Bay, Gulf of California; it differs in having less well developed radial threads on the surface, a variable character.

558. Florimetis dombei (Hanley, 1844) (Synonyms: **Scrobicularia producta** Carpenter, 1856; **Tellina excavata** Sowerby, 1867). Like *F. asthenodon*, this has a tapering posterior end. It differs from that species in the broad hinge, the oval adductor muscle scar, and the pallial sinus, which is confluent with the pallial line for a little less than half its length. A large specimen measures: length, 66 mm; height, 51 mm; diameter, 28 mm. Panama to Peru, mainly offshore in depths to 22 m.

Genus MACOMA LEACH, 1819

Lateral teeth wanting; shell elongate-quadrate, thin. The genus is much better represented in cold and temperate waters, and no species of *Macoma, s. s.,* occur in the Panamic province.

1. Shell with oblique sculpture..................................*Cymatoica*
   Shell smooth or sculpture not oblique..............................2
2. Posterior dorsal margin flangelike.........................*Rexithaerus*
   Posterior dorsal margin even, without flange.........................3
3. Posterior dorsal slope minutely granulate....................*Macoploma*
   Posterior dorsal slope smooth............................*Psammacoma*

Subgenus CYMATOICA DALL, 1890

Small, elongate, with undulating oblique sculpture.

559. Macoma (Cymatoica) undulata (Hanley, 1844) (Synonym: **Cymatoica occidentalis** Dall, 1890). The shell is small and thin, the posterior rather drawn out and flexed to the right. Length, 16 mm; height, 9 mm; diameter, 5 mm. The Gulf of California to Ecuador, offshore in 7 to 38 m.

Subgenus MACOPLOMA PILSBRY & OLSSON, 1941

Area of posterior slope near dorsal margin minutely granular.

560. Macoma (Macoploma) medioamericana Olsson, 1942. The shell is elongate and shaped much like the species of *Psammacoma*, but the posterior slope has concentric riblets and is dotted with granules. A large specimen measures: length, 101 mm; height, 54 mm; diameter, 24 mm. The Gulf of California to Caleta la Cruz, Peru, intertidally and offshore in depths to 80 m, on mud bottom.

Subgenus PSAMMACOMA DALL, 1900

The valves are elongate, convex, and thin, the posterior end markedly shorter. The pallial sinus is not confluent with the pallial line and is rounded and relatively short.

561. Macoma (? Psammacoma) carlottensis (Whiteaves, 1880) (Synonyms: **M. inflatula** Dall, 1897; **M. quadrana** Dall, 1916). Well known along the west North American coast, this small white clam has been dredged near Las Animas Island, Gulf of California, depth 1,490 to 1,550 m, a southward extension of range. Length, 16 mm; height, 11 mm. Aleutian Islands, Alaska, to Gulf of California.

562. Macoma (Psammacoma) elytrum Keen, 1958 (Synonym: **Tellina elongata** Hanley, 1844 [not Dillwyn, 1823]). A long, white shell, nearly smooth except for fine concentric lines of growth and some irregular oblique striations along the posterior umbonal ridge, with a depressed area on the lower middle

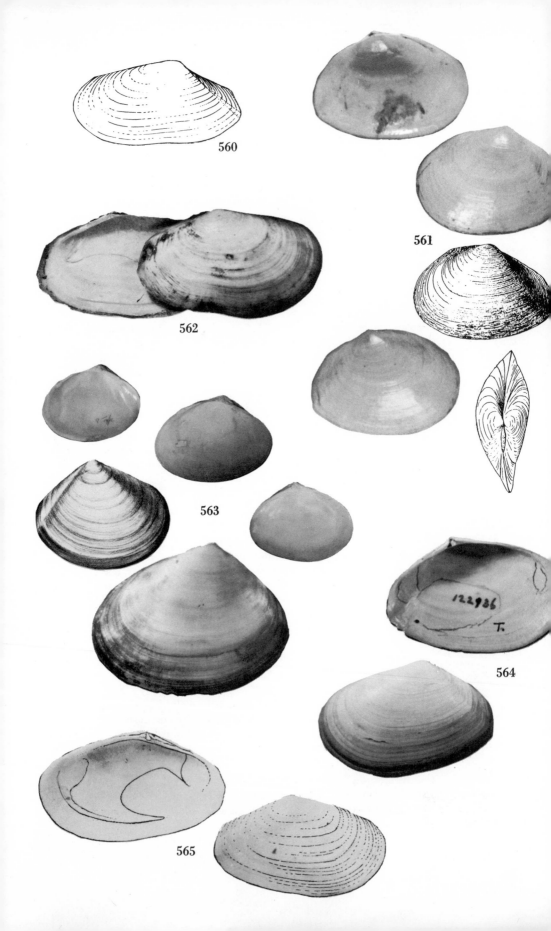

560

561

562

563

564

565

part of the anterior end. Length, 47 mm; height, 26 mm; diameter, 13 mm. Off-shore, Baja California and the Gulf of California south to Ecuador, in depths to 110 m.

563. Macoma (? Psammacoma) grandis (Hanley, 1844). Heavier and more solid than typical Psammacomas, this is also unusually short and trigonal. The shell is white under an olive periostracum and resembles *Florimetis* superficially, lacking the posterior flexure, the large pallial sinus, and the umbonal inflation. Length, 72 mm; height, 55 mm; diameter, 22 mm. Nayarit, Mexico, to Peru.

564. Macoma (Psammacoma) hesperus Dall, 1908. The thin white shell is more slender, with a more narrow posterior end than *M. elytrum*. It is known only from the type locality, Panama Bay, 333 m, but may be recognized now that a photograph is available and offshore collecting is less difficult. The principal differences between this and *M. lamproleuca* seem to be in the smaller size and the thinner texture. Length, 34 mm, height, 22 mm.

565. Macoma (? Psammacoma) lamproleuca (Pilsbry & Lowe, 1932) (Synonym: **M. parthenopa** Pilsbry & Lowe, 1932). In some ways this large *Macoma* resembles *M. elytrum*, but the dorsal margin slopes at a steeper angle, the shell is thicker, and the hinge is heavier. A large specimen measures: length, 73 mm; height, 41 mm; diameter, 21 mm. Santa Inez Bay, Gulf of California, to Peru, mostly offshore in depths to 90 m. Like *M. grandis*, this form is exceptionally heavy for a *Psammacoma*.

Macoma (Psammacoma) siliqua (C. B. Adams, 1852) (Synonyms: **Thracia carnea** Mörch, 1860; **M. panamensis** Dall, 1900). The posterior end is proportionately much shorter than in *M. (P.) elytrum*, and the shell is smaller. Length, 31 mm; height, 14 mm; diameter, 7 mm. Dall considered this form to be very close to *M. extenuata* Dall, 1900, of the Gulf of Mexico. Two geographic subspecies may be recognized:

566. Macoma (Psammacoma) siliqua siliqua. Shell surface smooth. Southern Mexico to Panama (type locality) in depths to 110 m.

567. Macoma (Psammacoma) siliqua spectri Hertlein & Strong, 1949. Shell surface with microscopically fine sculpturing that gives an iridescent sheen to fresh specimens. Gulf of California south along the West Mexican coast, in depths to 70 m.

Subgenus REXITHAERUS TRYON, 1869, *ex* CONRAD, MS

Shell thin, compressed, posterior area somewhat set off, its margin flangelike, with a sinuous posterior ventral margin.

568. Macoma (Rexithaerus) indentata Carpenter, 1864. Although better known as a California species, this form ranges north from La Paz and occurs in some abundance on the tidal flats of the northern end of the Gulf of California, east as far as Guaymas. At one time Dr. Harold Rehder evidently contemplated naming it as a distinct form (note in Steinbeck & Ricketts, 1941, p. 510), but the name was not validated. Length, 40 mm; height, 30 mm; diameter, 12 m.

Genus PSAMMOTRETA DALL, 1900
(SCROBICULINA DALL, 1900; SCHUMACHERIA COSSMANN, 1902)

The shell resembles that of *Psammacoma* but is shorter. The external ligament has a sunken lower portion or resilium separated off by a slight ridge. Two Panamic subgenera.

566

567

568

569

570

571

Subgenus **PSAMMOTRETA**, *s. s.*

Posterior end somewhat truncate, posterior margin lacking any flange.

569. Psammotreta (Psammotreta) aurora (Hanley, 1844) (Synonym:
? Tellina panamensis Philippi, 1844). The orange-red umbones of this form,
combined with the narrowly rectangular outline, make it easy to recognize. Length,
28 mm; height, 17 mm; diameter, 8 mm. The Gulf of California to Boca de Pan,
Peru, mostly offshore in depths to 33 mm. This is the type of the genus. A similar
species in the western Atlantic is *P. brevifrons* (Say, 1834).

570. Psammotreta (Psammotreta) mazatlanica (Deshayes, 1855). The
posterior end is more tapering than in any other of the West American Psam-
motretas, but it does not come to a point as in *P. (Ardeamya) columbiensis*. There
seems to be a gradational relationship to the next two species, and more material
may show they are all a single variable complex. Length of the holotype, which
is in the British Museum, 33 mm; height, 19 mm. Mazatlán (type locality) to
Ecuador.

571. Psammotreta (Psammotreta) pura (Gould, 1853). This has been
something of an enigma, perhaps because Gould's original figure did not show
the hinge. Although Gould gave the type locality correctly as Panama, Sowerby
(1867) figured the species as from "Vancouver Island," and later authors have
all failed to clarify its status. The type specimen, which, for unknown reasons, is
in the British Museum, bears Gould's type symbol, "Orig.," and matches the
figure well. It is a thin white shell, intermediate in proportions between *P. mazatlanica* and *P. viridotincta*, somewhat closer to the latter, though with the anterior
end longer. The hinge is more nearly parallel with the ventral margin than in
P. mazatlanica. Length, 25 mm; height, 16.5 mm. Known as yet only from the
type locality, Panama.

572. Psammotreta (Psammotreta) viridotincta (Carpenter, 1856) (Synonym: **Macoma pacis** Pilsbry & Lowe, 1932). The first figure of the hinge of
Carpenter's type specimen (Palmer, 1963) showed that it was not related to the
Tellina ochracea with which authors had confused it and that it was instead, as
Boss (1964) pointed out, a macomoid form lacking lateral teeth. Although somewhat variable, it is larger and relatively higher than *P. aurora*, the posterior end
more rounded. The umbones are white to yellowish salmon. Length, 49 mm;
height, 35 mm; diameter, 15 mm. The Gulf of California to Panama, mostly offshore in depths to 26 m. As mentioned under the discussion of the two previous
forms, this may prove to be part of a variable complex. However, Carpenter—
who must have had in hand the types of all three species—described *P. viridotincta*
as distinct, and he did not even cite *P. pura* in his discussion. If they prove synonymous, *P. pura* will take precedence as the oldest name.

Subgenus **ARDEAMYA** OLSSON, 1961

Posterior end pointed, posterior margin sinuous, flangelike.

573. Psammotreta (Ardeamya) columbiensis (Hanley, 1844). The thin,
compressed shell is a glossy white under a greenish-yellow periostracum. The anterior end is rounded and slightly longer than the posterior end, which has no
flexure. Length, 52 mm; height, 28 mm; diameter, 8 mm. Corinto, Nicaragua, to
Manta, Ecuador.

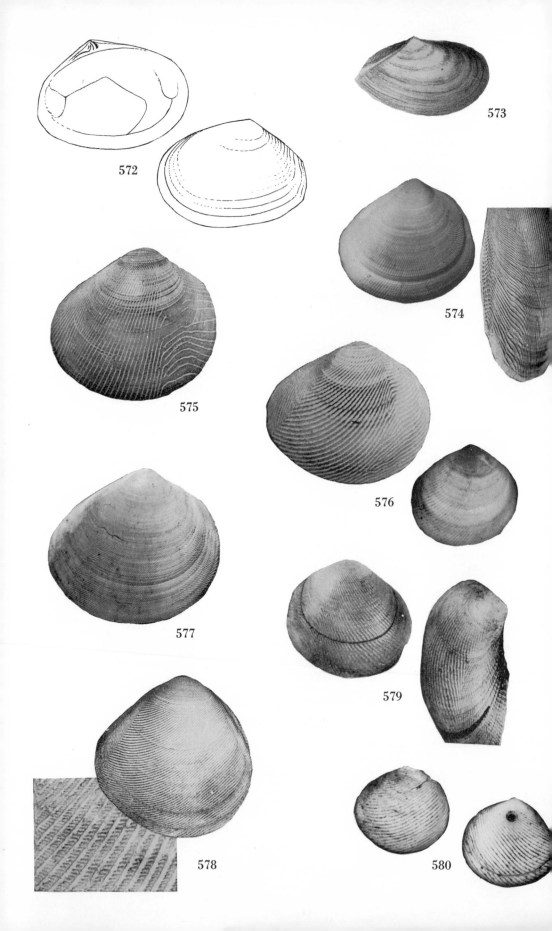

572

573

574

575

576

577

578

579

580

Genus STRIGILLA TURTON, 1822

Lenticular, mostly small shells, with fine, partially divaricate ribbing; tending toward pink coloration, especially on the umbones. Three Panamic subgenera.

Subgenus STRIGILLA, s. s.

Lacking zigzag lines on posterior area.

574. Strigilla (Strigilla) chroma Salisbury, 1934 (Synonym: **Tellina fucata** Gould, 1851 [not Hinds, 1845]). This has been misidentified by several authors as *S. costulifera* (Mörch). It is the commonest and widest-ranging Panamic species. The shell is relatively thick and convex, the anterior margin somewhat squared, so that it almost parallels the posterior margin. Color varies from white to pink or red, especially within. Length, 25 mm; height, 22 mm; diameter, 12 mm. Magdalena Bay, Baja California, to Ecuador. *S. pseudocarnaria* Boss, 1967, is a similar Atlantic species.

575. Strigilla (Strigilla) cicercula (Philippi, 1846) (Synonym: **S. maga** Mörch, 1860). Small-shelled, this form is white with a flush of bright pink on the umbones and with the divaricate sculpture more widely spaced on the anterior part. Length, 9.5 mm; height, 7.5 mm; diameter, 5 mm. Gulf of California to Ecuador. Afshar (1969) has made this species the type of a separate subgenus, *Roemerella* Afshar, 1969.

576. Strigilla (Strigilla) dichotoma (Philippi, 1846) (Synonym: **S. costulifera** Mörch, 1860). Although about the same shape and size as *S. cicercula*, this has the sculpture at a gentler angle and more crowded over the middle part of the valves. Along the lower part of the anterior end the ribs divide rather irregularly. Length, about 8.5 mm; height, 6.5 mm; diameter, 4.5 mm. Gulf of California to Ecuador.

577. Strigilla (Strigilla) disjuncta (Carpenter, 1856) (Synonym: "**S. sincera** Hanley, 1844" of authors [not of Hanley, an Australian species]). Compared with the last, this is thinner-shelled and flatter, with very fine markings, the anterior margin rounded. The lines of markings in the two valves do not agree at the edges—hence the name, meaning discrepant. The color is mostly white or white tinged with pink. A large specimen measures: length, 36 mm; height, 31 mm; diameter, 14 mm. This is one of the least common of the West American Strigillas. It ranges from Nicaragua to Peru, intertidally and to depths of 15 m. A similar form in the Caribbean is *S. gabbi* Olsson & McGinty, 1958.

578. Strigilla (Strigilla) ervilia (Philippi, 1846) (Synonym: **? Tellina lenticula** Philippi, 1846). Olsson (1961) interprets this as a medium-sized shell with a sharp posterior flexure and no flexure on the anterior slope; the ribs are close-set and even; the shape is ovate and nearly equilateral. Length, 11 mm; height, 10.8 mm; diameter, 6 mm. Baja California to Ecuador. *S. producta* Tryon, 1870, is a similar species in the Caribbean.

Subgenus PISOSTRIGILLA OLSSON, 1961

With one or more zigzag lines on posterior area.

579. Strigilla (Pisostrigilla) interrupta Mörch, 1860 (Synonym: **S. panamensis** Olsson, 1961). Pure white, the small plump shell has a sharp line of divarication of the sculpture on the posterior slope. Except for the color it is similar to *S. pisiformis* (Linnaeus, 1758) of the Caribbean. Length, large specimen, 8 mm; height, 7.5 mm; diameter, 5 mm. El Salvador to Ecuador.

Subgenus **SIMPLISTRIGILLA** OLSSON, 1961

Sculpture of unflexed simple diagonal lines.

580. Strigilla (Simplistrigilla) serrata (Mörch, 1860) (Synonym: **S. strata** Olsson, 1961). The sculpture is coarser and more even than in the other species of *Strigilla*, the riblets all running diagonally across the shell in one direction with no flexing. It is a white and rather solid little shell. Length, 7.3 mm; height, 6.5 mm. The type locality was not stated, though probably El Salvador or Nicaragua; the range extends south to Ecuador.

Genus **TELLIDORA** H. & A. ADAMS, 1856

Trigonal, with strong serrations along the dorsal margin.

581. Tellidora burneti (Broderip & Sowerby, 1829). The shell is white and shiny, with low concentric ribs. There is no other West American form with which it might be confused. *T. cristata* (Récluz, 1842) is a similar species in the Caribbean area. A large specimen of *T. burneti* measures: length, 49 mm; height, 40 mm; diameter, 7 mm. Baja California and throughout the Gulf of California, to Salango, Ecuador, in depths to 29 m.

Genus **TEMNOCONCHA** DALL, 1921
(**PSAMMOTHALIA** OLSSON, 1961)

Sculpture of fine oblique lines, especially on the anterior and central portions of the valves.

582. Temnoconcha cognata (C. B. Adams, 1852) (Synonyms: **Tellina concinna** C. B. Adams, 1852 [not Philippi, 1844]; **Psammobia casta** Reeve, 1857; **Tellina tenuilineata** Li, 1930). Previously assigned to *Tellina*, this actually is closer to *Macoma*, for there are no lateral teeth. The rectangular outline and the fine oblique sculpture are distinctive characteristics. The form named *"T. concinna"* is considered to be a smooth variant by Olsson (1961); actually the type lot shows faint and distant oblique striae. The shell is white, with a pale brownish or reddish tinge. Length, 43 mm; height, 30 mm. Mazatlán, Mexico, to Panama, offshore in depths of 8 to 73 m.

Family **DONACIDAE**

The members of this family have sturdy shells, mostly trigonal in shape, valves tightly closing, with a moderately large pallial sinus. The hinge has two cardinal teeth and one or more lateral teeth on either side. The ligament is external, on a small platform. Three Panamic genera.

Genus **DONAX** LINNAEUS, 1758

The internal margins of *Donax* are crenulate, reflecting the radial ribbing of the middle layer of the shell. Ribbing is well developed also on the surface of most forms. Living specimens of *Donax* are found on sandy beaches or in bays, where they burrow shallowly so near the surface they may be washed out by heavy waves. The active animals are able to dig back again very quickly, however. In areas of abundance they are sought as food. The type species of the genus *Donax* is *D. rugosus* Linnaeus, 1758, from west Africa. Panamic species of *Donax*, with only a few exceptions, do not seem to fit well into the subgeneric groups that have so far been named; thus, no attempt at full-scale subdivision will be made here.

The usage of many early authors (including Linnaeus) was to consider *Donax*

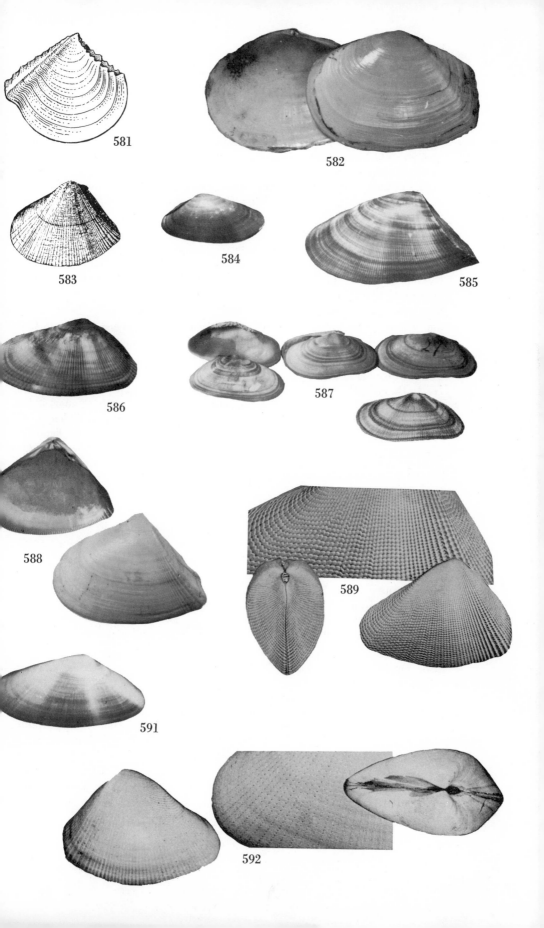

as a name of feminine gender. Actually the word "donax" in Latin is a masculine noun. Under the International Code of Zoological Nomenclature, the gender of the word in the classic language (whether Greek or Latin) determines the usage as a scientific name. This brings about what appear to be inconsistencies of spelling for those specific names that are adjectives and must have the endings changed. A compound inconsistency results for several of the specific names that end in -*fer* (masculine) or -*fera* (feminine). Authors who are unfamiliar with Latin grammar may assume that the masculine equivalent of the feminine -*fera* would be -*ferus*. Thus, they may write *D. dentiferus* to replace *D. dentifera*; technically, it should be *D. dentifer*.

583. Donax asper Hanley, 1845. The subcentral beaks and the evenly spaced ribs of the posterior slope, as well as the concentric roughening of the ribs, mark this species. The color is grayish white to purple. A large specimen measures: length, 35 mm; height, 26 mm; diameter, 17 mm. Tangola-Tangola Bay, southern Mexico, to Peru.

584. Donax californicus Conrad, 1837. The type locality is southern California, where the species is common in bays. South of Magdalena Bay, Baja California, it grades into *D. navicula*, which may prove to be identical with it. Length, large specimen, 27 mm; height, 12 mm; diameter, 8 mm.

585. Donax carinatus Hanley, 1843 (Synonyms: **?** **D. rostratus** C. B. Adams, 1852; **D. culminatus** Carpenter, 1857). The trigonal shape, strong sculpture, and sharp angulation setting off the posterior slope are identifying marks. The color is a purplish brown, purple within. A large specimen measures: length, 39 mm; height, 22 mm; diameter, 15 mm. Altata, Mexico, to Colombia, in depths to 24 m. Authors have differed in their attempts to identify Adams's *D. rostratus*. His type specimen being lost, it seems preferable to place the name in the synonymy of the only species that Adams cited for comparison.

586. Donax contusus Reeve, 1854 (Synonyms: **D. conradi** and **bitincta** Reeve, 1854). Like *D. punctatostriatus* and *D. culter*, this has the interspaces between the ribs punctate; that is, with fine pits. It is longer for its height than *D. punctatostriatus* and shorter than *D. culter*. Length, 30 mm; height, 16 mm; diameter, 11 mm. The southern part of the Gulf of California. *D. vellicatus* Reeve, 1854, is a similar Atlantic form.

587. Donax culter Hanley, 1845 (Synonyms: **Donax petallina** Reeve, 1854, and **D. petalina** Deshayes, 1855 [in part]; **Amphichaena gracilis** Mörch, 1860). For comparisons, seen under *D. contusus*, above. Length, 26 mm; height, 12 mm; diameter, 8 mm. Baja California, through the Gulf and south to Corinto, Nicaragua.

588. Donax dentifer Hanley, 1843 (Synonyms: **D. paytensis** Orbigny, 1845; **?** **D. assimilis** Hanley, 1845). This species resembles *D. asper*, but the shell is thinner, less acutely rounded anteriorly, with one to three raised ribs on the posterior slope coarser than the others. These ribs project as interlocking teeth at the posterior margin. Olsson (1961) has pointed out that Hanley's description of *D. assimilis* mentions such a tooth, although the figure in Reeve suggests a somewhat longer shell than *D. dentifer*. Length, 40 mm; height, 30 mm. Corinto, Nicaragua to Peru.

589. Donax ecuadorianus Olsson, 1961. A trigonal small shell with strong cancellate sculpture in somewhat oblique or wavering rows; white to pale yellow, with a few purplish stains, especially within. It seems to be closest to *D. obesus*, but the sculpture is more pronounced. Length, 19 mm; height, 13 mm; diameter, 9 mm. Panama to Ecuador.

590. Donax gouldii Dall, 1921. A Californian form, this has been reported as far south as Acapulco, but may have been a misidentification. It is more probably not to be found south of Magdalena Bay.

591. Donax gracilis Hanley, 1845. A polished shell, rather long and narrow, this has the beaks nearer the posterior end. The color is some shade of brown, with the interior brownish purple. A large specimen measures: length, 22.5 mm; height, 9.4 mm; diameter, 6 mm. Near Magdalena Bay, Baja California, through the Gulf and south to Negritos, Peru, intertidally and to depths of 24 m. An outer coast rather than a bay form. It differs from *D. navicula*, which it resembles, by being relatively a little longer and narrower, more compressed, with the posterior end more pointed.
See Color Plate XIII.

592. Donax mancorensis Olsson, 1961. The white to cream-colored shell usually is stained with purple inside; it is rather solid and trigonal in outline, the anterior end longer and somewhat pointed, the posterior end obliquely truncate. The surface is nearly smooth except for low, flattened radial riblets, between which may be small pits. Length, 25 mm; height, 18 mm; diameter, 13 mm. Manta, Ecuador, to Zorritos, Peru.

593. Donax navicula Hanley, 1845. Compared with *D. gracilis* the shell is more rounded ventrally and more rhomboidal in outline as well as more inflated. Length (large specimen), 21 mm; height, 10 mm; diameter, 7.8 mm. The Gulf of California to Panama, intertidally and to depths of 13 m. This species is so similar to the northward-ranging *D. californicus* that it may eventually be placed in synonymy, with a range extending north to southern California.

594. Donax obesulus Reeve, 1854, *ex* Deshayes, MS (Synonyms: **? D. granifera** Reeve, 1854; **D. curtus** Sowerby, 1866). A white shell, with dark purplish brown on the posterior end and on the anterior dorsal margin. The posterior slope is set off by a low carina or angle. Olsson (1961) attempted to identify *D. rostratus* C. B. Adams, 1852, with this (see under *D. carinatus*). Length, 20 mm; height, 15 mm. Central America to Peru.

595. Donax obesus Orbigny, 1845. The shell is small, subtriangular, and inflated. It is ornamented by radial grooves crossed by somewhat flexuous concentric grooves, forming a fine punctate pattern. The inner margin is crenulate, the color within dark purplish brown posteriorly, white on the outside. This form has a less abruptly truncate shell than *D. obesulus*, and the ridge that sets off the posterior slope is less angulate or almost rounded. A large specimen measures: length, 13 mm; height, 11 mm; diameter, 8.4 mm. Corinto, Nicaragua, to Paita, Peru.

596. Donax panamensis Philippi, 1849 (Synonyms: "**D. cayennensis** Lamarck" of Reeve, 1854 [not of Lamarck, an Atlantic species]; **D. reevei** and **sowerbyi** Bertin, 1881; **D. assimilis** of authors). The grayish to purplish shell resembles *D. asper* except that the beaks are more posterior and the shell longer. A large specimen measures: length, 41 mm; height, 27 mm; diameter, 17 mm. Mazatlán, Mexico, to Ecuador. This is cited by a number of authors as *D. assimilis*, which Olsson (1961) has shown to be probably a variant of *D. dentifer*.

593

594

595

596

597

597. Donax peruvianus Deshayes, 1855 (Synonyms: **D. radiatus** Valenciennes, 1827 [not Gmelin, 1791]; **D. aricana** Dall, 1909). Except that the interspaces lack pitted sculpture, this resembles *D. punctatostriatus*. It is common in Peru. Length, 34 mm; height, 20 mm. Manta, Ecuador, to Chile.

598. Donax punctatostriatus Hanley, 1843 (Synonym: **D. caelatus** Carpenter, 1857). Probably the most abundant of the Panamic donacids, this has a sturdy, fair-sized shell easily recognized, in spite of its color variants and variations in size and shape, by the pits in the interspaces between ribs. The majority of specimens are bright brown, violet-stained within. Typical specimens are higher than those of either *D. culter* or *D. contusus*, which also show some punctations in the interspaces. These three species may be allocated on the basis of the pitted sculpture to the subgenus *Chion* Scopoli, 1777, the type of which is *D. denticulatus* Linnaeus, 1758, from the West Indies. Length, 45 mm; height, 28 mm; diameter, 15 mm. San Ignacio Lagoon, Baja California, through the Gulf of California and south to Negritos, Peru.

599. Donax transversus Sowerby, 1825 (Synonym: **D. scalpellum** of authors, not of Gray, 1823). The slenderest of the Panamic species of *Donax*, this is easily recognized by its nearly smooth surface and short posterior end set off by a keel. The color is yellowish white with purple rays. A large specimen measures: length, 36 mm; height, 14.5 mm; diameter, 9 mm. Mazatlán, Mexico, to San Juan del Sur, Nicaragua. This species can be assigned to the subgenus *Machaerodonax* Römer, 1870, the type of which is a very similar-shaped form, *D. scalpellum* Gray, 1823.

Genus AMPHICHAENA PHILIPPI, 1847

Elongate, somewhat gaping posteriorly, inner margin crenulate only at ends, smooth along ventral margin; anterior end not narrowed; pallial sinus short; sculpture of fine radial striae, not ribs; hinge lacking posterior lateral tooth, anterior lateral teeth close to cardinals, cardinals compact, separated by a curved slot.

This genus has been assigned to Psammobiidae by many authors, but Olsson (1961) has suggested transfer to Donacidae, where he would even make it a subgenus. Anatomical studies are needed to establish its true position.

600. Amphichaena kindermanni Philippi, 1847 (Synonyms: **Donax petallina** Reeve, 1854, and **D. petalina** Deshayes, 1855, in part). *Amphichaena* is a genus with a single known species. One might mistake the shell for *Donax*, but the weak hinge, the marginal crenulations only on the anterior end, and the cylindrical form show that it is not. Most specimens are brownish, radially striped with white, the shell of a solid texture. Dimensions of a large specimen: length, 37 mm; height, 15 mm; diameter, 10 mm. Mazatlán, Mexico, to Guatemala.

Genus IPHIGENIA SCHUMACHER, 1817

The smooth inner margin distinguishes *Iphigenia* from *Donax*.

601. Iphigenia altior (Sowerby, 1833) (Synonym: **I. ambigua** Bertin, 1881). The shell is solid, rather large, subtrigonal, smooth, with a heavy, olive periostracum under which the color is yellowish or purplish white, the interior white and violet. Length, 68 mm; height, 51.5 mm; diameter, 31 mm. The Gulf of California to Tumbes, Peru, intertidally and to 24 m. Dr. S. S. Berry has recognized (*in litt.*) that the *I. altior* of authors comprises two distinct forms, separable on several points.

Family PSAMMOBIIDAE
(GARIDAE)

Shells inequilateral, slightly gaping, especially at posterior end; hinge with two small cardinal teeth; laterals weak to wanting; ligament on a nymph or platform; pallial sinus large. Two Panamic subfamilies.

Subfamily PSAMMOBIINAE

Oval to trapezoidal, equivalve; the gape small. Two Panamic genera.

Genus GARI SCHUMACHER, 1817
(PSAMMOBIA of authors, not of LAMARCK, 1818)

Rectangular, smooth to weakly sculptured, posterior end wider than anterior and gaping more widely; ligament external, sturdy.

No species of the subgenus *Gari, s. s.*, in which the shell is pointed posteriorly, compressed, and only slightly gaping, occur in the Panamic province.

Subgenus GOBRAEUS BROWN, 1844
(PSAMMOCOLA of authors, not of BLAINVILLE, 1824)

Smooth, posterior end somewhat truncate; gape moderate; pallial sinus large, partly confluent with pallial line.

602. Gari (Gobraeus) fucata (Hinds, 1844). Smooth, quadrate, with a greenish periostracum. Length, about 45 mm. The type locality is Magdalena Bay, Baja California. Apparently the species, if it is a valid one, ranges northward along the Baja California coast, like *G. regularis*, with which it does not seem to be synonymous. The relationship to Californian forms remains to be established. The type specimen is in the collection of the British Museum.

603. Gari (Gobraeus) helenae Olsson, 1961 (Synonym: **G. regularis** of authors, not of Carpenter, 1864). An elongate shell, with a slightly sinuous ventral margin, the greatest height in front of the midline. In color it is purplish, mottled with white, and there are several dark-colored radial markings. Length, 52 mm; height, 27 mm; diameter, 14 mm. The type locality is Panama. What has customarily been identified in the Gulf of California area as *G. regularis* is close to *G. helenae*, differing only by being relatively a little shorter for the height, with a less obliquely truncate posterior end, by having the ventral margin evenly arched, not sinuate in front of the junction with the posterior slope, and by having the color more mottled, without definite dark rays. Both forms show fine dichotomous ribbing on the posterior slope. Material from intermediate localities will establish whether the variation is clinal or may represent geographic subspecies. Accepting this form as *G. helenae* extends the range, from the Gulf of California through Panama to Colombia.

604. Gari (Gobraeus) species. A single specimen in the Stanford University collection collected by Ted Dranga in Costa Rica is here illustrated but left nameless. In outline it is close to *G. helenae*, but the color pattern is decidedly different, being of zigzag stripes in a roughly concentric series of bands. The ground color is white, the stripes reddish purple, with a suffusion of yellow along their edges, and there are flecks of the reddish color over the rest of the shell. Length, 40 mm; height, 21 mm; diameter, 12 mm. More material is needed to show whether there is variation in the color markings and to indicate the geographic range. The form may prove only to be a variant in the *G. helenae* complex.

PLATE IV · *Semele junonia* (Verrill)

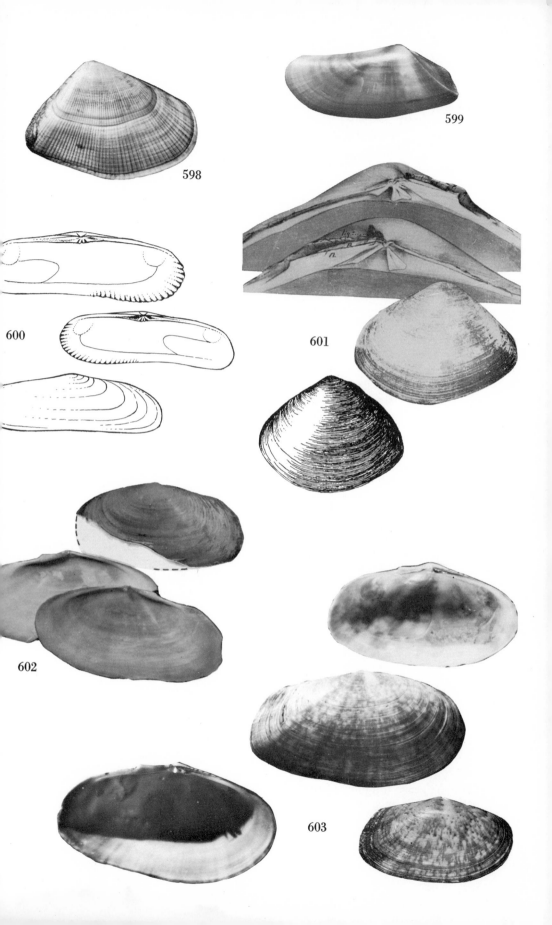

598

599

600

601

602

603

605. Gari (Gobraeus) lata (Deshayes, 1855). The shortest of the Garis for its height, this was described from Central America but seems to be more common in the southern part of its range, in Ecuador. Length, 60 mm; height, 40 mm.

606. Gari (Gobraeus) maxima (Deshayes, 1855). The color pattern is a distinguishing feature of this *Gari*, the shell being yellowish white rayed with brownish pink or purple. Length, 54 mm; height, 33 mm. The species has been reported intertidally from the Gulf of California to Colombia.

607. Gari (Gobraeus) panamensis Olsson, 1961. Resembling *G. maxima*, this is longer and lower, white-rayed with irregular bands of purplish brown. The shell is unusually thin in texture, with a varnish-like brown periostracum. Length, 55 mm; height, 32 mm; diameter, 16 mm. Panama.

608. Gari (Gobraeus) regularis (Carpenter, 1864). Palmer's (1958) figure of Carpenter's hitherto unillustrated type specimen shows it to be an elongate form with transverse lamellae on the posterior slope, unlike what has been identified in collections as *G. regularis*. The type locality is Cape San Lucas, but the range is probably northward along the outer coast of Baja California rather than southward into the Panamic province.

Genus HETERODONAX Mörch, 1853

Solid ovate shells, not gaping, with much variation in color markings. Hinge with two cardinal teeth in each valve, the anterior stronger; no laterals. In young specimens the hinge is disproportionately large, with markedly heavy teeth.

609. Heterodonax pacificus (Conrad, 1837) (Synonyms: **Tellina vicina** C. B. Adams, 1852; **Donax ovalina** Reeve, 1854, *ex* Deshayes, MS; **H. bimaculatus purpureus** and **salmoneus** Williamson, 1892). So similar are these shells to the Caribbean *H. bimaculatus* (Linnaeus, 1758) that most authors treat them as a single species, although a number of specific names have been given to the eastern Pacific form. There are small but consistent differences, as Dr. E. V. Coan (*in litt.*) points out: the Pacific form is larger, and the pallial sinus tends to meet the pallial line at a right angle, whereas in *H. bimaculatus* there is gradual merging, with a small gap between the two. Near high tide line on sandy shores of bays, shallowly buried. The white shell is variously striped or spotted with color, especially inside, where the pink or orange color may be concentrated in the two spots suggested in the scientific name Linnaeus gave it. The outside may be rayed with dull blue or gray, and many combinations of these colors occur. The largest specimens measure about 27 mm long and 22 mm high, but 15 mm is a more common length. Southern California to Panama. The Atlantic twin species, *H. bimaculatus*, ranges widely through the Caribbean.

Subfamily SANGUINOLARIINAE

Smooth, inequivalved as well as inequilateral; pallial sinus large.

Genus SANGUINOLARIA Lamarck, 1799

The generic name suggests a blood-red color, and this is a characteristic of the group, varying from bright pink to a reddish purple. The color, the thin, gaping, unsculptured shell, and lack of lateral hinge teeth or a posterior twist distinguish the genus from any of the tellens. One valve may be flatter than the other. Two Panamic subgenera.

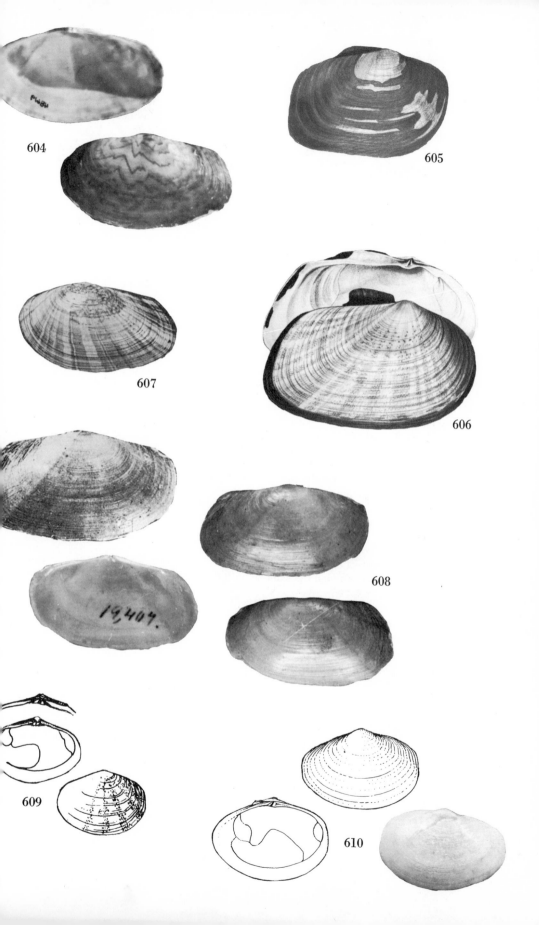

604

605

607

606

609

608

610

Subgenus SANGUINOLARIA, *s. s.*

Valves convex, only slightly unequal in size, rounded, the posterior end not markedly flexed; pallial sinus similar in the two valves, largely confluent with the pallial line.

610. Sanguinolaria (Sanguinolaria) ovalis Reeve, 1857 (Synonyms: **S. vespertina** Pilsbry & Lowe, 1932; **S. tenuis** Olsson, 1961). Authors have had some difficulty in recognizing this form, as the original figure was not satisfactory. The outline is almost intermediate between that of the two synonyms cited here, rounded at both ends, with low beaks. The color is mostly white, deepening on the umbones to pink in some specimens. Length, 22 mm; height, 14 mm. Tangola-Tangola Bay, Oaxaca, Mexico, to Ecuador.

611. Sanguinolaria (Sanguinolaria) tellinoides A. Adams, 1850 (Synonyms: **Tellina miniata** Gould, 1851; **S. purpurea** Deshayes, 1855). The color may vary from rose pink to almost purple or red (in the form called *S. purpurea*), laid on in darker and lighter bands, especially in the larger specimens. Large shells may be 90 mm in length. An average size would be: length, 75 mm; height, 45 mm; diameter, 22 mm. Gulf of California to Ecuador.

Subgenus PSAMMOTELLA HERRMANNSEN, 1852

Left valve flattened, less convex than right; posterior end of right valve flexed; pallial sinus discrepant (not the same in the two valves), only slightly confluent with pallial line.

612. Sanguinolaria (Psammotella) bertini Pilsbry & Lowe, 1932 (Synonyms: **T. rufescens** of authors, not of Gmelin, 1791; **T. hanleyi** Bertin, 1878 [not Dunker, 1853]). The colorful shell is dark rose red in zones, deeper toward the beak. The left valve has a somewhat pointed posterior end. A large specimen measures 92 mm in length and 47 mm in height, but the average size is more apt to be 60 mm in length and 30 mm in height. San Ignacio Lagoon, Baja California, through the Gulf and southward to Lobitos, Peru, intertidally. A similar species in the western Atlantic is *S. (P.) cruenta* [Lightfoot, 1786].

Family SOLECURTIDAE

The shells are rectangular to cylindrical and more than twice as long as they are high. The hinge is much the same as in the Psammobiidae. Two Panamic genera.

Genus SOLECURTUS BLAINVILLE, 1824

The distinguishing feature of *Solecurtus* is the oblique sculpturing, which cuts across the growth lines of the shell.

613. Solecurtus broggii Pilsbry & Olsson, 1941. A rare form, this may be recognized by the black periostracum and the oblique sculpturing on the posterior end only. It ranges from the Gulf of Chiriqui, Panama, to northern Peru, and was dredged at 64 to 73 m. Length, 84 mm; height, 33 mm. It is possible that more material will show this to be a synonym of *S. galapaganus* (Dall *in* Dall & Ochsner, 1928) from the Pleistocene of the Galápagos Islands. Dall's type was figured but not described, a smaller shell (length, 55 mm) but of about the same proportions.

614. Solecurtus guaymasensis Lowe, 1935. Also rare, this has a yellowish-brown periostracum and oblique sculpturing on both ends. A large specimen is 56 mm long and 23 mm high. Only dredged specimens have been reported, from 37 to 110 m. Cedros Island, Baja California, to Chiriqui, Panama.

611

612

613

614

615

616

617

Genus TAGELUS GRAY, 1847

The jackknife clams are mud-flat dwellers, cylindrical like the Solens but with the hinge near the middle of the dorsal margin, both ends rounded and, of course, gaping. The hinge has two small cardinal teeth. The pallial sinus is deep. A heavy periostracum is characteristic. Two Panamic subgenera.

Subgenus TAGELUS, *s. s.*

Medium-sized to large, without any internal rib or buttress.

615. Tagelus (Tagelus) affinis (C. B. Adams, 1852) (Synonym: **Solecurtus cylindricus** Sowerby, 1874). The shell is white under a yellowish-brown periostracum, relatively small. Length, 55 mm; height, 21 mm. The pallial sinus extends just beyond a line drawn vertically through the beaks. The Gulf of California to Panama, on mud flats or offshore to depths of 73 m. A similar species in the Atlantic is *T. (T.) plebius* [Lightfoot, 1786].

616. Tagelus (Tagelus) californianus (Conrad, 1837). A yellowish-white shell with rust-colored stains, with a dark periostracum that may be worn off from the central part, revealing in many specimens vertical scratchlike markings stained with darker color. The posterior dorsal margin is slightly sinuous, not sloping downward from the beaks. Larger than any other western *Tagelus*. The shell may measure as much as 120 mm in length; however, a length of 100 mm (four inches) and a height of 28 mm is more common. Compared with *T. affinis*, the shell is proportionately longer and thicker, and the pallial sinus is shorter, not extending past the beaks. It is a mud-flat species, occurring from Monterey, California, south to northern Mexico.

617. Tagelus (Tagelus) dombeii (Lamarck, 1818). The southern counterpart of *T. californianus*, this is narrowly elongate, rusty brown except for two narrow white rays extending obliquely down from the beak. The periostracum is coarse, wrinkled, dark brown. Length, 90 mm; height, 25 mm; diameter, 16 mm. Panama to Chile. It is the type species of *Solecurtellus* Ghosh, 1920, a subgenus that has not yet been adopted by authors but that may, according to studies of Dr. Eugene Coan, prove useful in future revisions of the group.

618. Tagelus (Tagelus) irregularis Olsson, 1961. This was named as a subspecies of *T. affinis*, from which it differs in being larger, thinner, and more inflated. Length, 80 mm; height, 27 mm; diameter, 20 mm. Ecuador.

619. Tagelus (Tagelus) longisinuatus Pilsbry & Lowe, 1932. The shell is exceptionally long for its height, and the long pallial sinus may extend through 60 percent of the length of the shell. Pilsbry and Lowe regarded the form as a subspecies of *T. affinis*. Length, 62 mm; height, 19 mm. Mazatlán to Oaxaca, Mexico.

620. Tagelus (Tagelus) peruanus (Dunker, 1861). Shortest and most broadly rectangular of the Panamic *Tagelus*, the shell is white with a dull olive-brown periostracum. Length, 42 mm; height, 18 mm; diameter, 12 mm. Nayarit, Mexico, to northern Peru.

621. Tagelus (Tagelus) violascens (Carpenter, 1857). This species is close in shape of the shell to *T. dombeii*. It is dark, tending toward violet staining, and the posterior end, at least in the type specimen—which is in the British Museum—is set off by a low radial ridge. The pallial sinus extends slightly beyond the midline of the shell. It may prove to be a synonym of *T. dombeii*.

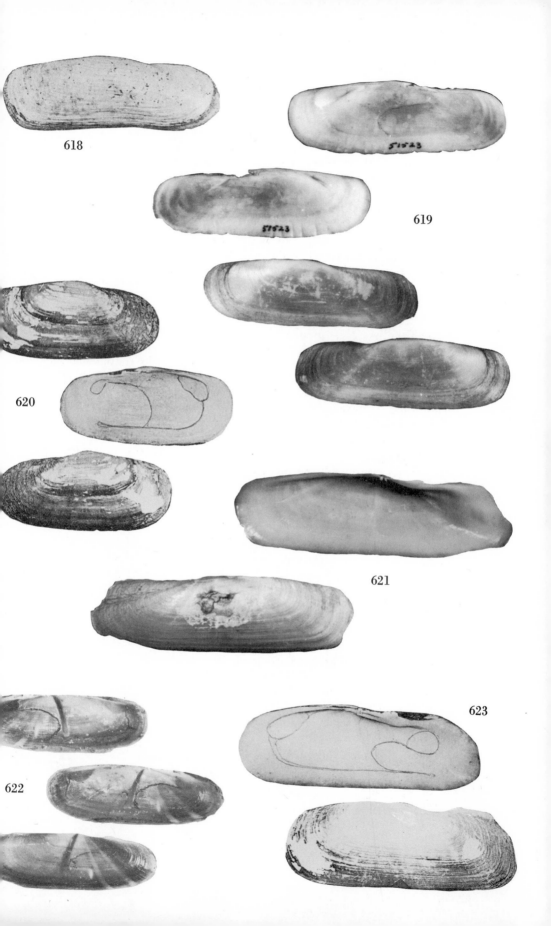

618

619

620

621

622

623

Length, 82 mm; height, 24 mm. The type locality is "Southwest Mexico." The *"Tagelus violascens"* of authors, as figured by Hertlein and Strong (1950), and refigured by Keen (1958, no. 472, p. 192), is a misidentification. Further study may well show this form to be an undescribed species that is restricted to the southern Mexican coast. The figure of Keen, 1958, is repeated here (no. 621*a*) for contrast with the true *T. violascens.*

Subgenus MESOPLEURA CONRAD, 1867

Interior reinforced by a radial rib, especially apparent in the young stages of growth. This rib often shows through to the outside as a color ray.

622. Tagelus (Mesopleura) bourgeoisae Hertlein, 1951. The smallest of the Panamic jackknife clams, this has a length of only 34 mm and a height of 11 mm. It is dark grayish purple to brownish purple, with an internal rib that shows through as a dark line and with two lighter-colored radial stripes slanting from the beaks to the posterior margin. The pallial sinus stops just short of the internal rib. Salina Cruz, Mexico, to Ecuador.

623. Tagelus (Mesopleura) peruvianus Pilsbry & Olsson, 1941. The periostracum is a dark olive-black, usually worn off the umbones, which show the light color and striping of the middle layer. Young shells are thin and show the internal rib of the subgenus *Mesopleura*; but older shells may be smooth and a little warped in the middle so that they gape at each end. The posterior dorsal margin has a flangelike extension that is characteristic of the species. Length, 57 mm; height, 21 mm. Diggs Point, Baja California, to Negritos, Peru.

624. Tagelus (Mesopleura) politus (Carpenter, 1857) (Synonyms: **Siliquaria carpenteri** Dunker, 1862; **S. nitidissima** Dunker, 1868). The shell is thin, almost translucent, and rayed with dark violet; the periostracum is glossy. Within, there is an internal rib, and the pallial sinus, rounded anteriorly, extends just to the ray. Length, 35 mm; height, 13 mm. Gulf of California to Panama.

Family SEMELIDAE

The shells of the Semelidae are rounded to ovate, mostly of medium size—some large—smooth or concentrically sculptured. The two most diagnostic features are the ligament, the major part of which is lodged in a capsule on the hinge plate back of the two small cardinal teeth, and the deep, widely rounded pallial sinus. The hinge also has well-developed lateral teeth in both valves. Although this is a family widely distributed in tropical seas, with more than 30 described species in the Panamic province alone, no one seems to have attempted a thoroughgoing revision, especially of the type genus *Semele*, by which grouping into subgenera might be achieved. Some of the genera, such as *Abra*, show resemblances to a related family, the Scrobiculariidae, in which lateral teeth and the posterior flexure are absent. The key is to genera.

1. Ligament partially external; shell rounded-trigonal.................*Abra*
   Ligament internal 2
2. Hinge with spoon-shaped resilifer projecting downward...........*Cumingia*
   Hinge with resilifer not markedly projecting......................... 3
3. Shell with posterior end pointed.............................*Leptomya*
   Shell with posterior end rounded................................... 4
4. Shells medium-sized to large, adults 20 mm long or more...........*Semele*
   Shells small, adults less than 10 mm long.....................*Semelina*

Genus SEMELE SCHUMACHER, 1817
(AMPHIDESMA of authors, not of LAMARCK, 1818)

One peculiar feature of *Semele* is the striate pattern of the area inside the pallial sinus, as though the surface were laid on in little radiating fibrous bundles. Another peculiarity is that the anterior end is the longer, the beaks being behind the midline.

Semele is one of many molluscan groups named after mythological figures. The original Semele was a Greek goddess; hence, the name should be pronounced as three syllables, not two: sem'-e-lee. Semele herself was a rather colorful figure, and it is not inappropriate that, if the name is to be used in Mollusca at all, it should be for this group, in which colors range from deep purple to brightest orange, subdued by tones of lavender, buff, and gray.

Juvenile specimens of *Semele* do not have the internal ligament strongly developed and may appear to be tellinid; two forms were, in fact, so described— *Tellina lamellata* and *T. regularis* Carpenter, 1857, from Mazatlán, Mexico. The types, in the British Museum, are too immature for species assignment, but they can be dismissed from Tellinidae.

625. Semele bicolor (C. B. Adams, 1852) (Synonyms: **Amphidesma striosum** and **ventricosum** C. B. Adams, 1852: **S. fucata** Mörch, 1860). A thin, orbicular shell with distinctive purple suffusion on a white ground. The surface is smooth near the beaks, but elsewhere shows fine, raised concentric lines and crowded microscopic radiating striae divergent posteriorly. Length, 21 mm; height, 19 mm; diameter, 11 mm. Gulf of California to Panama. Lowe reported it as rare on mud flats at La Paz.

626. Semele californica (Reeve, 1853, *ex* A. Adams, MS). Closely related to the larger and heavier *S. corrugata* (Sowerby, 1833) from Peru, this has a thinner shell, with concentric wrinkle-ridged ribs and dense, minute radial striae. Within, the color is yellow to almost a golden orange. Length, 40 mm; height, 36 mm; diameter, 14.5 mm. Only beach shells have been reported in the literature, from Magdalena Bay, Baja California, to the Gulf of California.

627. Semele corrugata (Sowerby, 1833). A large white shell, the interior yellowish orange, with a stain of dark red on the hinge. Length, 96 mm; height, 86 mm; diameter, 40 mm. It is a southern form, ranging from southern Ecuador to Chile.

628. Semele craneana Hertlein & Strong, 1949. A thin shell with coarsely corrugated concentric ribs and no radial striae, this otherwise resembles several of the Semeles in which the anterior end is rounded, the posterior short and set off by a faint angulation. Length, 38 mm; height, 29.5 mm; diameter, 13 mm. It has been dredged in the southern end of the Gulf of California and off Clarion Island in 82 to 90 m.

629. Semele elliptica (Sowerby, 1833). One of the larger forms, an average specimen of this measures 80 mm in length. The color is white to pinkish buff with a greenish periostracum, flushed with faint pink under the beaks within. Central America to Ecuador, mostly intertidally.

630. Semele flavescens (Gould, 1851) (Synonym: **Amphidesma proximum** C. B. Adams, 1852). Somewhat similar to *S. elliptica* in shape but higher, this has a moderately large, thin shell, grayish yellow on the outside, brighter within. In fresh specimens the surface is covered by a greenish periostracum. Length, 59 mm; height, 55 mm; diameter, 28 mm. Southern Baja California to Peru, intertidally.

631. Semele formosa (Sowerby, 1833). The unusual color pattern well merits the Latin adjective *formosa*—beautiful. On a creamy-white ground, interrupted pinkish-buff bands radiate out from the umbones, intersected by concentric bands of violet. Length, 55 mm; height, 45 mm; diameter, 22 mm. Gulf of California to Ecuador.

632. Semele guaymasensis Pilsbry & Lowe, 1932. Somewhat resembling *S. pulchra* (Sowerby), this form has much coarser sculpture, and incised radial sculpture is wanting on the posterior area. The color is light buff, faintly mottled or rayed with violet; the dorsal borders are dark purple. Length, 19 mm; height, 14.6 mm; diameter, 6.5 mm. Punta Peñasco, Sonora, Mexico, southward through the Gulf to La Paz, in 7 to 37 m.

633. Semele jovis (Reeve, 1853) (Synonym: **? Tellina barbarae** Boone, 1928). One of several large Semeles with a posterior flexure, this has extremely fine radial wrinkling on the close-set concentric ribbing. The coloring is characteristic: rose-fawn, the beaks red, with a white streak running through the middle of the umbonal areas, the interior rose and white. Length, 55 mm; height, 45 mm; diameter, 21 mm. Kino Bay, Sonora, Mexico, through the Gulf and south to Perlas Islands, Panama, on mud flats and offshore to depths of 27 m.
See Color Plate XIII.

634. Semele junonia (Verrill, 1870) (Synonym: "**Amphidesma rosea** Sowerby" of Reeve, 1863, not Sowerby, 1833). The most brilliantly colored of the West American Semeles, this rare shell is indeed beautiful. It is fairly large, with the umbones a rich orange red, shading to white near the margin. Length, 70 mm; height, 62 mm; diameter, 29 mm. Gulf of California, from La Paz to Guaymas, offshore in depths of a few meters.
See Color Plate IV.

635. Semele laevis (Sowerby, 1833). As the name suggests, this is a virtually smooth form, with only slight microscopic sculpture. The shell is white under a thin olive periostracum. Length, 68 mm; height, 53 mm; diameter, 26 mm. Guatemala to Ecuador in depths of 18 to 26 m.

636. Semele lenticularis (Sowerby, 1833). Although this resembles *S. bicolor* (Adams, 1852), with which Olsson (1961) would synonymize it, the outline is more circular, with the beaks lower and more nearly centrally placed. Normally white, the shell may be stained with yellow, purple, or brown; fine radial striae fan out in a surface patterning. Length, 25 mm; height, 21 mm; diameter, 12 mm. Barra de Navidad, Mexico (common), to Peru.

637. Semele pacifica Dall, 1915 (Synonym: **S. jaramija** Pilsbry & Olsson, 1941). Resembling *S. guaymasensis*, this differs in details of sculpture, having strong, well-developed concentric sculpture intersected with strong radial riblets on both the anterior and posterior portions, often in the right valve over the whole shell. The coloring is yellowish, with radial stripes of purple and purple spots along the dorsal margin. Length, 20 mm; height, 16.5 mm; diameter, 9 mm. Gulf of California to Panama, intertidally and to depths of 38 m. Fossil records are in the Pleistocene of Panama and the Pliocene of Ecuador.
See Color Plate XIII.

631

632

633

634

635

636

637

638. Semele pallida (Sowerby, 1833). Many of the southern species of *Semele* have counterparts in the northern part of the Panamic province. Two of the northern species that are very similar to this in outline are *S. paziana* Hertlein & Strong and *S. simplicissima* Pilsbry & Lowe. They differ in details of sculpture. *S. pallida* is ovate, with the posterior end somewhat truncate. The sculpture is of fine concentric ribs with smooth interspaces. Coloring of the white shell varies from a flush of salmon to pale orange or purple. Length, 26 mm; height, 19 mm. Ecuador.

639. Semele paziana Hertlein & Strong, 1949 (Synonym: **S. regularis** Dall, 1915 [not E. A. Smith, 1885]). A thin, delicate, mostly white shell, sculptured with spaced, low concentric lamellae between which are fine concentric lineations and no radial striation. A few specimens show a flush of orange within. Length, 22 mm; height, 17 mm; diameter, 6 mm. The Gulf of California, off La Paz, 18 to 55 m depth.

640. Semele pilsbryi Olsson, 1961. High for its length, the shell is trigonal with a rounded anterior end, sculptured with irregular radial riblets of varying width, the shell surface in addition being granular. In color it is a grayish white, flushed with apricot within except on the hinge, which is coral red. It may prove to be closest to *S. rupium*. Length, 46 mm; height, 42 mm. Known only from the type locality, Bucaro, Panama.

641. Semele pulchra (Sowerby, 1832) (Synonym: **? S. quentinensis** Dall, 1921). Most trigonal of the West American Semeles, the yellowish-gray shell has chevron-shaped markings and purple blotches; the interior is white below, purplish on the upper part. The outside is strongly sculptured concentrically with fine ribs; a few radial riblets cross the anterior slope. Length, 31 mm; height, 25 mm; diameter, 12.5 mm. The relationship to *S. quentinensis* is unclear. Perhaps only a single species is represented, with a range from southern California to Ecuador, on mud flats. If, however, two geographically separate subspecies are to be recognized, *S. pulchra pulchra* would be the southern form, ranging from Costa Rica southward.

642. Semele punctata (Sowerby, 1833). Much resembling the *S. decisa* (Conrad, 1837) from southern California and the west coast of Baja California, this is longer for the height. In addition to coarse concentric undulations, the sculpture is of fine radial striae in irregular rows. Length, 50 mm; height, 42 mm; diameter, 18 mm. Galápagos Islands.

643. ? Semele quentinensis Dall, 1921. As indicated in the discussion of *S. pulchra* above, this may prove to be a synonym or a subspecies rather than a distinct species, but more comparative material is needed before a decision can be made. The form has been considered by Hertlein and Strong (1949) to be smaller, lower, and less brightly colored than *S. pulchra*. Length, 24 mm; height, 19 mm; diameter, 8 mm. Ventura County, California, to Costa Rica, intertidally and to depths of 30 m.

Semele rosea (Sowerby, 1833). Two geographic subspecies can be recognized.

644. Semele rosea rosea. The shell is large and thin, with spaced concentric ribs and fine radial striations. It resembles *S. junonia* in this respect. The color is rose pink, shading into yellow, the beaks darker, and the interior of the same color. Fresh specimens show a thin brown periostracum. Length, 60 mm; height, 52 mm; Panama to Peru.

638

639

640

641

642

643

645

646

647

645. Semele rosea tabogensis Pilsbry & Lowe, 1932. Very similar in color and sculpture to *S. r. rosea*, this subspecies has a more ovate shell, with the anterior end proportionately longer. Length of a large specimen, 64 mm; height, 53 mm. Tangola-Tangola Bay, Mexico, to Panama.

646. Semele rupium (Sowerby, 1833) (Synonym: **S. floreanensis** Soot-Ryen, 1932). When not distorted by its nestling habit, this form shows a conspicuous furrow radiating from the beak, setting off the posterior slope. Length, about 30 mm. It has been definitely recorded only from the Galápagos Islands. A remarkably similar species in California is *S. rupicola* Dall, 1915.

647. Semele simplicissima Pilsbry & Lowe, 1932. The specific name is evidently intended to call attention to the smooth interspaces between the concentric ribs, in contrast to the striate spaces in *S. paziana*, the closest relative. A large shell measures 33 mm in length, 24 mm in height, 12.5 mm in diameter. Santa Inez Bay, Gulf of California, to Acapulco, Mexico; in depths of 22 to 110 m.

648. Semele sowerbyi Lamy, 1912 (Synonym: **Amphidesma purpurascens** Sowerby, 1833 [not *Venus purpurascens* Gmelin, 1791, a *Semele*]). More symmetrical than most Semeles, this has fine lamellar concentric sculpture. The color is of shades of violet and purple, and the beaks are marked with a ray of white. The inside is white, flushed with purple. It sometimes resembles *S. jovis* of the Mexican coast. Length, 47 mm; height, 36 mm; diameter, 13 mm. Panama to Ecuador.

649. Semele sparsilineata Dall, 1915. The specific name refers to the scattered oblique striae that intersect the fine, regular concentric riblets. The color is a dingy white with faint brownish-purple stains. A large specimen measures 25 mm in length, 21 mm in height, and 9 mm in diameter. Guaymas and Barra de Navidad, Mexico, to Panama, on mud flats and offshore.

650. Semele tortuosa (C. B. Adams, 1852) (Synonym: **Semele planata** Carpenter, 1856). The shell is dingy white, with a strong posterior flexure and somewhat irregular and wavering concentric ridges that are crossed by submicroscopic radiating striae. Length, 30 mm; height, 27 mm; diameter, 9 mm. Panama to Ecuador, known mostly by odd valves found in beach drift. Specimens resembling this species but with a flattened left valve have been collected in the Guaymas to Mazatlán area, Mexico (Shasky, *in litt.*).

651. Semele venusta (Reeve, 1853, *ex* A. Adams, MS). The quadrate shell is shiny, buff to flesh color, obscurely rayed with rose, the anterior end much the longer. Sculpture is of concentric grooving. Within, the shell is purple, edged with white. The pallial sinus is unusually long, about two-thirds the length of the shell. Length, 22 mm; height, 15 mm; diameter, 8.5 mm. Acapulco, Mexico, to west Colombia, probably intertidally as well as offshore in depths to 16 m.

652. Semele verrucosa Mörch, 1860 (Synonym: **S. margarita** Olsson, 1961). Because the type was unfigured, authors have misinterpreted the description and have illustrated for it a considerably different shell. True *S. verrucosa* is similar to *S. pacifica* Dall but more elongate, with broader concentric ribs and more nodose sculpture at both ends; it differs from *S. guaymasensis* by having well-developed radial ribbing posteriorly. In color it is pinkish buff with darker mottling. It is a small shell: length, 10 mm; height, 8 mm. Costa Rica to Panama.

653. Semele verruculastra Keen, 1966 (Synonym: **S. verrucosa** of authors, not of Mörch). The elongate-ovate shell is inequilateral, whitish blotched with

648

649

650

651

652

653

purple, the sculpture of close concentric ribs broken into scalelike projections at the ends, the whole surface with radial wrinkles. Length, about 50 mm; height, about 38 mm; diameter, about 17 mm. Gulf of California to Panama Bay; also in the fossil record in the Pleistocene, Gulf of California.

Genus ABRA LAMARCK, 1818

These are small, white, thin, nearly smooth shells. The ligament is partially external, with only a small portion in a little capsule back of the cardinal teeth.

654. Abra palmeri Dall, 1915. Relatively high for the length, the shell is inflated, with a silky surface and a thin, shiny, pale yellow periostracum. Length, 10 mm; height, 8 mm; diameter, 5.5 mm. Ballenas Lagoon, west coast of Baja California, through the Gulf and south to Panama Bay, offshore in depths of 30 to 165 m. Olsson (1961) has synonymized this with *A. tepocana*, although it is the better known and more widely distributed of the two. There seems sufficient difference in shape to justify separation.

655. Abra tepocana Dall, 1915. As in *Semele*, the anterior end is longer. The internal ligament is relatively large, and the lateral teeth are obsolete. Length, 8 mm; height, 6 mm; diameter, 3.5 mm. Off Puerto Peñasco to Cape Tepoca, Sonora, Mexico; in 26 m.

Genus CUMINGIA SOWERBY, 1833

The Cumingias also are white shells but thicker and more irregular in shape than the Abras. Like them, *Cumingia* has a pallial line that doubles back on itself so that it is confluent with the pallial sinus. The internal ligament is large; the cardinal teeth are small, the lateral teeth large in the right valve, weak or wanting in the left. The anterior end is short and rounded, the posterior longer and angulate.

656. Cumingia adamsi Olsson, 1961, *ex* Carpenter (Synonym: **C. adamsii** Carpenter, 1864, *nomen nudum*). The species is distinguished by a shell that is undeformed and thus of regular shape, the beaks nearly median, the posterior end somewhat pointed, the surface marked by fine and rather evenly spaced concentric lamellae, with finer radial striae between that fan out toward the margin. Length, 10 mm; height, 8 mm; diameter, about 4.5 mm. Panama to Ecuador.

657. Cumingia lamellosa Sowerby, 1833 (Synonyms: **C. coarctata** and **trigonularis** Sowerby, 1833; **C. similis** A. Adams, 1850; **C. moulinsii** De Folin, 1867). The nestling habit of this form—in sand, sponges, and rock crevices—leads to distortions that have tricked a number of workers into giving separate names to ecologic variants. The shell is thinner and more delicate than the northern *C. californica* Conrad, 1837. An undistorted specimen measures about 18 mm in length. San Martin Island, Baja California, southward through the Gulf of California to Peru, intertidally and to depths of 24 m.

658. Cumingia mutica Sowerby, 1833. A thinner, less distorted, more finely sculptured species than *C. lamellosa*. The anterior end is smoothly rounded, the posterior bluntly pointed, and the sculpture is of fine threadlike lines. Length, 20 mm; height, 16 mm; diameter, about 10 mm. Ecuador to Chile.

659. Cumingia pacifica (Dall, 1915). Described as an *Abra*, this has the hinge of *Cumingia*. The posterior end is slightly longer than the anterior. The surface is finely concentrically striate, giving it a silky look. Length, 9 mm; height, 9 mm; diameter, 3 mm. As yet the species has been reported only from the type locality, Guaymas, Mexico.

Genus LEPTOMYA A. ADAMS, 1864

Shell white, smooth, thin; pointed behind.

660. Leptomya ecuadoriana Soot-Ryen, 1957 (Synonym: **L. americana** Keen, 1958). Resembling a small *Macoma,* this is distinguished by the internal ligament. The shell is a creamy white, unsculptured, except for lines of growth and some weak granulations. Length, 33 mm; height, 23 mm; diameter, 12 mm. San Blas, Mexico, to northern Peru, on mud flats near mangrove swamps. This is the only American species of an otherwise western Pacific genus.

Genus SEMELINA DALL, 1900

The small elliptical shell has the posterior end much the shorter; the hinge has strong lateral teeth in the right valve, none in the left. There is one bifid cardinal tooth in each valve, and the resilifer is internal. Sculpture of regular concentric riblets.

661. Semelina subquadrata (Carpenter, 1857). The small shell, though not uncommon, has mostly escaped notice. Carpenter in describing it overlooked the large pallial sinus and did not stress sufficiently the strong concentric sculpture. He assigned it to the Montacutidae, where it has been listed as a *Mysella.* The color is mostly white, some specimens having a pinkish tinge. Length, 5.5 mm; height, 4 mm. Gulf of California to Colombia.

Superfamily SOLENACEA

Cylindrical to elongate, gaping at both ends; hinge weak, with one to three cardinals; laterals, if present, weak and laminar; pallial sinus short in most.

Family SOLENIDAE

Beaks terminal or nearly so; hinge with only one tooth in either valve; siphons fused; foot modified for rapid digging in sand. Two Panamic genera.

Genus SOLEN SCOPOLI, 1777

The shell is straight and tubular, with the beak terminal or nearly so, the anterior end either cut off squarely or slightly curved forward, one cardinal tooth in either valve. Two Panamic subgenera.

Subgenus SOLEN, s. s.

Cardinal teeth terminal; shell of same thickness throughout.

662. Solen (Solen) crockeri Hertlein & Strong, 1950. Proportionately to other razor clams, this is small and high; length, about 40 mm; height, 10.5 mm. The shell is white under a thin, light olive periostracum. The type was dredged in the Gulf of Fonseca, Nicaragua, in 9 to 29 mm depth.

663. Solen (Solen) mexicanus Dall, 1899. White, like *S. crockeri,* the shell is larger and relatively longer: length, 60 mm; height, 8.5 mm. It was described from the Gulf of Tehuantepec, southern Mexico, and later was dredged in Nicaragua.

664. Solen (Solen) oerstedii Mörch, 1860. Described from Costa Rica, it was not figured and it seems not to have been recognized since. Mörch stated that the umbonal area is red and that the shell is 69 mm long and 11 mm high. The type specimen is in the University of Copenhagen Zoological Museum and is figured

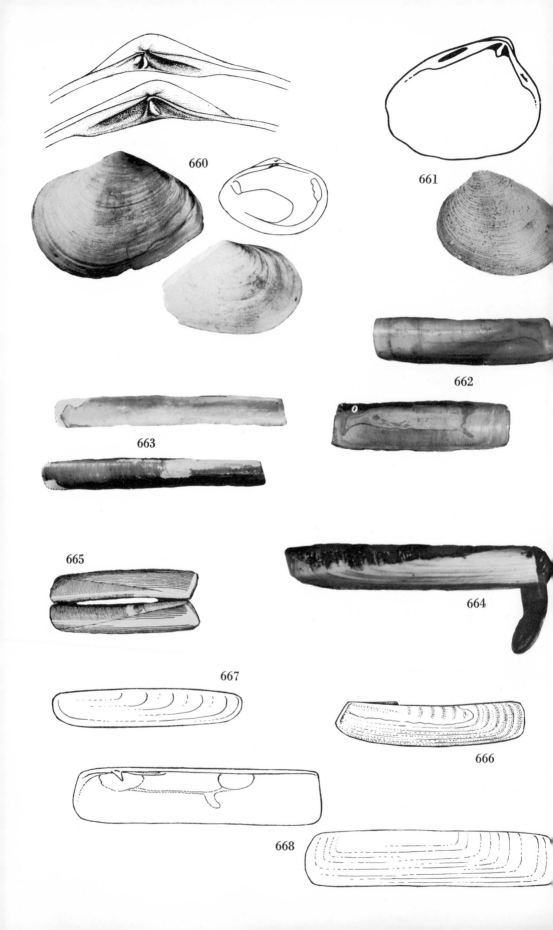

660

661

662

663

665

664

667

666

668

here. Its outline is closest to *S. rosaceus* except that the posterior end is more oblique. The thick periostracum mentioned by Mörch still is partially present, but the red color has faded to a dull yellow.

665. Solen (Solen) pazensis Lowe, 1935. The purple coloration of the periostracum concentrated in a triangle on each valve marks this species. The length is about 58 mm; height, 12 mm. Lowe collected the form intertidally at La Paz; it has since been dredged in 11 to 13 m in southern Mexico, in Tangola-Tangola Bay.

666. Solen (Solen) pfeifferi Dunker, 1862. Although the shell is banded with purple, the periostracum is a uniform brown color over the entire surface. With a maximum length of 52 mm and a height of 10 mm, this is slightly smaller than *S. pazensis*, which it most resembles. It is a more southern form, distributed from Tangola-Tangola Bay, Oaxaca, Mexico, to Ecuador, in 7 to 24 m.

667. Solen (Solen) rosaceus Carpenter, 1864. The pale rose umbones of this elongate shell distinguish it; possibly it is the northern representative of *S. oerstedii*. A large specimen measures 75 mm in length; a length of 48 mm with a height of 9 mm is more common. Santa Barbara, California, to Mazatlán, Mexico, intertidally on muddy shores, in burrows, to depths of 24 m.

Subgenus **SOLENA** Mörch, 1853
(**HYPOGELLA** Gray, 1854)

Cardinal teeth a little behind the thickened anterior margin of the shell.

668. Solen (Solena) rudis (C. B. Adams, 1852). Largest of the West American Solens, this has been confused by some authors with the Atlantic *S. obliquus* (Spengler, 1793), also known as *S. ambiguus* (Lamarck, 1818). Adams's type specimen was nearly 150 mm long, 30 mm high. He obtained many specimens from natives who were digging them for food. The shell is white under a rough brown periostracum that extends over the edges of the valves. Costa Rica south to northern Peru.

Genus **ENSIS** SCHUMACHER, 1817

The shell of *Ensis* is cylindrical, like that of *Solen*, but with a smooth arc or curve in its outline. In the left valve there are two cardinal teeth. The habits are similar to those of *Solen*.

669. Ensis californicus Dall, 1899. This has been confused with a more common form from southern California, *E. myrae* Berry, 1953. The true *E. californicus* has a smaller and more slender shell (length, about 56 mm; height, about 6.8 mm), with a rounded anterior end and the pallial line well removed from the margin. Magdalena Bay, Baja California, through the Gulf and south to Ecuador; intertidally and to depths of 50 m.

670. Ensis tropicalis Hertlein & Strong, 1955. Almost as straight as a *Solen*, the slender little grayish-brown shell measures about 50 mm in length and 6 mm in height. Puerto Peñasco, Sonora, Mexico (Skoglund collection); Perlas Islands, Panama (type locality) in 11 to 25 m, sandy bottom.

Order **MYOIDA**
(**ASTHENODONTA**)

Thin-shelled, burrowing forms with well-developed siphons; outline inequilateral; lunule and escutcheon weak to wanting.

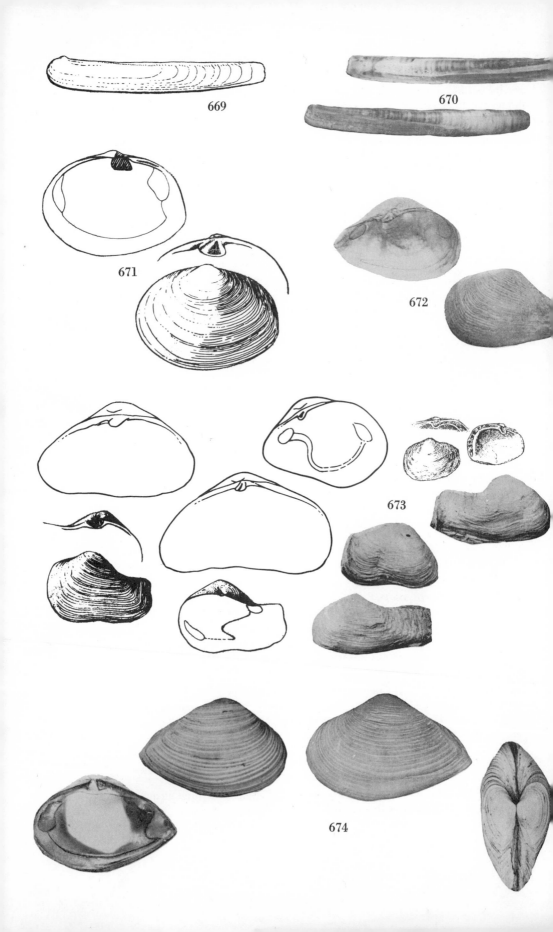

669

670

671

672

673

674

Suborder **MYINA**

Ligament external or with an internal resilium; hinge with small cardinal teeth, usually not more than one per valve.

Superfamily **MYACEA**

Ovate to elongate, valves subequal in size, porcelaneous to chalky, with a thin periostracum; hinge margin mostly without teeth, ligament mainly internal; valve margins smooth.

Family **MYIDAE**

Shells gaping at the ends, with a horizontally projecting spoon-shaped chondrophore or resilifer in one valve only. Pallial sinus mostly well developed. Three Panamic genera.

Genus **CRYPTOMYA** CONRAD, 1849

Rounded-quadrate, gaping at ends; resembling the northern group *Mya* in shape and chalky texture but lacking pallial sinus.

671. Cryptomya californica (Conrad, 1837) (Synonyms: **Mya tenuis** Philippi, 1887 [not Schroeter, 1802]; **C. magna** Dall, 1921). The smooth white shell is ovate-quadrate and looks like a short *Mya*, except for the pallial line, which is entire or has only the faintest hint of a sinus. Length, 25 mm; height, 18 mm. Alaskan Gulf to northern Peru, burrowing shallowly in sand, often in association with burrowing shrimp.

Genus **PLATYODON** CONRAD, 1837

Rounded-quadrate, gaping, inflated anteriorly; hinge as in *Mya*, with a chondrophore in one valve; sculpture fine, concentric. No species of *Platyodon, s. s.*, occur in the Panamic province.

Subgenus **AUSTROPLATYODON** OLSSON, 1961

Small shells, resembling *Platyodon* in outline.

672. Platyodon (Austroplatyodon) australis Olsson, 1961. The genus, which otherwise is confined in the Pacific to the California to Washington coast, is represented in the Panamic area by this one small form, with a pear-shaped, finely sculptured, somewhat translucent shell. Length, 5 mm; height, 3 mm. Ecuador.

Genus **SPHENIA** TURTON, 1822
(**TYLERIA** H. & A. ADAMS, 1854)

Small shells with the right valve edentulous and a small, shallow chondrophore in the left. The habit of nestling in cavities produces variations in shape. In general, Sphenias are white, with a broad, gaping posterior elongation and a shallow but wide pallial sinus.

673. Sphenia fragilis (H. & A. Adams, 1854) (Synonyms: **S. fragilis** Carpenter, 1857; **S. pacificensis** De Folin, 1867). The shell described by the Adams brothers as *Tyleria fragilis* proves, upon comparison of the type specimens in the British Museum, to be the same form that Carpenter described under the same specific name two years later. The Adams's shell was even more distorted than most Sphenias and seemed to have a series of compartments inside, formed by irregularities of growth. When not distorted the shell is somewhat pear-shaped. It is lustrous white under a yellow-green periostracum. Length, 8 mm; height,

4.5 mm; diameter, 4 mm. California, south through the Gulf of California to northern Peru; nestling in such cavities as worm burrows in other shells.

Family CORBULIDAE

Small, rather asymmetrical shells, mostly with concentric sculpture and a drawn-out posterior margin. The hinge has one principal cardinal tooth in the right valve that fits into a socket in the left, and the valves close tightly, with no gape. The pallial line is entire or slightly sinuate. Specimens found intertidally cling by a byssus under rocks or among gravel. A few groups have become adapted for existence in brackish to fresh water.

Genus CORBULA BRUGUIÈRE, 1797
(ALOIDIS MEGERLE VON MÜHLFELD, 1811)

The shells are sturdy and not only inequilateral but somewhat inequivalve, the left valve being the smaller, carrying the resilifer; there may be a blunt cardinal tooth in the right valve. Pallial sinus small to wanting.

The subgenera here assigned to *Corbula* all comprise marine forms, with the exception of *Panamicorbula*, which occurs in the brackish-water mud of mangrove swamps. Subgenus *Corbula, s. s.*, is not represented in the Panamic province.

1. Valves unlike in size, shape, and sculpture.....................*Varicorbula*
   Valves similar in size, shape, and sculpture..........................2
2. Posterior end pointed..3
   Posterior end truncate, not pointed................................5
3. Ventral margin striate.....................................*Serracorbula*
   Ventral margin smooth...4
4. Concentric sculpture of fine threads.......................*Caryocorbula*
   Concentric sculpture coarse, undulating.....................*Hexacorbula*
5. Posterior slope not set off by any angulation...............*Panamicorbula*
   Posterior slope set off by a keel...................................6
6. Shell thick, solid..*Juliacorbula*
   Shell thin, fragile ....*Tenuicorbula*

Subgenus CARYOCORBULA GARDNER, 1926

Valves slightly unequal in size, the right larger; hinge with a cardinal tooth in right valve, fitting into a socket in the left, ligament entirely internal, attached to a projecting, somewhat oblique chondrophore in the left valve and to two scars on a resilifer pit in the right; lateral teeth not present. Pallial line entire or with a very small sinus.

674. Corbula (Caryocorbula) amethystina Olsson, 1961. Large, solid, resembling *C. ovulata* but more rectangular, posterior end shorter, not twisted. Color varying from pink to violet or rose-purple. Length, 27 mm; height, 18 mm; diameter, 14 mm. Panama to Ecuador.

675. Corbula (Caryocorbula) luteola Carpenter, 1864. This southern Californian species has been reported by some authors as far south as La Paz. It is similar to *C. biradiata* but smaller and duller in color. Length, about 9 mm.

676. Corbula (Caryocorbula) marmorata Hinds, 1843. The best guide for recognition of this shell is the purplish-red spot under the beaks, combined with strong concentric and fine radial sculpture. Length, 7 mm; height, 5 mm; diameter, 3.5 mm. Sonora, Mexico, to Panama, dredged in depths of 13 to 27 m.

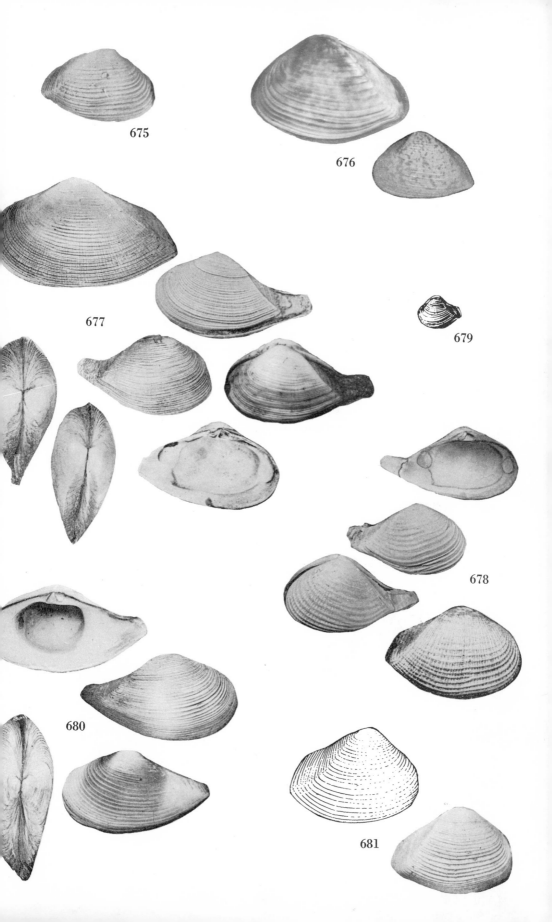

677. Corbula (Caryocorbula) nasuta Sowerby, 1833 (Synonyms: **C. fragilis** Hinds, 1843; **C. pustulosa** Carpenter, 1857). The spoutlike posterior end is a good characteristic of this species and also the fine pustular radial markings on the early part of the shell. The shell is whitish, thick, about 13 mm long, 7 mm high, and 6.5 mm in diameter. Magdalena Bay, Baja California, throughout the Gulf of California and south to Peru, mainly offshore in depths of 7 to 27 m.

678. Corbula (Caryocorbula) nuciformis Sowerby, 1833. As the name suggests, the shell is short, rounded, of nutlike shape, with a short, squarish posterior end. Length, 10 mm; height, 9.5 mm; diameter, 8 mm. Concepcion Bay, Gulf of California, to Ecuador, mostly offshore in depths of 11 to 90 m.

679. Corbula (Caryocorbula) obesa Hinds, 1843. The original figure shows a globose shell with irregular ribbing, about 9 mm long and 6 mm high. Its identity apparently must be judged on the basis of the description and figure, for the type material has not yet been located; it seems not to have been placed in the British Museum collection. Possibly this is the same form later named *C. porcella*. The range of both was reported by Dall as southern California to Panama.

680. Corbula (Caryocorbula) ovulata Sowerby, 1833. One of the largest West American Corbulas, this is fairly common in collections although usually represented by beach-worn material. The purplish-pink color of the umbones and interior, edged with white, and the oval shape are characteristic. A large shell may measure: length, 24 mm; height, 14 mm; diameter, 11.5 mm. The southern part of the Gulf of California to Peru, offshore in depths of 2 to 55 m.

681. Corbula (Caryocorbula) porcella Dall, 1916. An inflated, rather trapezoidal shell, ashy white in color, this may prove to be the same as *C. obesa* when specimens can be compared with Hinds's type. Length, 7.8 mm; height, 5 mm; diameter, 4.4 mm. Southern California to Panama, offshore in depths of 64 to 100 m.

682. Corbula (Caryocorbula) ventricosa Adams & Reeve, 1850. Adams and Reeve described the shell as "a very dull simple species, peculiar in form." It is one of the least common in collections. A large specimen reported by Hertlein and Strong (1950) measured: length, 13.6 mm; height, 9.8 mm; diameter, 8.2 mm. The Gulf of California to Panama, 51 to 90 m.

Subgenus **HEXACORBULA** OLSSON, 1932

Lunule and escutcheon not evident; sculpture of coarse undulations.

683. Corbula (Hexacorbula) esmeralda Olsson, 1961. The large shell has wavelike or undulating concentric sculpture, the ribs well spaced, finer growth lines showing in the interspaces. A depressed ray extends from the umbones to the ventral margin along the medial part. Length, 21 mm; height, 12 mm; diameter, 10 mm. Ecuador.

Subgenus **JULIACORBULA** OLSSON & HARBISON, 1953

With the posterior area well defined, set off by a ridge and often differently sculptured from the rest of the shell, which may have coarse, regular concentric riblets.

684. Corbula (Juliacorbula) bicarinata Sowerby, 1833 (Synonym: **C. alba** Philippi, 1846). The posterior end is abruptly truncate, with one sharp angulation running down from the umbo and a weaker one between it and the dorsal margin. The color is whitish, the sculpture of fine concentric threads. Length, 12 mm;

682

683

684

685

686

687

688

height, 8.5 mm; diameter, 7 mm. The Gulf of California to Ecuador, intertidally and to depths of 26 m.

685. Corbula (Juliacorbula) biradiata Sowerby, 1833 (Synonyms: **C. rubra** C. B. Adams, 1852; **C. polychroma** Gould & Carpenter, 1857). The two radiating white rays and the bright red color of the interior provide easy means of recognition. Length, 13 mm; height, 9 mm; diameter, 6 mm. Guaymas, Mexico, to Guayaquil, Ecuador, intertidally and to depths of 24 m.

686. Corbula (Juliacorbula) elenensis (Olsson, 1961). The ivory white shell resembles *C. biradiata* but is more convex and has more regular sculpture; it lacks the colored rays along the sides of the umbones, although there may be a white border on either side. Length, 17 mm; height, 12 mm; diameter, 8.5 mm. Ecuador and Peru. A similar species in the Atlantic is *C. (J.) knoxiana* (C. B. Adams, 1852).

687. Corbula (Juliacorbula) ira Dall, 1908. First illustrated by Olsson, 1961, this proves to be close to *C. bicarinata*, from which it differs by being longer, with a more elongate posterior end, and by having coarser sculpture. Length, 11 mm; height, 8 mm. The type locality is in deep water, 330 m, Panama Bay.

Subgenus PANAMICORBULA PILSBRY, 1932

The rounded shells show only a faint indication of a keel defining the posterior slope. They are inflated, thin, white, with a deep pit for the resilium in the right valve, a chondrophore in the left. The pallial sinus is small or wanting. The habitat is in mangrove swamps or near the mouths of streams, in mud.

688. Corbula (Panamicorbula) cylindrica (Morrison, 1946). Smaller and thinner than *C. inflata*, this has relatively low umbones and the beaks more anterior, so that the posterior end is longer and the shell appears to be cylindrical. Length, 22 mm; height, 16 mm; diameter, 16 mm. Panama to Ecuador.

689. Corbula (Panamicorbula) inflata (C. B. Adams, 1852) (Synonyms: **Potamomya aequalis** and **trigonalis** C. B. Adams, 1852; **? C. macdonaldi** Dall, 1912). The large white shell is moderately heavy, umbones high and prominent, the anterior end a little longer than the posterior. There is a thin yellow periostracum. Length of a large specimen, 27 mm; height, 22 mm; diameter, 19 mm. Mazatlán to northern Peru. This is the type species of *Panamicorbula*.

Subgenus SERRACORBULA OLSSON, 1961

The ventral margin is evenly serrate or striate, the edge thickened below the pallial line; adductor scars large and distinct.

690. Corbula (Serracorbula) tumaca (Olsson, 1961). The shell is brown, medium-sized, its surface with closely set concentric riblets. The crenulate ventral margin is distinctive. On tidal mud flats. Length, 12 mm; height, 7 mm. Panama to Peru.

Subgenus TENUICORBULA OLSSON, 1932

These are thin, white, strictly marine shells, trapezoidal in outline, both valves about the same size, the posterior area set off by an angle.

691. Corbula (Tenuicorbula) tenuis Sowerby, 1833 (Synonym: **C. glypta** Li, 1930). Here the posterior angulation is strong enough to be called a keel. The sculpture of the posterior slope is coarser than that of the rest of the shell. Length,

24 mm; height, 13 mm; diameter, 10.5 mm. A rare form, this has not been reported outside the Panama area.

Subgenus VARICORBULA GRANT & GALE, 1931

Left valve much smaller than right, umbones high and incurved, right valve inflated and with the posterior end somewhat set off by an angle.

692. Corbula (Varicorbula) speciosa Reeve, 1843 (Synonym: **C. radiata** Sowerby, 1833 [not Deshayes, 1824]). The right valve is much larger than the left. Both are whitish, conspicuously rayed with red. Length, 19 mm; height, 15 mm; diameter, 12.5 mm. Santa Inez Bay, Gulf of California, to Panama Bay.

Family SPHENIOPSIDAE

Small ovate-triangular shells with two hinge teeth in the right valve and a socket for the resilifer in the left valve; pallial sinus rounded.

Genus GRIPPINA DALL, 1912

Ovate, finely concentrically sculptured, with radial ridges bounding the lunule and escutcheon.

693. Grippina berryana Keen, new species. Small, white, trigonal, anterior margin rounded, posterior end bluntly pointed; hinge with a slotlike ligamental socket in the left valve, two strong cardinal teeth in the right, with a groove in front of them to receive the shelflike edge of the left valve; pallial sinus indistinct, apparently broad and shallow; shell surface with 25 or more low to somewhat undulating concentric ribs, ending at a radial ridge on the posterior slope. Length of holotype, 2.5 mm; height, 1.9 mm. Length of largest specimen seen, 2.5 mm. Type locality, northwest side of Bahía Salinas, Isla Carmen, Gulf of California, in 5 to 9 m. In outline the shell resembles a Tertiary eastern American and European genus, *Spheniopsis* Sandberger, 1863, but the hinge and sculpture show affinities to the type species of *Grippina*, *G. californica* Dall, 1912, which has a limited range in southern California and is a more quadrate form, not pointed posteriorly. At least three lots of *G. berryana* have been taken, ranging from the central part of the Gulf of California to Isla Espiritu Santo, near La Paz, Baja California, in depths to 90 m.

Superfamily GASTROCHAENACEA

Burrowing forms, the shells lying free within a linear cavity; hinge edentulous; the valves broadly gaping.

Family GASTROCHAENIDAE

The shells might at first glance be mistaken for Limas, but there are no ears, no radial sculpture. The two valves gape widely on the anterior ventral part and are held together by a small exterior ligament that rests on a platform. The burrowing habit protects the animal. Burrows are in soft rock, other shells, or coral heads, and extracting the shells unbroken is not easy.

Genus GASTROCHAENA SPENGLER, 1783
(CHAENA PHILIPSSON, 1788; ROCELLARIA BLAINVILLE, 1829)

The anterior end is narrowed, the posterior end flaring; some species have small shelly plates for the support of the muscles, extending into the shell cavity near the hinge.

694. Gastrochaena ovata Sowerby, 1834 (Synonyms: **?** **G. brevis** Sowerby, 1844; **? G. denticulata** Deshayes, 1857). The sculpture consists of concentric striae that follow the shape of the margin. Length, 10 mm; height, 4.8 mm. A large specimen measures 22 mm by 15 mm. The range is from San Diego, California, to Ecuador.

695. Gastrochaena rugulosa Sowerby, 1834. (Synonym: **? G. ecuadoriensis** Olsson, 1961). More slender than *G. ovata*, this has somewhat stronger sculpture on the posterior part. Galápagos Islands.

696. Gastrochaena truncata Sowerby, 1834. As the specific name suggests, the posterior end of the shell is squared off. The surface is nearly smooth except for a slight angulation. Length, 23 mm; height, 12 mm; diameter, 12 mm. Mazatlán, Mexico, to Panama, boring in large shells.

Superfamily HIATELLACEA

Quadrate to trapezoidal, valves slightly to widely gaping; hinge with one or two weak teeth; ligament on a nymph; pallial sinus mostly well developed.

Family HIATELLIDAE
(SAXICAVIDAE)

The members of this family, which is diversified in size, have the common characteristic of gaping valves, reflecting a boring or nestling habit, and a hinge with two small peglike cardinal teeth. Two Panamic genera.

Genus HIATELLA BOSC, 1801, *ex* DAUDIN, MS
(SAXICAVA FLEURIAU DE BELLEVUE, 1802)

The generic name literally means "little gaper," but in most specimens the gape is less apparent than the irregularity of outline. These are nestlers that float into some crevice as free-swimming larvae and then grow to fit the contours of the cavity, and no two shells are the same shape. Hence, there is difficulty in deciding whether only one worldwide species is present or whether several separate species can be isolated.

697. Hiatella arctica (Linnaeus, 1767) (Synonyms: **Saxicava acuta** and **initialis** De Folin, 1867). In uncrowded situations where it can develop to the full, one of these little shells is quadrangular in outline, the anterior end narrower, covered by a thin, yellowish periostracum that is raised into spines along a ridge from the umbones to the posteroventral margin. Old shells may have the teeth worn down to faint nubs, or the teeth may even be obsolete. The type locality of this species is the Arctic area of Norway, and it has been reported from the entire Arctic ranging southward to the tropics, on the west coast as far as Panama, intertidally and to depths of 390 m. Average length, about 20 mm. Part of the type material of *H. initialis* (De Folin) is in the British Museum; there is variation within the lots covering much the same range as in *H. arctica*.

698. Hiatella solida (Sowerby, 1834) (Synonyms: **Saxicava purpurascens** and **tenuis** Sowerby, 1834). Somewhat rectangular shells, proportionately longer than *H. arctica*, with a strong umbonal ridge setting off the posterior slope. The color is yellowish to brown or even grayish purple. Length, 28 mm; height, 20 mm. Nestling in crevices, Panama to Peru. The type was figured and discussed by Dell in 1964.

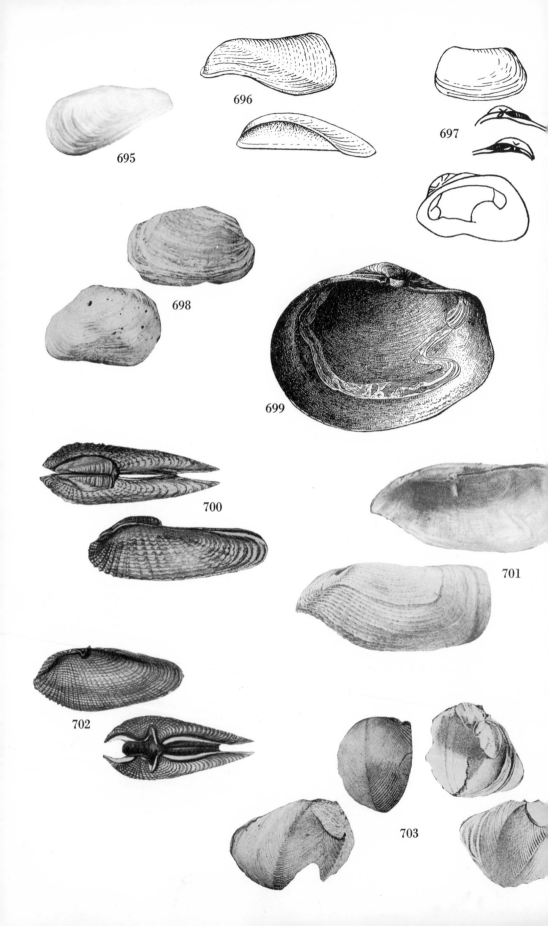

695

696

697

698

699

700

701

702

703

Genus **PANOPEA** Ménard, 1807
(**PANOPE** of authors, an unjustified emendation)

The Panopeas reach their greatest abundance and largest shell size on the north-west coast of North America. The valves are medium-sized to large, quadrate, gaping at both ends, the beaks subcentral. The ligament is large, on a nymph, and the pallial sinus is wide. In either valve there is one small cardinal tooth. Long siphons enable the animals to live some distance below the surface in muddy sand, a situation from which few enemies are able to dislodge them. Even man, the prime predator, finds the task no easy one; and he who does not know how to deal with shifting sand digs a big hole before reaching the clam. The clams themselves cannot dig to escape but can only contract as much as possible into the shell, though this is much too small to accommodate the large siphons. The common name gweduc, applied to the West American species of this genus and often mis-spelled as geoduck, is from a Nisqualli Indian word meaning, according to the Merriam-Webster dictionary, "dig deep."

699. Panopea globosa Dall, 1898. It seems strange to find an otherwise tem-perate-climate form so far south as the head of the Gulf of California, but there are similar forms elsewhere; for example, a rare *Panopea* that has been taken a few times along the Florida coast. *P. globosa* is not common. Mr. John Fitch found it at San Felipe in two to three feet of sandy mud. Head of the Gulf of California to off San Marcos Island in 60 m. Length of the type specimen, 160 mm (6 inches) ; height, 120 mm; diameter, 80 mm.

Suborder **PHOLADINA**

Shells adapted for burrowing in hard substrates such as peat, wood, or soft rock; soft parts greatly modified, foot truncate to circular, developed as a suction disc, mantle closed.

Superfamily **PHOLADACEA**

Shells inequilateral, the dorsal margin reflected anterior to beak, forming an at-tachment area for the anterior adductor muscle; hinge teeth lacking, but a small chondrophore and an internal ligament are usually present. A rocking motion of the valves when the animal is boring is facilitated by a pivotal point or ventral condyle, which develops opposite the umbones at the junction of the anterior slope and the main disk of the shell.

Family **PHOLADIDAE**

The pholads or piddocks have solved the problem of finding shelter by constructing their own, burrowing into wood, rock, or other shells. It was long supposed that they must use free acid secreted by special glands to do this, but experiments and laboratory observations have since demonstrated that the boring is done by the friction of the shell against the sides of the burrow and, in part, by keeping the cavity clean of algae so that the sea water can help in loosening rock-grains or wood cells. Slow rotation of the shell and the flushing action of ciliary currents gradually enlarge and deepen the burrow. Special structures have developed in the shell, such as a spoon-shaped calcareous apophysis, which curves into the shell from below the beak. This seems to serve as an added place of attachment for the necessarily powerful muscles used in rotation. Other special structures are the supplementary plates—protoplax, mesoplax, and metaplax—that develop along

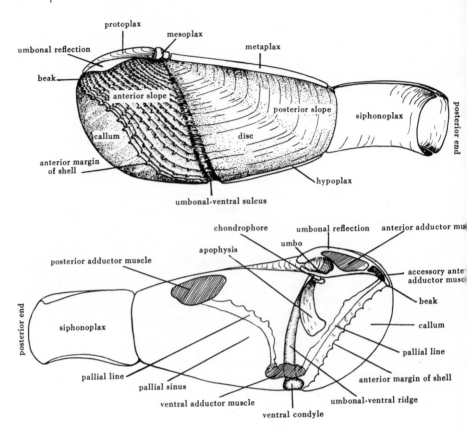

dorsal and ventral surfaces to fill in the wide gape between the valves. Lastly, there is the callum, a solid sheet of shell material that fills in the anterior gape after the shell has reached adult size and has ceased active growth.

Linnaeus recognized only one genus of this group, *Pholas*, but later workers have added many more, as the significance of differences in form and habitat became apparent. The latest revision of the family—at least of that part living in the Western Hemisphere—has been by Dr. Ruth Turner of the Museum of Comparative Zoology, Harvard. Her work forms the basis for the present summary, since it is by far the most comprehensive analysis anyone has undertaken. The key, which is to the genera within the four subfamilies, is adapted from one by Turner (1955).

1. Adult shell without a callum.. 2
   Adult shell with a callum... 5
2. Shell without apophyses....................................*Xylophaga*
   Shell with apophyses.. 3
3. Shell with only one dorsal plate............................*Barnea*
   Shell with more than one dorsal plate............................ 4
4. Shell with a largely horny protoplax, a mesoplax, but no metaplax; umbonal reflection not septate..............................*Cyrtopleura*
   Shell with a calcareous protoplax, mesoplax, and metaplax; umbonal reflection septate ....*Pholas*

5. Shell without apophyses ....*Jouannetia*
   Shell with apophyses.. 6
6. Dorsal plate one, a mesoplax...........................*Pholadidea*
   Dorsal plates two, a mesoplax and a metaplax...................... 7
7. Valves divided into three distinct areas; posterior area with overlapping
   horny plates; shell more than an inch in length..............*Parapholas*
   Valves divided into two areas; posterior area without horny plates; shell
   under one inch in length.. 8
8. With umbonal reflection closely appressed; callum extending between
   beaks and on either side of mesoplax; boring in shell and rock...*Diplothyra*
   With funnel-shaped pit below umbonal reflection; callum not extending between
   beaks; boring in wood.......................................*Martesia*

Subfamily PHOLADINAE

Adult gaping anteriorly; hypoplax and siphonoplax plates wanting. Three Panamic genera.

Genus PHOLAS LINNAEUS, 1758

Shells narrowly pointed anteriorly, protoplax divided lengthwise in two parts.

700. Pholas chiloensis Molina, 1782 (Synonyms: **P. chiloensis parva** Sowerby, 1834; **P. laqueata** Sowerby, 1849; **P. retifer** Mörch, 1860; **P. dilecta** Pilsbry & Lowe, 1932). The large size (length to 125 mm, or about 5 inches), thin, white shell, and reflected umbo with its series of vertical septal buttresses readily mark this species. Although it has a wide range (from the head of the Gulf of California to Chiloé Island, Chile), this pholad is comparatively rare in collections, known mostly by beach valves. It is said to bore into soft stone on Chiloé Island, the type locality. A similar Atlantic species is *P. campechiensis* Gmelin, 1791.

Genus BARNEA RISSO, 1826

The single supplementary plate on the dorsal side has led authors to separate this genus from *Pholas*, which has three plates.

701. Barnea subtruncata (Sowerby, 1834) (Synonym: **B. pacifica** Stearns, 1871). Sowerby's name for this species was overlooked until recent years because his type specimen has been lost and his description was poor, but Turner has shown that his is the earliest name for the species, which is better known as *B. pacifica*. Length, about 60 mm; height, 25 mm. Specimens of this adaptable piddock may be found in mud, clay, peat, or even soft rock or wood, from southern Oregon to Chile.

Genus CYRTOPLEURA TRYON, 1862

This genus, though resembling *Pholas*, does not have the reflected umbonal margin, and the supplementary plates are partially horny. The best-known species is the east American *C. costata*, the "Angel-wing."

702. Cyrtopleura crucigera (Sowerby, 1834). Rare throughout its range, *C. crucigera* is restricted to areas of soft stone. Length, about 38 mm; height, 17 mm. Guaymas, Mexico, to Ecuador.

Subfamily JOUANNETIINAE

Valves beaked; adult shell closed by a callum.

Genus JOUANNETIA DESMOULINS, 1828

The lack of apophyses inside the shell, together with production of a callum, are distinctive features. Two Panamic subgenera.

Subgenus JOUANNETIA, *s. s.*

With an internal shelf for the attachment of the posterior adductor muscles; margins of the siphonoplax smooth.

703. Jouannetia (Jouannetia) duchassaingi Fischer, 1862. A large shell, so rare it is known by only a few valves, these mostly broken. When complete it would be globose, valves unequal, the right larger. Length, 43 mm; height, 28 mm. Panama to Ecuador.

Subgenus PHOLADOPSIS CONRAD, 1849
(TRIOMPHALIA SOWERBY, 1849)

Without internal shelf or myophore; siphonoplax with a serrate or pectinate margin.

704. Jouannetia (Pholadopsis) pectinata (Conrad, 1849) (Synonym: **Triomphalia pulcherrima** Sowerby, 1849). In outline the shell is pear-shaped, with the siphonoplax of the right valve larger than that of the left, finely scalloped along the edge. The callum of the left valve is larger than that of the right. Relatively large (length to 51 mm, including callum and siphonoplax). Little is known about the habits; one report cites soft stone at low water as the habitat. Baja California to Peru; specimens in collections are very few.

Subfamily MARTESIINAE

Valves beaked; protoplax wanting. Four Panamic genera (see key for the Pholadidae).

Genus MARTESIA SOWERBY, 1824

These small pholads have, like the teredos, taken to wood as a substance in which to bore. The various species have become widely distributed, perhaps because they can drift from place to place in the logs they invade. The anterior callum, the pear-shaped outline, and the presence of supplementary plates distinguish them from *Teredo*.

705. Martesia striata (Linnaeus, 1758) (Synonyms: **Hiata infelix** Zetek & McLean, 1936; **M. intercalata** of authors [not the *M. intercalata* of Carpenter, 1857, which is *Penitella conradi* (Valenciennes, 1846), a California species]). A variable form, *M. striata* occurs in the tropical and subtropical Atlantic and Pacific, from South Carolina to Australia and Japan. Boring into wood in these warm waters, specimens may mature in one month and may grow to a length of 35 mm in four months. They can thus be very destructive to wharf piling, like the teredos or true shipworms. On the Pacific coast the species has been reported from Baja California and Sonora, Mexico, to Peru.

706. Martesia fragilis Verrill and Bush, 1898 (Synonym: **Pholadidea minuscula** Dall, 1908). A widespread species that can be distinguished only by the shape and sculpture of the mesoplax, which is depressed, has sharply keeled edges, and is strongly sculptured. It seems to be a more pelagic species, found mostly in floating seeds or nuts, on the west coast from Sonora, Mexico, to Panama.

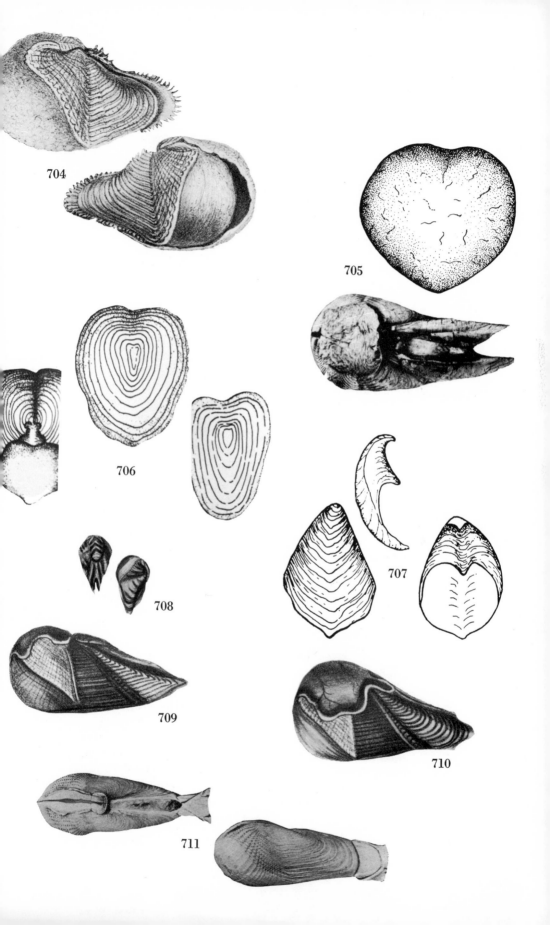

704

705

706

707

708

709

710

711

707. Martesia cuneiformis (Say, 1822) is primarily an Atlantic species, only once found at Balboa, Canal Zone, perhaps fortuitously. In this form the umbones are covered with callus, and the mesoplax has a central groove.

Genus DIPLOTHYRA TRYON, 1862

Although long considered a subgenus of *Martesia*, this group has come to be recognized as a separate genus on account of differences in the shape of the supplementary plates, in the umbonal reflection, in the ventral condyles, in a well-developed chondrophore and internal ligament, and in the habit of boring rock and shells rather than wood.

708. Diplothyra curta (Sowerby, 1834). Small shells, these may be found in shells and soft stone from the head of the Gulf of California to Ecuador. Length, about 10 mm. Specimens occur intertidally and have been dredged in 18 m of water.

Genus PARAPHOLAS CONRAD, 1848

Parapholas is distinguished from *Pholadidea* by having two dorsal supplementary plates and by the division of the valves into three well-marked regions.

709. Parapholas acuminata (Sowerby, 1834). It is a characteristic of *Parapholas* that a "chimney" forms over the posterior part of the shell, composed of grains of the bored rock that are agglutinated into a solid mass. Length, about 55 mm. Like *Pholadidea melanura*, this apparently occurs in abundance at certain offshore localities but is rare in collections. Baja California to Peru.

710. Parapholas calva (Sowerby, 1834) (Synonyms: **Pholas nana** Sowerby, 1834; **Parapholas bisulcata** Conrad, 1849). Relative to others of the genus, shells are small (about 45 mm long), differing from *P. acuminata* in being more rounded posteriorly, in having the plates of the posterior slope rounded, not angular, and by having a larger, lobed mesoplax. It has been reported from Guaymas, Mexico, to Ecuador, but is even rarer than *P. acuminata*.

Genus PHOLADIDEA TURTON, 1819

With one dorsal supplementary plate, the mesoplax. Unlike their relatives north of the border, which may be found in abundance in any soft sandstone or shale along shore, the tropical piddocks are rare, and apparently favor rock bottoms in several meters of water. Hence, although they may be locally abundant, they are seldom available to the collector. There are no species of *Pholas, s. s.*, in the Panamic province.

Subgenus HATASIA GRAY, 1851

Umbonal reflections closely appressed over the umbones, mesoplax with a basal portion, siphonoplax variable; a siphonal tube present.

711. Pholadidea (Hatasia) esmeraldensis (Olsson, 1961). Resembling *P. tubifera*, this is more slender, the posterior half of the valves longer and narrower, calcareous tube short. Length, 30 mm; height, 9.4 mm; diameter, 8.8 mm. Ecuador.

712. Pholadidea (Hatasia) melanura (Sowerby, 1834) (Synonym: **Penitella wilsonii** Conrad, 1849). The black periostracum and siphonoplax as well as the inflation of the posterior part of the shell are distinctive. Length is about 50 mm. During World War II, when new dredging was done to clear the western

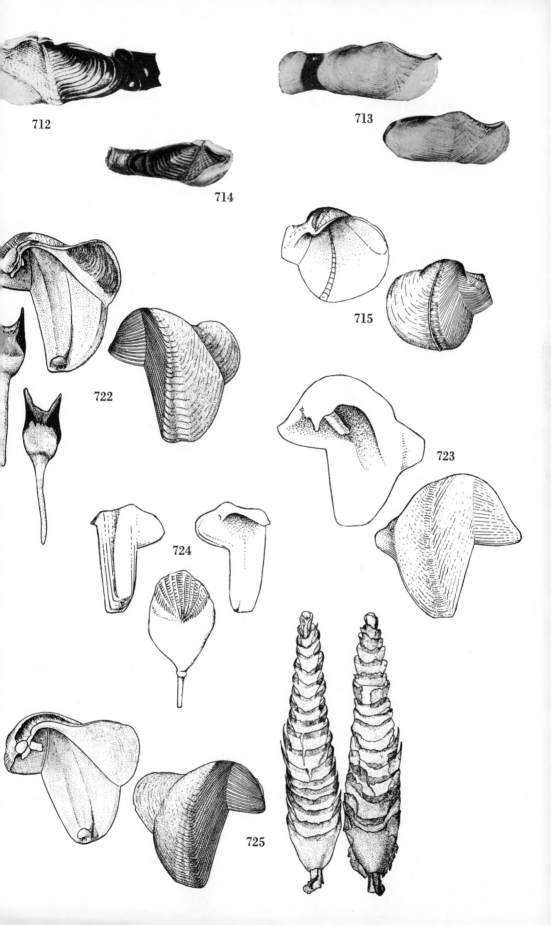

712

713

714

715

722

723

724

725

terminus of the Panama Canal, blocks of rock were dumped ashore that were riddled with the burrows of this form and of *Parapholas acuminata*. Unfortunately,
few collectors were there to reap the harvest, so the species remains a rare form
in collections. It has been recorded from Baja California to Ecuador, at widely
separated stations.

713. Pholadidea (Hatasia) quadra (Sowerby, 1834). Smaller and more
elongate than *P. melanura*, this is a thin and fragile shell with a horny brown
siphonoplax, swollen near the base and connected to a calcareous tube. Length,
including tube, 25 mm. Panama to Ecuador.

714. Pholadidea (Hatasia) tubifera (Sowerby, 1834). The all-white shell
and the conical tube at the end of the shell separate this from *P. melanura*. The
length is about 32 mm. Panama to Ecuador.

Subfamily XYLOPHAGINAE

Valves beaked and gaping anteriorly, without a callum.

Genus XYLOPHAGA TURTON, 1822

Resembling *Teredo* but without posterior gape; internal apophysis wanting.

715. Xylophaga mexicana Dall, 1908. The Xylophagas, like *Teredo*, burrow
in wood, except more shallowly. The type specimen of this species was taken from
a mud bottom off Acapulco, at a depth of 258 m, in submerged wood. The species
may prove identical with *X. globosa* Sowerby, 1835, which has a type locality
in Chile.

In a report on the deepwater dredgings of the Danish research vessel *Galathea*,
Knudsen (1961) documented the occurrence of a number of species of this genus
in plant remains (not always solid wood) on the sea floor, well offshore in the
Gulf of Panama:

716. Xylophaga aurita Knudsen, 1961. Gulf of Panama, 915 m.

717. X. concava Knudsen, 1961. Gulf of Panama, 3,260 to 3,660 m.

718. X. duplicata Knudsen, 1961. Gulf of Panama, 915 m.

719. X. obtusata Knudsen, 1961. Gulf of Panama, 915 m.

720. X. panamensis Knudsen, 1961. Gulf of Panama, 915 to 970 m.

721. X. turnerae Knudsen, 1961. Gulf of Panama, 915 m.

Family TEREDINIDAE

The teredos or shipworms are the most destructive of mollusks, and the damage
they have done to wooden structures—ships and wharf piling—over the centuries
is incalculable. The boring habit is not for the consumption of the wood (although
there is some evidence that commensal microorganisms in the digestive tract may
help in the digestion of a part of the fragments chipped off in burrow construction), but rather for protection. The animal grows rapidly for a few months, excavating its burrow, then settles down for a period of simple feeding and repro-
duction. Unlike the Pholadidae, in which the shell always encloses the soft parts,
the Teredinidae have shells that are reduced to efficient rasping tools at the anterior end of the long, soft body, which is encased in the burrow. The siphons, at
the posterior end, are tipped with calcareous structures called pallets. These may

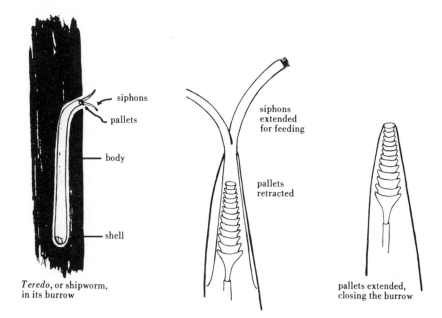

siphons

pallets

body

shell

Teredo, or shipworm,
in its burrow

siphons
extended
for feeding

pallets
retracted

pallets extended,
closing the burrow

be used to close the burrow or may be drawn back so that the tubular siphons can circulate seawater, which carries in suspension the microscopic particles of organic material that are the principal food of the mollusk. The text figure shows these structures diagrammatically.

Although the shipworms have been a recurrent problem to men for centuries, infiltrating any wooden structure placed in the sea, especially in the tropics and in the temperate zones, it has only been within the past few decades that serious study of them began. Sudden and widespread collapse of wharves in San Francisco Bay in the early 1900's launched the first major study on the West Coast. Laboratory and fieldwork have supplied much information, but much more must be done in devising methods of control, for the obvious precautions of pretreating pilings with toxic substances or of jacketing them only postpone but do not completely prevent destruction.

As knowledge increases, the classification of the group becomes more finely drawn. Early workers were content with a single genus, *Teredo*. Later, when a striking difference in the shape of the pallets was noted, a second genus, *Bankia*, was recognized. In *Teredo* the pallets are paddle-shaped; in *Bankia* they are segmented, looking rather like the rattles on a rattlesnake's tail. Although now two subfamilies and a number of genera and subgenera are accepted by the specialists, the pallet distinction remains basic. The shells themselves are surprisingly small in proportion to the length of the burrow. It may be added parenthetically that burrows do not intersect; the turning and twisting to avoid an adjacent burrow in a thickly infested piece of wood of course only riddles it faster.

A recent monograph on the Teredinidae by Turner (1966) lists all described species. It has illustrations of shells and pallets and notes on morphology, geographic distribution, and zoological classification. Having studied the soft parts, Turner recognizes as genera a number of groups that previously were considered subgenera. On the basis of shells alone even the genera cannot always be identified. The soft parts, however, show consistent distinctions. There are two Panamic subfamilies. Some 11 species not now recorded from the Panamic province but likely to occur there in time are listed following the descriptions of known species.

The review of this family has been prepared in consultation with Dr. Ruth Turner. Preliminary organization of data on Panamic teredinids was by Dr. Eugene V. Coan, whose assistance is also gratefully acknowledged.

Subfamily TEREDININAE

Pallets not segmented. Some forms brood their young. Three Panamic genera.

Genus TEREDO LINNAEUS, 1758

Pallets with a thin periostracum often extending beyond the calcareous portion. Siphons long and separate.

Subgenus TEREDO, s. s.

Pallets white, tipped with a yellow to brown periostracum.

722. Teredo (Teredo) bartschi Clapp, 1923. A Caribbean species with its type locality at Port Tampa, Florida, this has been reported at La Paz, Baja California (Turner, *in litt.*). The periostracum is thin, golden brown, and the shell measures: length, 4 mm; height, 4 mm.

Genus PSILOTEREDO BARTSCH, 1922

Pallets broad, almost entirely calcareous, with a short stalk. Blade concave on inner surface, convex and with a depression on the outer surface. Siphons united except at tip.

723. Psiloteredo miraflora (Bartsch, 1922) (Synonym: **T. healdi** Bartsch, 1931). The type locality is Mira Flores Lake, in the Panama Canal Zone. It is the only species of shipworm yet reported able to live in fresh water. Also recorded in the Caribbean. Length, 11.6 mm; height, 11.3 mm; diameter, 12 mm.

Genus UPEROTUS GUETTARD, 1770

Distal portion of blade with pronounced radiating ribs. Anterior and posterior shell slopes much reduced. Siphons united to the tip.

724. Uperotus panamensis (Bartsch, 1922). Type material of this species was dredged in Panama Bay in 93 m of water. Length, 3.3 mm; height, 4.6 mm; diameter, 5.5 mm.

Subfamily BANKIINAE

Pallets segmental, with a cone-in-cone structure. Larvae planktonic. Two Panamic genera.

Genus BANKIA GRAY, 1842

Pallets elongate, with numerous cones, which have calcareous bases, covered with a border of periostracum. This genus has been divided into a number of subgenera on the basis of pallet morphology. Turner (1966) suggests that these are of little significance because of transitional species. Three Panamic subgenera; subgenus *Bankia, s. s.*, is not represented in the Panamic province.

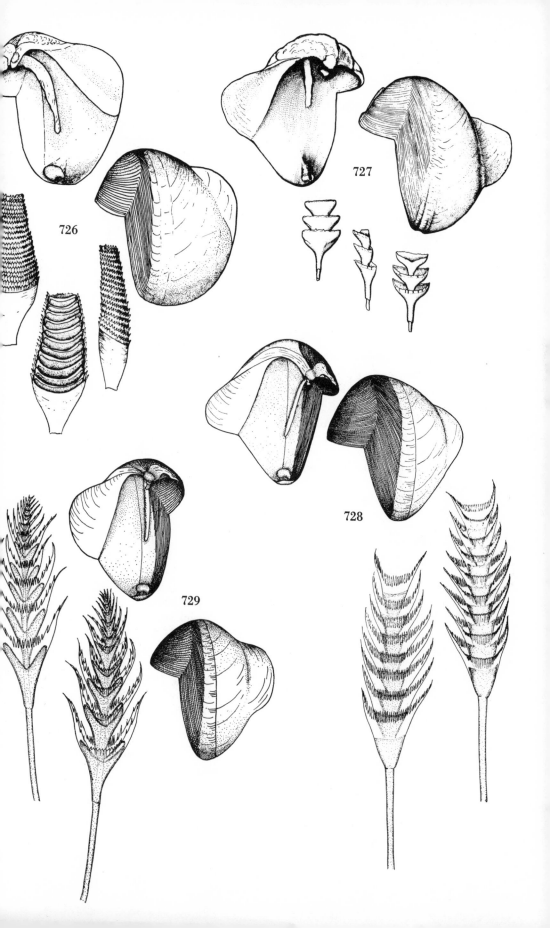

726

727

728

729

Subgenus **BANKIELLA** BARTSCH, 1921

Margins of cones smooth; lateral awns long and thin.

725. Bankia (Bankiella) gouldi (Bartsch, 1908) (Synonym: **B. mexicana** Bartsch, 1921). A species of the mangrove swamps, especially of west Mexico. Also in the Caribbean. Length of shell, 7 mm; height, 6.5 mm. Sinaloa, Mexico.

Subgenus **NEOBANKIA** BARTSCH, 1921

Margins of cones wide, even, serrate on both faces; no lateral awns.

726. Bankia (Neobankia) orcutti Bartsch, 1923. Small, white, tinged with pink. The pallets are finely fringed with periostracum. From Bocochibampo Bay, Sonora, Mexico, near Guaymas. Also in Asia.

727. Bankia (Neobankia) zeteki Bartsch, 1921. Cones more widely spaced than in the preceding. This is a species of the Canal Zone, the type locality being in the wood of the canal locks at Balboa. It is also reported at Puerto Armuelles, Panama. Length, 10 mm; height, 9.5 mm.

Subgenus **PLUMULELLA** CLENCH & TURNER, 1946

Margins of cones serrate on both faces, produced into lateral awns.

728. Bankia (Plumulella) cieba Clench & Turner, 1946. The pallets have elongate periostracal fringes and a broad tip. Diameter, about 3.5 mm. Balboa, Canal Zone.

729. Bankia (Plumulella) fimbriatula Moll & Roch, 1931 (Synonyms: **Teredo fimbriata** Jeffreys, 1860 [not DeFrance, 1828]; **B. canalis** Bartsch, 1944). The fringes are feather-like, as in the above, with the pallets becoming smaller toward the tip. This species has been taken in the Panama Canal, both on the Pacific and the Atlantic sides. It also occurs in the Atlantic. The bluish-white shell measures 6.2 mm long, 6.3 mm high, and 6.2 mm in diameter. Balboa, Panama Bay.

Genus **NAUSITORA** WRIGHT, 1864

Pallets elongate, with closely packed cones. Distal portion of blade with a papillose, calcareous covering. Valves relatively large.

730. Nausitora dryas (Dall, 1909) (Synonym: **Bankia jamesi** Bartsch, 1941). This is one of the largest members of the genus in Panamic waters. The pallets are club-shaped and tightly fused. The shell is pale pink stained with rust. Length, 12 mm; height, 12.2 mm; diameter, 13.6 mm. Nayarit, Mexico, to northern Peru.

731. Nausitora excolpa (Bartsch, 1922) (= ? **Teredo fusticulus** Jeffreys, 1860). The pallets of this form are elongate and feather-like. The type specimens were taken in cedar wood in the Gulf of California. Length, 3.8 mm; height, 4 mm; diameter, 4.5 mm. Mazatlán, Mexico, to Ecuador. If the suggested synonymy is correct, then the species also occurs in the Atlantic.

Other Teredinids Likely to Occur

Because of the facility of transport for the teredinids, a number of species reported in adjacent regions may eventually be detected in the Panamic province. The most likely candidates in the Teredininae (according to Turner, *in litt.*) are:

Teredo (Teredo) furcifera Von Martens in Semon, 1894. Caribbean.

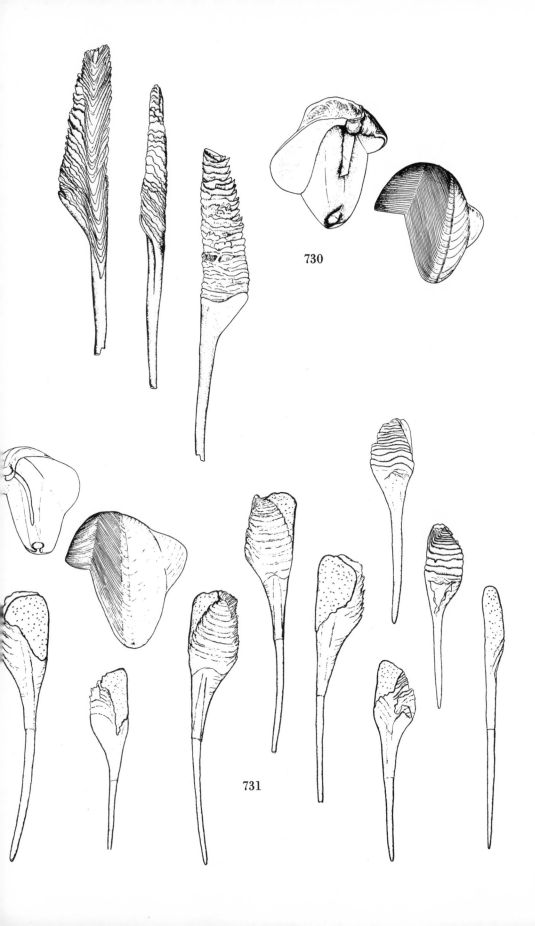

730

731

T. (T.) portoricensis Clapp, 1924. Puerto Rico.

Teredo (Zopoteredo) clappi Bartsch, 1923. Caribbean.

T. (Z.) fulleri Clapp, 1924. Caribbean.

T. (Z.) johnsoni Clapp, 1924. Caribbean, including eastern Panama.

Lyrodus affinis (Deshayes, 1863). Caribbean.

L. pedicellatus (Quatrefages, 1849). San Diego, California.

Teredothyra dominicensis (Bartsch, 1921). Eastern Panama.

The most likely candidates in the Bankiinae (again according to Turner) are:

Bankia (Neobankia) destructa Clench & Turner, 1946. Honduras and eastern Panama.

Bankia (Plumulella) fosteri Clench & Turner, 1946. Caribbean.

Nototeredo knoxi (Bartsch, 1917). Caribbean.

Subclass ANOMALODESMATA
(ANOMALODESMACEA)

Shells small to medium-sized, usually thin; muscle scars two, approximately equal in size; mantle lobes fused ventrally.

Order PHOLADOMYOIDA
(EUDESMODONTIDA)

Hinge margin thickened or enrolled, edentulous or with a ridgelike tooth in one valve, a socket in the other; lateral teeth absent; ligament behind beaks (wanting in some), often with a resilium that is supported or enclosed by a calcareous lithodesma.

These clams show rather a mixture of characteristics, having, on the one hand, shells with the nacreous inner lining that is the hallmark of the ancient stocks; yet having, on the other hand, the well-developed pallial sinus that elsewhere occurs only in the most advanced types, like the Veneridae and Tellinidae. Apparently here is a side branch among the bivalves, one that early took to a burrowing life—not, perhaps, an active burrowing as in the gaper and jackknife clams (Mactridae and Solenidae), but a more passive burrowing or a lying on one side below the surface of the sea floor. An inequivalve condition has resulted, and, in most forms, one valve is markedly flatter than the other. Although the hinge has no true teeth, various buttress and chondrophore arrangements strengthen weak edges and support the ligament.

Superfamily PANDORACEA

Medium-sized shells, the thin valves with or without periostracum. Soft parts with normal bivalve gills.

Family PANDORIDAE

Pallial sinus wanting; muscle scars joined by a faint line of minute scars.

Genus PANDORA BRUGUIÈRE, 1797

If odd shapes are any criterion, these are indeed Pandora's boxes. Some forms look like Turkish slippers done in pearl, others like hatchets, and still others like little fans. One wonders where the animal that formed the shells found space to live, for the right valve is flat, and the left valve only slightly arched. There is no hinge plate, but along the dorsal margin one to three radiating buttresses or laminae (called cardinal teeth, by courtesy, or, more accurately, crura) serve to protect the ligament and to act as a hinge. The key is to subgenera.

1. Lithodesma (calcified ligamental structure) wanting.............*Pandora*
   Lithodesma present 2
2. With teeth not joined by a deck.................................... 3
   With a bridge or deck between hinge teeth in left valve.............. 5
3. Left valve lacking a posterior lamina.......................*Heteroclidus*
   Left valve with a posterior lamina below the dorsal margin.............. 4
4. Right valve with three cardinal laminae......................*Clidiophora*
   Right valve with two cardinal laminae.........................*Pandorella*
5. Deck joining anterior tooth to anterior margin..................*Foveadens*
   Deck joining two radiating teeth..............................*Frenamya*

Subgenus PANDORA, *s. s.*

Smooth or with growth lines only; right valve with two crural or cardinal teeth, and there may be an obscure tooth in the left valve; lithodesma absent.

732. Pandora (Pandora) brevifrons Sowerby, 1835. An inequilateral shell somewhat more slender than *P. uncifera*, with a smoothly arched anterior margin. The posterior end is drawn out into a little spout, where the siphons protrude. Length, 25 mm; height, 10 mm. The type locality is Panama Bay, in 18 m, on a sandy bottom.

733. Pandora (Pandora) uncifera Pilsbry & Lowe, 1932. The distinctive hook-like curve along the anterior margin sets this apart. It is rather small (length, 12 mm), opaque to translucent white in color. The type locality is Acapulco, Mexico, in 37 m, but it has been taken from off Cape San Lucas and Guaymas, Gulf of California, to Costa Rica, in 13 to 27 m; shelly sand bottom.

Subgenus CLIDIOPHORA CARPENTER, 1864

Right valve with three crural laminae or teeth, posterior lamina separated from dorsal margin by a deep groove; left valve with three crura, the anteriormost nearly vertical, pointing toward anterior adductor scar.

734. Pandora (Clidiophora) arcuata Sowerby, 1835 (Synonyms: **P. claviculata** Carpenter, 1856; **C. cristata** Carpenter, 1864). This species is the type of the subgenus. Variations in shape account for the synonyms. The dorsal margin may be strongly concave or may be only slightly curved, and stoppages of growth with resumption at a different angle may result in serrations along the anterior margin, which led Carpenter to the separation of the form *P. cristata*. The commonest of the Panamic Pandoras, this has been variously misidentified in collections. The type specimen is illustrated here. The shell is white, lustrous, with slightly undulating concentric sculpture. Length up to 40 mm; height, about 22 mm. Scammon's Lagoon, Baja California, to northern Peru.

Subgenus FOVEADENS DALL, 1915

With three laminae or crura in either valve, the posteriormost two decked by a shelly bridge.

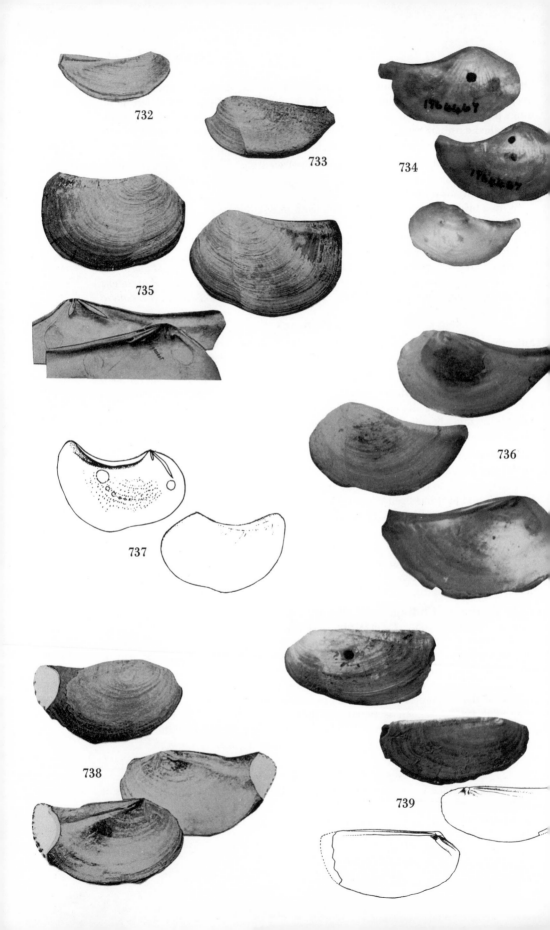

735. Pandora (Foveadens) panamensis Dall, 1915. The thin white shell has a single radiating riblet in the left valve close below the dorsal margin; except for faint undulations of the surface and growth lines, there is no other sculpture. Length, 18 mm; height, 11 mm. The type specimen was found, with other separated valves, on the beach at Panama; one worn right valve has since been dredged off El Salvador in 30 m. Olsson (1961) reports it as abundant in Panama, large specimens up to 26 mm in length.

Subgenus FRENAMYA IREDALE, 1930
(COELODON CARPENTER, 1864 [not AUDINET-SERVILLE, 1832])

Right valve with three, left valve with two crural laminae, the anterior left lamina tent-shaped; cavity of shell with radiating lines.

736. Pandora (Frenamya) radians Dall, 1915. A small white shell (length, 15 mm; height, 8 mm), this is the first American species of an otherwise Oriental subgenus. The center of the valve is marked by radiating lines. The type locality is Ballenas Lagoon, Baja California, in depths to 10 m. A specimen from western Colombia mentioned by Carpenter may extend the range southward. Olsson (1961) suggests synonymy with *P. radiata* Sowerby, but the type of that species seems instead to belong in *Pandorella*.

Subgenus HETEROCLIDUS DALL, 1903

Left valve with only one crural lamina, right with three.

737. Pandora (Heteroclidus) punctata Conrad, 1837. On the inside of the shiny white shell, tiny punctations like pinpricks mark the area of attachment of the soft parts. Length, 32 mm; height, 22 mm. British Columbia to the Gulf of California.

Subgenus PANDORELLA CONRAD, 1863
(KENNERLIA CARPENTER, 1864)

Resembling *Pandora, s. s.*, but with a lithodesma; right valve usually with radial markings exteriorly; hinge with two crural teeth, the anterior tooth not pointing toward the anterior adductor scar.

738. Pandora (Pandorella) cornuta C. B. Adams, 1852 (Synonym: **Clidiophora acutedentata** Carpenter, 1864). Authors have long been in doubt about this species, for it remained unillustrated until 1956. Olsson (1961) seems to have recognized what its correct placement should be, although it does not show the characteristic radial marking on the right valve. The shell is ovate, smooth, the anterior end short, rounded. The holotype is badly chipped posteriorly, and the supposed "horns" are purely accidental. Length, about 20 mm; height, about 12 mm. Panama.

739. Pandora (Pandorella) granulata Dall, 1915. Along the dorsal surface of the greenish shell are elevated threads that are delicately beaded. Length, 8.5 mm; height, 4 mm. Southern California to Guaymas, Mexico, on muddy bottoms, to depths of 33 m.

740. Pandora (Pandorella) radiata Sowerby, 1835 (Synonym: **Kennerlyia convexa** Dall, 1915). The shell is small, inequilateral, with the posterior end somewhat drawn out into a spout. The right valve is nearly flat, the left convex, especially in the central part. Radial rays show up clearly on the right valve, and

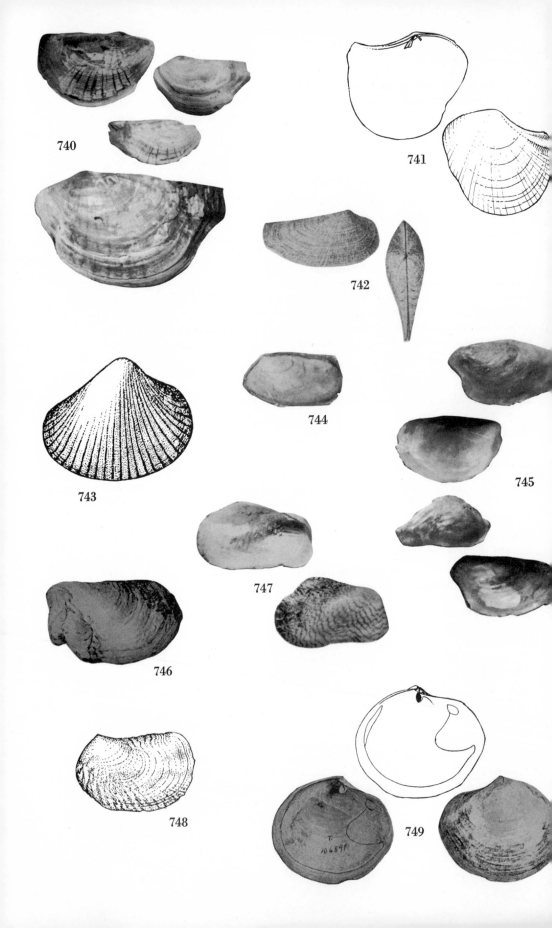

740

741

742

743

744

745

746

747

748

749

there is a thin brown periostracum. Length (holotype) 13 mm; height, 7 mm; length of Dall's holotype of *P. convexa*, 21 mm; height, 12 mm, diameter, 3.5 mm. Ballenas Lagoon, outer coast of Baja California, to Mazatlán, Mexico, in depths to 140 m.

741. Pandora (Pandorella) rhypis Pilsbry & Lowe, 1932. Fine radiating riblets like the spokes of a fan distinguish this rather large Pandora, the length of which is about 25 mm. The type locality is El Salvador, in 80 m.

Family LYONSIIDAE

The Lyonsias have rather thin shells, variable in size and shape, with a well-developed periostracum. The valves are more nearly equal than in some other families, very inequilateral, convex, and somewhat gaping. There is no hinge except for a small internal ligament. Interiorly, the shells are pearly, with a small pallial sinus. Two Panamic genera.

Genus LYONSIA TURTON, 1822

Elongate-ovate, inequilateral, posterior end longer and attenuate; slightly gaping posteriorly; sculpture of radial striations and raised threads.

742. Lyonsia gouldii Dall, 1915 (Synonym: **Osteodesma nitidum** Gould, 1853 [not Fabricius, 1798]). The slender shell has the pearly surface ornamented by fine raised radial lines. Length is about 16 mm. San Diego, California, to Acapulco, Mexico, mainly taken by dredging in 7 to 73 m.

743. L. panamensis Dall, 1908. Deep water, Gulf of Panama, 1,017 m.

Genus ENTODESMA PHILIPPI, 1845

Periostracum coarse. Lithodesma large. The type species is *E. chilense* Philippi, 1845, which is confined to the southern Chilean coast. It has been considered a synonym of *Anatina cuneata* Gray, 1828 (from Arica, in northern Chile), by some authors, but Gray's species is described as having an adherent greenish periostracum that overlaps the edges of the shell, and Dall plausibly suggests that it is probably a *Lyonsia*, whereas *E. chilense* looks much like *E. inflatum* except that the posterior part is flatter and more rounded. Three Panamic subgenera.

Subgenus ENTODESMA, s. s.

Periostracum uniformly colored; surface of shell smooth.

744. Entodesma (Entodesma) inflatum (Conrad, 1837) (Synonym: **Lyonsia diaphana** Carpenter, 1856). Like other Entodesmas, this is a nestler, and the shell is therefore subject to much variation in shape. The anterior end is short, the whole shell about 20 mm in length. The tough, yellowish periostracum will, on drying, crack the shell to pieces unless it has been protected by a coating of lubricant such as vaseline. *L. diaphana* Carpenter, 1856, originally described from Mazatlán, Mexico, is probably the young of this species. California to Guayaquil, Ecuador, from shallow water to a depth of 37 m. Lowe reported the species at Punta Peñasco, Sonora, Mexico, on the reef, among compound ascidians.

Subgenus AGRIODESMA DALL, 1909

Periostracum coarse, horny; shells distorted by growth in rock crevices; lithodesma large.

745. Entodesma (Agriodesma) brevifrons (Sowerby, 1834). Small, the anterior end broadly gaping, periostracum thin, horn-colored. Length, 18 mm; height, 9 mm. In sandy mud, Santa Elena, Ecuador, depth 11 to 15 m.

746. Entodesma (Agriodesma) sechuranum Pilsbry & Olsson, 1935. Moderately large, irregularly shaped, because of nestling, pearly within, periostracum dark-colored, beaks small and terminal. Antero-ventral margin with a gape. Length, 68 mm; height, 41 mm; diameter, about 35 mm. Panama to Peru.

Subgenus **PHLYCTICONCHA** BARTSCH & REHDER, 1939
(**PHLYCTIDERMA** BARTSCH & REHDER, 1939 [not DALL, 1899])

Periostracum with zigzag color pattern and pitted sculpture.

747. Entodesma (Phlycticoncha) lucasanum (Bartsch & Rehder, 1939). The irregularly shaped white shell is overlain by a periostracum that is yellowish, marked with light to dark brown zigzag lines. The surface of the young shell is finely dotted with white pustules. The type, which was evidently immature, is only 18.5 mm long. It was dredged at a depth of 11 to 18 m off Cape San Lucas, Baja California. Larger specimens (length, 28 mm; height, 20 mm) have been found intertidally at Puerto Peñasco, Sonora, Mexico, and on the Oaxaca coast of Mexico.

748. Entodesma (Phlycticoncha) pictum (Sowerby, 1834). The shell is brownish with black radial or slightly zigzag markings, gaping along part of the antero-ventral margin. Length, 21 mm; height, 13 mm; diameter, 9 mm. The type locality is Guayaquil, Ecuador. The specimen described and figured by Olsson (1961) as *E. picta* from the Philadelphia Academy collection seems instead to be a *Lyonsia inflata*, perhaps part of Conrad's missing original lot; it was purchased from a British dealer in 1856.

Family **PERIPLOMATIDAE**

With an internal ligament lodged in a spoon-shaped resilifer in either valve, buttressed by a shelly ridge on the inside of the shell. Shell material translucent white with a nacreous sheen inside.

Genus **PERIPLOMA** SCHUMACHER, 1817

Thin, right valve more convex than left and overlapping it; surface granular, beaks opisthogyrate; resilium in a spoon-shaped chondrophore in each valve, anterior muscle scar long and narrow, posterior small and crescentic; pallial sinus short, rounded.

Subgenus **PERIPLOMA**, *s. s.*

Valves smooth or with only microscopic sculpture.

749. Periploma (Periploma) carpenteri Dall, 1896. Nearly orbicular, this has prominent umbones that are nearly at the midline. The surface granulation is dense and has no definite pattern. Length, 49 mm; height, 42 mm. It is not likely to be found intertidally, as the type was dredged in Panama Bay at a depth of 380 m. Hertlein and Strong (1946) report a possible second specimen dredged at a depth of 5 m off El Salvador.

750. Periploma (Periploma) discus Stearns, 1890. The specific name is well chosen, for the flat left valve is much like a disk in shape. The beaks are nearly central or slightly nearer the posterior end, as in *P. altum*. The surface ornamentation is of fine granules that are arranged in radial rows on parts of the shell.

Length, 25 to 35 mm. Monterey, California, to El Salvador, on mud flats and in shallow water at least to a depth of 9 m.

751. Periploma (Periploma) lagartilla Olsson, 1961. The shell is thin, rather small, subcircular, the beaks slightly nearer the posterior end, right valve more convex than the left; the surface is minutely granular, visible only under magnification, in a concentric patterning. There is a thin light-colored periostracum. The outline is close to that of *P. stearnsii* except that it is relatively shorter and the posterior end is more produced. The surface granulations are much finer. Length, 23 mm; height, 18 mm; diameter, 8 mm. Panama.

752. Periploma (Periploma) lenticulare Sowerby, 1834 (Synonyms: **Anatina alta** C. B. Adams, 1852; ? **P. excurva** Carpenter, 1856). Because the type was unfigured, this has long been supposed by authors to be a synonym of *P. planiusculum*. Study of the type lot in the British Museum (figured here) shows that *P. lenticulare* is instead a prior name for *P. altum*, the type of which itself remained unfigured for over a century. The type lot consists of a single valve of a mature shell and a pair that is smaller. The type locality is Isla Muerte, Ecuador. The shell is subcircular, the beaks at the midline. The surface is nearly smooth, with fine granules in concentric rows. Length of Sowerby's larger specimen, 31 mm; height, 24 mm. Length of Adams's type, 50 mm; height, 44 mm (? Mazatlán, Mexico); Panama to Ecuador. *P. carpenteri* is very similar in outline except that the umbones are higher; surface granulations apparently are coarser.

753. Periploma (Periploma) planiusculum Sowerby, 1834 (Synonyms: **P. argentaria** Conrad, 1837; **P. obtusa** Hanley, 1842; **P. papyracea** Carpenter, 1856). Quadrate in shape, the shell is noticeably longer than high and is more sturdy than in the other species. The surface is smooth or with irregularly arranged pustules. Length, about 50 mm; height, 35 mm. Southern California to Peru, from the extreme low tide level to depths of a few meters.

754. Periploma (Periploma) stearnsii Dall, 1896. Like *P. discus* this is an orbicular form, but the surface granules are larger and more definitely arranged in radial rows. It is more compressed than *P. discus* and the twisted portion or rostrum at the posterior end of the valves is wider. Length, 47 mm; height, 39 mm. Point Fermin, head of the Gulf of California, to Cape San Lucas, Baja California, in depths of 15 to 44 m.

755. Periploma (Periploma) teevani Hertlein & Strong, 1946. Nearest to *P. planiusculum* in form, this differs by being relatively higher for the length, the surface having radial rows of pustules. Length, 23 mm; height, 19 mm. The type specimen was dredged in 55 m depth, mud bottom, off Oaxaca, Mexico.

Subgenus ALBIMANUS PILSBRY & OLSSON, 1935
Valve ribbing more than microscopic; ribbing radial.

756. Periploma (Albimanus) pentadactylus Pilsbry & Olsson, 1935. Three radial ribs that project at the margin are the unique feature of the species. The shell is white, and its outline, like a small outstretched hand, is referred to in both the subgeneric and the specific names. Length, 21 mm; height, 15 mm. Western Nicaragua to Los Santos province, Panama.

Subgenus HALISTREPTA DALL, 1904
Valve ribbing more than microscopic; ribbing concentric. This subgenus was known only from a single species in southern California until recent years.

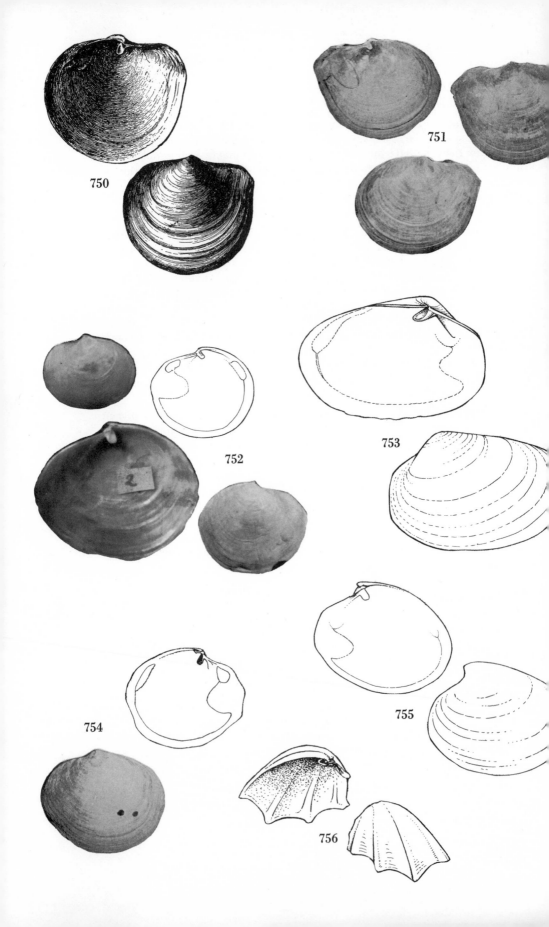

750

751

752

753

754

755

756

757. Periploma (Halistrepta) myrae Rogers, 1962. Except for three low radial ribs the sculpture consists of undulating concentric ribbing, with the posterior area set off by coarser sculpture. The shell is very thin, white, with microscopic surface granulation. Length, 20 mm; height, 16 mm. Dredged off Carmen Island, Baja California, depth 27 to 45 m.

Family THRACIIDAE

The ligament is lodged in a resilifer in one valve (rarely is there a resilifer in both) or is external. Within, the pallial sinus is well marked, wide if not deep. The key is to genera.

1. Shell surface smooth, not granular or wavy........................*Bushia*
   Shell surface granular or wavy..................................... 2
2. Sculpture of oblique undulations or granulations..............*Cyathodonta*
   Sculpture granular, but without pattern............................ 3
3. Ligament entirely internal ....*Asthenothaerus*
   Ligament external or mainly so................................*Thracia*

Genus THRACIA SOWERBY, 1823

Smooth, not nacreous-textured, inequivalve, the right valve larger; surface granular in most forms, hinge edentulous, resilium in an obliquely directed chondrophore; posterior end broadly truncate and set off by a low ridge; pallial line with a sinus.

758. Thracia anconensis Olsson, 1961. An ovate shell with apparently opisthogyrate beaks, sculpture concentric and irregular, somewhat stronger anteriorly. The entire shell surface has coarse granulations. The pallial sinus is short. Length, 34 mm; height, 22 mm; diameter, about 13 mm. Ecuador, described from a single left valve.

759. Thracia colpoica Dall, 1915. The shell is whitish, with a posterior area set off by a weak rib, a rough but not markedly granular surface, and a short pallial sinus. Length, 17 mm; height, 16 mm; diameter, 8.7 mm. Dredged, Gulf of California off La Paz and Topolobampo, Mexico, depths of 7 to 165 m. Olsson (1961) reports the species as common on the mud flat at the mouth of the Tumbes River, Peru.

760. Thracia curta Conrad, 1837. Since specimens of this species nestle in rock cavities, the shells may become distorted. The finely granulose surface is a good guide in recognition. Length, 26 mm; height, 22 mm; diameter, 14 mm. *T. curta* has been reported from southern Alaska to Ecuador.

761. Thracia squamosa Carpenter, 1856. This species may be distinguished not only by the densely granulose surface but also by the more regular shape and by the presence of a vertical scar under the beaks marking the attachment of what is called a lithodesma, a calcareous reinforcement of the internal ligament. Length, 30 mm; height, 18 mm; diameter, 10 mm. Magdalena Bay, Baja California, through the Gulf at least as far as the Sonoran coast of Mexico.

Genus ASTHENOTHAERUS CARPENTER, 1864

One of the few genera described from west coast material and later found in another province, this is comprised of only a few species—one in the Panamic province, two in the Caribbean. The shells are much like a small *Thracia* in appearance,

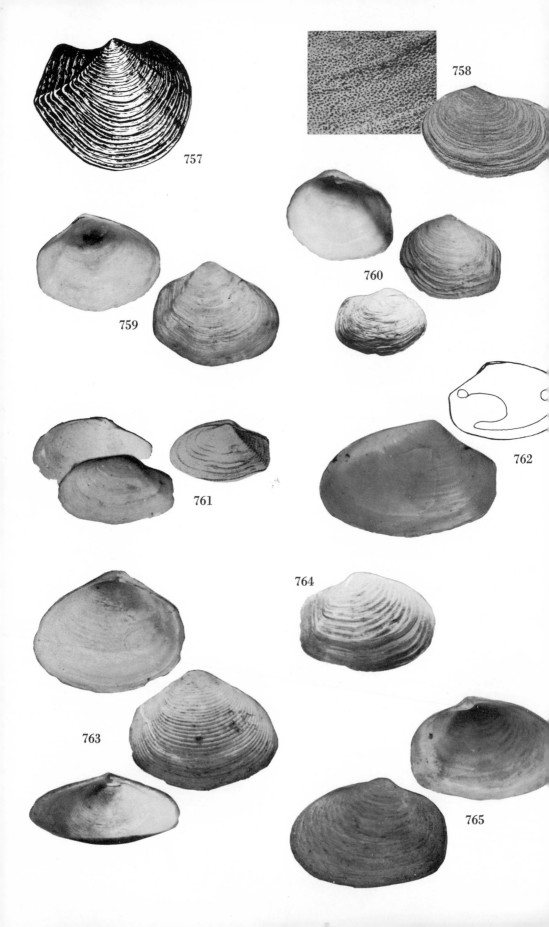

757

758

759

760

761

762

763

764

765

earthy in texture, with the surface finely granular. The ligament is internal, lodged just under the beaks and supported by a calcified structure, an ossicle, which looks like a long-winged butterfly poised under the hinge.

762. Asthenothaerus villosior Carpenter, 1864 (Synonym: **Thracia diegensis** Dall, 1915). Perhaps because specimens look so like an immature *Thracia*, few collectors have recognized this species. Length, 10 mm; height, 6 mm; diameter, 4 mm. San Pedro, California, to Cape San Lucas; specimens labeled *T. diegensis* are not uncommon in collections, taken in muddy parts of San Diego Bay, California.

Genus BUSHIA DALL, 1886

The ligament is external, and the underside of the beaks is filled with shelly material, though the hinge is edentulous. In texture the shell is porcelaneous.

763. Bushia panamensis Dall, 1890. First described from material dredged in Panama Bay, 93 m, mud, this species has not been reported again. Length, 14 mm; height, 11 mm; diameter, 8 mm. The present figure is of the holotype.

Genus CYATHODONTA CONRAD, 1849

Shell with undulating sculpture or oblique rows of granules; ligament internal, on a chondrophore.

764. Cyathodonta dubiosa Dall, 1915. This species can be distinguished from the larger and commoner *C. undulata* by the greater relative convexity of the left valve and by the surface granulation, which is in irregular concentric lines, not radially distributed. Length, 20 mm; height, 16 mm. San Pedro, California, to Champerico, Guatemala, in depths to 26 m.

765. Cyathodonta lucasana Dall, 1915. The surface granulations are arranged as in *C. dubiosa*, but the shell is more ovate, longer for its height, and the beaks are nearer the posterior end. Length, 21 mm; height, 14 mm. La Paz, Gulf of California, to Port Guatulco, Mexico; on mud flats, 40 to 45 cm (16 to 18 inches) deep.

766. Cyathodonta undulata Conrad, 1849 (Synonyms: **C. undulata peruviana** and **? C. tumbeziana** Olsson, 1961). The obliquely rippled surface with its radiating rows of granules is distinctive, but if one overlooks the granules and the hinge, one could confuse the shell with *Mactra thracioides*. Large specimens may be as much as 50 mm long. The average size is 40 mm in length, 31 mm in height. The subspecies named by Olsson on the basis of greater proportionate height seems to fall within the range of variation of the species. Puerto Peñasco and Guaymas, Sonora, Mexico, and the southern part of the Gulf of California to Peru, in depths of 3 to 110 m. Also reported as fossil in Baja California and Peru. The latter record may prove to be of *C. tumbeziana*, a more truncate form, with finer ribbing, which ranges from Tumbes, Peru, southward, that Olsson regards—perhaps rightly—as specifically distinct. A similar species from the Caribbean is *C. magnifica* (Jonas, 1850). *C. undulata* is the type species of the genus *Cyathodonta*.

Superfamily POROMYACEA
(SEPTIBRANCHIA)

Gills muscular, forming a septum-like sheet, so profound a modification that some authors have even regarded the group as a separate subclass. Instead of being a series of flaplike or straplike structures for the circulation of water, the gills are fused and evidently assist in the management of food. Unlike other bivalves, these animals are carnivorous. F. R. Bernard of the Biological Station, Nanaimo, British Columbia, has made a study of stomach contents, which he finds include protozoa and fragments of larger invertebrates. Notes on distribution and classification have been supplied by Mr. Bernard, whose assistance is here gratefully acknowledged.

Family POROMYIDAE

These are mostly deepwater forms, at least in the Panamic province. The shell is thin, fragile, nacreous within, trigonal to ovate.

Genus POROMYA FORBES, 1844

Small, inflated; surface granular. Two Panamic subgenera. Subgenus *Poromya, s. s.*, with an internal ligament or cartilage, has not yet been recorded from the Panamic province.

Subgenus CETOCONCHA DALL, 1886

With an external ligament; shell surface polished, granules minute.

767. Poromya (Cetoconcha) perla Dall, 1908. Gulf of Panama to Ecuador, in 3,073 to 3,435 m.

768. Poromya (Cetoconcha) scapha (Dall, 1902). Shell thin, equivalve, with many rows of tubercles. Length, 13 mm; height, 8 mm. Off Cocos Island, Gulf of Panama, in 183 m—unusually shallow for this genus.

769. P. (C.) smithi (Dall, 1908). Off Acapulco, Mexico, in 3,400 m.

Subgenus DERMATOMYA DALL, 1889

Surface not granular; hinge strong; pallial line with a small sinus.

770. Poromya (Dermatomya) equatorialis Dall, 1908. Gulf of Panama to Ecuador, in 1,356 to 3,060 m.

771. P. (D.) mactroides Dall, 1889. Type of subgenus. Southern Baja California to Chile, in 637 to 915 m.

Family CUSPIDARIIDAE

Shells lacking any nacreous luster, the outline somewhat like that of *Lyonsia*, except that the posterior end is elegantly drawn out into a spout. The ligament is internal, in a lithodesma or calcified structure. The pallial line is indistinct, not showing any sinus.

These are offshore forms, never found intertidally, rarely even as beach drift. Some occur in very deep water—3,000 to 4,000 meters (1,500 to 2,000 fathoms). The key is to genera.

1. Shell surface smooth....................................*Cuspidaria*
   Shell surface with sculpture.. 2
2. Sculpture pustulose, irregular ....*Plectodon*
   Sculpture radial, regular 3
3. Radial ribs few; hinge edentulous.............................*Myonera*
   Radial ribs several; hinge with teeth.........................*Cardiomya*

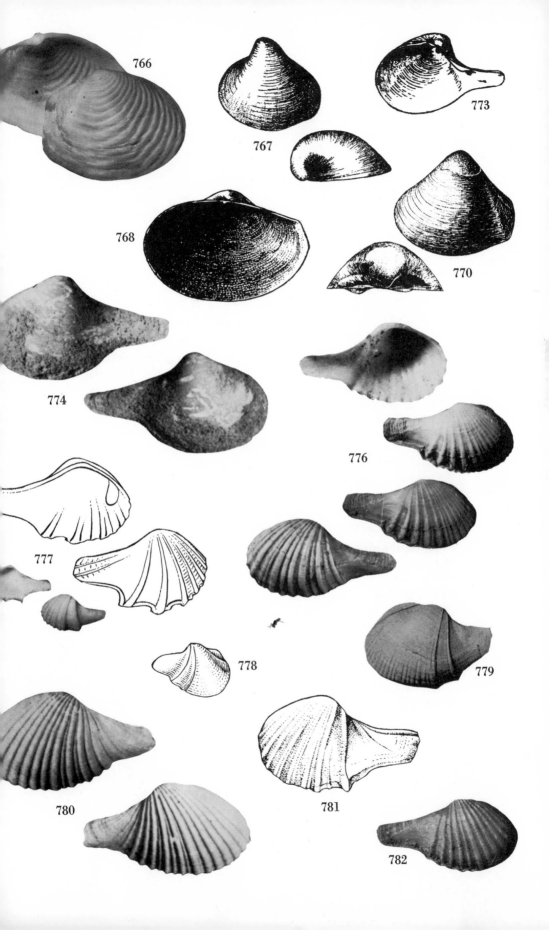

Genus **CUSPIDARIA** NARDO, 1840
(**NEAERA** GRIFFITH & PIDGEON, 1834 [not ROBINEAU-DESVOIDY, 1830])

Nearly equivalve but markedly inequilateral, posterior end rostrate to spoutlike; surface smooth; hinge with one posterior lateral tooth.

772. Cuspidaria chilensis Dall, 1908. Costa Rica to Chile, in 1,239 to 1,896 m.

773. Cuspidaria panamensis Dall, 1908. Baja California to Panama Bay, in 915 to 1,280 m.

774. Cuspidaria parapodema Bernard, 1969. Smooth, inflated, polished within. Length, 15 mm; height, 9 mm. Southern California to the Gulf of California in 53 to 275 m.

775. Cuspidaria patagonica (E. A. Smith, 1885). Southern Baja California and Panama Bay south to Chile and east to the South Atlantic, in 300 to 3,050 m.

Genus **CARDIOMYA** A. ADAMS, 1864

With radial ribs on the main part of the shell, the rostrum or spout smooth; generally in shallower water than *Cuspidaria*.

776. Cardiomya californica (Dall, 1886). Proportionately longer than *C. pectinata*, with more ribs (16 to 20). Length, 7 mm; height, 3.6 mm. British Columbia to southern Baja California and the Galápagos Islands, in 15 to 640 m.

777. Cardiomya costata (Sowerby, 1834) (Synonym: **Cuspidaria dulcis** Pilsbry & Lowe, 1932). Shell white, with a thin buff periostracum; some of the ribs of the left valve twinned, right valve with 7 to 8 simple ribs. Length, 7.5 mm; height, 4.5 mm. Southern California and the northern part of the Gulf of California to Guayaquil, Ecuador, and the Galápagos Islands, in 4 to 84 m; also in the western Atlantic.

778. Cardiomya didyma (Hinds, 1843). Except for two prominent ribs in front of the rostrum, the shell is smooth. Length, about 7 mm. The northern part of the Gulf of California to Panama, in 18 to 48 m.

779. Cardiomya ecuadoriana (Olsson, 1961). The small, plump shell has two heavier ribs and about 20 fine radial riblets. Length, 8.5 mm; height, 5.5 mm. Galeras, Ecuador (type locality).

780. Cardiomya isolirata Bernard, 1969. The slightly inequivalve shell has a grayish periostracum over the 12 to 20 rounded and uniform ribs. Length, 8 mm; height, 4.8 mm. Southern California to the central part of the Gulf of California, in 55 to 183 m.

781. Cardiomya lanieri (Strong & Hertlein, 1937). One or two larger ribs jut out at the posterior ventral margin of the shell. Length, about 5 mm. San Quintín Bay, Baja California, through the Gulf of California and south to Ecuador and the Galápagos Islands, in 15 to 238 m.

782. Cardiomya pectinata (Carpenter, 1864). Although reported as ranging from Alaska to Panama, this has not been recognized in recent years south of Cedros Island, Baja California.

783. Cardiomya planetica (Dall, 1908). Bering Sea to Cedros Island, Baja California, in 135 to 2,400 m. Records from the Galápagos Islands probably represent misidentifications.

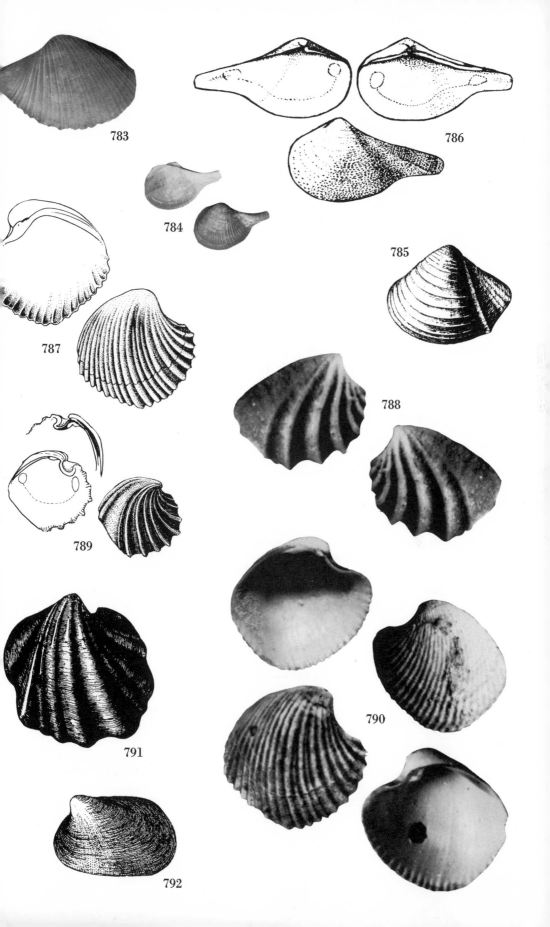

783

786

784

785

787

788

789

791

790

792

784. Cardiomya pseustes (Dall, 1908). Gulf of Panama, in 2,320 m.

Genus MYONERA DALL, 1886

Edentulous, with a vertical or posteriorly directed fossette.

785. Myonera garretti Dall, 1908. Gulf of Panama, in 1,650 m.

Genus PLECTODON CARPENTER, 1864

The hinge has both an anterior and a posterior lateral tooth, and the surface of the shell is studded with fine granules.

786. Plectodon scaber Carpenter, 1864. Not only is the pustular surface distinctive but there is a tinge of pink on the umbones of the otherwise dingy white shell. One of the largest representatives of the family in the eastern Pacific. Length, about 24 mm. Catalina Island, California, through the Gulf of California and south to Panama and the Galápagos Islands, in 20 to 250 m.

Family VERTICORDIIDAE

Heart-shaped, inflated, nacreous within, radially ribbed; hinge weak, with one or two low teeth. Three Panamic genera.

Genus VERTICORDIA SOWERBY, 1844, *ex* WOOD, MS

Small, lustrous, with strong, curved, radial ribs; ligament internal, in a large lithodesma. The type species is from the Pliocene of England. Two Panamic subgenera.

Subgenus VERTICORDIA, *s. s.*
(TRIGONULINA ORBIGNY, 1846)

Hinge with one conical tooth in right valve, no laterals.

787. Verticordia (Verticordia) aequacostata A. Howard, 1950. Finer-ribbed than the more common *V. ornata*. Length, 2 to 4 mm. Off Catalina Island, California, to Isla Angel de la Guarda, Gulf of California, in 165 to 190 m.

788. Verticordia (? Verticordia) hancocki Bernard, 1969. Inequilateral, subquadrate, compressed, surface granular. Length, 2 mm; height, 1.4 mm. Gorgona Island, Colombia, in 73 to 110 m. Bernard placed this in the subgenus *Trigonulina*, which he regards as distinct from *Verticordia, s. s.*

789. Verticordia (Verticordia) ornata (Orbigny, 1846). No one has yet differentiated eastern Pacific from western Atlantic specimens (the type locality is Jamaica). Both have the same number of ribs (8 to 9) on the anterior and central part of the shell; both are about 2 to 4 mm in length and height. Catalina Island, California, through the Gulf of California, south to Peru and the Galápagos Islands, in 18 to 168 m; also in the Caribbean. This is the type of *Trigonulina*, a subgenus that is doubtfully distinct from *Verticordia, s. s.*

Subgenus HALIRIS DALL, 1886

Globose; left valve with a small cardinal tooth and a lateral.

790. Verticordia (Haliris) spinosa Bernard, 1969. The shell has 25 to 30 ribs that may be somewhat spinose toward the ventral margin; right valve with a lateral tooth, left valve with a peglike cardinal and a lateral socket. Length, 7 mm; height, 6.5 mm. Off the tip of Baja California, in 275 m.

Genus HALICARDIA DALL, 1895
(HALICARDISSA DALL, 1913)

This is a deepwater genus, the shells of which resemble *Verticordia* in shape but with fewer ribs and much larger dimensions, length being as much as 40 mm. They are surprisingly sturdy. The presumed differences of *Halicardissa* are insignificant, according to Bernard (*in litt.*). The type species of *Halicardia* occurs in the Atlantic.

791. Halicardia perplicata (Dall, 1890). This species, the type of *Halicardissa*, ranges from off British Columbia and Oregon to the Galápagos Islands (type locality) in depths of 1,110 to 1,486 m.

Genus LYONSIELLA G. SARS, 1872, *ex* M. SARS, MS

Quadrangular, inflated shells, with the beaks toward the anterior end; the left valve with one weak tooth on the hinge; the right with none. The pallial line is without a sinus.

792. Lyonsiella magnifica Dall, 1913 (Synonym: **L. magnifica** Dall, 1923). A pearly white shell under a pale olive periostracum, with the outer surface minutely granular. The shell itself is very thin and one valve is larger than the other. Length, 25 mm; height, 16.5 mm; diameter, 16 mm. Dall's two descriptions cite slightly different type localities, in fairly deep water (115 m or more) at the southern end of the Gulf of California.

PELECYPOD DATA RECEIVED TOO LATE FOR INCLUSION IN THE TEXT

A paper by Knudsen (1970), received too late for more than minimal recognition in the text, makes a number of additions and corrections to the deepwater pelecypod fauna of the Panamic province.

Range extension:

31. Nuculana agapea (Dall, 1908). Gulf of California to Ecuador.

New species:

48a. Malletia (Neilo) cuneata (Jeffreys, 1876). Gulf of Panama, 2,950 to 3,190 m. Also Atlantic.

Name change and range extension:

97. Bathyarca orbiculata (Dall, 1881) (Synonym: **Arca nucleator** Dall, 1908). Off southern California to the Gulf of Panama in depths to 2,320 m; also western Pacific and Atlantic.

Name change:

104. Limopsis dalli Lamy, 1912 (Synonym: **L. compressus** Dall, 1896, not Nevill, 1874). Regarded by Knudsen as a subspecies of *L. pelagica* Smith, 1885.

New species:

148a. Modiolus abyssicola Knudsen, 1970. Gulf of Panama, 3,270 to 3,670 m.

New genus and species:

Genus DACRYDIUM TORELL, 1859

Shell minute, smooth, hinge crenate to striate; resilium internal.

155a. Dacrydium panamensis Knudsen, 1970. Acapulco, Mexico, to Panama, 3,270 to 3,670 m.

New species:

187a. Cyclopecten graui Knudsen, 1970. Gulf of Panama, 3,270 to 3,670 m.

189a. Cyclopecten neoceanicus (Dall, 1908). Gulf of Panama to southwest of Galápagos Islands, 3,270 to 3,670 m.

New genus and species:

Genus **KELLIELLA** M. SARS, 1870

Shell minute, rounded-ovate; hinge with two teeth in each valve.

265a. Kelliella galatheae Knudsen, 1970. Gulf of Panama, 2,950 to 3,570 m.

New species:

653a. Abra californica Knudsen, 1970. Off southern Baja California, 3,480 to 3,518 m.

Range extension:

767. Poromya (Cetoconcha) perla Dall, 1908. Gulf of California to Ecuador, to 3,518 m.

New species:

772a. Cuspidaria haasi Knudsen, 1970. Off Central America, 3,570 m.

774a. Cuspidaria parkeri Knudsen, 1970. Off Gulf of California, 2,770 to 2,817 m.

785a. Myonera mexicana Knudsen, 1970. Off west Mexico, 3,529 to 3,557 m.

New genus and species:

Genus **POLICORDIA** DALL, BARTSCH & REHDER, 1938

Shell ovate, with fine radial ribs; edentulous; ligament in a groove.

792a. Policordia alaskana (Dall, 1895). Alaska to west Mexico, in depths to 3,570 m.

 # CLASS GASTROPODA

Class Gastropoda

The Gastropoda ("stomach-footed ones") are, in general, free-moving animals that creep about on a flat, disklike foot. The shell alone is an unsatisfactory basis for classification, as the same form (for example, a limpet-like shape) may appear in what are obviously unrelated stocks. The radula seems to be the best single guide to relationships, but the gills (ctenidia), heart, and nervous system are also useful. One does not, of course, have to make an anatomical study of every specimen in order to discover a name for it. The shells are very convenient when it comes to identification—after the classification or pigeonholing has been worked out by the zoologist.

A unique feature of the gastropods is the phenomenon called torsion. This takes place within the first few hours of the embryonic snail's life, when the bilaterally symmetrical larva twists so that all the organs behind the "neck" are reversed in position and the mantle cavity moves up and forward, over the head. Part of the paired organs (heart, gills, kidney) cease developing, and the little snail begins to be asymmetrical and spirally coiled. Torsion forms a good basis for separation of two subclasses.

Subclass PROSOBRANCHIA
(PROSOBRANCHIATA, STREPTONEURA)

Torsion brings the water-circulating organ (ctenidium or gill) to the front, which is emphasized in the subclass name; it also causes a crossing of previously parallel nerve cords, alluded to in the alternative name Streptoneura.

Order ARCHAEOGASTROPODA

As was pointed out for pelecypods, the earliest mollusks seem to have had nacreous shells; hence, this kind of material in a modern stock may be construed as indicating ancient lineage. Early gastropods also had a round aperture, or if there was a notch it was at the top, marking the position of the excurrent or anal siphon, and the gills were paired.

Modern archaeogastropods have the round aperture and a few have the notch; but the nacreous shell texture may be partly or entirely replaced by a porcelaneous texture. The shells are cap-shaped to turbinate, conical, or globose. Various modifications of the gills or breathing apparatus are evident. In food habits, the animals are predominantly herbivorous, browsing on microscopic algae that they scrape from the rocks with the radula; some have been found also to include sponges in their diet.

The text for the Archaeogastropoda has been prepared, with minor changes to bring it into uniformity with the remainder of the book, by Dr. James H. McLean of the Los Angeles County Museum of Natural History, whose doctoral thesis dealt with problems in the classification of some Archaeogastropod families. His assistance in providing both the text and the illustrations is gratefully acknowledged. He prefers to follow, in the sequence of systematic units above species level, a systematic order rather than the nominate-group and alphabetical order adopted as the general pattern in this book.

Superfamily PLEUROTOMARIACEA

Here seem to be the most direct descendants of those ancient stocks that had a slit in the outer lip, but the only tropical American species retaining the nacreous shell lining are in the genus *Haliotis*, a few specimens of which have been taken in the offshore waters of the Panamic province. All the rest of the Panamic Pleurotomariacea have progressed to porcelaneous shell texture. The spiral shape is maintained in *Haliotis* and the minute gastropods of the family Scissurellidae. The rest of the group lose the spiral shape early in the life of the individual, and the adult becomes limpet-like or conical, with the slit preserved as a perforation at the apex or as a channel or groove that connects the apex to the margin. The radula is rhipidoglossate ("with fan tongue"); in each row of teeth there are a limited number of central and lateral teeth, but the marginals are indefinitely numerous, somewhat resembling the ribs of a fan.

Family HALIOTIDAE

Shell ear-shaped, depressed, spire excentric, low; aperture occupying most of underside, inner lip rimlike; no operculum.

Genus HALIOTIS LINNAEUS, 1758

Shell with nacreous interior; rounded to elongate, few-whorled, with a row of holes marking the position of the selenizone or slit band. Animal with an epipodium, a sensory ridge around the edge of the foot, bearing a series of tentacles.

The type species of the genus is the Indo-Pacific *H. asinina* Linnaeus, 1758, the soft parts of which are twice as wide as the shell and much too large ever to be completely retracted beneath it.

Subgenus PADOLLUS MONTFORT, 1810

Shell large enough to cover the soft parts. Most of the species of *Haliotis* other than *H. asinina* fall within *Padollus*.

1. **Haliotis (Padollus) dalli** Henderson, 1915 (Synonym: "**H. pourtalesii**" of Dall, 1890, not of Dall, 1881, a Caribbean species). Shell relatively flat, with a raised mid-dorsal rib and corresponding interior channel between the row of holes and the margin; sculpture of flat threads separated by deeply incised lines; five holes open on the selenizone or slit band. Color, pale brick-red, flecked or mottled with greenish white. Length, 25 mm; width, 19 mm. Gorgona Island, Colombia, to Galápagos Islands, Ecuador (type locality), depth 46 to 64 m. Dall's description of the Caribbean form was from memory, the original specimen having been destroyed in the Chicago fire in 1871; rediscovery of the species led to the realization that the Pacific form is distinct.

2. Haliotis (Padollus) roberti McLean, 1970. Smaller than *H. dalli* but with the same coloration. The shell is more inflated, with thicker and more separated spiral cords; the mid-dorsal ridge reduced to a trace, with only a shallow channel below the row of openings. Length, 18 mm; width, 14 mm; height, 6.5 mm. A single lot taken in 1938 off Cocos Island, Costa Rica, in 73 to 86 m.

Family SCISSURELLIDAE

Minute turbinate shells, umbilicate; porcelaneous except for a thin nacreous layer; lip with a slit or foramen; operculum multispiral. The species of the eastern Pacific were reviewed by McLean (1967). Two Panamic genera.

Genus SCISSURELLA ORBIGNY, 1824

Shell with an open slit on outer lip, its trace on earlier whorls preserved as a slit band or selenizone. No species of subgenus *Scissurella, s. s.,* occur in the Panamic province.

Subgenus ANATOMA WOODWARD, 1859

Spire elevated, selenizone at the periphery, sculpture cancellate. Two recently described species are the first records of the group in the Panamic province.

3. Scissurella (Anatoma) epicharis McLean, 1970. Upper part of the whorls only slightly rounded, the suture closely adjacent to the selenizone, the edges of which project markedly. Sculpture finely cancellate. Height, 1.3 mm; diameter, 1.9 mm. Isabela Island, Galápagos Islands, depth 22 m.

4. Scissurella (Anatoma) keenae McLean, 1970. Whorls rounded, suture impressed; sculpture finely cancellate, the axial and spiral ridges of nearly equal strength. Height, 2.0 mm; diameter, 2.0 mm. Cape San Lucas, Baja California, and north in the Gulf of California to Raza and San Pedro Nolasco Islands, in 73 to 146 m.

Genus SINEZONA FINLAY, 1927
(**SCHISMOPE** of authors, not of JEFFREYS, 1856; **CORONADOA** BARTSCH, 1946)

Minute, with the slit closed at the apertural margin.

5. Sinezona rimuloides (Carpenter, 1865) (Synonym: **Coronadoa simonsae** Bartsch, 1946). Sculpture is of strong axial folds. The shell is white. Height, 0.8 mm. Common in gravel, from the Farallon Islands, California, to Guaymas, Gulf of California, and south to Iquique, Chile, intertidally and offshore to a depth of 30 m. The genus *Coronadoa* was based on juvenile specimens less than 0.5 mm in height, at which stage the foramen or opening is not developed.

Superfamily FISSURELLACEA

Shell porcelaneous, conical, having a spiral protoconch; with a perforation, slit, notch, or emargination for the passage of the exhalant water current. Upper side of foot with epipodial tentacles.

Family FISSURELLIDAE

Popularly known as keyhole limpets, from the apical opening that occurs in many forms; water that has passed over the gills is channeled through this opening or a slit near the margin.

The family is divisible into two main groups: those in which the apex of the shell is preserved in adults and the fissure is an indentation on the margin or a perforation on the anterior slope, and those in which the perforation occurs at the summit of the shell and the apical area is absorbed. The first group is the more primitive, comprising the subfamily Emarginulinae. The second group includes two subfamilies, the Fissurellidinae and the Fissurellinae, which are separated by a major difference in the radula. The subfamily distinctions differ from the usage of authors previous to McLean (1966). Genera may be keyed as follows:

1. With an open slit or marginal indentation............................ 2
   With a perforation or orifice..................................... 5
2. Slit band not defined.......................................*Hemitoma*
   Slit band well defined from anterior margin to apex................. 3
3. Apex remote from the posterior margin....................*Emarginula*
   Apex at posterior end, close to posterior margin...................... 4
4. Without an internal septum.......................................*Nesta*
   With an internal septum at posterior end........................*Zeidora*
5. Perforation on anterior face................................... 6
   Perforation at apex... 7
6. Internal septum present at apex............................*Puncturella*
   Internal septum lacking.......................................*Rimula*
7. Callus truncate posteriorly within apex........................*Diodora*
   Callus not truncate... 8
8. Shell relatively large, over 40 mm in length...................*Stromboli*
   Shell relatively small, under 40 mm in length........................ 9
9. Profile of shell conical.....................................*Fissurella*
   Profile of shell low, platelike...............................10
10. Sculpture of fine concentric ridges only....................*Leurolepas*
    Sculpture both radial and concentric......................*Lucapinella*

Subfamily EMARGINULINAE

The apex remains intact in the adult shell. In the more primitive genera there is a slit at the anterior margin; in the more advanced genera the margin is sealed and the slit located on the anterior slope. The rachidian tooth of the radula is broad and rectangular, lacking cusps. The shells are usually white externally and mostly to be found offshore.

Genus EMARGINULA LAMARCK, 1801

Apex posterior, elevated above the margin of the shell; slit and its slit band (selenizone) narrow, bordered by raised ridges; sculpture cancellate.

6. Emarginula tuberculosa Libassi, 1859. With numerous ribs, every fourth rib slightly larger at the margin. Length, 13 mm; height, 7 mm. A species of wide distribution, known in the eastern Pacific from Isabela Island, Galápagos Islands, to Octavia Rocks and Port Utria, Colombia, in 80 to 180 m; also in the western and eastern Atlantic.

7. Emarginula velascoensis Shasky, 1961. Small, white, with relatively few radial ribs; sculpture coarsely cancellate. Length, 5.3 mm; width, 3.7 mm; height, 2.7 mm. Angel de la Guarda Island to Cape San Lucas, Gulf of California, in 73 to 550 m.

Genus HEMITOMA SWAINSON, 1840

Sturdy shells with a slight anterior notch and a distinct internal groove running toward the apex. Two Panamic subgenera.

Subgenus HEMITOMA, *s. s.*

Apex nearly central, slightly inclined posteriorly, sculpture radial, primary ribbing strong.

8. Hemitoma (Hemitoma) natlandi Durham, 1950 (Synonyms: **H. scrippsae** Durham, 1950; **H. chiquita** Hertlein & Strong, 1951). Flattened or moderately elevated, the shell has six primary ribs of nearly equal size, projecting conspicuously at the margin, with two low secondary ribs between the anteriormost strong rib and the next pair; the posterior rib is doubled. Height is variable, and on some specimens the primary ribs may be nodular. Length, 12 mm; width, 8 mm. On rocky bottoms at depths of 8 to 45 m, Barra de Navidad, Jalisco, Mexico, to Port Utria, Colombia. The type locality is Coronado Island, Gulf of California, Pleistocene. The form named *H. scrippsae* falls within the range of variation of the species, and *H. chiquita* represents the juvenile stage of growth.

Subgenus MONTFORTIA RÉCLUZ, 1843

Apex strongly inclined posteriorly, sculpture both radial and concentric; three anterior ribs much stronger than the other primary ribs.

9. Hemitoma (Montfortia) hermosa Lowe, 1935. The chalky white shell is horn-colored within, sculptured with uneven, nodulous ribs. Variations in outline range from shells that are broad and only moderately elevated to those that are high and narrow; some show fine cancellate sculpture in the early stages. Length, 15 mm; width, 12 mm; height, 8.5 mm. On gravel bottoms, in depths of 18 to 73 m; from Magdalena Bay, Baja California, to Guaymas, Sonora, and the Tres Marias and Revillagigedo Islands and Banderas Bay, Mexico.

Genus NESTA H. ADAMS, 1870

Apex at the posterior end; slit open at margin, slit band or selenizone broad, running the length of the dorsal surface; posterior margin thickened, forming a narrow shelf. Only three species are known, one in the Red Sea, one in the Caribbean, and a newly described one in the eastern Pacific.

10. Nesta galapagensis McLean, 1970. Shell small, white, sculptured with fine concentric and radial ribs that form squarish cancellations. Length, 5.5 mm; width, 3.3 mm; height, 1.7 mm. A single specimen dredged at Isabela Island, Galápagos, depth 145 to 180 m.

Genus PUNCTURELLA LOWE, 1827

Conical white shells with a fissure on the anterior slope near the apex; interior with a septum posterior to the fissure. Primarily a cool-water group, the genus is only sparsely represented in the Panamic province. Two Panamic subgenera.

Subgenus PUNCTURELLA, *s. s.*

Fissure not preceded by a double anterior rib; roof of the mantle entire, not divided.

11. Puncturella (Puncturella) punctocostata Berry, 1947 (Synonym: **P. ralphi** Berry, 1947). Smallest member of the genus in the eastern Pacific, this has a high conical shell with weakly beaded ribs and minute white punctations between. The fissure is elongate, and the septum is uncurved. Length, 4.3 mm; width, 3 mm; height, 2.8 mm. Monterey Bay, California, to Guadalupe Island, Baja California; one specimen dredged at Carmen Island, Gulf of California, in 37 m. Described from the Pleistocene of southern California.

Subgenus CRANOPSIS A. ADAMS, 1860

Fissure preceded by a double anterior rib; roof of mantle split from the margin to the fissure.

12. Puncturella (Cranopsis) expansa (Dall, 1896). A large, thin, oval shell with numerous fine radial ribs. Length, 32 mm; width, 26 mm; height, 10 mm. In deep water from southwest of Ensenada, Baja California, to Panama Bay and the Galápagos Islands, 1,015 to 2,712 m.

Genus RIMULA DEFRANCE, 1827

Small, white shells with cancellate sculpture, an elongate fissure midway on the anterior slope, and a slit band (selenizone) bordered by raised ridges.

13. Rimula mexicana Berry, 1969 (Synonym: **R. astricta** McLean, 1970). The fragile shell has an elongate basal margin, cancellate sculpture, the apex near the posterior margin, and the fissure drawn out into a long tapered slit anteriorly. Length, 6.4 mm; width, 3.9 mm; height, 2 mm. Specimens have been dredged at Tortuga Island, Guaymas, and San Esteban Island, Gulf of California, depth, 30 to 80 m.

Genus ZEIDORA A. ADAMS, 1860

Fragile and translucent shells with the apex at the posterior end, the slit band (selenizone) running the length of the dorsal surface, with an open slit at the anterior margin. Internally there is a broad septum at the apex, much as in *Crepidula*. Deepwater forms, with two species in the Caribbean.

14. Zeidora flabellum (Dall, 1896). Extremely fragile, the shell of the only known specimen was somewhat crushed in the dredge but in good enough condition so that a drawing could be made reconstructing its original outline. Length, 10 mm. Off Clarion Island, Revillagigedo Islands, Mexico, in 840 m.

Subfamily FISSURELLIDINAE

Apex absorbed in mature shells, perforation at summit. Color patterns simple, mostly rays of gray or reddish gray on a white ground. Radula like that of the Emarginulinae, rachidian tooth broad, without cusps. Three Panamic genera.

Genus DIODORA GRAY, 1821
(DIADORA of authors, unjustified emendation)

Sculpture cancellate; callus inside apex squared posteriorly; muscle scars with hooked ends.

15. Diodora alta (C. B. Adams, 1852). The shell is relatively high, with coarse cancellate sculpture. It is grayish white, rayed with dark gray or black, and the

perforation is small and oval. Length, 13 mm; width, 10 mm; height, 8 mm. Head of the Gulf of California to Paita, Peru.

16. Diodora digueti (Mabille, 1895) (Synonym: **D. constantiae** Kanakoff, 1953). Distinguished from the similar and more common *D. inaequalis* by an oval or nearly circular orifice, the sides of the shell tending to be convex, whereas in *D. inaequalis* they are flat. Also, the internal callus in *D. digueti* is broader. The whitish shells are rayed and the internal callus bordered with gray. Length, 24 mm; width, 14 mm; height, 9 mm. San Ignacio Lagoon, Baja California, and throughout the Gulf of California, south to Salinas, Ecuador. The juvenile apex or spur persists just posterior to the perforation until the shell reaches about 7 mm in length. The species occurred in southern Californian embayments in the Late Pleistocene (recorded as *D. inaequalis* by Arnold [1903]), and the immature stage was described as a separate species, *D. constantiae*.

17. Diodora fontainiana (Orbigny, 1841) (Synonyms: **Fissurella aspera** Sowerby, 1835 [not Rathke, 1833]; **Fissuridea asperior** Dall, 1919 [new name]). Relatively large, with coarse cancellate sculpture and a large oval perforation. Length, 28 mm; width, 22 mm. Pacasmayo to Islay, Peru (type locality); not reported in recent years.

Diodora granifera (Pease, 1861). Described from the Hawaiian Islands and otherwise an Indo-Pacific form; reported as living on Clipperton Island.

18. Diodora inaequalis (Sowerby, 1835) (Synonyms: **Rimula mazatlanica** Carpenter, 1857; **Fissurella pluridentata** Mabille, 1895). The orifice tends to be tripartite, broad in the middle and constricted at both ends, becoming oval in some older specimens. Shell proportions are variable. Coloring is tan with gray or brown rays, the callus within bordered by gray. The anterior end of the shell is usually more narrowed than in *D. digueti*. Length, about 27 mm; width, 16 mm; height, 8 mm. Head of the Gulf of California to Santa Elena Peninsula and the Galápagos Islands, Ecuador. Carpenter's type, named as a *Rimula*, proves, upon study, to be a juvenile *Diodora*, probably *D. inaequalis*.
See Color Plate XII.

19. Diodora panamensis (Sowerby, 1835). Small, relatively high and straight-sided, with an oval base; orifice oval, nearly circular. The color markings are concentric dark streaks in a zigzag pattern. Length, 18 mm; width, 11 mm. La Union, El Salvador, to Gulf of San Miguel, Panama, in depths of 9 to 18 m. One of the least common Panamic species of the genus.

20. Diodora pica (Sowerby, 1835). Orifice unusually small and nearly circular, about one-third the distance from the anterior margin. The margins on either side of the shell are nearly parallel; outline evenly rounded, not narrowed anteriorly. The shell is white around the orifice, greenish gray at the margins, with an irregular and abrupt line marking off the two areas. Length, 25 mm; width, 15 mm; height, 6.5 mm. San Juan del Sur, Nicaragua, to Santa Elena Peninsula, Ecuador. Like *D. panamensis*, this is of limited distribution and uncommon occurrence.

21. Diodora punctifissa McLean, 1970. Small, high, the sides raised relative to the ends, with an extremely small orifice, the apex remaining intact in the mature shell. Internal callus broad. Length, 9.5 mm; width, 7 mm; height, 6 mm. Known only from eight dead specimens dredged off Wenman Island, Galápagos Islands, in 180 m depth. The species is unique in retaining the apex in the adult.

22. Diodora pusilla Berry, 1959. Smallest of the eastern Pacific Diodoras, this is also the tallest. The orifice is round and is nearly central. Coloring is of grayish-brown rays on a whitish shell. Length, 5 mm; width, 3.6 mm; height, 4 mm. Head of the Gulf of California to Panama, on gravel bottoms, 9 to 27 m.

23. Diodora saturnalis (Carpenter, 1864). Distinguished from *D. inaequalis* by being thicker-shelled, broader, and higher for the length; marginal crenulations are coarser. The orifice is about one-third the distance back from the anterior margin, small, slotlike, three-lobed, rounded in older specimens. The margin is nearly in one plane, or only slightly arched upward at the sides. The callus is bordered with gray, as in *D. inaequalis*. Length, 24 mm; width, 17 mm; height, 10 mm. Carpenter chose the name because the "hole resembles the telescopic appearance of Saturn when the rings are reduced to a line." San Ignacio Lagoon, Baja California, throughout the Gulf of California and south to Salinas and the Galápagos Islands, Ecuador.

Genus LUCAPINELLA PILSBRY, 1890

Apex subcentral, the perforation or orifice large, sculpture of imbricate radial ribs. The shell is crenulate at the ends within, the posterior margin slightly elevated. Soft parts are too large to be completely concealed by the shell. West American species have been reviewed by McLean (1967).

24. Lucapinella aequalis (Sowerby, 1835). Orifice large, oblong, the sides of both the orifice and the shell nearly parallel. Coloring is of gray rays on a whitish ground, the sculpture showing up as imbricate only near the margins. Length, 23 mm; width, 13 mm; height, 4 mm. Port Guatulco, Mexico, to Santa Elena Peninsula, Ecuador.

25. Lucapinella callomarginata (Dall, 1871). Relatively large, with coarse, strongly imbricate ribs and a markedly elongate aperture; rayed with gray on a whitish ground. Length, 23 mm; width, 13 mm; height, 6 mm. Morro Bay, California, to Magdalena Bay, Baja California.

26. Lucapinella crenifera (Sowerby, 1835). The ribbing is coarse and the orifice elongate. Compared with *L. callomarginata*, this has more conspicuous ribbing and a narrower orifice. The color pattern is of reddish or gray rays on a white ground. Length, 20 mm; width, 11.5 mm; height, 5 mm. Salinas, Ecuador, to Independencia Bay, Peru, and the Galápagos Islands. Previously allocated to *Diodora* by some authors (for example, Keen, 1958), this is probably the basis of the southern record of *L. callomarginata* by others (Dall, 1909).

27. Lucapinella eleanorae McLean, 1967. The orifice is oval and proportionately smaller than in the other species. The shell is thin, reddish buff in color, with radiating bands of gray. Length, 18 mm; width, 11 mm; height, 3 mm. An offshore form, on rock and gravel bottoms in 18 to 37 m; Guaymas, Mexico, to Santa Elena Peninsula, Ecuador.

28. Lucapinella milleri Berry, 1959. Smallest of Panamic members of the genus, this has an elongate, oval orifice, relatively larger than the others. The whitish shell is rayed with reddish or gray. Length, 9 mm; width, 5 mm; height, 2 mm. Intertidally and offshore to depths of 18 m. Head of the Gulf of California to Pearl Islands, Panama Bay.

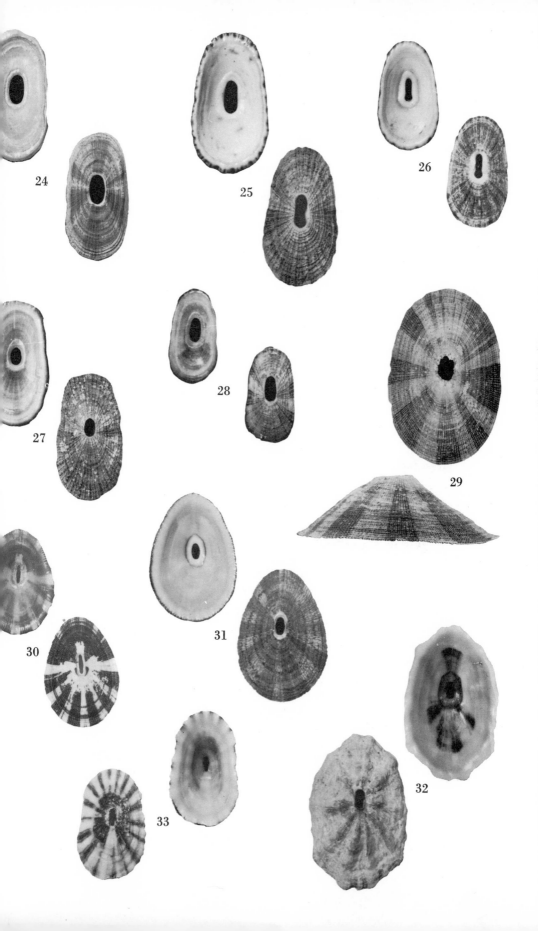

Genus STROMBOLI BERRY, 1954

With a relatively narrow callus only slightly truncate posteriorly (not, as in *Diodora*, squarely cut off); muscle scar with hooked ends. Orifice large and oval, as in the Californian *Megathura*, but the soft parts are relatively much smaller. Only one species is known.

29. Stromboli beebei (Hertlein & Strong, 1951). The sculpture of the relatively large shell is finely cancellate, and the orifice is large and oval. Coloring is of dark gray on a white ground, the interior white. Length, 68 mm; width, 46 mm; height, 17 mm. Punta Abreojos, Baja California, through the southern end of the Gulf of California (type locality near Gorda Bank) to Bahía Concepción, and south to La Plata Island, Ecuador, in 66 to 183 m.

Subfamily FISSURELLINAE

Perforation at summit, apex absorbed in mature shells; color patterns variegated, mostly of red or green rays. Rachidian tooth of radula narrow, similar in size and shape to adjacent laterals. Two Panamic genera.

Genus FISSURELLA BRUGUIÈRE, 1789

Sculpture of radial ribbing. Chiefly intertidal animals, the Fissurellas live attached to rocks, often where there is heavy surf. The soft parts of the animal are small enough so that the shell may be clamped firmly to the rock.

The genus reaches its most prolific development on the west coast of South America, in the temperate regions of Peru and Chile, where numbers of species and abundant large specimens occur. Two Panamic subgenera; there are no Panamic species that can be assigned to *Fissurella, s. s.*, the type species of which is West Indian.

Subgenus CLYPIDELLA SWAINSON, 1840

Shells depressed, with elevated ends, the orifice elongate and anterior to the midline. Two other species occur in the Caribbean.

30. Fissurella (Clypidella) morrisoni McLean, 1970. The ends of the shell are slightly elevated, the anterior end narrowed. The orifice is greatly elongate, and the radial ribbing is extremely fine. Outside, the shell is rayed with white and purplish brown; the inside is light green, with the external color pattern showing through. Length, 16 mm; width, 11 mm; height, 3 mm. Four specimens are known from Bahía Honda and San José Island, Pearl Islands, Panama.

Subgenus CREMIDES H. & A. ADAMS, 1854

Medium-sized shells with radial ribbing and a crenulate margin, lacking any internal dark border. Most of the Caribbean and tropical eastern Pacific species are assignable to this subgenus.

31. Fissurella (Cremides) asperella Sowerby, 1835 (Synonym: **F. ostrina** Reeve, 1850). The shell is flat-sided, markedly narrowed anteriorly, with sculpture of broad, flat ribbing that is fairly even-surfaced. Pink and white rays of color are flecked with gray, especially in specimens from Peru. The orifice is elongate-oval, faintly tripartite. Inside, the shell is pale green, the callus with a gray-green border. Length, 28 mm; width, 18 mm; height, 7 mm. Manzanillo, Mexico, to Lobos Island, Peru.

32. Fissurella (Cremides) decemcostata McLean, 1970. A large, flattened form, most specimens with 10 broad, rounded, nodular ribs projecting beyond the margin of the shell; the exterior usually is eroded or covered with coralline or other algae. The orifice is relatively small, oblong, and the interior of the shell is pale green, its callus outlined with reddish purple, the adjacent area with some irregular reddish-purple staining. Length, 28 mm; width, 18 mm; height, 6 mm. On flat exposed reefs, intertidally, Mazatlán to Puerto Angel, Oaxaca, Mexico. The species resembles *F. rubropicta* in having reddish staining on the inside.

33. Fissurella (Cremides) deroyae McLean, 1970. Smallest of the Panamic Fissurellas, this has fine irregular ribbing. It is white with pink rays, the interior pale green, banded concentrically with darker green, the external pattern of white rays showing through; callus area bordered with pink, orifice oval and tripartite. Length, 15 mm; width, 10 mm; height, 5 mm. Galápagos Islands.

34. Fissurella (Cremides) gemmata Menke, 1847 (Synonyms: **F. alba** Carpenter, 1857 [not Philippi, 1845]; **F. tenebrosa** Sowerby, 1863 [not Conrad, 1833]). The sturdy white shell has coarse, nodular ribs, alternate ribs often marked with gray; interior cream-colored, the callus bordered with gray. Some specimens have finer and more numerous ribs, but the sculpture is in any case rough-textured. Length, 35 mm; height, 11 mm. Mazatlán to Puerto Angel, Oaxaca, Mexico, on flat exposed rocks. Menke's type specimen has not been available for study; it seems from the description to be immature and probably worn.

35. Fissurella (Cremides) longifissa Sowerby, 1863. The orifice, as the specific name suggests, is of unusual length. The greenish shell is rayed with brown, surface ribbing broad and flattened, grooves between being bright pink, these pink lines persisting after the surface sculpture is eroded away. Within, the shell is dark green, its callus encircled with a pink line, the only mainland species to show a pink-bordered callus. Length, 28 mm; width, 19 mm; height, 8 mm. San Juan del Sur, Nicaragua, to La Plata Island, Ecuador.

36. Fissurella (Cremides) macrotrema Sowerby, 1834. The apical orifice is large, elongate, and beveled. The shell is pale green, finely sculptured with radial cords and grooves, with broad rays of brown. The inside is pale green, the callus not bordered. Resembling *F. virescens*, this form has finer sculpture and a distinctive color pattern. Length, 32 mm; width, 22 mm; height, 12 mm. Galápagos Islands.

37. Fissurella (Cremides) microtrema Sowerby, 1835 (Synonyms: **F. chlorotrema** and **humilis** Menke, 1847; **F. rugosa** of authors, not of Sowerby, 1835). Variable, with an irregular outline, depressed, narrow anteriorly usually with three anterior ribs that are stronger than the rest, projecting at the margin; the entire shell with rough and scaly irregular ribbing. Orifice relatively small, tripartite in young specimens, oval in mature shells. Interior greenish white, the callus area outlined with gray or yellow-orange. Color pattern variegated, mostly reddish- or greenish-rayed, major ribs light-colored. On exposed rocks, usually covered with tufts of algae. Length, 32 mm; width, 22 mm; height, 8 mm. Cape San Lucas and the southern part of the Gulf of California through Mazatlán, Mexico, and south to Salinas, Ecuador.

38. Fissurella (Cremides) nigrocincta Carpenter, 1856. Resembling *F. gemmata*, this has a smoother shell that lacks nodes on the broad ribs, which are rayed with gray and tan alternately, the grooves between lighter-colored. Interior

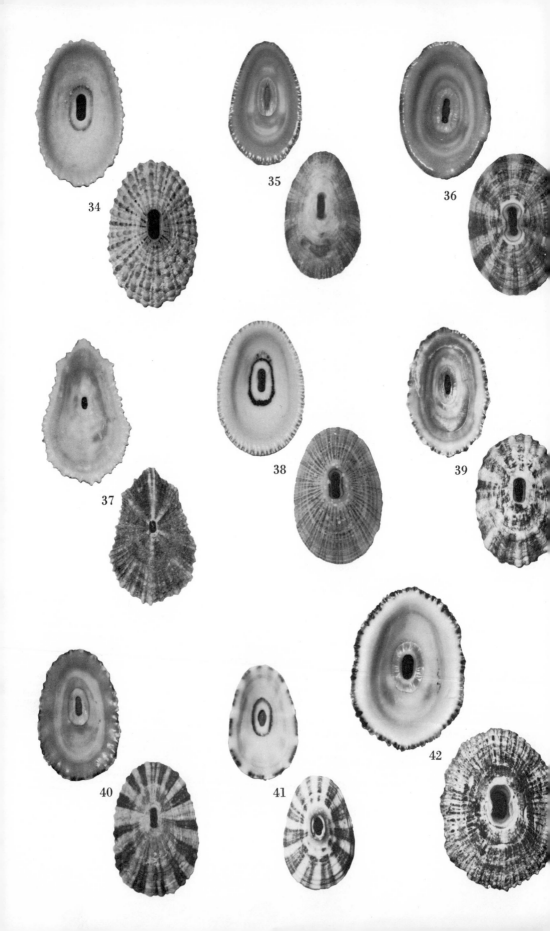

white, the callus bordered by black, and the adjacent area stained with gray. Length, 28 mm; width, 19 mm; height, 10 mm. Mazatlán, Mexico, to Salina Cruz, Oaxaca, Mexico.

39. Fissurella (Cremides) obscura Sowerby, 1835. Moderately elevated, the shell has strong, rounded ribs that project slightly at the margin and interspaces with fine radial striae. The color is reddish to dark brown or black, the ribs usually lighter than the interspaces. Interior solid green, the callus bordered with red. Length, 25 mm; width, 17 mm; height, 7 mm. Galápagos Islands; abundant.

40. Fissurella (Cremides) rubropicta Pilsbry, 1890. Moderately elevated, with coarse radial ribbing rayed with black and red. Within, the shell is pale green, shading to yellow or lavender in young specimens, the callus bordered by dark red; in mature specimens the interior becomes white, with irregular staining of red or purple around the callus. Length, 30 mm; width, 18 mm; height, 9 mm. Lagoon Heads, outer coast of Baja California, through the Gulf of California as far north as Tiburon Island; south to Oaxaca, Mexico. The largest specimens occur on the outer coast of Baja California; in the Cape San Lucas area this form occurs interspersed with *F. microtrema*.

41. Fissurella (Cremides) spongiosa Carpenter, 1857. Relatively small, lacking radial ribs but with fine radial grooves, rayed with gray on a white ground, some specimens gray with darker rays. The interior is white, the gray pattern showing through, the callus bordered by a reddish-brown line. Length, 16 mm; width, 10 mm; height, 6 mm. Mazatlán to Salina Cruz, Oaxaca, Mexico, more abundant in the southern part of its range.

42. Fissurella (Cremides) virescens Sowerby, 1835 (Synonyms: **F. nigropunctata** Sowerby, 1835; **Megatebennus cokeri** Dall, 1909). Largest of the Panamic Fissurellas, this species is easily recognized by the greenish color and uniform green interior. The internal callus is broad and flat. Young specimens are rayed with red, the color eroded away in mature shells. Length, 50 mm; width, 38 mm; height, 18 mm. Intertidally, Mazatlán, Mexico, to Lobos Island, Peru and the Galápagos Islands. Specimens from Panama and Colombia usually show black on the elevated radial ridges; this is the color form named *F. nigropunctata*. Both to the north and to the south the normal color pattern replaces it.

Genus LEUROLEPAS McLEAN, 1970

Sculpture wanting, shell nearly smooth except for raised growth lines; animal too large for the soft parts to be retracted under the shell; orifice large.

43. Leurolepas roseola McLean, 1970. The shell is pink, with radiating bands and flecks of tan. Internal callus narrow, margin rounded. Length, 11 mm; width, 7 mm; height, 3 mm. Offshore on rocky bottoms, in 10 to 20 m. Espíritu Santo Island, Gulf of California, to La Plata Island, Ecuador.

Superfamily PATELLACEA

Limpets have a conical or cap-shaped shell, the muscle scar horseshoe-shaped and open at the anterior end. The radula is docoglossate, differing from that of all other archaeogastropods in having a reduced number of teeth, with not more than three marginal teeth instead of the relatively large number that occur in rhipidoglossate forms. The families differ in the structure of the radula and gills.

Family PATELLIDAE

The Patellidae lack ctenidia (gills) altogether, having instead a series of leaflike structures along the mantle margin called the branchial cordon, which functions as a secondary gill. The radula has a central or rachidian tooth that may be either functional or vestigial.

Subfamily PATELLINAE

Radula with three lateral teeth.

Genus ANCISTROMESUS DALL, 1871

Shell massive, interior porcelaneous and opaque, not iridescent; radula with a functional rachidian tooth. A monotypic genus.

44. Ancistromesus mexicanus (Broderip & Sowerby, 1829) (Synonym: **Patella maxima** Orbigny, 1841). The dull white shell is generally eroded and shows a few low angles or obsolete ridges. Immature shells have ribs projecting beyond the margins and have yellowish interiors, the shell more narrowed anteriorly than in acmaeid species. The animal is black, mottled with white. The largest of living limpets, attaining a length of 120 to 150 mm or more. Mazatlán, Mexico, to Paita, Peru, on surf-beaten rocks at the low-water line. Overcollecting of larger specimens for food could become a major threat to the species along the more accessible parts of its range.

See Color Plate XIII.

Family ACMAEIDAE

The Acmaeidae differ from the Patellidae chiefly in having only a single ctenidium, a feather-shaped gill in the mantle cavity. Some genera have a branchial cordon, others do not. The rachidian tooth of the radula is lacking. There are three pairs of laterals; the marginals in this family are two at most and may be lacking altogether.

The acmaeid limpets are conspicuous on the west coasts of North and South America, both in number of species and in number of individuals. They occur mostly on rocks in the intertidal zone, although some species are characteristically subtidal and a few—chiefly northern species—occur on algae or marine grasses. The animals are entirely herbivorous. A majority of them have a more or less permanent home base to which they return after feeding; as a result, the edges of the shell will often be found to conform with the irregularities of the site of attachment.

American authors have customarily placed nearly all species in the genus *Acmaea* Eschscholtz, in Rathke 1833, and have tended to ignore the available subgenera, which are founded chiefly upon differences in the radula. A revised classification of the Panamic species, based upon characters of the radula and the presence or absence of a branchial cordon, is presented here. The genus *Acmaea*, s. s., is not represented in the Panamic province; it is a northern group of white or pink-rayed forms, chiefly sublittoral in habit, with a radula specially adapted for browsing upon coralline algae. More information on the northern species is given by McLean (1969).

A workable key to genera cannot be based upon shell characters, although generic separation on anatomical characters is simple.

1. Branchial cordon present...*Scurria*
   Branchial cordon lacking.. 2
2. Marginal teeth of radula well formed, two......................*Patelloida*
   Marginal teeth none or one....................................... 3
3. Marginal tooth of radula rudimentary, single...................*Collisella*
   Marginal teeth lacking.......................................*Notoacmea*

Genus **COLLISELLA** DALL, 1871
(**NOMAEOPELTA** BERRY, 1958)

Shell with apex well anterior to center, posterior slope convex; color variegated. Sculpture of heavy to fine radial ribbing. A branchial cordon is lacking. Marginal teeth of the radula are represented by a single pair of flaplike structures (often called uncini) that probably are vestigial remnants of marginal teeth. This group is represented in Japan and on the western coasts of North and South America; a few species occur in the western Atlantic.

45. Collisella acutapex (Berry, 1960). Relatively small and variable, generally with an elevated apex and 12 to 14 prominent whitish ribs that project at the margin; the interspaces smooth or finely ribbed. Ribs irregular and undulating, marked with a reticulate pattern of fine brown lines that tend to fuse and coalesce. Interior bluish white, external pattern tending to show through. Length, 16 mm; width, 13 mm; height, 7 mm. Mid-intertidal zone, head of the Gulf of California, south to Guaymas on the east and Cerralvo Island on the west. Specimens from San Felipe have more numerous white ribs, with solid black penciling in the interspaces. This has been confused with *C. mitella*, a more southern species with finer ribbing that does not project at the margin; it most resembles the Californian species *C. conus* (Test, 1945), from which it differs in lacking the solid brown central area inside the shell.

46. Collisella atrata (Carpenter, 1857). Relatively large, sculptured with strong whitish ribs that project at the margin; interspaces with fine radial ridges, rayed with black. The apex usually eroded, exposing some black banding; interior whitish, with irregular brownish clouding of the central area and black checkering at the margin. Length, 48 mm; width, 41 mm; height, 15 mm. Common at the midtide line, Magdalena Bay, Baja California, and north in the Gulf of California to Kino Bay, Sonora, Mexico. Records of the species south to Mazatlán and Acapulco have not been confirmed in recent years.

47. Collisella dalliana (Pilsbry, 1891). The shell of Dall's limpet is low, its dorsal surface arching smoothly to the apex, the brown surface sculptured with narrow but finely imbricate riblets that are flecked with radially arranged white maculations. The inside is bluish white, with a brown central stain, the outer color pattern showing through the thin shell. Length, 46 mm; width, 32 mm; height, 6.5 mm. Northwestern Gulf of California, from Puerto Peñasco to San Francisquito Bay. The unusually active animals retreat to protected positions on undersides of boulders during daylight hours and when the tide is out; consequently, the shell is uneroded and the margin smooth. The sensory tentacles along the mantle margin, present in all acmaeids, are especially well developed. It is the type species of *Nomaeopelta* Berry, a taxon based chiefly on the marked development of these tentacles; other anatomical features are characteristic of *Collisella*.

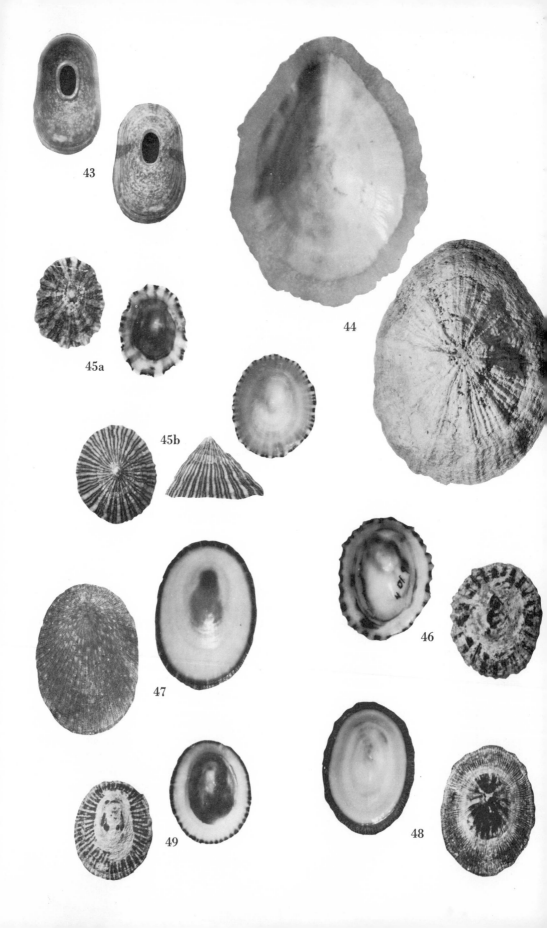

43

44

45a

45b

46

47

48

49

48. Collisella discors (Philippi, 1849) (Synonyms: **Acmaea mutabilis** Philippi, 1851; **Patella aenigmatica** Mabille, 1895). Sculpture of fine, even, radial ribbing, usually worn around the apex; ribs whitish, interspaces black-rayed, surface erosion revealing irregular dark areas. The inner margin of the shell solid black or gray with black rays, interior whitish with irregular staining of bluish gray. Shells are variable in height, some relatively flat, others high-arched. Length, 48 mm; width, 37 mm; height, 14 mm. On rocky reefs at the mid-tide to low-tide levels, Cape San Lucas, Tres Marias, and Revillagigedo Islands, and on the mainland from Mazatlán to Banderas Bay, Mexico. Records in southern Mexico are in need of confirmation.

49. Collisella mitella (Menke, 1847) (Synonym: **Patella navicula** Reeve, 1854). With fine, even, white ribbing and blackish interspaces, the ribs only slightly crenulating the margin, some specimens almost entirely black. Internal border checkered, interior bluish white, with darker central stains. Length, 13 mm; width, 10 mm; height, 6 mm. Mazatlán, Mexico, to Cabo Corriente, Colombia, not common south of Mexico. Mid-intertidal zone, often attached to *Chiton articulatus*. Southern specimens tend to have fewer ribs.

50. Collisella pediculus (Philippi, 1846) (Synonym: **Patella corrugata** Reeve, 1854). The shell is low, with about ten stout rounded ribs that project at the margins, one at each end and four along the sides; the color is yellowish white, with some black between the ribs in the early growth stages; irregular brown markings within. Length, 30 mm; width, 22 mm; height, 7 mm. On surf-exposed rocks at the low-tide line, Espíritu Santo Island, Gulf of California, to Port Utria, Colombia.

51. Collisella stanfordiana (Berry, 1957) (Synonyms: **"Acmaea" goodmani** Berry, 1960; **"A." concreta** Berry, 1963). Although the shell of this variable form resembles that of *Scurria mesoleuca*, the animal differs in lacking the branchial cordon, and the apex of the shell is closer to the anterior margin. Shells vary from thin and low to moderately thick and high; sculpture of fine radial striae, with white flecking or streaking, the interior a bright blue-green. This is the only New World *Collisella* with the interior green. As in *C. dalli*, the animal is active. The shell margin is even, not fitting any particular site on the rock. Length, 28 mm; width, 31 mm; height, 10 mm. Head of the Gulf of California south to Guaymas and Espíritu Santo Island, Mexico.

52. Collisella strigatella (Carpenter, 1864) (Synonym: **Acmaea paradigitalis** Fritchman, 1960). The shell is medium in size, moderately elevated, nearly smooth-surfaced. The apex is usually eroded, with a color pattern of irregularly forked black stripes on a white ground, some specimens with darker and lighter maculations and streaks. Internally the shell is a glossy bluish white, irregularly brown-stained in the central area, some of the outer pattern showing through. Length, 19 mm; width, 14 mm; height, 7 mm. On partially protected sloping surfaces in the mid-intertidal zone, British Columbia to Cape San Lucas, north in the Gulf of California to Loreto and Guaymas. Specimens from Guaymas tend to have a slightly more bluish cast than those from Cape San Lucas, the type locality. Californian specimens were given the name *A. paradigitalis* by Fritchman; this may qualify as a geographic subspecies, although morphologic differences are slight.

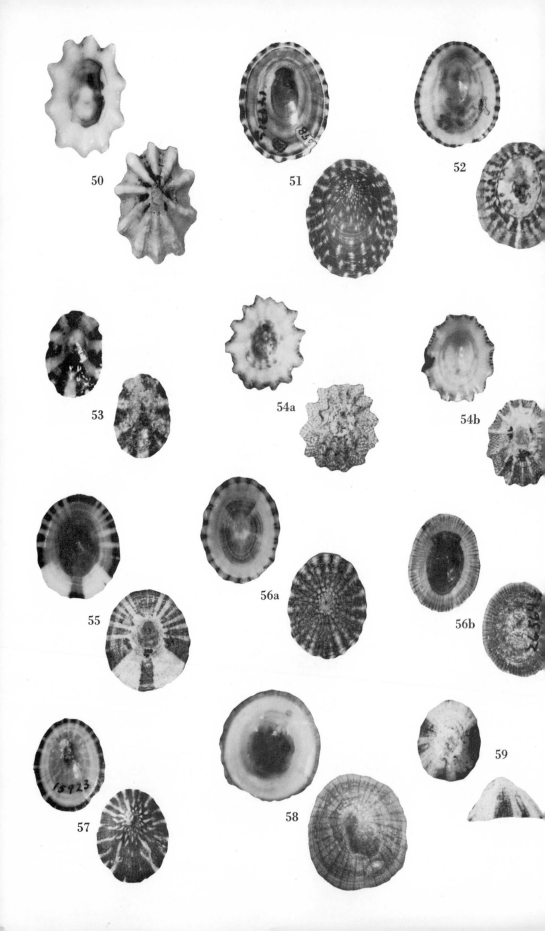

50

51

52

53

54a

54b

55

56a

56b

57

58

59

15923

53. Collisella strongiana (Hertlein, 1958). Shell small, low, with irregular radial ridges and bands of dark gray and greenish white. The surface is usually eroded; uneroded specimens show a fine reticulate network of light brown lines, the external color pattern plainly showing on the inside. Length, 9 mm; width, 6 mm; height, 2 mm. Length of a large specimen, 15 mm. Gulf of California from the head south to Puerto Libertad on the east and Cape San Lucas on the west, in the mid-intertidal zone.

54. Collisella turveri (Hertlein & Strong, 1951) (Synonym: **Acmaea turveri fayae** Hertlein, 1958). The shell is thin and unusually low, sculptured with 10 to 14 low radial ribs that project at the margin, intersected by fine concentric striae, the interspaces lacking the ribbing of *C. acutapex*. The apical area is usually worn, exposing a stellate pattern of shell layers. The outermost area of the shell is marked with a fine reticulate pattern of brown lines on white. The interior is bluish white, with yellow and brown staining in the center, the border usually dark. Length, 18 mm; width, 11 mm; height, 4 mm. Head of the Gulf of California south to Guaymas and Bahía de Los Angeles. Specimens from San Felipe have been given the name *C. turveri fayae* (Hertlein, 1958). They are consistently more elevated and the ribs are narrower.

Genus NOTOACMEA IREDALE, 1915

Notoacmea differs principally from *Collisella* in a small but consistent structural feature of the radula—absence of the marginal teeth or uncini. Shells are not heavily ribbed, in contrast to most of the species of *Collisella*. The genus is widely distributed: it occurs in South Australia and New Zealand, where *Collisella* is not present; in Japan; and on both coasts of North and South America. As in *Collisella*, the branchial cordon is lacking. Shell colors are variegated and the species are chiefly intertidal in habitat. Iredale's original spelling of *Notoacmea* is unfortunate, but it is now too late to emend it to *Notoacmaea*.

55. Notoacmea biradiata (Reeve, 1855) (Synonyms: **Acmaea vernicosa** Carpenter, 1865; **A. fonsecana** Pilsbry & Lowe, 1932). Nearly circular in outline, with a moderately elevated apex; surface finely striate radially. Color may be predominantly white with darker rays, or the darker rays may cover the greater part of the shell. Many specimens show two broad white posterior rays. Darker shells have a bluish interior; in the lighter ones the exterior pattern shows through. Length, 15 mm; width, 13 mm; height, 6.5 mm. Gulf of Fonseca, El Salvador, to Panama Bay. Although the range is shared with *N. subrotundata*, this species is distinguished by having a shell that is higher and also by the different color pattern. Reeve's cited locality was "China Seas," where no such shell has since been taken, nor has the species been recognized anywhere else. His figure is a perfect match for this eastern Pacific species. The earlier name is here utilized.

56. Notoacmea fascicularis (Menke, 1851) (Synonyms: **Patella opea** Reeve, 1854; **Nomaeopelta myrae** Berry, 1959). The shell is low and sculptured with minute radial ribbing; color reddish brown to white with thin, dark brown lines, some specimens with tan maculations near the lines. Interior bluish white, color pattern showing through on young shells, interiors of mature shells white, with brown staining in the central area. Specimens from southern populations are relatively large, often uniformly reddish brown or white. Length, 25 mm; width, 18 mm; height, 7 mm; large southern specimens reach a length of 37 mm. Mazatlán, Mexico, to Playas del Coco, Costa Rica. There is some indication that this is a complex involving more than one species; some populations have the three lateral teeth of nearly equal size, while in others the outer lateral is reduced. As yet no shell characters have been found that correlate with the radular differences.

57. Notoacmea filosa (Carpenter, 1865). The shell is low and broadly oval, with minute radial and concentric sculpture. It is gray-brown with irregular white rays, the inside bluish white with a broad checkered border, the outside colors showing through. Length, 18 mm; width, 13 mm; height, 3 mm. La Libertad, El Salvador, to Gorgona Island, Colombia, and the Galápagos Islands, on the undersides of rocks at low tide.

58. Notoacmea subrotundata (Carpenter, 1865). Resembling *N. filosa* but more rounded—nearly circular in outline. The surface is brownish, the radial ribs tending to be slightly darker. Some specimens show faint whitish rays near the apex; in others the narrow white rays extend to the margins. The interior is dark gray-blue with irregular brown staining in the central area, the margin dark, white rays showing through. Length, 18 mm; width, 17 mm; height, 4.5 mm. La Libertad, El Salvador, to Panama, on undersides of rocks in the mid-intertidal zone, higher than *N. filosa* where the two occur together.

Genus PATELLOIDA QUOY & GAIMARD, 1834

The radula in this genus differs from all other acmaeids in having two pairs of elongate, probably functional, marginal teeth. This is primarily an Australian, New Zealand, and Japanese group, where the species rival the eastern Pacific Collisella in numbers and size. In the New World, it is represented by a single eastern Pacific and a single western Atlantic species.

59. Patelloida semirubida (Dall, 1914). The shell is small, with a high apex. Sculpture is of fine radial and concentric riblets, the color whitish red with red or pink rays, some with chevron-shaped markings of pink. Length, 7 mm; width, 6 mm; height, 4.5 mm. Head of the Gulf of California to Panama, on rocks at low tide, more abundant offshore, to depths of 10 m. Superficially this resembles the Californian *Acmaea rosacea* Carpenter, 1864, but that species does not show the chevron-shaped markings. *Patelloida pustulata* (Helbling, 1799) is a related Caribbean species that attains a much larger size.

Genus SCURRIA GRAY, 1847

In addition to the single ctenidium there is a branchial cordon—a secondary gill structure consisting of leaflike flaps located just inside the mantle margin. The branchial cordon is continuous around the entire margin. In *Lottia*, a related Californian genus, it is missing in the head area. The radula is similar to that of *Collisella*, having the vestigial marginal remnants or uncini. The genus is best developed on the western coast of South America, in the Peruvian province.

60. Scurria mesoleuca (Menke, 1851) (Synonyms: **Patella diaphana** Reeve, 1854; **P. floccata, striata** [sp. 99, not 58], and **vespertina** Reeve, 1855). Although the shell resembles that of *Collisella stanfordiana*, especially in its blue-green interior, the branchial cordon is an important diagnostic feature that shows it to be *Scurria*. Shells have fine radial ribbing and are highly variable. Color patterns differ somewhat from those of *C. stanfordiana*; some specimens have white rays extending to the edge of the shell, and on others there may be large areas with white mottling. Length, 34 mm; width, 28 mm; height, 8 mm. Common in protected rocky areas intertidally, southern Baja California from Cerralvo Island to Cape San Lucas; Mazatlán, Mexico, to Santa Elena Peninsula and the Galápagos Islands, Ecuador.

61. Scurria stipulata (Reeve, 1855). This resembles *S. mesoleuca* in having the interior bright green, but it differs in the ribbing, which is irregular, of rounded, broad rays, the ribs separated by darkly pigmented grooves. Shells are irregular in outline, some of the larger ribs projecting more than others. The exterior is greenish black, faintly rayed with white, the interior intensely blue-green, the entire central area usually a solid chestnut-brown. Length, 30 mm; width, 23 mm; height, 9 mm. Gulf of Fonseca, El Salvador, to Gorgona Island, Colombia, intertidally in rocky areas.

Superfamily TROCHACEA

Globose to conical shells with several whorls, interior nacreous. Radula rhipido-glossate, with numerous marginal teeth. Animal having a series of epipodial tentacles around the upper side of the foot.

Family TROCHIDAE

The top shells are conical, with a round (entire) aperture, the inner layer of the shell nacreous. The horny operculum is multispiral and nearly circular. Many genera have a characteristic radula, which affords a good basis for subfamily grouping; apertural differences, such as presence or absence of a columellar tooth, prove to be less satisfactory as criteria. Four of the available subfamily groupings, based on radular differences, are recognized here. The key is to the genera in these subfamilies.

1. Aperture smooth, lacking columellar plicae or denticles................ 2
   Aperture with columellar plicae or denticles.......................... 5
2. Sculpture bluntly nodose...................................*Bathybembix*
   Sculpture not bluntly nodose.. 3
3. A beaded cord present, bordering umbilicus....................*Solariella*
   No beaded cord bordering umbilical area.............................. 4
4. Sculpture of fine beading...................................*Calliostoma*
   Sculpture wanting, shell smooth...................................*Gaza*
5. Columella with a fold on the inner wall............................. 6
   Columella with one or more denticles at base........................ 7
6. Small, under 5 mm in height...............................*Mirachelus*
   Large, over 10 mm in height...................................*Turcica*
7. Inner lip reflected over umbilicus............................*Monilea*
   Inner lip not reflected over umbilicus.........................*Tegula*

Subfamily MARGARITINAE

In this subfamily, regarded as the most primitive, shells are relatively thin, especially in the pigmented outer layer. Most of the species are to be found offshore. The central and lateral teeth of the radula are elongate and have numerous cusps.

Genus BATHYBEMBIX CROSSE, 1893

Deepwater trochids with thin shells. The well-spaced axial and spiral ribs are nodose at the intersections. Two Panamic subgenera.

Subgenus BATHYBEMBIX, *s. s.*

Large forms, mature shells not umbilicate.

62. Bathybembix (Bathybembix) bairdii (Dall, 1889) (Synonym: **Solariella oxybasis** Dall, 1890). Shell large, covered with a yellow-green periostracum. Height, 43 mm; diameter, 37 mm. Bering Sea, Alaska, to the Gulf of Tehuantepec, Mexico, in 457 to 915 m.

63. Bathybembix (Bathybembix) macdonaldi (Dall, 1890). Larger than *B. bairdii*, with nodes only at the periphery; periostracum dark brown. Height, 75 mm; diameter, 66 mm. Manta, Ecuador, to Coquimbo, Chile, in 730 to 1,100 m.

Subgenus SOLARICIDA DALL, 1919

Smaller than *Bathybembix, s. s.*; mature shells umbilicate. Occurring only at abyssal depths in the Panamic province.

64. Bathybembix (Solaricida) ceratophora (Dall, 1896). With two cords per whorl, squarely nodose. Height, 28 mm; diameter, 24 mm. San Diego, California, to El Salvador, in 1,460 to 2,740 m.

65. Bathybembix (Solaricida) equatorialis (Dall, 1896). The nodes are more closely spaced and less strongly projecting than in *B. ceratophora*. Height, 21 mm; diameter, 20 mm. San Diego, California, to Panama, 1,830 to 2,190 m.

66. Bathybembix (Solaricida) galapagana (Dall, 1908). Spiral cords more numerous than in the preceding species. Height, 17 mm; diameter, 15 mm. Known from a single specimen, 2,490 m. Galápagos Islands.

Genus MIRACHELUS WOODRING, 1928

Small white shells with coarse cancellate sculpture, resembling *Turcica* in having a columellar fold.

67. Mirachelus galapagensis McLean, 1970. The small white shell has three spiral cords per whorl, the subsutural cord weakly developed, crossed by axial ribs that form square cancellations. Height, 3.6 mm; diameter, 2.9 mm. Cocos Island, off Costa Rica, to Galápagos Islands, Ecuador, in 91 to 183 m.

Genus SOLARIELLA WOOD, 1842

Deeply umbilicate shells in which the umbilicus is usually bordered by a beaded cord. The radula is diagnostic, with lateral teeth two to four in number, cusped only on the outer side. All species occur offshore, some in relatively deep water.

68. Solariella diomedea Dall, 1919. Whitish with brown flecking, with about five or six rounded spiral cords per whorl, beaded on the upper part of the whorl. Height, 7 mm; diameter, 6.5 mm. Cocos Island, off Costa Rica, to Galápagos Islands, Ecuador (type locality), in 37 to 128 m.

69. Solariella elegantula Dall, 1925. Relatively small, with a metallic luster and brown mottling. Spiral sculpture of a keel at the shoulder and one at the periphery, umbilical wall with axial ribbing. Diameter and height, 5.5 mm. Guaymas, Mexico, to Octavia Bay, Colombia, 37 to 90 m.

70. Solariella nuda Dall, 1896. The gray-white shell is smooth except for axial and spiral sculpture on the first two whorls. Height, 15 mm; diameter, 19 mm. Vancouver Island, British Columbia, to Clarion Island, Mexico, 366 to 915 m.

71. Solariella peramabilis Carpenter, 1864. Characterized by three to four spiral cords per whorl and fine axial lamellae. At the northern end of its range in fairly shallow water, this species occurs at depths greater than 180 m in the Gulf of California. Forrester Island, Alaska, to the Gulf of California.

72. Solariella tavernia Dall, 1919. Small, with fine axial and spiral sculpture. Height, 3 mm; diameter, 4 mm. Galápagos Islands, in 1,160 m.

73. Solariella triplostephanus Dall, 1910. There are three strong keels per whorl; base and umbilicus show fine spiral ribbing. Fresh shells have a brassy luster and are mottled with brown. Height, 6.5 mm; diameter, 7.5 mm. Cedros Island, Baja California, north in the Gulf of California to Angeles Bay, south to Santa Elena Bay, Ecuador, 18 to 55 m.

Genus TURCICA A. ADAMS, 1854

Resembling *Bathybembix* in having nodular sculpture. The columella has a twisted fold, resembling that of *Tegula*, but the radula shows closer affinity to *Bathybembix*.

74. Turcica admirabilis Berry, 1969. With numerous noded spiral cords on the whorls and base, the periphery defined by two strong cords, a channel between. Height, 38 mm; diameter, 33 mm. Cedros Island, Baja California, through the Gulf of California to Tiburon Island, and south to Panama Bay, at depths of 50 to 180 m. The Californian species *T. caffea* (Gabb, 1865) is smaller, with a single cord at the periphery. A Pliocene form, *T. brevis* Stewart, 1941, also from California, is larger, with coarser nodes than *T. admirabilis*.

Subfamily CALLIOSTOMATINAE

Radula with the first marginal tooth exceptionally large and thick; central and lateral teeth with numerous cusps.

Genus CALLIOSTOMA SWAINSON, 1840

The generic name (pronounced Callios'-toma) literally means "beautiful mouth," referring to the brilliant nacre of the interior. The shells are either flat-sided or with rounded whorls, mostly with fine beading and color markings. The base may have a slight umbilical depression; rarely is it deep. The aperture is round, with no folds or teeth on the columella.

75. Calliostoma aequisculptum Carpenter, 1865. The shell is low-spired, with convex whorls and an umbilical pit. Numerous spiral cords are sharply beaded throughout. Color is consistent, dark pink with brown markings on the spiral cords. Height, 18 mm; diameter, 25 mm. In rocky areas just below the low-tide line, Mazatlán to Acapulco, Mexico.

76. Calliostoma antonii (Koch in Philippi, 1843) (Synonym: **Trochus lima** Philippi, 1850). More conical than *C. aequisculptum*, having similar sculpture, the whorls slightly convex below the suture, with less evidence of an umbilical pit. Color varies from mottled gray to dark orange. Height, 22 mm; diameter, 24 mm. Gulf of Fonseca, El Salvador, to Cabo Blanco, Peru; intertidally in rocky areas.

77. Calliostoma bonita Strong, Hanna & Hertlein, 1933. The tan shell is maculated with brown, sculptured with smooth spiral threads on which are scattered bright brown dots. The shoulder is concave, and there is one strong carination; the base has a bright purple channel bordering the inner lip. Height, 23 mm; diameter, 24 mm. Mazatlán to Acapulco, Mexico, 37 to 73 m.

78. Calliostoma eximium (Reeve, 1843) (Synonym: **Trochus versicolor** Menke, 1851). The shell is light yellowish pink, or grayish with nearly smooth spiral ribs irregularly checkered with black, white, and deep red. The base is glossy, with well-spaced spiral cords. Height, 25 mm; diameter, 22 mm. Scammons Lagoon, Baja California, through the Gulf of California and south to Ecuador; offshore to 40 m, occasionally at low tide on sandbars.

79. Calliostoma fonkii (Philippi, 1860). The whorls are flat-sided and have three main spiral cords, the two uppermost beaded. Height, 17 mm; diameter, 16 mm. A single specimen of this south Peruvian species has been dredged at 457 m at Santa Maria Island, Galápagos Islands.

80. Calliostoma gordanum McLean, 1970. The shell is grayish white with light tan maculations and an unmarked base. The numerous spiral cords are finely beaded, the periphery defined by two cords with an intercalary thread between. Height and diameter, 20 mm. San Francisquito Bay to Gorda Bank, off Cape San Lucas in 128 m (type locality), Baja California, Mexico.

81. Calliostoma iridium Dall, 1896. The whorls are flat-sided, with two strong, beaded cords at the periphery and one below the suture, the area between these nearly smooth on the early whorls but showing fine beaded cords on the later whorls. Yellowish pink with radiating brown flammules, bronze pink iridescence strong on the early whorls. Height, 21 mm; diameter, 19 mm. Gulf of Panama, 232 to 280 m.

82. Calliostoma jacquelinae McLean, 1970. Lacking a color pattern but highly opalescent with lavender and green, the shell has two strong peripheral cords, fine cording on the rest of the whorl and the base. Height, 11 mm; diameter, 10 mm. Santa Cruz Island, Galápagos Islands, 146 m.

83. Calliostoma keenae McLean, 1970. The whorls are convex, the spiral cording unbeaded until the fourth whorl. later whorls with numerous finely beaded cords and interspaces of nearly equal width. The color is drab green or yellow with brown flammules. Height, 15 mm; diameter, 15 mm. Laguna Beach, California, to Cape San Lucas and the Revillagigedo Islands, Mexico, 55 to 110 m.

84. Calliostoma leanum (C. B. Adams, 1852) (Synonym: **Trochus macandreae** Carpenter, 1857). A relatively small form, reddish brown with lighter mottling. Whorls flat on the shoulder but with a rounded base. The spiral cording is broad and thick, with low beading, the interspaces narrow. The three cords above the suture are the broadest. Height, 12 mm; diameter, 11 mm. Tiburon Island, Gulf of California, to Santa Elena Bay, Ecuador, and the Galápagos Islands, on rocks at and just below lowest tide level.

85. Calliostoma marshalli Lowe, 1935 (Synonyms: **C. gemmuloides** and **angelenum** Lowe, 1935). The shell of this variable species runs to olive, tan, or reddish brown, with darker radiating flammules. The spiral ribs are well marked, and the beading is sharp and spinose. Height, 14 mm; diameter, 13.5 mm. Head of the Gulf of California south to Guaymas and Magdalena Bay, at low tide and offshore. *C. gemmuloides* is a slender variant and *C. angelenum* is a reddish variant with slightly less spinose beads.
See Color Plate XIII.

86. Calliostoma mcleani Shasky & Campbell, 1964. Similar to *C. leanum* but with more numerous, more finely beaded spiral cords. The color is a dull orange or yellow, with only faint indications of the mottling of *C. leanum*. Height, 11 mm;

diameter, 10 mm. Cabo Tepoca, Sonora, Mexico, to La Plata Island, Ecuador; on rocky bottoms offshore at depths greater than 9 m.

87. Calliostoma nepheloide Dall, 1913. The olive brown shell is marked with darker flammules and has flat-sided whorls with finely beaded spiral cords, the last whorl slightly convex; base defined by two elevated cords. Height, 25 mm; diameter, 22 mm. Point Abreojos to Cape San Lucas, Baja California; Mazatlán, Mexico, to Panama, not common, 73 to 128 m.

88. Calliostoma palmeri Dall, 1871. This species resembles *C. bonita* in general form and color pattern, including the purple channel bordering the inner lip, but it has a lower spire, reaches a larger size, has more subdued color markings, and the basal and peripheral carinations are not as strong. The peripheral carination is higher on the whorl than in *C. bonita*. No intergradation is known and the two species are regarded as distinct. Height, 24 mm; diameter, 28 mm. Head of the Gulf of California south to Guaymas, from the low intertidal level to 45 m depth.

89. Calliostoma rema Strong, Hanna & Hertlein, 1933. The shell is low-spired, with a polished surface and spiral sculpture resembling that of *C. eximium*. The umbilicus may be open, partially obstructed, or completely filled with callus. In specimens with an open umbilicus, it is bordered by a rounded carina ending in a small tooth at the columella. Specimens from a single locality may show different stages of umbilical filling, which tends to be more complete in larger shells and in those toward the south end of the range. Height, 14 mm; diameter, 19 mm. Mazatlán, Mexico, to Santa Elena Bay, Ecuador, in 18 to 45 m.

90. Calliostoma sanjaimense McLean, 1970. Resembling *C. iridium* in proportion, this differs in having strong beading on the early whorls. Yellow brown with lighter and darker markings on the cords, the early whorls highly iridescent. Height, 20 mm; diameter, 18 mm. Known only from the San Jaime Bank, off Cape San Lucas, 137 m.

91. Calliostoma santacruzanum McLean, 1970. The whorls are flat-sided and coarsely beaded; the base has fine spiral ribs and is defined by a strong peripheral cord and narrow open umbilicus. The tan shell has broad brownish maculations and shows yellow and green iridescence on the early whorls. Height and diameter, 7 mm. Known from a single, probably immature, specimen, dredged in 45 mm, Santa Cruz Island, Galápagos Islands.

92. Calliostoma veleroae McLean, 1970. The shell has a markedly concave outline; spiral cording is even, and the base is defined by a greatly thickened cord. There is a shallow depression in the umbilical area. The color is yellowish tan, with darker flammules. Height, 16 mm; diameter, 17 mm. Gulf of Panama, 100 m. Known from a single specimen with a broken outer lip.

Subfamily MONODONTINAE

Sturdy shells with variegated color patterns, usually with one or more denticles at the base of the columella. The rachidian tooth of the radula does not have a pointed overhanging cusp, but the lateral and marginal teeth are well cusped.

Genus TEGULA LESSON, 1835

Shells medium-sized, variously colored, globose to conic in form and smooth to strongly ribbed, always with a tooth at the base of the columella. Three Panamic subgenera.

Subgenus TEGULA, *s. s.*

Comprises a single species. The shell is nonumbilicate, with broad spiral cords that are coarsely beaded. Relationship to the other groups of *Tegula* is indicated by the radula.

93. Tegula (Tegula) pellisserpentis (Wood, 1828) (Synonyms: **Trochus elegans** Lesson, 1835; **T. strigilatus** Anton, 1839). The large, solid shell has zigzag markings of black; a smooth area of the base near the aperture has a greenish tinge in young specimens. Height, 45 mm; diameter, 37 mm. Gulf of Fonseca, El Salvador, to Gorgona Island, Colombia, intertidally on boulders at mid-tide level.

Subgenus CHLOROSTOMA SWAINSON, 1840
(OMPHALIUS PHILIPPI, 1847)

Some members of this group are umbilicate; others are imperforate at maturity. Most are large forms with obliquely rugose sculpture. Primarily it is a temperate-zone group, well represented in Japan, California, and Chile. One species occurs in the Gulf of California.

94. Tegula (Chlorostoma) rugosa (A. Adams, 1853) (Synonym: **Omphalius rufotinctus** Carpenter, 1857). The shell is heavy and dull gray, variegated with brown, black, or red streaks; the whorls are roughly and irregularly sculptured with oblique folds and some spiral threads. The umbilicus is deep, white within. Height, 36 mm; diameter, 33 mm. Head of the Gulf of California south to Concepcion Bay and Guaymas, abundant on rocks in the upper intertidal zone.

Subgenus AGATHISTOMA OLSSON & HARBISON, 1953

Small to medium-sized shells, with spiral sculpture, open umbilicus, base of columella with one to three denticles, the largest at the termination of a strong spiral cord bordering or passing into the umbilical cavity. Numerous species are represented in the tropical and subtropical Panamic and Caribbean provinces.

95. Tegula (Agathistoma) bergeroni McLean, 1970. Of small to medium size, this has a finely noded carination midway on the whorl. Base with well-spaced spiral threads, umbilical area white, callus tongue and umbilical wall greenish; denticles three. The yellow shell has axial bands of brown, extending uninterruptedly across the base. Height, 13 mm; diameter, 14 mm. Playas del Coco, Costa Rica, to Cabo Corrientes, Colombia, intertidally and offshore to 18 m.

96. Tegula (Agathistoma) cooksoni (E. A. Smith, 1877). The low-spired shell is small, sculpture consisting of fine striae. Whorls evenly rounded or slightly angulate. With a single columellar denticle. Color gray or brown, mottled with white or pink, umbilical area white. Height, 9 mm; diameter, 10 mm. Intertidally in rocky areas, Galápagos Islands. This resembles the West Indian *T. fasciata* (Born, 1778), type species of the subgenus *Agathistoma*, but is much smaller.

97. Tegula (Agathistoma) corteziana McLean, 1970 (Synonym: **T. globulus** of authors, not of Carpenter, 1857). The gray-brown shell is relatively small, the whorls evenly rounded, and the sutures deeply impressed. The spiral cords are even and prominent and are weakly mottled with yellow, having fine spiral striae in the interspaces. There is a single columellar tooth; the umbilical area is greenish white. Height, 12 mm; diameter, 13 mm. Abundant on rocks at low tide, north end of the Gulf of California to Cerralvo Island, southeastern Gulf of California, to Guaymas, Sonora.

98. Tegula (Agathistoma) corvus (Philippi, Mar. 1850) (Synonym: **Trochus impressus** Philippi, Dec. 1850, *ex* Jonas, MS). Relatively small, the gray-brown shell may show some axial markings of white, its base usually with diagonal white markings and the umbilical area green. Spiral sculpture of about four low cords per whorl, the base nearly smooth except for weak spirals, area in advance of the aperture eroded. Height, 13 mm; diameter, 13 mm. Paita to Lobos Island, Peru, intertidally. This name has been used erroneously for *T. globulus* and *T. corteziana*.

99. Tegula (Agathistoma) eiseni Jordan, 1936 (Synonyms: **Tegula ligulata** of authors, not of Menke, 1850; **T. (A.) mendella** McLean, 1964). The brownish shell is relatively large, whorls evenly rounded, spiral ribs strong, nodular, with alternating dark and light areas; base usually eroded in advance of the aperture, lip lirate, umbilical area white. Height, 25 mm; diameter, 22 mm. Los Angeles County, California, to the outer entrance of Magdalena Bay, Baja California, intertidally to 18 m. This Californian-province species has been confused with certain Panamic forms.

100. Tegula (Agathistoma) felipensis McLean, 1970. Of medium size, the yellow-green shell is axially banded with brown. Sculpture of strong spiral ribs, with a strong carination midway on the whorl, basal cords strong with narrower interspaces, umbilical area green. Height, 15 mm; diameter, 17 mm. Known only from the extreme north end of the Gulf of California, where it is abundant at San Felipe and uncommon at Puertecitos and Puerto Peñasco. Related to *T. mariana*, which has not been collected at San Felipe. Both species occur at Puertecitos and Puerto Peñasco, and there is no evidence of intergradation.

101. Tegula (Agathistoma) globulus (Carpenter, 1857). Smaller than *T. corteziana*, this has the spiral cords more numerous and the interspaces lacking the fine spiral striae. Shells are gray to light brown, often mottled with lighter areas, spiral cords of base marked with light and dark areas, umbilical area green. Height, 9 mm; diameter, 10 mm. Topolobampo, Sinaloa, and Tres Marias Islands to Acapulco, Mexico, abundant intertidally.

102. Tegula (Agathistoma) ligulata ligulata (Menke, 1850). The shell is brown with white flecking, and the umbilical area is white. The spiral cords are strongly developed. Cording above the midwhorl carination may be heavily noded in axial rows; entire shell with fine spiral striae. Height, 15 mm; diameter, 17 mm. The typical subspecies is known only from the Mazatlán area, at low tide and to depths of 9 m in rocky areas.

103. Tegula (Agathistoma) ligulata mariamadre Pilsbry & Lowe, 1932. Differs from the typical subspecies in having faintly indicated spiral cording, but the pattern and arrangement of the cording corresponds to that of the typical subspecies. Height, 16 mm; diameter, 18 mm. Cape San Lucas area east to Cabo Pulmo; Tres Marias Islands (type locality); Banderas Bay to Manzanillo, Mexico, at low tide and in the sublittoral zone. Since the two forms have not been collected together and the differences are strictly quantitative, they are regarded as geographic subspecies. This was originally described as a subspecies of *T. mariana*.

104. Tegula (Agathistoma) maculostriata (C. B. Adams, 1845). Relatively small, similar to *T. globulus* but having a lower spire and lacking all traces of green in the columellar area. Height, 7 mm; diameter, 9 mm. Cocos Island, Costa Rica. Pilsbry and Vanatta (1902) first pointed out the occurrence of this other-

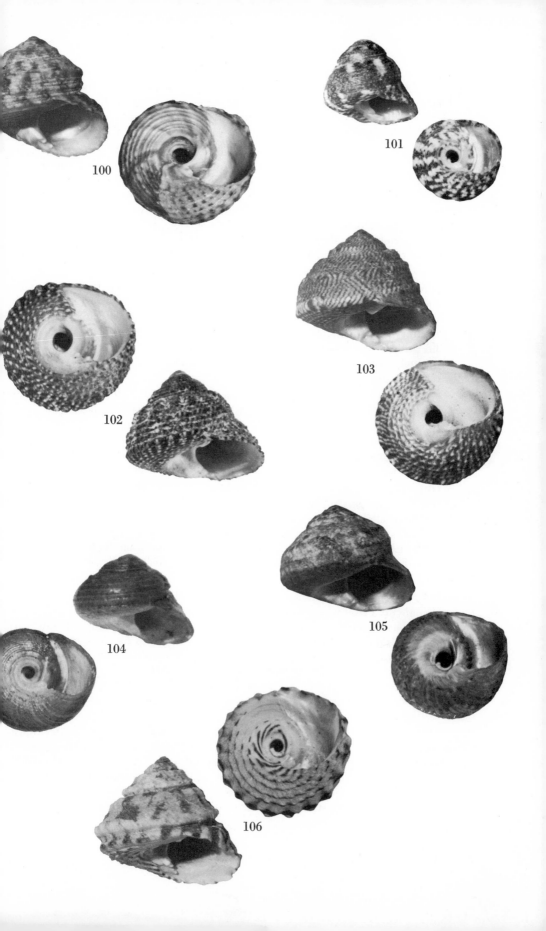

wise Caribbean species—known from Jamaica and the Bahama Islands—in the eastern Pacific only on Cocos Island. Material of this species has been collected at Cocos Island by other expeditions. It seems morphologically inseparable from Caribbean populations.

105. Tegula (Agathistoma) mariana Dall, 1919 (Synonyms: **Trochus coronulatus** of authors, not of C. B. Adams, 1852; **Omphalius turbinatus** Pease, 1869 [not of A. Adams, 1859]). The light brown or gray shell is variegated with lighter and darker areas, the umbilical area green. Spiral sculpture is variable, but the shell has a noded carination midway on the whorl, the basal area nearly smooth, with a waxen appearance. Height, 14 mm; diameter, 8 mm. The distribution is discontinuous: abundant from the head of the Gulf of California to Guaymas and La Paz; less common in the Gulf of Panama, with one record from Cabo Blanco, Peru.

106. Tegula (Agathistoma) melaleucos (Jonas, 1844). A large, conical form with deeply impressed sutures, the periphery defined by two strong keels, base with about five narrow, raised cords. Cream-colored, with diagonal brown banding, spiraling in toward the umbilicus on the base. Umbilical area white, aperture with strong internal lirae. Height, 25 mm; diameter, 24 mm. La Libertad, Ecuador, to northern Peru. This is larger and more conical than *T. rubroflammulata*.

107. Tegula (Agathistoma) panamensis (Philippi, 1849) (Synonym: **Omphalius smithii** Tapparone-Canefri, 1874). The large shell is globose, with evenly rounded whorls, suture not deeply impressed. The fine spiral ribs are nearly smooth, ground color reddish brown, streaked or mottled with greenish white, umbilical area whitish or slightly tinted with green. Height, 15 mm; diameter, 19 mm. La Libertad, El Salvador, to Paita, Peru, at low tide.

108. Tegula (Agathistoma) picta McLean, 1970, *ex* Dall, MS. This resembles *T. verrucosa*, but the sculpture is more subdued, particularly on the base, which has a smooth glossy appearance and has about five low cords, with broad, smooth interspaces. Color also differs and consists of pinkish-tan axial streaks on a greenish-white ground that fall into a radiating pattern on the base. Height, 19 mm; diameter, 24 mm. Intertidally, Cape San Francisco, Ecuador, to Talara, Peru.

109. Tegula (Agathistoma) rubroflammulata (Koch in Philippi, 1843) (Synonym: **Trochus coronulatus** C. B. Adams, 1852). Similar to *T. melaleucos* in having two peripheral carinations and a deeply impressed suture, with a strongly noded spiral cord below, this differs in being smaller and having a low spire, the sides of the whorls convex. Base with strong spiral cording, umbilical area white, aperture lirate within. Color is whitish, radiately striped with pink color bands, not spiraling into the umbilicus as in *T. melaleucos*. Height, 17 mm; diameter, 18 mm. On sand flats near rock and offshore on sand to 18 m. Head of the Gulf of California to Port Utria, Colombia.

110. Tegula (Agathistoma) snodgrassi (Pilsbry & Vanatta, 1902) (Synonym: **Tegula (Chlorostoma) barkeri** Bartsch & Rehder, 1939). Small to medium in size, the shell has the whorls evenly rounded and the spiral cording regular. Color is dark brown, occasionally reddish brown, with broad radiating areas of grayish white, the umbilical area green, callus tongue an intense green. Height, 15 mm; diameter, 14 mm. Intertidally, Galápagos Islands.

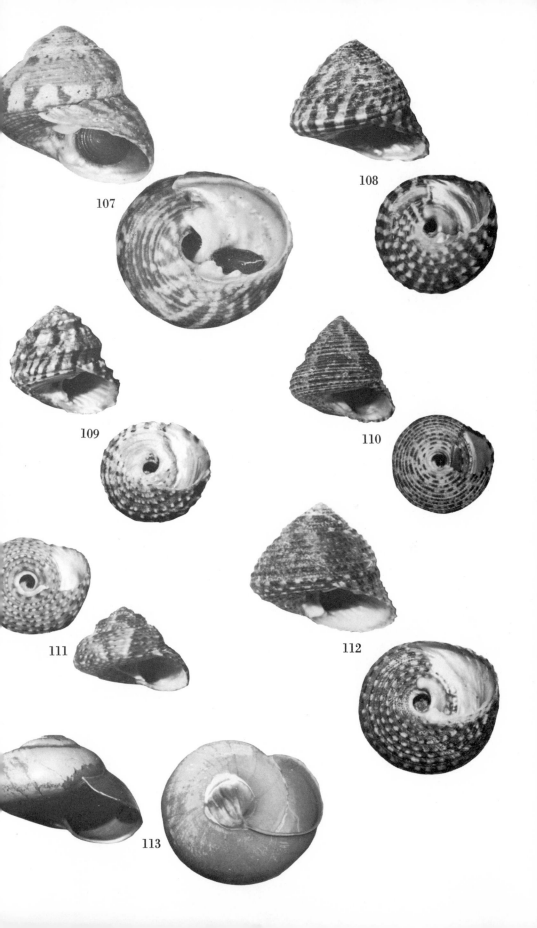

107

108

109

110

111

112

113

111. Tegula (Agathistoma) verdispira McLean, 1970. This resembles *T. globulus* in having rounded whorls and even spiral cording; it is lower-spired, has more deeply impressed sutures, and more pronounced spiral cording, particularly on the early whorls. It is characterized by the broad and well-defined spiral platform descending within the umbilicus, colored bright green. Height, 9 mm; diameter, 12 mm. Tres Marias Islands, Mexico, with *T. globulus*. A single specimen in the Los Angeles County Museum collection has been found with *T. corteziana* from Los Frailes, Baja California.

112. Tegula (Agathistoma) verrucosa McLean, 1970 (Synonyms: **Trochus reticulatus** Wood, 1828 [not Sowerby, 1821]; **Tegula byroniana** of authors, not Wood, 1828). Large and sturdy, whorls evenly rounded, base angulate, sutures weakly impressed. The spiral cords are coarsely but evenly beaded; fine spiral striae are apparent between the cords of the body whorl and base; interspaces of the basal cords are narrower than the cords, not broad as in *T. picta*. Height, 19 mm; diameter, 22 mm. La Libertad, El Salvador, to Sechura Bay, Peru, but not common south of Panama.

Subfamily UMBONIINAE

The umbilicus of the mature shell tends to be covered or partially obscured by callus, although some genera assigned to this group do not show this character at all. Of more importance is the structure of the central and lateral teeth of the radula, which lack the overhanging cusps of the other groups. Two Panamic genera.

Genus GAZA WATSON, 1879

Relatively large, globular shells, umbilicus partially or completely obscured by callus in mature shells. All occur in moderately deep water.

113. Gaza rathbuni Dall, 1890. The yellow-green shell is highly opalescent on the early whorls, the surface with fine spiral striae. The umbilicus is fully sealed with callus in mature specimens. Height, 32 mm; diameter, 28 mm. Galápagos Islands, 717 to 915 m.

Genus MONILEA SWAINSON, 1840

Sturdy shells with rounded whorls and spiral sculpture, the inner lip recurved toward the umbilicus. An Indo-Pacific group except for the one American species.

114. Monilea patricia (Philippi, 1851) (Synonym: **? Monilea kalisoma** A. Adams, 1853). The apex is pointed and the sutures are channeled; color tan, with light and darker markings on the spiral cords, the inner lip reflected slightly across the umbilicus. Height, 17 mm; diameter, 20 mm. The original locality for both names was indefinite and an eastern Pacific occurrence was doubted until four specimens were dredged by a California Academy of Sciences expedition from the vicinity of Corinto, Nicaragua.

Family SKENEIDAE

Small to minute, the white shells lacking interior nacre; operculum corneous, multispiral; aperture circular, umbilicus or umbilical chink present; radula rhipidoglossate, lateral teeth with cusps. No other family combines the features of rhipidoglossate radula, lack of nacre, and multispiral operculum. The key is to genera.

1. With surface sculpture, either spiral or axial or both.................. 2
   Without surface sculpture...................................... 4
2. Axial sculpture lacking.............................*Haplocochlias*
   Axial and spiral sculpture present................................ 3
3. Axial sculpture predominating.............................*Brookula*
   Spiral sculpture predominating..........................*Parviturbo*
4. Peritreme incomplete ....*Ganesa*
   Peritreme complete ....*Granigyra*

Genus BROOKULA IREDALE, 1912

Inflated, shells glassy in texture; sculpture of axial ribs and spiral threads; minute. No species of the subgenus *Brookula*, s. s., occur in the Panamic province.

Subgenus VETULONIA DALL, 1913

Umbilicate; outer lip reflected, thickened.

115. Brookula (Vetulonia) galapagana (Dall, 1913). Height, 2.2 mm; diameter, 3.4 mm. Galápagos Islands, in 1,160 m.

Genus GANESA JEFFREYS, 1883

Whorls rounded; smooth; peritreme incomplete; minute shells, mostly from deep water.

116. Ganesa panamensis Dall, 1902. Height, 4.7 mm; diameter, 4.3 mm. Gulf of Panama, in 1,870 m.

Genus GRANIGYRA DALL, 1889

Whorls rounded; smooth; peritreme complete. Minute; described from deep water.

117. Granigyra filosa (Dall, 1919). Height, 2.3 mm; diameter, 2.5 mm. Galápagos Islands, 1,160 m.

118. Granigyra piona (Dall, 1919). Height, 2 mm; diameter, 2 mm. Galápagos Islands, 1,160 m.

Genus HAPLOCOCHLIAS CARPENTER, 1864

Globose shells with fine spiral sculpture, narrow umbilicus, and thickened outer lip.

119. Haplocochlias cyclophoreus Carpenter, 1864. A sturdy yellow-white shell, the early whorls darker, the umbilicus nearly closed. Height, 5 mm; diameter, 5.2 mm. Magdalena Bay, Cape San Lucas, and Espíritu Santo Island, Baja California; Tres Marias Islands and south to Banderas Bay, Jalisco, Mexico. Intertidally to 10 m.

120. Haplocochlias lucasensis (Strong, 1934). Much smaller than *H. cyclophoreus*, this is also higher-spired. Height, 1.7 mm; diameter, 1.6 mm. San Pedro Nolasco Island and Guaymas, south to Cape San Lucas and Socorro Island, Revillagigedo Islands, Mexico. Intertidally to 10 m.

Genus PARVITURBO PILSBRY & McGINTY, 1945

Small, globose shells with a narrow umbilicus, sculpture of strong spiral and axial threads.

121. Parviturbo acuticostatus (Carpenter, 1864) (Synonym: **Fossarus angiolus** Dall, 1919). Penultimate whorl with two strong spiral keels, some speci-

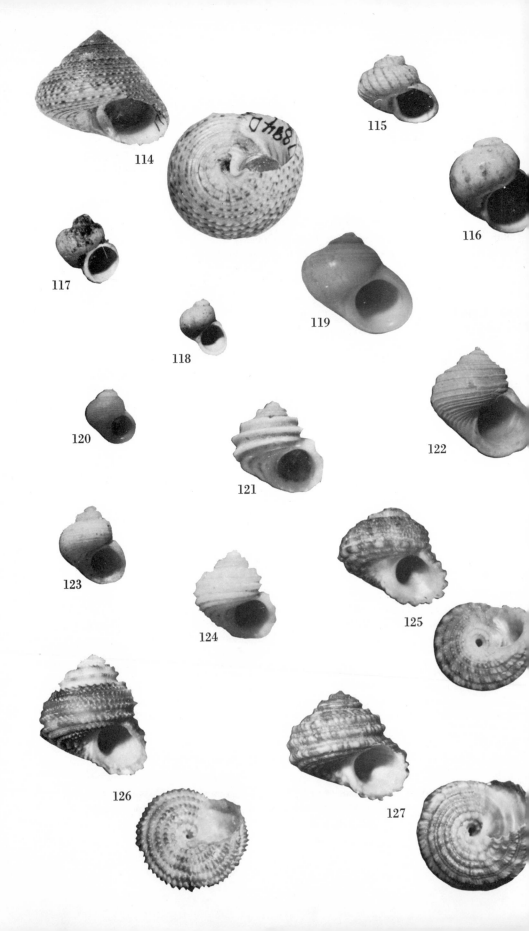

mens with coarse cancellate sculpture on the upper part of the whorl. Height, 2.5 mm; diameter, 2.5 mm. On gravel bottoms, Monterey, California, to Cape San Lucas and Cerralvo Island, Gulf of California, in 3 to 30 m.

122. Parviturbo concepcionensis (Lowe, 1935). Relatively large, this has four main cords on the penultimate whorl, interstitial spiral threads on the final whorl. Height, 5.6 mm; diameter, 5.6 mm. Head of the Gulf of California to Panama, in 15 to 35 m.

123. Parviturbo erici (Strong & Hertlein, 1939). With fine spiral sculpture; umbilicus bordered by a pair of exceptionally strong basal cords. Height, 2.1 mm; diameter, 1.8 mm. Guaymas, Mexico, to Gorgona Island, Colombia, 10 to 50 m.

124. Parviturbo stearnsii (Dall, 1918) (Synonym: **Liotia heimi** Strong & Hertlein, 1939). With three strong spiral cords on the penultimate whorl, usually with strong axial ribs on the uppermost cord. Height, 2.2 mm; diameter, 2.2 mm. Asuncion Island, outer coast of Baja California, north in the Gulf of California to Guaymas; south to Port Utria, Colombia, 3 to 30 m on gravel bottoms.

Family LIOTIIDAE

Small, umbilicate, the shells nacreous within; operculum multispiral, corneous on the inner surface, the outer surface with calcareous granules or beads in a radiating pattern and with bristles between the beads near the outer edges. Radula rhipidoglossate, rachidian tooth uncusped, lateral teeth cusped, marginals numerous. The liotiid operculum is distinctive. Two genera occur in the tropical eastern Pacific. The key is to genera and subgenera.

1. Shell white; sculpture imbricate only........................*Macrarene*
   Shell colored; sculpture both beaded and imbricate.................... 2
2. Spiral cords numerous, finely beaded........................*Otollonia*
   Spiral cords few, coarsely beaded.................................. 3
3. Base with coarsely beaded spiral cords........................*Arene, s. s.*
   Base flat, usually with radiating ridges......................*Marevalvata*

Genus ARENE H. & A. ADAMS, 1854

Small to minute, with variegated color patterns; spiral cords beaded or fluted into short spines, umbilicus deep, the umbilical wall with one or more spiral cords; suture descending in the last whorl, lip usually thickened. Inner surface of the operculum with a glossy corneous layer, outer surface concave, studded with beads in a radiating pattern, scanty bristles at the outer edge. Three Panamic subgenera.

Subgenus ARENE, *s. s.*

Sculpture of coarsely beaded spiral cording on the body whorl and base, the periphery rounded or stellate.

125. Arene (Arene) adusta McLean, 1970. The tan-colored shell is mottled with white, its umbilical area lighter, the outer lip slightly thickened. Distinguished from *A. lurida* mainly by having a strongly developed basal carination, especially prominent in juvenile shells. Height, 4.8 mm; diameter, 6.1 mm. Isla Partida and Espíritu Santo Island (type locality), Gulf of California, intertidally, with *A. (A.) luria*.

126. Arene (Arene) echinata McLean, 1970. Largest of the eastern Pacific Arenes, with rounded whorls and fluted scales on the spiral cords, the shell reddish brown, with irregular lighter areas. Height, 8.5 mm; diameter, 8.0 mm. Galápagos Islands.

127. Arene (Arene) ferruginosa McLean, 1970. Brick red, with lighter markings on the cords, this has three major beaded cords, a beaded subsutural cord, with interstitial threads on the final whorl, only the uppermost cord beaded on the first three whorls. Height, 6.0 mm; diameter, 6.9 mm. Occurring in deeper water than the other species—46 to 91 m—north of Acapulco, Mexico, to Bahía Honda, Panama; dwarf specimens have been taken off the Galápagos Islands and Cocos Island.

128. Arene (Arene) guttata McLean, 1970. The whitish shell is spotted with pink on the evenly beaded spiral cords; the whorls are rounded in profile. Height, 4.5 mm; diameter, 5.0 mm. On rocks at low tide, Galápagos Islands.

129. Arene (Arene) hindsiana Pilsbry & Lowe, 1932. A relatively large form, with a thickened outer lip. The basal cord is strong, with fluted scales. The tan shell is usually mottled with dark gray, often with a dark spiral band on the base. Height, 5.3 mm; diameter, 7.7 mm. At low tide and to depths of 10 mm in rocky areas, Mazatlán to Manzanillo and the Tres Marias Islands, Mexico.

130. Arene (Arene) lurida (Dall, 1913) (Synonym: **Liotia carinata** of authors, not of Carpenter, 1857). The whorls are rounded and the spiral cords are of even strength, finely beaded. Color uniform gray or reddish brown. Height, 5.9 mm; diameter, 6.5 mm. On rocks, at low tide, western side of the Gulf of California; Angel de la Guarda Island to Cerralvo Island. Populations from Carmen and Cerralvo Islands tend toward gray, those from Espíritu Santo Island to reddish brown.

131. Arene (Arene) olivacea (Dall, 1918) (Synonyms: **Liotia olivacea** var. **litharia** Dall, 1918; **A. winslowae** Pilsbry & Lowe, 1932). The whorls are rounded, although the peripheral carination is slightly more prominent than the others; this cord tends to develop fluted spines, especially on the early whorls. The shells are uniform gray, with some lighter flecking. Height, 5.5 mm; diameter, 6.2 mm. Intertidally, San Juan del Sur, Nicaragua, to Santa Elena Peninsula, Ecuador.

132. Arene (Arene) socorroensis (Strong, 1934). With three keels on the final whorl, the lower or basal keel the strongest, base with a concave area adjacent to this cord; the basal cords descending into the umbilicus. Spiral cords on shoulder finely beaded. Shells are brown or pink; height, 3.5 mm; diameter, 4.5 mm. Intertidally to 3 m on rocks, Cape San Lucas area north to Margarita Island and Cabo Pulmo; Socorro Island, Mexico.

133. Arene (Arene) stellata McLean, 1970. The uniformly pinkish-tan shell is low-spired, with two strongly elevated keels, on which are fluted scales, the area between the keels nearly smooth; shoulder and basal areas rounded and finely beaded; umbilical area with three prominent beaded cords. Height, 2.9 mm; diameter, 4.8 mm. Type lot of five specimens, none taken alive, from off Espíritu Santo Island, near La Paz, Gulf of California, in 9 m.

Subgenus MAREVALVATA OLSSON & HARBISON, 1953

Base flat, with weak spiral lirae or radial ridges extending from the crinkled cord bordering the umbilicus; mature specimens lacking spiral cords in the umbilicus.

134. Arene (Marevalvata) balboai (Strong & Hertlein, 1939). The shell is gray or mottled with tan and cream, sculpture consisting of three elevated spiral cords, occasionally with finer interstitial cords; base flat, cord bordering the umbilicus fluted, some specimens with radiating ridges across the base. Height, 3.8 mm; diameter, 4.0 mm. Head of the Gulf of California to La Libertad, Ecuador, in 18 to 37 m. The radial ridges across the base are stronger than those in the type species of the subgenus, *A. (M.) tricarinata* (Stearns, 1872), from the Caribbean.

Subgenus OTOLLONIA WOODRING, 1928

Spiral cords numerous and finely beaded; umbilicus of immature shells broadly open, closely constricted in mature specimens and bounded by a massive cord. The type species is from the Miocene of Jamaica, and the similarity to *Arene* has not previously been recognized.

135. Arene (Otollonia) fricki (Crosse, 1865) (Synonym: **Liotia rammata** Dall, 1918). The whorls are rounded, sculpture consisting of fine spiral cords, three slightly more prominent on the final whorl, cording in the subsutural and umbilical areas beaded. Color variable, usually a mottling of white, gray, red, brown, or black; a frequently occurring color form is purplish red with white axial markings. Height, 4.7 mm; diameter, 5.5 mm. The most abundant species, common on rocky bottoms, 3 to 20 m, head of the Gulf of California to Santa Elena Peninsula, Ecuador.

Genus MACRARENE HERTLEIN & STRONG, 1951

Relatively large, low-spired, white-shelled forms, sculptured with fine axial fimbriations (especially evident in juvenile specimens) and coarse axial and spiral riblets; periphery strongly keeled, usually with stellate projections. The operculum is markedly concave and has numerous narrow whorls, with dense tufts of bristles, especially at the margin; under magnification, fine calcareous beads are discernible between the tufts. Four species are known in the eastern Pacific; all occur offshore.

136. Macrarene californica californica (Dall, 1908) (Synonym: **Liotia pacis** Dall, 1908). Mature shells have a stellate periphery and a finely fimbriate and pitted basal ridge bordering the broad umbilicus. Juvenile shells have strong cancellate sculpture. Height, 20 mm; diameter, 26 mm. Guadalupe and Cedros Islands (type locality), south to Cape San Lucas and La Paz, Gulf of California, on rocky bottoms, in 37 to 183 m. The name *Liotia pacis* was based on a half-grown specimen.

Macrarene californica cookeana (Dall, 1918) (Synonym: **M. coronadensis** Stohler, 1959). More northern in distribution, ranging from southern California to Punta Banda, Baja California. The holotype was only 3 mm in diameter, but growth series connect this juvenile form to the adult named as a separate species— *M. coronadensis*—here regarded as a northern subspecies that has somewhat stronger spiral ridges on the base in mature shells.

137. Macrarene farallonensis (A. G. Smith, 1952). The whitish shell has a stellate periphery and a strongly noded basal cord. The shoulder is relatively smooth in the early whorls; in some specimens it develops a weak spiral cord, and an additional cord may appear between the basal and peripheral cords. Height,

9 mm; diameter, 12 mm. Dredged at depths of 64 to 137 m, Farallon Islands, California, to Cape San Lucas; juvenile specimens have been dredged near Angel de la Guarda and Raza Islands, Gulf of California.

138. Macrarene lepidotera McLean, 1970. Differing from *M. farallonensis* by having four cords rather than a single spiral between the stellate periphery and the main basal cord. Specimens also tend to be smaller and to have more numerous axial ribs. In none of the other species are fluted scales developed on the basal cords; hence the name, meaning scaly. Height, 7 mm; diameter, 10 mm. Socorro, Clarion, and San Benedicto Islands, Revillagigedo group, off west Mexico, in 73 to 238 m.

139. Macrarene spectabilospina Shasky, 1970. Characterized by prominent upturned spines that mark the periphery, this differs from the other species also in having a prominent basal ridge with more deeply impressed square pits on either side. There is an added spiral cord on the shoulder slope; on the base the one or two cords tend to become obsolete in mature specimens. Height, 10 mm; diameter, 14 mm. The Gulf of Tehuantepec, Mexico, in 55 to 75 m.

Family TURBINIDAE

Sturdy turbinate or top-shaped shells of few whorls, the aperture rounded, nacreous within. The operculum is calcareous and thickened, with a corneous layer on its inner surface. Three subfamilies are recognized here, based upon opercular and radular characters. The key is to genera and subgenera, omitting *Turbo*, *Astraea*, and *Homalopoma, s. s.*, subgenera that do not occur in the Panamic province.

1. Operculum with a multispiral pattern internally........................ 2
   Operculum with a paucispiral pattern internally...................... 3
2. Shell whitish in color....................................*Cantrainea*
   Shell red, with whitish nodes on the cords...................*Panocochlea*
3. Operculum round, nucleus central................................. 4
   Operculum ovate, nucleus lateral................................. 7
4. Mature shells umbilicate.................................*Chaenoturbo*
   Mature shells imperforate.. 5
5. Operculum pustulose, with a spiral ridge.....................*Callopoma*
   Operculum smooth or with a shallow central pit...................... 6
6. Shell sculpture of fluted scales or nodes...................*Marmarostoma*
   Shell sculpture of smooth spiral ribbing.....................*Taeniaturbo*
7. Operculum surface smooth...............................*Pomaulax*
   Operculum strongly ridged 8
8. Ridges three ....*Megastraea*
   Ridges two on operculum..................................*Uvanilla*

Subfamily HOMALOPOMATINAE

Small shells, operculum multispiral on the inner surface, outer surface slightly concave, thickened, with a paucispiral pattern; radula with an uncusped rachidian, five cusped lateral teeth, and numerous small marginals.

Genus HOMALOPOMA CARPENTER, 1864

Small shells with a slight denticle at the base of the columella, sculpture chiefly spiral over the body whorl and base, juvenile shells umbilicate, adult shells usually imperforate. *Homalopoma, s. s.*, does not occur in the Panamic province, though there are a number of shallow-water species in the Californian province. Two Panamic subgenera.

Subgenus CANTRAINEA JEFFREYS, 1883

White-shelled, with spiral ribs on the upper part of the whorl and a smooth base. Chiefly an Atlantic group, occurring in moderately deep water.

140. Homalopoma (Cantrainea) panamense (Dall, 1908). The whitish shell is relatively large, with a concave shoulder, three angulate cords at the periphery, and fine spiral striae throughout; operculum unknown. Height, 9.5 mm; diameter, 11.0 mm. Off Point Abreojos and Cape San Lucas, Baja California, to the Gulf of Panama, 380 to 1,150 m.

Subgenus PANOCOCHLEA DALL, 1908

Reddish shells, base smooth or with fine spiral ribbing, spiral cords of body whorl slightly nodose. Although *Panocochlea* was originally proposed as a subgenus of the trochid genus *Clanculus* Montfort, 1810, it is here employed for a group of *Homalopoma* species occurring in moderately deep water in the Panamic province.

141. Homalopoma (Panocochlea) clippertonense (Hertlein & Emerson, 1953). Relatively large and low-spired with even spiral ribs, the reddish shell is elegantly beaded with white nodes, base smooth. Height, 8.6 mm; diameter, 10.5 mm. Described from Clipperton Island, this has also been taken at San Benedicto Island, Cape San Lucas, and San Pedro Nolasco Island, Gulf of California, 135 to 360 m. The operculum is as yet unknown.

142. Homalopoma (Panocochlea) grippii (Dall, 1911). Shell red, spire elevated, ribs variable in strength and spacing, usually with a more prominent spiral rib at the midposition on the whorl; ribs sculptured with small white nodes, base smooth or finely striate, area in advance of the aperture with a thin glaze of callus. Operculum as in *Homalopoma, s. s.* Height, 7 mm; diameter, 8 mm. Santa Rosa Island, California, to Cape San Lucas and the Revillagigedo Islands, Mexico, 75 to 180 m.

143. Homalopoma (Panocochlea) rubidum (Dall, 1908). Resembling *H. clippertonense* but with a flatter spire and fewer spiral cords per whorl. The upper surface is pink, and the cords are less strongly beaded. Height, 7 mm; diameter, 13 mm. Known from two specimens, Gulf of Panama, 330 to 475 m. This is the type species of *Panocochlea*; the operculum is unknown.

Subfamily TURBININAE

Shells relatively large, with rounded whorls and entire aperture; operculum thick, its nucleus nearly central. The second marginal tooth of the radula is exceptionally large.

Genus TURBO LINNAEUS, 1758

Variously colored shells with whorls rounded and base convex. The operculum is calcareous, with a thin corneous layer on the inside that shows it is paucispiral, with a few rapidly expanding whorls; the outer surface may be smooth or sculptured. Many of the subgeneric divisions of *Turbo* are based on opercular differ-

ences. Adult shells are usually imperforate, but juvenile shells have a narrow umbilicus bordered by a beaded cord. Early sculpture consists of two strong peripheral keels. No species of subgenus *Turbo, s. s.*, occur in the Panamic province.

Subgenus CALLOPOMA GRAY, 1850

Operculum granular, with a strong spiral central ridge, a deep pit, and a marginal band of pustular ribs and deep grooves. Adult shells imperforate. The keels of the juvenile shells are slightly stellate. This is strictly an eastern Pacific group.

144. Turbo (Callopoma) fluctuosus Wood, 1828 (Synonyms: **T. fokkesi** Jonas, 1843; **T. assimilis** and **tessellatus** Kiener, 1847–48; **T. fluctuatus** Reeve, 1848). A variable form; in some specimens the spiral ribs are weak and nearly smooth, in others the ribs are strongly developed and nodular. Under a closely adherent periostracum, the color is greenish brown with white and darker brown markings along the spiral cords, often in a chevron pattern. Operculum as described for the subgenus, the central area white, the marginal area green. In some specimens from the outer coast of Baja California and from Mazatlán, the central spiral of the operculum is shallow, as in *T. saxosus*; in specimens from the Gulf of California the central pit is deep, showing two whorls. Height, 62 mm; diameter, 65 mm. The range is discontinuous; abundant from Cedros Island, outer coast of Baja California, through the Gulf and south to Banderas Bay and the Tres Marias Islands, Mexico; uncommon in the southern part of the range, from La Plata Island, Ecuador, to Paita, Peru, in rocky areas at low tide. This is the type of the subgenus.

145. Turbo (Callopoma) funiculosus Kiener, 1847–48 (Synonym: **Callopoma fluctuatum** var. **depressum** Carpenter, 1856). Lower-spired and with less tabulate whorls than *T. fluctuosus*, the shell is heavier and the area of the flaring lip below the columella is more pronounced. Operculum similar to that of *T. fluctuosus*, with a deeply penetrating central spiral ridge. Height, 56 mm; diameter, 55 mm. This is fairly common at the Revillagigedo Islands, Mexico, and has been collected but rarely in the Cape San Lucas area.

146. Turbo (Callopoma) saxosus Wood, 1828 (Synonyms: **T. nitzschii** Anton, 1839; **T. venustus** Philippi, 1845). Differing from *T. fluctuosus* by having fine imbricate or scaly axial sculpture, which is never displayed in *T. fluctuosus*. Periphery emphasized by fluted nodes. The operculum is similar to that of *T. fluctuosus*; in some populations, the spiral rib penetrates the central pit, in others it does not. The brownish-green shell may have axial markings of white. Height, 45 mm; diameter, 40 mm. The range is less broad than previously reported: San Juan del Sur, Nicaragua, to Paita, Peru, in rocky areas at low tide.

Subgenus CHAENOTURBO McLEAN, 1970

Mature shells umbilicate, operculum granular, with a thick spiral ridge and deep pit in the center, lacking grooves along the outer edge. The two keels of the juvenile shell weakly stellate. Based on a single West American species.

147. Turbo (Chaenoturbo) mazatlanicus Pilsbry & Lowe, 1932. Smallest of the West American Turbos and the only one with an open umbilicus, the shell has spiral sculpture and some fine axial striae. Coloration is variable, from greenish brown to orange, with lighter axial maculations, base speckled. Height, 15 mm; diameter, 15 mm. Cape San Lucas area and Mazatlán, Mexico, to Port Utria, Colombia, offshore on rocky bottoms, to depths of 37 m.

Subgenus MARMAROSTOMA SWAINSON, 1829

Mature shells imperforate, sculpture nodulose or spiny, operculum convex, smooth or finely striate or granular along the outer edge. The spiral carinae of juvenile shells of the two eastern Pacific species are not stellate as in *Callopoma* or *Chaenoturbo*.

148. Turbo (Marmarostoma) scitulus (Dall, 1919) (Synonym: **T. agonistes** Dall & Ochsner, 1928). Half-grown shells resemble those of *T. squamiger*, but fully mature specimens have broadly spaced, strong spines at the shoulder and a nearly smooth base. Color, usually purple, resembling that of coralline algae, with some white mottling. The operculum is unusual in having a shallow, darkly stained central pit that appears when the shell reaches a diameter of about 6 mm. Height, 29 mm; diameter, 31 mm. Galápagos Islands, 37 to 73 m. Dall described a juvenile specimen 2 mm in diameter; this matches the nuclear whorls of larger specimens, showing the reddish-brown flecking that occurs on the nucleus. No such marking occurs on the nucleus of *T. squamiger*. *T. agonistes* was described from a Pliocene fossil specimen for which the operculum was unknown; only in recent years have mature living specimens been found.

149. Turbo (Marmarostoma) squamiger Reeve, 1843 (Synonym: **T. pustulata** Reeve, 1843). Mature sculpture in this species consists of nodes or fluted scales on the spiral ribbing, particularly strong at the shoulder and on the base. The whitish operculum is smooth and has a shallow channel near the outer edge. Color greenish brown, variegated with white. Juvenile shells white. Height, 40 mm; diameter, 38 mm. Bahía San Luis Gonzaga, Gulf of California, to Paita, Peru, offshore on rocky bottoms, to 50 m. Although Reeve's material was supposed to have come from the Galápagos Islands, this form has not been collected there in recent years; thus, the original locality may be in error.

Subgenus TAENIATURBO WOODRING, 1928

Sculpture of spiral ribs, operculum nearly smooth.

150. Turbo (Taeniaturbo) magnificus Jonas, 1844. Nearly smooth, except for weak ribbing, the brown shell is marbled with white and purplish brown. Operculum smooth except for faint granules and an incised groove along the outer edge. Height, 50 mm; diameter, 48 mm. Manta, Ecuador, south to Callao, Peru, living below low tide line in rocky areas, to depths of 25 m.

Subfamily ASTRAEINAE

In this subfamily the operculum is oval or elongate rather than round, the nucleus of the operculum near the outer edge of the aperture. Radula as in Turbininae, the second marginal tooth larger than the rest.

Genus ASTRAEA RÖDING, 1798

Moderately large, conical shells, mostly imperforate with flat base. As in *Turbo*, the subgenera are distinguished primarily upon external sculpture of the operculum. Three Panamic subgenera; no species of subgenus *Astraea, s. s.*, occur in the Panamic province.

Subgenus MEGASTRAEA McLEAN, 1970

Exceptionally large shells characterized by three spinose ridges on the operculum and a thick, fibrous periostracum. The type species, *A. undosa* (Wood, 1828), is common in southern California, ranging south along the outer coast of Baja

California to Punta Abreojos. The two species in the subgenus are members of the subtropical Californian province.

151. Astraea (Megastraea) turbanica (Dall, 1910) (Synonyms: **A. petrothauma** Berry, 1940; **A. rupicollina** Stohler, 1959). This differs from the familiar *A. undosa* in having two strong peripheral ridges in mature shells rather than one. The lowermost ridge of the operculum is curved and nearly lacking spines, in contrast to that of *A. undosa*, which is uncurved, thick, and spiny. Height, 150 mm; diameter, 135 mm. Much less common than *A. undosa*, this has been found by divers and dredged on rocky bottoms, usually near kelp, at depths of 15 to 65 m, from the Coronado Islands, near San Diego, to the vicinity of Magdalena Bay, Baja California. Dall's specimen was live-collected but immature; Berry's name was based upon lower Pleistocene specimens from Los Angeles County. Large living specimens remained unknown until described by Stohler, 1959.

Subgenus POMAULAX GRAY, 1850

The operculum is smooth-surfaced, with a shallow groove running the length of the upper margin. Sculpture rugose, with strong spiral cording on the base; color reddish, under a thin periostracum.

152. Astraea (Pomaulax) gibberosa (Dillwyn, 1817) (Synonyms: **Trochus inaequalis** Martyn, 1784 [nonbinomial]; **T. diadematus** Valenciennes, 1846; **T. ochraceus** Philippi, 1846; **Turbo rutilus** C. B. Adams, 1852; **A. inaequale** var. **depressum** Dall, 1909; **Pachypoma magdalena** Dall, 1910; **P. lithophorum** Dall, 1910; **A. inaequalis montereyensis** Oldroyd, 1927; **A. guadalupeana** Berry, 1957). The large number of synonyms attests to the variability of this species. Some specimens have a rounded, others a stellate periphery, and some have a relatively concave base, but all comprise a single species. Height, 47 mm; diameter, 55 mm. Queen Charlotte Islands, British Columbia, south to Magdalena Bay area, Baja California, intertidal at the north end of the range, but at depths greater than 17 m at the south end.

153. Astraea (Pomaulax) spirata (Dall, 1911). Although described as a variety of *A. inaequale* [= *A. gibberosa*], this proportionally taller form is outside the range of variation of that species. Sculpture of the reddish-brown shell consists of oblique ridges, with strong spiral ridges on the base. Height, 37 mm; diameter, 30 mm. Not collected in recent years. All specimens known are beach-worn shells; the original locality was "Gulf of California," but there are single beach-worn shells from Sinaloa, Mexico, and the Gulf of Nicoya, Costa Rica, in the U.S. National Museum.

Subgenus UVANILLA GRAY, 1850

Small to medium-sized shells, the operculum in this subgenus has two smooth or granular ridges. The periostracum is relatively thin; sculpture consists of diagonal ridges, the concave base with fine spiral ribs. Limited to the tropical eastern Pacific.

154. Astraea (Uvanilla) babelis (Fischer, 1874). Resembling *A. unguis* in having the blunt spines along the base, this is smaller, proportionally taller and has coarser spiral sculpture on the base; one of the outermost ribs on the base has coarse beading. The upper ridge of the operculum is thick and broadly curving, the lower ridge relatively small. Height, 24 mm; diameter, 26 mm. The known range is limited, Cape San Francisco to the Gulf of Guayaquil, Ecuador, at low tide in rocky areas. Although Fischer described the species as of an unknown locality, Dall (1909) recognized it as Ecuadorian, probably because of the beading on the basal cords shown in the original illustration.

155. Astraea (Uvanilla) buschii (Philippi, 1844). Under a thin, fibrous periostracum, the greenish-brown shell is variegated with brown, green, or white. The periphery bears triangular spines and the base has fine lamellar sculpture and a single strong spiral rib. Height, 37 mm; diameter, 50 mm. Corinto, Nicaragua, to Paita, Peru, in rocky areas at low tide.

156. Astraea (Uvanilla) olivacea (Wood, 1828) (Synonyms: **Trochus brevispinosus** Valenciennes, 1846; **T. erythropthalmus** Philippi, 1848; **T. (Calcar) melchersi** Menke, 1851). Somewhat larger than *A. buschii*, this lacks the stellate projections at the periphery but has similar sculpture on the base, consisting of a single spiral rib. The most striking feature is the brilliant spot of reddish orange in the umbilical pit, bordered by dark brown or black; the rest of the shell is greenish brown under a fibrous periostracum. Diameter, 65 mm; height, 55 mm. Cape San Lucas and La Paz, Baja California; Mazatlán to Salina Cruz, Oaxaca, Mexico, at low tide and just offshore in rocky areas.
See Color Plate V.

157. Astraea (Uvanilla) unguis (Wood, 1828) (Synonyms: **Trochus digitatus** Deshayes, 1839; **T. amictus** and **chemnitzii** Valenciennes, 1846; **T. multipes** Philippi, 1850). The shell is variegated with brown, having long, curved, blunt spines on the periphery. Base lighter-colored, with numerous fine axial ribs of even size. The outer lip extends through about half the circumference of the shell, facilitating firm attachment on wave-exposed rocks. Height, 50 mm; diameter, 63 mm. The distribution is apparently discontinuous; abundant from Guaymas to Acapulco, Mexico; uncommon at Santa Elena, Ecuador, in rocky areas at low tide and just offshore. This is the type species of the subgenus.

Family PHASIANELLIDAE

Smooth, ovate shells lacking interior nacre, operculum calcareous, inner side paucispiral, nucleus near the basal margin. Color patterns variegated.

Genus TRICOLIA RISSO, 1826
(EUCOSMIA CARPENTER, 1864 [not STEPHENS, 1829]; EULITHIDIUM PILSBRY, 1898; USATRICOLIA HABE, 1956)

Small to minute forms, shells smooth or with fine spiral threads; base with a narrow umbilical chink; operculum radially striate along the outer margin. Radula with rachidian tooth elongate-oval, uncusped. Eastern Pacific species were reviewed by Strong (1928), and some were discussed by Robertson (1958) in his monograph on western Atlantic species.

158. Tricolia cyclostoma (Carpenter, 1864). Minute, early whorls finely striate; nucleus white, third whorl brown, last whorls smooth, glossy; whitish, with brown maculations or spots. Height, 2.6 mm; diameter, 2.1 mm. Dredged, 18 to 37 m. Puertecitos to Cape San Lucas, Baja California, and the Revillagigedo Islands, Mexico.

159. Tricolia diantha McLean, 1970. Minute, with fine spiral striae throughout. Color, pink with red spotting or streaking, some specimens with white maculations. Height, 2.1 mm; diameter, 1.7 mm. Galápagos Islands, 18 to 70 m.

160. Tricolia perforata (Philippi, 1848) (Synonyms: **? Phasianella perforata striulata** Carpenter, 1857; **P. mazatlanica** Strong, 1928). Largest of Panamic Tricolias, this always has red or brown spiral lines descending from the suture and a mottling of pink, tan, light green, or gray, the periphery with lighter markings. Height, 5 mm (maximum, 7 mm); diameter, 3 mm. Intertidally and just offshore in rocky areas, Mazatlán, Mexico, to Paita, Peru (type locality). Carpenter's type specimen is juvenile and not well preserved, thus not positively identifiable. The "*P. perforata* Philippi" of Reeve, 1862, is unfortunately a misidentification; Reeve's figure was copied by Strong, 1928.

161. Tricolia phasianella (Philippi, 1849) (Synonym: **Turbo phasianella** C. B. Adams, 1852). Larger than *T. substriata*, this has strong spiral sculpture throughout and is slightly angulate at the base; bright color patterns are variegated with reddish brown, white, and tan, the operculum usually dark-colored. Height, 4 mm (maximum, 6 mm); diameter, 3.1 mm. Nicaragua to Santa Elena Peninsula, Ecuador, intertidally and just offshore; also reported on Clarion Island, west Mexico.

Tricolia pulloides (Carpenter, 1865) (Synonyms: **Phasianella pulloides** Carpenter, 1864, *nomen nudum*; **P. compta** var. **elatior** Carpenter, 1865). Although this common Californian species has been cited as ranging south to Cape San Lucas, its southernmost verified locality is Asuncion Island, outer coast of Baja California.

162. Tricolia substriata (Carpenter, 1864). Relatively small, with fine spiral striae throughout, the shells are mottled with brown and white in various patterns. Height, 2.9 mm; diameter, 2.3 mm. Catalina Island, California, to Cape San Lucas, Baja California, and north in the Gulf of California to Puerto Libertad, Sonora, Mexico.

163. Tricolia umbilicata (Orbigny, 1840) (Synonym: **Phasianella minima** Philippi, 1860). Minute, smooth, lacking spiral sculpture; umbilical chink relatively large. Specimens from the northern end of the range tend to be mottled with red and white, those from the southern end to be slate gray. Height, 1.8 mm; diameter, 1.5 mm. Paita, Peru, to Chimba, Chile.

164. Tricolia variegata (Carpenter, 1864) (Synonyms: **Phasianella punctata** Carpenter, 1864 [not *T. punctata* Risso, 1826]; **Eulithidium typicum** Dall, 1908, unnecessary new name for *T. variegata*; **T. carpenteri** Dall, 1908, unnecessary new name for *P. punctata* Carpenter; [not preoccupied by *Phasianella variegata* Lamarck, 1822]). Relatively large, this invariably has a color pattern of dark dots upon an otherwise reddish brown, gray, or yellowish ground. Height, 4.2 mm (maximum height, 5.5 mm); diameter, 3.3 mm. In rocky areas, Santa Maria Bay, outer coast of Baja California, to the head of the Gulf of California, intertidally and just offshore.

Superfamily NERITACEA

Nearly globular, few-whorled, mostly sturdy shells; spire low; inner layer not nacreous; operculum calcareous, paucispiral, with excentric nucleus.

Family NERITIDAE

The nerites have a long geologic record. Now mainly tropical in distribution, they occupy the niche along tropic shores that the littorines or periwinkles do in the north, clinging to the rocks near high-tide line. The genus *Nerita* is predominantly marine, only a few forms having adapted to somewhat brackish water. In *Neritina* and *Theodoxus*, however, there are species that live not only in estuaries but even in rivers miles from the sea. One might suspect—and radular studies confirm—that from such an old and adaptable stock some of the present-day land and freshwater molluscan groups descended.

The shells are low-spired, nearly globular in shape, and apparently to take full advantage of the inner space, the animal resorbs the walls of the inner whorls, which is unusual in the Gastropoda. The outer lip may be thickened near the aperture or toothed within, and the inner lip not only may be toothed but may be spread out as a mass of callus or even as a sort of deck, so that the shell comes to look like a slipper limpet. The operculum is calcareous and oddly shaped, with a little clawlike process or apophysis that fits under the edge of the columella and enables a tight fit when the animal withdraws into the shell. This effectively seals in moisture during periods when the animal is out of the water—as at the time of low tide—and permits life in the upper splash zone of the shore. The marine forms are solid-shelled, the freshwater ones thin-shelled. None are nacreous. Three Panamic genera.

Genus NERITA LINNAEUS, 1758

Shell solid, last whorl large, operculum ribbed, with apophysis. Two Panamic subgenera; no species of subgenus *Nerita, s. s.*, occur in the Panamic province.

Subgenus RITENA GRAY, 1858

The spire moderately elevated, the inner lip with several teeth, and the callus area irregularly folded.

165. Nerita (Ritena) scabricosta Lamarck, 1822 (Synonyms: **N. ornata** Sowerby, 1823; **N. fuscata** Menke, 1829; **N. papilionacea** Valenciennes, 1832; **N. deshayesii** Récluz, 1841; **N. multijugis** Menke, 1847). The shells are among the largest of the genus, dark gray, with spiral ribs that are rough-surfaced and somewhat irregularly spaced. In the southern part of the range specimens tend to be smaller, the shells more globose, more regularly ribbed. These have been given the subspecific name *N. s. ornata*. However, there are enough exceptions to this tendency to make the division seem unjustified. Like the littorines, these snails may be found on sunbaked rocks in the splash zone, wetted only at high tide. Height of an average specimen, 45 mm; diameter, 38 mm. Punta Pequeña, outer coast of Baja California, to Ecuador.

Subgenus THELIOSTYLA MÖRCH, 1852

The shell is low-spired, the inner lip finely denticulate, the callus area granular.

166. Nerita (Theliostyla) funiculata Menke, 1851 (Synonym: **N. bernhardi** Récluz, 1855). Smaller than *N. scabricosta*, this has irregular wrinkles and pustules on the callus area. Diameter, 20 mm; height, 15 mm. Baja California through the Gulf and southward to Peru, also the Galápagos Islands, on rocks intertidally.

Genus NERITINA LAMARCK, 1816

Shell smaller and thinner than *Nerita*, the outer lip thin, the inner lip smooth or finely toothed. No species of subgenus *Neritina, s. s.*, occur in the Panamic province.

Subgenus **CLYPEOLUM** RÉCLUZ, 1842
(**ALINA** RÉCLUZ, 1842, not RISSO, 1826)

Aperture large, flaring, the upper expansion larger than the lower; opercular peg long, slender; fluviatile, living in and near river mouths.

167. Neritina (Clypeolum) latissima Broderip, 1833 (Synonyms: **N. globosa** Broderip, 1833 [not Eichwald, 1830]; **N. intermedia** Sowerby, 1833; **N. cassiculum** Sowerby, 1836; **N. fontaineana** Orbigny, 1840; **N. guayaquilensis** Sowerby, 1849; **N. pilsbryi** Tryon, 1888). The widely expanded outer lip of this shell makes it look a little like a limpet; there is some variation in the amount of flare. The color is a uniform olive brown, with a lacing of light spots bordered by black showing through the dark periostracum; there may be one or two color bands, and worn shells may have a flush of pink. The form *N. cassiculum* was described without locality, but later authors who have discussed it seem agreed it is West American. It has been compared to the Caribbean *N. punctulata* Lamarck, 1816, but Russell in 1941, in a monograph of the Caribbean Neritidae, compares it instead to *N. latissima*; as the points of difference he cites do not hold up for suites of specimens in the Stanford collection, the form is here regarded as a synonym. Possibly further study, especially comparison of radulas as well as shells, will show that at the north and south ends of its range, *N. latissima* has a less flaring shell, lower-spired and somewhat heavier (especially in Ecuador)—in which case the name *N. l. cassiculum* would be available as a subspecific name for the northern form, *N. l. fontaineana* for the southern one. A large specimen measures: diameter, 30 mm; height, 38 mm. Acapulco, Mexico, to Guayaquil, Ecuador, in tidepools and running streams at river mouths.

Genus **THEODOXUS** MONTFORT, 1810

Shells smooth, opercular peg weak or wanting; mainly estuarine to freshwater in distribution. No species of subgenus *Theodoxus, s. s.,* occur in the Panamic province.

Subgenus **VITTOCLITHON** H. B. BAKER, 1923

Operculum smooth, with a small blunt peg, as contrasted to the European *Theodoxus, s. s.,* in which the peg is absent.

168. Theodoxus (Vittoclithon) luteofasciatus Miller, 1879 (Synonyms: **Neritina picta** Sowerby, 1832 [not Eichwald, 1830]; **N. picta guttata** Miller, 1879 [not Récluz, 1841]; **N. usurpatrix** Crosse & Fischer, 1892; **N. picta albescens** and **nigrofuscata** Miller, 1879; **N. picta serta** and **subnigra** Von Martens, 1901). A small, shiny shell, variously patterned with lines and spots of color, but with brown predominating around the aperture. The apertural callus is golden chestnut. Diameter, 11 mm; height, 12 mm. The Gulf of California to Puerto Pizarro, Peru, on the margins of mangrove swamps and on mud flats.

Family **PHENACOLEPADIDAE**

Conical or cap-shaped shells with apex turned backward and near posterior margin; surface smooth to radially sculptured.

Genus **PHENACOLEPAS** PILSBRY, 1891

The muscle scar is horseshoe-shaped and opens anteriorly, as in the Acmaeidae. There is no operculum.

169. Phenacolepas malonei Vanatta, 1912 (Synonym: **P. magdalena** Dall, 1918). Shell white, with radial striae and concentric rows of granules that give it a beaded appearance. Length, 15 mm; width, 14 mm; height, 4.5 mm; the apex, 3.5 mm from the posterior margin. Magdalena Bay, Baja California, to the northern end of the Gulf of California and east at least to Guaymas, Sonora, Mexico.

170. Phenacolepas osculans (C. B. Adams, 1852) (Synonym: **Scutellina navicelloides** Carpenter, 1857). The apex is so near the edge or even overhangs it in such a way that the edge appears to be a shelf inside. Length, 12.5 mm; width, 9 mm; height, 3 mm. The northern end of the Gulf of California to Panama.

171. ? Phenacolepas puntarenae (Mörch, 1860). This was named as a *Lepeta*, but the type specimen in the University of Copenhagen museum, a juvenile shell, seems not to be a patellacean. The faint muscle scars are more those of *Phenacolepas*. The conical shell has fine and somewhat twinned radial ribs and a blunt apex that appears to be slightly coiled. It does not match closely either of the two other described species, for the apex is somewhat farther from the posterior margin. Length of the type, 6 mm. Type locality, Puntarenas, Costa Rica.

Family TITISCANIIDAE

Marine snails without either a shell or an operculum, anatomically related to the Neritidae; with a pallial cavity and ctenidium.

Genus TITISCANIA BERGH, 1890

White, slug-shaped, with two thin, long tentacles; roof of the mantle cavity forming an oval area anteriorly about one-fourth of the body length; with a row of dense white glands on both sides of the back; ctenidium more or less projecting from the pallial cavity.

172. Titiscania limacina (Bergh, 1875). A nonpelagic, free-living snail, easily confused by collectors with the nudibranchs, although completely unrelated. Length of a preserved specimen, 6 to 12 mm; width, 3 mm; length of tentacles, 1.5 mm. The species was described from the western Pacific but has been recorded by Marcus and Marcus (1967) at Puerto Lobos, Sonora, Mexico, and on the Pacific coast of the Panama Canal Zone.
See Color Plate XV.

Superfamily COCCULINACEA

Resembling the Patellacea but radula rhipidoglossate; apex posteriorly directed.

Family COCCULINIDAE

Shells small, cap-shaped, round to oblong. Mostly deepwater.

Genus COCCULINA DALL, 1882

Shells limpet-like, smooth to weakly sculptured, white.

173. Cocculina agassizii Dall, 1908. Gulf of Panama in 1,015 m.

174. C. diomedae Dall, 1908. Gulf of Panama in 1,870 m.

175. C. nassa Dall, 1908. Gulf of Panama in 2,320 m.

Family LEPETELLIDAE

Shell small, conical, apex central or slightly posterior, not spiral; aperture rounded to oval.

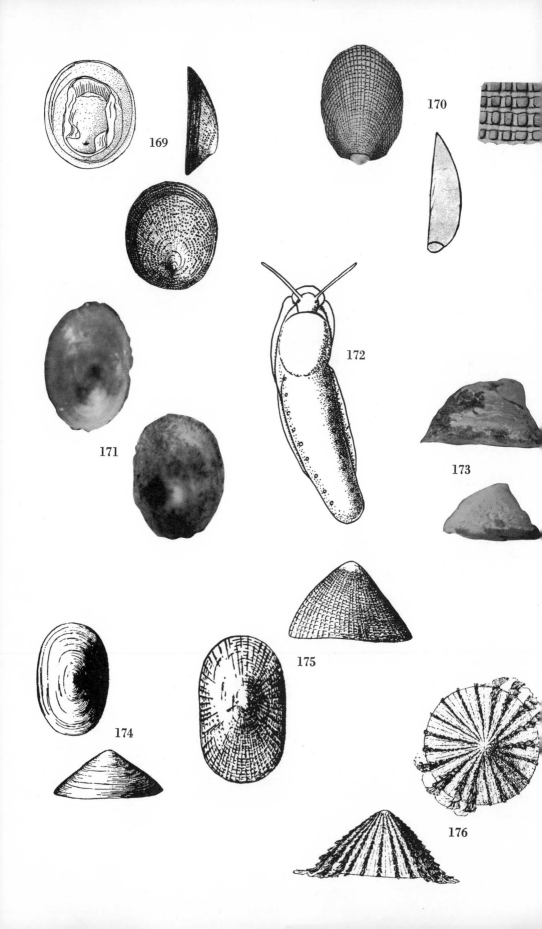

Genus **BATHYSCIADIUM** DAUTZENBERG & FISCHER, 1900

Steeply conical, with radial striae.

176. Bathysciadium pacificum Dall, 1908. Sechura Bay, Peru, on a squid beak (type locality), 4,060 m.

Order **MESOGASTROPODA**
(**CTENOBRANCHIA**)

A ctenidium is present in all of the Mesogastropoda; it is monopectinate—that is, the gill is reduced to a single row of filaments. Shells (except in one family) are porcelaneous in texture. In most of the more primitive forms the aperture is entire, correlated with herbivorous food habits, but there are exceptions. In the more advanced forms the aperture has an anterior notch and the animals may be carnivorous. All but a few families have a taenioglossate radula, with a few pairs of marginal teeth, two lateral teeth, and one central or rhachidian tooth in each row. A parasitic habit, developed in some groups, may lead to loss or degeneration of the radula.

Superfamily **SEGUENZIACEA**

Shells with turbinate coiling, a strong notch in the posterior part of the aperture, and an anterior notch at the end of the columella.

Family **SEGUENZIIDAE**

Inner layer nacreous, as in the Archaeogastropoda, but radula taenioglossate, as in most Mesogastropoda. Authors are not yet in complete agreement on the correct systematic placement of this family.

Genus **SEGUENZIA** JEFFREYS, 1876

Whorls inflated, sculptured with several spiral keels and fine axial striations. Mostly deepwater forms.

177. Seguenzia occidentalis Dall, 1908. Off Acapulco, Mexico, in 1,210 m.

178. S. stephanica Dall, 1908. Off Mazatlán, Mexico, in 1,825 m.

Superfamily **LITTORINACEA**

Ovate to turbinate, small to medium-sized; smooth or with weak to nodose sculpture, never spiny; aperture entire; operculum horny, paucispiral.

Family **LITTORINIDAE**

The littorines or periwinkles are among the commonest marine snails of the world, and few are the areas that have none clinging to the rocks between midtide and high-tide line, often so high they receive only an occasional splash of seawater. They are true inhabitants of the littoral—the seashore—and, like the nerites, have probably provided the ancestral stock for some of the present-day land snails. The operculum is not calcareous as in the nerites, but is a solid little horny scale that effectively blocks the aperture and keeps enough moisture inside the shell when the animal is retracted so that the gill does not become dry. The littorines mostly have small shells, porcelaneous in texture, and the animals feed upon microscopic plant material that they rasp from the surface of the rocks. The radula, as in many of the families in Mesogastropoda, is taenioglossate ("ribbon-tongued"). Three Panamic genera.

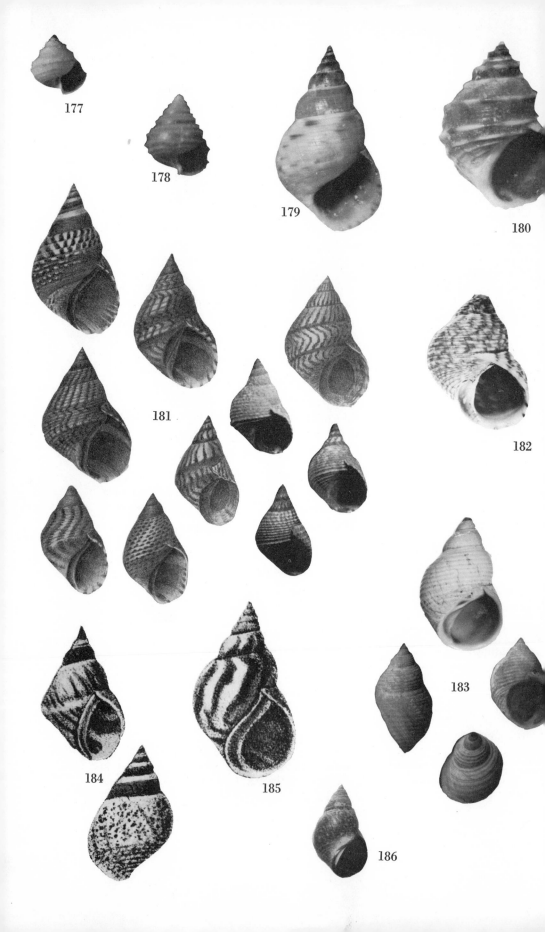

Genus **LITTORINA** FÉRUSSAC, 1822

Surface smooth or only weakly sculptured.

Although a number of subgeneric divisions have been proposed, authors have not generally utilized them, for they are based mainly on slight differences of shell morphology and on differences in radula, reproductive organs, and spawning habits. Rosewater (1970), in a paper published too late for full consideration here, has recommended the following allocations of eastern Pacific species: to *Littorina, s. s., L. planaxis*; to *Littoraria* Griffith & Pidgeon, 1834, *L. pullata, schmitti*, and *zebra*; to *Littorinopsis* Mörch, 1876, *L. aberrans, fasciata, modesta, paytensis*, and *varia*; and to a new subgenus, *Austrolittorina* Rosewater, 1970, *L. araucana*, *aspera, penicillata*, and *peruviana*.

179. Littorina aberrans Philippi, 1846. A species of the mangrove swamps, this is probably only subspecifically different from the wide-ranging and very variable tropical species *L. scabra* (Linnaeus, 1758), which has scarcely distinguishable subspecies in the western Pacific and the Caribbean. The shell is white to light brown in color, darker within the aperture, nearly smooth except for fine spiral striae. Length, 18 mm. Type locality, Panama; the probable range is from Puerto Peñasco, near the head of the Gulf of California, to Ecuador.

180. Littorina albicarinata McLean, 1970. Small, turbinate, dark-colored, with two thin, strong, rounded carinae; base with a few incised spiral lines; axial sculpture lacking; a small umbilical chink in young specimens, a slight groove outside the callus area in adults. Spire grayish brown, columella white, aperture dark brown, body whorl with brown maculations. Height (maximum), 4.7 mm; width, 3.2 mm. Cape San Lucas to the northern end of the Gulf of California and eastward to Guaymas, Sonora, Mexico, intertidally, in crevices near barnacles, often associated with a superficially similar small *Fossarus*.

Littorina araucana Orbigny, 1840 (Synonym: **L. thersites** Reeve, 1857). A small, ovate form, with fine spiral grooving and a brown aperture. Although reported from Nicaragua to Chile (type locality, Valparaiso), this probably does not occur north of Peru.

181. Littorina aspera Philippi, 1846 (Synonyms: **? L. glabrata** Philippi, 1846; **L. parvula** Philippi, 1849; **L. apicina** Menke, 1850; **L. dubiosa** C. B. Adams, 1852; **L. philippii** Carpenter, 1857; **L. penicillata** Carpenter, 1864; **L. philippii alba, latistrigata**, and **subsuturalis** Von Martens, 1901). As evidenced by the number of synonyms, this is a variable form. Early workers even identified some of the variants with *L. ziczac* (Gmelin, 1791) from the Caribbean or *L. neritoides* (Linnaeus, 1758) from the Indo-Pacific. The shell is white to light brown, mostly with flattened spiral sculpture, but one or more of the spiral riblets near the periphery may be larger, giving it a somewhat angulate appearance; color markings are of wavering or zigzag axial stripes in shades of brown, with occasional bluish spots. Separation of the species into geographic subspecies seems difficult, for within a single area one may find variants that bridge the extremes between *L. aspera*—usually identified as the larger, browner, more coarsely sculptured form—and *L. dubiosa, L. philippii*, or *L. penicillata*, all smaller and with brighter axial markings. However, it is possible that careful work will demonstrate the desirability of recognizing more than one species within this complex. Height, to 16 mm (average about 11 mm); diameter to 10 mm (average, 7 mm). Manuela Lagoon, Baja California, through the Gulf of California and south at least to Ecuador, possibly to northern Peru.

182. Littorina fasciata Gray, 1839. The banded littorine is one of the largest forms of the genus, golden brown, banded in oblique axial lines with darker brown. Length, 30 mm; diameter, 23 mm. Magdalena Bay, Baja California, through the Gulf and south to Ecuador.

183. Littorina modesta Philippi, 1846 (Synonyms: **L. conspersa** and **puncticulata** Philippi, 1847; **L. albida** Philippi, 1848). The shell is white, dotted with reddish brown; sculpture is of well-incised spiral grooves; the aperture is orange-brown within, the columella broad and somewhat excavated. Height, 16 mm; diameter, 10 mm. Ensenada, Baja California, to Ecuador. The erroneous type locality of "Sitka, Alaska" has cast doubt on the validity of the name *L. modesta*; syntypic material in the British Museum proves its identity with what had long been known as *L. conspersa*.

184. Littorina paytensis Philippi, 1847. Except for the angulate periphery, this is close to *L. modesta* and may prove to be a southern subspecies. Height, 12 mm. Ecuador to Peru.

185. Littorina peruviana (Lamarck, 1822). Most boldly marked of all eastern Pacific littorines, the large shell has irregular axial stripes of purplish brown and white. Although reported from Central America by early workers, it apparently is restricted to the coast south of Ecuador; there is one record from the Galápagos Islands.

Littorina planaxis Philippi, 1847. A Californian species reported on the offshore islands as far south as Socorro Island; one lot in the Stanford University collection extends the range to Cocos Island, off Panama.

186. Littorina pullata Carpenter, 1864. Color, blackish or purplish brown, with very fine spiral lines of a lighter shade, some specimens showing an obscure checkered design. Length, 8 mm; diameter, 5.5 mm. Southern Baja California to Panama.

187. Littorina schmitti Bartsch & Rehder, 1939. The shell is ovate, with a pattern of fine spiral dots and dashes, the dots being lighter, the dashes darker or almost black. The Hawaiian species *L. pintado* Wood, 1828, is closely related; in fact, this may be an eastern Pacific representative of the *L. pintado* stock, known as yet only on the offshore islands. Length, 18 mm; diameter, 12 mm. Clipperton Island.

188. Littorina varia Sowerby, 1832 (Synonyms: **L. variegata** Souleyet, 1852; **L. costulata** Tryon, 1887, *ex* Souleyet vernacular). The varied littorine resembles *L. fasciata* in shape but has interrupted axial markings and has more variable spiral ridges. The color is predominantly grayish white, spotted with buff and brown, especially on the spiral ribs. Length, 30 mm; diameter, 20 mm. Panama southward to Ecuador and perhaps as far as Peru.

189. Littorina zebra Donovan, 1825 (Synonym: **L. pulchra** Sowerby, 1832). The zebra littorine is a large, solid, low-spired form, one of the showiest of the littorines. The terra-cotta-colored shell is obliquely banded with brown stripes, and the spiral lines are very fine and even. It is the type of the subgenus *Littoraria* Griffith & Pidgeon, 1834. Length, 32 mm; diameter, 25 mm. Costa Rica to Buenaventura Bay, Colombia.

Genus NODILITTORINA VON MARTENS, 1897

Spirally nodose; operculum oval and paucispiral.

190. Nodilittorina galapagiensis (Stearns, 1892) (Synonym: **Littorina atyphus** Stearns, 1893*b*). The small shell has several rows of strong nodes, especially on the body whorl. It is dark brown, with a darker aperture. The operculum has about five turns around a subcentral nucleus. The columella is broad and concave, the anterior end of the canal narrowed. Height, 6 mm; diameter, 4 mm. Galápagos Islands to Ecuador.

Genus PEASIELLA NEVILL, 1884 (? 1885)

Low conic, with peripheral carina; base perforate or umbilicate; operculum concentric; multispiral.

191. Peasiella roosevelti Bartsch & Rehder, 1939. This seems to be the first record of the genus in the eastern Pacific. The shell is like a small *Tegula* in shape, except for the peripheral carina. The general coloration is brownish, mottled and checkered with bluish-white bands and somewhat zigzag axial stripes. The columella is pale brown, as is the inside of the outer lip. Height, 3.2 mm; diameter, 3 mm. James Island, Galápagos. There is a strong resemblance to the widespread Pacific species *P. tantilla* (Gould, 1849), which occurs as far east as Hawaii.

Family LACUNIDAE

Small shells with a chink along the margin of the columella; animals vegetarian.

Genus LACUNA TURTON, 1827

Resembling *Littorina* but with an umbilical chink. The little snails cluster on seaweeds rather than on rock and are more common in temperate than in tropical seas. Only a few species have been reported south of the Californian province.

192. Lacuna succinea Mörch, 1860. The shell is unusually solid for a *Lacuna* and resembles a small *Littorina*, lacking color. The umbilical chink is very narrow, hardly apparent. Length, 6 mm. Puntarenas, Costa Rica. It is possible that this may prove to be an *Iselica*, family Pyramidellidae.

Superfamily RISSOACEA

Small to minute shells variously sculptured and shaped. Few of the species exceed a length of 5 mm, and all require a microscope for proper identification. Therefore, they will not be treated in detail here. Useful references are cited for each generic or family group. The original generic allocation is shown in square brackets. Supplementary or later figures are cited, especially for those species not adequately illustrated by the original author. The key is to families.

1. Shells ovate to cylindrical, height greater than diameter. 2
   Shells lenticular, diameter greater than height. 6
2. Apical whorls shed in adult stage; shell cylindrical. . . . . Truncatellidae
   Apical whorls retained in adult; shell ovate-conic. 3
3. Shells smooth 4
   Shells with varying amounts of sculpture. 5
4. Base imperforate, with a wide inner lip callus. . . . . Assimineidae
   Base narrowly umbilicate, without callus. . . . . Rissoellidae
5. Aperture with a small anterior canal. . . . . Rissoinidae
   Aperture rounded anteriorly. . . . . Rissoidae
6. Apertural margin entire; apical whorls tilted at an angle. . . . Cyclostremellidae
   Apertural margin somewhat interrupted by body whorl; axis of apical
     whorls not tilted. . . . . Vitrinellidae

Family RISSOIDAE

Small shells, spire mostly longer than aperture; smooth to sculptured; aperture entire; operculum thin and simple, without apophysis in most forms. Useful reference: Baker, Hanna & Strong, 1930a. Five Panamic genera.

1. Base umbilicate, inner lip shelflike........................*Amphithalamus*
   Base not umbilicate... 2
2. Shell smooth except for apical pitting.........................*Barleeia*
   Shell variously sculptured....................................... 3
3. Aperture angulate above.......................................*Onoba*
   Aperture rounded, not angulate above............................ 4
4. Conic, few-whorled, aperture entire...........................*Alvinia*
   Cylindrical, many-whorled, aperture slightly detached......*Nannoteretispira*

Subfamily RISSOINAE

Sculpture well developed, axial ribs often predominating and stronger than growth lines.

Genus ALVINIA MONTEROSATO, 1884
(ALVANIA of authors, not of RISSO, 1826)

Sculpture cancellate, axial ribs weaker on basal cords; aperture relatively large. Useful reference: Bartsch, 1911h. Two Panamic subgenera.

Subgenus ALVINIA, s. s.

Axial ribs as strong as spiral ribs on upper parts of whorls; suture descending evenly.

193. Alvinia (Alvinia) albolirata (Carpenter, 1864) [*Rissoa*]. Gulf of California.

194. A. (A.) clarionensis (Bartsch, 1911) [*Alvania*] (Synonym: ? "**Alvania lirata**" of Bartsch 1911, not of Carpenter, 1857). Clarion Island; probably to southern end of the Gulf of California.

195. A. (A.) electrina (Carpenter, 1864) [*Diala*]. Cape San Lucas.

196. A. (A.) galapagensis (Bartsch, 1911) [*Alvania*]. Galápagos Islands.

197. A. (A.) gallegosi (Baker, Hanna & Strong, 1930) [*Alvania*]. Cape San Lucas.

198. A. (A.) granti (Strong, 1938) [*Alvania*]. Tres Marias Islands, Mexico.

199. A. (A.) halia (Bartsch, 1911) [*Alvania*]. Galápagos Islands.

200. A. (A.) herrerae (Baker, Hanna & Strong, 1930) [*Alvania*]. Cape San Lucas to Tres Marias Islands.

201. A. (A.) hoodensis (Bartsch, 1911) [*Alvania*]. Hood Island, Galápagos Islands.

202. A. (A.) ima (Bartsch, 1911) [*Alvania*]. Galápagos Islands.

203. A. (A.) inconspicua (C. B. Adams, 1852) [*Rissoa*]. Panama.

204. A. (A.) ingrami (Hertlein & Strong, 1951) [*Alvania*]. Port Guatulco, Mexico.

205. A. (A.) lara (Bartsch, 1911) [*Alvania*]. Galápagos Islands.

PLATE V · *Astraea olivacea* (Wood)

206. A. (A.) lucasana (Baker, Hanna & Strong, 1930) [*Alvania*]. Cape San Lucas, Baja California.

207. A. (A.) monserratensis (Baker, Hanna & Strong, 1930) [*Alvania*]. Monserrate Island, Gulf of California.

208. A. (A.) nemo (Bartsch, 1911) [*Alvania*]. Galápagos Islands.

209. A. (A.) perlata (Mörch, 1860) [*Alvania*]. El Salvador. Figured, Keen, 1966.

210. A. (A.) profundicola (Bartsch, 1911) [*Alvania*]. Galápagos Islands.

211. A. (A.) tumida (Carpenter, 1857) [*Alvania*]. Mazatlán, Mexico, to Panama. Figured, Keen (1968).

212. A. (A.) veleronis (Hertlein & Strong, 1939) [*Alvania*]. Panama.

Subgenus LAPSIGYRUS BERRY, 1958

Interspaces between spiral ribs finely cancellate; last whorl descending; outer lip thickened.

213. Alvinia (Lapsigyrus) contrerasi (Jordan, 1936) [*Alvania*]. Magdalena Bay, Baja California (type of subgenus).

214. A. (L.) milleriana (Hertlein & Strong, 1951) [*Alvania*]. Ballena Bay, Costa Rica.

215. A. (L.) mutans (Carpenter, 1857) [*Alaba*]. Mazatlán, Mexico. Figured, Keen (1968).

216. A. (L.) myriosirissa (Shasky, 1970) [*Lapsigyrus*]. Mazatlán, Mexico.

Subfamily ANABATHRINAE

Relatively high-spired, smooth or with spiral cords. Two Panamic genera.

Genus AMPHITHALAMUS CARPENTER, 1865

Inner margin of aperture separated from body whorl and shelflike. Useful reference: Bartsch (1911g).

217. Amphithalamus inclusus Carpenter, 1865. California to Gulf of California (type of genus).

218. A. stephensae Bartsch, 1927. Magdalena Bay to La Paz, Baja California.

219. A. trosti Strong & Hertlein, 1939. Panama.

Genus NANNOTERETISPIRA HABE, 1961

Minute, smooth, cylindrical, aperture subquadrate.

220. Nannoteretispira kelseyi (Bartsch, 1911) [*Nodulus*]. Southern California to Nayarit, Mexico.

Subfamily BARLEEINAE

Sculpture reduced to pitting, on nuclear whorls only.

Genus BARLEEIA CLARK, 1853

Shell ovate-conic, whorls inflated. Operculum with a peg. Useful reference: Bartsch (1920a).

221. Barleeia alderi (Carpenter, 1857) [*Jeffreysia*]. Gulf of California to Tres Marias Islands. Figured, Keen (1968).

222. B. bentleyi Bartsch, 1920. Southern California to Cape San Lucas.

223. B. carpenteri Bartsch, 1920. Cape San Lucas.

224. B. orcutti Bartsch, 1920. Magdalena Bay to Gulf of California.

225. B. polychroma (De Folin, 1870) [*Rissoa*]. Panama.

226. B. zeteki Strong & Hertlein, 1939. Panama.

Subfamily CINGULINAE

Slender, subcylindrical, sculpture weak to wanting.

Genus ONOBA H. & A. ADAMS, 1852

Cylindrical, with fine spiral striae and faint axial riblets.

227. Onoba fortis Pilsbry & Olsson, 1941. Ecuador.

228. O. ? fusiformis (Carpenter, 1857) [*Aclis*]. Mazatlán, Mexico. Figured, Keen (1968).

Family ASSIMINEIDAE

These small snails are characteristically intertidal, living in bays and salt marshes beneath vegetation and drift debris. No species of the family are known to live below low-tide level, where superficially similar shells may represent several other families of the Rissoacea.

The review of this family is compiled from notes supplied by Drs. Dwight W. Taylor, J. P. E. Morrison, and Eugene V. Coan, whose assistance is gratefully acknowledged.

Genus ASSIMINEA FLEMING, 1828
(SYNCERA of authors)

The smooth conical shells are 3 to 5 mm long, with a uniform glossy chestnut color when fresh. They lack obvious sculpture or color pattern. The inner lip is appressed to the preceding whorl, leaving no umbilical chink. The operculum is simple and paucispiral, its nucleus subcentral. Living specimens examined under a hand lens reveal a unique feature of the family: the eyes are on short ocular peduncles, and tentacles seem to be lacking. Useful literature: Bartsch, 1920; Keen and Coan in Moore (in press).

229. Assiminea californica (Tryon, 1865) (Synonyms: **Jeffreysia translucens** Carpenter, 1864, of authors, *nomen nudum*; **J. translucens** Carpenter, 1866; **Syncera magdalenensis** Bartsch, 1920). The shell is stoutly conical, with a rounded body whorl, convex spiral whorls, and an impressed suture. Length, 3.8 mm; diameter, 2.3 mm; number of whorls, 5. Southern British Columbia (Puget Sound) to Cape San Lucas and the upper Gulf of California. The radula has three basal cusps on both sides of the central plate, which suggests that the species may be assignable to *Assiminea* (*Assiminea*).

230. Assiminea compacta (Carpenter, 1864). The type specimen is in rather poor condition but shows a few-whorled shell with a blunt spire. The type, which is in the U.S. National Museum, was figured by Palmer (1963). Length, 1.3 mm; diameter, 1 mm. Cape San Lucas (type locality). Morrison (*in litt.*) has identified specimens from the Galápagos Islands as being *A. compacta*, which is a considerable range extension.

231. Assiminea dubiosa (C. B. Adams, 1852). The shell is smaller and narrower in outline than *A. californica*, with smaller aperture and a more flat-sided spire. Length, 2.1 mm. Panama.

Family RISSOELLIDAE
(JEFFREYSIIDAE of authors)

Shells thin and nearly transparent, ovate, with a narrow umbilical chink; operculum with a projecting internal plate.

Genus RISSOELLA J. E. GRAY, 1847
(JEFFREYSIA FORBES & HANLEY, 1850)

Smooth, ovate, whorls inflated. Intertidal forms, clinging to seaweed. Although new evidence makes clear that *Jeffreysia* has slight priority as a generic name over M. E. Gray's 1850 validation of *Rissoella*, it seems better in the interests of stability to accept the 1847 introduction of the name, although it had been rejected on a technicality by some workers. Useful reference: Bartsch (1920*a*).

232. Rissoella anguliferens (De Folin, 1870) [*Rissoa*]. Panama.

233. R. bifasciata (Carpenter, 1857) [*Jeffreysia*]. Cape San Lucas to Mazatlán. Figured, Keen (1968).

234. R. ? conica (De Folin & Périer, 1870) [*Rissoa*]. Panama.

235. R. excolpa Bartsch, 1920. Gulf of California.

236. R. johnstoni Baker, Hanna & Strong, 1930. Cape San Lucas.

237. R. ? paupercula (C. B. Adams, 1852) [*Cingula*]. Panama.

238. R. tumens (Carpenter, 1857) [*Jeffreysia*]. Cape San Lucas to Mazatlán. Figured, Keen (1968).

Family RISSOINIDAE

Ovate to cylindrical, mostly sturdy shells. Aperture crescentic, narrow above. Operculum with an apophysis or peg on the inner surface.

Subfamily RISSOININAE

Slender, sculpture present, mostly axial, spiral in some, or axial with spiral threads in interspaces. Aperture tending to develop an anterior spout, inner lip sinuous. Useful reference: Bartsch (1915).

Genus RISSOINA ORBIGNY, 1840

Axial ribs well developed, at least on spire. Four Panamic subgenera.

Subgenus RISSOINA, *s. s.*

Spire regularly coiled; outer lip angulate but not notched.

239. Rissoina (Rissoina) adamsi Bartsch, 1915. Nicaragua to Panama.

240. R. (R.) alarconi Hertlein & Strong, 1951. Costa Rica.

241. R. (R.) allemani Bartsch, 1931. Panama.

242. R. (R.) axeliana Hertlein & Strong, 1951. Puerto Guatulco, Mexico.

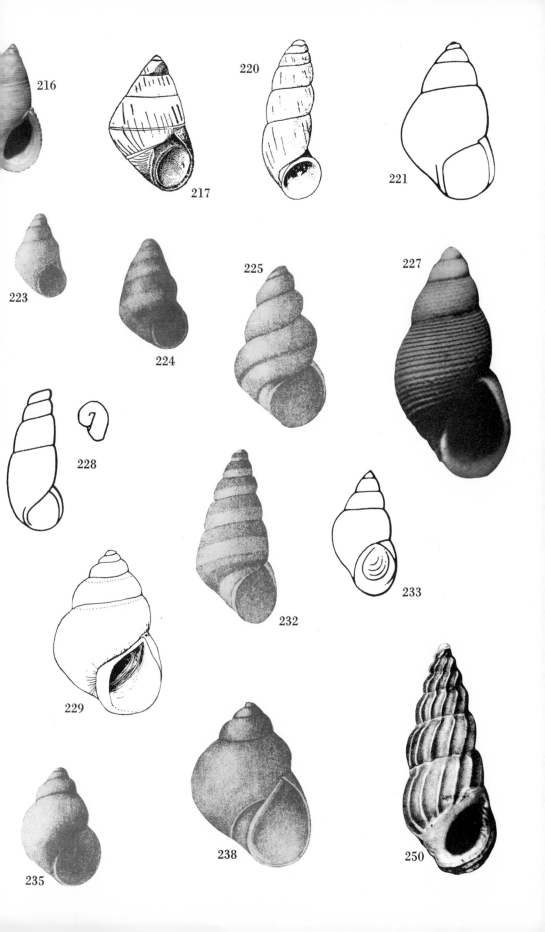

243. R. (R.) bakeri Bartsch, 1902. Cape San Lucas.

244. R. (R.) barthelowi Bartsch, 1915. Gulf of California to Mazatlán, Mexico. Length, 7 mm.

245. R. (R.) basilirata Baker, Hanna & Strong, 1930. Gulf of California.

246. R. (R.) burragei Bartsch, 1915. Gulf of California.

247. R. (R.) cancellata Philippi, 1847. Ecuador and Peru.

248. R. (R.) clandestina (C. B. Adams, 1852) [*Rissoa*]. Panama. Figured, Turner (1956).

249. R. (R.) dina Bartsch, 1915. Galápagos Islands. Length, 7 mm.

250. R. (R.) effusa Mörch, 1860. Central America. Figured, Keen (1966).

251. R. (R.) excolpa Bartsch, 1915. Gulf of California to Mazatlán, Mexico.

252. R. (R.) expansa Carpenter, 1865. Guaymas to Acapulco, Mexico. Length, 9 mm.

253. R. (R.) favilla Bartsch, 1915. Cape San Lucas, Baja California. Length, 7 mm.

254. R. (R.) firmata (C. B. Adams, 1852) [*Rissoa*]. (Synonym: **R. scalariformis** C. B. Adams). Panama. Figured, Turner (1956).

255. R. (R.) fortis (C. B. Adams, 1852) [*Rissoa*]. Nicaragua to Panama and the Galápagos Islands. Figured, Turner (1956). Length, 7.5 mm.

256. R. (R.) gisna Bartsch, 1915. Nicaragua to Panama. Length, 7 mm.

257. R. (R.) hartmanni Jordan, 1936. Pleistocene, Magdalena Bay, Baja California.

258. R. (R.) helena Bartsch, 1915. Peru. Length, 7 mm.

259. R. (R.) histia Bartsch, 1915. Gulf of California.

260. R. (R.) Inca Orbigny. 1840. Galápagos Islands to Chile. Length, 8 mm. (type of genus).

261. R. (R.) io Bartsch, 1915. Galápagos Islands. Length, 9 mm.

262. R. (R.) janus (C. B. Adams, 1852) [*Rissoa*]. Panama. Figured, Turner (1956).

263. R. (R.) laurae De Folin, 1870. Panama.

264. R. (R.) mazatlanica Bartsch, 1915. Gulf of California. Length, 6 mm.

265. R. (R.) melanelloides Baker, Hanna & Strong, 1930. Cape San Lucas, Baja California.

266. R. (R.) mexicana Bartsch, 1915. Gulf of California.

267. R. (R.) nereina Bartsch, 1915. Outer coast of Baja California.

268. R. (R.) peninsularis Bartsch, 1915. Cape San Lucas, Baja California. Length, 6 mm.

269. R. (R.) porteri Baker, Hanna & Strong, 1930. Gulf of California.

270. R. (R.) scalariformis (C. B. Adams, 1852). (See *R. (R.) firmata*.)

271. R. (R.) stricta Menke, 1850. Cape San Lucas through the Gulf of California and south to Tres Marias Islands. Length, 9 mm.

272. R. (R.) townsendi Bartsch, 1915. Gulf of California. Length, 5 mm.

273. R. (R.) woodwardi Carpenter, 1857. Gulf of California to Mazatlán, Mexico. Figured, Keen (1968).

274. R. (R.) zeltneri (De Folin, 1867). Bahía San Luis Gonzaga, Gulf of California, to Panama.

Subgenus FOLINIA CROSSE, 1868

Axial ribs sinuous; aperture with a slight posterior notch.

275. Rissoina (Folinia) ericana Hertlein & Strong, 1951. Puerto Guatulco, Mexico.

276. R. (F.) signae Bartsch, 1915 (Synonym: **Rissoa insignis** De Folin, 1867 [not Adams & Reeve, 1850]). Panama.

Subgenus SULCORISSOINA KOSUGE, 1965

Sculpture of incised spiral grooves; axial ribs weak or wanting.

277. Rissoina (? Sulcorissoina) berryi Baker, Hanna & Strong, 1930. Cape San Lucas, Baja California.

278. R. (? S.) lapazana Bartsch, 1915. La Paz, Gulf of California. Length, 6 mm.

279. R. (? S.) lirata (Carpenter, 1857) [*Rissoa*]. Mazatlán, Mexico. Figured, Keen (1968).

280. R. (? S.) stephensae Baker, Hanna & Strong, 1930. Cape San Lucas, Baja California.

Subgenus TIPHYOCERMA BERRY, 1958

Apical whorls with axial riblets; coiling somewhat irregular.

281. Rissoina (Tiphyocerma) preposterum (Berry, 1958) [*Tiphyocerma*]. Puerto Peñasco, Sonora, Mexico.

Genus RISSOINA, s. l. (?)

282. ? Rissoina infrequens (C. B. Adams, 1852) [*Rissoa*]. Panama. Length, 6 mm. Allocated to genus *Pliciscala* De Boury, 1887 (family Epitoniidae) by Bartsch, 1915, without comment as to evidence. Family and genus remain uncertain, although the figure of the type given by Turner (1956) suggests rissoinid affinities.

Family TRUNCATELLIDAE

The family name is apt: the shells are characteristically truncate. As the snail matures, the early whorls are broken off, and the new, blunt apex is sealed with a shelly plug.

The review of this family has been prepared by Dr. Dwight W. Taylor, whose assistance is gratefully acknowledged.

Genus TRUNCATELLA RISSO, 1826

The adult shell is turriform, having an abruptly truncate apex, nearly flat-sided convex whorls, and an impressed suture. The young snails with the original apex may have 8 whorls and a length of 7 mm, but adults are usually only 5 mm long with 4 whorls. Sculpture consists of weakly sinuous ribs about as wide as their interspaces, varying within a colony from prominent to obsolete. The operculum is thin, simple, and paucispiral, the nucleus close to the base of the columella.

Truncatella is semiterrestrial, semimarine. The animals live beneath vegetation or drift close to high-water mark. Here they are rarely covered by the sea, yet they are not found inland beyond the strand line.

283. Truncatella bairdiana C. B. Adams, 1852. The shell is coarsely ribbed, with a thick, heavy crest behind the lip and a heavy internal thickening of the aperture. Length, 5 mm; width, 1.9 mm; whorls, four. Panama.

284. Truncatella californica Pfeiffer, 1857. Less coarsely ribbed than the preceding, sometimes nearly smooth. The crest may be absent, but if present is weak, and the aperture has no thick internal callus. Length, 5 mm; width, 2 mm; whorls, four and a quarter. Southern California to the upper Gulf of California.

Family CYCLOSTREMELLIDAE

Small to minute, lenticular; initial whorls set at an angle, *i.e.*, heterostrophic; later whorls smooth to striate, widely umbilicate; apertural margin complete, thin.

Genus CYCLOSTREMELLA BUSH, 1897

Planispirally coiled; suture with a channel, aperture triangular-ovate, wider below, with a sinus at suture; operculum oval, paucispiral. Useful reference: D. Moore (1966).

285. Cyclostremella orbis (Carpenter, 1857) [*Vitrinella*]. Mazatlán, Mexico. Figured, Keen (1968).

Family VITRINELLIDAE

Small to minute, mostly discoidal or low-spired; shell material porcelaneous to glassy; operculum horny, multispiral, circular.

Panamic vitrinellids have been reviewed and the shells well illustrated by Pilsbry & Olsson (1945, 1952). More work on the soft parts is needed to clear up confusion with such other families as the Liotiidae and Fossaridae. Mostly small forms less than 5 mm in diameter. Since all require study under a microscope, detailed treatment is not given here. Original generic allocations are shown in square brackets. Subsequent figures are cited for those not adequately illustrated in the original description. Two Panamic subfamilies. The key is to genera.

1. Umbilical area wide, not obscured by callus........................ 2
   Umbilical area partially or entirely filled with callus.................10
2. Periphery with a finely serrate keel..........................*Episcynia*
   Periphery not serrate... 3
3. Suture marked by a row of beads...................*Discopsis* (*Alleorus*)
   Suture smooth 4
4. Columellar lip widened at some part by a callosity............*Solariorbis*
   Columellar lip even, not widened at any point...................... 5
5. Shell smooth or nearly so, without conspicuous ribs...........*Vitrinella*
   Shell with well-developed sculpture.............................. 6
6. Periphery rounded ....*Parviturboides*
   Periphery emphasized by a keel.................................. 7
7. Height nearly equal to diameter.........................*Aorotrema*
   Height only about one-half as much as diameter...................... 8

8. Spiral keel one, confined to periphery......................*Vitrinorbis*
   Spiral keels more than one above and below periphery................. 9
9. Keels moderate, uppermost not markedly high............*Cyclostremiscus*
   Keels strong, uppermost one forming a low collar..............*Lydiphnis*
10. Callus pad smooth... 11
    Callus pad sculptured or divided................................ 13
11. Callus pad distributed over most of base...................*Teinostoma*
    Callus pad small, central..................................... 12
12. Shell porcelaneous, base with radial ribs.....................*Anticlimax*
    Shell glassy in texture, base smooth........................*Vitridomus*
13. Callus pad two-pronged.....................................*Panastoma*
    Callus pad grooved.......................................*Woodringilla*

Subfamily VITRINELLINAE

Mostly lenticular, variously sculptured; umbilicate. Eight Panamic genera.

Genus VITRINELLA C. B. ADAMS, 1852

Lenticular, glassy to opaque-white, thin, fragile, smooth or only weakly sculptured, umbilicus wide. Two Panamic subgenera.

Subgenus VITRINELLA, s. s.

Umbilicus bordered by a spiral thread, periphery rounded.

286. Vitrinella (Vitrinella) goniomphala Pilsbry & Olsson, 1952. Ecuador.

287. V. (V.) modesta C. B. Adams, 1852. Panama. Figured, Pilsbry & Olsson (1945).

288. V. (V.) ? naticoides Carpenter, 1857. Mazatlán, Mexico. Figured, Keen (1968).

Subgenus VITRINELLOPS PILSBRY & OLSSON, 1952

Discoidal, few-whorled, periphery carinate in some; umbilicus smooth, no spiral thread at its margin.

289. Vitrinella (Vitrinellops) bifilata Carpenter, 1857. Mazatlán, Mexico. Figured, Keen (1968).

290. V. (V.) campylochila Pilsbry & Olsson, 1952. Panama.

291. V. (V.) dalli (Bartsch, 1911) [*Cyclostremella*]. Gulf of California.

292. V. (V.) fortaxis Pilsbry & Olsson, 1952. Panama.

293. V. (V.) guaymasensis Durham, 1942. Guaymas, Mexico.

294. V. (V.) lucasana (Baker, Hanna & Strong, 1938) [*Delphinoidea*]. Gulf of California.

295. V. (V.) magister Pilsbry & Olsson, 1952. Colombia.

296. V. (V.) margarita Pilsbry & Olsson, 1952. Panama.

297. V. (V.) martensiana (Hertlein & Strong, 1951) [*Scissilabra*]. Corinto, Nicaragua.

298. V. (V.) multispiralis Pilsbry & Olsson, 1952. Ecuador to Peru.

299. V. (V.) ponceliana De Folin, 1867. Panama Bay.

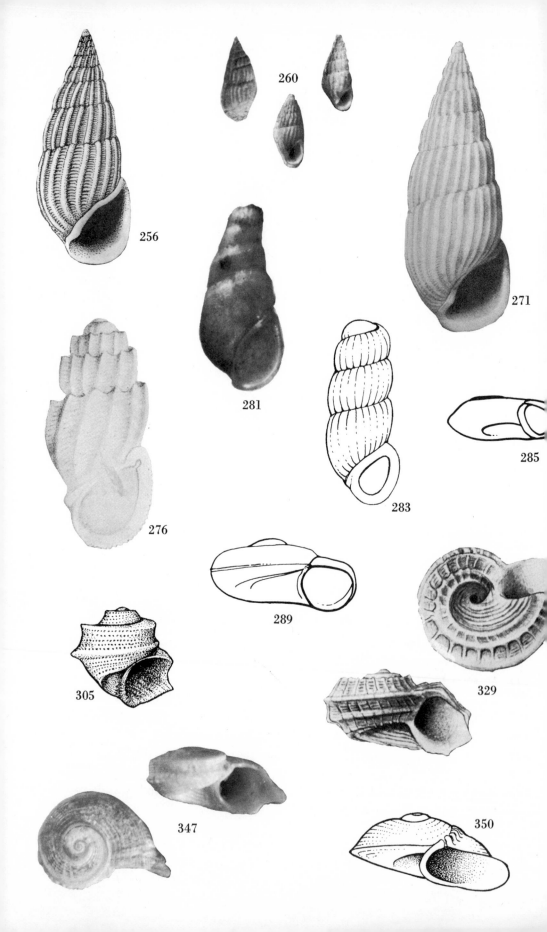

256

260

271

281

283

285

276

289

329

305

347

350

300. V. (V.) proxima Pilsbry & Olsson, 1952. Panama.

301. V. (V.) stephensae (Baker, Hanna & Strong, 1938) [*Delphinoidea*]. Tres Marias Islands.

302. V. (V.) subquadrata Carpenter, 1857. Mazatlán. Figured, Keen (1968).

303. V. (V.) tiburonensis Durham, 1942. Gulf of California.

304. V. (V.) zonitoides Pilsbry & Olsson, 1952 (type of subgenus). Panama.

Genus **AOROTREMA** SCHWENGEL & MCGINTY, 1942

With two strong keels; sculpture pitted; umbilicus present.

305. Aorotrema humboldti (Hertlein & Strong, 1951) [*Cyclostremiscus*]. Gulf of California to Costa Rica.

Genus **CYCLOSTREMISCUS** PILSBRY & OLSSON, 1945

Lenticular, with several spiral keels and weaker axial ribs. Three Panamic subgenera.

Subgenus **CYCLOSTREMISCUS**, *s. s.*

Apertural margin even and relatively thin.

306. Cyclostremiscus (Cyclostremiscus) adamsi (Bartsch, 1911) [*Cyclostrema*]. Panama.

307. C. (C.) azuerensis Pilsbry & Olsson, 1952. Panama to Ecuador.

308. C. (C.) bailyi (Hertlein & Strong, 1951) [*Circulus*]. Corinto, Nicaragua.

309. C. (C.) balboa Pilsbry & Olsson, 1945. Panama.

310. C. (C.) baldridgae (Bartsch, 1911) [as *Cyclostrema baldridgei*]. Gulf of California.

C. (C.) bartschi Strong & Hertlein, 1939 [not Mansfield, 1930]. (See *C. (C.) veleronis.*)

311. C. (C.) bifrontia (Carpenter, 1857) [*Vitrinella*]. Mazatlán, Mexico. Figured, Keen (1968).

312. C. (C.) cerrosensis (Bartsch, 1907) [*Circulus*]. Southern California to Gulf of California.

313. C. (C.) coronatus (Carpenter, 1857) [*Vitrinella*]. Mazatlán, Mexico. Figured, Keen (1968).

314. C. (C.) colombianus Pilsbry & Olsson, 1945. Colombia.

315. C. (C.) cosmius (Bartsch, 1907) [*Circulus*]. Ecuador.

316. C. (C.) diomedeae (Bartsch, 1911) [*Circulus*]. Panama Bay.

C. (C.) exigua (C. B. Adams, 1852). (See *C. (C). trigonatus.*)

317. C. (C.) gallo Pilsbry & Olsson, 1945. Colombia.

318. C. (C.) glyptobasis Pilsbry & Olsson, 1952. Panama.

319. C. (C.) glyptomphalus Pilsbry & Olsson, 1952. Pleistocene, Panama.

320. C. (C.) gordanus (Hertlein & Strong, 1951) [*Cyclostrema*]. Gulf of California.

321. C. (C.) janus (C. B. Adams, 1852) [*Vitrinella*]. Panama. Figured, Turner (1956).

322. C. (C.) lirulatus (Carpenter, 1857) [*Vitrinella*]. Mazatlán. Figured, Keen (1968).

323. C. (C.) lowei (Baker, Hanna & Strong, 1938) [*Cyclostrema*]. Cape San Lucas.

324. C. (C.) madreensis (Baker, Hanna & Strong, 1938) [*Circulus*]. Tres Marias Islands.

325. C. (C.) major Olsson & Smith, 1952. Guaymas, Mexico, to Panama.

326. C. (C.) nodosus (Carpenter, 1857) [*Vitrinella perparva*, var.]. Panama and Mazatlán. Figured, Keen (1968).

327. C. (C.) nummus Pilsbry & Olsson, 1952. Panama.

328. C. (C.) ornatus (Carpenter, 1865) [*Vitrinella*]. Mazatlán, Mexico. Figured, Keen (1968).

329. C. (C.) panamensis (C. B. Adams, 1852) [*Vitrinella*]. Mexico to Panama. Figured, Pilsbry & Olsson (1945).

330. C. (C.) parvus (C. B. Adams, 1852) [*Vitrinella*]. Mexico to Panama. Figured, Pilsbry & Olsson (1945).

331. C. (C.) pauli Pilsbry & Olsson, 1952. Panama. (Figured, Pilsbry & Olsson (1945) as *C. bartschi*.)

332. C. (C.) perparvus (C. B. Adams, 1852) [*Vitrinella*]. Panama. Figured, Pilsbry & Olsson (1945).

333. C. (C.) peruvianus Pilsbry & Olsson, 1945. Tumbes, Peru.

334. C. (C.) planospira Pilsbry & Olsson, 1945. Panama.

335. C. (C.) psix Pilsbry & Olsson, 1952. Ecuador.

336. C. (C.) solitarius Hertlein & Allison, 1968. Clipperton Island.

337. C. (C.) spiceri (Baker, Hanna & Strong, 1938) [*Cyclostrema*]. Gulf of California.

338. C. (C.) spiritualis (Baker, Hanna & Strong, 1938) [*Delphinoidea*]. La Paz, Baja California.

339. C. (C.) taigai (Hertlein & Strong, 1951) [*Circulus*] ["*C. faigai*" spelling error, Pilsbry & Olsson, 1952.] Corinto, Nicaragua.

340. C. (C.) tenuisculptus (Carpenter, 1864) [*Vitrinella*]. Mazatlán, Mexico. Figured, Keen (1968).

341. C. (C.) tricarinatus (C. B. Adams, 1852) [*Vitrinella*]. Panama. Figured, Pilsbry & Olsson (1945).

342. C. (C.) trigonatus (Carpenter, 1857) [*Vitrinella*]. (Synonym: **V. exigua** C. B. Adams, 1852, not of Philippi, 1845). Guaymas, Mexico, to Panama. Figured, Pilsbury & Olsson (1945).

343. C. (C.) valvatoides (C. B. Adams, 1852) [*Vitrinella*]. Panama. Figured, Pilsbry & Olsson (1945).

344. C. (C.) veleronis (Strong & Hertlein, 1947) [*Cyclostrema*] [new name for *C. bartschi*]. Panama.

345. ? C. (C.) verreauxii (Fischer, 1857) [*Adeorbis*]. "California." Figured, Pilsbry & Olsson (1945).

346. C. (C.) xantusi (Bartsch, 1907) [*Cyclostrema*]. Cape San Lucas, Baja California, to Panama.

Subgenus **MIRALABRUM** PILSBRY & OLSSON, 1945

Outer lip with a spur.

347. Cyclostremiscus (Miralabrum) planospiratus (Carpenter, 1857) [*Vitrinella*]. Mazatlán, Mexico. Figured, Keen (1968).

348. C. (M.) unicornis (Pilsbry & Olsson, 1945) [*Miralabrum*]. Ecuador.

Subgenus **PACHYSTREMISCUS** OLSSON & McGINTY, 1958

Apertural margin thickened.

349. Cyclostremiscus (Pachystremiscus) pachynepion Pilsbry & Olsson, 1945. Colombia.

Genus **DISCOPSIS** DE FOLIN & PÉRIER, 1870

With a single prominent peripheral keel. Subgenus *Discopsis, s. s.,* is not represented in the Panamic province.

Subgenus **ALLEORUS** STRONG, 1938

Suture beaded, carina smooth.

350. Discopsis (Alleorus) deprellus (Strong, 1938) [*Alleorus*]. Gulf of California.

Genus **EPISCYNIA** MÖRCH, 1875

Periphery carinate, keel finely beaded under a deciduous periostracum; last whorl slightly descending.

351. Episcynia bolivari Pilsbry & Olsson, 1946. Colombia to Peru; Pleistocene, Panama. Figured, Pilsbry & Olsson (1952).

352. Episcynia medialis Keen, new species. The small white shell has $5\frac{1}{2}$ whorls, rounded above, with a well-marked peripheral carina set off by a slight groove above it; spiral sculpture of a few faint lirae, axial sculpture of growth lines only, sloping backward from the suture in a smooth curve, some of the lines at regular intervals being heavier; the last whorl falls slightly below the beaded carination, which is minutely denticulate; the umbilicus is open, with vertical sides, the base set off from the umbilicus by a sharp angle but not a carina; outer lip slightly reflected near the juncture with the columellar lip. Diameter (holotype), 3.1 mm; height, 1.9 mm. Type locality, off Cabo Haro, Guaymas, Mexico, in 18 m. A second specimen measuring 2 mm in diameter, with $3\frac{1}{2}$ glassy whorls, was collected at Banderas Bay, Mexico. The species is closest to *E. bolivari* Pilsbry & Olsson, 1946, differing in the smaller size and the deflection of the final whorl below the suture, which is absent in *E. bolivari*; from *E. nicholsoni* it differs in its relatively greater height, from the Californian *E. devexa* Keen, 1946, in smaller size and lesser deflection of the body whorl.

353. Episcynia nicholsoni (Strong & Hertlein, 1939) [*Circulus*]. Panama.

Genus LYDIPHNIS MELVILL, 1906

Depressed, with one to three prominent keels, apertural margin angulate. No species of subgenus *Lydiphnis, s. s.,* occur in the Panamic province.

Subgenus CYMATOPTERYX PILSBRY & OLSSON, 1946

Discoidal, middle keel broadly expanded and wavy.

354. Lydiphnis (Cymatopteryx) cincta (Carpenter, 1857) [*Vitrinella*]. Mazatlán, Mexico. Figured, Keen (1968).

355. L. (C.) cymatotropis Pilsbry & Olsson, 1945. Ecuador.

356. L. (C.) mariae (Baker, Hanna & Strong, 1938) [*Cyclostrema*]. Tres Marias Islands.

357. L. (C.) strongi Pilsbry & Olsson, 1952. Ecuador.

Genus PARVITURBOIDES PILSBRY & McGINTY, 1950

Resembling the archaeogastropod genus *Parviturbo* in outline, with spiral sculpture predominating over axials; umbilicus small.

358. Parviturboides clausus (Pilsbry & Olsson, 1945) [*Parviturbo*]. Ecuador.

359. P. copiosus (Pilsbry & Olsson, 1945) [*Parviturbo*]. Guaymas, Mexico, to Ecuador.

360. P. decussatus (Carpenter, 1857) [*Vitrinella*]. Mazatlán, Mexico. Figured, Keen (1968).

361. P. germanus (Pilsbry & Olsson, 1945) [*Parviturbo*]. Colombia to Ecuador.

362. P. monile (Carpenter, 1857) [*Vitrinella*]. Mazatlán, Mexico. Figured, Keen (1968).

363. P. monilifer (Carpenter, 1857) [*Vitrinella*]. Mazatlán, Mexico. Figured, Keen (1968).

Genus SOLARIORBIS CONRAD, 1865

Depressed, spirally striate to smooth, umbilicus with a spiral ridge ending in a lobe on inner lip. Four Panamic subgenera.

Subgenus SOLARIORBIS, s. s.

Periphery rounded, aperture grooved at upper angle; umbilical wall wide.

364. Solariorbis (Solariorbis) allomphalus Pilsbry & Olsson, 1952. Colombia.

365. S. (S.) ametabolus Pilsbry & Olsson, 1952. Tumbes, Peru.

366. S. (S.) bailyanus Pilsbry & Olsson, 1952. Panama.

367. S. (S.) bakeri (Strong & Hertlein, 1939) [*Circulus*]. Panama.

368. S. (S.) concinnus (C. B. Adams, 1852) [*Vitrinella*] (Synonym: **Cyclostrema mccullochae** Strong & Hertlein, 1939). Panama. Figured, Pilsbry & Olsson (1945).

369. S. (S.) gibraleonis Pilsbry & Olsson, 1952. Panama to Ecuador.

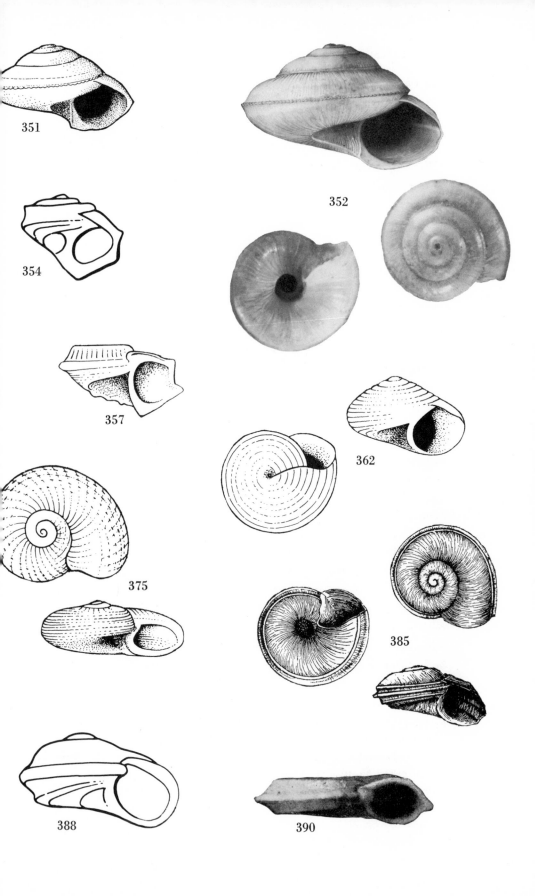

351

352

354

357

362

375

385

388

390

370. S. (S.) hambachi (Strong & Hertlein, 1939) [*Delphinoidea*]. Panama.

371. S. (S.) hannai (Strong & Hertlein, 1939) [*Delphinoidea*]. Panama.

372. S. (S) hypolius Pilsbry & Olsson, 1952. Tumbes, Peru.

S. (S.) mccullochae. (See *S. concinnus.*)

373. S. (S.) miguelensis Pilsbry & Olsson, 1952. Panama.

374. S. (S.) millepunctatus (Pilsbry & Olsson, 1945) [*Cyclostremiscus*]. Colombia.

375. S. (S.) minutus (C. B. Adams, 1852) [*Vitrinella*]. Panama. Figured, Pilsbry & Olsson (1945).

376. S. (S.) narinensis Pilsbry & Olsson, 1952. Colombia.

377. S. (S.) pyricallosus (Carpenter, 1857) [*Globulus*]. Mazatlán. Figured, Keen (1968).

378. S. (S.) regularis (C. B. Adams, 1852) [*Vitrinella*]. Panama. Figured, Pilsbry & Olsson (1945).

Subgenus EULEREMA PILSBRY & OLSSON, 1952

Aperture not grooved at upper angle.

379. Solariorbis (Eulerema) pellucidus Pilsbry & Olsson, 1952. Ecuador (type of subgenus).

Subgenus HAPALORBIS WOODRING, 1957

With one or more strong spiral keels, especially at periphery; umbilical wall narrow, lobe small.

380. Solariorbis (Hapalorbis) carianus Pilsbry & Olsson, 1952. Colombia.

381. S. (H.) carinatus (Carpenter, 1857) [*Teinostoma*]. Mazatlán, Mexico. Figured, Pilsbry & Olsson (1952).

382. S. (H.) carinulatus (Carpenter, 1857) [*Vitrinella*]. Mazatlán, Mexico. Figured, Keen (1968).

383. S. (H.) ditropis Pilsbry & Olsson, 1952. Colombia.

384. S. (H.) exquisitus Pilsbry & Olsson, 1952. Ecuador.

385. S. (H.) liriope (Bartsch, 1911) [*Circulus*]. Gulf of California (type of subgenus).

386. S. (H.) pacificus Pilsbry & Olsson, 1952. Panama.

387. S. (H.) seminudus (C. B. Adams, 1852) [*Vitrinella*]. Panama. Figured, Pilsbry & Olsson (1945).

Subgenus SYSTELLOMPHALUS PISBRY & OLSSON, 1941

Early whorls with oblique axial ribs, last whorl smooth above, spirally ribbed below, umbilical border overhanging.

388. Solariorbis (Systellomphalus) annulatus (Carpenter, 1857) [*Vitrinella*]. Mazatlán, Mexico. Figured, Keen (1968).

389. S. (S.) elegans Pilsbry & Olsson, 1952. Guaymas, Mexico, to Tumbes, Peru (type of subgenus).

Genus **VITRINORBIS** PILSBRY & OLSSON, 1952

Discoidal, periphery carinate; aperture oblique; base widely umbilicate.

390. Vitrinorbis callistus Pilsbry & Olsson, 1945. Ecuador.

391. V. galloensis Pilsbry & Olsson, 1945. Colombia.

Subfamily **TEINOSTOMATINAE**

Base partially to entirely sheathed with callus. Five Panamic genera.

Genus **TEINOSTOMA** A. ADAMS, 1851

Lenticular, smooth or weakly spirally striate. Four, perhaps five, Panamic subgenera.

Subgenus **TEINOSTOMA**, *s. s.*

Spire nearly enclosed by last whorl; shell relatively large.

392. Teinostoma (Teinostoma) gallegosi Jordan, 1936. Magdalena Bay, Baja California. Figured, Durham (1950).

393. T. (T.) politum A. Adams, 1851. Ecuador. Figured, Pilsbry & Olsson (1952). Type of genus.

394. T. (T.) ultimum Pilsbry & Olsson, 1945. West Mexico. Probably a synonym of *T. gallegosi.*

Subgenus **ESMERALDA** PILSBRY & OLSSON, 1952

Spirally striate; upper edge of aperture arched forward.

395. Teinostoma (Esmeralda) concavaxis Pilsbry & Olsson, 1945. Panama to Colombia.

396. T. (E.) esmeralda Pilsbry & Olsson, 1945. Ecuador

397. T. (E.) imperfectum Pilsbry & Olsson, 1945. Panama.

Subgenus **IDIORAPHE** PILSBRY, 1922

Suture concealing spire, as in *Teinostoma, s. s.,* but deviating abruptly in its final turn; peristome rounded; shell thick.

398. Teinostoma (Idioraphe) narina Pilsbry & Olsson, 1945. Colombia.

Subgenus **PSEUDOROTELLA** FISCHER, 1857

Umbilical callus not bounded by a cord; callus of inner lip thin.

399. Teinostoma (Pseudorotella) americanum Pilsbry & Olsson, 1945. Ecuador.

400. T. (P.) amplectans Carpenter, 1857. Mazatlán, Mexico. Figured, Keen, 1968.

401. T. (P.) cecinella Dall, 1919. Magdalena Bay, Baja California. Probably = *T. amplectans.*

402. T. (P.) ecuadorianum Pilsbry & Olsson, 1941. Ecuador (Pleistocene) to Tumbes, Peru. Figured, Pilsbry & Olsson (1952).

403. T. (P.) herbertianum Hertlein & Strong, 1951. Costa Rica.

404. T. (P.) lampetes Pilsbry & Olsson, 1952. Ecuador.

405. T. (P.) lirulatum (Carpenter, 1857) [*Globulus*]. Mazatlán, Mexico. Figured, Keen (1968).

406. T. (P.) millepunctatum Pilsbry & Olsson, 1945. Ecuador.

407. T. (P.) ochsneri Strong & Hertlein, 1939. Panama.

408. T. (P.) pallidulum (Carpenter, 1857) [*Globulus*]. Mazatlán, Mexico. Figured, Keen (1968).

409. T. (P.) percarinatum Pilsbry & Olsson, 1945. Sechura Bay, Peru.

410. T. (P.) soror Pilsbry & Olsson, 1945. Ecuador.

411. T. (P.) substriatum Carpenter, 1857. Mazatlán, Mexico. Figured, Keen (1968).

412. T. (P.) sulcatum (Carpenter, 1857) [*Globulus*]. Mazatlán, Mexico. Figured, Keen (1968). Possibly a synonym of *T. (P.) tumens*.

413. T. (P.) tumens (Carpenter, 1857) [*Globulus*]. Mazatlán, Mexico. Figured, Keen (1968).

414. T. (P.) zacae Hertlein & Strong, 1951. Costa Rica.

Genus TEINOSTOMA, Subgenus uncertain

415. Teinostoma hemphilli Strong & Hertlein, 1939. Panama.

416. T. invallatum (Carpenter, 1864) [*Ethalia*]. Monterey, California, to Gulf of California.

417. T. rarum Pilsbry & Olsson, 1945. Panama to Colombia.

418. T. supravallatum (Carpenter, 1864) [*Ethalia*]. Monterey, California, to Baja California.

Genus ANTICLIMAX PILSBRY & McGINTY, 1946

Spirally striate, with wavelike radial ribs on base. No species of subgenus *Anticlimax, s. s.*, occur in the Panamic province.

Subgenus SUBCLIMAX PILSBRY & OLSSON, 1950

Umbilicus nearly or entirely closed by a callus pad united to the columella.

419. Anticlimax (Subclimax) willetti Hertlein & Strong, 1951. Costa Rica.

Genus PANASTOMA PILSBRY & OLSSON, 1945

As yet known only from the Pleistocene of western Panama; solid; strongly spirally striate, inner lip with a two-pronged callus.

420. Panastoma azulense Pilsbry & Olsson, 1945. Pleistocene, Panama.

Genus VITRIDOMUS PILSBRY & OLSSON, 1945

Thin and fragile, with fine spirals, periphery bluntly carinate, umbilical callus small.

421. Vitridomus fragilis Pilsbry & Olsson, 1945. Ecuador.

422. V. nereidis Pilsbry & Olsson, 1945. Ecuador (type of genus).

Genus WOODRINGILLA PILSBRY & OLSSON, 1951

Solid, with spiral sculpture; umbilical callus grooved.

393

400

411

419

422

423

424

425

423. Woodringilla glyptulus Pilsbry & Olsson, 1951. Ecuador (type of genus).

? Family CHORISTIDAE

Small to medium-sized, inflated, umbilicate, smooth; operculum paucispiral, with central nucleus.

Genus CHORISTES CARPENTER in DAWSON, 1872

Shell thin, spire somewhat tabulate, slightly elevated; aperture entire; operculum concave.

424. Choristes carpenteri Dall, 1896. In deep water, off west Mexico to the Bay of Panama, 2,700 to 3,440 m. Figured, Dall (1908).

Superfamily ARCHITECTONICACEA

The apex is blunt because the larval stages have the type of coiling known as ultra-dextral or hyperstrophic (each whorl rising above the periphery of the previous one instead of falling below as in normal helical coiling), and the larval shell can be seen only through the umbilicus, which in most forms is broadly open. The larval shells may even appear to be sinistrally coiled. Because of a long pelagic larval life, many of the species have a wide distribution and may occur in more than one marine province.

This superfamily has been revised in consultation with Dr. Robert Robertson of the Academy of Natural Sciences of Philadelphia, whose assistance is here gratefully acknowledged. His review of *Heliacus* was facilitated by the loan of material from the Los Angeles County Museum's collection.

Family ARCHITECTONICIDAE

Generally low-spired shells, trochoid in outline, sculptured with spiral ribs that may be evenly beaded, the umbilicus usually marked by a heavier cord. The operculum is horny and has a peg projecting into the foot. Two Panamic genera.

Genus ARCHITECTONICA RÖDING, 1798
(SOLARIUM LAMARCK, 1799)

Umbilicus bounded by a beaded ridge; operculum flat and few-whorled. Two Panamic subgenera.

Subgenus ARCHITECTONICA, *s. s.*

Spire low, gently convex, the suture hardly apparent.

425. Architectonica (Architectonica) nobilis Röding, 1798 (Synonyms: **Solarium granulatum** Lamarck, 1816; **S. granosum** Valenciennes, 1832; **S. quadriceps** Hinds, 1844; **S. verrucosum** Philippi, 1849; **A. valenciennesi** Mörch, 1860). The noble sundial shell varies from grayish brown to a bright flesh color, with bands of light brown spots. Diameter, 32 mm; height, 18 mm. Magdalena Bay, Baja California, through the Gulf of California and southward to Peru, on tideflats and to depths of 37 m. The species occurs in both the eastern and western Atlantic as well as the eastern Pacific, and as yet no authors have pointed out features that clearly distinguish the populations. Should anyone succeed in finding consistent differences, there are abundant names available for the Panamic representative.

Subgenus DISCOTECTONICA MARWICK, 1931
(ACUTITECTONICA HABE, 1962)

Spire flatter than in *Architectonica, s. s.*, the suture generally falling below the periphery and the aperture thus not entirely enclosing the strong peripheral cord.

426. Architectonia (Discotectonica) placentalis (Hinds, 1844). A small, flat form with a single sharp peripheral keel. In color it is creamy white, mottled with buff or brown. Diameter, 12 to 15 mm. Magdalena Bay, Baja California (type locality), to Guaymas, Mexico (Shasky collection). A twin species in the Atlantic is *A. (D.) peracuta* (Dall, 1889).

Genus HELIACUS ORBIGNY, 1842
(TORINIA GRAY, 1847)

Smaller than *Architectonica*, with more reticulate and stronger sculpture. The distinctive operculum is a spiral, horny, pagoda-like structure of numerous turns, with a fringed edge.

427. Heliacus architae (Costa, 1844) (Synonym: **H. panamensis** Bartsch, 1918). The sculpture is strongly reticulate, the base at all stages convexly angulate in profile, with a major cord at the crest; the spire has normally four cords per whorl, the second from the periphery narrowest, six to eight spiral cords on base; two prominent keels at the periphery. Umbilicus wide and markedly stepped. Protoconch discoidal, smaller than in other species of *Heliacus*. Diameter (large specimen), 13 mm; height, 5 mm; most specimens about 6 mm in diameter. Southern California, through the Gulf of California and south to Ecuador. Also in the Atlantic and Mediterranean.

428. Heliacus bicanaliculatus (Valenciennes, 1832) (Synonyms: **Euomphalus radiatus** Menke, 1851; **H. chiquita** Pilsbry & Lowe, 1932). Like small trochids in shape but with regular spiral ribbing and a color pattern of brown and white spots. There are two furrows or channels on the inner lip, reflections of spiral ribs in the umbilicus. The spiral fringed operculum is a conspicuous feature. Spire normally with three cords per whorl in addition to the upper peripheral keel. Diameter, 15 mm; height, 12 mm. La Paz, Baja California, through the Gulf of California and south to Panama and the Galápagos Islands, on zoanthids (colonial sea anemones) at extreme low tide. A similar species in the western Pacific is *H. variegatus* (Gmelin, 1791), and a like one in the Atlantic is *H. cylindricus* (Gmelin, 1791), of which *H. cyclostoma* (Menke, 1830) is a synonym.

429. Heliacus caelatus (Hinds, 1844) (Synonym: **H. planispira** Pilsbry & Lowe, 1932). Resembling *H. mazatlanicus* but distinguished by having the three outermost cords of each spire whorl replaced by strong radial ribs, traces of the three cords remaining on the early whorls. The subsutural noded cord persists. On the base are normally only five noded spiral cords; periphery with two keels. Protoconch discoidally coiled. Diameter of a large specimen, 9 mm; height, 3 mm. Concepcion Bay and Guaymas, Gulf of California, to Chiapas, Mexico, in the eastern Pacific; wide-ranging in the western Pacific (type locality). Woodring in 1957 made *H. planispira* the type of a subgenus, *H. (Astronacus)*.

430. Heliacus mazatlanicus Pilsbry & Lowe, 1932. Spire varying from nearly flat to roundly convex, with an anal keel or angle on the protoconch, spire whorls normally with four nodose cords on each; periphery with two prominent keels, an

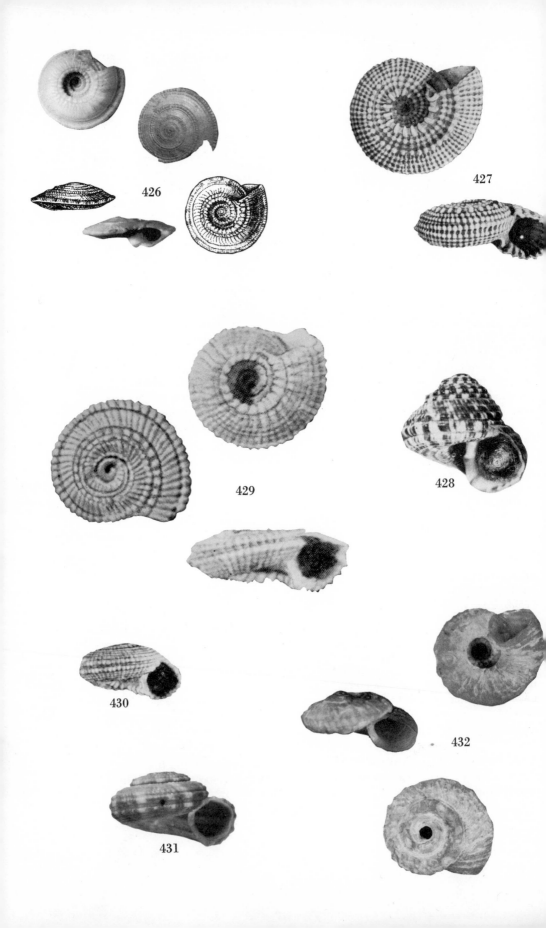

426

427

429

428

430

431

432

intercalary thread or minor cord not always developed; base with seven noded spiral cords and threads, the first and third from the periphery the narrowest, often absent; base convexly angulate only in juvenile stages. Color varying from nearly uniform creamy white to dark brown, usually with variegation or banding. Diameter (large specimen), 10 mm; height, 6.3 mm; average-sized specimens about 6 mm in diameter. San Felipe, near the head of the Gulf of California, to Ecuador and the Galápagos Islands; the most common of the small *Heliacus*. A similar form in the Atlantic has been cited as *H. bisculatus* (Orbigny, 1842).

431. Heliacus perrieri (Rochebrune, 1881) (Synonym: ? "**H. infundibuliformis strigatus** [Hanley, 1863]" of Hertlein & Allison, 1968). Resembling *H. bicanaliculatus* in size but with a much wider umbilicus, lower spire, and a channeled suture; whorls circular in profile, with wide spiral cords alternating with much smaller cords or threads. Diameter, about 12 mm; height, about 5 mm. Mazatlán and Panama; also in the Caribbean and eastern Atlantic (type locality). A similar species in the western Pacific is *H. infundibuliformis* (Gmelin, 1791), incorrectly cited by some authors under the name *H. crenellus* (Linnaeus, 1758), a *nomen dubium*. The record by Hertlein and Allison (1968) of *H. infundibuliformis strigatus* from Clipperton Island probably represents *H. perrieri* instead.

432. ? Heliacus radialis (Dall, 1908). Although live-taken, this small white shell from very deep water is too corroded for satisfactory assignment. Diameter, 9 mm. Gulf of Panama, 2,320 m.

Superfamily TURRITELLACEA

Low- to high-spired sturdy shells with several whorls, a round aperture, and a multispiral horny operculum.

Family TURRITELLIDAE

The tower shells are among the most slender-spired of the gastropods. One wonders how the animal can carry so delicately turned a shell without breaking it, but perhaps that is because animal and shell are safely buried just under the surface of the sea floor, where the animal feeds on the detritus—fine bits of seaweed and other broken fragments—that settles like dust. Most species live beyond the low-tide line.

The shells are somewhat variable—even within the course of growth of one individual—in the expression of the spiral ribs. Axial sculpture is weak. The aperture is round and is closed by a remarkable operculum, a flat or saucer-shaped multispiral coil that is thin and membranous at the edge, supported by flexible bristles somewhat prolonged beyond the margin. This operculum is so flexible it may be withdrawn well within the shell when the animal retreats. Two Panamic subfamilies.

Subfamily TURRITELLINAE

Coiling evenly spiral throughout all whorls.

Genus TURRITELLA LAMARCK, 1799

The Turritellas have tightly coiled shells. Although usually living just below the sea floor, the animal is capable of crawling about. These are gregarious mollusks, tending to form large colonies where they do occur.

433. Turritella anactor Berry, 1957. The handsome shell is yellowish gray, heavily suffused and streaked with purplish brown. It resembles *T. gonostoma* in size but differs in the concave outline of the whorls and the beveled keel just above the suture. Length, 122 mm; diameter, 30 mm. The type locality is San Felipe, near the head of the Gulf of California; to Puerto Peñasco, Sonora, Mexico.

434. Turritella banksi Reeve, 1849. This is flat-whorled, with a strong spiral cord at the periphery. The cord, which forms a sutural band on the spire, is checkered with darker brown on the buff to brown shell. Height, 48 mm; diameter, 16 mm. Guaymas, Mexico, to Ecuador.

435. Turritella broderipiana Orbigny, 1840. Confused with *T. gonostoma* by some authors, this is a southern species, largest of West American members of the genus; it has been made the type of the subgenus *Broderiptella* Olsson, 1964. The whorls are flat-sided and nearly smooth, creamy white with fine brown-speckled axial stripes. Length, 180 mm; diameter, 30 mm. The type locality is Paita, Peru.

436. Turritella cingulata Sowerby, 1825 (Synonym: **T. tricarinata** King & Broderip, 1835). The shell is ivory white with three raised spiral ribs that are dark brown. Length, about 50 mm. Manta, Ecuador, to Chile.

437. Turritella clarionensis Hertlein & Strong, 1951. The shell is white, marked with light brown. Its distinguishing features are the two spiral ribs above and below the suture separated by a nearly smooth central area on which the sinuous growth lines are apparent. The shell is wider for its height than either *T. radula* or *T. mariana*, species with somewhat similar ribbing. Height, 56 mm; diameter, 16.5 mm. Off Angel de la Guarda Island, Gulf of California, to Panama, in depths of 70 to 100 m.

438. Turritella gonostoma Valenciennes, 1832 (Synonyms: **T. punctata** and **marmorata** Kiener, 1843; **"T. goniostoma"** of authors, an unjustified emendation). In coloring the shell varies from light gray to dark purplish brown, mottled with white, and the sculpture varies from several spiral cords per whorl, with impressed sutures, to almost smooth and flat-sided, the latter variant approaching in form the larger and heavier South American *T. broderipiana*. Young specimens always have a spiral rib in the middle of the whorl. Height, 115 mm; diameter, 22 mm. Gulf of California to Ecuador.

439. Turritella lentiginosa Reeve, 1849. Though considered a synonym of *T. gonostoma* by most authors, this can be separated readily in any large suite of specimens. The sculpture is coarser than in *T. gonostoma*, with two to three cords strongly developed on the lower part of each whorl. Dimensions are similar to those in *T. gonostoma*; height, 115 mm; diameter, 22 mm. Gulf of California.

440. Turritella leucostoma Valenciennes, 1832 (Synonyms: **T. tigrina** Kiener, 1843–44; **T. cumingii** Reeve, 1849; **T. dura** Mörch, 1860). More delicate-appearing than *T. lentiginosa*, this has a light buff coloring mottled and striped with reddish brown. Each whorl is contracted just above the suture, giving it a characteristic profile. Height, 115 mm; diameter, 20 mm. Cedros Island, Baja California, south through the Gulf of California to Panama, in depths of as much as 40 m; common as dead shells, but not common living.

441. Turritella mariana Dall, 1908. As Hertlein and Strong (1955) have shown, it is probable that this will prove to be the same as *T. radula* Kiener. It

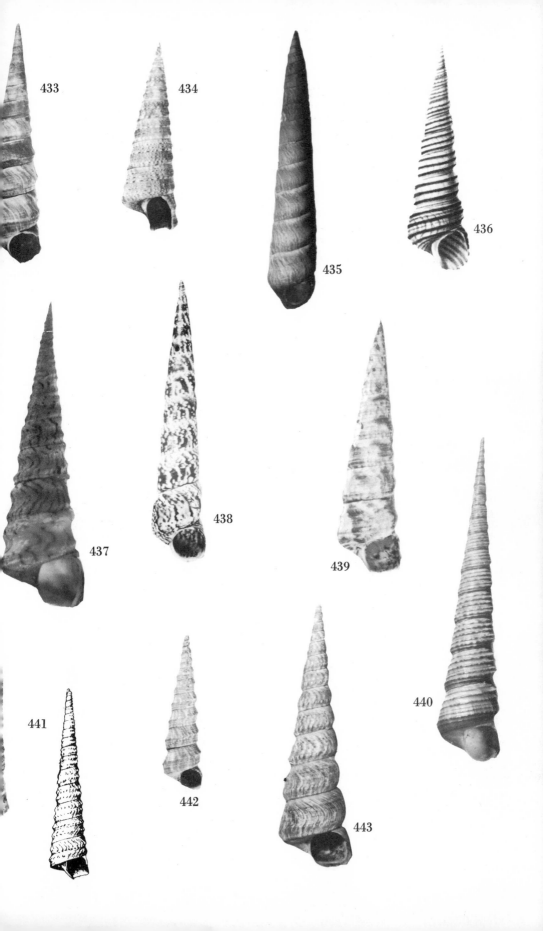

resembles *T. clarionensis* in having two heavy spiral ribs just above and just below the suture, but here the area between is sculptured with several finely beaded spiral ribs. Dall's holotype specimen was a young individual only 25 mm in height. A mature specimen measures: height, 62 mm; diameter, 12.5 mm. Cedros Island, Baja California, through the southern part of the Gulf of California and south to Colombia, in 22 to 150 m.

442. Turritella nodulosa King & Broderip, 1832 (Synonym: **T. papillosa** Kiener, 1843–44). In color the shell is gray to buff, with somewhat irregular axial stripes. The two to four principal spiral ribs are beaded. Height, 32 mm; diameter, 8 mm. Magdalena Bay, Baja California, through the southern part of the Gulf of California and south to Ecuador in 4 to 170 m.

443. Turritella parkeri McLean, 1970. Resembling *T. anactor* in having one main beveled keel just above the suture, the shell is smaller and thinner, with a deeply indented growth line trace, the single main keel being the only one apparent in the early whorls. Color, tan with irregular brown streaks along the growth lines. Additional spiral cording is variable in size and number of riblets. Base nearly smooth. Height, 48 mm; diameter, 14 mm. Espíritu Santo Island, Gulf of California, in depths of 82 to 145 m. In contrast to other Turritellas, which are relatively shallow-water forms, this occurs well offshore.

444. Turritella radula Kiener, 1843–44. As indicated above, under the discussion of *T. mariana*, this may be the earlier name for the species. The original figure is reproduced here. The type specimen measures 80 mm in height and probably came from Ecuador, which was designated by Hertlein and Strong (1955) as type locality. Merriam (1941) recorded the range as "Acapulco, Mexico, to Ecuador," 20 to 110 m.

445. Turritella rubescens Reeve, 1849. With the early whorls resembling those of *T. nodulosa*, the shell gradually shows a change of sculpture by adding more spiral carinas that are less and less elevated; the mature whorls are almost flat. The color pattern is of reddish-brown irregular flecks and blotches. Length, 57 mm; diameter, 15 mm. San Francisco Island, Gulf of California, to Gorgona Island, Colombia, offshore, mostly in depths of 27 to 55 m.

446. Turritella willetti McLean, 1970 (Synonym: "**T. sanguinea** Reeve" of Shasky, 1961 [not of Reeve, 1849]). Early whorls convex, with fine spirals and impressed suture; later whorls flat-sided, with about 14 cords of irregular size, the suture somewhat beveled; base rounded, with numerous faint spiral cords. In color the shell is whitish with brown maculations. Length, 61 mm; diameter, 14 mm. Sonora to Sihuatanejo, Guerrero, Mexico, offshore in depths of 27 to 70 m.

Subfamily VERMICULARIINAE

Coiling regular in early whorls, becoming loose or disjunct in adult. Two Panamic genera.

Genus VERMICULARIA LAMARCK, 1799

Although confused with the Vermetidae by many collectors, these shells are easily recognized by the difference in coiling of the early whorls, which form a smoothly tapering cone, like the early whorls of a *Turritella*, the whorls evenly and tightly joining. Later, the coiling may become lax or disjunct and irregular. Some specimens may be cemented to rocks and stones in the gravelly sand of their habitat, but others seem to be able to move about to some extent.

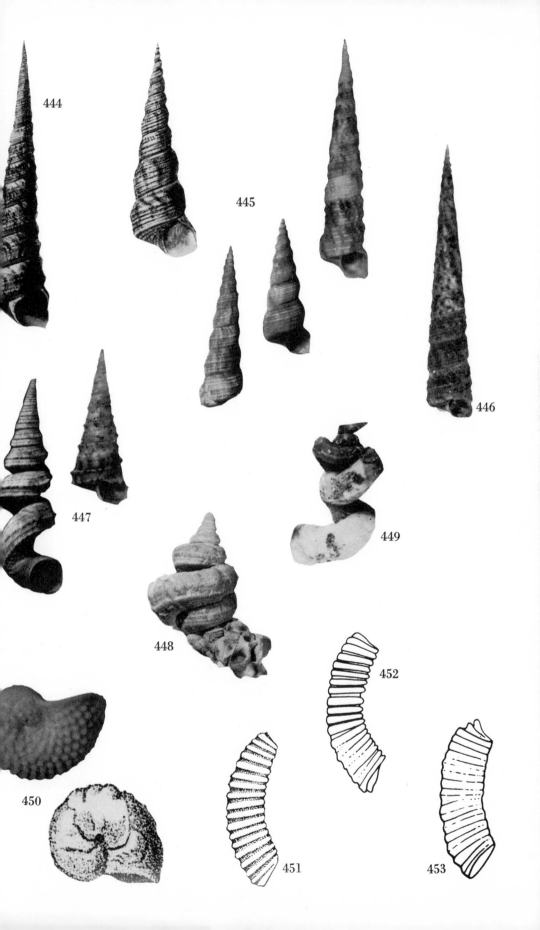

447. Vermicularia frisbeyae McLean, 1970. The shell is white with irregular brown spots along the growth lines and on the base. Early whorls of the shell are remarkably large and turritelloid in appearance, with two spiral keels. After about 12 turns the shell may develop disjunct coiling, and the last whorls may be only weakly spiral, the spiral keels showing only as angulations. Length of a mature coil, 61 mm; diameter of aperture, 11 mm. Bahía de Tenacatita, Jalisco, Mexico, to El Salvador, in depths of 33 to 110 m. This is an offshore form that closely resembles a *Turritella* and in the juvenile stage could readily be misidentified. A similar species in the Atlantic is *V. fargoi* Olsson, 1951, which, however, lives on mud flats.

Vermicularia pellucida (Broderip & Sowerby, 1829). Whether there is only a single variable intertidal *Vermicularia* in the eastern Pacific or whether there are two or more is a question not yet resolved. An additional complication is that the earliest available name is based on material described without known locality. One might rather arbitrarily recognize two subspecies.

448. Vermicularia pellucida pellucida. Coarsely sculptured, the shell is moderately heavy and solid in texture. Length, about 65 mm; diameter, 12 mm. Probably restricted to the coast south of Panama.

449. Vermicularia pellucida eburnea (Reeve, 1842) (Synonym: **Vermetus fewkesi** Yates, 1890). The shell is thin, and some specimens are elegantly white, but the irregular coiling detracts from their beauty except for the symmetrical apex, which often is well preserved. Length of an average specimen, about 50 mm; diameter of tube, 8 mm. This seems to be the form that occurs in southern California and is common from the Gulf of California to Panama, although the type locality was given as "South America." The shells are anchored among gravel and small stones, intertidally.

Genus STEPHOPOMA Mörch, 1860

The nuclear whorls are low but regularly coiled, with conspicuous nodes, especially above the periphery. The operculum is set with long, branched bristles.

450. Stephopoma pennatum Mörch, 1860 (Synonym: **S. p. bispinosum** Mörch, 1861). Shells may occur singly, in a tight little coil firmly cemented on one side to the substrate, or, when crowded, as intertwined masses. The shell is dark brown, lighter near the aperture, somewhat quadrangular or pentagonal in section, weakly ribbed, with irregular growth striae. The initial whorls are flat, resembling a small *Heliacus*, lighter-colored than the adult shell, studded with radial rows of pustules. Opercular bristles are long and dark brown in color. Specimens have been found at several localities in Panama by Eugene Bergeron, associated mostly with the vermetid *Tripsycha tulipa*, often partially walled in under the outer edge of the vermetid coils, some of the young Stephopomas even attaching to the adult vermetid in the manner of young vermetids. Diameter of an isolated individual, about 7 mm; of the end of the tube, 2 mm; length of a single tube, about 20 mm. Nicaragua to Peru.

Family CAECIDAE

Small shells in which the spiral nuclear whorls are shed when the teleoconch begins to form and are replaced in most species by an apical plug. Under stones and in crevices, intertidally and offshore.

Although genera have been named on the basis of shell characters, they are far from being clear-cut, and many species are assigned with difficulty because they seem to be on the borderline between two genera. Because few specimens reach a length of 5 mm and all must be studied under the microscope for identification, the described species are merely listed here, without detailed treatment. Original generic assignments are cited in brackets. Useful references: Carpenter (1858–59); De Folin (1867); Bartsch (1920); Strong & Hertlein (1939); Keen (1968).

1. Surface of shell smooth.....................................*Fartulum*
   Surface of shell sculptured....................................... 2
2. Sculpture entirely of rings encircling the tube...................... 3
   Sculpture not entirely of rings.................................. 4
3. With rings distantly spaced and rather coarse....................*Caecum*
   With rings closely spaced and fine.........................*Micranellum*
4. Sculpture of both longitudinal ridges and rings...........*Elephantanellum*
   Sculpture of longitudinal ridges only.....................*Elephantulum*

Genus CAECUM FLEMING, 1813

With well-developed annulations.

451. Caecum bahiahondaense Strong & Hertlein, 1939. Panama.

452. C. clathratum Carpenter, 1857. Mazatlán to Tres Marias Islands, Mexico. Figured, Keen (1968).

453. C. compactum Carpenter, 1857. Mazatlán, Mexico. Figured, Keen (1968).

454. C. diminutum C. B. Adams, 1852. Panama. Figured, Turner (1956).

455. C. eburneum C. B. Adams, 1852. Panama. Figured, Turner (1956).

456. C. elongatum Carpenter, 1857. Mazatlán. Figured, Keen (1968).

457. C. farcimen Carpenter, 1857. Mazatlán. Figured, Keen (1968).

458. C. firmatum C. B. Adams, 1852. Mazatlán, Mexico, to Panama. Figured, Turner (1956).

459. C. laqueatum C. B. Adams, 1852. Panama. Figured, Turner (1956).

460. C. mirificum De Folin, 1867. San Miguel, Pacific Ocean.

461. C. monstrosum C. B. Adams, 1852. Panama. Figured, Turner (1956).

462. C. paradoxum De Folin, 1867. Panama.

463. C. parvum C. B. Adams, 1852. Panama. Figured, Turner (1956).

464. C. pygmaeum C. B. Adams, 1852. Panama. Figured, Turner (1956).

465. C. quadratum Carpenter, 1857. Mazatlán. Figured, Keen (1968).

466. C. richthofeni Strong & Hertlein, 1939. Panama.

467. C. semilaeve Carpenter, 1857. Mazatlán. Figured, Keen (1968).

468. C. subimpressum Carpenter, 1857. Cape San Lucas to Mazatlán. Figured, Keen (1968).

469. C. uncinatum De Folin, 1867. Perlas Islands, Panama.

470. C. undatum Carpenter, 1857. Mazatlán to Panama. Figured, Keen (1968).

Genus ELEPHANTANELLUM BARTSCH, 1920

With faint annulations crossed by stronger longitudinal riblets.

471. Elephantanellum carpenteri Bartsch, 1920. Magdalena Bay to Gulf of California. Pleistocene. Unfigured.

472. E. heptagonum (Carpenter, 1857) [*Caecum*]. Bahía San Luis Gonzaga, Gulf of California, to Panama. Figured, Keen (1968).

473. E. liratocinctum (Carpenter, 1857) [*Caecum*]. Bahía San Luis Gonzaga, Gulf of California, to Panama. Figured, Keen (1968).

474. E. subconicum (Carpenter, 1857) [*Caecum*]. Mazatlán. Figured, Keen (1968).

475. E. subobsoletum (Carpenter, 1857) [*Caecum*]. Mazatlán. Figured, Keen (1968).

476. E. tenuiliratum (Carpenter, 1857) [*Caecum*]. Mazatlán. Figured, Keen (1968).

Genus ELEPHANTULUM CARPENTER, 1857

Sculpture of longitudinal ridges only.

477. Elephantulum abnormale (Carpenter, 1857) [*Caecum*]. Mazatlán. Figured, Keen (1968).

478. E. insculptum (Carpenter, 1857) [*Caecum*]. Mazatlán. Figured, Keen (1968).

479. E. obtusum (Carpenter, 1857) [*Caecum*]. Mazatlán. Figured, Keen (1968).

480. E. subspirale (Carpenter, 1857) [*Caecum*]. Mazatlán. Figured, Keen (1968).

Genus FARTULUM CARPENTER, 1857

Smooth, relatively small; apical plug cap-shaped.

481. Fartulum bakeri Bartsch, 1920. Magdalena Bay to Gulf of California. Unfigured.

482. F. dextroversum (Carpenter, 1857) [*Caecum*]. Mazatlán. Figured, Keen (1968).

483. F. glabriforme (Carpenter, 1857) [*Caecum*]. Mazatlán. Figured, Keen (1968).

484. F. laeve (C. B. Adams, 1852) [*Caecum*]. Mazatlán to Panama. Figured, Turner (1956).

485. F. reversum (Carpenter, 1857) [*Caecum*]. Mazatlán. Figured, Keen (1968).

486. F. teres (Carpenter, 1857) [*Caecum*]. Mazatlán. Figured, Keen (1968).

Genus MICRANELLUM BARTSCH, 1920

Relatively large, with finer annulations than *Caecum*.

487. Micranellum corrugulatum (Carpenter, 1857) [*Caecum*]. Mazatlán. Figured, Keen (1968).

488. M. elongatum (Carpenter, 1857) [*Caecum*]. Mazatlán. Figured, Keen (1968).

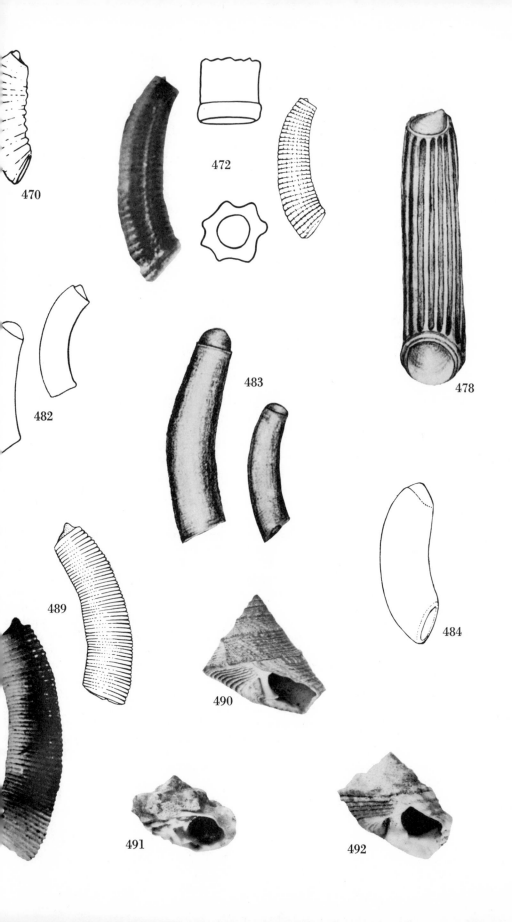

489. M. lohri Strong & Hertlein, 1939. Panama.

Family MODULIDAE

Turbinate, somewhat flat-topped shells with spiral to nodose sculpture; umbilicus narrow.

Genus MODULUS POTIEZ & MICHAUD, 1838
(As MODOLUS; emended, GRAY, 1842)

Resembling the turban shells of the subgenus *Tegula* (*Agathistoma*), but entirely porcelaneous and with a grooved inner lip that ends in a sharp tooth. The operculum is horny and multispiral.

490. Modulus catenulatus (Philippi, 1849) (Synonym: **M. trochiformis** Eydoux & Souleyet, 1852). The white shell is dotted with reddish brown and sculptured with even spiral cords, wider above the angulate periphery. Diameter and height, 17 mm. Gulf of California to Ecuador, on mud flats. A similar species in the Caribbean is *M. carchedonius* (Lamarck, 1822).

491. Modulus cerodes (A. Adams, 1851). Conspicuous nodes at the ends of radial ribs encircle the periphery of the white shell, and there are some irregular markings of red in the interspaces of the intersecting spiral ribs. Diameter, 15 mm; height, 12 mm. The Gulf of California to Panama, on mud flats. The type locality was cited by Adams as "Mozambique," an obvious error.

492. Modulus disculus (Philippi, 1846) (Synonym: **M. dorsuosus** Gould, 1853). The periphery is marked by one or two rows of small tubercles, and the aperture is violet within. Otherwise the shell is white, checkered with chestnut brown at the suture and periphery. Diameter, 15 mm; height, 13 mm. The Gulf of California to Panama, on mud flats; fairly common. *M. modulus* (Linnaeus, 1758) is a similar Atlantic species.

Family VERMETIDAE

The worm gastropods are among the most puzzling of shells, and authors have confused them at every classificatory level—from species through genus and family to phylum, for many tube-building annelid worms have been described as vermetid gastropods and vice versa. The distinction between the vermetids and the worms is clear-cut: the gastropods have a three-layered shell that is glossy within and that begins with a tightly and spirally coiled embryonic shell, whereas the tube-building worms have a two-layered shell that is dull-surfaced within and that begins with a single tubular chamber. The Vermetidae have been confused with the Vermicularias (here treated as a subfamily of Turritellidae though ranked by some authors as a family), but the nuclear whorls are very different. In the Vermetidae the nuclear whorls, which the little snail has when it emerges from the egg capsule, are two or three in number, with an aperture twisted forward and upward. Immediately upon emergence from the protection of the parent's tubular shell, the young vermetid finds a suitable place for starting life and attaches the shell to the substrate, beginning to coil its adult whorls at a right angle to the nuclear whorls. In changing the plane of coiling, the Vermetidae show some relationship to the Architectonicidae. The vermicularias continue the regular spiral Turritella-like coiling and gradually become lax or unwound.

Several genera of the Vermetidae are recognized, but much work needs to be done to evaluate the significance of the differences one sees. The classification given here is still a provisional one, although arrived at only after examination of hundreds of specimens, both west coast and exotic. Useful reference: Keen (1961).

Genus VERMETUS DAUDIN, 1800

The type species of *Vermetus* is a west African form not very similar to any other species except to a couple in the Mediterranean. It adheres to rocks and makes rather loose and irregular clusters of tubes. The tubes are dark gray-brown in color, white and smooth within. The operculum is a small, partly coiled horny plate, not as large as the diameter of the aperture. Although one cannot with certainty assign species from other parts of the world to *Vermetus* in the strict sense, the name is convenient for use in the broad sense, when one cannot be certain of the correct placement of a form, especially if the nuclear whorls and the operculum are missing. It is practically impossible to give a correct identification for a vermetid on the basis of a part of the conch, especially of the beach-worn or broken fragments that find their way into most collections—some of which even have been made type specimens. For identification of the vermetids it is necessary to have the entire specimen or colony, with a part of the base to which it was attached, and the specimens should not be brushed or cleaned (this destroys newly hatched young specimens that might be lightly attached in crevices) and should not be placed in any preservative, such as formalin, that might damage the fragile nuclear whorls. Quick drying in hot sunshine is a good technique, or preservation in neutral alcohol.

Subgenus THYLAEODUS MÖRCH, 1860
(BIVONIA of authors)

Buff to brown shells of moderate to small diameter, with strongly cancellate sculpture; feeding-tube scars (abandoned remnants of former vertical tubes) present on most specimens; operculum less than three-fourths the diameter of the aperture, with a central spiral lamina rising free from the disk.

493. Vermetus (Thylaeodus) contortus (Carpenter, 1857). This species is type of the subgenus. The color of the shell is a warm wax-brown, and the sculpture is of longitudinal threads evenly beaded at the intersections of cross-threads, the whorls rounded in section. Internal spiral laminae are completely lacking. The two nuclear whorls are conic, moderately inflated, pinkish brown in color. The diameter of the operculum is about one-half that of the aperture. Length of a mature coil 15 to 20 mm; diameter of the tube, about 2 mm. Topolobampo to Mazatlán, Mexico.

494. Vermetus (Thylaeodus) indentatus (Carpenter, 1857). The longitudinal sculpture is heavier and less evenly beaded than in *V. (T.) contortus*, of which it was considered a variety by Carpenter. Coiling seems to be less regular, and there is a tendency toward angulation of the whorls by strong development of some of the lirae. The color is dark brown, almost black in some specimens. Length of coil, about 20 mm; diameter of tube, 2 mm. Gulf of California, from Puerto Peñasco to Cape San Lucas and Mazatlán, Mexico.

Genus PETALOCONCHUS LEA, 1843

The type species of this genus is a Miocene fossil, from the Caribbean. It is a moderately large shell, attached only in the early stages, the coiling being rather

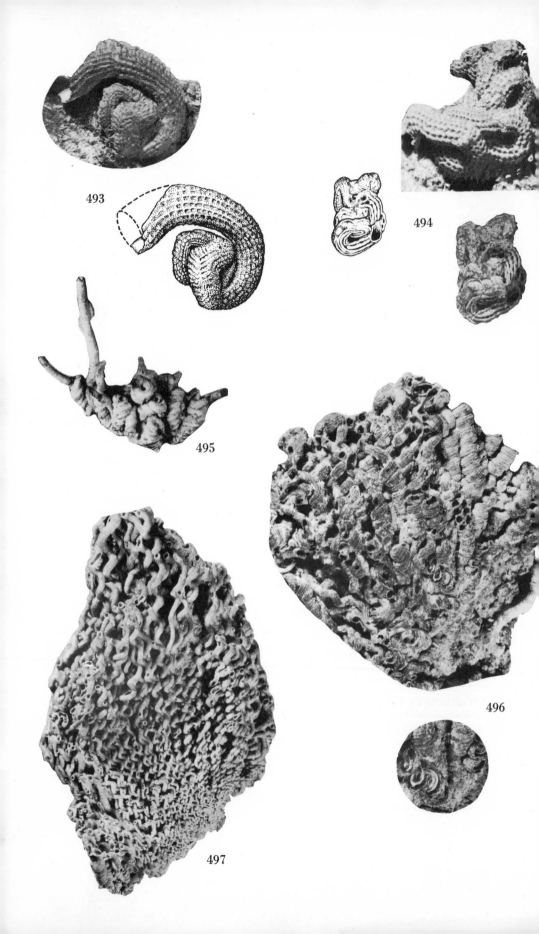

493

494

495

496

497

regular, in the form of a hollow cylinder ending in one long and nearly straight turn. Inside, in the medial whorls, there is a complex structure called a spiral lamina—two or more spiral calcareous plates that diverge from the columella or hang down from the upper margin of the tube. The function of these structures is not known. The only species in the Panamic province that might be assigned to *Petaloconchus*, on the basis of manner of coiling, is *Tripsycha tripsycha*, which, however, lacks the internal laminae. This illustrates the problems one encounters in trying to classify and identify the vermetids.

Subgenus MACROPHRAGMA CARPENTER, 1857

Shells rather small, nuclear whorls cylindrical, three- to four-whorled, postnuclear whorls mostly cancellately sculptured, tightly coiled but firmly attached to the substrate throughout life except for temporary vertical feeding tubes, the scars of which show on many whorls; spiral laminae well developed inside the middle whorls; operculum thin, horny, spirally coiled, in two parts, appearing to be a coil inside a flat outer plate. The numerous living species of the tropics seem to belong here rather than in *Petaloconchus* in the strict sense.

495. Petaloconchus (Macrophragma) complicatus Dall, 1908. The sculpture is irregular, more of wrinkles than of ribs, the color a pale brown. Length of coiled portion about 16 mm, of the erect part 27 mm, the diameter of tube at aperture, 2.3 mm. Two internal lamellae are present. The type locality is near Cocos Island, off Panama, in 120 m. A white form that may prove to be distinct has been taken at three stations in the Galápagos Islands, 110 to 275 m.

496. Petaloconchus (Macrophragma) flavescens (Carpenter, 1857). In dense clusters, this resembles *P. innumerabilis*, but it has tight corkscrew spirals with beaded ribs. The type locality was erroneously cited as Sicily, but the type specimen in the British Museum compares precisely with west Mexican material. Length of a single coil, about 25 mm; diameter of tube, 2 mm. Bahía San Luis Gonzaga to Mazatlán, Mexico.

497. Petaloconchus (Macrophragma) innumerabilis Pilsbry & Olsson, 1935. In color these shells vary from terra-cotta to bright brown. The animals are able to live in dense colonies in which growth may be so rapid that there is no time for coiling. Diameter at aperture, 1.5 mm; length of coil, 100 mm. Internal laminae are well developed. Off Mazatlán, Mexico, to Boca Pan, Tumbes, Peru.

498. Petaloconchus (Macrophragma) macrophragma Carpenter, 1857. As Carpenter picturesquely described it, "The shell is of small diameter; when growing freely taking a tolerably regular spiral, like a Turritella squeezed sideways, the whirls enlarging very slowly, and resembling a winding staircase. It is known when fresh by its lustrous purple-brown colour and absence of pits on the surface." Diameter at aperture, 5 mm; length of coil, 25 to 50 mm. Baja California to Panama.

Genus TRIPSYCHA KEEN, 1961

Relatively large shells, the coiled portions lacking feeding-tube scars; nuclear whorls of two to four turns; operculum nearly as large as aperture, multispiral, concave, spiral lamina appressed. Mainly solitary shells, not forming contorted masses even when crowded. Two Panamic subgenera.

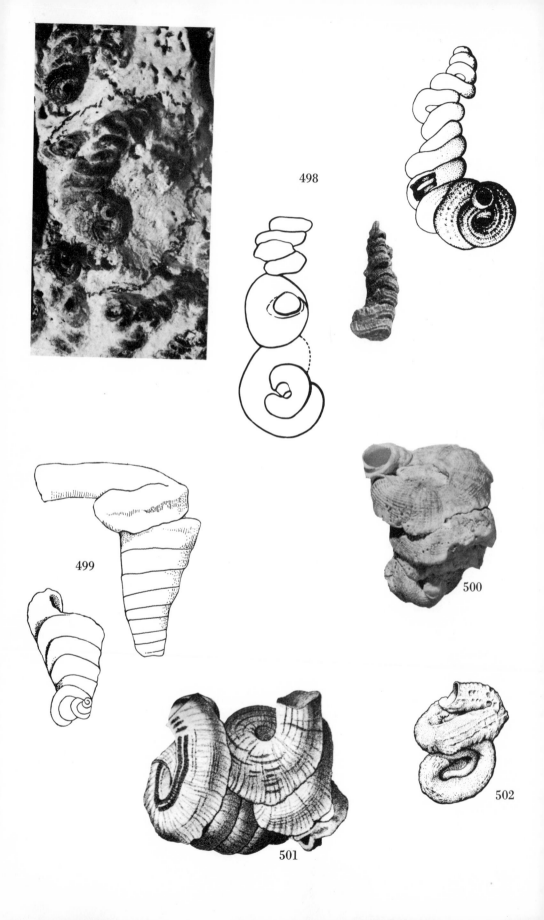

498

499

500

501

502

Subgenus TRIPSYCHA, *s. s.*

White shells of moderate size, early whorls firmly attached to substrate, coiling in a tight spiral; later whorls not attached but coiling in a hollow cone, last volution disjunct or with lax coiling; nuclear whorls unusually large, with up to four turns.

499. Tripsycha (Tripsycha) tripsycha (Pilsbry & Lowe, 1932). This is the type of the genus. The coiling pattern is that of *Petaloconchus, s. s.*—a gradually increasing hollow cone, with a lax final whorl, but the inside of the shell entirely lacks the spiral laminae so characteristic of *Petaloconchus*. Also, the operculum is a concave disk without the secondary coiled lamina of the *Petaloconchus* operculum. The nuclear whorls are large but more the shape of those of *Serpulorbis*. Hence, this is an anomalous species. Individuals seem to prefer a solitary life, and when two or more young attach near each other, they attempt to grow in opposite directions. One spectacular cluster in the Stanford collection, donated by Mr. A. Sorensen, who found it at Guaymas, has eight individuals entwined yet obviously trying to avoid each other. Diameter at aperture, 10 mm; length of coil, 80 mm. Puerto Peñasco, Sonora, to Mazatlán, Mexico.

Subgenus EUALETES KEEN, 1971

Cream-colored to brown shells of moderate to large size, whorls cemented to substrate throughout but tending to have regular, tight spirals, with the outer edge of each volution appressed to substrate; nuclear whorls of about two turns, more globose than in *Tripsycha, s. s.*; operculum as large as aperture, concave, brown, with six to eight or more volutions, the edge of each slightly upturned but not forming a spiral lamina.

This subgenus has the same relationship to *Tripsycha* that *Macrophragma* does to *Petaloconchus* in that instead of the whorls forming a hollow cone they may be cemented throughout most of their length and may thus have what Carpenter described as the "Turritella-squeezed-sideways" pattern of coiling. The contour of the aperture may be affected by the splaying out of the edges of the coil as they are appressed to the substrate on the one side and the center of the coil on the other, so that the aperture may appear to be crescentic rather than circular.

500. Tripsycha (Eualetes) centiquadra (Valenciennes, 1846) (Synonyms: **Vermetus peronii** Valenciennes, 1846 [not Chenu, 1844]; **? Bivonia sutilis** Mörch, 1862). The yellow-brown shell, with punctations between the low ribs, is loosely coiled, with a flattened margin to each whorl. The operculum is thin and concave, with six to eight turns, the edges of successive growth increments turned up, giving an appearance of being dichotomous. Diameter of the aperture about 6 mm; length of an average coiled specimen, about 75 mm. Gulf of California to southern Mexico. Type of the subgenus.

501. Tripsycha (Eualetes) tulipa (Chenu, 1843, *ex* Rousseau, MS) (Synonyms: **Vermetus angulatus** and **panamensis** Chenu, 1844, *ex* Rousseau, MS; **V. effusus** Chenu, 1844, *ex* Valenciennes, MS). Shell firmly attached to substrate throughout life, last whorl not becoming disjunct or lax. Color ivory white to light brown, often with variegations or blotches of color. Margins of coils markedly appressed, surface of whorls nearly smooth, not punctate. Nuclear whorls relatively small, nearly globose; operculum large, of several whorls, concave, thin, brown. Diameter of aperture of a large specimen, 16 mm; length of a coiled specimen, about 60 mm. Numerous individuals have been collected at several localities in Panama Bay by Eugene Bergeron, frequently associated with *Stephopoma pennatum*. The specimens tend to be solitary rather than colonial.

Genus SERPULORBIS SASSI, 1827
(ALETES CARPENTER, 1857)

The mollusks of this genus are mucus feeders and have dispensed with an operculum. Each individual, whether solitary or in a colony, extrudes a film of mucus into the water to entrap floating debris and swimming microscopic animals. When the net is full, the animal pulls it in and devours it, catch and all. A whole colony, with tangled nets, must pull simultaneously or one individual may get more than a fair share. Tubes moderately large, the coil tending to be a flat spiral; with feeding-tube scars; sculpture nodose to scaly.

502. Serpulorbis eruciformis (Mörch, 1862). The type specimen of this species was first figured by Keen in 1961, and the species is still not well recognized. The specimen has three whorls, with a beaded sculpture pattern. The shell is pale lavender, spotted with yellow and white. It has a few ribs running the length of the tube, three of which are noded, and of these the outer two are stronger. The diameter at the aperture of the type specimen is 7 to 8 mm. Length of the coil measures about 30 mm. The type was apparently from the Gulf of California, on a *Crucibulum*.

503. Serpulorbis margaritaceus (Chenu, 1844, *ex* Rousseau, MS) (Synonym: **Vermetus margaritarum** Valenciennes, 1846). The brownish tubes either are regularly wound into a flat spiral or may be nearly straight, but they are always firmly attached to rocks or to other shells. Small scales roughen the spiral ribs. Diameter of a large coil, 70 mm, of an average coil, 30 mm; diameter of tube, 10 to 13 mm. Gulf of California to southern Mexico; the commonest of the larger vermetids on the Mexican coast.

504. Serpulorbis oryzata (Mörch, 1862). The shell is loosely coiled while it remains attached, but after a few whorls it ceases to be fixed to anything. The tube then arches in a broad curve or becomes almost straight. The surface is delicately pebbled, with here and there a larger node. The nuclear whorls and operculum are yet unknown, so that generic assignment remains tentative; some shells suggest a relationship to *Tripsycha*. A specimen in the Stanford University collection measures: length, 250 mm (10 inches); diameter of aperture, 15 mm. Guaymas to Acapulco, Mexico.

Genus DENDROPOMA MÖRCH, 1861
(SPIROGLYPHUS DAUDIN, 1800, of authors)

These small shells embed a part of the whorl in the substrate, into which they have corroded a channel. The nuclear whorls are two to three in number, brown, rapidly increasing in size, the outer lip of the aperture drawn forward into a rounded lobe. Subsequent whorls have sinuous growth lines and one or more rows of nodes along the periphery. The operculum is as wide as the aperture, thick, reddish in color, with a thickened central button and a smooth marginal band on the inside.

505. Dendropoma lituella (Mörch, 1861). This is the type species of the genus. It was described from a colony on a California abalone, but the animals

503

504

505

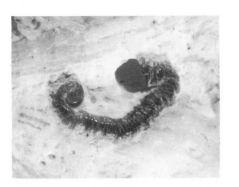

seem also capable of attachment to rock. Variations are protean. On rock the vermetid cannot corrode a trench, and the shell becomes more solid, with a heavy carina on the outer margin of the coil and a fairly regular flat coil, especially in uncrowded situations. On shells, the ability to corrode a trench lets the whorls sink below the surface so that only a series of concave growth striae appear. The shells are mostly white or grayish with no color markings. The carina in rock-based specimens may be fluted or crested with imbricate scales. Diameter of coil, 7 to 12 mm; diameter of aperture, 1.5 to 3 mm. Southern California to the southern end of the Gulf of California at La Paz, Baja California.

There are probably several other species of *Dendropoma* to be described, but as yet material is insufficient to justify additional naming.

Superfamily CERITHIACEA

Elongate, many-whorled shells, with the anterior canal more or less well developed, usually somewhat twisted. Mostly marine, a few groups adapted to brackish-water conditions.

Family CERITHIIDAE

Marine forms, small to medium-sized. A conservative classification is adopted here, treating as subfamilies several groups ranked as families by some authors. A useful summary of the species in the northern part of the Panamic province was given by Baker, Hanna & Strong (1938). The key is to the twelve genera recognized here.

1. Coiling of shell sinistral.................................*Triphora*
   Coiling of shell dextral.. 2
2. Aperture with a spoutlike to slotlike anterior canal.................. 3
   Aperture entire or nearly so, without evident anterior canal........... 9
3. Suture not impressed; whorls flat-sided............................. 4
   Suture impressed; whorls rounded in profile or angulate.............. 5
4. Sculpture cancellate, spiral ribs heavily beaded................*Eumetula*
   Sculpture of spiral ribs only..................................*Seila*
5. Anterior canal spoutlike, not well developed.....................*Bittium*
   Anterior canal well developed, slotlike............................. 6
6. Sculpture of even spiral grooves, axial cancellation only on earliest
      whorls ....*Liocerithium*
   Sculpture not evenly spiral, axial nodes present...................... 7
7. Height of adult specimens more than 10 mm...................*Cerithium*
   Height of adult specimens 10 mm or less............................ 8
8. Shell short for its width, few-whorled......................*Cerithiopsis*
   Shell slender, with more than 8 whorls........................*Metaxia*
9. Aperture triangular, widest below..........................*Diastoma*
   Aperture rounded, not triangular................................. 10
10. With rounded low varices, irregularly spaced....................*Alaba*
    Without varices 11
11. Columella smoothly arched below............................*Alabina*
    Columella obliquely cut off at lower end........................*Litiopa*

Subfamily CERITHIINAE

Sturdy shells, variously colored, the anterior canal well developed.

Genus CERITHIUM BRUGUIÈRE, 1789

Sculpture of coarse to fine often irregular spiral ribs, beaded to nodose; irregular varices often present; operculum ovate, paucispiral. No Panamic species is at present considered to be assignable to *Cerithium, s. s.*

Subgenus OCHETOCLAVA WOODRING, 1928

Sculpture of beaded spiral ribs; outer lip ascending near aperture; posterior canal of aperture relatively long; anterior canal short.

506. Cerithium (Ochetoclava) gemmatum Hinds, 1844 (Synonym: **Clava californica** Dall, 1919). Yellowish white, marbled and spotted with brown, sculpture of beaded spiral ribs; columella with a fold near the base. Height, 34 mm; diameter, 12 mm. Magdalena Bay, Baja California, through the Gulf of California and south to Ecuador, mostly offshore. Type of the subgenus.

Subgenus THERICIUM MONTEROSATO, 1890

Sculpture of irregular spiral threads and ribs, variously noded or beaded by weak axial lirae; suture even, not ascending at aperture; posterior canal of aperture small to wanting; anterior canal narrow to open.

507. Cerithium (Thericium) adustum Kiener, 1841. A short, stout, nearly black shell, speckled with white, this is a southern form closly related to *C. maculosum*. The whorls are nearly smooth or with a few fine nodes on the spire. Height, 43 mm; diameter, 22 mm. Mazatlán, Mexico, to Ecuador, more common in the southern part of its range, in sand among rocks at low tide, and even in tide pools filled by freshwater seepages.

508. Cerithium (Thericium) browni (Bartsch, 1928) (Synonym: **C. stercusmuscarum exaggeratum** Pilsbry & Lowe, 1932). Possibly a subspecies of the more northern *C. stercusmuscarum*, this differs in being a little smaller, wider for its height, with color markings that are coarser. The shell is gray, with blotches of dark brown and white, especially on a row of knobs just below the suture, the base having additional rows of smaller nodules, unevenly color-spotted. Length, 23 mm; diameter, 15 mm. Type locality, Guayaquil, Ecuador. The form from Panama named by Pilsbry and Lowe as a subspecies seems to be only a local color variant. The range of the species probably is Panama to Ecuador. The name *C. billeheusti* has been misapplied in some collections to the Ecuadorean *C. browni*. Actually, *C. billeheusti* Petit, 1851, is an Indo-Pacific form originally described without known locality.

509. Cerithium (Thericium) gallapaginis Sowerby, 1855, *ex* A. Adams, MS (Synonym: **C. galapaginis** A. Adams, 1863). A small, black form, like a slender dwarf *C. menkei* but with stronger sculpture; spiral grooves are crossed by close-set axial riblets, which give the shell a somewhat beaded texture. Length, 13 mm; diameter, 8 mm. El Salvador to Panama and the Galápagos Islands.

510. Cerithium (Thericium) maculosum Kiener, 1841 (Synonym: **C. alboliratum** Carpenter, 1857). Sometimes confused with *C. adustum*, this has a more slender outline, and the distribution is more northern. The shell is bluish gray, speckled with white. Height to 50 mm; diameter to 23 mm. Magdalena Bay, Baja California, through the Gulf of California and south to Mazatlán and Tres Marias Islands, Mexico. Carpenter's *C. alboliratum* seems to be the juvenile stage of *C. maculosum*.

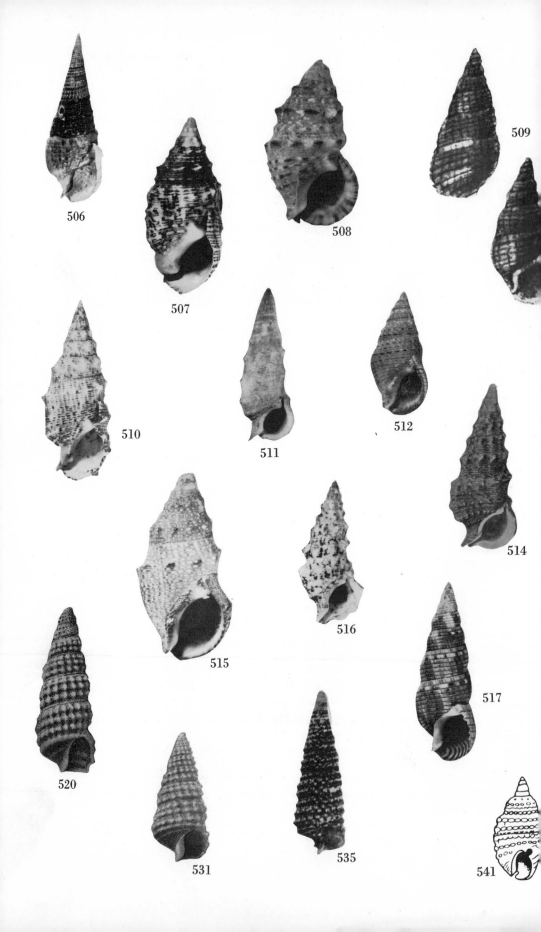

506

507

508

509

510

511

512

514

515

516

517

520

531

535

541

511. Cerithium (Thericium) mediolaeve Carpenter, 1857. Because Carpenter named this as a variety of *C. famelicum*—a synonym of *C. uncinatum*—and because his type specimen was not figured, authors assumed it to be a synonym also. However, a photograph of the type shows (Keen, 1968) not only that the shell is smoother than *C. uncinatum*, with subdued color markings, but that it is much more slender. Its recurved canal and the placement of the sculpture relates it more to *C. nicaraguense*, but it differs in being smaller and more slender. Length, 25 mm; diameter, 10 mm. Type locality, Mazatlán, Mexico.

512. Cerithium (Thericium) menkei Carpenter, 1857 (Synonym: **C. interruptum** Menke, 1851 [not Lamarck, 1804]). The rather small shell is dark gray to brown with a whitish margin to the aperture and white on the end of the columella. The sculpture is of fine spiral ribs, two or three larger and with low nodes. Length, 16 mm; diameter, 9 mm. Gulf of California to Ecuador, under rocks at low tide.

513. Cerithium (Thericium) nesioticum Pilsbry & Vanatta, 1906. This Hawaiian form has been reported on Clipperton Island.

514. Cerithium (Thericium) nicaraguense Pilsbry & Lowe, 1932. The shell has some resemblance to *C. (Ochetoclava) gemmatum* in outline but has no internal fold on the columella; color and sculpture are more subdued. The anterior canal is well developed and recurved. Height, 26.5 mm; diameter, 11 mm. Nicaragua to Panama and Ecuador, intertidally and offshore to 37 m.

515. Cerithium (Thericium) stercusmuscarum Valenciennes, 1833 (Synonyms: "**C. ocellatum** Bruguière" of authors, not of Bruguière, 1792; **C. irroratum** Gould, 1851). The bluish-gray to brownish shell is finely speckled with white and has a well-marked spiral row of pointed tubercles below the suture. Height, 25 mm; diameter, 11 mm. Baja California to Peru; common on sand flats and in estuaries. In the southern part of its range this species and *C. browni* are difficult to separate.

516. Cerithium uncinatum (Gmelin, 1791) (Synonym: **C. famelicum** C. B. Adams, 1852). The white shell is mottled with brown spots that are divided by fine white spiral lines. The spiral sculpture consists of a row of small nodes just below the suture and a row of larger ones at the periphery of each whorl. Height, 29 mm; diameter, 12 mm. The Gulf of California to Guayaquil, Ecuador, mostly offshore, in depths to 37 m.

Genus LIOCERITHIUM TRYON, 1887

Slender, columnar or cylindrical, early whorls cancellate, later whorls smooth except for well-incised spiral grooves.

517. Liocerithium judithae Keen, new species (Synonyms: **Cerithium incisum** Sowerby, 1855 [not Hombron & Jacquinot, 1854]; **C. sculptum** of authors, not of Sowerby, 1855). Relatively slender, the shell is smooth except for cancellate early whorls and, on later whorls, regularly spaced, deeply incised spiral grooves. The spiral cords thus formed are wider on the upper parts of the whorls, narrowing anteriorly, the upper two or three cords lighter-colored to white, marked with regularly spaced squarish black dots. The shell is olive green to dark gray in color. The oval aperture has an anterior notch but no distinct anterior canal; strong lirations mark the inside of the outer lip. Young specimens may be cancellately sculptured and streaked or spotted with large reddish brown and white blotches. Height of a large specimen, 20 mm; diameter, 6 mm. Not uncommon under rocks between tides from Magdalena Bay, Baja California, through the Gulf of California at least to Mazatlán, Mexico. The species is here described as new because other alternatives involve acceptance of names of doubtful application. Sowerby (1855) described *C. incisum* as being from Australia and the Philippines, but his illustration was a good representation of the west Mexican form. The name *incisum* had already been used and thus was invalid. He described at the same time a *C. curtum* that had similar coloring but was much shorter, not citing a locality but implying that it came from the same area as *C. incisum*. This name also was preoccupied. It was replaced by Bayle in 1880 with the name *C. eurus*. Bayle, however, cited the locality as "Virginia." In 1921 Dall suggested that Sowerby's figures of *C. sculptum* (described in the same work, without known locality) might be of the West American species. Later authors have followed Dall's precedent. However, Sowerby's figures of this shell show a very short, broad form, and his description mentions axial sculpture. One concludes, therefore, that this fairly common west Mexican species has had up to now no properly documented name. Although Tryon in 1887 suggested synonymy of "*C. curtum*" [= *C. eurus*] with "*C. incisum*," this has not been adopted by subsequent authors, and the name seems best regarded as a *nomen dubium*, unidentifiable. *C. judithae*, which is distinguished by the smooth spiral ribbing and the columnar outline, thus becomes the type and sole recognized species of *Liocerithium*.

Subfamily CERITHIOPSINAE

Small shells with a slight to flaring anterior notch and well-developed sculpture. Because most specimens require study under a microscope, detailed treatment will not be given here for these and the next three subfamilies. Species are listed by name, and some useful references are cited.

Genus CERITHIOPSIS FORBES & HANLEY, 1851

Columnar shells with a small aperture, anterior canal evident; generally brown in color, sculpture of several rows of nodes or beads. Useful references: Bartsch (1911e); Baker, Hanna & Strong (1938).

518. Cerithiopsis adamsi Bartsch, 1911. Panama.

C. albonodosa Carpenter, 1857. Mazatlán, Mexico. (See *C. tuberculoides*.)

519. C. anaitis Bartsch, 1918 (Synonym: **C. helena** Bartsch, 1917 [not Boettger, 1901]). Figured, Bartsch (1917).

520. C. aurea Bartsch, 1911. Cape San Lucas. Length, 7 mm.

521. C. bicolor Bartsch, 1911. Galápagos Islands.

522. C. bristolae Baker, Hanna & Strong, 1938. Cape San Lucas.

523. C. cerea Carpenter, 1857. Mazatlán. Figured, Keen (1968).

524. C. curtata Bartsch, 1911. Panama to Galápagos Islands.

525. C. destrugesi (De Folin, 1867) [*Cerithium*]. Panama (possibly the Philippine Islands). Length, 6 mm.

526. C. eiseni Strong & Hertlein, 1939. Panama.

527. C. galapagensis Bartsch, 1911. Indefatigable Island, Galápagos Islands.

528. C. gissleri Strong & Hertlein, 1939. Panama.

529. C. guanacastensis Hertlein & Strong, 1951. Costa Rica.

530. C. guatulcoensis Hertlein & Strong, 1951. Puerto Guatulco, Mexico.

531. C. halia Bartsch, 1911. Baja California.

C. helena Bartsch, 1917 [not Boettger, 1901]. (See *C. anaitis*.)

532. C. infrequens (C. B. Adams, 1852) [*Triphoris*]. Panama. Figured, Turner (1956).

533. C. kinoi Baker, Hanna & Strong, 1938. Cape San Lucas.

534. C. lohri Jordan, 1936. Magdalena Bay (Pleistocene).

535. C. montezumai Strong & Hertlein, 1939. Panama Bay.

536. C. neglecta (C. B. Adams, 1852) [*Cerithium*]. Panama to Galápagos Islands. Figured, Turner (1956).

537. C. oaxacana Hertlein & Strong, 1951. Puerto Guatulco, Mexico.

538. C. perrini Hertlein & Strong, 1951. Puerto Guatulco, Mexico.

539. C. porteri Baker, Hanna & Strong, 1938. Gulf of California.

540. C. pupiformis Carpenter, 1857. Mazatlán, Mexico. Figured, Keen (1968).

541. C. sorex Carpenter, 1857. Mazatlán, Mexico. Figured, Keen (1968).

542. C. subgloriosa Baker, Hanna & Strong, 1938. Gulf of California.

543. C. tuberculoides Carpenter, 1857 (Synonym: **?** **C. albonodosa** Carpenter, 1857). Mazatlán, Mexico. Figured, Keen (1968).

Genus BITTIUM GRAY, 1847

Sturdy small shells with nodose sculpture; aperture with only a slight anterior notch, not a canal. Useful reference: Bartsch (1911*f*).

544. Bittium arenaense Hertlein & Strong, 1951. Arena Bank, Gulf of California.

545. B. cerralvoense Bartsch, 1911. Cerralvo Island, Gulf of California. Length, 8 mm.

546. B. decussatum (Carpenter, 1857) [*Cerithiopsis*]. Mazatlán. Figured, Keen (1968).

547. B. mexicanum Bartsch, 1911. Gulf of California. Length, 6 mm.

548. B. nicholsi Bartsch, 1911. Gulf of California. Length, 6 mm.

549. B. nitens Carpenter, 1864. Cape San Lucas. Figured, Palmer (1963). Length, 6 mm.

550. B. panamense Bartsch, 1911. Panama. Length, 14 mm.

551. B. peruvianum (Orbigny, 1841) [*Cerithium*]. Paita, Peru. Figured, Keen (1966). Length, 8 mm.

Genus EUMETULA THIELE, 1912
(**EUMETA** MÖRCH, 1868 [not WALKER, 1855]; **LASKEYA** IREDALE, 1918)

Spire high, evenly tapering, periphery of body whorl angulate. Useful reference: Bartsch (1911*c*).

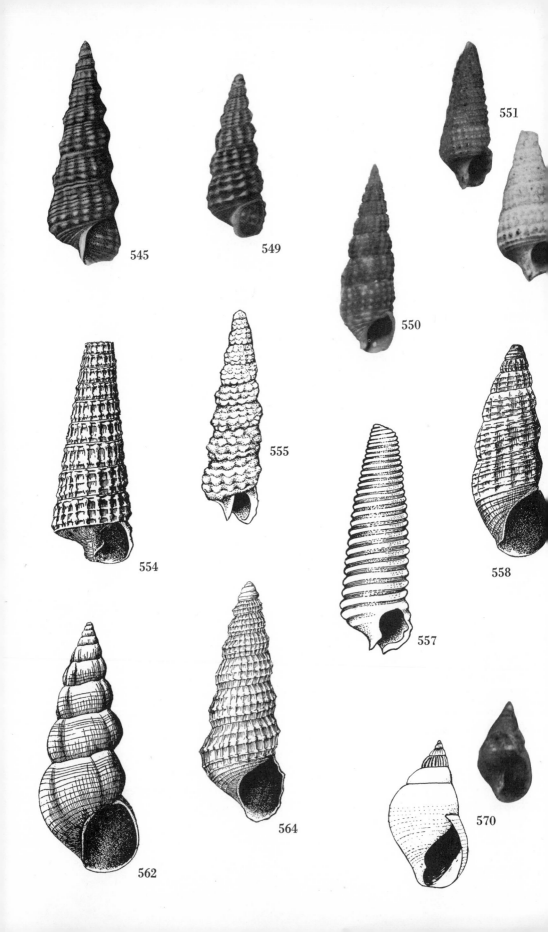

545

549

551

550

555

554

557

558

562

564

570

552. Eumetula bimarginata (C. B. Adams, 1852) [*Cerithium*]. Panama. Figured, Turner (1956).

553. E. eucosmia Bartsch, 1911. Galápagos Islands.

554. E. intercalaris Carpenter, 1865 [*Cerithiopsis*]. Guacomayo, Guatemala. Figured, Bartsch (1911).

Genus METAXIA MONTEROSATO, 1884

Slender, columnar, sculpture nodose, whorls inflated, suture indented. Useful reference: Baker, Hanna & Strong (1938).

555. Metaxia convexa (Carpenter, 1857) [*Cerithiopsis*]. Gulf of California to Panama.

556. M. diadema Bartsch, 1907. Gulf of California.

Genus SEILA A. ADAMS, 1861

Slender, high-spired, without axial sculpture. Useful reference: Baker, Hanna & Strong (1938).

557. Seila assimilata (C. B. Adams, 1852) [*Cerithium*] (Synonyms: **Cerithiopsis kanoni** and **moreleti** De Folin, 1867). Gulf of California to Panama. Figured, Turner (1956).

Subfamily DIASTOMATINAE

Small shells with round to triangular aperture. Three Panamic genera.

Genus DIASTOMA DESHAYES, 1850

Aperture triangular, wider anteriorly, with an angular notch. Useful reference: Bartsch (1911*d*).

558. Diastoma chrysalloidea Bartsch, 1911. Gulf of California.

Genus ALABA H. & A. ADAMS, 1853

Aperture circular to quadrangular; sculpture weak to wanting; varices present on one or more whorls. Useful reference: Bartsch (1910).

559. Alaba guayaquilensis Bartsch, 1928. Ecuador.

560. A. interruptelineata Pilsbry & Lowe, 1932. Nicaragua to Panama.

561. A. jeannettae Bartsch, 1910. Gulf of California. (? = *A. supralirata*.)

562. A. supralirata Carpenter, 1857. Gulf of California to Panama. Figured, Keen (1968).

Genus ALABINA DALL, 1902

Ovate-conic shells, high-spired, aperture rounded; sculpture varying from axial to spiral. Useful reference: Bartsch (1911*b*).

563. Alabina crystallina (Carpenter, 1864) [*Fenella*]. Cape San Lucas. Figured, Palmer (1963).

A. diomedeae Bartsch, 1911. (= *A. excurvata*.)

564. A. effusa (Carpenter, 1857) [*Alvania*]. Mazatlán. Figured, Keen (1968).

565. A. excurvata (Carpenter, 1857) [*Alvania*] (Synonym: **A. diomedeae** Bartsch, 1911) Gulf of California to Mazatlán. Figured, Keen (1968).

566. A. ignati Bartsch, 1911. Baja California.

567. A. monicensis Bartsch, 1911. Southern California to Bahía San Luis Gonzaga, Gulf of California.

568. A. occidentalis (Hemphill, 1894) [*Eulimella*]. Baja California to Tres Marias Islands. Figured, Bartsch (1911).

569. A. veraguaensis Strong & Hertlein, 1939. Panama.

Subfamily LITIOPINAE

Small, pelagic; spire with axial riblets, body whorl smooth; columella truncate below.

Genus LITIOPA RANG, 1829

Characters as for the subfamily.

570. Litiopa melanostoma Rang, 1829 (Synonym: **L. divisa** Carpenter, 1856). Pelagic, tropics. Figured, Keen (1968).

Subfamily TRIPHORINAE

Resembling the Cerithiopsinae but with sinistral coiling.

Genus TRIPHORA BLAINVILLE, 1828
(TRIPHORIS, TRIFORIS of authors)

Adults with spoutlike anterior canal. Useful references: Bartsch (1907); Baker (1926); Baker & Spicer (1935).

571. Triphora adamsi Bartsch, 1907. Nicaragua to Galápagos Islands.

572. T. alternata C. B. Adams, 1852. Panama. Figured, Turner (1956).

573. T. chamberlini Baker, 1926. Gulf of California. Length, 7 mm.

574. T. chathamensis Bartsch, 1907. Galápagos Islands.

575. T. contrerasi Baker, 1926. Gulf of California to Tres Marias Islands. Length, 8 mm.

576. T. cookeana Baker & Spicer, 1935. Gulf of California.

577. T. dalli Bartsch, 1907. Tres Marias Islands to Panama. Length, 6.5 mm.

578. T. escondidensis Baker, 1926. Gulf of California.

579. T. evermanni Baker, 1926. Gulf of California.

580. T. excolpa Bartsch, 1907. Cape San Lucas, Baja California, to Guacomayo, Guatemala.

581. T. galapagensis Bartsch, 1907. Galápagos Islands.

582. T. hannai Baker, 1926. Gulf of California. Length, 8 mm.

583. T. hemphilli Bartsch, 1907. Baja California. Length, 5 mm.

584. T. inconspicua C. B. Adams, 1852. Mazatlán, Mexico, to Panama.

585. T. johnstoni Baker, 1926. Gulf of California.

586. T. marshi Strong & Hertlein, 1939. Panama.

587. T. oweni Baker, 1926. Gulf of California.

588. T. palmeri Strong & Hertlein, 1939. Panama.

589. T. panamensis Bartsch, 1907. Panama. Length, 8.7 mm.

590. T. pazensis Baker, 1926. Gulf of California.

591. T. peninsularis Bartsch, 1907. Gulf of California.

592. T. postalba Bartsch, 1907. Galápagos Islands (variety of *T. galapagensis*).

593. T. slevini Baker, 1926. Gulf of California.

594. T. stearnsi Bartsch, 1907. Gulf of California.

595. T. stephensi Baker & Spicer, 1935. Gulf of California.

596. T. unicolor Bartsch, 1907. Galápagos Islands (variety of *T. galapagensis*).

597. T. vanduzeei Baker, 1926. Gulf of California. Length, 5 mm.

Family PLANAXIDAE

Although the shape of the shell suggests the Littorinidae, there is a slight notch in the anterior end of the aperture, and the shells are thicker and heavier, with a fibrous periostracum.

Genus PLANAXIS LAMARCK, 1822

Small, brownish, with spirally grooved sculpture.

598. Planaxis obsoletus Menke, 1851 (Synonyms: **P. acutus** Menke, 1851 [not Krauss, 1848]; **P. nigritella** Forbes, 1852). The shell is chocolate brown, polished, under a thin periostracum, engraved with narrow spiral lines except on the middle of the body whorl. Height, 12 mm; diameter, 6.5 mm. The Gulf of California to Salina Cruz, Mexico.

599. Planaxis planicostatus Sowerby, 1825 (Synonyms: **Buccinum planaxis** Wood, 1828; **P. canaliculatus** Duval, 1840; **P. circinatus** Lesson, 1842). Also dark chocolate in color, this has almost regular spiral ribs and grooves throughout but is lighter-colored inside. A large specimen measures: height, 28 mm; diameter, 16 mm. Mazatlán, Mexico, to Peru, under stones near the high-tide line.

Family POTAMIDIDAE

Shells brownish to horn-colored, with anterior canal less strongly developed than in the Cerithiidae. Operculum subcircular, with central nucleus and several to many turns. Habitat: brackish-water estuaries. Three Panamic genera.

Genus BATILLARIA BENSON, 1842

Whorls flat or weakly convex; aperture with a short columellar margin; base truncate, canal short. Operculum subcircular, closely coiled.

600. Batillaria mutata (Pilsbry & Vanatta, 1902). Thin, with rather inflated whorls, this has fine spiral sculpture crossed by coarser axial riblets, some specimens with a beaded cord just below the suture. The shell is dark-colored, with a reddish band near the periphery and vague whitish markings. Length, 14 mm; diameter, 5.6 mm. The original material was collected in a salt tide pool in a mangrove swamp, Albemarle (Isabela) Island, Galápagos. A similar species in the Atlantic is *B. minima* (Gmelin, 1791).

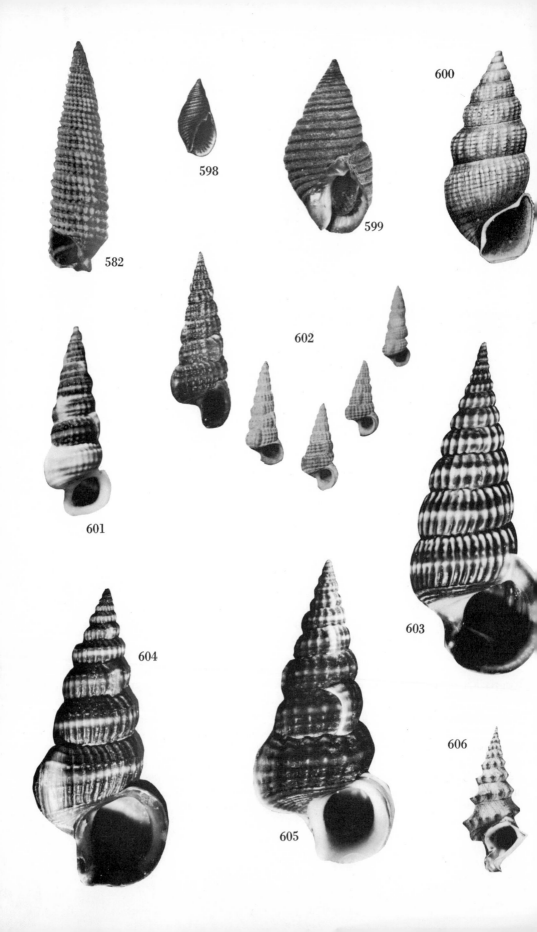

Genus CERITHIDEA SWAINSON, 1840

The shells are thin in texture, axially ribbed, wide-mouthed, with a rudimentary anterior canal (a mere notch in some forms), and with a multispiral horny operculum. The animal lives in brackish mud or entirely out of the water on reeds and twigs.

601. Cerithidea albonodosa Gould & Carpenter, 1857. The brown shell has regular varices of white and a white peripheral stripe. The species has been considered a variant of *C. californica* (Haldeman, 1840) or of *C. mazatlanica*, but Berry (1956) has shown that it is clearly distinct and occupies a separate ecologic niche. Height, 30 mm; diameter, 12 mm. Specimens may be found in great abundance at the midtide level on tidal flats in the Gulf of California, especially in the upper part of the Gulf. San Ignacio Lagoon, Baja California, to Guaymas, Mexico.

602. Cerithidea mazatlanica Carpenter, 1857 (Synonym: "**Cerithium hegewischii** Philippi" of authors, not of Philippi, 1848). Philippi's name of *C. hegewischii* was revived by Bequaert (1942) for this, although Philippi's figure seems closer to *C. valida*. However, since Philippi's cited locality was "Mexico," there is no proof that the name applies to any west Mexican species. It seems best, therefore, to use Carpenter's name until Philippi's type specimen can be critically studied. The species much resembles the Californian *C. californica* (Haldeman, 1840), but is smaller and more slender. The shell is dark brown to black and is sculptured with a network of axial and spiral ribs that rise into rounded beads where they cross. Height, 27 mm; diameter, 9 mm. In the Gulf of California, where astronomical numbers of individuals occur, the species occupies a slightly higher niche on the tide flats than *C. albonodosa*, as Berry (1956) has shown. San Ignacio Lagoon, Baja California, through the Gulf and south to Panama.

603. Cerithidea montagnei (Orbigny, 1839) (Synonym: **Cerithium reevianum** C. B. Adams, 1852). The most elegant of the Panamic horn shells, its bright brown surface ornamented only with curved axial ribs. Height, 35 mm; diameter, 17 mm. San Ignacio Lagoon, Baja California, through the Gulf and south to Ecuador, on mud flats and especially in mangrove swamps, where the snails cluster on the mangrove roots.

604. Cerithidea pulchra (C. B. Adams, 1852) (Synonyms: **Cerithium varicosum** Valenciennes, 1832 [not Defrance, 1817]; **Cerithidea solida** Gould & Carpenter, 1857). The axial ribs are almost as numerous as in *C. montagnei*, but they are finely beaded by spiral sculpture, there are well-marked varices on the spire, and the periostracum is velvety. The shell is widest for its height among the Panamic species: height, 31 mm; diameter, 17 mm. Panama to Guayaquil, Ecuador, in muddy sand or mangrove swamps at high-tide level.

605. Cerithidea valida (C. B. Adams, 1852) (Synonyms: **Cerithium varicosum** Sowerby, 1834 [not Valenciennes, 1832] [= **Cerithium fortiusculum** Bayle, 1880, and **Cerithidea aguayoi** Clench, 1934, new names]; **? Cerithium delectum** Reeve, 1865, *ex* A. Adams, MS; **Potamides meta** Li, 1930). In sculpture this somewhat resembles *C. mazatlanica*, but it is sturdier and wider for the height; it differs from *C. pulchra* by its coarser sculpture. Height, 40 mm; diameter, 19 mm. Gulf of California to Ecuador, on mud flats.

Genus RHINOCORYNE VON MARTENS, 1900

Turreted, with a relatively long anterior canal and an adherent periostracum. Operculum thin, horny, circular, with five to seven turns, its nucleus central.

606. Rhinocoryne humboldti (Valenciennes, 1832) (Synonyms: **Cerithium lamarckii** Valenciennes, 1832; **C. pacificum** Sowerby, 1833). The shell is grayish to chocolate brown, spirally striate, the whorls with an acute spinose spiral rib. Occasionally specimens have the rib doubled, with two rows of spines; this is the form named as *R. lamarckii*. Height, 37 mm; diameter, 20 mm. Sonora, Mexico, to Chile, in estuaries and offshore to 27 m.

Superfamily STROMBACEA

Shells thick and solid, with a relatively large body whorl; aperture narrow, notched at both ends; operculum narrow, too small to close the aperture.

Family STROMBIDAE

The mollusks of this family have a narrow, peculiarly arched foot, the greatly elongate posterior portion of which (metapodium) bears at its end the relatively small operculum. The foot facilitates an unusual mode of locomotion for a gastropod. Instead of the normal gliding motion there is a scrambling or leaping, with the pointed operculum used as a sort of claw. Many collectors have been startled on picking up the shell to find that the animal inside is adept at kicking in an attempt to escape. The development of the eyes is noteworthy, too—eyes more complex and apparently keener than in most other gastropod groups, situated at the ends of long, stout stalks. The "stromboid notch" is an embayment of the outer lip that marks the position of one eye when the animal is feeding, the other eye protruding along the anterior canal. It would be tempting to suppose that such eyes would be useful for searching prey, but actually they are defensive devices, to detect predators. The strombs are herbivorous, grazing on bottom deposits and fine, filamentous algae.

Genus STROMBUS LINNAEUS, 1758

The shells are stout and solid, with a spire of numerous turns, a long aperture, and an anterior notch in the aperture. The outer lip becomes thickened and extended or even winglike with age. Immature shells are much lighter in weight, with thin, sharp, easily broken outer lips, and look more like cones than strombs. The operculum is claw-shaped, with serrations along one edge. All of the known species are tropical in distribution, most of them living in the intertidal zone or below low-tide line. Useful reference: Emerson (1963). Three Panamic subgenera.

Subgenus STROMBUS, *s. s.*

Shell smooth except for a row of nodes at the periphery.

607. Strombus (Strombus) gracilior Sowerby, 1825. A much smaller shell than *S. galeatus*, this has a higher spire armed with blunt spines. The color is a yellowish brown with a lighter central band, the aperture white, edged with orange-brown. The periostracum is thin, horn-colored or varnished in appearance. Length, 75 mm (3 inches); width, 50 mm (2 inches). The Gulf of California to Peru, on sand flats and in muddy lagoons, offshore in depths to 45 m. A closely similar species in the Atlantic is the type of the genus—the so-called "fighting conch," *S. pugilis* Linnaeus, 1758.

Subgenus **LENTIGO** JOUSSEAUME, 1886
Body whorl with several spiral rows of low, irregular nodes.

608. Strombus (Lentigo) granulatus Swainson, 1822. About the size of *S. gracilior*, this is easily distinguished by its more slender spire and the strongly developed tubercles both on the spire and on the body whorl. Characteristic granulations develop inside the outer lip of mature specimens. The color varies, mostly being of brown spots on a whitish or violet-tinged background. The periostracum is thin, light-colored, and velvety. The northern end of the Gulf of California to Ecuador, on exposed beaches of rock and sand, but mostly offshore in depths to 75 m.

Subgenus **TRICORNIS** JOUSSEAUME, 1886
Body whorl with low spiral ridges bounded by incised lines.

609. Strombus (Tricornis) galeatus Swainson, 1823 (Synonyms: ? *S. crenatus* Sowerby, 1825; **S. galea** Wood, 1828). This is probably the heaviest-shelled if not the largest West American gastropod. Young shells are variegated with brown on white, banded or blotched in orange-yellow. Mature shells are ivory white with a darker spire. A heavy, brown periostracum covers the body whorl, and the spire, when not eroded, shows axial ribs and spiral striae. The aperture changes from white in immature specimens to brown and dull orange in adults. Because of the large size, the animal is a favored article of food among some Mexicans, and large heaps of the shells may be seen along the Gulf of California coast. The operculum of this species may serve as base for at least two different species of *Crepidula*. The spire of an otherwise attractive large specimen often being corroded and worm-eaten, the collector should examine all specimens carefully and retain only those in good condition, placing the others back in the water at the site where collected. Length, 190 mm (7½ inches); width, 125 mm (5 inches). The northern end of the Gulf of California to Ecuador, just below low-tide line. A similar Atlantic species is *S. goliath* Sowerby, 1842.

610. Strombus (Tricornis) peruvianus Swainson, 1823. When fully mature the shell has a pronounced "wing" on the outer lip, and the aperture is a brilliant orange color, with a series of folds or ridges on the upper part of the columella and another along the outer lip. Exteriorly the shell is tan or brown, the low spire variegated with brown and white, under a thick, dark brown periostracum. Though not as large or heavy a shell as *S. galeatus*, this may reach a length of 150 mm or more. The females are said to be larger than the males. Tres Marias Islands, Mexico, to northern Peru, at or near low-tide mark, in tide pools.
See Color Plate XVI.

Superfamily **EPITONIACEA**
(**PTENOGLOSSA**)
Slender, high-spired, with numerous whorls; axially ribbed.

Family **EPITONIIDAE**
(**SCALARIIDAE**)
The wentletraps or spiral-staircase shells have a classic beauty of form and a bewildering complexity of species. The pre-Linnaean name of *Scala*—staircase—has been applied to them by many European authors, but the first valid generic name was *Epitonium*, which literally means "braced up," and refers to the peg

used in tuning a violin. Although this is a longer word than *Scala*, it still is shorter than the common name, wentletrap, so readily adopted by amateurs who insist they cannot remember scientific names. "Wentletrap" actually is the Dutch word for spiral staircase; but probably few of the people who use it are aware of its meaning.

Although the shells have a simple or entire aperture (which in most gastropods indicates herbivorous feeding habits), the evidence mounts that these are carnivorous snails. The purple color of the soft parts in some—which may even leave a purple stain on the shells—and the reduced radula suggest this, and it was confirmed when in 1956 Dr. Gunnar Thorson observed a specimen of a Californian *Opalia* inserting its proboscis or feeding tube into the base of a sea anemone. That the Epitoniidae are usually found in association with anemones and corals is therefore no coincidence if it is upon these that they feed. Mostly they are in shallow water, at depths of less than 15 meters, although there are some that occur at greater depths.

The shells are mostly slender and conical, of numerous whorls, with a round aperture and regular axial ribs or costae. A few are smooth or at least lack regular costae. Significant variations to be observed are the number and form of the costae; the amount of spiral ribbing; the shape of the base; and the relative proportions of the outline. Color is generally unimportant, for practically all specimens are white; a few are brown. The operculum is horny and few-whorled.

Authors have grouped the species of the Epitoniidae into numerous genera, subgenera, and smaller divisions they have called "sections," a category not accepted in modern nomenclature. Under the conservative treatment adopted here, only five genera, some of which are divided into subgenera, are recognized, as shown in the keys to the units. Many of the named species have remained unillustrated, but photographs of types are now available. Even with these, not every specimen can be identified with confidence, because the limits of variation that one should expect are not yet worked out for many forms.

Revision of this family in the eastern Pacific has been an especial interest of Helen (Mrs. Joseph) DuShane, of Whittier, California. She has generously supplied information on synonymies and new data on distribution and morphology, and has provided a number of the illustrations. This aid is gratefully acknowledged.

1. With lamellar costae (often incorrectly called varices) . . . . *Epitonium*
   Without lamellar costae (simple axial ribbing may be present) 2
2. Aperture with anterior margin almost forming a notch . . . . *Alora*
   Aperture evenly rounded anteriorly 3
3. With somewhat cancellate sculpture, both spiral and axial . . . . *Amaea*
   With axial or spiral sculpture only, not both 4
4. Outer lip sharp-edged, not thickened . . . . *Acirsa*
   Outer lip thickened . . . . *Opalia*

Genus **Epitonium** Röding, 1798
(**Scala** of authors; **Scalaria** Lamarck, 1801)

With regular axial costae; interspaces smooth or only weakly sculptured.

1. Spiral ribbing present between costae 2
   Spiral ribbing not present between costae 3
2. Shell white . . . . *Asperiscala*
   Shell brown . . . . *Pictoscala*

3. Base of body whorl set off by a strong ridge.................*Cirsotrema*
   Base of body whorl not set off by a spiral ridge...................... 4
4. Body whorl large; spire low.............................*Sthenorytis*
   Body whorl not markedly large; spire slender, tapering................ 5
5. Costae coronate or spinous at periphery....................*Hirtoscala*
   Costae smoothly arched, no pointed spines at periphery........*Nitidiscala*

Subgenus **ASPERISCALA** DE BOURY, 1909

With spiral sculpture between axial costae.

611. Epitonium (Asperiscala) acapulcanum Dall, 1917 (Synonyms: **E. xantusi** Dall, 1917; **E. keratium** Dall, 1919; **E. strongi** Bartsch, 1928; **E. slevini** Strong & Hertlein, 1939). Ridges between costae are evenly spaced and prominent, continuing sometimes onto the face of the slightly reflected costae; suture moderately deep, inner lip lying close to the costae, its anterior end with a strong wrinkled fasciole behind the lip; number of costae per whorl tending to be greater toward the southern end of the geographic range. Height, 5 to 10 mm; diameter, 2.5 to 4.5 mm; costae, 11 to 15. Magdalena Bay, Baja California, throughout the Gulf of California and south to the Galápagos Islands.

612. Epitonium (Asperiscala) billeeanum (DuShane & Bratcher, 1965). White, with a yellowish-buff periostracum, the whorls rapidly enlarging; convex, thin, and fragile, with cancellate sculpture; outer lip thin, flaring on the inner margin; umbilicus small. Height, 5 to 15 mm; diameter, 3.5 to 9 mm. Cape San Lucas, Baja California, throughout the Gulf of California and south to Galápagos Islands, Ecuador.
See Color Plate XIV.

613. Epitonium (Asperiscala) canna Dall, 1919 (Synonym: **E. reedi** Bartsch, 1928). Fresh shells have a brown flush between costae. Height, 9 to 20 mm; diameter, 5 to 9 mm; costae, six. Magdalena Bay, Baja California, through the Gulf of California and south to Ecuador.

614. Epitonium (Asperiscala) centronium (Dall, 1917). Small, slender; suture deep; costae with a spine between suture and periphery; spiral sculpture weak between whorls; aperture rounded. Height, 4.5 mm; diameter, 2 mm; costae, nine. Gulf of California.

615. Epitonium (Asperiscala) cookeanum Dall, 1917. Small, solid; costae compact, smooth; spiral sculpture of fine uniform threads between costae. Height, 9 mm; diameter, 4 mm; costae, ten to eleven. Mainly a southern Californian species, reported intertidally as far south as Cape San Lucas, Baja California.

616. Epitonium (Asperiscala) emydonesus Dall, 1917 (Synonyms: **E. imperforatum** Dall, 1917; **E. manzanillense** Hertlein & Strong, 1951). Small, fragile, suture deep, spiral threads same width as interspaces, costae not coalescing at parietal lip but riding under the face of the lip. Height, 4.5 mm; diameter, 2.5 mm; costae, fifteen to twenty-one. Gulf of California to the Galápagos Islands, Ecuador.

617. Epitonium (Asperiscala) eutaenium (Dall, 1917) (Synonym: **E. vivesi** Hertlein & Strong, 1951). With numerous rounded whorls, suture deep; costae reflected, with coronations or points at the shoulder; spiral sculpture of fine threads, definite on early whorls, fading on later whorls; lip margin wide,

reflected, a spine at the top. Height, 4.5 to 11 mm; diameter, 2 to 4 mm; costae, seven to eight. Gulf of California to the Galápagos Islands, Ecuador.

618. Epitonium (? Asperiscala) gaylordianum Lowe, 1932. Height, 28.5 mm; diameter, 9.5 mm; costae, twelve to fourteen. Puertecitos, Gulf of California, to Panama. This may prove to be not an *Asperiscala* but a synonym of *E. (Nitidiscala) gradatum* (Sowerby, 1844) (DuShane, *in litt.*).

619. Epitonium (Asperiscala) habeli Dall, 1917 (Synonym: **E. kelseyi** Baker, Hanna & Strong, 1930). Space between fine ridges about three times the width of ridge; suture and umbilicus deep; ridges evanescent on later whorls; sculpture in fresh specimens of rounded and twisted tubercles. Height, 7.5 to 10 mm; diameter, 4 to 6 mm; costae, fourteen to seventeen. Gulf of California to the Galápagos Islands.

620. Epitonium (Asperiscala) huffmani DuShane & McLean, 1968. The brown shell has about fifty narrow white axial costae and a few spiral riblets. Height, 11 mm; diameter, 7 mm. Gulf of California to Salinas, Ecuador, and the Galápagos Islands, intertidally.

621. Epitonium (Asperiscala) indistinctum (Sowerby, 1844) (Synonyms: **Acirsa albemarlensis** Dall & Ochsner, 1928; **E. chalceum** Olsson & Smith, 1951). Costae thicker at sutures, spirals threadlike, base imperforate. Height, 24 mm; diameter, 7 mm; costae, twenty-six to thirty-two. San Blas, Mexico, to Galápagos Islands. Resembling *Amaea* but lacking the spiral thread near the top of the aperture that is characteristic of *Amaea*.

622. Epitonium (Asperiscala) longinosanum DuShane, 1970*b*. Whorls brown, costae white, spirals threadlike, a sharp spine on shoulder of each whorl; suture deep. Height, 16.5 mm; diameter, 5.5 mm; costae, eleven. Outer coast of southern Baja California.

623. Epitonium (Asperiscala) lowei Dall, 1906. Height, 7 to 17 mm; diameter, 4 to 9 mm; costae, twenty-seven to thirty-two. Southern California to Punta Abreojos, Baja California.

624. Epitonium (Asperiscala) macleani DuShane, 1970*b*. Fragile, with convex whorls and deep suture; axial costae low, of differing strength, spiral ribs fine, numerous. Height, 7.3 mm; diameter, 5 mm; costae, thirty-two. Cerralvo Island, Baja California, Mexico.

625. Epitonium (Asperiscala) minuticostatum (De Boury, 1912) (Synonyms: **E. onchodes** and **pacis** Dall, 1917; **E. clarki** T. Oldroyd, 1921 [Pleistocene]; **E. cedrosense** Jordan & Hertlein, 1926 [Pliocene]; **E. nesioticum** Dall & Ochsner, 1928 [Pleistocene]). Spiral ridges strong, evenly spaced, costae reflected, umbilicus partially concealed by inner lip, which is free from costae; aperture oval. Some specimens have wide costae or true varices at irregular intervals. Height, 13 to 35 mm; diameter, 5 to 17 mm; costae, thirteen to twenty. California to the Galápagos Islands.

626. Epitonium (Asperiscala) regularis (Carpenter, 1857). It is possible that *E. (A.) acapulcanum* Dall may prove to be a synonym; the outline is relatively a little more slender than would be indicated by Carpenter's stated dimensions. However, the syntype lot of *E. regularis* in the British Museum (label not in Carpenter's handwriting) comprises shells that are even more slender, with more costae. Carpenter's stated dimensions: height, 7 mm; diameter, 3.3 mm; costae, ten to twelve. Acapulco, Mexico, to Panama (type locality).

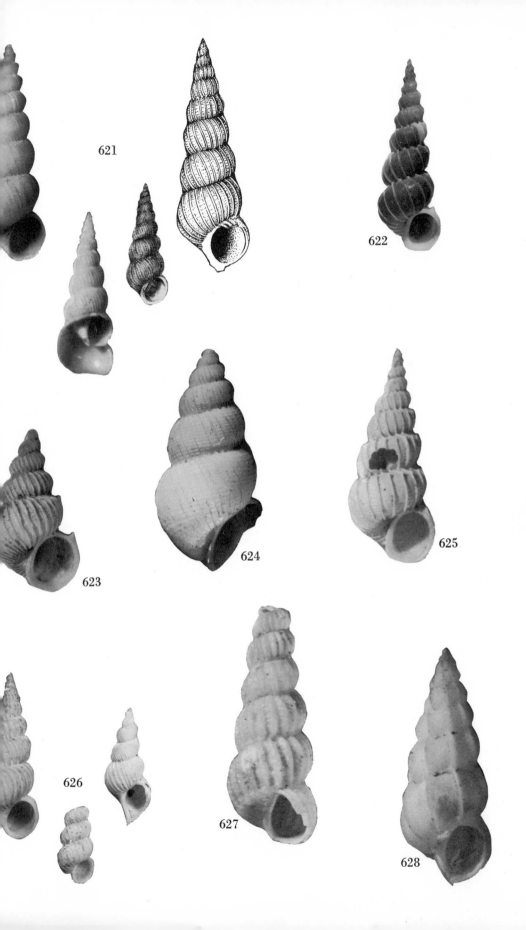

627. Epitonium (Asperiscala) rhytidum Dall, 1917. Whorls spirally sculptured, rounded; costae rounded; base with strong cord. Height, 3.5 mm; diameter, 2 mm; costae, eighteen to nineteen. Galápagos Islands, Ecuador.

628. Epitonium (Asperiscala) thylax Dall, 1917. Costae nearly parallel with axis, lacking spines; spiral threads fine, aperture oval; base imperforate. Height, 6 to 7 mm; diameter, 2.5 to 3 mm; costae, eight. Bahía Concepcion and Pulmo Reef, Gulf of California, to Mazatlán, Mexico, and south to Panama.

629. Epitonium (Asperiscala) tinctorium Dall, 1919. Costae slightly reflected, wider at crossing of deep suture; spiral threads fine and close; a brown spiral band in front of suture. Height, 7 mm; diameter, 3 mm; costae, eleven to twelve. Gulf of California to Panama.

630. Epitonium (Asperiscala) venado (Olsson & Smith, 1951). Resembling *E. (A.) acapulcanum* but with spiral ridges finer, more uneven, and more widely spaced, interspaces five to six times as wide as ridges; upper third of each whorl without ridges; base not umbilicate; a wrinkled fasciole behind inner lip anteriorly. Height, 10 mm; diameter, 5 mm; costae, thirteen. Venado Beach, Panama.

631. Epitonium (Asperiscala) walkerianum Hertlein & Strong, 1951. Whorls rounded; axial ribs low, suture distinct but not deep; spiral threads sharp, well defined; aperture nearly circular. Height, 3.7 to 8 mm; diameter, 1.2 to 2 mm; costae, twenty. San Felipe, Gulf of California, to Nicaragua.

632. Epitonium (Asperiscala) zeteki Dall, 1917. Spiral sculpture threadlike, more widely spaced on later whorls, nearly obsolete on body whorl. Height, 6 mm; diameter, 3 mm; costae, eleven. Panama.

Subgenus CIRSOTREMA MÖRCH, 1852

With thickened costae and spiral ribs; basal disk well marked.

633. Epitonium (Cirsotrema) togatum Hertlein & Strong, 1951. Whorls tabulate; suture deep; ribs strongly retractive, with close axial striae; a spiral cord on base; aperture circular. Height, 37.5 mm; diameter, 14 mm; costae, twenty. Guaymas, Sonora, Mexico, and Outer Gorda Bank, Baja California, south to the Galápagos Islands.

634. Epitonium (Cirsotrema) vulpinum (Hinds, 1844) (Synonym: **C. pentedesmium** Berry, 1963). The specific name, meaning "of a fox," probably refers to the coloring, which varies from reddish brown to pale orange or frosty white, with spiral threads a deeper brown. The shell is slender, spirally striate, suture deep, costae heavy and rounded, with occasional true varices. Height, 3.2 to 14 mm; diameter, 1.5 to 5 mm; costae, nine. Gulf of California to Panama.

Subgenus HIRTOSCALA MONTEROSATO, 1890

Costae reflected, with a pointed spine below the suture; otherwise similar to *Nitidiscala*.

635. Epitonium (Hirtoscala) mitraeforme (Sowerby, 1844). Whorls convex; suture deep; costae not expanded over whorls; aperture rounded. Height, 18 mm; diameter, 8 mm; costae, twelve. One live-taken specimen in the Shasky collection, from the Gulf of Tehuantepec, Mexico, depth 9 to 27 m, may be this species.

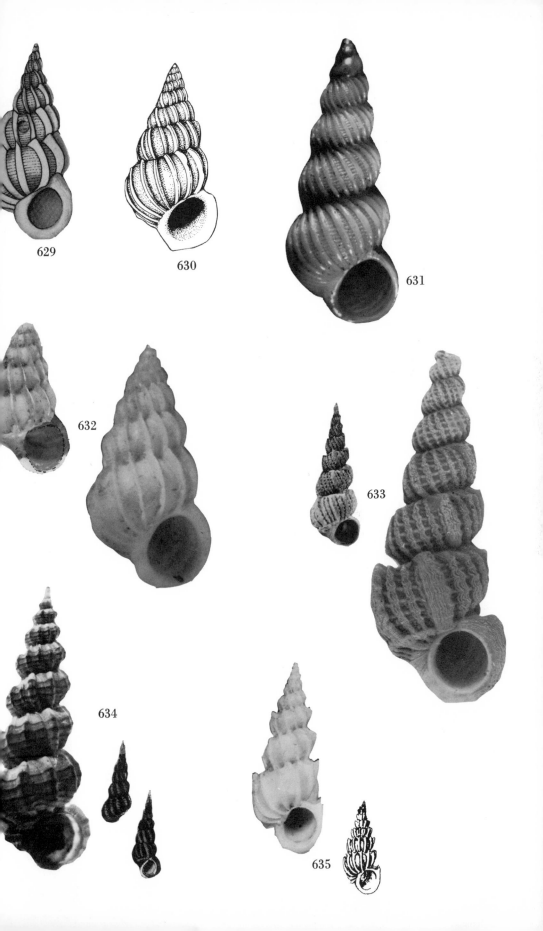

629

630

631

632

633

634

635

636. Epitonium (Hirtoscala) reflexum (Carpenter, 1856). Whorls nearly disjunct; costae strongly reflected; aperture round. Height, 10 to 15 mm; diameter, 4 to 5 mm; costae, five. Throughout the Gulf of California and south to Manzanillo, Mexico.

637. Epitonium (Hirtoscala) replicatum (Sowerby, 1844) (Synonyms: **E. bialatum** Dall, 1917; **E. wurtsbaughi** Strong & Hertlein, 1939; **E. oerstedianum** Hertlein & Strong, 1951). Short, white, some specimens with a brown spot between costae; costae strongly reflected; aperture oval, its margin broad. Height, 6.5 to 15 mm; diameter, 4 to 10 mm; costae, seven to eight. Gulf of California to the Galápagos Islands.

Subgenus NITIDISCALA DE BOURY, 1909

Shell smooth except for axial costae.

638. Epitonium (Nitidiscala) aciculinum (Hinds, 1844). Brown with smooth white costae; whorls convex. Height, 8.5 mm; costae, ten. Central America.

639. Epitonium (Nitidiscala) bakhanstranum Keen, 1962 (Synonym: **E. apiculatum** Dall, 1917 [not Dall, 1889]). Whorls rapidly enlarging, costae sharp, high, thin, reflected, with an angle at the shoulder; suture deep; outer lip reflected; aperture oval. Height, 9 to 16 mm; diameter, 3.5 to 6 mm; costae, eight. Gulf of California to Panama.

640. Epitonium (Nitidiscala) barbarinum Dall, 1919. Shell white; suture deep; anterior surfaces of costae flat; aperture oval. Height, 7.5 to 19 mm; diameter, 4.5 to 7.5 mm; costae, eleven to twelve. Southern California (type locality) and the northern part of the Gulf of California.

641. Epitonium (Nitidiscala) callipeplum Dall, 1919. Whorls rounded, suture deep; costae thin, low, narrow; surface polished; aperture oval. Height, 9 to 11 mm; diameter, 4 to 5 mm; costae, eleven. Magdalena Bay, Baja California, and Mazatlán, Mexico.

642. Epitonium (Nitidiscala) colpoicum Dall, 1917. Whorls glassy, rapidly enlarging, well rounded; suture deep, with pitlike cavities; costae thin, sharp, not reflected; lip buttressed by ends of costae. Height, 6.5 to 10 mm; diameter, 4.5 to 6 mm; costae, eight to ten. Gulf of California to Mazatlán, Mexico, possibly to the Galápagos Islands.

643. Epitonium (Nitidiscala) columnella Dall, 1917. Small, whorls enlarging rapidly; costae high, thin, spinose at shoulder; aperture round. Height, 3 mm; diameter, 2 mm; costae, eleven. Panama.

644. Epitonium (Nitidiscala) compradora Dall, 1917. Small, costae thin, sharp, erect. Height, 4 mm; diameter, 2.5 mm; costae, thirteen. Punta Abreojos, Baja California, to Gulf of California.

645. Epitonium (Nitidiscala) cumingii (Carpenter, 1856) (Synonym: **E. gissleri** Strong & Hertlein, 1939). Whorls glassy, not as rapidly enlarging as in the similar *E. (N.) colpoicum*; costae slightly reflected, with coronations on shoulders of whorls; terminal costa free from body whorl, supported by anterior ends of preceding costae. Height, 6.5 to 9 mm; diameter, 4 to 5.5 mm; costae, eight to nine. Gulf of California to Panama.

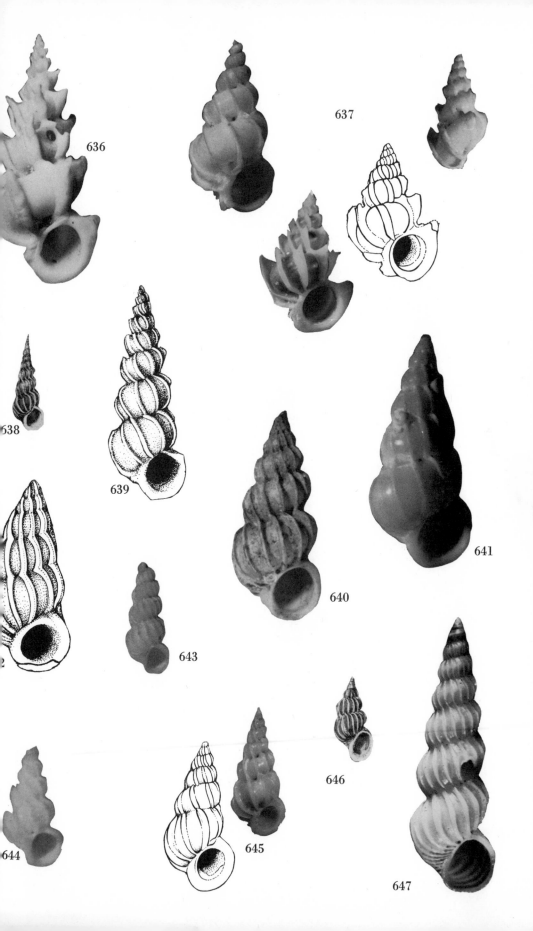

636

637

638

639

640

641

643

644

645

646

647

646. Epitonium (Nitidiscala) curvilineatum (Sowerby, 1844) (Synonym: **E. imbrex** Dall, 1917). Costae numerous, curved; whorls few; aperture oval. Height, 16 mm; diameter, 7.5 mm; costae, six to sixteen. Central America.

647. Epitonium (Nitidiscala) durhamianum Hertlein & Strong, 1951. Elongate-conic, suture moderately deep; costae low, rounded, slightly reflected, fusing at the suture; aperture oval. Height, 5.7 mm; diameter, 1.8 mm; costae, sixteen. Puertecitos, Gulf of California, to Nicaragua. Possibly this will prove to be a synonym of *E. (N.) curvilineatum* Sowerby (DuShane, *in litt.*).

648. Epitonium (Nitidiscala) elenense (Sowerby, 1844) (Synonyms: **Scalaria raricostata** Carpenter, 1857 [not Wood, 1828]; **S. carpenteri** Tapparone-Canefri, 1876; **E. phanium** Dall, 1919). Small, suture distinct; costae not continuous from whorl to whorl, low, rounded; aperture oval. Height, 3.2 to 8 mm; diameter, 1.5 to 3 mm; costae, six to eight. Magdalena Bay, Baja California, and Concepcion Bay, Gulf of California, to Ecuador.

649. Epitonium (Nitidiscala) gradatum (Sowerby, 1844). Outline somewhat turreted, suture deep; growth lines sometimes evident; costae slightly reflected; aperture oval. Height, 16 to 32 mm; diameter, 6 to 16 mm; costae, twelve to fourteen. Puertecitos, Gulf of California, to Ecuador.

650. Epitonium (Nitidiscala) hancocki DuShane, 1970a. Glassy in texture, costae numerous, a sharp spine at the shoulder; suture deep; umbilicus small; whorls rounded; outer lip reflected. Resembling *E. (A.) kelseyi* Baker, Hanna & Strong, 1930, in outline. Height, 13 mm; diameter, 5.2 mm; costae, twenty-one. Galápagos Islands.

651. Epitonium (Nitidiscala) hexagonum (Sowerby, 1844) (Synonym: **E. propehexagonum** Dall, 1917). Whorls rounded; suture deep; costae strong, oblique, thick; aperture round. Height, 5 to 28 mm; diameter, 2 to 11 mm; costae, six (rarely seven). Magdalena Bay, Baja California, and Concepcion Bay, Gulf of California, south to Panama.

652. Epitonium (Nitidiscala) hindsii (Carpenter, 1856). Whorls numerous, rounded; suture unusually deep; costae with an angle at the shoulder; aperture oval and free from body whorl. Height, 26 mm; diameter, 10 mm; costae, eight. Panama.

653. Epitonium (Nitidiscala) obtusum (Sowerby, 1844) (Synonyms: **Scalaria suprastriata** and **tiara** Carpenter, 1857). Solid, short, costae angular above shoulder, with a spine on early whorls; some specimens with spiral striae on early whorls; inner lip spirally twisted. Height, 11 to 25 mm; diameter, 8.5 to 10 mm; costae, eight to fifteen. Southern Baja California and Mazatlán, Mexico, to Ecuador.

654. Epitonium (Nitidiscala) pazianum Dall, 1917 (Synonyms: **E. cylindricum** and **musidora** Dall, 1917). Whorls convex, suture very deep; costae thin, with a spine at the shoulder; aperture round. Height, 4 to 20 mm; diameter, 1.5 to 9 mm; costae, nine to fifteen. Southern California to Peru.

655. Epitonium (Nitidiscala) politum (Sowerby, 1844) (Synonyms: **E. appressicostatum** Dall, 1917; **E. implicatum** Dall & Ochsner, 1928 [Pleistocene]; **E. pedroanum** Willett, 1932). Thin, slender, polished, costae nearly obsolete on body whorl, which is slightly angulate at the shoulder; whorls round-

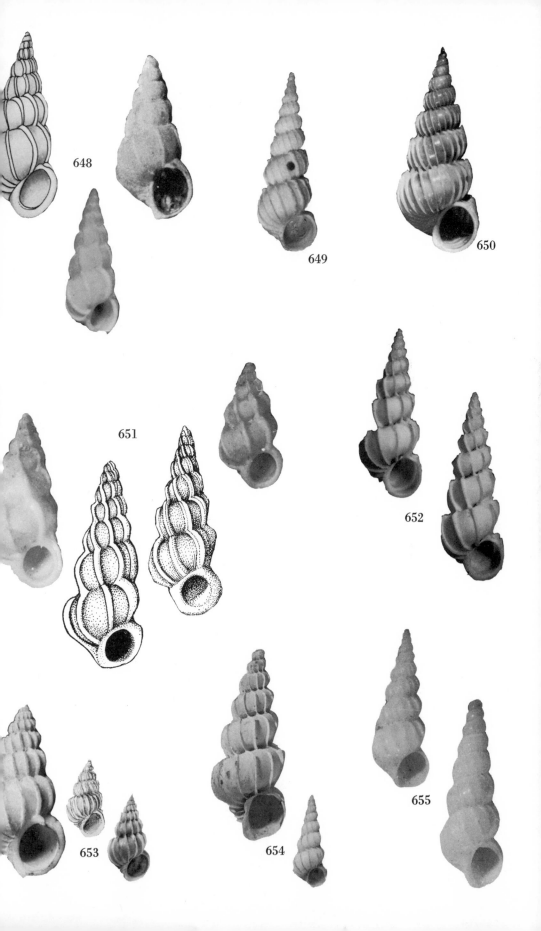

648

649

650

651

652

653

654

655

ed; outer lip thin. Height, 9 to 16 mm; diameter, 3.5 to 5 mm; costae, nine to fifteen. San Pedro, California, to Ecuador and the Galápagos Islands.

656. Epitonium (Nitidiscala) roberti Dall, 1917. Solid, some specimens with a small spine on shoulder of early whorls; aperture oval; base with a simulated disk defined by a slender spiral cord that does not cross over the costae. Height, 12 mm; diameter, 6.5 mm; costae, twelve to fourteen. Southern Gulf of California to Panama. Similar to *E. (N.) obtusum* except for the basal cord.

657. Epitonium (Nitidiscala) shyorum DuShane & McLean, 1968. The whorls have a distinctive steplike contour. Height, 12 mm; diameter, 4 mm; costae, eight to nine. Gulf of California to La Libertad, Ecuador.

658. Epitonium (Nitidiscala) statuminatum (Sowerby, 1844) (Synonym: **E. strongianum** Lowe, 1932). Short, thick, costae heavy and broad, obliquely continuous from whorl to whorl. Height, 11 to 19 mm; diameter, 5 to 9 mm; costae, five. Mazatlán, Mexico, to Peru, in depths to 37 m.

659. Epitonium (Nitidiscala) subnodosum (Carpenter, 1856). Height, 35 mm; diameter, 12 mm; costae, fourteen to sixteen. Panama. The holotype seems to have been lost.

660. Epitonium (Nitidiscala) tabogense Dall, 1917. Height, 3 mm; diameter, 1.5 mm; costae, eleven. Panama Bay. The holotype is a juvenile specimen probably too young for positive identification with adult topotype material.

661. Epitonium (Nitidiscala) willetti Strong & Hertlein, 1937. Whorl rounded; suture deep; costae reflected, thin, sharp, wide at shoulder, curved on base, flattening near the aperture; aperture circular, lip thin. Height, 3.2 mm; diameter, 1.6 mm; costae, eighteen to twenty. Guaymas, Sonora, to Manzanillo, Colima, Mexico, in depths of 22 to 31 m.

662. Epitonium (Nitidiscala) zephyrium Dall, 1917 (Synonym: **E. basicum** Dall, 1917). Costae cordlike, some specimens with a narrow brown band below suture and crossing the suture; basal ridge strong; aperture oval. Height, 11.5 to 15 mm; diameter, 6 to 7 mm; costae, nine to eleven. Range reported by Dall as San Diego, California, to Panama. Probably restricted to southern California (DuShane, *in litt.*).

Subgenus PICTOSCALA DALL, 1917

Shell dark brown, with fine spiral striations and numerous white costae.

663. Epitonium (Pictoscala) purpuratum Dall, 1917. With a few heavy white varices in addition to costae. Color darker below basal cord; lip white, heavy. Height, 10 mm; diameter, 4.5 mm; costae, eighteen. Panama (type locality), the holotype a beach specimen.

Subgenus STHENORYTIS CONRAD, 1863

With a large body whorl and short spire; axial costae well developed.

664. Epitonium (Sthenorytis) dianae (Hinds, 1844) (Synonym: **E. (S.) paradisi** Hertlein & Strong, 1951). The type locality is Gulf of Nicoya, Costa Rica, but because some later authors erroneously considered the species to be Indo-Pacific, it long went unrecognized. The holotype is a small specimen, 8 mm long; height, 8 to 35 mm; diameter, to 26.5 mm; costae, nine to ten. Outer Gorda Bank, Cape San Lucas, Baja California, in 82 to 145 m, southward to Costa Rica.

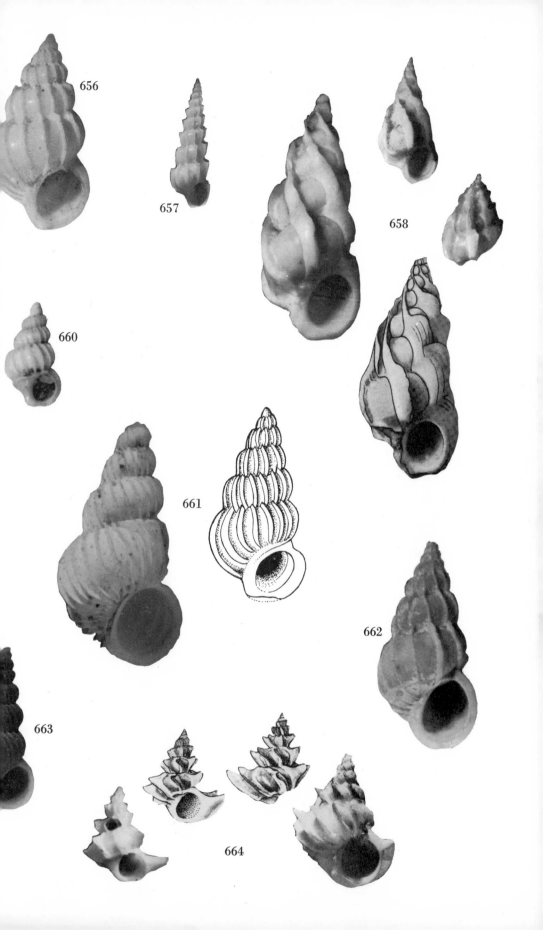

665. Epitonium (Sthenorytis) turbinum Dall, 1908 (Synonym: **E. (S.) toroense** Dall, 1912). Although described as a fossil from the Pliocene of Toro Point, Canal Zone, Panama, *E. toroense* seems indistinguishable from the living form; it was, however, reported by Hanna and Hertlein in 1927 from the Pliocene of Monserrate Island, Gulf of California. Dall's figure of the type (Dall, 1925, pl. 18, fig. 5) matches well his original figure for *E. turbinum*. Height, 35 to 40 mm; diameter, 30 mm; costae, ten to twelve. San Pedro Nolasco Island, Gulf of California, to Galápagos Islands (type locality) in 110 to 550 m.

Genus ACIRSA MÖRCH, 1857

Spiral sculpture of fine striae; no axial costae; basal disk indistinct.

666. Acirsa cerralvoensis (DuShane, 1970*b*). Light brown, with about 16 pale brown spiral ridges on body whorl; suture moderately impressed. Height, 12 to 15 mm; diameter, 3 to 5 mm. Angel de la Guarda Island to Cape San Lucas, Baja California.

667. Acirsa menesthoides (Carpenter, 1864). Sculpture of fine spiral threads. Height, 4 mm; diameter, 1 mm. Cape San Lucas, Baja California.

668. Acirsa murrha (DuShane, 1970*b*). Sculpture of about 19 sinuous axial ridges and numerous spiral cords; suture moderately impressed; base defined by a low ridge; aperture oval. Height, 19 mm; diameter, 7 mm. Secas Islands, Panama. The specific name refers to the porcelain-like texture of the shell.

Genus ALORA H. ADAMS, 1861

Sculpture of both spiral and axial riblets; no basal disk; aperture with a slight trace of an anterior notch. Useful reference: Keen (1969).

669. Alora gouldii (A. Adams, 1857) (Synonym: **? Recluzia insignis** Pilsbry & Lowe, 1932). The sculpture is strongly cancellate; periostracum present, dull buff in color. Height, 15 to 37 mm. Off Manzanillo, Mexico, depth 31 m (Shy collection, 1968), to Panama (type locality).

Genus AMAEA H. & A. ADAMS, 1853

Sculpture cancellate, of coarse and fine axial riblets and spiral ridges; suture impressed; outer lip simple or with a thickened edge; basal ridge present. As yet no species of subgenus *Amaea, s. s.,* has been recognized in the eastern Pacific.

Subgenus SCALINA CONRAD, 1865
(FERMINOSCALA DALL, 1908)

Conic to slender, sculpture markedly different below basal ridge; base with or without an umbilical chink.

670. Amaea (Scalina) brunneopicta (Dall, 1908) (Synonym: **Eglisia nebulosa** Dall, 1919). Slender, with cancellate sculpture; yellowish brown, with a darker brown peripheral band; lip white and thin; aperture oval. Height, 15 to 43 mm; diameter, 5 to 12 mm. Cedros Island, Baja California (type locality) and the Gulf of California, south to the Galápagos Islands, Ecuador.

671. Amaea (Scalina) deroyae DuShane, 1970*a*. Axial and spiral sculpture foliaceous; suture deep; basal ridge strong; outer lip thin, with about seven crenulations. Height, 23 mm; width, 7.3 mm. Gulf of California to Galápagos Islands. A similar species in the Atlantic is *A. (S.) retifera* (Dall, 1889).

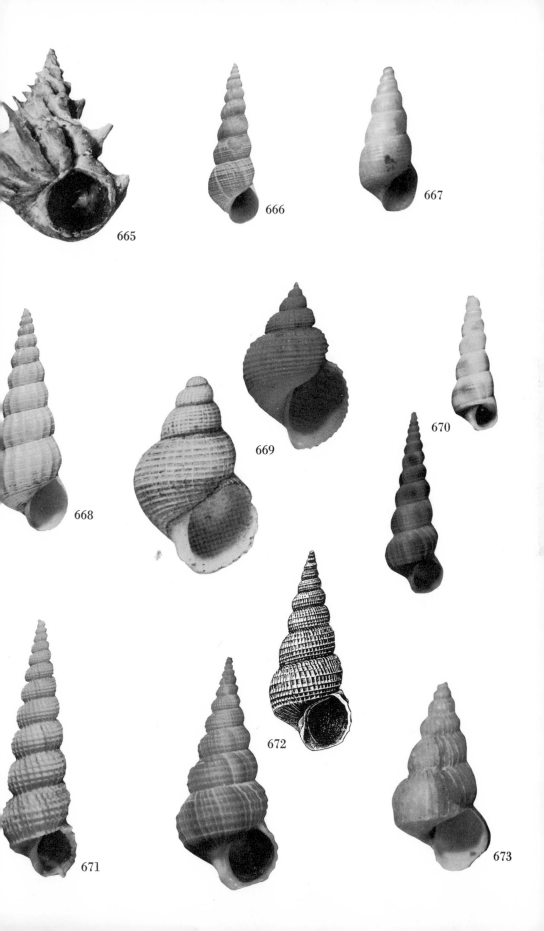

665

666

667

668

669

670

671

672

673

672. Amaea (Scalina) ferminiana (Dall, 1908) (Synonym: **Scala weigandi** Böse, 1910). Sculpture cancellate, whorls rounded, suture moderately deep, shell brown with darker streaks, lip white and thin, aperture brown. Larger and wider in proportion than *A. (S.) brunneopicta*. Height, 23 to 61 mm; diameter, 9 to 23 mm. Gulf of California south to Colombia.

673. ? Amaea (Scalina) pompholyx (Dall, 1890). Sculpture reticulate, with about 32 costae on last whorl; suture distinct; shell white with a yellow periostracum. Height, 14 mm; diameter, 7.6 mm. Galápagos Islands, 1,485 m.

674. Amaea (Scalina) tehuanarum DuShane & McLean, 1968. Spire profile somewhat convex, not evenly tapering; outer lip heavy. Height, 40 mm; diameter, 15 mm; ribs, thirty-eight to forty. Gulf of California to Gulf of Tehuantepec, Mexico, in depths to 73 m.

Subgenus AMAEA, *s. l.*

675. Amaea contexta DuShane, 1970. Whorls and ribs numerous, the low spiral ribs coarse; suture deep; basal ridge well defined; shell dark brown in color; aperture oval; base without umbilicus. Height, 15.5 mm; diameter, 4.5 mm; ribs, twenty-four. Manzanillo to Petatlan Bay, Mexico, in depths of 10 to 20 m.

Genus OPALIA H. & A. ADAMS, 1853

Axial sculpture of strong ribs but not of regular and laminar costae; spiral sculpture various. Outer lip thickened by the last rib.

1. Spiral sculpture of clear-cut ribs; base not set off by a spiral ridge. .*Nodiscala*
   Spiral sculpture weak, of fine punctations; base set off by a spiral ridge. . . . 2
2. Suture deep, angulate below, crenulated by upper margin of ill-defined
   axial ribs . . . . *Dentiscala*
   Suture moderate, lacking angulation, not crenulate. . . . .*Opalia, s. s.*

Subgenus OPALIA, *s. s.*

Suture line evenly spiral, not deeply impressed, whorls only slightly inflated to nearly flat in profile.

676. Opalia (Opalia) exopleura (Dall, 1917). Outline somewhat resembling that of a littorinid, apex acute; spirally punctate sculpture over entire surface, with a prominent thread at periphery of whorls; aperture thickened at margin. Height, 3.5 mm; diameter, 2 mm. Cape San Lucas, Baja California.

Subgenus DENTISCALA DE BOURY, 1886

Basal disk bounded by a keel, the shell surface usually with a thin, chalky outer layer; axial ribs crenulating the suture, sometimes partly obsolete; spiral sculpture of fine threads, with rows of pits between.

677. Opalia (Dentiscala) colimana (Hertlein & Strong, 1951). Small, elongate-conic, the suture deep; sculpture of strong rounded ribs, rounded interspaces wider than the ribs, the ribs terminating at the basal disk, spiral striations alternating with rows of fine pits across the axial ribs; color brown or white; outer lip thick, with a slightly raised inner edge. Height, 7.6 to 12 mm; diameter, 2.8 to 3 mm; ribs, eight to ten. Santa Cruz, Nayarit, to Manzanillo, Colima, Mexico.

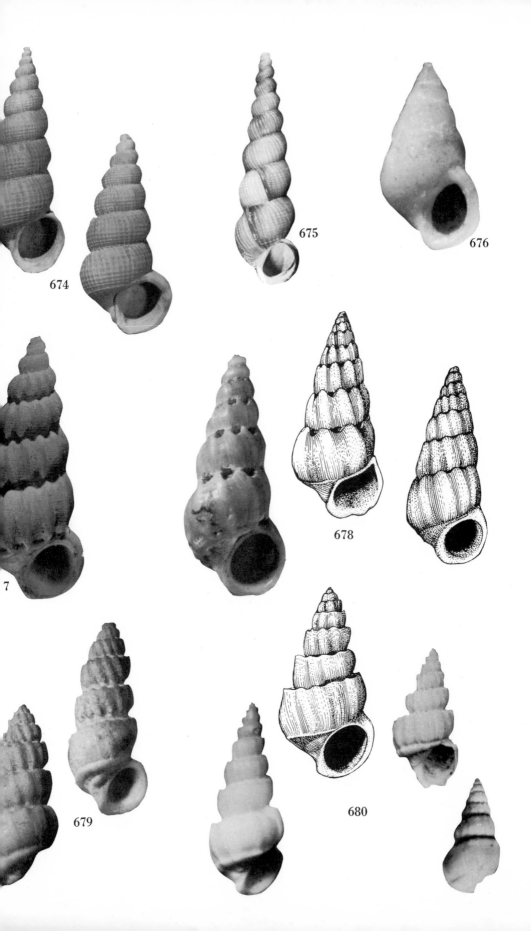

674

675

676

7

678

679

680

678. Opalia (Dentiscala) crenatoides (Carpenter, 1864) (Synonym: **Epitonium golischi** Baker, Hanna & Strong, 1930). Relatively large, texture dull; ribs retractive, obsolete on the lower whorls of some specimens, separated by deep depressions at the suture; with spiral and vertical threads over entire surface, pits at their intersections. Height, 13 to 16 mm; diameter, 6 to 8 mm; ribs, ten. Southern California to the Galápagos Islands. No other eastern Pacific *Dentiscala* has a double row of pits, one above and one below the basal cord.

679. Opalia (Dentiscala) diadema (Sowerby, 1832). Spire profile somewhat convex. Height, 13.5 mm; diameter, 6 mm; ribs, ten. Ecuador to the Galápagos Islands.

680. Opalia (Dentiscala) funiculata (Carpenter, 1857) (Synonyms: **O. insculpta** Carpenter, 1864; **? Scala gereti** De Boury, 1913; **Dentiscala crenimarginata** and **nesiotica** Dall, 1917). Suture deep, coronated by ribs that tend to become obsolete on the last whorl; spiral sculpture of fine threads alternating with rows of pits on fresh specimens; a heavy varix at aperture and sometimes another on spire, indicating a resting period. Long confused with *O. diadema*, this is distinguished by its evenly tapering spire. Height, 10.5 to 17 mm; diameter, 5 to 8.5 mm; ribs, twelve to twenty. Southern California and through the Gulf of California to Panama. The type of *O. insculpta* is from the Pleistocene of southern California.

Subgenus NODISCALA DE BOURY, 1889

Surface with a spongy punctate outer layer, the axial ribs faint over the whorls, crenulating the suture.

681. Opalia (Nodiscala) bullata Carpenter, 1864 (Synonyms: **O. tremperi** Bartsch, 1927; **Epitonium ordenanum** Lowe, 1932). Small, white or brown; suture distinct, crenulated by the ribs; lip heavy; aperture oval. Height, 8 mm; diameter, 2.5 mm; ribs, fifteen to twenty. Southern California to Nicaragua.

682. Opalia (Nodiscala) espirita (Baker, Hanna & Strong, 1930). Milk-white elongate-conic shells, the whorl shoulders slopingly concave, ribs crenulating the suture; spiral cords numerous, crossing the ribs, with pits between the spiral cords; base long, aperture oval; lip with continuous callus. Height, 6 mm; diameter, 2.3 mm; ribs, fifteen. Gulf of California to Chiapas, Mexico.

683. Opalia (? Nodiscala) mazatlanica Dall, 1908. Small, slender, white or brown; suture distinct, crenulated by the ribs; aperture heavy, oval. Height, 7 to 12 mm; diameter, 2 to 4 mm; ribs, fifteen to twenty. Mazatlán, Mexico (type locality), possibly south to Nicaragua. The status of this species is doubtful, for the type is badly worn (DuShane, *in litt.*).

684. Opalia (? Nodiscala) mexicana Dall, 1908. Slender, with deep pits at suture between ribs; sculpture of fine punctations; aperture round. Height, 11 mm; diameter, 4 mm; ribs, nine. Acapulco, Mexico.

685. Opalia (Nodiscala) sanjuanensis (Lowe, 1932) (Synonym: **Dentiscala clarki** Olsson & M. Smith, 1951). Whorls well rounded, suture distinct; entire surface with faint spiral punctate lines; aperture oval and thick. Height, 10 to 12 mm; diameter, 3.5 to 4.5 mm; ribs, nine to fifteen. Gulf of California to Nicaragua.

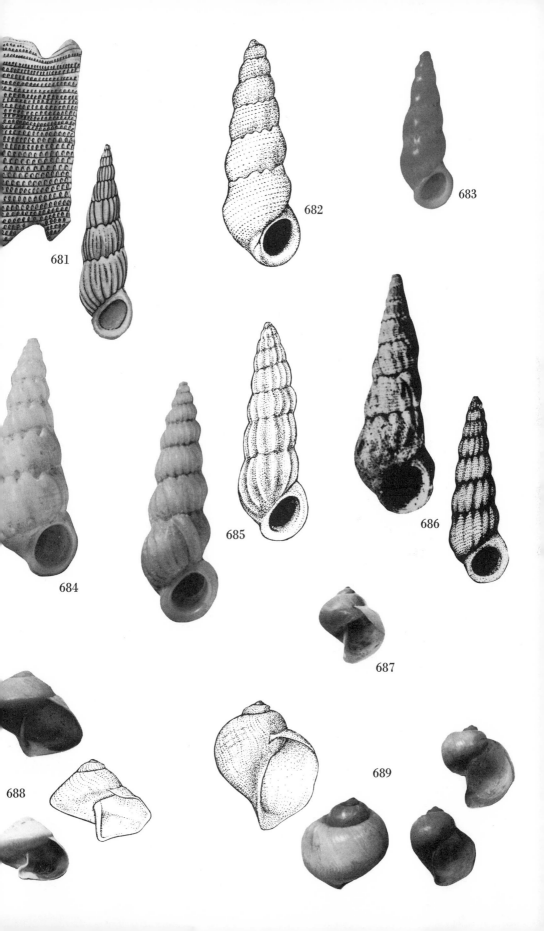

686. Opalia (Nodiscala) spongiosa Carpenter, 1864 (Synonym: **O. reti-porosa** Carpenter, 1864). Whorls with sloping shoulders, ribs ending in rounded knobs at suture; color, yellowish to dark brown; surface with spiral punctations in rows over all whorls; lip thickened, aperture oval. Height, 7 to 12.5 mm; diameter, 2.5 to 5 mm; ribs, fifteen to twenty. Southern California south throughout the Gulf of California to the Galápagos Islands.

Family JANTHINIDAE

Members of this family are pelagic, floating on the surface of the sea, buoyed up by a raft of air bubbles. The float, which is constructed by the foot, entraps bubbles of air in a mucilaginous secretion and serves also as a place of attachment for the eggs. The shell is thin, globose, whitish or purplish. There is no operculum. Two Panamic genera.

Genus JANTHINA RÖDING, 1798
(IANTHINA, an emendation)

Because they are floating organisms, the purple snails are widely distributed, and one species may range through great distances. They are gregarious snails, so that when onshore winds and currents start them washing ashore, vast numbers may be stranded along miles of coast. For the beachcombing collector this is a feast, but it is likely to be followed by a famine that may last for years. Tow-nets behind a boat are a much surer means of making a collection of these and other pelagic gastropods.

The Janthinas are purple, at least on the basal half of the shell, which, when the animal is floating, is the part uppermost. The outer lip may have a sinuous margin, a bending back of the growth lines, that marks the point of attachment of the float.

Nomenclature of the species is confused, in part because of the widespread geographic distribution. Three species seem to be recognizable in the Panamic province, but what names to use is not easy to decide. Dr. S. S. Berry (1958) suggests that the following are probably correct:

687. Janthina globosa Blainville, 1822 (Synonym: **J. umbilicata** Orbigny, 1840; "**J. exigua** Lamarck" of authors, not of Lamarck, 1816). The shell is light purple, banded at the suture with lighter color. The diameter varies from about 6 to 19 mm. The range is throughout the tropical Pacific and Atlantic oceans.

688. Janthina janthina (Linnaeus, 1758) (Synonyms: **J. fragilis** Lamarck, 1801; **J. striulata** and **J. s. contorta** Carpenter, 1857; **J. carpenteri** Mörch, 1860). This is a larger and lower-spired form than *J. globosa*. A large specimen may attain a diameter of 40 mm. The color is a uniform purplish or violet below the periphery, lavender or white above. As with the other species, the range is throughout the warmer parts of the Pacific and Atlantic oceans.

689. Janthina prolongata Blainville, 1822 (Synonyms: **J. globosa** Swainson, 1823 [not Blainville, 1822]; **J. decollata** Carpenter, 1857). In this species the aperture is proportionately large, the columella twisted, the sutures well marked. The coloring is much as in *J. janthina*, and there is variation in the depth of the sinus on the outer lip. Height, 25.5 mm; diameter, 22 mm. The geographic range is the same as for *J. janthina*.

Genus RECLUZIA PETIT, 1853

Shell white, with a periostracum.

690. Recluzia palmeri (Dall, 1871). Dall at first thought this was a freshwater snail, a *Lymnaea*, for the type was picked up some distance back from the shore near the head of the Gulf of California. Height, 22.5 mm; diameter, 13 mm. Few additional specimens have been reported.

Superfamily EULIMACEA
(GYMNOGLOSSA)

Small, smooth, ovate, mostly slender shells. The animals are parasitic in habit, on seastars, sand dollars, and other Echinodermata. A review of Panamic species was published by Bartsch in 1917, which is still the most comprehensive report available, although its generic nomenclature must be somewhat amended under modern taxonomic rules. Shell size is given for forms more than 5 mm in length and figures cited for those species not illustrated by the original author. Two Panamic families, the Aclididae with a single species.

Family EULIMIDAE

Whorls flattened, suture not indented. The key is to genera.

1. Shell not conspicuously glossy; suture evident..................*Cythnia*
   Shell polished; suture mostly indistinct............................. 2
2. Apex mucronate, with minute pointed tip........................... 3
   Apex evenly tapering.. 5
3. Outline globose ....*Stilifer*
   Outline ovate to conic.. 4
4. Spire bluntly rounded, outline ovate.....................*Hypermastus*
   Spire cylindrical, outline conic...........................*Mucronalia*
5. Base umbilicate .... *Niso*
   Base not umbilicate... 6
6. Outline ovate ....*Turveria*
   Outline conic to slenderly tapering.............................. 7
7. Periphery with a keel..............................*Scalenostoma*
   Periphery rounded, not keeled.................................. 8
8. Slender, many-whorled; aperture elongate......................*Eulima*
   Blunt, relatively few-whorled; aperture short...................... 9
9. Whorls inflated ....*Sabinella*
   Whorls flat-sided 10
10. Inner lip smoothly appressed to body whorl......................*Balcis*
    Inner lip slightly elevated from body whorl.................*Eulimostraca*

Genus EULIMA RISSO, 1826
(STROMBIFORMIS of authors, not of DA COSTA, 1778)

Shell very slender, the aperture long and pointed above. Because these are small shells and sculpture must be studied under a microscope, detailed treatment will not be given here. Most species were illustrated when described; figures are cited for those that were not or that have been refigured in modern literature.

691. Eulima acuta (Sowerby, 1834) [*Leiostraca*]. W. Panama. Figured, Bartsch (1917).

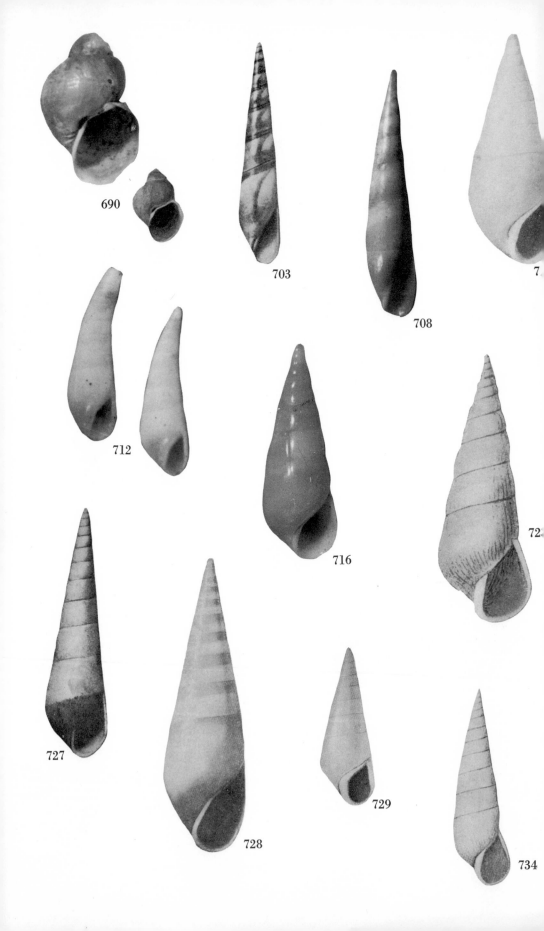

692. E. barthelowi (Bartsch, 1917) [*Strombiformis*]. Santa Maria Bay, Baja California. Length, 5 mm; diameter, 1.3 mm.

693. E. burragei (Bartsch, 1917) [*Strombiformis*]. Gulf of California.

694. E. elegantissima De Folin, 1867. Panama Bay. Figured, Bartsch (1917).

695. E. fuscostrigata Carpenter, 1864. Cape San Lucas. Bartsch (1917).

696. E. healeyi (Strong & Hertlein, 1939) [*Strombiformis*]. Panama. 5 to 16 m.

697. E. hemphilli (Bartsch, 1917) [*Strombiformis*]. Baja California.

698. E. hua (Bartsch, 1926) [*Strombiformis*]. Ecuador.

699. E. inca (Bartsch, 1926) [*Strombiformis*]. Ecuador.

700. ? E. involuta (Carpenter, 1865) [*Mucronalia*]. Mazatlán. Figured, Keen (1968).

701. E. jaculum (Pilsbry & Lowe, 1932) [*Strombiformis*]. Mazatlán.

702. E. lapazana (Bartsch, 1917) [*Strombiformis*]. Gulf of California. Length, 7.8 mm; diameter, 1.3 mm.

703. E. panamensis (Bartsch, 1917) [*Strombiformis*]. Panama Bay. Length, 6 mm; diameter, 1.3 mm.

704. E. paria (Bartsch, 1926) [*Strombiformis*]. Ecuador.

705. E. proca De Folin, 1867. Panama Bay (?). Figured, Bartsch (1917).

706. E. salsa (Bartsch, 1926) [*Strombiformis*]. Ecuador.

707. E. townsendi (Bartsch, 1917) [*Strombiformis*]. La Paz, Gulf of California. Length, 11 mm; diameter, 2.5 mm.

708. E. varians Sowerby, 1834. Ecuador to Panama. Figured, Bartsch (1917). Length, 13 mm, height, 2.5 mm.

Genus **BALCIS** GRAY, 1847
(**MELANELLA** of authors)

Shell ovate-conic, aperture ovate, not markedly pointed; white.

709. Balcis abreojosensis (Bartsch, 1917) [*Melanella*]. Baja California.

710. B. adamantina (De Folin, 1867) [*Eulima*]. Panama. Figured, Bartsch (1917).

711. B. baldra (Bartsch, 1917) [*Melanella*]. Baja California, Tres Marias Islands. Length, 5 mm; diameter, 2 mm.

712. B. bipartita (Mörch, 1859) [*Eulima*]. El Salvador. Figured, Keen (1966).

713. B. capa (Bartsch, 1926) [*Melanella*]. Ecuador.

714. B. corintonis Hertlein & Strong, 1951. Corinto, Nicaragua.

715. B. cosmia (Bartsch, 1917) [*Melanella*]. Baja California.

716. B. dalli (Bartsch, 1917) [*Melanella*]. Gulf of California. Length, 20 mm; diameter, 7.5 mm.

717. B. drangai Hertlein & Strong, 1951. Port Guatulco, Mexico.

718. B. elenensis (Bartsch, 1926) [*Melanella*]. Ecuador.

719. B. elodia (De Folin, 1867) [*Eulima*]. Panama. Figured, Bartsch (1917).

720. B. falcata (Carpenter, 1865) [*Eulima*]. Acapulco, Mexico. Figured, Bartsch (1917). Length, 7.6 mm; diameter, 2.2 mm.

721. B. gibba (De Folin, 1867) [*Eulima*]. Panama. Figured, Bartsch (1917).

722. B. gracillima (Reeve, 1865) [*Eulima*]. Guatemala (? east coast).

723. B. hastata (Sowerby, 1834) [*Eulima*]. Ecuador. Figured, Bartsch (1917). Length, 20 mm; diameter, 6 mm.

724. B. hemphilli (Bartsch, 1917) [*Melanella*]. Baja California to Mazatlán. Length, 8 mm; diameter, 3 mm.

725. B. iota (C. B. Adams, 1852) [*Eulima*]. Panama. Figured, Bartsch (1917).

726. B. lastra (Bartsch, 1917) [*Melanella*]. Baja California.

727. B. linearis (Carpenter, 1857) [*Leiostraca*]. Mazatlán. Figured, Keen (1968).

728. B. mexicana (Bartsch, 1917) [*Melanella*]. Gulf of California. Length, 6 mm; diameter, 2 mm.

729. B. ochsneri (Bartsch, 1917) [*Melanella*]. Galápagos Islands. Length, 9 mm; diameter, 3 mm.

730. B. olssoni (Bartsch, 1924) [*Melanella*]. Ecuador.

731. B. panamensis (Bartsch, 1917) [*Melanella*]. Panama Bay.

732. B. parva (Reeve, 1866) [*Eulima*]. Mazatlán.

733. B. producta (Carpenter, 1864) [*Leiostraca*]. Mazatlán. Figured, Keen (1968).

734. B. pusilla (Sowerby, 1834) [*Eulima*]. Ecuador. Length, 8 mm; diameter, 2 mm. Figured, Sowerby (1854).

735. B. recta (C. B. Adams, 1852) [*Eulima*]. Panama. Length, 9 mm; diameter, 2.3 mm. Figured, Bartsch (1917).

736. B. retexta (Carpenter, 1857) [*Leiostraca*]. Mazatlán. Figured, Keen (1968).

737. B. solitaria (C. B. Adams, 1852) [*Eulima*]. Panama. Figured, Turner (1956).

738. B. tia (Bartsch, 1926) [*Melanella*]. Ecuador.

739. B. townsendi (Bartsch, 1917) [*Melanella*]. Gulf of California.

740. B. yod (Carpenter, 1857) [*Leiostraca*]. Guaymas to Mazatlán, Mexico. Figured, Keen (1968).

Genus CYTHNIA CARPENTER, 1864

Shell small, transparent, with an umbilicus set off by a low ridge; operculum of concentric spirals. Parasitic on seastars.

741. Cythnia asteriaphila Carpenter, 1864. Cape San Lucas.

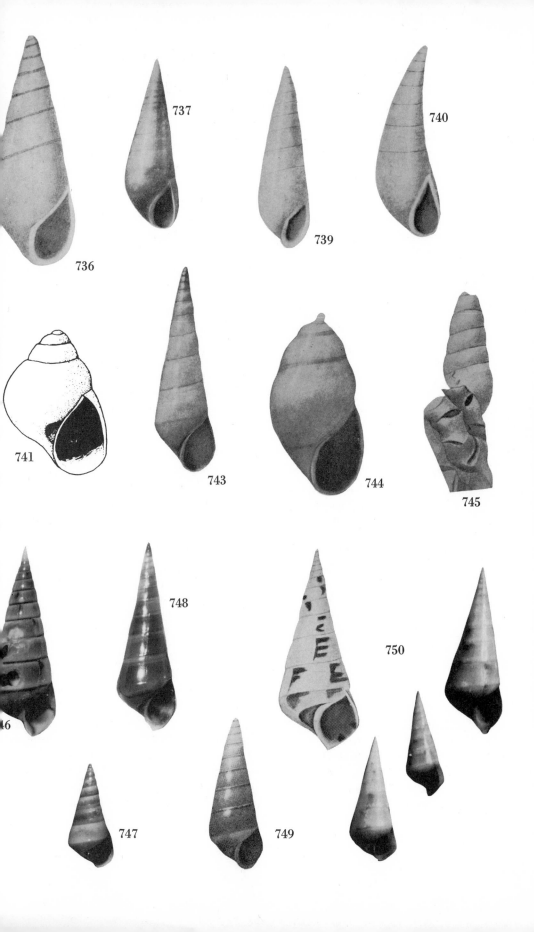

736

737

739

740

741

743

744

745

748

750

747

749

Genus EULIMOSTRACA BARTSCH, 1917

Shell slender, aperture pear-shaped; with a brown band below the suture. Conic, whorls flattened, inner lip slightly elevated from body whorl; color markings present.

742. Eulimostraca bartschi Strong & Hertlein, 1937. Mazatlán.

743. E. galapagensis Bartsch, 1917. Galápagos Islands.

Genus HYPERMASTUS PILSBRY, 1899
(LAMBERTIA SOUVERBIE, 1879, not ROBINEAU-DESVOIDY, 1863)

Apex mucronate (with a small pointed tip), shell rounded-ovate, inner lip of aperture appressed to the base of the body whorl.

744. Hypermastus cookeanus (Bartsch, 1917) [*Lambertia*]. Baja California to Mazatlán.

Genus MUCRONALIA A. ADAMS, 1860

Apex mucronate, spire cylindrical, inner lip not appressed to base.

745. ? Mucronalia bathymetrae (Dall, 1908) [*Stilifer*]. Panama Bay. 3,245 m.

Genus NISO RISSO, 1826

Slender, base widely and deeply umbilicate; color markings present on most species. Useful reference: Emerson (1965). Two Panamic subgenera.

Subgenus NISO, *s. s.*

Whorls flat, smoothly tapering; color markings in spiral bands.

746. Niso (Niso) aeglees Bush, 1895. Similar in coloring and markings to *N. lomana*, but smaller and proportionately more slender, especially at the apex. Length, 21 mm; diameter, 9 mm. Galápagos Islands, in 170 to 200 m. Also in the western Atlantic (type locality), recognized by McLean (*in litt.*) as occurring also in the eastern Pacific.

747. Niso (Niso) baueri Emerson, 1965. Small, with nuclear whorls proportionally smaller than in other species; broadly conic, spire white, later whorls brownish tan with white bands below suture and at the slightly angulate periphery of the body whorl. Length, 3.1 mm; diameter, 1.3 mm. Guaymas, Mexico, to Los Coronados Islands, Gulf of California (type locality), in depths of 37 to 80 m.

748. Niso (Niso) emersoni McLean, 1970. Similar to *N. interrupta* but more slender, with dark nuclear whorls. The periphery of the reddish-brown shell is defined by a low ridge, slightly darker in color, with a light band below that shows through the outer lip. Length, 12 mm; diameter, 4 mm. Off Chiapas, Mexico, to San José Point, Guatemala (type locality), in 13 to 20 m.

749. Niso (Niso) hipolitensis Bartsch, 1917. Less sturdy than *N.* (*N.*) *baueri*, umbilicus narrower; tip of spire white, base white with a broad brown band in the middle, anterior half of the aperture white. Length, 3.1 mm; diameter, 1.2 mm. San Diego, California, to Bahía de los Angeles, Gulf of California.

750. Niso (Niso) interrupta (Sowerby, 1834). The color pattern is of irregular brown blotches, the nuclear whorls being light and the inner lip and umbilicus dark. Length, 18 mm; diameter, 8 mm. Gulf of California near San Pedro Martír Island to Guayaquil, Ecuador, in depths to 55 m.

751. Niso (Niso) lomana Bartsch, 1917. With a chestnut-brown band at the periphery, bordered by lighter areas above and below, and a brown band along the basal ridge, the suture channeled. Length, 20 mm; diameter, 9 mm. Cedros Island, Baja California, to Angel de la Guarda and Tiburon Islands, Gulf of California, in 9 to 175 m. Described from the Pleistocene of southern California, the type a fragmentary specimen.

752. Niso (Niso) splendidula (Sowerby, 1834). Probably the largest-sized species in the entire family, this has a white shell with brown flecks on either side of two spiral bands of brown. Described from a single fine specimen dredged off Ecuador by Hugh Cuming in the early 1830's, the species remained rare in collections until recent years, when a number of additional specimens have been taken. Length, 38 mm; diameter, 17 mm. Off Angel de la Guarda Island, Gulf of California, to Guayaquil, Ecuador, in 10 to 110 m.

See Color Plate XIV.

Subgenus NEOVOLUSIA EMERSON, 1965
(VOLUSIA ADAMS, 1861, not ROBINEAU-DESVOIDY, 1830)

Profile of whorls angulate; suture indented below a spiral ridge.

753. Niso (Neovolusia) excolpa Bartsch, 1917. The shell is flesh-colored, with fine rows of brown flecks. Length, 17.5 mm; diameter, 6.5 mm. Gulf of California to Panama, in depths to 48 m.

754. Niso (Neovolusia) imbricata (Sowerby, 1834). White, with a band of vertical reddish stripes. Length, 21 mm. Clarion Island, off west Mexico, to Santa Elena (type locality) and the Galápagos Islands, Ecuador, in 11 to 15 m. Also reported as a fossil from the Pliocene of Ecuador. The species is the type of the subgenus.

Genus SABINELLA MONTEROSATO, 1890

Conic, whorls somewhat inflated, aperture large, with outer lip flaring, inner lip not appressed to base.

755. Sabinella chathamensis Bartsch, 1917. Galápagos Islands.

756. S. meridionalis Bartsch, 1917. Galápagos Islands.

757. S. opalina (De Folin, 1867) [*Eulima*]. Panama.

Genus SCALENOSTOMA DESHAYES, 1863

Periphery with an acute keel that continues up spire as a sutural cord.

758. Scalenostoma babylonia (Bartsch, 1912) [*Odostomia*]. Baja California.

759. S. rangii (De Folin, 1867) [*Chemnitzia*]. Panama. Figured, Dall & Bartsch (1909).

Genus STILIFER BRODERIP, 1832

Apex mucronate (initial whorls forming a narrow cylindrical tip); shell otherwise inflated to nearly globular, inner lip not appressed to base of body whorl. Parasitic on Echinodermata. Two Panamic subgenera.

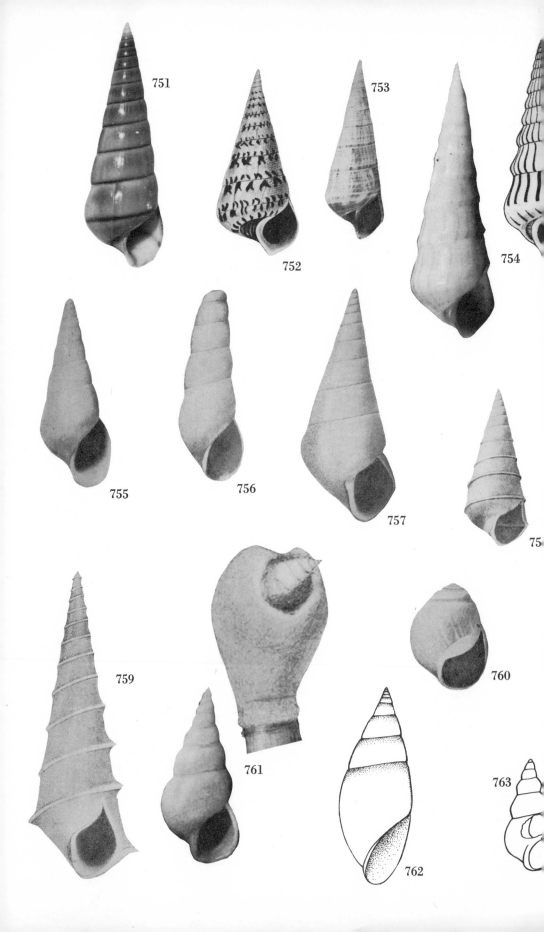

751

752

753

754

755

756

757

758

759

760

761

762

763

Subgenus **STILIFER**, *s. s.*

Body whorl globose. Parasitic on seastars.

760. Stilifer (Stilifer) astericola Broderip, 1832. Galápagos Islands.

Subgenus **PELSENEERIA** KOHLER & VANEY, 1908
(**ROSENIA** SCHEPMAN, 1913 [not WAAGEN & WENTZEL, 1887])

Body whorl ovate to globose, mucronate tip longer than in *Stilifer, s. s.* Parasitic on sea urchins, the type species from deep water, eastern Atlantic. American species forming cavities in echinoid spines, possibly meriting subgeneric separation. First recorded in the eastern Pacific by Shasky (1968).

761. Stilifer (Pelseneeria) nidorum (Pilsbry, 1956). Small, ovate, white. Length, about 4 mm. Gulf of California (type locality, Caribbean). *Aclis tumens* Carpenter, 1857, may also prove to belong here.

Genus **TURVERIA** BERRY, 1956

Ovate-conic, the aperture ovate; a brown color band below the suture.

762. Turveria encopendema Berry, 1956. Sonora, Mexico, on *Encope*, a large sand dollar.

Family **ACLIDIDAE**

Small, high-spired shells, similar in shape to the Eulimidae but not polished, whorls more inflated.

Genus **ACLIS** LOVÉN, 1846

Spire whorls slender, smoothly tapering.

763. Aclis tumens Carpenter, 1857. Mazatlán, Mexico. Figured, Keen (1968).

Superfamily **HIPPONICACEA**

Limpet-shaped to spiral but few-whorled shells, mostly with strong cancellate sculpture.

Family **HIPPONICIDAE**

Shells cap-shaped, not spirally coiled in adult.

Genus **HIPPONIX** DEFRANCE, 1819
(**AMALTHEA** SCHUMACHER, 1817 [not RAFINESQUE, 1815]; **HIPPONYX**
of authors, emend.)

The horse hoof limpets are thick, obliquely conical, not spiral, the apex somewhat posterior and directed backward; sculpture is of radial ribs variously roughened. The muscular impression is horseshoe-shaped, opening anteriorly. In some species a calcareous base (in effect, an attached operculum) is secreted, on which the animal rests. The shell is white in most species, with a dark-colored fibrous periostracum.

Morrison (1965) has suggested that several generic or subgeneric groups can be recognized on the basis of the embryonic whorls. He has been able to allocate several Panamic species, but because data on initial whorls are not yet available for all of the species, a subdivision of *Hipponix* is not attempted here.

764. Hipponix delicatus Dall, 1908. The principal points of difference between this shell and *H. panamensis* are the fine lamellar sculpturing and the lack of a

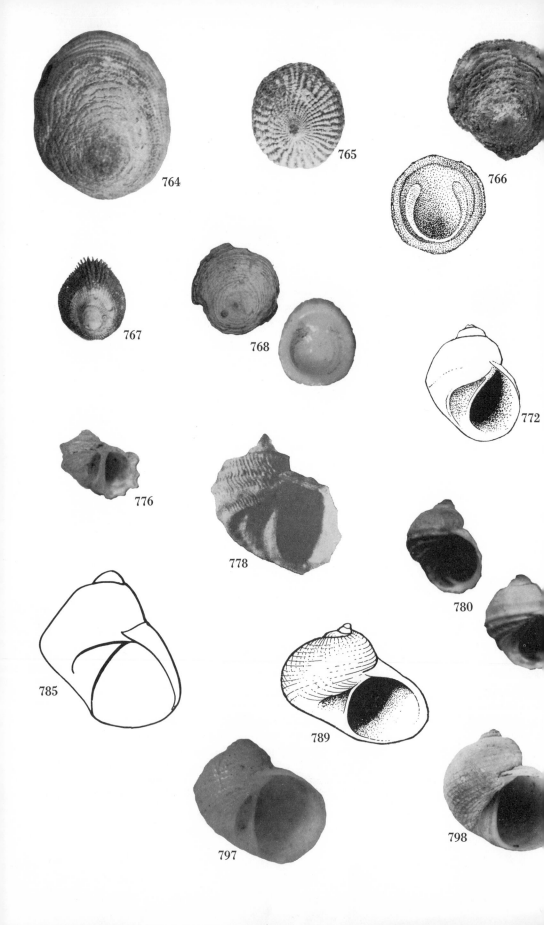

764

765

766

767

768

772

776

778

780

785

789

797

798

flattened lip edge. These features may be a result of the habitat—deep water rather than intertidal rocks. Length, 12 mm; width, 10 mm; height, 6 mm. Panama Bay, 333 m.

765. Hipponix grayanus Menke, 1853 (Synonym: **H. radiatus** Gray, 1835 [not Blainville, 1824]). This resembles *H. pilosus* but has fewer periostracal hairs and more coarsely beaded ribs. The shell is brown-tinged within. Major diameter, 11 mm; height, 5 mm. Mazatlán, Mexico, to Ecuador. Morrison allocates this to *Cochlear* Mörch, 1877.

766. Hipponix panamensis C. B. Adams, 1852 (Synonyms: **Patella antiquata** Linnaeus, 1767, of authors; **H. serratus** Carpenter, 1856; **H. fimbriatus** Bartsch & Rehder, 1939). A low conoidal form with the apex near the posterior margin. The surface of the white shell seems to be made up of many close-set lamellae, the interstices being filled with shreds of dark brown periostracum. Linnaeus's species was long considered to have wide distribution in both the Atlantic and the Pacific. Species differences for so simple a shell are difficult to detect, especially because attachment of the shell to irregular surfaces may result in no two individuals being precisely the same shape. Using a name bestowed on Panamic material seems preferable, however, until a thorough study of soft parts as well as shells from other areas demonstrates the true relationships. Major diameter, about 25 mm; height, about 11 mm. Gulf of California to Peru, intertidally. Morrison allocates this species to *Antisabia* Iredale, 1927.

767. Hipponix pilosus (Deshayes, 1832) (Synonym: **Hipponyx barbatus** Sowerby, 1835). The most unmistakable species of the genus, this is fairly regular in shape, with regular radial ribs between which the yellow-brown periostracum is developed like rows of fine bristles. Major diameter, 23 mm; height, 11 mm. Gulf of California to Ecuador. Morrison allocates the species to *Pilosabia* Iredale, 1929, and suggests that the specific name *Patella trigona* Gmelin, 1791, may be an earlier synonym; Gmelin's name was based on a short description of a shell from an unknown locality and on the citation of a figure in a nonbinomial work.

768. Hipponix planatus Carpenter, 1857. A small, very flat form that seems to have a smoother and more central apex than *H. panamensis* and less well-developed periostracum. Major diameter, 14 mm; height, 3 mm. Mazatlán, Mexico, to Panama.

Family FOSSARIDAE

Spirally coiled, few-whorled, with a wide umbilicus. Two Panamic genera.

Genus FOSSARUS PHILIPPI, 1841

Small shells, with spiral sculpture predominating. Spire normally tapering, mostly not set off by change of sculpture or color. Because these are small shells and sculpture must be studied under a microscope, detailed treatment will not be given here, but references to available figures will aid in identification. There is no single useful reference on the group, which has been neglected by authors.

769. Fossarus abjectus (C. B. Adams, 1852) [*Adeorbis*]. Panama. Figured, Turner (1956).

770. F. angiostoma (C. B. Adams, 1852) [*Littorina*]. Panama. Figured, Turner (1956).

771. F. angulatus Carpenter, 1857. Mazatlán. Figured, Brann (1966).

772. F. atratus (C. B. Adams, 1852) [*Littorina*]. Panama. Figured, Turner (1956).

773. F. excavatus (C. B. Adams, 1852) [*Littorina*]. Panama. Figured, Turner (1956).

774. F. foveatus (C. B. Adams, 1852) [*Littorina*]. Panama. Figured, Turner (1956).

775. F. guayaquilensis Bartsch, 1928. Ecuador. Figured, Bartsch (1928).

776. F. lucanus Dall, 1919. Cape San Lucas. Unfigured.

777. F. mediocris De Folin, 1867. Panama. Figured, Tryon (1887).

778. F. megasoma (C. B. Adams, 1852) [*Littorina*]. Panama. Figured, Turner (1956).

779. F. parcipictus Carpenter, 1864. Cape San Lucas. Figured, Palmer (1963).

780. F. porcatus (Philippi, 1845) [*Littorina*]. Galápagos Islands. Height, 6 mm.

781. F. purus Carpenter, 1864. Cape San Lucas. Figured, Palmer (1963).

782. F. saxicola (C. B. Adams, 1852) [*Litiopa*]. Panama. Figured, Turner (1956).

783. F. tuberosus Carpenter, 1857. Mazatlán. Figured, Keen (1968).

Genus **MACROMPHALINA** COSSMANN, 1888
(**MEGALOMPHALUS** BRUSINA, 1871, of authors, in part; **DISCOPSIS** of authors, not of DE FOLIN, 1870; **CHONEBASIS** PILSBRY & OLSSON, 1945)

Small, apex smooth, pointed, set at an angle to later whorls, usually brown in color; body whorl rapidly enlarging. Useful references: Pilsbry & Olsson (1945; 1952).

784. Macromphalina argentea (Bartsch, 1918) [*Discopsis*]. Panama.

785. M. cryptophila (Carpenter, 1857) [*Vanicoro*]. Mazatlán, Mexico. Figured, Keen (1968).

786. M. dipsycha (Pilsbry & Olsson, 1945) [*Chonebasis*]. Colombia.

787. M. equatorialis (Pilsbry & Olsson, 1945) [*Chonebasis*]. Ecuador.

788. M. hancocki (Strong & Hertlein, 1939) [*Megalomphalus*]. Panama.

789. M. hypernotia Pilsbry & Olsson, 1952. Zorritos, Peru.

790. M. immersiceps (Pilsbry & Olsson, 1945) [*Chonebasis*]. Colombia.

M. panamensis (Bartsch, 1918) [*Discopsis*]. (See *M. scabra*.)

791. M. peruvianus (Pilsbry & Olsson, 1945) [*Chonebasis*]. Zorritos, Peru.

792. M. philippii Pilsbry & Olsson, 1945 [*Chonebasis*]. Colombia.

793. M. recticeps (Pilsbry & Olsson, 1945) [*Chonebasis*]. Ecuador.

794. M. scabra (Philippi, 1849) [*Adeorbis*] (Synonym: **Discopsis panamensis** Bartsch, 1918). Panama. Figured, Pilsbry & Olsson (1945).

795. M. souverbiei (De Folin, 1867) [*Sigaretus*]. Panama.

796. M. symmetrica (Pilsbry & Olsson, 1945). [*Chonebasis*]. Ecuador.

Family VANIKORIDAE

Rather small white shells, umbilicate, with a wide aperture, sculptured with narrow spiral ribs and less regular axial ribs. Apical whorls not tilted at an angle. A tropical group, with most of its species in the Pacific.

Genus VANIKORO QUOY & GAIMARD, 1832
(MERRIA GRAY, 1839)

The name *Vanikoro* has been rejected by some authors who feel it was not properly proposed as a Latin name. However, this is debatable, and long usage seems to justify its adoption in preference to *Merria*.

797. Vanikoro aperta (Carpenter, 1864). White, solid, apex smooth, brown; later whorls deeply cancellate, with strong spiral cords and regular axial riblets; last whorls rapidly enlarging, sculpture reduced to irregular threads; umbilicus large, bounded by a heavy rib. Height, 10.6 mm; diameter, 11.4 mm. Cape San Lucas to Guaymas, Sonora, Mexico.

798. Vanikoro galapagana Hertlein & Strong, 1951. Shell larger and less heavily sculptured than *V. aperta*; umbilicus small. Diameter, 13.5 mm. Galápagos Islands.

Superfamily CALYPTRAEACEA

Conical or limpet-like to few-whorled, with entire aperture or only a small anterior notch; a fibrous to bristly periostracum present in some. Mostly sedentary, subject to many distortions of shape and sculpture, thus tending to be variable in form.

Family CALYPTRAEIDAE

Conical, with an internal shelf, deck, or lamina for the support of the soft parts. The key is to Panamic genera and subgenera.

1. Apex marginal 2
   Apex central or subcentral 3
2. Shelf decklike . . . . *Crepidula*
   Shelf detached along part of one side . . . . *Crepipatella*
3. Spiral suture evident 4
   Spiral suture inconspicuous 5
4. Outer surface smooth or nearly so . . . . *Calyptraea, s. s.*
   Outer surface with oblique ribbing . . . . *Trochita*
5. Internal shelf with free and pointed ends . . . . *Cheilea*
   Internal shelf cuplike, entire, no pointed ends 6
6. Shelf free along most of its margin . . . . *Crucibulum, s. s.*
   Shelf attached along its right side and partially flattened . . . . *Dispotaea*

Genus CALYPTRAEA LAMARCK, 1799

The shell is conical, its apex central and spiral, the aperture basal, having a spiral diaphragm that is twisted at the margin to form a false umbilicus. Two Panamic subgenera.

Subgenus **CALYPTRAEA**, *s. s.*

Sculpture weak, shell nearly smooth.

799. Calyptraea (Calyptraea) conica Broderip, 1834 (Synonyms: **C. sordida** Broderip, 1834; **C. aspersa** C. B. Adams, 1852). The surface of the thin shell is a little roughened, as if it had once been granular and had been worn almost smooth. The ground color is yellowish white, on which are spots of chestnut-brown that show through to the bluish-white interior. Diameter, 24 mm; height, 14 mm. Magdalena Bay, Baja California, through the Gulf of California and southward to Ecuador, mostly offshore in depths to 37 m.

800. Calyptraea (Calyptraea) lichen Broderip, 1834. This may prove to be a form of *C. mamillaris*. The shell is low, smooth, and white, with a few scattered spots or lines of reddish brown. Diameter, 18 mm. The type locality is Guayaquil, Ecuador.

801. Calyptraea (Calyptraea) mamillaris Broderip, 1834 (Synonyms: ? **C. lamarckii** Menke, 1847; **C. regularis** C. B. Adams, 1852; **Galerus sinensis fuscus** Carpenter, 1856; **Trochita solida** Reeve, 1859). The shell is white and smooth, with a flush of purple or brown on the spire. The outline may vary from smoothly conic, as in the illustration, to elevated, with contracted sutures, as in the form named *solida* by Reeve, which evidently had been attached to a small convex base, such as the body whorl of a spiral shell, and which therefore was obliged to increase its relative height to gain space within. Diameter, 35 mm; height, 15 mm. Magdalena Bay, Baja California, through the Gulf and southward to Paita, Peru, offshore in depths to 37 m; not uncommon on mud flats, attached to rocks and other shells.

802. Calyptraea (Calyptraea) subreflexa (Carpenter, 1856). Very similar to *C. mamillaris*, this is yellowish white in color, with fine diagonal lines ending in prickles. Diameter, 16 mm; height, 4 mm. Gulf of California to Acapulco, Mexico.

Subgenus **TROCHITA** SCHUMACHER, 1817

Oblique spiral ribbing well developed.

803. Calyptraea (Trochita) spirata (Forbes, 1852). Considered by some authors a northern subspecies of *C. (T.) trochiformis*, this Mexican form seems separable on several counts—size, color, sculpture, as well as distribution. It is large, coarsely ribbed, whitish or grayish outside, dark brown within. Diameter, 60 mm; height, 20 mm. Mazatlán to the Gulf of Tehuantepec, Mexico. Specimens may be found clinging tenaciously to the most surf-beaten rocks on exposed coasts, especially at Manzanillo.

See Color Plate XIV.

804. Calyptraea (Trochita) trochiformis (Born, 1778) (Synonyms: "**Patella trochiformis** Gmelin 1791" of authors; **Trochus radians** Lamarck, 1816; **P. trochoides** Dillwyn, 1817; **Trochita spiralis** Schumacher, 1817; **Calyptraea peruviana** Deshayes, 1830; **C. araucana** Lesson, 1830). A nearly conical white shell, with slightly inflated whorls strongly and obliquely ribbed in most specimens, although there is some variation. Unworn shells show a light brown periostracum. Diameter, 50 mm; height, 15 mm. Manta, Ecuador, to Valparaiso, Chile. Rehder (1943) has discussed the complex nomenclatural history of this species.

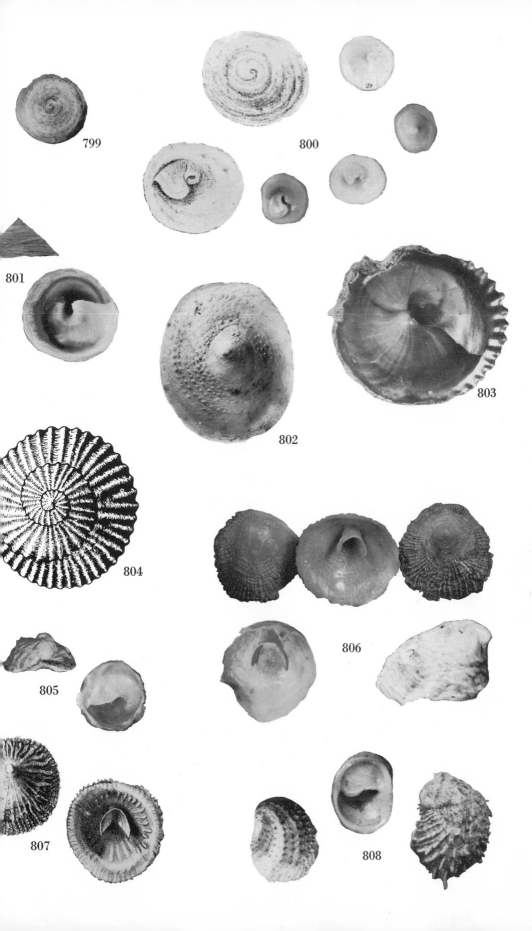

799

800

801

802

803

804

805

806

807

808

805. Calyptraea (Trochita) ventricosa (Carpenter, 1857). Named as a sub-species of *C. spirata*, this is probably more distinctive as to color than for the difference of outline noted by Carpenter. The inside is white, not brown, in which character it is closer to *C. (T.) trochiformis*. The whorls are angulate below the suture, which gives the shell a somewhat pagoda-like outline. Diameter, 19 mm; height, 6 mm. Mazatlán, Mexico.

Genus CHEILEA MODEER, 1793
(MITRULARIA SCHUMACHER, 1817)

The internal support is attached only at its base and hangs free within the cavity of the shell. The shell itself is conical but rather irregular in form, with a some-what spiral and nearly central apex.

806. Cheilea cepacea (Broderip, 1834) (Synonyms: **Patella equestris** Linnaeus, 1758, of authors; **Calyptraea cornea** Broderip, 1834; **Calyptraea plan-ulata** C. B. Adams, 1852; **Calyptraea alveolata** and **tortilis** Reeve, 1858). The white shell is irregularly conic, with minute radial ridges. Some authors regard the species as identical with the Caribbean *C. equestris* (Linnaeus, 1758), which is a similar or twin species. Sculpture varies, and a number of names have been given, some of which may prove to apply to recognizably distinct forms: *C. cornea* is nearly smooth, though this may be a matter of wear; *C. alveolata* has spinose radial riblets; and *C. tortilis* has scaly or almost frilled secondary concentric sculpture. Diameter, 20 mm; height, 12 mm. Puerto Peñasco, Gulf of California to Chile, on dead shells, from extreme low-tide level to depths of 20 m.

807. Cheilea corrugata (Broderip, 1834) (Synonym: **Calyptraea varia** Broderip, 1834). The outer surface is coarsely spirally ribbed, the ribs scaly near the margin. In color, the shell may be white or light yellowish brown. About 40 mm in diameter. The Gulf of California to Peru. The type specimen was taken by Cuming under stones dredged from a depth of 25 m. Shasky and Campbell (1964) found it in colonies under rocks at 8 to 10 m.

Genus CREPIDULA LAMARCK, 1799

The shells of the slipper limpets are low, oval, with the apex at or near the margin. Very young shells are spiral, but these spiral whorls are soon lost and the shell becomes flat, with a deck or shelf across about one-half of the aperture. Although individuals can move about, they customarily remain attached to whatever spot the young shell settles on, and great variations in shape may result. Those that live inside the apertures of dead gastropod shells may be smooth and curved back-ward to fit the contour of the shell, whereas specimens from the same brood lodg-ing on the outside of the shell may become rough and distorted. In some species the individuals cluster together, perhaps even stacked one upon the back of an-other. Berry (1950) has reviewed some of the problems in identification.

808. Crepidula aculeata (Gmelin, 1791) (Synonyms: **Calyptraea echinus** and **hystrix** Broderip, 1834; **? C. bilobata** Carpenter, 1857). The spiny slip-per has the apex spirally curved to one side and the surface of the shell covered with spiral rows of spines, with much variation in coarseness or fineness. In some specimens there is a spiral ray or two of brown, and the interior may be spotted with brown. The septum or deck is white, slightly notched at the sides and in the center. Length, 15 mm; width, 12 mm; height, 5.5 mm. Very large specimens may be as much as 40 mm long. California through the Gulf and south to Val-

809

810

811

812

813

814

816

817

818

819

820

paraiso, Chile. The same species is reported from the Caribbean and even from the western Pacific. Lowe reported it on or under rocks or on dead clam shells intertidally.

809. Crepidula arenata (Broderip, 1834). The shell is yellowish white, dotted and streaked—especially near the margin—with reddish brown, the internal septum white, with a slight indentation near the center and an evident muscle impression at the right edge inside (as the shell is held with the septum upward). The species is close to *C. excavata*, from which it differs in having the septum less deeply sunken, and *C. onyx*, which has two curves in the edge of the septum. Length, 33 mm; width, 24 mm; height, 10 mm. Scammon's Lagoon, Baja California, through the Gulf and south to Chile, on beaches but mostly offshore in depths to 100 m.

810. Crepidula excavata (Broderip, 1834). In color this is yellowish brown with dark spots on the earlier part and the rest striped with brown. The septum is shaped like that of *C. arenata* but is sunken well below the margin. One amusing test for determining specimens of *C. excavata* is to see that they will hang on the tip of a lead pencil inserted into the little cavity between the base of the septum and the outer margin of the shell. Length, 33 mm; width, 24 mm; height, 12 mm. Baja California throughout the Gulf and south to Panama, on other shells, especially *Polinices*.

811. Crepidula incurva (Broderip, 1834). A small shell, dark brown exteriorly and nearly black within, except for the white septum; the shell is ribbed, even when attached to a smooth surface. Length, 19 mm; width, 13 mm; height, 10 mm. Baja California through the Gulf and south to Paita, Peru, on shells, intertidally and offshore to 18 m.

812. Crepidula lessonii (Broderip, 1834). The ovate shell is white streaked with reddish-brown lines, constructed of frilled concentric laminae; an occasional specimen may be entirely white or brownish within. Length, 23 mm; width, 15 mm; height, 7 mm. Head of the Gulf of California to Paita, Peru, intertidally, under stones.

813. Crepidula marginalis (Broderip, 1834). The low, oval shell is yellowish brown, concentrically striped. The interior is brownish black, with a white septum. Actually, this may be only a color form of *C. onyx*. Length, 28 mm; width, 21 mm; height, 8 mm. El Salvador to Ecuador. Lowe reported finding these limpets attached to the valves of the mussel *Mytella guyanensis*, on mud flats.

814. Crepidula onyx Sowerby, 1824 (Synonyms: **C. cerithicola** C. B. Adams, 1852; **C. lirata** Reeve, 1859). The shell is rather thick and rough, fresh specimens having a shaggy periostracum; the color is flesh to brown, with a white septum. Length, 38 mm; width, 25 mm; height, 15 mm. Southern California to Chile, intertidally on dead shells. A worker in South America, J. J. Parodiz, in an account of South American Crepidulas published in 1939, has reported *C. onyx* as common in the western South Atlantic between latitudes 42° and 47° South, both intertidally and offshore to depths of 70 m.

815. Crepidula perforans (Valenciennes, 1846) (Synonym: **C. explanata** Gould, 1853). This well-known Californian form is reported as occurring as far south as Panama, living in abandoned pholad holes. The adult shell, compressed at the sides to fit the hole, assumes a characteristic concave form. The Panamic form, however, is probably a situs or environmental expression of *C. striolata* and not the true *C. perforans*.

816. Crepidula rostrata C. B. Adams, 1852. This is more strongly and roughly ribbed than *C. incurva*, which it somewhat resembles, and the beak rises straight over the margin instead of overhanging it in a dunce-cap point. Length, 10 mm; width, 6 mm; height, 5 mm. Panama.

817. Crepidula striolata Menke, 1851 (Synonyms: **Calyptraea squama** Broderip, 1834 [not Deshayes, 1830]; **Crepidula nivea** C. B. Adams, 1852; **C. nebulata** Mabille, 1895). This has been confused with the Californian *C. perforans* under the name of *C. nivea*. Unfortunately, it seems that the best-known name for this form was proposed later than two others, the first a homonym and the second not accompanied by an illustration. Menke's description, though brief, seems to fit one of the variants of the species, and therefore the name *C. striolata* becomes the earliest valid name. The shell is low, ovate, white, some specimens showing a few radial riblets on the dorsal surface. Others are smooth. Most specimens are white, with a shaggy yellow periostracum and with a flush of brown near the apex, which is marginal. Length, 31 mm; width, 21 mm; height, 8 mm. The Gulf of California to Panama, attached to the underside of stones.

818. Crepidula uncata Menke, 1847. A relative of the Californian *C. adunca*, this is lower and lighter-colored, with flecks of brown on a terra-cotta shell and with the interior a mottled light purplish brown except for the deck, which is a transparent white. Length, 10 mm; width, 8 mm; height, 7 mm. The species seems to be restricted to the Gulf of California area.

Genus CREPIPATELLA LESSON, 1830

Internal deck detached along a part of one side.

819. Crepipatella dorsata (Broderip, 1834). The white shell has a few irregular radial rays of brownish color on the outer edges. Although reported sparingly from north of the equator, it seems rather to be a southern species, the type locality being Ecuador. Diameter, 15 mm. Tres Marias Islands, Mexico, to Chile.

820. Crepipatella lingulata Gould, 1848. The nearly circular shell is white, with a somewhat glossy texture. It is well known in the northern part of its range, but has been confused by some workers with the southern *C. dorsata*. Diameter, 20 mm; height, 5 mm. Alaska to southern Mexico and perhaps as far south as Panama.

Genus CRUCIBULUM SCHUMACHER, 1817

The internal shelf or septum is concave and principally attached near the apex of the shell, standing more or less free in later stages of growth—hence the popular name "cup-and-saucer limpet." Because of the sessile habits of these snails, the shells may be distorted by growth in cramped quarters or on irregular surfaces; sculpture and shape may be variable. Two Panamic subgenera.

Subgenus CRUCIBULUM, s. s.

Septum attached along a narrow part of the right side, the edge or margin mostly free.

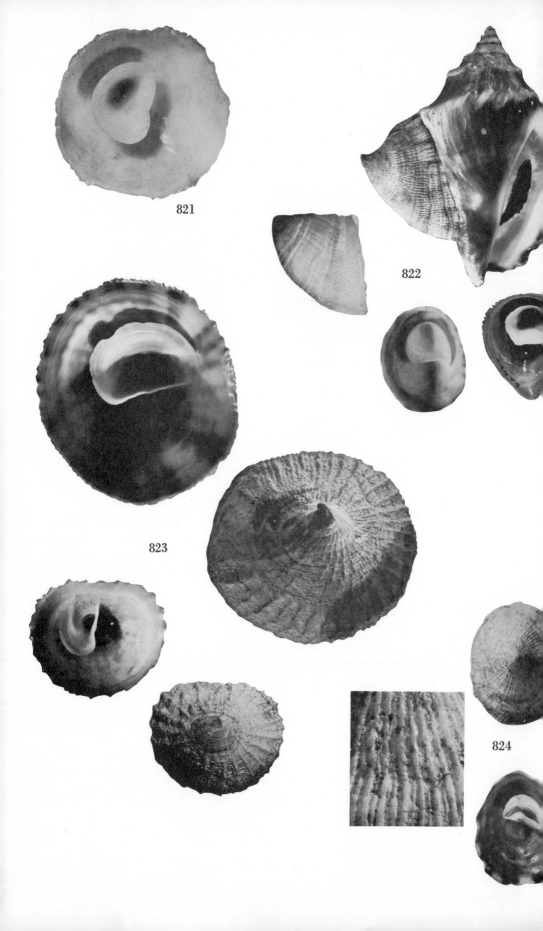

821

822

823

824

821. Crucibulum (Crucibulum) cyclopium Berry, 1969. One of the largest-sized members of the genus, with heavy radial ribs; internal septum large, free except at apex, with a ridge on its outside right margin, like a narrow seam. Resembling *C. umbrella* but larger, inner surface less lustrous. Diameter, 66 mm; height, 22 mm. Magdalena Bay, Baja California, Mexico, to Playas del Coco, Costa Rica (type locality, Manzanillo, Mexico); intertidally and to 6 m, on undersides of boulders.

822. Crucibulum (Crucibulum) lignarium (Broderip, 1834) (Synonyms: ? **Calyptraea tenuis** Broderip, 1834; **? Calyptraea gemmacea** Valenciennes; **Calyptraea trigonalis** Adams & Reeve, 1850). The outline varies from trigonal to almost rounded, depending on the shape of the base to which the animal was attached (often other shells); the color is tawny, with radiating lines of darker color and fine radial riblets that are slightly coarser in front, the internal septum well developed. Diameter, 25 mm; height, 12 mm. Intertidally, on stones or dead shells, from the Gulf of California to Ecuador. The species has been confused with *C. quiriquinae* (Lesson, 1830), a form that apparently occurs only south of Peru.

823. Crucibulum (Crucibulum) monticulus Berry, 1969. The medium-sized shell is rather thin, whitish stained with brown, with widely spaced narrow radial ribs that give the margin a serrate appearance; cup a little flattened in front and attached on the right side rather more widely than in others of the subgenus. Diameter, 29 mm; height, 17 mm. An offshore form, from the Mazatlán area to the Gulf of Tehuantepec, Mexico, in depths of 64 to 183 m.

824. Crucibulum (Crucibulum) personatum Keen, 1958 (Synonym: **Calyptraea radiata** Broderip, 1834 [not Deshayes, 1830]). Although superficially resembling *C. spinosum*, this is easily separable on the basis of sculpture, which is of fine radial threads lacking spines, and of the placement of the septum, partially flattened and well removed from the margin. The shell is white, with a subcentral apex that is not recurved. Diameter, 25 mm; height, 12 mm. Guaymas, Mexico, to Panama.

825. Crucibulum (Crucibulum) scutellatum (Wood, 1828) (Synonyms: **Calyptraea imbricata** Sowerby, 1824 [not Fischer, 1807]; **Calyptraea rugosa** Lesson, 1830 [not Borson, 1825]; **Calyptraea maculata** Broderip, 1834; **Crucibulum corrugatum** Gould & Carpenter, 1857; **C. imbricatum broderipii** Carpenter, 1857). Although variable, like others of the family, this can usually be recognized by the brownish color and the coarse, scaly, radial ribs, which may be somewhat latticed by concentric sculpture. The internal cup is attached along one side as well as at the apex. Major diameter, 50 mm; height, 16 mm. Cedros Island, Baja California, through the Gulf and southward to Ecuador, on stones or other shells on intertidal mud flats to offshore depths of 27 m.

826. Crucibulum (Crucibulum) spinosum (Sowerby, 1824) (Synonyms: **Patella peziza** Wood, 1828; **Calyptraea tubifera** Lesson, 1830; **Calyptraea hispida** Broderip, 1834). The shell is yellowish white, with or without curved purplish to brownish rays. The apex is recurved, its tip to one side, like a little stocking-tailed cap. Oblique wrinkles roughen the surface, overlain by more or less well-developed rows of spines, some even tubular. The internal cup is white and attached along most of one side. Diameter, 20 mm; height, 7.5 mm. California and southward through the Gulf to Tomé, Chile, on stones and dead shells, intertidally, and offshore in depths to 55 m.

825

826

827

828

827. Crucibulum (Crucibulum) umbrella (Deshayes, 1830) (Synonym: **Calyptraea rudis** Broderip, 1834). The principal point of difference between this and *C. scutellatum* is that the internal cup is attached only at the apex and stands free in the middle of the shell. Diameter, 55 to 60 mm; height, 15 to 20 mm. Intertidally, Gulf of California to Panama.

Subgenus DISPOTAEA SAY, 1824

Cup attached by the right side to the wall of the shell, the area of attachment varying somewhat; cup generally flattened. The type species is a fossil form from the Miocene of the eastern United States. Authors are not in complete agreement on the identification of this species; thus the status of the subgenus remains somewhat insecure. It is utilized here because there does seem to be a distinctive group of species with a flattened cup.

828. Crucibulum (Dispotaea) concameratum Reeve, 1859 (Synonym: ? **C. castellum** Berry, 1963). Described from an unknown locality, this had not been recognized until recent years, when tropical West American offshore material became more readily available. The white shell is beautifully latticed by radial and concentric sculpture, and the spaces between successive layers of growth tend to be developed as deep pits. The cup is flattened, as in *C. serratum*. Length, 11 mm; width, 10 mm; height, 7 mm. Fairly common in dredgings throughout the southern end of the Gulf of California and south to Acapulco, Mexico, in depths to 7 to 90 m.

829. Crucibulum (Dispotaea) pectinatum Carpenter, 1856. Fresh shells are bright orange-brown, which tends to fade to a dull grayish brown. The radial ribs extend beyond the margin of the thin shell; successive periods of growth are marked by deep pits, so that shells may come to have a pagoda-like profile. The cup is flattened, as in *C. concameratum*. Length of a large specimen, 25 mm; width, 23 mm; height, 14 mm. Offshore, in depths to 27 m. Mazatlán, Mexico, to Peru (type locality).

830. Crucibulum (Dispotaea) serratum (Broderip, 1834). This is a thinner-shelled and lower form than the others, with a white shell marked by one or more chestnut stripes, the apex sharp and toward the posterior end. The inner cup is distinctive, being flattened down against the outer shell instead of being rounded and attached only along one side. About 20 mm in maximum diameter, with a height of about 10 mm. Corinto, Nicaragua, to Muerte Island, Ecuador.

831. Crucibulum (Dispotaea) subactum Berry, 1963. Resembling *C. lignarium* in miniature, this has relatively coarser riblets with somewhat nodose sculpture on them and is grayish white in color. The cup is slightly flattened, not as much so as in *C. concameratum*, placed nearer the margin than in other species. The attachment area of the cup is wider. Length, 11 mm; width, 9 mm; height, 6.5 mm. Off Sinaloa, Mexico, in depths of 46 to 64 m.

Family CAPULIDAE

Sessile forms, attached to other invertebrates, especially to Pectens; cap-shaped, usually with a periostracum but lacking any internal shelf or deck. Three Panamic genera.

Genus CAPULUS MONTFORT, 1810

Cap-shaped, with a pointed apex and weak to obsolete radial sculpture.

829

830

831

832

833

834

835

832. Capulus sericeus J. & R. Burch, 1961. The oval shell has a slightly spiral apex twisted to the left; surface with fine radiating lines and a velvety periostracum extending beyond the margin. Attached near the byssal opening of *Pecten sericeus*. Length, 15 mm; width, 12 mm; height, 6 mm. Gulf of California (type locality, off Guaymas, Mexico), in depths to 180 m.

833. Capulus ungaricoides (Orbigny, 1841). Though shaped like *C. sericeus*, this species has a higher, narrower apex, more strongly incurved. The shell is very thin, with well-developed radial riblets and an adherent periostracum that still is present on the holotype, which is in the British Museum of Natural History in London, more than a century after the specimen was collected. Length, 14 mm; width, 13 mm. Type locality, Paita, Peru.

Genus CYCLOTHYCA STEARNS, 1890

Small, with one or two spiral coils, whorls rapidly enlarging.

834. Cyclothyca corrugata Stearns, 1890. The shell is white, mottled with gray, sculptured by irregular ribs. It seems not to have been collected again since Stearns described it. Diameter, 4.5 mm; height, 2.5 mm. West coast of Nicaragua. The shell has a striking resemblance to the New Zealand genus *Zelippistes* Finlay, 1927, as illustrated and discussed by Dell in 1964. There being no subsequent American records, one wonders if Stearns's material could have been incorrectly labeled with respect to locality.

Genus THYCA H. & A. ADAMS, 1854

White, without periostracum, apex a little spiral; septum weak. Parasitic on seastars. No species of subgenus *Thyca, s. s.*, occurs in the Panamic province.

Subgenus BESSOMIA BERRY, 1959

Apex submarginal, larval coil of three whorls; septal shelf sculptured.

835. Thyca (Bessomia) callista Berry, 1959. The body whorl of the shell is not spirally twisted, and the larval coils are embedded; the last whorl is nodulose, with crenulations of the margin in front. Length, 7.4 mm; diameter, 5.4 mm; height, 3.1 mm. On seastars, Bahía San Carlos, Sonora, Mexico.

Family TRICHOTROPIDIDAE

Spirally coiled, aperture triangular, with a small anterior canal.

Genus CERITHIODERMA CONRAD, 1860

Sculpture cancellate; spire of several whorls; aperture with a narrow umbilical chink. Although the genus is mainly comprised of Eocene species from the southeastern United States, some deepwater Pacific forms have been assigned to it.

836. Cerithioderma pacifica Dall, 1908. Panama Bay, in 2,320 m.

Family XENOPHORIDAE

Shell trochoid in outline, whorls flat-sided, tending to overhang at the periphery; texture mostly thin and brittle. Foot and operculum used as in the Strombidae for a leaping rather than gliding motion.

Genus XENOPHORA FISCHER DE WALDHEIM, 1807

The carrier shells have the unusual habit of picking up bits of debris from the sea floor and cementing them in place along the periphery of the growing whorl.

In time the whole shell comes to look like a pile of rubble, which may serve as a protective device. Collectors have long marveled at the selectiveness and, one might almost say, artistry of the otherwise dim-witted snail, for each individual, once starting with a given type of building material, tends to continue with that, whether it be stones, round bivalve shells like *Glycymeris*, or long snail shells like *Terebra*. Of course this may be coincidental, a result of the most common material available at the site where the carrier shell lives. Stripped of its decorations, the shell looks like a thin *Trochus*, but the inside is not nacreous and the horny operculum has the nucleus at one side, not central as in the Trochacea. There are only a few species of carrier shells, all tropical in distribution.

837. Xenophora robusta Verrill, 1870. The shell is white, the inside of the aperture brown, the operculum brown, horny, and pear-shaped. Diameter, 105 mm; height, 59 mm. The species seems to be confined to the west Mexican coast from Guaymas and La Paz in the Gulf of California to the Gulf of Tehuantepec, mostly offshore in depths of 45 to 50 m. A similar species in the Atlantic is *X. conchyliophora* (Born, 1780).

Superfamily **ATLANTACEA**
(**HETEROPODA**)

Pelagic mollusks, swimming with the foot uppermost; shells thin, nearly or entirely obsolete; soft parts large and in most forms not contained within the shell. Coiling of the shell involute or planorboid rather than helical. Several genera are widely distributed in tropical and temperate seas. Shells rarely drift ashore, and the heteropods must be collected offshore in tow nets. Only a few forms have been reported in the Panamic province as yet, but there are a number in adjacent areas that may be expected to be found with more intensive collecting. References: Tesch (1949); Coan (1965). Lists and needed information on this group were assembled and organized by Dr. Eugene V. Coan, whose assistance is gratefully acknowledged. Some hitherto unpublished data on distribution were generously provided by Roger Seapy, graduate student at the University of California, Irvine, indicated here as "Seapy, *in litt.*"

In the following summary, localities are cited only for those species that have been definitely recorded in the Panamic province. Other species listed are those that presumably might be expected.

Family **ATLANTIDAE**

Shell present, laterally compressed, large enough to contain soft parts of the animal; operculum present. Perhaps three Panamic genera.

Genus **ATLANTA** LESUEUR, 1817

Shell dextral, keel narrow; aperture oval.

838. Atlanta brunnea Gray, 1850 (Synonym: **A. fusca** Souleyet, 1852). Panama Bay.

839. A. gaudichaudi Souleyet, 1852.

840. A. helicinoides Souleyet, 1852.

841. A. inclinata Souleyet, 1852.

842. A. inflata Souleyet, 1852.

843. A. lesueuri Souleyet, 1852.

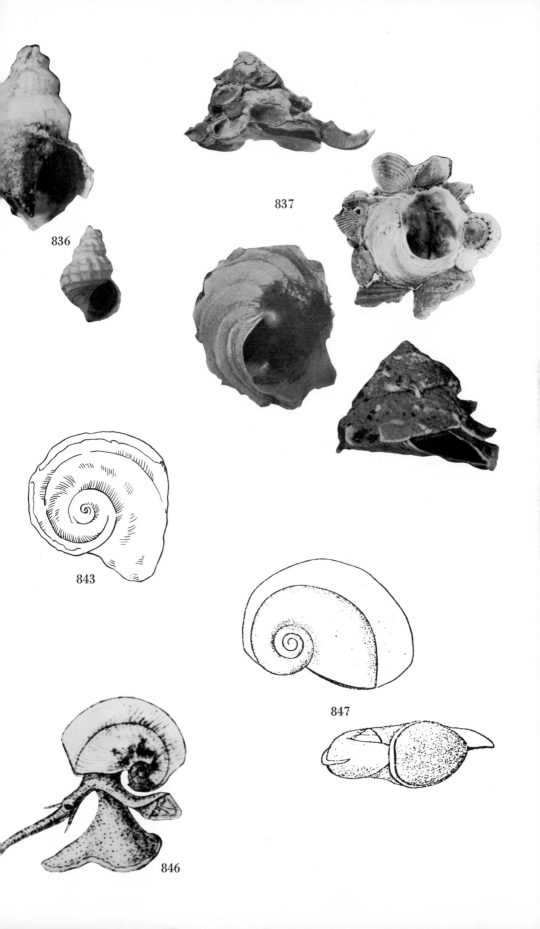

836

837

843

847

846

844. A. peroni Lesueur, 1817.

845. A. turriculata Orbigny, 1836.

Genus OXYGYRUS BENSON, 1835

Shell cartilaginous; keel relatively broad, continuing to periphery of aperture; aperture triangular.

846. Oxygyrus keraudrenii (Lesueur, 1817). Panama Bay.

Genus PROTATLANTA TESCH, 1908

Shell calcareous but transparent, with a few spiral lines on both sides; operculum round, with apical spiral nucleus.

847. Protatlanta souleyeti (Smith, 1888).

Family CARINARIIDAE

Shell small to medium-sized, covering part or most of the visceral nucleus; swim-fin large; no operculum. Three Panamic genera.

Genus CARINARIA LAMARCK, 1801

Shell covering visceral nucleus; body cylindrical; swim-fin with a sucker on posterior edge.

848. Carinaria cithara Benson, 1835.

849. C. cristata (Linnaeus, 1766).

850. C. galea Benson, 1835.

851. C. japonica Okutani, 1955 (Synonym: **? C. latidens** Dall, 1919, *nomen dubium*).
  See Color Plate XV.

Genus CARDIAPODA ORBIGNY, 1835

Shell minute, coiled, with winglike flanges at the edges of the aperture; sucker reduced in size.

852. Cardiapoda placenta (Lesson, 1830). Panama Bay.

Genus PTEROSOMA LESSON, 1827

Shell small, flattened; body of animal also flattened, with dorsal crest and with an attenuated tail.

853. Pterosoma planum (Lesson, 1827). Panama Bay.

Family PTEROTRACHEIDAE

No shell; body cylindrical, visceral nucleus small, unstalked, in a groove; swim-fin anterior to visceral nucleus. Two Panamic genera.

Genus PTEROTRACHEA FORSKÅL, 1775

Pre-oral tentacles absent; a short tail present; a sucker in the middle of the swim-fin margin.

854. Pterotrachea coronata Forskål, 1775. Gulf of California (Seapy, *in litt.*).

855. P. hippocampus Philippi, 1836.

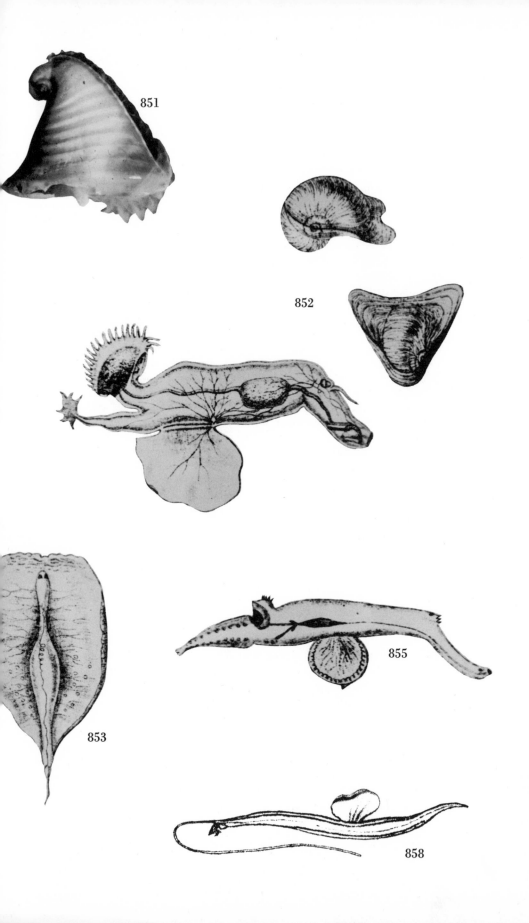

856. P. minuta Bonnevie, 1920.

857. P. scutata Gegenbaur, 1855.

Genus FIROLOIDA LESUEUR, 1817

Pre-oral tentacles present, tail much reduced; sucker on anterior border of swim-fin.

858. Firoloida demarestia Lesueur, 1817 (Synonym: **F. desmaresti** of authors). Gulf of California (Seapy, *in litt.*).

Superfamily NATICACEA

The moon snails are one of the few exceptions to the rule that the gastropods with rounded apertures are herbivorous, for these snails do a better job of drilling holes in clams than the so-called "drill shells." The latter, in fact, do more chipping than drilling, but the moon snails can produce as neat a beveled hole—with no other tool than the radula—as a skilled workman could with a battery of steel bits. The moon snail plows about just beneath the surface of the sea floor hunting its prey—clam or other snail—which, when found, is enveloped in the massive foot (a foot too large in many species to be withdrawn into the shell), and the rasping file of the radula is applied with a circular motion to some convenient part of the shell. Early workers assumed that the drilling process may be supplemented by the action of free acid which the animal was presumed to secrete in special glands, but research has revealed no such acid, and evidence has been found for the purely mechanical action of the radula. Drill holes made in soft material have shown the actual marks of the radular teeth. It is possible that, as in the muricid snails, enzymatic action may dissolve the organic cement of the prey's shell, so that the radula can more effectively scrape away the loosened grains of shell material. Occasionally a victim may escape the grip of the predator, or the snail may weary of the task before the hole has reached a vulnerable part of the prey. Now and then a shell is found with not just one hole but two or three or even a row of them, the last of which presumably hit the mark.

Not only are the feeding habits of the moon snails unusual, but the egg capsules are also. The female when ready to spawn deposits the eggs in a collar-shaped structure made of sand grains cemented together with mucus, which is molded into its peculiar shape over the margin of the aperture.

Family NATICIDAE

With a short spire and large body whorl; aperture semicircular to oval, angular above, broadly rounded below, with more or less callus on the inner lip; operculum horny to calcareous, paucispiral. The key is to genera and subgenera.

1. Operculum calcareous, with spiral grooves; umbilical area with a funicle
   or ridge of callus at or below middle of inner lip.................... 2
   Operculum horny or absent; umbilicus wanting or with callus entering
   from upper part of inner lip...................................... 4
2. Shell with straplike axial ribs intersected by fine spiral striae...... *Stigmaulax*
   Shell smooth or axially wrinkled near suture........................ 3
3. Umbilicus narrowed, the funicle reduced to a narrow twisted vertical col-
   umn along columellar lip......................................*Lunaia*
   Umbilicus open, with a central ridge or funicle ending in a callus lobe
   near middle of inner lip......................................*Natica, s. s.*

4. Umbilical area filled with callus...............................*Neverita*
   Umbilical area not filled with callus.............................. 5
5. Shell ear-shaped, spire low.....................................*Sinum*
   Shell not ear-shaped, spire pointed............................... 6
6. Surface with fine spiral grooves.............................*Eunaticina*
   Surface smooth except for growth lines........................*Polinices*

Genus NATICA SCOPOLI, 1777

The shells of the Naticas are sturdy, porcelaneous, mostly smooth or polished, with bold color patterns in shades of brown, orange, or gray. The funicle in the umbilicus is a distinguishing feature—a riblike structure that is built up from the little bulge of callus on the outer side of the inner lip just below the middle of the lip. Specialists divide the genus into a number of subgenera, some of which, like *Stigmaulax*, may even be treated as genera in order that smaller subdivisions may be recognized. A conservative classification (three subgenera) is adopted here, however, pending a much-needed revision of the family by some ambitious student.

Subgenus NATICA, *s. s.*

Natica in the restricted sense is a tropical group, but there is a closely related subgenus, *Cryptonatica* Dall, 1892 (or *Tectonatica* Sacco, 1890, as some workers prefer to call it), in the Arctic area. The shell is relatively smooth, not ribbed, with a calcareous operculum that is ovate and flat.

859. Natica (Natica) brunneolinea McLean, 1970. A thin shell, yellowish above, whitish below, with fine axial grooves below the suture that fade out above the periphery, which is marked by irregular vertical brown penciled lines. Operculum white, with two deep grooves at the outer edge. Height, 11 mm; diameter, 10.5 mm (holotype); larger specimens may be 46 mm in height, 42 mm in diameter. Galápagos Islands.

860. Natica (Natica) caneloensis Hertlein & Strong, 1955. On an ashy brown ground there are two rows of chestnut spots, one at the shoulder and one at about the middle of the whorl with, in some specimens, one or two fainter rows toward the base of the shell. The operculum has four ribs and five grooves, the widest being second from the edge. It is a thinner shell than the more common *N. chemnitzii* and differs from *N. grayi* in the higher spire and more impressed sutures. Height, 27 mm; diameter, 24 mm. Mazatlán, Mexico, to Manta, Ecuador.

861. Natica (Natica) chemnitzii Pfeiffer, 1840 (Synonyms: "**N. marochiensis** Gmelin, 1791," of authors; **N. pritchardi** Forbes, 1852). Most abundant and most variable as to color markings of the Panamic moon snails, this has been considered inseparable from the widespread Indo-Pacific species *N. marochiensis*. Modern students take the view that it is better to give separate names to forms in isolated geographic areas until it can be proved that they are identical with those of other regions. Lumping together what may be only superficially similar species under a single ill-defined name may obscure significant distinctions. Hence, the name *N. chemnitzii* is being adopted for the Panamic moon snails that closely resemble and seem to have the same range of variations as the Indo-Pacific, Caribbean, and west African forms. The shell is grayish blue or grayish yellow, with four to five bands of arrow-shaped markings that are either brownish or nearly white, and the interior is dark brown, banded. The operculum is white and smooth

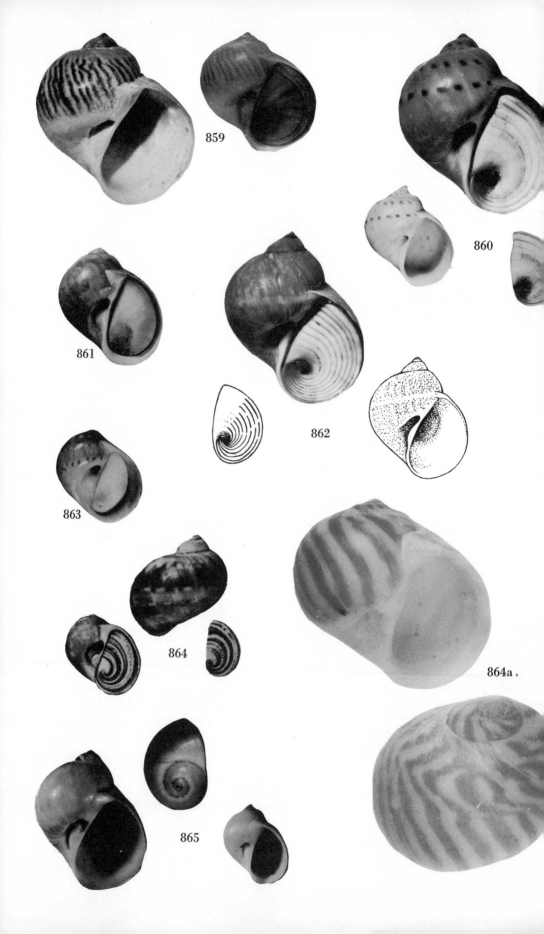

859

860

861

862

863

864

864a

865

except for a roughened area near the nucleus. Height, 33 mm; diameter, 31 mm. Magdalena Bay, Baja California, through the Gulf and south to northern Peru; intertidally on mud flats.

See Color Plate XIV, Figs. 3 and 6.

862. Natica (Natica) colima Strong & Hertlein, 1937. A thin, pale brown shell with two lighter spiral bands, one just below the suture, the other at the periphery. The operculum has eight deep, square grooves. Height, 21 mm; diameter, 18 mm. The type locality is off the coast of Colima, west Mexico, in 95 m.

863. Natica (Natica) grayi Philippi, 1852 (Synonyms: **N. depressa** Gray, 1839 [not Sowerby, 1812]; **N. catenata** Philippi, 1853). In color, this is light brown, with four whitish concentric bands marked with dark brown spots. The operculum is white, nearly flat, with a single narrow groove near the outer edge. Long known to collectors as *N. catenata*. Height, 18 mm; diameter, 15 mm. Magdalena Bay, Baja California, through the Gulf and south to the Galápagos Islands and Manta, Ecuador, mostly offshore in depths to 37 m.

864. Natica (Natica) idiopoma Pilsbry & Lowe, 1932. The low-spired shell is dull brown, with four spiral bands of darker brown alternating with whitish spots, lighter around the umbilicus. The operculum has three principal spiral ribs, but the outer one is doubled and the two parts are bridged by a thin layer of radial rods. Because of the unusual operculum, this species was made type of *Glyphepithema* Rehder, 1943, but there seem to be transitional forms that relate it to *Natica, s. s.* Height, 12 mm; diameter, 11 mm. San Juan del Sur, Nicaragua (type locality); Galápagos Islands.

864a. Natica inexpectans Olsson, 1971. Somewhat resembling *N. brunneolinea* in color marking but with broader stripes that may break up into zigzag patterns on the back of the body whorl; also, the spire is relatively lower. Height, 14 mm; diameter, 15.8 mm. Gulf of Panama, dredged.

865. Natica (Natica) othello Dall, 1908. The spire is somewhat pointed, as in *N. colima*, but the coloring is a more uniform brown except for a lighter zone below the periphery. The brown periostracum tends to flake off readily. The umbilicus is nearly filled by the large funicle. On the operculum there is a single groove near the margin. Height, 25 mm; diameter, 21 mm. An offshore form, ranging from the Gulf of Tehuantepec to Panama Bay (type locality) in depths to 85 m.

866. Natica (Natica) scethra Dall, 1908. The shell is brown, with three rather vague paler bands. On the white operculum are two grooves near the outer edge. Height, 17 mm; diameter, 16 mm. Panama Bay, 280 m.

867. Natica (Natica) sigillata McLean, 1970. The globose shell is narrowly umbilicate; it has a thin yellowish periostracum and is smooth except for growth lines; umbilicus lacking a funicle. The color is chestnut-brown with tent-shaped markings of white; operculum with four raised ridges at the outer edge, the two inner ridges broader. Height, 10 to 19 mm; diameter, 10 to 19 mm. Gulf of California, off Carmen Island, to Galápagos Islands, in depths to 90 m.

868. Natica (Natica) unifasciata Lamarck, 1822. Although this form would have been unrecognizable from Lamarck's original description, a figure by Delessert in 1841 of a specimen from Lamarck's collection has helped authors to identify it as West American. The shell is dark gray or even almost black, with a single

866

867

868

870

869

871

872

874

873

875

band of yellowish white just below the suture; it is relatively large, solid in texture. Height, 40 mm; diameter, 35 mm. Costa Rica to northern Peru.

Subgenus LUNAIA BERRY, 1964

Thin, inflated, nearly smooth; umbilicus narrow, funicle reduced to a narrow vertical ridge of callus along inner lip edge; operculum calcareous, with several marginal grooves.

869. Natica (Lunaia) lunaris (Berry, 1964). The shell has a somewhat horny texture, light yellowish brown in color, with a paler zone around the umbilicus and columella, bordered by a darker zone of brown. The umbilicus is remarkably narrow, and the inner lip somewhat overhangs it. The calcareous operculum is smooth except for a few incised grooves near the outer margin. Height, 28 mm; diameter, 26 mm. Dredged off Sonora, Mexico, in depths to 45 m.

Subgenus STIGMAULAX MÖRCH, 1852

The axial ribs are much more pronounced than in *Natica, s. s.,* and they may be intersected by fine spiral sculpture, especially in the grooves.

870. Natica (Stigmaulax) broderipiana Récluz, 1844 (Synonym: **N. iostoma** Menke, 1847). Orange to yellowish brown in color, this has three white bands checkered with dark brown spots, and the inside of the aperture may be tinted with violet. The operculum has one coarse medial rib and four to five small outer ribs. Height, 27 mm; diameter, 25 mm. Cedros Island, Baja California, through the Gulf and south to Lobitos, Peru, mostly offshore in depths to 55 m.

871. Natica (Stigmaulax) elenae Récluz, 1844 (Synonym: **N. excavata** Carpenter, 1856). The white shell is marked with reddish-brown close-set but irregular lines running from suture to base. The operculum has the middle ridge grooved near the end, and the smaller outer spirals are rather coarse and irregular. Height, 32 mm; diameter, 32 mm. Magdalena Bay, Baja California, to Santa Elena, Ecuador, mostly offshore in depths to 37 m.

Genus EUNATICINA FISCHER, 1885

These shells resemble *Sinum* but are higher-spired and less ear-shaped. The spiral sculpture varies in degree of expression.

872. Eunaticina heimi Jordan in Hertlein, 1934. The white shell is rather coarsely spirally grooved. Height, 10 mm; diameter, 7 mm. Although described as a Pleistocene fossil, it has been found living from southern Baja California, Mexico, to Ecuador. *Sigatica semisulcata* (Gray, 1839), from the Caribbean, has similar sculpture but is lower-spired, with a wider umbilicus. Whether *E. heimi* should be referred to *Sigatica* Meyer & Aldrich, 1886, the type of which is an Eocene fossil from the eastern United States, or to *Eunaticina*, the type species of which is a Recent form from the western Pacific, is a matter for future research.

Genus POLINICES MONTFORT, 1810

The horny operculum that distinguishes the group from *Natica* is paucispiral, thin, and ovate, with the nucleus at the larger end. In another contrast to *Natica*, the umbilicus does not have a true funicle, for the major part of the callus is at the upper edge of the inner lip. Several minor subdivisions recognized by specialists are grouped here in two subgenera and *Polinices, s. l.*

Subgenus **POLINICES**, *s. s.*

With the umbilical area partially open below.

873. Polinices (Polinices) bifasciatus (Griffith & Pidgeon, 1834). The shell is fairly solid, light brown to tan-colored, lighter near the suture and around the umbilicus, with two narrow white bands. The umbilicus is partly covered above by the brown columellar callus. Height, 40 mm; diameter, 34 mm. Gulf of California to Panama, on mud flats between tides.

See Color Plate XIV.

874. Polinices (Polinices) caprae (Philippi, 1852) (Synonym: **P. clarki** M. Smith, 1950). A thin white shell, with clouded markings of brown, especially along the inner lip. Height, 29 mm; diameter, 26 mm. Mazatlán, Mexico, to Panama, intertidally.

875. Polinices (Polinices) galapagosus (Récluz, 1844). Lowest-spired among the eastern Pacific species, this is nearly globular, brown, especially inside the aperture, with a lighter zone below the suture. Height, 24 mm; diameter, 22 mm. Galápagos Islands.

876. Polinices (Polinices) helicoides (Gray, 1825) (Synonyms: **Natica patula** Sowerby, 1824 [not J. Sowerby, 1822]; **N. glauca** Lesson, 1830; **N. bonplandi** Valenciennes, 1832). This species is much better known under the name of *P. glauca*, often incorrectly attributed to Von Humboldt. The thin shell is low-spired, the last whorl relatively large, buff to light brown in color, the surface somewhat striate. The umbilicus is partly overhung by a tongue of callus. Height, 23 mm; diameter, 48 mm. Magdalena Bay, Baja California, through the Gulf and south to Callao, Peru, mostly beyond the low-tide line and offshore in depths to 37 m. This has been made the type of the subgenus *Hypterita* Woodring, 1957.

877. Polinices (Polinices) intemeratus (Philippi, 1853) (Synonym: **Natica alabaster** Reeve, 1855). This is closely related to *P. uber* but differs by being a little larger and by having a lower, less pointed spire. The shell is white, with a thick callus pad that resembles the funicle of *Natica* but is connected to the callus from the upper wall of the aperture. Height, 40 mm; diameter, 37 mm. Cedros Island, Baja California, to Panama, in depths to 333 m.

878. Polinices (Polinices) limi Pilsbry, 1931. The most globose of the species related to *P. uber*. Pilsbry named it as a subspecies of the Peruvian *P. rapulum* (Reeve, 1855). The aperture is relatively small and the callus area narrow. The shell is smooth, pale brown, with a white aperture. Height, 39 mm; diameter, 31.5 mm. Panama, offshore in 4 to 13 m.

879. Polinices (Polinices) otis (Broderip & Sowerby, 1829) (Synonyms: **Natica salongonensis** Récluz, 1844; **N. fusca** Sowerby, 1883 [not Weinkauff, 1868]; **Ruma subfusca** Dall, 1919). The shell is conic-ovate, yellowish white or light gray, with a buffy-orange band at the suture. Below, it is chestnut brown in the open umbilicus and the aperture. Height, 40 mm; diameter, 38 mm. The Gulf of California to Ecuador and the Galápagos Islands. Rare on mud flats at extreme tides, commoner offshore.

880. Polinices (Polinices) panamaensis (Récluz, 1844) (Synonym: **Natica amiculata** Philippi, 1849). Large and solid, with a rounded angulation below the suture, so that the profile of the shell appears nearly vertical-sided as contrasted to the rounding in most species. The umbilicus is small and the callus pad

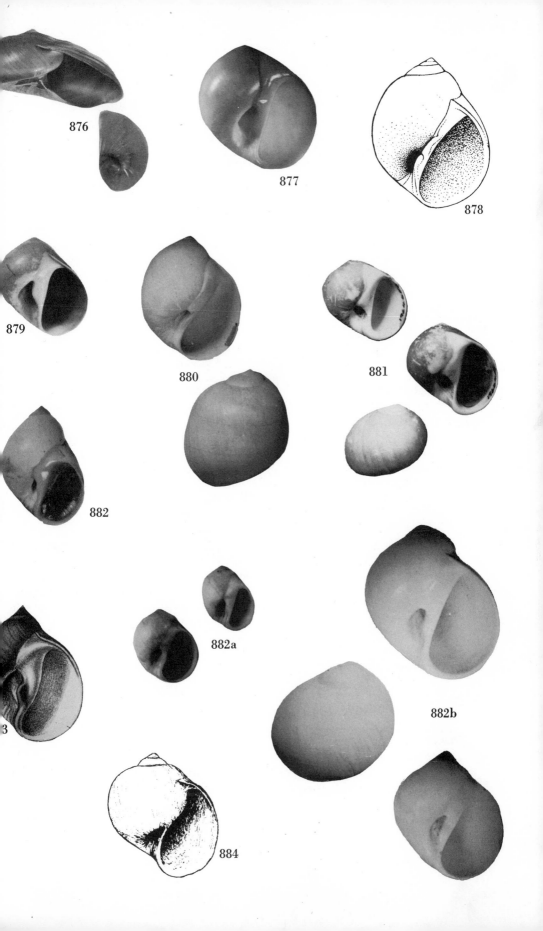

876

877

878

879

880

881

882

882a

882b

884

of the upper part of the inner lip very broad. The shell is pure white. Height, 44 to 54 mm; diameter, 40 to 47 mm. Panama to northern Peru, mostly offshore in depths to about 45 m.

881. Polinices (Polinices) ravidus (Souleyet, 1852). The low-spired shell is white under a rusty brown periostracum. The callus pad of the upper part of the inner lip is broad and heavy. Height, 32 mm; diameter, 30 mm. Panama to Paita, Peru (type locality).

882. Polinices (Polinices) uber (Valenciennes, 1832) (Synonyms: **Natica virginea** Récluz, 1850; **N. ovum** Menke, 1850). The degree of variability to admit within this species has been a matter of debate among conchologists. Present practice is to recognize *P. uber* as a central form among a number of related species of white *Polinices*—namely, *P. cora* (Orbigny, 1840), *P. dubius* (Récluz, 1844), and *P. rapulum* (Reeve, 1855), from western South America; *P. lacteus* (Guilding, 1834) and *P. uberina* (Orbigny, 1842); from the Caribbean area; and *P. intemeratus, P. limi*, and *P. unimaculatus*, of the Panamic province. In typical *P. uber* the spire is of moderate height and the umbilical opening is only slightly closed, with a somewhat reflected callus on the inner lip that does not form a funicle. An average specimen of *P. uber* measures: height, 19 mm; diameter, 16 mm. An exceptionally large shell measures 51 mm in height and 42 mm in diameter. The range is from Scammon's Lagoon, Baja California (possibly even San Diego, California), throughout the Gulf of California and south to Paita, Peru; not uncommon both intertidally and offshore in depths of 4 to 90 m.

883. Polinices (Polinices) unimaculatus (Reeve, 1855). Without the operculum this form might be mistaken for a *Natica, s. s.*, because of the development of a funicular ridge. However, this ridge is connected to a heavy callus pad in the upper wall of the aperture, which is characteristic of *Polinices*. The shell is light-colored outside, with a brown spot on the funicular ridge and a brown aperture. Height, 35 mm; diameter, 28 mm. Mazatlán to Panama.

Subgenus POLINICES, *s. l.*

Several species have been described from deep water in the eastern Pacific, assigned to the subgenus *Euspira* Agassiz, 1839. The type species of this group is from the Eocene of Europe. Other authors prefer to use *Lunatia* Gray, 1847, for the Recent species that are somewhat intermediate in callus development between *Polinices, s. s.*, and *P.* (*Neverita*). Precise placement must await careful revision of the family.

884. Polinices agujanus Dall, 1908. Panama to northern Peru, in 980 to 3,055 m.

885. P. crawfordianus Dall, 1908. Mazatlán, Mexico, to Aguja, Peru, in 1,000 to 1,860 m.

886. P. litorinus Dall, 1908. Galápagos Islands, in 1,485 m.

887. P. pardoanus Dall, 1908. Panama Bay to the Galápagos Islands, in 1,620 to 2,690 m.

Subgenus NEVERITA RISSO, 1826

Callus area extending from the apertural wall across the umbilicus, nearly or entirely filling it.

885

887

888

889

890

891

892

894

893

895

888. Polinices (Neverita) recluzianus (Deshayes, 1839). Shell chunky, light brown, with a whitish base. Height, 56 mm; diameter, 53 mm. More common in southern California but ranging throughout the Gulf of California to Tres Marias Islands, Mexico, on sandbars intertidally, where the snail preys vigorously on clams.

Genus SINUM RÖDING, 1798
(SIGARETUS LAMARCK, 1799)

A periostracum may or may not be present on the low, ear-shaped shell. The sculpture is of fine spiral grooves, and the color varies from brown to white. The animal is much too large to retract into the shell.

889. Sinum cymba (Menke, 1828) (Synonym: **Sigaretus concavus** Lamarck, 1822, of authors). Shell relatively large and heavy, with coarse spiral sculpture, the apex dark, the aperture chestnut-brown. The umbilicus is somewhat narrowed. Height, 44 mm; diameter, 52 mm. Ecuador to Chile. Lamarck's species proves to be west African in distribution.

890. Sinum debile (Gould, 1853) (Synonym: **S. pazianum** Dall, 1919). The shell is white, somewhat stained with rusty yellow. Height, 10 mm; diameter, 28 mm. Baja California to Panama, on tide flats at extreme low tide, and offshore to depths of 48 m. The form described as *S. pazianum* proves to be only the juvenile stage of *S. debile*.

891. Sinum grayi (Deshayes, 1843) (Synonym: **S. cortezi** J. & R. Burch, 1964). Although long confused with *S. concavum* (which was thought by authors to be the Peruvian form now known as *S. cymba* but which was actually a west African species), *S. grayi* has been recognized as a distinct species only in recent years. Compared with the South American *S. Cymba* it is smaller and brighter-colored, with finer sculpture. The apex and base are lighter-colored than the medial part of the shell. The spire is pointed, and the umbilicus is wide. Height, 39 mm; diameter, 44 mm. Most specimens have been taken in shrimp trawls, from the Guaymas area of Mexico to Panama Bay, depths probably ranging from 25 to 45 m.

892. Sinum noyesii Dall, 1903. The shell is mottled purple-brown above and white below, with two brown bands inside the aperture. The soft parts are mahogany-colored and envelop the shell when the animal is moving about. Major diameter, 36 mm; minor diameter, 26 mm; height, 10 mm. Nicaragua to Panama. Lowe (1932) reported it on mud flats among small stones and coarse gravel, in company with *Protothaca grata*, about three inches below the surface.

893. Sinum sanctijohannis (Pilsbry & Lowe, 1932). The shell is fairly solid and white throughout. The sculpture is of fine spiral lines rendered wavy by low radial folds. Height, 16.5 mm; major diameter, 27 mm; minor diameter, 23 mm. The type locality is San Juan del Sur, Nicaragua.

Superfamily LAMELLARIACEA

This superfamily and the next (Triviacea) form a transition between the Naticacea and the Cypraeacea. The shells of the Lamellariacea resemble those of *Sinum* in the Naticacea, but the soft parts and especially the larval development reveal differences. The Lamellariacea share with the Triviacea a type of larval shell called the echinospira larva, which occurs in no other gastropods except *Capulus* in the Calyptraeacea. The echinospira larval shell is two-walled: in effect, one shell inside another, the inner one fitting tightly over the back part of the tiny snail, the outer shell surrounding it and being ornamented with one or more keels. In the Lamellariacea the coiling of this larval shell is nautiloid or involute. The radula, nervous system, and manner of feeding link the Lamellariacea with the Triviacea. All members of both superfamilies are carnivorous, preying on tunicates or ascidians.

Family LAMELLARIIDAE

Shells few-whorled, with a periostracum; outer lip sharp-edged, never denticulate, inner lip smooth, slightly reflected.

Genus LAMELLARIA MONTAGU, 1815

The shell is internal and in most species concealed or enclosed by the living tissues of the animal. Hence, it is thin and white, with little sculptural ornamentation. The living animal resembles a sea slug or nudibranch, although very different anatomically. The favorite or perhaps the sole food of the lamellarians is tunicates or compound ascidians, the colonies of which many of them come to resemble so closely that they are effectively concealed. Authors have reported that lamellarians also may feed on sponges, but this characteristic remains to be demonstrated in laboratory work.

894. Lamellaria diegoensis Dall, 1885. Polished, white, and translucent, the shell has a malleated surface texture, as though it has been beaten with a tiny hammer. The soft parts are described as dark, even velvety black. There are records in the literature of red animals, perhaps a confusion with *Pleurobranchus digueti*, although it is possible that, in this species as in some others, color is not constant. Diameter, 16 mm. Southern California to Sonora, Mexico, apparently living on ascidians offshore.

895. Lamellaria inflata (C. B. Adams, 1852). The shell is rather solid for the genus, partly or entirely white, the spiral columella showing within to the apex. Diameter, 11 mm. Puertecitos and La Paz, Gulf of California, to Panama (type locality). A specimen collected by Gale Sphon at La Paz at a depth of 18 m had a light yellow body marked with brown.

896. Lamellaria orbiculata Dall, 1871. Shell of two and a half whorls, inflated, oblique, spire somewhat elevated, columella rather thick. Length, 9 mm; width, 8 mm; height, 6 mm. Alaska to the Gulf of California. Mrs. G. E. MacGinitie reported specimens at Puerto Peñasco, Sonora, Mexico, under large flat rocks, with encrusting sponges and tunicates that the whitish animal closely resembled.

897. Lamellaria perspicua (Linnaeus, 1758) (Synonym: ? **L. p. mopsicolor** E. Marcus, 1958). The shell is white, thin, its growth lines distinct; smooth and shiny inside, with three whorls, the apex only slightly prominent; columella open, revealing the apex from below. Length, males about 15 mm, females 18 mm; width, males 10 mm, females 19 mm; height, males 9 mm, females 8 mm. The species, which has been reported worldwide except in polar regions, was recorded at Puerto Peñasco, Sonora, Mexico, by Marcus and Marcus (1967). They report the animals to be translucent, whitish, the male with a pinkish hue, the female with a brownish hue, glandular inclusions on the skin opaque white in the male, glassy in the female. The animal of the specimen identified by them as *L. p. mopsicolor* was dark bluish gray, with dark cutaneous bosses surrounded by lighter rings.

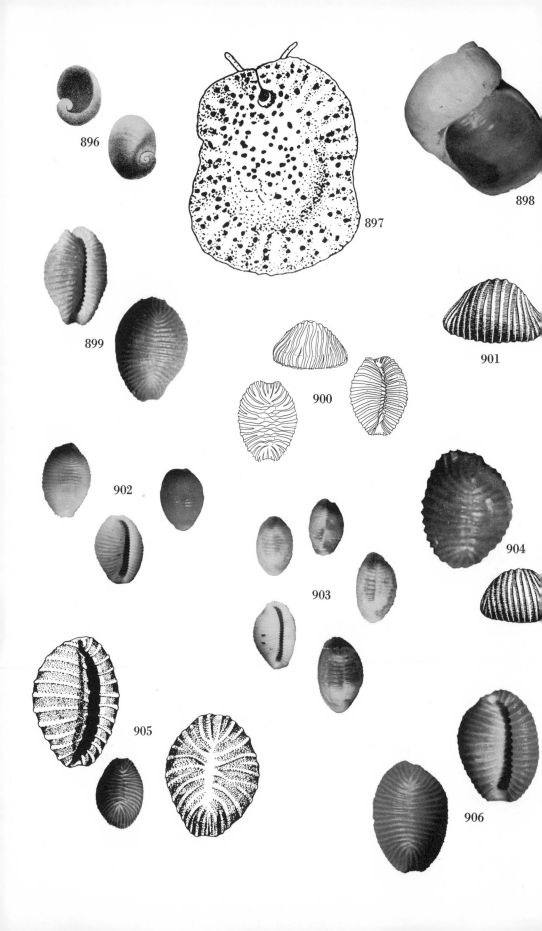

898. Lamellaria sharonae Willett, 1939. The shell is naticoid in form, relatively higher than in the other species, thin, imperforate, white under a pinkish periostracum that later appears greenish; columella brown, set off by a groove; surface with very fine spiral striations. Height, 7.5 mm; diameter, 5.5 mm. The soft parts, which completely conceal the shell, are purplish brown, lighter below, with six low radiating ridges marked by darker spots. Southern California to the northern end of the Gulf of California.

See Color Plate XVI. Cate & Cate (1962) discuss Willett's type specimens.

Superfamily TRIVIACEA

For many years combined with the Cypraeacea because of the shape of the shells, this group was later transferred to the Lamellariacea because of the similarity of larval development and mode of feeding. More recently, Schilder (1966) has recommended that it be accorded separate status because its differences from the Lamellariacea are as great as the differences from the Cypraeacea. The Triviacea have an echinospira larva, but the coiling of this larval shell is helicoid, not nautiloid. From the Cypraeacea the group is distinguished by having a short pedal ganglion and semilunar osphradium (in the Cypraeacea the pedal ganglion is long and the osphradium trifid). The Triviacea resemble the Cypraeacea in having teeth along the margins of the aperture and in the manner of coiling, but the shells are in general more strongly sculptured, often with well-developed spiral ribbing.

Family TRIVIIDAE

Globose to ovate, aperture the length of the shell or nearly so, spire low. Two Panamic subfamilies.

Subfamily TRIVIINAE

Spire concealed in adult, spiral ribbing well developed; fossula connected with the dorsal wall in front; aperture roughly parallel to the axis of coiling.

Genus TRIVIA BRODERIP, 1837

Color markings may occur as spots, but there are no color bands. *Trivia, s. s.,* comprises small, white, finely ribbed forms unlike any on the west coast; the Panamic species are allocated to other subgenera.

1. Inner lip rising to a point posteriorly................................... 2
   Inner lip not elevated posteriorly................................... 3
2. Shell colored, dorsal furrow present.........................*Cleotrivia*
   Shell white, dorsal furrow weak or wanting...................*Dolichupis*
3. Shell light-colored, dorsum with darker spots.....................*Niveria*
   Shell dark-colored, color spots, if present, not darker...............*Pusula*

Subgenus CLEOTRIVIA IREDALE, 1930

Small, ovate, unspotted, ribs narrow; inner lip pointed posteriorly.

899. Trivia (Cleotrivia) rubescens (Gray, 1833). Most of the records of *T. rubescens* in the literature must be discounted, for authors have confused two or three species. The original description compares well with the type lot, which is in the British Museum: ovate-globose, pale brown, thin, unspotted, ribs narrow, sharp, continuing over dorsum, outer lip narrow and flexed. The posterior end of the inner lip is conspicuous. Length, 8 mm; diameter, 5 mm. Tres Marias Islands, Mexico, to the Galápagos Islands (type locality); offshore in depths to 55 m.

Subgenus DOLICHUPIS IREDALE, 1930

Unspotted, ends of shell somewhat produced, dorsal furrow weak, inner lip pointed posteriorly.

900. Trivia (Dolichupis) acutidentata (Gaskoin, 1836). Although the type specimen was broken before it could be illustrated, the description has been considered by Schilder and Tomlin (1931) sufficient for recognition, for specimens from near the type area have been found that match it well. The ribs on the dorsum of the dull white shell are beaded but are not interrupted by a dorsal furrow. The aperture is conspicuously narrow and slotlike. Length, about 7 mm. Known only from Ecuador.

901. Trivia (Dolichupis) paucilirata (Sowerby, 1870) (Synonyms: **Cypraea buttoni** Melvill, 1900; **T. panamensis** Dall, 1902). Solem (1963) has shown that Sowerby's type in the British Museum, although described from an unknown locality, seems to match precisely that of Melvill, from an unknown West American locality, and also that of Dall, from Panama. Compared with *T. acutidentata* it is more slender, with a wider aperture and narrower outer lip. The shell is pure white. Length, about 6 mm; width, about 4 mm. Panama, dredged in 33 m, to Santa Elena, Ecuador, intertidally under rocks.

Subgenus NIVERIA JOUSSEAUME, 1884

Ends of shell blunt, ribs fine, with granular interspaces; dorsum with a long furrow and one or more darker spots of color.

902. Trivia (Niveria) maugeriae (Sowerby, 1832, *ex* Gray, MS) (Synonym: **T. maugeri** of authors, an emendation). The shell was described as oval, thin, pellucid, rose-colored with darker ends and a central dorsal blotch, the ribs thin, numerous, rather close, the interstices roughened, the dorsal groove narrow, the lip white, with sharp teeth. The type material in the British Museum matches this description well. Length, largest shell, 16 mm; width, 11 mm. Galápagos Islands.

903. Trivia (Niveria) pacifica (Sowerby, 1832, *ex* Gray, MS). The shell is pinkish in color, with dark gray dorsal spots. A similar Atlantic species is *T. (N.) quadripunctata* (Gray, 1827). Length, 9 to 11 mm; width, 6 to 7 mm. Outer coast of Baja California at Pescadero Point, through the southern end of the Gulf and south to Mancora, Peru, and the Galápagos Islands (type locality).

See Color Plate XVI.

Subgenus PUSULA JOUSSEAUME, 1884

Ends blunt, dorsal furrow evident in most species, often with beading of the interrupted ribs; rib interspaces granular; inner lip not projecting posteriorly.

904. Trivia (Pusula) atomaria Dall, 1908. Smallest of the eastern Pacific Trivias, this has a pinkish-brown cast that seems to be the result of pigmentation at two levels, the lower olive, the outer bright pink. The ribs continue across the dorsum almost without interruption, smooth, their interspaces minutely granular; the ends of the shell are rounded, not produced. Length, 3.2 mm; width, 2.6 mm; height, 2.2 mm. Panama Bay, offshore in 33 m.

905. Trivia (Pusula) californiana (Gray, 1827) (Synonyms: **Cypraea californica** Sowerby, 1832, *ex* Gray, MS; **? T. elsiae** Howard & Sphon, 1960). The coffee-bean shell is the only member of the genus that ranges as far north as central California. The common name describes its appearance well, as it is a dark purplish brown and about the shape and size of a coffee bean, measuring 10 mm in length. The range is from California to Acapulco, Mexico. The form described as *T. elsiae* seems to be a juvenile *Trivia* that comes close to *T. californiana* in size and number of ventral ribs; the dorsum, however, is at this stage smooth.

906. Trivia (Pusula) fusca (Sowerby, 1832, *ex* Gray, MS) (Synonyms: **Cypraea rufescens** Sowerby, 1832, *ex* Gray, MS [not Gmelin, 1791]; **T. galapagensis** Melvill, 1900; **T. pulloidea** Dall & Ochsner, 1928; "**T. occidentalis** Schilder, 1922" of Schilder, 1931*b*). The identification of this species has been a problem to authors, because the original figure was small and the type lot in the British Museum had meanwhile been incorrectly relabeled *T. rubescens*. Sowerby's figured specimen, refigured here, matches his description well: ovate, dark brown, with pale dorsal streak, extremities slightly produced; ribs small, rather close, some of them not continuing over the edge of the lips beneath. It also matches well the figures given by Solem (1963) for Melvill's type. Worn specimens (as most of Melvill's specimens were) lose the ribbing on the dorsum and appear to be smooth; this has contributed to the misinterpretation of the species. Length, about 8 mm; width, about 6 mm. Type locality, Galápagos Islands. The range seems to include also Panama and Ecuador on the mainland, but most of the records of *T. fusca* in the literature are misidentifications.

907. Trivia (Pusula) myrae Campbell, 1961. The shell is minute, ovately globular, dark purplish brown, with the right side and the ends margined; dorsal sulcus shallow, crossed by a few ribs; ribs narrow and sharp, interspaces minutely granular. Length, 4.8 mm; width, 3.6 mm. Northern end of the Gulf of California, from off Monserrate Island, Baja California, northward and eastward to Guaymas, Sonora, Mexico, in depths to 55 m.

908. Trivia (Pusula) radians (Lamarck, 1811) (Synonyms: **Cypraea rota** Weinkauff, 1881; **T. sanguinea circumdata** Schilder, 1931). This is the largest of the West American members of the genus; the shell is pinkish, with brown spots on the dorsum. Length, 21 mm; diameter, 15 mm. Magdalena Bay, Baja California, to Ecuador, under rocks, intertidally.

909. Trivia (Pusula) sanguinea (Sowerby, 1832, *ex* Gray, MS). The shell is purplish brown, with a red spot in the middle of the dorsum, whitish at each end, the ribs also whitish. Length, 12 mm. The Gulf of California to Ecuador, not uncommon offshore but rare intertidally, under rocks. Authors have confused this form with some others, failing to notice the stain of bright red along the dorsal furrow; this color still is apparent in the type lot at the British Museum, even after more than 140 years.

910. Trivia (Pusula) solandri (Sowerby, 1832, *ex* Gray, MS) (Synonym: **T. solanderi** of authors, an emendation). This resembles *T. radians*, but the color is more brownish and the shell averages smaller, with a few less transverse ribs. Length, 16 mm; width, 12 mm. Southern California throughout the Gulf of California and south to Peru; fairly common in beach drift, living intertidally under rocks.

Subfamily ERATOINAE

Spire evident, not concealed in adult; aperture long, at an angle to the axis of coiling, narrowed at one end, anterior canal open; anterior margin of the fossula free; shells smooth except for denticulations of the apertural margins. In outline the shells are somewhat pear-shaped; many are brightly colored. The animal has a long proboscis, which it can insert into the cavity of a tunicate colony, the radular teeth then rasping away bits of tunicate flesh.

Genus ERATO Risso, 1826

The generic name is derived from a Greek word meaning "pleasing," which is appropriate for these smooth and rather shiny little shells. They resemble the Marginellas, and early authors tended to confuse the two, failing to notice that the Eratos lack any strong columellar folds. There may be a line of weak denticles, however, along the border of the inner lip. Two Panamic subgenera; no species of subgenus *Erato*, *s. s.*, occur in the Panamic province.

Subgenus ERATOPSIS Hörnes & Auinger, 1880

The terminal ridges at the anterior end of the columella are decidedly oblique.

911. Erato (Eratopsis) oligostata Dall, 1902. The tiny shell is greenish white, flushed with rose-purple at the anterior end. The type specimen measures 3.3 mm in length and was dredged at 33 m in Panama Bay.

Subgenus HESPERERATO Schilder, 1932

The terminal ridges at the end of the columella are weak and not oblique.

912. Erato (Hespererato) columbella Menke, 1847 (Synonym: **E. leucophaea** Gould, 1853). The shell is orange-brown to gray-brown, with a white lip. Length, about 7 mm. Monterey, California, and southward through the Gulf of California to Panama; intertidally, under rocks.

913. Erato (Hespererato) galapagensis Schilder, 1933. Named as a subspecies of *E. marginata* Mörch, this is more probably a southern relative of *E. columbella*. The shell is smooth, pinkish-flesh-colored to brown, with the early whorls reddish, and the base, the right margin, and the suture line white. Some specimens have a green tinge at the anterior end. Length, 4.6 mm; width, 2.3 mm. Albemarle (Isabela) Island, Galápagos Islands.

914. Erato (Hespererato) panamensis Carpenter, 1856 (Synonym: ? **E. marginata** Mörch, 1860). The grayish-pink shell is more uniformly colored than *E. columbella*, its outer lip not white; the spire is lower and the body whorl more inflated than *E. maugeriae* Sowerby, 1832, from the West Indies, which it otherwise very much resembles. Because neither Carpenter's type material nor Mörch's was figured until recent years, authors have not been able to recognize the species; Carpenter's type lot has, in fact, received several other identifications. Length, 7.5 mm; diameter, 5 mm. The type locality is Panama.

915. Erato (Hespererato) scabriuscula Sowerby, 1832 (Synonyms: **Marginella cypraeola** Sowerby, 1832; **M. granum** Kiener, 1843). The shiny grayish-brown little shell is easily recognized by the minute granules on the surface. The length is about 8 mm and the range from the southern end of the Gulf of California to Ecuador, intertidally.

Superfamily CYPRAEACEA

Colorful shells, mostly with the spire concealed and the aperture long and narrow.

Family CYPRAEIDAE

Cowrie shells, because of their high polish and beauty of coloring, have been favored by primitive people in tropical areas for personal adornment and today

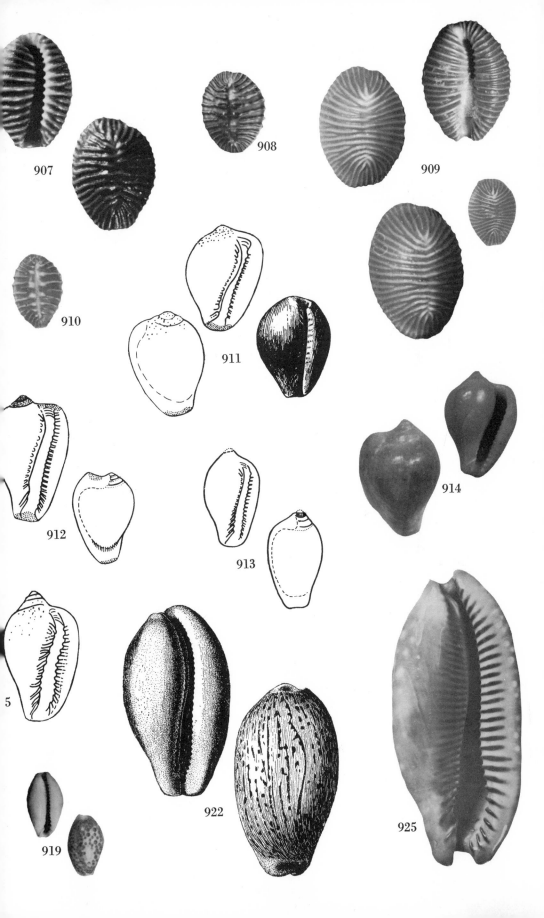

907

908

909

910

911

912

913

914

5

919

922

925

are highly prized as curios or souvenirs by persons who otherwise show little interest in shells. They are the prime favorites of many collectors.

The shell of an adult cowrie differs greatly in appearance from that of a young specimen. Immature shells (see text figure) are fragile, somewhat polished, with an elongate body whorl, short spire, and sharp lip. The color pattern shows a tendency to vague spiral bands. As the shell nears maturity, the outer lip turns abruptly in toward the columella, and both lips begin to show signs of teeth. The shell also becomes stronger and heavier, and the color pattern begins to take on the adult character. Fully mature specimens have a narrow aperture with well-developed teeth (in most species), a distinctive color pattern, and a fairly sturdy shell. A heavy layer of enamel is deposited by the mantle upon the surface of the shells, often leading the uninitiated to assume that the shell has been varnished or polished, and this layer of enamel almost or entirely conceals the spire of the shell. In some species the enamel is heavily deposited about the sides of the shell (known as the margins), forming callus deposits. The margins may show color markings called "lateral spots." The term "base" is commonly used to describe the flattened part of the adult shell where the aperture and teeth are located, and the opposite side is known as the dorsum. Periostracum is completely lacking.

Because of the great variability in the size of fully mature shells of the same species, there has long circulated a story that the animal of *Cypraea* is able to dissolve its shell and resecrete another larger one within a short period of time. This ability would appear to be entirely legendary, although the animal is able to dissolve partially or resorb the early whorls, making more room for its body.

The animal of *Cypraea* is as striking and colorful as its shell, with fringed or warty-appearing filaments on the mantle, two lobes of which envelop the shell from either side, meeting along an irregular line down the center of the dorsum. This mantle tissue protects the shell from wear and also deposits layers of glaze or enamel, the result being a highly polished or glossy texture. The animal is very shy and ordinarily completely withdraws into its shell on being disturbed. Professor Ostergaard observed at the Marine Laboratory in Hawaii that the female cowrie remains with the newly laid egg mass for some days, until the young are released, and that she will attempt to discourage, by menacing postures and movements, any intruder from disturbing her "nest." He also studied the microscopic larval forms of some of the Hawaiian species and found that at an early stage they possess a well-developed operculum.

Cowries show a preference for warm, shallow water and are most numerous on the coral reefs of the western Pacific and Indian oceans. On the western American coast, where coral reefs are poorly developed, they are found living under stones or in pockets of rocky reefs, at or below low water.

Because the family has long held attraction for collectors and specialists, the variations within it have been intensively studied and a complex terminology developed for the shell features, shown here in a text figure (adapted from Schilder, 1938). In all systematic work it is a human tendency to become more impressed with differences the more deeply one goes in surveying a group. Thus, perhaps inevitably, specialists such as Dr. and Mme. Schilder have divided and redivided the Linnaean genus *Cypraea* into numerous genera and subgenera that are based on what seem to the nonspecialist to be very slight differences. One school of conservatives would prefer to keep the many species within a single genus, *Cypraea*. Another would maintain the one genus but recognize several to

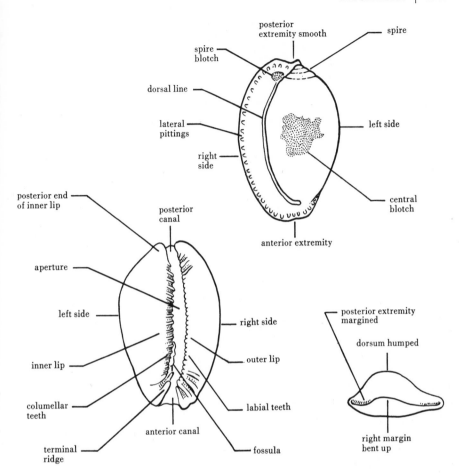

many subgenera. As a practical expedient, that is the course adopted here; bracketed notations in the text point out the ranking that is given to the various subgeneric units in the Schilder classification when this differs significantly. Theoretically it is probable that a group such as the Cypraeidae—which has been evolving since Cretaceous time—would have diverged considerably. Some of the studies of the living animals, however, have not confirmed the subdivisions that shell characters seem to suggest. Therefore, it seems that more work is needed before a finely drawn classification can be adopted.

Of around 170 living species, only seven are really native to the Panamic marine province. Several Indo-Pacific forms have managed to extend their geographic range to the outlying islands, especially to Clipperton, and one has been reported on the mainland. These Indo-Pacific species will merely be cited here; a detailed discussion of them has been published by Cate (1969). Other useful review papers on the Panamic Cypraeidae are: Ingram (1947, 1951); Schilder & Schilder (1938–39, 1969); Bakus (1968).

The review herein has been prepared in consultation with Crawford N. Cate of Los Angeles, who has generously supplied notes on morphology and distribution as well as classification; his assistance is gratefully acknowledged.

Genus CYPRAEA LINNAEUS, 1758

Spire concealed in adult by the glaze that covers the body whorl; aperture the length of the shell, narrow, denticulate on either side. No species of subgenus *Cypraea*, s. s., occur in the Panamic province.

Subgenus ARABICA JOUSSEAUME, 1884

Sturdy shells, the teeth well developed; dorsal markings of brown lines and spots. Regarded as a subgenus of *Mauritia* Troschel, 1863, by Schilder (1969).

916. Cypraea (Arabica) depressa Gray, 1824. Clipperton Island to western Pacific.

917. C. (A.) maculifera Schilder, 1932. Clipperton Island to western Pacific.

918. C. (A.) scurra indica Gmelin, 1791. Clipperton Island to western Pacific.

Subgenus EROSARIA TROSCHEL, 1863

Aperture wider anteriorly, arched posteriorly; upper margins of shell excavated and pitted; apertural teeth darker than the interspaces; margins with well-developed callus. Regarded as a genus under the subfamily Erosariinae by Schilder (1969).

919. Cypraea (Erosaria) albuginosa Gray, 1825 (Synonym: **E. a. nariaeformis** Schilder, 1930). Shell somewhat pear-shaped, with raised sides thickened by callus; base convex, narrow; teeth numerous and fine; dorsum buff, margins and part of base grayish lavender; dorsal surface marked with small white spots and small brown rings with light gray centers. Length, 29 mm; width, 17 mm; height, 13 mm. Distribution discontinuous: central part of the Gulf of California and Socorro Island to Manzanillo, Mexico; and Cocos Island to Panama and Ecuador, including the Galápagos Islands.
    See Color Plate XV.

920. Cypraea (Erosaria) caputserpentis Linnaeus, 1758. Clipperton Island and Indo-Pacific.

921. Cypraea (Erosaria) helvola Linnaeus, 1758. Clipperton Island and Indo-Pacific.

Subgenus LURIA JOUSSEAUME, 1884

Cylindrical, without clear-cut margins, the ends set off by color spots; apertural teeth lighter than the interspaces. Regarded as a genus of the tribe Luriini by Schilder (1969).

922. Cypraea (Luria) isabellamexicana Stearns, 1893. The pale brown dorsum has darker brown streaks, the margins lighter and the base and interior white. Ends with a conspicuous orange and brown spot. Apertural teeth small and numerous. Length, 38 mm; width, 21 mm; height, 17 mm. Guaymas and La Paz, Gulf of California, to Panama and the Galápagos Islands, Ecuador. A similar species in the Indo-Pacific, *C. isabella* Linnaeus, 1758, is slightly smaller and less brightly colored.

Subgenus LYNCINA TROSCHEL, 1863

Somewhat pear-shaped, arched posteriorly; teeth numerous; sides smooth, with callus. Regarded as a genus in the Tribe Cypraeini by Schilder (1969).

923. Cypraea (Lyncina) schilderorum (Iredale, 1939). Clipperton Island to Indo-Pacific.

924. Cypraea (Lyncina) vitellus Linnaeus, 1758. Clipperton Island to Indo-Pacific.

Subgenus MACROCYPRAEA SCHILDER, 1930

Apertural teeth long, columellar folds short and cross-striate; shell large and somewhat cylindrical, lacking anterior callus. Regarded as a subgenus of *Trona* Jousseaume, 1884, by Schilder (1969).

925. Cypraea (Macrocypraea) cervinetta Kiener, 1843. The dorsum is rich yellowish brown, with pale grayish spots toward the margins. The base is lighter, the columella of many specimens having a brown blotch. The shell is purplish within, and the numerous apertural teeth are dark brown. Immature specimens are unspotted, with four brown bands. Length, 40 to 115 mm; width, 15 to 52 mm. Puerto Peñasco, Sonora, Mexico, to Paita, Peru, and the Galápagos Islands. Two West Indian species are similar: *C. (M.) zebra* Linnaeus, 1758, and *C. (M.) cervus* Linnaeus, 1771.

See Color Plate VI.

Subgenus MONETARIA TROSCHEL, 1863

Small, dorsum yellow, with longitudinal growth lines. Regarded as a genus of the subfamily Erosariinae by Schilder (1969).

926. Cypraea (Monetaria) moneta Linnaeus, 1758. Clipperton, Cocos, and the Galápagos Islands westward to the Indo-Pacific.

Subgenus PSEUDOZONARIA SCHILDER, 1927

Aperture nearly central and straight, canals strong and deep, columella smooth; color markings tending to fuse into elongate stripes. Regarded as a subgenus of *Zonaria* Jousseaume, 1884, subfamily Erroneinae, by Schilder (1969).

927. Cypraea (Pseudozonaria) arabicula (Lamarck, 1811). The color of the dorsal surface varies from light buff to greenish blue, speckled and streaked with light or dark brown. The dorsum is humped, with rather sharp angulate margins, the lateral areas spotted, and the base nearly flat. Juvenile specimens are blue-gray, finely speckled with brown, with three broad bands, and the humped outline is apparent. Perhaps the most distinctive feature is the fine, numerous, sharply chiseled apertural teeth. About 26 mm in length; 16 mm in width; 13 mm in height. Gulf of California, from Guaymas to Pulmo Reef, south to the Galápagos Islands and Peru. Type species of the subgenus *Pseudozonaria*.

See Color Plate XV.

928. Cypraea (Pseudozonaria) nigropunctata Gray, 1828 (Synonyms: **C. irina** Kiener, 1843; **C. gemmula** Weinkauff, 1881; **C. massauensis** Schilder, 1922). The coloring resembles that of *C. (P.) arabicula* except that the lateral spots are smaller and much more numerous. The shell is longer and more quadrate. The teeth are intermediate in size between those of *C. arabicula* and *C. robertsi*. Length, 28 mm; width, 15 mm; height, 12 mm. Galápagos Islands; rarely in Ecuador and northern Peru.

929. Cypraea (Pseudozonaria) robertsi (Hidalgo, 1906) (Synonym: **C. punctulata** Gray, 1824 [not Gmelin, 1791]). Broad, deltoid, with a heavy callus on both margins, not as humped dorsally or as sharply margined as *C. arabicula*.

927

928

929

931

933

935

936

934

The margins are flesh-colored to purplish brown, with small, dark brown spots. The base and teeth are white, the interior violet. Teeth are blunt and coarse, the aperture narrow. Length, 23 mm; width, 14 mm; height, 11 mm. Nicaragua to Paita, Peru, and the Galápagos Islands, most abundant at Panama.

Subgenus TALOSTOLIDA IREDALE, 1931

Pear-shaped, margins thick, angled on the right side; dorsum with diffused brown markings; teeth fine and numerous. Regarded as a synonym of *Bistolida* Cossmann, 1920, subfamily Erroneinae, by Schilder (1969).

930. Cypraea (Talostolida) rashleighana Melvill, 1888. An Indo-Pacific species reported on Cocos Island off Costa Rica.

931. Cypraea (Talostolida) teres pellucens Melvill, 1888. With three narrow but vague color bands near the middle of the shell and a large central blotch, the margins with a few spots. Length, 23 to 26 mm; width, 12 to 15 mm; height, 9.5 to 10 mm. Living on corals, Secas Island and Bahía Honda, Panama Bay; Clipperton Island, the Hawaiian Islands, and westward to the Indo-Pacific. Emerson and Old (1968) report this as the first Indo-Pacific cowrie species to be taken living on the mainland of the Americas.

932. Cypraea (Talostolida) teres teres Gmelin, 1791. An Indo-Pacific form reported on Clipperton and the Galápagos Islands; perhaps these records will prove to be misidentifications of *C (T.) teres pellucens*.

Subgenus ZONARIA JOUSSEAUME, 1884

Pear-shaped, aperture arched anteriorly; teeth lacking on the posterior canal. Regarded as a genus in the subfamily Erroneinae by Schilder (1969).

933. Cypraea (Zonaria) annettae Dall, 1909 (Synonyms: **C. ferruginosa** Kiener, 1843 [not Gmelin, 1791]; **C. sowerbyi** Kiener, 1843 [not Gray, 1832]). Dorsal surface heavily spotted with brown, with faint brown banding, the sides well rounded and spotted, the base varying from flesh-colored to brown, the teeth lighter or nearly white; interior purple. Young specimens lighter, weakly spotted, with five brown bands. Two subspecies have been recognized.
See Color Plate XV.

Cypraea (Zonaria) annettae annettae Dall, 1909. Length, 40 mm; width, 23 mm; height, 18 mm. The range seems to be restricted to the Gulf of California, south only to Pulmo Reef, southeastern Baja California; intertidally under rocks. A variant that is smaller, wider, paler, cream-colored within, with finer teeth, occurs in the Concepcion Bay area; it may eventually prove to be distinct enough for separation (Cate, 1961).

934. Cypraea (Zonaria) annettae aequinoctialis (Schilder, 1933). The shell is slightly larger and rounder, with heavier, coarser teeth and a somewhat more diffuse color pattern. A nacre-like deposit free of color and spotting occurs above the anterior terminal collar. Length, 44 mm; width, 28 mm; height, 22 mm. Panama Bay to Peru.

Family OVULIDAE
(AMPHIPERATIDAE)

The Ovulas or egg shells are distinguished from the cowries by several important anatomical features and two shell differences—the lack of color bands or spots and the involute or inwardly curving spire. These mollusks are carnivorous, and each species seems to confine itself to a single type of coelenterate prey: stony corals (either solitary or reef type) ; gorgonians or leather corals; or others. The colors of the shells tend to blend with those of the host, but there are exceptions, and this is even more true of the mantle, which is exposed in life, spreading out over the shell. Two Panamic subfamilies.

Subfamily OVULINAE

Surface smooth or nearly so. Two Panamic genera.

Genus CYPHOMA Röding, 1798
(ULTIMUS Montfort, 1810; BINVOLUTA Schlüter, 1838; CARINEA Swainson, 1840)

With a transverse ridge across the dorsum; aperture long and slotlike; lips thickened and slightly roughened within.

935. Cyphoma emarginatum (Sowerby, 1830) (Synonym: **? C. marginata** Chénu, 1859). Whitish to flesh-colored, the shell has the dorsal ridge rather sharply defined, the ends of the shell narrowed and drawn out. Fine striae occur on the dorsal surface of some specimens. Length, 15 mm; width, 8 mm. Agua de Chale, near San Felipe, Baja California, southward and eastward through Sonora, Mexico, to Ecuador. Two similar species in the western Atlantic are *C. gibbosum* (Linnaeus, 1758) and *C. intermedium* (Sowerby, 1828).
See Color Plate XVIII.

Genus SIMNIA Risso, 1826
(NEOSIMNIA Fischer, 1884)

Shell smoothly oval, tapering at both ends, more slender than *Cyphoma*, the dorsal surface lacking any heavy horizontal ridge. Outer lip thickening with age. *Neosimnia* was distinguished on the basis of the thick lip from *Simnia*, which was thought to have a thin lip, but this proves not to be a constant difference. Specialists therefore are reverting to the use of the older name, *Simnia*. The preferred coelenterate hosts are the gorgonians—sea whips and sea fans.

A California collector, Faye Howard, observed specimens of *Simnia* in Mexico. They are not confined to a single host colony but can move about. Leaving the host, they drift up to the surface of the water, where with the foot they attach to the surface film and ride, suspended upside down. After a time they peel the foot loose from the film, retract into the shell, and settle down to the bottom. If no new host colony is encountered, they repeat the upward movement and try again. Experimenting with a half-dozen specimens in an aquarium, she found that two made a transfer from an old to a new host, a distance of 15 inches, on the first try; others had to make more than one attempt, but all succeeded.

A very conservative classification is adopted here, for several taxonomic problems have yet to be resolved. Critical study of the type material in the British Museum will be necessary, for photographs show that most of the type lots are composite. Thus, judicious selection of lectotypes will be needed to refine the concepts for several of the available older names. It may well be that shell features such as are pointed out by Cate (1969) will prove useful in separating species. More work also is needed on the question of whether these mollusks are selective in their choice of gorgonian species and whether both the shell and the

PLATE VI · *Cypraea cervinetta* Kiener (The Little Deer Cowry)

soft parts are affected by the coloration of the host. The synonymies suggested here are therefore highly tentative. Notes and figures are supplied for several forms that may ultimately prove to be distinct.

936. Simnia aequalis (Sowerby, 1832) (Synonyms: **Ovula variabilis** C. B. Adams, 1852; **O. livida** Reeve, 1865; **O. vidleri** Sowerby, 1881; **S. quaylei** Lowe, 1935; **Neosimnia vidleri tyrianthina** Berry, 1960). The color varies from white to lavender-rose or orange, with the anterior and posterior ends somewhat darker, yellow or orange. The dorsum is usually smooth but may be finely striate centrally, with transverse (i.e., spiral) ridges or grooves developing toward the ends of the shell. The anterior half of the columella may be smooth or may have a low carina. Shells in the northern end of the Gulf of California tend to have the columellar carina less evident (the form named *S. quaylei*), but there are exceptions, as in *N. v. tyrianthina*, with a perceptible carina. Length, to 22 mm; width, to 8 mm. Monterey, California, through the Gulf of California and south to Panama and the Galápagos Islands.

See Color Plate XVIII.

937. Simnia avena (Sowerby, 1832). Lavender-brown in color, the shell is small, wide for its length, with transverse (spiral) ridges or striae over the entire dorsum. Length, 12 mm; width, 6 mm. Southern Baja California to Panama (type locality). Three of the four specimens in the type lot of *S. livida* (Reeve, 1865) are specimens of *S. avena*, but since Reeve figured the fourth specimen, which seems to be an *S. aequalis*, his name is here listed in the synonymy of that species.

938. Simnia inflexa (Sowerby, 1832). The posterior flexure is unusually strong in the type lot of this species. No columellar carina is evident, and the dorsum of what seems to be Sowerby's figured specimen has regular transverse (spiral) ridges. The type locality is the Gulf of Dulce. Length, 17 mm; width, 6 mm. Central America, probably from Costa Rica to Panama.

939. Simnia rufa (Sowerby, 1832) (Synonyms: **? Ovula neglecta** C. B. Adams, 1852; **Ovulum subrostratum** of authors, not of Sowerby, 1849 [an Atlantic species]; **Ovula californica** Reeve, 1865, *ex* Sowerby, MS). Color may be variable; the most distinctive shell feature is a readily visible ridge along the entire length of the columella, especially well shown in the type lot of the *S. californica* of Reeve. The dorsum is nearly smooth or with a few low spiral ridges near the ends of the shell. Length, 15 to 17 mm; width, 5 to 6 mm. Southern California to Ecuador. The original figured specimen of Sowerby has not in recent years been positively identified; thus, the concept rests upon two lots of specimens in the British Museum that represent supposed synonyms.

Subfamily **EOCYPRAEINAE**
(**JENNERIINAE** of authors)

Surface sculptured with nodes or ribs. Two Panamic genera.

Genus **JENNERIA** JOUSSEAUME, 1884
(**PUSTULARIA** of authors, not of H. & A. ADAMS, 1854)

Dark-colored shells with finely striate surface markings, the dorsum studded with nodes. Although the shells resemble Triviinae, in the Lamellariacea, the soft parts are nearer to the Ovulidae: they lack the echinospira larva, they have three-

937

938

939

940

941

943

945

942

946

branched rather than two-branched osphradia (sensory organs in the mantle cavity), and D'Asaro (1969) has shown that the egg capsules are similar to those of *Simnia* and *Cyphoma*. The spire is always involute, even in juvenile stages of growth.

940. Jenneria pustulata [Lightfoot, 1786]. This species can be confused with no other living species on earth. The dorsum is covered with bright orange pustules, each of which is encircled by a dark-colored ring, the dorsal surface itself being a glistening gray with fine spiral lines. On the base the apertural teeth extend to the margins as white ridges on a dark brown ground. The interior is violet. Length, 25 mm; width, 15 mm. The northern end of the Gulf of California to Ecuador, in and near masses of stony coral.

See Color Plate XVIII.

Genus PSEUDOCYPRAEA SCHILDER, 1927

Outer lip lirate, inner lip with fine furrows and some denticles anteriorly; spire involute.

941. Pseudocypraea adamsonii (Sowerby, 1832). Small, white, with brown flecks; having numerous low ribs, no dorsal furrow; apertural edges slightly arched. Length, 20 mm; width, 12 mm. Galápagos Islands and west to Japan and the western part of the Indian Ocean.

Superfamily TONNACEA

Generally thin shells with a low spire and large aperture; anterior canal short; periostracum thin or wanting.

Family TONNIDAE

The tun shells have much thinner and more capacious shells than the Cassididae, lacking true varices and axial sculpture. The family is confined to the tropics. No species of the type genus, *Tonna* Brunnich, 1771, occur in the Panamic province. Two Panamic subfamilies.

Subfamily TONNINAE

Large, inflated, with broad spiral ribs; operculum wanting; anterior notch somewhat constricted.

Genus MALEA VALENCIENNES, 1832

The shell is porcelaneous, fairly thick, with the outer lip reflected and toothed along its inner edge. Sculpture is of wide, smooth, flat, spiral ribs. The inner lip has a heavy callus that rises into a few low folds on either side of a deep columellar excavation.

942. Malea ringens (Swainson, 1822) (Synonyms: **Dolium dentatum** Barnes, 1824; **D. personatum** Menke, 1828; **M. latilabris** and **crassilabris** Valenciennes, 1832; **D. plicosum** Deshayes & Milne-Edwards, 1843). One of the largest of Panamic shells, a full-grown specimen of the grinning tun is a fine sight. The white shell is delicately flecked with yellow spots under a thin periostracum, and a large specimen may be 150 to 240 mm (6 to 9 inches) in height. Average diameter is about 100 mm. Puerto Peñasco, Mexico, to Paita, Peru. Large specimens were taken by Lowe under ledges of rock at extreme tides at Mazatlán and smaller specimens on sand bars in Costa Rica and Panama.

Subfamily OOCORYTHINAE

Thin, more or less inflated shells, resembling the Tonnidae and some Cassididae but with the anterior notch open as in the Ficidae. Deep water, mostly in the tropics.

Genus OOCORYS FISCHER, 1884

Sculpture of narrow, well-spaced spiral ridges; whorls inflated; mostly low-spired; inner lip with a thin callus and parietal shield; operculum present, paucispiral, with a raised apical nucleus; anterior canal ending in a shallow spout.

943. Oocorys elevata Dall, 1908. Galápagos to Peru, in 4,090 m.

944. O. pacifica (Dall, 1896). Panama Bay, in 2,130 to 2,190 m.

945. O. rotunda Dall, 1908. Panama Bay, in 3,050 m.

Family CASSIDIDAE
(CASSIDAE of authors)

The helmet shells are solid, porcelaneous, globose to triangular, the spire flat to only moderately high. Varices present in most. The inner lip has a wide callus that may be raised or shelflike, its surface smooth to ridged or granulate. The operculum is horny, shaped like a half-moon, sometimes with radial ribs, the nucleus at the middle of the inner margin. That the animals are carnivorous is indicated by the siphonal notch in the anterior end of the aperture, characteristic of active predators. Some cassids are known to feed on sea urchins, drilling a hole through the plates. Useful reference: Abbott (1968). Three Panamic genera; the key includes also subgenera.

1. Shell cylindrical, aperture more than two-thirds the length of the shell 2
   Shell ovate, aperture less than two-thirds the length 5
2. Shell thin, large, aperture with a posterior notch . . . . *Cypraecassis*
   Shell solid, medium-sized to small, aperture lacking any posterior notch 3
3. Anterior canal recurved, with a siphonal fasciole . . . . *Levenia*
   Anterior canal simple, lacking a siphonal fasciole 4
4. Sculpture regularly cancellate . . . . *Cancellomorum*
   Sculpture irregularly nodose . . . . *Morum, s. s.*
5. Outer lip with a few blunt spines anteriorly . . . . *Casmaria*
   Outer lip smooth anteriorly . . . . *Semicassis*

Genus CASSIS SCOPOLI, 1777

Shells ovoid to trigonal, with a relatively narrow aperture; varices present, mostly well developed. Some forms in the East Indies and the West Indies reach a large size. No members of *Cassis, s. s.*, occur now in the Panamic province, but there are fossil records in southeastern California. Three Panamic subgenera.

Subgenus CYPRAECASSIS STUTCHBURY, 1837

Spire short, aperture narrow; a single varix only, at outer lip, no varix scars on spire. The type species is the Indo-Pacific *Cassis rufa* (Linnaeus, 1758), a showy orange-brown form. Abbott (1968) regards this as a full genus.

946. Cassis (Cypraecassis) tenuis Wood, 1828 (Synonym: **C. massenae** Kiener, 1835). The thin shell is brown, mottled with lighter and darker spots. Height, 110 mm; diameter, 50 mm. La Paz, Gulf of California, to Ecuador; not uncommon as beach shells but apparently living only offshore.

Subgenus LEVENIA GRAY, 1847

Spire low, aperture narrow, outer lip not reflected outward, the upper part turned slightly inward. Abbott (1968) ranks this as a subgenus of *Cypraecassis*.

947. Cassis (Levenia) coarctata Sowerby, 1825. A smaller and more solid shell than *C. tenuis*, this has somewhat similar color markings. The aperture is contracted posteriorly. Height, 45 mm; diameter, 25 mm. Head of the Gulf of California to Ecuador; mostly living offshore but not uncommon in beach drift. The type species of the subgenus.
See Color Plate XVI.

Subgenus SEMICASSIS MÖRCH, 1825
(TYLOCASSIS WOODRING, 1928)

Spire more pointed and higher than in *Cassis, s. s.*, the aperture wider, whorls more inflated; varices inconspicuous; inner lip callus raised, shelflike, sculptured. Ranked as a subgenus of *Phalium* Link, 1807, by Abbott (1968), with *Tylocassis* a coordinate subgenus, *Semicassis* having the sculpture of the lip callus a series of ridges or folds, *Tylocassis* with the ridges broken up into granules.

948. Cassis (Semicassis) centiquadrata (Valenciennes, 1832) (Synonyms: "*C. abbreviata* Lamarck" of authors, not of Lamarck, 1822; **C. doliata** Valenciennes, 1832). The white shell has regular spiral ribs, and the face of the lip callus is granular. Height, 55 mm; diameter, 35 mm. The Gulf of California to the Galápagos Islands and Lobitos, Peru, living in sand at very low water. A similar Atlantic form, which is the type of *Tylocassis*, is *C. granulata* (Born, 1780).
See Color Plate XVI.

Genus CASMARIA H. & A. ADAMS, 1853

Smooth and polished, with a few blunt spines on the lower edge of the outer lip; no varices, except for the thickened outer lip.

949. Casmaria vibexmexicana (Stearns, 1894). The glossy light brown surface is marked with fine spiral pencil lines of dark brown and somewhat wavy axial stripes of brown. The rim of the outer lip and the anterior notch are brown-spotted. The inner lip is smooth and white, with a yellow spot on either side of the canal. Average height, 44 mm; width, 25 mm; height of a large specimen, 70 mm. San José Island and La Paz, Baja California, southward to the Tres Marias Islands, Mexico. This was at first identified with the Indo-Pacific species *C. erinaceus* (Linnaeus, 1758)—under the name *Cassis vibex* (Linnaeus, 1758)—a form that is slightly more slender and not so brightly colored. A similar Caribbean species is *Casmaria atlantica* Clench, 1944.
See Color Plate XVI.

Genus MORUM RÖDING, 1798
(LAMBIDIUM LINK, 1807; ONISCIA SOWERBY, 1824)

The anterior end of the aperture looks as though it had been broken, because it lacks the turned-back edge of most related forms. The young shell has even been mistaken for a cone by unsuspecting collectors. Two Panamic subgenera.

Subgenus MORUM, *s s.*

Cone-shaped, spire low; sculpture of a few blunt nodes at periphery.

947

948

951

949

950

952

953

950. Morum (Morum) tuberculosum (Reeve, 1842, *ex* Sowerby, MS) (Synonyms: **Oniscia lamarckii** Deshayes in Lamarck, 1844 [not Lesson, 1840]; **M. xanthostoma** A. Adams, 1854). The cylindrical shell has about five concentric rows of blunt tubercles; it is brown, dotted with white, the interior white or saffron-yellow. Length, 17 mm; diameter, 10 mm. The outer coast of Baja California through the Gulf and south to Mancora, Peru, under rocks at extreme low tide. A similar species in the western Atlantic is *M. oniscus* (Linnaeus, 1767).

Subgenus CANCELLOMORUM EMERSON & OLD, 1963

Spire somewhat elevated, sculpture cancellate, of broad axial ribs and narrowed spiral threads.

951. Morum (Cancellomorum) veleroae Emerson, 1968. Looking like a cross between a stromb and a harp shell, the weakly cancellate shell is grayish white, with four spiral bands of light brown on the body whorl and some irregular patches of brown on the spire and between the bands. The aperture is white, the lirations of the inner lip white, edged with light apricot. Across the front of the shell the inner lip callus is raspberry red, fading to a vivid pinkish lavender, studded with white pustules. Length, 36 to 53 mm; width, 21 to 30 mm. Cocos Island, Costa Rica, to Galápagos Islands, dredged at depths of 55 to 90 m. A twin species in the western Atlantic is *M. (C.) dennisoni* (Reeve, 1842), which, although long known, still is regarded as a rare shell.

Family FICIDAE

The fig shells are thin, elegantly curved, about the size and shape of a large fig, finely sculptured with spiral and concentric riblets. There is no operculum. The animal has a long siphon and glides about on a large foot, from the sides of which the mantle folds up over the shell in wide lobes. The family is tropical in distribution.

Genus FICUS RÖDING, 1798
(PYRULA LAMARCK, 1799)

The generic name *Ficus* is the Latin name for fig and is one of the few nouns ending in *-us* that are feminine in gender. *Venus* is another. The shape of the shell is reminiscent of fruit, for the synonymous name of Lamarck, *Pyrula*, means "little pear."

952. Ficus ventricosa (Sowerby, 1825) (Synonym: **Bulla decussata** Wood, 1828). The shell is creamy tan to gray in color, mottled with brown spots along the spiral ridges. The smoothly polished aperture is bluish lavender within. Height, 90 mm (3½ to 4 inches); diameter, 50 mm. Magdalena Bay, Baja California, throughout the Gulf of California and south to Negritos, Peru. Not uncommon as beach shells but living offshore. A similar Atlantic species is *F. communis* (Röding, 1798).

See Color Plate XV.

Superfamily CYMATIACEA

With an anterior canal, well-developed sculpture, and regular varices.

Family CYMATIIDAE

The triton shells constitute a group widespread in the tropics, some of the species even, apparently, occurring in two or more oceans. The free-swimming larval shells may drift great distances before settling to the bottom. They are slender and several-whorled, projecting as a sharp tip on the spire of well-preserved adult shells, and may be so differently colored and differently sculptured from the adult as to deceive even veteran workers into bestowing separate names upon them. The adult shells are sturdy and porcelaneous, mostly with a fibrous or hairy periostracum, under which the shell may be brightly colored. Varices are characteristic of the family, most genera exhibiting a regular pattern of varices occurring at a constant angular distance apart, such as 120°, 270°, etc. The operculum is horny, ovate, with the nucleus near the anterior end of the aperture. In most species the inner lip is extensively wrinkled, and varices of the outer lip are formed by an inward folding, as a sort of gutter that is then filled with shell material.

The nomenclature as well as the distribution of the tritons has been a matter of much confusion and disagreement among workers. One of the anomalies of distribution is that some Caribbean species resemble those of the western Pacific more than they do their nearer neighbors in the Panamic province. Many generic and subgeneric names have been proposed, some of which seem to be useful. Two Panamic genera; the key is to subgenera.

1. Inner-lip callus forming a rectangular or shield-shaped area; suture at varying angles .... *Distorsio*
   Inner-lip callus confined to the columellar area; suture regularly spiral.... 2
2. Whorls shouldered 3
   Whorls rounded 5
3. Without varices except at outer lip.......................... *Linatella*
   With varices on spire.. 4
4. Massive, surface nearly smooth under a flaky periostracum.... *Cymatium, s. s.*
   Medium-sized, surface with fine spiral ribs, periostracum slight, of thin hairs .... *Turritriton*
5. Canal long, narrow, recurved.............................. *Gutturnium*
   Canal short to moderate in length, straight.......................... 6
6. Spiral sculpture of wide primary ribs with wide, nearly smooth interspaces .... *Monoplex*
   Spiral sculpture of alternating and rather irregular ribs of two sizes.... *Septa*

Genus CYMATIUM Röding, 1798
(LOTORIUM Montfort, 1810; NYCTILOCHUS Gistel, 1848)

The generic name *Triton* Montfort, 1810, though used by many of the older writers, cannot be retained, for there is an earlier *Triton* in Amphibia. Authors who have refused to accept Röding's generic name *Cymatium* have adopted one or another of the several synonyms, adding to the nomenclatural confusion that surrounds the whole group. Six Panamic subgenera.

Subgenus CYMATIUM, *s. s.*

Massive, varices heavy; canal short.

953. Cymatium (Cymatium) tigrinum (Broderip, 1833). Largest of the West American tritons, this is also one of the least common. The shell is bright yellowish brown, tinged with dark brown between the spiral ribs. In a fully mature specimen the outer lip expands to make the body whorl very wide, and the shell may attain a length of 110 to 155 mm. La Paz, Gulf of California, south to Panama; rare, from extreme low tide to depths of 20 m. *C. femorale* (Linnaeus, 1758) is a similar species in the Atlantic.

Subgenus GUTTURNIUM MÖRCH, 1852

Spire with varices; body whorl inflated, anterior canal elongate; periostracum velvety, with rows of longer hairs.

954. Cymatium (Gutturnium) amictoideum Keen, new species (Synonym: "C. (G.) amictum" of authors, not of Reeve, 1844). The shell, with its long canal and slender outline, bears a remarkable resemblance to the western Pacific *C. amictum* as figured by Reeve, but the holotype in the British Museum proves that Reeve's species has much coarser sculpture and less tapering whorls. Thus, the eastern Pacific species is in need of a name, here supplied, the word chosen being intended to point out the similarity. The shell is medium-sized, elegantly tapered, with less than one varix per whorl, the varices appearing at irregular intervals on the spire, with low axial nodes between, decreasing in number with growth, about four on the body whorl; spiral sculpture of irregular threads; shell white under a grayish-yellow periostracum, which is sparsely studded with long brown hairs somewhat irregularly disposed in axial rows; aperture ovate, with about six large denticles on the outer lip, one heavy knob or boss at the posterior end of the inner lip, and five or six raised lirae simulating folds on the lower part of the columella; anterior canal long, narrow, and nearly straight, appearing to be slightly recurved in profile view. Length of holotype, 53 mm; diameter, 23.5 mm, maximum. Type locality, 27 to 55 m off the northwestern end of San José Island, Panama Bay (dredged by Captain R. G. Shaver). Distribution: off Angel de la Guarda Island, Gulf of California, to Panama Bay, in depths of 30 to 100 m.

Subgenus LINATELLA GRAY, 1857

No varices on spire, one on outer lip.

955. Cymatium (Linatella) wiegmanni (Anton, 1839) (Synonyms: **Triton chemnitzii** Reeve, 1844, *ex* Gray; **T. perforatus** Conrad, 1849). Clench and Turner (1957) suggest that there is a prior name for this form—*Triton cynocephalum* Lamarck, 1816, proposed for a species of unknown locality. Kiener in 1842 erroneously (as we now realize) cited as *T. cynocephalum* a Caribbean form that does not match Lamarck's original figure, and his misidentification has been accepted for the past century. Lamarck's figure has more resemblance to the Panamic form, but there still are several points of discrepancy. Hence, it seems advisable to continue the use of the name *C. wiegmanni* for the Panamic species. Even should Lamarck's type specimen, if extant, prove to be identical with the Panamic form, his name, long used in another connection, could be considered a *nomen oblitum*, and under the International Rules on Zoological Nomenclature a petition could be entered to have the name suppressed. In color the shell is yellowish brown, the spiral ribs a little darker than the interspaces; there are a few nodes on the shoulders of the whorls. About 75 mm in length. San Ignacio Lagoon, Baja California, through the Gulf of California and south to Peru. Rare living, though not uncommon as beach shells.

Subgenus MONOPLEX PERRY, 1811

Varices widely spaced; spiral sculpture of uniformly wide ribs.

956. Cymatium (Monoplex) parthenopeum keenae (Beu, 1970) (Synonym: **Murex parthenopeus** Von Salis, 1798, of authors). Although *C. parthenopeum* has long been considered to have cosmopolitan distribution in the Mediterranean, Caribbean, and eastern and western Pacific, Beu (1970) has shown

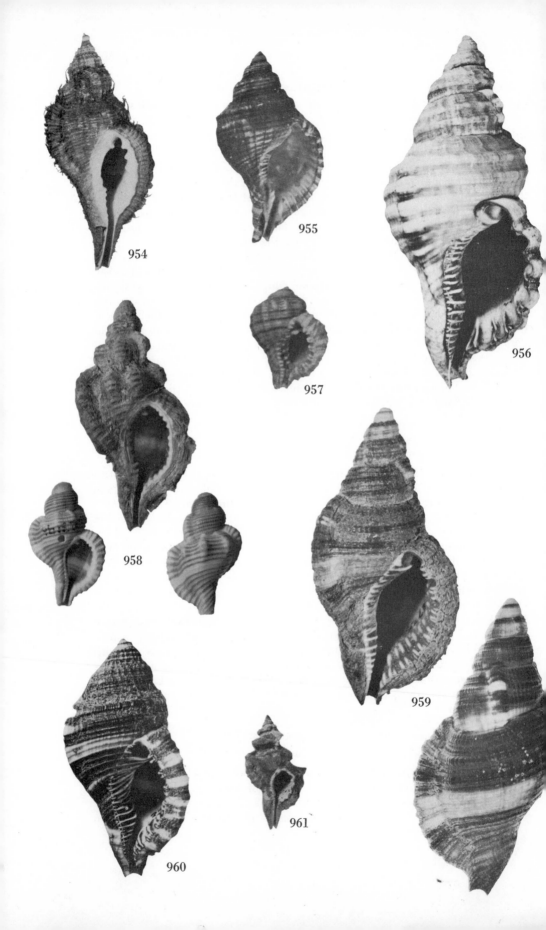

that there are consistent geographic differences. The eastern Pacific form differs by having seven spiral ridges per whorl, more than in any of the other species or subspecies he recognizes. Beach-worn shells show heavy spiral and axial ribbing, but specimens taken alive have shells that are completely obscured by a heavy, fibrous, almost woolly, brown periostracum. A mature specimen from Mexico measures 80 mm in length and 45 mm in diameter. La Paz, Gulf of California, to the Galápagos Islands, intertidally and offshore.

957. Cymatium (? Monoplex) lignarium (Broderip, 1833). The solid little shell is yellowish brown with dark brown revolving bands. On the orange-colored columella are a double series of small white nodes. The outer lip is stained with orange, the tuberculations white. Length, about 25 mm; diameter, 15 mm. The Gulf of California to Negritos, Peru.

Subgenus **SEPTA** PERRY, 1810
(**LAMPUSIA** SCHUMACHER, 1817)

Varices widely spaced; spiral sculpture of ribs alternating in size.

958. Cymatium (Septa) lineatum (Broderip, 1833). A yellowish-brown shell with chestnut-brown spiral ribs, the columella and aperture dark brown with white lirations and teeth. Length, 56 mm. Galápagos Islands. A similar species in the Atlantic is *C. (S.) krebsii* (Mörch, 1877).

959. Cymatium (Septa) pileare (Linnaeus, 1758) (Synonyms: **Triton aquatilis** Reeve, 1844; **T. martinianum** Orbigny, 1845; **Litiopa effusa** C. B. Adams, 1850; **T. intermedius** Pease, 1869; **T. veliei** Calkins, 1878; **C. vestitum insulare** Pilsbry, 1921; **Dissentoma prima** Pilsbry, 1945; **C. vestitum** of west coast authors, in part). The synonymy for this species has been worked out by Emerson and Old (1963), who show that two species of *Septa* (previously called *Lampusia*) occur in the eastern Pacific. The *Cymatium pileare* complex is wide-ranging, from the western Pacific to the western Atlantic; thus, its occurrence in the Panamic province is plausible. It is a relatively slender form, with appressed sutures, the spire equal to or longer than the aperture. The inside of the aperture is red or rusty brown, with white plications and teeth. Length, 70 to 90 mm; width, 35 to 45 mm. Cedros Island, Baja California, through the Gulf of California and south to Panama; also in the western Pacific and the Caribbean. Houbrick and Fretter in 1969 reported this species as feeding on bivalves in Hawaii, anesthetizing their prey with an acid fluid.

960. Cymatium (Septa) vestitum (Hinds, 1844) (Synonym: **Triton vestitum senior** C. B. Adams, 1852). The shell is light brownish under a bristly periostracum. The spire is relatively low, shorter than the aperture, and the suture is channeled; whorls are more inflated than those of *C. pileare*, and the aperture is white with dark brown stains between the plications and teeth. Length, 50 to 80 mm; width, 30 to 40 mm. Manzanillo, Mexico, to Panama and the Galápagos Islands, possibly to northern Peru.

Subgenus **TURRITRITON** DALL, 1904

With shouldered whorls and deeply indented suture.

961. Cymatium (Turritriton) gibbosum (Broderip, 1833) (Synonym: ? **C. adairense** Dall, 1910). There are four varices per whorl on the yellowish-brown shell. The flat shoulders of the whorls and the spiral-staircase outline are characteristic. Length, about 40 mm; diameter, 23 mm. Sonora, Mexico, to the Galápagos Islands and Peru, on and under rocks at low tide. *C. adairense* may qualify as a subspecies that is smaller, with coarser sculpture, confined to the northern part of the Gulf of California. Whether it is validly different remains to be determined.

Genus DISTORSIO RÖDING, 1798

The flaring callus and the wavering suture set this genus apart from the other members of the family. Emerson and Puffer (1953) have published a useful summary of species, synonymies, and distribution. No species of *Distorsio, s. s.*, occur in the Panamic province.

Subgenus RHYSEMA CLENCH & TURNER, 1957

Callus plate narrower and less detached from body whorl than in *Distorsio, s. s.*

962. Distorsio (Rhysema) constricta (Broderip, 1833). Compared with *D. decussata*, the shell is shorter and thicker, with a shorter canal and only one fold on the upper part of the inner lip, with two or three tubercles above it. Length, 40 mm; diameter, 25 mm. The range is Tiburon Island, Sonora, Mexico, south to Mancora, Ecuador. A similar Atlantic species is *D. clathrata* (Lamarck, 1816), type of the subgenus.

963. Distorsio (Rhysema) decussata (Valenciennes, 1832) (Synonyms: **D. ridens** of authors, not of Reeve, 1844; **D. cancellinus** of authors, not of Roissy, 1805). This has a comparatively slender and thin shell with a long anterior canal and two folds on the inner lip near the posterior end of the aperture, one emerging at the angle and one a little lower. Length, 55 mm; diameter, 30 mm. Cape Tepoca and Guaymas, Sonora, Mexico, south to Manta, Ecuador, offshore in depths to 82 m.

Family BURSIDAE

Spire bluntly pointed; sculpture nodose; whorls laterally compressed; anterior canal slotlike, with a thickened edge.

Genus BURSA RÖDING, 1798

The frog shells lack the fuzzy periostracum of the Cymatiidae and have mostly two varices per whorl. The aperture has a deep slot for the posterior siphon, resembling the anal notch of the Turridae, but the growth lines do not bend back around it.

964. Bursa caelata (Broderip, 1833) (Synonym: **B. c. louisa** Smith, 1948). The chestnut-brown shell has spiral rows of blackish-brown nodes that are strongest on the periphery; the aperture is whitish. Length, 50 mm; diameter, 30 mm. Baja California and the Gulf of California south of Guaymas, Mexico, southward to Peru; reported on Socorro and the Galápagos Islands. Intertidally under stones and in coral masses offshore to depths of a few meters.

965. Bursa calcipicta Dall, 1908. Light brown under a chalky and finely reticulate outer coating, this is a nodosely sculptured form with a somewhat tubular posterior notch. The inside of the aperture is white, flushed with pink. Length, 44 mm; diameter, 25 mm. Tenacatita Bay, Jalisco, Mexico, 9 to 11 m, to Panama Bay (type locality), intertidally to 120 m, and La Plata, Ecuador.

Bursa granularis (Röding, 1798). Reported on Clipperton Island. It is an Indo-Pacific species, observed feeding on annelid worms in Hawaii.

966. Bursa nana (Broderip & Sowerby, 1829) (Synonym: **? Ranella albo-fasciata** Sowerby, 1841). A typical specimen is nearly smooth except for a row of tubercles at the shoulder of the whorl. The color is purplish to reddish brown, with a white band below the periphery and with white on the inner and outer lips of the aperture, the brown banding showing within. Some authors have regarded *B. n. albofasciata* as a subspecies. It has several spiral rows of granules in addition to the one at the shoulder, but there is much variation between the two extremes; the more sculptured specimens seem to predominate at the southern end of the range, especially south of Panama. Length, 40 mm; diameter, 24 mm. Guaymas, Mexico, to Ecuador, offshore, in depths to 37 m.

967. Bursa sonorana Berry, 1960. Although named as a subspecies of the Californian species *B. californica* (Hinds, 1843), the Sonoran frog shell has a number of points of difference: it is smaller, the spire is higher and more conic, the nodes are smaller and sharper. The shell is white with a brown band and some brown staining. Height, 99 mm; diameter, 62 mm. In beach drift, Cape Tepoca (Mrs. Paul Skoglund, collector) and from shrimp boats in the Guaymas area, Sonora, Mexico (type locality). Further collecting will be necessary to show whether there is any overlap in range and shell characters with the more northern *B. californica*. For convenience, *B. sonorana* is treated as a distinct species here, but it may well prove not to be separable.

Family COLUBRARIIDAE

Although resembling the Cymatiidae, these shells are smaller and more slender, with smoother and more polished surfaces. Varices are somewhat irregularly spaced. The operculum is horny, its nucleus apical.

Genus COLUBRARIA SCHUMACHER, 1817

Elongate shells with a short anterior canal. Sculpturing may be granular, cancellate, or of sinuous axial ribs. Varices are present, not regularly spaced and not continuous from whorl to whorl. Inner lip with a wide callus plate, smooth, not wrinkled as in the Cymatiidae. The suture rides up on previous whorl at the varix. Two Panamic subgenera.

Subgenus COLUBRARIA, *s. s.*

Sculpture regularly cancellate, at least on spire; body whorl shorter than spire (i.e., less than half the length of the shell).

968. Colubraria (Colubraria) jordani Strong, 1938. Most slender and high-spired of the Panamic Colubrarias, this has a light brown shell with two spiral rows of darker spots, each whorl with one or two varices. Length, 35 mm; maximum diameter, 10 mm. Socorro Island, Revillagigedo group, off west Mexico.

969. Colubraria (Colubraria) lucasensis Strong & Hertlein, 1937 (Synonyms: **? Triton reticulatus** Sowerby, 1833 [not Blainville, 1829]; **"C. soverbii"** of authors, not of Reeve, 1844). Shell brownish, with whorls more inflated than in the other species, the nuclear whorls glassy, the remainder cancellately sculptured and spotted or streaked with darker and lighter color. Length, 27 mm; diameter, 10.5 mm. Off Cape San Lucas, Baja California, in 37 to 45 m. Guaymas, Sonora, Mexico, and south to Panama and the Galápagos Islands. Emerson and D'Attilio (1966) have discussed the status of this species.

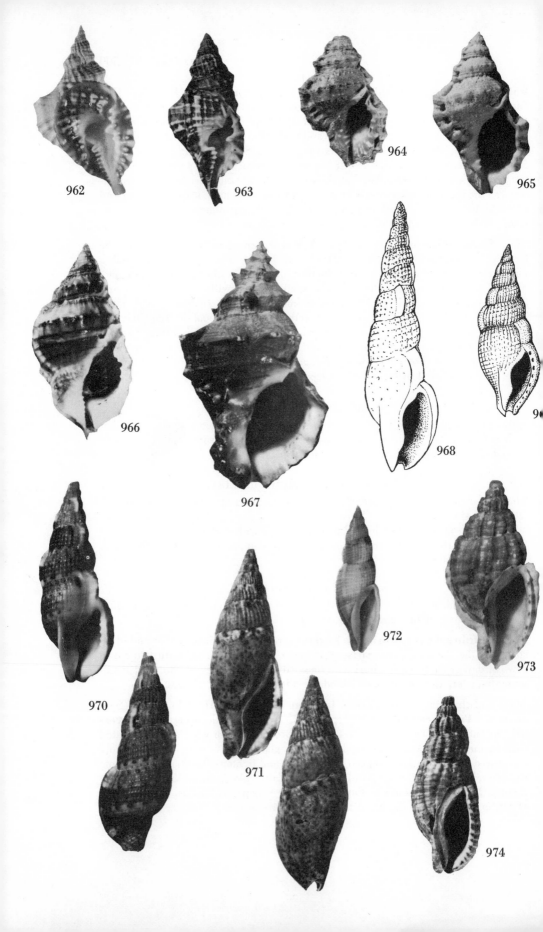

962

963

964

965

966

967

968

96

970

971

972

973

974

970. Colubraria (Colubraria) ochsneri Hertlein & Allison, 1968. Most colorful of the West American species, the shell is light brown with a darker medial concentric band, above and below which on the body whorl are small brown spots in a spiral row; varices one per whorl, lighter in color but with brown spots corresponding to the bands. Sculpture regularly cancellate; outer lip with paired lirae within, inner lip with a broad, smooth, white callus. Length, 30 mm; width, 11 mm. Clipperton Island. Perhaps closest to the central Pacific species *C. nitidulus* (Sowerby, 1833), but differing in a number of features (smaller size, wider aperture, stubbier outline, and coarser sculpture).

971. Colubraria (Colubraria) procera (Sowerby, 1832). Named as a *Columbella* and thought to be from Panama, the immature type specimen, which was unillustrated, lay forgotten in the British Museum collection for over a century. In 1967 a living specimen—fortunately an adult—was taken by a California collector, Lawrence Thomas, who had found it while diving in southern Mexico. It matched a photograph of Sowerby's type then available. Thus, identification of the species could be made, and it could be assigned to its proper genus, *Colubraria*. The mature shell is a somewhat mottled reddish brown, nearly smooth except for some cancellate sculpture on the spire and a few irregularly spaced varices. Just below the suture is an indistinct color band of lighter and darker spots, and the outer lip, lirate within, has two brown spots on its edge. The ovate aperture and the callus of the inner lip are ivory white to buff. Sowerby's type specimen, which was collected by Hugh Cuming (who at that time had not done much work north of Panama and who therefore had probably acquired the shell from some seaman) is faded, though not beach-worn; it now is a lavender-gray, with brown markings. The species is the largest of the West American forms and rivals the Indo-Pacific *C. soverbii* (Reeve, 1844) for size. Length, 75 mm; diameter, 28 mm. Since 1967 other collectors have taken specimens, all in the general vicinity of Manzanillo, Mexico, under rocks at depths to about 18 m. Also known as a Pleistocene fossil from Oaxaca, Mexico.

972. Colubraria (Colubraria) siphonata (Reeve, 1844) (Synonyms: **C. aphrogenia** Pilsbry & Lowe, 1932; **C. panamensis** and **perla** M. Smith, 1947). Campbell (1961) has supplied a good description of the species: the slender, fusiform shell is dark bluish gray to light brown, spotted with orange-brown on whorls; with whitish areas on the varices; the two whorls of the protoconch are dark purple to light gray in the middle and are followed by six subsequent convex whorls with eight to ten spiral cords intersected by narrow axial ribs; varices small, irregularly spaced, about two per whorl; aperture narrow, contracted into a canal anteriorly; lip finely denticulate. Length, 20 to 26 mm; diameter, 10 to 13 mm. Off Puertecitos, Baja California, and Guaymas, Mexico, to Panama Bay, in depths to 18 m.

Subgenus TRITONOHARPA DALL, 1908

Sculpture mainly axial, not regularly cancellate; aperture more than one-half the shell length.

973. Colubraria (Tritonoharpa) vexillata (Dall, 1908). A white shell with six to eight brown bands; varices sharp, with several axial ribs between; spiral bands ending at small denticles in the outer lip. Length, 15 mm; width, 7.5 mm. Galápagos Islands, 1,460 m. Type species of *Tritonoharpa*.

974. Colubraria (? Tritonoharpa) xavieri Campbell, 1961. Relatively solid, with coarsely sculptured whorls and somewhat sinuous axial ribs crossed but not rendered cancellate by fine concentric striae; varices approximately two per whorl; color a medium brown with some darker and lighter areas, varices lighter; spiral lines in pairs, ending at small denticles in the outer lip; aperture purplish brown within. Length, 27 mm; width, 10 mm. An offshore form, dredged off Guaymas, Sonora, Mexico, 180 m. The whorls are more cylindrical and the anterior canal relatively shorter than in *C. vexillata*, and the axial ribs differ in number and arrangement.

Order NEOGASTROPODA
(STENOGLOSSA)

With minor exceptions the Neogastropoda are all carnivorous, and their feeding habits are reflected in the modifications of the soft parts, especially the radula, which may be rachiglossate ("spiny-tongued") or toxoglossate ("bow-tongued"). The radula has been dispensed with altogether in some, when there is a ready source of soft food, such as coral flesh. A long inhalant siphon and consequent anterior notch or canal are characteristic.

Superfamily MURICACEA

The radula is rachiglossate, which means that it is narrow, with three teeth in each transverse row, a central or rachidian tooth, and a pair of laterals. The teeth may be simply sharp-edged or may be variously cusped. All the Muricacea are carnivorous.

Family MURICIDAE

The rock shells reach the climax among the gastropods in complexity of spines, from symmetrical, slender ranks, as in the Venus's-comb, to intricately frilled varices. Varices—periodic thickenings of the outer lip—are characteristic of the family, the shape being determined by the convolutions of the edge of the mantle. The operculum is horny and oval, its nucleus at the anterior end (apical) in many, but there are some with a "purpuroid" type, in which the nucleus has moved laterally to the right. The radula seems to be the best single character for consistently dividing the various groups, especially the subfamilies. In all the Muricidae each row of radular teeth has a central tooth flanked on either side by a hook-shaped unicuspate lateral. The central tooth has several cusps, and it is the number and arrangement of these that is significant. Egg capsules are also distinctive structures, but useful in classification at the generic rather than the higher levels. Dr. Gunnar Thorson of the Marine Biological Laboratory at Helsingör, Denmark, generously made available for study his unparalleled collection of egg capsules.

Members of this family have been called "drill shells," but mostly they prefer to chip the edges of clam shells in getting at the juicy meat rather than to drill in the manner that *Natica* or *Polinices* do, and the holes a muricid makes are rougher, less even, and less finely beveled than those of the moon snails. Fruitful studies on the mechanism of boring have been made by Dr. Melbourne Carriker at the Marine Biological Laboratory, Woods Hole, Massachusetts, during the 1960's. Experimenting with muricid species he found that drilling is in large part facilitated by the use of enzymes to soften the organic cement of the prey's shell, so that the radula teeth may then scrape it away grain by grain. Many of the muricid species have developed a tooth on the outer lip that seems to be useful as a lever in holding or wedging apart the valves of mussels or clams. Sorensen (1943) has an illuminating description of the pink-mouthed and the black murexes at work.

Because the shells are striking in appearance, they early attracted the attention of collectors and describers. Misinterpretations of some of the illustrations and errors in stated locality have complicated the modern problem of attaching correct names, and the result has been no little confusion in the literature. Differences of ideas among authors as to the degree of subdivision of the Linnaean genus *Murex* that is justifiable also have led to confusion, many authors not taking variations into account and others ignoring or being unaware of previous work. The classification adopted here is something of a compromise between the extremes of splitting and lumping. It has been benefited to no small degree by the published work of Dr. Emily Vokes (1964 on), Dr. William K. Emerson (1959 on), Mr. Anthony D'Attilio (1963 on), and Dr. George Radwin (1970 on), as well as by correspondence and advice from them, which is gratefully acknowledged. Classification of the Muricidae continues to be revised because of new insights into relationships, made possible by fresh collections of live-taken material that have recently become increasingly available. Parallel development of forms that look alike but that come from basically unrelated stocks, through the rapid evolution of the family during the Tertiary, has added complications to the problems of interpreting kinship, and it is to be expected that still further adjustments will be required before a really stable scheme is worked out. Five subfamilies are here recognized in the Panamic province, only one of which, the Typhinae —with tubular open spines—can be defined on shell characters alone. Because basic differences for the other four involve a study of the radula, no key to subfamilies is practicable here.

Subfamily MURICINAE

Medium-sized to large shells, varices well developed in regular series of three, four, six, or more per whorl. The operculum has an apical nucleus in most species of the subfamily, but there are exceptions. The radula is regarded as the most stable basis for assessing relationships; it has a central tooth with five unequal-sized cusps, the middle and outer ones larger than those between. The egg capsules tend to occur in large masses, individual capsules with a platform above and below that is welded to adjacent capsules in a distinctive network pattern, reminding one of modular structures in modern architecture. Several specific names that are doubtfully to be assigned to the Muricinae are discussed in the Appendix. The ten Panamic genera may be keyed as follows:

1. Varices winglike, broad...................................... *Pterynotus*
   Varices not winglike or flaring, tending to be spinose................... 2
2. Anterior canal long, straight, tubular............................. *Murex*
   Anterior canal short to medium in length, somewhat bent............... 3
3. Spiral sculpture dominant, of regular raised ribs................. *Murexsul*
   Spiral sculpture intersected by axial ribs or spines.................... 4
4. Varices three per whorl...................................... *Phyllonotus*
   Varices more than three per whorl................................. 5
5. Shells small to medium-sized, adults less than 50 mm in height.......... 6
   Shells large, adults more than 50 mm in height....................... 9

6. Spiral rib ends foliose, the ribs intricately webbed together by growth lamellae .... *Murexiella*
   Spiral ribs not webbed by growth lamellae........................... 7
7. Spire relatively low; operculum with lateral nucleus.......... *Homalocantha*
   Spire pointed; operculum with apical nucleus......................... 8
8. Columella granular or with low folds; outer layer of shell not chalky....
   .... *Muricopsis*
   Columella smooth; outer layer somewhat chalky.................. *Paziella*
9. Posterior end of aperture with a raised anal notch at the suture on all varices .... *Hexaplex*
   Posterior end of aperture lacking a raised notch, all varices soldered firmly and smoothly at suture.. *Muricanthus*

Genus MUREX LINNAEUS, 1758

Linnaeus included in this genus many species that now would be assigned to other families. Fixation of the type species as the Indo-Pacific *M. tribulus* Linnaeus, 1758, restricts the concept in the strict sense to forms in which the shell has a long and nearly straight anterior canal and varices that are mostly three in number, more or less spinose. Members of the genus *Murex* so defined are tropical in distribution, but some of the related genera have become adapted to life in colder waters. Several subgenera of *Murex* have been erected, only one of which, *Murex, s. s.*, is represented in the eastern Pacific.

Subgenus MUREX, *s. s.*

Spire low, body whorl inflated, anterior canal long, straight, nearly closed; varices and canal with spines.

975. Murex (Murex) elenensis Dall, 1909 (Synonym: **M. plicatus** Sowerby, 1834 [not Gmelin, 1791]). A creamy to yellowish-white shell, with relatively short and stubby spines, longer on the canal. Aperture tinged with lavender within. Length, about 75 mm. Scammon's Lagoon, Baja California, throughout the Gulf of California and south to Ecuador.

Murex (Murex) recurvirostris Broderip, 1833 (Synonym: **M. nigrescens** Sowerby, 1841). The spines are fewer in number than in *M. elenensis* or almost wanting. In color this varies from whitish to terra-cotta, with two to three broad brown bands that are most evident within the aperture. Length, about 50 mm; diameter, 25 mm. The species has a clinal range of variation in expression of spines that is close to being geographically separable, so much so that three subspecies might be recognized:

976. Murex (Murex) recurvirostris recurvirostris Broderip, 1833. Spines sparsely developed or entirely wanting on varices and anterior canal. Southern Mexico to Ecuador.

977. Murex (Murex) recurvirostris lividus Carpenter, 1857. A few stubby spines at the base of anterior canal and one at the shoulder of each varix; interspaces with low, concentrically striate nodes, shell uniformly colored, mostly light buff. Gulf of California to Mazatlán, offshore in depths to 55 m.

978. Murex (Murex) recurvirostris tricoronis Berry, 1960 (Synonyms: **M. funiculatus** Reeve, 1845 [not Defrance, 1827]; **M. lividus** Carpenter of authors, not Carpenter, 1857). Spines three per varix, some spines at base of

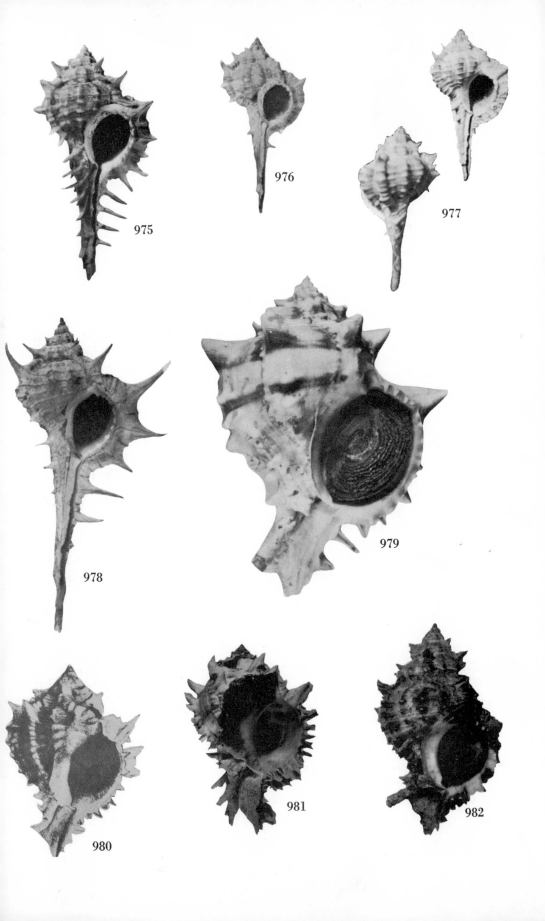

canal but none on canal (as distinguished from *M. elenensis*); interspaces with a tendency toward fine red or brown spiral lines. Cedros Island (type locality), Baja California, to the southern end of the Gulf of California, offshore, in depths to 75 m.

Genus HEXAPLEX PERRY, 1810

The type species, *Murex cichoreum* Gmelin, 1791, is Indo-Pacific in distribution. It shares with American species the tendency toward pink coloring around the aperture and an anal notch that stands free from the suture at each varix. The group is very close to the American *Phyllonotus*, and many workers would prefer to allocate the American species to the latter. However, there are some consistent points of difference, such as the more numerous varices in *Hexaplex* (five and a half to seven per whorl, as compared with three to three and a half in *Phyllonotus*) and the more flaring lip callus in *Phyllonotus*, that seem to justify the separation of a group of West American species as *Hexaplex*. Figures of the radulas of the two suggest that there is a very close relationship, and the fossil record indicates that it may have been even closer in the past.

979. Hexaplex brassica (Lamarck, 1822) (Synonym: **Murex ducalis** Broderip, 1829). The cabbage murex is white to light brown or pinkish, with three brown bands. The edges of the varices and margin of the aperture are bright pink. The operculum is exceptional in having the nucleus nearly central, with coarse concentric laminae surrounding it. The varices are not so frilly as are those of other species of *Hexaplex*. Length, 150 to 200 mm (6 to 8 inches). Guaymas, Mexico, to Peru; rare on tide flats, commoner offshore in depths to 55 m.

980. Hexaplex erythrostomus (Swainson, 1831) (Synonyms: **Murex bicolor** Valenciennes, 1832 [not Risso, 1826]; **? M. rhodocheilus** King, 1832; **M. hippocastanum** Philippi, 1845 [not Linnaeus, 1758]). The pink-mouthed murex is probably the most abundant member of the family in this province. The white shell is dull-surfaced on the outside but polished and bright pink within. Now and then one finds a specimen in which the pink is reduced to a faint trace around the edge of the lip. About 100 mm (4 inches) in length. Both Sorensen (1942, 1943) and Lowe (1933) have commented upon the feeding habits of these snails, which prey upon such large clams as *Megapitaria squalida* at extreme low-tide level and can be taken abundantly in offshore traps. Egg capsules, described by Faye Wolfson (1968), are triangular or tongue-shaped, in a network pattern, with the appearance of a series of chevron-shaped modules. The Gulf of California to Peru.
See Color Plate VII.

981. Hexaplex regius (Swainson, 1821). The royal murex is the brightest-colored of the three Panamic species of *Hexaplex*. The shell itself is white with a bright pink aperture, as in *H. erythrostomus*, but the columellar lip and the wall above it are brown, shading from light brown in some areas to almost black in others. The varices being somewhat reflected and doubled, the pink lining is brought to the surface in stripes, and the upper edges of the brown wash on the front of the body whorl remain as spots between varices on the whorls of the spire. Length, 100 to 125 mm. The southern part of the Gulf of California to Peru, on tide flats at low tide.
See Color Plate VII.

Genus PHYLLONOTUS SWAINSON, 1833

Resembling *Hexaplex* but with fewer varices (mostly three to three and a half) per whorl; varices with subdued spines or none, and with less frilling; pink coloration of aperture mostly wanting; inner lip callus somewhat more flaring. Egg capsules as in *Hexaplex*.

Because of the closeness of relationship, it would seem logical to make one group a subgenus of the other, and if *Phyllonotus* had nomenclatural priority, *Hexaplex* probably would be regarded as a subgenus of it. Subordinating *Phyllonotus* under *Hexaplex*, however, would be unacceptable to many workers. The conservative course, adopted here, is to regard both as genera. *Phyllonotus* was thought to be a group confined to the Western Atlantic until the shrimp boats began bringing in offshore Eastern Pacific material.

982. Phyllonotus peratus Keen, 1960 (Synonym: **P. peratus decoris** Keen, 1960). Medium-sized, grayish brown, with traces of underlying color bands of brown below suture and on base; areas between varices lighter, intervarical nodes with darker spots; aperture white, suffused along the margin with creamy yellow, showing four dark bands on the outer lip that continue above the aperture as a brown blotch. Aperture ovate, anterior canal nearly closed, recurved. The variant named *P. p. decoris* probably is only a color form, although it seems to be slightly shorter and wider and to be more northern in distribution. Length, 57 to 69 mm; width, 38 to 41 mm. Gulf of Tehuantepec to Panama, in depths to 77 m.

Genus HOMALOCANTHA MÖRCH, 1852

Spire short; suture somewhat tabulate; varices four to ten, frilled, tending to have the spines expanded at their ends; operculum with a lateral nucleus.

983. Homalocantha multicrispata (Dunker, 1869) (Synonyms: **Murex crispus** Broderip, 1833 [not Lamarck, 1803]; "**M. tortuosus** Sowerby, 1833" of authors [in synonymy; not Borson, 1821]). The blunt spines are finely frilled along their edges, and the light brown shell has its larger ribs and fronds chocolate brown. Length, 50 mm; width, 30 mm. Ecuador to northern Peru, in depths to 90 m.

984. Homalocantha oxyacantha (Broderip, 1833) (Synonyms: **Murex varicosus** Sowerby, 1841 [not Brocchi, 1814]; **M. stearnsii** Dall, 1918). With about eight varices crossed by scaly spiral ribs, the shell has the spines of the lip varix especially long but sharp-pointed, not splayed out as in western Pacific species. Color is grayish white, the spine tips of the last few varices tinged with light to medium brown. The spire is low and somewhat tabulate. Length, 45 mm; diameter, 31 mm. Manzanillo, Mexico, to southern Ecuador; Pleistocene, Magdalena Bay, Baja California. The species has been misidentified as *Murex melanamathos* Gmelin, 1791, by some authors, but that form turns out—as D'Attilio (1967) has shown—to be west African; it has a higher spire, less marked spiral sculpture, and spines on the area just below the suture, wanting in *H. oxyacantha*. The form named *H. stearnsii* by Dall is a variant with heavy spiral ribs.

Genus MUREXIELLA CLENCH & PÉREZ FARFANTE, 1945

The varices are (with a few exceptions) four to six in number, composed of a row of spines intricately netted or webbed together by fine layers of frills; ante-

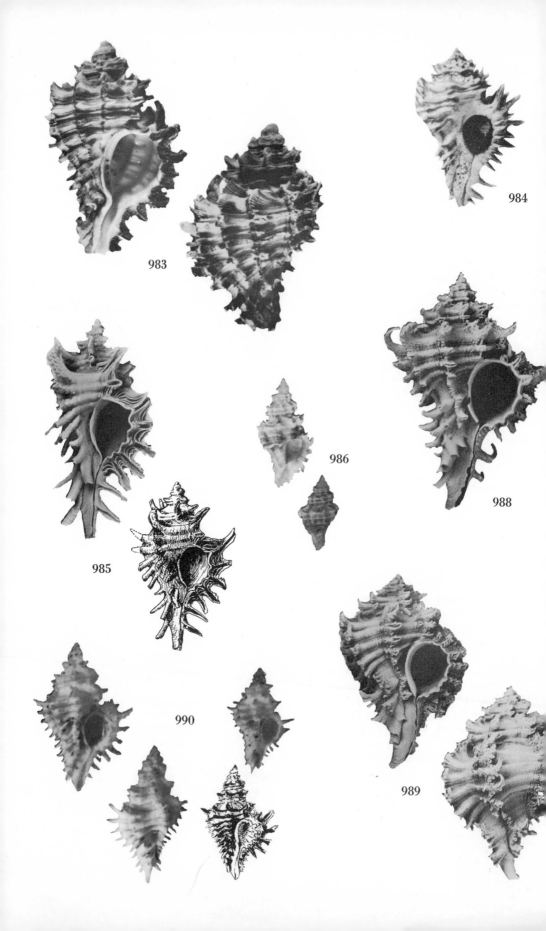

983

984

985

986

988

990

989

rior canal moderately broad; operculum with a subapical nucleus. The type species is western Atlantic, and its relationship to West American species has only recently been recognized (Vokes, 1964; 1970).

985. Murexiella diomedaea (Dall, 1908). The small reddish-brown shell has four to five spinose varices. Length, 29 mm; diameter, 15 mm. Reported only from the type locality, Panama Bay, in 155 m depth.

986. Murexiella dipsacus (Broderip, 1833) (Synonym: **Murex peruvianus** Sowerby, 1841 [not Lamarck, 1816]). There are about ten varices per whorl, with recurved spines at the shoulders that reminded Broderip of the hooks of the teasel-head. The shell is light brown, with a band of white on the peripheral spines and another on the base, where another row of spines sets off the pillar from the body whorl. This may be a southern form of *M. lappa*. Length, 25 mm; width, 15 mm. Salinas, Ecuador (type locality) to Peru.

987. ? Murexiella exigua (Broderip, 1833). The type specimen seems to have been lost. The figure suggests a juvenile shell that is now probably unrecognizable. Length, 12 mm; width, 8 mm. Salango, Ecuador.

988. Murexiella humilis (Broderip, 1833). This species has been generally misidentified because a specimen other than the type was later figured by Sowerby. The type seems to have been lost, but a specimen that matches Sowerby's original figure has been found by Ruth Purdy and illustrated by Vokes (1970); it is refigured here. The color is white, with two indistinct brown bands, one above the aperture, the other across the base. Spiral ribs are well spaced, the anterior canal straight and relatively narrow. Length, 32 mm; width, 22 mm. Sonora, Mexico, to Santa Elena (type locality), Ecuador, intertidally and offshore to depths of 15 to 33 m.

989. Murexiella keenae Vokes, 1970. This is the form commonly mistaken for *M. humilis*. It is a chunky pinkish-brown shell with about five varices, intricately webbed, having fine sculpture in the interspaces between the spiral ribs. The anterior canal is broader than in *M. humilis* and somewhat curved. Length, 35 mm; diameter, 24 mm. Nicaragua to Ecuador, intertidally and offshore to depths of 33 m.

990. Murexiella lappa (Broderip, 1833). The shell is light brown, with rather elongate spines on the upper part of the lip varix. Length, 30 mm. Magdalena Bay, Baja California, through the southern part of the Gulf of California and south to Ecuador, mostly offshore in depths to 37 m.

991. Murexiella laurae Vokes, 1970. The shell is chestnut-brown, with an outline reminiscent of that of *M. lappa*. In contrast to *M. keenae*, the intervarical spaces are smooth, not ribbed. Length, 20 mm; width, 13.5 mm. Manzanillo to Acapulco, Mexico, in 30 to 45 m.

992. Murexiella minuscula (M. Smith, 1947). Named as a subspecies of *Murex vittatus*, it was said to be easily separated by the much smaller size, more slender shape, and pinched appearance on the back of the last whorl. The figure shows a sturdy shell with a broad dark band across the upper half of the whorls, white below, with an unusually large siphonal fasciole. Length, 18.5 mm; width, 11.5 mm. Panama Bay.

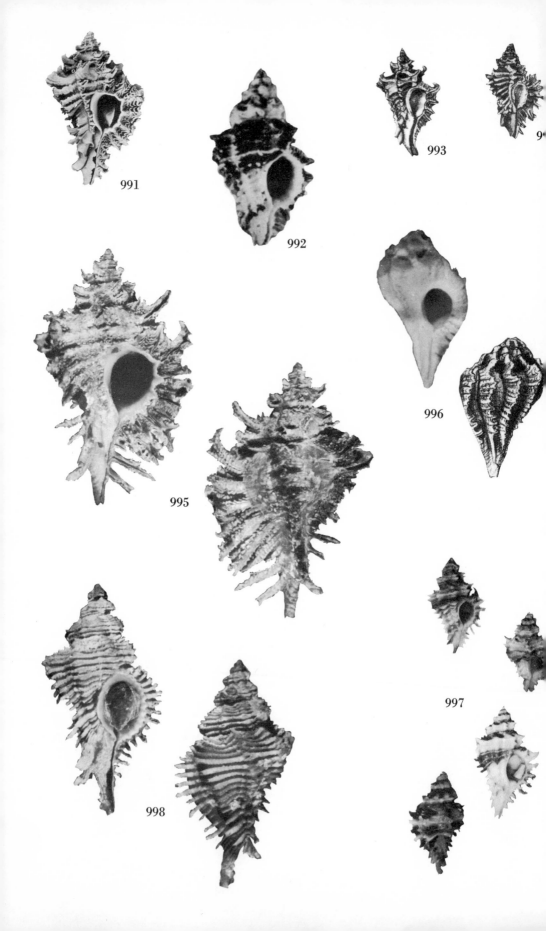

991

992

993

9

995

996

997

998

993. Murexiella perita (Hinds, 1844). The rather slender shell is brown, with darker markings on some of the ribs. Length, 23 mm; diameter, 15 mm. Magdalena Bay, Baja California, to Manzanillo, Mexico, 13 to 95 m depth.

994. Murexiella radicata (Hinds, 1844). The five varices of this species are delicately spinose. The shell is light brown or whitish. Length, about 20 mm. Guaymas, Mexico, to Panama, under rocks at extreme tide or offshore to depths of 20 m.

995. Murexiella radwini Emerson & D'Attilio, 1970. The shell is pinkish violet in color, with a violaceous aperture and flesh-toned varices, which consist of long recurved spines, intricately webbed. The spines at the anterior end are long and free of the webbing. The outline is somewhat reminiscent of that of *M. laurae* Vokes. Length, 34 mm; diameter, 22 mm. Tagus Cove, Isabela Island, Galápagos Islands, depth 100 m.

996. Murexiella santarosana (Dall, 1905) (Synonym: **Murex fimbriatus** A. Adams, 1853 [not Brocchi, 1814]). A Californian form not positively reported south of Cedros Island, Baja California, except that Adams's type specimen was said to have come from the Gulf of California. A photograph of the apertural view of Adams's holotype in the British Museum confirms the suspicion of authors that his form is identical with Dall's. Length, 40 mm.

997. Murexiella vittata (Broderip, 1833). The seven varices are made up of short spines. The shell itself is waxy white, spirally banded with black. About 20 mm in length. The Gulf of California to Guayaquil, Ecuador, and the Galápagos Islands, intertidally and to depths of 20 m.

Genus **MUREXSUL** IREDALE, 1915

With eight varices and strong spiral sculpture that ends in delicate fronds on the varices; spire relatively high, as in the Ocenebrinae, but radula and operculum as in the Muricinae; outer lip smooth within. The type species, *M. octogonus* (Quoy & Gaimard), is southern Pacific in distribution, and the genus is only recently recognized in the eastern Pacific.

998. Murexsul jacquelinae Emerson & D'Attilio, 1969. Pinkish violet in color, the small shell has conspicuous spiral ridges, with hollow fronds at the varices, especially on the outer lip; the aperture is oval, rosy violet within. Length, 20 to 26 mm; width, 11 to 16 mm. Galápagos Islands, 90 m. The shell somewhat resembles *Murexiella lappa* but is not color-banded as that is, and it lacks webbing between spines.

Genus **MURICANTHUS** SWAINSON, 1840

Although the position of the anal siphon is marked by a slight ridge on the body whorl, there is no looping back of the outer lip, and the varices are smoothly cemented to the shell at the suture line. The shells are mostly white with brown or black markings, with 5 to 14 varices per whorl, large, and solid.

999. Muricanthus ambiguus (Reeve, 1845) (Synonym: **Murex melanoleucus** Mörch, 1860). This is perhaps not a distinct species but a transitional form between the southern *M. radix* and the northern *M. nigritus*. However, specimens are easily separable both in form and geographic distribution. The varices are less numerous than those of *M. radix*, the shell is larger, and it is not so definitely pear-shaped. Compared with *M. nigritus* there are more varices and the

999

1000

1001

canal and spire are not so long in proportion to the diameter of the shell. Large specimens may measure as much as 175 mm (7 inches) in length. The average length is about 100 mm. Southern Mexico to Panama.

1000. Muricanthus callidinus Berry, 1958 (Synonym: **Murex nitidus** of Reeve, 1845, not of Broderip, 1833). Somewhat smaller than *M. ambiguus*, to which it is closely related, this form has spinose rather than elaborately frilled varices. The shell is white, with brown stripes of variable width. Length of a large specimen, 99 mm (4 inches); maximum diameter, 86 mm (3½ inches). Guatemala to Costa Rica.

1001. Muricanthus nigritus (Philippi, 1845). This name has been attributed to Meuschen, 1787, by some writers, but that author is now considered not to have conformed to the principles of binomial nomenclature; thus, his names have no standing. Besides, what he had was apparently the shell later named *M. radix* by Gmelin. The black murex is clearly distinct from *M. radix*, having a much higher spire and longer canal. Sorenson (1943) has illustrated a growth series that shows how the young shells, which are nearly pure white, become progressively more clouded with black as they increase in size to a maximum length of about 150 mm (6 inches). He also discussed food habits. Black murexes prey especially upon clams, chipping the edges of the valves in order to insert the proboscis and suck out the rich juice. *M. nigritus* seems to be confined to the Gulf of California, where it is fairly common; Lowe (1935) reported it as abundant on reefs in the northern part of the Gulf, feeding on *Cerithium stercusmuscarum*. Egg capsules, which are laid singly, not in netted masses, are vase-shaped, attached at the slender base by a small platform.

1002. Muricanthus princeps (Broderip, 1833) (Synonym: **? Murex norrisii** Reeve, 1846). The shell is somewhat biconic in shape, with five to eight varices, whitish, with the ribs and spines tinged brown. About 125 mm in length. The southern part of the Gulf of California to Peru, intertidally on rocks at extreme low tide and offshore in shallow water.

1003. Muricanthus radix (Gmelin, 1791). The root murex is a somewhat pear-shaped shell, white, with numerous black-spined varices. The spines are imbricating or overlapping, like tipped-up roof tiles. It is a solid-feeling shell for its size, in contrast to the northern *M. nigritus*, which is much larger and thinner. Length, 100 mm (4 inches). Panama to southern Ecuador. This species has been fixed as the type of the genus by the International Commission on Zoological Nomenclature (Opinion 886, 1969).

Genus **MURICOPSIS** BUCQUOY, DAUTZENBERG & DOLLFUS, 1892

Biconic, somewhat resembling *Murexsul* but less strongly ribbed spirally; outer lip strongly denticulate within; lip varix with scaly surface marking; columella slightly roughened at its base by granulations or lirae.

1004. Muricopsis armatus (A. Adams, 1854) (Synonyms: **Muricidea squamulata** Carpenter, 1865; "**Muricopsis squamulifera** Pfeiffer" of M. Smith, 1939). A photograph of Adams's type specimen, supplied by the British Museum (Natural History), shows that it is the form long known under Carpenter's later name. The sculpture is finely scaly, the shell being a dull yellowish white. About 25 mm in length. Throughout most of the Gulf of California; intertidally, under rocks at extreme low tide.

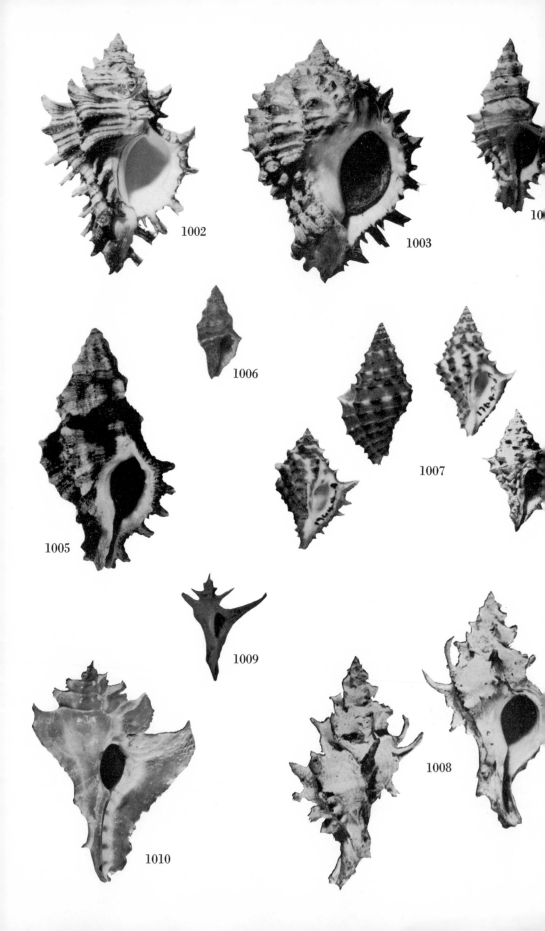

1002

1003

10

1006

1005

1007

1009

1010

1008

1005. Muricopsis jaliscoensis Radwin & D'Attilio, 1970. Light yellowish brown in color, the shell has five spinose varices; it is less slender in outline than *M. zeteki* and with blunter spines; outer lip with seven denticles. Length, 26 to 27 mm; diameter, 14 to 16 mm. Banderas Bay, Jalisco, to Manzanillo, Colima, Mexico, offshore in depths of 5 to 22 m.

1006. Muricopsis pauxillus (A. Adams, 1854). Smallest member of the genus, this has seven varices. The shell is dark gray, with whitish spiral ribs. About 12 mm in length. Restricted to the southern part of the Gulf of California, especially around Mazatlán, Mexico, under rocks at extreme low tides. The apex in juvenile shells is flattened, as in the Caribbean genus *Risomurex*, but other shell features place the species in *Muricopsis*.

1007. Muricopsis zeteki Hertlein & Strong, 1951 (Synonyms: **Murex aculeatus** Wood, 1828 [not Lamarck, 1822]; **M. dubius** Sowerby, 1841, new name [not Dillwyn, 1817]). The shell is white, with brown on spines and varices, a tinge of brown on the columella of some specimens, the aperture bluish white. Length, about 23 to 30 mm; diameter, about 15 mm. Puertecitos, near the head of the Gulf of California, to Guayaquil, Ecuador.

Genus PAZIELLA JOUSSEAUME, 1880

Medium-sized, biconic, the varices dissected into spines; aperture smooth within, anterior canal nearly closed, with short spines near its base. The type species is from the Caribbean. *Poirieria* Jousseaume, 1880, an Indo-Pacific group, is similar but lacks spines on the canal; Vokes (1970) regards *Paziella* as a subgenus of *Poirieria*.

1008. Paziella galapagana (Emerson & D'Attilio, 1970). The dull white shell somewhat resembles *Muricopsis* in outline; spines on the varices are long and recurved. The anterior canal is nearly closed and is slightly bent to the right. Some specimens show a striate chalky outer layer, as in most of the Aspellinae. Length, 47 mm; diameter, 25 mm. Galápagos Islands, at depths of 100 to 200 m. The species is assigned to *Paziella* with some hesitation, for the development of the spines and of the chalky layer differ somewhat from typical members of that group.

Genus PTERYNOTUS SWAINSON, 1833
(PTERONOTUS of authors, unjustified emendation)

Varices three, winglike, their surfaces ornamented by growth striae and spiral ribbing; operculum with apical, i.e. basal, nucleus. Two Panamic subgenera; no species of subgenus *Pterynotus*, *s. s.*, occur in Panamic province.

Subgenus CALCITRAPESSA BERRY, 1959

Varices reduced to long, pointed spines with a central groove on the anterior face; siphonal canal long, closed; operculum with apical (basal) nucleus.

1009. Pterynotus (Calcitrapessa) leeanus (Dall, 1890). One might call this an economy-model muricine, for the varices are stripped down to a minimal diameter, and there is no wasted shell material expended in frills or ornament. The color is off-white to light grayish tan. Length, 50 to 65 mm; overall width, 55 to 83 mm. Best known from off the west coast of Baja California, the species ranges into the Gulf of California as far as San Jose Island, in depths to 110 m; also known from the Pleistocene of southern California.

Subgenus PURPURELLUS JOUSSEAUME, 1880

Siphonal canal wide, tending to have a curved interruption of the lip varix between the body whorl and the end, so that the anterior portion of the canal has a lobelike flange; surface ornament of the varices less developed than in the Oriental *Pterynotus, s. s.*; coloration tending to brown bands or blotches; operculum with subcentral lateral nucleus; radula typically muricine (*fide* Emerson & D'Attilio, 1969).

1010. Pterynotus (Purpurellus) macleani Emerson & D'Attilio, 1969. Shell wide for its height, thin, fragile-appearing, tending toward pink coloration, with brown blotches beginning at each varix but fading out between varices; outer lip varix only weakly interrupted along anterior canal. Length, 22 to 29 mm; width, 13 to 22 mm. Off Carmen Island, Gulf of California, to Panama Bay, in depths of 22 to 45 m.

1011. Pterynotus (Purpurellus) pinniger (Broderip, 1833) (Synonym: **Centrifuga inezana** Durham, 1950). A well-developed specimen of this species is a coveted prize of collectors. The shell is white, with bands of brown irregularly developed, absent entirely in some, showing only faintly through the surface layer of others, but when fully expressed beginning at each varix as a strong, wide stripe that may persist across to the next varix or may fade out midway. The edges of the varices are wavy, and the outer lip shows well the interruption that is characteristic of the West African type species of the subgenus *P. (P.) gambiensis* (Reeve, 1845). Authors are now generally agreed that *P. (P.) inezana* (Durham), described as a Pleistocene fossil in the Gulf of California, is a synonym, for dredging has brought to light specimens at a number of localities along the Central American coast. Length, about 50 mm; width, about 25 mm. San Marcos Island and Guaymas, Gulf of California, to Ecuador, in depths to 82 m.

Subfamily ASPELLINAE

Varices not foliose, mostly inconspicuous, low, few, rarely with spines; sculpture generally subdued; shell surface often with a thin chalky layer; operculum muricoid, with apical nucleus; radula with central tooth having three larger cusps, with two smaller ones between. The narrow anterior canal is short to moderate in length in most species; aperture with a smooth to dentate outer lip. The key is to genera.

1. Outer surface of shell with a chalky texture......................*Aspella*
   Outer surface of shell not markedly chalky.......................... 2
2. Varices two per whorl... 3
   Varices not two per whorl....................................... 4
3. Sculpture subdued, not conspicuously cancellate................*Eupleura*
   Sculpture of regular spiral and axial ribs, cancellate...........*Phyllocoma*
4. Sculpture reduced to fine spiral threads, mostly on base..........*Attiliosa*
   Sculpture stronger than fine threads, over entire shell................... 5
5. Axial ribs several (about ten) per whorl....................*Calotrophon*
   Axial ribs few (eight or less) ....*Favartia*

Genus ASPELLA MÖRCH, 1877

Small shells with a rather chalky surface texture, finely spirally striate, with two to three (rarely six) major varices per whorl; the canal short, a little twisted, aperture relatively small.

1. Varices two per whorl.....................................*Aspella, s. s.*
   Varices more than two per whorl or irregular........................ 2
2. Varices three per whorl, with smaller ones between................... 3
   Varices not in a pattern of three major, three minor................. 4
3. Spiral ribs numerous (eight or more) ; outer lip smooth within...*Dermomurex*
   Spiral ribs few (about five) ; outer lip denticulate..............*Trialatella*
4. Spire low, shell pyriform....................................*Maxwellia*
   Spire elevated, outline biconic...........................*Gracilimurex*

Subgenus ASPELLA, s. s.

With two principal varices per whorl, having one or more low costae between. The type species, *A. anceps* (Lamarck, 1822), is wide-ranging in the tropics, occurring in both the Atlantic and the western Pacific.

1012. Aspella (Aspella) hastula (Reeve, 1844). Resembling *A. pyramidalis* in shape, the shell differs in being spirally beaded and in being brown with white varices. Length, 14 mm; width, 7 mm. Described from an unknown locality but recognized by Dr. Emily Vokes (*in litt.*) as from the Galápagos Islands.

1013. Aspella (Aspella) pyramidalis (Broderip, 1833). The spire has several tapering whorls, with a suture that undulates over the ends of a few concealed axial ribs. The shell is white, some specimens with spiral lines of brown. Length, 18 mm; width, 9 mm. Mazatlán, Mexico, to Panama and the Galápagos Islands.

Subgenus DERMOMUREX MONTEROSATO, 1890

With three principal varices per whorl, alternating with smaller axial ridges diagonally ascending the spire.

1014. Aspella (Dermomurex) indentata (Carpenter, 1857) (Synonym: **A. perplexa** Keen, 1958). Fresh shells have a worn look. Worn shells may show brown spots or bands on a whitish surface. The varices are of nearly equal size, six to seven per whorl. Length, 34 mm; diameter, 15 mm. Mazatlán, Mexico (type locality), exact range uncertain, possibly south to Panama. The type specimen has a heavy lip varix that contracts the aperture; a specimen in the Los Angeles County Museum collection that may have come from Panama has a similar aperture.

1015. Aspella (Dermomurex) myrakeenae Emerson & D'Attilio, 1970. With six varices crossed by strong spiral cords, sculpture weak between; surface chalky, striately sculptured; shell white, the aperture with a flaring white border, about six spiral color bands showing within, which underlie the chalky layer and may show to the outside on worn shells; anterior canal short, recurved, open; operculum with basal nucleus. Length, 21 to 25 mm; diameter, 13 to 15 mm. La Paz, Baja California, to Sihuatanejo, Guerrero, Mexico (type locality, Banderas Bay, Nayarit). This has been confused with *A. indentata* and was figured as a variant of *A. perplexa* by Keen, 1958.

1016. Aspella (? Dermomurex) obeliscus (A. Adams, 1853). The shell is ovate-pyramidal in outline, with several somewhat granular spiral cords that are spotted with reddish brown, especially on the three principal varices. The anterior canal is narrowed and somewhat recurved. Length, 33 mm; diameter, 14 mm. Mazatlán area, Mexico, to Masachapa, Nicaragua. The species was described from an unknown locality, later said to be West Indian. As Emerson and D'Attilio

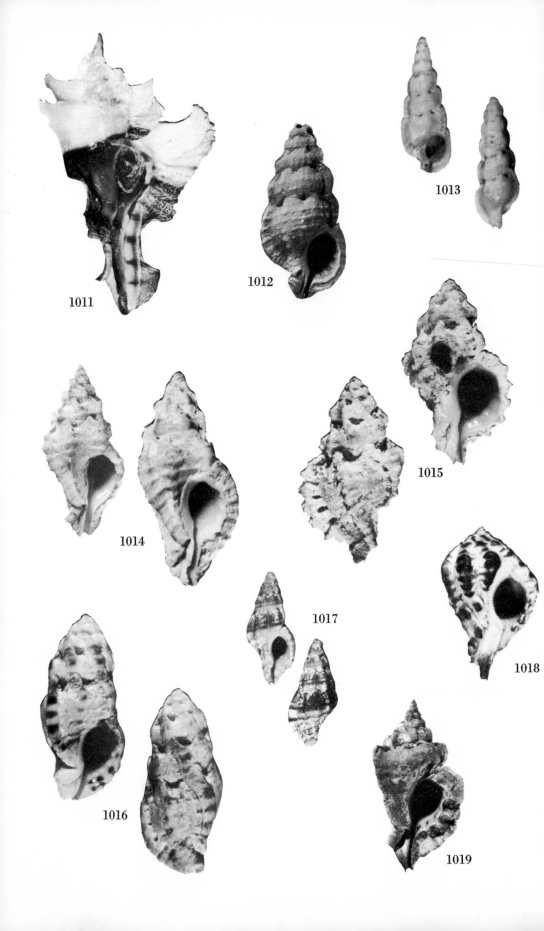

1011

1012

1013

1015

1014

1017

1018

1016

1019

(1970) have pointed out, material from the eastern Pacific matches the original illustration better than does that from the Caribbean.

Subgenus GRACILIMUREX THIELE, 1929

Varices inconspicuous, somewhat irregularly distributed; with color bands of brown and white, especially on spire.

1017. Aspella (Gracilimurex) bakeri Hertlein & Strong, 1951 (Synonym: **Murex bicolor** Thiele, 1929 [not Risso, 1826]). The varices are not conspicuous on this shell, which is white, with broad bands of pale brown on the spire and with some narrower ones below. Length, 18 mm; diameter, 8 mm. Gulf of California.

Subgenus MAXWELLIA BAILY, 1950

Small, solid shells with globose spire and rather short canal, the varices six to seven in number, reflected and joined diagonally to varices of previous whorls, with deep pits behind them at the suture. The relationship to *Aspella* has been suggested on the basis of the radula (E. Vokes, *in litt.*) The type species is *A. (M.) gemma* (Sowerby, 1879) from southern California.

1018. Aspella (Maxwellia) angermayerae Emerson & D'Attilio, 1965. Resembling the Californian *Maxwellia gemma* (Sowerby, 1879), the shell has seven rounded axial ribs that are black under the chalky white surface layer and a few spiral riblets. Length, 21 mm; width, 14 mm. Galápagos Islands, dredged in depths to 18 m.

Subgenus TRIALATELLA BERRY, 1964

Spire whorls with six varices, body whorl with three, the varices spirally ridged, especially on the faces; aperture denticulate.

1019. Aspella (Trialatella) cunninghamae (Berry, 1964). The shell is white, with about six brown denticles on the inner edge of the outer lip; the most conspicuous feature is the grooving of the varices by the spiral sculpture. Length, 17 mm; width, 10 mm. Puerto San Carlos, Guaymas, Mexico, in depths of 27 to 64 m.

Genus ATTILIOSA EMERSON, 1968

Small to medium-sized, rather solid shells, with reduced sculpture, mostly of fine and slightly scaly spiral riblets on the lower part of the whorls; outer lip lirate within, inner lip with a well-developed margin; siphonal fasciole large, canal short, not closed; nucleus of operculum apical (basal), as in the Muricinae, the margin within thickened. Radula with five cusps on central tooth, the two intermediate teeth smaller.

1020. Attiliosa carmen (Lowe, 1935). Small, cream-colored, with a few flecks of brown on the upper part of each whorl; with two or three strong spiral ribs, the peripheral rib bluntly nodose. Length, about 10 mm; diameter, 5 mm. The Gulf of California, in depths of 15 to 40 m. The sloping whorl profile and narrow anterior canal separate this from the similarly marked *A. rufonotata*.

1021. Attiliosa incompta (Berry, 1960). The white shell is generally so encrusted with lime that the sculpture cannot be seen. The shell is sturdy, with blunt nodes, especially at the periphery of the whorls. Placement of the species in the Muricacea has been a matter of some uncertainty; the original allocation was to Coralliophilidae. When it was made type of a new genus, *Attiliosa*, this was assigned to a spot in the Thaididae, near *Cronia*. It does not fit neatly anywhere, but seems least inharmonious in the Aspellinae. Length, 35 mm; width, 20 mm. The Gulf of California, from Puerto Peñasco, Sonora, Mexico, southward to the Secas Islands, Panama, in depths of 18 to 80 m.

1022. Attiliosa rufonotata (Carpenter, 1864). The small white shell has a few low cords at the periphery of the whorls, with dots of red just below. Length, 12 mm; width, 9 mm. Southern end of the Gulf of California, from Cape San Lucas and Carmen Island to the Tres Marias Islands and the mainland Mexican coast, in depths to 40 m.

Genus **CALOTROPHON** HERTLEIN & STRONG, 1951
(**HERTLEINELLA** BERRY, 1958)

Solid, fusiform, lacking periostracum but with surface layer chalky; whorls shouldered, axial sculpture of strong costae; canal open, siphonal fasciole broad; columella smooth, outer lip lirate within.

1023. Calotrophon turritus (Dall, 1919) (Synonyms: **Trophon (Calotrophon) bristolae** Hertlein & Strong, 1951; **Hertleinella leucostephes** Berry, 1958). The sturdy white shell has a few irregular color bands of brown. Its surface layer in well-preserved specimens is chalky, as in *Aspella*, but this layer is readily lost, and the sculpture shows as fine imbricating scales on spiral riblets rather irregularly disposed. Length, 39 mm; width, 20 mm. Off Baja California and in the southern end of the Gulf of California, in depths to 145 m. The synonymy has been worked out by McLean and Emerson (1970).

Genus **EUPLEURA** H. & A. ADAMS, 1853

The two varices on opposite sides of the whorls are reminiscent of the Cymatiidae, with which the group was affiliated by early workers, but the structure of the radula shows that it belongs in the Muricidae. Strong denticles occur inside the outer lip.

1024. Eupleura muriciformis (Broderip, 1833) (Synonyms: **Ranella plicata** and **triquetra** Reeve, 1844; **E. m.** var. **unispinosa** and var. **limata** Dall, 1890). The shell varies in color from waxy white to grayish black in irregular patches or areas. Variations in shape and in expression of the varices have led to repeated naming, but attempts to separate the forms into clear-cut subspecies on the basis of distribution or shell features have not been successful as yet, because of the numerous intergradations. *E. plicata* applies to a form with wide, rounded varices, and *E. triquetra* to one that has the last varix drawn upward and outward into a triangular shape. Length, 38 mm; maximum diameter, 24 mm. Cedros Island, Baja California, throughout the Gulf and southward to Ecuador, intertidally and offshore. Lowe (1935) reported it as not rare on reefs, feeding on *Cerithium*. A similar Atlantic species is *E. caudata* (Say, 1822).

1025. Eupleura nitida (Broderip, 1833). A much smaller form than *E. muriciformis*, this is dark brown or purplish, banded or spotted with white, the varices narrow and platelike. Length, 21 mm; maximum diameter, 15 mm. Mazatlán, Mexico, to Panama; rare, at extreme low tides but commoner offshore.

1026. Eupleura pectinata (Hinds, 1844). A flattened and rather shiny brown shell with a long canal and with the varices ornamented by several finger-like spines. Length, 38 mm; diameter, 19 mm. Southern Mexico to Panama Bay.

1020

1021

1022

1023

1024

1025

1026

1027

1028

1029

1030

Genus FAVARTIA JOUSSEAUME, 1880

Shell relatively short and broad, varices four to six, diagonally ascending the spire; anterior canal short, bent, broad; sculpture of spiral riblets intersected by axial striae to form a reticulate pattern, elaborately festooning the varices, often with pits between ribs. Axial ribs somewhat undulating, low; spire mostly blunt.

1027. Favartia erosa (Broderip, 1833). The color varies from dull grayish to brownish, with a worn look caused by the fine surface sculpture. Length, 20 mm; diameter, 10 mm. Mazatlán, Mexico, to Panama, under stones at extreme low tide and offshore. The species has been allocated to *Aspella*, but its strong coloration and the irregular varices suggest removal from there.

1028. Favartia incisa (Broderip, 1833) (Synonym: "**Tritonalia margaritensis** Dall" of M. Smith, 1931). The chunky little shell has about seven axial ribs that rise as low swellings overridden by narrow spiral ribs. The shell normally is white, but the spiral ribs may be touched here and there with brown. The siphonal fasciole is broad, the anterior canal nearly closed. Length, 32 mm; width, 20 mm. There is a somewhat worn ivory white specimen of this form in the British Museum collection, received from C. B. Adams in the 1850's from Panama and identified by Carpenter as *"Rhizochilus nux = Purpura osculans* C. B. Adams"— both erroneous determinations. Three identical-appearing specimens in the Stanford University collection were taken by Henry Hemphill at Magdalena Bay, Baja California, probably prior to 1900; a note in W. H. Dall's handwriting suggests they are not west coast, but comparison with the figure of the holotype of *F. incisa* and with the twin species from the Caribbean, *F. cellulosa nucea* (Mörch, 1850), confirms their identity as Panamic. Thus, the range is Magdalena Bay to Ecuador and the Galápagos. The form was confused with *Maxwellia gemma* by some early workers.

1029. Favartia peasei (Tryon, 1880) (Synonym: **Murex foveolatus** Pease, 1869 [not Hinds, 1844]). Specimens that match the original figure well for shape and sculpture have been taken during the 1960's by several collectors in the Gulf of California. These specimens are pinkish brown, with a white band on the lower part of the body whorl; the outer lip is dentate within, the canal is narrow but not closed, the siphonal fasciole broad. Length, about 15 mm. Gulf of California, from La Paz to the coast of Sonora, Mexico, offshore to 37 m.

Genus PHYLLOCOMA TAPPARONE-CANEFRI, 1881
(CRASPEDOTRITON DALL, 1904)

Similar in appearance to some Colubrariidae but with the radula of the Muricidae. Shell with well-developed spiral ribs crossed by almost equal axial ribs. Varices strong. Inner-lip callus flaring.

1030. Phyllocoma scalariformis (Broderip, 1833). The shell is white, elegantly sculptured with spiral and axial ribs. Length, 30 mm; diameter, 14 mm. Guaymas, Mexico, in 10 m, to the Galápagos Islands.

Subfamily OCENEBRINAE

Small to medium-sized shells, varices three or more per whorl; sculpture tending to be scaly; anterior canal narrow or closed into a tube in most forms; outer lip often developing denticles within, some groups having in addition a tooth on the anterior margin; operculum with lateral nucleus; radula having the central tooth with two or three larger cusps in the middle, flanked by several smaller ones; in addition, there may be two large cusps at the outer ends of the tooth.

The type genus of the subfamily, *Ocenebra* Gray, 1847 (*Tritonalia* of authors), is better represented in temperate than in tropical seas; there seems, in fact, to be only one species of *Ocenebra* in the Panamic province, all the rest so assigned in the first edition of this book now being transferred to other genera as a result of new information and revisionary work during the last decade. The key is to genera.

1. With regular varices, three or more per whorl. 2
   Without varices or with varix-like laminae only on last whorl. 4
2. Outer lip entire, without a tooth at anterior end. . . . . *Pteropurpura*
   Outer lip with a tooth on its anterior margin. 3
3. Anterior canal broad to recurved; outer lip varix solid to scaly. . .*Ceratostoma*
   Anterior canal narrowed, straight; outer lip varix thin. . . . .*Pterorytis*
4. Spiral sculpture of fine and regular lirae. . . . .*Ocenebra*
   Spiral sculpture irregular or wanting. . . . .*Vitularia*

Genus **OCENEBRA** GRAY, 1847

(**OCINEBRA**, unjustified emendation; **TRITONALIA** FLEMING, 1828, suppressed, ICZN Opinion 886, 1969)

With spiral and axial sculpture; varices, if present, irregular.

1031. Ocenebra buxea (Broderip, 1833). The tawny brown shell is marked with four to five lighter spiral ridges that form elongate nodes over the low axial ribs, especially at the shoulder; the spiral cords are scaly. The specific name *buxea*, meaning boxwood, evidently refers to the color and the fine texture, especially of slightly worn specimens. Length, 30 mm; diameter, 16 mm. Reports of occurrence at Magdalena Bay, Baja California, probably are in error; the range seems rather to be from Pacosmayo, Peru, southward to Chile. A Californian species with which it may have been confused is *Ocenebra poulsoni* Carpenter, 1864. Some authors have assigned the species *O. buxea* to the Buccinidae on account of the *Cantharus*-like outline, but the operculum, with its lateral nucleus, indicates muricid affinities.

Genus **CERATOSTOMA** HERRMANNSEN, 1846

(**PURPURA** of authors, not of BRUGUIÈRE, 1789)

Varices three or more per whorl, ridgelike to broad; sculpture spiral, not strongly scaly; anterior canal closed, moderately long; outer lip entire, with a well-developed apertural tooth anteriorly. The generic name has been placed on the Official List by the International Commission on Zoological Nomenclature (Opinion 704, 1964), the type species being the Californian *C. nuttalli* (Conrad, 1837). Although species of the northern Pacific have developed the largest shells, there are a number of tropical forms. The subgeneric name *Microrhytis* Emerson, 1959, may prove useful for these. Its type species is from the Miocene of eastern Oaxaca, Mexico. In the eastern Atlantic, the genus *Jaton* Pusch, 1837, shows a remarkable resemblance to *Ceratostoma*.

1032. Ceratostoma fontainei (Tryon, 1880) (Synonym: **Murex monoceros** Orbigny, Dec. 1841 [not Sowerby, June 1841]). Very similar to *C. lugubre*, this may prove to be only a southern variant. The shell is more sturdy, and the varices more solid-appearing. The type, in the British Museum, shows a small tooth on the outer lip. Length, 36 mm; width, 23 mm. Paita, Peru (type locality).

1033. Ceratostoma lugubre (Broderip, 1833). The grayish-white shell is banded with brown on the upper and middle part of the whorls, and the end of the anterior canal may be brown. There are seven varices per whorl, the lip varix of even width, not indented. The type lot in the British Museum consists of four specimens, only the one labeled holotype now retaining the minute tooth on the outer lip, though in the other specimens it shows up at previous varices. Length, about 30 mm; width, 20 mm. The type locality is Costa Rica, but the range remains to be worked out, modern collectors not having recorded the species. It is not the *Acanthina lugubris* of the outer coast of Baja California.

1034. Ceratostoma monoceros (Sowerby, 1841). Primarily a species of the outer coast of Baja California; the shell is easily confused with *Pteropurpura erinaceoides* if one fails to notice the tooth on the outer lip. There are three low varices per whorl, with nodes between. Most specimens are grayish white, but an occasional one shows a suffusion of brown, especially on newly formed sections of the shell. Length, 60 mm; width, 30 mm. The range and relationship to the next species remain to be worked out. Specimens in the Stanford University collection are from San Ignacio Lagoon and Magdalena Bay, Baja California.

1035. Ceratostoma unicorne (Reeve, 1849). Described without locality, the species remains open to question. Reeve's figure, reproduced here, shows a heavy brown shell that resembles the *C. nuttalli* (Conrad, 1837) of southern California but with a more massive tooth on the outer lip. Some specimens in the Stanford University collection from Magdalena Bay come close to matching the figure for size and color. Length, 45 to 55 mm; width, 23 to 29 mm. The species has been reported (E. Vokes, *in litt.*) at La Paz, Gulf of California.

Genus PTEROPURPURA JOUSSEAUME, 1880

Varices winged, as in some species of *Ceratostoma*, but much more highly ornamented, tending to have fine scales and surface foliations of the anterior faces; operculum with lateral nucleus; no apertural tooth on outer lip. Two Panamic subgenera.

Subgenus PTEROPURPURA, *s. s.*

Varices with smoothly curved or sinuous outer margin, not dissected into spines.

1036. Pteropurpura (Pteropurpura) erinaceoides (Valenciennes, 1832) (Synonym: **Murex californicus** Hinds, 1844). The entire surface of the light brown shell is covered by fine scales. Length, 50 mm; maximum diameter, 28 mm. Southern California through the Gulf to Guaymas, Mexico; not uncommon among rocks at low tide.

Subgenus CENTRIFUGA GRANT & GALE, 1931

Varices dissected into elongate spinelike lobes.

1037. Pteropurpura (Centrifuga) centrifuga (Hinds, 1844) (Synonym: **Pterynotus swansoni** Hertlein & Strong, 1951). A yellowish-white shell with varices that are produced into flat spines, especially at the shoulders of the whorls. Collection of material at all stages of growth by shrimp trawlers has demonstrated that *P. swansoni* is the adult form of *P. centrifuga*. Length of an average adult specimen, 59 mm; width, 49 mm. Magdalena Bay, Baja California, and through the Gulf of California southward to Panama, in depths of 64 to 180 m. Emerson (1960) succeeded in locating Hinds's long lost type specimen.

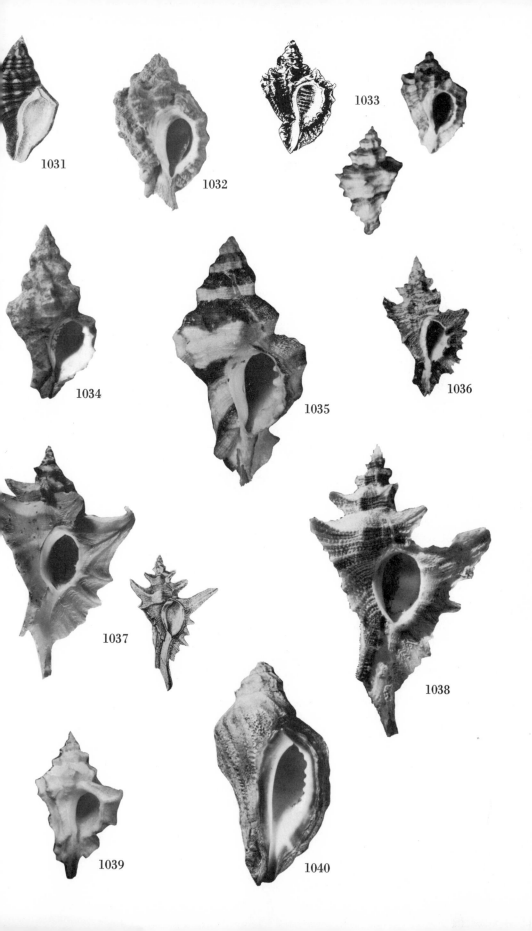

1031

1032

1033

1034

1035

1036

1037

1038

1039

1040

1038. Pteropurpura (Centrifuga) deroyana Berry, 1968. The ivory-white shell resembles *P. (C.) centrifuga* in outline, but the margins of the varices are less dissected, with only one or two small digitations or spines, and the surface sculpture of the varices is more foliaceous and scaly. Length, 33 mm; width, 21 mm. Galápagos Islands, Ecuador, in 100 m.

Genus **PTERORYTIS** CONRAD, 1862
(**PTERORHYTIS** of authors; **NEURARHYTIS** OLSSON & HARBISON, 1953)

Slender to biconic, mostly low-spired; varices bladelike and foliated, four to six (rarely more) in number; outer lip with an indentation at the site of the well-developed apertural tooth; anterior canal closed. The genus is closely related to *Ceratostoma*, and further study may relegate it to subgeneric status under *Ceratostoma*; the subgenus *Neurarhytis* was proposed for species having four varices and a smaller apertural tooth. Emerson (1959) has reviewed the Atlantic species.

Like several other molluscan genera—called paciphiles by Woodring (1966)—this had a wide distribution in the Atlantic during the Tertiary (the type species is from the Miocene of Virginia) but has become extinct there, surviving on the Pacific side of the Isthmus of Panama. Dr. Emily Vokes (*in litt.*) has recognized its occurrence in the eastern Pacific. Other examples of paciphile genera include *Northia* and *Trajana* in the Buccinacea and *Harvella* in the Pelecypoda.

1039. Pterorytis hamatus (Hinds, 1844). A light yellow shell with a brown canal and six varices per whorl. The lip varix is broad and thin, its indentation at the level of the apertural tooth deep and wide. The apertural tooth is relatively small. Length, 30 mm; width, 18 mm. Guayaquil, Ecuador (type locality).

Genus **VITULARIA** SWAINSON, 1840

With varices irregular in shape and development, mostly vaulted and sharp-edged, not scaly; surface sculpture of fine vermiculations.

1040. Vitularia salebrosa (King & Broderip, 1832) (Synonym: **V. s. extensa** M. Smith, 1947). The rich terra-cotta color of this shell is one of the best means of recognition. The surface texture may vary from almost smooth, with only a single lip varix, to finely obliquely wrinkled, with several varices, perhaps lapped one within another. The canal is open, and several large denticles stud the inside of the outer lip. Length, 60 mm; diameter, 35 mm. Cedros Island, Baja California, La Paz and Guaymas, Gulf of California, southward to Gorgona Island, Colombia, under rocks at extreme low tide. The subspecific name given by Maxwell Smith to a low-spired variant from Panama seems not to be significant geographically.

Subfamily **TROPHONINAE**

These are shells that are thin for their size, with either regular or irregular, mostly nonscaly varices or axial laminae that are thin-edged and vaulted to smoothly rounded. The outer lip is smooth within, the operculum in most species with the nucleus near the middle of the outer edge. The radula is of the muricine pattern except for two additional cusps at the outer ends. Egg capsules paddle-shaped. Two Panamic genera.

Genus **TROPHON** MONTFORT, 1810

Most of the members of this genus are deepwater forms, with white shells variously sculptured. Three Panamic subgenera.

Subgenus **ACANTHOTROPHON** HERTLEIN & STRONG, 1951

Axial sculpture of simple spines, largest at periphery; area between suture and periphery nearly smooth, sloping downward.

1041. Trophon (Acanthotrophon) carduus (Broderip, 1833). Formerly placed in the *Coralliophila* because of its scaly sculpture, this species has been found in recent years to have a radula, which apparently would place it in the Trophoninae. Although the sculpture is more elaborate than that of *T. (A.) sorenseni*, there is a general resemblance in shape and size. The sculpture of the white shell is relatively coarse, with one or two longer scales or spines near the aperture. Length, 25 mm; diameter, 16 mm. Mazatlán, Mexico, to Peru, occurring mostly offshore.

1042. Trophon (Acanthotrophon) sentus (Berry, 1969). The white shell resembles that of *T. (A.) sorenseni* but is smaller, with more numerous and stouter spines, which are paired in two rows at the periphery. Length, 20 mm; diameter, 15 mm. Academy Bay, Santa Cruz Island, Galápagos Islands.

1043. Trophon (Acanthotrophon) sorenseni Hertlein & Strong, 1951. The shell is grayish white, with a single row of vertical spines at the periphery. Length, 31 mm; diameter, 14 mm. Southern end of the Gulf of California, offshore in depths to 110 m. This is the type of the subgenus.

Subgenus **AUSTROTROPHON** DALL, 1902

Axial sculpture of vaulted spines forming winglike varices.

1044. Trophon (Austrotrophon) cerrosensis Dall, 1891. This and the related Californian species, *T. (A.) catalinensis* I. Oldroyd, 1927, have been thought by many authors to belong to the genus *Forreria* rather than to *Trophon*. However, *Forreria*, as typified by the Californian *F. belcheri* (Hinds, 1844), has a large, heavy shell with a deep spiral furrow at the base of the body whorl very different from the thin, delicate shells of *Austrotrophon*, of which *T. cerrosensis* is type. The shell is flesh-colored, fading to white in some specimens near the apex of the thin varices, with about ten vaulted varices per whorl. About 40 mm in length and 24 mm in diameter. Off Cedros Island, Baja California, to Acapulco, Mexico, mostly offshore in depths to 88 m.

1044a. Trophon (Austrotrophon) panamensis (Olsson, 1971). Height, 32 mm; diameter, 27 mm. Gulf of Panama, dredged.

1045. Trophon (Austrotrophon) pinnatus Dall, 1902. The light yellowish shell has numerous wide lamellar varices that show a few fine spiral striae. The specific name *pinnatus* refers to the winglike effect that this gives, especially on the outer lip. Length, 70 mm; width, 55 mm. Magdalena Bay, Baja California, in depths of 37 to 135 m.

Subgenus **ZACATROPHON** HERTLEIN & STRONG, 1951

Coiling nearly disjunct, suture falling below periphery of earlier whorls; whorls tabulate, with small vertical spines at the shoulder.

1046. Trophon (Zacatrophon) beebei Hertlein & Strong, 1948. Pinkish brown in color, the shell is loosely coiled and bears a row of guttered spines on the shoulder of the whorl. Length, 52 mm; diameter, 23 mm. In 90 to 110 m depth, southern part of the Gulf of California.

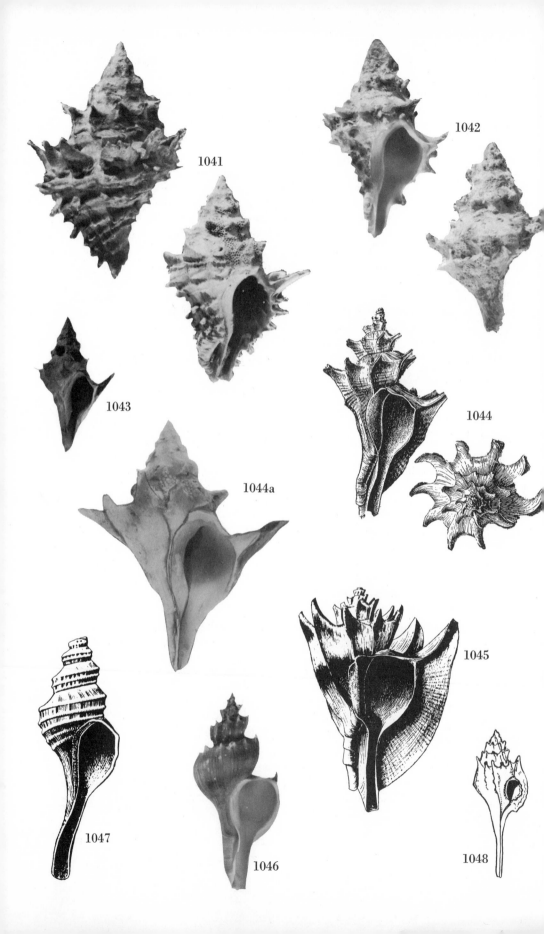

1041

1042

1043

1044

1044a

1045

1046

1047

1048

Genus TROPHONOPSIS BUCQUOY, DAUTZENBERG & DOLLFUS, 1882

Canal long, narrow; sculpture of both spiral and axial ribs, cancellate in early whorls. Mostly cold-water forms; one species has been recorded in the Panamic province, in deep water.

1047. Trophonopsis panamensis (Dall, 1902) [*Boreotrophon*]. Off Panama Bay, in 2,320 m. (Figured, Dall, 1925.)

Subfamily TYPHINAE

With tubular spines at the periphery, either between or on the varices, the final tube opening into the aperture. The radula resembles that in the Ocenebrinae but with three larger cusps and between each pair of these three smaller cusps. The key, adapted from Keen (1944), is to the genera and subgenera of the Panamic Typhinae.

1. Tubes per whorl two.....................................*Distichotyphis*
   Tubes per whorl more than two...................................... 2
2. Tubes per whorl three... 3
   Tubes per whorl four.. 4
3. Tube openings at tips of spines on varices.................*Tripterotyphis*
   Tubes separate from but just behind varices...................*Pterotyphis*
4. Canal slender and long (half the entire length).............*Haustellotyphis*
   Canal short and wide (less than half the entire length).................. 5
5. Tubes forming part of a subdued varix....................*Cinclidotyphis*
   Tubes separate from varices....................................... 6
6. Tubes reinforced on posterior side by solid remnant of previous lip varix;
      pillar area broad.....................................*Typhisopsis*
   Tubes free, not reinforced by varix remnant; pillar area narrow....*Talityphis*

Genus TYPHIS MONTFORT, 1810

The type species of *Typhis* is an Eocene fossil. During the Tertiary the genus radiated widely, although never abundantly, since only a relatively few specimens of each species have been found. In spite of rarity, the group developed a number of distinctive forms.

Subgenus HAUSTELLOTYPHIS JOUSSEAUME, 1880

Canal long and slender.

1048. Typhis (Haustellotyphis) cumingii Broderip, 1833. The dark brown shell has elegant white varices tinged with lighter brown. Length, 20 mm; width, 10 mm. Acapulco, Mexico, to Guayaquil, Ecuador, intertidally and to depths of 13 m.

Subgenus TALITYPHIS JOUSSEAUME, 1882

Varices winglike, broad, four per whorl.

1049. Typhis (Talityphis) latipennis Dall, 1919. In color this is a buffy white, with tinges of brown. The tubes have a backward slant. Length, 30 mm; maximum diameter, 24 mm. Gulf of California to Panama Bay, in depths to 45 m.

Subgenus TYPHISOPSIS JOUSSEAUME, 1880

Tubes reinforced behind by a buttress; pillar broad.

1050. Typhis (Typhisopsis) clarki Keen & Campbell, 1964 (Synonym: **T. quadratus** Hinds, 1843, of authors). Light brown to white, lighter on the lip varix and above and below the aperture, the shell has some touches of darker brown, especially on the end of the canal and the tubes, which are turned sharply backward, each soldered to a remnant of a previous varix. Length, 20 mm; width, 12 mm. Off Puerto Peñasco, northern end of the Gulf of California to Panama Bay (type locality), intertidally and to a depth of 16 m.

1051. Typhis (Typhisopsis) coronatus Broderip,. 1833 (Synonyms: **T. quadratus** Hinds, 1843; **T. martyria** Dall, 1902). The original figure of *T. coronatus* showed an immature individual; thus, for years the species remained unrecognized. The tube is appressed to the preceding varix and buttressed by the remnant of the former lip, which makes a bridge from the later to the earlier whorl. A large adult may be 40 mm in length. Magdalena Bay, Baja California, through the Gulf to Puerto Peñasco, Mexico, and south to Santa Elena, Ecuador, mostly offshore, in depths to 33 m; apparently in colonies, for several specimens may be taken in a single dredge haul at a suitable location.

1052. Typhis (Typhisopsis) grandis A. Adams, 1855. Although sometimes cited as a synonym of *T. coronatus*, the species has proved to be distinct, as shown by Shasky and Campbell (1964). The shell is more massive, with a broader pillar area, is proportionately wider for the height, more strongly ribbed spirally, and the lip varix is more widely flaring. Length of a large specimen, 39 mm; width, 23 mm. Dredged near Guaymas, Sonora, Mexico, at depths to 15 m.

Genus CINCLIDOTYPHIS DuShane, 1969

Varices four per whorl, expressed only as axial swellings, barely apparent, the tubes emerging from posterior ends of the varices, folded backward along suture line; sculpture well developed, of regular axial and spiral riblets. Related to a western Pacific and Caribbean group, *Siphonochelus* Jousseaume, 1880, but distinguished by the strong cancellate sculpture and the emergence of the tubes from the varices along the suture line, without elevation.

1053. Cinclidotyphis myrae DuShane, 1969. The shell is creamy white, with rounded and somewhat inflated whorls, its most distinctive feature being the appressed tube openings at the suture. The anterior canal is rather more open than in most members of the subfamily. Length, 13 mm; width, 6.5 mm. Sayulita Bay, Nayarit (Gale Sphon collection) to Tenacatita Bay, Jalisco, Mexico (type locality, DuShane collection), intertidally, from hermit crabs.

Genus DISTICHOTYPHIS KEEN & CAMPBELL, 1964

Varices two per whorl; tubes forming part of the varix; anterior canal long, closed.

1054. Distichotyphis vemae Keen & Campbell, 1964. The shell is translucent white, the apex of the only known specimen missing and sealed off from within. Height, 8 mm; diameter, 8 mm. Gulf of Panama, in 1,856 m. This is the maximum depth record for the subfamily. Most Muricidae, in fact, occur in water depths of less than 200 m.

Genus PTEROTYPHIS JOUSSEAUME, 1880

Varices three per whorl, mostly broad and flaring.

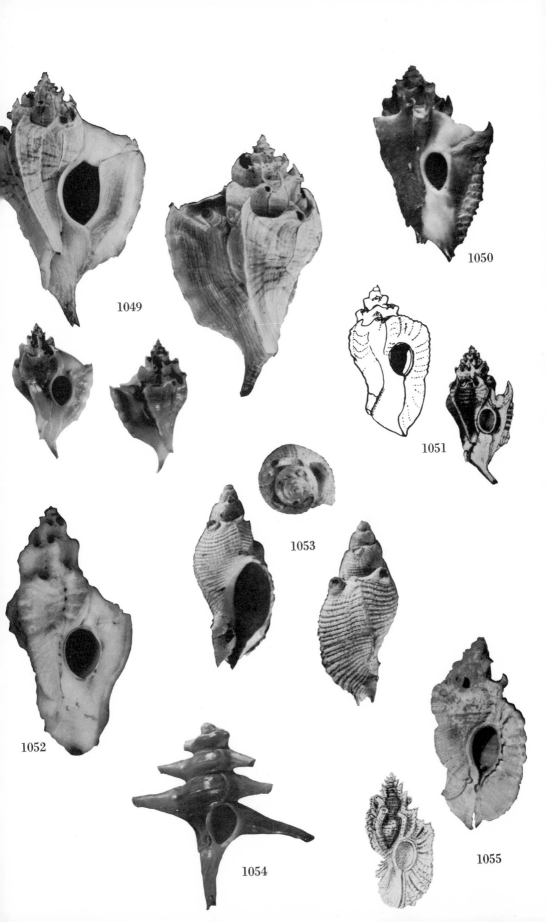

1049

1050

1051

1052

1053

1054

1055

Subgenus PTEROTYPHIS, s. s.
(TRIGONOTYPHIS JOUSSEAUME, 1882)

Tubes separate from varices, formed immediately before the construction of the varix.

1055. Pterotyphis (Pterotyphis) fimbriatus (A. Adams, 1854) (Synonym: ? **Murex jamrachi** Von Martens, 1861). Long known only from the holotype, which is in the British Museum, the species has been rediscovered in the 1960's. A showy shell, it is basically yellowish white, with a flush of pink and brown on the body whorl and part of the lip varix, which is notably broad. Length, 20 mm; width, 10 mm. The type locality is "Gulf of California," but Mrs. Faye Howard has found a specimen at Barra de Navidad, Jalisco, Mexico, in beach drift, and another has come to light from Cuastecomate, near Manzanillo, Colima, Mexico (Los Angeles County Museum collection).

Subgenus TRIPTEROTYPHIS PILSBRY & LOWE, 1932

Tubes formed as a part of the varix, emerging at the tip.

1056. Pterotyphis (Tripterotyphis) arcana DuShane, 1969. An ivory-white and brown shell, with the color irregularly distributed in blotches on the body whorl and varices. The shell has less regular sculpture than previously described species, only two major spiral cords, with a few minor spirals above and below. Length, 15 mm; width, 8 mm. Mazatlán, Mexico, to Banderas Bay, Mexico, intertidally on rocks.

1057. Pterotyphis (Tripterotyphis) fayae Keen & Campbell, 1964. Fine spiral riblets distinguish the species, about 22 on the body whorl. The shell is white, irregularly suffused with brown blotches; stronger color in rib interspaces gives the impression of color banding in some specimens. The outer lip varix is relatively narrow. Length, 20 mm; width, 11 mm. Cape San Lucas, Baja California, and the mainland of west Mexico from Barra de Navidad, Jalisco, to Bahía de Audencia, Colima, intertidally. Also known in the Pleistocene of Oaxaca, Mexico.

1058. Pterotyphis (Tripterotyphis) lowei (Pilsbry, 1931). A creamy white shell with a few spiral ribs, intersected by low axial riblets that form a somewhat cancellate pattern, with deeper pits at the edge of the varix just above the aperture. Length, 15 mm; width, 9 mm. Cape San Lucas, Baja California; Guaymas, Sonora, Mexico, and southward to Panama, intertidally and offshore at shallow depths. A similar species in the Atlantic is the type of the subgenus, *P. (T.) triangularis* (A. Adams, 1855) (*Murex cancellatus* Sowerby, 1841, preoccupied).

Family CORALLIOPHILIDAE
(MAGILIDAE of authors)

This family, obviously related to the Muricidae, is comprised of a number of unlike-appearing genera that have in common the habit of preying upon corals, some forms even going so far as to attach themselves to the coral permanently, abandoning spiral growth, and adding to the aperture only as needed to keep pace with the surface of the coral; those that do not keep pace may be distorted in form by near-sedentary life in coral colonies. As with some other parasitic gastropods that have found an easy food source, these coral lovers have lost the radula. The shell surface is sculptured with many spiral rows of overlapping scales. The operculum is like that of the next family, the Thaididae, with a lateral nucleus. A conservative classification is adopted here, authors not being in agreement upon fine points of subdivision. The key is to Panamic genera and subgenera.

1. With a peripheral keel made up of overlapping spines.........*Babelomurex*
   Without a peripheral keel, although periphery may be angulate........... 2
2. Spire low, aperture as long as shell.................................. 3
   Spire evident, aperture shorter than shell length...................... 4
3. Sculpture weak ....*Quoyula*
   Sculpture of well-developed axial and spiral threads............*Coralliobia*
4. Axial sculpture well-developed, of regular ribs..............*Pseudomurex*
   Axial sculpture weak or very irregular.....................*Coralliophila*

Genus CORALLIOPHILA H. & A. ADAMS, 1853

Aperture with a well-defined anterior canal. Two Panamic subgenera.

Subgenus CORALLIOPHILA, s. s.

Sculpture dominantly spiral, with only faint axial ribs or none; canal spoutlike. Egg capsules blade-shaped, folded.

1059. Coralliophila (Coralliophila) californica (A. Adams, 1855). Because Adams's material was not adequately figured, authors have synonymized the name with other earlier ones. However, the type lot in the British Museum seems distinctive. The outline is not unlike that of *C. parva*, but the shell is larger, its sculpture much finer, with imbricate spiral ribs, which end at the aperture as a fringe of spines, and a few low axial ribs not well defined. The canal is almost as long and slender as that of *C. (Pseudomurex) orcuttiana*, but the suture is not impressed. The color is a uniform white, inside and out. Length, 30 mm; width, 17 mm. Type locality "Gulf of California." It is possible that the form may prove not to be West American.

1060. Coralliophila (Coralliophila) macleani Shasky, 1970. The white shell has a short and slightly tabulate spire, the body whorl somewhat globose; sculpture is variable, mostly of spiral cords alternating in size, the strongest at the shoulder, with weak axial ribs on the spire that fade out on the body whorl. Outer lip crenulate at the edge and flaring; columella straight, the anterior canal open and a little recurved; operculum yellow to brown, its nucleus excentric. Length, 18 to 24 mm; diameter, 12 to 16 mm. Puerto Peñasco, Sonora, to Mazatlán, Mexico, on the bases of white gorgonians, depth 2 to 10 m. A similar species in the Atlantic is *C. caribaea* Abbott, 1958.

1061. Coralliophila (Coralliophila) neritoides (Lamarck, 1816) (Synonyms: **Purpura neritoidea** Sowerby, 1832 [not Linnaeus, 1758] of authors; **P. violacea** and **diversiformis** Kiener, 1836). The type species of the genus, this has a wide distribution in the western Pacific, where it is common on coral reefs. The sturdy white shell is bright pinkish violet within. There may be considerable variation in the height of the spire, the inflation of the whorls, and the outline. Length, 20 to 25 mm; width, 10 to 15 mm. Eastern Pacific records: Clarion, Clipperton, and Galápagos Islands.

Subgenus PSEUDOMUREX MONTEROSATO, 1872

Sculpture reticulate, axial ribs strong, spiral sculpture scaly; outer lip with digitate outer margin, somewhat lirate within; siphonal fasciole small.

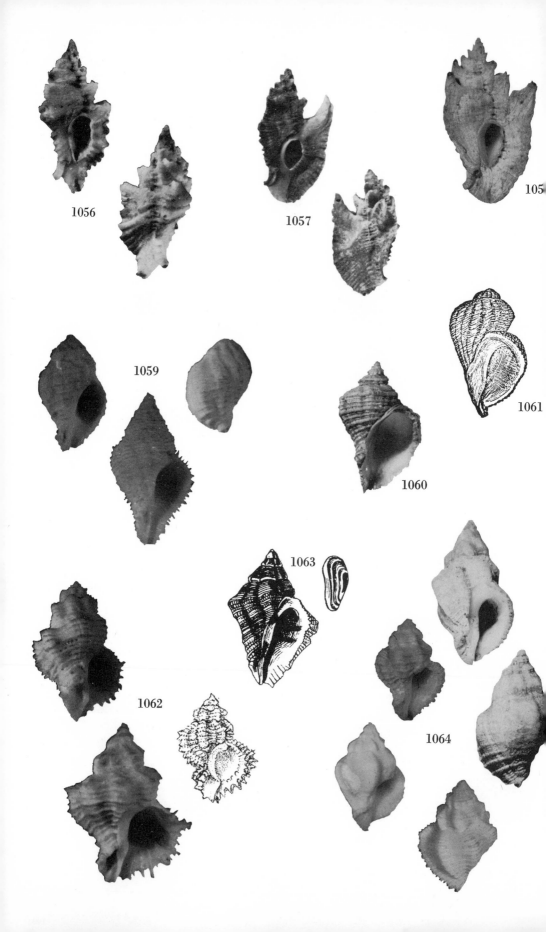

1056

1057

105

1059

1060

1061

1062

1063

1064

1062. Coralliophila (Pseudomurex) aspera (A. Adams, 1855). The original figure, reproduced here, shows spines and a large siphonal fasciole. No specimens have been reported in recent years. Length, 30 mm; diameter, 24 mm. The type lot, in the British Museum, was said to be from the Gulf of California. It may well prove not to be West American.

1063. Coralliophila (Pseudomurex) costata (Blainville, 1832) (Synonyms: "**Purpura diadema** Lamarck" of authors, a fictitious name; **P. foveolata** C. B. Adams, 1852). The grayish-white shell is biconic in shape, with a tendency toward a peripheral keel. Length, 25 mm; diameter, 17 mm. Baja California to Panama, on rocks between tides or offshore.

1064. Coralliophila (Pseudomurex) nux (Reeve, 1846) (Synonym: ? **Purpura osculans** C. B. Adams, 1852). This is a rather small, solid, white shell, with a few oblique axial ribs. The type specimen of Adams's *P. osculans* has been lost. It was described as having tumid whorls and an impressed suture, which suggests resemblance to *C. nux.* Length, 20 mm; diameter, 13 mm. Baja California south to Ecuador and the Galápagos Islands, under rocks at low tide.

1065. Coralliophila (Pseudomurex) orcuttiana Dall, 1919 (Synonyms: **Ocenebra sloati** and **O. s. hambachi** Hertlein, 1958). Although illustrated (M. Smith, 1939), the species remained unrecognized and was redescribed from other material. The shell is pure white, with axial ribs stronger on the spire than on the body whorl, fine spiral sculpture, and a long anterior canal. Length, 22 mm; width, 11 mm. Magdalena Bay, Baja California, to Gulf of Tehuantepec, Mexico, in depths to 95 m.

1066. Coralliophila (Pseudomurex) parva (E. A. Smith, 1877). This form has been confused with *C. nux,* though a comparison of specimens shows considerable difference, the main point of similarity being the cutaway effect of the outer lip where it joins the canal. *C. parva* is buffy white, with irregular spiral ribs that are cordlike as they ride over the curved axial ribs. Height, 10 to 24 mm; diameter, 7 to 15 mm. The Gulf of California to the Galápagos Islands. It is possible that the mainland form may prove to be separable from *C. parva,* the type of which is a small shell only 10 mm in length, type locality Galápagos Islands.

1067. Coralliophila (Pseudomurex) squamosa (Broderip, 1833). The sculpture is relatively coarse, and the columella is brownish in color. About 28 mm in length and 19 mm in diameter. The species was described from Peru, but specimens have been reported in Panama.

Genus LATIAXIS SWAINSON, 1840

With a strong peripheral keel. No species of subgenus *Latiaxis, s. s.,* occurs in the Panamic province.

Subgenus BABELOMUREX COEN, 1922

Spire moderate to high; sculpture scaly, with a peripheral keel made up of vaulted or imbricating spines; umbilicus small to medium-sized.

1068. Latiaxis (Babelomurex) hindsii Carpenter, 1857 (Synonym: **Trophon muricatus** Hinds, 1844 [not Montagu, 1803]). At the shoulder of the whorl the peripheral keel of the grayish to yellowish-white shell is made up of loosely overlapping spines that may either curve upward or slope outward. Length, about 20 mm; diameter, about 15 mm. Puertecitos, Baja California, to Panama, mostly offshore. *L. (B.) oldroydi* (I. Oldroyd, 1929) is a larger form from southern California, and an Atlantic species with a more elaborate keel is *L. (B.) dalli* Emerson & D'Attilio, 1963. Palmer (1958) has suggested that a return to Hinds's specific name might be made, since it is not a primary homonym of *Murex muricatus* Montagu; however, the latter is the type of *Trophonopsis*, which is treated as a subgenus of *Trophon* by some authors, and therefore confusion could result from secondary homonymy. Carpenter's replacement name has been well established, and retention of it seems preferable.

1069. Latiaxis (Babelomurex) santacruzensis Emerson & D'Attilio, 1970. The shell is flushed with a pale ochre color exteriorly, the aperture with rosy pink. The suture is deeply indented below a row of about ten long, hollow, upward-curving spines. Sculpture is of spiral cords that are scaly-surfaced, and the ends of the cords crenulate the outer lip of the aperture; the inner lip is somewhat flaring and free. Length, 34 mm; width, 34 mm. Galápagos Islands, in depths of 134 to 150 m. The closest relative of this striking form seems to be *L. (B.) dalli* Emerson & D'Attilio, 1963, from the Atlantic.

Genus CORALLIOBIA H. & A. ADAMS, 1853

Bulbous shells, spire almost concealed; outer lip distorted by attachment to the surface of coral heads, the inner lip broadly flaring; sculpture of body whorl cancellate.

1070. Coralliobia cumingii (H. & A. Adams, 1864). Described from "California?" and based upon a specimen that has apparently since been lost, the species has gone unrecognized. It was described as a *Campulotus* and compared to *Magilus*. Specimens that match the original description well are in the collection of the Los Angeles County Museum, taken alive on the coral *Pavona* at Gorgona Island, Colombia. The shell is white, with a yellowish-brown cancellate outer layer, the sculpture rising to prickles at the intersection of spiral and axial riblets; the anterior canal is narrowly spoutlike. Length, 21 mm; width, 13 mm. It closely resembles *Coralliobia robillardi* (Lienard, 1870) from Mauritius, a species reported on Clipperton Island by Bartsch and Rehder (1939). Thus, *C. cumingii* probably will be found on the offshore islands from Clipperton southward through the Galápagos.

Genus QUOYULA IREDALE, 1912

The aperture of the rather globose shell is almost entire; that is, there is scarcely any indication of an anterior canal. The outer lip is thin and smooth, bent to fit the coral to which the shell is attached.

1071. Quoyula madreporarum (Sowerby, 1834) (Synonym: **Purpura monodonta** Quoy & Gaimard, 1833 [not Blainville, 1832, *ex* Quoy & Gaimard, MS]). This is one of the few species widely distributed in the Indo-Pacific that has reached the mainland shores of West America. The animals feed upon corals and therefore can be found only in and around coral heads. The shell is white on the outside, pink to purple within. Length, 13.5 mm; diameter, 10 mm. The southern part of the Gulf of California and Tres Marias Islands, Mexico, to Panama and on the offshore islands, intertidally among corals.

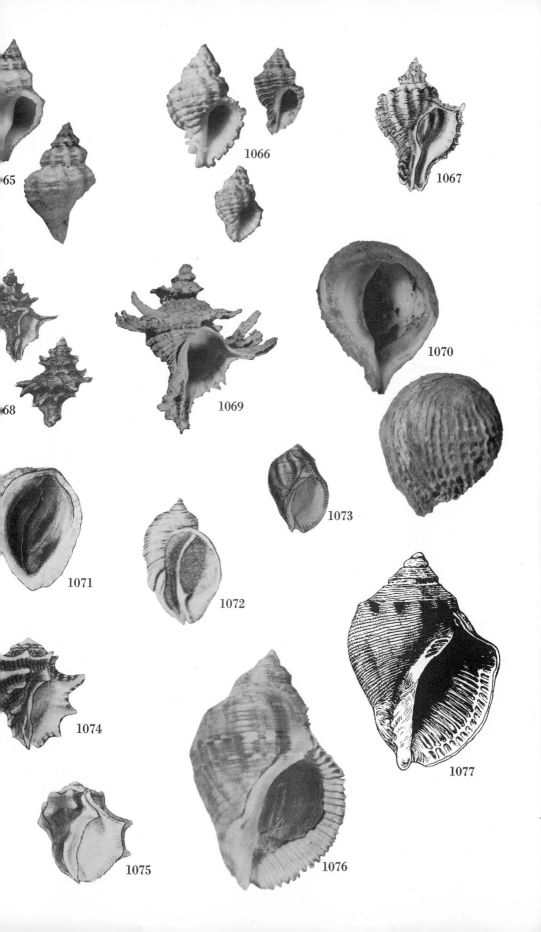

65

1066

1067

68

1069

1070

1071

1072

1073

1074

1075

1076

1077

1072. Quoyula monodonta (Blainville, 1882, *ex* Quoy & Gaimard, MS). It is possible that only one species of *Quoyula* is represented. Authors seem not to be in agreement on the significance of a columellar tooth at the anterior end of the canal that appears on an occasional specimen. If it is a variation that comes and goes in any population, then *Q. madreporarum* will fall as a synonym, for *Q. monodonta* has priority. Specimens with the columellar tooth were found at Acapulco, Mexico, by Mrs. G. Riley Hettick in 1959.

Family THAIDIDAE
(*Nom. conserv.*, ICZN Op. 886, 1969; PURPURIDAE, THAISIDAE)

The family name Thaididae has been conserved as of more frequent usage than the technically older Purpuridae, and the corrected spelling Thaididae has been made official by the International Commission on Zoological Nomenclature.

The thaids are active predators, creeping about over rocks in search of their prey—chiefly mussels and barnacles. The shells are not so varied in form as those of the Muricidae or so individually variable as those of the Coralliophilidae. They are medium-sized, rather solid, lacking varices but with knobby or slightly spinose sculpture. The anterior canal is short and the notch broadly open. A horny operculum has a heavy and polished ridge along the inside of the outer edge, its nucleus at about the middle of this ridge. Egg capsules are columnar or on a short stalk. The key is to genera in the three Panamic subfamilies.

1. With a spine at the lower edge of the outer lip 2
   Without a spine at the anterior end of the outer lip 3
2. With no siphonal fasciole or only a small one . . . . *Acanthina*
   With a conspicuous siphonal fasciole . . . . *Neorapana*
3. Shell small (length under 25 mm), slender, biconic . . . . *Morula*
   Shell more than 25 mm in adults, not slender 4
4. Columella with a raised ridge near its center . . . . *Cymia*
   Columella smoothly arcuate, without ridges 5
5. Sculpture, if nodose, with larger peripheral rows . . . . *Thais*
   Sculpture nearly evenly nodose over entire shell 6
6. Aperture contracted by denticles along both lips . . . . *Drupa*
   Aperture not contracted by denticles 7
7. Aperture reddish-orange within . . . . *Purpura*
   Aperture light buff within . . . . *Xanthochorus*

Subfamily THAIDINAE

Aperture not strongly dentate within, on only one lip if at all; siphonal fasciole mostly narrow. Four Panamic genera.

Genus THAIS Röding, 1798

Wherever there are rocks one may expect to find these sturdy mollusks crawling about. Although they are apparently capable of using the radula to drill holes in the shell of the prey, they do not always resort to this, for there is some evidence that a secretion from well-developed glands is used to narcotize or immobilize the prey. This secretion is a milky fluid when first extruded, but it turns to purple on exposure to air. Some of the Muricidae also secrete such a fluid, which was the basis for the Tyrian dye of the Mediterranean peoples.

Subgeneric divisions are not easy to distinguish, and some assignments are rather arbitrary.

1. Suture roughened by an area of fluted scales....................*Thaisella*
   Suture smoothly appressed.. 2
2. Columellar lip expanded, with oblique raised ridges..............*Tribulus*
   Columellar lip not expanded, smooth.............................. 3
3. Shell thick and heavy; spire low, rounded......................... 4
   Shell not conspicuously thick; spire either flat or elevated.............. 5
4. Shell with nodes or spines...............................*Mancinella*
   Shell smooth ....*Vasula*
5. Mature shells moderately large (over 50 mm in length); outer lip with a
   lirate or threaded margin within.........................*Stramonita*
   Mature shells less than 50 mm in length; outer lip smooth or dentate but
   not regularly threaded....................................*Thais, s. s.*

Subgenus THAIS, *s. s.*

Columella flattened and somewhat excavated.

1073. Thais (Thais) callaoensis (Gray, 1828). A short, solid little shell, light brown in color, with a large aperture. Length, 28 mm; diameter, 21 mm. The range is from Panama and the Galápagos Islands to Peru.

Subgenus MANCINELLA LINK, 1807

Ovate, solid, outer lip often lirate; with siphonal fasciole.

1074. Thais (Mancinella) speciosa (Valenciennes, 1832) (Synonyms: **Purpura centiquadra** Duclos, 1832; **P. triserialis** Blainville, 1832). It is ironic that three names should have been given to this species in a single year; Valenciennes' name apparently has priority by a few weeks or months, for the publication is cited by Duclos in May 1832, when his own *P. centiquadra* was published, and Blainville's work is known to have appeared later than May. The white shell is easily recognized by its spiral bands of brown squares; it has a yellowish aperture. Length, 36 mm; diameter, 30 mm. Magdalena Bay, Baja California, through the Gulf and south to Peru, on rocks between tides.

1075. Thais (Mancinella) triangularis (Blainville, 1832) (Synonym: **Purpura carolensis** Reeve, 1846). Although resembling *T. speciosa*, the light brown shell lacks the square dots, and the two rows of nodes at and below the shoulder are of equal size, whereas the upper one is larger in *T. speciosa*. Length, 30 mm; diameter, 26 mm. Cape San Lucas, Baja California, through the Gulf and south to Peru, intertidally, on rocks.

Subgenus STRAMONITA SCHUMACHER, 1817

Spire higher than in *Thais, s. s.*, columella less flattened; outer lip lirate within.

1076. Thais (Stramonita) biserialis (Blainville, 1832) (Synonym: **Purpura haematura** Valenciennes, 1846). Although a recent monograph (Clench, 1947) would make this and the similar Caribbean form synonymous with the eastern Atlantic *T. (S.) haemostoma* (Linnaeus, 1767), there is a slight color difference and a form difference that would seem to be enough to warrant separation. The Panamic province specimens do not have the orange-red coloration of true *T. haemostoma* but are, like the Caribbean specimens, more of a terracotta brown on the inner margin of the aperture and still lighter inside. None of the specimens examined in west coast collections had the rounded or knobby nodes on the shoulder of the whorls evident in the Mediterranean and west African shells. The exterior of the shell is dark gray and may vary from nearly smooth to roughly corded, the color pattern varying from plain to mottled with light and dark brown. Mature specimens may attain a length of more than 75 mm (3 inches), with a diameter of 50 mm. Cedros Island, Baja California, through the Gulf and south to Chile, on rocks, intertidally; also the Galápagos Islands.

1077. Thais (Stramonita) chocolata (Duclos, 1832). This differs from *T. biserialis* in color, being a uniform brown, the columella tinged with orange, the interior bluish or yellowish. It is of about the same size and ranges from Ecuador southward to Peru.

1078. Thais (Stramonita) delessertiana (Orbigny, 1841) (Synonym: **Purpura blainvillei** Deshayes, 1844). Higher-spired than the other two species, this is a uniform brown on the outside, bluish within the aperture, and the shell is rather thin. Length, 34 mm; diameter, 22 mm. Ecuador to Paita, Peru.

Subgenus THAISELLA CLENCH, 1947

With remnants of anal notch of previous whorls as flutings below suture.

1079. Thais (Thaisella) kiosquiformis (Duclos, 1832). A gray or brown shell, now and then marked with white, the aperture brown or brown-and-white-banded. The frilled suture and the angulate outline are characteristics easy to recognize. Length, 43 mm; diameter, 26 mm. Magdalena Bay, Baja California, throughout the Gulf and south to Peru. It is a common form in mangrove swamps, where the animals feed on attached oysters, but specimens may also be found on rocks in muddy areas.

Subgenus TRIBULUS H. & A. ADAMS, 1853

With oblique ridges on columella and lower inner lip; outer lip lirate within.

1080. Thais (Tribulus) planospira (Lamarck, 1822) (Synonym: **Haustrum pictum** Perry, 1811 [suppressed by the International Commission on Zoological Nomenclature as a Rejected and Invalid Name (Op. 886, 1969)]). The yellowish-brown shell is white within except for reddish ribs inside the outer lip and on the columella. Across the deep channel in the middle of the columella is an oblique black rib, a most unusual form of marking. Length, 53 mm; diameter, 43 mm. Cape San Lucas, Gulf of California, to Peru, on rocks at extreme low tide.

Subgenus VASULA MÖRCH, 1860

Thick, solid, smooth, columella rounded; no siphonal fasciole.

1081. Thais (Vasula) melones (Duclos, 1832) (Synonym: **Purpura crassa** Blainville, 1832). Another easily recognizable form, this very thick shell is nearly globular, black, with yellow and white patches. Length, 48 mm; diameter, 34 mm. The Gulf of Tehuantepec, Mexico, to Callao, Peru, and the Galápagos Islands.

Genus ACANTHINA FISCHER DE WALDHEIM, 1807
(MONOCEROS SOWERBY, 1827 [not LACÉPÈDE, 1798])

The shell has a long thornlike tooth on the outer lip just above the canal, which may be used by the animal as a wedge in preventing closure of the valves of its prey.

1078

1079

1081

1082

80

1084

1083

1085

1086

1087

1082. Acanthina angelica I. Oldroyd, 1918. This species resembles the Baja Californian *A. lugubris*, but shells are smaller and more slender, the spire higher, and the coloration less subdued, the aperture being white within. Length, 39 mm; diameter, 24 mm. The species seems to be restricted in distribution to the Gulf of California. Egg capsules are described by Wolfson (1970).

1083. Acanthina brevidentata (Wood, 1828) (Synonyms: **Purpura cornigera** Blainville, 1832; **P. ocellata** Kiener, 1835–36; **Monoceros maculatum** Gray, 1839). A grayish-black shell with white spots on the spiral nodules that encircle the whorls; the inner margin of the outer lip dark brown or black. Length, 30 mm. Mazatlán, Mexico, to Paita, Peru, on rocks between tides.

1084. Acanthina lugubris (Sowerby, 1822) (Synonyms: **Monoceros cymatum** Sowerby, 1825 [*nomen nudum*]; **M. armatum** and **denticulatum** Wood, 1828). Because this species ranges outside the area covered by this report, it could have been omitted here but is cited for comparison. The shell is an olive-drab color without and within, mottled by patches of brown or black on the rough, dull exterior. The aperture is a uniform shiny yellowish brown, smooth except for elongate denticles near the margin of the outer lip. Length, 44 mm; diameter, 31 mm. The species ranges from southern California to Magdalena Bay, Baja California; it has also been reported in the Galápagos Islands; not uncommon on rocks between tides.

1085. Acanthina tyrianthina Berry, 1957. The color of the shell is buffy gray, checkered with blackish brown, the interior tawny brown. This is somewhat intermediate between *A. angelica* and *A. lugubris*, having a much lower spire than the former and more shouldered whorls than the latter; it is smaller than either. Length, 27 mm; diameter, 18 mm. Magdalena Bay, Baja California, to the southern part of the Gulf of California.

Genus CYMIA MÖRCH, 1860
(CUMA SWAINSON, 1840 [not MILNE-EDWARDS, 1828])

The shell is rough-textured, with deeply incised spiral grooves and one row of knobby spines at the shoulder, which slopes downward from the suture. The siphonal fasciole is conspicuous, composed of the ends of successive stages in growth of the twisted canal.

1086. Cymia tecta (Wood, 1828). The dark grayish-brown shell is heavy and solid for its size, with a brownish to yellowish aperture in which the external sculpture shows as raised lines. Length, 48 mm; diameter, 33 mm. Costa Rica to Ecuador, on rocks between tides.

Genus PURPURA BRUGUIÈRE, 1789
(PLICOPURPURA COSSMANN, 1908; PATELLIPURPURA DALL, 1909)

Although the generic term *Purpura* was applied by the earliest writers on shells to marine snails from which purple dye could be extracted, there has been some difference of opinion among more modern workers as to the application of the name within the framework of our system of zoological nomenclature. An Opinion of the International Commission on Zoological Nomenclature (Op. 886, 1969) clears *Purpura* for use in a long-accustomed and appropriate sense.

The Tyrian dye actually was obtained in the main from species of the Muricidae—which, like *Purpura*, have the mucous gland partially modified for the secretion of a purple fluid that perhaps is useful to the mollusk in subduing prey. The Indians of the West Central American area had learned of the properties of the dye in *Purpura*, and being better conservationists than the Mediterranean peoples—who obtained it by crushing the animals, shells and all—they had found a way to assure a continuous supply. As Clench (1947) describes it: "Carrying skeins of loosely twisted cotton thread, the natives collect individual specimens of *Purpura*, blow upon them, and, as the mollusk recedes into the shell exuding a milky froth, they dab this froth upon the cotton thread. The shell is then placed back on the rock or in a tide pool and used again. . . . The milky froth turns purple and the threads are thus dyed."

1087. Purpura columellaris (Lamarck, 1822) (Synonym: **Haustrum dentex** Perry, 1811 [suppressed by the International Commission on Zoological Nomenclature as a Rejected and Invalid Name (Op. 886, 1969)]). The shell is grayish brown, thick and heavy for its size, with an orange-brown aperture, the outer lip of which is studded with teeth. The columella also has a couple of raised nodes near its center. About 59 mm in length and 42 mm in diameter; a large specimen, 79 mm in length and 49 mm in diameter. The southern part of the Gulf of California to Chile. Lowe (1932) reported it as occurring on surf-beaten rocks with *P. pansa* but as being much rarer and difficult to distinguish. The type of *Plicopurpura*, which could become a subgenus for species with nodose sculpture.

1088. Purpura pansa Gould, 1853. This has been called *P. patula* (Linnaeus, 1758) by authors, but the West American form is easily separable from true *P. patula*, which is the Caribbean twin species, by the white in the aperture as well as by size. The dark gray shell has a bright salmon-brown aperture, especially bright-colored along the border and on the columella, with a curved area of dark brown above the columella and a white inner margin to the columella. The operculum is rather small for the size of the aperture and closes it only when the animal is deeply withdrawn into the shell. A large specimen may be nearly 100 mm (4 inches) in length, but the average is about 64 mm; the average diameter is 41 mm. Magdalena Bay, Baja California, through the south end of the Gulf and south to Colombia and the Galápagos Islands, common on rocks in exposed locations.

Subfamily DRUPINAE

Small to medium-sized, the aperture strongly dentate, with teeth on both lips.

Genus DRUPA RÖDING, 1798
(RICINULA LAMARCK, 1816; SISTRUM MONTFORT, 1810)

Low-spired, aperture contracted by large denticles.

1089. Drupa albolabris (Blainville, 1832) (Synonym: **D. ricinus** [Linnaeus, 1758] of authors, not of Linnaeus). A white shell with conspicuous black nodes and a white aperture. Length, 26 mm. Principally Indo-Pacific, occurring only on offshore islands, especially Clipperton Island.

Genus MORULA SCHUMACHER, 1817

Small, biconic, with nodose spiral ribs and apertural teeth mainly on the outer lip.

Subgenus MORULA, *s. s.*

Nodes relatively coarse.

1090. Morula (Morula) aspera (Lamarck, 1816). Chunky, somewhat rounded, with a violet aperture and paired black spiral lines on a grayish-white ground. Length, 20 mm. An Indo-Pacific species that occurs on Clipperton Island.

1091. Morula (Morula) uva (Röding, 1798). The white shell has the nodes aligned spirally and vertically; inner lip teeth not conspicuous. Length, 24 mm; width, 16 mm. Another western Pacific species with a foothold in the Panamic province, on Guadalupe and Clipperton islands. It is the type of the genus.

Subgenus MORUNELLA EMERSON & HERTLEIN, 1964

Surface finely nodose; outer lip sharp-edged, thickened within by a lirate ridge studded with small denticles at the ends of the lirae.

1092. Morula (Morunella) ferruginosa (Reeve, 1846). Somewhat resembling *Muricopsis zeteki*, this form may be distinguished by its more slender outline, by the lack of varices and foliaceous sculpture, by the rounded, not elongate, nodes, and by the brown and blue spotting of the aperture. The shell is dark gray to black, obscurely banded with white, the nodes black. Length, 21 to 26 mm; width, 11 to 16 mm. Magdalena Bay, Baja California, through the Gulf of California and south along the Sonoran coast of Mexico at least as far as Guaymas, intertidally under rocks. It is probable that this and the next species and other New World Morulas will prove to be generically rather than subgenerically separable. According to A. D'Attilio (*in litt.*), their principal point in common with the Indo-Pacific Morulas is the nodose sculpture.

1093. Morula (Morunella) lugubris (C. B. Adams, 1852) (Synonyms: **Ricinula rugosoplicata** Baker, 1891; **Fusinus orcutti** Dall, 1915; **Cantharus exanthematus** Dall, 1919). Smaller than *M.* (*M.*) *ferruginosa* and smoother, the cinnamon-brown shell is marked with dark brown on the rows of nodes. Apertural teeth are not strongly developed. Length, 15 mm; diameter, 7 mm. San Diego, California, to Panama, intertidally under rocks and offshore to 40 m. This is the type species of the subgenus.

Subfamily RAPANINAE

Few whorled; aperture large, smooth within or with lirae or only small denticles but with a tendency to develop a tooth on the outer margin near the canal; siphonal fasciole wide. Radula lacking marginal cusps, resembling the Muricidae in the apparent reduction in number of cusps but with small serrations of the cusp margins. Operculum with a lateral nucleus. Egg capsules columnar.

Genus NEORAPANA COOKE, 1918

Formerly placed with *Acanthina* because of the tooth on the outer lip, these shells differ in larger size, stronger sculpture, and large siphonal fasciole as well as in the pattern of the radula.

1094. Neorapana grandis (Sowerby, 1835, *ex* Gray, MS) (Synonym: **Purpura grayi** Kiener, 1835–36). The interspaces between the brown scaly-edged spiral ribs are lighter-colored and deeply excavated, forming a convoluted pattern of the outer lip. The aperture is yellowish within. Length, 60 to 90 mm; width, 45 to 65 mm. Galápagos Islands.

1095. Neorapana muricata (Broderip, 1832) (Synonym: **Purpura truncata** Duclos, 1833). Largest member of the genus, this has a yellowish to brownish

shell, somewhat pink within, the spire low, the first rib below the suture not nodose but standing out as a strong ridge. A large specimen may be 100 mm in length, an average one 55 mm, with a diameter of 51 mm. Guaymas, Mexico, to Ecuador.

1096. Neorapana tuberculata (Sowerby, 1835). This may be only a sub-species of *N. muricata*, but it seems distinct. The spire is higher and the first rib is broken up into nodes. Also, the shell is smaller and more slender, about 50 mm long and 40 mm in diameter. The species seems to be restricted in distribution to the Gulf of California area, from Cape San Lucas throughout the Gulf to Mazatlán; fairly common on rocks between tides.

Genus **XANTHOCHORUS** FISCHER, 1888

Thick, heavy, last whorl rounded, spire low; axial costae numerous; outer lip dentate within.

1097. Xanthochorus broderipii (Michelotti, 1841) (Synonyms: **Murex horridus** Broderip, 1833 [not Brocchi, 1814]; **M. boivinii** Kiener, 1843). A sturdy brownish-gray shell with numerous varix-like costae, the aperture buffy within and studded with a row of small teeth. Length, 45 mm; width, 22 mm. Panama to Chile. Michelotti's name for Broderip's preoccupied *M. horridus* has been overlooked by most authors and was only recently noted by Dr. Emily Vokes (*in litt.*).

Superfamily **BUCCINACEA**

Mostly sturdy shells with well-developed but not spinose sculpture. Aperture with an anterior canal or notch, inner lip mostly smooth. Radula rachiglossate.

Family **BUCCINIDAE**

Most buccinids have a somewhat tapering spire, fairly large body whorl, and short canal. With few exceptions the animals are carnivorous, but they may be scavengers rather than active predators, content to feed on dead fish and other scraps. The operculum is ovate, thin, horny, its nucleus apical.

The type genus, *Buccinum*, is limited to cold water, mainly of the northern hemisphere. Early authors, however, used the generic name as a catchall for all forms with a short anterior canal. Most of the temperate to tropical species so named now are placed in other genera. The so-called buccinids from warmer waters and those assigned to the related cold-water family Neptunidae are much in need of revision.

The sequence of genera adopted here departs from strictly alphabetical order for convenience in comparing such closely similar warm-water groups as *Cantharus* and *Solenosteira*. The key is to genera.

1. Outer lip varicose. 2
   Outer lip somewhat thickened but not varicose. 4
2. Anterior canal open in adult; spire not tapering. . . . .*Bailya*
   Anterior canal partially closed in adult; spire tapering. 3
3. Outer lip varix wide, somewhat fluted. . . . .*Neoteron*
   Outer lip varix narrow, rounded. . . . .*Trajana*
4. Spire slender, distinctly longer than aperture. 5
   Spire equal to or shorter than aperture. 6
5. Sculpture becoming obsolete on later whorls. . . . .*Northia*
   Sculpture on body whorl and spire equally strong. . . . .*Phos*
6. Sculpture finely and evenly cancellate throughout. . . . .*Metula*
   Sculpture coarsely cancellate or unevenly cancellate. 7

Genus BAILYA SMITH, 1944

Small, with well-developed regular axial ribs and fine spiral threads under a persistent periostracum. The operculum is smaller than the aperture, its nucleus terminal.

1098. Bailya anomala (Hinds, 1844) (Synonym: **Fusus bellus** C. B. Adams, 1852 [not Conrad, 1833]). The axial ribs are strong, ten to twelve per whorl, and the shell is yellowish brown in color. Type species of the genus. Length, 15 mm; diameter, 8 mm. Guaymas, Mexico, to Nicaragua, intertidally. A similar species in the Atlantic, *B. parva* (C. B. Adams, 1852), has been shown to have a buccinid radula.

Genus CADUCIFER DALL, 1904

Originally considered to be a member of the Cymatiidae, as was *Bailya*, this genus has been transferred to the Buccinacea because of the radula. The outer lip is thickened rather than varicose. The shell is ovate, the spire in complete specimens bluntly tapering. In many of the species complete specimens are rare, the animal having sealed off the early whorls from within, after which they easily break away. Sculpture is both axial and spiral, the latter tending to be of straplike cords. The posterior part of the aperture has at least two heavy teeth that define a tunnellike slot for the posterior siphon; both the inner and outer lips may be wrinkled or denticulate also.

Subgenus CADUCIFER, s. s.

Apex almost invariably truncate and sealed from within. No Panamic species seem referable to this subgenus, the type of which is a western Pacific form.

Subgenus MONOSTIOLUM DALL, 1904

Apex present in most specimens, bluntly pointed. The type species, *C. (M.) swifti* (Tryon, 1880), is from the Caribbean.

1099. Caducifer (Monostiolum) biliratus (Reeve, 1846) (Synonym: **Fusus apertus** Carpenter, 1857). The ovate shell is light brown with darker mottling; each of the spiral ribs is doubled, as Reeve's specific name suggests. Length, 22 mm; width, 11 mm; Gulf of California, in depths of 7 to 146 m, to Galápagos Islands (type locality).

1100. Caducifer (Monostiolum) cinis (Reeve, 1846) (Synonym: **C. thaleia** Pilsbry & Lowe, 1932). The sturdy shell has a bluntly tapering spire and about 20 axial ribs cut by raised spiral cords; it is brown with white spots that may form bands. The aperture has somewhat flaring lips, wrinkled to denticulate, and a narrow canal anteriorly, a rounded notch posteriorly. Length, 28 mm; width, 10 mm. Jalisco, Mexico, to Cocos and the Galápagos Islands.

1101. Caducifer (Monostiolum) crebristriatus (Carpenter, 1856) (Synonym: **C. tabogensis** Pilsbry & Lowe, 1932). The sculpture is of strong spiral cords with relatively wide interspaces, with only faint axials, the shell white, densely spotted with reddish brown. The outer lip has a weak varix; the inside of the aperture is smooth except for two low teeth near the posterior siphonal notch. Length, 18 mm. Barra de Navidad, Jalisco, Mexico, to Panama (type locality).

1102. Caducifer (Monostiolum) nigricostatus (Reeve, 1846). Described as olive yellow in color, the shell has several spiral ribs beaded with black, the axial beading more evident and numerous than in *C. biliratus*. Length, 19 to 25 mm; diameter, 9 to 12 mm. Jalisco, Mexico, in 9 to 18 m, to Panama (type locality).

1103. Caducifer (Monostiolum) pictus (Reeve, 1844) (Synonym: **? C. pervaricosa** Dall & Ochsner, 1928). The bluntly tapering shell is distantly ribbed axially and closely ribbed spirally, spotted with dark brown to reddish brown, the aperture white within. Axial sculpture is somewhat sinuous and the spiral riblets fine and even; the outer lip varix is weak. Length, 18 mm; width, 6 mm. Galápagos Islands. Comparison of a photograph of Reeve's type with the figure of the Pleistocene shell described by Dall and Ochsner reveals no significant difference in size or sculpture.

Genus **CANTHARUS** RÖDING, 1798

(**PISANIA, POLLIA, PSEUDONEPTUNEA, TRITONIDEA** of authors)

The type species, *C. tranquebaricus* (Gmelin, 1791), is from the Indian Ocean, a solid, short shell with a rapidly tapering spire, impressed sutures, low axial folds, and numerous irregular spiral ribs. The siphonal fasciole is large and the aperture ovate, with a short canal. The horny operculum has an apical nucleus.

Various groupings of the species that fit into this general pattern have been attempted, but description of the differences is not easy. A sound classification awaits studies of the soft parts, especially of the radula. Some of the Panamic species can be assigned to the subgenus *Gemophos*; the rest must for the time being remain in *Cantharus* without subgeneric allocation.

Subgenus **CANTHARUS**, *s. l.*

Several species that bear an obvious relationship to *Cantharus* but that do not fit readily into groups at present recognized for the eastern Pacific are here listed as *Cantharus* in the broad sense. *Pseudoneptunea* Kobelt, 1882, cited for some of these, has been shown not to be appropriate for American species.

1104. Cantharus panamicus (Hertlein & Strong, 1951). The color is light brown, variegated, the tubercles darker, with a vague band of brown around the base of the last whorl. The shell is relatively thin. Length, 39 mm; diameter, 25 mm. Guaymas, Sonora, Mexico, to Panama, in depths to 73 m. [See note, p. 853.]

1105. Cantharus rehderi Berry, 1962 (Synonym: **Solenosteira elegans** Dall, 1908 [not Griffith & Pidgeon, 1834]). The shell resembles some of the small Cymatiidae in outline, lacking, however, any varices. The periostracum is thin and reveals the axial and spiral sculpture, which forms rather even elongate nodes. One or

1101

1102

1103

1104

1105

1106

1107

1108

1109

1110

1111

two rows of the nodes are more prominent and may be touched with white. Length, 46 mm; width, 28 mm. Cedros Island, Baja California, and the southern end of the Gulf of California, south through Panama Bay (type locality) to Peru, mostly offshore in depths of 46 to 333 m.

1106. Cantharus shaskyi Berry, 1959. In sculpture the shell resembles both *C. panamicus* and *C. rehderi*, but it is more brightly colored than either. A fresh specimen is pinkish buff, with white raised lines where the spiral ribs cross the axials and a few maculations of darker and lighter color. Length, 44 mm; width, 25 mm. From shrimp boats, Gulf of California, probably from south of Guaymas, in moderate depths.

Subgenus GEMOPHOS OLSSON & HARBISON, 1953

Shell solid, with a bluntly tapering spire; whorls rounded, having both axial and spiral sculpture; periostracum present in most species of the subgenus; siphonal fasciole weak to wanting.

1107. Cantharus (Gemophos) berryi McLean, 1970. Relatively small, yellowish white under a thin, closely adherent periostracum; axial ribs brown near periphery, with a lighter band below the suture. The species shares the habit of *Solenosteira* of depositing eggs on the outside of the shell, but otherwise it is closer to *Cantharus (Gemophos)*. Length, 21 mm; diameter, 11.7 mm. Banderas Bay, Jalisco, Mexico, 18 to 27 m.

1108. Cantharus (Gemophos) elegans (Griffith & Pidgeon, 1834, *ex* Gray, MS) (Synonyms: **Buccinum insignis** Reeve, 1846; **Pisania aequilirata** Carpenter, 1857 [not *Solenosteira elegans* Dall, 1908]). Having a rather longer canal and smaller siphonal fasciole than most species of the genus, this is a brown shell variegated with white, especially on the nodose part of the ribs. The periostracum is fibrous and olive brown in color. About 47 mm in length, 24 mm in diameter. Magdalena Bay, Baja California, throughout the Gulf and south to Peru, on rocks, intertidally.

1109. Cantharus (Gemophos) gemmatus (Reeve, 1846) (Synonym: **Buccinum mutabile** Valenciennes, 1846 [not Linnaeus, 1758]). A brownish shell under an olive green periostracum, the spiral ribs touched with dark brown at their nodose junctions with the axial ribs. The aperture is white. About 35 mm in length. Mazatlán, Mexico, to Ecuador, on rocks between tides. This is the type species of the subgenus.

1110. Cantharus (Gemophos) janellii (Kiener, 1835–36). The low-spired sturdy gray shell is shaped much like that of *C. sanguinolentus*—with which it has been confused—but the aperture is ringed with a solid band of dark brown or purplish black within; the axial ribs are knobby at the periphery and white-spotted. Length, about 25 mm; width 20 mm. Galápagos Islands to Paita, Peru (type locality).

1111. Cantharus (Gemophos) lautus (Reeve, 1846). Under a thin brown periostracum like that of *C. vibex*, the shell is strikingly marked with black bars on the axial ribs and a few touches of yellowish brown. Length, 30 mm; width, 18 mm. Port Guatulco, Mexico, to Ecuador (type locality).

1112. Cantharus (Gemophos) pagodus (Reeve, 1846). Sculpture is subdued: the whorls rounded, axial ribs low, and spiral cords not deeply incised. The

PLATE VII · *Upper left* · *Hexaplex erythrostomus* (The Pink-mouthed Murex)
Center and lower left · *Hexaplex regius* (The Royal Murex)

periostracum is persistent, somewhat velvety, mottled in shades of dark brown. Compared to *C. vibex*, the shell is larger, with more axial ribs but with the ribs ill-defined. Length, 45 mm; diameter, 25 mm. Mazatlán, Mexico, to Panama.

1113. Cantharus (Gemophos) pastinaca (Reeve, 1846). A robust, rather wide form, prominently sculptured, with a decided shoulder and tunnel-like slot at the upper end of the aperture for the posterior siphon. About 27 mm in length. Confined to the coast of Panama.

1114. Cantharus (Gemophos) ringens (Reeve, 1846). The wide slot for the posterior siphon gives the aperture of this shell the appearance of a grinning mouth—hence the name, *ringens*. The shell is, like *C. pastinaca*, brownish under a smooth, drab brown periostracum. About 28 mm in length. Mazatlán, Mexico, to Ecuador, on rocks, intertidally.

1115. Cantharus (Gemophos) sanguinolentus (Duclos, 1833) (Synonyms: **Pollia haemastoma** Gray, 1839; **? Columbella apthaegera** Lesson, 1842; **Buccinum janelii** Valenciennes, 1846 [not *Purpura janellii* Kiener, 1835–36]). Colored much like the other species, this has a distinctive red margin to the aperture, dotted with white pustules. About 25 mm in length. The outer coast of Baja California through the southern part of the Gulf of California to Guaymas, Mexico, and south to Ecuador; not uncommon around masses of coral.

1116. Cantharus (Gemophos) vibex (Broderip, 1833) (Synonyms: **Triton turbinelloides** Griffith & Pidgeon, 1834, *ex* Gray, MS; **C. v. marjoriae** M. Smith, 1944). With shouldered whorls and seven axial ribs, the shell is brown variegated with white under a rough brown periostracum; the aperture is white. About 35 mm in length and 20 mm in diameter. Magdalena Bay, Baja California, and along the Central American coast to Panama, intertidally and offshore to 22 m. The color form named *C. v. marjoriae* has a yellow-tinged aperture.

Genus **SOLENOSTEIRA** DALL, 1890
(**HANETIA** of authors, not of JOUSSEAUME, 1880)

Though considered a subgenus of *Cantharus* by many authors, *Solenosteira* is treated as a genus here, placed next to *Cantharus* for comparison. The name was long thought to be a synonym of *Hanetia*, but that genus has now been shown (Emerson, 1968) to be muricid. The Solenosteiras have in general larger shells than *Cantharus*, with a longer anterior canal. The periostracum is well developed and may show specifically distinct growth patterns. Beneath it the shells are white, rarely with any touches of color. A habit shared only occasionally by species of other genera is that of depositing the egg masses on the outside of the shell; many specimens may be completely encrusted with the chaffy, wheat-shaped capsules.

1117. Solenosteira anomala (Reeve, 1847) (Synonyms: **Pyrula lignaria** Reeve, 1847; **Neptunea anceps** H. & A. Adams, 1853). One of Reeve's two figured specimens, in the British Museum, is refigured here. It is relatively slender, with strong spiral cords and a pinched-in groove below the suture; the siphonal fasciole is large and the anterior canal well developed; two or three of the spiral cords are stronger than the rest, but they do not stand out as peripheral nodes. Reeve's other specimen, which had a lower spire, was considered by Dall to be more characteristic of the species. Unfortunately, that specimen seems to have been lost. Length, 40 mm; width, 21 mm. Baja California southward, possibly to Ecuador, intertidally and offshore to depths of 73 m.

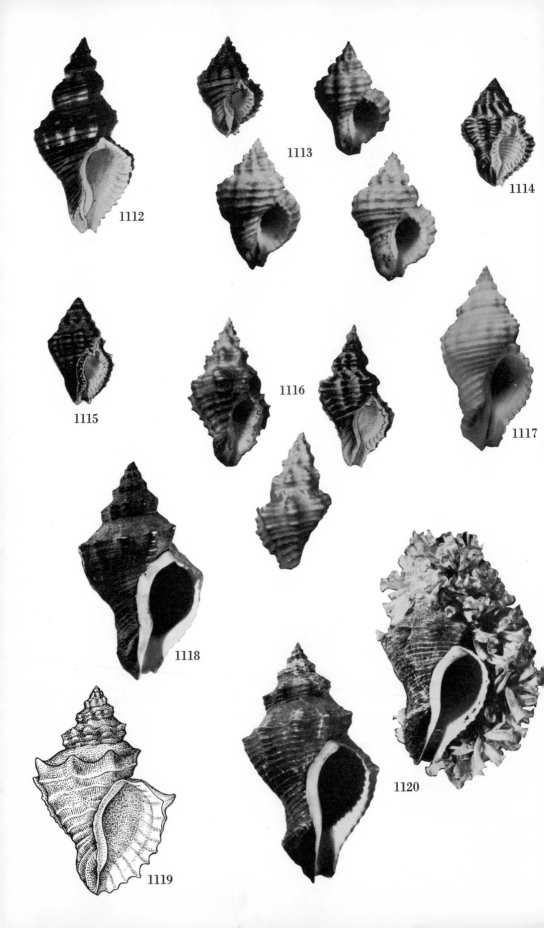

1112

1113

1114

1115

1116

1117

1118

1119

1120

1118. Solenosteira capitanea Berry, 1957. The brown periostracum is distinctive in being laid on in vertical flakes, with scattered bristly hairs; the periphery of the shell is slightly nodose. Length, 56 mm; diameter, 33 mm. San Felipe, near the head of the Gulf of California (type locality).

1119. Solenosteira fusiformis (Blainville, 1832) (Synonyms: **Fusus purpuroides** Orbigny, 1839; **Purpura dorbignyi** Reeve, 1846). Under a fibrous brown periostracum the shell is white, mottled with rusty brown. The shape is somewhat like that of *Thais kiosquiformis*, but the operculum is buccinoid. Length, about 50 mm. Panama to Peru, intertidally, in rock crevices. This species was made the type of the subgenus *Fusinosteira* Olsson, 1932. The large nodes of the peripheral keel are distinctive.

1120. Solenosteira gatesi Berry, 1963. The medium-sized shell is acutely conic and turreted, with about ten axial ribs crossed by numerous spiral cords, rising to points at the periphery; periostracum buffy brown, scurfy, finely spinulose, aligned with the growth striae; interior of the aperture bluish white. Length, 44 mm; diameter, 25 mm. Guaymas, Sonora, Mexico, to Mazatlán (type locality), in depths to 27 m.

1121. Solenosteira macrospira Berry, 1957. Very close to *S. pallida*, this differs in being slightly shorter for its width, with a stronger siphonal fasciole and more subdued spiral ribs; also, it has a groove just below the suture that sets off a collar-like cord, as in *S. anomala*. The periostracum is fringed. Length, 46 mm; diameter, 28 mm. Near the head of the Gulf of California (type locality, San Felipe).

1122. Solenosteira mendozana (Berry, 1959) (Synonym: **Fusus turbinelloides** Reeve, 1858 [not Grateloup, 1853]). Largest of the Solenosteiras, this is an offshore form with a strongly carinate periphery, which the axial ribs cross as sharp points; the siphonal fasciole is larger than in *S. capitanea*, and the periostracum lacks the scattered spinelike spicules. Length, 62 mm; diameter, 41 mm. Magdalena Bay, Baja California, 18 to 46 m (type locality) to Paita, Peru (Te Vega, 1968).

1123. Solenosteira pallida (Broderip & Sowerby, 1829) (Synonym: **? Buccinum modificatum** Reeve, 1846). The white shell has a densely pilose or velvety periostracum, with moderately coarse spiral ribs and low axial undulations. Length, about 40 mm; diameter, 28 mm. The exact range remains to be worked out now that a number of distinct species have been sorted out from the *"Hanetia pallida"* of authors. Unfortunately, the type specimen is lost. A specimen from the type lot, in the Gray collection at the British Museum, is figured here; it may be considered a lectotype or replacement type. It resembles *S. macrospira* but is more slender, with stronger sculpture, and it lacks a clear-cut subsutural groove.

Genus ENGINA GRAY, 1839

Shells small to medium-sized, biconic, with denticles on both sides of the aperture. Color variability has complicated recognition of the named species. The assistance of Dr. James H. McLean in construing some of the synonymies is gratefully acknowledged.

? Engina contracta (Reeve, 1846). For a discussion of this enigmatic species, see the Appendix.

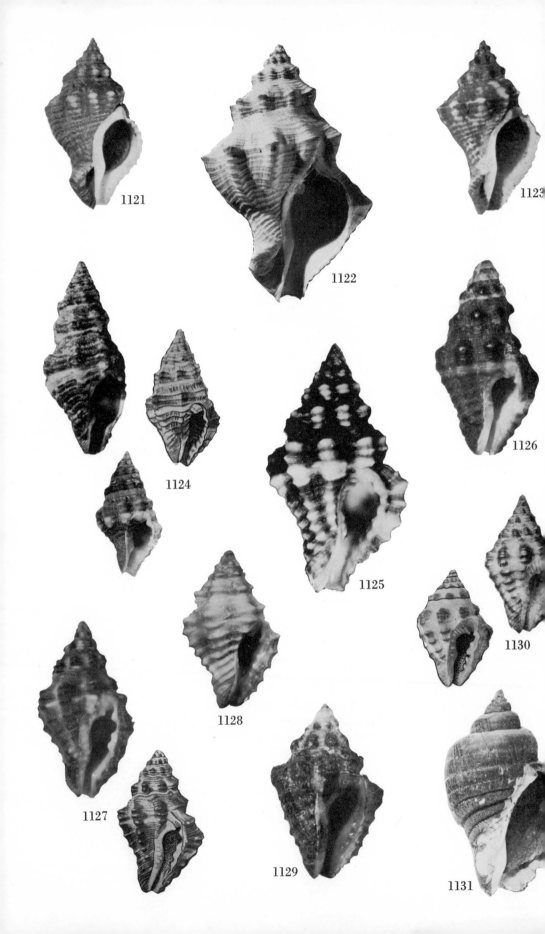

1124. Engina fusiformis Stearns, 1894 (Synonyms: **E. solida** Dall, 1917; **E. senae** and **senae multa** Pilsbry & Lowe, 1932). Strong growth lines form axial striations between the spiral ribs of the white-banded brown shell. There are six to seven massive axial folds in the majority of specimens; the form named by Pilsbry and Lowe as a subspecies may have eight. Stearns's type specimen, though immature, was unusually large and high-spired, and because it was unillustrated, authors have overlooked the name. Length, 11 to 12 mm; diameter, 6 to 6.5 mm. Puertecitos, near the head of the Gulf of California, to Acapulco, Mexico.

1125. Engina jugosa (C. B. Adams, 1852). The light brown shell, with a fibrous periostracum, is studded with white dots on the intersections of axial and spiral ribs and has a white band just below the periphery of the last whorl. Length, about 23 mm; diameter, 12 mm. Head of the Gulf of California to Ecuador and the Galápagos Islands, in depths to 37 m.

1125a. Engina macleani Olsson, 1971. Height, 22 mm; diameter, 9.8 mm. Gulf of Panama, dredged.

1126. Engina mantensis Bartsch, 1928. Probably related to *E. maura*, the dark brown form has a yellow band just below the periphery of the shouldered whorls. The spire is higher than in *E. maura*. Length, 17 mm; diameter, 9 mm. Manta, Ecuador.

1127. Engina maura (Sowerby, 1832) (Synonyms: **Ricinula carbonaria** Reeve, 1846; **E. panamensis** Bartsch, 1931). A nearly black shell with a bluish-white aperture, this has the axial ribs rounded and the spirals alternating in size. Some specimens may have orange and yellow mixed into the plain black and white. The spire tends to be turreted. Length, 12 to 20 mm; diameter, 8 to 17 mm. Barra de Navidad, near Manzanillo, Mexico, to Ecuador.

1128. Engina pulchra (Reeve, 1846) (Synonyms: **? Columbella livida** Sowerby, 1832 [*nomen oblitum*]; **Ricinula reeviana** C. B. Adams, 1852 [not to be confused with *Engina reevei* Tryon, 1883]). The light brown to buff shell has two thin white spiral ribs at the periphery, a brown band between, a grayish band above, and a yellow band below; the aperture is bluish white tinged with brown on the columella. Length, 15 mm; diameter, 11 mm. Panama to Ecuador, intertidally, under stones. For a discussion of the name *Columbella livida*, see the Appendix.

1129. Engina pyrostoma (Sowerby, 1832) (Synonyms: **Ricinula crocostoma** and **forticostata** Reeve, 1846; **E. earlyi** Bartsch & Rehder, 1939). The black shell has an orange aperture; its outline is more angular than in most other species of *Engina*, with a distinct shoulder at the periphery. Length, 12 to 15 mm; diameter, 8 to 10 mm. Galápagos Islands.

1130. Engina tabogaensis Bartsch, 1931. A yellowish shell with black nodes at the intersections of the ribs and a yellow-orange aperture. It differs from *E. pyrostoma* in its lighter and warmer color and its more rounded whorls. Length, about 13 mm; diameter, 9 mm. Guaymas, Mexico, to Panama.

Genus MACRON H. & A. ADAMS, 1853

Axial sculpture wanting, spiral sculpture of more or less deeply incised grooves. Siphonal fasciole well developed; periostracum velvety and adherent. Endemic to West America, mostly confined to western Baja California.

1131. Macron aethiops (Reeve, 1847) (Synonyms: **Purpura trochlea** Gray, 1839, of authors [not *Buccinum trochlea* Bruguière, 1798]; **M. kellettii** A. Adams, 1854). The shell is white under a dark green periostracum. Between the strongly ribbed form *M. aethiops* and one ribbed only on the base, *M. kellettii*, there is a complete gradation. E. A. Smith (1903), who examined the type specimens, showed that both forms constitute a single species, but he mistakenly thought that the name *trochlea* should apply. Whether *M. orcutti*, from Magdalena Bay, Baja California, is only a smaller and smoother form remains to be seen. Length, about 60 mm. Common on the outer coast of Baja California, rare in the Gulf of California.

1132. Macron orcutti Dall, 1918. Shell flesh-colored under a brown periostracum, aperture white, flushed with pale purple; suture not channeled; shell surface finely spirally striate, outer lip lirate within, inner lip with a white callus. Length, 36 mm; diameter, 21 mm. Magdalena Bay, Baja California.

Genus **METULA** H. & A. ADAMS, 1853
(**ANTEMETULA** REHDER, 1943)

Long, oval shells with finely cancellate sculpture and a smooth aperture. Although there is a difference of opinion among authors about the type species of *Metula*, the strongest arguments seem to be on the side of conservatism—regarding the type as *Buccinum metula* Hinds, by hidden tautonymy (based on the mention by H. and A. Adams of "*Buccinum* sp. Hinds" when they proposed the name *Metula*). Otherwise, the first subsequent selection of a type species was of one that, by present standards, may not even belong in the same family.

1133. Metula amosi Vanatta, 1913. A pinkish flesh-colored shell with faint brown markings. Length, 38 mm; diameter, 13 mm. Guaymas, Mexico, to Panama, mostly offshore in depths to 18 m.

1134. Metula metula (Hinds, 1844) (Synonym: **Metula hindsii** H. & A. Adams, 1853). A smaller and more brightly colored shell than *M. amosi*, this is whitish with four spiral rows of brown spots. Length, about 17 mm; diameter, 4 mm. Reported only from Panama, offshore in depths of several meters.

Genus **NEOTERON** PILSBRY & LOWE, 1932

Small, spire acute, base and suture deeply channeled; outer lip with a flaring, flattened varix. This was named as a subgenus under *Hindsia* A. Adams, 1851, a genus not now considered appropriate for eastern Pacific species. *Neoteron* seems closest to *Trajana*, but definite placement must await live-taken material and study of the radula.

1135. Neoteron ariel (Pilsbry & Lowe, 1932). The general color of the shell is dull wine-brown, with a lighter spire. The flaring lip varix is fluted by fine spiral threads between which are finer axial lirae, much as in the muricid genus *Aspella*. The structure of the fossa or groove setting off the base and the manner of closure of the anterior canal is not muricid but is closer to that of the Nassariidae in the Buccinacea. Length, 10 mm; diameter, 5.6 mm. Corinto, Nicaragua. The type has been well figured by Vokes (1969).

Genus **NORTHIA** GRAY, 1847

Smooth, high-spired shells, with a siphonal fasciole and fossa. In spite of the superficial appearance of relationship to the Nassariidae by the presence of a

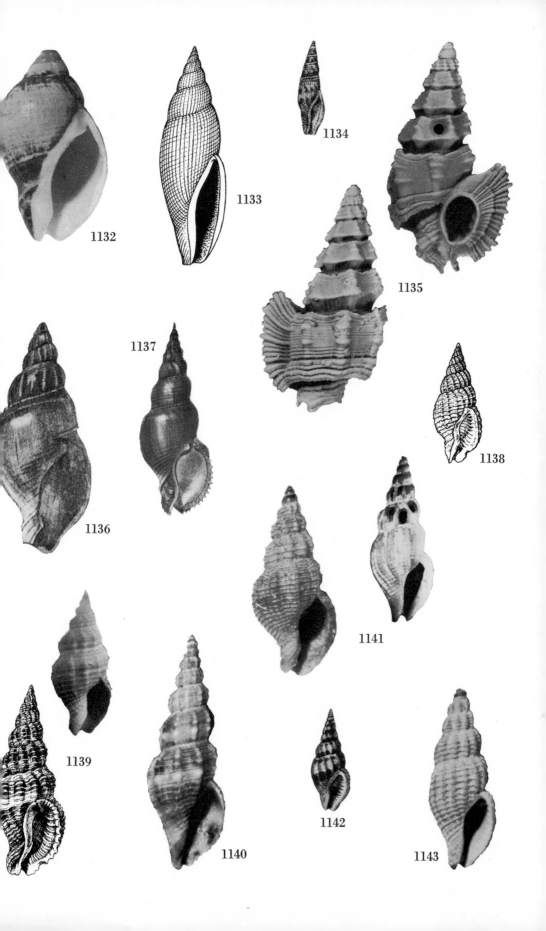

fossa or channel surrounding the anterior canal of the shell, the animal has a buccinid radula, as was shown by Thiele (1929) and confirmed by Dr. A. A. Olsson (*in litt.*).

1136. Northia northiae (Griffith & Pidgeon, 1834, *ex* Gray, MS). A shiny olive-brown shell with several spire whorls and a few denticles on the outer edge of the outer lip of the aperture. Length, 50 mm; diameter, 22 mm. Southern Mexico to Panama. Field evidence has suggested to collectors the presence of two distinct species, but museum workers have been slow to accept this. *N. northiae*, according to its original illustration, is the short form, with a relatively stubby spire and smoother aperture. Differences in range and habitat remain to be worked out; that of *N. northiae* includes at least Mazatlán to Oaxaca, Mexico.

1137. Northia pristis (Deshayes in Lamarck, 1844) (Synonym: **Buccinum serratum** Kiener, 1834, not Brocchi, 1814). The original figure by Kiener shows the elongate form that has been lumped with *N. northiae*. The spire has numerous turns, the aperture has strong prickles on the leading edge of the outer lip, and the body whorl shows a tendency toward tabulation below the suture. Length, 65 mm; diameter, 25 mm. Mazatlán, Mexico, to Ecuador.

Genus PHOS MONTFORT, 1810

Slender shells with a tapering spire, a rather short aperture, and strong axial sculpture predominant over weaker spiral ribs. A shallow fossa, somewhat as in the Nassariidae, may occur. Four Panamic subgenera; none of the eastern Pacific species can be assigned to *Phos*, *s. s.*, the type of which, *P. senticosus* (Linnaeus, 1758), is Indo-Pacific. Assistance by Dr. James H. McLean of the Los Angeles County Museum in the revision of this group is gratefully acknowledged.

Subgenus ANTILLOPHOS WOODRING, 1928

Anterior canal wide, with only a slight fasciolar fold; nuclear whorls keeled.

1138. Phos (Antillophos) veraguensis Hinds, 1843 (Synonyms: **P. biplicatus** Carpenter, 1856; **P. alternatus** Dall, 1917). The shell is nearly a uniform brown, without banding. Distinctive characteristics are the rounded whorls and evenly cancellate sculpture. Length, 25 mm; diameter, 12 mm. Bahía San Luis Gonzaga, Gulf of California, south to Port Utria, Colombia, offshore in depths of 37 to 402 m.

Subgenus CYMATOPHOS PILSBRY & OLSSON, 1941

Shell sturdy, with a small, smooth nucleus, aperture weakly lirate within, siphonal fasciole limited by a simple thread, not a furrow.

1139. Phos (Cymatophos) crassus Hinds, 1843 (Synonym: ? **P. clarki** M. Smith, 1944). The shell is compact, heavy, ivory buff in color, some specimens with brownish color bands. The anterior canal is slightly recurved, and there is a denticle on the inner lip adjacent to the posterior end in many mature specimens. Length, 45 to 50 mm; diameter, 20 to 21 mm. The Gulf of Tehuantepec, Mexico, to Panama, in depths of about 40 m. In the northern part of the range, specimens tend to be more slender, thinner, with stronger color bands, as in the original figure by Hinds. Southern specimens tend to be relatively shorter and wider. Possibly further collecting will justify separation of two subspecies, in which case the name *P. clarki* will be available for the southern form.

1140. Phos (Cymatophos) dejanira (Dall, 1919) (Synonym: **P. mexicanus** Dall, 1917 [not Böse, 1906]). Resembling the more widely distributed *P. gaudens*, this is larger, with fewer axial ribs and more broadly spaced nodes on the axial ribs, a node-free band at the periphery, the surface with a somewhat chalky appearance. There are three smooth nuclear whorls rather than the usual four. Length of a large specimen, 28 mm. Gulf of California, from the northern end southward to La Paz.

1141. Phos (Cymatophos) fusoides (C. B. Adams, 1852) (Synonym: **Fusus porticus** Dall, 1915). Narrow, raised spiral cords are characteristic of this species, especially strong on the base. There are three smooth nuclear whorls. Postnuclear sculpture shows only three spiral cords, with low, broad cancellations. Maximum length, 24 mm. Panama.

1142. Phos (Cymatophos) gaudens Hinds, 1844. A pale, shiny shell with widely spaced ribs, a dark brown band near the suture, white nodes on the ribs, and a darker stain on the area of the siphonal fasciole. The outer lip is nearly smooth within. There are four smooth nuclear whorls. Length, 25 mm; diameter, 8.5 mm. Southern end of the Gulf of California, Escondido Bay, southward to La Libertad, Ecuador, offshore in depths of 18 to 110 m.

1143. Phos (Cymatophos) minusculus Dall, 1917. The narrow raised spiral bands are dark-colored, which is the most characteristic feature of this small, cancellately sculptured form. Length, 12 mm; diameter, 5 mm. The southern part of the Gulf of California to Port Utria, Colombia, offshore in depths of 27 to 86 m.

Subgenus METAPHOS OLSSON, 1964

Nucleus larger than in *Cymatophos*, having spiral striations and irregular axial riblets that are thickened, some slightly, others strongly; remainder of shell with strong axial ribs that tend to form an angulate periphery.

1144. Phos (Metaphos) articulatus Hinds, 1844 (Synonyms: **P. turritus** A. Adams, 1853; **P. cocosensis** Dall, 1896). Most variable of the West American species of *Phos* in outline, specimens from a single lot varying from slender and weakly shouldered to robust and strongly shouldered, all, however, having here and there oddly thickened ribs, resembling varices. The shell is white, clouded with brown, with brown and white spots on the axial ribs. Length, 46 mm; diameter, 19 mm. Tortuga Island, Gulf of California, to Lobos Island, Peru, offshore in depths of 37 to 232 m.

1145. Phos (Metaphos) laevigatus (A. Adams, 1851) (Synonym: **P. chelonia** Dall, 1917). Long unrecognized because it was given an erroneous African locality, this species has been shown by Emerson (1967) to be Panamic instead. The shell is ivory to buff, tinged with purplish brown on the body whorl, the anterior canal, and the lowermost spiral riblets, with a weak mottling of brown along the suture; aperture white within, purplish tinged; operculum crescentic in shape, its nucleus apical. Length, 26 to 42 mm. Galápagos Islands.

Subgenus STROMBINOPHOS PILSBRY & OLSSON, 1941

Columella with fine lirae; nucleus of shell smooth; radula with rachidian teeth multicuspate.

1146. Phos (Strombinophos) cumingii Reeve, 1846. The light brown shell has a few interrupted spiral lines of darker color; axial ribs are low and wider-

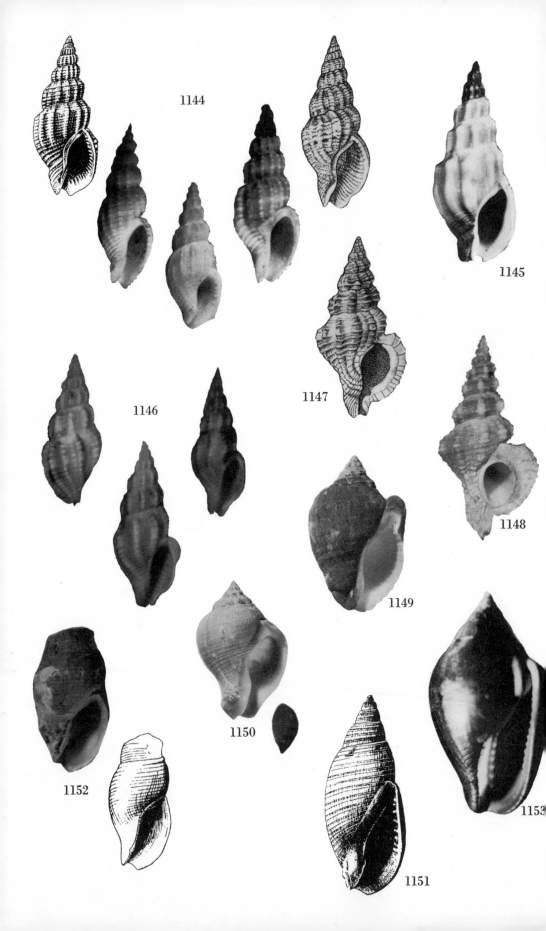

1144

1145

1146

1147

1148

1149

1150

1151

1152

1153

spaced than in other species. There is a broad and shallow stromboid notch on the outer lip that gives a characteristic shape to the aperture, the lip varix being especially thick at the upper part. The nucleus consists of four smooth whorls. Length, 32 mm; diameter, 12 mm. Bahía Isla Grande, Mexico, to La Libertad, Ecuador, in 18 to 27 m. The species has generally been overlooked because it was thought to be a synonym of *P. articulatus*, from which it differs in the columellar lirae as well as in sculpture. An erroneous date of 1859 or 1860 has been cited for it, but Iredale in 1922 showed that the name was first proposed by Reeve in 1846.

Genus TRAJANA GARDNER, 1948
(HINDSIA of authors, not of A. ADAMS, 1851)

Spire elevated, with a small smooth nucleus; axial ribs varicose but not spinose; aperture elliptical, smooth within, anterior canal closed, short, bounded above by a shallow fossa; outer lip with a varix; suture incised to channeled. The genus was named in the Muricidae, but the radula proves to be buccinoid, with a multi-cuspate central and strongly hooked laterals. Vokes (1969), who suggested an affinity with the Nassariidae, has reviewed the distribution of the genus, a "paci-phile" group widespread in the Atlantic Tertiary, surviving only in the eastern Pacific.

1147. Trajana acapulcana (Pilsbry & Lowe, 1932). The color is a chestnut-brown with a darker band below the suture and light buff between the ribs and on the base. The holotype measures: length, 17.5 mm; diameter, 9.7 mm. Aca-pulco, Mexico (type locality) to Gulf of Tehuantepec, 37 to 73 m. The type speci-men was immature and did not have the anterior canal closed; adult material later collected shows that the canal normally is closed except for a narrow chink at the anterior end.

1148. Trajana perideris (Dall, 1910). Lighter in color than *T. acapulcana*, with fewer ribs, the shell is yellowish white with a basal and in some specimens a sutural band of brown. There are six rounded axial ribs and six equally spaced spiral threads. Length, 29 mm; diameter, 12 mm. Gulf of California, from La Paz, Baja California, to Guaymas, Sonora, Mexico, in 37 to 55 m.

Genus TRIUMPHIS GRAY, 1857

Subglobose shells resembling *Cantharus* in general outline but body whorl lacking axial ribs; spire with low cancellate sculpture, which is reduced with growth, with spiral lirae tending to be more persistent. Aperture with a well-developed posterior siphonal notch. Two Panamic subgenera.

Subgenus TRIUMPHIS, s. s.

Siphonal fasciole weak to wanting; anterior canal hardly evident; posterior siphonal notch deep, often forming an earlike expansion; a strong hump forming behind the aperture along the uneven suture, which is slightly ascending near the apertural lip.

1149. Triumphis (Triumphis) distorta (Wood, 1828) (Synonym: **Colum-bella triumphalis** Duclos, 1843). The body whorl is nearly smooth except for the earlike extension of the aperture and a few low spiral threads near the base. Under a tightly adherent, smooth, greenish-brown periostracum, the shell is white, mottled and striped with brown. About 40 mm in length. La Union, El Salvador, to Ecuador, intertidally, on rocks in mud flats.

Subgenus NICEMA WOODRING, 1964

Siphonal fasciole evident; anterior canal short but distinct; posterior siphonal notch not forming an expansion; spiral sculpture only on basal part of body whorl.

1150. Triumphis (Nicema) subrostrata (Wood, 1828) (Synonym: **? Fusus lapillus** Broderip & Sowerby, 1829). This shell bears a remarkable resemblance to *Cancellaria solida* but of course lacks the columellar folds. It is a muted flesh-orange in color, with a low spire on which can be seen a few axial and spiral riblets. Length, 38 mm; diameter, 25 mm. It has been reported as fairly common on the mud flats at San Blas, Mexico, but is otherwise rare, ranging south to Colombia.

Genus TRUNCARIA ADAMS & REEVE, 1848

The shell is ovate in outline, with a wide aperture, flaring below, and with the columella obliquely cut off anteriorly.

1151. Truncaria brunneocincta (Dall, 1896). The compact shell is bright pinkish, with narrow and distant brown spiral lines and a few brown spots near the suture, sculptured with fine spiral riblets and lines. Length, 31.5 mm; diameter, 13 mm. Dredged in Panama Bay at a depth of 102 m.

Genus VOLUTOPSIUS MÖRCH in RINK, 1857

Nuclear whorls large, spire whorls inflated; aperture longer than spire, with a broad and shallow anterior notch. A cold-water group with one deepwater record in the tropical Pacific.

1152. Volutopsius amabilis Dall, 1908. Panama Bay, 2,320 m.

Family COLUMBELLIDAE

The dove shells are mostly small, though rarely microscopic in size, with boldly marked, smooth or ribbed shells. The animals are not as timid as many mollusks are and may be seen plowing about on the floor of a tide pool at low tide—especially in the tropics—searching for the minute organisms on which they feed. Some have developed the picturesque habit of spinning a thread of mucus, which is attached to a bit of seaweed or even to the surface film of a quiet tide pool, and the little owner of the shell then dangles and swings on its elastic support. As the animals live in abundance in the intertidal area, the shells may be cast ashore by the tides and thus form a considerable part of the fine beach drift on the shores of the world, for these are adaptable mollusks, and not a few species or even genera have moved out of the tropics and into the temperate or boreal seas.

The arrangement of material for this family has been done in consultation with Dr. George E. Radwin, of the San Diego Natural History Museum, whose doctoral thesis has been a study of the family, especially as it occurs in the Atlantic. His studies of the radula and other diagnostic features indicated the need for a number of reassignments of American species. The key is to genera.

1. Aperture with both a posterior and an anterior canal.................. 2
   Aperture with anterior canal only.................................. 4
2. Shell with reticulate sculpture.............................*Zetekia*
   Shell smooth or with axial sculpture only........................... 3

3. Shell smooth; posterior canal slotlike, at suture . . . . *Bifurcium*
   Shell with axial lirae; posterior canal spoutlike, at periphery of whorl
   . . . . *Microcithara*
4. Outer lip of mature shell smooth within 5
   Outer lip of mature shell denticulate 9
5. Shell cone-shaped, aperture long . . . . *Parametaria*
   Shell not cone-shaped, aperture moderate to short 6
6. Pillar with an oblique fold . . . . *Mazatlania*
   Pillar with normal spiral striations only 7
7. Shell subcylindrical in outline, aperture short . . . . *Aesopus*
   Shell biconic, aperture about half the length of the shell 8
8. Sculpture of spiral ribs only . . . . *Cosmioconcha*
   Sculpture of both axial and spiral ribs . . . . *Decipifus*
9. Spire slender and tapering, outer lip thickened 10
   Spire not conspicuously tapering, outer lip not markedly thickened 18
10. Shell small (less than 15 mm long) . . . . *Ruthia*
    Shell medium-sized (more than 15 mm long) . . . . *Strombina*
11. With axial ribs predominating on the spire 12
    Without axial sculpture on the spire 13
12. Spiral ribs wider than the interspaces . . . . *Anachis*
    Spiral ribs narrower than spaces between them . . . . *Nassarina*
13. Spire short; shell thick and heavy, moderately large . . . . *Columbella*
    Spire as long as aperture; shell thin, rather small . . . . *Mitrella*

Genus **COLUMBELLA** LAMARCK, 1799
(**PYRENE** RÖDING of authors, not of RÖDING, 1798)

The Lamarckian name *Columbella* was long used for all of the dove shells; then, when it was realized that Röding's *Pyrene* had priority, *Columbella* went into an eclipse (except for use as a base of the family name) and was more or less restricted to the Atlantic *C. mercatoria*. However, recent studies by Dr. George Radwin (in press) have shown that on the basis of the radula and sculpture of the early whorls, the western Pacific species of typical *Pyrene* are markedly different from the eastern Pacific species that have been allocated to it. The latter are, in spite of sculptural differences, more closely related to the type species of *Columbella*. Therefore, on his advice, the generic name *Columbella* is revived for the supposed Pyrenes of the eastern Pacific.

The shells are medium-sized to relatively large, smooth or with spiral sculpture, many with a dense periostracum that obscures the bold striped or spotted markings. Apertures may be contracted or strongly dentate within, and the anterior canal is shallow. Some members of the genus, notably the type species, *Columbella mercatoria* (Linnaeus, 1758), are herbivorous, grazing on algae—in contrast to the carnivorous habits of most of their relatives.

1153. Columbella aureomexicana (Howard, 1963) (Synonym: **Pyrene aureola** Howard, 1963 [not Duclos in Chenu, 1846]). Resembling *C. fuscata* but with the apertural margin orange-brown; inner lip with a rectangular area of callus posteriorly and with numerous fine, even denticulations anteriorly. Length, 21 mm; diameter, 12 mm. Cedros Island, Baja California, through the Gulf of California to Topolobampo, Sinaloa, Mexico. Shasky (1970) is of the opinion that *C. luteola* Kiener, 1841, may prove to be a prior name for this. Kiener's type has not been restudied. It was from an unknown locality, and the only comparisons were with a Mediterranean species.

1154. Columbella castanea Sowerby, 1832. Chestnut-brown, the shell is spotted with white, its aperture tinged with orange. Length, 21 mm; diameter, 12 mm. Galápagos Islands.

1155. Columbella fuscata Sowerby, 1832 (Synonyms: **C. gibbosa** Valenciennes, 1832; **C. meleagris** and **nodalina** Kiener, 1840; **? C. luteola** Kiener, 1841; **C. pallescens** Wimmer, 1880). The chestnut-brown shell is dotted and irregularly spotted with white. Just below the suture is a band of triangular white markings. The shell is covered with a thin, smooth, light olive periostracum, and the aperture is lavender-colored, fading to white; the callus of the posterior part of the inner lip is weakly developed, and the denticles of the anterior part of the inner lip are few or of uneven size. Length, 20 mm; diameter, 12 mm. Magdalena Bay, Baja California, through the southern end of the Gulf of California and south to Peru.

1156. Columbella haemastoma Sowerby, 1832. Dark brown, shading from chestnut to chocolate, this has white blotches zigzagging around the shoulder and at the base, and the aperture is orange-colored. Length, about 25 mm. Magdalena Bay and the southern part of the Gulf of California to Ecuador and the Galápagos Islands, under rocks between tides.

1157. Columbella labiosa Sowerby, 1822 (Synonym: **C. venilia** Duclos, 1846). Under a thin, smooth, olive periostracum the shell is ash-gray, with many narrow brown spiral lines. The margin of the aperture is white. About 22 mm in length. Nicaragua to Ecuador, on sides and tops of rocks, exposed to surf, according to Lowe (1932).

1158. Columbella major Sowerby, 1832 (Synonym: **C. gibbosa** Duclos, 1840 [not Valenciennes, 1830]). Many authors regard this as a synonym of *C. strombiformis*. The only distinction seems to be in the periostracum, which has spiral sculpture on the shoulder only, not all over; the white markings of the shell tend to be dotted rather than made up of zigzag lines, but this is a variable character. Length, 27 mm; diameter, 16 mm. The southern end of the Gulf of California to Peru, under rocks between tides.

1159. Columbella paytensis Lesson, 1830. Stout, with a short spire and narrow aperture; brown in color, flecked with white, aperture white to purple within. Length, about 25 mm. Ecuador and the Galápagos Islands to northern Peru.

1160. Columbella socorroensis Shasky, 1970. The ovate shell has about 18 axial ribs on the spire whorls, the later whorls smooth except for some spiral grooves anteriorly; the color is chocolate-brown and bluish white, with markings that vary from irregular cloudings to rectangular or square dots in a checkerboard pattern. The outer lip is thickened at the edge and has 12 to 14 lirae within; an inner lip callus is evident. Length, 18.7 mm; width, 9.2 mm. Socorro Island, off Colima, Mexico, depth 2 m.

1161. Columbella sonsonatensis (Mörch, 1860) (Synonyms: **C. festiva** Kiener, 1841 [not Laborde, 1830]; **C. lucasana** Dall, 1916). Mörch's name for this species was a "forgotten name" for almost the requisite fifty years but was revived just short of that time. The shells are small, least in size among the West American members of the genus, smooth, white around the suture, spotted and

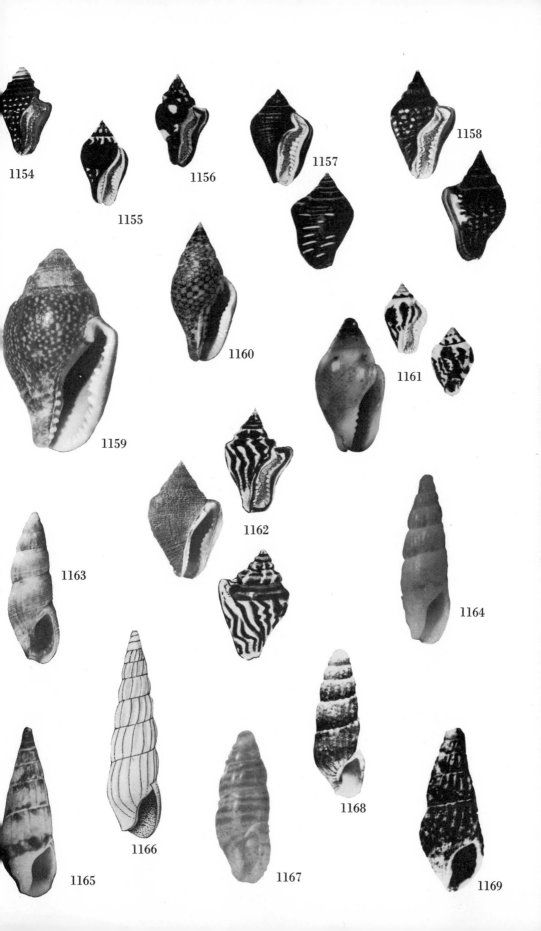

1154

1155

1156

1157

1158

1159

1160

1161

1162

1163

1164

1165

1166

1167

1168

1169

streaked with brown and white, the aperture white. Length, about 9 mm. Although Dall's replacement name implies occurrence at Cape San Lucas, modern records suggest that it does not occur north of southern Mexico and is commoner southward as far as Panama. Mörch's type locality is El Salvador.

1162. Columbella strombiformis Lamarck, 1822 (Synonyms: **? Columbella buccinoides** Lesson, 1842 [not Sowerby, 1833]; **C. bridgesii** Reeve, 1858). Largest of the West American Columbellas, this has a swollen body whorl, with a ridge at the upper part that forms a sort of spout at the aperture. Under a shaggy olive periostracum the shell is dark reddish brown, spotted and marked with zigzag lines and triangular blotches of white. The aperture is white or tinted with orange. About 30 mm in length. The northern end of the Gulf of California south to Peru, under rocks between tides.

Genus AESOPUS GOULD, 1860

The short aperture and small size are distinguishing features. Two Panamic subgenera.

Subgenus AESOPUS, *s. s.*

Sculpture weak to wanting, mainly spiral.

1163. Aesopus (Aesopus) sanctus Dall, 1919. The whitish shell measures about 5 mm in length and 1.7 mm in diameter. Its range is from southern California to the Gulf of California.

Subgenus ITHIAESOPUS OLSSON & HARBISON, 1953

Sculpture of straight, rounded axial riblets, interspaces smooth or with weak spiral threads.

1164. Aesopus (Ithiaesopus) arestus Dall, 1919. Resembling the next species but more slender, with a longer, narrower aperture. Length, 7 mm; diameter, 2 mm. Magdalena Bay, Baja California.

1165. Aesopus (Ithiaesopus) eurytoides (Carpenter, 1864). The shell is white to mottled brown, about 7 mm long and 2.8 mm in diameter. Southern California to Panama.

1166. Aesopus (Ithiaesopus) fredbakeri Pilsbry & Lowe, 1932. Pale buff in color, clouded with dull brown, this shell has several smooth axial ribs. It is about 7.5 mm long and 2 mm in diameter. Chiapas, Mexico, to Nicaragua (type locality).

1167. Aesopus (Ithiaesopus) fuscostrigatus (Carpenter, 1864). The whitish shell has three or four interrupted spiral bands of brown color spots. Length, 3.5 mm; diameter, 1.2 mm. San Hipolito Point, Baja California, to Cape San Lucas.

1168. Aesopus (Ithiaesopus) osborni Hertlein & Strong, 1951. Smallest of the west coast species, the shell is only 3 mm in length. Off Port Guatulco, Mexico.

1169. Aesopus (Ithiaesopus) subturritus (Carpenter, 1864) (Synonym: **Anachis petravis** Dall, 1908). Color markings are somewhat as in *A. fuscostrigatus* but axial ribs are stronger. Length, 4 mm; diameter, 1.6 mm. Southern California to Tres Marias Islands, Mexico.

Genus ANACHIS H. & A. ADAMS, 1853

Distinguished by strong axial ribs on the spire that tend to fade toward the base of the body whorl; spiral sculpture various. The outer lip has teeth within in most forms, the inner lip smooth or nearly so. The notch at the anterior end of the aperture is deep, but the anterior canal is short to absent. A heavy periostracum that obscures the color pattern is developed in a few species. Coloring runs to shades of brown, banded with yellow or white. The genus comprises more species than any other columbellid group in the Panamic province. The key is to the five recognized subgenera.

1. Suture well indented to tabulate.............................*Anachis, s. s.*
   Suture only slightly impressed.................................... 2
2. Axial sculpture predominant over entire shell......................... 3
   Axial sculpture not predominant over entire shell..................... 4
3. Relatively large, adults mostly more than 6 mm in length.......*Costoanachis*
   Relatively small, adults less than 6 mm in length..............*Parvanachis*
4. Axial ribs intersected by spiral threads of nearly equal strength, surface
      somewhat cancellate ....*Glyptanachis*
   Axial ribs on spire only, shell mainly with fine spiral threads.........*Zafrona*

Subgenus ANACHIS, *s. s.*

Relatively large, suture well indented to tabulate; spire blunt; ribs heavy; anterior canal indistinct; columellar lip with wrinkles or denticles.

1170. Anachis (Anachis) lyrata (Sowerby, 1832). An elegantly marked and colored shell—yellowish with two bands of brown spots on the ribs and a white aperture. Length, about 20 mm. Costa Rica to Panama, under rocks, intertidally. A similar species in the Atlantic is *A. terpsichore* (Sowerby, 1822).

1171. Anachis (Anachis) scalarina (Sowerby, 1832) (Synonym: **A. pachyderma** Carpenter, 1857). This is the type species of the genus *Anachis*. The shell is brown with white bands at the shoulder and middle of the whorl. About 20 mm in length. Mazatlán, Mexico, to Panama. It is possible that *A. pachyderma* may prove to be a northern subspecies of the more southern *A. scalarina*, distinguished by smaller size (length about 17 mm) and heavier periostracum.

Subgenus COSTOANACHIS SACCO, 1890

Spire more tapering than in *Anachis, s. s.*, suture less indented, whorls tending to be flat, not tabulate; a short anterior canal in some species; columellar edge smooth or finely denticulate.

1172. Anachis (Costoanachis) adelinae (Tryon, 1883). The white shell has a checkerboard pattern of brown spots on the base that is characteristic; there is also a broken brown band at the periphery. Length, 15 mm. Magdalena Bay, Baja California, through the Gulf to the Sonoran coast of Mexico.

1173. Anachis (? Costoanachis) berryi Shasky, 1970. Slender, with about 22 axial ribs per whorl slightly noded at the suture, with fine spiral threads between, stronger anteriorly; color light yellowish tan with scattered brown triangular blotches; outer lip with three denticles and one or two lirae, columella with two to three faint lirae; anterior canal short and recurved. Length, 9 mm; width, 3.3 mm. Pulmo Reef, Baja California, in 1 to 3 m depth. Somewhat resembling *A. gracilis* but more slender and less striate.

1174. Anachis (Costoanachis) boivini (Kiener, 1841). The dark gray shell is speckled with white, the shoulder with a row of nodes and the spire with axial ribs; around the aperture is an edging of brown. Length, about 20 mm; diameter, 13 mm. Nicaragua to Colombia, intertidally.

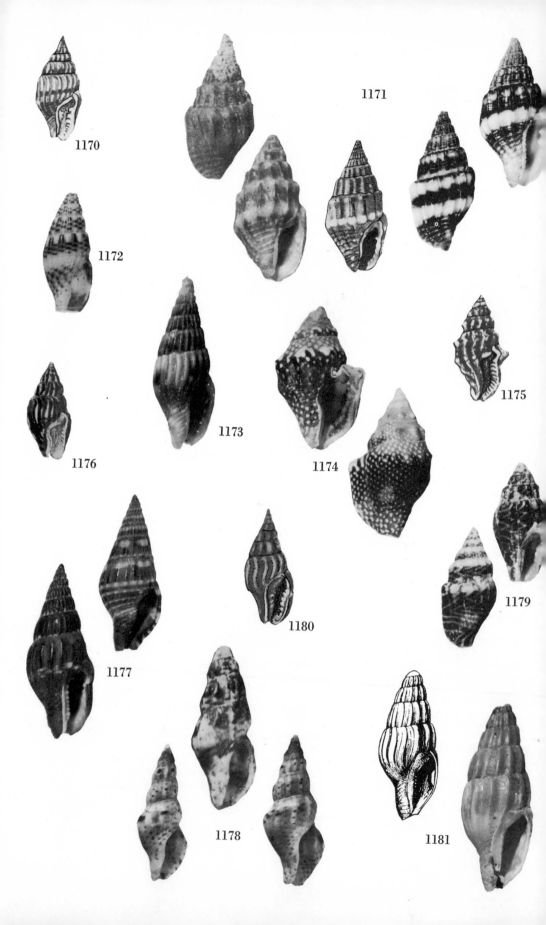

1170

1171

1172

1173

1174

1175

1176

1177

1178

1179

1180

1181

1175. Anachis (Costoanachis) coronata (Sowerby, 1832). The shell is yellowish white, mottled or striped with brown. Length, 14 mm; diameter, 6 mm. The outer coast of Baja California through the Gulf and south to Panama.

1176. Anachis (Costoanachis) costellata (Broderip & Sowerby, 1829). The shell is bright brown, with darker bands and spots, about 18 mm in length, with ribs that are close-set, narrow, and slightly curved. Mazatlán, Mexico, to Paita, Peru.

1177. Anachis (Costoanachis) decimdentata Pilsbry & Lowe, 1932. A brownish-black shell with a spiral band of white spots. The length is 13.5 mm; diameter, 5.5 mm, for the type specimen, which came from Nicaragua. Depth, 27 m.

1178. Anachis (? Costoanachis) fayae Keen, new species (Synonym: **A. bartschii** of authors, not of Dall, 1918). The ivory white shell has widely spaced and somewhat sinuous axial ribs crossed by very fine spiral lirae; there are about nine axial ribs on the body whorl. The nucleus consists of two smooth whorls; the next two are light brown with fine axial threads. The spire and body whorl show two interrupted color bands of flecks and spots, one with larger spots just below the suture and at the periphery, the other—of smaller spots in a checkerboard pattern—on the base. Anterior part of body whorl showing only spiral sculpture. Outer lip thin, white, without teeth. Length (holotype), 7 mm; diameter, 2.9 mm. Guaymas, Sonora, Mexico (type locality), to Mazatlán. About 200 specimens have been found in the Guaymas area by Mrs. Faye Howard and others. This seems to be the species figured by Baker, Hanna, and Strong (1938) (copied by Keen, 1958) as *A. bartschii*, although their specimen, from San Marcos Island, Gulf of California, has darker color bands and a slightly broader spire. Distinguished from other species by the fine spiral sculpture and the relatively few axial ribs.

1179. Anachis (Costoanachis) fluctuata (Sowerby, 1832) (Synonyms: **Columbella suturalis** Griffith & Pidgeon, 1834; **C. costata** Duclos, 1840 [not Valenciennes, 1832]; **C. fluctuosa** Duclos in Chenu, 1846). Under a yellowish periostracum the shell is whitish, with zigzag brown markings and curved axial ribs. About 18 mm in length. Nicaragua to Ecuador.

1180. Anachis (Costoanachis) fulva (Sowerby, 1832). The long narrow axial ribs on this reddish-brown shell fade out on the lower part and are replaced by spiral ribs. The shell averages about 23 mm in length and ranges from Nicaragua to Ecuador.

1181. Anachis (Costoanachis) fusidens (Dall, 1908). A deepwater form from 550 m, Galápagos Islands.

1182. Anachis (? Costoanachis) gilva (Menke, 1847). Described as being brownish black, with a white line up the spire. Length, about 16 mm. The species was unfigured, and the type has not since been studied. Carpenter thought the form might be a synonym of *A. coronata*. Pilsbry and Lowe in 1932 figured a specimen as *A. gilva* that is instead *A. scalarina*. The species remains of doubtful status. Type locality, Mazatlán, Mexico.

1183. Anachis (? Costoanachis) gracilis (C. B. Adams, 1852). Pale brown in color, the shell has a row of darker spots just below the suture and at the periphery a white band broken by arrow-shaped brown spots. Length, 7.5 mm; diameter, 3 mm. Ballena Bay, Costa Rica, to Panama. The axial ribs are not as well developed as in other species of *Anachis*.

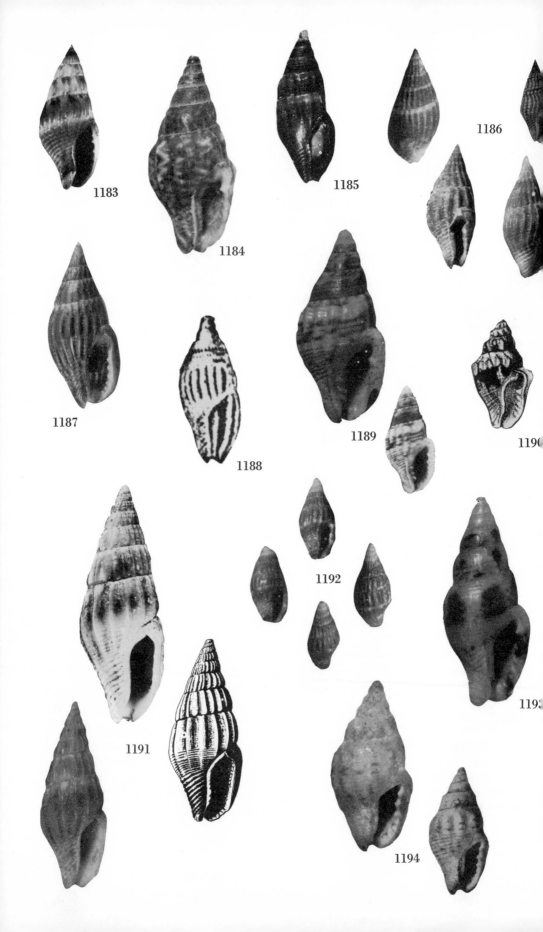

1183

1184

1185

1186

1187

1188

1189

1190

1191

1192

1193

1194

1184. Anachis (Costoanachis) hannana Hertlein & Strong, 1951. Described as a variant of *A. coronata*, this is brighter-colored and has less conspicuous nodes at the periphery. Length, 14 mm; width, 6.3 mm. The distribution seems to be restricted to the west coast of Baja California, from Scammon's Lagoon to Cape San Lucas.

1185. Anachis (Costoanachis) moesta (C. B. Adams, 1852). A smooth area around the middle of the last whorl of this dark brown shell is a distinctive feature. Length, about 8 mm; diameter, about 2.8 mm. El Salvador to Panama.

1186. Anachis (Costoanachis) nigricans (Sowerby, 1844). The shell is dark brown to blackish, with a row of granules just below the suture that may be dotted with white, and there is also a light band around the middle of the shell. About 8 mm in length. Gulf of California to Panama and the Galápagos Islands, under rocks between tides.

1187. Anachis (Costoanachis) nigrofusca Carpenter, 1857. A dark brown shell with wavy light-colored lines. Length, about 10 mm; diameter, 4.3 mm. The species seems to be confined to the west Mexican coast between Mazatlán and Manzanillo.

1188. Anachis (Costoanachis) oerstedi Von Martens, 1897. Described as having a brownish-black shell with a dark aperture, the upper part of the whorls with narrow, straight, rather widely spaced axial ribs that fade out sharply below the periphery. Length, 5 mm. Bocorones, Costa Rica. Not again recorded by collectors; possibly a synonym of *A. spadicea*.

1189. Anachis (Costoanachis) ritteri Hertlein & Strong, 1951. Resembling *A. (Zafrona) incerta*, this is larger, with more irregularity in the finely penciled spiral lines, which may vary from almost entire absence to a maximum of six strong bands of reddish-brown lines on the white shell. Length, 7.4 mm. Along the west Mexican coast in the Gulf of Tehuantepec area, in depths to 13 m.

1190. Anachis (Costoanachis) rugosa (Sowerby, 1832) (Synonym: **Columbella bicolor** Kiener, 1841). The upper part of the shell is whitish, the lower part gray to drab brown, clouded irregularly with dark brown. A spiral row of blunt tubercles encircles the upper part of the whorl. About 20 mm in length. Lowe (1932) reported that "this common and rather variable species is found on top of rocks, on exposed ledges, or on mud flats, clinging to small stones or dead shells." Nicaragua to Ecuador.

1191. Anachis (? Costoanachis) sanfelipensis Lowe, 1935. A flesh-colored shell with blotches of light brown between the axial ribs and wavy axial lines of brown on the base. Length, 17 mm; diameter, 6.5 mm. The species is restricted in range to the northern end of the Gulf of California, in the lower tidal zone.

1192. Anachis (? Costoanachis) spadicea (Philippi, 1846). The type material in the British Museum, well figured by Reeve, shows a rather slender brownish-gray shell resembling *A. nigricans* in shape but lacking the nodes at the posterior ends of the ribs; also, it has a lighter band around the middle of the whorls, the pillar is lighter, and the apical whorls are nearly white; it resembles *A. nigrofusca* in size but is much more slender. Length, 9 mm; diameter, 3.5 mm. Mazatlán to Acapulco, Mexico. The figure by Pilsbry and Lowe (1932), copied by Keen, 1958, is apparently—like their *A. gilva*—of a variant of *A. scalarina*.

1193. Anachis (Costoanachis) teevani Hertlein & Strong, 1951. The yellow-ish-white shell has patches of reddish-brown color of varying size and faint axial and spiral sculpture. Length, 8 mm; diameter, 3.5 mm. Dredged off Acapulco, Mexico, in 50 m depth.

1194. Anachis (Costoanachis) treva Baker, Hanna & Strong, 1938. The flesh-colored shell has a spiral band of brown spots at the periphery, a row of smaller dots on the base, and a narrow band of fine brown axial lines next to the suture. Length, 9.5 mm; diameter, 4.5 mm. Dredged off Tres Marias Islands, Mexico.

1195. Anachis (Costoanachis) varia (Sowerby, 1832) (Synonyms: **Columbella veleda** Duclos, 1846; **C. encaustica** Reeve, 1858). There is a white band at the suture and on the periphery of this brown shell, which resembles *A. scalarina* in size and coloring but has a smooth, not shouldered, suture, and a rounded, not angulate, periphery. In some specimens the spire especially may have almost cancellate sculpture because of the well-developed spiral riblets. About 18 mm in length and 9 mm in diameter. Sonora, Mexico, to Panama.

1196. Anachis (Costoanachis) varicosa (Gaskoin, 1852) (Synonyms: **Columbella macrostoma** Reeve; *ex* Anton, MS; **C. valida** Reeve, 1859). Long confused with *A. costellata* but now considered distinct, the ribs on the last whorl being straighter, more widely spaced, and a bit more nodose at their upper ends. The shell is white, under a buff periostracum, marked with blackish brown in two bands. Some specimens resemble *A. rugosa* in coloring but have much finer ribs. Length, 15 mm; diameter, 8 mm. Panama to Peru.

1197. Anachis (Costoanachis) vexillum (Reeve, 1858). A yellowish-brown shell with wavy axial stripes of dark brown, for which the Latin name *vexillum*—flag—is appropriate. About 15 mm long and 6 mm in diameter. The species seems to be confined to the Gulf of California area, from the head of the Gulf as far south as Acapulco, Mexico, under rocks intertidally.

Subgenus **GLYPTANACHIS** PILSBRY & LOWE, 1932

With strong spiral sculpture almost equal in strength to axial ribs, forming a beaded surface pattern; inner lip smooth to wrinkled; anterior canal short.

1198. Anachis (Glyptanachis) atramentaria (Sowerby, 1844). Blue-black, the shell is sparsely dotted with white and is densely grooved spirally; the aperture is deep violet within. Length, 10 mm. Galápagos Islands.

1199. Anachis (Glyptanachis) hilli Pilsbry & Lowe, 1932. The shell has a marbled pattern of black, yellow, and white. There are about 20 axial ribs crossed by strong spiral sculpture. About 8.5 mm in length and 4 mm in diameter. Northern end of the Gulf of California to Nicaragua, intertidally. The type species of the subgenus *Glyptanachis*.

1200. Anachis (Glyptanachis) lentiginosa (Hinds, 1844) (Synonyms: **Columbella tessellata** C. B. Adams, 1852; **C. guatemalensis** and **pertusa** Reeve, 1859; **A. tabogensis** Bartsch, 1931). Although Adams felt that the form he named was broader, comparatively, than Hinds's, a photograph of the type lot of the latter, in the British Museum, confirms the identity that later authors suspected. Shells are greenish or brownish black, spotted with white on the quadrate granules of the surface. Length, 7 to 12 mm; diameter, 3.3 to 5 mm. Guatemala to Panama, under rocks at low tide.

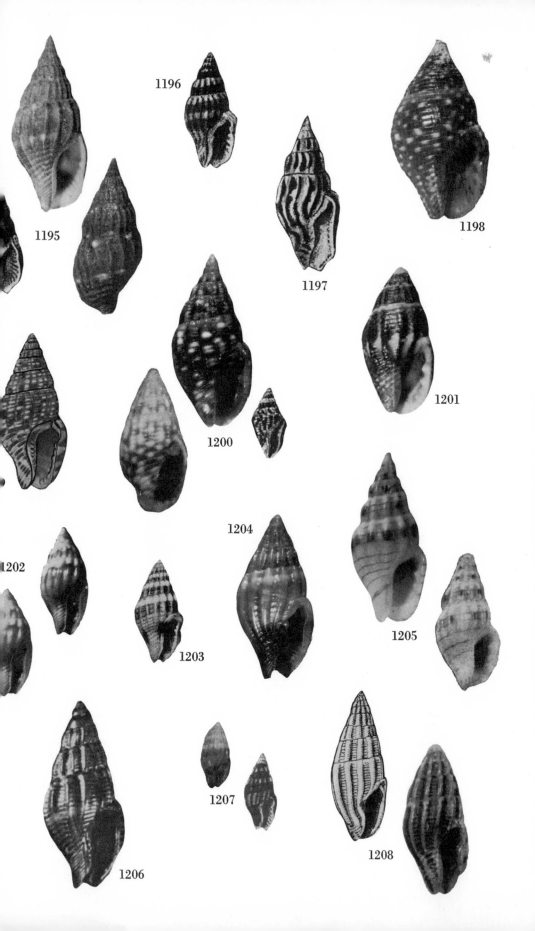

1195

1196

1197

1198

1200

1201

1202

1203

1204

1205

1206

1207

1208

1201. Anachis (Glyptanachis) rugulosa (Sowerby, 1844). The yellowish shell is closely dotted with dark brown except for a light band in the middle of the whorl; the aperture is chestnut brown. Length, 13 mm. Panama to Ecuador and the Galápagos Islands.

Subgenus PARVANACHIS RADWIN, 1968

Small, stout, with body whorl inflated, anterior canal short or wanting, outer lip thickened. Radula of type species—from the Caribbean—showing a peculiar hook on the outer cusps of the lateral tooth.

1202. Anachis (Parvanachis) albonodosa (Carpenter, 1857). The shell is light brown with a greenish tint, marked with white spots at the ends of axial ribs, just below the suture. The outer lip of the type specimen at the British Museum is contracted and thin, perhaps because it is not fully mature. Length, 3 mm; diameter, 1.1 mm. Mazatlán, Mexico.

1203. Anachis (Parvanachis) dalli Bartsch, 1931. The shell is pale yellow with two zones of interrupted brown spots and one of white. The type specimen is 6 mm in length and 3 mm in diameter and came from Panama.

1204. Anachis (Parvanachis) diminuta (C. B. Adams, 1852) (Synonym: **A. rufotincta** Carpenter, 1857). One of the smallest members of the genus, the little brown shell has a darker brown area at the canal and measures about 4 mm in length and 2 mm in diameter. The outer coast of Baja California, near Point Abreojos, to Panama. The type specimen of Carpenter's species proves upon inspection to be a faded specimen of *A. diminuta*.

1205. Anachis (Parvanachis) gaskoini Carpenter, 1857 (Synonyms: **Columbella taeniata** Philippi, 1846 [not Link, 1807]; **A. bartschii** Dall, 1918; not *A. gaskoini* of Grant & Gale, 1931, or of Keen, 1958). One of the most attractive of the small columbellids, this has had an unfortunate nomenclatural history and was not accurately figured until 1968. The shell is white, with a row of brown spots on the axial ribs at the periphery, bounded by fine golden-brown lines on either side; on the spire not every rib is spotted, but the double row of spiral lines carries across. The base has four or five more of these golden brown lines. Length, about 6 mm; diameter, about 2 mm. Bahía San Luis Gonzaga, Gulf of California, southward at least to Manzanillo, Mexico, and probably farther south; one specimen in the British Museum collection is reported to be from Callao, Peru, which needs confirmation.

1206. Anachis (Parvanachis) guerreroensis Strong & Hertlein, 1937. This is uniformly brown except for a lighter central band. Length, 4.2 mm; diameter, 1.9 mm. It is known only from off Acapulco, Mexico, in depths of 37 to 80 m.

1207. Anachis (Parvanachis) milium (Dall, 1916) (Synonym: **Columbella parva** Sowerby, 1844 [not Lea, 1841]). Dall's name being a replacement for a homonym, the type specimen must be Sowerby's material, which is at the British Museum. In the type lot are four specimens, only one of which is the form usually so identified. It is illustrated here and selected as lectotype. The shell is white above, with a dark brown band at the periphery of the body whorl and another broader one on the base, the space between being light brown. Length, 4.8 mm; width, 2.1 mm. The type locality is Monte Cristi (Manta), Ecuador, and the northern end of the range is probably not much north of Panama, most specimens previously identified as *A. parva* proving to be of other species.

1208. Anachis (Parvanachis) pardalis (Hinds, 1843) (Synonyms: **Columbella sulcosa** of authors, not of Sowerby, 1832; **A. carmen** Pilsbry & Lowe, 1932). The shell is dark grayish brown, with lighter or sometimes bright yellow axial ribs. Length, about 9 mm; diameter, 3.5 mm. Nicaragua to Colombia. Hinds assigned the species to *Clavatula*, a turrid genus, and had rather a poor figure, which did not show the aperture. The type lot in the British Museum, illustrated here, proves identical with the columbellid later renamed by Pilsbry and Lowe.

1209. Anachis (Parvanachis) pygmaea (Sowerby, 1832) (Synonyms: **Mitrella elegantula** Mörch, 1860; **Columbella deshayesii** De Folin, 1867). Small white shells with two or three rows of brown spots. Length, 6 mm. The southern part of Baja California through the Gulf of California and south to Ecuador, under rocks between tides.

1210. Anachis (Parvanachis) reedi Bartsch, 1928. The shell is a warm golden brown, with a narrow band of brown at the suture and a broad band of much darker brown on the anterior third of the base; banding also shows in the aperture. Length, 5.3 mm; width, 2.8 mm. Ecuador. Three of the four specimens in the type lot of *A. "parva"* (Sowerby) in the British Museum are this form; oddly, he figured the fourth specimen, thus fixing the species concept, and the present form remained nameless for nearly a century. The pattern of banding is almost the opposite of that in *A. "parva"* (now *A. milium*).

1211. Anachis (Parvanachis) sinaloa Strong & Hertlein, 1937. Very close to *A. guerreroensis*, this was considered by the authors to be distinct because of the color differences and more slender form. The shell is relatively slender for the group and of a uniform light shade of brown. Length, 4.2 mm; diameter, 1.8 mm. Off Mazatlán, Mexico, in 22 m.

1212. Anachis (Parvanachis) strongi Bartsch, 1928. The shell is brown, its posterior half even darker chestnut brown, the aperture dark but paler within. It is relatively wider for its height than *A. milium*. Some specimens of what seems to be this stubby form, in the British Museum collection, from western Colombia, have been labeled *"A. parva"* (i.e. *milium*). They show a lighter band just below the suture and another just above the aperture. One may suspect there is variation in coloring. Length, 6 mm; diameter, 3 mm. Colombia to Ecuador (type locality, Guayaquil).

Subgenus ZAFRONA IREDALE, 1916

Axial sculpture weak to obsolete on body whorl, present as fine riblets on spire; sculpture mainly of fine spiral threads.

1213. Anachis (Zafrona) incerta (Stearns, 1892). The white shell is banded or mottled with reddish brown. Length, 6 mm; diameter, 2.7 mm. Guaymas, Mexico, to Panama and the Galápagos Islands, under rocks and offshore to depths of 13 m.

Genus ANACHIS, s. l.

1214. Anachis phanea Dall, 1919. The description of this unfigured species mentions a white shell with no other color markings than a brown line on the base. The type specimen, which was 9 mm in length and 3.5 mm in diameter, apparently is lost. The type locality is Salina Cruz, Mexico.

Genus BIFURCIUM FISCHER, 1884

With a slotlike posterior canal and a tapering spire.

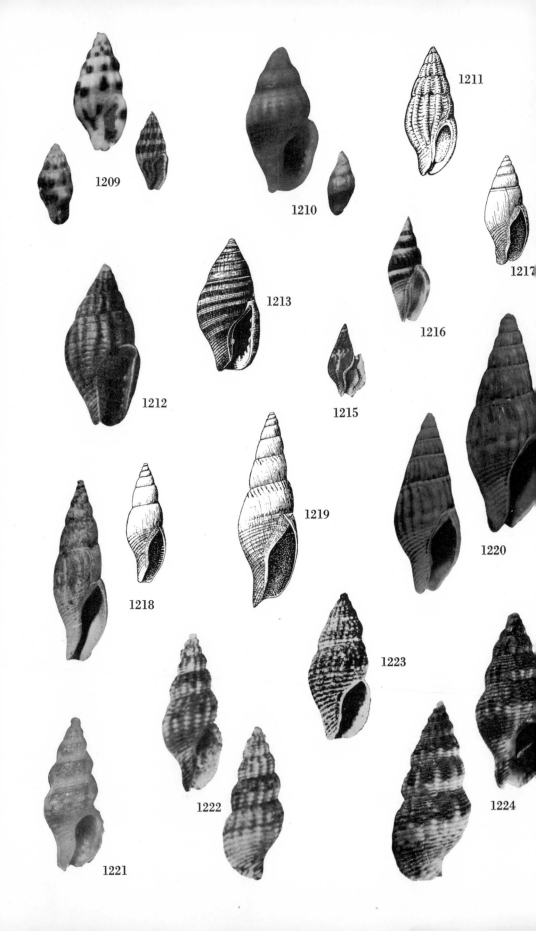

1215. Bifurcium bicanaliferum (Sowerby, 1832) (Synonym: **Columbella bicanaliculata** Duclos, 1835–40). A smooth shell, pale brown, with wavy axial lines of dark brown, the outer edge of the lip brown, the inside of the aperture lavender. About 12 mm in length. The southern part of the Gulf of California to Ecuador and the Galápagos Islands, mostly offshore.

Genus COSMIOCONCHA DALL, 1913

The shell is nearly smooth except for spiral ribs on the lower part. One spiral furrow below the suture sets off a collar-like band. There is also a thickening behind the edge of the outer lip, the inside of the lip being smooth or with a few lirations but no denticles.

1216. Cosmioconcha modesta (Powys, 1835) (Synonym: **Strombina laevistriata** Li, 1930). Type species of the genus, this is a brown shell shading to yellow on the spire, with fine spiral ribs over much of the surface, especially on the lower part of the body whorl. Length, 24 mm; diameter, 10 mm. El Salvador to Ecuador, on mud flats among rocks at extreme low tide.

1217. Cosmioconcha palmeri (Dall, 1913). Of a uniform pale brown, this shell is nearly smooth but has an impressed line below the suture and spirals on the base. Length, 19 mm; diameter, 9 mm. The head of the Gulf of California to Acapulco, Mexico, in depths to 45 m.

1218. Cosmioconcha parvula (Dall, 1913). A slender, pale olive shell. Length, about 20 mm. Off La Paz, Gulf of California, in 205 m.

1219. Cosmioconcha pergracilis (Dall, 1913). Brownish flammules form an obscure band below the suture of the whitish shell. Length, 25 mm; diameter, 7 mm. Off west Mexico in 106 m.

1220. ? Cosmioconcha rehderi (Hertlein & Strong, 1951). The shell is nearly a uniform brown except for mottling in the middle part of the whorls. Besides being small, this differs from other members of the genus in having axial ribs. Hertlein and Strong, in fact, assigned the species to *Anachis*. The sutural collar, however, as pointed out by Dr. George Radwin (*in litt.*), suggests that it may be a *Cosmioconcha*. Final decision must await study of the radula. Length, 8.5 mm; diameter, 3.2 mm. Guerrero, Mexico, to Ecuador, offshore to depths of 48 m.

Genus DECIPIFUS OLSSON & McGINTY, 1958

Sculpture of low, narrow riblets finely beaded by spiral threads; axial sculpture becoming obsolete on the base. Aperture with a small and indistinct posterior canal, the outer lip simple, the columella straight, slightly twisted at the tip of the pillar but lacking any fasciole; anterior canal scarcely differentiated.

1221. Decipifus dictynna (Dall, 1919). The whitish shell has a few spiral cords that are nodulous at the periphery, where they are intersected by about ten low axial threads that are riblike above the periphery but fade out below. Length, 3.8 mm; diameter, 2 mm. Cape San Lucas, Baja California.

1222. Decipifus gracilis McLean, 1959. Slender, with slightly sinuous axial ribs crossed by about seven flat-surfaced spiral cords. The color is buff, variegated with dark brown and bluish-green mottling, brown on the four upper cords of each whorl and the pillar, and with many of the beads white. Length, 8.2 mm; diameter, 3.7 mm. Bahía de los Angeles, Baja California, to Guaymas, Sonora, Mexico (type locality), under rocks at a depth of about 2 m.

1223. Decipifus lyrta (Baker, Hanna & Strong, 1938). The color is yellowish brown, with a darker brown band at the periphery. The shell is less slender than *D. gracilis*, and the rather more straight axial ribs become obsolete farther up the body whorl. Length, 9 mm; diameter, 4 mm. Northern part of the Gulf of California, from San Felipe and Bahía San Luis Gonzaga to Angel de la Guarda Island, intertidally and offshore to a few meters depth. The species is perhaps the closest of West American forms to the type of the genus, *D. sixaolus* Olsson & McGinty, 1958, from the Atlantic coast of Panama.

1224. Decipifus macleani Keen, new species. Color varying from buff to brown with three light bands, one just below the suture, one at the top of the aperture, and one on the base; axial ribs about 20 per whorl, cut by seven to eight spiral cords, forming beads at the intersections; the base with a checkerboard color pattern; aperture quadrangular, with a very short anterior canal; axial ribs dying out at periphery and on some specimens not developed at all, spirals then showing as flat-surfaced or straplike cords. Whorls about seven in number, the nucleus of two smooth turns, the next with fine axial riblets; spiral beading well developed by the fifth whorl. Length, 8 mm; diameter, 3.1 mm. Puertecitos (type locality, where 96 specimens were found in 1962 by Dr. James McLean) to Bahía San Luis Gonzaga (Helen DuShane, 14 specimens), in the northwestern part of the Gulf of California, intertidally. The form is closest to *D. lyrta*, with color banding and with weaker axial ribs and more numerous spiral cords.

1225. ? Decipifus serratus (Carpenter, 1857). The type lot in the British Museum consists of three juvenile specimens. They show some resemblance to *Anachis* (*Glyptanachis*) *hilli*, but have the more delicate sculpturing and indented sutures of *Decipifus*. The shell is tan, mottled with grayish-brown spots. The largest specimen measures in length, 3.5 mm; diameter, 2 mm. Mazatlán, Mexico.

Genus **MAZATLANIA** DALL, 1900
(**EURYTA** ADAMS, 1853 [not GISTL, 1848])

Like a small *Terebra* in shape, with an oblique cord on the pillar resembling a fold, and with a rather short aperture. Radula columbelloid.

1226. Mazatlania fulgurata (Philippi, 1846) (Synonyms: **Terebra arguta** Gould, 1853; **Euryta fulgurans** of authors, a spelling error). The slender, tapering whitish shell has zigzag brown axial lines. Height, 14 mm; diameter, 4 mm. Mazatlán, Mexico, to Nicaragua; common locally.

1227. Mazatlania hesperia Pilsbry & Lowe, 1932. The buff-colored shell is streaked with interrupted zigzag lines of brown, and the impressed sutures give the whorls a convex outline. Length, 15 mm; diameter, 5 mm. The Gulf of California.

Genus **MICROCITHARA** FISCHER, 1884

Axial riblets are conspicuous at the periphery of the whorls, fading out below; at maturity the angulate shoulder of the body whorl is emphasized by the spoutlike posterior canal; the outer lip may be strongly lirate within.

1228. Microcithara cithara (Reeve, 1859). The white shell is covered with a reticulate pattern of chocolate brown; just below the suture is a band of alternating brown and white spots. The surface is nearly smooth except for the row of axial lirae at the periphery. Length, 11 mm; diameter, 5 mm. This was described

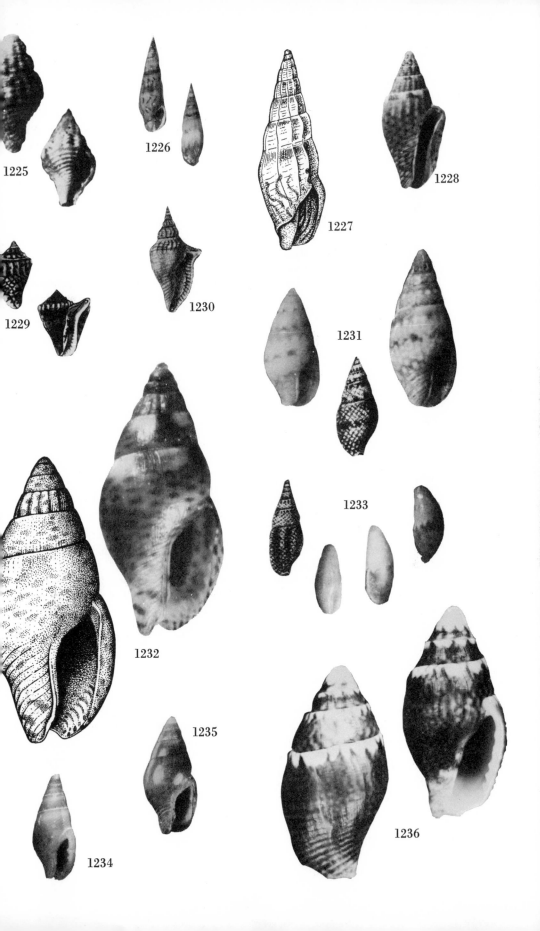

1225

1226

1227

1228

1229

1230

1231

1232

1233

1234

1235

1236

without known locality and remained unrecognized until 1969, when Robert Bullock of Harvard University collected several specimens at San Carlos, Panama, and thus was able to add another species to the already rich Panamic fauna.

1229. Microcithara harpiformis (Sowerby, 1832) (Synonym: **Columbella citharula** Duclos, 1840). A reticulate mottling of gray or brownish black resembles that in *M. cithara*, but the spire is lower and the spout larger on the thickened outer lip. There is a thin periostracum. Length to 20 mm; diameter, 19 mm. El Salvador to Panama, under rocks at extreme low tide. This is the type of the genus.

1230. Microcithara uncinata (Sowerby, 1832). An olive-brown shell, freckled with lighter dots that are larger around the upper part of the whorls and interrupted by reddish-brown lines; the aperture is lavender. Length, 11 mm; diameter, 6 mm. Tres Marias Islands, Mexico, to Central America, mostly offshore in depths to 37 m. This has been assigned to *Bifurcium* by some authors but seems instead to belong here, being intermediate in outline between the other two species, with a moderately tapering spire sculptured with somewhat curved axial lirae.

Genus MITRELLA RISSO, 1826
(ALIA, ASTYRIS, NITIDELLA of authors)

Somewhat resembling *Cosmioconcha*, these are mostly smaller and more compact. Spiral sculpture is generally restricted to the anterior end near the base; axial sculpture is, with few exceptions, entirely absent. The outer lip is dentate in adult specimens.

1231. Mitrella baccata (Gaskoin, 1852) (Synonyms: **Columbella cervinetta** and **C. c. obsoleta** Carpenter, 1857). The yellowish shell is latticed with brown lines and light circular spots, the markings forming one to three bands on the body whorl, with a flush of grayish purple on some. Length, about 6 mm. Magdalena Bay, Baja California, to Nicaragua, under rocks intertidally. This species shares with the next an exceptional development of axial sculpture—the spire shows axial ribs on the third or fourth turns, which disappear abruptly on later whorls.

1232. Mitrella caulerpae Keen, new species. Shell small, stubby, rather broad for its length, nearly biconic in outline; nuclear whorls three, the first white, the second and third brown, smooth, the fourth with about 12 to 14 axial ribs; last two to three whorls smooth, with silky growth lines; ground color of shell buff, the ribs whitish, later whorls marked with about five spaced brown blotches just below the suture, alternating with buff to white areas, anterior to which spiral rows of elongate brown dots develop in a somewhat quincunx pattern; base with six to twelve spiral threads, fading out above; outer lip thickened or with a varix, denticulate within; inner lip smooth except for two low folds at anterior end; operculum small, diamond-shaped, the nucleus apparently apical. Length of holotype, 4.6 mm; width, 2.3 mm; average length, of 13 paratypes, 4.4 mm, width, 2.0 mm; largest specimen seen, length, 6.3 mm. Distinguished from *M. baccata* by its color pattern; from *M. dorma*, which it otherwise resembles (except for the axial ribbing of the spire), by smaller size and white areas below the suture. Type locality, Puerto Ballandra, about 10 miles northeast of La Paz, in sand from among the holdfasts of the green alga *Caulerpa*, 14 specimens, collected by Allyn G. Smith in 1960. Another specimen taken by Dr. Donald Shasky on a *Spondylus* in 3 to 5 m depth, at Punta Diablo, near La Paz, Baja California, Mexico.

1233. Mitrella delicata (Reeve, 1859). The smooth and shiny shell is yellowish white, with a close network of orange-red lines. About 13 mm in length, more slender than *M. ocellata*, a similarly marked form. Guaymas, Mexico, to Panama, under rocks, intertidally. Adults easily lose the slender tip of the shell, so that most specimens appear truncate. A complete specimen is more cylindrical than any other West American *Mitrella*.

1234. Mitrella densilineata (Carpenter, 1864). Coloring and outline are very similar to those of *M. lalage*, and the two may prove to be not more than subspecifically different. A photograph of one of Carpenter's type specimens (reproduced here) shows that the markings are of microscopically fine oblique lines that are not broken up into dots. Length, 6 mm; diameter, 2.4 mm. Cape San Lucas, Baja California.

1235. Mitrella dorma Baker, Hanna & Strong, 1938. In color this is bright chestnut-brown with a faint microscopic network of lighter lines. Length, 6 mm; diameter, 3 mm. Throughout the Gulf of California. Some specimens may be unicolored; others show a pattern of fine light dots between the reticulations of the bright brown markings, as if there were two levels of pigmentation.

1236. ? Mitrella electroides (Reeve, 1858). According to the original description, the shell is thin, reddish brown, with a row of white dots below the suture. The photograph of the type in the British Museum shows what appears to be a chunky and rather solid shell. Length, 12 mm; width, 6 mm. Described as from Guayaquil, Ecuador, but not reported by modern authors.

1237. Mitrella elegans (Dall, 1871) (Synonym: **? M. guttata baileyi** Bartsch & Rehder, 1939). Described as resembling *M. dichroa* (Sowerby, 1844) from the western Atlantic, the shell is marked with brown and white rectangular bars. As the photograph of the type specimen shows, there is a fine pattern of white dots on the base and a scattering of them on some of the dark blotches. The dot pattern resembles that of *M. guttata*, but the spire is more bluntly tapering and the shell is wider for the height. A specimen that closely matches the photograph of the type is in the DuShane collection, found along the coast of Nayarit, Mexico. Length, 7 mm; diameter, 2.8 mm. Mexico and Galápagos Islands (?) to Panama (type locality). The form named as a subspecies of *M. guttata* by Bartsch and Rehder has the color pattern of the Atlantic *M. ocellata* (Gmelin, 1791); in outline it is more similar to *M. elegans* than to *M. guttata*, but it lacks the rectangular color markings.

1238. Mitrella granti Lowe, 1935. The color is a dark brown, somewhat suffused with yellow. Length, 9.4 mm; diameter, 3.4 mm. Northern end of the Gulf of California (type locality, San Felipe).

1239. Mitrella guttata (Sowerby, 1832) (Synonyms: **Buccinum ocellatum** Gmelin, 1791, of authors [not of Gmelin]; **B. cribrarium** of authors [not of Lamarck, 1822]). Long thought to be indistinguishable from the western Atlantic *M. ocellata*, this has been shown by Faye Howard (1963) to have a consistently different color pattern. The Atlantic species is uniformly dark brown with an overall pattern of white dots. The eastern Pacific species, *M. guttata*, has a lighter band around the body whorl just below the periphery. In most specimens the tip or apex of the shell is broken off, but when present it is slender and acutely tapering, in contrast to the blunt apex of *M. elegans*. Length, about 14 mm, diameter, 4 mm, for

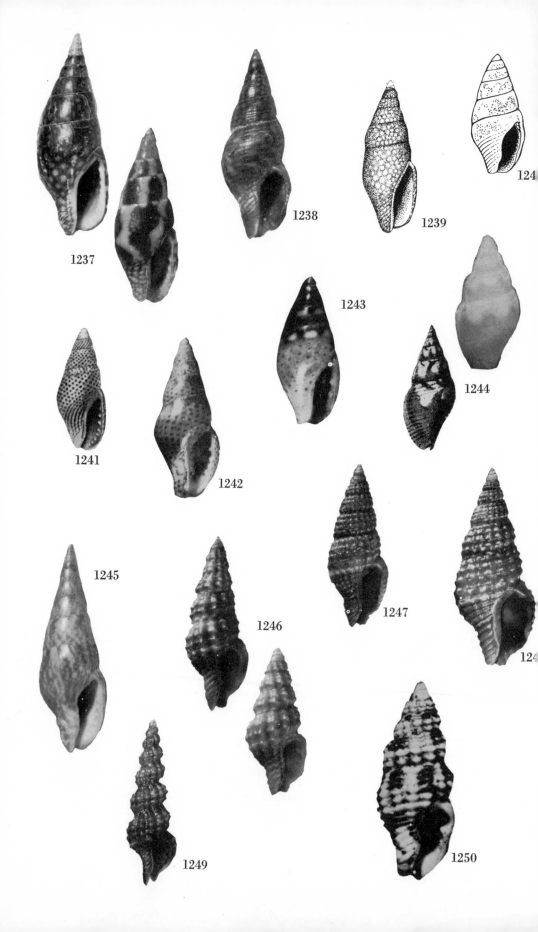

1237

1238

1239

124

1243

1244

1241

1242

1245

1246

1247

124

1249

1250

a large specimen. Magdalena Bay, Baja California, through the Gulf of California and south to Panama, under stones intertidally.

1240. Mitrella harfordi Strong & Hertlein, 1937. The minute shell is whitish, with a central band of irregular brown markings. Length, 3.4 mm; diameter, 1.8 mm. Dredged off Acapulco, Mexico. The range is to Panama Bay, in depths of 13 to 80 m.

1241. Mitrella lalage Pilsbry & Lowe, 1932. The dull buff shell is evenly dotted with brown spots that may fall into curved stripes near the aperture. The type specimen is 5.5 mm in length and 2.3 mm in diameter and was taken at Mazatlán. The range is from the southern part of the Gulf of California to Oaxaca, Mexico.

1242. Mitrella millepunctata (Carpenter, 1864). Also a light brown shell with orange-brown dots, this somewhat resembles *M. lalage*, with coarser spots that do not show a tendency to fall into curved rows. Length, 6.5 mm; diameter, 2.8 mm. Restricted to the southern part of Baja California and the Gulf of California.

1243. Mitrella pulchrior (C. B. Adams, 1852). On a pale-yellowish ground are fine dots of reddish brown, somewhat as in *M. lalage*; in addition a row of rectangular spots forms a band below the suture. About 5 mm in length and 2.5 mm in diameter. Reported only in Panama, under stones at low-water mark.

1244. Mitrella santabarbarensis (Gould & Carpenter, 1857) (Synonym: **Columbella reevei** Carpenter, 1864). The shell has more spiral grooves than most Mitrellas. It is white, clouded or spotted with brown, with a row of spots below the suture. About 8 mm in length. The southern part of the Gulf of California to Central America.

1245. Mitrella xenia (Dall, 1919). Largest West American member of the genus, this has a white shell painted with brown in irregular clouds or streaks, the whorls flat-sided, the canal short, slightly recurved. Length, 18 mm; diameter, 6 mm. Collected at Cape San Lucas, Baja California, by W. H. Dall. It is possible that this may prove to be a *Strombina*.

Genus NASSARINA DALL, 1889

Small shells with both axial and spiral sculpture; whorls more rounded and sutures more deeply indented than in *Anachis*; spiral cords narrower than their interspaces. The subgenera in the Panamic province may be keyed as follows:

1. With outer lip contracted to form an anterior canal.................... 2
   With outer lip not contracted anteriorly............................ 3
2. Sculpture reticulate, axial ribs and spiral cords nearly equal........*Cigclirina*
   Sculpture of undulating axial ribs, stronger than spiral cords........... 4
3. Spire tapering, whorls not more than 8....................*Nassarina, s. s.*
   Spire elongate, whorls as many as 10........................*Radwinia*
4. Axial ribs about equal in strength to spiral cords..............*Steironepion*
   Axial ribs spaced, more prominent than spiral cords...........*Zanassarina*

Subgenus NASSARINA, s. s.

Sculpture relatively coarse; apex with about one and a half smooth, inflated whorls; aperture with the posterior part shelflike, from the columellar margin across to the junction with the outer lip, anteriorly with a distinct canal. Type species, *N. bushii* Dall, 1889, from the Caribbean.

1246. Nassarina (Nassarina) vespera Keen, new species. Shell light brown, banded with ivory-white, one band below suture, one just above the aperture, one on base. Distinctive features are the shape of the upper part of the aperture and the spiral sculpture, which consists of four cords, the uppermost weak; the whorls have a somewhat angulate profile, with the suture indented under the lowest cord and the area below the suture convex outward to the first cord. Axial ribs about twelve, beaded at their intersections with the spiral cords. Aperture reflecting the exterior sculpture as lirations, especially on the outer lip, which has one strong denticle above and a few lirations below it; anterior canal distinct. Base and pillar with about nine spiral cords, no axials. Length of holotype, 7.5 mm (range, in other material, 6 to 7.5 mm); width of holotype, 3 mm (range, 2.5 to 3 mm). Teacapán, Sinaloa (Berry collection); Manzanillo, Colima (California Academy of Sciences collection), 55 m; Puerto Guatulco, Oaxaca (Los Angeles Museum), 45 to 55 m; Port Parker, Costa Rica, 27 m (type locality; California Academy of Sciences collection). This is the first record of the subgenus *Nassarina, s. s.*, on Pacific shores—as implied by the specific name *vespera*, an adjective meaning "western." The shelflike edge of the inner lip does not show as conspicuously in the specimens from Sinaloa as in the other three lots, especially in the type lot from Costa Rica, but in other respects the three lots are similar in form.

Subgenus **CIGCLIRINA** WOODRING, 1928

Sculpture evenly reticulate; apex blunt, of one and a half smooth whorls; aperture with a moderately developed anterior canal.

1247. Nassarina (Cigclirina) helenae Keen, new species. Buff to pinkish brown, with a lighter band below the suture, showing just below the top of the aperture on the body whorl; spire slender and acutely tapering; sculpture of beaded cords, about four spiral cords per whorl, of nearly equal strength, and 18 to 20 axial riblets; spire of two to three nearly smooth nuclear whorls, followed by about six cancellate whorls; aperture relatively narrow, inner lip somewhat detached from body whorl along the pillar edge; outer lip with a broad posterior sinus and a few denticles within; anterior canal distinct. Length of holotype, 7.5 mm (range, 7 to 8 mm); diameter of holotype, 2.8 mm (range, 2.7 to 2.8 mm). Puertecitos (DuShane collection), 7 m; Puerto Peñasco (McLean collection); Guaymas (type locality), 45 m; Banderas Bay (Stanford collection; Manzanillo, Colima (California Academy of Sciences collection), 55 m; Puerto Guatulco, Oaxaca, Mexico (Los Angeles County Museum), 45 m. In color this is very similar to the next, with the same range of variations in development of bands. Possibly the two may prove to be geographic subspecies. The ranges overlap at Puerto Guatulco, without evident intergradation. The northern form, *N. helenae*, is consistently more slender, smaller, with a relatively shorter, narrower aperture; it has more axial ribs per whorl and finer beading.

1248. Nassarina (Cigclirina) perata Keen, new species. Buff to purplish or pinkish brown, with four strong spiral cords per whorl and about 14 to 17 axial ribs, beaded at the intersections, lighter color bands just below the suture and in the middle of the whorls. Spire bluntly tapering, of three smooth nuclear whorls and six cancellate subsequent whorls; aperture somewhat flaring, inner lip not firmly appressed at its outer edge to the pillar, leaving a narrow chink; base with about ten spiral cords, axial sculpture obsolete. Length of holotype, 9.1 mm; diameter, 3.8 mm (average length of other material, 8.5 mm). Puerto Guatulco,

Oaxaca, 45 m (Los Angeles County Museum and California Academy of Sciences), Puerto Videra, Chiapas, Mexico, 37 to 45 m (type locality) (Dr. Donald Shasky collection) to Santa Elena, Ecuador, 16 m (Los Angeles County Museum). A larger and more sturdy shell than *N. helenae*, with a more southern distribution. Specimens are consistently wider for the height, with coarser beading and fewer axial ribs. The specific name (an adjective meaning "of the west") is selected to call attention to the first records of the subgenus *Cigclirina* in the Pacific. The type species, *C. sigma* Woodring, 1928, is from the Miocene of Jamaica.

Subgenus RADWINIA SHASKY, 1970

Slender, spire elongate, with three smooth nuclear whorls, up to seven additional teleoconch whorls, having nodose axial sculpture; suture indented; outer lip varicate, with two denticles inside the margin, aligned with the varix; anterior canal slightly recurved.

1249. Nassarina (Radwinia) tehuantepecensis (Shasky, 1970). The high-spired shell is brown, fusiform, the spire longer than the aperture; the columella is faintly lirate, the anterior canal open. Length, 9 mm; width, 2.8 mm. Off Chiapas, Gulf of Tehuantepec, Mexico, in 27 to 55 m. The sculpture is similar to that in *Nassarina*, s. s., but the spire is more slender, with two or more additional whorls, and the shelflike upper margin of the aperture is less clearly developed. The radula is distinctive in having hooked marginal teeth (otherwise unknown in the family); the rachidian tooth is uncusped, as in other columbellids.

Subgenus STEIRONEPION PILSBRY & LOWE, 1932
(PSAROSTOLA REHDER, 1943)

Sculpture reticulate; nuclear whorls three, tending to develop a carina; no anterior contraction of the outer lip.

1250. Nassarina (Steironepion) hancocki Hertlein & Strong, 1939. Small, slender, with 14 vertical ribs crossed on each whorl by two larger spiral cords, the intersections strongly beaded, and a few lesser cords; white with irregular patches of brown. Length, 4 mm; diameter, 1.5 mm. Galápagos Islands (Pleistocene, type locality); living, Salinas, Ecuador (Stanford collection).

1251. Nassarina (Steironepion) melanosticta (Pilsbry & Lowe, 1932). Small, buff-colored, with deep pits between the intersections of the axial and spiral ribs; irregular brown nodes on the intersections; outer lip and columella weakly denticulate. Length, 4.4 mm; width, 1.7 mm. Guaymas, Mexico, to Nicaragua (type locality).

1252. Nassarina (Steironepion) tincta (Carpenter, 1864) (Synonyms: "**Pleurotoma lineolata** Reeve" of authors [not of Reeve, 1846]; **Mangelia fredbakeri** Pilsbry, 1932; [not *Anachis tincta* of authors]). Small, white to buff, with flecks of brown forming vague bands; aperture rectangular, lip thickened in the middle. Length, about 5 mm. Cedros Island, Baja California, through the southern end of the Gulf of California to Guaymas and south to Banderas Bay, Mexico. Carpenter's type has been recently detected at the U.S. National Museum (Radwin, *in litt.*) and proves to be the form better known from Pilsbry's description. Previous guesses as to the identity of *A. tincta* have been unsuccessful, and the figure by Baker, Hanna, and Strong, 1930 (copied by Keen, 1958) was far from the mark, for the specimen turns out to be a worn *Mitrella baccata*.

The corrected range of *Nassarina (Steironepion) tincta* is based on material in the Los Angeles County Museum and Stanford University collections.

Subgenus ZANASSARINA PILSBRY & LOWE, 1932

Nuclear whorls two to three, conic, not inflated; axial ribs well spaced; inner lip not set off or shelflike; aperture not contracted into an anterior canal.

1253. Nassarina (Zanassarina) anitae Campbell, 1961. The fusiform shell is brown with a peripheral orange-brown band that colors the interspaces and alternately the nodes on the axial ribs; axial ribs about twelve, narrower than their interspaces; aperture with shallow posterior sinus and five denticles on the outer lip. Length, 10 mm; width, 4.5 mm. Off Puerto Peñasco and Guaymas (type locality), Sonora, Mexico, in depths to 55 m.

1254. Nassarina (Zanassarina) atella Pilsbry & Lowe, 1932. The shell is grayish white with lighter spiral cords. Length, 6.8 mm; diameter, 3 mm. Agua de Chale, northwestern Gulf of California, to Guaymas and Acapulco (type locality), Mexico.

1255. Nassarina (Zanassarina) conspicua (C. B. Adams, 1852). On the white ground are bands of brownish orange and some brown spots. Length, about 5 mm; diameter, 2.5 mm. Only a single specimen, from Panama, has been reported.

1256. Nassarina (Zanassarina) cruentata (Mörch, 1860) (Synonyms: **Columbella humerosa** Carpenter, 1865; **Mangilia whitei** Bartsch, 1928; **N. xeno** and **N. x. albipes** Pilsbry & Lowe, 1932). The white shell has a relatively short, broad aperture, with a curving posterior sinus; axial ribs about seven in number, crossed by six to seven spiral cords, which bear brown flecks of color at irregular intervals; axial ribs obsolete on base. Length, 6.5 mm; width, 3 mm. Mazatlán, Mexico, to Ecuador.

1257. Nassarina (Zanassarina) pammicra Pilsbry & Lowe, 1932. The slender black shell has a few pale spots near the middle of the whorls. Length, 4.4 mm; diameter, 1.9 mm. Bahía San Luis Gonzaga, Gulf of California, to western Nicaragua (type locality).

1258. ? Nassarina (Zanassarina) penicillata (Carpenter, 1864). More slender shelled than other species of the subgenus, with sinuous narrow axial ribs crossed by inconspicuous spiral cords and a few brown spiral lines. Length, 5 mm; width, 2 mm. Southern California to Cape San Lucas, Baja California.

1259. Nassarina (Zanassarina) poecila Pilsbry & Lowe, 1932. Dark brown blotches mark the yellowish-white shell above the middle of the whorl, and there are some small dark spots on the base. Length, 5 mm; diameter, 2.1 mm. The fine sculpture gives an impression of scaliness to the surface. Mazatlán, Mexico, to Nicaragua. This is the type of the subgenus.

1260. Nassarina (Zanassarina) whitei (Bartsch, 1928). The chunky flesh-colored shell has irregular spots and streaks of dark brown or rusty yellow. The rather even axial and spiral cords intersect as elongate nodes. The posterior sinus is wide, and the thickened outer lip has six denticles. Length, 7.7 mm; width, 3.4 mm. Guayaquil, Ecuador. This is not the same form as the *N. (Z.) whitei* (Bartsch, 1928) that falls as a synonym of *N. cruentata*. The present species was described as an *Anachis*; thus, primary homonymy does not exist, and fortunately there is a prior name for the second species.

Genus PARAMETARIA DALL, 1916
(META REEVE, 1859 [not KOCH, 1835])

Conical shells with a low spire, long aperture, and smooth outer lip.

1261. Parametaria dupontii (Kiener, 1849–50) (Synonyms: **Conus concinnus** Broderip, 1833 [not Sowerby, 1821]; **Strombus dubius** Sowerby, 1842 [not Swainson, 1823]; **C. concinnulus** Crosse, 1858; **Meta cedonulli** Reeve, 1850; **P. dupontiae** of authors, an unjustified emendation). Resembling a cone, this differs in the form of the operculum and radula and also in having a thinner shell, which is white, variously marked with yellow, orange, or brown in irregular blotches and lines. About 28 mm long and 16 mm wide. A variable form as to height and outline of spire; it is probable that careful work will demonstrate the need to divide what seems to be one species into two or more. The Gulf of California to Tres Marias Islands, west Mexico.

1262. Parametaria macrostoma (Reeve, 1858, *ex* Anton, MS). A narrower form than *P. dupontii*, marked with dark reddish brown; it is more shiny and has a rounded shoulder, with the suture riding upward near the aperture. Length, about 21 mm; width, 11 mm. Manzanillo, Mexico, to Panama, intertidally under stones. The species was named as a *Meta* and is not to be confused with *Columbella macrostoma* Reeve, 1859, which is an *Anachis*. The type specimen in the British Museum matches Reeve's figure well except that it is slightly smaller; two other specimens on the mount are *P. dupontii*. The species is very similar to one in the Atlantic, *P. ovulata* (Lamarck, 1822).

Genus RUTHIA SHASKY, 1970

Small shells, with a tapering spire longer than the aperture; sculpture of a few axial ribs cut into nodes by spiral lirae, the ribs not fading out on the lower parts of the whorls; aperture with a lirate outer lip; columella smooth; anterior canal short.

1263. Ruthia ecuadoriana Shasky, 1970. The shell is yellow, somewhat turreted in form, with about seven axial ribs per whorl, having three nodes per rib on the later whorls. Length, 14.5 mm; diameter, 4.8 mm. Off Cape San Francisco, Ecuador, at a depth of 4 m.

1264. Ruthia mazatlanica Shasky, 1970. Smaller than *R. ecuadoriana*, this is dark grayish brown, with about six axial ribs per whorl except on the last, where a terminal rib forms a lip varix; ribs with two rows of cream-colored nodes on spire, body whorl with six to eight light-colored nodes. Length, 12 mm; diameter, 3.9 mm. Intertidally and in shallow water offshore at Mazatlán, Sinaloa, Mexico. Type species of the genus.

Genus STROMBINA MÖRCH, 1852
(STROMBOCOLUMBUS COSSMANN, 1901)

The Strombinas have slender shells with an elongate and tapering spire. On the body whorl there is a thickening behind the outer lip, and on the inner side of the outer lip there may be a heavy ridge or a few teeth. Sculpture varies from microscopically fine to rather coarse nodes. The anterior canal mostly is long and well developed or even recurved. Except for one Caribbean species, the genus is confined to the west coast of the Americas, thus almost qualifying as what Woodring

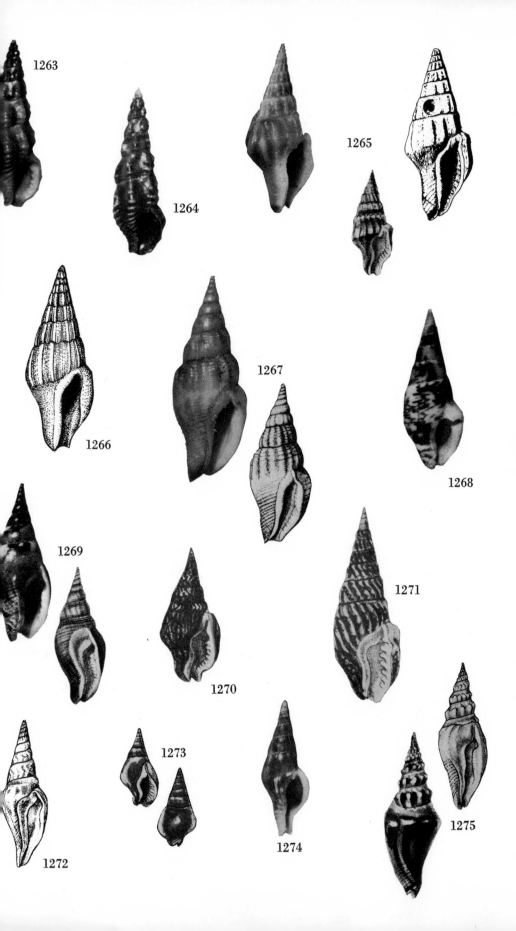

1263

1264

1265

1266

1267

1268

1269

1270

1271

1272

1273

1274

1275

(1966) would term a "paciphile," for during the Tertiary, from Miocene to Pliocene time, it was widespread in both the western Atlantic and the eastern Pacific. Two Panamic subgenera.

Subgenus STROMBINA, *s. s.*

Back of body whorl with a thickening or hump.

1265. Strombina (Strombina) angularis (Sowerby, 1832) (Synonym: **S. subangularis** Lowe, 1935). A yellowish-white shell stained with brown. The strong axial ribs fade out at the angulate periphery and are replaced by spiral striae. Length, 32 mm; width, 11.5 mm. Gulf of California to Panama, offshore in depths to at least 37 m.

1266. Strombina (Strombina) bonita Strong & Hertlein, 1937. The shell is not so large as *S. angularis* and has a few more axial ribs; also, it is not so strongly angulate. Length, 19 mm; diameter, 7.5 mm. Dredged in the southern part of the Gulf of California, in 37 to 45 m depth.

1267. Strombina (Strombina) carmencita Lowe, 1935. This also resembles *S. angularis* but has weaker ribs and a more rounded periphery. Length, 30 mm; diameter, 11 mm. Off Carmen Island, Gulf of California, to Panama, in depths to 75 m.

1268. Strombina (Strombina) clavulus (Sowerby, 1834). A small yellowish-white shell with reticulate markings of brown. The outer lip is greatly thickened, with a slot for the anal siphon that resembles a turrid notch. Length, 23 mm. Mazatlán, Mexico, to Panama, at extreme low tide or offshore in depths to 37 m.

1269. Strombina (Strombina) colpoica Dall, 1916 (Synonym: **Columbella subulata** Sowerby, 1847 [not Duclos, 1840]). Except for a slight shoulder, the shell has a rounded contour to the whorls. The periostracum is yellowish, stained with light brown. A large specimen may be 35 mm in length. The Gulf of California to Panama.

1270. Strombina (Strombina) dorsata (Sowerby, 1832) (Synonym: **Columbella nasuta** Menke, 1851). The shell is yellowish white, marked with wavy or zigzag lines. The thickening of the lip results in a hump on the back of the shell. About 25 mm in length. The head of the Gulf of California to Ecuador, mostly offshore in depths to 37 m.

1271. Strombina (Strombina) elegans (Sowerby, 1832). This is of the group of *S. angularis* but is larger and less angulate, with straighter axial ribs. About 37 mm in length. The west coast of Central America.

1272. Strombina (Strombina) fusinoidea Dall, 1916 (Synonym: **Columbella fusiformis** Hinds, 1844 [not Anton, 1839]). In coloring and shape this form resembles *S. recurva*, but the shoulder is more rounded, without nodes. Length, about 50 mm; diameter, 18 mm. Santa Maria Bay, Baja California, to Panama.

1273. Strombina (Strombina) gibberula (Sowerby, 1832). Callus thickenings on the back and side give the shell a humpbacked look. They are white against the yellowish-white surface of the shell itself, which is reticulately marked with brown lines. About 14 mm in length. Magdalena Bay, Baja California, through the Gulf and south to Peru, intertidally and in depths to 100 m.

1274. Strombina (Strombina) hirundo (Gaskoin, 1852). The shell is whitish, freckled with many orange-brown lines, slender, and shining, without any callus thickening on the body whorl. On the outer lip there is a single strong denticle. About 18 mm in length and 6.6 mm in diameter. Southern Baja California to Cape Pasado, Ecuador, mostly offshore.

1275. Strombina (Strombina) lanceolata (Sowerby, 1832). A slender, light-colored shell somewhat resembling *S. recurva* but more elongate, with a shorter canal. Length, 31 mm; width, 11 mm. Ecuador and the Galápagos Islands. This is the type species of the genus.

1276. Strombina (Strombina) lilacina Dall, 1916. A short, stumpy white shell, with strong spiral ribs and weak axial sculpture, the spire tinged with violet. Length, 23.5 mm. Gulf of California to Manzanillo, Mexico, mostly offshore, in depths to 37 m.

1277. Strombina (Strombina) maculosa (Sowerby, 1832). The slender shell is white, reticulated with brown, the whorls ornamented with a row of tubercles at the shoulder. About 25 mm in length. The northern part of the Gulf of California to Panama, on mud flats and offshore to depths of 37 m.

1278. Strombina (Strombina) marksi Hertlein & Strong, 1951. A light brown shell with a few irregular white spots below the suture, and with faint axial sculpture on the spire, the spiral ribs developed on the anterior part of the body whorl. Length, 24 mm; diameter 9.5 mm. Dredged in the southern part of the Gulf of California, where the species occurs in depths of 70 to 100 m.

1279. Strombina (Strombina) paceana Dall, 1916. A dark brown shell with a silky periostracum, the spire smooth and tapering, the aperture narrow and edentulous. On the body whorl a row of small pustules occurs just below the suture. Length, 37 mm; diameter, 10.5 mm. Scammon's Lagoon, Baja California, to the Gulf of California.

1280. Strombina (Strombina) pavonina (Hinds, 1844) (Synonym: **Columbella haneti** Petit de la Saussaye, 1850). The yellowish-white shell has wavy stripes and patches of dark brown, and the canal is narrowed and recurved. About 22 mm in length. Mazatlán, Mexico, to Panama, mostly offshore in depths to 37 m.

1281. Strombina (Strombina) pulcherrima (Sowerby, 1832). Yellowish brown in color, the shell has spiral ridges tinged with brown and a distinctive outline. Length, 23 mm. Tres Marias Islands, Mexico, to Costa Rica, offshore in depths to 37 m.

1282. Strombina (Strombina) recurva (Sowerby, 1832) (Synonym: **Drillia limonetta** Li, 1930). A buff-colored shell with fine axial ribbing on the spire and a row of tubercles at the shoulder of the body whorl. The outer surface of the shell in many specimens has a network of fine irregular grooves. About 30 mm in length and 12 mm in diameter. San Ignacio Lagoon, Baja California, and south along the coast of Mexico to Lobitos, Peru, mostly offshore in depths to 37 m.

1283. Strombina (Strombina) sinuata (Sowerby, 1875). A flesh-colored shell with a peculiarly flaring lip, brown within but edged with white. Length, about 18 mm; diameter, 8 mm. By error the type locality was given as "Upper California." Specimens in the Stanford collection from San José, Guatemala, dredged at a depth of 18 m, match the original figure well; the range includes also El Salvador and probably Panama.

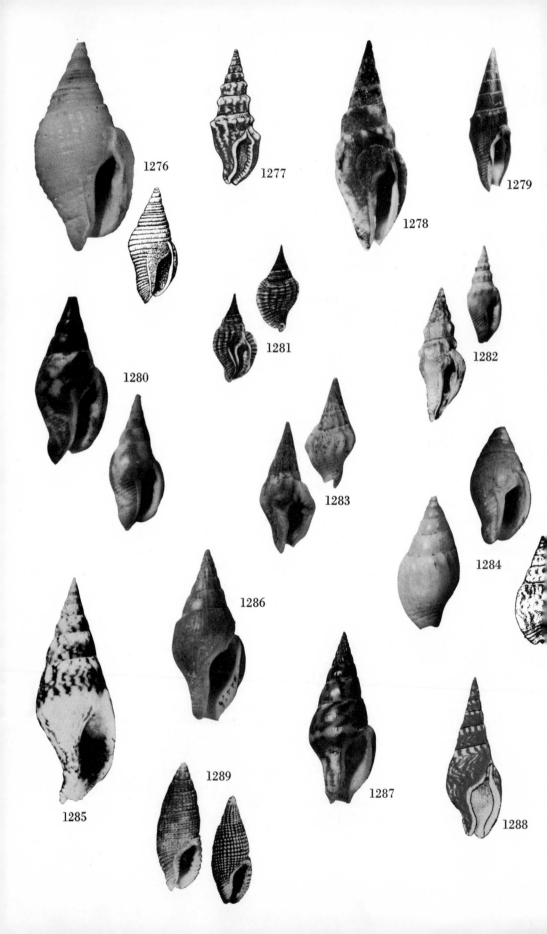

1284. Strombina (Strombina) solidula (Reeve, 1859). The rather solid white shell has a few spiral grooves and is a little puckered on the back just below the suture, streaked or marbled with brown, the interior of the aperture and the ends of the shell tinged with violet. Length, 15 mm. Reported only in the southern end of the Gulf of California.

Subgenus COTONOPSIS OLSSON, 1942

Back of body whorl lacking any thickening or hump. Subgenus based on a fossil form from the Tertiary of Central America.

1285. Strombina (Cotonopsis) deroyae Emerson & D'Attilio, 1969. Slender and smoothly tapering, the shell is medium brown in color, with one or two broad bands of zigzag markings below the suture and on the base, the first four spire whorls with eleven axial ribs; the anterior canal is somewhat recurved. Length (holotype), 49 mm; width, 27 mm. Galápagos Islands, depth, 158 m.

1286. Strombina (? Cotonopsis) edentula Dall, 1908. The whitish shell has obscure brown cloudings and an ill-defined pale peripheral band, under a smooth, pale yellow periostracum. Shorter than *S. deroyae*, with a less recurved anterior canal, the first four spire whorls with fourteen to fifteen axial ribs. Length, 34 mm; width, 13.5 mm. Off the tip of Baja California to El Salvador, in depths to 110 m.

1287. Strombina (Cotonopsis) mendozana Shasky, 1970. With about 20 axial ribs on the early spire whorls, the white shell has straight or zigzag bands of brown and a white zone near the center of the body whorl; some specimens tend toward yellowish brown or even solid yellow. Length, 22.5 mm; width, 9 mm. Smaller than the other species of the subgenus, it is distinguished from *S. edentula* by having several lirae in the outer lip and from *S. deroyae* by the shorter anterior canal and narrower aperture. Gulf of Fonseca, El Salvador, in 37 to 110 m.

1288. Strombina (? Cotonopsis) turrita (Sowerby, 1832) (Synonym: **S. elegans** Li, 1930 [not Sowerby, 1832]). The smooth brown shell has zigzag markings that form a spotted band just below the suture; the aperture is smooth, lacking teeth or lirae along the lip, white within. Length, 35 mm. Guatemala to Ecuador, mostly offshore in depths to 25 m.

Genus ZETEKIA DALL, 1918

Small, subcylindrical, suture not indented; sculpture reticulate; aperture with a broad but shallow posterior notch and one or more columellar denticles. Because of the posterior sinuosity of the aperture, the genus has been assigned to the Turridae; however, a similar apertural feature occurs among others of the Columbellidae, such as in the subgenus *Steironepion* of the genus *Nassarina* and in some species of *Anachis*.

1289. Zetekia gemmulosa (C. B. Adams, 1852) (Synonym: **Z. denticulata** Dall, 1918). The small brown shell has granular sculpture and a bullet-shaped outline. Length, about 6 mm. Panama. Type of the genus. Transferred to the Columbellidae on the advice of Dr. Donald Shasky, who has collected specimens in Panama, and Dr. James McLean, who rejects the group from the Turridae.

Family MELONGENIDAE

Pear-shaped, solid, dark-colored or banded with lighter color.

Genus **MELONGENA** SCHUMACHER, 1817
(**GALEODES** RÖDING, 1798 [not OLIVIER, 1791])

Shells with a short spire, an oval aperture, and a short open canal. The columella is smooth, the outer lip simple, and the operculum clawlike, with an apical nucleus. One or more rows of blunt spines may develop on the shoulders of the whorls.

1290. Melongena patula (Broderip & Sowerby, 1829). The bright brown shell is banded with pale yellow or white, and the aperture may be yellowish to pinkish. Young shells are more spindle-shaped, but the smooth columella gives the clue that they are not fasciolarids. A large adult shell may be as much as 250 mm long (10 inches), but most are 150 to 200 mm. According to Tryon (1881), the animal has a yellow foot spotted with brown, rather square in shape; the head is long and narrow, and the anterior siphon is chestnut-brown in color. On sand and mud flats from the northern part of the Gulf of California to Panama. *M. melongena* (Linnaeus, 1758) is a similar species in the western Atlantic.

Family **NASSARIIDAE**

The Nassa or basket shells are mostly scavengers, greatly attracted by the smell of decaying—or, for that matter, freshly killed—flesh. The speed with which a quiescent group of these snails, concealed in the sand of an aquarium, can surface and converge on a bit of fish dropped into the tank is convincing proof of their sensitivity. Plowing along the sea floor or a tidal mud flat on their unusual-shaped foot, which is divided into two lobes, they leave a characteristic track that may lead to a small mound of sand under which the snail is buried. The Nassariidae are at their best in subtropical to tropical waters, and the variations they exhibit have been the despair of many a student. One such worker, F. P. Marrat, who lived before the days when a sound theoretical explanation for variation had been formulated or methods devised for dealing with such problems, first tried to name every variant he found (which often meant giving a specific name to every specimen) and later came to the opposite conclusion, that the family consists of a single immensely variable species. The truth, of course, lies somewhere in between.

Subdivision of the family in the Panamic province, if one is conservative, is not difficult. Most of the species can be assigned to *Nassarius*. In some other regions of the world there is greater generic diversity.

Genus **NASSARIUS** DUMÉRIL, 1805
(**NASSA** LAMARCK, 1799 [not RÖDING, 1798]; **ALECTRION** MONTFORT, 1810, of authors)

The distinctive shell feature of the genus is the fossa, a deep groove on the base of the shell around the siphonal fasciole and the slotlike recurved canal. Sculpture is apt to be cancellate, the spire tapering, the aperture rounded and lirate within. The operculum is ovate and small for the size of the aperture, has an apical nucleus, and is notched along the edges in some species.

The type of the genus is a Mediterranean species, *N. mutabilis* (Linnaeus, 1758). Subgeneric divisions are difficult to define, and although several are available, only two, *Arcularia* and *Pallacera*, are distinguished here; other species are assigned to *Nassarius, s. l.* Much more work needs to be done on this family.

Genus **NASSARIUS**, *s. l.*

1291. Nassarius angulicostis (Pilsbry & Lowe, 1932a) (Synonyms: **Nassa nodifera** of authors [not of Powys, 1835]; **Nassa nodicinctus** of authors [not

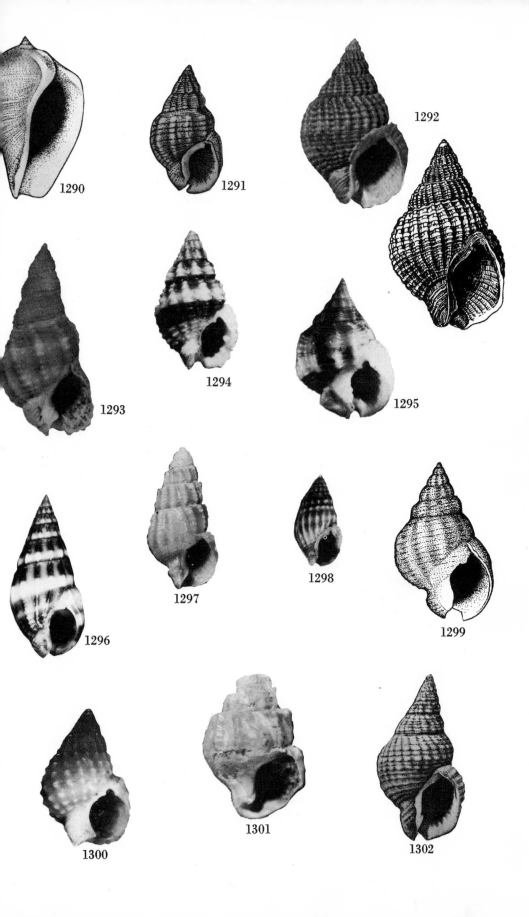

1290

1291

1292

1293

1294

1295

1296

1297

1298

1299

1300

1301

1302

A. Adams, 1852]). Resembling the more common *N. versicolor* but with several fine brown spiral lines and, in some specimens, dark grayish to brownish spots irregularly scattered. Length, 13 mm; diameter, 7 mm. The northern part of the Gulf of California to Panama, intertidally and offshore to depths of 37 m.

1292. Nassarius catallus (Dall, 1908) (Synonym: **Nassa polistes** Dall, 1917). The sculpture is sharply reticulate on this pale brown, slightly dark-banded shell. Length, 14 to 24 mm; diameter, 8 to 15 mm. An offshore form, occurring from Baja California to Peru and the Galápagos Islands, in.depths of 37 to 333 m.

1293. Nassarius cerritensis (Arnold, 1903). A buff to light brown shell with faint and irregular color spots or bands of brown; slender, thin-shelled, the spire with well-marked cancellate sculpture, the axial ribs becoming more and more widely spaced on later whorls. Length, 17 to 20 mm; diameter, 8 to 10 mm. Outer coast of Baja California through the Gulf of California to Bahía San Luis Gonzaga and Guaymas, in depths of 35 to 55 m. Described as a Pleistocene fossil in southern California.

1294. Nassarius collarius (C. B. Adams, 1852). The dingy white little shell has one or more broad bands of brown color and sharply granular sculpture. Length, 10 mm; diameter, 5.5 mm. San José, Guatemala, to Panama Bay (type locality).

1295. Nassarius corpulentus (C. B. Adams, 1852). A nearly globular shell, mostly white, with bands of fine spiral lines of brown. Specimens sometimes have such wide bands as to appear brown with a white central band. Length, 22 mm; diameter, 16 mm. Guaymas, Mexico, to Ecuador, mostly below low-tide level.

1296. Nassarius exilis (Powys, 1835) (Synonym: **Nassa panamensis** C. B. Adams, 1852). The brown shell is banded in buff, with smooth axial ribs between which are fine spiral threads. Length, 15 mm; diameter, 6 mm. Panama to Paita, Peru.

1297. Nassarius exsarcus (Dall, 1908) [described as *Alectrion (Tritia) exsarcus*], is from deep water. Length, 9 mm; diameter, 4.5 mm. Off the Galápagos Islands, depth 366 m.

1298. Nassarius fontainei (Orbigny, 1841). Considered a synonym of *N. exilis* by Tryon and others, this proves upon inspection of the type in the British Museum to be distinct. It has a more stubby shell than *N. exilis*, with a broader spire and more axial ribs (about 18 on the body whorl), which end at the suture in a row of nodes. The color is ivory or buff, with three brown bands. Length, 20 mm; diameter, 11 mm. Type locality, Paita, Peru.

1299. Nassarius gallegosi Strong & Hertlein, 1937. Resembling *N. catallus*, this is pale brown, darker on the back of the body whorl. The holotype measures 21 mm in length and 13 mm in diameter. Guaymas, Sonora, Mexico, to Salango Island, Ecuador, in depths of 18 to 110 m.

1300. Nassarius gemmulosus (C. B. Adams, 1852). Whitish, tinged with yellow or light brown, and finely striped with brown, the distinctive feature of the shell is the even beading of the surface by the intersection of axial and spiral ribs as nodes. The dimensions of the holotype are: length, 6 mm; diameter, 4 mm. Punta Piaxtla, Sinaloa, Mexico, to Panama.

1301. Nassarius goniopleura (Dall, 1908). From deep water off the Galápagos Islands, depth 1,194 m.

1302. Nassarius guaymasensis (Pilsbry & Lowe, 1932*a*). Shell cream-colored, with two very faint brown bands. This is another member of the deepwater group of *N. catallus* and *N. gallegosi*. Length, 17 mm; diameter, 9.5 mm. Puerto Peñasco, intertidally, to off Guaymas, Mexico, in 37 m.

1303. Nassarius howardae Chace, 1958. Resembling *N. catallus* in sculpture and outline but with the apex more acute. The shell is an evenly colored ivory white, without banding, and the outer lip is more flaring than in *N. catallus*. Length, 28 mm; diameter, 13.5 mm. San Felipe (type locality), near the head of the Gulf of California, to Guaymas, Mexico, in depths of 18 to 27 m. Although it might be taken for an unbanded color form of *N. guaymasensis*, it is larger and more slender, with finer sculpture.

1304. Nassarius insculptus (Carpenter, 1864) (Synonym: **N. i. gordanus** Hertlein & Strong, 1951). With fine and regularly spaced spiral cords, crossed by a few low axial ribs, especially on the spire. Length, 22 mm; diameter, 11.5 mm. Point Arena, California, to the southern part of the Gulf of California, in depths to 135 m. The form named as a subspecies, *N. i. gordanus*, was thought to have more distinct axial sculpture on the body whorl and to be limited to the Gulf of California area; however, this variant also occurs offshore in southern California.

1305. Nassarius limacinus (Dall, 1917). A small, glistening, yellowish-white shell with flecks of brown. Length, about 5.3 mm; diameter, 3 mm. Puerto Peñasco, near the head of the Gulf of California, intertidally, to Sinaloa, Mexico.

1306. Nassarius miser (Dall, 1908). Another of the group of deepwater species resembling *N. catallus*. The shell is solid, pale straw-colored, with reticulate sculpture, beaded at the intersections of the ribs. Length, 20 mm; diameter, 11 mm. Cedros Island, Baja California, through the Gulf of California and south to Panama, 150 to 590 m.

1307. Nassarius nodicinctus (A. Adams, 1852). Similar to *N. angulicostis*, with which it has been confused, this has a more angulate outline, with the suture strongly indented or tabulate. Length, 21 mm; diameter, 11 mm. Galápagos Islands.

1308. Nassarius onchodes (Dall, 1917). This was described as a "small, short, acute, swollen, evenly reticulate, whitish shell with a channeled suture, three smooth nuclear and three reticulate whorls, a heavily thickened outer lip with 7 or 8 denticles..." Length, 6 mm; width, 5 mm. Baja California, through the southern end of the Gulf of California to Panama, in 18 to 55 m.

1309. Nassarius pagodus (Reeve, 1844) (Synonyms: **Buccinum decussatum** Kiener, 1841 [not Linnaeus, 1758]; "**Nassa angulifera** A. Adams" of authors; **N. canescens** C. B. Adams, 1852; **N. p. acuta** Carpenter, 1857 [not Say, 1822]). The mottled or banded light brown shell is rather thin for its size, with an angulate outline, as implied in the specific name. Length, 22 mm; diameter, 15 mm. Magdalena Bay, Baja California, through the Gulf of California and south to Ecuador. Carpenter's subspecies may prove to be a distinct species, but the name will have to be replaced. His type specimen, in the British Museum, has fewer axial ribs than *N. pagodus* and lacks color banding (Keen, 1968, figure).

1310. Nassarius planocostatus (A. Adams, 1852). Ash-colored, closely and finely flat-ribbed, ribs sometimes obsolete on the body whorl. Length, 20 mm. Paita, Peru.

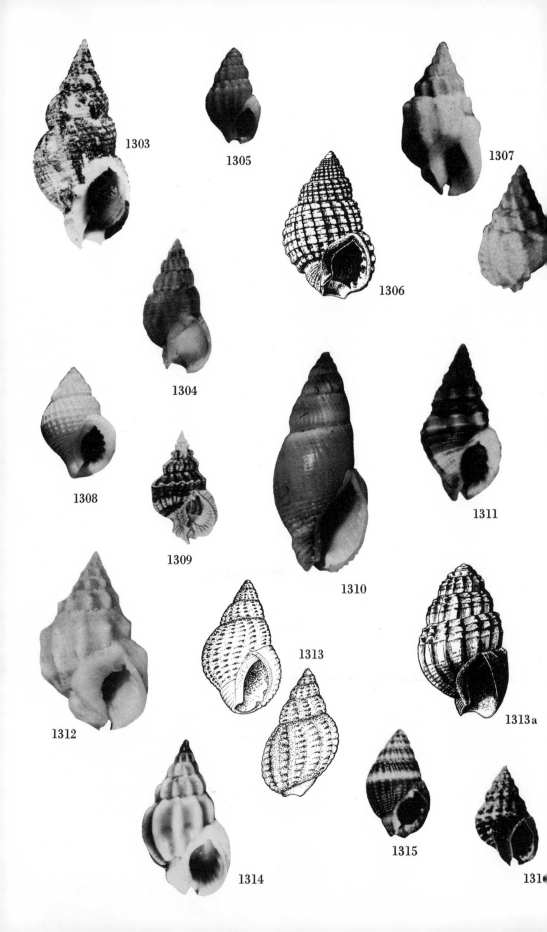

1303

1305

1307

1306

1304

1308

1309

1310

1311

1312

1313

1313a

1314

1315

131

1311. Nassarius scabriusculus (Powys, 1835) (Synonym: **Nassa stimpsoniana** C. B. Adams, 1852). A slender grayish-brown shell, this has the spiral ribs of two sizes, the larger ones reddish or whitish where they cross the axial ribs. Length, 18 mm; diameter, 8 mm. San José, Guatemala, to Panama, intertidally and offshore in shallow water.

1312. Nassarius shaskyi McLean, 1970. Resembling both *N. pagodus* and *N. corpulentus*, this differs from both in having nearly obsolete spiral sculpture on the final whorl, except for two series of equally projecting nodes, with the area between slightly concave. The shell is whitish, some specimens showing faint brown banding toward the base. Length, 23 mm; diameter, 14 mm. Isla Partida, Gulf of California (latitude 28°53′N), to Gorgona Island, Colombia.

1313. Nassarius taeniolatus (Philippi, 1845) (Synonym: **Buccinum nucleolus** Philippi, 1846). Though small, this is an easy form to recognize, for the white shell always has a reddish-brown band at the suture. There may or may not be other bands below. Length, 6 mm; diameter, 3 mm. Puertecitos and Acapulco, Mexico, to Chile.

1313a. Nassarius townsendi (Dall, 1890). Short, stout, yellowish white. Length, 10 mm; diameter, 6 mm. Off the Galápagos Islands, in deep water, 1,485 m.

1314. Nassarius versicolor (C. B. Adams, 1852) (Synonyms: **Nassa rufocincta** A. Adams, 1852; **Nassa glauca, proxima, striata,** and **striatula** C. B. Adams, 1852; **Nassa crebristriata** Carpenter, 1857; **Nassa lecadrei** De Folin, 1867). The numerous synonyms testify to the variability of this species. Basically it has a white shell with coarse axial folds and fine spiral threads, tinged with yellowish brown and marked with fine spiral brown lines. Length, 12 mm; diameter, 5.6 mm. Magdalena Bay, Baja California, throughout the Gulf and south to Paita, Peru, intertidally and offshore to depths of 46 m.

1315. Nassarius wilsoni (C. B. Adams, 1852). The dark brown shell has a whitish band above the suture. It has been considered a synonym of *N. (Arcularia) complanatus*, but the figure of the holotype (Turner, 1956) shows it to be distinct. Length, 8 mm; diameter, 5 mm. Gulf of Fonseca, El Salvador, to Panama.

Subgenus ARCULARIA LINK, 1807

The lip callus is spread across the apertural face of the shell, and a varix may be formed either as a hump on the back of the shell or at the left side.

1316. Nassarius (Arcularia) bailyi (Pilsbry & Lowe, 1932*b*). The shell is purplish gray to black, with an interrupted white band at the shoulder and with the border of the aperture a brownish orange. It differs from *N. (A.) luteostoma* in size and in the spacing of the spiral ribs. Length, 14 mm; diameter, 9.2 mm. Mazatlán, Mexico, to Champerico, Guatemala.

1317. Nassarius (Arcularia) complanatus (Powys, 1835). The lip callus is not well developed in this form, but the varix at the left side suggests that the species belongs in *Arcularia*. The grayish-brown shell is banded with white, and the margin of the outer lip is also white. Length, 10 mm; diameter, 5 mm. Gulf of Fonseca, El Salvador, to Panama.

1318. Nassarius (Arcularia) iodes (Dall, 1917). Callus well developed, white; shell light-colored, with a violet-brown band in the middle of the whorl, and feebly sculptured. Length, 8 mm; diameter, 5 mm. Head of the Gulf of California to Mazatlán, Mexico.

1319. Nassarius (Arcularia) luteostoma (Broderip & Sowerby, 1829) (Synonym: **Nassa nodulifera** Carpenter, 1857, *ex* Philippi, MS). The bright yellow apertural callus in this species is distinctive, with the brown of the body whorl showing through above the aperture. Length, 22 mm; diameter, 16 mm. Gulf of California to Tumbes, Peru, intertidally.

1320. Nassarius (Arcularia) moestus (Hinds, 1844) (Synonyms: **Nassa moesta brunneostoma** Stearns, 1893; **Nassa leucops** Pilsbry & Lowe, 1932). Very similar to *N. (A.) complanatus*, this is proportionately wider. The last whorl is smooth near the aperture. Color of the callus may vary from white around the margin to bright brown, the latter color form having been considered a subspecies by Stearns. Length, 10 to 15 mm; diameter, 6 to 9 mm. Head of the Gulf of California to Guaymas, Mexico.

1321. Nassarius (Arcularia) tiarula (Kiener, 1841) (Synonyms: **Nassa coronula** of authors, not of A. Adams, 1852; **Nassa complanata major** Stearns, 1894). This seems to be the Panamic representative (a subspecies, according to some workers) of the Californian *N. tegula* (Reeve, 1853). The color is buff to brown, with a fillet of dark brown just below or between the nodes of the shoulder. Length, 16 mm; diameter, 10 mm. Throughout the entire Gulf of California and south to Panama. Type of *Phrontis* H. & A. Adams, 1853, which might become a subgenus of *Arcularia* if that group is ever accorded generic rank.

See Color Plate XVIII.

Subgenus PALLACERA WOODRING, 1964

Nuclear whorls small, fossa deep, siphonal fasciole strongly inflated, with sinuous axial swellings; lower part of the columella bordered by a wide smooth platform.

1322. Nassarius (Pallacera) myristicatus (Hinds, 1844). Grayish brown, with a lighter canal and peripheral band, the intersections of spiral and axial ribs touched with bright brown. Length, 28 mm; diameter, 16 mm. Nicaragua to Panama, intertidally and offshore to depths of 37 m. The species has been made type of *Pallacera*, a group that Woodring (1964) regards as a genus, with an extensive range during the Tertiary of the Central American area but now restricted to one Pacific species and a twin species in the Atlantic, *N. (P.) guadelupensis* (Petit, 1842).

Family FASCIOLARIIDAE

Medium-sized to large, with a well-developed, long anterior canal and moderate to high spire. A periostracum may be present. The large operculum is horny and claw-shaped. Central teeth in each transverse row of the rachiglossate radula are narrow, the lateral teeth wider and multicuspate. Placement of several taxa in the family, especially in the genera *Latirus* and *Leucozonia*, is on the advice of Mr. Robert Bullock of Harvard University and Dr. James H. McLean of Los Angeles County Museum; their assistance is gratefully acknowledged. The key is to the genera in the two Panamic subfamilies.

1. Outer lip with a long, sharp tooth anteriorly..................*Opeatostoma*
   Outer lip smooth or nearly so, with no large tooth..................... 2
2. Columella smooth, without evident folds........................*Fusinus*
   Columella with one or more folds.................................. 3

3. Adult shells large, more than 100 mm in height................*Fasciolaria*
   Adult shells medium-sized, less than 100 mm in height................. 4
4. Base without a color band; radula with a small cusp on inner side of
   lateral teeth ....*Latirus*
   Base with a color band; radula lacking a small cusp on inner side of
   lateral teeth ....*Leucozonia*

Subfamily FASCIOLARIINAE

Columella with one or more low folds; periostracum, if present, smooth, not bristly or hairy; shell material densely porcelaneous, solid in texture. Four Panamic genera.

Genus FASCIOLARIA LAMARCK, 1799

Anterior canal well developed. The type species, *F. tulipa* (Linnaeus, 1758), is Caribbean; it has a smoothly tapering spire. No Pacific species resemble it closely enough to be placed in the subgenus *Fasciolaria, s. s.*

Subgenus PLEUROPLOCA FISCHER, 1884

Heavy shells, with a tendency to develop nodose sculpture at the periphery of the whorls. The type species, *F. (P.) trapezium* (Linnaeus, 1758), is Indo-Pacific in distribution.

1323. Fasciolaria (Pleuroploca) granosa Broderip, 1832. The nodose-shouldered heavy, dark shell is covered by a tightly adherent, almost black periostracum. Within, the aperture is brownish orange to bluish gray. Length, 125 to 175 mm (5 to 7 inches). The Gulf of California to Peru. Lowe (1932) reported the species as living on mud flats below low-water mark. Of the three Panamic species of *Pleuroploca*, this is most similar to the type of the subgenus.

1324. Fasciolaria (Pleuroploca) princeps Sowerby, 1825 (Synonym, **F. aurantiaca** Sowerby, 1828 [not Lamarck, 1816]; **F. reevei** "Jonas" Philippi, 1850). One of the largest of the Panamic province gastropods, this has an orange-brown shell with a smooth, dark brown periostracum. The columella and inside of the aperture are orange, with close-set spiral lines of red. The operculum has about five furrows along its length and some irregular diagonal ribs. An average specimen measures 150 to 225 mm (6 to 9 inches) in length and a large one nearly 300 mm (12 inches). Sorensen (1943) reported that these conchs prey upon *Hexaplex erythrostomus*. Gulf of California to Peru, offshore. Strangely, no one seems to have proposed a subgenus for this and its twin species, *F. (P.) gigantea* Kiener, 1840 (? = *F. papillosa* Sowerby, 1825) from the Caribbean, although large size, biconic outline, and unusual operculum would seem to qualify them for separation. Instead, the custom seems to be to use *Pleuroploca* as a generic name for them, although neither is close to the type of that group.

1325. Fasciolaria (Pleuroploca) salmo (Wood, 1828) (Synonym: **F. valenciennesi** Kiener, 1840). The shell is light orange or flesh-colored to yellowish, with a thin yellowish-brown periostracum and a bright salmon-colored aperture; the whorls are rounded, not nodose as in *F. granosa*. Length, 100 to 125 mm. Acapulco, Mexico, to Panama, mostly offshore.

Genus LATIRUS MONTFORT, 1810

Medium-sized shells, mostly slender but sturdy, with canal and spire of about equal length, an adherent periostracum, and a few small folds on the lower part

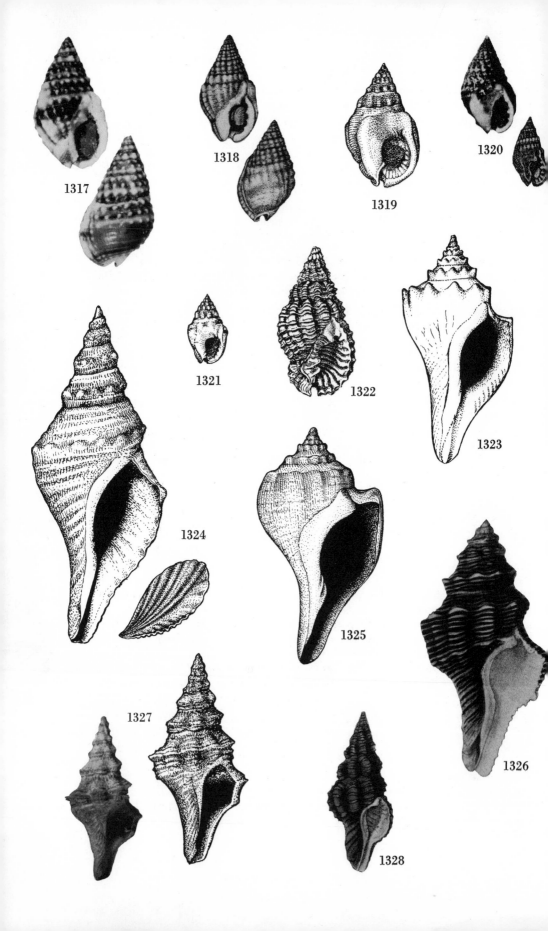

1317   1318   1319   1320

1321   1322   1323

1324   1325   1326

1327   1328

of the columella. The lateral teeth of the radula have a small cusp on the inner edge, next to the central teeth.

1326. Latirus candelabrum (Reeve, 1847). Large for the genus, with somewhat turreted whorls, this has axial sculpture of six to seven ribs forming large compressed nodes at periphery, spiral sculpture alternating in size; shell white under a dark periostracum. Length, 85 mm; diameter, 43 mm. Punta Cotuda, Colombia, in 9 to 13 m (*Te Vega* Expedition, 1968), to Santa Elena, Ecuador (type locality).

1327. Latirus centrifugus (Dall, 1915). The dark-colored shell has strong spiral sculpture rising to sharp nodes or low spines at the periphery, with a lighter band below the periphery. The columella shows some faint folds. Length, 22 mm; diameter, 10 mm. Galápagos Islands, depth 60 m. Bullock (*in litt.*) suggests reallocation of this species, which has been considered a *Fusinus*, because of its similarity to the Caribbean *L. distinctus* A. Adams, 1855.

1328. Latirus concentricus (Reeve, 1847) (Synonym: **Turbinella spadiceus** Reeve, 1847). Dull orange-brown in color, the shell has spiral ribs that are somewhat darker, with a frilled suture line. Length, 47 mm; diameter, 21 mm. Guaymas, Mexico, to Ecuador. *L. spadiceus* was described without locality but was recognized by Melvill (1891) as occurring at Panama; the figure suggests a finely ribbed form of *L. concentricus*.

1329. Latirus hemphilli Hertlein & Strong, 1951a. The color is yellowish, under a dark brown periostracum, the interior of the aperture white. Length, 69 mm; diameter, 24 mm. As Hertlein and Strong show, this is probably the *L. spadiceus* Reeve of authors, not Reeve. The range of *L. hemphilli* is from off Santa Margarita Island, Magdalena Bay, Baja California, and from Carmen Island, Gulf of California, to Pearl Islands, Panama Bay, in depths to 27 m.

1330. Latirus mediamericanus Hertlein & Strong, 1951a (Synonyms: **Murex acuminatus** Wood, 1828 [not Pennant, 1777]; **"Turbinella acuminata"** of authors; **Turbinella castanea** Reeve, 1847 [not Gray, 1839]). A chestnut-brown, nearly smooth shell, with a few spiral threads on the spire and the canal. Length, 55 mm; diameter, 20 mm; length of a large specimen, 98 mm. Manzanillo, Mexico, to Santa Elena, Ecuador (*Te Vega* Expedition, 1968).

1331. Latirus praestantior Melvill, 1892 (Synonym: **Turbinella gracilis** Reeve, 1847 [not Anton, 1839]). Reeve's figure shows an orange-brown shell with seven low and rounded axial ribs, crossed by numerous fine spiral lines, the suture undulating over the ends of the ribs; the anterior canal is unusually long. Length, 57 mm; diameter, 21 mm; length of a large specimen, 80 mm. Although Reeve cited the locality as unknown, the species was recognized as from west Central America by Melvill (1891). Specimens that match Reeve's figure well have been taken in recent years from off Puertecitos, Guaymas, and Carmen Island, Gulf of California, in 10 to 20 m. Melvill's type specimen was described as from Mauritius. However, R. Bullock (*in litt.*), who has studied a color photograph of it, considers the form conspecific with Reeve's. Whether the locality Mauritius is in error or whether this is a wide-ranging species in the Pacific remains to be determined.

1332. Latirus rudis (Reeve, 1847). White under a brownish-olive periostracum, the shell has subangular whorls with about ten low axial ribs crossed by a few spiral cords, the one at the periphery being heaviest. Length, 44 mm; diameter, 20 mm. Described without locality, it was identified as West American by later authors but incorrectly assigned to *Leucozonia*, where it was confused with the unrecognized *L. knorrii*. Specimens in the collection of the Los Angeles County Museum that match the figure and description well are from localities in Panama Bay.

1333. Latirus sanguineus (Wood, 1828) (Synonym: **Turbinella varicosa** Reeve, 1847). The slender, light buff shell has rounded whorls, with about a dozen brown axial ribs, the fine spiral sculpture showing only as striae in the interspaces. Length, 58 mm; diameter, 21 mm. Galápagos Islands, Ecuador.

1334. Latirus socorroensis Hertlein & Strong, 1951*b* (Synonym: **L. clippertonensis** Hertlein & Allison, 1968). The shell is yellowish white under an orange-brown periostracum. It differs from *L. concentricus* by having an angulate periphery, with the area between suture and periphery smooth. The type specimen of *L. socorroensis* measures 39 mm in length, 17 mm in diameter, whereas the type of *L. clippertonensis* is 51 mm in length. Sculptural differences between the two forms being very slight, it would seem that size might be a matter of variation. Revillagigedo group, off west Mexico, to Clipperton Island, off west Central America, in depths to 42 m.

1335. Latirus tumens Carpenter, 1856. Although the type specimen was figured by Melvill (1891), authors have in general overlooked this, and the species remained an enigma to west coast collectors until the 1960's. The whorls of the medium-sized shell are inflated and sturdy, with about eight broad axial ribs crossed by numerous incised spiral grooves. The shell is whitish under a dark greenish-brown periostracum. Length, 68 mm; diameter, 33 mm. Mazatlán, Mexico, to Panama (type locality).

Genus LEUCOZONIA GRAY, 1847

The shell resembles that of *Latirus* in being sturdy and compact, with a few low folds on the columella. The outer lip tends to develop a small tooth anteriorly, but this may be weak to wanting. The principal difference is in the radula, the lateral teeth of which lack the small cusp on the inner edge next to the central tooth (R. Bullock, *in litt.*). Color banding of the base of the body whorl may also be characteristic.

1336. Leucozonia cerata (Wood, 1828). The yellowish-brown shell has lighter-colored or white nodes on the axial ribs; the periostracum is dark brown. Length, about 50 mm; diameter, 26 mm. The southern part of the Gulf of California, from Guaymas south to Panama and the Galápagos Islands. Lowe (1932) reported that the animal is brilliant reddish salmon color; on rocks at very low tide.

1337. Leucozonia knorrii (Reeve, 1847) (Synonyms: "**L.** [*Latirus*] **rudis** Reeve" of authors, not *Turbinella rudis* Reeve, 1847; **Lagena californica** A. Adams, 1852). The shell is dark brown, lacking periostracum, with a lighter band below the suture, the axial sculpture showing only as blunt nodes at the periphery and giving a somewhat angulate outline to the whorls; anterior canal short, the tooth of the outer lip weak to wanting. A similar species in the Atlantic, *L. nassa* (Gmelin, 1791) (better known as *L. cingulifera* [Lamarck, 1816]), has the lighter band much more conspicuous, yellow to white, and has a longer anterior canal. Length, 45 mm; diameter, 22 mm. Honduras (type locality) [i.e. Gulf of Fonseca] to Panama, probably to Ecuador.

1338. Leucozonia tuberculata (Broderip, 1833) (Synonym: **Turbinella tubercularis** Griffith & Pidgeon, 1834). Elaborately sculptured, the shell is ivory-white to buff, with a nodose peripheral carina studded with brown; several smaller spiral ribs are beaded. Length, 45 mm; diameter, 26 mm. Galápagos Islands.

Genus OPEATOSTOMA BERRY, 1958

Spire short, body whorl relatively large and somewhat shouldered; the outer lip with a long, slender tooth.

1339. Opeatostoma pseudodon (Burrow, 1815) (Synonyms: **Monoceros cingulatum** Lamarck, 1816; **M. angulatum** Rogers, 1913). It is unfortunate that Lamarck's long-used and appropriate name *Leucozonia cingulata* must disappear, but this species is so different from the type of *Leucozonia* that it should be removed from that genus. Authors have overlooked the earlier specific name given by Burrow because they assumed his *pseudodon* dated from the second edition of his book, in 1825; but, as Dr. S. S. Berry has shown, actually it was validated in the rare first edition, in 1815, a year ahead of Lamarck's *cingulata*. Burrow evidently had a decrepit specimen with a broken outer lip and was misled into proposing for the specific name a term that means "false tooth"—this for a shell that normally has perhaps the longest apertural tooth developed by any gastropod. One wonders if the spinelike tooth is used—as a similar one in Acanthinas has been observed to be—for wedging apart the plates of barnacles or the valves of clams while the meat within is being devoured. The shell, under a yellowish-brown periostracum, is white. Several dark brown, smooth spiral ridges give it the look of having been turned in a lathe. Length, about 42 mm, diameter, 31 mm. Cape San Lucas, Baja California, through the Gulf of California and southward to Peru; among rocks at low tide.

Subfamily FUSININAE

Columella without folds; shell texture somewhat chalky; periostracum tending to be fibrous.

Genus FUSINUS RAFINESQUE, 1815
(FUSUS of authors, not of HELBLING, 1779)

The spindle shells have spire and canal almost equally long. Ribbing is both spiral and axial. Most of the shells are white or light buff in color. The operculum is horny, with an apical nucleus. Subgeneric grouping is at present rather artificial and in need of revision. There are a number of unnamed groups as well as unnamed species. Three subgenera are recognized here; some of the species that cannot readily be assigned to these are grouped as *Fusinus, s. l.*

Subgenus FUSINUS, *s. s.*

Fusiform, large; apical whorls evenly tapering.

1340. Fusinus (Fusinus) dupetitthouarsi (Kiener, 1840) (Synonyms: **Fusus funiculatus** Lesson, 1842; **Fusus dupetitthouarsi** vars. **aplicatus** and **nodosus** Grabau, 1904). Beneath the greenish-yellow periostracum the shell is white. This is one of the longest forms in the genus, a large specimen being as much as 250 mm (10 inches) long. The outer coast of Baja California, throughout the Gulf and south to Ecuador, on mud flats and sandbars at extreme tide and offshore to 55 m.

1341. Fusinus (Fusinus) irregularis (Grabau, 1904). Resembling *F. (F.)* *dupetitthouarsi* but more slender, anterior canal relatively longer and more sinuous; tip of canal within and edge of the outer lip tinged with purple. Length, 145 mm (large specimen, 235 mm); diameter, 33 mm. Outer coast of Baja California, mainly Cedros Island to San Juanico Bay, in 22 to 42 m, less commonly to the southern end of the Gulf of California.

Subgenus **APTYXIS** TROSCHEL, 1868

Anterior canal relatively short.

1342. Fusinus (Aptyxis) cinereus (Reeve, 1847) (Synonyms: **Fusus taylorianus** Reeve, 1848; **Fusus tumens** Carpenter, 1857; **Pisania elata** Carpenter, 1864; **Fusinus cinereus coronadoensis** and **F. c. sonoraensis** Lowe, 1935). The shell is dull gray, with one or two of the spiral ribs touched with brown. Length, about 30 mm; diameter, 13 mm. Gulf of California, on rocks in mud flats.

1343. Fusinus (Aptyxis) felipensis Lowe, 1935. A small brown form with a purplish aperture, this has ten to eleven axial ribs that fade out on the body whorl. Length, 19 mm; diameter, 7.7 mm. San Felipe (type locality), near the head of the Gulf of California.

Subgenus **BARBAROFUSUS** GRABAU & SHIMER, 1909

Apical whorls compressed, button-like.

1344. Fusinus (Barbarofusus) colpoicus Dall, 1915. The anterior canal is conspicuously twisted; narrow axial ribs thirteen, with eight varices on last whorl; spiral threads sharp. Length, 66 mm; diameter, 18 mm. Guaymas, Sonora, Mexico, to Gorda Bank, southern end of the Gulf of California, in depths of 110 to 165 m. Resembling the southern Californian *F. (B.) barbarensis* (Trask, 1855), but with more and narrower axial ribs.

Subgenus **FUSINUS**, *s. l.*

Several subdivisions, based on differences in the apical whorls, may be represented among the following species.

1345. Fusinus allyni McLean, 1970. The large, thin shell is light in weight, with a yellowish periostracum, the inflated whorls rounded except for a peripheral carina. Length, 88 mm; diameter, 34 mm. Galápagos and Cocos Islands, in 128 to 146 m.

1346. Fusinus ambustus (Gould, 1853) (Synonyms: **F. hertleini** Lowe, 1935, with color forms **F. h. albescens** and **F. h. brunneocincta**). The shell is yellowish, shaded with brown, especially between the axial ribs, of which there are eight to eleven. Length, 45 mm; diameter, 18 mm. The Gulf of California, from La Paz to Mazatlán, Mexico. Lowe (1932) reported specimens spawning on the mud flats in January.

1347. Fusinus fragilissimus (Dall, 1908). Off Ecuador, depth 2,877 m.

1348. Fusinus fredbakeri Lowe, 1935. Cream-colored to light brown, the shell has about seven whorls and twelve axial ribs per whorl. Length, 38 mm; diameter, 15.5 mm. San Felipe, near the head of the Gulf of California, southward along the Sonoran coast to Mexico.

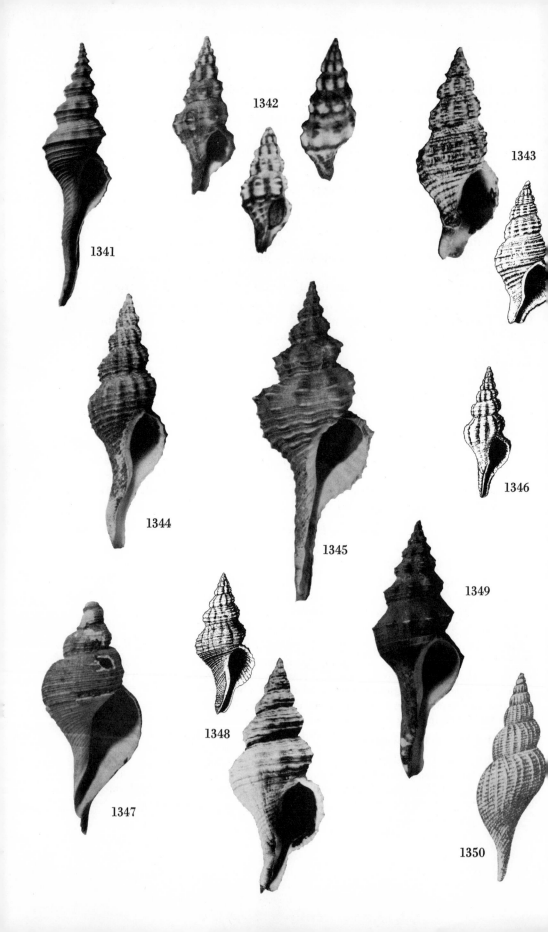

1341

1342

1343

1344

1345

1346

1347

1348

1349

1350

1349. Fusinus panamensis Dall, 1908. The white shell has an olive periostra-cum, eight whorls, nine to ten axial ribs. Length, 75 to 150 mm; diameter, 26 to 50 mm. West Mexico to Ecuador, in depths of 35 to 275 m, specimens not uncommon in shrimp-trawling hauls.

1350. Fusinus rufocaudatus (Dall, 1896). Deep water, Acapulco to Panama Bay, in 2,320 to 3,616 m.

1351. Fusinus zacae Strong & Hertlein, 1937. Dark brown, lighter between ribs and on the base, with eight angulate whorls and eight axial ribs, spiral sculpture weak except for the peripheral cord. Length, 52 mm; diameter, 20 mm. Off Cape San Lucas, Baja California, in 35 to 400 m.

Superfamily VOLUTACEA

Columella with more or less well developed columellar folds, the anteriormost plications being the strongest. Radula rachiglossate but in some forms reduced from the normal 3 teeth per transverse row to one sharply tricuspate tooth, the central.

Family VOLUTIDAE

The volutes, like the cowries, are among the shell aristocracy, for a number of species are so rare that good specimens command high prices. The shells are somewhat spindle-shaped and run to shades of brown and orange, often gaily painted with symmetrical patterns in darker shades of brown. They reach their climax of abundance in the western Pacific. The western Atlantic has a number of species. Only a few kinds occur in the eastern Pacific, none of the type genus, *Voluta*. Four Panamic genera, three in deep water only.

Genus LYRIA GRAY, 1847

The aperture is narrow, with a thickened outer lip and with the columellar callus roughened above the columellar folds. The anterior folds are conspicuously larger than the posterior. No species of subgenus *Lyria, s. s.*, occur in the Panamic province.

Subgenus ENAETA H. & A. ADAMS, 1853

The outer lip has a blunt tooth within.

1352. Lyria (Enaeta) barnesii (Gray, 1825 [as **Voluta barnsii**; emend. Carpenter, 1864]) (Synonym: **V. harpa** Barnes, 1824 [not Mawe, 1823]). The ovate shell has a pale flesh or grayish color, marked with brown spots that may form diffused bands. Length, 32 mm; diameter, 16 mm. The southern part of the Gulf of California to Peru, offshore in depths to 37 m. *L. (E.) guildingi* (Sowerby, 1844) is a similar species in the Caribbean.

1353. Lyria (Enaeta) cumingii (Broderip, 1832) (Synonyms: **Voluta cyl-leniformis** Sowerby, 1844; **E. pedersenii** Verrill, 1870). With a more pointed spire than *L. barnesii* and more nodose ribbing, this is similarly colored. Length, 30 mm; diameter, 17 mm. Magdalena Bay, Baja California, through the Gulf of California and south to Peru, in sandy mud intertidally. The form named as *L. pedersenii* seems to be only an immature stage of *L. cumingii*.

Genus ADELOMELON DALL, 1906

Large, spire longer than canal; smooth; columellar folds few, somewhat weak.

1354. Adelomelon benthalis (Dall, 1896). Deep water, Panama Bay, in 3,054 m.

Genus CALLIOTECTUM DALL, 1890

With a short canal and a well developed periostracum; operculum with a spiral nucleus, curved.

1355. Calliotectum vernicosum Dall, 1890. Deep water, Ecuador and Peru, depths 741 to 1,036 m.

Genus TRACTOLIRA DALL, 1896

Columellar lip with a single strong fold.

1356. Tractolira sparta Dall, 1896. Deep water, Acapulco, Mexico, to Panama Bay, in 3,054 to 3,084 m.

Family HARPIDAE

The harps are a group widespread in the western Pacific but represented by a single species in the Panamic province. Unlike a related family, the Vasidae, the harps have not dispersed into the Caribbean or the Atlantic. They are among the most beautiful of shells, the surface somewhat shiny, rhythmically decorated with ribs and delicate scalloped patterns in brown and rose. The aperture is wide, with a notch rather than a canal and, in most species, with a brightly colored callus area extending across the columellar margin and pillar. This is the only family among the Volutacea that has no folds on the columella; also, there is no operculum.

Genus HARPA RÖDING, 1798

With the characters of the family.

1357. Harpa crenata Swainson, 1822 (Synonyms: **H. scriba** Valenciennes, 1832; **H. rivoliana** Lesson, 1834). Rather dull beside the western Pacific forms, this is grayish, with inconspicuous axial ribs and fine brownish pencilings. Mexican fishermen in recent years have taken a number of specimens from off the Mazatlán-Manzanillo area, but most of these seem to be dead shells brought up in the trawling nets. Length, 82 mm; diameter, 52 mm. Magdalena Bay, Baja California, through the southern end of the Gulf of California and south to Gorgona Island, Colombia.

See Color Plate XI.

Family TURBINELLIDAE (ICZN Opinion 489, 1957)
(XANCIDAE)

The shells are mostly sturdy and fusoid in outline, with weak to strong columellar folds, not as clear-cut or as numerous as in the Volutidae. The family name Xancidae (based on *Xancus* Röding, 1798) had been in use for some years, but in 1957 the International Commission on Zoological Nomenclature accepted a petition—against considerable professional opposition—to suppress *Xancus* in favor of the later *Turbinella* Lamarck, 1799. Rehder (1967) has summarized the problems involved.

Genus SURCULINA DALL, 1908
(PHENACOPTYGMA DALL, 1918)

Slender, with a long and open anterior canal, weak columellar folds, and a turreted spire. Rehder (1967) has shown that the radula indicates volutid rather

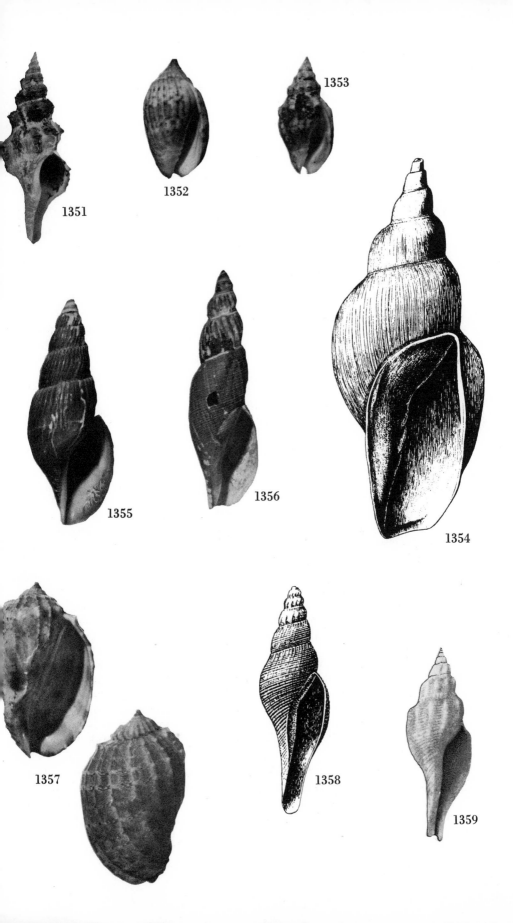

1351

1352

1353

1355

1356

1354

1357

1358

1359

than turrid affinities. Hitherto the species, which are deepwater in distribution, had been allocated to the Turridae, although authors have not been in agreement on the exact placement within that family.

1358. Surculina blanda (Dall, 1908). Cocos Island, 1,953 m. Type of the genus.

1359. S. galapagana (Dall, 1919). Off the Galápagos Islands, 1,159 m.

Family OLIVIDAE

Like the cowries, the olives have a brilliantly polished shell because two wide lobes of the foot and mantle fold back over the shell when the animal is moving about, and the living tissue protects the delicate surface of the shell from being dulled. The mantle edge also has an unusual threadlike extension that lies along the suture, called a filament, which probably has a sensory function. Most olives prefer a sandy substrate, where they plow about just beneath the surface, leaving a characteristic trail. They are carnivorous animals, and Olsson (1956) reports that members of one genus, *Oliva*, may prey upon those of another, *Olivella*. He also points out that there are more species of *Olivella* in the American tropics than in any other part of the world. *Oliva* has its metropolis in the western Pacific. The key is to genera and (in *Oliva*) subgenera.

1. Spire low, aperture long, shells mostly large, solid; operculum wanting..... 2
   Spire pointed, shells mostly small to medium-sized; operculum present..... 3
2. Body whorl uniformly colored, no incised line medially..........*Oliva, s. s.*
   Body whorl with a wide band below a medial incised line........*Strephonella*
3. Surface smooth but not glazed, shells medium-sized..............*Agaronia*
   Surface highly glazed, shells small............................*Olivella*

Genus OLIVA BRUGUIÈRE, 1789

The cylindrical shells have a long aperture and a channeled suture on the low spire. The inner lip is wrinkled but does not have a wide callus area. Useful reference: Zeigler and Porreca (1969)—illustrations of all the species in color. Two Panamic subgenera.

Subgenus OLIVA, *s. s.*

Shell uniformly colored over entire body whorl except on columellar area.

1360. Oliva (Oliva) incrassata [Lightfoot, 1786] (Synonym: **O. angulata** Lamarck, 1811). The thick shell is, in the adult, angularly swollen above the middle, resulting in a distinctive outline. The coloration is a dove-gray to brown, with fine zigzag pattern of markings on a creamy white ground, tinted with pink on the columellar area. Young individuals do not have the angle developed at the shoulder but may be distinguished from *O. spicata*, when alive, by the fawn-colored soft parts spotted with brown (Gifford, 1951). Length, 55 mm; diameter, 30 mm. A large specimen may be 75 mm (3 inches) or more in length. Magdalena Bay, Baja California, throughout the Gulf of California and south to Peru, the preferred habitat being the outer side of sandspits at extreme low-tide level.
See Color Plate XVII.

1361. Oliva (Oliva) julieta Duclos, 1835 (Synonym: "**O. timorea**" of authors [not **O. timoria** Duclos, 1835]; **O. pantherina** Philippi, 1848; **O. mariae** Ducros de Saint Germain, 1857; **O. graphica** and **porcea** Marrat in Sowerby,

1360

1361

1362

1363

1364

1365

1368

1367

1366

1369

1870). The outline resembles that of *O. incrassata*, but the shell is smaller, with less of an angle at the shoulder and a rather uniform color pattern of blurred, triangular reddish-brown spots on a white ground. Some authors have regarded this as a subspecies of *O. spicata*. Length, 50 mm; diameter, 33 mm. Matenchen, Mexico, to Peru.

See Color Plate XVII.

1362. Oliva (? Oliva) kaleontina Duclos, 1835. The shell is white to bluish brown, variegated with reddish brown, the spots larger just below the suture; columella and aperture are purplish white. The outer lip tends to be longer than the columella. Length, about 33 mm. The southern part of the Gulf of California to Ecuador and Peru, more common in the southern part of the range.

1363. Oliva (Oliva) polpasta Duclos, 1833 (Synonyms: **O. callosa** Li, 1930; **O. davisae** Durham, 1950). Because this resembles *O. spicata* in its color markings, some authors have regarded it as a subspecies, but it is distinguished by the more pear-shaped outline, the uniform dull gray background spotted with brown, and the finer liration of the columellar wall. *Oliva davisae*, described as a Pleistocene fossil, represents the extreme trend toward low spire and width in the upper part of the body whorl. Length, 36 mm; diameter, 20 mm. Magdalena Bay, Baja California, through the Gulf of California and south to Ecuador.

See Color Plate XVII.

1364. Oliva (Oliva) porphyria (Linnaeus, 1758) (Synonyms: **O. panamensis** Montfort, 1810; **O. porphyracea** Perry, 1811). The tent olive is one of the few West American mollusks to have been named by Linnaeus. So striking is the pattern of the shell that specimens were early collected by explorers and sailors and carried back to Europe. The crowded chestnut-brown zigzag markings on a flesh-colored ground and the size of the shell are unique among olives. More than one person, having a specimen of this in hand and reading of the rare western Pacific cone, *Conus gloriamaris* Chemnitz, 1777, has concluded that it must be, by its very beauty, that rarity. Lowe (1933) reported finding live specimens in sand intertidally. The range is from the Gulf of California to Panama.

See Color Plates VIII and XVII.

1365. Oliva (Oliva) spicata (Röding, 1798) (Synonyms: **Porphyria arachnoidea** and **litterata** Röding, 1798; **O. araneosa** and **venulata** Lamarck, 1811; **O. obesina, oriola** [not Lamarck, 1811], and **pindarina** Duclos, 1835; **O. subangulata** Philippi, 1848; **O. cumingii** Reeve, 1850; **O. melchersi** Menke, 1851; **O. intertincta** Carpenter, 1857; **O. fuscata, oblonga, punctata,** and **violacea** Marrat in Sowerby, 1870; **O. hemphilli** and **perfecta** Johnson, 1911; **O. rejecta** Burch & Burch, 1962; **O. ionopsis** Berry, 1969). Most variable of the West American olives, as the long list of synonyms attests, this is basically of a buff-olive color, flecked with small irregular brown dots. The spire is fairly high. Many color forms have been named; for example, *O. cumingii* is plain yellow to brown, perhaps with two or three wide brown bands; *O. fuscata* is an almost uniformly colored dark brown with V-shaped markings that show faintly through the glaze. All the color forms can be found in nearly any large population of *O. spicata*; thus, they cannot well be accorded subspecific status. One puzzling possible exception is the form named as *O. rejecta*, in which only slight color differences can be seen (such as the light purple columella instead of white) and slight difference of outline (more slender, spire more tapering). Indian artisans have

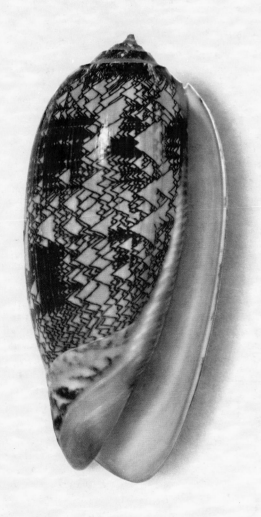

PLATE VIII · *Oliva porphyria* (Linnaeus)

noted, however, that such shells shatter when worked, whereas normal *O. spicata* specimens do not. A difference in mineralogical composition was demonstrated by Donohue and Hardcastle (1962). Whether the difference is genetic or whether it may in some way be environmentally induced remains for further investigation. As Zeigler and Porreca (1969) point out, even if the form proves valid, the name *O. violacea* for it would take precedence; *O. ionopsis* Berry is similar. Length, about 45 mm; diameter, about 23 mm. Throughout the Gulf of California and south to Panama. The animal is buff-colored and marked with short brown wavy lines (Gifford & Gifford, 1951).

1366. Oliva (Oliva) spendidula Sowerby, 1825. On a creamy ground there are fine dots and triangles of cinnamon brown that blend together to form two somewhat interrupted bands. The columella is tipped with dull pink. Length, 46 mm; diameter, 25 mm. Tres Marias Islands, west Mexico, to Panama; not common, rare north of Panama.

Subgenus STREPHONELLA DALL, 1909

Small, low-spired shells with an incised line around the middle forming a wide band between it and the fasciole below that differs in coloration and texture.

1367. Oliva (Strephonella) undatella Lamarck, 1810 (Synonyms: **O. tenebrosa** Wood, 1828; **O. nedulina** Duclos, 1835). White to yellowish in ground color, this form has varied markings of brown. The columellar callus is thick but not wide, with folds along its entire extent. Length, 12 mm; diameter, 6 mm. The size suggests affinity to *Olivella*, but the radula is that of *Oliva*, according to Dr. A. A. Olsson (*in litt.*). Magdalena Bay, Baja California, through the Gulf and south to Ecuador, intertidally. The species is type of the subgenus *Strephonella*.

Genus AGARONIA GRAY, 1839

Thin, unglazed shells, with a flaring aperture, pointed spire, and subdued coloring. The soft parts are grayish to buff in color, and there is an operculum.

1368. Agaronia murrha Berry, 1953. The shell is smaller than that of either of the other two species, porcelain-white outside, brown within. Length, 36 mm; diameter, 15 mm. Corinto, Nicaragua.

1369. Agaronia propatula (Conrad, 1849) (Synonym: **A. hiatula** of authors, not *Voluta hiatula* Gmelin, 1791). Though long grouped with *A. testacea*, this seems to be a distinct form, resembling but not identical with the true *A. hiatula* of the eastern Atlantic. The shell is shorter and wider than that of *A. testacea*, more uniformly colored bluish gray, though some specimens show faint zigzag lines of brown. The spire is proportionately shorter and more bluntly tapering. Length, about 42 mm; diameter, 20 mm. Southern Mexico to Ecuador.

1370. Agaronia testacea (Lamarck, 1811) (Synonyms: **A. reevei** Mörch, 1860; **Oliva philippi** and **testacea griseoalba** Von Martens, 1897). The general color is a soft blue-gray or yellow-gray with brownish wavy axial lines, the base and apex brown, the columella white, and the interior of the aperture violet-brown. According to the Giffords (1951), the animal itself is grayish to buffy. They found these shells living the highest up of any mollusks on the beach. Length, 50 mm; diameter, 20 mm. The head of the Gulf of California to Peru.

See Color Plate XVII.

Genus OLIVELLA SWAINSON, 1840

The olivellas are dainty little shells, more slender and with a higher spire than the olives. The operculum is a thin horny scale not large enough to close the aperture. Papers by Olsson (1956) and by Burch and Burch (1963) are useful in listing species and subgeneric groups. Olsson's key is here abridged for Panamic subgenera.

1. Columellar callus stopping at end of aperture.................*Minioliva*
   Columellar callus extending above end of aperture.................... 2
2. Columellar wall partly removed from within, leaving any lirations sharply cut off at inner end... 3
   Columellar wall not excavated from within, lirations continuing to inside undiminished 4
3. Lirations present on columellar wall.......................*Olivella, s. s.*
   Lirations wanting, columellar wall smooth....................*Pachyoliva*
4. Columellar callus with one tongue-shaped central area................. 5
   Columellar callus with uniform lirations or a single basal fold.......... 7
5. Shell more than 10 mm long; markings fine.................*Dactylidella*
   Shell less than 10 mm long; markings coarse........................ 6
6. Outer lip smooth; columellar lip without any toothlike projection above
   .... *Dactylidia*
   Outer lip lirate within; columellar lip with a tooth at its upper end...*Niteoliva*
7. Columella ending in a single fold, without lirations..............*Callianax*
   Columellar callus with uniform lirations........................... 8
8. Lirations numerous, along most of columellar edge...........*Lamprodoma*
   Lirations few, short, not extending widely on callus.............*Zanoetella*

Subgenus OLIVELLA, *s. s.*

Lirations of columellar wall sharply cut off from within.

1371. Olivella (Olivella) alba (Marrat in Sowerby, 1871) (Synonym: **Oliva miriadina** Duclos, 1835; **O. myriadina** of authors, unjustified emendation). White shells, some specimens with a dark apex, with a widely grooved suture. The species has been established as West American by Olsson (1956). Length, about 8.5 mm. Margarita Bay, Baja California, to Acapulco, Mexico; possibly to Guatemala.

1372. Olivella (Olivella) altatae Burch & Campbell, 1963. A short-spired shell with a color pattern of brown zigzag lines and two dark bands just below the suture; the fasciolar band has brown and white alternating rectangular spots; pillar with seven lirae on columellar lip. Length, 8.5 mm. Altata to San Blas, Mexico. This can be confused with *O. morrisoni* because of the similar color pattern; the columella and the geographic range differ.

1373. Olivella (Olivella) aureocincta Carpenter, 1857. With about ten to twelve fine lirae on the columella; body whorl with two golden yellow bands, one just below the suture, the other just above the fasciole, between which are wavering-to-zigzag diagonal lines. Length, 8 mm. Mazatlán, Mexico (type locality). The species remained unfigured until recently (Keen, 1968) and thus is not well known. Dr. S. S. Berry (*in litt.*) reports having specimens.

1374. Olivella (Olivella) bitleri Olsson, 1956. The coloration is variable, mostly in shades of dark brown, and much resembles that of the common *O. (Lam-*

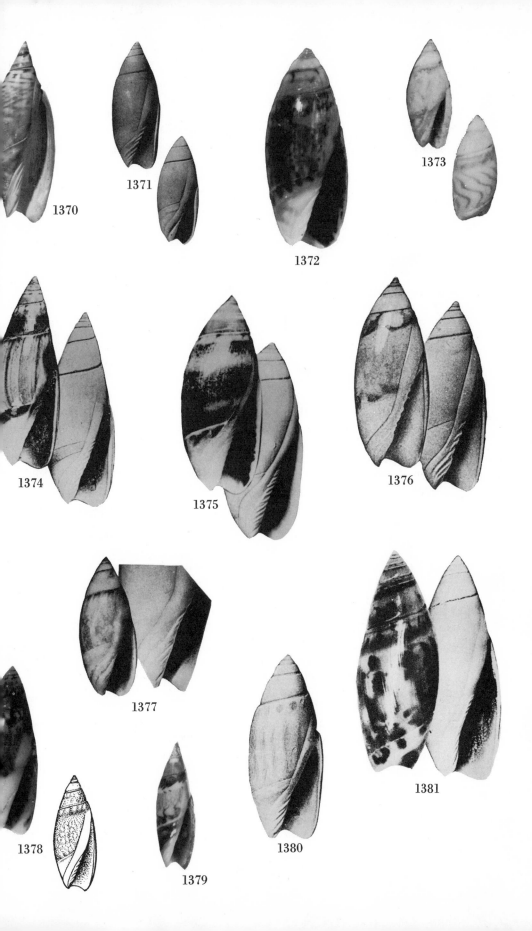

1370

1371

1372

1373

1374

1375

1376

1377

1378

1379

1380

1381

prodoma) *volutella*, from which this species is distinguished by the cutting off of the inner ends of the columellar lirations. Length, 15 mm; diameter, 5.8 mm. The species was described from Panama Bay. Specimens in the Stanford University collection extend the range to the Galápagos Islands.

1375. Olivella (Olivella) broggii Olsson, 1956. A grayish or purplish-brown shell mottled with small dots in ill-defined bands, resembling *O. tergina* but more ovate; pillar with small lirae. Length, 17.5 mm. Northwestern Peru, from Zorritos southward.

1376. Olivella (Olivella) cocosensis Olsson, 1956. Although similar to *O. gracilis*, this is much smaller, the spire shorter, and the whorls with more callus. Length, 10 mm; diameter, 4 mm. The type locality is Cocos Island, off Panama, but specimens in the Stanford University collection extend the range northward to Nicaragua.

1377. Olivella (Olivella) dama (Wood, 1828, *ex* Mawe, MS) (Synonyms: **Oliva purpurata** Swainson, 1831; **O. lineolata** Gray, 1839). This is the type species of the genus. The shell is white, variably marked with faint zigzags of brownish or grayish color, the apex of the spire and the aperture being violet. About 20 mm in length and 9 mm in diameter. The head of the Gulf of California, where it associates with *Oliva incrassata* on the outer sides of sandspits, along the west Mexican coast at least as far south as Mazatlán.

See Color Plate XVII.

1378. Olivella (Olivella) fletcherae Berry, 1958. Frequently misidentified in collections as *O. anazora* or *O. gracilis*, the slender, shiny little shell is marked with bright yellowish brown on a whitish or creamy ground. The inner lip callus does not reach the suture, and the columella has five to six small folds. Length, about 10 mm; diameter, 3.5 mm. Restricted to the northern part of the Gulf of California, from San Luis Gonzaga Bay to southern Sonora, Mexico.

1379. Olivella (Olivella) gracilis (Broderip & Sowerby, 1829) (Synonyms: **O. versicolor** Marrat in Sowerby, 1871; **O. gracilis gaylordi** Ford, 1895). A slender, white shell with brown zigzag markings in two bands, some specimens also with a brown basal area. The pillar is deeply excavated, and the columellar edge has six to seven folds. Length, 20 mm; diameter, 7 mm. Small specimens may be confused with *O. (Dactylidella) anazora*, which, however, has a single large columellar fold instead of several small ones. Guaymas, Mexico, to Panama. Burch and Burch (1963), who have studied type specimens of both the cited synonyms, report that although they appear to be distinct, no similar specimens have been found in recent years by collectors in this area; thus the two names remain of doubtful application.

1380. Olivella (Olivella) rehderi Olsson, 1956. A dull white or cream-colored shell, like *O. alba* but thinner. The pillar has four to five folds on the columellar edge. Length, 7.7 mm; diameter, 3.3 mm. Dredged in Panama Bay at a depth of 55 m.

1381. Olivella (Olivella) riverae Olsson, 1956. The shell is deep mahogany brown on a gray to yellow base, brown within, the callus white in some specimens. The pillar is a low ridge, finely lirate at the lower end of the columellar lip. The holotype measures 12.7 mm in length, 5.1 mm in diameter. Gulf of Nicoya, Costa Rica, to Peru.

1382. Olivella (Olivella) sphoni Burch & Campbell, 1963. Resembling *O. fletcherae* but with twelve lirae on pillar; white mottled with gray, with two brown bands and numerous brown blotches, the posterior band of the fasciole white with small brown dots; outer lip light brown within. Length, 14 mm. Guaymas (type locality) to Nicaragua, 18 to 37 m.

1383. Olivella (Olivella) steveni Burch & Campbell, 1963 (Synonym: ? *O. steveni campbelli* Burch & Campbell, 1963). The color is gray with two bands below the suture joined by narrow vertical bars; outer lip dark brown within; pillar with a narrow plate having eight to ten folds; fasciolar band white with brown bars. Length, 10 mm. Northern end of the Gulf of California near San Felipe (type locality) to Puertecitos, intertidally. The named subspecies, distinguished only by white rather than gray color, more intense color bands, and fasciolar bands with brown blotches, occurs at Guaymas, Sonora, 18 to 37 m.

1384. Olivella (Olivella) tergina (Duclos, 1835) (Synonym: **O. salinasensis** Bartsch, 1928). A narrowly ovate shell with a color pattern of gray or yellow blotches, and arrow-shaped white markings edged with brown. The small nucleus projects conspicuously from the rounded spire. Pillar with six to nine strong lirae; aperture short. About 16 mm in length. Magdalena Bay, Baja California, and Mazatlán, Mexico, to Zorritos, Peru.

1385. Olivella (Olivella) walkeri Berry, 1958. A whitish shell with ivory-yellow spire and fasciole and indistinct brown markings, especially below the suture. A fold bounding the canal has two smaller ones below and several above; inner lip callus heavy, wide, sharply bounded, extending to whorl above; spire short, only one-third the shell length. Length, 10.5 mm; diameter, 5.2 mm. Near Guaymas, Mexico.

Subgenus CALLIANAX H. & A. ADAMS, 1853
Columella smooth except for one fold anteriorly.

1386. Olivella (Callianax) intorta Carpenter, 1857. Smoothly inflated, columella ending in a simple fold, no lirations on inner lip; callus extending above aperture but not to suture; shell elliptical, with a short and pointed spire. The color is grayish, indistinctly banded below the suture with olive, the body whorl with irregular lines and spots of purplish brown. Length, 13 mm. Magdalena Bay, Baja California, to the southern end of the Gulf of California.

Subgenus DACTYLIDELLA WOODRING, 1928
Relatively large; columella with a tongue-shaped central area.

1387. Olivella (Dactylidella) anazora (Duclos, 1835). This species is type of the subgenus. The shell is shaped like an *Agaronia*, with a high, slender spire. The color is mostly brownish, many specimens with a fine, zigzag pattern of darker lines, the fasciole solid brown. The columella has a strong keel or fold at the end. Length, 18 mm; diameter, 7.4 mm. The Gulf of California to northern Peru (type locality).

1388. Olivella (Dactylidella) cymatilis Berry, 1963. A gray and white shell with brown markings much as in *O. anazora*, but more robust in form, with lower spire, paler color, finer and more distinct zigzag lineation. Length, 18 mm; diameter, 7.2 mm. Magdalena Bay, Baja California.

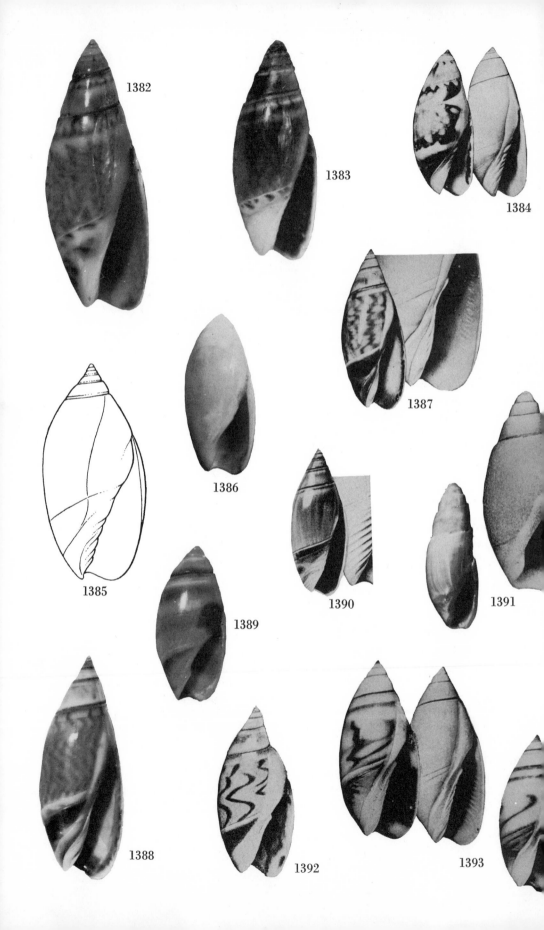

1382

1383

1384

1385

1386

1387

1388

1389

1390

1391

1392

1393

Subgenus DACTYLIDIA H. & A. ADAMS, 1853

Relatively small shells; outer lip smooth within.

1389. Olivella (? Dactylidia) zonalis (Lamarck, 1811) (Synonym: **? Oliva salasia** Duclos, 1835). Columella with one plait, as in *Callianax*, but shell much smaller. Body whorl yellow to pink-tinged, with three rusty-red spiral bands and a chestnut blotch at the upper end of the inner lip, the lower brown band at the base ending in a basal notch. Length, 4 to 7 mm. Puertecitos, near the head of the Gulf of California, to Acapulco, Mexico. The assignment to subgenus *Dactylidia* has been suggested by Olsson (*in litt.*).

Subgenus LAMPRODOMA SWAINSON, 1840

Columella with numerous lirations along its entire length.

1390. Olivella (Lamprodoma) volutella (Lamarck, 1811) (Synonym: **Oliva razamola** Duclos, 1835). Though the color varies, the commonest pattern is a uniform brown with a white fasciole. The distinguishing feature is the structure of the pillar, with fine lirations encircling the columella and running across the fasciole but not up on the face of the whorl. Length, 19 mm; diameter, 6 mm. Central America to Ecuador; most common at Panama, where shells are abundant on the mud flats.

Subgenus MINIOLIVA OLSSON, 1956

Columellar callus stopping at the upper end of the aperture.

1391. Olivella (Minioliva) inconspicua (C. B. Adams, 1852) (Synonym: "**Oliva myriadina** Duclos" of authors). A minute white shell with a turreted spire. Length, about 3.5 mm. Chiapas, Mexico (Shasky), to Panama (type locality).

Subgenus NITEOLIVA OLSSON, 1956

Outer lip lirate within; shells small.

1392. Olivella (Niteoliva) morrisoni Olsson, 1956. The ivory-white little shell has a heavy callus at the center of the inner lip and a smooth outer lip; the zigzag markings are widely spaced and dark brown. Length, 9 mm; diameter, 4 mm. San Blas, Mexico (Stanford University collection; Mrs. W. C. Frisbey, collector) to Panama Bay (type locality).

1393. Olivella (Niteoliva) peterseni Olsson, 1956. A gray shell with brown zigzag lines on a white or yellow base; callus white, fasciole doubled; lirae of inner lip more numerous than in *O. morrisoni*. Length, 13 mm. Zorritos, Peru.

Subgenus PACHYOLIVA OLSSON, 1956

Columellar wall entirely smooth.

1394. Olivella (Pachyoliva) columellaris (Sowerby, 1825). Type of the subgenus, this has a heavy callus, short, knoblike spire, and a large body whorl. The spire and fasciole are white and the body whorl has three broad brown or bluish-gray spiral bands. About 14 mm long. Nicaragua to Peru.

1395. Olivella (Pachyoliva) semistriata (Gray, 1839) (Synonyms: **Oliva attenuata** Reeve, 1850; **Olivina semisulcata** Gray, 1858; **Oliva affinis** Marrat in Sowerby, 1871). This has a higher and more slender spire than *O. columellaris*, and a faint series of vertical striae at the upper margin of the body whorl is distinctive. Length, 15 mm; diameter, 7 mm. The Gulf of California to northern Peru.

Subgenus ZANOETELLA OLSSON, 1956

Columellar lirations uniform in size but few in number.

1396. Olivella (Zanoetella) zanoeta (Duclos, 1835) (Synonym: **O. guayaquilensis** Bartsch, 1928). The *O. zonalis* of authors (not of Lamarck), a white shell with two purple-brown spiral bands. It is the type of the subgenus. About 17 mm in length. The northern part of the Gulf of California to Ecuador. The locality was originally cited, by error, as Japan.

See Color Plate XVII.

Family VASIDAE

Only one species of this widespread tropical group related to the sacred chank shells of India occurs in the Panamic province. Seeing the shell from the back, one might easily confuse it with a large *Thais*, but the heavy columellar folds in the aperture reveal its relationship to the volutes and miters. A thick periostracum covers the surface. The operculum is horny and clawlike, with an apical nucleus.

Genus VASUM RÖDING, 1798

Spire short, aperture relatively long and narrow; outline low biconic, with strong nodose sculpture, especially at the periphery.

1397. Vasum caestus (Broderip, 1833) (Synonym: **? Turbinella ardeola** Valenciennes, 1832 [? *nomen oblitum*]). Very thick and heavy for its size, the shell is white under an adherent fibrous brown periostracum. Length, 90 mm; diameter, 70 mm. A closely related form, *V. muricatum* (Born, 1778), occurs in the Caribbean. La Paz and Guaymas, Gulf of California, to Manta, Ecuador. In sand under rocks at extreme low tide. If Valenciennes' material can be proved to have come from the Panamic province, his specific name would have priority; however, it may well have been an erroneous locality record of an Indo-Pacific species.

Family MARGINELLIDAE

The marginellas are active mollusks, probably chiefly predaceous, with smooth shells, white or brightly colored, high- to low-spired, always having several columellar folds. Most are small, although on the west coast of Africa, where the group seems to attain its metropolis, shells of several species measure more than 25 mm (one inch) in length. One South American species and one in Brazil may have shells 125 mm (5 inches) long. The family is mainly tropical to subtropical in distribution. Frequency estimates are based on total number of lots in major collections.

The text for this family has been prepared by Dr. Eugene V. Coan and Mr. Barry Roth, whose assistance is gratefully acknowledged. Useful references: Tomlin (1917), Coan (1965), Coan and Roth (1966), Graham (1966), Roth and Coan (1968), Marcus & Marcus (1968), Ponder (1970). The key is to the genera within the two Panamic subfamilies.

1. Spire high, a little less than half total length.................*Dentimargo*
   Spire low to concealed... 2
2. Aperture with a distinct anterior notch, visible from dorsal (back) view.... 3
   Aperture with anterior end rounded................................. 4

3. Shell small to minute, less than 6 mm long......................*Granula*
   Shell medium-sized, over 6 mm long..........................*Persicula*
4. Small to medium-sized, over 6 mm long, usually colored................ 5
   Small to minute, less than 6 mm long, white......................... 6
5. Shell more than 10 mm long, globose to ovate-elongate.............*Prunum*
   Shell less than 10 mm long, cylindrical.......................*Volvarina*
6. Outer lip smooth within; spire low but not concealed by outer lip....*Cysticus*
   Outer lip denticulate within, its posterior end concealing spire.....*Granulina*

Subfamily MARGINELLINAE

Shells small to large (1 to 90 mm); spire medium to low or concealed; surface polished, white to brightly colored. Recent anatomical work demonstrates that some members of this subfamily have no radula, while most have a single rachidian plate with from 8 to 40 cusps. Eventual division into two subfamilies seems probable. Four Panamic genera (see key, above).

Genus PRUNUM HERRMANNSEN, 1852

Ovate to elongate, with a medium to low spire, and a callus pad on body whorl near posterior end of aperture.

Subgenus PRUNUM, s. s.

Spire low; outer lip not greatly thickened, smooth within; anterior canal shallow.

1398. Prunum (Prunum) curtum (Sowerby, 1832). Ashy yellow with mottled white spiral banding, the shell is tinged with dull orange outside the outer lip and along the anterior margin. Shorter than *P. sapotilla*, with heavier callus around the aperture, more divergent columellar folds, and a flattened base to the columella. Length, 21 mm. Manta, Ecuador, to Iquique, Chile, in 2 to 20 m, on sand; not common.

1399. Prunum (Prunum) sapotilla (Hinds, 1844) (Synonym: **Marginella evax** Li, 1930). Ashy yellow, sometimes with faint spiral banding, tinged with dull orange outside the outer lip. The largest Panamic marginellid; shells may reach a length of 27 mm. Bahía Honda to Bella Vista and Isla Pedro Gonzales, Panama, intertidally and to 60 m, on mud; not common. A twin or homologous species (an "analogue" of authors) in the Caribbean is *P. (P.) prunum* (Gmelin, 1791).

1400. Prunum (Prunum) woodbridgei (Hertlein & Strong, 1951). The shell is slate-gray, with two faint darker bands and with a trace of orange on the apex, the outside of the outer lip, along the suture, and between the columellar folds. Aperture brown within. Length, 12 mm. Known only from San José, Guatemala, in 23 m; rare. A similar Caribbean species is *P. (P.) apicinum* (Menke, 1828).

Genus DENTIMARGO COSSMANN, 1899
(VOLVARINELLA HABE, 1951)

Small, biconic, generally with a strong denticle on the posterior part of the outer lip and a series of smaller ones anteriorly.

1401. Dentimargo eremus (Dall, 1919) (Synonym: **Marginella anticlea** Dall, 1919; first revisers: Coan & Roth, 1966). White, circled by two narrow brown bands. The height of the spire varies somewhat. Length, 5 mm. Near the

1394

1395

1396

1397

1398

1399

1400

1401

1402

1403

1404

Galápagos Islands, in 80 to 1,300 m; rare. A twin or homologous species in the Caribbean is *D. aureocinctus* (Stearns, 1873).

Genus PERSICULA SCHUMACHER, 1817

Medium-sized to small, with a flat spire that is sometimes concealed by callus, and with a strong anterior notch; most are brightly marked with spots or lines of color.

1402. Persicula accola Roth & Coan, 1968 (Synonym: **Voluta porcellana** Gmelin, of authors, not of Gmelin, 1791). Pale yellowish tan, with spiral rows of brown rectangular spots and a brown tinge along the edge of the outer lip. Length, 13 mm. Bahía Montijo, Isla Coiba, and Isla Jicaron, Panama, intertidally; rare. The true *P. porcellana* (Gmelin) is from the Caribbean and is a twin or homologous species.

1403. Persicula bandera Coan & Roth, 1965. The yellowish-tan shell is ringed by 15 to 17 brown lines, and there is brown along the outer lip. Length, 12.5 mm. Collected at Bahía Banderas, Mexico, in 3 to 10 m, in mud; rare. A twin or homologous species in the Caribbean is *P. multilineata* (Sowerby, 1846).

1404. Persicula hilli (M. Smith, 1950). A grayish, solid shell with a brown band at the suture and a few indistinct bands on the body whorl. The outer lip and the columella are touched with brown. Length, 15 mm. Bahía Chamela, Mexico, in 30 to 80 m; rare.

1405. Persicula imbricata (Hinds, 1844) (Synonyms: **Marginella vautieri** Bernardi, 1853; **P. dubiosa** Dall, 1871; **P. adamsiana** Pilsbry & Lowe, 1932). White to yellowish, with reddish-brown dashes of varying size arranged in spiral bands and, to a lesser degree, in axial columns. Bands of heavier dashes may divide the shell into thirds. Brown markings may also appear on the columella, around the spire, and outside the outer lip. Length, 11 mm. Cabo San Lucas, Baja California, and Tenacatita, Jalisco, Mexico, to La Libertad, Ecuador, and the Galápagos Islands, in 16 to 60 m, in rubble; not common. This species is variable, populations differing conspicuously from one another, but none of these seems isolated enough to warrant subspecific recognition. A related Caribbean species is *P. muralis* (Hinds, 1844) (? = *P. maculosa* [Kiener, 1834]).

1406. Persicula phrygia (Sowerby, 1846). White or pale straw-colored, with reddish-brown markings in the shape of horseshoes arranged in rows both spirally and axially, largest and darkest on the periphery and just above the anterior canal. Length, 6.5 mm. Bahía Magdalena and Puertecitos, Baja California, to Panama and the Galápagos Islands, intertidally and to 20 m; not common. A similar Caribbean species is *P. swainsoniana* (Petit, 1851).

Genus VOLVARINA HINDS, 1844
(HYALINA of authors)

Small, cylindrical to conic, without an anterior notch. The spire may be low to slightly elevated.

Subgenus VOLVARINA, *s. s.*

Cylindrical, spire low, outer lip smooth or only weakly denticulate within.

1407. Volvarina (Volvarina) .... Roth & Coan, MS. Small, polished, translucent white, with three spiral orange-brown bands; outer lip white, thickened along edge, color bands showing in aperture; columella with four equally spaced sharp folds. Length, 5.7 mm; width, 3 mm. Type locality, Isla Pinta (Abingdon), Galápagos Islands, Ecuador, intertidally, rare; also on Isla Genovesa, Galápagos. Distinguished from *V. taeniolata taeniolata* by the smaller size and from *V. t. rosa* by the color banding.

1408. Volvarina (Volvarina) taeniolata taeniolata Mörch, 1860 (Synonyms: **Marginella varia** Sowerby of authors, not of Sowerby, 1846; **M. californica** Tomlin, 1916; **M. californica parallela** Dall, 1918). White or yellowish, with three brown bands more or less evident. Specimens from the southern end of the Gulf of California may be pale, with a relatively high spire. Length, 9 mm. Point Conception, California, to Salinas, Ecuador, intertidally and to 40 m; not common in the Panamic province. A twin or homologous Caribbean species is *V. avena* (Kiener, 1834, *ex* Valenciennes, MS).

1409. Volvarina (Volvarina) taeniolata rosa (Schwengel, 1938). This subspecies differs from the typical form by having the shell bright pink in color, with the spiral banding only faintly developed. Length, 8.3 mm. Galápagos Islands, intertidally and to 5 m.

Subfamily CYSTISCINAE

Shells minute to small (0.5 to 5 mm); spire low to concealed; surface polished, white; outer lip thickened but not reflected. Radula present as narrow rachidian plates with few cusps. Three Panamic genera.

Genus CYSTISCUS STIMPSON, 1865

Small shells, with a low but unconcealed spire and a smoothly rounded anterior end without a notch. The outer lip is smooth within.

1410. Cystiscus palantirulus Roth & Coan, 1968. The pear-shaped shell is about 3 mm long, shouldered, with the anterior end twisted to the left as seen in dorsal view. Length is 1.6 times the diameter. Isla Monserrate to Cabo San Lucas, Baja California, in 6 to 80 m; rare. Similar to the Californian *C. jewettii* (Carpenter, in Gould & Carpenter, 1857).

1411. Cystiscus politulus (Dall, 1919) (Synonyms: **Marginella regularis** Carpenter of authors, not of Carpenter, 1864; **Hyalina myrmecoon** Dall, 1919; first revisers: Roth & Coan, 1968). Less than 3 mm long, elongate and not shouldered. The anterior end is evenly rounded, and the length is 1.8 times the diameter. Santa Barbara, California, to Cabo San Lucas, Pulmo Reef, and Isla Cerralvo, Baja California, intertidally and to 60 m; not common. A similar Caribbean species is *C. bocasensis* (Olsson & McGinty, 1958), the only member of the genus thus far reported from the western Altantic.

Genus GRANULA JOUSSEAUME, 1875
(KOGOMEA HABE, 1951)

This genus differs from *Cystiscus* in having a distinct notch at the anterior end, often most easily seen in dorsal view. The outer lip frequently is denticulate within.

1412. Granula . . . . Roth & Coan, MS. Minute, white or colorless under a bluish-white glaze, polished, surface with narrow raised ridges trending somewhat obliquely across body whorl, fainter anteriorly. Length, 2.6 mm; width, 1.6 mm. Isla del Coco (Cocos Island), Costa Rica, depth 25 m, not common.

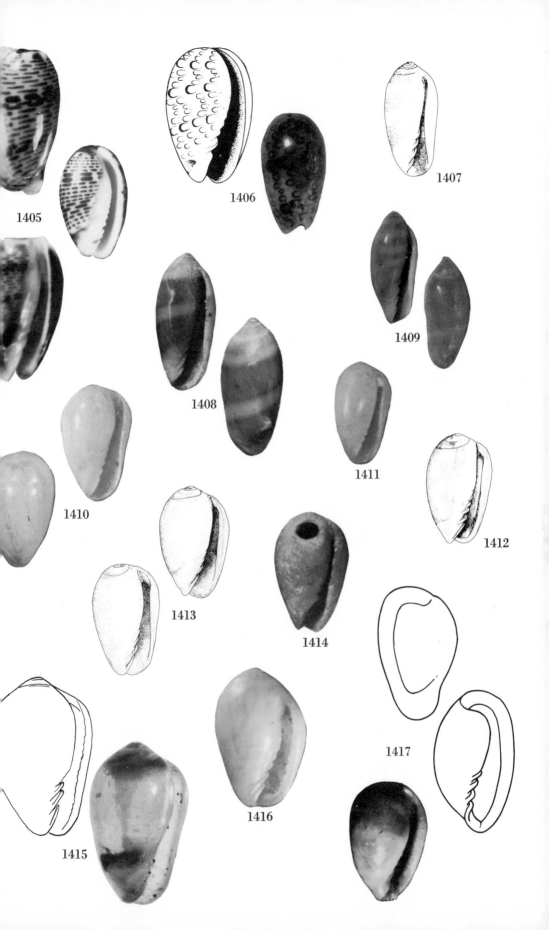

1405

1406

1407

1408

1409

1410

1411

1412

1413

1414

1415

1416

1417

Distinguished from other small marginellids by the surface ornamentation, which is uncommon in this family. The species has been misidentified as *Marginella minor* by some authors.

1413. Granula . . . . Roth & Coan, MS. Minute, white, polished, and without sculpture; elongate-ovate, with a very low spire that is evenly dome-shaped; outer lip smooth within, thickened at edge, especially posteriorly; columella with five folds, inner lip slightly granular near the folds; siphonal notch broad. Length, 3.3 mm; width, 2 mm. Tagus Cove (type locality), Isla Isabela (Albemarle), Galápagos Islands, intertidally, rare; also on three other islands of the Galápagos group.

1414. Granula minor (C. B. Adams, 1852). Most conical of the Panamic species, with a low spire. The outer lip is conspicuously denticulate within. Length, 2.3 mm; width, 1.5 mm. Puerto Parker, Costa Rica, to La Libertad, and the Galápagos Islands, Ecuador, intertidally and to 120 m depth; not common.

1415. Granula polita (Carpenter, 1857) (Synonym: **Marginella morchii** Redfield, 1870, new name for **Gibberula coniformis** Mörch, 1860 [not *M. coniformis* Sowerby, 1850]). A thin shell with small teeth scarcely obvious on the outer lip. Length, 2 mm. White's Point, San Pedro, California, through the Gulf of California, to Costa Rica, possibly to Panama, intertidally and to 100 m; not common. The Caribbean *G. lavalleeana* (Orbigny, 1841) resembles this species.

1416. Granula subtrigona (Carpenter, 1864) (Synonyms: **Marginella regularis** Carpenter, 1864; **M. oldroydae** Jordan, 1926). Solid, with a thickened and denticulate outer lip and a fairly high spire. Length, 3.5 mm. Monterey, California, to Cabo San Lucas and La Paz, Baja California, intertidally and to 100 m; not common.

Genus GRANULINA JOUSSEAUME, 1888
(CYPRAEOLINA CERULLI-IRELLI, 1911; MEROVIA DALL, 1921)

Spire concealed by body whorl; outer lip smooth or denticulate within. At present, it is possible to distinguish morphologically only one wide-ranging species in the eastern Pacific.

1417. Granulina margaritula (Carpenter, 1857) (Synonym: **Volutella pyriformis** Carpenter, 1864). The long aperture and concealed spire are diagnostic. Length, 2.5 mm. Alaska, along the entire west coast, through the Gulf of California to Panama and the Galápagos Islands, intertidally and to 110 m; not uncommon. A related Caribbean species is *G. ovuliformis* (Orbigny, 1841).

Superfamily MITRACEA

Ovate to elongate, mostly sturdy shells, with or without periostracum; columella with folds. The radula is rachiglossate, but the presence of a poison gland in most of them points toward relationship with the Conacea.

Family MITRIDAE

The miter shells, which in the Indo-Pacific province run to bright colors and numerous species, are, in the Panamic province, mostly of somber hues and reduced numbers. They are offshore forms mainly, and the coveted large ones must be obtained by dredging. The uppermost folds on the columella are heaviest. Useful references: Sphon (1961, 1969). Assistance in the organization of material by Dr. Eugene V. Coan is gratefully acknowledged. Cernohorsky (1970) has discussed the systematics of the family. The key is to the genera and subgenera in the three subfamilies of the Panamic Mitridae.

1. Sculpture of raised spiral ribs, narrower than the interspaces.....*Subcancilla*
   Sculpture not of ribs that are narrower than interspaces................ 2
2. Aperture short, flaring anteriorly...............................*Isara*
   Aperture nearly half the shell length............................ 3
3. Shell brightly colored, with irregular blotches.................*Mitra, s. s.*
   Shell not brightly colored, any color present in spiral bands.............. 4
4. Shell smooth or nearly so, with a conspicuous black periostracum...*Atrimitra*
   Shell sculptured, periostracum inconspicuous........................ 5
5. Outer lip smooth within; shell medium-sized..................*Strigatella*
   Outer lip lirate within; shell small..............................*Thala*

Subfamily MITRINAE

Smooth to cancellately sculptured or pitted; radula with a small rachidian tooth, laterals elongate, multicuspid.

Genus MITRA LAMARCK, 1798 (Sept.)
(MITRA RÖDING, 1798 [Dec.] of authors)

Medium-sized to large, smooth or with spiral grooves or pits; outer lip smooth within; columella with three to seven folds. Four, possibly five, Panamic subgenera; see key, above.

Subgenus MITRA, s. s.

Shells light colored, with bright color markings of regular to irregular blotches.

1418. Mitra (Mitra) papalis (Linnaeus, 1758). An Indo-Pacific species reported on Clipperton Island.

Subgenus ATRIMITRA DALL, 1918

Medium-sized to large shells, mostly smooth, some with spiral grooves or punctations; periostracum persistent, shell usually dark. *Fuscomitra* Pallary, 1900, may prove to be an older name for this group; *Episcomitra* Monterosato, 1917, also may be congeneric; the types of both are Mediterranean species, whereas the type of *Atrimitra* is West American.

1419. Mitra (Atrimitra) belcheri Hinds, 1843. Largest and most sought-after of the miters of the eastern Pacific, *M. belcheri* is a handsome form with broad, flat spiral ribs and narrow, deeply cut grooves. The shell is white under a dark, greenish-black periostracum. A large specimen measures, length, 120 mm; diameter, 30 mm (record size, length, 148 mm; diameter, 49 mm). Off Magdalena Bay, Baja California, and Mazatlán, Mexico, to Panama, in depths of 37 m or more.

1420. Mitra (Atrimitra) fultoni E. A. Smith, 1892. The black shell—which otherwise resembles *M. idae* Melvill, 1893, from California, type of the subgenus—is distinguished by its sculpture of regularly spaced punctations in incised spiral grooves, the pits lining up vertically along well-marked growth lines. Length, 36 mm; diameter, 13 mm. Baja California, from San Martin Island, on the outer coast, to Puertecitos, Gulf of California (Sphon, 1961); also Pleistocene of southern California.

1421. Mitra (Atrimitra) swainsonii Broderip, 1836 (Synonyms: **M. mexicana** Dall, 1919; **M. zaca** Strong, Hanna & Hertlein, 1933). Under a thin black periostracum the buff to ivory-white shell is nearly smooth, with only a few spiral threads on the shoulders and base of the whorls. In size it comes close to *M. belcheri*, but it is usually more slender. Large specimens tend to develop tabulation on the upper part of the whorls; for a time it seemed two species might be recognized, the tabulate *M. swainsonii* in Ecuador and *M. zaca*, a more slender northern form. However, larger suites of dredged material from Panama northward blur such a distinction. *M. mexicana* proves to be the juvenile stage. Length, 130 to 135 mm; diameter, 27 to 36 mm. Guaymas, Mexico, to Ecuador, in depths of 5 to 73 m.

Subgenus ISARA H. & A. ADAMS, 1853

High-spired, the aperture short, flaring anteriorly; spirally striate.

1422. Mitra (Isara) effusa Broderip, 1836. The spiral sculpture is of fine threads somewhat alternating in size. The shell is orange-brown under a chestnut-brown periostracum. Length, 33 mm; diameter, 9 mm. The Gulf of Tehuantepec, Mexico, to Ecuador and the Galápagos Islands.

Subgenus STRIGATELLA SWAINSON, 1840

Medium-sized shells, smooth to spirally striate or with upper whorls cancellate; aperture about one-half shell length, anterior canal short. Some of these species may prove to belong elsewhere in the Mitridae when the radulas are studied.

1423. Mitra (? Strigatella) crenata Broderip, 1836. A small brown shell with slightly raised threadlike spiral ribs crossed by a few low axial striae. Length, 9 to 11 mm; diameter, 4 to 4.5 mm. Guaymas, Mexico, to Ecuador (type locality), in depths of 5 to 55 m.

1424. Mitra (Strigatella) gausapata Reeve, 1845. Nearly black below, the shell has a band of yellowish white between the suture and the slightly angulate shoulder, with three or four fine spiral lines of the dark color penciled on it; spiral sculpture is stronger on the base than on the rest of the shell, and axial sculpture is of rather undulating ill-defined ribs that die out on the base. Length, 19 mm; diameter, 7 mm. Galápagos Islands.

1425. Mitra (Strigatella) inca Orbigny, 1841. Because of its southern type locality, this species was only recently recognized in the Panamic province. The shell is grayish brown, its spiral ribs evenly beaded at the intersections of axial sculpture that dies out on the base. Length, 28 mm; diameter, 12 mm. Guaymas, Mexico, to Paita, Peru (type locality), in depths to 18 m. Cernohorsky (1970) considers *M. nucleola* Lamarck, 1811, an earlier name; Lamarck's species is more likely the Caribbean *M. barbadensis* (Gmelin, 1791).

1426. Mitra (Strigatella) lens Wood, 1828 (Synonyms: **Tiara foraminata** Broderip, 1836; **M. dupontii** Kiener, 1839). Bluish gray to brownish under a dark brown periostracum, this rather coarse-looking shell is probably the commonest miter of the province. The deep, square pits in one or two spiral rows are distinctive features. Length, 40 mm; diameter, 18 mm. The head of the Gulf of California to Peru, in sand and under rocks at extreme low tide.

1427. Mitra (Strigatella) semigranosa Von Martens, 1897. Somewhat resembling *M. inca*, the grayish-brown shell has evenly cancellate spire whorls, but

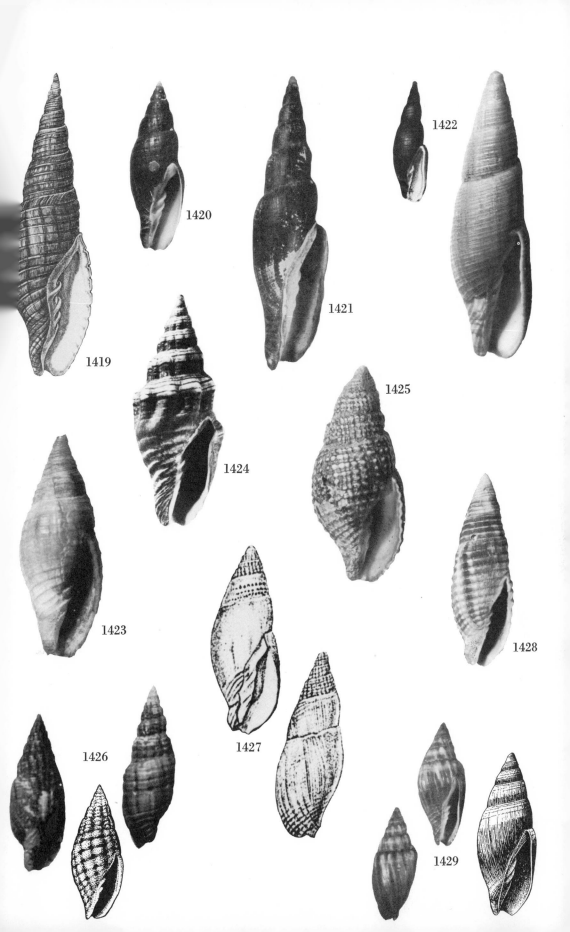

1419

1420

1421

1422

1424

1425

1423

1427

1426

1428

1429

sculpture dies out on the body whorl. The aperture is white within. Length, 45 mm; diameter, 16 mm. Ecuador to Chile.

1428. Mitra (? Strigatella) sphoni Shasky & Campbell, 1964. Larger than *M. crenata*, though resembling it, this has well-developed spiral ribs wider than their interspaces. It lacks any axial striae and is tan under a dark brown periostracum. Length, 23 mm; diameter, 8 mm. Off Guaymas, Sonora, Mexico, 2 to 31 m.

1429. Mitra (Strigatella) tristis Broderip, 1836 (Synonyms: **M. dolorosa** Dall, 1903; **M. salinasensis** Bartsch, 1928). The names *tristis* and *dolorosa* suggest sadness. The color is a drab olive-brown, with a lighter band on the upper part of the whorl and a few low axial ribs at the shoulder that fade out on the lower part of the body whorl. The supposed differences between *M. tristis* and the presumably larger and more slender northern *M. dolorosa* are within the limits of variation of the species and not consistent. Length, 21 to 28 mm; diameter, 8 to 10 mm. The northern end of the Gulf of California to Ecuador and the Galápagos Islands. Cernohorsky (1970) suggests that the name *Mitra olivacea* Anton, 1839, may prove to be another synonym; also *M. jousseaumiana* Mabille, 1898, described as from "California."

Genus MITRA, *s. l.*

Several Panamic species having strongly developed axial and spiral sculpture do not fit readily into any of the named mitrid subgenera, most of which are based on Indo-Pacific type species. Before a special subgenus is namd to accommodate them, studies of the soft parts and especially of the radula are required. The name *Scabricola*, formerly used, is inappropriate, for the type species of that group proves to be a member of another subfamily, the Imbricariinae.

1430. Mitra lignaria Reeve, 1844. Somewhat resembling *M. (Strigatella) lens*, the shell is more slender, has regular spiral sculpture and strong axial ribs, but no pits at the intersections. It is yellowish orange under a dark periostracum. Length, 48 mm; diameter, 16 mm. Guaymas, Mexico, to Ecuador, offshore in depths to 26 m.

1431. Mitra marshalli Bartsch, 1931. A dark brown shell with shouldered whorls and rather coarsely cancellate sculpture, its spiral ribs cut into elongate nodes. Length, 14 mm; diameter, 6 mm. Panama.

1432. Mitra muricata Broderip, 1836. Although the shell has shouldered whorls, the spire is more slender than that of *M. marshalli*. It is brown under a very dark brown periostracum. Length, 23 mm; diameter, 10 mm. Central America to Ecuador and the Galápagos Islands, mostly offshore to depths of 11 m.

1433. Mitra rupicola Reeve, 1844. The shell is pinkish brown under an olive-brown periostracum. The spiral ribs are narrow and evenly spaced, rising to long nodes on the axial folds. Length, 35 mm; diameter, 15 mm. Panama to Ecuador, intertidally, at extreme low tides, and offshore.

Subfamily IMBRICARIINAE

Smooth to spirally grooved; radula with a relatively large central tooth, laterals variable, relatively few-cusped.

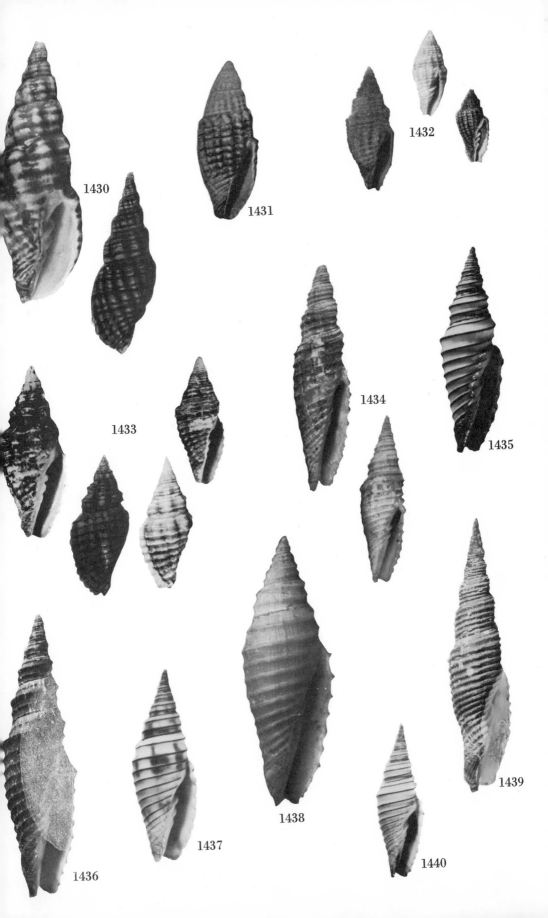

Genus **SUBCANCILLA** OLSSON & HARBISON, 1953
(**TIARA** of authors, not of SWAINSON, 1831)

Biconic to slender, with raised spiral ribs narrower than their interspaces; axial sculpture reduced to lirations between spiral ribs; periostracum thin and light-colored or absent. The species of this group have been ranked in the subgenus *Tiara*, but unfortunately the earliest type designation for this taxon removes it to another setting. The name *Cancilla* Swainson, 1840, based on an Indo-Pacific species, could take its place. However, so long as a name change from *Tiara* is required, it seems preferable to adopt *Subcancilla*, which is based on an eastern Pacific species. Differences from *Mitra* in shell and soft parts are sufficient to warrant elevation of the unit to generic rank.

1434. Subcancilla attenuata (Broderip, 1836). The shell is long and gently tapering, with two bands of darker color, a broad one on the middle of the spire whorls and a narrower one about halfway down the aperture. There are four spiral ribs per whorl, the uppermost weak, so that the whorls have a slightly shouldered appearance. The type lot in the British Museum shows some variation in the depth of color. Length, 39 mm; diameter, 10 mm. Panama Bay, 37 m.

1435. Subcancilla calodinota (Berry, 1960). Slightly wider than *S. hindsii*, this has brighter color, brownish gray, with a pinkish-brown spot in the aperture. The three upper intercostal spaces are white, the one below the suture brown. Height, 27 mm; diameter, 8.7 mm. Guaymas to the Gulf of Nicoya, Costa Rica, offshore in depths of 18 to 46 m.

1436. Subcancilla directa (Berry, 1960). Similar to *S. attenuata*, with more angulate whorls; the columellar folds steep and the spiral cords dark. A fawn-colored periostracum is present that is unusually persistent. Length, 32 mm; diameter, 8 mm. Off Guaymas, Sonora, Mexico, 37 to 91 m.

1437. Subcancilla erythrogramma (Tomlin, 1931) (Synonym: **Mitra lineata** Broderip, 1835 [not Schumacher, 1817]). The whitish shell not only has the spiral ribs brown but also has brown axial markings here and there as vague stains and streaks. Length, 25 mm; diameter, 9 mm. Bahía San Luis Gonzaga to Colombia, offshore in depths of 18 to 37 m.

1438. Subcancilla funiculata (Reeve, 1844). Nearest to *S. sulcata* in rib pattern, both having white spiral cords, this form differs in outline; the aperture is long and narrow, nearly three-fourths the length of the shell. Length, 26 mm; diameter, 9 mm. Near the southern end of the outer coast of Baja California to Guatemala, in depths of 14 to 24 m.

1439. Subcancilla gigantea (Reeve, 1844, *ex* Swainson, MS). The specific name is well chosen, for this is the largest of the Panamic Subcancillas. It is a buffy white with a thin greenish-brown periostracum and strong spiral ribs. Length, 70 mm; diameter, 16 mm. Panama to Ecuador.

1440. Subcancilla hindsii (Reeve, 1844). The spiral ribs are almost keel-like, especially one in the middle of the whorl on the spire and on the shoulder of the body whorl. Some specimens show fine vertical striations; others are smooth. The shell is pinkish white under an olive periostracum, the spiral ribs brown. Length, 38 mm; diameter, 11 mm. Santa Margarita Island, Baja California, and the head of the Gulf of California southward to Ecuador, mostly offshore, in depths of 11 to 51 m. Though considered identical with *M. gigantea* by some authors, this is clearly separable.

1441. Subcancilla lindsayi (Berry, 1960). Having the general coloring of *S. hindsii*, the shell is more shouldered, with a sinuous outline; the spirals are pale brown. Length, 23 mm; diameter, 8 mm. Puerto Peñasco to Guaymas, Sonora, Mexico, in depths of 18 to 37 m.

1442. Subcancilla malleti (Petit de la Saussaye, 1852). Axial striae are especially well developed, and an incised spiral line midway between the spiral ribs emphasizes them. The shell itself is rosy brown with an olive periostracum, the aperture flesh-colored. Length, 24 mm; diameter, 10 mm. The species was described without known locality but has been recognized at Guaymas, Sonora, Mexico (Berry, 1964).

1443. Subcancilla phorminx (Berry, 1969). The slender shell has shouldered whorls and flat, widely spaced spiral cords, with fine axial striae. The shell is white, the spiral cords yellowish brown. The area between suture and shoulder is set off by a color band of yellow or light brown. Length, 40 mm; diameter, 10.5 mm. Off Rio Bolsa, near Acapulco, Guerrero, Mexico, in 65 to 80 m.

1444. Subcancilla sulcata (Swainson in Sowerby, 1825) (not preoccupied by *Voluta sulcata* Gmelin, 1791) (Synonym: **? M. haneti** Petit de la Saussaye, 1852). The shell varies in color from white with a few brown spots to light brown, under a brown periostracum. The spiral ribs are not colored as brightly as in *S. hindsii*, and the shell is less tapering. Length, 21 to 32 mm; diameter, 8 to 12 mm. The southern part of the Gulf of California to Ecuador. Type species of *Subcancilla*.

Subfamily VEXILLINAE

Axial sculpture well developed, especially on spire; outer lip lirate within; periostracum, if present, inconspicuous.

Genus THALA H. & A. ADAMS, 1853
(MICROMITRA BELLARDI, 1888; MITROMICA BERRY, 1958)

Small, spire as long as or longer than aperture; sculpture cancellate. The genus has been reviewed by Sphon (1969), especially as to eastern Pacific records.

1445. Thala gratiosa (Reeve, 1845) (Synonyms: **Mitra solitaria** of authors, not of C. B. Adams, 1852; **M. nodocancellata** Stearns, 1890). Small, black, somewhat cylindrical in outline, sculpture cancellate, aperture narrow, with a clearly set-off anterior canal, outer lip lirate within, anal sulcus more evident in older specimens. Length, 11.5 mm; diameter, 3 mm. Throughout the Gulf of California and south to Panama and the Galápagos Islands (type locality).

1446. Thala jeancateae Sphon, 1969. The rather tapering spire is white, the base below the middle of the aperture brown, the outer lip white within. The anterior canal is slightly recurved and the aperture moderately narrow. Length, 9.4 mm; diameter, 3.8 mm. Albemarle (Isabela) Island, Galápagos, in 91 to 110 m.

1447. Thala solitaria (C. B. Adams, 1852). The shell is black, some specimens with white nodes, larger and less cylindrical than *T. gratiosa*. Aperture moderately narrow, canal slightly recurved. A subsutural band is present. Length, 17 mm; diameter, 6 mm. Banderas Bay, Jalisco, Mexico, to Panama (type locality) and the Galápagos Islands; intertidally, under rocks.

Family CANCELLARIIDAE

The members of this family have (as the name suggests) cancellate sculpture on at least some part of the shell: narrow axial and spiral ribs intersecting to form a latticework. The columella develops a few strong folds. The aperture is moderately large, the anterior canal short. Olsson (1970) has shown that the radula, which had by some authors been considered toxoglossate or like that of the Turridae (because the teeth seem to be brushlike and marginal), is really a complex, unique structure. The usual stenoglossate ribbon is lacking; instead, the radula is made up of several parts, with long anterior and posterior filaments that may represent modified teeth. Olsson proposes a new name, Nematoglossa, for this group, which he would thus consider a separate suborder. Nothing is known about the food habits of the group, but Olsson suggests, from the radular structure, that the animals must feed on soft-bodied microorganisms, which could be loosened from the sea floor by action of the filaments. The cancellariids are mostly offshore forms; thus, the study of their living habits will be difficult.

An early paper by Jousseaume, in 1887, laid the groundwork for the modern classification of the Cancellariidae. Marks (1949) has summarized the classification and has indicated some characteristics of the shell that seem useful in subdividing especially the West American representatives, but work still remains to be done on the classification. Three genera, *Cancellaria*, *Perplicaria*, and *Trigonostoma*, are recognized here, with subgenera, especially in *Cancellaria*. Mr. Richard E. Petit of Ocean Drive Beach, South Carolina, specialist on the family, has advised on synonymies and reallocations of species in the present revision, and his assistance is gratefully acknowledged. The key is to genera and subgenera.

1. Columella straight or bent to left; diameter of body whorl less than or equal to length of aperture.. 2
   Columella bent toward outer lip; diameter of body whorl greater than length of aperture...16
2. Outline slender, diameter half the shell height or less.................. 3
   Outline ovate to globose, diameter more than half the height............ 6
3. With irregular varices in addition to axial ribs....................*Narona*
   Without varices, axial ribbing regular.............................. 4
4. Spire equal to aperture in length; columella pointed............*Hertleinia*
   Spire longer than aperture; columella blunt......................... 5
5. Inner lip with a wide callus....................................*Aphera*
   Inner lip not markedly calloused............................*Perplicaria*
6. Base umbilicate, with a wide siphonal fasciole....................... 7
   Base imperforate or with only a slight umbilical chink.................13
7. Spire blunt, rounded ....*Ovilia*
   Spire pointed but not elevated..................................... 8
8. Periostracum well developed, in tufts at periphery...............*Agatrix*
   Periostracum wanting 9
9. Spire low, aperture nearly circular.................................10
   Spire moderately high, aperture narrow to crescentic..................11
10. Margins of aperture white....................................*Bivetiella*
    Margins of aperture brightly colored........................*Bivetopsia*

11. Sculpture cancellate, whorls rounded....................*Cancellaria, s. s.*
    Sculpture strongly clathrate, whorls shouldered........................12
12. Aperture not flaring anteriorly, wider above midline..............*Bivetia*
    Aperture flaring anteriorly, wider below midline.................*Solatia*
13. Inner lip without callus.......................................*Massyla*
    Inner lip with well-developed callus..............................14
14. Body whorl smooth ....*Pyruclia*
    Body whorl sculptured...15
15. Axial ribs numerous, spirals moderately heavy...................*Euclia*
    Axial ribs few, widely spaced, spirals weak.....................*Sveltia*
16. Whorls disjunct, not in contact at suture.....................*Extractrix*
    Whorls not disjunct, suture in contact............................17
17. Axial ribs spirally corded, varix-like........................*Olssonella*
    Axial sculpture laminar to faint, not spirally corded or varix-like...*Ventrilia*

Genus CANCELLARIA LAMARCK, 1799

Whorls rounded to angulate at shoulder but not concave; coiling tight; aperture rounded to ovate. Fourteen Panamic subgenera; see key, above.

Subgenus CANCELLARIA, s. s.

Sculpture evenly reticulate; early whorls cancellate.

1448. Cancellaria (Cancellaria) albida Hinds, 1843. A slender white shell with a few brown lines. An average specimen measures 21 mm in length, 11 mm in diameter. The range is from off Maldonado Point, Mexico, to the Bay of Guayaquil, Ecuador, mostly offshore in depths to 128 m.

1449. Cancellaria (Cancellaria) darwini Petit, 1970. A cream-colored shell with evenly spaced brown spiral ridges. Axial sculpture is fine and weak, somewhat cancellate or beaded at intersections of spiral riblets. Columella with three plaits, strongest above; aperture white. Length, 19 mm; diameter, 10 mm. Academy Bay, Isla Santa Cruz, Galápagos Islands, depth 170 to 200 m.

1450. Cancellaria (Cancellaria) decussata Sowerby, 1832. Reddish brown to fawn-colored, the shell may have a band of white about the middle of the somewhat globose body whorl. Length, 27 mm; diameter, 17 mm. Magdalena Bay, Baja California, through the southern part of the Gulf of California and south to Cape Pasado, Ecuador, offshore in depths to 80 m.

1451. Cancellaria (Cancellaria) gemmulata Sowerby, 1832 (Synonym: **C. emydis** Dall & Ochsner, 1928). The general color is light brownish with two darker brown bands, one at the middle, the other at the base of the last whorl. Length, 17 mm; diameter, 12 mm. Gulf of California to Panama Bay and the Galápagos Islands, offshore in depths to 73 m.

1452. Cancellaria (Cancellaria) obesa Sowerby, 1832 (Synonym: **C. acuminata** Sowerby, 1832). The somewhat globose white shell is obscurely banded by three broad areas of brown. Length, 48 mm; diameter, 35 mm. A slender variant form measures 45 mm in length, 27 mm in diameter. Cedros Island, Baja California, through the Gulf of California and south to Ecuador, mostly offshore in depths to 90 m (it has been taken intertidally at Puertecitos near the head of the Gulf of California).

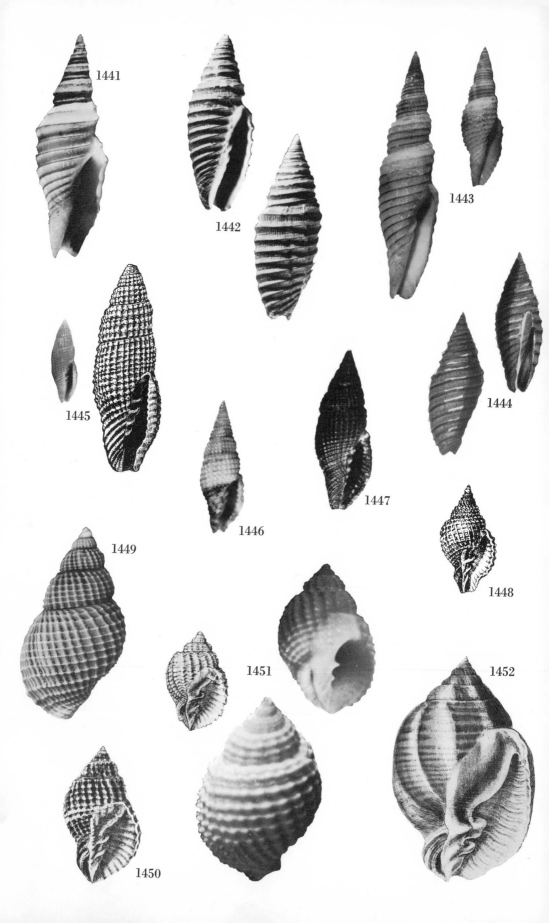

1453. Cancellaria (Cancellaria) ovata Sowerby, 1832. Although synonymized with *C. obesa* by some authors, this seems to be distinct. The spire is lower and the color darker, tending toward reddish brown. Length, 40 mm; diameter, 33 mm. Ecuador.

1454. Cancellaria (Cancellaria) urceolata Hinds, 1843. The shells vary in color from creamy white to light brown, with the whorls somewhat shouldered and the sides of the body whorl flattened below the suture. Length, 23 mm; diameter, 15 mm. Magdalena Bay, Baja California, throughout the Gulf of California and south to Cape Pasado, Ecuador, offshore in depths to 73 m.

1455. Cancellaria (Cancellaria) ventricosa Hinds, 1843 (Synonym: **C. affinis** C. B. Adams, 1852). A white shell, somewhat suffused with brown, with inflated whorls that are slightly shouldered below the suture. Length, 34 mm; diameter, 20 mm. Magdalena Bay, Baja California, south to Panama, offshore in depths to 180 m.

Subgenus AGATRIX PETIT, 1967

Spire turreted, with widely shouldered whorls; sculpture reticulate; periostracum present, bristle-like, in tufts at the periphery.

1456. Cancellaria (Agatrix) deroyae Petit, 1970. A chalky white to light brown form with a large body whorl, shouldered above; sculpture of axial riblets overridden by straplike spiral cords, interspaces smooth, with elongate nodes at the summits of the ribs; axial riblets sinuous. Aperture ovate, outer lip thin, smooth within or showing some lirations well removed from the edge. Columella with three plaits, the anteriormost forming a keel at the border of the widely rounded siphonal canal. Length, 16 mm; diameter, 10.7 mm. Academy Bay, Isla Santa Cruz, Galápagos Islands, depth, 150 m.

1457. Cancellaria (Agatrix) strongi Shasky, 1961. The small shell is olive-brown, with a thin, darker-colored periostracum that is tufted at the shoulder of the whorls. The anterior canal is short, the aperture somewhat trigonal. Length, 18 mm; diameter, 11 mm. Gulf of California, from Punta Arena, Baja California, to Guaymas, Sonora, Mexico, offshore in depths of 37 to 165 m.

Subgenus APHERA H. & A. ADAMS, 1854

Shell somewhat cylindrical in outline, with the aperture flaring anteriorly and with a heavy callus on the body whorl.

1458. Cancellaria (Aphera) tessellata Sowerby, 1832. The white shell is banded and spotted in spiral rows with dark brown, and the sculpture is rather granular at the intersections of spiral and axial ribs. Length, 25 mm; diameter, 10 mm. Southern end of the Gulf of California to Peru. This is the type of *Aphera*. The species *C. oblonga* Sowerby, 1832, which has been cited as Panamic, proves upon study of material at the British Museum to be an Indo-Pacific form. A second *Aphera* has been described, however, in the fossil record—*C. (A.) wigginsi* Emerson & Hertlein, 1964, from Pleistocene beds on Isla Monserrate, Gulf of California.

See Color Plate XVIII.

Subgenus BIVETIA JOUSSEAUME, 1887

With strongly clathrate sculpture; aperture somewhat pyriform, with a posterior canal.

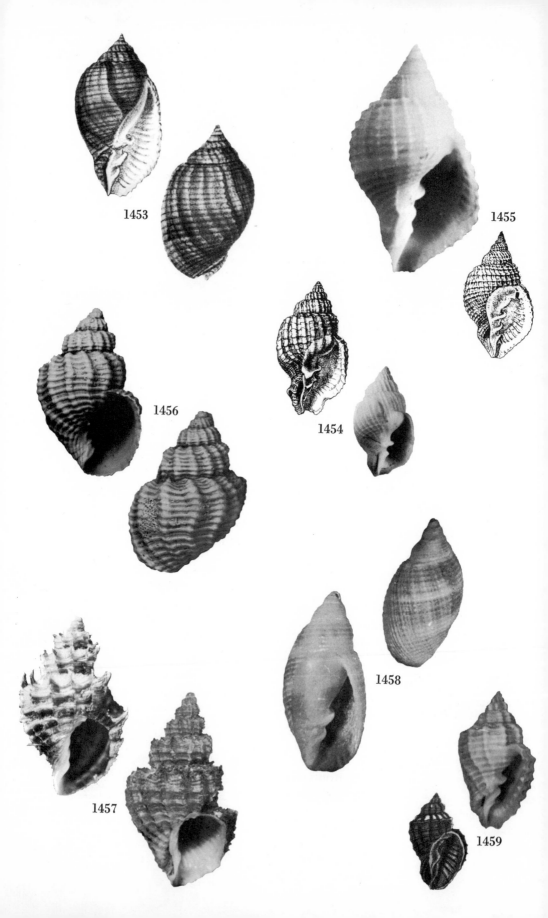

1453

1455

1456

1454

1457

1458

1459

1459. Cancellaria (Bivetia) cremata Hinds, 1843 (Synonym: **C. affinis** Reeve, 1856 [not C. B. Adams, 1852]). This is very similar to *C. indentata* and may prove to be a synonym. The specimens of the type lot at the British Museum have slightly coarser axial sculpture. Length, 26 mm; diameter, 15 mm. West Mexico to Central America.

1460. Cancellaria (Bivetia) indentata Sowerby, 1832 (Synonym: **B. mariei** Jousseaume, 1887; "**B. mariae**" of authors). The synonymy of *Bivetia* has been worked out by Petit (1968), the type species hitherto having been an enigma; it now proves to be a synonym of *C. indentata*, which thus becomes type of the subgenus. The shell of *C. indentata* has a somewhat angulate shoulder to the whorls; the color is reddish brown, and there are about 20 strong axial ribs. On the inner lip a marked indentation sets off the columellar folds. Length, 27 mm; diameter, 16 mm. Santa Inez Bay, Gulf of California, to Cape Pasado, Ecuador, offshore in depths to 110 m.

1461. Cancellaria (Bivetia) jayana Keen, 1958. This species, described in the subgenus *Narona*, now seems better situated in *Bivetia*, which it resembles in the sculpturing. The spire is higher than in *C. indentata*. When fresh, the shell is flesh-colored, fading to white. Length, 20 mm; diameter, 10 mm. Off Sonora, Mexico, to Panama, in depths to 75 m.

Subgenus **BIVETIELLA** WENZ, 1943

Whorls stout, inflated, spire low; shell umbilicate.

1462. Cancellaria (Bivetiella) pulchra Sowerby, 1832. Whitish with bands and lines of dark brown, the shell has a wide umbilicus and is slightly shouldered, with blunt spines. Length, 35 mm; diameter, 26 mm. Guaymas, Mexico, to Ecuador, offshore in depths to 55 m.

Sugenus **BIVETOPSIA** JOUSSEAUME, 1887

Whorls shouldered, body whorl wide, aperture short.

1463. Cancellaria (Bivetopsia) chrysostoma Sowerby, 1832. The white shell is sparsely dotted and lined with brown, with a row of dots on the umbilical ridge; the columella and, in some specimens, the margin of the outer lip are orange-colored. Length, 25 mm; diameter, 18 mm. Panama to Peru, offshore.

1464. Cancellaria (Bivetopsia) haemostoma Sowerby, 1832. Very similar in form to *C. chrysostoma*, of which it has been considered to be a subspecies, the shell has the columella and lip entirely orange-red. The ribs are smaller and more widely spaced; a few specimens lack the apertural coloring. Length, 23 mm; diameter, 17 mm. Galápagos Islands.

Subgenus **EUCLIA** H. & A. ADAMS, 1854

The body whorl with swollen axial ribs and a tendency to form nodes on the shoulder, the outer lip feebly lirate within to smooth.

1465. Cancellaria (Euclia) balboae Pilsbry, 1931. Resembling the more common *C. cassidiformis*, this has more numerous axial ribs, blunter spines, and a thicker shell. Length, 40 mm; diameter, 25 mm. Restricted to Panama. The range seems to be from Mexico (questionably at Guaymas, more positively identified at San Blas and in the Gulf of Tehuantepec area) to Panama, mostly in depths of 18 to 53 m.

1460

1461

146

1463

1464

1465

1466

1467

1468

1469

1466. Cancellaria (Euclia) cassidiformis Sowerby, 1832. The shell is flesh-colored to orange-brown, with a white band on the lower part of the body whorl. Length, 40 mm; diameter, 28 mm. The northern part of the Gulf of California to Peru, intertidally at extreme low tides but mostly offshore in depths to 37 m.

Subgenus HERTLEINIA MARKS, 1949

Slender, cancellately sculptured shells with one stronger spiral rib below the suture; aperture long, with a sinus on the outer lip near the anterior end.

1467. Cancellaria (Hertleinia) mitriformis Sowerby, 1832 (Synonyms: **C. uniplicata** Sowerby, 1832; **C. sowerbyi** Crosse, 1861; "**C. mitraeformis**" of authors, spelling error [not preoccupied by *Voluta mitraeformis* Brocchi, 1814]). A purplish-brown shell with well-developed cancellate sculpture. The sinus in the outer lip gives it the appearance of being cut or broken away. Length, 35 mm; diameter, 14 mm. Intertidal crab specimens not uncommon in Panama, but occurring mostly offshore in depths to 37 m, Panama to Peru. This is the type of the subgenus.

Subgenus MASSYLA H. & A. ADAMS, 1858

Ovate, low-spired, sculpture of low spiral threads; aperture contracted anteriorly, with a truncate columella.

1468. Cancellaria (Massyla) corrugata Hinds, 1843. Nearly smooth, the brownish shell has fine spiral threads, especially on the anterior half, with faint axials that are hardly more than growth striae. Length, 21 mm; diameter, 12 mm. Guaymas, Mexico, to Guayaquil, Ecuador (type locality), in depths to 31 m.

Subgenus NARONA H. & A. ADAMS, 1854

Sculpture of swollen axial ribs, some resembling varices; early whorls with bicarinate sculpture; shell slender, spire tapering.

1469. Cancellaria (Narona) clavatula Sowerby, 1832 (Synonym: ? **C. elata** Hinds, 1843). The spiral sculpture is much weaker than the axial on this brown shell. There is a narrow white band at the shoulder and one on the middle of the body whorl. About 22 mm in length and 10 mm in diameter. Mazatlán, Mexico, to Paita, Peru, offshore in depths to 110 m. This is the type species of the subgenus. Hinds's type specimen of *C. elata* in the British Museum is a juvenile shell with a defective outer lip, faded but showing some brownish-yellow color, strong axial ribbing, and fine spiral riblets; it is about 16 mm in height.

1470. Cancellaria (Narona) exopleura Dall, 1908. This is described as larger than *C. clavatula*, with one more whorl and more axial ribs that, however, are less rounded, and prominent; also, the whorls do not increase as rapidly or have a tabulate shoulder. Length, 26 mm; diameter, 12 mm. Panama (type locality) to Paita, Peru, in depths to 128 m.

Subgenus OVILIA JOUSSEAUME, 1887

Low-spired, with incised spiral sculpture crossed in the interspaces by fine axial lirae; umbilicate, anterior end of aperture narrowed or spoutlike.

1471. Cancellaria (Ovilia) cumingiana Petit de la Saussaye, 1844. Spire higher than in *C. obtusa*. Shell of solid texture, yellowish to light brown, the spiral cords of even size, axial ribs irregular, low, rounded. Length (larger of two syntypes in the British Museum), 46 mm; diameter, 31 mm. Type locality unknown; modern material from west Mexico (shrimp boats working out of Guaymas).

1472. Cancellaria (Ovilia) obtusa Deshayes, 1830 (Synonym: **C. cumingiana subobtusa** Crosse, 1863). Nearly globose, spire whorls inflated; spiral cords strong; color yellow-orange, lighter on the base. Length, 29 mm; diameter, 24 mm. Original locality unknown. Crosse's supposed subspecies, thought to be intermediate in dimensions, actually is closer to *C. obtusa*. It is from Peru. Rous (1908) reported a specimen from Panama. Thus the range seems to be Panama Bay to Paita, Peru. The two species may later prove nonseparable.

Subgenus PYRUCLIA Olsson, 1932

Sculpture almost lacking on the body whorl, the outer lip lirate within.

1473. Cancellaria (Pyruclia) bulbulus Sowerby, 1832. This orange-colored shell has a longer canal and less tabulate whorls than *C. solida*. *C. bulbulus* measures 33 mm in length, 20 mm in diameter. Nicaragua to Panama, offshore in depths to 37 m.

1474. Cancellaria (Pyruclia) solida Sowerby, 1832. With a lower spire and more inflated whorls than *C. bulbulus*, this shell superficially looks remarkably like *Triumphis subrostrata*, although, of course, it is not at all closely related. Length, 28 mm; diameter, 20 mm. The Gulf of California to Peru, mostly offshore, in depths to 37 m.

Subgenus SOLATIA Jousseaume, 1887

With strong cancellate to clathrate sculpture, the upper part of the aperture contracted to form a posterior canal and the lower part flaring, sharply contracted to the anterior canal, which is short.

1475. Cancellaria (Solatia) buccinoides Sowerby, 1832. Coloring varies from tawny brown to dark chocolate, with a white band on the middle of the body whorl. Length, about 40 mm. Nicaragua to Chile, mostly offshore.

Subgenus SVELTIA Jousseaume, 1887

Axial ribs widely spaced, crossed by fine spiral threads; base not umbilicate; spire moderately high and pointed.

1476. Cancellaria (Sveltia) centrota Dall, 1896. The well-developed axial sculpture rises to spines at the shoulder, unusual in this family. The shell is pinkish white to grayish, with subdued spiral sculpture. Length, 35 mm; diameter, not including the spines, 20 mm. Gorda Banks, off Baja California, to Cocos Island (type locality), in depths to 120 m.

Genus CANCELLARIA, s. l.

Two species of *Cancellaria* have been reported from deep water. They were placed with some hesitation in the subgenus *Merica*, H. & A. Adams, 1854, which otherwise does not occur in the eastern Pacific.

1477. Cancellaria io Dall, 1908. Described as white or pink under an olive periostracum; length, 43 mm. Panama Bay, 589 m.

1478. Cancellaria microsoma Dall, 1908. Length, 6.5 mm. Off Acapulco, 1,207 m.

1470

1471

1472

1473

1474

1475

1476

1477

1478

1479

Genus PERPLICARIA DALL, 1890

Slender, few-whorled, rather cylindrical shells with weak spiral sculpture and a few columellar folds. The type species is a fossil form, from the Pliocene of Florida.

1479. Perplicaria clarki M. Smith, 1947. A brownish-yellow shell with rectangular spots of lighter color on the last two whorls, this has indistinct spiral ribs and a single thickened varix a little behind the outer lip. There are three folds upon the columella, with granulations on the surface around them, and the anterior notch is only faintly indicated. The type specimen was a juvenile only 16 mm in length. Additional material (Olsson & Bergeron, 1967) shows the adult length to be at least 33 mm; diameter, 9 mm. Banderas Bay, Jalisco, Mexico, to Panama Bay, intertidally.

Genus TRIGONOSTOMA BLAINVILLE, 1827

Low-spired shells with loose coiling, a wide concave shoulder to the whorls, and a trigonal aperture, the columella bent toward the right. Three Panamic subgenera; no species of subgenus *Trigonostoma, s. s.*, have been reported from the Panamic province.

Subgenus EXTRACTRIX KOROBKOV, 1955

Whorls so loosely coiled that there is no suture line; columellar folds faint.

1480. Trigonostoma (Extractrix) milleri Burch, 1949. A chalky white to straw-colored shell so loosely wound that adjacent whorls do not touch. The columellar folds are weak and hardly evident. The paratype here figured measures 19 mm in length and 14 mm in maximum diameter. Costa Rica.

Subgenus OLSSONELLA PETIT, 1970

Suture deeply indented, outer lip arched, internally lirate; axial ribs varix-like, crossed by corded spiral riblets; umbilicus present in varying degrees; aperture ovate-trigonal.

1481. Trigonostoma (Olssonella) campbelli Shasky, 1961. Small, spire turreted, the shell dark brown, with six axial ribs crossed by fine spiral threads; aperture brown within, the outer lip finely lirate; columella with two folds; base deeply umbilicate. Length, 16 mm; diameter, 9 mm. Gulf of California, from Bahía San Luis Gonzaga and Guaymas, Sonora (type locality), southward to the Gulf of Tehuantepec, in depths of 18 to 91 m. Similar to *T. (O.) funiculatum* but darker in color, the suture deeper, the aperture more trigonal.

1482. Trigonostoma (Olssonella) funiculatum (Hinds, 1843). The white shell, tinged with orange-brown, has, according to the original figure, more rounded whorls than other species of *Trigonostoma*. Length, about 20 mm. Magdalena Bay, Baja California, through the southern part of the Gulf of California and south to Panama, offshore.

Subgenus VENTRILIA JOUSSEAUME, 1887

More tightly coiled than either *Trigonostoma, s. s.*, or *T. (Extractrix)*, but with spire whorls tending to overhang or slope outward at suture; umbilicus well developed.

1483. Trigonostoma (Ventrilia) breve (Sowerby, 1832). A white shell flecked with brown between the axial ribs and in the umbilicus. Length, 20 mm;

1480

1481

1482

1483

1484

1485

1486

1487

1488

diameter, 15 mm. Nicaragua to Panama intertidally at extreme low tides and offshore to depths of a few meters.

1484. Trigonostoma (Ventrilia) bullatum (Sowerby, 1832). The rather inflated shell is ornamented with low ridges and blunt nodes. It is whitish, tinged with brown in vague bands. Length, 30 mm; diameter, 25 mm. Puerto Peñasco, Sonora, Mexico, to Panama, offshore, in depths of 33 to 82 m.

1485. Trigonostoma (Ventrilia) elegantulum M. Smith, 1947. Apparently rather similar to *T. breve* but smaller, with more angulate whorls, this is chalky white flecked with dark brown, with a flesh-colored nucleus. Length, 10.5 mm; diameter, 11.5 mm. Cuastecomate, Mexico (Los Angeles County Mus. coll.), to Perlas Islands, Panama Bay.

1486. Trigonostoma (Ventrilia) goniostoma (Sowerby, 1832) (Synonym: **Cancellaria rigida** Sowerby, 1832). The shell is white, variously lined and spotted with brown. The shoulder is deeply excavated and the umbilicus wide. Length, 25 mm; diameter, 17 mm. The head of the Gulf of California to Panama, intertidally at extreme low tides and offshore to depths of a few meters.

1487. Trigonostoma (Ventrilia) pygmaeum (C. B. Adams, 1852). Smallest member of the genus, this waxy white shell has brown spots on the ribs and shoulders of the whorls. It is closest to *T. breve,* but has the spiral ribs rather than the axials most prominent on the whorls of the spire. Length, 8 mm; diameter, 6 mm. Panama.

1488. Trigonostoma (Ventrilia) tuberculosum (Sowerby, 1832). This differs from *T. bullatum* in the more clear-cut spiral and axial sculpture and the sharper nodes on the shoulders of the whorls. Length, 40 mm. Panama to Peru, offshore in depths to 13 m.

Superfamily CONACEA

Shells with a slotlike constriction, a notch, or sinuous growth lines below the suture; radula mostly with harpoon-like teeth, a poison gland being present in most groups.

Family CONIDAE

This family, as interpreted by most malacologists, consists of but one very large genus, *Conus,* though there have been some not very successful attempts in recent years to split it into a number of separate genera. Probably the genus can with some profit be divided into a number of subgenera—as here—but the dividing lines between the natural groups are so tenuous that correct assignment to subgenus is not always easy. A catalog by Tomlin (1937) lists some 2,700 named species. About half of these names apply to fossil forms and many others fall as synonyms. Thus, the actual number of recognizable living species is much smaller—probably about 400.

Very few other shells in any part of the world could be easily confused with a cone; and probably only the species of the genus *Parametaria* (family Columbellidae) in the Panamic region might be. The shell of a typical cone is broad at the top of the body whorl and tapers gradually to a narrow base. The spire is steplike or turreted in some forms. Ornamentation with blunt nodes at the shoulder, known as coronations, is frequent, or there may be beads. The aperture is long, narrow, and mostly of uniform width. The outer lip is thin and sharp and is very

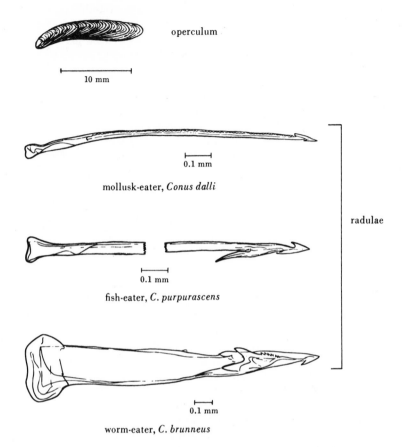

operculum

10 mm

mollusk-eater, *Conus dalli*

0.1 mm

radulae

fish-eater, *C. purpurascens*

0.1 mm

worm-eater, *C. brunneus*

0.1 mm

easily broken or chipped, as the collector soon learns, and as is attested by the frequency with which marked growth flaws occur. Sculpture is mainly of revolving grooves on the base of the shell. A cancellate or pustulose type of sculpture occurs in a few species.

Many cones have a well-developed periostracum, so thick that it may obscure the color pattern beneath. For this reason, some collectors remove it by using a caustic solution. Serious collectors leave it on most of their specimens, for it may be a useful guide in separating otherwise similar species.

The operculum of a typical cone is clawlike or unguiculate, with a terminal nucleus, and it is much smaller than the aperture of the shell. Some species are said to be inoperculate; perhaps because of the small size, the operculum in these has been overlooked.

Although other mollusks such as some Muricacea are able to narcotize prey, and although there is a poison gland in some of the Mitracea as well as in others of the Conacea, the cones have become specialists in the art of using the venom in capturing live food, even killing such agile animals as fish. The venom may be potent enough to cause human fatalities. In all species of the genus the teeth of the radula have become harpoon-like, with a hollow shaft that is connected to the poison gland and a barbed tip that can be ejected forcibly into the prey when the prey comes close to the end of the proboscis. Nybakken (1970*b*) has shown that the form of the radular teeth differs according to food preference, and three types of radular structure can be recognized, depending on whether the prey is fish, annelid worms, or other mollusks. In some of the species of the Indo-Pacific province the "sting" given an unwary collector may be very painful, and a number of deaths have been recorded. The venom is a nerve poison not unlike that of the rattlesnake. Though no reports of injury from cones in American waters are known, it is advisable to handle the larger specimens with care, not holding the shell of a live cone so that the aperture is in contact with the hand.

Most of the species of *Conus* are confined to tropical seas, but a few venture into temperate waters, such as those of the Mediterranean and the California coast. They are known from both deep and shallow water. About 30 species occur on the West American coast—few in comparison to the Philippines, where there are more than 150 species.

Useful references: Emerson & Old (1962); Hanna & Strong (1949); Hanna (1963); Nybakken (1967; 1969; 1970). The paper by Hanna has a key to the species but does not attempt a subgeneric classification.

Genus **CONUS** LINNAEUS, 1758

Characters of the family. The review of this genus (and family) has been done in consultation wth Dr. James Nybakken of Moss Landing Marine Laboratories, who not only has studied the soft parts of Panamic cones and made radular preparations but has also observed the living animals both in the field and in aquaria. He has generously loaned figures from one of his papers for use here, assistance that is gratefully acknowledged. Classification on the basis of the radula, as he points out, proves not to be consonant with a classification based on shell characters, although the radular differences may often enable the separation of puzzling species complexes. Obviously, much more remains to be done to reconcile softpart and shell discrepancies. The subgeneric allocations utilized here, based as they are on shell characters alone, may well prove to be more a matter of convenience than of genetic relationships. The structure of the radula correlates with food habits, as shown in the text figures, but food habits do not necessarily correlate with shell form.

1. Spire with knobs or coronations.............................*Conus, s. s.*
   Spire smooth or with weak beading................................. 2
2. Sculpture of strong spiral cords, especially anteriorly.............*Asprella*
   Sculpture inconspicuous or wanting................................. 3
3. Shoulder of whorls sharply angulate to keeled.................*Leptoconus*
   Shoulder of whorls not sharply angulate............................ 4
4. Color markings of triangular spots and fine lines...............*Cylindrus*
   Color markings, if present, not triangular......................... 5
5. Color pattern of rows of fine dots........................*Ximeniconus*
   Color pattern not of fine dots.................................... 6
6. Spire low to flattened.......................................*Lithoconus*
   Spire moderately elevated, not markedly low........................ 7
7. Shells small, less than 25 mm in length....................*Stephanoconus*
   Shell medium-sized to large, in adults to more than 25 mm long.......... 8
8. Outline narrowly elongate, shoulder rounded..................*Chelyconus*
   Outline pear-shaped, adults large............................*Pyruconus*

Subgenus CONUS, *s. s.*

Spire low, with a tendency to develop nodes or coronations.

1489. Conus (Conus) brunneus Wood, 1828 (Synonyms: **C. bartschi** Hanna & Strong, 1949; **C. andrangae** Schwengel, 1955). The color is in general a rich dark brown, with still darker revolving lines and white blotches on the spire or in a central band. The spire is low, mostly somewhat eroded, with well-developed coronations on the body whorl. Characteristic coloring of the aperture is a dull grayish white, brown at the outer lip, yellowish-tinted on the anterior canal. The periostracum is thick, smooth, and light horn-colored. The soft parts are a deep purple-red or wine color (Nybakken, *in litt.*). Length, about 56 mm; diameter, 37 mm. Magdalena Bay, Baja California, through the Gulf of California and south to Manta, Ecuador; although reported intertidally, it is more common offshore (Nybakken, *in litt.*).

1490. Conus (Conus) chaldeus (Röding, 1798) (Synonym: **C. vermiculatus** Lamarck, 1810). An Indo-Pacific immigrant, marked with wavy bands, that has been found on Clipperton Island and in the Galápagos Islands.

1491. Conus (Conus) diadema Sowerby, 1834 (Synonyms: **C. prytanis** Sowerby, 1882; **C. d. pemphigus** Dall, 1910). The shell is plain chestnut-brown, with a light buff central band, lacking the white blotches of *C. brunneus*, with which it has been confused. The interior of the aperture is purple, not gray, and the animal is orange-colored; the radula is clearly distinct from that of *C. brunneus*. Length, 34 mm; diameter, 20 mm. Isla Monserrate and southward in the Gulf of California and on the offshore islands of west Mexico to the Galápagos Islands, Ecuador; on rocky ledges intertidally (Nybakken, *in litt.*). The form named *C. d. pemphigus* Dall, 1910, from the Tres Marias Islands, has slightly raised spiral rows of pustules.

1492. Conus (Conus) ebraeus Linnaeus, 1758. The hebrew cone is a widespread Indo-Pacific species, one of the few that seem to have gained a foothold on the American continent. The rather small white shell is boldly marked with squarish black spots under a somewhat thin, smooth, light-colored periostracum. The aperture is white. Length, 32 mm; diameter, 20 mm. A shallow-water form, this has been taken by a number of collectors on Clipperton Island and the Galápagos group; Mr. Ted Dranga found it in Costa Rica. The related *C. chaldeus* (Röding, 1798) [*C. vermiculatus* Lamarck, 1810], with wavy bands, another Indo-Pacific immigrant, has also been found on the Galápagos and Clipperton islands.

1493. Conus (Conus) gladiator Broderip, 1833 (Synonym: **C. tribunis** Crosse, 1865). On a white or grayish background are light brown zones or blotches and darker spiral lines. The spire is low and the coronations weakly developed. Within, the aperture is white, some brown showing through. The periostracum is thick, rough, and dark brown. Length, 27 mm; diameter, 22 mm. Magdalena Bay to Guaymas, Gulf of California, south to Ecuador and the Galápagos Islands, intertidally, on rock ledges.

1494. Conus (Conus) princeps Linnaeus, 1758 (Synonyms: **C. regius** Hwass in Bruguière, 1792 [not Gmelin, 1791]; **C. lineolatus** Valenciennes, 1832; **C. apogrammatus** Dall, 1910). A striking and easily distinguished shell, of an orange or pink color with dark brown wavy axial stripes that extend up onto the low and well-coronated spire. The aperture is the color of the outside but

1489

1491

1492

1493

1494

1494a

1495

1496

1497

1498

1499

1500

1501

without the stripes. Also distinctive is the periostracum, which is heavy and dark with wide-spaced bristly spiral lines. Two color varieties may be recognized— C. princeps lineolatus Valenciennes, 1832 (fig. 1494a), with the axial stripes replaced by numerous hairlines, and C. p. apogrammatus Dall, 1910, solid orange or pink in color—but they are not true subspecies. Length, 55 mm; diameter, 33 mm. The head of the Gulf of California to Ecuador, intertidally on rocky ledges.

1495. Conus (Conus) tiaratus Sowerby, 1833, ex Broderip, MS (Synonyms: **C. inconstans** Smith, 1877; **C. roosevelti** Bartsch & Rehder, 1939). A variable form, resembling C. miliaris Hwass in Bruguière, 1792, of the Indo-Pacific, with which it has been synonymized by some authors, this cone has a background color of light to dark brown, or even of flesh or pink. The spire may be nearly flat to well elevated, and spiral striation varies in degree of expression. Constant characteristics are a large brownish-purple blotch inside the outer lip and a smaller one below, the two separated by a gray band. Coronations of the spire are as strong as in C. brunneus, to which some authors consider it closely related. Length, 30 mm; diameter, 18 mm. Mexico to Ecuador, mainly on the offshore islands in shallow water.

Subgenus ASPRELLA SCHAUFUSS, 1869

Spire somewhat nodose; body whorl with incised grooves and spiral cords, especially anteriorly.

1496. Conus (Asprella) arcuatus Broderip & Sowerby, 1829 (Synonym: **C. borneensis** Adams & Reeve, 1848). The carina at the shoulder of the body whorl is sharp and the area below the suture gently curved. Under a thin, lemon-yellow periostracum the shell is white, with three zones of brown markings. Spiral riblets are evident near the anterior end but fade out toward the upper part of the whorl. The spire has regularly spaced nodules, especially in the early whorls; on some specimens the spiral grooves of the body whorl may be present throughout, in most only below the middle of the body whorl. Length, 35 mm; diameter, 15 mm. Santa Inez Bay, Gulf of California, to Octavia Bay, Colombia, in depths to 50 m.

1497. Conus (Asprella) emersoni Hanna, 1963. On a clay-colored ground are reddish-brown markings forming three spiral bands, the lower ones made up of lines and spots irregularly spaced, and there are some dots and dashes between the spiral grooves. Spire whorls are finely beaded, the beads weaker toward the aperture; body whorl with shallow spiral grooves, stronger toward the base. Length, 43 mm; diameter, 18.5 mm. Off Los Frailes, Cape San Lucas, Baja California, depth, 550 m.

Subgenus CHELYCONUS MÖRCH, 1852

Narrowly conic, the shoulder rounded.

1498. Conus (Chelyconus) californicus Reeve, 1844, ex Hinds, MS (Synonyms: **C. ravus** Gould, 1853; **C. dealbatus** A. Adams, 1853; **C. c. fossilis** T. Oldroyd, 1921). The California cone is small, dull gray or tan in color, and of a rounded shape, under a heavy, rough, dark brown periostracum. It is a common shore-loving species of southern California and ranges sparingly into the Panamic province, one specimen in the Stanford University collection being from La Paz. Length, about 30 mm.

1499. Conus (Chelyconus) orion Broderip, 1833 (Synonym: **C. drangai** Schwengel, 1955). On an off-white ground are bold markings of dark brown in three ill-defined spiral bands, the anterior end bright apricot in color. Length, 33 mm; diameter, 17 mm. Guaymas, Mexico, to Manta, Ecuador, in depths to 30 m. Because the type was not figured in Reeve's monograph on *Conus*, authors have been slow to recognize this form, and it was needlessly named a second time.

1500. Conus (Chelyconus) purpurascens Sowerby, 1833, *ex* Broderip, MS (Synonyms: **C. regalitatis** Sowerby, 1834; **C. comptus** Gould, 1853; **C. p.** var. **rejectus** Dall, 1910). The shell, though variable in form and color, can be distinguished by its broad outline and decided purplish cast. Color patterns may be of revolving brown lines or brilliant combinations of violet, dark purple, and brown, in bands and blotches, or the shell may be nearly unicolored. The aperture is a light grayish blue within, purple along the outer lip edge, and the periostracum is somewhat like that of *C. princeps* in being heavy and studded with revolving lines of bristles. Length, 50 mm; diameter, 28 mm. Range, Magdalena Bay, Baja California, and the entire Gulf of California south to Ecuador; common, especially southward, in tide pools and on rocky ledges. Nybakken (1967) has shown that this is a fish-eating species and has published a series of photographs that show the cone stinging and ingesting a small fish.

1501. Conus (Chelyconus) vittatus Hwass in Bruguière, 1792 (Synonym: **? C. henoquei** Bernardi, 1860). Ground color varies from white through orange and lilac to light brown, with fine spiral lines of dark brown on most specimens. A dark brown central band is broken by white spots and stripes. On the spire are numerous broad, dark brown flame-shaped markings separated by white or light brown zones. Spiral threads, slightly raised and marked with brown, circle the body whorl. The aperture is white, the periostracum so thick and rough as to conceal the markings of the shell. Length, 35 mm; diameter, 22 mm. Santa Inez Bay and Guaymas, Gulf of California, south to Manta, Ecuador, mostly offshore in depths of 18 to 50 m.

Subgenus **CYLINDRUS** DESHAYES, 1824
(**CYLINDER** MONTFORT, 1810 [not VOET, 1796])

Spire smooth, shoulder rounded; color pattern of fine white triangles on a dark ground.

1502. Conus (Cylindrus) dalli Stearns, 1873. Dall's cone, like other members of the textile cone group, has a color pattern reminiscent of fabric, in soft browns and yellows and darker brown bands, the whole interspersed with white triangular markings. The aperture is tinged with soft rose or peach color. A thin periostracum is so smooth and transparent that the colors show through. Length, 50 mm; diameter, 27 mm. Guaymas, Mexico, to Panama and the Galápagos Islands, in depths of 15 m; it may also be taken intertidally, especially on the offshore islands. Nybakken (1968) has shown that *C. dalli* preys on other mollusks.

See Color Plate IX.

1503. Conus (Cylindrus) lucidus Wood, 1828 (Synonyms: **C. reticulatus** Sowerby, 1834 [not Born, 1778]; **C. loomisi** Dall & Ochsner, 1928). The color pattern resembles the crackle design on old china or a brown web wrapped around the pale violet body whorl of the shell. Irregular brown blotches form a central band, and there are flamelike markings on the spire. The aperture is pale to deep violet. The periostracum is horn-colored, rather thick, with a silky appearance. Length, 50 mm; width, 27 mm. Magdalena Bay, Baja California, through the Gulf of California and south to Ecuador.

Subgenus LEPTOCONUS SWAINSON, 1840

Spire fairly high, concave, the shoulder sharply keeled to angular; body whorl nearly smooth; aperture slightly wider at base, the anal slit deep.

1504. Conus (? Leptoconus) gradatus Wood, 1828. *ex* Mawe, MS (Synonym: "**C. gradatus** Mawe, 1823" of authors, *nomen nudum*). The original figure shows a shell with a high and turreted spire; the type locality is "California." Until the type is detected and studied, it seems wisest to regard the species as indeterminate and to utilize for the complex of which it might be a member a name based on well-figured type material.

1505. Conus (Leptoconus) poormani Berry, 1968. The shell is similar in outline to that of *C. recurvus*, thin, with a white aperture; coloring is of brown blotches irregularly spaced on a white background, and there is a conspicuous medial white band. The periostracum is tan to light brown, thick, with a fringe on the shoulder. Length, 40 mm; diameter, 22 mm. Offshore, Sonora, Mexico, to Octavia Bay, Colombia, in depths of 55 to 70 m.

1506. Conus (Leptoconus) recurvus Broderip, 1833 (Synonyms: **C. incurvus** Sowerby, 1833; **C. emarginatus** Reeve, 1844; **C. scariphus** Dall, 1910). The shape of the shell is distinctive in that the sides of the body whorl are somewhat concave, with a "pinched-in" appearance toward the anterior end. The spire is moderately high, concave in outline, each whorl with a concave furrow or channel. The profile of the outer lip is markedly curved. The coloration also is distinctive—dark brown flamelike irregular axial stripes that may break up into blotches on a glistening white ground. The aperture is white, the periostracum thin, smooth, and dark horn-colored. The shell is thin for its size, easily fractured. Length, 50 mm; diameter, 22 mm. Magdalena Bay and the Gulf of California south to Colombia, offshore at depths of 35 to 145 m. Although the shells may sometimes resemble those of the more near-shore *C. regularis*, the radula is clearly distinct (Nybakken, *in litt.*).

1507. Conus (Leptoconus) regularis Sowerby, 1833 (Synonyms: **C. monilifer** Broderip, 1833; **C. syriacus** Sowerby, 1833; **C. angulatus** A. Adams, 1854; **C. magdalenensis** Bartsch & Rehder, 1939; **C. gradatus thaanumi** and **recurvus helenae** Schwengel, 1955). It may be that this will prove to be superseded by *C. gradatus* if that name can be given proper documentation. The variability of the species complex is attested by the number of synonyms, and some authors would add to this complex *C. scalaris* and *C. dispar*. The most common form is low-spired, the background color ivory-white, profusely marked with spiral rows of rectangular brown spots and axial streaks of purplish brown, the aperture purplish brown within. An overlay of tan may obscure the color pattern in some specimens (such as in the color form named *thaanumi*). Other variations may be higher-spired, and there is wide difference in the amount and distribution of the color markings. An average specimen measures: length, 59 mm; diameter, 32 mm. Magdalena Bay, Baja California, and through the Gulf of California, south to Panama, possibly to Peru, intertidally and offshore, mostly in depths of 5 to 90 m.

1502

1503

1505

1506

1507

1508

1509

1510

1508. Conus (Leptoconus) scalaris Valenciennes, 1832 (Synonym: ? **C. dispar** Sowerby, 1833). The shell is slender, with a high, turreted spire slightly concave in profile. Color markings trend to yellows, irregularly clouded and not in squares or in stripes. The variant *C. dispar* has a lower and straighter spire. Because of the intergradation with the *C. regularis* complex, some authors regard this as not a separable species. Length, 47 mm; diameter, 17 mm. Cedros Island, Baja California, through the Gulf of California and south to Acapulco, Mexico, in depths of 15 to 145 m.

1509. Conus (Leptoconus) virgatus Reeve, 1849 (Synonyms: **C. cumingii** Reeve, 1849 [not Reeve, 1848]; **C. sanguinolentus** Reeve, 1849 [not Quoy & Gaimard, 1834]; **C. signae** Bartsch, 1937). On a pinkish-buff ground color are dark brown axial stripes. There are some brown spiral lines, and a very fine wavy spiral sculpture that imparts a peculiar silky texture to the shell. Length, 33 mm; diameter, 17 mm. Cedros Island, Baja California, through the Gulf of California and south to Ecuador, intertidally and offshore in depths to 100 m.

Subgenus LITHOCONUS Mörch, 1852

Shells medium- to large-sized; spire flat or low, nearly smooth or with some spiral ridges near base; apertural margins nearly parallel.

1510. Conus (Lithoconus) archon Broderip, 1833 (Synonym: **C. sanguineus** Kiener, 1849). The color pattern consists of large, irregular, light to dark reddish-brown blotches on a white ground, under a thin, dark brown periostracum. The aperture is white. The spire is low, sharply pointed, and deeply concave, and the body whorl is smooth except for growth lines and a few spiral threads near the base. Length, 63 mm; diameter, 34 mm. The Gulf of California to Panama, mostly offshore in depths of 26 to 400 m.

1511. Conus (Lithoconus) fergusoni Sowerby, 1873 (Synonyms: **C. xanthicus** Dall, 1910; **C. chrysocestus** Berry, 1968). One of the larger Panamic cones, this escaped the notice of the early monographers. Adult specimens are pure white, with no banding. Young shells show color, however, being bright yellow to orange, obscurely banded with white under a tenacious velvety periostracum. Length of a large specimen, 150 mm; width, 87 mm. Turtle Bay [Bahía Bartolomé], Baja California, through the Gulf of California and south to Santa Elena, Ecuador, and the Galápagos Islands, intertidally; more common offshore in depths to 165 m.

1512. Conus (Lithoconus) tessulatus Born, 1778 (Synonyms: **C. tesselatus** of authors; **C. edaphus** Dall, 1910). A widespread tropical species, this, like *C. ebraeus,* is one of the few Indo-Pacific immigrants that have found their way to the West American mainland. The shell is a soft yellowish-flesh color with spiral rows of darker and rather rectangular spots. A small specimen measures 21 mm in length, 13 mm in diameter. The range on the Pacific coast is on the west Mexican coast, especially the offshore islands; thence it ranges to Hawaii and Japan and the south seas to Australia and east Africa; rare on the Pacific coast.

Subgenus PYRUCONUS Olsson, 1967

Pear-shaped, large, smooth; radular teeth small.

1513. Conus (Pyruconus) patricius Hinds, 1843 (Synonym: **C. pyriformis** Reeve, 1843). The color of the massive shell varies from off-white (especially in

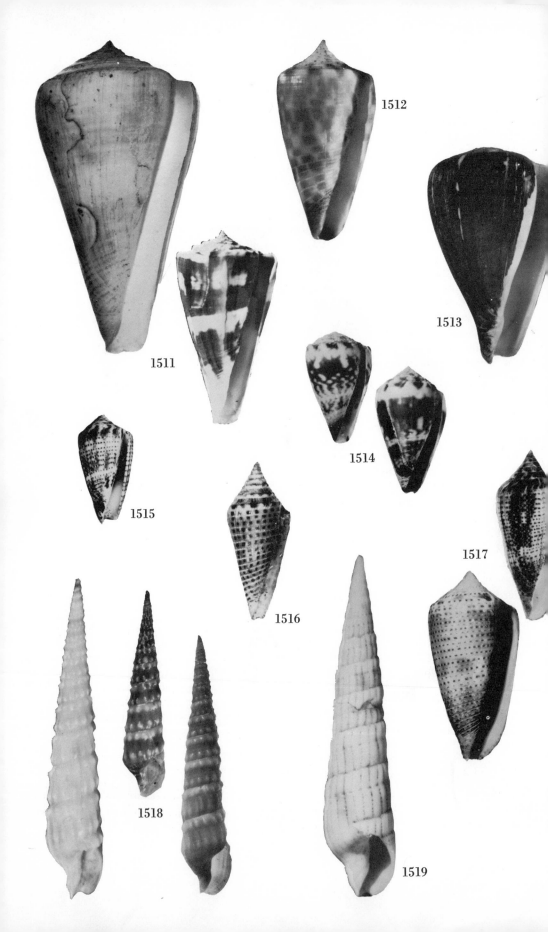

1511

1512

1513

1514

1515

1516

1517

1518

1519

beach-worn shells) to a dull orange, the aperture being white with a tinge of violet-brown at the base. The heavy, brown periostracum has a distinctive pattern of interlaced ridges. About 70 mm in length, but a very large one in the Stanford collection measures 140 mm in length (nearly six inches) and 90 mm in diameter. The species has long been known under the name given by Reeve, but Hinds's name was published a month earlier. Nicaragua south to Ecuador; not uncommon at low tide. This species is the type of the subgenus.

Subgenus STEPHANOCONUS MÖRCH, 1852

Spire low to moderately high, conic, knobbed; body whorl with spiral ridges; aperture narrow above, with a small anal slit.

1514. Conus (Stephanoconus) nux Broderip, 1833 (Synonym: **C. pusillus** Gould, 1853 [not Lamarck, 1810]). Smallest and perhaps the commonest of the Panamic cones, this is like a small *C. ebraeus* but with reddish-brown markings arranged in indistinct bands. There is a purple blotch on the anterior tip, and the aperture commonly shows two purple bands on a white ground. Coronations of the spire are weak. The periostracum is thin and horn-colored. The foot of the animal is pink. Length, 22 mm; diameter, 14 mm. Magdalena Bay, Baja California, and the entire Gulf of California, south to Ecuador, intertidally on rocky ledges.

Subgenus XIMENICONUS EMERSON & OLD, 1962

Fusiform, with a color pattern of closely placed spiral rows of dots, with or without pustules between rows.

1515. Conus (Ximeniconus) perplexus Sowerby, 1857. The ground color may vary from grayish white to pale violet, overlain by numerous spiral rows of brown dashes or dots, some of which may be raised into pustules or ridges, especially near the base of the shell. Axial stripes or blotches may appear on spire and body whorl. The aperture is pale bluish violet, darker within, the periostracum thin, smooth, and light-colored. Length, 26 mm; diameter, 16 mm. Magdalena Bay, Baja California, throughout the Gulf of California and south to Ecuador, mostly on sandbars but also recorded offshore to depths of 37 m. In outline and markings this species is similar to *C. ximenes*, but the radula is strikingly dissimilar (Nybakken, *in litt.*), being more like that of *C. lucidus*, which, on the basis of shell form, is placed in subgenus *Cylindrus*.

1516. Conus (Ximeniconus) tornatus Sowerby, 1833, *ex* Broderip, MS (Synonyms: **C. catenatus** Sowerby, 1878 [not Sowerby, 1850]; **C. concatenatus** Sowerby, 1887 [not Kiener, 1849]; **C. desmotus** Tomlin, 1937). On a whitish background, dark brown rectangular dots or irregular blotches of lighter brown run in spiral rows. Most specimens are heavily beaded or pustular on straplike spiral ribs that are separated by incised furrows, but some may be completely smooth. The periostracum is very thin, smooth, and light brown. Height, 23 mm; diameter, 10 mm. Cedros Island, Baja California, through the Gulf of California and south to Ecuador, offshore in water less than 37 m.

1517. Conus (Ximeniconus) ximenes Gray, 1839 (Synonyms: **C. interruptus** Broderip & Sowerby, 1829 [not Wood, 1828]; **? C. mahogani** Reeve, 1843; **C. exaratus** Reeve, 1844). This is the type of the subgenus. Compared with *C. perplexus* the shell is narrower, with a higher spire, a row of brown dots above and another below the suture. Axial stripes are few or wanting. The aperture is purple within, except in the form considered by some authors to be a distinct species

(*C. mahogani*), which seems rather to be a color variant that is smaller, a little narrower, with a white aperture, but that is not distinguishable on the basis of the radula (Nybakken, *in litt.*). Length, 40 mm; diameter, 20 mm. Throughout the Gulf of California and south to Panama or perhaps to Peru, intertidally and offshore in depths to 90 m. Although the shells of *C. ximenes* and *C. perplexus* can be distinguished with certainty only by use of statistical measures (Wolfson, 1962), the radulas are not at all close (Nybakken, *in litt.*). Thus, they should be considered distinct species.

Family TEREBRIDAE

The auger shells resemble the Turritellas in being long and slender and many-whorled, but here the resemblance ends. The Terebras have a short anterior canal or notch and a narrow aperture; the sculpture is more strongly axial than spiral; there are one or two low folds on the columella; and the operculum has a terminal nucleus. These mollusks are almost exclusively inhabitants of tropical waters, and there is evidence that at least some of them paralyze prey by the use of a poison gland. Little is known about the anatomy of the family. A report by Rudman (1969) shows that in some of the species there is a radula with harpoon-shaped teeth and a poison gland, as in the Conidae; in others either the radular teeth or the poison gland or both may be wanting. Observations are needed on the feeding habits of such forms. Accounts by divers who have collected terebras indicate that they prefer a fine, sandy sea floor, where they move about just under the surface, hidden under a small mound of sand.

Although a number of subdivisions of the family have been attempted on shell characters, the only one that seems practicable for Panamic species is recognition of two genera, *Terebra* and *Hastula*. Subgeneric divisions that have been proposed for *Terebra* turn out to be artificial and not very useful, because of the many exceptions that occur. The species will therefore be given in simple alphabetical order here, but any subgeneric taxa that have Panamic species as types will be mentioned. Cernohorsky (1969) has listed the type specimens in the British Museum.

The review of the family has been prepared by Twila (Mrs. Ford) Bratcher, of Hollywood, California, and Robert D. Burch, of Downey, California. Their text was so nearly in conformity with the style adopted for this book that only minor editorial changes have been made; one change in synonymy is discussed under *Hastula*. They also have provided most of the illustrations. Their contribution to our understanding of the family is gratefully acknowledged.

Genus TEREBRA BRUGUIÈRE, 1789

Many-whorled, slender, sturdy shells, with an anterior notch but no anterior canal, and with one or two folds on the columella. Sculpture usually low, of axial riblets and weak spiral cords, often with a subsutural band or sulcus. Operculum horny, its nucleus terminal.

1518. Terebra adairensis Campbell, 1964. The slender, noded shell resembles *T. roperi* but is more solid, somewhat larger, and is bluish gray or tan rather than brown, with the nodes on the periphery of the body whorl not so sharp. The aperture is elongate and the columella without plications. Length, 29.5 mm; diameter, 5.7 mm. Santa Maria Bay, Baja California, to Sinaloa, Mexico, 7 to 24 m.

1519. Terebra affinis Gray, 1834. The ivory-colored shell shows pinkish-brown blotches; the subsutural band is flat, the whorls and band with broad, flat axial

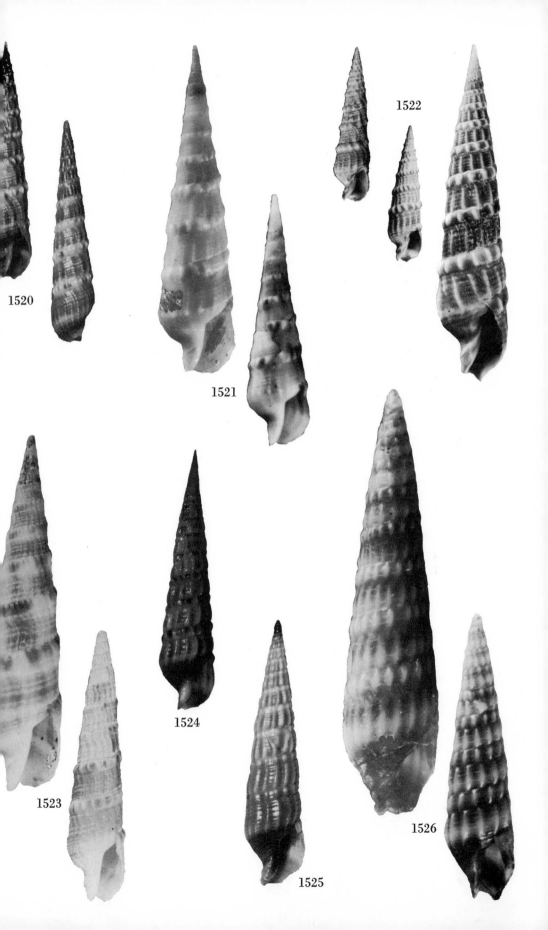

1520

1521

1522

1523

1524

1525

1526

ribs separated by punctations. The aperture is elongate and the columella straight, with a faint plication. Length, 30 mm; diameter, 5 mm. Intertidally at San Luis Gonzaga Bay, Gulf of California; otherwise the species is western Pacific in distribution.

1520. Terebra albocincta Carpenter, 1857. The dark brown shell has a white subsutural band spotted with brown and a white line at the periphery. The axial ribs often overhang the suture. Separable from *T. variegata* by the longer protoconch, of three and a half whorls, the weaker spiral sculpture, fewer axial ribs, and less strong plications of the columella. The aperture is moderately elongate, and the columella has two plications. Length, 46 mm; diameter, 10 mm. Mazatlán, Mexico (type locality), to Guatemala, in 3 to 46 m.

1521. Terebra allyni Bratcher & Burch, 1970. Pale rust-brown mottles the cream-colored shell. Many specimens show a single row of nodes at the periphery; a few have several rows of small nodes. Separable from *T. intertincta* by the nodes at the periphery; those above the suture being formed by more obvious ribbing, by the less concave outline, and by the brown color. The aperture is elongate and the columella straight, without plication. Length, 39 mm; diameter, 8.8 mm. Baja California to Jalisco, Mexico, mostly at island locations, intertidally and offshore in depths to 20 m.

1521a. Terebra argosyia Olsson, 1971. Length, 77.4 mm; diameter, 15 mm. Isla La Plata, Ecuador.

1522. Terebra armillata Hinds, 1844 (Synonyms: **T. albicostata** Adams & Reeve, 1850; **T. marginata** Deshayes, 1857). The shell usually is brown, with cream-colored axial ribs and cream-colored nodes on the prominent subsutural band, light tan or yellowish in some specimens. The aperture is quadrate, the columella recurved, with two plications, the posterior of which forms the keel of the siphonal fasciole. Resembling *T. variegata* but with fewer ribs, rougher sculpture, and an unspotted subsutural band. Length, 64 mm; diameter, 14.9 mm. Santa Maria Bay, outer coast of Baja California, to Peru, intertidally and offshore to depths of 73 m.

1523. Terebra balaenorum Dall, 1908 (Synonym: **T. pulchella** Deshayes, 1857 [not Röding, 1798]). The straw-colored shell has brownish blotches, the early whorls straight-sided, later whorls concave. The body whorl is unusually long, the aperture elongate, the columella straight, with no plication. Length, 45.5 mm; diameter, 9.8 mm. Outer coast of Baja California to Jalisco, Mexico, in depths of 22 to 117 m.

1524. Terebra berryi Campbell, 1961. The shiny grayish-white shell with yellowish blotches has sharper ribs and a less prominent subsutural band than *T. variegata*. The aperture is elongate, the columella curved, with one faint plication. Length, 57.1 mm; diameter, 7.6 mm. Puertecitos, Gulf of California, to Costa Rica, in depths of 2 to 37 m.

1525. Terebra brandi Bratcher & Burch, 1970. The slender bluish shell has a shiny dark brown nucleus. The evenly spaced spiral grooves do not cross the slightly curved axial ribs. Aperture moderately elongate, columella curved, with a faint plication. Length, 24 mm; diameter, 4.6 mm. Gulf of California to Peru, intertidally and offshore to 18 m.

1526. Terebra bridgesi Dall, 1908 (Synonym: **T. dushanae** Campbell, 1964). The small shiny brown shell has whitish nodes and ribs and darker brown

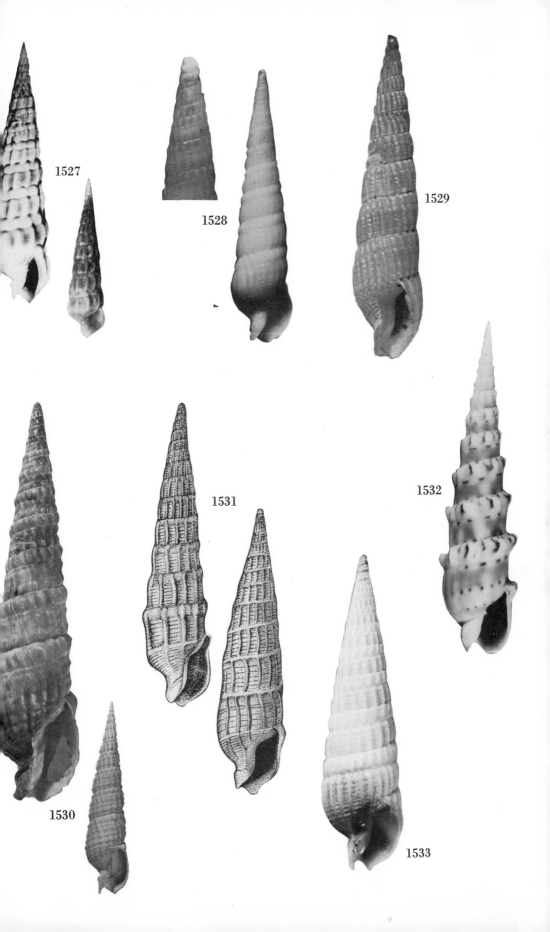

1527

1528

1529

1530

1531

1532

1533

areas on the whorls and below the periphery. The aperture is semiquadrate, the columella straight, with no plication. Length, 10.5 mm; diameter, 2.5 mm. Gulf of California to Panama.

1527. Terebra brunneocincta Pilsbry & Lowe, 1932 (Synonym: **T. varicosa** Hinds, 1844 [not Gmelin, 1791]). Color varies from pale grayish brown to warm reddish or yellowish brown. The subsutural band is flat, often of a lighter color, with the same light color at the periphery of the body whorl. The arched ribs have fine spiral striae in the interspaces. The aperture is quadrate and the columella recurved, with no plications. Length, 42 mm; diameter, 7.5 mm. Manzanillo, Mexico, to Guatemala, in 4 to 20 m depth. This has been confused with *T. larvaeformis* and *T. puncturosa*, from which it is distinct.

1528. Terebra churea Campbell, 1964. The slender white shell has curved axial ribs that continue over the subsutural band. The wider interspaces are evenly divided by four or five spiral cords not crossing the ribs. The aperture is slightly quadrate and the columella recurved, with a faint plication. Length, 14.4 mm; diameter, 3.2 mm. The type locality is Guaymas, Sonora, Mexico.

1529. Terebra corintoensis Pilsbry & Lowe, 1932. The slender shell resembles *T. elata* but is smaller, more slender, and has more axial ribs. The color varies from light bluish brown to tan with brown blotches. The aperture is elongate and the columella straight, with no plications. Length, 13.5 mm; diameter, 3.2 mm. Baja California to Ecuador, in 13 to 18 m depth.

1530. Terebra cracilenta Li, 1930 (Synonym: **T. tenuis** Li, 1930). Although noded and light flesh-colored, this may be separated from *T. tuberculosa* by the elongate aperture. In most specimens the nodes are formed by impressed lines cutting the curved axial ribs, but in some specimens they are more elevated. The aperture is elongate and the columella straight, with a faint plication. Length, 56.5 mm; diameter, 12.3 mm. Tres Marias Islands, Mexico, to Ecuador, in 18 to 37 m depth.

1531. Terebra crenifera Deshayes, 1859 (Synonym: **T. ligyrus** Pilsbry & Lowe, 1932). The color varies from white to tan. Axial ribs are sharp and widely spaced, usually starting from a node on the subsutural band. Wide interspaces are crossed by fine spiral lines. The aperture is quadrate and the columella straight, with no plication. Length, 36.5 mm; diameter, 7 mm. Southern California to Ecuador; intertidally and offshore to depths of 110 m; otherwise, western Pacific in distribution.

1532. Terebra crenulata (Linnaeus, 1758) (Synonyms: **T. maculata** Perry, 1811 [not Linnaeus, 1758]; **T. fimbriata** Deshayes, 1857; **T. interlineata** Deshayes, 1859). The flesh-colored shell has small brown dots on the whorls and brown lines between the nodes on the subsutural band. The aperture is quadrate and the columella straight, with a faint plication. Length, 57.1 mm; diameter, 12.9 mm. Revillagigedo group, off Mexico, in 73 m; mainly a western Pacific species.

1533. Terebra dislocata (Say, 1822) (Synonyms: **T. rudis** Gray, 1834; **T. petiti** Kiener, 1839). The color of this shiny shell may be white, bluish gray or tan, often with a lighter subsutural band. There are many ribs on the heavy convex band, and the remainder of the whorl is sharply ribbed, with spiral lines in the interspaces. The aperture is elongate and the columella recurved, with one plica-

tion. Length, 45 mm; diameter, 10.4 mm. Redondo Beach, California, to Panama, intertidally and to 18 m depth; also in the western Atlantic (type locality).

1534. Terebra dorothyae Bratcher & Burch, 1970. The slender brown shell may have a lighter subsutural band and a lighter stripe at the periphery of the body whorl. Axial ribs are broken into definite rounded nodes. Larger and less slender than *T. adairensis* and *T. roperi*, which have only two rows of nodes per whorl. The noded *T. cracilenta* and *T. tuberculosa* are larger and heavier, with early sculpture of elongate nodes rather than the ribs of *T. dorothyae*. Another noded form, *T. glauca*, is less uniformly so; it is not of a dull brown color. The aperture is elongate, the columella slightly curved, with no plication. Length, 36 mm; diameter, 7.7 mm. Gulf of California to the Galápagos Islands, in depths of 13 to 90 m.

1535. Terebra elata Hinds, 1844 (Synonyms: **T. belcheri** E. A. Smith, 1873 [not Philippi, 1852]; **T. guayaquilensis** E. A. Smith, 1880; **T. ira** Pilsbry & Lowe, 1932). The color of the shell varies from white to cinnamon-brown, sometimes with bands or blotches of brown or yellow. The whorls are axially ribbed, spiral cords sometimes crossing the ribs. The aperture is elongate and the columella straight, with no plications. Length, 41.1 mm; diameter, 9 mm. Santa Maria Bay, Baja California, to the Galápagos Islands; intertidally and to depths of 90 m.

1536. Terebra formosa Deshayes, 1857 (Synonyms: **T. incomparabilis** Deshayes, 1859; **T. pachyzona** Mörch, 1860). The shell is white with squarish brown spots, three rows on the body whorl, two on the others. The subsutural band occupies about two-thirds of the whorl except on the body whorl of adult specimens. The aperture is elongate and the columella recurved. Length, 75 mm; diameter, 16 mm. Manzanillo, Mexico, to Panama, in depths to 7 m.

1537. Terebra frigata Hinds, 1844 (Synonym: **T. galapagina** Dall & Ochsner, 1928). The white shell has slightly concave whorls. Ribs on the whorl join nodes on the band, making the ribs appear swollen at the posterior end. Punctations between the ribs mark the subsutural band. The aperture is elongate and the columella straight, with no plication. Length, 29.9 mm; diameter, 6.9 mm. Galápagos Islands, in depths of 4 to 82 m.

1538. Terebra glauca Hinds, 1844 (Synonyms: **T. aspera** Hinds, 1844 [not Bosc, 1801]; **T. radula** Hinds, 1844 [not Gravenhorst, 1807]; **T. petiveriana** Deshayes, 1857). This is one of the most variable of the Panamic terebras, with variations of sculpture, outline, and color. The color ranges from blackish brown, with a white subsutural band, to uniform rosy beige. Constant features are the long slender nucleus, the first few flat postnuclear whorls with axial ribs extending from suture to suture, and short, recurved, very broad anterior canal. The axial ribs are cut into nodes by spiral lines. The aperture is elongate and the columella almost straight, with plications of varying prominence (two in some, obsolete in others). Length, 45.3 mm; diameter, 13.3 mm. Baja California to Ecuador; intertidally and to a depth of 6 m.

1539. Terebra grayi E. A. Smith, 1877 (Synonym: **T. gracilis** Gray, 1834 [not Lea, 1833]). The shell varies in color from whitish to deep brown with whitish ribs. The brown form resembles *T. armillata* but has fewer ribs, not broken into beads by spiral sculpture, and smaller, sharper nodes on the subsutural band. The aperture is elongate and the columella curved, with one plication. Length, 48

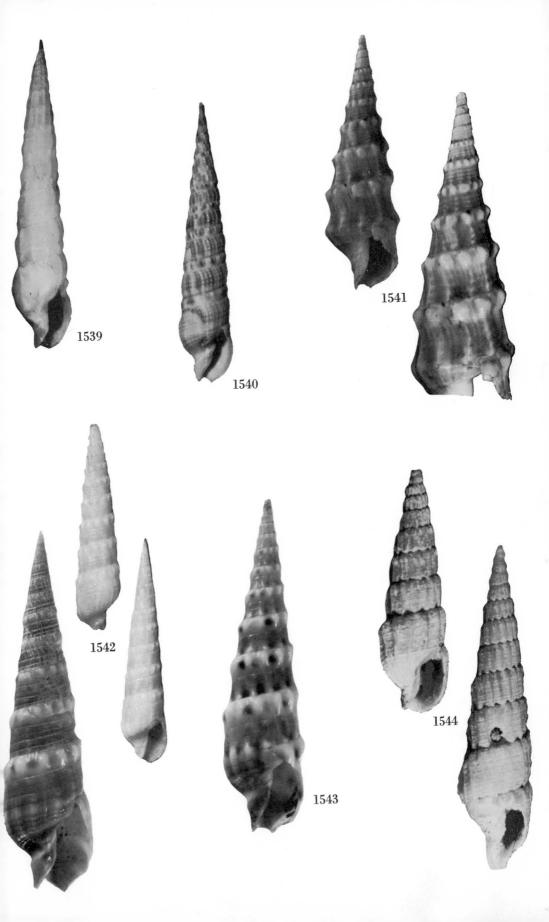

1539

1540

1541

1542

1543

1544

mm; diameter, 11.3 mm. Bahía de Adair, Sonora, Mexico, to the Galápagos Islands, in depths of 5 to 80 m. The species was described, erroneously, as from west Africa and has therefore been overlooked by authors; it has not been collected there subsequently. A photograph of the type in the British Museum (Natural History) makes identification possible as a Panamic form.

1540. Terebra hancocki Bratcher & Burch, 1970. The large, shiny, beige-colored shell may be marked with blotches of purplish gray, gold, reddish brown, or a combination of all three. The surface has a rough, filelike appearance caused by spiral cords crossing axial ribs. Longer than *T. variegata*, with a more slender nucleus and rougher sculpture; smaller than *T. dislocata*, with less heavy plications on the columella. The aperture is semiquadrate, the columella with two plications that are heavier and more prominent than in any other West American species of *Terebra*. Length, 75.3 mm; diameter, 15.5 mm. Tres Marias Islands, Mexico, to Ecuador; intertidally and offshore to a depth of 90 m.

1541. Terebra hertleini Bratcher & Burch, 1970. The small white shell is shouldered below the suture. Cords cross obsolete ribs that begin in large nodes at the suture. Obsolete ribs on the body whorl end in small nodes at the periphery. The aperture is quadrate, the columella straight, with one weak plication. Length, 11.8 mm; diameter, 5 mm. Galápagos Islands, in depths of 5 to 45 m.

1542. Terebra hindsii Carpenter, 1857 (Synonym: **T. malonei** Vanatta, 1925). The color varies from flesh to pinkish or bluish gray. There are white nodes on the subsutural band, with brown flecks between. Below, at the periphery, is another row of faint nodes, often scarcely discernible. This may be separated from *T. tiarella* by its smaller size, smaller nodes, and more widely spaced spiral sculpture. The aperture is quadrate and the columella recurved, with a faint plication. Length, 31 mm; diameter, 7 mm. Baja California to Colombia, intertidally.

1543. Terebra intertincta Hinds, 1844. The shell is flesh to gray in color, with prominent white nodes and brown spots between the nodes and on the subsutural band. Another row of faint nodes occurs anterior to the band, and the axial ribs become enlarged to form a row posterior to the suture and at the periphery of the body whorl. The shell is larger, more heavily noded, and the whorls have a more convex look than in *T. hindsii*. The aperture is quadrate and the columella straight, with one plication. Length, 36 mm; diameter, 11 mm. Baja California to Ecuador; intertidally and to a depth of 37 m. Erroneously described as from west Africa.

1544. Terebra iola Pilsbry & Lowe, 1932. The type of the subgenus *Microtrypetes* Pilsbry & Lowe, 1932. The shell is uniformly light brown in color, with continuous axial ribs that swell slightly anterior to the suture. Interspaces are wider than the axial ribs and are faintly crossed with fine spiral lines. The aperture is elongate, the columella curved, with a faint plication. Length, 19 mm; diameter, 5 mm. Santa Maria Bay, outer coast of Baja California, to Mazatlán, Mexico, intertidally and offshore to 55 m.

1545. Terebra jacquelinae Bratcher & Burch, 1970. The shiny cream-colored shell has very concave whorls. The axial ribs fade at the center of the whorl and become large nodes at both ends, those anterior to the suture being slightly more prominent. The interspaces are almost smooth. The aperture is elongate and the columella straight, with a rounded plication. Length, 36.9 mm; diameter, 10.8 mm. Galápagos Islands, in depths of 4 to 37 m.

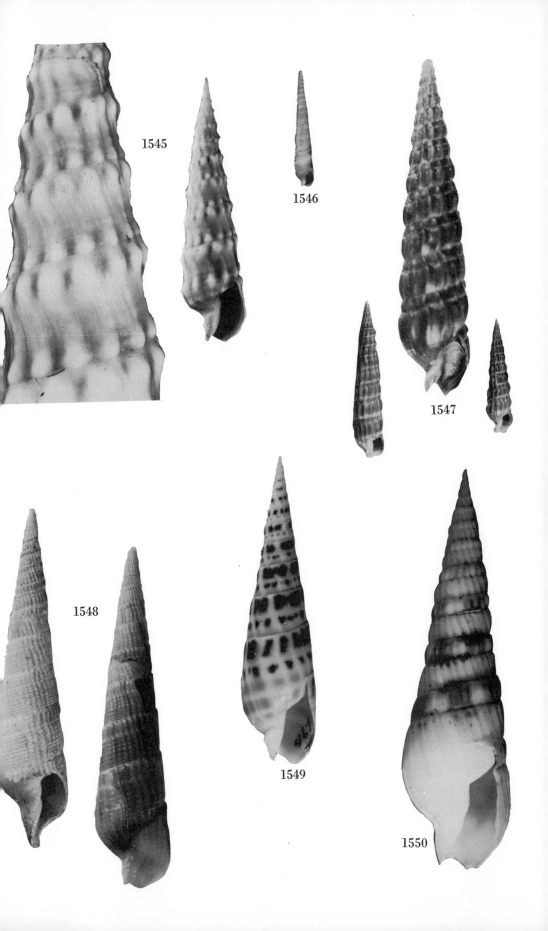

1546. Terebra laevigata Gray, 1834 (Synonym: **T. stylus** Dall, 1908). Color pale yellowish brown; axial sculpture reduced to curved lines of growth; spiral sculpture appearing to form a double subsutural band. Length, 29 mm; diameter, 7 mm. Panama Bay (type locality of *T. stylus*); otherwise western Pacific in distribution.

1547. Terebra larvaeformis Hinds, 1844 (Synonym: **T. isopleura** Pilsbry & Lowe, 1932). The color varies from pinkish white to black. Axial ribs are widely spaced, curved, and extend from suture to suture. The subsutural band is marked by an incised line that is in the interspaces on early whorls, later cutting through the ribs. There are faint spiral lines in the interspaces. The aperture is elongate and the columella straight, with no plication. Length, 40.4 mm; diameter, 8.4 mm. Santa Maria Bay, Baja California, to Ecuador, in depths of 5 to 73 m.

1548. Terebra lucana Dall, 1908. The straw-colored shell has flat-sided whorls slightly shouldered below the suture. A flat subsutural band is marked by deep punctations between the numerous axial ribs. The aperture is elongate and the columella recurved, with one plication. Length, 85.5 mm; diameter, 13.5 mm. Cedros Island, Baja California, to Peru, in depths of 11 to 275 m.

1549. Terebra maculata maculata (Linnaeus, 1758). The shell varies from shiny dark cream to buff color, ornamented with irregular squarish brown spots on the flat subsutural band and one row of smaller brown spots on the whorl. The axial ribs of early sculpture fade in later whorls. The aperture is elongate and the columella short, with one oblique fold. Length over 300 m. This form has been collected at Cocos Island, Costa Rica, but is mainly western Pacific in distribution.

1550. Terebra maculata roosevelti Bartsch & Rehder, 1939. The shell resembles *T. maculata maculata* in color and early sculpture, but the regular, straight, slightly slanted axial ribs continue through the body whorl of adult specimens. The subspecies also is smaller in size. The aperture is elongate and the columella recurved, with one oblique plication. Known only from Socorro Island, Mexico, 18 to 37 m.

1551. Terebra mariato Pilsbry & Lowe, 1932. A small, dull-brown-colored shell with the apex lighter. Axial ribs straight, about eighteen to the last whorl. Length, 7.3 mm. Collected at Montijo Bay, Panama.

1552. Terebra montijoensis Pilsbry & Lowe, 1932. The slender shell is pale brown marbled with cinnamon-brown, the later whorls whitish below the suture. Sculpture is rather coarse, of fifteen axial ribs on the last whorl, four spiral grooves on the spire and eight additional on the base. Length, 15 mm; diameter, 3.3 mm. Nicaragua to Ecuador.

1553. Terebra ninfae Campbell, 1961. The translucent amber shell has a light purplish-brown stripe at the subsutural band and another below the periphery of the body whorl. This form is one of the smallest of the genus. It may be distinguished from *T. bridgesi* by the smaller size and by the definite swelling of the axial ribs at the anterior end into a teardrop shape, overhanging the suture. The aperture is elongate, the columella straight, with no plication. Length, 6.2 mm; diameter, 2 mm. Manzanillo to Chiapas, Mexico, in depths of 27 to 55 m.

1554. Terebra ornata Gray, 1834. The stubby shell is beige in color, with squarish brown spots, four rows on the body whorl and two or three on others. The body whorl is short. The aperture is elongate and the columella recurevd with

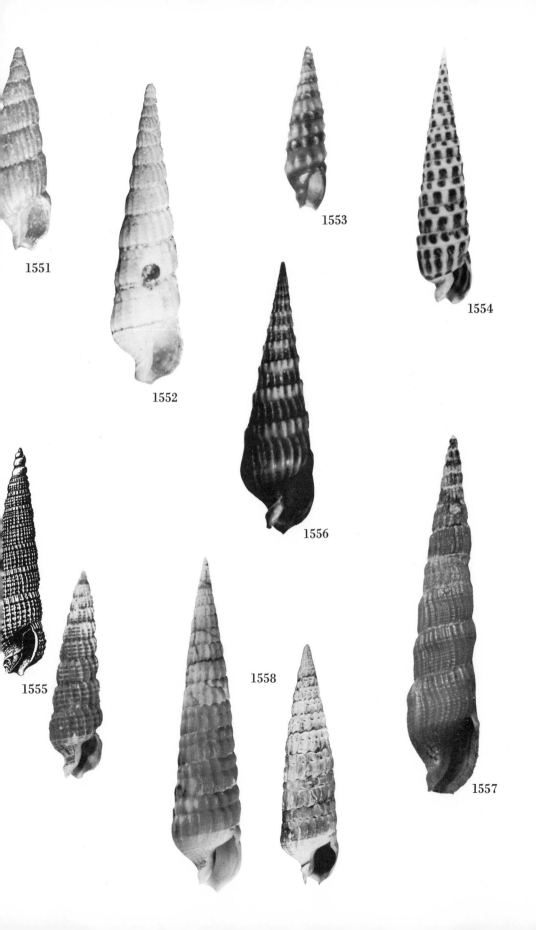

1551

1552

1553

1554

1555

1556

1557

1558

two plications, the anterior one forming a sharp ridge. Length, 82 mm; diameter, 18.6 mm. San Luis Gonzaga Bay, Gulf of California, to the Galápagos Islands, intertidally to 82 m.

1555. Terebra panamensis Dall, 1908. The brownish shell has convex whorls and cancellate sculpture. The shell layer containing the sculpture often erodes or breaks away in the early whorls, as in the figure, exposing the smooth underlayer. The aperture is elongate and the columella recurved, with two plications. Length, 22 mm; diameter, 6 mm. The west coast of Baja California to Panama, in depths of 88 to 280 m.

1556. Terebra plicata Gray, 1834. The shiny cream-colored shell has interspaces wider than the axial ribs and faint, finely punctate spiral lines crossing the ribs. The early whorls usually have a violet tint. The aperture is elongate, the columella straight, with two plications. Length, 70 mm; diameter, 18 mm. Nicaragua to the Galápagos Islands, in depths of 5 to 128 m. Common in the Galápagos, rare elsewhere.

1557. Terebra polypenus Pilsbry & Lowe, 1932. The shell is walnut-brown, darker below the suture and on the base, the nuclear whorls white. Length, 14.1 mm; diameter, 3 mm. Dredged off Mazatlán, Mexico, in 37 m depth.

1558. Terebra puncturosa Berry, 1961. This whitish shell resembles *T. larvaeformis*, but has more numerous and straighter axial ribs, with punctations between that mark the subsutural band. Spiral sculpture is almost microscopic. The aperture is elongate, the columella slightly recurved with no plication. Length, 76.3 mm; diameter, 12.3 mm. Santa Maria Bay, Baja California, to Peru, in depths of 4 to 90 m.

1559. Terebra purdyae Bratcher & Burch, 1970. The shiny slender shell is pale cream-colored, with inconspicuous tan blotches. Axial ribs are crossed by spiral cords, giving it a finer cancellate sculpture than the larger *T. panamensis*. The aperture is elongate and the columella straight, with a faint plication. Length, 20 mm; diameter, 3 mm. Panama to Peru, in depths of 15 to 146 m.

1560. Terebra robusta Hinds, 1844 (Synonyms: **T. lingualis** Hinds, 1844; **T. loroisi** Guérin-Méneville, 1854; **T. insignis** Deshayes, 1857; **T. macrospira** Li, 1930; **T. dumbauldi** Hanna & Hertlein, 1961). The yellowish-white shell has brown spots that coalesce in some specimens to form axial stripes. The early whorls are heavily sculptured. Outline varies, some specimens being slender, others obese. The aperture is elongate and the columella slightly recurved, with one plication. Length, 140 mm; diameter, 34 mm. Santa Maria Bay, outer coast of Baja California, to the Galápagos Islands, intertidally and offshore to 90 m. The broad form named *T. dumbauldi* seems commoner in the southern part of the range, but outline varies considerably in all populations studied. Olsson (1967) regarded *T. robusta* as distinctive enough to be made type of a new subgenus, which he named *Panaterebra* Olsson, 1967.

1561. Terebra roperi Pilsbry & Lowe, 1932. The slender shiny brown shell has two rows of nodes, those on the subsutural band being rounded, the anterior ones more elongate. The outer lip is almost transparent, the aperture elongate, and the columella curved, with two faint plications. Length, 29 mm; diameter, 4 mm. Concepcion Bay, Gulf of California, to Ecuador, in depths of 4 to 31 m.

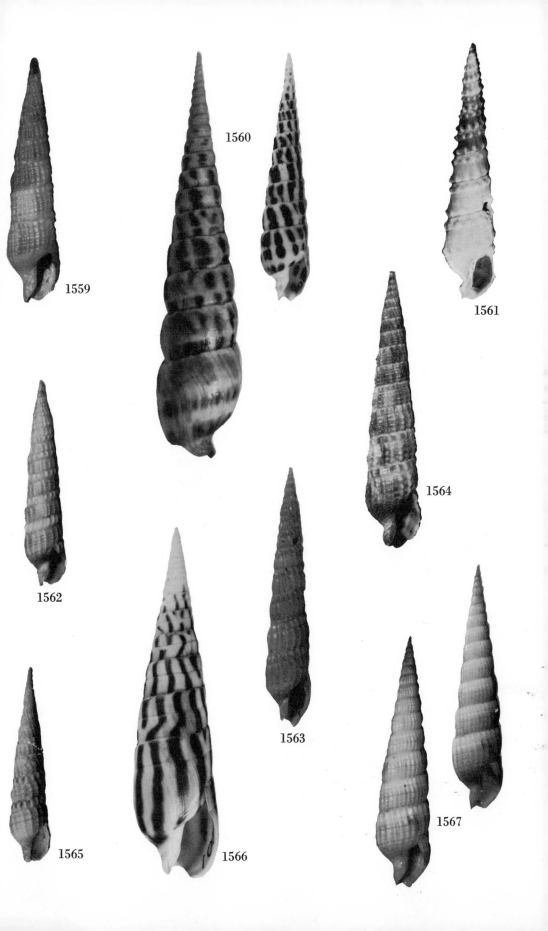

1559

1560

1561

1562

1564

1563

1565

1566

1567

1562. Terebra rufocinerea Carpenter, 1857. The flesh-colored shell with brown flecks has sharp, almost triangular nodes on the subsutural band. The remainder of the whorl is broken into smaller nodes by spiral grooves cutting through axial ribs. The aperture is quadrate, the columella recurved, with one well-defined plication. Length, 35.1 mm; diameter, 9.1 mm. Santa Maria Bay, Baja California, to Guerrero, Mexico, in depths of 5 to 37 m.

1563. Terebra sanjuanensis Pilsbry & Lowe, 1932. This small brownish shell has a spiral groove marking the subsutural band and another posterior to the suture that continues around the periphery of the body whorl. The aperture is elongate, the columella straight, without plication. Length, 16 mm; diameter, 2.9 mm. Costa Rica to Panama, in depths of 9 to 27 m.

1564. Terebra shyana Bratcher & Burch, 1970. The slender shell is usually cancellate, dark buff in color, with darker fulvous blotches. Separable from *T. purdyae* by the darker color and smaller size and by having the axial ribs continuing anterior to the periphery of the body whorl. The shell is not as broad as in *T. panamensis*, the cancellate sculpture finer. The aperture is elongate, the columella curved, with a slight plication. Length, 36 mm; diameter, 7.5 mm. Outer coast of Baja California to Manzanillo, Mexico, in depths of 3 to 80 m.

1565. Terebra specillata Hinds, 1844. The whitish shell with blotches of cinnamon-brown has flat-sided whorls and a slightly convex subsutural band. It is shouldered below the suture, and the body whorl is long. The aperture is elongate, the columella straight, with no plication. Length, 57 mm; diameter, 11 mm. Santa Maria Bay, outer coast of Baja California, to Ecuador, in depths of 9 to 90 m.

1566. Terebra strigata Sowerby, 1825 (Synonyms: **Buccinum elongatum** Wood, 1828; **T. flammea** Lesson, 1832 [not Lamarck, 1822]; **T. zebra** Kiener, 1839). This is the heaviest of the Panamic terebras and has a broad apical angle. The shell is white with dark brown wavy axial stripes. It has a very long body whorl. The aperture is elongate and the columella straight, without plications. Length, 120 mm; diameter, 39.2 mm. The Gulf of California to the Galápagos Islands, intertidally and offshore to depths of 55 m.

1567. Terebra stohleri Bratcher & Burch, 1970. The small sturdy shell has the color and appearance of ivory, with faintly darker blotches. The slightly curved axial ribs are about equal to the interspaces and are crossed by evenly spaced spiral grooves. The aperture is elongate, the columella straight, without plication. Length, 21.4 mm; diameter, 5.1 mm. Pulmo Reef, Baja California, to Socorro Island, Mexico, in depths of 4 to 42 m.

1568. Terebra subnodosa Carpenter, 1857. This species is regarded as indeterminate, for the holotype in the British Museum (Natural History) is a beachworn specimen that shows early damage. Keen (1968) suggested that it might be a synonym of *T. intertincta*.

1569. Terebra tiarella Deshayes, 1857 (Synonym: **T. fitchi** Berry, 1958). The white or reddish-brown shell has a row of large nodes below the suture on otherwise flat whorls. Axial ribs cross the whorls in early sculpture but fade out in later whorls. Spiral sculpture is of numerous fine lines. The aperture is elongate, the columella curved, with two weak plications. Length, 34 mm; diameter, 8 mm. Outer coast of Baja California.

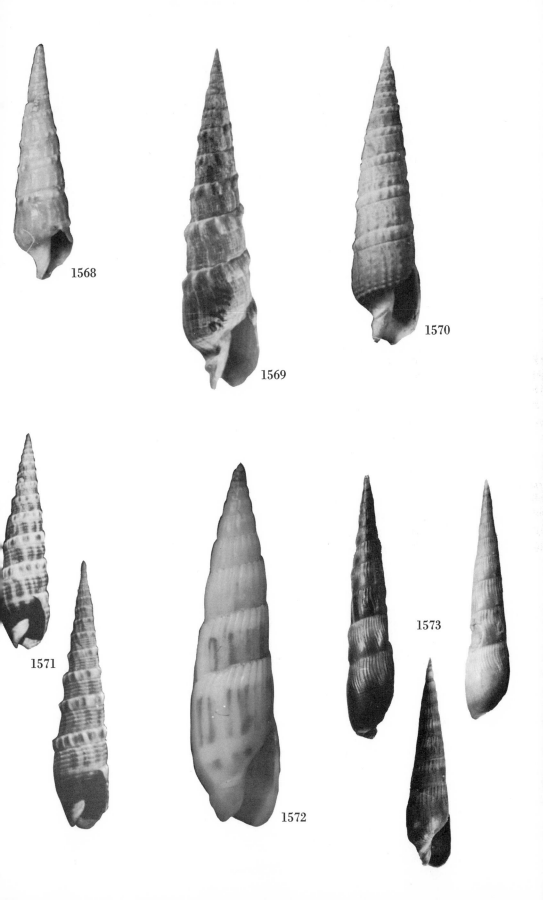

1568

1569

1570

1571

1572

1573

1570. Terebra tuberculosa Hinds, 1844. The color of this shiny shell varies from gray to tan, the early whorls often having a bluish color. There are one to three rows of nodes below the nodose subsutural band. The aperture is slightly quadrate, the columella straight, with a faint plication. Length, 68 mm; diameter, 14.3 mm. Santa Maria Bay, Baja California, to Guatemala; intertidally and to depths of 45 m.

1571. Terebra variegata Gray, 1834 (Synonyms: **T. africana** Griffith & Pidgeon, 1833, *ex* Gray, MS [*nomen oblitum*, ICZN petition pending]; **T. hupei** Lorois, 1857; **T. melia** Pilsbry & Lowe, 1932). The most common of the Panamic terebras. The shell typically is flat-sided, grayish, with a brown-and-white-spotted subsutural band and a white peripheral band that shows through the aperture as a stripe. Streaks and spots of color mark the surface. Spiral sculpture may be almost absent or may cut the curved axial ribs into beadlike nodes, as in the purplish-brown, more slender form. The aperture is quadrate, the columella recurved, with two plications. Length, 87 mm; diameter, 19.3 mm. Santa Maria Bay, Baja California, to Ecuador; intertidally and to depths of 110 m.

Genus HASTULA H. & A. ADAMS, 1853

Shell smooth and shiny except for fine wrinkles below the suture on spire whorls. Thiele (1929) has pointed out differences in the soft parts, especially in the location of the eyes, from those of *Terebra*, and Rudman (1969) has given a further summary.

1572. Hastula albula (Menke, 1843) (Synonyms: **Terebra casta** Hinds, 1844; **T. mera** Hinds, 1844; **T. incolor** Deshayes, 1859; **T. bipartita** Deshayes, 1859; **T. philippiana** Deshayes, 1859; **T. medipacifica** Pilsbry, 1921; **T. medipacifica melior** Pilsbry, 1921). The shell is white, tan, or pale brown, with a light stripe below the suture and at the periphery. Most specimens are ribbed, though some have smooth whorls, with the ribs fading out below the suture. This Indo-Pacific species has been collected at several localities in the Revillagigedo Islands off Mexico. The aperture is elongate and the columella straight, without plication. Length, 27.9 mm; diameter, 6 mm. Revillagigedo Islands, in depths of 7 to 110 m.

1573. Hastula luctuosa (Hinds, 1844). The slender shell is gray or tan, with a dark-spotted lighter band at the suture; some specimens are uniform blackish brown. Axial riblets are fine and fade out rapidly; spiral striae are microscopic. The aperture is elongate, the columella straight, with no plication. Length, 41.2 mm; diameter, 8 mm. Tenacatita Bay, Mexico, to Ecuador, intertidally and offshore to depths of 27 m. Bratcher and Burch (*in litt.*) regard this as indistinguishable from the twin species in the Atlantic, *H. cinerea* (Born, 1778), populations on both coasts apparently having the same range of variations. However, Morrison (1968) argues that in *H. luctuosa* the microscopic surface pattern of growth increments is slightly coarser, the radular teeth are slightly longer and narrower, and both dark and light color phases (which he observed living together in Panama Bay) are darker than their western Atlantic counterparts.

Family TURRIDAE
(PLEUROTOMIDAE; TURRITIDAE of authors)

A posterior sinus or notch along the upper part of the outer lip is the hallmark of this family. The notch may be faint, hardly more than a slight deflection of the

CLAVINAE

vestigial central,
comblike to
compressed lateral,
solid marginal

PSEUDOMELATOMINAE

large central,
massive marginal

ZONULISPIRINAE
BORSONIINAE
MITROLUMNINAE
CLATHURELLINAE
MANGELIINAE
DAPHNELLINAE

hollow marginal only,
barbed or unbarbed,
basal portion variously
shaped

TURRICULINAE
CRASSISPIRINAE
STRICTISPIRINAE

marginal only, duplex or
modified wishbone-shaped,
distal limb severed

TURRINAE

marginal only,
wishbone-shaped

growth lines in some. Mostly, however, it is a V-shaped or U-shaped slot that marks the position of the anal siphon. Shells are in general slender and fusiform, but there are exceptions. Much work remains to be done on the soft parts. In a review of the family, Powell (1966) could cite only one major published study on turrid anatomy (others have since appeared). More is known about the radulas, as will be discussed by Dr. McLean, below. Differences in radular structure and in the buccal apparatus no doubt must reflect marked differences in mode of feeding. According to Powell (1966), all turrid groups have in the buccal area a well-developed poison gland.

The turrids are abundant in numbers of species, but individuals are apt to be thinly scattered, probably as a corollary of their predatory habits. They are not confined to the tropics but may occur in temperate to cold or deep water. Some have been recorded at abyssal depths. In general they are less common intertidally than they are offshore.

Nineteenth-century workers tried to group the known species under a few generic divisions, especially under *Pleurotoma* Lamarck, 1799. This name, meaning "notch-side," was well chosen and long used; nearly a century later, however, malacologists realized that *Turris* Röding, 1798, had priority. Use of only one or a few generic names yields simplicity in indexing, but it compounds the difficulty of making identifications, for in a large group such as this one it brings together an impossibly large number of species of unlike appearance. The opposite trend

CHARACTERS OF THE SUBFAMILIES OF THE TURRIDAE

Subfamily	Radular teeth			Operculum	Earliest apical whorls	Columellar folds	Parietal callus	Position of sinus
	Central	Lateral	Marginal					
Pseudomelatominae	Large	None	Solid	Present	Smooth	None	None	Shoulder
Clavinae	Vestigial	Broad, comblike	Solid	Present	Smooth or carinate	None	Present	Shoulder
Turrinae	Large, vestigial, or absent	None	Solid, wishbone	Present	Smooth	None	None	Periphery
Turriculinae	Large, vestigial, or absent	None	Solid, wishbone or duplex	Present	Smooth	None	None	Shoulder
Crassispirinae	Rarely present	None	Solid, duplex	Present	Smooth or weakly carinate	None	Present	Shoulder
Strictispirinae	None	None	Solid	Present	Smooth	None	Present	Shoulder
Zonulispirinae	None	None	Hollow, mostly barbed	Present	Smooth	None	Present	Shoulder
Borsoninae	None	None	Hollow, rarely barbed	Either present or absent	Smooth	Either present or absent	None	Shoulder
Mitrolumninae	None	None	Hollow, no barbs	None	Smooth	Present	None	Suture, shallow
Clathurellinae	None	None	Hollow, no barbs	None	Usually carinate	None	Present	Shoulder
Mangelinae	None	None	Hollow, rarely barbed	None	Smooth, sub-carinate, or cancellate	None	Either present or absent	Shoulder
Daphnellinae	None	None	Hollow, no barbs	None	Usually diagonally reticulate	None	Either present or absent	Suture

of splitting has developed during the present century, to which the turrids lend themselves well. Some 550 generic and subgeneric names are now available, and the end is not in sight.

Dr. James H. McLean of the Los Angeles County Museum has prepared the text for this family. What he began as a simple list of named species has developed, as he got deeper into the subject, into a complete revision of the family in the eastern Pacific, with descriptions of a number of new taxa (McLean, 1971; McLean & Poorman, 1971; Shasky, 1971). He has studied and photographed type material in all of the major museums of the United States and has made radular mounts for more than 70 of the species. The accompanying illustrations are provided by him, except for those photographed at the British Museum. Dr. McLean's introductory remarks, which follow, include acknowledgments of the assistance he received from colleagues.

The monographic study of worldwide turrid genera by Dr. A. W. B. Powell (1966) has laid the groundwork for the present revision. Clues to relationships in turrids are afforded chiefly by radular characters. In addition to radular studies of my own I have benefited from those of two persons: Dr. J. P. E. Morrison of the United States National Museum, who has made available for comparative study his numerous drawings of eastern Pacific turrid radulae that formed the basis for his summary published in 1966; and Mrs. Virginia Orr Maes of the Philadelphia Academy of Natural Sciences, whose advice and suggestions have been of constant aid, though it should not be inferred that she is in agreement with all of the conclusions reached.

A comprehensive collection of eastern Pacific turrids has been available at the Los Angeles County Museum of Natural History (LACM). In addition to the Museum's holdings, the Hancock collection, resulting from expeditions of the Allan Hancock Foundation and now on long-term loan to the Museum, has proved rich in turrids. Recent fieldwork has also yielded material. Collections at the United States National Museum, the Philadelphia Academy of Natural Sciences, and the Pacific coast museums were studied and the type specimens photographed. Several private collections also were available for study, notably those of Helen DuShane, Leroy Poorman, and Donald R. Shasky.

The radula proves to be the best single character for subdivision of the turrids, though its use must take into account a combination of other characters as well. Basically, there are two kinds of turrid radulae. In the first, the teeth are arranged on a ribbon that has a strong basal membrane, the teeth consisting of central, laterals, and marginals (many have lost all but the marginal teeth). The marginal teeth are solid in structure, some having a wishbone shape; and in some the distal limb of the wishbone seems to have become detached, producing what is called a duplex tooth. This type of radula occurs in the subfamilies Pseudomelatominae, Clavinae, Turrinae, Turriculinae, Crassispirinae, and Strictispirinae. In the second radular type, the marginal teeth are hollow and truly toxoglossate, used singly as hypodermic needles to paralyze the prey, as in the family Conidae. Here the teeth are also on a ribbon, but the basal membrane is weak. In some the teeth are hollow tubes that may be barbed at the tip, others are unbarbed, whereas in some the teeth may resemble partially rolled leaves. The subfamilies in this second group are the Zonulispirinae, Borsoniinae, Mitrolumninae, Clathurellinae, Mangeliinae, and Daphnellinae.

Previous authors have offered widely different subfamily arrangements in the family Turridae. Thiele (1929) recognized three subfamilies, one comprising genera with solid marginal teeth, the other two those with the hollow marginal teeth. Powell (1942, 1966) has offered classifications that employ a number of subfamilies but rely chiefly on shell characters. He would accept within a single subfamily disparate radular types, such as those with solid and those with hollow marginal teeth. Morrison (1966) acknowledged the distinction between the solid and hollow marginal teeth to the point of suggesting separation at the family level, employing a few subfamilies to account for further radular distinctions. Family separation, however, is perhaps premature, because so little is known about the anatomy and functional morphology of the group as a whole. The course adopted here has been to use still more subfamilies, defined both on radular features and shell characters. No subfamily, however, contains genera with fundamentally dissimilar radulae. But shell characters are not always clear-cut, and somewhat similar forms may appear in different subfamilies.

The alphabetical arrangement followed elsewhere in this book is adopted here only at the species level. The sequence for subfamilies, genera, and subgenera is instead what is considered to be systematic relationship. Subfamilies are in sequence from those with combinations of less specialized or primitive features to those that are more highly modified or advanced. Characters considered advanced are hollow marginal teeth, loss of operculum, and pronounced development of parietal callus. In each subfamily, however, the nominate genus is treated first. Related genera are then grouped sequentially, i.e. in systematic order.

Because shell characters alone are so poor a guide to the classification, a key to the subfamilies is impracticable. Instead, the tabular summary on p. 688 is given, showing the combination of characters that comprise the twelve subfamilies recognized here.

Subfamily PSEUDOMELATOMINAE

Posterior sinus on the shoulder, operculum leaf-shaped, with a terminal nucleus. Rachidian tooth of radula rectangular in shape, with a large projecting cusp; marginal tooth massive, tapered to a sharp point.

The radula has a superficial resemblance to that of the muricids, as pointed out by Powell (1966), who placed these genera in the Turriculinae because of the similar shape of the sinus. Morrison (1966) considered the radula sufficiently different to justify rejection from the family. However, recognition as a subfamily in the Turridae is given here because of the presence of a poison gland.

1. Canal short, less than half the length of aperture...........*Pseudomelatoma*
   Canal elongate, equal to length of aperture............................ 2
2. Periostracum thin, yellowish...............................*Hormospira*
   Periostracum thick, brownish black........................*Tiariturris*

Genus PSEUDOMELATOMA DALL, 1918
(LAEVITECTUM DALL, 1919)

Medium-sized, elongate, body whorl narrow, tapering to a short and shallowly notched anterior canal; adult shell with a subsutural collar ornamented by nodes above the shallow and smoothly rounded posterior sinus; lip not thickened; sculpture of a few oblique axial ribs.

1574. Pseudomelatoma penicillata (Carpenter, 1864) (Synonyms: **Drillia moesta** Carpenter, 1864; **D. eburnea** Carpenter, 1865; **Pleurotoma digna** E. A. Smith, 1877; **D. moesta** var. **maculata** Williamson, 1905). Typical specimens are grayish yellow under brown periostracum, with zigzag markings that may be broken up into dots between the low axial ribs and nodes. Length, 35 mm; diameter, 11 mm. Santa Barbara, California, southward along the Baja California coast to Magdalena Bay, intertidally and offshore. Records from the Gulf of California need confirmation. The type of the genus.

Genus HORMOSPIRA BERRY, 1958

Tall, slender, shoulder concave, with a slightly swollen subsutural fold; sculpture of fine spiral striae; periphery nodular, sinus moderately deep, lip arcuate; anterior canal elongate. Periostracum thin, light brown. Only one species known.

1575. Hormospira maculosa (Sowerby, 1834). Cream-colored, with a row of white nodes at the periphery and irregular dots and flecks of brown that may form axial stripes. Length, 40 mm; diameter, 13 mm. Head of the Gulf of California to Guayaquil, Ecuador, intertidally and in depths to 30 m, on sand bottoms.

Genus TIARITURRIS BERRY, 1958

Large, spire elongate, with narrow oblique axial ribs; later whorls with broad ribs that rise to nodes at the periphery, shoulder concave, sinus moderately deep. Periostracum thick, dark-colored.

1576. Tiariturris libya (Dall, 1919). The large shell has few peripheral nodes on the final whorl; early whorls with slanting axial ribs, pillar finely striate. Under a greenish-black periostracum the shell is buff white, painted with irregular zigzag brown axial markings. Length, 47 mm; diameter, 17 mm. Magdalena Bay and Cape San Lucas to Tenacatita Bay, Mexico, in 35 to 90 m. *T. ochsneri* (Anderson & Martin, 1914) is a similar Californian Miocene species.

1576a. Tiariturris spectabilis Berry, 1958. Similar in sculpture and periostracum to *T. libya* but reaching a larger size, with more early whorls, spire outline concave. The irregular zigzag pattern tends to coalesce into a brown band below the periphery. Length, 77 mm; diameter, 23 mm. Angel de la Guarda Island, Gulf of California, to Gulf of Fonseca, El Salvador, in 35 to 90 m. This is the type species of the genus.

Subfamily CLAVINAE

Shells of moderate to large size, slender, high-spired; anterior canal short and truncate or moderately elongate; posterior sinus well developed, deep, often subtubular; shoulder smooth or concave, lacking a raised subsutural thread. Apical whorls smooth or carinate.

The radula is distinctive, its dentition somewhat primitive, with the central tooth a single cusp, laterals broad and comblike, marginals long and flattened. Central and lateral teeth may be reduced in some or lost altogether, but marginal teeth ratain the same appearance.

This is the Drilliinae of Morrison (1966); Powell (1966) included in the Clavinae a number of genera that are here allocated elsewhere, especially some with heavy subsutural collars. Generic criteria useful in recognizing eastern Pacific Clavinae are nuclear whorls, which may be smooth or carinate; length of anterior canal and whether it is nearly straight or bent at an angle to the edge of the outer lip; back of the last whorl, whether having thickened axial ribbing or obsolete sculpture; and rim of the anal sinus, whether raised or flush. These shell characters are consistent within the genera. Relative strength of axial and spiral sculpture is less reliable. The key is to genera and, in *Drillia*, subgenera. Fourteen Panamic genera.

1. Second protoconch whorl carinate.................................. 2
   Second protoconch whorl smooth................................ 6
2. Carination arising from first whorl.........................*Calliclava*
   Carination arising on second whorl............................. 3
3. Aperture truncate, anterior canal lacking....................*Elaeocyma*
   Aperture moderately elongate into anterior canal..................... 4
4. Canal not at angle to edge of outer lip......................*Leptadrillia*
   Canal at angle to edge of outer lip............................. 5
5. Back of last whorl with strong node..........................*Imaclava*
   Back of last whorl smooth, rounded..........................*Kylix*
6. Canal elongate, at angle to edge of outer lip..................*Agladrillia*
   Canal not elongate, not at angle to edge of lip...................... 7
7. Ends of ribs forming beads in subsutural area...............*Globidrillia*
   Ends of ribs not forming beads in subsutural area................... 8
8. Parietal callus weak, sinus sutural................................ 9
   Parietal callus strong, sinus outward-projecting.....................10
9. Whorls rounded ....*Bellaspira*
   Whorls angulate .... *Spirotropis*
10. Axial sculpture of peripheral nodes........................*Splendrillia*
    Axial sculpture of elongate ribbing...............................11
11. Pillar twisted to right.....................................*Cerodrillia*
    Pillar not twisted to right......................................12
12. Sinus rim not projecting, nearly flush.........................*Iredalea*
    Sinus rim spoutlike, projecting beyond whorl outline.................13
13. Anterior canal moderately long..........................*Syntomodrillia*
    Anterior canal short...14
14. Spiral sculpture of strong ribbing................*Drillia (Clathrodrillia)*
    Spiral sculpture of incised striae........................*Drillia, s. s.*

Genus CALLICLAVA McLEAN, 1971

Small to medium-sized, body whorl relatively short; axial ribs weak across the shoulder, tending to form nodes at the periphery, spiral sculpture of incised grooves; shell surface glossy, often with brown or pink banding; anterior canal short, deeply notched; mature outer lip preceded by a thickened axial rib one-quarter turn back; protoconch large, strongly carinate from the beginning. The lateral tooth of the radula is nearly square and has few cusps, rather than elongate with many cusps as in other clavine genera. Distinctive shell characters are the short canal and completely carinate protoconch.

1577. Calliclava aegina (Dall, 1919). Small for the genus, the shell is brownish, often with a whitish film on the early whorls; axial sculpture of strong knobs at the periphery, spiral striae well developed. Length, 13 mm; diameter, 5 mm. Bahía de los Angeles, to Ceralvo Island, Gulf of California, in 15 to 30 m, on sandy bottoms.

1578. Calliclava albolaqueata (Carpenter, 1865) (Synonym: **Pleurotoma cretata** E. A. Smith, 1888). The ivory-white shell is sturdy, with prominent, slanting axial ribs, the subsutural area smoothly concave, base with a constricted profile, spiral sculpture particularly strong on the base. Length, 24 mm; diameter, 9 mm. San Juan del Sur, Nicaragua, to Panama Bay, at low tide and offshore on muddy bottoms.

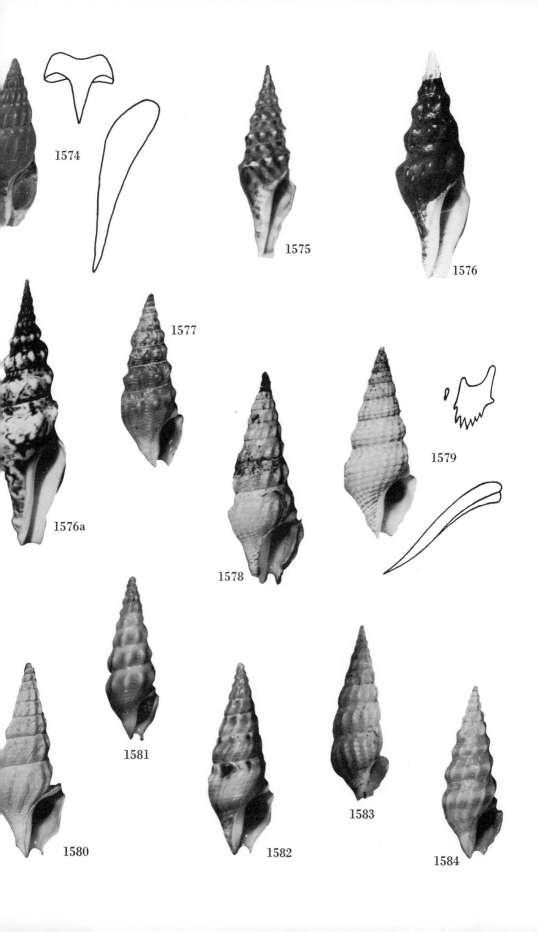

1574

1575

1576

1577

1576a

1578

1579

1580

1581

1582

1583

1584

1579. Calliclava alcmene (Dall, 1919) (Synonym: **Kylix turveri** Hertlein & Strong, 1951). The deeply incised spiral sculpture and the closely spaced axial ribbing give a square beaded effect to the sculpture. Length, 19 mm; diameter, 7.4 mm. Tiburon Island to Espíritu Santo Island, Gulf of California, 40 to 70 m. Dall's taxon was based upon an immature specimen.

1580. Calliclava craneana (Hertlein & Strong, 1951). The dull white shell has a slight pink cast. The axial ribs tend to run from suture to suture, particularly on early whorls; spiral sculpture weak, stronger toward base. Smaller than *C. albolaqueata*, with a more evenly curved base. Length, 21 mm; diameter, 8 mm. Chacahua Bay, Oaxaca, Mexico, to Bahía Honda, Panama, 20 to 70 m.

1581. Calliclava jaliscoensis McLean & Poorman, 1971. Smallest member of the genus, the polished shell varies from white to dark pink and is characterized by relatively few axial ribs forming knobs at the periphery; the spiral incisions evident, except on the shoulder. Length, 12 mm; width, 4 mm. Banderas Bay, Mexico, to Puerto Culebra, Costa Rica, 10 to 50 m.

1582. Calliclava lucida McLean & Poorman, 1971. Not as strongly shouldered as most members of the genus, the relatively small shell is highly polished, white to light pink, with irregular chestnut mottling in the shoulder area; incised spiral sculpture strong, the axial ribs weak on the final whorl. Length, 16 mm; width, 6 mm. Southern tip of Baja California, from Arena Bank to Cape San Lucas, 20 to 70 m, on sandy bottoms.

1583. Calliclava pallida (Sowerby, 1834). The slender shell is white, with numerous sinuous axial ribs that extend across the rounded whorls and base. The most characteristic feature is the relative height of the spire, the aperture being only one-fourth to one-fifth the height of the shell. Length, 25 mm; diameter, 7.5 mm. Gulf of Tehuantepec to Costa Rica, 10 to 25 m. This species has been misunderstood by previous authors; Reeve illustrated a shell with a proportionately larger aperture (possibly *C. albolaqueata*). A syntype specimen is illustrated here; a specimen in the Shasky collection is the only additional specimen yet recognized as matching the type lot.

1584. Calliclava palmeri (Dall, 1919). Relatively small, the strongly shouldered shell has a glossy surface, a variable number of axial ribs (nine to seventeen), and rather weak development of the incised spiral sculpture. The color pattern is characteristic, the shoulder area having a faint brown band and two more bands extending across the base; the area adjacent to the prominent white knob on the back of the last whorl is brown. Length, 15 mm; diameter, 5 mm. Head of the Gulf of California to Puertecitos and Guaymas, Mexico, on sand flats at low tide and to depths of 20 m. This is the type species of the genus.

1585. Calliclava rhodina McLean & Poorman, 1971. Color is distinctive: this is the only eastern Pacific turrid with a dark pink ground, the subsutural area yellow-brown. Axial ribs are lighter-colored across the shoulder, and the swollen axial rib on the last whorl is stained brown. Length, 20 mm; width, 6.5 mm. Isla Grande Bay to Tangola-Tangola Bay, Oaxaca, Mexico, in 20 to 30 m.

1586. Calliclava subtilis McLean & Poorman, 1971. The small pinkish-white shell has deeply incised spiral sculpture similar only to that of *C. alcmene*, from which it differs in having fewer and more broadly spaced axial ribs, a more constricted base, and a more defined anterior canal. Length, 15 mm; diameter, 5 mm. Bahía Honda to Jicarita Island, Panama, 20 to 55 m, sandy bottom.

Genus ELAEOCYMA DALL, 1918

Medium-sized to moderately large, the overall appearance is similar to that of *Calliclava*, with a short anterior canal and incised spiral striae, but the large protoconch is two-whorled, the first smooth, the second developing a low carination. The mature lip is not preceded by a thickened rib. The radula resembles that of other clavine genera, with an elongate, many-cusped lateral tooth. The type species is the southern Californian *Elaeocyma empyrosia* (Dall, 1899).

Elaeocyma has the short canal and carinate second nuclear whorl of *Cymatosyrinx* Dall, 1889, but the type species and other related species of *Cymatosyrinx* from the Florida Pliocene have a beaded subsutural cord and lack spiral sculpture altogether; no eastern Pacific species are related.

1587. Elaeocyma amplinucis McLean & Poorman, 1971. The glossy shell has an especially large protoconch; axial sculpture consists of broad, sinuous white ribs with brown interspaces. A groove defines the shoulder area, but the body whorl is essentially free of spiral sculpture. Length, 20 mm; diameter, 7 mm. Isla Santiago, Galápagos Islands, 30 m.

1588. Elaeocyma arenensis (Hertlein & Strong, 1951). The white shell has a large final whorl and an unusually long spire, with an almost concave profile. Length, 45 mm; diameter, 14.5 mm. Off the southern tip of Baja California, Arena and Gorda Banks, 80 to 100 m.

1589. Elaeocyma melichroa McLean & Poorman, 1971. Shell of moderate size, whorls rounded, axial sculpture of broad, weak folds, the wide shoulder area defined by a groove; body whorl with well-spaced spiral incisions; ground color flesh or tan, with irregular brown mottling on the shoulder. Length, 25 mm; width, 9 mm. Galápagos Islands, 40 to 90 m.

1590. Elaeocyma ricaudae Berry, 1969. Relatively large, the whitish shell is frequently banded with brown on the shoulder and across the base, the peripheral area white. This is closest to the type species of the genus, the southern Californian *E. empyrosia* (Dall), and may prove to be a southern subspecies of it. It differs in being more robust, with a lower spire. Length, 29 mm; diameter, 13 mm. Outer coast of Baja California, Point Abreojos to Boca Soledad, in 25 to 55 m.

1591. Elaeocyma splendidula (Sowerby, 1834). This has a concave shoulder and an angulate periphery, the spiral striae broadly spaced across the body whorl. Flesh-colored, darker between the axial ribs. Length, 28 mm; diameter, 10 mm. Galápagos Islands, in depths to 140 m.

Genus KYLIX DALL, 1919

Protoconch as in *Elaeocyma*, the first whorl bulbous, with a lateral tip, the second with a low-set peripheral carina. Shoulder somewhat concave, the axial ribs sinuous, subdued at the shoulder. Spiral sculpture usually deeply incised. The anterior canal is elongate, set at an angle to the outer lip. The axial ribs increase in number on the final whorl and on the back of the last whorl become obsolete. Lighter-colored beads occur adjacent to the suture.

Distinctive characters of this group are the combination of carinate second whorl of the protoconch, elongate canal, smooth back of the last whorl, and subsutural beading.

1592. Kylix alcyone (Dall, 1919). Relatively small and slender, the spiral incisions superficial and broadly spaced on the dull-surfaced white shell. Length,

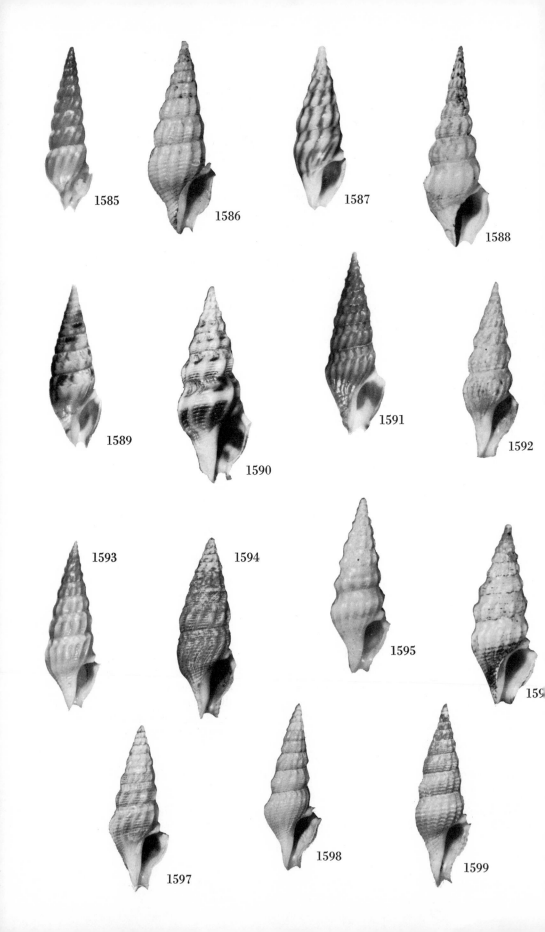

1585

1586

1587

1588

1589

1590

1591

1592

1593

1594

1595

159

1597

1598

1599

15 mm; diameter, 4.5 mm. Off Cabo Lobos, Gulf of California, 139 m. This is the type species of the genus, known only from the type specimen, which lacks the nuclear whorl and a mature lip.

1593. Kylix contracta McLean & Poorman, 1971. The highly polished shell ranges from pure white to flesh-colored, the back of the last whorl brown and the subsutural beading prominent. Aperture and canal short relative to the height of the shell; axial ribs and spiral incisions relatively numerous. Length, 15 mm; diameter, 5 mm. Tenacatita Bay to Guatulco Bay, Mexico, in 20 to 50 m. This is frequently confused with *Agladrillia pudica*, which has a different nucleus and a much longer canal.

1594. Kylix halocydne (Dall, 1919). The largest and most robust member of the genus, the shell pale or dark flesh-colored. The axial ribbing is sinuous and the spiral striae strong on the shoulder. Length, 21 mm; diameter, 7 mm. Chiefly a species of the Californian province: Ventura County, California, to Magdalena Bay, Baja California, 30 to 60 m.

1595. Kylix hecuba (Dall, 1919). The strongly shouldered white shell may be faintly tinted with pink or may have faint flesh-colored banding. The surface is glossy and the weak spiral striae are relatively closely spaced. Length, 15 mm; diameter, 5 mm. Head of the Gulf of California to Concepción Bay on the west and Puerto Peñasco, Sonora, Mexico, on the east, in 10 to 25 m. This is frequently found with *Calliclava palmeri*, which has a shorter canal.

1596. Kylix ianthe (Dall, 1919). The shell surface is highly polished, the whorls strongly tabulate, and the spiral sculpture deeply incised except on the shoulder. White, often with tan banding, back of last whorl brown. Length, 20 mm; diameter, 6.5 mm. Sonoran coast of Mexico from Puerto Peñasco to Rio San Ignacio, 13 to 50 m.

1597. Kylix impressa (Hinds, 1843). The shell is rose-pink, with lighter axial ribs and weak spiral grooves, the back of the last whorl brown. Length, 11 mm; diameter, 4 mm. Gulf of Papaguayo, Costa Rica, to Perlas Islands, Panama, 10 to 30 m.

1598. Kylix panamella (Dall, 1908). Small, the white to flesh-colored glassy-surfaced shell with numerous axial ribs and deeply incised spiral sculpture. Length, 14 mm; diameter, 5 mm. Banderas Bay, Mexico, to Panama Bay, 30 to 250 m.

1599. Kylix paziana (Dall, 1919). The shell is pale brown, frequently with a darker brown band below the periphery. The number of axial ribs is variable and the spiral sculpture deeply incised, the shoulder area often with numerous spiral striae. Less tabulate than *K. ianthe*. Length, 17 mm, diameter, 6 mm. Central and southern Gulf of California, Tiburon Island to Cape San Lucas, in 20 to 80 m.

1600. Kylix rugifera (Sowerby, 1834) (Synonym: **Drillia albemarlensis** Pilsbry & Vanatta, 1902). The lighter-colored axial ribs are broad and strong on this brownish glossy-surfaced shell, spiral incisions broadly spaced, appearing not to override the axial ribs; ribs obsolete on back of last whorl, but brown and white vertical striping persisting. Length, 13 mm; width, 5 mm. Galápagos Islands, in 20 to 55 m. This resembles the Galapagan species of *Elaeocyma*, but the angulation of the lip leading to the elongate canal and features of the back of the last whorl are those of *Kylix*.

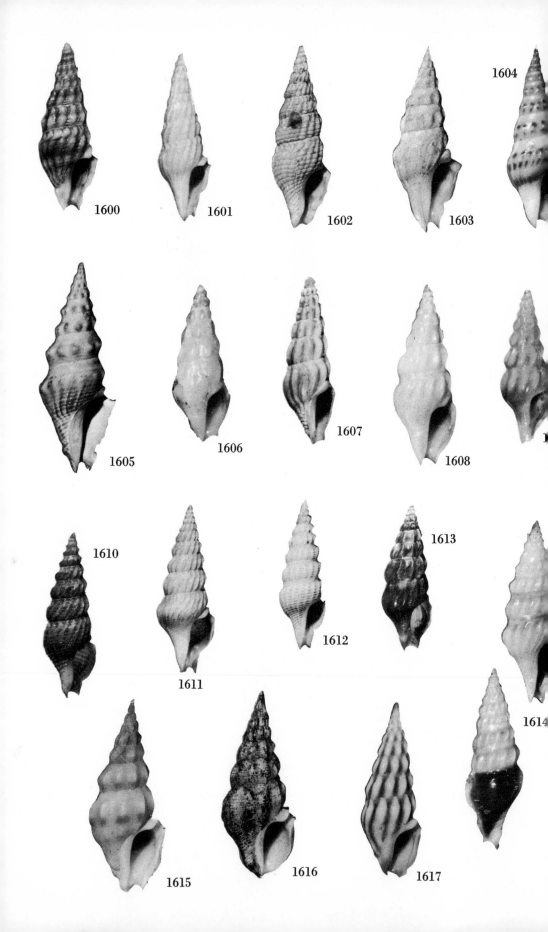

1600   1601   1602   1603   1604

1605   1606   1607   1608

1610   1611   1612   1613   1614

1615   1616   1617

1601. Kylix woodringi McLean & Poorman, 1971. The shell is white with a faint flesh-colored tint on back of last whorl; axial ribs numerous and narrow, spiral sculpture very fine but sharply incised. Length, 16 mm; width, 5.6 mm. Bahía Honda, Panama, to Gorgona Island, Colombia, in 20 to 65 m, on sandy bottoms.

1602. Kylix zacae Hertlein & Strong, 1951. The spiral sculpture of this brownish-white shell is deeply incised and the axial ribs are broad, the interspaces narrow, the resultant sculpture broken into narrow rectangular segments. Length, 14.5 mm, diameter, 5 mm. Central Gulf of California from Tiburon Island to Santa Inez Bay, Mexico, in 10 to 40 m. Sculpture texture is similar to that of *Elaeocyma alcmene*, which has a different protoconch and a shorter canal.

Genus IMACLAVA BARTSCH, 1944

Relatively large, with a strongly noded periphery. As in *Cymatosyrinx* and *Kylix*, the nuclear tip is lateral and the second nuclear whorl is carinate. The anterior canal is elongate and at an angle to the outer lip, as in *Kylix*, but the back of the last whorl has a massive thickened rib one-third of a whorl back from the lip. No subsutural beading on the last whorl. *Imaclava* has been considered a subgenus of *Clathrodrillia*, but that group has a different protoconch and a shorter canal. *Imaclava* therefore seems to warrant full generic recognition.

1603. Imaclava asaedai (Hertlein & Strong, 1951). The shell is whitish, with relatively few axial ribs, which rise to a crest at the periphery and extend halfway across the base; spiral sculpture is reduced to microscopic striae. Length, 27 mm; diameter, 10 mm. Arena Bank, off Cape San Lucas, Baja California (type locality), to La Libertad, Ecuador, in depths to 82 m.

1604. Imaclava pilsbryi Bartsch, 1950. The whitish shell has a pale lavender tint, with relatively low axial ribs, forming nodes on the spire and reaching their crest at the periphery of the final whorl, the crests with a series of brown dots above and below, a brown band extending across the base, the broad varix on the last whorl brown-stained. Length, 28 mm; diameter, 9 mm. Puerto Peñasco, Sonora, Mexico, to Panama, in 10 to 70 m.

1605. Imaclava unimaculata (Sowerby, 1834) (Synonyms: **Clavus pembertoni** Lowe, 1935; **I. ima** Bartsch, 1944). Largest member of the genus, the axial ribs form strongly projecting nodes and the spiral sculpture is well developed across the base. The color pattern is similar to that of *I. pilsbryi*, the back of the last whorl with a broad brownish stain. Length, 47 mm; diameter, 16 mm. Head of the Gulf of California to Puerto Utria, Colombia, in 20 to 70 m. This is the type species of the genus (as *I. pembertoni*).

Genus LEPTADRILLIA WOODRING, 1928

Small, glassy white or faint flesh-colored shells with axial ribs that extend from suture to suture, spiral sculpture absent except for fine threads on the canal; canal moderately long, but not at an angle to the edge of the outer lip; lip preceded by a slightly thickened rib; nucleus of two bulbous whorls, the second whorl with a low peripheral carination.

The type species is from the Jamaican Miocene and there are two living Caribbean species.

1606. Leptadrillia elissa (Dall, 1919). Axial ribs extend from suture to suture, uninterrupted in the shoulder area; spiral sculpture lacking. Length, 10 mm; diameter, 3.5 mm. Chacahua Bay, Mexico, to Panama Bay, in 45 to 90 m.

1607. Leptadrillia firmichorda McLean & Poorman, 1971. The pinkish-white shell has the same general appearance as *L. elissa* but has a more prominent nucleus, and the axial ribs turn toward the columella on the base, separated by deeply incised channels. Length, 9 mm; diameter, 3 mm. Panama Bay to Puerto Utria, Colombia, in 20 to 90 m.

1608. Leptadrillia quisqualis (Hinds, 1843) (Synonym: **Cymatosyrinx hespera** Dall, 1919). The white and polished shell differs from the others in having a concave shoulder in which the axial ribs are subdued; spiral sculpture is lacking altogether. Length, 9 mm; diameter, 3.5 mm. Nicaragua to Panama Bay, on muddy bottoms to depths of 93 m.

Genus SYNTOMODRILLIA WOODRING, 1928

Sculpture as in *Leptadrillia* but protoconch whorls completely smooth, lacking a carination. Several species occur in the Caribbean.

1609. Syntomodrillia vitrea McLean & Poorman, 1971. White or faintly flesh-colored, axial ribs extend from suture to suture, the whorls are rounded, the canal long and sculptured with fine spiral striae. Length, 9 mm; diameter, 3 mm. Off Taboga Island, Panama Bay, in 5 to 10 m.

Genus AGLADRILLIA WOODRING, 1928

This genus greatly resembles *Kylix* in having a long anterior canal, set at an angle to that of the outer lip, deeply incised spiral sculpture, and axial ribs that are obsolete upon the back of the last whorl, with a tendency in some species toward subsutural beading. The nuclear characters differ, however. Instead of the bulbous, carinate nucleus of Kylix, the nucleus of *Agladrillia* consists of three smooth whorls, the tip small and immersed in the center.

1610. Agladrillia badia McLean & Poorman, 1971. The shell is uniformly brown except for the white nucleus and siphonal fasciole. Axial ribs are numerous and slanting, spiral incisions deep and broadly spaced. Height, 19 mm; diameter, 5.5 mm. Galápagos Islands, in 130 to 200 m.

1611. Agladrillia flucticulus McLean & Poorman, 1971. The dull-surfaced shell is whitish on the spire and canal with a light brown band below the periphery and brown on the back of the last whorl. Axial ribs are sinuous and numerous and the spiral striae numerous but not deeply incised. Height, 19 mm; diameter, 7 mm. Rio Balsa, Sinaloa, to the Gulf of Tehuantepec, Mexico, in 40 to 70 m.

1612. Agladrillia gorgonensis McLean & Poorman, 1971. The glistening white shell is strongly tabulate, has a faint flush of pink, and the spiral sculpture is sharply incised, not overriding the axial ribs. Height, 12.5 mm; diameter, 4 mm. Off Gorgona Island, Colombia, in 70 to 110 m.

1613. Agladrillia plicatella (Dall, 1908). Smallest member of the genus, the shell is brownish with relatively few axial ribs, the spiral striae faint. Height, 9 mm; diameter, 3.3 mm. Espíritu Santo Island, Gulf of California, to Panama Bay, in 55 to 165 m.

1614. Agladrillia pudica (Hinds, 1843). Most striking member of the genus, this has an exceptionally long anterior canal. The entire shell is white except for the back of the last whorl, which is dark reddish brown; spiral striae strongly developed only on the canal. Height, 21 mm; diameter, 7 mm. Concepción Bay, Gulf of California, to Lobos Island, Peru, 20 to 70 m.

Genus **DRILLIA** GRAY, 1838
(**DOUGLASSIA** BARTSCH, 1934)

Medium- to large-sized shells, white or pink-banded, with strong axial ribs, weaker on the shoulder area, spiral sculpture mostly weak. Anterior canal short or slightly drawn out but not at an angle to the edge of the outer lip. The parietal callus next to the sinus is strong, giving the sinus an outward-projecting aspect. The nucleus consists of two to two and a half smooth whorls, its tip small and immersed. On the back of the last whorl there is a thickened rib one-third to one-fourth whorl away from the edge of the salient lip, which is crenulated at its edge by spiral sculpture. The smooth nucleus, relatively straight lip edge, and rimmed sinus are the most characteristic features of *Drillia*. As in *Calliclava*, there is variation in relative strengths of axial and spiral sculpture. Here, subgenera prove useful for grouping the smooth and the sculptured forms. Two Panamic subgenera.

Subgenus **DRILLIA**, *s. s.*

Spiral sculpture lacking or consisting of broadly spaced incised lines, with a tendency toward interrupted pink banding in some species. The type species is *D. umbilicata* Gray, 1838, from west Africa.

1615. Drillia (Drillia) acapulcana (Lowe, 1935). Shell pale flesh-colored, the relatively few axial ribs strongest at the periphery, marked with pink below; spiral striae increasing in strength toward the base. Length, 23 mm; diameter, 10 mm. Tiburon Island, Gulf of California, to Santa Elena Bay, Ecuador, in 20 to 60 m.

1616. Drillia (Drillia) aerope (Dall, 1919). Resembling *D. acapulcana* in general appearance but without color spots. Spiral sculpture of widely spaced, deeply incised grooves. Length, 24 mm; diameter, 9.5 mm. Angel de la Guarda Island, Gulf of California, to Cape San Lucas and Santa Maria and Magdalena Bays, Baja California, in 20 to 55 m.

1617. Drillia (Drillia) albicostata (Sowerby, 1834) (Synonyms: **Pleurotoma roseobasis** Pilsbry & Vanatta, 1902 [not Smith, 1888]; **P. roseotincta** Dall, 1923, new name [not Montrouzier, 1872]; **P. testudinis** Pilsbry & Vanatta, 1923, new name; **Cymatosyrinx zeteki** Dall & Ochsner, 1928 [Pleistocene]). Axial ribs white, stout, strongest at the shoulder; interspaces and base of aperture pinkish brown, spiral sculpture lacking. Length, 28 mm; diameter, 10 mm. Galápagos Islands, 10 to 70 m.

1618. Drillia (Drillia) clavata (Sowerby, 1834) (Synonym: **D. decenna** Dall, 1908). A relatively small form with a shell that is glossy, faintly pink-banded on mainland specimens, Galápagos Island specimens more strongly banded; axial ribs low, spiral striae microscopically fine. Length, 13 mm; diameter, 5 mm. Puerto Culebra, Costa Rica, to Xipixapi, Ecuador, and the Galápagos Islands, in 20 to 70 m.

1619. Drillia (Drillia) cunninghamae McLean & Poorman, 1971. Resembling *D. roseola* in general proportions but with a less tabulate shoulder and narrower ribs. The white shell has a chalky surface, fine spiral striae. Length, 32 mm; diameter, 13 mm. Sonora, Mexico, from Rio Tastiota to Guaymas, 40 to 70 m.

1620. Drillia (Drillia) inornata McLean & Poorman, 1971. Shell medium-sized, with a chalky white surface; axial ribs relatively few, constricted at the base,

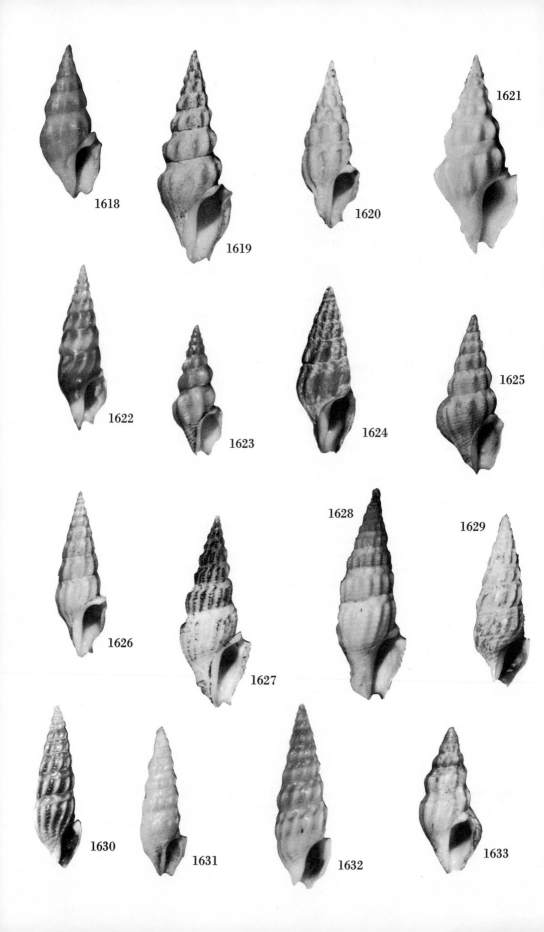

most strongly developed at the shoulder; shoulder area with a rounded subsutural swelling. Length, 21 mm; diameter, 9 mm. Angel de la Guarda Island, Gulf of California, to the Gulf of Dulce, Costa Rica, in 50 to 100 m.

1621. Drillia (Drillia) roseola (Hertlein & Strong, 1955) (Synonym: **Pleurotoma rosea** Sowerby, 1834 [not Quoy & Gaimard, 1833]). A white or rose-colored shell with a markedly tabulate shoulder, this has massive rounded ribs and a recurved anterior canal; spiral sculpture lacking. Length, 31 mm; diameter, 13 mm. Head of the Gulf of California, to Santa Elena Bay, Ecuador, in 10 to 55 m.

1622. Drillia (Drillia) sinuosa McLean & Poorman, 1971. A slender form with a highly polished surface and microscopic spiral striae; axial ribs narrow, low, slanting; reddish brown with lighter banding across the base, axial ribs and anterior canal white. Length, 20 mm; diameter, 6 mm. Isla Santa Cruz, Galápagos Islands, in 150 m.

1623. Drillia (Drillia) tumida McLean & Poorman, 1971. Relatively small, pinkish white, resembling *D. roseola*, but with only six markedly tumid axial ribs and with spiral striae throughout, deeply incised on the anterior canal. Length, 11.3 mm; diameter, 4.4 mm. Tiburon Island, Gulf of California, to Bahía Honda, Panama, in 35 to 70 m.

1624. Drillia (Drillia) valida McLean & Poorman, 1971. Whorls nearly flat-sided, with strong, slanted axial ribs slightly impressed in the subsutural area. The surface is chalky and finely striate. On fresh specimens the areas between the ribs are pinkish brown, as in *D. albicostata*. Height, 30 mm; diameter, 12 mm. Cedros Island to Cape San Lucas, Baja California, in 90 to 140 'm.

Subgenus **CLATHRODRILLIA** DALL, 1918

Axial and spiral sculpture nearly equal in strength.

1625. Drillia (Clathrodrillia) allyniana (Hertlein & Strong, 1951). The shell has a markedly concave shoulder, the spiral sculpture strongly incised, inner lip raised away from a twisted fasciole. Length, 21 mm; diameter, 8 mm. Santa María Bay and Arena Bank off Cape San Lucas, Baja California, to Santa Cruz Bay, Oaxaca, Mexico, in 40 to 90 m.

1626. Drillia (Clathrodrillia) berryi McLean & Poorman, 1971. Relatively slender, with a weakly defined shoulder; the axial ribs are numerous and the fine spiral striae are deeply incised. White, with a narrow yellow-brown band below the periphery; the thickened rib on the back of the last whorl has a reddish-brown stain. Length, 23 mm; diameter, 8 mm. Tovari, Sonora, Mexico, to La Plata Island, Ecuador, in 50 to 165 m.

1627. Drillia (Clathrodrillia) salvadorica (Hertlein & Strong, 1951). A relatively large form in which the spiral sculpture is incised, broadly spaced, and overrides the narrow axial ribs; siphonal fasciole thick, the aperture twisted to the right. Length, 29 mm; diameter, 11 mm. Tiburon Island, Gulf of California, to La Libertad, El Salvador, in 20 to 45 m.

1628. Drillia (Clathrodrillia) walteri (M. Smith, 1946). This resembles the type species of *Clathrodrillia*, from the West Indies, *D. (C.) gibbosa* (Born, 1778), but the shoulder is not as concave and the spiral sculpture not as coarse. The white shell is tinted with pink near the aperture. Height, 37 mm; diameter, 18 mm. Perlas Islands, Panama.

Genus GLOBIDRILLIA WOODRING, 1928

Small, high-spired, glossy-surfaced shells with broadly spaced incised spiral sculpture and numerous low, slanted axial ribs that extend across the base, are interrupted at the shoulder, and terminate at the suture as a row of beads. The aperture and canal are short, the columella and outer lip slanted to the left, stromboid notch barely perceptible. Nucleus smooth, its tip immersed, of one and a half to three rounded whorls.

In addition to the Panamic species, there are two Californian species, *G. arbela* (Dall, 1919) from Scammon's Lagoon, and *G. hemphilli* (Stearns, 1871) in southern California. The type species is from the Jamaican Miocene.

1629. Globidrillia ferminiana (Dall, 1919). Attaining a larger size than the other species. The early whorls of this yellowish-brown shell are usually covered with a whitish film; spiral sculpture is relatively weak. Length, 14 mm; diameter, 5 mm. Tiburon Island to Point San Fermin, Baja California, 20 to 45 m.

1630. Globidrillia micans (Hinds, 1843) (Synonym: **Elaeocyma aeolia** Dall, 1919). The horn-colored shell has sinuous white axial ribs; the incised spiral striae are evenly spaced. This species has a nucleus of three whorls. Length, 10 mm; diameter, 3 mm. Head of the Gulf of California to Santa Elena Bay, Ecuador, in 10 to 30 m.

1631. Globidrillia paucistriata (E. A. Smith, 1888) (Synonym: **Elaeocyma attalia** Dall, 1919). Shell glossy white; axial ribs weak on the base, increasing to their maximum size at the periphery; spiral striae well developed only on the base. Height, 8.5 mm; diameter, 3 mm. Smith's species was described from "California," but specimens are known from Espíritu Santo Island, southern Gulf of California, and Mazatlán, Mexico.

1632. Globidrillia strohbeeni (Hertlein & Strong, 1951). The shell is pale flesh-colored, with faint brown banding across the base and a darker brown band connecting the beads in the shoulder area. Length, 11.5 mm; diameter, 3.5 mm. Tiburon Island, Gulf of California, to Cape San Lucas, Baja California, in 20 to 40 m.

Genus CERODRILLIA BARTSCH & REHDER, 1939

Shell small, axial ribs running from suture to suture, angulate at the periphery, spiral sculpture of fine cords across the base; anterior end truncate, columella twisted toward the right; nucleus of two to three smooth whorls; shell brownish or flesh-colored with lighter axial ribs.

Three Recent West Indian species have two smooth nuclear whorls; the eastern Pacific species have three.

1633. Cerodrillia asymmetrica McLean & Poorman, 1971. Smaller than *C. cybele*; axial ribs on this pale flesh-colored shell are more numerous and more slender, the three-whorled nucleus appearing disproportionately large. Length, 6.1 mm; width, 2.7 mm. Galápagos Islands, 20 to 40 m.

1634. Cerodrillia cybele (Pilsbry & Lowe, 1932). The waxen brown shell has relatively few whitish axial ribs, spiral sculpture of fine striae on the canal only. Length, 12 mm; diameter, 5 mm. Tiburon Island, Gulf of California, to Santa Elena Bay, Ecuador, in 10 to 100 m.

Genus SPLENDRILLIA HEDLEY, 1922

Small, glossy-surfaced shells with a smooth, concave shoulder area, axial sculpture of strong, projecting peripheral nodes, which may be obsolete on the final whorl, spiral sculpture lacking or of fine striae; sinus narrow, with a heavy deposit of parietal callus; protoconch two-whorled, smooth; shells colored, frequently banded.

1635. Splendrillia academica McLean & Poorman, 1971. The shell is blue-gray, with white nodes and brown spots between the nodes, concave in the shoulder area below a rounded subsutural collar; canal deeply striate. Height, 13.7 mm; diameter, 5.3 mm. Galápagos Islands, in 20 to 60 m.

1636. Splendrillia arga McLean & Poorman, 1971. Resembling *S. lalage*, but with larger protoconch and axial sculpture lacking on the final whorl. The white or faintly flesh-colored shell differs from the others in lacking all traces of color banding. Length, 9.3 mm; diameter, 3.2 mm. Port Guatulco, Oaxaca, Mexico, in 15 m.

1637. Splendrillia bratcherae McLean & Poorman, 1971. Pale flesh-colored with lighter peripheral nodes and a yellowish-brown band below the nodes; on some specimens there is faint noding in the shoulder area. The aperture is slightly twisted to the right; spiral sculpture absent. Length, 12.2 mm; diameter, 4.1 mm. Tiburon Island, Gulf of California, to La Paz, Baja California, in 10 to 40 m.

1638. Splendrillia lalage (Dall, 1919) (Synonym: **Elaeocyma baileyi** Berry, 1969). Smaller than *S. bratcherae* and with a shorter spire; axial nodes fewer and lower. The shoulder area is darkly banded, and there are two narrow bands across the base. Height, 8.5 mm; diameter, 3.3 mm. Southern tip of Baja California: from Muertos Bay to Cape San Lucas, in 20 to 40 m. Although Dall's type specimen is bleached and beach-worn, a trace of the banding pattern remains.

Genus IREDALEA OLIVER, 1915
(BREPHODRILLIA PILSBRY & LOWE, 1932)

Small glossy-surfaced shells with low axial ribs running suture to suture; sinus narrow, with a heavy parietal callus; nucleus large, of three and a half smooth whorls. Powell (1966) showed that *Brephodrillia* is inseparable from the Indo-Pacific *Iredalea*.

1639. Iredalea ella (Pilsbry & Lowe, 1932). The shell is cinnamon-brown, with a narrow white band at the periphery, the numerous axial ribs running somewhat diagonally. Length, 6 mm; diameter, 2.3 mm. Guaymas, Mexico, to Panama Bay, on gravel bottoms near rocks, just below low-tide line.

1640. Iredalea perfecta (Pilsbry & Lowe, 1932). The glossy flesh-pink shell has relatively few axial ribs and three somewhat spotted narrow brown bands, especially apparent on the axial ribs. Length, 5.8 mm; diameter, 2.2 mm. Barra de Navidad, Jalisco, Mexico, to La Plata Island, Ecuador, in 20 to 70 m, on sand bottoms.

Genus BELLASPIRA CONRAD, 1868

Small to medium-sized shells with axial ribs that are best developed at the periphery; anterior end sharply truncate, anal sinus narrow, closely constricted, its slot walls directed apically; nucleus of two smooth whorls. Confused with the Mangeliinae by authors because of an error in figuring the type species, which

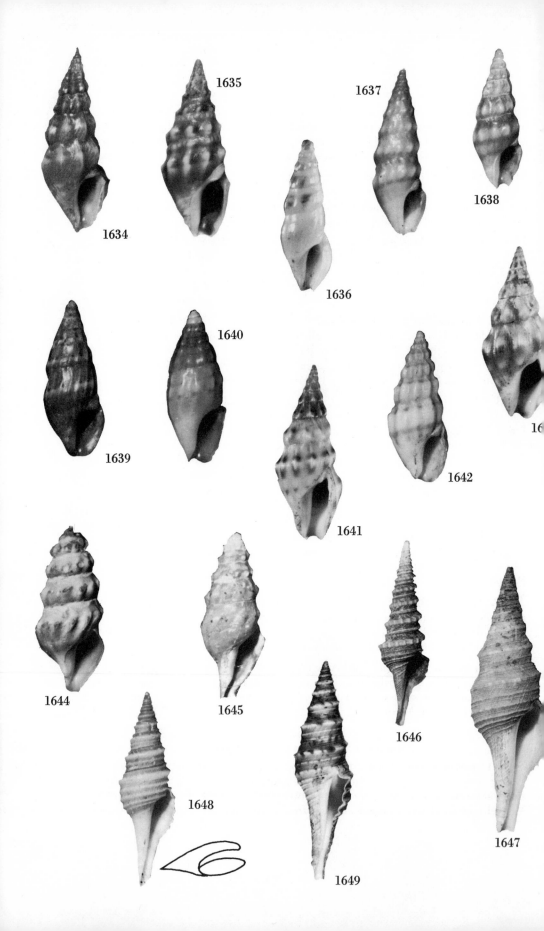

1634

1635

1636

1637

1638

1639

1640

1641

1642

1644

1645

1646

1647

1648

1649

is a Miocene form from Virginia. This genus was reviewed by McLean & Poorman (1970).

1641. Bellaspira acclivicosta McLean & Poorman, 1970. Largest-shelled of the three species, this has strong axial ribs, spiral sculpture of microscopic threads, and is flesh-colored, with three rows of brown spots along the ribs. Length, 18 mm; diameter, 7 mm. Guaymas to Manzanillo, Mexico, on gravel bottoms near rocky areas, in 20 to 40 m.

1642. Bellaspira clarionensis McLean & Poorman, 1970. The shoulder is strongly tabulate and the spiral sculpture is relatively coarse in this pinkish-white form; there is a dark pink band on the shoulder and another across the base. Length, 13.5 mm; diameter, 6 mm. Clarion Island, Revillagigedo Islands, Mexico, in 50 to 80 m. This is closest among Panamic species to the Californian *B. grippi* (Dall, 1908), but is larger, with more numerous spiral striae and a narrower shoulder area.

1643. Bellaspira melea Dall, 1919. A stout rose-pink shell with six to eight low axial ribs continuous up the spire, spiral striae microscopic. Periphery with a darker pink band; subsutural area whitish. Length, 17 mm; diameter, 7 mm. Tepoca Bay, Sonora, to Puerto Utria, Colombia, 20 to 70 m.

Genus SPIROTROPIS G. O. SARS, 1878

Whorls strongly carinate, canal elongate, nucleus of two smooth whorls. Deep water.

1644. Spirotropis genilda (Dall, 1908) [*Mangilia*]. Length, 10.5 mm; diameter, 4.6 mm. Gulf of Panama, 1,855 to 2,323 m.

1645. Spirotropis laodice (Dall, 1919) [*Mangilia*]. Length, 7.7 mm; diameter, 3.5 mm. Off Manta, Ecuador, 733 m.

Subfamily TURRINAE

Posterior sinus on or very near the peripheral keel; operculum leaf-shaped, with terminal nucleus; radula with a pair of marginal teeth that are either wishbone-shaped or the duplex type, with the distal limb severed.

1. Slit band smooth . . . .*Polystira*
   Slit band beaded. 2
2. Anterior canal short . . . .*Cryptogemma*
   Anterior canal long. . . . .*Gemmula*

Genus GEMMULA WEINKAUFF, 1875

Posterior sinus slotlike, deep and narrow, the slit band, which represents its former positions, regularly beaded. Radula of the wishbone type, with or without a rachidian tooth. A number of deepwater eastern Pacific species described in this genus are now assigned elsewhere.

1646. Gemmula hindsiana Berry, 1958 (Synonym: **Pleurotoma gemmata** Reeve, 1843, *ex* Hinds, MS [not Conrad, 1835]). The brown shell has a slit band ornamented with white beads. Length, 18 mm; diameter, 5.3 mm. Magdalena Bay, Baja California; Angel de la Guarda Island, south to Puerto Utria, Colombia, in 40 to 70 m. This is the type of the genus. Unfortunately, the specific name *gemmata* proved to have been preoccupied.

Genus POLYSTIRA WOODRING, 1928
(PLEUROLIRIA GREGORIO, 1890, of authors)

Slender, with a high spire and a long anterior canal; sculpture of several strong spiral cords and numerous smaller threads. Sinus shallow, on the peripheral carina, radula of the wishbone type. A tropical American group, the type species of which, *P. albida* (Perry, 1811), is Caribbean. The group *Pleuroliria* probably represents the early Tertiary ancestral stock.

1647. Polystira nobilis (Hinds, 1843). One of the largest and finest of West American turrids, this white shell may attain a length of 100 mm and a diameter of 32 mm. The uppermost carina is conspicuously elevated, with the result that the whorls are shouldered. Head of the Gulf of California to Panama, to 165 m.

1648. Polystira oxytropis (Sowerby, 1834) (Synonyms: **Pleurotoma albicarinata** Sowerby, 1870; **Pleuroliria artia** and **parthenia** Berry, 1957). Smaller and more slender than *P. nobilis*, this does not have a broad tabulation of the upper part of the whorls. The shell is creamy white to buff, the main carination white, and the axial growth striae between the carinae are strongly developed. Length, 43 mm; diameter, 13 mm. Cedros Island, Baja California, north in the Gulf of California to Tepoca Bay, Sonora, and south to La Libertad, Ecuador, offshore in depths to 110 m.

1649. Polystira picta (Reeve, 1843, *ex* Beck, MS) (Synonym: **Pleurotoma rombergii** Mörch, 1857). Slightly larger than *P. oxytropis*, the white shell has flecks of brown on the spiral cords. It is more slender than *P. nobilis*. Length, 57 mm; diameter, 15 mm. Off Consag Rock, near the head of the Gulf of California, to Puerto Utria, Colombia, in depths of 20 to 70 m.

Genus CRYPTOGEMMA DALL, 1918

Small to medium-sized, with a short, twisted anterior canal; rounded nodules occur along the peripheral slit band of the sinus. Radula of the modified wishbone type. In deep water.

1650. Cryptogemma benthima (Dall, 1908) [*Gemmula*]. Length, 28 mm; diameter, 12 mm. Gulf of Panama to Ecuador and the Galápagos Islands, in 1,485 to 2,490 m. This is the type species of the genus.

1651. C. eldorana (Dall, 1908) [*Gemmula*]. Length, 8 mm; diameter, 5 mm. Gulf of Panama to Galápagos Islands, 1,485 to 2,320 m.

1652. C. polystephanus (Dall, 1908) [*Pleurotomella*]. Length, 25 mm; diameter, 12 mm. Galápagos Islands, 1,159 m.

Subfamily TURRICULINAE

Shells of moderate to large size, fusiform, the spire high, the anterior canal long and often somewhat flexed. Posterior sinus usually deep, J-shaped or U-shaped, above the periphery and often occupying most of the shoulder slope. Operculum leaf-shaped. Radula with a pair of wishbone-shaped marginal teeth that may in some be of the modified or duplex type, with the distal part detached; a central tooth present in some genera, absent in others. Closely related to the Turrinae but with the sinus above rather than at the periphery. The key is to the eight Panamic genera.

Genus FUSITURRICULA WOODRING, 1928

Shell of moderate size, fusiform, the spire high, the anterior canal long and somewhat flexed. Axial ribs regular and evenly spaced but with irregular swollen ribs on the final whorl; early spire whorls with two to three keels; anal sinus deep, with a nearly vertical angle at the suture. Radula of the modified wishbone or duplex type, thus differing from the Indo-Pacific genus *Turricula* Schumacher, 1817, which has the wishbone-shaped marginals.

1653. Fusiturricula andrei McLean & Poorman, 1971. The shell is pinkish brown, with darker areas between the axial ribs and on the shoulder. This differs from *F. armilda* in having more numerous axial ribs and spiral cords, a straighter canal, and a more tabulate shoulder. Length, 32 mm; diameter, 10 mm. Isla Santa Cruz, Galápagos Islands, in 90 to 180 m.

1654. Fusiturricula armilda (Dall, 1908) (Synonyms: **Turris dolenta** and **fusinella** Dall, 1908). The shell is flesh-colored, with lighter-colored nodes and some white banding along the axial ribs. The early whorls are bicarinate, but the spiral threads increase in number and the axial ribs become indistinct on fully mature specimens, with every third to fifth rib swollen on the final whorl. Length, to 40 mm; diameter, to 14 mm. Cabo Lobos, Sonora, Mexico, south through the Gulf of California to Panama, in depths of 46 to 280 m. This is the type species of the genus (as *T. fusinella*). The name *T. fusinella* was based on an immature specimen, that of *F. armilda* on a half-grown specimen, and that of *T. dolenta* on a fully mature specimen.

Genus COCHLESPIRA CONRAD, 1865
(ANCISTROSYRINX DALL, 1881)

Canal long; peripheral keel spinose. Radula with a central tooth and marginals of the modified wishbone type. Powell (1969) has shown that the type of *Cochlespira* —*Pleurotoma cristata* Conrad, 1847, from the Oligocene of Mississippi—is a species so similar to the type of *Ancistrosyrinx* that the latter must fall as a synonym.

1655. Cochlespira cedonulli (Reeve, 1843). Most decorative of the American members of the family, this is a light brown shell, smooth except for the upcurved spines of the keel. Large specimens may be as much as 35 mm in length. Off Tib-

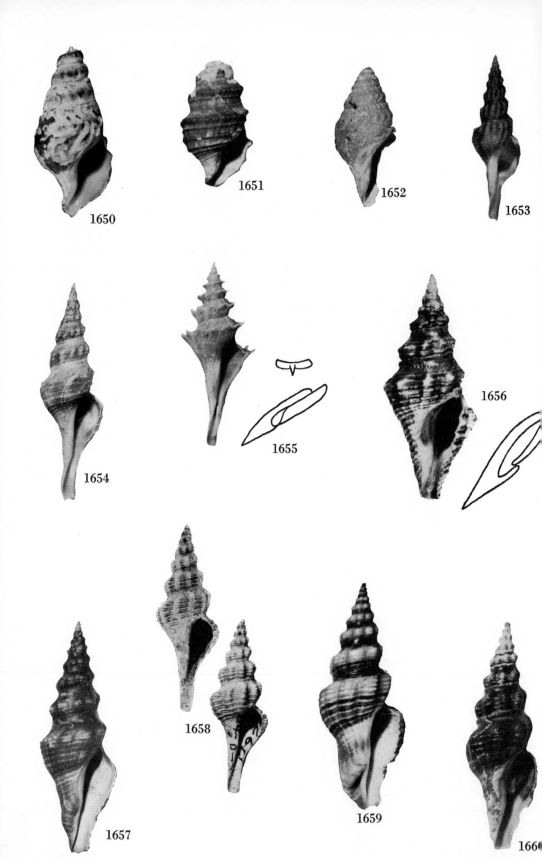

1650

1651

1652

1653

1654

1655

1656

1657

1658

1659

1660

uron Island, Gulf of California, to Puerto Utria, Colombia, in depths of 55 to 275 m; also Galápagos Islands.

Genus KNEFASTIA DALL, 1919

Large, robust, biconic to fusiform shells with coarse sculpture; suture somewhat appressed; operculum ovate, with terminal nucleus; mostly with a thick periostracum. Anal sinus deep, in the middle of the shoulder slope. Radula with a vestigial central tooth and marginals of the wishbone type with the distal limb detached.

1656. Knefastia dalli Bartsch, 1944. Slender to robust, this has an olive brown to dark brown periostracum, the shell pinkish or gray-brown, the spiral ribs somewhat lighter, aperture gray, often stained with brown. The spiral cords are strong across the base, somewhat pustular on the pillar; the margin of the outer lip is crenulate but the inside of the lip is smooth. Length, 61 mm; diameter, 25 mm. Head of the Gulf of California to Guaymas and La Paz, at low tide and just offshore on sandy and rocky bottoms. Exceptionally robust specimens (as the holotype) occur on mud flats in La Paz Bay, but the range of variation includes more slender specimens at the same locality, the slender forms prevalent throughout the rest of the range. Slender specimens have frequently been misidentified as *K. funiculata*.

1657. Knefastia funiculata (Kiener, 1840, *ex* Valenciennes, MS). This resembles *K. tuberculifera* but has a less prominent peripheral carination and a different color pattern. The lip is thin and smooth within. The shell is flesh-colored under a thin, brown periostracum, the smooth shoulder darker brown, with some narrow faint brown banding across the base, the columella and aperture a bright yellow-orange. Length, 53 mm; diameter, 18 mm. Mazatlán to Salina Cruz, Mexico, offshore to depths of 50 m.

1658. Knefastia howelli (Hertlein & Strong, 1951). This resembles *K. walkeri* but has a few more axial ribs and more numerous spiral cords. Known only from the bleached, immature holotype specimen, the lip of which is broken back and obscures the sinus. Length, 31 mm, diameter, 11 mm. Off Judas Point, Costa Rica, 77 to 112 m. Originally described as a *Fusiturricula*.

1659. Knefastia olivacea (Sowerby, 1833). Under an olive-green periostracum the shell is tawny orange, often with narrow bands of brown across the base. Sculpture of fine spiral ribs overriding the low, slanted axial folds, especially at the rounded periphery. The outer lip is lirate within, and large specimens tend to have a twisted siphonal fasciole. Length, 62 mm; diameter, 23 mm. Guaymas, Mexico, to La Libertad, Ecuador, at low tide and to depths of 50 m. This is the type species of *Knefastia*.

1660. Knefastia princeps Berry, 1953. With relatively few and massive axial ribs, seven on the last whorl, overridden by thin, narrow spiral cords. Height, 70 mm; diameter, 22 mm. Outer coast of Baja California, Cedros Island to Santa Maria Bay, in 30 to 45 m.

1661. Knefastia tuberculifera (Broderip & Sowerby, 1829). The axial sculpture is sharply nodose at the periphery, and the spiral sculpture is faint. The shell is yellow-orange under an olive-green periostracum and is characterized by two narrow brown bands across the base. The outer lip is faintly lirate within. Length,

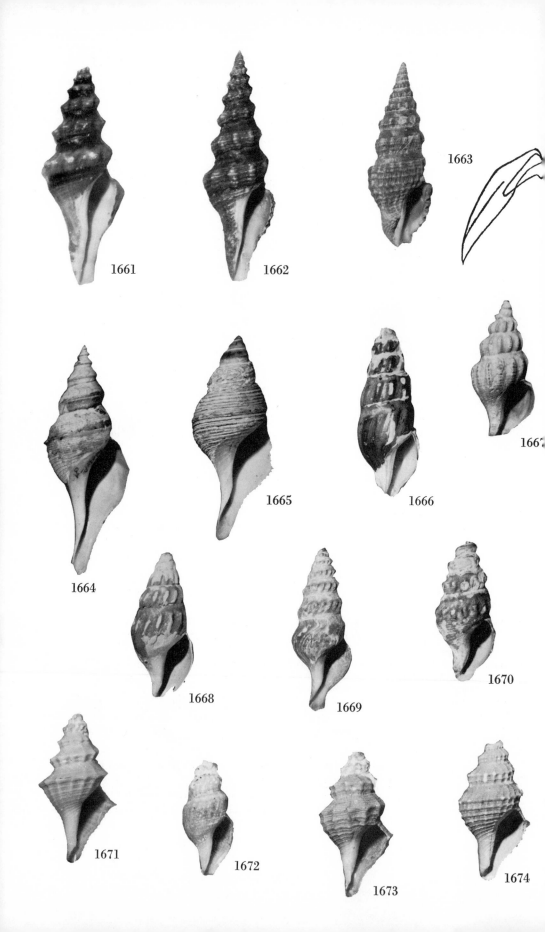

65 mm; diameter, 24 mm. Head of the Gulf of California to Banderas Bay, Mexico, in 20 to 50 m.

1662. Knefastia walkeri Berry, 1958. The relatively thin shell has a shiny yellowish-brown periostracum and two spiral color bands, a dark brown one below the periphery and a lighter but wider one on the base. The axial ribs are exceptionally narrow and are overridden by numerous spiral cords. Length, 67 mm; diameter, 23 mm. Angel de la Guarda Island, Gulf of California, to Isabela Island, off Mazatlán, Mexico, in 40 to 70 m.

Genus PYRGOSPIRA McLEAN, 1971

Shell of medium size, high-spired, with tabulate whorls, shoulder concave, periphery with vertical nodes; sculpture across the body whorl coarsely clathrate, anterior canal short, stromboid notch deep. Mature specimens have a slight deposition of columellar callus next to the sinus. Radula with marginal teeth only; main arm of the tooth massive, distal limb small. An Atlantic representative of this genus is *P. ostrearum* (Stearns, 1872).

1663. Pyrgospira obeliscus (Reeve, 1843) (Synonyms: **Clathrodrillia aenone** Dall, 1919; **Crassispira tomliniana** Melvill, 1927; **Clathrodrillia nautica** Pilsbry & Lowe, 1932). The shell is yellow-orange under a brown periostracum. Shoulder concave, with a single low thread. On mature specimens the siphonal fasciole is recurved and the outer lip margin crenulated by the spiral sculpture. Length, 35 mm; diameter, 12 mm. Consag Rock, near the head of the Gulf of California, to Puerto Utria, Colombia, in depths to 40 m on sand bottoms. Reeve's type specimen was from an unknown locality but proves to represent this common eastern Pacific species. This is the type species of the genus.

Genus AFORIA DALL, 1889
(IRENOSYRINX DALL, 1908)

Large, thin, elongate, anterior canal flexed, long. Deep water.

1664. Aforia goodei (Dall, 1890 [*Leucosyrinx*]) (Synonyms: **L. goodei** var. **persimilis** Dall, 1890; **Irenosyrinx persimilis leonis** Dall, 1908; **L. persimilis blanca** Dall, 1919; **L. amycus** Dall, 1919). Queen Charlotte Sound, British Columbia, to southern Chile, in 1,220 to 1,950 m. The type species of *Irenosyrinx*.

Genus STEIRAXIS DALL, 1896

Large, with fine spiral sculpture and a long, bent anterior canal; sinus broad, without sutural fasciole. Deep water.

1665. Steiraxis aulaca (Dall, 1896) [*Pleurotoma*]. Acapulco, Mexico, to Panama, in 3,240 to 3,440 m. This is the type species of the genus.

Genus LEUCOSYRINX DALL, 1889

Large, thin, with a turreted spire and a slightly bent anterior canal; anal sinus occupying entire area above periphery, notch broad and shallow. Deep water. Some of the species listed here eventually may be allocated elsewhere.

1666. ? Leucosyrinx clionella Dall, 1908. Length, 35 mm; diameter, 12 mm. Gulf of Panama to Ecuador, in 733 to 935 m.

1667. ? L. equatorialis (Dall, 1919) [*Lora*]. Length, 13 mm; diameter, 6 mm. Manta, Ecuador, to Patagonia, 223 to 733 m.

1668. ? L. esilda (Dall, 1908) [*Pleurotomella*]. Length, 23 mm; diameter, 10 mm. Gulf of Panama, 1,335 m.

1669. L. exulans (Dall, 1890) [*Pleurotoma*]. Length, 32 mm; diameter, 13 mm. Ecuador and Galápagos Islands, 1,159 to 2,070 m.

1670. L. herilda (Dall, 1908) [*Gemmula*]. Length, 18 mm; diameter, 7 mm. Gulf of Panama, 3,050 m.

Genus **ANTICLINURA** THIELE, 1934

(**CLINUROPSIS** THIELE, 1929 [not VINCENT, 1913])

Deepwater shells of moderate size, periphery strongly angulate, canal elongate. There is no apparent sinus. The closest affinity seems to be to *Marshallena* Allan, 1926, a deepwater Indo-Pacific group, which was placed by Powell (1969) in the Turriculinae.

1671. Anticlinura monochorda (Dall, 1908) [*Clinura*]. Length, 11.5 mm; diameter, 6.5 mm. Gulf of Panama, 1,870 m. Type of the genus.

1672. ? A. movilla (Dall, 1908) [*Mangilia*]. Length, 5 mm; diameter, 2 mm. Off Acapulco, Mexico, 1,390 m.

1673. A. peruviana (Dall, 1908) [*Clinura*]. Length, 9 mm; diameter, 4 mm. Northern Peru, 1,896 m.

1674. A. serilla (Dall, 1908) [*Gemmula*]. Length, 8 mm; diameter, 4 mm. Gulf of Panama, 2,323 m.

Subfamily **CRASSISPIRINAE**

Medium-sized to large shells with well-developed parietal callus bordering the sub-tubular sinus; usually with a narrow but evident fold below the suture, the shoulder area otherwise sculptured only by growth lines; body whorl with axial ribs and spiral cords. Operculum leaf-shaped, with a terminal nucleus. Protoconch initially smooth-whorled, then developing axial riblets. Radula of the duplex type, with a pair of elongate marginals only and a narrow, much smaller accessory plate superimposed on the larger main member. A rachidian radula tooth is present in only one group, a subgenus of *Crassispira*.

Morrison (1966) recognized the subfamily Crassispirinae, but his limits are different from those in the present revision. Powell (1966) allocated these genera to the Clavinae. The key is to generic and subgeneric units in the eight Panamic genera.

1. Back of last whorl with a thickened axial rib behind the lip 2
   Back of last whorl lacking such a rib 12
2. Shoulder concave, subsutural cord wanting . . . .*Carinodrillia*
   Shoulder with a subsutural cord 3
3. Ground color tan . . . .*Buchema*
   Ground color dark brown or black 4
4. Axial ribs strong across base, spiral sculpture fine 5
   Axial and spiral sculpture weak on base, but forming strong tubercles at
     intersections 9

5. Sinus shallow, lip thick at edge..............................*Striospira*
   Sinus deep, lip thin at edge....................................... 6
6. Sinus with open slot leading toward suture.......................... 7
   Sinus lacking open slot.. 8
7. Subsutural cord smooth...............................*Crassispira, s. s.*
   Subsutural cord beaded.......................................*Burchia*
8. Parietal callus a rounded tubercle...........................*Crassiclava*
   Parietal callus angular, spurlike..........................*Crassispirella*
9. Shell over 25 mm in length..10
   Shell under 25 mm in length.......................................11
10. Sinus lacking an open slot leading toward the suture...........*Gibbaspira*
    Sinus with an open slot leading toward the suture..............*Glossispira*
11. Sinus deep ....*Dallspira*
    Sinus shallow ....*Monilispira*
12. Lip lacking massive terminal varix.................................13
    Lip with massive terminal varix....................................14
13. Subsutural cord lacking......................................*Doxospira*
    Subsutural cord prominent....................................*Hindsiclava*
14. Sinus entrance broadly open............................*Miraclathurella*
    Sinus nearly sealed with abutting callus.............................15
15. Anterior canal elongate, lip edge thin...................*Lioglyphostoma*
    Anterior canal short, lip edge thick.........................*Maesiella*

Genus **CRASSISPIRA** SWAINSON, 1840
(ICZN Opinion 754, 1965)

Medium- to large-sized dark-colored shells, some with banding of lighter colors, fusiform but with anterior end truncate, with a narrow, raised subsutural fold, shoulder usually concave; back of last whorl with a thickened axial rib behind lip.

The Crassispiras have their center of abundance in the tropical American region, and a relatively large number of species occur. The nine subgenera employed here are based chiefly upon the structure of the sinus, secondarily on sculpture. A number of species previously regarded as *Crassispira* are here assigned to other genera, some even to other subfamilies.

Subgenus **CRASSISPIRA**, *s. s.*

Relatively large and slender; periostracum thick, dark brown; aperture elongate, axial and spiral sculpture subdued across the base; sinus moderately deep; parietal wall with a callus tubercle; posterior part of the aperture narrowed above the sinus into a vertical slot.

1675. Crassispira (Crassispira) incrassata (Sowerby, 1834) (Synonym: **Pleurotoma bottae** Kiener, 1839–40, *ex* Valenciennes, MS). The shell surface is glossy black, aperture and parietal callus pad whitish, lip edge crenulate, thickened by a swollen rib. Head of the Gulf of California, to the Gulf of Guayaquil, Ecuador, at low tide and offshore to depths of 35 m, on sandy bottoms. The type species of the genus is *C. bottae*, which is here regarded as a synonym of *C. incrassata*.

1676. Crassispira (Crassispira) maura (Sowerby, 1834) (Synonyms: **Turricula nigricans** Dall, 1919; **Drillia inaequistriata** Li, 1930 [cited as *inter-*

striata by Pilsbry, 1931]; **C. perla** M. Smith, 1947). Largest and most elongate member of the subgenus, the shell is brown with a dark periostracum; axial ribs are nodular on the spire and slant obliquely on the base; the spiral sculpture across the base is relatively coarse. Length, 70 mm; diameter, 17 mm. Bahía de los Angeles, Gulf of California, to Santa Elena Bay, Ecuador, on sandy and muddy bottoms to depths of 55 m. Immature specimens frequently have a pale basal periostracum, as in Dall's taxon, *T. nigricans*.

Subgenus GLOSSISPIRA McLEAN, 1971

Relatively large and solid; periostracum thin; subsutural cord and base with coarse tubercles in a clathrate pattern; sinus narrow, shallow; posterior part of the aperture narrowed above the sinus into a vertical slot; lip edge forming a projecting tongue between the sinus and the vertical slot. Monotypic (known only by a single species).

1677. Crassispira (Glossispira) harfordiana (Reeve, 1843) (Synonyms: **Pleurotoma corrugata** of authors, not of Sowerby, 1834; **C. adamsiana** Pilsbry & Lowe, 1932). Shell dark gray, fusiform, with small spots of white at the upper ends of the axial ribs and occasional white blotching on the shoulder. Length, 35 mm; diameter, 15 mm. Concepción Bay, Baja California, to San Francisco Bay, Ecuador, at low tide and offshore to depths of 30 m, on muddy bottoms.

Subgenus BURCHIA BARTSCH, 1944

Relatively large, slender, high-spired shells; axial sculpture consisting of low ribs reappearing as beads in the immediate subsutural area; spiral sculpture of fine cording, increasing in strength toward the base. Sinus with a posteriorly directed slot and a glossy parietal tubercle. The sinus is broader and not as deep as that of *Crassispira, s. s.*

1678. Crassispira (Burchia) semiinflata (Grant & Gale, 1931) (Synonym: **Pseudomelatoma redondoensis** T. Burch, 1938). The glossy-surfaced periostracum is black, although some specimens show brown between the axial ribs. Length, 48 mm; diameter, 14 mm. Santa Barbara, California, to Turtle Bay, Baja California, in 10 to 30 m, on gravel bottoms. The largest member, and the type species, of the subgenus.

1679. Crassispira (Burchia) unicolor (Sowerby, 1834) (Synonyms: **C. erebus** Pilsbry & Lowe, 1932; **C. tangolaensis** Hertlein & Strong, 1951). Yellowish brown under a dark periostracum; axial ribs sinuous and spiral sculpture consisting of fine striae on the base, subsutural beading prominent. Length, 24 mm; diameter, 9 mm. Head of the Gulf of California to Santa Elena Bay, Ecuador, in sandy areas near rocks at low tide and offshore to 20 m depth.

Subgenus CRASSICLAVA McLEAN, 1971

Relatively large and slender, aperture elongate, the closely adherent periostracum dark brown, lighter between the axial ribs; sinus moderately deep, bordered on the parietal wall by a callus tubercle. The radula of the two species differs from that of all the other members of the subfamily in having a moderately large rachidian tooth in addition to the duplex marginal teeth.

1680. Crassispira (Crassiclava) cortezi Shasky & Campbell, 1964. Axial ribs nodulous on the spire but elongate across the base; spiral sculpture lacking

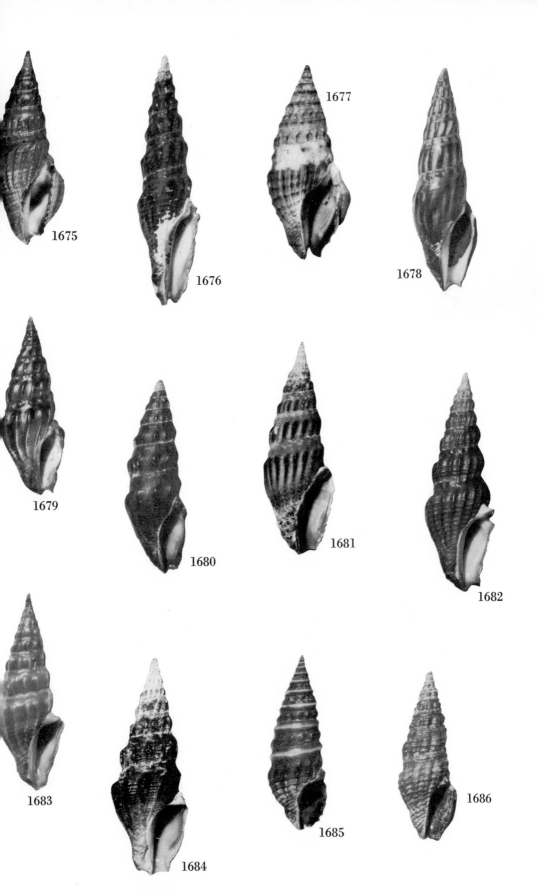

1675

1676

1677

1678

1679

1680

1681

1682

1683

1684

1685

1686

on the spire, increasing in strength toward the base. Length, 26 mm; diameter, 9 mm. Gulf of California: Guaymas to Carmen Island, on gravel near rocks, in 10 to 20 m.

1681. Crassispira (Crassiclava) turricula (Sowerby, 1834) (Synonyms: **Pleurotoma corrugata** Sowerby, 1834; **P. sowerbyi** Reeve, 1843). Shell orange-colored, with dark ribs, under an adherent periostracum; spiral sculpture producing fine beads on crossing the narrow axial ribs. Length, 36 mm; diameter, 12 mm. San Hipolito Point, outer coast of Baja California, north in the Gulf of California to Guaymas, and south to Santa Elena Bay, Ecuador, in 20 to 55 m. This is the type species of the subgenus.

Subgenus CRASSISPIRELLA BARTSCH & REHDER, 1939

Sinus deep, laterally directed, parietal callus directed downward, layered in mature specimens, but not tightly constricting the sinus opening. Shoulder area with subsutural thread, otherwise concave; base evenly rounded below the shouldered periphery, with axial and spiral ribbing.

1682. Crassispira (Crassispirella) ballenaensis Hertlein & Strong, 1951. The dark brown shell has a rounded periphery; axial ribs are markedly narrow, crossed by evenly spaced spiral cords, producing a reticulate sculpture pattern. Described originally as a subspecies of *C. turricula*, from which it differs in sculpture and in sinus structure. Length, 33 mm; diameter, 11 mm. Banderas Bay, Mexico, to the Gulf of Nicoya, Costa Rica, in 20 to 90 m.

1683. Crassispira (Crassispirella) brujae Hertlein & Strong, 1951. The shell is yellow-orange under a periostracum that varies from black to brown; the axial ribs are nearly smooth on the spire but intersected by spiral sculpture across the base, the sinus walls projecting on mature specimens. Length, 29 mm; diameter, 9 mm. San Francisco Island, Gulf of California, to Puerto Utria, Colombia, in 30 to 80 m.

1684. Crassispira (Crassispirella) chacei Hertlein & Strong, 1951. Relatively large; outline tabulate, spiral cords fine and of even strength and spacing, the sinus opening becoming constricted in mature shells. The shell is yellow-orange under a thin brown periostracum, darker along the axial ribs. Length, 33 mm; diameter, 11 mm. Tiburon Island, Gulf of California, to Arena Bank, Baja California, in 40 to 100 m.

1685. Crassispira (Crassispirella) discors (Sowerby, 1834). This is closely related to *C. rustica* but smaller and characterized by a prominent yellow subsutural cord; axial and spiral ribs forming tubercles at intersections. Length, 21 mm; diameter, 8 mm. Guaymas, Mexico, to La Plata Island, Ecuador, on gravel bottoms in rocky areas, low tide to 25 m.

1686. Crassispira (Crassispirella) epicasta Dall, 1919. The small shell is dark brown to black, with a broad dark orange band at the periphery, sculptured with prominent spiral cords that override the strong axial ribs. In addition there are fine axial and spiral striae. Length, 13 mm; diameter, 4 mm. Banderas Bay, Mexico, to Puerto Utria, Colombia, on gravel bottoms in rocky areas, intertidally to 40 m.

1687. Crassispira (Crassispirella) rugitecta (Dall, 1918). Dark brown, with a yellow band at the periphery; axial and spiral sculpture strong, the spiral

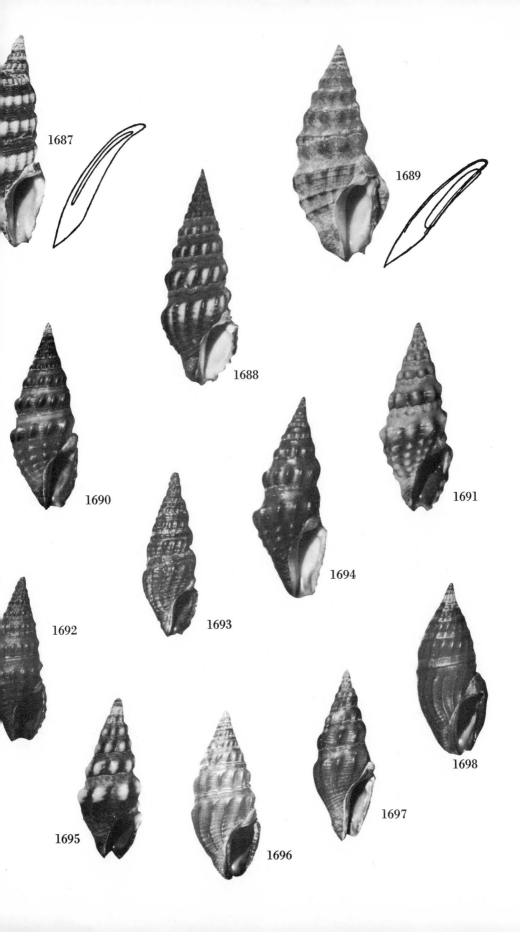

1687

1689

1688

1690

1691

1693

1694

1692

1695

1696

1697

1698

striae especially prominent throughout the shoulder area. Length, 30 mm; diameter, 10 mm. Type species of the subgenus, known chiefly from the outer coast of Baja California as far north as Cedros Island, and in the Gulf of California at Angel de la Guarda and Espíritu Santo Islands, in 40 to 110 m, gravel bottoms.

1688. Crassispira (Crassispirella) rustica (Sowerby, 1834). The brown shell has darker axial ribs and a lighter band that shows up in the interspaces of the axial ribs at the periphery; axial ribs across the base rendered almost spinose by the spiral sculpture. Length, 33 mm; diameter, 12 mm. San Luis Island, Gulf of California, to Manta, Ecuador, occurring at low tide in Panama and offshore to depths of 20 m.

Subgenus GIBBASPIRA McLEAN, 1971, *ex* BARTSCH, MS

Moderately large shells, the subsutural cord swollen and bluntly noded, the shoulder concave below; shell spirally striate throughout, base with strong axial and spiral sculpture, noded at intersections, axial sculpture terminating above in white-tipped nodes at the periphery; sinus deep, its slot walls laterally directed, with callus tubercles above and below. The suture descends and then rises on the last whorl, giving a lateral twist to the shell.

1689. Crassispira (Gibbaspira) rudis (Sowerby, 1834) (Synonym: **Drillia albovallosa** Carpenter, 1857). The laterally distorted shell is gray, the upper sides of the main peripheral nodes whitish. The spire erodes to reveal a white band on the early whorls. Length, 27 mm; diameter, 16 mm; specimens from Mexico somewhat smaller. Mazatlán, Mexico, to Santa Elena Peninsula, Ecuador, intertidally and offshore on rocky bottoms. This is the type species of the subgenus. A related Caribbean species is *C. dysoni* (Reeve, 1846).

Subgenus DALLSPIRA BARTSCH, 1950

Subsutural cord weakly developed, sinus deep, its slot walls apically directed, bordered within by a curved parietal callus. Periphery with strongly projecting nodes; axial and spiral sculpture subdued across the base but forming strong tubercles at intersections.

1690. Crassispira (Dallspira) abdera (Dall, 1919) (Synonyms: **Dallspira dalli** and **lowei** Bartsch, 1950). Shells dredged offshore are dull waxen in color, those collected intertidally are dark brown with paler zones along the shoulder and across the base. The subsutural thread is not pronounced, and the peripheral nodes are elongate and somewhat slanted, the tubercles of the base usually lighter in color. Length, 17 mm; diameter, 7 mm. San Juan del Sur, Nicaragua, to Santa Elena Peninsula, Ecuador, at low tide and to depths of 33 m. This was described as an *Elaeocyma*, from a light-colored offshore specimen. The type species of the subgenus is the synonymous *Dallspira dalli*, based on a dark-colored intertidal specimen.

1691. Crassispira (Dallspira) bifurca (E. A. Smith, 1888) (Synonym: **C. flavonodosa** of authors, not Pilsbry & Lowe, 1932). The brown shell has massive yellow peripheral tubercles, the shoulder area gray, the subsutural cord yellow, broad and undulating; sinus deep, the parietal callus nearly obstructing the opening. This is often confused with *Pilsbryspira nymphia*, but the two species are different in sinus structure. Length, 18 mm; diameter, 6.5 mm. Head of the Gulf of California, Mexico, to Santa Elena Peninsula, Ecuador, at low tide and offshore in rocky areas. E. A. Smith's specimen was from an unknown locality;

PLATE IX · *Conus dalli* Stearns (Dall's Cone)

the holotype of the Pilsbry & Lowe species is a worn specimen of *C. eurynome*, although paratypes do represent the intended species.

1692. Crassispira (Dallspira) cerithoidea (Carpenter, 1857). Relatively small, uniformly dark brownish black; the subsutural cord weak, the periphery noded, and the base with coarse tubercles. The shell is finely striate throughout. Length, 15 mm; diameter, 7 mm. Mazatlán to Barra de Navidad, Mexico, on rocky bottoms offshore to depths of 20 m.

1693. Crassispira (Dallspira) coelata (Hinds, 1843) (Synonyms: **Pleurotoma caelata** of authors, unjustified emendation; **Drillia hanleyi** Carpenter, 1857). Most slender of the Crassispiras, the shell is dark brown, slanted axial ribs numerous, and spiral threads increasing in strength toward the base. The subsutural cord is narrow and raised. Length, 11 mm; diameter, 4 mm. Mazatlán, Mexico, to Panama Bay, at low tide on gravel in rocky areas and offshore to depths of 35 m.

1694. Crassispira (Dallspira) erigone Dall, 1919. This resembles *C. abdera* but is larger, has a more elongate aperture, and the axial ribs are nearly vertical at the periphery; base with strong tubercles. Color, orange-brown to dark brown. Length, 22 mm; diameter, 8 mm. Guaymas, Mexico, to Cape San Francisco, Ecuador, in 20 to 50 m.

1695. Crassispira (Dallspira) eurynome Dall, 1919 (Synonym: **C. flavonodosa** Pilsbry & Lowe, 1932). A small member of the subgenus, this resembles *C. bifurca* in being brightly colored, with yellowish nodes on a dark ground, the shoulder area gray. The peripheral nodes are more elongate than those of *C. bifurca* and the basal nodes fewer and less prominent. Length, 12 mm; diameter, 5 mm. Mazatlán, Mexico, to San Juan del Sur, Nicaragua and Panama Bay, at low tide and to depths of 20 m, in rocky areas.

1696. Crassispira (Dallspira) martiae McLean & Poorman, 1971. A small form resembling *C. coelata* in general sculptural features, but much broader and with a less prominent subsutural cord, the shoulder area adjacent to the subsutural cord showing pronounced spiral striae, walls of sinus slot nearly vertical, the slot thus apically directed. Length, 10.7 mm; diameter, 4.7 mm. Panama Bay, on gravel bottoms near rocks, intertidally.

Subgenus **STRIOSPIRA** BARTSCH, 1950
(**ADANACLAVA** BARTSCH, 1950)

Medium sized; sinus relatively shallow and open, indenting the massive thickened lip, the walls of the sinus slot apically directed; parietal edge of sinus with a tubercle glazed over by callus; base evenly rounded below the shouldered periphery, with axial ribs and finer spiral sculpture. The suture rises on the last whorl, increasing the length of the aperture and parietal callus. The entire radular ribbon is unusually small, which suggests that it may have limited use.

1697. Crassispira (Striospira) adana (Bartsch, 1950). Relatively small, yellowish brown in color, shoulder deeply concave, with only a faint subsutural cord but with well-developed spiral striae. As in *C. nigerrima* and *C. tepocana*, the parietal callus extends toward the apex. Length, 14 mm; diameter, 5.5 mm. Mazatlán, Mexico, to Panama Bay, in 30 to 55 m. This is the type species of *Adanaclava*.

1698. Crassispira (Striospira) coracina McLean & Poorman, 1971. This has the general appearance of *C. xanti* and *Strictispira stillmani*, with an elevated, subsutural cord and sinuous axial ribs, but the sinus structure differs from that in both. The shallow sinus is narrow and laterally directed, its upper edge on a continuous curve of the inner lip; the outer lip has a massive appearance. The glazed tubercle characteristic of *Striospira* is lacking. Length, 15 mm, diameter, 6.5 mm. El Pulmo, Baja California, to Panama Bay, at low tide and offshore in gravel near rocks.

1699. Crassispira (Striospira) kluthi E. K. Jordan, 1936 (Synonyms: **Clavatula luctuosa** Hinds, 1843, not *Pleurotoma luctuosa* Orbigny, 1842; **S. lucasensis** and **tabogensis** Bartsch, 1950). Glossy black, the shoulder smooth except for a faint subsutural cord; periphery weakly noded; base with fine spiral striae, axial ribbing variable, frequently obsolete. Length, 16 mm; diameter, 6 mm. Cedros Island, outer coast of Baja California and throughout the Gulf of California, south to Salango Island, Ecuador, at low tide on mud and sand flats and to depths of 50 m. Type of the subgenus (as *S. lucasensis*). The familiar early name of Hinds, *C. luctuosa*, is preoccupied by that of Orbigny, a West Indian *Crassispira* of the subgenus *Crassispirella*.

1700. Crassispira (Striospira) nigerrima (Sowerby, 1834) (Synonym: **Pleurotoma cornuta** Sowerby, 1834). The glossy black shell is larger than *C. kluthi* and is characterized by a massive deposition of callus projecting toward the apex; the shoulder has a single low but prominent cord, and the axial ribs are slightly noded at the periphery. Length, 20 mm; diameter, 8 mm. Cape San Lucas, Baja California, to Santa Elena Bay, Ecuador, at low tide on mud flats and to depths of 60 m.

1701. Crassispira (Striospira) tepocana Dall, 1919. Resembling *C. nigerrima* in the structure of the sinus but larger and more slender, the surface brown and rough rather than glossy black, the shoulder area with fine spiral threading rather than a single prominent cord. Length, 24 mm; diameter, 8 mm. Angel de la Guarda Island and Cabo Tepoca, Gulf of California, to Santa Elena Bay, Ecuador, in 30 to 70 m.

1702. Crassispira (Striospira) xanti Hertlein & Strong, 1951. A stout black form resembling *C. kluthi* but with a strong subsutural cord; axial ribs are narrow; spiral striae strong over the entire surface. Larger than *Strictispira stillmani*, this lacks the deep sinus and blue-gray periostracum of that species. Length, 16 mm; diameter, 6 mm. Punta Lobos, Sonora, Mexico, to Santa Elena Bay, Ecuador, in 20 to 55 m. On some specimens the axial ribs may be obsolete, but the prominent subsutural cord distinguishes this species from *C. kluthi*.

Subgenus MONILISPIRA BARTSCH & REHDER, 1939

Medium- to small-sized, sinus relatively shallow, its slot apically directed, parietal callus of moderate thickness, layered down toward the sinus; outer lip not greatly thickened; spiral sculpture relatively strong, forming brightly colored tubercles on overriding the axial ribs; subsutural cord of moderate strength, surface finely spirally striate throughout. This group has been confused with *Pilsbryspira*, in which there is truly toxoglossate radular dentition and a different sinus structure.

1703. Crassispira (Monilispira) appressa (Carpenter, 1864). Most slender member of the subgenus, this has strong axial ribbing, the peripheral nodes elongate. The subsutural cord is swollen and nodular. Specimens from the northern part of the range are gray with whitish markings on the nodes and tubercles; those from the southern area tend to have orange cording, the ground color of

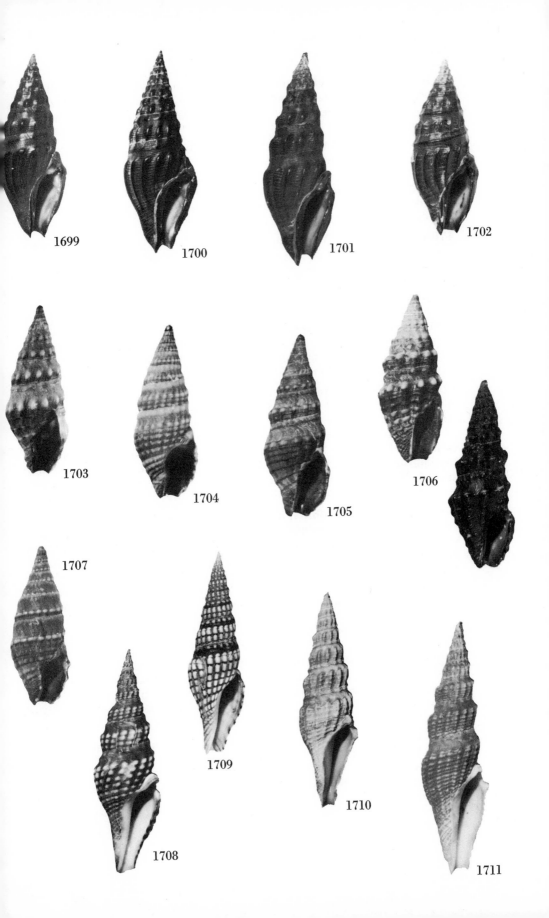

the shoulder and pillar gray and of the periphery brown. Length, 7.5 to 15 mm; diameter, 3 to 5 mm. Western side of the Gulf of California, Puertecitos to Cape San Lucas and the Tres Marias Islands, Mexico, at low tide and offshore, in rocky areas.

1704. Crassispira (Monilispira) currani McLean & Poorman, 1971. Relatively small, the color pattern is characteristic, the spiral cords being dark orange upon a gray ground, the subsutural band broad and uniformly orange. Length, 11 mm; diameter, 4 mm. Sayulita, Nayarit, to Banderas Bay, Jalisco, Mexico, at low tide and offshore, in gravel near rocks. A red filter was used to bring out the dark orange pattern in the photograph.

1705. Crassispira (Monilispira) monilifera (Carpenter, 1857). Upon a gray ground there is a peripheral row of projecting yellowish-white to orange tubercles, the basal cording a faint muddy yellow or orange, the subsutural cord narrow and faintly colored, the shoulder area often a lighter shade of gray, some specimens with a gray band across the base. Sculpture is more subdued that that of *C. pluto*, where the two occur together. Length, 18 mm; diameter, 6.5 mm. Western side of the Gulf of California from San Luis Gonzaga Bay to Ceralvo Island; on the mainland from Mazatlán to Manzanillo, Mexico, at low tide and offshore in rocky areas. This is the type species of the subgenus. It should not be confused with *Pilsbryspira nymphia*, which it superficially resembles.

1706. Crassispira (Monilispira) pluto Pilsbry & Lowe, 1932. Largest and most coarsely sculptured member of the subgenus, the gray-black shell has clathrate sculpture across the base and a broad, undulating subsutural cord. Typical specimens have the upper half of the main peripheral nodes yellow or white, but variant specimens occur at Puertecitos and Bahía San Luis Gonzaga in which the basal tubercles are also brightly marked. Length, 19 mm; diameter, 7 mm. Head of the Gulf of California to Guaymas on the east and Cape San Lucas on the west, at low tide in rocky areas.

1707. Crassispira (Monilispira) trimariana (Pilsbry & Lowe, 1932). Characterized by having twice as many peripheral tubercles as any of the other Monilispiras, the fusiform shell is gray on the shoulder and black on the base, with only the slightest indentation at the shoulder, the subsutural cord but faintly indicated. The main peripheral tubercles are gray-white, the basal tubercles only faintly colored. Length, 11 mm; diameter, 4 mm. Santa Cruz, Nayarit, to Banderas Bay and the Tres Marias Islands, Mexico, at low tide and just offshore in rocky areas.

Genus HINDSICLAVA HERTLEIN & STRONG, 1955
(TURRIGEMMA BERRY, 1958)

Light-colored shells with thin olivaceous periostracum, straight-sided, with a high spire and an elongate aperture; shoulder concave and smooth but for a narrow raised subsutural cord; sculpture coarsely reticulate, sinus broad and deep, callus at the upper extremity forming a projecting tongue in mature shells. Radula of the duplex type.

1708. Hindsiclava andromeda (Dall, 1919) (Synonym: **Turrigemma torquifer** Berry, 1958). Mature shells tan to tawny brown, with coarse, white-surfaced tubercles across the body whorl. Length, 39 mm; diameter, 12 mm. Western side of the Gulf of California: Angel de la Guarda Island to Cape San Lucas, in 40 to 160 m.

1709. Hindsiclava hertleini Emerson & Radwin, 1969. Sculptured with squarish white nodules, separated by axial and spiral channels, the spiral channels and the smooth shoulder darker-colored; the surface gives the illusion of high relief. Fully mature specimens with the projecting spur are as yet unknown. Length, 37 mm; diameter, 12 mm. Galápagos Islands, in 70 to 100 m.

1710. Hindsiclava militaris (Reeve, 1843, *ex* Hinds, MS) (Synonyms: **Turricula dotella** and **notilla** Dall, 1908). The slanting axial ribs are more strongly elevated than the spiral cords and are finely nodulous at the intersections. The number of axial ribs and the size at maturity vary greatly; few specimens show the mature development of the lip with the projecting spur. Length, 30 to 40 mm; diameter, 9 to 11 mm. San Luis Gonzaga Bay, Gulf of California, to Cabita Bay (Cabo Corrientes), Colombia, in depths of 20 to 55 m. This is the type species of the genus.

1711. Hindsiclava resina (Dall, 1908). Largest member of the genus, this resembles *H. militaris* but has finer and more numerous lines of spiral cords and a straighter basal profile, lacking any constriction on the pillar. Length, 55 mm; diameter, 15 mm. The broken type specimen was dredged at 589 m in the Gulf of Panama and presumably originated in shallower water; one fully mature specimen is in the LACM collection from 82 m off Puerto Utria, Colombia.

Genus DOXOSPIRA McLEAN, 1971

Relatively large, fusiform, with elongate canal, axial sculpture strong on the early whorls; shoulder concave and smooth, lacking a subsutural cord; sinus broad and shallow, the parietal callus at the upper extremity forming a spur as in *Hindsiclava*. Radula of the duplex type. This appears most closely related to *Hindsiclava* on the basis of sinus structure but differs in the overall fusiform shape and lack of the subsutural cord. Only a single species known.

1712. Doxospira hertleini Shasky, 1971. Relatively large, with numerous spiral cords below an unusually broad, smooth shoulder area; three spiral cords are prominent on the early whorls but do not form a distinct carination. Yellow-white, with irregular narrow banding of brown below the periphery. Length, 42 mm; diameter, 17 mm. Judas Point, Costa Rica, to La Plata Island, Ecuador, 20 to 75 m.

Genus BUCHEMA COREA, 1934

Moderate- to small-sized, high-spired, with tabulate whorls, anterior end truncate, aperture narrow, lip with distinct stromboid notch, sinus deep, U-shaped, upward-directed, bordered within by a curved callus pad; axial sculpture of massive ribs, overridden by narrow spiral cords, subsutural cord a thickened fold adjacent to the suture. Light colored; dark areas between axial ribs. Radula of the duplex type. The type and several other species are known from the Caribbean.

1713. Buchema granulosa (Sowerby, 1834) (Synonym: **Clathrodrillia callianira** Dall, 1919). Spiral ribs are rather crowded and irregular on the relatively small shell, and an inflated cord adjoins the suture above the concave shoulder. Some specimens are uniformly grayish white, others with dark areas between the axial ribs, and occasional specimens are dark across the entire base. Length, 19 mm; diameter, 6 mm. Puertecitos, Gulf of California, to La Libertad, Ecuador, in 20 to 55 m.

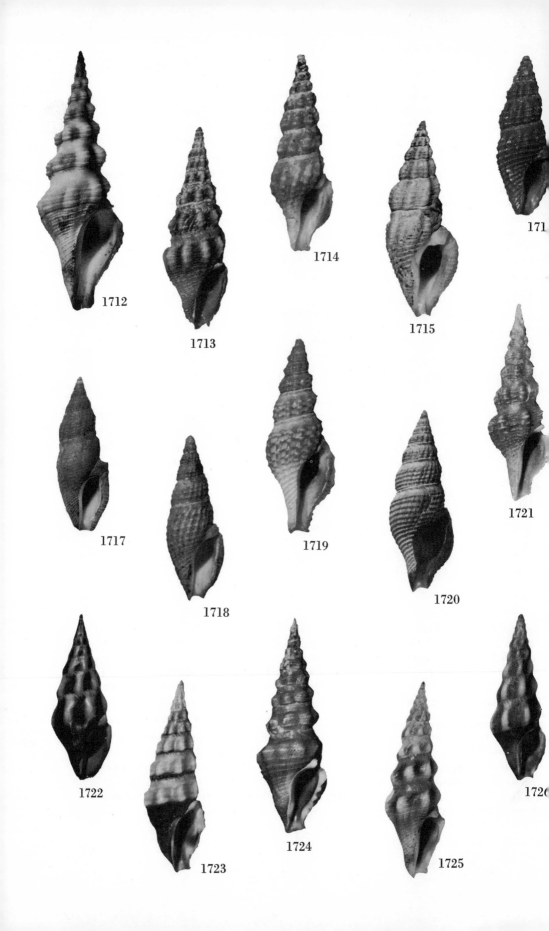

1712

1713

1714

1715

171[]

1717

1718

1719

1720

1721

1722

1723

1724

1725

1726

Genus LIOGLYPHOSTOMA WOODRING, 1928
(GLYPHOSTOMOPS BARTSCH, 1934)

Medium- to small-sized shells with rounded whorls, shoulder not deeply concave, subsutural cord weakly developed or wanting; first two nuclear whorls smooth, rounded, third nuclear whorl subcarinate; mature sculpture of strong sinuous axial ribs, overridden by spiral cords; aperture elongate, twisted to the left, outer lip with a massive varix and a thin, projecting edge; sinus deep, subtubular, vertically walled, parietal callus pad of mature specimens rounded, sinus entrance nearly sealed by callus projecting from the lip. Operculum with terminal nucleus, radula of the duplex crassispirine type.

Lioglyphostoma has traditionally been regarded as closely related to *Glyphostoma*, but its radula and operculum with terminal nucleus removes it from that association. The subcarinate third whorl of the nucleus, constricted sinus rim, and twisted canal are the characteristic features. The type species is from the Miocene of Jamaica.

1714. Lioglyphostoma ericea (Hinds, 1843) (Synonyms: **Glyphostoma sirena** Dall, 1919; **Clathurella erminiana** Hertlein & Strong, 1951). Light brown in color and weakly shouldered, axial ribs are slanting, crossed by two to three spiral cords per whorl, forming elongate whitish beads at intersections. Length, 16 mm; diameter, 6 mm. Tiburon Island, Gulf of California, to Gorgona Island, Colombia, and the Galápagos Islands, 40 to 120 m.

1715. Lioglyphostoma rectilabrum McLean & Poorman, 1971. This lacks all traces of a subsutural cord; axial ribs are strong on the early whorls of the white shell, spiral cords and striae strong throughout; mature lip greatly thickened. Length, 16 mm; diameter, 6 mm. Guaymas, Mexico, in 30 to 80 m.

Genus MAESIELLA McLEAN, 1971

Small to medium-sized, whorls rounded, shoulder not deeply concave, subsutural cord a narrow raised thread; first two nuclear whorls smooth, rounded; strong diagonal axial ribs arise on the third nuclear whorl, persist for half a turn and abruptly cease, replaced by weaker vertical ribs and spiral cords; mature sculpture of sinuous axial ribs (obsolete on final whorl in some species), crossed by spiral cords and microscopic spiral striae, aperture elongate but not drawn into an anterior canal, outer lip thickened by a massive varix, stromboid notch shallow; sinus deep, the opening nearly obstructed by downward growth of the lip between the sinus and body whorl. Operculum with terminal nucleus; radula of the duplex type.

1716. Maesiella hermanita (Pilsbry & Lowe, 1932). The brown shell has coarse axial and spiral sculpture and fine, microscopic spiral striae, subsutural cord moderately developed. Length, 9.3 mm; diameter, 3.4 mm. Guaymas, Mexico, to Puerto Utria, Colombia, on gravel bottoms, 30 to 40 m.

1717. Maesiella maesae McLean & Poorman, 1971. A small, white to tan, robust form with a thin subsutural thread, spiral sculpture of cords and threads, axial sculpture weakly developed, parietal callus massive, nearly closing the sinus. Length, 9.2 mm; diameter, 3.3 mm. Puertecitos to Guaymas, Mexico, on gravel bottoms, in 15 to 30 m. The type species of the genus.

1718. Maesiella punctatostriata (Carpenter, 1856) (Synonym: **Crassispira solitaria** Pilsbry & Lowe, 1932). Shell grayish white to dark brown with sinuous axial ribs and spiral cords of nearly equal strength, a weak subsutural cord and fine spiral striae. Length, 18 mm; diameter, 6 mm. Mazatlán, Mexico, to Panama Bay. One specimen in the Shasky collection from Mazatlán is whitish, fully mature at a length of 12.6 mm.

Genus MIRACLATHURELLA WOODRING, 1928

Medium-sized shells with rounded whorls, subsutural cord prominent, first nuclear whorl rounded, next two whorls with a strong keel on the lower part of the whorl; mature sculpture of axial ribs and spiral cords giving a clathrate effect below the smooth, concave shoulder; canal elongate, straight; sinus deep, bordered by parietal callus but not constricted as in *Lioglyphostoma*. Operculum with terminal nucleus, radula of the duplex type. The type species, which resembles *M. mendozana*, is from the Miocene of Jamaica.

1719. Miraclathurella bicanalifera (Sowerby, 1834) (Synonyms: **Pleurotoma variculosa** Sowerby, 1834; **P. gracillima** Carpenter, 1856; **Lioglyphostoma acapulcanum** Pilsbry & Lowe, 1932). Early whorls with three to four spiral cords below the concave shoulder, spiral cords numerous and more closely spaced on the pillar; axial ribs closely spaced, obsolete on the base, beaded at intersections with the spiral cords. Yellowish white to brown, the pillar and the beaded tips of the ribs lighter in color. Adult size variable; length, 11 to 23 mm; diameter, 3.7 to 7.5 mm. San Luis Gonzaga Bay, Gulf of California, to Puerto Utria, Colombia, in 20 to 70 m.

1720. Miraclathurella mendozana Shasky, 1971. Lower-spired and broader than *M. bicanalifera*, with rounded whorls; axial ribs nearly obsolete on the final whorl, the spiral sculpture strong and beaded; uniformly brown. Length, 16 mm; diameter, 5.3 mm. Espíritu Santo Island, Gulf of California, to Cupica Bay, Colombia, in 40 to 130 m.

Genus CARINODRILLIA DALL, 1919

Shells of moderate size, shoulder concave, subsutural cord wanting; axial ribs on the early whorls relatively few, tending to be continuous from whorl to whorl; spiral sculpture weak or obsolete; periphery rounded; anterior canal of moderate length or relatively short; sinus deep, the slot walls directed apically, the parietal callus well developed but not tending to constrict the sinus entrance; first two nuclear whorls smooth, third whorl slightly subcarinate. Shells have a waxen surface texture, ground color light to dark brown, the axial ribs lighter in color. The radula differs from that of other crassispirine genera in having the main member more elongate, with more parallel sides, and the accessory plate less developed.

Most of the species previously regarded as *Carinodrillia* are here assigned to *Compsodrillia*, which has barbed, toxoglossate teeth, stronger spiral sculpture, and a more laterally directed sinus slot. In addition to having radular differences, *Carinodrillia* lacks the strong spirals and subsutural cord of *Buchema*, although sinus structure is much the same in the two genera.

1721. Carinodrillia adonis Pilsbry & Lowe, 1932. The periphery is rounded, the spiral cords relatively well developed; color, cinnamon-brown, the summits of the ribs whitish where crossed by the spiral cords. Length, 26 mm; diameter, 8 mm. Tiburon Island, Gulf of California, to Santa Elena Bay, Ecuador, and the Galápagos Islands, in 30 to 80 m.

1722. Carinodrillia alboangulata (E. A. Smith, 1882). Intercostal and basal areas are dark brown to black, the axial ribs lemon-yellow with yellow and black flecks in the subsutural area; anterior end truncate, spiral sculpture lacking except for striae on the pillar. Length, 19 mm; diameter, 7 mm. Mazatlán, Mexico, to Cape San Francisco, Ecuador, at low tide and to depths of 30 m, on gravel bottoms in rocky areas. Described without locality data and previously unillustrated, this has recently been recognized in the Panamic fauna.

1723. Carinodrillia dichroa Pilsbry & Lowe, 1932. The base is brown, the upper half of the whorl lemon-yellow. Spiral sculpture is moderately developed, the anterior canal more elongate than in *C. alboangulata*. Length, 26 mm; diameter, 9 mm. Puertecitos, Gulf of California, Mexico, to Santa Elena Bay, Ecuador, in 10 to 40 m, on gravel bottoms near rock.

1724. Carinodrillia halis (Dall, 1919). Unlike the other members of the genus, this has the periphery carinate, a single prominent spiral cord forming a low keel; base with numerous spiral cords. Buff-colored, the elongate nodes lighter in color than the rest of the shell. Length, 31 mm; diameter, 10 mm. Puertecitos, near the head of the Gulf of California, to Santa Elena Peninsula, Ecuador, in 20 to 55 m. This is the type of the genus. Dall's holotype was immature, and the species has been confused with the smaller *Compsodrillia albinodata*. *Carinodrillia halis* differs from *Compsodrillia alcestis* in having a different sinus structure, more waxen surface, weaker spiral cords, and a shorter canal.

1725. Carinodrillia hexagona (Sowerby, 1834) (Synonym: **Clathrodrillia pilsbryi** Lowe, 1935). Spiral cords are weakly developed, the six axial ribs per whorl forming massive rounded nodes at the periphery. Greenish brown, with lighter ribs and with dark brown between the ribs at the periphery. Length, 35 mm; diameter, 12 mm. Head of the Gulf of California, Mexico, to La Libertad, Ecuador, at low tide and to depths of 40 m, on sandy bottoms.

1726. Carinodrillia lachrymosa McLean & Poorman, 1971. Smaller and more slender than *C. alboangulata*, with a more elongate aperture. The color pattern differs—the axial ribs are white on a light brown ground, the base set off by being darker. Length, 16 mm; diameter, 5.5 mm. Guaymas, Sonora, to Barra de Navidad, Jalisco, Mexico, on gravel in rocky areas in 10 to 40 m.

Subfamily STRICTISPIRINAE

Dark-colored shells of moderate size, sculpture both axial and spiral, shoulder concave, with a well-marked subsutural cord; parietal callus well developed, partially obstructing the entrance to the deep, laterally directed sinus slot. Operculum leaf-shaped, its nucleus terminal. Radula with marginal teeth only, the teeth solid and massive, elbow-shaped, with a projecting collar-like structure on the inside.

Two genera with a distinctive and hitherto unrecorded radular pattern are grouped as a subfamily here. The radula in *Cleospira* somewhat resembles that of the Pseudomelatominae, except that it lacks any rachidian tooth. On the basis of the shell characters there are some disparities, for the members of the genus *Strictispira* resemble the Crassispirinae in having axial ribs predominating, whereas *Cleospira* resembles the Zonulispirinae in having spiral sculpture predominant. Anatomical studies can perhaps resolve the problems of the true relationships of this group.

Genus STRICTISPIRA McLEAN, 1971

Medium-sized, dark brown or black shells, with the appearance of *Crassispira*; posterior sinus deep, the slot laterally directed, the parietal callus so large on its lower side as nearly to seal off the sinus. Marginal tooth of the elbow type, with an inner flange, outer profile with about a 90-degree curve.

The sculpture resembles that of *Crassispira (Crassispirella)* and *C. (Striospira)*, but the constricted sinus is distinctive.

1727. Strictispira ericana (Hertlein & Strong, 1951). Brown under a thin, usually abraded periostracum, the shell is high-spired, the outline tabulate; spiral striae even, axial ribs variable in number. Resembling *Crassispira (Striospira) tepocana* but differing in sinus structure. Length, 22 mm; diameter, 7.5 mm. Puertecitos, near the head of the Gulf of California, to Santa Elena Bay, Ecuador, in 10 to 80 m. Originally regarded as a small form with limited distribution, this proves to be variable and wide-ranging. Type species of the genus.

1728. Strictispira stillmani Shasky, 1971. Ovate-biconic in shape; black under a thin blue-gray periostracum, the spiral cords increasing in strength toward the base. Resembling *Crassispira (Striospira) xanti* and *C. (S.) coracina* but differing in sinus structure. The blue-gray periostracum is distinctive. Length, 15 mm; diameter, 6 mm. Head of the Gulf of California to Panama Bay in sandy areas near rocks, intertidally and offshore. *Strictispira ebenina* (Dall, 1890) is a related Caribbean species.

Genus CLEOSPIRA McLEAN, 1971

Dark-colored, with a tabulate profile and truncate anterior end; axial ribs strong, crossed by broad spiral cords and narrow, deeply incised grooves; shoulder concave, with a narrow subsutural cord and fine spiral striae; axial ribs particularly numerous on the early whorls; sinus slot deep, laterally directed, parietal callus extending downward, nearly sealing off the sinus on mature shells; outer lip with a stromboid notch well developed. Radula of the elbow type, with an inner flange, the outer profile with a curve amounting to about 45 degrees. This differs from *Compsodrillia* by having much denser axial ribbing on the early whorls. It is in this character also similar to *Pyrgospira* and *Tiariturris*, but the parietal callus is more developed. One species known.

1729. Cleospira ochsneri (Hertlein & Strong, 1949) (Synonym: **Pleurotoma bicolor** Sowerby, 1834 [not Risso, 1826]). The base and the shoulder area are waxen brownish black, the periphery banded with yellow, darker in the grooves. Length, 18 mm; diameter, 6 mm. Galápagos Islands, at low tide and offshore in rocky areas. Also reported at Panama, but the record needs to be confirmed.

Subfamily ZONULISPIRINAE

Shells medium-sized to large, resembling the Crassispirinae in having well-developed parietal callus about the sinus and usually a narrow, elevated subsutural fold; sculpture of axial ribs and spiral cords. Unlike the Crassispirinae, these have the spiral sculpture usually stronger on the base. Operculum leaf-shaped, with a terminal nucleus; protoconch initially smooth-whorled or subcarinate, then developing axial riblets. Radula of marginals only that are hollow tubes, usually strongly barbed at the tip. This group differs from most toxoglossate subfamilies in having a fully developed operculum, which is usually either wanting or vestigial in the other toxoglossate subfamilies. The key is to generic and subgeneric units in the four Panamic genera.

1. Sinus slot laterally directed, constricted............................ 2
   Sinus with open slot, the walls directed toward the suture................ 4
2. Early whorls with spiral cording only......................*Zonulispira*
   Early whorls with axial ribbing or beading.......................... 3
3. Canal long, the spiral cords of base not beaded..............*Compsodrillia*
   Canal short, the spiral cords of base beaded.............*Pilsbryspira, s. s.*
4. Anterior canal short, truncate..........................*Nymphispira*
   Anterior canal elongate.................................*Ptychobela*

Genus ZONULISPIRA BARTSCH, 1950

Biconic shells with a short, truncate aperture; ground color black or brown, with raised yellowish spiral cords. Sculpture on early whorls consisting of the projecting subsutural and peripheral cords, weak axial ribs appearing by the fourth whorl; sinus deep, entrance constricted by the parietal callus. Radula of hollow marginals.

1730. Zonulispira chrysochildosa Shasky, 1971. The relatively large shell is robust, with yellow-orange spiral cords, forming elongate tubercles on the axial ribs and continuous yellow cords on the pillar. Length, 21 mm; diameter, 8 mm. Panama Bay to Cape San Francisco, Ecuador; intertidally, near rocks.

1731. Zonulispira grandimaculata (C. B. Adams, 1852) (Synonyms: **Crassispira dirce** Dall, 1919; **Z. reigeni** Bartsch, 1950). The black or brownish shell has white spots of various sizes diffused outward from the nodes where the main spiral keel crosses the low axial ribs; in some specimens the pattern is faint, in others absent. Length, 26 mm; diameter, 9 mm. Head of the Gulf of California to Santa Elena Peninsula, Ecuador. Larger than *Z. zonulata*. In some localities both species may occur together, at low tide and offshore, in sand near rocks.

1732. Zonulispira zonulata (Reeve, 1843) (Synonym: **Pleurotoma cincta** Sowerby, 1834 [not Lamarck, 1822]). This resembles *Z. grandimaculata*, but is smaller and has a shorter aperture and a different color pattern. The shell is black, the main peripheral cord yellow-white and of uniform width; a second, equally prominent, yellow band occurs on one of the basal cords. Length, 18 mm; diameter, 6.5 mm. Mazatlán, Mexico, to Panama Bay, intertidally and offshore on rocky bottoms. This is the type species of the genus.

Genus COMPSODRILLIA WOODRING, 1928

Slender, turreted shells of moderate size, with relatively few axial ribs on the early whorls, axial ribs overridden by prominent spiral cords; periphery usually carinate; anterior canal moderately long, sinus deep, the parietal callus expanded above the inner lip, tending to constrict the sinus entrance; protoconch smooth, the third whorl subcarinate in some species. Radula of rolled marginals, usually barbed at the tip. Most shells are unicolored, some with axial color patterns, a few spirally banded. Mainly occurring offshore on soft bottoms, less commonly intertidally in rocky areas.

Many of the species of *Compsodrillia* were originally described in *Carinodrillia*, a genus now assigned to the Crassispirinae because the type species proved to have solid marginal teeth. Although the type of *Compsodrillia*, from the Jamaican

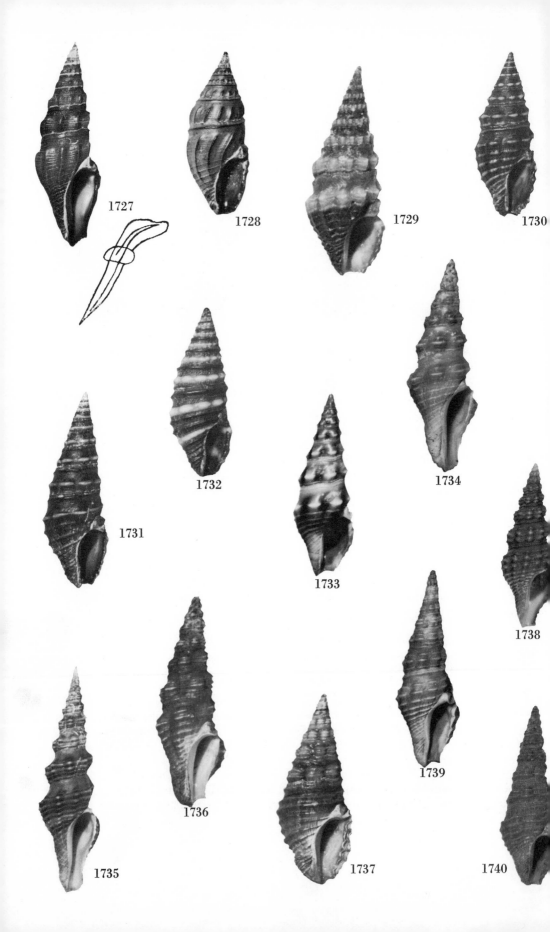

Miocene, is a slender form, the species that seem to belong here range from slender to robust.

1733. Compsodrillia albonodosa (Carpenter, 1857) (Synonym: **Carinodrillia halis soror** Pilsbry & Lowe, 1932). Small, with a single raised peripheral cord and relatively few spiral cords on the base, with no spiral striations between. Some specimens are brown with white nodes at the periphery; others are yellowish brown with the white color diffused above and below the nodes. Rib interspaces are darker brown. Length, 17 mm; diameter, 6 mm. Head of the Gulf of California to Banderas Bay, Mexico, on sand flats at low tide and to depths of 20 m.

1734. Compsodrillia alcestis (Dall, 1919). A pale yellowish-brown shell, darker along the axial ribs, with two to three major spiral cords, one forming an undulating keel. The pillar is wide and the canal slightly recurved, twisted to the right. Length, 35 mm; diameter, 12 mm. Guaymas, Mexico, to Puerto Utria, Colombia, in 20 to 90 m.

1735. Compsodrillia bicarinata (Shasky, 1961). Relatively large, high-spired, characterized by two sharply elevated spiral cords per whorl, overriding the strong axial ribs; color, tan with dark brown bands between the whitish spiral cords. Length, 48 mm; diameter, 15 mm. Guaymas, Mexico, to La Plata Island, Ecuador, in 40 to 110 m.

1736. Compsodrillia duplicata (Sowerby, 1834). The pinkish buff or tan shell has a massive appearance, with a thick pillar, the aperture appearing offset because of the twisted fasciole. Axial ribs are few and are in alternating alignment up the spire. There are three prominent spiral cords on the penultimate whorl, and the periphery is rounded. Length, 34 mm; diameter, 12 mm. San Luis Gonzaga Bay, Gulf of California, to Santa Elena Bay, Ecuador, in 20 to 45 m.

1737. Compsodrillia excentrica (Sowerby, 1834). Uniformly orange under a brown periostracum, this greatly resembles *C. duplicata* but has a shorter, more twisted aperture and stronger axial and spiral sculpture, the subsutural cord being especially well developed. Length, 27 mm; diameter, 12 mm. Guaymas, Mexico, to Santa Elena Bay, Ecuador, intertidally and to depths of 30 m, in gravel near rocks.

1738. Compsodrillia gracilis McLean & Poorman, 1971. Small, slender, tan-colored, with rounded whorls; three spiral cords per whorl are beaded in white on crossing the narrow axial ribs. Length, 15 mm; diameter, 5 mm. Galápagos Islands, in 100 to 200 m.

1739. Compsodrillia haliplexa (Dall, 1919). Tall, slender, light to dark brown, with the suture not deeply channeled. There are three spiral cords per whorl, the centermost slightly carinate; interspaces with fine spiral striae. Highly variable, some are mature at half the usual size; some have the intersection of the cords and axial ribs light, others dark. Length, 28 mm; diameter, 9 mm. Magdalena Bay through the Gulf and south to Santa Elena Bay, Ecuador, in 10 to 55 m.

1740. Compsodrillia jaculum (Pilsbry & Lowe, 1932). Small, slender, reddish brown, with two prominent spiral cords per whorl and spiral striae throughout. This resembles the small forms of *C. haliplexa*, but the shell is darker and the aperture shorter. Length, 19 mm; diameter, 5.5 mm. Tenacatita Bay, Mexico, to Piñas Bay, Panama, in 10 to 40 m.

1741. Compsodrillia olssoni McLean & Poorman, 1971. A relatively small but robust form with a marked peripheral carination, as in *C. alcestis*, but differing from this and other species in having a short aperture and relatively few spiral cords on the pillar. The shell is tan, with lighter spiral cords and a whitish pillar. Length, 17 mm; diameter, 6.5 mm. Guaymas, Mexico, to Santa Elena Bay, Ecuador, 10 to 70 m.

1742. Compsodrillia opaca McLean & Poorman, 1971. The medium-sized brown shell is banded with yellow or white below the periphery, the periostracum usually obscuring the color band. The periphery is rounded, with three strong spiral cords. Length, 27 mm; diameter, 9.5 mm. Cedros Island, outer coast of Baja California, to Cape San Lucas, in 95 to 140 m, also off Angel de la Guarda Island, Gulf of California, in 99 to 124 m.

1743. Compsodrillia thestia (Dall, 1919). Relatively small, ground color tan, the raised cording lighter in color. Length, 14 mm; diameter, 5 mm. Head of the Gulf of California to Guaymas and Puertecitos, sandy bottoms to 20 m.

1744. Compsodrillia undatichorda McLean & Poorman, 1971. The shell is tan, with a strongly tabulate profile; the numerous spiral cords are sharply raised, and the axial ribs undulate across the otherwise smooth shoulder. Length, 21 mm; diameter, 8.5 mm. Galápagos Islands, in 80 to 150 m.

Genus PILSBRYSPIRA BARTSCH, 1950

Medium- to small-sized, dark-colored shells, with broad color bands or colored beading, spiral sculpture predominating over axial sculpture, basal cords with well-spaced, strong tubercles; aperture short, truncate, pillar twisted; sinus deep, parietal callus thickened above inner lip. Radula as in *Zonulispira*, marginal teeth usually barbed at the tip.

This group has been confused with *Monilispira*, which is of a different radular type. *Pilsbryspira* has a deep, laterally directed sinus, whereas *Monilispira* may be recognized by its shallow sinus. The type species of *Pilsbryspira* is *P. pilsbryi* Bartsch, 1950, a synonym of the Caribbean *P. albomaculata* (Orbigny, 1842).

In the typical subgenus, the sinus is similar to that in *Zonulispira*, constricted by callus, whereas in *P. (Nymphispira)* there is an open slot.

Subgenus PILSBRYSPIRA, s. s.

Sinus opening constricted by anteriorly directed callus.

1745. Pilsbryspira (Pilsbryspira) albinodata (Reeve, 1843). With a strong, slightly undulating subsutural cord, periphery noded by a pair of narrow cords; base with strong tubercles; color, brown with yellow-orange nodes. Length, 27 mm; diameter, 11 mm. Zihuantanejo, Guerrero, Mexico, to Panama Bay, at low tide and offshore to depths of 30 m.

1746. Pilsbryspira (Pilsbryspira) amathea (Dall, 1919). The gray-brown shell is mottled with white, especially on the upper part of the whorl. Shoulder area broad, sculptured with a prominent subsutural cord. Height, 23 mm; diameter, 9 mm. Santa Cruz, Nayarit, to Acapulco, Guerrero, Mexico, at low tide in rocky areas.

1747. Pilsbryspira (Pilsbryspira) aterrima (Sowerby, 1834) (Synonyms: **Pleurotoma maura** Kiener, 1840, *ex* Valenciennes, MS [not Sowerby, 1834]; **P. atrior** C. B. Adams, 1852). The shell is glossy black, with a strong subsutural cord and a narrow orange band above the blunt peripheral tubercles; base with two tuberculate spiral cords. A color variant from Panama has the entire shoulder and periphery yellow-orange. Length, 20 mm; diameter, 8.5 mm. Mazatlán, Mexico, to Santa Elena Bay, Ecuador, at low tide in rocky areas.

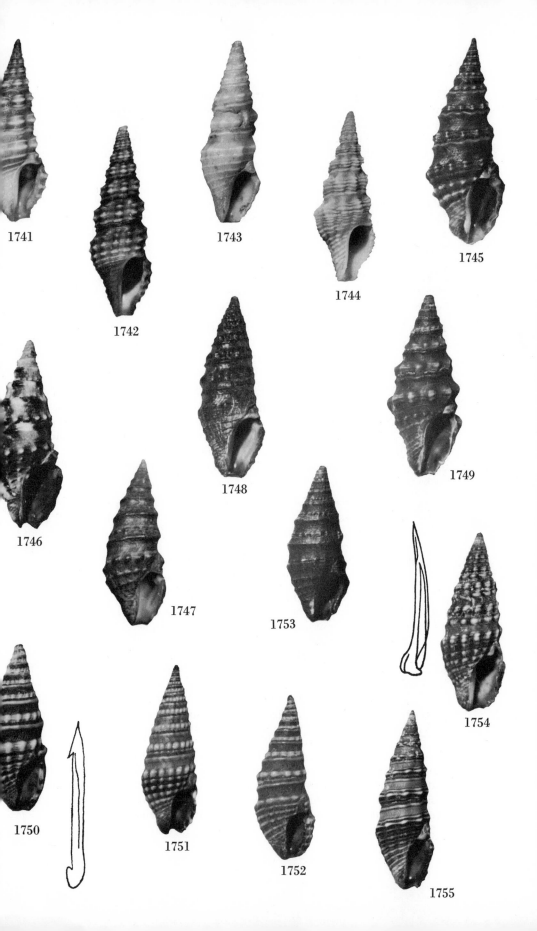

1741

1742

1743

1744

1745

1746

1747

1748

1749

1750

1751

1752

1753

1754

1755

1748. Pilsbryspira (Pilsbryspira) atramentosa (E. A. Smith, 1882) (Synonym: **Crassispira fonseca** Pilsbry & Lowe, 1932). The dark brown shell has a peripheral row of nodules connected by two spiral lirations. As in *P. aterrima* there is a thin, faint yellow line above the periphery, but the shell is more slender, brown rather than black, with more numerous spiral cords on the base. Length, 18 mm; diameter, 7 mm. Gulf of Fonseca, El Salvador, to Panama Bay, at low tide and offshore.

1749. Pilsbryspira (Pilsbryspira) aureonodosa (Pilsbry & Lowe, 1932). This has a weaker subsutural cord than *P. albinodata* and is smaller, but it has a similarly beaded periphery, with a double cord and strong tubercles on the base. Specimens in the northern part of the range are brown with orange nodes, whereas in the southern part of the range at Panama some are banded with gray-green, others grading into the typical coloration. Length, 22 mm; diameter, 9 mm. Mazatlán, Mexico, to Panama Bay, at low tide and to depths of 40 m, on gravel bottoms near rock.

1750. Pilsbryspira (Pilsbryspira) collaris (Sowerby, 1834) (Synonym: **Crassispira nephele** Dall, 1919). The biconic shell is black, with the projecting sculpture yellow, subsutural cord strong, peripheral nodes connected by two spiral cords. Length, 15 mm; diameter, 6 mm, frequently mature at 10 mm length. Mazatlán, Mexico, to Manta, Ecuador, intertidally.

1751. Pilsbryspira (Pilsbryspira) garciacubasi Shasky, 1971. This resembles *P. collaris* but is more slender, with orange rather than yellow beading, a less projecting subsutural cord, and a gray shoulder area with microscopic spiral striae. The sinus is deep and the parietal callus well developed. Length, 17 mm; diameter, 6 mm. Banderas Bay to Acapulco, Mexico, at low tide and offshore on rocky bottoms.

1752. Pilsbryspira (Pilsbryspira) loxospira (Pilsbry & Lowe, 1932). The peripheral and basal tubercles are elongate and connected on the last whorl; ground color gray-black throughout, sculptured with fine spiral striae. The suture drops and then rises again on the last whorl, giving the last whorl and aperture a characteristic tilted effect. Length, 14 mm; diameter, 5 mm. Mazatlán, Mexico, to Port Parker, Costa Rica, on rocky bottoms offshore to depths of 20 m.

1753. Pilsbryspira (Pilsbryspira) melchersi (Menke, 1851). Like *P. aterrima*, this has an orange line above the periphery, but the shell is smaller, with more deeply incised spiral striae throughout, the peripheral tubercles being not as bluntly nodose but instead connected by two or three fine cords; spiral cords on the pillar more pronounced. This is the more abundant of the two at Mazatlán, whereas at Panama Bay both are common and readily separable. At other localities, the differences between the two are not pronounced; thus, it is possible they may prove not to be separate species. Length, 14 mm; diameter, 6.5 mm. Mazatlán, Mexico, to Santa Elena Bay, Ecuador, at low tide and offshore, in rocky areas.

Subgenus **NYMPHISPIRA** McLEAN, 1971

Sinus deep, not constricted as in *Pilsbryspira, s. s.,* but having an open slot with walls directed toward the suture. Subsutural cord not strongly developed, shoulder concave, periphery noded, the nodes representing the summits of axial ribs that are more evident on the base than in the typical subgenus.

1754. Pilsbryspira (Nymphispira) arsinoe (Dall, 1919). The brown shell has yellowish nodes and a gray shoulder area; the subsutural cord is thick and undulating and there are fine spiral striae throughout. Length, 17 mm; diameter, 7.5 mm. Outer coast of Baja California: San Bartolomé Bay to Santa Maria Bay, in 10 to 40 m, on gravel bottoms.

1755. Pilsbryspira (Nymphispira) bacchia (Dall, 1919). The subsutural cord and the basal cords are yellow-white, with dark brown between, shoulder slate-gray, a narrow brown band between the suture and the subsutural cord, entire surface finely striate. Length, 18 mm; diameter, 6 mm. Gulf of California: Guaymas to Cape San Lucas, on gravel bottoms near rocks to depths of 20 m.

1756. Pilsbryspira (Nymphispira) nymphia (Pilsbry & Lowe, 1932). Characterized by large white peripheral tubercles, the subsutural cord and the basal tubercles bright orange, the shoulder slate-gray, and the basal areas between tubercles brown. Entire shell finely striate; the mature lip with deep sinus and a sutural slot. Length, 18 mm; diameter, 7 mm. Head of the Gulf of California to Guaymas and Cape San Lucas, intertidally in gravel near rocks. This is the type species of the subgenus. The shell should not be confused with the similar-appearing *Crassipira (Dallspira) bifurca* or with *C. (Monilispira) monilifera*, in both of which the sinus differs.

Genus PTYCHOBELA THIELE, 1925

Moderately large shells, with strong axial and spiral sculpture; shoulder concave, lacking a subsutural cord, the sinus broad, with an open slot next to the suture and a blunt pad of parietal callus. Marginal teeth of radula barbed. This is chiefly an Indo-Pacific group, incorrectly synonymized by Powell (1966) under *Inquisitor*, which has a radula of duplex type.

1757. Ptychobela lavinia (Dall, 1919 [pl. 1, fig. 5, not 6]). Shell yellowish white, with numerous spiral cords, canal elongate, slightly recurved. Length, 49 mm; diameter, 16 mm. This species is enigmatic. The original specimen is from an unknown locality, but it was presumed by Dall to have come from Mexico. A second specimen was collected by Olsson from an Indian grave at Mancora, Peru. This is clearly a representative of an otherwise Indo-Pacific group and the localities might therefore be suspect. However, the specimens are not identifiable with any known Indo-Pacific species, and the species is provisionally regarded as a member of the eastern Pacific fauna.

Subfamily BORSONIINAE

Shell fusiform in outline; columella smooth or with one to three plicae, sinus at the shoulder or periphery, broad, U-shaped, not bordered by heavy callus; operculum lacking, vestigial, or fully developed, with a terminal nucleus; radula with marginals only, which are slightly curved, minutely barbed or smooth at the tip, and expanded basally.

Powell's (1966) concept of this group included only the inoperculate forms with strong columellar plicae. However, all degrees of development of the operculum and the columellar folds are represented in some genera. The subfamily is here envisioned as including fusiform shells resembling *Turricula* but lacking strong parietal callus, and having true toxoglossate radular dentition, the marginal teeth unbarbed or minutely barbed at the tip. The key is to generic and subgeneric units in the four Panamic genera.

1. Mature shells over 10 mm in length............................... 2
   Mature shells under 10 mm in length............................. 4
2. Anterior canal equal to length of aperture................*Cruziturricula*
   Anterior canal shorter than aperture............................. 3
3. Columella with one or two strong plaits..................*Borsonella, s. s.*
   Columellar plication weak or lacking......................*Borsonellopsis*
4. Early whorls axially costate............................*Microdrillia*
   Early whorls not axially costate...............................*Taranis*

Genus BORSONELLA DALL, 1908

Small to medium-sized, with angulate whorls, some with axial ribbing; columella typically with one, occasionally two nearly horizontal folds; outer lip thin, arcuate; sinus deep. U-shaped, in the shoulder sulcus; shell whitish under an olive periostracum. Marginal teeth of radula not barbed at the tip. Two Panamic subgenera.

Subgenus BORSONELLA, s. s.

Columellar fold prominent, operculum lacking. Several species, including the type species, occur in moderate depths (100 to 200 m) off California. The Panamic species are mostly from deep water.

1758. Borsonella (Borsonella) abrupta McLean & Poorman, 1971. Yellow-brown under a thin brown periostracum; subsutural area swollen, undulating; vertical axial ribs terminate abruptly at the shoulder; columellar fold strong. Length, 22 mm; diameter, 7.5 mm. Isla Santa Maria, Galápagos Islands, in 457 m.

1759. Borsonella (Borsonella) agassizii (Dall, 1908) [*Borsonia*] (Synonym: **Borsonia hooveri** Arnold, of Dall, 1908, not Arnold, 1903). Length, 23 mm; diameter, 11 mm. Mazatlán, Mexico, to Ecuador and the Galápagos Islands, in 1,485 to 2,320 m.

1760. Borsonella (Borsonella) galapagana McLean & Poorman, 1971. Characterized by having two columellar plicae; the whorls are tabulate, with a slightly concave shoulder, the axial ribbing of the early whorls obsolete on the final whorl; base with fine spiral striae. Length, 18 mm; diameter, 7 mm. Isla Santa Cruz, Galápagos Islands, in 170 to 200 m.

1761. Borsonella (Borsonella) saccoi (Dall, 1908) [*Borsonia*]. Length, 14 mm; diameter, 7 mm. Gulf of Panama, 590 m.

Subgenus BORSONELLOPSIS McLEAN, 1971, ex BARTSCH, MS

Shell broad and angulate, with nodes at periphery; columellar fold weak or lacking; operculum vestigial. Deep water.

1762. Borsonella (Borsonellopsis) callicesta (Dall, 1902) [*Pleurotoma*] (Synonyms: **Gemmula esuriens** and **esuriens** var. **pernodata** Dall, 1908; **Antiplanes amphitrite** and **amphitrite** var. **beroe** Dall, 1919; **Cryptogemma cymothoe** and **eidola** Dall, 1919). Length, 19 mm; diameter, 8 mm. Santa Barbara, California, to Gulf of Panama and the Galápagos Islands, in 700 to 4,000 m.

1763. Borsonella (Borsonellopsis) erosina (Dall, 1908) [*Leucosyrinx*]. Length, 28 mm; diameter, 12 mm. Gulf of Panama, 3,058 m. The type species of the subgenus.

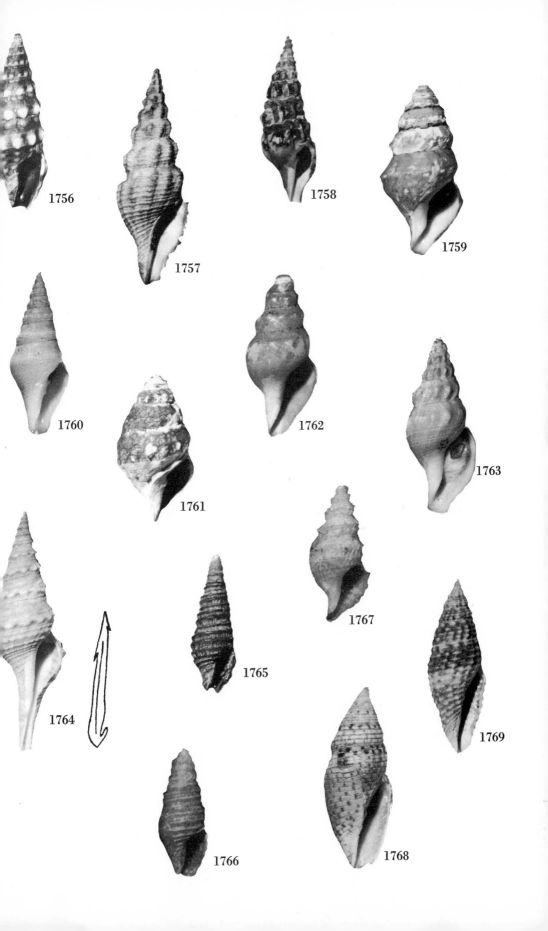

1756

1757

1758

1759

1760

1761

1762

1763

1764

1765

1766

1767

1768

1769

Genus CRUZITURRICULA MARKS, 1951

Axial sculpture weak and irregular; early spire whorls with a single keel; anal sinus a deep slot, its upper edge abutting obliquely against the suture. Radula with marginal teeth expanded at the base, barbed at the tip and halfway up the shaft. A group with a tropical American fossil record dating from the Eocene, only a single Recent species surviving.

1764. Cruziturricula arcuata (Reeve, 1843) (Synonym: **Turricula panthea** Dall, 1919 [pl. 1, fig. 6, not 5]). The white shell is sculptured with spiral ridges, the main peripheral cord with brown streaks between the undulations of the peripheral cord. Length, to 62 mm; diameter, to 19 mm. Guatulco Bay, Oaxaca, Mexico, to Puerto Utria, Colombia, in depths of 30 to 90 m. Because of an error in numbering of the original illustration of Dall's species, this has been confused with *Ptychobela lavinia.*

Genus MICRODRILLIA CASEY, 1903

Small, slender, anterior end truncate, protoconch with as many as five axially ribbed whorls; postnuclear sculpture of strong spiral cords and incremental growth lines; sinus on the shoulder, between the strong subsutural cord and uppermost keel; columella smooth or plicate; radula with awl-shaped marginal teeth having an expanded base. Widely distributed since Eocene time. The subfamily allocation of this and the next genus is tentative.

1765. Microdrillia tersa Woodring, 1928. Small, slender, spire longer than aperture, columella with two plicae deep within; posterior sinus broad, deep. Length, 5.5 mm; diameter, 2 mm. Panama Bay, in 22 m. A single specimen in the LACM collection is slightly larger than the type, which measures 3.8 mm in length, but otherwise the specimen matches Woodring's species, described from the Miocene of Jamaica.

1766. Microdrillia zeuxippe (Dall, 1919). Small, brownish white, with spiral cords and arcuate axial striae. This differs from *M. tersa* in having a broader and more concave shoulder, the spiral cords across the base being more numerous, and the columella smooth. Length, 5 mm; diameter, 2 mm. Galápagos Islands and Cocos Island, Costa Rica, in 20 to 110 m.

Genus TARANIS JEFFREYS, 1870

Small, biconic, with a shallow sinus at the peripheral keel.

1767. Taranis panope Dall, 1919. Length, 4.5 mm; diameter, 2.3 mm. Deep water off Manta, Ecuador, 733 m.

Subfamily MITROLUMNINAE

Relatively small shells; axial and spiral sculpture of nearly equal strength; columella with folds or plicae that may be pronounced or nearly obsolete; sinus either not apparent or consisting of a shallow indentation next to the suture. Marginal teeth of radula expanded at the base, lacking barbs, slightly constricted below the tip. Operculum wanting.

These genera were grouped in the Borsoniinae by Powell (1966) as the "mitromorphid genera." They differ from the Borsoniinae in lacking a deep U-shaped sinus; they are usually smaller, the sculpture tending to be clathrate, and they are characteristic of shallower water. Four Panamic genera.

1. Shell under 10 mm in length. 2
   Shell over 10 mm in length. 3
2. Columellar plicae present. . . . . *Cymakra*
   Columellar plicae lacking. . . . . *Mitromorpha*
3. Shell slender, axial ribs continuous whorl to whorl. . . . . *Diptychophlia*
   Shell biconic, axial ribs not continuous whorl to whorl. . . . . *Mitrolumna*

Genus MITROLUMNA BUCQUOY, DAUTZENBERG & DOLLFUS, 1883
(ARIELIA SHASKY, 1961)

Medium-sized, biconic, resembling *Mitromorpha* but with two well-formed columellar plications; sinus represented by a shallow indentation in the outer lip near the suture. Members of the genus also occur in the Mediterranean and on the west coast of Africa. Emerson and Radwin (1969) have discussed the systematic position of the group.

1768. Mitrolumna keenae Emerson & Radwin, 1969. The shell is sculptured by fine incised grooves; white with spots of chestnut-brown, darkest just below the suture, brown-tinged in the interspaces below, the color spots randomly arranged in the interspaces but tending to form alternating rows of dots on the spiral ribs. Length, 17 mm; diameter, 6.5 mm. Tagus Cove, Isabela Island, Galápagos Islands, in 75 to 100 m.

1769. Mitrolumna mitriformis (Shasky, 1961). The narrowly fusiform shell has reticulate sculpture with fine beads at the intersections of the axial and spiral ribs; outer lip lirate within, sinus shallow. Buff-white with banding and mottling of brown. Length, 17 mm; diameter, 6 mm. Bahía de los Angeles to Gorda Point, Gulf of California, in 22 to 165 m; also tentatively identified from the Galápagos Islands. This is the type species of *Arielia*.

Genus MITROMORPHA CARPENTER, 1865

Small, biconic, columellar plicae wanting; sculpture of flat-topped spiral cords, anterior end truncate, aperture narrow, more than half the length of the shell, lip thick and evenly denticulate. Known in the fossil record from the Miocene of France and the Caribbean.

1770. Mitromorpha carpenteri Glibert, 1954 (Synonym: **Daphnella filosa** Carpenter, 1864 [not *Columbella filosa* Dujardin, 1837, a *Mitromorpha*]). A sturdy, dark brown shell with a white protoconch, with even spiral ribbing and a periostracum that shows up as incremental striae; columella smooth, outer lip with fine lirations. Length, 8.5 mm; diameter, 4 mm. Monterey, California, to Panama and the Galápagos Islands, in rocky areas at low tide and offshore to 20 m. This is the type species of the genus.

Genus CYMAKRA GARDNER, 1937

Small, biconic, sinus not apparent; sculptured with prominent spiral cords and axial ribs that are strongest on early whorls; protoconch smooth, obtuse; sutures deeply channeled, columella with two strong folds, outer lip lirate within. Although Powell (1966) assigned this to the Mitridae, it is clearly a mitrolumnine turrid, resembling *Mitromorpha* except that the columellar plaits are strongly developed as in *Mitrolumna*. The type is from the Florida Miocene. Recent species occur in the western Atlantic and eastern Pacific.

1771. Cymakra baileyi McLean & Poorman, 1971. The shell is tan with light and darker markings. Narrow spiral cords override the axial ribs; columellar plaits and the lirae of the outer lip are especially prominent. Length, 4.7 mm; diameter, 2.1 mm. West side of the Gulf of California, from Bahía de los Angeles to Cape San Lucas, in 20 to 60 m.

1772. Cymakra granata McLean & Poorman, 1971. Spiral sculpture with fine granular beading; lip thickened and lirate within; columella with two weak but distinct plicae; color, light brown, with white mottling and a white pillar. This resembles the Californian *C. gracilior* (Tryon, 1884) but is more ovate, with less inflated whorls. Length, 6.5 mm; diameter, 3 mm. Off Monserrate Island, Gulf of California, in 40 to 70 m.

Genus DIPTYCHOPHLIA Berry, 1964

Small, slender, tall-spired, aperture narrow, axial ribs continuous whorl to whorl, columellar plicae and sinus as in *Mitrolumna*. Only one species known.

1773. Diptychophlia occata (Hinds, 1843). The slender shell has six axial ribs that are continuous from whorl to whorl, overridden by four narrow spiral cords; fine tufts of periostracum extend above and below the spiral cords, increasing their apparent width. The shell is tan, with a white aperture. Length, 16 mm; diameter, 4 mm. Guaymas, Mexico, to Puerto Utria, Colombia, in 30 to 55 m.

Subfamily CLATHURELLINAE

Small to moderately large, with axial and spiral sculpture, sinus broad and deep, encircled with callus; anterior canal notched; inner and outer lips smooth or heavily denticulate. The radula consists of long, slender, slightly curved marginals, unbarbed, and with evenly swollen bases; none having the hilted dagger tooth of the Mangeliinae. Many of the genera grouped here have the protoconch carinate between the smooth nuclear tip and the mature sculpture; some, such as the Californian genus *Crockerella* Hertlein & Strong, 1951, lack such carination.

This group of genera has usually been regarded as mangeliine, but it differs in radular features and in having a deep subtubular sinus and deeply notched anterior canal. The size in some genera far exceeds that of mangeliine genera. The key includes the two subgeneric units of *Glyphostoma*.

1. Length under 10 mm.. 2
   Length over 10 mm... 3
2. Sinus rim not projecting.....................................*Clathurella*
   Sinus spoutlike, projecting.................................*Nannodiella*
3. Inner and outer lips not denticulate.................*Strombinoturris*
   Inner and outer lips denticulate.................................... 4
4. Surface glossy ....*Euglyphostoma*
   Surface microscopically granular.................*Glyphostoma, s. s.*

Genus CLATHURELLA Carpenter, 1857

Small, solid, sculpture of strong axial ribs with narrow, overriding spiral cords, sinus deep, setting off the thickened outer lip; parietal callus weakly developed, lip denticulate, columella weakly lirate; protoconch keeled for at least two whorls. Entire surface minutely granular.

1774. Clathurella maryae McLean & Poorman, 1971. High-spired, with deeply impressed sutures and a constricted pillar, the anterior canal somewhat elongate; nucleus with three carinate whorls; sculpture coarsely clathrate, with more numerous spiral cords on the shoulder; whitish, darker on the base. Length, 7.5 mm; diameter, 2.9 mm. Off Cedros Island, outer coast of Baja California, and north in the Gulf of California to Puertecitos; also Galápagos Islands, in 30 to 140 m.

1775. Clathurella rava (Hinds, 1843) (Synonyms: **Defrancia serrata** Carpenter, 1856; **Philbertia telamon** Dall, 1919). Bright yellow, with narrow brown bands at the shoulder and below the periphery, pillar tipped with brown; columella with three folds, lip with five denticles. Length, 9.4 mm; diameter, 3.9 mm, often mature at half this size. Concepción Bay, Baja California, to Panama, at low tide in rocky areas. This is the type species of the genus.

1776. Clathurella rigida (Hinds, 1843) (Synonym: **C. affinis** Dall, 1871). Purplish brown, with a whitish band just below the periphery; spiral cords are narrow and override the massive, closely spaced axial ribs, which undulate the suture; aperture narrow, strongly denticulate on both sides. Length, 8 mm; diameter, 3 mm. Puertecitos, near the head of the Gulf of California, to Perlas Islands, Panama, at low tide and offshore, in rocky areas.

Genus NANNODIELLA DALL, 1919

Small, slender to biconic; anterior canal short, slightly notched, sinus rim projecting, spoutlike; aperture smooth to denticulate, axial and spiral sculpture nearly equal, forming rectangular interspaces, protoconch large, third and fourth whorls carinate.

1777. Nannodiella fraternalis (Dall, 1919). Whitish, with faint banding of brown; aperture brown, weakly denticulate in mature specimens. Shoulder with numerous spiral striae; periphery with three strong cords. Length, 4.8 mm; diameter, 1.8 mm. Puertecitos, Gulf of California, to Gorgona Island, Colombia, in 20 to 70 m.

1778. Nannodiella nana (Dall, 1919). Smaller than *N. fraternalis*, lacking apertural denticulation, with coarser sculpture, the shoulder lacking spiral ribbing, and having only two strong peripheral cords. Whitish, with darker subsutural band and base. Length, 3.5 mm; diameter, 1.5 mm. San Luis Gonzaga Bay, Gulf of California, to Gorgona Island, Colombia, in 20 to 70 m. This is the type species of the genus.

Genus GLYPHOSTOMA GABB, 1872
(RHIGLYPHOSTOMA WOODRING, 1970)

Medium-sized to moderately large, high-spired, canal elongate, deeply notched; parietal callus strongly developed and raised, columella and outer lip with denticles or ridges; protoconch keeled for two turns. All species occur offshore on sand bottoms.

Shells of *Glyphostoma*, s. s., have microscopic granular sculpture, whereas shell surface in the subgenus *Euglyphostoma* is glossy. *Rhiglyphostoma*, intended as a subgenus for small forms with weak apertural dentition, seems not to have clear limits.

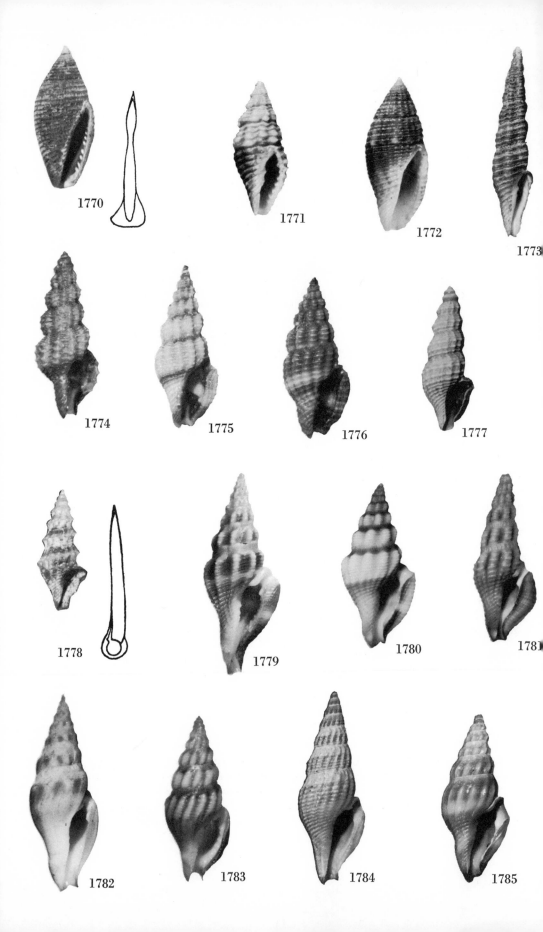

1770   1771   1772   1773

1774   1775   1776   1777

1778   1779   1780   1781

1782   1783   1784   1785

Subgenus **GLYPHOSTOMA**, *s. s.*

Shell surface microscopically granular. The type species is from the Miocene of the Dominican Republic.

1779. Glyphostoma (Glyphostoma) bayeri Olsson, 1971. The shoulder is strongly concave, the canal unusually long, and the outer lip markedly thickened and projecting; tan with reddish-brown squares between the axial ribs; rib surfaces usually white, a narrow white band below the periphery; spiral sculpture fine and even, overriding the narrow axial ribs. Length, 18 to 26 mm; diameter, 7 to 10 mm. Gulf of Nicoya, Costa Rica, to Puerto Utria, Colombia, in 30 to 80 m.

1780. Glyphostoma (Glyphostoma) myrae Shasky, 1971. Relatively broad and low-spired, with fine spiral ribbing overriding the broad axial folds; reddish brown on the concave shoulder and on the base, the broad peripheral area yellowish white. Length, 13 mm; diameter, 7 mm. Jicarita Island, Panama, to Gorgona Island, Colombia, in 40 to 60 m.

1781. Glyphostoma (Glyphostoma) neglecta (Hinds, 1843) (Synonyms: **Defrancia intercalaris** Carpenter, 1856; **Clathurella aurea** Carpenter, 1857; **G. adana** and **C. adria** Dall, 1919; **Lioglyphostoma armstrongi** Hertlein & Strong, 1951; **G. myrakeenae** Olsson, 1964). High-spired and slender, the concave shoulder with irregular lines of growth, the broad axial ribs fading on the base and crossed by narrow, raised spiral cords, beaded on the pillar. Yellowish brown, often banded with darker brown on the base, the narrow spaces between the axial ribs often darker brown. With marked variations in size, sculpture, and color. Length, 10 to 16 mm; diameter, 4 to 5.5 mm. Head of the Gulf of California to Santa Elena Peninsula, Ecuador, in 20 to 50 m.

1782. Glyphostoma (Glyphostoma) pustulosa McLean & Poorman, 1971. Large, whitish, with spotting of brown between the blunt peripheral nodes; finely striate, entire surface microscopically pustulose; sinus closely appressed to the body whorl, the parietal callus pad with four lirations. Length, 23 mm; diameter, 8.5 mm. Galápagos Islands, 20 to 40 m.

1783. Glyphostoma (Glyphostoma) scobina McLean & Poorman, 1971. More slender and with more numerous slanting ribs than *G. myrae*, spiral sculpture of fine striae, the shoulder and base banded in brown, rib surfaces white, with darker interspaces. Length, 13 mm; diameter, 5 mm. Galápagos Islands, 90 to 200 m, also from Cocos Island, Costa Rica.

1784. Glyphostoma (Glyphostoma) thalassoma (Dall, 1908). Large and slender, axial ribs low and numerous, the spiral cording strong and uniform. Light brown, with a broad white band below the periphery, intersections of the cords whitish. Length, 23 mm; diameter, 8 mm. Gulf of California, from Cape Lobos, Sonora, to Tiburon and Angel Islands, in 70 to 140 m.

Subgenus **EUGLYPHOSTOMA** WOODRING, 1970

Shell surface glossy, spiral sculpture tending to fade out on the final whorl.

1785. Glyphostoma (Euglyphostoma) candida (Hinds, 1843) (Synonym: **G. partefilosa** Dall, 1919). Brown, with a whitish band below the periphery and another on the pillar; shell surface glossy, with smooth, oblique ribs on the upper part of the whorl, pillar with deeply incised striae only. Length, 15 mm; diameter, 6 mm. Cedros Island, Baja California, and north in the Gulf of California to Cabo

Tepoca, Sonora, south to La Plata Island, Ecuador, in 40 to 90 m. This is the type species of the subgenus (as *G. partefilosa*).

1786. Glyphostoma (Euglyphostoma) immaculata (Dall, 1908). Lower-spired than *G. candida*, with low spiral cords over the entire surface, white with a flush of pink on the final whorl. Length, 10 mm; diameter, 5 mm. Gulf of Panama, in 280 m; known only from the original two specimens.

Genus STROMBINOTURRIS HERTLEIN & STRONG, 1951

Large, slender, with a high spire and elongate, notched anterior canal; sinus deep and rounded, with a rimmed margin and parietal callus pad, lip with a varix, the outer edge thin and projecting; surface microscopically granular, protoconch with a small, two-whorled, keeled phase. This was originally allocated to the Columbellidae, more recently by Powell (1966) to the Clavinae, but it has the radula and other features of *Glyphostoma*, differing by lacking the denticulation. Only one species known.

1787. Strombinoturris crockeri Hertlein & Strong, 1951 (Synonym: "**Pleurotoma stromboides** Sowerby" of Reeve, 1843, not of Sowerby, 1832). Yellowish brown, the aperture and callus white; shoulder smoothly concave, periphery with strong nodes, the fine axial and spiral cords of base forming tubercles at intersections. Length, 48 mm; diameter, 15 mm. Santa Inez Bay, southwestern Gulf of California, to Panama, in 25 to 110 m.

Subfamily MANGELIINAE

Small shells, mostly slender, with several whorls, the anterior canal relatively short, not deeply notched; sinus shallow to moderately deep; outer lip with a varix, some forms denticulate within. Operculum absent in tropical and subtropical groups; radula of short and broad marginals, resembling partially rolled leaves, many with an upcurved spur at the base of the tooth.

The small turrids have been difficult to study, and authors have tended not to use for them the generic taxa that are available in the literature or not to notice that they are lumping together forms that are morphologically dissimilar under the frequently employed name *Mangelia* Risso, 1826. The type species of this genus, however, is European and relatively large in size, with rounded axial ribs that extend from whorl to whorl, lacking a strongly indented posterior sinus. No Panamic species is closely comparable; thus none qualifies as true *Mangelia*. Instead, the eastern Pacific species seem referable to several genera, the type species of which are in the main from the western Atlantic, Tertiary to Recent. Generic criteria found useful in this revision are depth and direction of the posterior sinus, sculpture of the protoconch, presence or absence of lip dentition, relative size, and sculpture. The sequence adopted here is not alphabetical but in the order of increasing depth and complexity of the sinus and lip dentition. The key is to genera and subgenera. Fourteen Panamic genera.

1. No apparent sinus. 2
   Sinus well developed. 6
2. Whorls rounded, pillar twisted to left. *Glyptaesopus*
   Whorls angulate, canal straight, elongate. 3
3. Spiral sculpture frosted or finely beaded. 4
   Spiral sculpture not frosted, or microscopically beaded. 5
4. Protoconch relatively small . . . . *Granoturris*
   Protoconch relatively large. . . . . *Kurtzina*

5. Last whorl of protoconch with cancellate sculpture.........*Kurtziella, s. s.*
   Last whorl of protoconch with numerous fine axials.........*Rubellatoma*
6. Sinus at the suture, parietal callus lacking.......................... 7
   Sinus on the shoulder, with some parietal callus..................... 9
7. Axial sculpture obsolete on the base....................*Cacodaphnella*
   Axial sculpture not obsolete on the base........................ 8
8. Aperture less than half the shell length..................*Notocytharella*
   Aperture more than half the shell length.....................*Tenaturris*
9. Inner side of outer lip smooth........................... 10
   Inner side of outer lip denticulate............................ 16
10. Sculpture coarsely clathrate............................. 11
    Sculpture not coarsely clathrate.............................. 13
11. Shell with raised subsutural cord.........................*Thelecythara*
    Shell lacking raised subsutural cord............................ 12
12. Shell over 8 mm in length................................*Acmaturris*
    Shell under 6 mm in length..............................*Platycythara*
13. Sculpture frosted (with a finely beaded surface).................*Kurtzia*
    Sculpture not frosted... 14
14. Usually over 10 mm in length...........................*Euclathurella*
    Usually under 10 mm in length................................ 15
15. Spiral sculpture of incised grooves....................*Agathatoma, s. s.*
    Spiral sculpture of raised riblets.........................*Vitricythara*
16. Lip with four or more low denticles.......................*Ithycythara*
    Lip with single denticle at base of sinus.......................... 17
17. Whorls rounded, shell more than 10 mm in length...........*Bellacythara*
    Whorls angulate, shell less than 10 mm in length.............*Pyrgocythara*

Genus **KURTZIELLA** DALL, 1918

Aperture elongate, sculpture of low axial ribs with wider interspaces, periphery angulate; spiral sculpture of dense lirae; lip varix no thicker than a regular axial rib, sinus a slight inflection of the outer lip on the shoulder slope. Most species are sand dwellers.

Four subgenera are distinguished on details of the nucleus and spiral sculpture.

Subgenus **KURTZIELLA**, *s. s.*

Protoconch of three regularly increasing whorls, the last with fine reticulate sculpture, spiral striae not frosted or finely beaded.

1788. Kurtziella (Kurtziella) antiochroa (Pilsbry & Lowe, 1932) (Synonym: **Mangelia cymatias** Pilsbry & Lowe, 1932). With relatively few axial ribs, the spiral cord at the periphery swollen where it overrides the axial ribs; the slender shell is pale yellow to tawny brown, with a brown band below the periphery and on the canal. Length, 9.2 mm; diameter, 3 mm. Head of the Gulf of California to La Libertad, Ecuador, in 10 to 50 m.

1789. Kurtziella (Kurtziella) plumbea (Hinds, 1843) (Synonyms: **Mangelia acuticostata** var. **subangulata** Carpenter, 1857; **M. angulata** Carpenter, 1864 [not Reeve, 1846]; **M. sulcata** Carpenter, 1865; **M. hecetae** Dall & Bartsch, 1910; **M. alesidota, hebe, oenoa,** and **tersa** Dall, 1919; **M. barbarensis** Oldroyd, 1924 [new name for *M. angulata*]; **M. wrighti** Jordan, 1936). Markedly variable in strength of the peripheral angulation, this generally has more axial

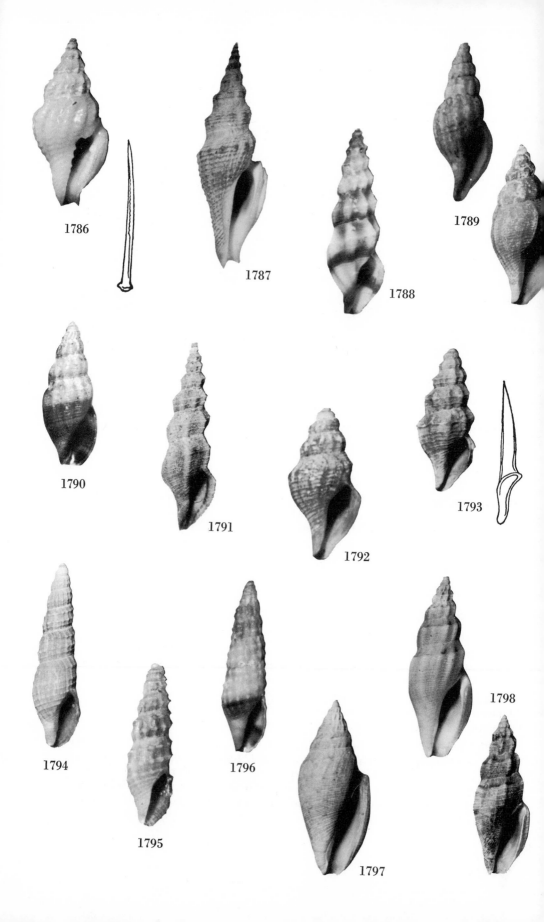

1786

1787

1788

1789

1790

1791

1792

1793

1794

1795

1796

1797

1798

ribs than *K. antiochroa*; the shell is tan, some specimens with narrow, irregular brown bands. Length, 8 to 11 mm; diameter, about 4 mm. British Columbia to Mazatlán, Mexico, and throughout the Gulf of California, in 10 to 50 m.

Subgenus **RUBELLATOMA** BARTSCH & REHDER, 1939

Third whorl of protoconch with numerous slanting axial riblets, body whorl only weakly angulate.

1790. Kurtziella (Rubellatoma) powelli Shasky, 1971. Relatively small, the white or tan shell broadly banded with brown below the periphery. The large nucleus with fine axial ribbing is characteristic; the anterior end is more truncate than the other Kurtziellas. Length, 5.1 mm; diameter, 2.0 mm. Head of the Gulf of California to Santa Elena Peninsula, Ecuador, on sand flats at low tide.

Subgenus **GRANOTURRIS** FARGO, 1953

Spire high, strongly angulate, with fine spiral threads crossed by incremental axials; minutely beaded, giving a frosted effect. Protoconch small, its third whorl carinate and crossed by numerous axial riblets.

1791. Kurtziella (Granoturris) antipyrgus (Pilsbry & Lowe, 1932). Relatively large, the grayish-white shell is somewhat darker between the ribs and at the tip of the canal; the angulate whorls and frosted sculpture are characteristic. Length, 10.5 mm; diameter, 3.2 mm. San Luis Gonzaga Bay, Gulf of California, to Gorgona Island, Colombia, in 20 to 70 m.

Subgenus **KURTZINA** BARTSCH, 1944

Shell relatively small, with a large nucleus, nuclear tip immersed, third nuclear whorl with cancellate sculpture, the intersections of which are beaded; spiral sculpture showing a weak frosted effect.

1792. Kurtziella (Kurtzina) beta (Dall, 1919). The brown shell is relatively broad, the spiral cords even, with nearly equal interspaces. Length, 5 mm; diameter, 2 mm. Farallon Islands, central California, to Santa Maria Bay, Baja California, in 45 to 110 m. This is the type species of the subgenus.

1793. Kurtziella (Kurtzina) cyrene (Dall, 1919). The tan shell more slender than in *K. beta*, with a proportionately smaller protoconch, alternate spiral cords on the base thinner. Length, 4.2 mm; diameter, 1.2 mm. Head of the Gulf of California to San Francisco Bay, Ecuador, in 10 to 70 m.

Genus **GLYPTAESOPUS** PILSBRY & OLSSON, 1941

Slender, tall, whorls flat-sided or rounded rather than angulate; sinus wanting, lip not thickened; third whorl of nucleus with fine cancellate sculpture, later whorls with both spiral and axial riblets. The shells resemble *Terebra*, and, like that group, these mollusks live on sand flats, burrowing just beneath the surface. Radwin (1968) has shown that the radula has detached awl-shaped teeth and that the genus therefore does not belong in the Columbellidae as originally assumed (it was named as a subgenus of *Aesopus*).

1794. Glyptaesopus oldroydi (Arnold, 1903) (Synonym: **Mangelia cetolaca** Dall, 1908 [unnecessary new name]; not to be confused with *Mangelia oldroydi* Arnold, 1903). Three prominent spiral threads per whorl are minutely beaded by the narrow axial ribs, entire surface finely striate throughout; uniformly grayish or brownish white. Length, 8.5 mm; diameter, 2.5 mm. Described from the Pleistocene of southern California; living from Point Abreojos to Magdalena Bay, outer coast of Baja California, in 5 to 10 m. Originally described in the genus *Columbella*, this is not a homonym of Arnold's *Mangelia oldroydi*.

1795. Glyptaesopus phylira (Dall, 1919) (Synonym: **Philbertia amyela** Dall, 1919). Two prominent spiral cords per whorl form sharply projecting tubercles when crossed by the narrow axial ribs, surface finely striate throughout. The shell is uniformly tan or with a brown band below the suture and a tinge of brown on the columella. Length, 7 mm; diameter, 2 mm. Concepción Bay, Gulf of California, to San Francisco Bay, Ecuador, on sand flats at low tide and to 20 m.

1796. Glyptaesopus xenicus (Pilsbry & Lowe, 1932). In addition to the two main bluntly beaded spiral cords per whorl, there is a third, subsutural cord that is also bluntly beaded. White, with a narrow brown band below the suture on later whorls. Whorls are flatter than in the other two species. Length, 7.5 mm; diameter, 2.2 mm. Concepción Bay, Gulf of California, to Acapulco, Mexico. The type of the genus.

Genus TENATURRIS WOODRING, 1928

Relatively large, somewhat cylindrical, shoulder slightly concave, aperture as long as or longer than the spire, axial ribs sinuous, not forming continuous ridges up the spire, spiral sculpture of fine striae; sinus broad and deep, parallel to body whorl, bordered on the inside with minimal or no callus; nucleus smooth, the third whorl developing axial riblets and then spiral striae; lip thin-edged, with a slight indication of a sinuosity or stromboid notch, but thickened behind by a final varix. The type species is from the Jamaican Miocene.

1797. Tenaturris concinna (C. B. Adams, 1852) (Synonym: **Cithara sinuata** Carpenter, 1856). Ovate-biconic, the axial ribs nearly obsolete upon the final whorl, the shoulder slightly convex. White, with faint flecking of reddish brown, especially on the shoulder. Length, 11 mm; diameter, 4.3 mm. Tres Marias Islands, Mexico, to Santa Elena Bay, Ecuador, intertidally and to 20 m.

1798. Tenaturris merita (Hinds, 1843) (Synonyms: **Cithara fusconotata** Carpenter, 1864; **Cytharella nereis** Pilsbry & Lowe, 1932). The basal outline of the shell is concave; the shoulder is smooth and somewhat concave, a variable number of axial ribs fade out toward the base. A brown line defines the shoulder, back of the last whorl often clouded with brown. Length, 9 mm; diameter, 3.5 mm. Head of the Gulf of California to Santa Elena Peninsula, Ecuador, intertidally and to 10 m depth, in rocky areas. The extremes of variation in this species are represented by Hinds's type specimen, with relatively few axial ribs, and that of the Pilsbry and Lowe synonym, with numerous ribs.

1799. Tenaturris verdensis (Dall, 1919) (Synonym: **Cytharella burchi** Hertlein & Strong, 1951). Resembling *T. concinna* but larger, with stronger axial ribs; an occasional rib more massive than the others; tan with brown flecking, often brown on the back of the last whorl. Length, to 16.5 mm; diameter, to 6.3 mm. Guaymas, Mexico, to Cape San Lucas, Baja California, in 10 to 80 m.

Genus NOTOCYTHARELLA HERTLEIN & STRONG, 1955

Medium-sized, cylindrical, spire height greater than aperture length, whorls rounded, sculptured with spiral striae and axial ribs that extend from suture to suture; nucleus smooth, the third whorl with fine cancellate sculpture; lip thickened, sinus flush to body whorl, with a thin glaze of parietal callus.

1800. Notocytharella striosa (C. B. Adams, 1852) (Synonyms: **Pleurotoma exigua** C. B. Adams, 1852; **Cytharella niobe** Dall, 1919; **C. hastula** Pilsbry & Lowe, 1932). The shell is slender to moderately inflated and may be uniformly white or variously striped with brown. Length, 7.6 mm; diameter, 2.8 mm. Concepción Bay, Gulf of California, to Santa Elena Peninsula, Ecuador, at low tide and to depths of 10 m. This is the type species of the genus (as *N. niobe*) and the only known species.

Genus CACODAPHNELLA PILSBRY & LOWE, 1932

Slender, the spire nearly twice as long as the aperture, nucleus developing fine clathrate sculpture; body whorl with coarse clathrate sculpture, base with fine spiral striae only, sinus as in *Tenaturris* and *Notocytharella*, flush with the body whorl. Only one known species. Originally allocated to the Daphnellinae; sinus and nuclear characters are those of the Mangeliinae.

1801. Cacodaphnella delgada Pilsbry & Lowe, 1932. The shell is buff-colored, with faint banding of brown. Length, 8 mm; diameter, 2.5 mm. San Juan del Sur, Nicaragua; known only from the type specimen.

Genus EUCLATHURELLA WOODRING, 1928

Shells moderately large for the subfamily, cylindrical, whorls rounded, shoulder tabulate, axial ribs sinuous, flexed across the shoulder, spiral sculpture of fine striae; sinus broad and deep, bordered on the inside with thickened parietal callus; aperture elongate; outer lip with a sinuosity or stromboid notch that is broad and shallow, lip thickened by a massive final varix; nuclear tip small, with fine axial ribs beginning on the second nuclear whorl. This resembles *Tenaturris*, differing in having thickened parietal callus and axial ribs that extend across the shoulder. The type species is widespread in the Caribbean Miocene.

1802. Euclathurella acclivicallis McLean & Poorman, 1971. The pale flesh-colored shell is slender, with numerous axial ribs that extend over the sharply tabulate shoulder. Length, 13.2 mm; width, 4.5 mm. Galápagos Islands, in depths to 150 m.

1803. Euclathurella carissima (Pilsbry & Lowe, 1932). Axial ribs strong and sinuous, sharply curved at the shoulder; shell sculptured with fine spiral striae throughout, the parietal callus exceptionally thick. Color pale cinnamon-pink, with spots of brown on the ribs. Length, 10 mm; diameter, 3.5 mm. San Luis Gonzaga Bay, Gulf of California, to Manzanillo, Mexico, intertidally to 10 m in rocky areas.

Genus ACMATURRIS WOODRING, 1928

Shells moderately large for the subfamily, whorls rounded, with coarsely clathrate sculpture and fine spiral striae; protoconch smooth at first, the third whorl weakly keeled and sculptured with axial riblets; outer lip strongly varicose, anterior end drawn into a short canal; sinus slot diagonally directed on the shoulder, moderately deep. The type species and two others are known from the Miocene of Jamaica.

1804. Acmaturris ampla McLean & Poorman, 1971. The whorls are strongly shouldered, the clathrate sculpture beaded at intersections of the axial ribs and

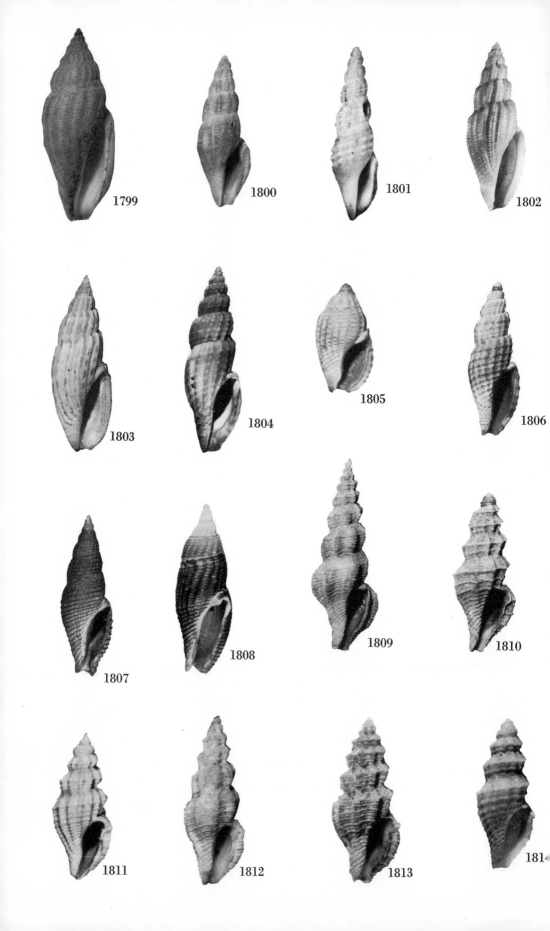

1799

1800

1801

1802

1803

1804

1805

1806

1807

1808

1809

1810

1811

1812

1813

181

spiral cords; tan, banded with brown. Length, 9.9 mm; diameter, 3.2 mm. Known from a single dead specimen, Gulf of Guayaquil, Ecuador, 64 m.

Genus PLATYCYTHARA WOODRING, 1928

Shells relatively small, with narrow raised axial and spiral ribbing, interspaces with microscopically beaded, frosty-appearing spiral striae; nucleus smooth, developing curved axials on the third whorl, sinus moderately deep, the slot diagonally directed, on the shoulder slope, parietal callus thick, outer lip with a varix and a shallow sinuosity or stromboid notch near the base, pillar slightly twisted to the left. The type species is from the Miocene of Jamaica.

1805. Platycythara curta (Dall, 1919). The small shell is broadly inflated and looks foreshortened, with raised cancellate sculpture. This is the smallest eastern Pacific member of the family; length, 2.3 mm; diameter, 1.5 mm. Panama Bay, 53 m, known only from the original specimen.

1806. Platycythara electra (Dall, 1919). The basic color of the shell is purplish brown, with the raised axial and spiral sculpture contrasting in white, the minutely beaded spiral striae visible only under high magnification. Length, 5.5 mm; diameter, 2 mm. Guaymas, Mexico, to Panama Bay, 20 to 70 m.

Genus THELECYTHARA WOODRING, 1928

Small to medium-sized, aperture elongate and twisted to the left, sculpture reticulate, the immediate subsutural spiral cord thickened and separated from the rest; sinus on the shoulder, the inner lip callus somewhat constricting the sinus opening; nuclear whorls smooth, elongate.

1807. Thelecythara dushanae McLean & Poorman, 1971. The brown shell is taller-spired and with more impressed sutures than *T. floridana*; the spiral cording is stronger than the axial ribbing, the terminations of the cords imparting a serrate edge to the lip. Length, 8.4 mm; diameter, 2.9 mm. Guaymas, Mexico, to Panama Bay, in 20 to 40 m.

1808. Thelecythara floridana Fargo, 1953. The axial ribs are sinuous at the shoulder and fade on the base, the spiral cords not overriding the ribs on the upper part of the whorl but increasing in strength toward the base; brown, the protoconch whitish. Length, 8 mm; diameter, 3 mm. Mazatlán, Mexico, to Panama Bay. Specimens from Mazatlán average about 10 mm in length. Described as a fossil, from the Pliocene of Florida.

Genus KURTZIA BARTSCH, 1944

Whorls angulate as in *Kurtziella*, but with a diagonally directed, shallow to moderately deep sinus, which may be bordered within by a low callus fold; lip thinedged, strengthened behind by a large final varix; sculpture of axial ribs and frosted spiral cords; third whorl of protoconch with fine cancellate sculpture, the axial ribs more numerous than on later whorls.

1809. Kurtzia aethra (Dall, 1919). Largest member of the genus; the shell is pale brown, the frosted spiral striae dense and fairly even, the axial ribs broadly spaced, the canal slightly twisted to the left. Length, 12 mm; diameter, 4.5 mm. Cabo Tepoca, Sonora, to Chiapas, Mexico, in 20 to 70 m.

1810. Kurtzia arteaga (Dall & Bartsch, 1910) (Synonyms: **Mangelia arteaga** var. **roperi** Dall, 1919; **K. gordoni** Bartsch, 1944). Sculptured with fine frosted striae throughout, early whorls with a second spiral cord below the main peripheral cord, canal with additional strong cords, sinus relatively shallow. Length, 9 mm; diameter, 3 mm. Vancouver Island, British Columbia, throughout the Gulf of California and south to the Gulf of Tehuantepec, Mexico, in 20 to 90 m. This is the type species of the genus.

1811. Kurtzia elenensis McLean & Poorman, 1971. Relatively small, the frosted striae visible only under high magnification; spiral cords are evenly spaced below the peripheral angulation, the clathrate postnuclear sculpture particularly apparent. Length, 4.4 mm; diameter, 1.9 mm. Gulf of Guayaquil, Ecuador, 65 m.

1812. Kurtzia ephaedra (Dall, 1919). Whitish with faint bands of brown across the shoulder and some brown flecks on the axial ribs, the frosted spiral striae varying slightly in thickness. Length, 8.3 mm; diameter, 3.2 mm. Panama Bay to Secas Islands, Panama, to depths of 30 m.

1813. Kurtzia granulatissima (Mörch, 1860) (Synonym: **Philbertia aegialea** Dall, 1919). The brown shell is medium-sized, but with much coarser spiral striae than in other species, the frosted effect pronounced. Length, 7 mm; diameter, 3 mm. Magdalena Bay, Baja California, throughout the Gulf of California and south to the Gulf of Nicoya, Costa Rica, in 20 to 40 m on gravel bottoms in rocky areas.

1814. Kurtzia humboldti McLean & Poorman, 1971. Similar to *K. arteaga* but smaller and more slender, with a stronger development of the major spiral cords below the periphery. Length, 5.2 mm; diameter, 2.0 mm. Galápagos Islands, 20 to 40 m.

Genus AGATHOTOMA COSSMANN, 1899

Slender, small to moderate-sized shells; sculptured with axial ribs tending to be continuous from whorl to whorl, periphery rounded rather than angulate, usually shouldered; spiral sculpture of fine striae; outer lip thickened, sinus deep, the slot walls laterally directed; nuclear whorls smooth, with deeply impressed sutures. Two Panamic subgenera.

Subgenus AGATHOTOMA, s. s.

Spiral sculpture of fine, incised striae.

1815. Agathotoma (Agathotoma) aculea (Dall, 1919). Aperture about one-third the shell length. Axial ribs are light-colored, sinuous, not continuous from whorl to whorl; spiral striae fine and regular. Height, 6 mm; diameter, 2 mm. Known as yet from two broken specimens, one from Cape San Lucas (type locality), the other from Magdalena Bay, Baja California.

1816. Agathotoma (Agathotoma) alcippe (Dall, 1918) (Synonyms: **Pleurotoma parilis** E. A. Smith, 1888 [not Edwards, 1860]; **Cytharella euryclea** and **pyrrhula** Dall, 1919). The whitish shell is variously banded with narrow or broad bands of brown, axial ribs tending to be continuous, the shoulder strongly tabulate, spiral striae fine and even. Mature specimens from 5 to 11 mm in height; diameter, 2.5 to 5 mm; larger specimens showing thickened varices that represent earlier resting stages. Head of the Gulf of California to Santa Elena Peninsula, Ecuador, at low tide and to depths of 20 m in gravel near rocks.

1817. Agathotoma (Agathotoma) camarina (Dall, 1919). White, hexagonal; axial sculpture of six to seven continuous, rounded ribs. Length, 6 mm; diameter, 2.5 mm. Galápagos Islands.

1818. Agathotoma (Agathotoma) finitima (Pilsbry & Lowe, 1932). Cinnamon-brown, with six to eight ribs on the last whorl, the somewhat irregular spiral striae increasing in strength toward the base. Length, 6 mm; diameter, 2.3 mm. San Juan del Sur, Nicaragua, to Salinas, Ecuador, at low tide.

1819. Agathotoma (Agathotoma) neglecta (C. B. Adams, 1852) (Synonyms: **Defrancia despecta** H. & A. Adams, 1853, unnecessary new name; **Cytharella phryne** Dall, 1919). Like *A. alcippe* this is variable, but typically has continuous axial ribs and fine, even spiral striae; it differs, however, in having a rounded rather than tabulate shoulder. Tan, variously banded with brown, usually with a narrow band at the shoulder, a broad median band and brown at the anterior end of the pillar. Length, 6 mm; diameter, 2.3 mm. Tres Marias Islands, Mexico, to Panama. Dall's *C. phryne* is a broad-shelled variant.

1820. Agathotoma (Agathotoma) quadriseriata (Dall, 1919). This resembles *A. neglecta*, particularly in the pattern of banding, differing in having a proportionately shorter aperture and stronger spiral cords. Length, 5 mm; diameter, 2 mm. Originally said to range from the Gulf of California to Acapulco, Mexico; however, the Acapulco specimens have not since been detected. The type specimen and all others known are from localities (mostly unspecified) in the Gulf of California; the only verified locality is Bahía Concepción (LACM collection).

1821. Agathotoma (Agathotoma) stellata (Mörch, 1860) (Synonyms: **Cytharella hippolita** Dall, 1919; **C. taeniornata** Pilsbry & Lowe, 1932). Characterized by the narrow brown color bands on a buff-colored shell, the bands separated and outlined by spiral grooves. Axial ribbing is variable, the last whorl with five to eight continuous or discontinuous ribs. Length, 5.7 mm; diameter, 2 mm. San Hipolito Point, outer coast of Baja California, through the Gulf of California, and south to Santa Elena Peninsula, Ecuador. Mörch's specimen prabably came from Costa Rica or Nicaragua.

Subgenus VITRICYTHARA FARGO, 1953

Slender, the spiral riblets rounded, interspaces deeply incised; nucleus glassy, sutures impressed. Although Powell (1966) considered this group a subgenus of *Pyrgocythara*, it seems closer to *Agathotoma*, because the whorls are flat-sided and shouldered rather than angulate.

1822. Agathotoma (Vitricythara) klasmidia Shasky, 1971. Spiral cords are fine, alternating in thickness and overriding the axial ribs. Length, 5 mm; diameter, 1.9 mm. Puertecitos, Gulf of California, Mexico, to Panama Bay, in 20 to 40 m.

1823. Agathotoma (Vitricythara) secalis Shasky, 1971. The narrow, rounded spiral cords are broadly spaced, overriding the more prominent axial cords; parietal callus nearly blocking the sinus entrance. Length, 7 mm; diameter, 2.3 mm. Mazatlán, Mexico, to Panama Bay, 15 to 30 m.

Genus PYRGOCYTHARA WOODRING, 1928

Small to medium-sized, whorls rounded to angulate, not shouldered; axial ribs extending suture to suture and across the base, usually not continuous from whorl

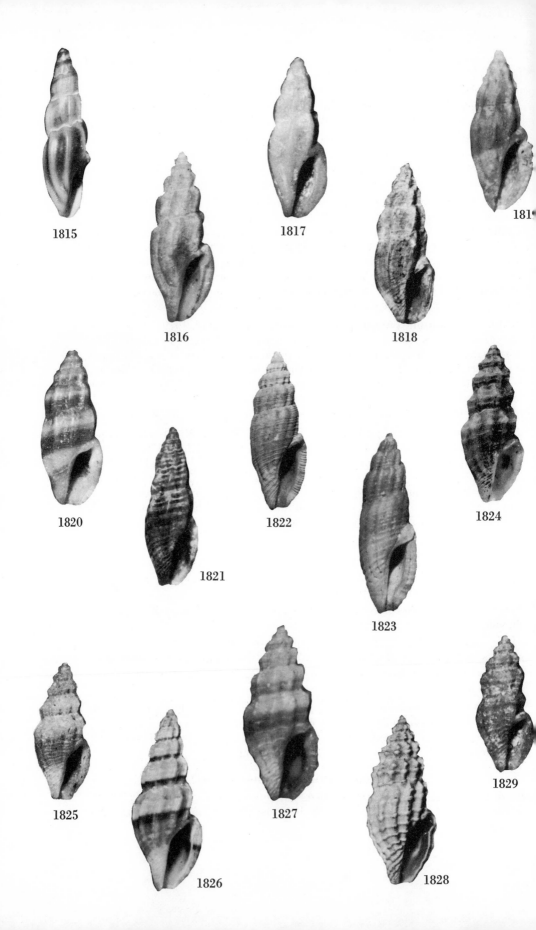

1815

1816

1817

1818

181

1820

1821

1822

1823

1824

1825

1826

1827

1828

1829

to whorl; overridden by spiral striae or cords; outer lip with a strong varix, sinus deep, slot walls diagonally directed, with parietal callus on the inside and a thickened tubercle projecting inward on the lip below the sinus; protoconch smooth at first, then developing fine axial riblets.

1824. Pyrgocythara angulosa McLean & Poorman, 1971. The yellowish-brown shell is whitish on the axial ribs and pillar, darker between the ribs at the periphery; axial ribs are prominent and strongly angulate, spiral cords broad and weakly defined, pillar with finely incised spiral cords. Length, 4.5 mm; diameter, 2.0 mm. Bahía Guasimas, Sonora, Mexico, on mud flats at low tide.

1825. Pyrgocythara danae (Dall, 1919) (Synonym: **Crockerella pederseni** Hertlein & Strong, 1951). The whitish shell has a faint brown band between the axial ribs just below the strongly angulate periphery; spiral striae are finely incised, giving a slight frosted effect. Length, 4.8 mm; diameter, 1.9 mm. Head of the Gulf of California to Guaymas, Sonora, and Agua Verde Bay, Baja California, Mexico, 10 to 30 m.

1826. Pyrgocythara emersoni Shasky, 1971. Relatively large; shoulder white, a narrow brown band below the periphery, a broad light band on the base, pillar white. Axial ribs are curved on the shoulder, spiral sculpture weakly developed. Length, 7.8 mm; diameter, 3.2 mm. Head of the Gulf of California to Guaymas and Puertecitos, at low tide in rocky areas.

1827. Pyrgocythara fuscoligata (Carpenter, 1856). The whorls are tabulate, the axial ribs strongly developed, crossed by broadly spaced, narrow spiral cords. Yellowish white, with a broad peripheral band of brown. Length, 7.5 mm; diameter, 3 mm. Panama Bay, at low tide.

1828. Pyrgocythara hamata (Carpenter, 1865) (Synonym: **Mangelia cerea** Carpenter, 1865). The shell is flesh-tinted orange, with narrow axial ribs crossed by equal-sized spiral cords, the intersections nodose and lighter in color. Length, 6.2 mm; diameter, 2.6 mm, Gulf of Fonseca, El Salvador, to Panama Bay. Although Dall (1919) speculated that Carpenter's material might prove to be a fossil from California, the species had been correctly cited by Carpenter as from Panama.

1829. Pyrgocythara helena (Dall, 1919). Smallest eastern Pacific member of the genus, the shell is angulate on the early whorls, rounded on the final whorl; brown, with clathrate sculpture, intersections of the ribs whitish. Length, 3.5 mm; diameter, 1.3 mm. San Luis Gonzaga Bay, Gulf of California, to Acapulco, Mexico, to 40 m depth.

1830. Pyrgocythara melita (Dall, 1919) (Synonym: **Crockerella hilli** Hertlein & Strong, 1951). Small and slender, the peripheral angulation strong and high on the whorl, the spiral cords broad and weak. The pillar is brown, and there is a brown band just below the suture. Length, 5.2 mm; diameter, 2 mm. Head of the Gulf of California to Banderas Bay, Mexico, in depths to 40 m. Specimens maturing at a length of about 4 mm were regarded by Hertlein and Strong as representing a separate species.

1831. Pyrgocythara phaethusa (Dall, 1919). A stout form that looks foreshortened, variable in sculpture, but usually with a brown band just below the periphery. Length, 3.2 to 4.2 mm; diameter, 1.6 to 2.1 mm. La Libertad, Sonora, to the Gulf of Tehuantepec, Mexico, to depths of 40 m.

1832. Pyrgocythara scammoni (Dall, 1919). Large for the genus; early whorls angulate, later whorls rounded, axial ribs numerous, overridden by spiral cords having nearly equal interspaces. The yellowish shell is banded with brown just below the periphery. Length, 7 mm; diameter, 2.5 mm. Scammon's Lagoon, outer coast of Baja California (type locality), Magdalena Bay and the north end of the Gulf of California as far as Tiburon Island, at low tide.

1833. Pyrgocythara subdiaphana (Carpenter, 1864). The shell is relatively slender, the final whorl rounded, with numerous low axial ribs crossed by fine spiral striae; translucent white, with a broad brown band on the base. Length, 5 mm; diameter, 2 mm. Cape San Lucas area, Baja California. A specimen figured by Dall was erroneously stated to be the type, which is illustrated here for the first time.

Genus BELLACYTHARA McLean, 1971

Relatively large and slender, with rounded whorls and an elongate canal; axial sculpture of low, rounded ribs crossed by numerous fine cords; lip thickened, sinus shallow, with a sharply pointed tubercle below; nucleus nodosely carinate on the third whorl. Resembling some genera in the Clathurellinae, except that the sinus is not as deep and the protoconch and lip tubercle differ.

1834. Bellacythara bella (Hinds, 1843). The yellowish shell is banded with light brown, the pillar dark reddish brown. Length, 15 mm; diameter, 5 mm. Mazatlán, Mexico, to San Francisco Bay, Ecuador, in 10 to 30 m. This is the type and only known species in the genus.

Genus ITHYCYTHARA Woodring, 1928

Small, hexagonal shells with six continuous axial ribs per whorl; periphery angulate, spiral sculpture of fine striae; sinus moderately deep, bordered below by a tubercle and a series of smaller denticles within the edge of the outer lip; nucleus smooth at first, developing axial riblets by the third whorl.

1835. Ithycythara penelope (Dall, 1919). Glassy white; the periphery of the last whorl only weakly angulate, the spiral striae numerous and evenly distributed. Guaymas, Mexico, to Panama Bay and the Galápagos Islands, in 20 to 70 m. The only known eastern Pacific member of the Mangeliinae with a dentate outer lip.

Subfamily DAPHNELLINAE

Shells small to moderately large, sinus close to suture, shaped like a reversed L, the outer lip descending vertically from the suture and produced tangentially forward; mature sculpture usually cancellate, operculum wanting; protoconch elongate, with several rounded, frequently diagonally cancellate whorls; radula having awl-shaped marginals, with the tip constricted, resembling a candle flame. In some genera the sinus is rendered subtubular by a parietal tubercle, and on some the diagonal sculpture of the protoconch is replaced by axial or cancellate riblets. Eleven genera are represented in the Panamic province, seven of which occur in shallow water.

1. Shells with mottled or banded color pattern........................... 2
   Shells uniformly whitish or brown............................... 8
2. Protoconch diagonally reticulate.................................... 3
   Protoconch not diagonally reticulate............................... 6

Genus **DAPHNELLA** HINDS, 1844
(**EUDAPHNE** BARTSCH, 1931 [not REUSS, 1922];
EUDAPHNELLA BARTSCH, 1933)

Ovate-cylindrical shells, with rounded whorls and fine reticulate sculpture, axial ribs tending to become obsolete on the final whorl; anterior end truncate, lip broadly flaring, slightly thickened in some species; protoconch diagonally reticulate. Widely distributed in the tropics.

1836. Daphnella allemani (Bartsch, 1931) (Synonym: **D. thalia** Schwengel, 1938). The shell is whitish, mottled with brown. Axial sculpture fades out on the final whorl, leaving only fine, numerous spiral riblets. The riblets become increasingly strong toward the base, and the aperture is squarely truncate at the anterior end. Length, 14 mm; diameter, 5 mm. Panama Bay and the Galápagos Islands.

1837. Daphnella bartschi Dall, 1919. Slender, tapering anteriorly, yellowish white, with brown flames and spots; surface fine and rasplike, intersections of the axial and spiral sculpture raised; lip slightly thickened in mature specimens. Length, 12.5 mm; diameter, 4.5 mm. Magdalena Bay area, Baja California; San Luis Gonzaga Bay, Gulf of California, to Barra de Navidad, Jalisco, Mexico, and the Galápagos Islands, to depths of 20 m.

1838. Daphnella gemmulifera McLean & Poorman, 1971. Thin-shelled, with an inflated, almost tabulate final whorl; axial sculpture strong on the early whorls, lost on the final whorl, on which the sculpture consists of finely beaded spiral cords. Light brown, mottled with white and darker brown. Length, 13 mm; diameter, 5.5 mm. Isla Santa Cruz, Galápagos Islands, in 200 m.

1839. Daphnella mazatlanica Pilsbry & Lowe, 1932 (Synonym: **D. panamica** Pilsbry & Lowe, 1932). With relatively coarse axial and spiral sculpture, the axial sculpture persisting on the last whorl; aperture flaring anteriorly; lip slightly thickened in mature specimens; color russet, mottled with white. Length, 15.5 mm; diameter, 6 mm. Head of the Gulf of California to Manta, Ecuador, and the Galápagos Islands, to depths of 40 m.

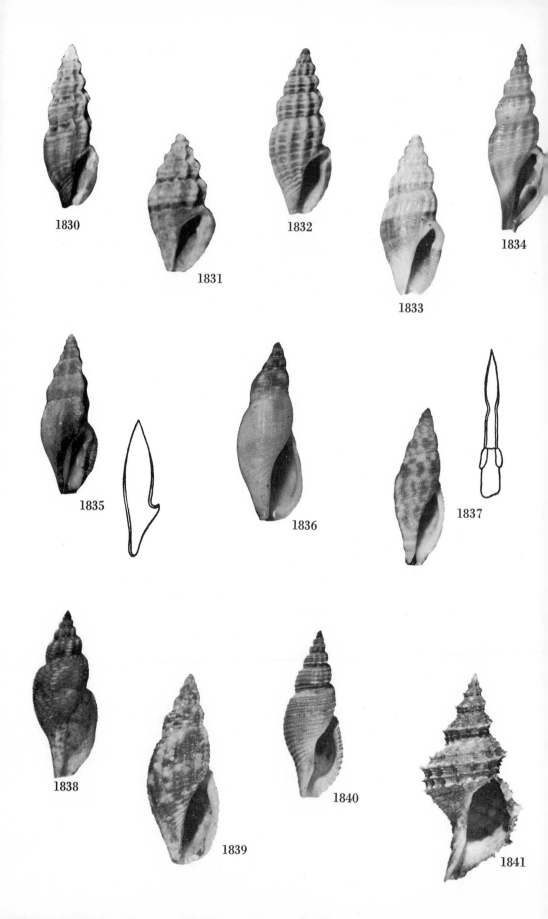

1830

1831

1832

1833

1834

1835

1836

1837

1838

1839

1840

1841

1840. Daphnella retusa McLean & Poorman, 1971. Yellowish white, except for faint mottling of brown and a prominent white spiral cord on the early whorls; whorls rounded, the anterior end tapered, lip moderately thick in mature specimens. No axial sculpture on the final whorl. There are fewer spiral cords than on *D. allemani*, with which this has been confused. Length, 13 mm; diameter, 5 mm. San Luis Gonzaga Bay, Gulf of California, to Secas Islands, Panama, in 20 to 55 m.

Genus RIMOSODAPHNELLA COSSMANN, 1915

Medium-sized, fusiform, with high spire and elongate anterior canal; sculptured with narrow axial ribs and overriding spiral cords; protoconch diagonally reticulate, lip edge thin, sinus deep, anal fasciole lacking spiral sculpture but with raised, axial growth lines. The type and several species are known from the Italian Pliocene.

1841. Rimosodaphnella deroyae McLean & Poorman, 1971. Sculpture spinose at the intersections of the narrow axial ribs and spiral cords; surface with fine imbricate striae between the major spiral cords. Early whorls strongly carinate. Length, 26 mm; diameter, 14 mm. Isla Santa Cruz, Galápagos Islands, in 150 to 200 m.

1842. Rimosodaphnella sculpta (Hinds, 1843). High-spired, the yellowish-brown shell has relatively flat-sided whorls, a constricted pillar, and an elongate canal; the narrow axial ribs fade upon the pillar. Height, 15 mm; diameter, 6.6 mm. Known only from the original specimen from Panama Bay, 13 m depth. *R. gracilispira* (E. A. Smith, 1879) is a related Japanese species having more inflated whorls.

Genus PHILBERTIA MONTEROSATO, 1884

Shells small to medium-sized, high-spired, with rounded whorls, sculpture cancellate, anterior canal short, at an angle to the outer lip; lip thickened, internal denticulation corresponding to spiral sculpture, sinus deeply excavated, bounded by thickened lip callus, parietal callus wanting; first whorl of protoconch spirally lirate, succeeding whorls diagonally cancellate. The type species, *P. philberti* (Michaud, 1830), is European.

1843. Philbertia doris Dall, 1919 (Synonym: **Clathurella crebriforma** Shasky & Campbell, 1964). Small, with deeply impressed sutures, and the aperture drawn out into a short, slightly twisted anterior canal, the pillar profile constricted, outer lip thickened. Light brown, several of the peripheral cords white, the base dark brown; occasional specimens uniformly light or dark brown. Length, 8 mm; diameter, 3 mm, often mature at smaller size. Head of the Gulf of California to Panama Bay, in 10 to 50 m, on gravel bottoms near rock.

1844. Philbertia shaskyi McLean & Poorman, 1971. Small, ovate-biconic, having prominent axial ribs that undulate the suture, with fine spiral ribs overriding the axials; outer lip thickened. The shell is tan-colored, with flecks of brown and white, the protoconch dark brown. Length, 5.7 mm; diameter, 2.5 mm. Off Monserrate Island and Guaymas, Gulf of California, in 40 to 70 m, gravel bottoms.

Genus TRUNCADAPHNE MCLEAN, 1971

Small, sculptured with heavy axial ribs that are overridden by narrow, elevated spiral cords; anterior end truncate, columella twisted; sinus laterally directed, bordered by heavy parietal callus; columella smooth, outer lip with a strong varix, lirate within; protoconch of three diagonally cancellate whorls. Only a single species known. The Indo-Pacific genus *Pseudodaphnella* Boettger, 1895, similar in form, has latticed sculpture, not diagonally cancellate, on the nuclear whorls.

1845. Truncadaphne stonei (Hertlein & Strong, 1939). Whitish, with a brown protoconch and brown flecks on the axial ribs. Length, 4 mm; diameter, 1.8 mm. Galápagos Islands, in 60 to 110 m; described as a Pleistocene fossil.

Genus KERMIA OLIVER, 1915

Shell small, elongate-cylindrical, sculptured with narrow axial ribs and spiral cords, beaded at intersections; sinus deep, U-shaped, bordered by parietal callus; outer lip greatly thickened, crossed by the spiral cords; protoconch with a smooth nucleus followed by three whorls with narrow, curved axial riblets. An Indo-Pacific group with one species in the eastern Pacific.

1846. Kermia informa McLean & Poorman, 1971. The surface of the shell is chalky white; the early whorls with two prominent spiral cords, the spiral cords of the base extending around the massive, thickened lip, crenulating the margin. Length, 6.7 mm; diameter, 2.5 mm. Galápagos Islands, 20 m.

Genus VEPRECULA MELVILL, 1917

Small, high-spired, with narrow axial ribs, crossed by spiral cords of equal strength, spinose at intersections; canal elongate, pillar flexed; sinus sutural; deep, parietal callus lacking; lip smooth within, protoconch with thin axial ribs and microscopic spiral threads, not diagonally cancellate. Widely distributed in the Indo-Pacific.

1847. Veprecula tornipila McLean & Poorman, 1971. Small, with a short, markedly twisted anterior canal; yellowish brown, the nuclear whorls and pillar brown. Length, 3.6 mm; diameter, 1.8 mm. Galápagos Islands, in 145 to 180 m.

Genus MICRODAPHNE MCLEAN, 1971

Minute, with coarse clathrate sculpture, spinose at intersections, pillar slanted to the left, canal moderately elongate, sinus sutural, deep, not bordered by parietal callus; outer lip with a heavy denticulate varix that reduces the aperture to a narrow opening; pillar smooth within; protoconch of three dark-colored whorls, with thickened, slanting axial folds. Only one species known.

1848. Microdaphne trichodes (Dall, 1919) (Synonym: **Pleurotoma hirsutum** De Folin, 1867 [not Bellardi, 1848]). Yellowish brown, with a dark brown protoconch and a white band below the periphery, pillar whitish. Length, 4 mm; diameter, 2 mm. Puertecitos, head of the Gulf of California, to Gorgona Island, Colombia, and the Galápagos Islands, 10 to 70 m. This species probably is widely distributed in the Indo-Pacific; it was illustrated by Maes in 1967 from Cocos-Keeling Atoll in the Indian Ocean.

Genus XANTHODAPHNE POWELL, 1942

Shells medium-sized, ovate, thin, the surface nearly smooth or with fine axial growth lines and spiral striae; aperture broad, with an arcuate outer lip; columella nearly straight; canal short, sinus sutural, wide and deep. The protoconch of the type species, from New Zealand, and of most others is unknown, but in the eastern

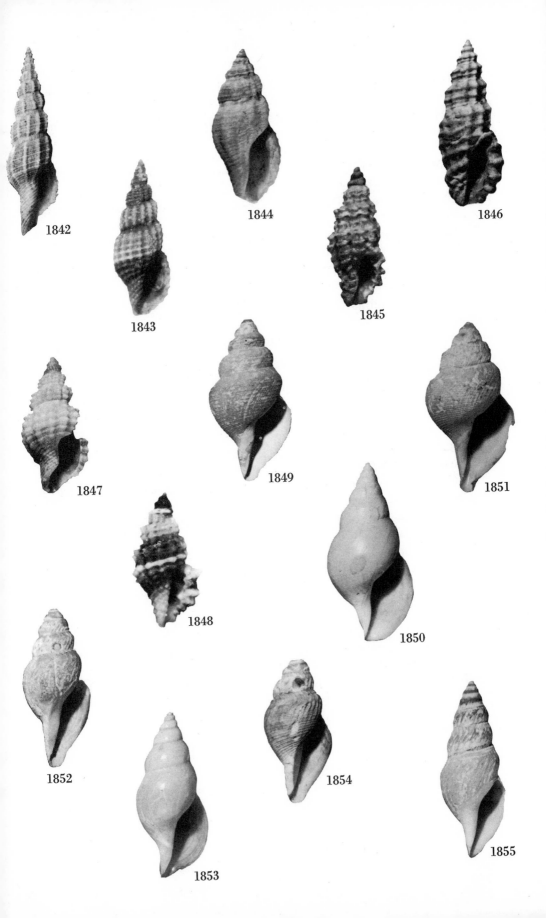

1842

1843

1844

1845

1846

1847

1848

1849

1850

1851

1852

1853

1854

1855

Pacific *X. imparella*, it is diagonally reticulate. The genus is widely distributed at abyssal depths.

1849. Xanthodaphne agonia (Dall, 1890) [*Pleurotomella*] (Synonym: **P. agonia** var. **altina** Dall, 1908). Length, 17 mm; diameter, 8 mm. Galápagos Islands and Cocos Island, 1,159 to 1,850 m.

1850. X. argeta (Dall, 1890) [*Pleurotoma*]. Length, 43 mm; diameter, 20 mm. Galápagos Islands, 1,592 m.

1851. X. egregia (Dall, 1908) [*Pleurotoma*]. Length, 23 mm; diameter, 12 mm. Peru, 4,060 m.

1852. X. encella (Dall, 1908) [*Mangilia*]. Length, 11 mm; diameter, 5 mm. Cocos Island, off Costa Rica, 1,950 m.

1853. X. imparella (Dall, 1908) [*Daphnella*]. Length, 13 mm; diameter, 5.5 mm. Gulf of Panama, 2,320 m.

1854. X. sedillina (Dall, 1908) [*Mangilia*]. Length, 8 mm; diameter, 4 mm. Gulf of Panama, 2,320 m.

1855. X. suffusa (Dall, 1890) [*Pleurotomella*]. Length, 31 mm; diameter, 11 mm; Galápagos Islands, 1,485 m.

Genus PLEUROTOMELLA VERRILL, 1873

Biconic to ovate, whorls shouldered, sculptured with sinuous axial ribs, sutural sinus deep, of the "reversed-L" type; anterior canal flexed. Widely distributed in deep water.

1856. Pleurotomella dinora Dall, 1908. Length, 15 mm; diameter, 7 mm. Galápagos Islands, 1,485 m.

1857. P. enora (Dall, 1908) [*Mangilia*]. Length, 9.5 mm; diameter, 4.2 mm. Off Ecuador, 2,074 m.

1858. P. hermione (Dall, 1919) [*Mangilia*]. Length, 8 mm; diameter, 4 mm. Galápagos Islands, 1,485 m.

1859. P. orariana (Dall, 1908) [*Clathurella*] (Synonym: **P. oceanida** Dall, 1919, same specimen). Length, 12 mm; diameter, 5 mm. Gulf of Panama, 2,320 m.

1860. P. parella Dall, 1908. Length, 41 mm; diameter, 13 mm. Ecuador, 2,074 m.

Genus PHYMORHYNCHUS DALL, 1908

Large, thin-shelled, spirally ridged, with glossy brown periostracum, sinus as in *Pleurotomella*; protoconch unknown.

1861. Phymorhynchus castaneus (Dall, 1895) [*Pleurotomella*]. Length, 53 mm; diameter, 23 mm. Galápagos Islands, 1,592 m. Type of the genus.

1862. P. cingulatus (Dall, 1890) [*Pleurotomella*]. Length, 73 mm; diameter, 30 mm. Galápagos Islands, 2,412 m.

1863. P. clarinda (Dall, 1908) [*Pleurotomella*]. Length, 39 mm; diameter, 18 mm. Gulf of Panama, 3,242 m.

1864. P. speciosus Olsson, 1971. Length, 39 mm; diameter, 18 mm. Gulf of Panama, 3,200 m. [See note, p. 853.]

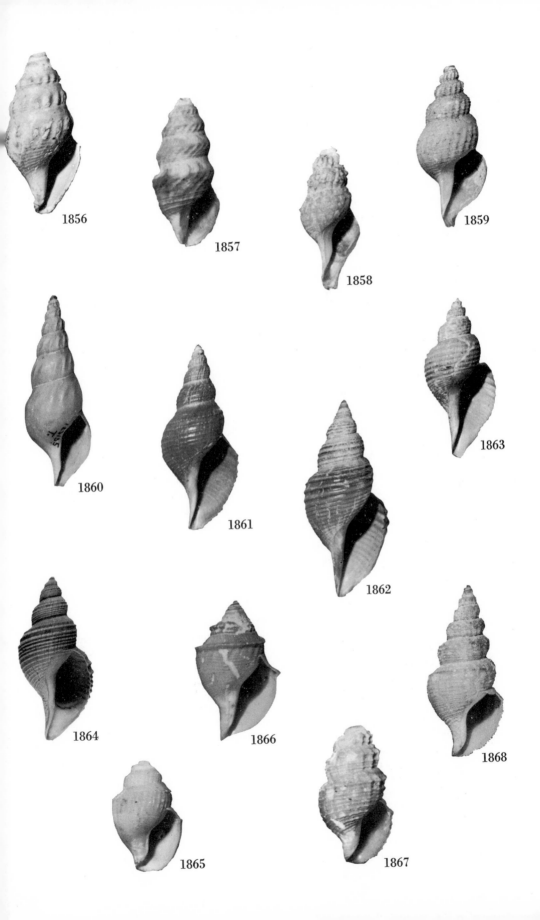

1856

1857

1858

1859

1860

1861

1862

1863

1864

1865

1866

1867

1868

Genus **GYMNOBELA** VERRILL, 1884

Ovoid, spire short, whorls tabulate, sinus weak. Widely distributed in deep water. The protoconch of the Atlantic type species is diagonally reticulate; sculpture not as yet observed in eastern Pacific species.

1865. Gymnobela brachis (Dall, 1919) [*Lora*]. Length, 4.5 mm; diameter, 3.3 mm. Galápagos Islands, 1,485 m.

1866. G. isogonia (Dall, 1908) [*Pleurotomella*]. Length, 11 mm; diameter, 7.7 mm. Gulf of Panama, 1,870 m.

1867. G. vicella (Dall, 1908) [*Gemmula*]. Length, 8.5 mm; diameter, 4.5 mm. Gulf of Panama, 2,323 m.

1868. G. xylona (Dall, 1908) [*Pleurotomella*]. Length, 27 mm; diameter, 12 mm. Galápagos Islands, 2,490 m.

Unassigned and Probably Indeterminate Panamic Turridae

1869. Mangelia striosa acuticostata Carpenter, 1857. The type was unfigured and has been lost. The species was compared to *M. striosa*, of which it may be a synonym.

1870. Pleurotoma adusta Sowerby, 1834. Type locality, Monte Cristi, Ecuador. Unfigured; not cited by Reeve (1843) in his monograph on *Pleurotoma*. Possibly the type is in the British Museum, but it has not been detected.

1871. Clavatula aspera Hinds, 1843. Type material not found. The figure shows strong denticles on the outer lip, none on the columella. No Panamic material has been seen that matches the figure. Possibly the original locality, "Guayaquil and North Coast of New Guinea," was incorrect in including Ecuador.

1872. Lachesis craticulata Mörch, 1860. The type locality is Costa Rica. The type should be in the museum of the University of Copenhagen but has not yet been detected. It was unillustrated, and the description is of little aid.

1873. L. perlata Mörch, 1860. Same status as *L. craticulata*.

1874. Philbertia dione Dall, 1919. The type at the U.S. National Museum has been lost; it was unillustrated.

Subclass **OPISTHOBRANCHIA**
(**OPISTHOBRANCHIATA** of authors; **EUTHYNEURA**)

Although torsion takes place in all gastropods, its effects are not equally apparent in all subclasses. In the opisthobranchs a subsequent detorsion may occur, varying in degree in different groups. Nerves are no longer crossed (hence the alternative subclass name Euthyneura); and the gill is no longer a forward-pointing two-branched plume but may become thickened, turned to one side, moved backward, or even lost altogether and its function taken over by other body tissues.

Throughout the long geologic history of this subclass there has been a tendency toward loss of shell, and a parallel tendency to revert to bilateral symmetry. The shells tend to be thin—bubble-like in some groups, scalelike and embedded in the dorsal surface in others—or lost entirely, as in the nudibranchs.

The animals are mostly active carnivores. Some burrow beneath the surface of the sea floor and move about in search of prey; others are parasitic; still others graze on sessile invertebrates such as hydroids; and there are those that are able to swim vigorously or to float at the sea's surface.

Order ENTOMOTAENIATA

With columellar plaits or other laminae within the aperture. Laminae are most strikingly developed in an extinct superfamily, the Nerineacea.

Superfamily PYRAMIDELLACEA

Columellar folds are present, but the most distinctive feature is the heterostrophic coiling of the many-whorled shells. The early whorls are ultradextral or hyperstrophic and thus apparently sinistral. When normal coiling begins, it is in a different plane, and thus the nuclear whorls seem to rest on or are embedded in the next whorl at a right angle. The radula is obsolete, the operculum few-whorled, oval, and with an apical nucleus.

Although long classed near the Eulimidae in Prosobranchia, this order has been transferred to the Opisthobranchia by modern systematists because of similarities in coiling of the larval shell and in several anatomical features.

Family PYRAMIDELLIDAE

Most members of the family are minute, and a microscope is required for their study. As with other groups of microscopic forms, the treatment here is abridged to lists of named forms and illustrations of sample species.

Useful references for the numerous forms in this family are: Dall & Bartsch (1909); Bartsch (1912); Baker, Hanna & Strong (1928); Pilsbry & Lowe (1932); Strong & Hertlein (1939); Strong (1949); Hertlein & Strong (1951); Brann (1966); Keen (1968). The key is to genera and, in *Pyramidella*, subgenera (the many subgenera of *Odostomia* and *Turbonilla* are keyed separately).

1. Columellar folds three 2
   Columellar folds fewer than three 6
2. Base umbilicate . . . . *Pyramidella, s. s.*
   Base not umbilicate 3
3. Surface smooth except for growth lines 4
   Surface sculptured 5
4. Periphery with a groove or sulcus . . . . *Longchaeus*
   Periphery not sulcate . . . . *Voluspa*
5. Base without cords . . . . *Pharcidella*
   Base with cords that may enter the aperture . . . . *Triptychus*
6. Shell slender, many-whorled, cylindrical . . . . *Turbonilla*
   Shell relatively short, few-whorled, ovate-conic 7
7. Columellar folds strongly developed . . . . *Odostomia*
   Columellar folds obsolete or only weakly developed 8
8. With one entering basal cord (not a columellar fold) . . . . *Peristichia*
   Without basal cords 9
9. Surface strongly sculptured . . . . *Iselica*
   Surface smooth . . . . *Syrnola*

Genus PYRAMIDELLA LAMARCK, 1799

Shell conic, elongate, the whorls somewhat inflated; columella with one to three folds; outer lip more or less strongly lirate within. Shells small but not minute (mostly more than 5 mm in length).

Subgenus PYRAMIDELLA, *s. s.*

Base of shell umbilicate.

1875. Pyramidella (Pyramidella) bairdi Dall & Bartsch, 1909. Around the periphery of the milk-white shell is a narrow yellow band. Length, 5.1 mm; diameter, 1.7 mm. Restricted to the Gulf of California.

1876. Pyramidella (Pyramidella) hancocki Strong & Hertlein, 1939. The slender shell is dull brown in color, with a distinct sulcus on the periphery. Length, 8.8 mm; diameter, 2.8 mm. Dredged in Panama Bay at a depth of 5 to 16 m.

Subgenus LONGCHAEUS MÖRCH, 1875

1877. Pyramidella (Longchaeus) adamsi Carpenter, 1864. The early whorls are white, the later ones mottled with dark brown. Length, 11.3 mm; diameter, 3.8 mm. Southern California through the Gulf of California and south to southern Mexico.

1878. Pyramidella (Longchaeus) bicolor Menke, 1854. In this species the early whorls are white, the next pinkish-tinted, and in the last the tint deepens to rose-purple. Length, 9.8 mm; diameter, 3.5 mm. The range is indefinite, along the Central American coast, probably from southern Mexico to Costa Rica.

1879. Pyramidella (Longchaeus) conica C. B. Adams, 1852. The early whorls are flesh-colored, the later ones light brown, some specimens having a flesh-colored varix. Length, 13 mm; diameter, 4.3 mm. Confined to the Panama coast.

1880. Pyramidella (Longchaeus) elenensis Bartsch, 1924. Pinkish white with a lighter zone in the middle of each whorl; suture denticulate and somewhat channeled. Length, 6.4 mm; diameter, 2.5 mm. Santa Elena Bay, Ecuador.

1881. Pyramidella (Longchaeus) mazatlanica Dall & Bartsch, 1909. The horn-colored shell has a darker band midway of the whorls, bordered below by a faint light line. Length, 11 mm; diameter, 3.5 mm. Southern California through the Gulf and south to southern Mexico, mostly offshore in depths to 26 m.

Subgenus PHARCIDELLA DALL, 1889

Shell faintly spirally striate, with obscure vertical ribbing, especially at the suture.

1882. Pyramidella (Pharcidella) achates (Gould, 1853). A white shell with irregular spots of rust, deepest on the upper third of the whorl. Length, 11.6 mm; diameter, 4.4 mm. Probably restricted to the southern part of the Gulf of California south to Mazatlán, Mexico.

1883. Pyramidella (Pharcidella) ava Bartsch, 1926. The shell is stout, pupoid, the axial ribbing fine and faint; an incised spiral groove falling below the periphery. Outer lip with seven spiral lamellae within. Length, 4.1 mm; diameter, 1.6 mm. Santa Elena Bay, Ecuador.

1884. Pyramidella (Pharcidella) hastata (A. Adams in Sowerby, 1854). Pale yellowish to flesh-colored, the shiny shell is irregularly clouded with light brown. Length, 11.5 mm; diameter, 4.6 mm. The species has been reported at Acapulco, Mexico.

1885. Pyramidella (Pharcidella) magdalenensis Bartsch, 1917. Pale horn-yellow, the shell has fine axial ribbing; outer lip with four denticles within. Length, 5.8 mm; diameter, 2.1 mm. Magdalena Bay, Baja California, depth 25 m.

1886. Pyramidella (Pharcidella) moffati Dall & Bartsch, 1906 (Synonym: **Obeliscus clavulus** Adams in Sowerby, 1854 [not Beck, 1838]). The white

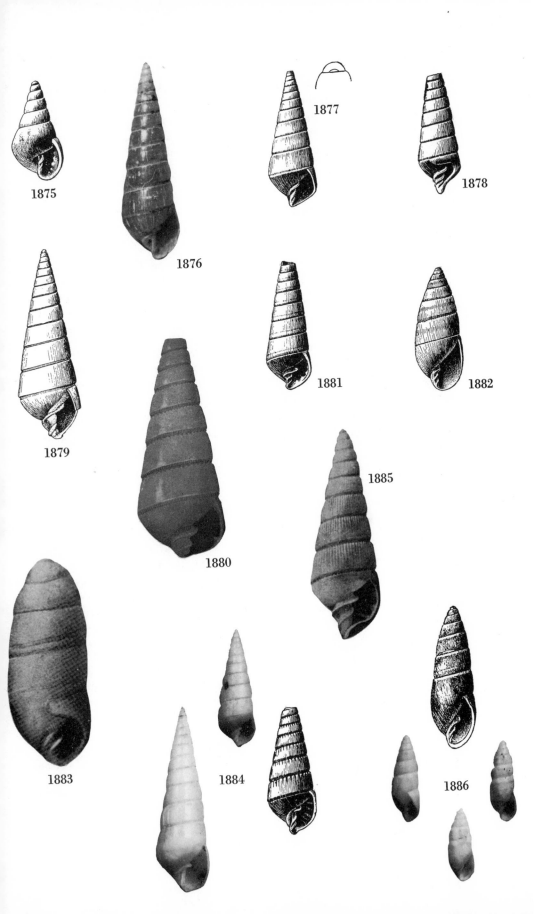

1875

1876

1877

1878

1879

1880

1881

1882

1883

1884

1885

1886

shell is clouded with rust-brown and has an ear-shaped aperture. Length, 11.5 mm; diameter, 4 mm. Acapulco, Mexico.

1887. Pyramidella (Pharcidella) panamensis Dall & Bartsch, 1909. A horn-yellow shell with the last whorl suffused with pale rose-purple. Length, 8.8 mm; diameter, 3 mm. Dredged in Panama Bay in 91 m depth.

Subgenus VOLUSPA DALL & BARTSCH, 1904

Shell surface smooth.

1888. Pyramidella (Voluspa) auricoma Dall, 1889. Yellowish white in color, the shell has fine golden-yellow spiral lines. Length, 10.6 mm; diameter, 3.8 mm. The Gulf of California to Mazatlán, Mexico.

1889. Pyramidella (Voluspa) linearum Pilsbry & Lowe, 1932. The pale pink shell is covered with fine brown spiral lines that are interrupted by some pinkish axial streaks. Length, 16 mm; diameter, 5.3 mm. Puerto Peñasco to Acapulco, Mexico, to depths of 37 m.

Genus ISELICA DALL, 1918
(ISAPIS H. & A. ADAMS, 1854 [not DOUBLEDAY, 1847])

Small, ovate, with cancellate sculpture; base umbilicate; columella with a low, blunt fold or ridge. Many workers have placed the genus in the Fossaridae.

1890. Iselica fenestrata (Carpenter, 1864) [*Isapis*]. Height, 6 mm; diameter, 4 mm. Puget Sound, Washington, to the Gulf of California.

1891. I. kochi Strong & Hertlein, 1939. Height, 1.5 mm. Bahía Honda, Panama, in 5 to 15 m.

1892. I. maculosa (Carpenter, 1857) [*Fossarus*]. Height, 4 mm; diameter, 3 mm. Mazatlán, Mexico.

1893. I. ovoidea (Gould, 1853) [*Narica*]. Height, 8 mm; diameter, 5 mm. Mazatlán, Mexico.

Genus ODOSTOMIA FLEMING, 1813

Minute shells (mostly under 5 mm in length), these are short-conic to ovate, variously sculptured, with one fairly strong fold on the columella. A number of subgenera are recognized, some (notably *Chrysallida, Miralda,* and *Evalea*) being considered distinctive enough to be ranked as genera by many workers. About 100 species have been described for the Panamic province. The Odostomias for which food habits are known have proved to be parasitic upon other mollusks—oysters, scallops, and slipper limpets—and also upon other invertebrates, especially polychaete worms. Other species are apt to be of similar habit. The degree of host specificity—i.e., whether a given *Odostomia* will live upon more than one host, and which species are associated with which involuntary hosts—is yet to be worked out, especially for tropical American species. Useful references are cited in the discussion of *Pyramidella*.

1. Early postnuclear whorls loosely coiled, later ones most tightly coiled
   . . . .*Lysacme*
   Postnuclear whorls evenly coiled throughout. 2
2. Varices present . . . .*Salassiella*
   Varices wanting 3

3. Axial sculpture present.. 4
  Axial sculpture absent 15
4. Axial sculpture not crossed by spirals ....*Salassia*
  Axial sculpture intersected by at least some spirals 5
5. Spiral sculpture weak 6
  Spiral sculpture equal to or stronger than axials 8
6. Spiral threads equally distributed ....*Pyrgulina*
  Spiral threads not equally distributed 7
7. Spiral threads stronger above base, weak on base ....*Besla*
  Spiral threads on base only ....*Egila*
8. Spiral sculpture equal to axial, reticulate 9
  Spiral sculpture stronger than axial...............................11
9. Sculpture reticulate on spire, spiral on base.................*Chrysallida*
  Sculpture reticulate throughout 10
10. Intersections cusped, texture rough ....*Haldra*
  Intersections not conspicuously nodose ....*Ividella*
11. Axials present only on upper parts of whorls 12
  Axials present as fine lirations between spirals 14
12. Spiral lirations stronger, and one or more crenulate, anteriorly .....*Miralda*
  Spiral lirations equally distributed below 13
13. Whorls rounded ....*Evalina*
  Whorls tabulate ....*Ivara*
14. Base umbilicate ....*Iolaea*
  Base not umbilicate...*Menestho*
15. Surface with microscopic spiral striae ....*Evalea*
  Surface polished, not showing spiral striae.........................16
16. Peritreme continuous ....*Heida*
  Peritreme interrupted by body whorl..............................17
17. Body whorl with a peripheral keel.........................*Eulimastoma*
  Body whorl rounded...18
18. Spire whorls inflated..*Amaura*
  Spire whorls flattened.................................*Odostomia, s. s.*

Subgenus ODOSTOMIA, s. s.

Small, smooth, with microscopically fine striae; whorls flattened.

1894. Odostomia (Odostomia) mammillata Carpenter, 1857. Height, 1.1 mm. Mazatlán, Mexico. Figured, Brann (1966), Keen (1968).

Subgenus AMAURA MÖLLER, 1842

Relatively large, inflated, sculpture reduced to microscopic striae.

1895. Odostomia (Amaura) subturrita Dall & Bartsch, 1909. Height, 6.9 mm. Santa Barbara, California, to Todos Santos Bay, Baja California.

Subgenus BESLA DALL & BARTSCH, 1904

Sculpture of three strong spiral ribs crossed by axials; base spirally ribbed.

1896. Odostomia (Besla) convexa Carpenter, 1857. Height, 2.4 mm. Mazatlán and Bahía San Luis Gonzaga, Mexico. Figured, Brann (1966), Keen (1968). Type of the subgenus.

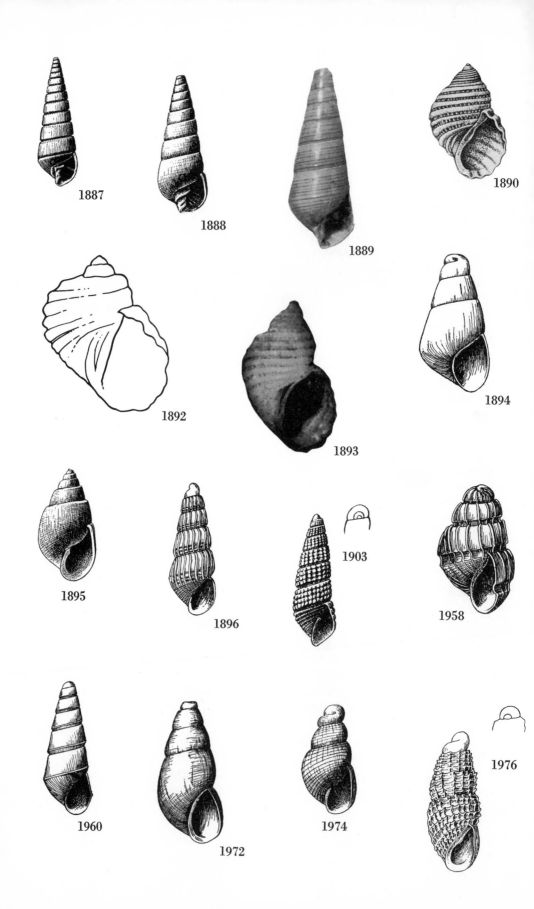

1887

1888

1889

1890

1892

1893

1894

1895

1896

1903

1958

1960

1972

1974

1976

1897. O. (B.) excolpa Bartsch, 1912. Head of the Gulf of California.

Subgenus **CHRYSALLIDA** CARPENTER, 1857

Axial ribs crossed by spirals of equal weight, intersections nodulous; base with spiral cords only.

1898. Odostomia (Chrysallida) acrybia Dall & Bartsch, 1909. Point Abreojos, Baja California.

1899. O. (C.) ata Bartsch, 1926. Santa Elena Peninsula, Ecuador.

1900. O. (C.) audax Baker, Hanna & Strong, 1928. Cape San Lucas, Baja California.

1901. O. (C.) benthina Dall & Bartsch, 1909 (Synonym: **O. (C.) oblonga** Carpenter, 1857 [not Macgillivray, 1848]). Mazatlán, Mexico, taken from *Spondylus*.

1902. O. (C.) capa Bartsch, 1926. Santa Elena Peninsula, Ecuador.

1903. O. (C.) clathratula (C. B. Adams, 1852) [*Chemnitzia*]. Height, 2.8 mm. Panama. Figured, Turner (1956).

1904. O. (C.) collea Bartsch, 1926. Santa Elena Peninsula, Ecuador.

1905. O. (C.) communis (C. B. Adams, 1852) [*Chemnitzia*]. Panama. Figured, Turner (1956).

1906. O. (C.) contrerasi Baker, Hanna & Strong, 1928. Gulf of California.

1907. O. (C.) corintoensis Hertlein & Strong, 1951. Corinto, Nicaragua, in beach drift.

1908. O. (C.) costaricensis Hertlein & Strong, 1951. Near Port Parker, Costa Rica, dredged in 22 to 27 m, shelly mud.

1909. O. (C.) deceptrix Dall & Bartsch, 1909. San Hipolito Point (type locality) and Point Abreojos, Baja California.

1910. O. (C.) defolinia Dall & Bartsch, 1909 (Synonym: **Noemia angusta** De Folin, 1872 [not *Chrysallida angusta* Carpenter, 1864]). Margarita Island, Panama Bay.

1911. O. (C.) defolinia contracta (De Folin, 1872) [*Noemia*]. No locality or figure given.

1912. O. (C.) defolinia difficilis Dall & Bartsch, 1909 (Synonym: **Noemia angusta** var. **ovata** De Folin, 1872 [not *Chrysallida ovata* Carpenter, 1857]). No locality or figure given.

1913. O. (C.) effusa Carpenter, 1857. Mazatlán and Cape San Lucas, Mexico. Figured, Brann (1966), Keen (1968).

1914. O. (C.) eiseni Jordan, 1936. Magdalena Bay, Baja California. Pleistocene.

1915. O. (C.) era Bartsch, 1927. Magdalena Bay, Baja California.

1916. O. (C.) eugena Dall & Bartsch, 1909. San Pedro, California, to San Hipolito Point, Baja California.

1917. O. (C.) evermanni Jordan, 1936. Magdalena Bay, Baja California. Pleistocene.

1918. O. (C.) excelsa Dall & Bartsch, 1909. Panama Bay.

1919. O. (C.) fasciata Carpenter, 1857. Mazatlán, Mexico. Figured, Brann (1966), Keen (1968).

1920. O. (C.) guatulcoensis Hertlein & Strong, 1951. Near Port Guatulco, Mexico, dredged in 13 m, sand and shell bottom.

1921. O. (C.) hipolitensis Dall & Bartsch, 1909. San Hipolito Point, Baja California.

1922. O. (C.) inconspicua (C. B. Adams, 1852) [*Cingula* (?)]. Panama. Figured, Turner (1956).

1923. O. (C.) lapazana Dall & Bartsch, 1909. Off La Paz, Baja California, in 48 m.

1924. O. (C.) licina Dall & Bartsch, 1909. Manuela Lagoon, Baja California.

1925. O. (C.) limbaughi Hertlein & Allison, 1968. Clipperton Island.

1926. O. (C.) loomisi Dall & Bartsch, 1909. Panama Bay.

1927. O. (C.) melitta Bartsch, 1924. Santa Elena, Ecuador.

1928. O. (C.) nodosa Carpenter, 1857. Mazatlán, Mexico. Figured, Brann (1966), Keen (1968).

1929. O. (C.) olssoni Bartsch, 1924. Santa Elena, Ecuador.

1930. O. (C.) ooniscia Dall & Bartsch, 1909 (Synonym: **O. (C.) ovulum** Carpenter, 1857 [not *Parthenia ovulum* Lea, 1845]). Mazatlán, Mexico; specimens taken from *Spondylus* and *Chama*. Figured, Brann (1966), Keen (1968).

1931. O. (C.) ovata Carpenter, 1857. Mazatlán, Mexico. Figured, Brann (1966), Keen (1968).

1932. O. (C.) pacha Bartsch, 1926. Santa Elena Peninsula, Ecuador.

1933. O. (C.) paupercula (C. B. Adams, 1852) [*Cerithium*] (Synonym: "**Cingula paupercula** Adams" of Dall and Bartsch, 1909, not of Adams, 1852, which is a member of the Rissoacea [*see under Rissoella*]). Panama. Figured, Turner (1956).

1934. O. (C.) promeces Dall & Bartsch, 1909. Todos Santos Bay, Baja California.

1935. O. (C.) proxima (De Folin, 1872) [*Noemia*]. Margarita Island, Panama Bay. Figured, Dall & Bartsch (1909).

1936. O. (C.) pulchra (De Folin, 1872) [*Noemia*]. Margarita Island, Panama Bay. Figured, Dall & Bartsch (1909).

1937. O. (C.) quilla Bartsch, 1926. Santa Elena Peninsula, Ecuador.

1938. O. (C.) reedi Bartsch, 1928. Ecuador.

1939. O. (C.) reigeni Carpenter, 1857. Mazatlán, Mexico. Figured, Brann (1966), Keen (1968).

1940. O. (C.) rinella Dall & Bartsch, 1909. Panama.

1941. O. (C.) rotundata Carpenter, 1857. Mazatlán, Mexico. Figured, Brann (1966), Keen (1968).

1942. O. (C.) salinasensis Bartsch, 1928. Ecuador.

1943. O. (C.) santamariensis Bartsch, 1917. Santa Maria Bay, Baja California.

1944. O. (C.) sorenseni Strong, 1949. Mazatlán, Mexico.

1945. O. (C.) swetti Strong & Hertlein, 1939. Off Taboga Island, Panama, in 5 to 15 m.

1946. O. (C.) talama Dall & Bartsch, 1909. Scammon's Lagoon, Baja California.

1947. O. (C.) taravali Bartsch, 1917. Santa Maria Bay, Redondo Point, and Magdalena Bay, Baja California.

1948. O. (C.) telescopium Carpenter, 1857. Mazatlán. Figured, Brann (1966), Keen (1968).

1949. O. (C.) terissa Pilsbry & Lowe, 1932. San Juan del Sur, Nicaragua.

1950. O. (C.) torrita Dall & Bartsch, 1909 (Synonym: **O. (C.) communis** (C. B. Adams) of Carpenter, 1857, not *Chemnitzia communis* C. B. Adams, 1852). Mazatlán, Mexico, on *Chama* and *Spondylus*. Type of the subgenus.

1951. O. (C.) trimariana Pilsbry & Lowe, 1932. Tres Marias Islands, Gulf of California.

1952. O. (C.) tyleri Dall & Bartsch, 1909. Panama Bay.

1953. O. (C.) vira Bartsch, 1926. Santa Elena Peninsula, Ecuador.

1954. O. (C.) virginalis Dall & Bartsch, 1909 (Synonym: "**Evalea graciliente** Carpenter" of Keep, 1888 [not Monterosato, 1884]). San Pedro, California, to western Baja California, at Point Abreojos, San Hipolito Point, and Todos Santos Bay (type locality).

1955. O. (C.) vizcainoana Baker, Hanna & Strong, 1928. La Paz, Baja California (type locality) in 7 m; also from Puerto Escondido, Agua Verde Bay, Espíritu Santo, and San José Island, Gulf of California. Pleistocene, Magdalena Bay.

1956. O. (C.) woodbridgei Hertlein & Strong, 1951. Costa Rica.

1957. O. (C.) zeteki Bartsch, 1918. Panama.

Subgenus EGILA DALL & BARTSCH, 1904

With strong axial ribs from apex to base; periphery with a sulcus.

1958. Odostomia (Egila) lacunata Carpenter, 1857. Height, 1 mm. Mazatlán, Mexico. Figured, Brann (1966), Keen (1968).

1959. O. (E.) poppei Dall & Bartsch, 1909. Point Abreojos, Baja California.

Subgenus EULIMASTOMA BARTSCH, 1916
(TELLODA HERTLEIN & STRONG, 1951)

Slender, smooth; periphery with a carina or keel.

1960. Odostomia (Eulimastoma) dotella Dall & Bartsch, 1909. Height, 2.3 mm. Southern end of the Gulf of California (type locality, Cerralvo Island), in 16 to 18 m.

1961. O. (E.) subdotella Hertlein & Strong, 1951. Port Parker, Costa Rica (type locality) to Corinto, Nicaragua, in 22 to 27 m.

Subgenus EVALEA A. ADAMS, 1860
Surface with fine, incised spiral lines.

1962. Odostomia (Evalea) donilla Dall & Bartsch, 1909. San Pedro, California (type locality), to Todos Santos Bay, Baja California.

1963. O. (E.) gallegosiana Hertlein & Strong, 1951. Puerto Guatulco, Mexico.

1964. O. (E.?) granadensis Dall & Bartsch, 1909. Panama Bay, 115 m.

1965. O. (E.) isthmica Strong & Hertlein, 1939. Bahía Honda, Panama (type locality), 5 to 15 m; Taboga Island, Panama.

1966. O. (E.) lucasana Dall & Bartsch, 1909. Cape San Lucas, Baja California.

1967. O. (E.) martinensis Strong, 1938. San Martin Island, Baja California.

1968. O. (E.) minutissima Dall & Bartsch, 1909. San Diego, California (type locality), to Point Abreojos, Baja California.

1969. O. (E.) palmeri Bartsch, 1912. Head of the Gulf of California.

1970. O. (E.) parella Dall & Bartsch, 1909. Near Galápagos Islands, 1,160 m.

1971. O. (E.) socorroensis Dall & Bartsch, 1909. Socorro Island, Mexico.

1972. O. (E.) tenuis Carpenter, 1857. Mazatlán, Mexico. Figured, Brann (1966), Keen (1968).

1973. O. (E.) valeroi Bartsch, 1917. Magdalena and Santa Maria bays, Baja California.

Subgenus EVALINA DALL & BARTSCH, 1904
Axial sculpture near sutures only; spiral cords numerous, fine.

1974. Odostomia (Evalina) intermedia (Carpenter, 1857) [*Chemnitzia*]. Height, 1.4 mm. Mazatlán. Figured, Brann (1966), Keen (1968).

1975. O. (E.) tehuantepecana Hertlein & Strong, 1951. Near Port Guatulco, Mexico, in 13 m, sand and shell bottom.

Subgenus HALDRA DALL & BARTSCH, 1904
Axials and spirals raised, equal, giving a roughened appearance.

1976. Odostomia (Haldra) photis Carpenter, 1857. Height, 1.2 mm. Mazatlán, Mexico. Figured, Brann (1966), Keen (1968).

Subgenus HEIDA DALL & BARTSCH, 1904
Smooth, peritreme continuous; aperture resembling Rissoidae.

1977. Odostomia (Heida) panamensis Clessin, 1900. Height, 3 mm. Panama. Figured, Dall & Bartsch (1909).

Subgenus IOLAEA A. ADAMS, 1867
(IOLE A. ADAMS, 1860 [not BLYTH, 1844]; IOLINA BAILY, 1948, unnecessary replacement)
Umbilicate; spiral cords predominating, axials showing between.

1978. Odostomia (Iolaea) amianta Dall & Bartsch, 1907. San Pedro, California, to Point Abreojos, Baja California.

1979. O. (I.) delicatula Carpenter, 1864. Height, 2.3 mm. Cape San Lucas, Baja California. Figured, Dall & Bartsch (1909).

1980. O. (I.) eucosmia Dall & Bartsch, 1909 (Synonym: **Oscilla insculpta** Keep, 1888, *ex* Carpenter, MS [not De Kay, 1843]). San Pedro, California, to Todos Santos Bay, Baja California.

Subgenus **IVARA** DALL & BARTSCH, 1903

Spirals even-sized, axials weak; whorls tabulate.

1981. Odostomia (Ivara) terricula Dall & Bartsch in Arnold, 1903. (Corrected to *turricula* by Dall and Bartsch, 1909.) Height, 4 mm. San Pedro, California (type locality), to Todos Santos Bay, Baja California. Type of the subgenus.

Subgenus **IVIDELLA** DALL & BARTSCH, 1909

With laminar spiral and axial ribs over entire surface.

1982. Odostomia (Ividella) contabulata (Mörch, 1860) [*Rissoina*]. Sonsonate, El Salvador. Figured, Keen, 1966.

1983. O. (I.) mariae Bartsch, 1928. Quaternary, from a well core near Vichayal, Peru.

1984. O. (I.) mendozae Baker, Hanna & Strong, 1928. Cape San Lucas, Baja California (type locality).

1985. O. (I.) navisa Dall & Bartsch, 1907. Scammon's Lagoon, Baja California. Type of the subgenus.

1986. O. (I.) notabilis (C. B. Adams, 1852) [*Rissoa*] (Synonyms: **Cingula turrita** C. B. Adams, 1852; **Parthenia quinquecincta** Carpenter, 1857; **O. orariana** Dall & Bartsch, 1909). Height, 2 mm. Mazatlán, Mexico, to Panama.

1987. O. (I.) pedroana Dall & Bartsch, 1909. San Pedro, California (type locality), to Scammon's Lagoon, Baja California.

1988. O. (I.) ulloana Strong, 1949. La Paz, Baja California.

Subgenus **LYSACME** DALL & BARTSCH, 1904

Spire whorls loosely coiled just below apex; base lirate.

1989. Odostomia (Lysacme) clausiformis Carpenter, 1857. Height, 4 mm. Mazatlán, Mexico. Figured, Brann (1966), Keen (1968). Type of the subgenus.

Subgenus **MENESTHO** MÖLLER, 1842

Spiral cords of even width, axials incremental only; no umbilicus.

1990. Odostomia (Menestho) aequisculpta Carpenter, 1864. Cape San Lucas. Figured, Dall & Bartsch (1909).

1991. O. (M.) callipyrga Dall & Bartsch, 1904 (Synonym: **Odetta elegans** De Folin, 1871 [not *Odostomia elegans* A. Adams, 1860]). Margarita Island, Panama. Figured, Dall & Bartsch (1909).

1992. O. (M.) ciguatonis Strong, 1949. Gulf of California.

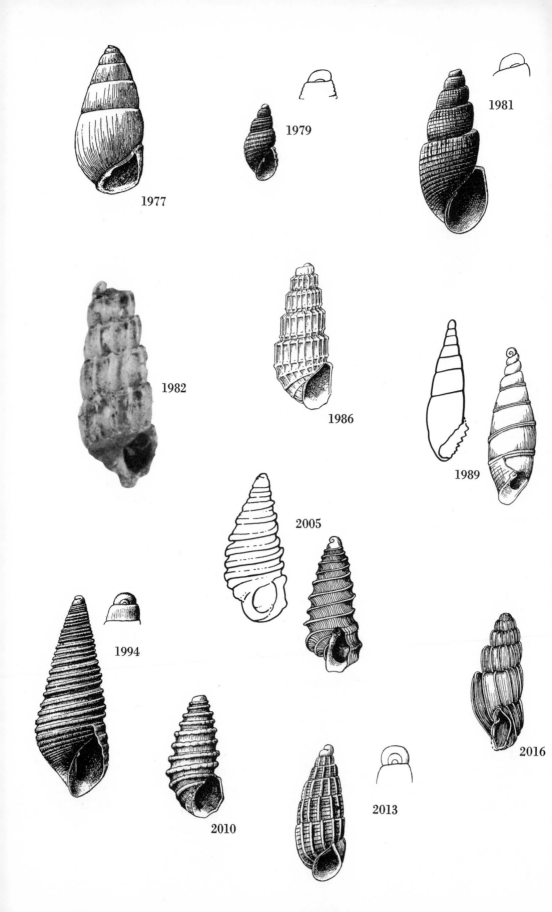

1977

1979

1981

1982

1986

1989

2005

1994

2010

2013

2016

1993. O. (M.) gloriosa Bartsch, 1912. San Diego, California (type locality) to San Hipolito Point, Baja California.

1994. O. (M.) grammatospira Dall & Bartsch, 1903. Height, 5.3 mm. Cape San Lucas, Baja California.

1995. O. (M.) grijalvae Baker, Hanna & Strong, 1928. Cape San Lucas, Baja California.

1996. O. (M.) navarettei Baker, Hanna & Strong, 1928. Amortajada Bay, San José Island, Gulf of California.

1997. O. (M.) nicoyana Hertlein & Strong, 1951. Near Port Parker, Costa Rica, in 20 to 27 m, sandy and shelly mud.

1998. O. (M.) recta (De Folin, 1872) [*Odetta*]. Margarita Island, Panama Bay. (Unfigured; type broken before it could be illustrated.)

1999. O. (M.) sublirulata Carpenter, 1857. Mazatlán. Figured, Brann (1966), Keen (1968).

2000. O. (M.) ziziphina Carpenter, 1857. Mazatlán. Figured, Brann (1966), Keen (1968).

Subgenus MIRALDA A. ADAMS, 1864

With strong spiral keels, the anterior two or three crenulate.

2001. Odostomia (Miralda) aepynota Dall & Bartsch, 1909. San Pedro, California (type locality), to Cape San Lucas, Baja California.

2002. O. (M.) aepynota planicosta Baker, Hanna & Strong, 1928. Cape San Lucas.

2003. O. (M.) armata Carpenter, 1857. Mazatlán. Figured, Brann (1966), Keen (1968).

2004. O. (M.) azteca Strong & Hertlein, 1939. Bahía Honda, Panama (type locality) in 5 to 15 m.

2005. O. (M.) exarata Carpenter, 1857. Height, 6 mm. Mazatlán. Figured, Brann (1966), Keen (1968).

2006. O. (M.) galapagensis Dall & Bartsch, 1909. Near Galápagos Islands, in 73 m on coral sand bottom.

2007. O. (M.) hemphilli Dall & Bartsch, 1909. San Pedro, California, to San Hipolito Point (type locality) and Point Abreojos, Baja California.

2008. O. (M.) porteri Baker, Hanna & Strong, 1928. Gulf of California.

2009. O. (M.) rhizophorae Hertlein & Strong, 1951. Near Corinto, Nicaragua, in 22 to 27 m, on mangrove leaves.

2010. O. (M.) terebellum (C. B. Adams, 1852) [*Cingula* (?)]. Height, 2.2 mm. Panama, in calcareous sand. Figured, Turner (1956).

Subgenus PYRGULINA A. ADAMS, 1863

With strong axials from apex to base, fine spirals between.

2011. Odostomia (Pyrgulina) herrerae Baker, Hanna & Strong, 1928. Gulf of California.

2012. O. (P.) mara Bartsch, 1926. Santa Elena Peninsula, Ecuador.

2013. O. (P.) marginata (C. B. Adams, 1852) [*Chemnitzia*]. Height, 2.8 mm. Panama. Figured, Turner (1956).

Subgenus SALASSIA DE FOLIN, 1885

Pupiform, whorls tabulate; with axial ribs, no varices.

2014. Odostomia (Salassia) gabrielensis Baker, Hanna & Strong, 1928. San Gabriel Bay, Espíritu Santo Island, Gulf of California (type locality), Monserrate Island and San José Island, Gulf of California; La Paz, Baja California.

2015. O. (S.) hertleini Strong, 1938. Off Maria Madre, Tres Marias Islands, in 18 to 45 m.

2016. O. (S.) scalariformis Carpenter, 1857. Height, 2.8 mm. Mazatlán, Mexico. Figured, Brann (1966), Keen (1968).

2017. O. (S.) tropidita Dall & Bartsch, 1909 (Synonym: **Salassia carinata** De Folin, 1872 [not *Scalenostoma carinata* Deshayes, 1863; not *O. carinata* H. Adams, 1873]). Perlas Island, Panama Bay. Type of the subgenus.

Subgenus SALASSIELLA DALL & BARTSCH, 1909

Pupiform; axial ribs strong, with irregular varices.

2018. Odostomia (Salassiella) laxa Dall & Bartsch, 1909. Height, 4.3 mm. San Diego, California, to Scammon's Lagoon, Baja California (type locality). Type of the subgenus.

Subgenus Uncertain

2019. Odostomia gallegosi Hertlein, 1934. Pleistocene, Maria Magdalena Island.

Genus PERISTICHIA DALL, 1889

Strongly sculptured, resembling *Triptychus*, but with only one basal cord, which continues into the aperture, and no columellar folds.

2020. Peristichia hermosa (Lowe, 1935) [*Pyramidella*]. Length, 7 mm. San Felipe, Baja California.

2021. P. pedroana (Dall & Bartsch, 1909) [*Odostomia*]. Length, 6.7 mm. Southern California to southern Baja California.

Genus SYRNOLA A. ADAMS, 1860

Slender, smoothly tapering, somewhat resembling the prosobranch genus *Eulima*; columellar folds weakly developed; outer lip with reinforcing lirations or lamellae.

2022. Syrnola collea (Bartsch, 1926) [*Pyramidella*]. Length, 3.4 mm. Santa Elena Bay, Ecuador.

Genus TRIPTYCHUS MÖRCH, 1875

Sculpture nodose, fairly strong; base with two to three cords that enter the aperture and one or more columellar folds; outer lip lirate within.

2023. Triptychus olssoni (Bartsch, 1926) [*Pyramidella*]. Length, 5 mm; diameter, 1.6 mm. Santa Elena Bay, Ecuador.

Genus **TURBONILLA** RISSO, 1826

More slender and tapering than the Odostomias, these are also minute shells, rarely exceeding 5 mm in length, though there are numerous whorls. The single columellar fold is so weak that it is not readily seen. Illustrations of a few subgenera of the Turbonillas are given here.

The genus holds the record for number of species in the Panamic province, over 200 having been described. Useful references are listed above (in the discussion of the Pyramidellidae). Several subgenera have been recognized, although dividing lines are not always clear. The key, which is to Panamic subgenera, uses the characters suggested in an early key by Dall & Bartsch (1909).

1. With a basal keel or cord. . . . . *Asmunda*
   Without basal keel. 2
2. Varices present in addition to normal axial ribbing. . . . . *Mormula*
   Varices not present; axial ribbing uniform. 3
3. Columella with a strong fold. . . . . *Ugartea*
   Columellar fold not evident in apertural view. 4
4. Spiral sculpture wanting or of microscopic striae. 5
   Spiral sculpture present, stronger than striae. 7
5. Axial ribs weak, on early whorls only. . . . . *Ptycheulimella*
   Axial ribs strong. 6
6. Axial ribs between sutures only, not on base. . . . . *Chemnitzia*
   Axial ribs both on spire whorls and on base. . . . . *Turbonilla, s. s.*
7. Axial sculpture reduced to fine threads. . . . . *Careliopsis*
   Axial sculpture stronger than fine striae. 8
8. Axial sculpture present as weak ribs. . . . . *Cingulina*
   Axial ribs strong. 9
9. Spiral markings fine and narrow. 10
   Spiral markings of strongly incised lines. 11
10. Aperture suboval . . . . *Pyrgolampros*
    Aperture subquadrate . . . . *Strioturbonilla*
11. Summits of whorls rounded, not shouldered. . . . . *Pyrgiscus*
    Summits of whorls strongly shouldered. 12
12. Spiral sculpture as strong as axials, intersections subnodulous; axials not
    continuing on base. . . . . *Bartschella*
    Spiral sculpture finer than axials; axials continuing across base. . . . *Dunkeria*

Subgenus **TURBONILLA**, *s. s.*

Axially ribbed throughout, no spiral sculpture.

2024. Turbonilla (Turbonilla) axeli Bartsch, 1924. Santa Elena Bay, Ecuador.

2025. T. (T.) centrota Dall & Bartsch, 1909 (Synonym: **Chemnitzia acuminata** C. B. Adams, 1852). Length, 2.8 mm. Panama (type locality) to Cape San Lucas, Baja California. Figured, Turner (1956).

2026. T. (T.) ima Dall & Bartsch, 1909. Gulf of Panama.

2027. T. (T.) lucana Dall & Bartsch, 1909. Cape San Lucas, Baja California.

2028. T. (T.) prolongata Carpenter, 1857. Mazatlán, Mexico. Figured, Keen (1968).

2029. T. (T.) salinasensis Bartsch, 1928. Guayaquil Bay, Ecuador.

Subgenus ASMUNDA DALL & BARTSCH, 1904

Base with a cord entering aperture above the columella.

2030. Turbonilla (Asmunda) churia Bartsch, 1926. Santa Elena, Ecuador.

2031. T. (A.) turrita (C. B. Adams, 1852) [*Chemnitzia*]. Length, 4.7 mm. Panama. Figured, Turner (1956).

Subgenus BARTSCHELLA IREDALE, 1916
(**DUNKERIA** of DALL & BARTSCH, 1909, not 1904; not of CARPENTER, 1857)

Axial and spiral sculpture subequal, intersections nodulous.

2032. Turbonilla (Bartschella) excolpa Dall & Bartsch, 1909. Length, 3.7 mm. Gulf of California.

2033. T. (B.) semela Bartsch, 1924 (spelled *semele* in plate explanation). Santa Elena, Ecuador.

2034. T. (B.) subangulata (Carpenter, 1857) [*Dunkeria*]. Mazatlán, Mexico. Type of subgenus.

2035. T. (B.) vestae Hertlein & Strong, 1951. Corinto, Nicaragua.

Subgenus CARELIOPSIS MÖRCH, 1875

Axial sculpture very faint, hardly more than growth striae.

2036. Turbonilla (Careliopsis) beltiana Hertlein & Strong, 1951. Corinto, Nicaragua.

2037. T. (C.) hannai Strong, 1938. Tres Marias Islands, Mexico.

2038. T. (C.) israelskyi Strong & Hertlein, 1939. Bahía Honda, Panama, in depths of 5 to 15 m.

2039. T. (C.) stenogyra Dall & Bartsch, 1909. Length, 5.5 mm. San Hipolito Point, Baja California (type locality), through the Gulf of California and south to Bahía Honda, Panama.

Subgenus CHEMNITZIA ORBIGNY, 1839

Axial ribs not extending onto base.

2040. Turbonilla (Chemnitzia) aculeus (C. B. Adams, 1852) [*Chemnitzia*]. Panama. Figured, Turner (1956).

2041. T. (C.) aepynota Dall & Bartsch, 1909. San Martin Island, Baja California.

2042. T. (C.) amortajadensis Baker, Hanna & Strong, 1928. Amortajada Bay, San José Island (type locality) to La Paz, Gulf of California.

2043. T. (C.) houseri Dall & Bartsch, 1909. Galápagos Islands.

2044. T. (C.) kelseyi Dall & Bartsch, 1909. Southern California to the Gulf of California.

2045. T. (C.) muricata (Carpenter, 1857) [*Chemnitzia*]. Mazatlán, Mexico. Figured, Keen (1968).

2046. T. (C.) nicarasana Hertlein & Strong, 1951. Corinto, Nicaragua.

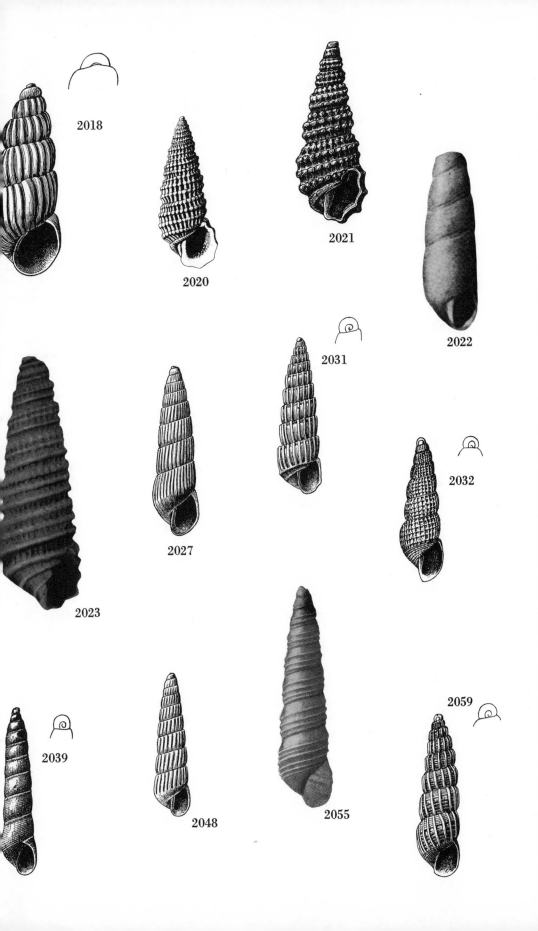

2018

2020

2021

2022

2031

2027

2032

2023

2039

2048

2055

2059

2047. T. (C.) oenoa Bartsch, 1924. Santa Elena Bay, Ecuador.

2048. T. (C.) paramoea Dall & Bartsch, 1909 (Synonym: **Chemnitzia similis** C. B. Adams, 1852 [not De Koninck, 1843]). Length, 5.9 mm. Panama. Figured, Turner (1956).

2049. T. (C.) rimaca Bartsch, 1926. Santa Elena Bay, Ecuador.

2050. T. (C.) sinaloana Strong, 1949. Mazatlán, Mexico.

2051. T. (C.) theone Bartsch, 1924. Santa Elena Bay, Ecuador.

Subgenus **CINGULINA** A. ADAMS, 1860

Axial ribs present but faint.

2052. Turbonilla (Cingulina) academica Strong & Hertlein, 1939. Bahía Honda, Panama, in depths of 5 to 15 m.

2053. T. (C.) evermanni Baker, Hanna & Strong, 1928. San José Island, Gulf of California.

2054. T. (C.) realejoensis Hertlein & Strong, 1951. Corinto, Nicaragua.

2055. T. (C.) urdeneta Bartsch, 1917. Length, 5.9 mm. Santa Maria Bay, near Magdalena Bay (type locality), outer coast of Baja California, to Espíritu Santo Island, Gulf of California.

Subgenus **DUNKERIA** CARPENTER, 1857
(**PYRGISCULUS** MONTEROSATO, 1884)

Axial ribs predominating over spirals and continuing across base.

2056. Turbonilla (Dunkeria) andrewsi Dall & Bartsch, 1909. Panama.

2057. T. (D.) cancellata (Carpenter, 1857) [*Chemnitzia*]. Mazatlán, Mexico.

2058. T. (D.) clippertonensis Hertlein & Allison, 1968. Clipperton Island.

2059. T. (D.) eucosmia Dall & Bartsch, 1909. Length, 4.8 mm. La Paz, Gulf of California.

2060. T. (D.) festiva De Folin, 1867. Margarita Island, Panama Bay. Figured, Dall and Bartsch (1909).

2061. T. (D.) genilda Dall & Bartsch, 1909. Panama Bay.

2062. T. (D.) guilleni Bartsch, 1917. Santa Maria Bay, near Magdalena Bay, Baja California.

2063. T. (D.) hipolitensis Dall & Bartsch, 1909. San Hipolito Point, Baja California.

2064. T. (D.) monilifera Dall & Bartsch, 1909. Gulf of California.

2065. T. (D.) paucilirata (Carpenter, 1857) [*Chemnitzia*]. Mazatlán, Mexico. Type of subgenus (designation of Dall and Bartsch, 1904).

2066. T. (D.) sedillina Dall & Bartsch, 1904. La Paz, Gulf of California.

2067. T. (D.) utuana Hertlein & Strong, 1951. Costa Rica, depth 27 m.

Subgenus **MORMULA** A. ADAMS, 1864

With irregular varices in addition to regular axial ribbing.

PLATE X · *Stenoplax magdalenensis* (Hinds) (The Magdalena Chiton)

2068. Turbonilla (Mormula) coyotensis Baker, Hanna & Strong, 1928. Concepción Bay, Gulf of California.

2069. T. (M.) guatulcoensis Hertlein & Strong, 1951. Puerto Guatulco, Mexico.

2070. T. (M.) heterolopha Dall & Bartsch, 1909. Southern California to southern Baja California.

2071. T. (M.) ignacia Dall & Bartsch, 1909. San Ignacio Lagoon, Baja California.

2072. T. (M.) inca Bartsch, 1926. Santa Elena Bay, Ecuador.

2073. T. (M.) major (C. B. Adams, 1852) [*Chemnitzia*]. Length, 9.6 mm. Panama. Figured, Turner (1956).

2074. T. (M.) phalera Dall & Bartsch, 1909. Panama Bay.

2075. T. (M.) santosana Dall & Bartsch, 1909. Todos Santos Bay, southern Baja California.

2076. T. (M.) scammonensis Bartsch, 1912. Scammon's Lagoon, Baja California.

2077. T. (M.) sebastiani Bartsch, 1917. Magdalena Bay, Baja California.

2078. T. (M.) vesperis Pilsbry & Lowe, 1932. Acapulco, Mexico.

2079. T. (M.) viscainoi Bartsch, 1917. Magdalena Bay, Baja California.

Subgenus PTYCHEULIMELLA SACCO, 1892

Axial sculpture on early whorls only.

2080. Turbonilla (Ptycheulimella) abreojensis Dall & Bartsch, 1909. Punta Abreojos, Baja California.

2081. T. (P.) magdalinensis Bartsch, 1927 (correctly spelled *magdalenensis* on plate explanation). Magdalena Bay, Baja California.

2082. T. (P.) obsoleta (Carpenter, 1857) [*Eulimella*]. Length, 2 mm. Mazatlán, Mexico.

2083. T. (P.) penascoensis Lowe, 1935. Length, 10.4 mm. Punta Peñasco, Sonora, Mexico.

2084. T. (P.) portoparkerensis Hertlein & Strong, 1951. Costa Rica, in 22 m.

Subgenus PYRGISCUS PHILIPPI, 1841

Axial and spiral sculpture strong; whorls evenly convex, not shouldered above.

2085. Turbonilla (Pyrgiscus) alarconi Strong, 1949. San Luis Gonzaga Bay to Concepción Bay, Gulf of California.

2086. T. (P.) almejasensis Bartsch, 1917. Magdalena Bay, Baja California.

2087. T. (P.) amandi Strong & Hertlein, 1939. Bahía Honda, Panama, in 5 to 15 m.

2088. T. (P.) amiriana Hertlein & Strong, 1951. Port Parker, Costa Rica.

2089. T. (P.) angusta (Carpenter, 1864) [*Chrysallida*]. Cape San Lucas, Baja California.

2090. T. (P.) annettae Dall & Bartsch, 1909. Manta, Ecuador.

2091. T. (P.) aripana Strong, 1949. Puerto Escondido, Gulf of California.

2092. T. (P.) auricoma Dall & Bartsch, 1903. Scammon's Lagoon, Baja California.

2093. T. (P.) aya Bartsch, 1926. Santa Elena Bay, Ecuador.

2094. T. (P.) ayamana Hertlein & Strong, 1951. Port Parker, Costa Rica.

2095. T. (P.) azteca Baker, Hanna & Strong, 1926. San Luis Gonzaga Bay, Gulf of California.

2096. T. (P.) baegerti Bartsch, 1917. Magdalena Bay, Baja California.

2097. T. (P.) bartolomensis Bartsch, 1917. San Bartolomé Bay, Baja California.

2098. T. (P.) bartonella Strong & Hertlein, 1939. Bahía Honda, Panama, depth, 5 to 15 m.

2099. T. (P.) beali Jordan, 1936. Magdalena Bay, Baja California (Pleistocene).

2100. T. (P.) biolleyi Hertlein & Strong, 1951. Costa Rica to Nicaragua.

2101. T. (P.) cabrilloi Bartsch, 1917. Magdalena Bay, Baja California.

2102. T. (P.) callipeplum Dall & Bartsch, 1909. Panama Bay.

2103. T. (P.) ceralva Dall & Bartsch, 1909. La Paz (type locality) to Cerralvo Island, Gulf of California.

2104. T. (P.) chinandegana Hertlein & Strong, 1951. Corinto, Nicaragua.

2105. T. (P.) cholutecana Hertlein & Strong, 1951. Corinto, Nicaragua.

2106. T. (P.) cinctella Mörch, 1859. Sonsonate, Guatemala. Figured, Keen (1966).

2107. T. (P.) cochimana Strong, 1949. Puerto Escondido, Gulf of California.

2108. T. (P.) colimana Hertlein & Strong, 1951. Manzanillo, Mexico.

2109. T. (P.) collea Bartsch, 1926. Santa Elena Bay, Ecuador.

2110. T. (P.) cora (Orbigny, 1840) [*Chemnitzia*]. Paita, Peru.

2111. T. (P.) corsoensis Bartsch, 1917. Magdalena Bay, Baja California.

2112. T. (P.) cortezi Bartsch, 1917. Magdalena Bay, Baja California.

2113. T. (P.) craticulata Mörch, 1859. Punta Arenas, Costa Rica, in 55 m. Figured, Keen (1966).

2114. T. (P.) crickmayi Strong & Hertlein, 1939. Bahía Honda, Panama, in 5 to 15 m.

2115. T. (P.) dina Dall & Bartsch, 1909. Panama Bay.

2116. T. (P.) domingana Hertlein & Strong, 1951. Santa Inez Bay, Gulf of California, in depths of 7 to 24 m.

2117. T. (P.) ekidana Hertlein & Strong, 1951. Port Parker, Costa Rica.

2118. T. (P.) evadna Bartsch, 1924 (spelled *evadne* on plate explanation). Santa Elena Bay, Ecuador.

2119. T. (P.) favilla Dall & Bartsch, 1909 (Synonym: **Chemnitzia coelata** Carpenter, 1865 [not Gould, 1861]). "Probably Panama."

2120. T. (P.) flavescens (Carpenter, 1857) [*Chemnitzia*]. Mazatlán, Mexico.

2121. T. (P.) garthi Strong & Hertlein, 1939. Taboga Island, Panama, 5 to 15 m.

2122. T. (P.) gordoniana Hertlein & Strong, 1951. Corinto, Nicaragua.

2123. T. (P.) gracillima (Carpenter, 1857) [*Chemnitzia*]. Mazatlán, Mexico.

2124. T. (P.) gruberi Hertlein & Strong, 1951. Corinto, Nicaragua.

2125. T. (P.) guaicurana Strong, 1949. La Paz, Gulf of California.

2126. T. (P.) guanacastensis Hertlein & Strong, 1951. Port Parker, Costa Rica.

2127. T. (P.) halidoma Dall & Bartsch, 1909. Gulf of California, from La Paz (type locality) to San Luis Gonzaga Bay.

2128. T. (P.) histias Dall & Bartsch, 1909. Off Baja California.

2129. T. (P.) indentata (Carpenter, 1857) [*Chrysallida*]. Mazatlán, Mexico.

2130. T. (P.) intia Bartsch, 1926. Santa Elena Bay, Ecuador.

2131. T. (P.) jewetti Dall & Bartsch, 1909. Southern California to southern Baja California.

2132. T. (P.) johnsoni Baker, Hanna & Strong, 1928. Gulf of California.

2133. T. (P.) kaliwana Strong, 1949. San Luis Gonzaga Bay, Gulf of California.

2134. T. (P.) lamna Bartsch, 1917. Magdalena Bay, Baja California.

2135. T. (P.) lara Dall & Bartsch, 1909. La Paz (type locality) to Cerralvo Island, Gulf of California.

2136. T. (P.) larunda Dall & Bartsch, 1909. La Paz (type locality) to Concepcion Bay, Gulf of California.

2137. T. (P.) lazaroensis Bartsch, 1917. Magdalena Bay, Baja California.

2138. T. (P.) lepta Dall & Bartsch, 1909. La Paz, Gulf of California.

2139. T. (P.) macbridei Dall & Bartsch, 1909. La Paz (type locality) to Puerto Peñasco, Gulf of California, to depths of 18 m.

2140. T. (P.) macra Dall & Bartsch, 1909. Punta Abreojos, Baja California.

2141. T. (P.) madriella Strong, 1938. Maria Madre, Tres Marias Islands, Mexico.

2142. T. (P.) mara Bartsch, 1926. Santa Elena Bay, Ecuador.

2143. T. (P.) mariana Bartsch, 1917. Santa Maria Bay, Magdalena Bay area, Baja California.

2144. T. (P.) marshalli Dall & Bartsch, 1909. La Paz, Gulf of California.

2145. T. (P.) mayana Baker, Hanna & Strong, 1928. Gulf of California.

2146. T. (P.) melea Bartsch, 1924. Santa Elena Bay, Ecuador.

2147. T. (P.) nicoyana Hertlein & Strong, 1951. Port Parker, Costa Rica.

2148. T. (P.) nuttalli Dall & Bartsch, 1909. "South America."

2149. T. (P.) otnirocensis Hertlein & Strong, 1951. Corinto, Nicaragua.

2150. T. (P.) ottomoerchi Hertlein & Strong, 1951. Corinto, Nicaragua.

2151. T. (P.) ozanneana Hertlein & Strong, 1951. Corinto, Nicaragua.

2152. T. (P.) pequensis Dall & Bartsch, 1909. Punta Abreojos, Baja California.

2153. T. (P.) pericuana Strong, 1949. Concepción Bay, Gulf of California.

2154. T. (P.) porteri Baker, Hanna & Strong, 1928. Gulf of California.

2155. T. (P.) recta Dall & Bartsch, 1909. Southern California to southern Baja California (Point Abreojos, type locality).

2156. T. (P.) rhizophorae Hertlein & Strong, 1951. Corinto, Nicaragua.

2157. T. (P.) rima Bartsch, 1926. Santa Elena Bay, Ecuador.

2158. T. (P.) sanctorum Dall & Bartsch, 1909. Length, 5.6 mm. La Paz (type locality) to Concepcion Bay, Gulf of California.

2159. T. (P.) sealei Strong & Hertlein, 1939. Bahía Honda, Panama, depths 5 to 15 m.

2160. T. (P.) shimeki Dall & Bartsch, 1909. Galápagos Islands.

2161. T. (P.) stonei Strong & Hertlein, 1939. Bahía Honda, Panama, in 5 to 15 m.

2162. T. (P.) striosa (C. B. Adams, 1852) [*Chemnitzia*]. Panama. Figured, Turner (1956).

2163. T. (P.) subula Mörch, 1859. Los Bocorones Islands, near Punta Arenas, Costa Rica.

2164. T. (P.) sulacana Hertlein & Strong, 1951. Port Parker, Costa Rica.

2165. T. (P.) superba Dall & Bartsch, 1909. La Paz (type locality) to Concepcion Bay, Gulf of California.

2166. T. (P.) tecalco Bartsch, 1917. Magdalena Bay, Baja California.

2167. T. (P.) tehuantepecana Hertlein & Strong, 1951. Puerto Guatulco, Mexico.

2168. T. (P.) templetonis Hertlein & Strong, 1951. Port Parker, Costa Rica.

2169. T. (P.) tenuicula (Gould, 1853) [*Chemnitzia*] (Synonyms: (?) **Chemnitzia terebralis** and **unifasciata** Carpenter, 1857; **C. crebrifilata** and **subcuspidata** Carpenter, 1864). Southern California through the Gulf of California to Mazatlán, Mexico, *fide* Dall & Bartsch, 1909.

2170. T. (P.) tenuilirata (Carpenter, 1857) [*Chemnitzia*]. (Considered a synonym of the otherwise more northern species *T. (P.) tenuicula* by Dall & Bartsch, 1909.) Mazatlán, Mexico. Figured, Keen (1968).

2171. T. (P.) terebralis (Carpenter, 1857) [*Chemnitzia*]. (Considered a synonym of the otherwise more northern *T. (P.) tenuicula* by Dall and Bartsch, 1909). Mazatlán, Mexico. Figured, Keen (1968).

2172. T. (P.) tia Bartsch, 1926. Santa Elena, Ecuador.

2173. T. (P.) tolteca Baker, Hanna & Strong, 1928. Gulf of California.

2174. T. (P.) ulloa Bartsch, 1917. Santa Maria Bay, near Magdalena Bay, Baja California.

2175. T. (P.) ulyssi Hertlein & Strong, 1951. Corinto, Nicaragua.

2176. T. (P.) unifasciata (Carpenter, 1857) [*Chemnitzia*]. (Considered a synonym of the otherwise more northern *T. (P.) tenuicula* by Dall & Bartsch, 1909.) Mazatlán, Mexico. Figured, Keen (1968).

2177. T. (P.) vivesi Hertlein & Strong, 1951. Santa Inez, Gulf of California.

2178. T. (P.) wetmorei Strong & Hertlein, 1937. Sinaloa; near Mazatlán, Mexico.

2179. T. (P.) yolettae Hertlein & Strong, 1951. Santa Inez, Gulf of California, in 7 to 24 m.

2180. T. (P.) zacae Hertlein & Strong, 1951. Port Parker, Costa Rica.

Subgenus **PYRGOLAMPROS** SACCO, 1892

Axial ribs strong, spiral cords fine; aperture somewhat ovate.

2181. Turbonilla (Pyrgolampros) francisquitana Baker, Hanna & Strong, 1928. San Francisquito Bay, Baja California.

2182. T. (P.) gonzagensis Baker, Hanna & Strong, 1928. San Luis Gonzaga Bay (type locality) to La Paz, Gulf of California.

2183. T. (P.) meanguerensis Hertlein & Strong, 1951. Length, 5.6 mm. Gulf of Fonseca, El Salvador.

2184. T. (P.) pazensis Baker, Hanna & Strong, 1928. La Paz (type locality) to Concepción Bay, Gulf of California.

2185. T. (P.) soniliana Hertlein & Strong, 1951. Puerto Guatulco, Mexico.

Subgenus **STRIOTURBONILLA** SACCO, 1892

Axial ribs strong, spiral cords fine; aperture subquadrate in outline.

2186. Turbonilla (Strioturbonilla) affinis (C. B. Adams, 1852) [*Chemnitzia*]. Panama. Figured, Turner (1956).

2187. T. (S.) asuncionis Strong, 1949. Asuncion Island, Baja California.

2188. T. (S.) ata Bartsch, 1926. Santa Elena Bay, Ecuador.

2189. T. (S.) caca Bartsch, 1926. Santa Elena Bay, Ecuador.

2190. T. (S.) calvini Dall and Bartsch, 1909. La Paz, Gulf of California.

2191. T. (S.) capa Bartsch, 1926. Santa Elena Bay, Ecuador.

2192. T. (S.) c-b-adamsi (Carpenter, 1857) [*Chemnitzia*]. Mazatlán, Mexico. (Erroneously cited as the Californian *T. (S.) buttoni* Dall & Bartsch, 1909, by Baker, Hanna & Strong, 1928, *fide* Strong, 1949.)

2193. T. (S.) chalcana Baker, Hanna & Strong, 1928. Gulf of California.

2194. T. (S.) contrerasiana Hertlein & Strong, 1951. Puerto Guatulco, Mexico.

2195. T. (S.) cookeana Bartsch, 1912. Gulf of California.

2073

2082

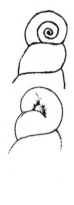

2083

2183

2150

2158

2212

2226

2196. T. (S.) corintoensis Hertlein & Strong, 1951. Corinto, Nicaragua.

2197. T. (S.) cowlesi Strong & Hertlein, 1939. Bahía Honda, Panama, in 5 to 15 m.

2198. T. (S.) doredona Bartsch, 1917. Magdalena Bay, Baja California.

2199. T. (S.) evagone Bartsch, 1924. Santa Elena Bay, Ecuador.

2200. T. (S.) galapagensis Dall & Bartsch, 1909. Galápagos Islands.

2201. T. (S.) galianoi Dall & Bartsch, 1909. Southern California to Cape San Lucas (type locality), Baja California.

2202. T. (S.) gracilior (C. B. Adams, 1852) [*Chemnitzia*]. Length, 6 mm. Panama. Figured, Turner (1956).

2203. T. (S.) haleyi Strong & Hertlein, 1939. Taboga Island, Panama Bay, in 5 to 15 m.

2204. T. (S.) hua Bartsch, 1926. Santa Elena Bay, Ecuador.

2205. T. (S.) imperialis Dall & Bartsch, 1909. Panama.

2206. T. (S.) masayana Hertlein & Strong, 1951. Corinto, Nicaragua.

2207. T. (S.) mcguirei Strong & Hertlein, 1939. Taboga Island, Panama, in 5 to 15 m.

2208. T. (S.) mexicana Dall & Bartsch, 1909. Off Baja California to San José Island, Gulf of California.

2209. T. (S.) nahuana Baker, Hanna & Strong, 1928. Gulf of California.

2210. T. (S.) nahuatliana Hertlein & Strong, 1951. Corinto, Nicaragua.

2211. T. (S.) nicaraguana Hertlein & Strong, 1951. Corinto, Nicaragua.

2212. T. (S.) nicholsi Dall & Bartsch, 1909. San Luis Gonzaga Bay to La Paz, Gulf of California.

2213. T. (S.) nychia Bartsch, 1924. Santa Elena Bay, Ecuador.

2214. T. (S.) oaxacana Hertlein & Strong, 1951. Puerto Guatulco, Mexico.

2215. T. (S.) panamensis (C. B. Adams, 1852) [*Chemnitzia*]. Panama. Figured, Turner (1956).

2216. T. (S.) pazana Dall & Bartsch, 1909. La Paz (type locality), Gulf of California.

2217. T. (S.) phanea Dall & Bartsch, 1909. La Paz (type locality) and Cerralvo Island, Gulf of California.

2218. T. (S.) redondoensis Bartsch, 1917. Magdalena Bay area, Baja California, in 25 m.

2219. T. (S.) santamariana Bartsch, 1917. Santa Maria Bay, Magdalena Bay, Baja California.

2220. T. (S.) schmitti Bartsch, 1917. Point Abreojos (type locality) to Cape San Lucas, Baja California, and to San José Island, Gulf of California.

2221. T. (S.) smithsoni Dall & Bartsch, 1909. Cape San Lucas (type locality), Baja California.

2222. T. (S.) stearnsii Dall & Bartsch, 1903. Southern California Pliocene; stated also to be living, Gulf of California.

2223. T. (S.) stephanogyra Dall & Bartsch, 1909. Panama Bay.

2224. T. (S.) thyne Bartsch, 1924. Santa Elena Bay, Ecuador.

2225. T. (S.) undata (Carpenter, 1857) [*Chemnitzia*]. Mazatlán, Mexico.

Subgenus UGARTEA BARTSCH, 1917

Cylindrical, with a strong columellar fold.

2226. Turbonilla (Ugartea) juani Bartsch, 1917. Length, 4.1 mm. Magdalena Bay, Baja California.

Subgenus and Status Uncertain

2227. Turbonilla (?) gibbosa (Carpenter, 1857) [*Chemnitzia*]. Mazatlán, Mexico. Allocated to subgenus *Pyrgolampros* with some doubt by Dall & Bartsch, 1909; regarded as indeterminate by Keen, 1968, probably not pyramidellid.

Order CEPHALASPIDEA
(BULLOMORPHA; TECTIBRANCHIATA)

Head and body with shieldlike lobes; a shell present, although sometimes internal and rudimentary.

Superfamily ACTEONACEA

Relatively unspecialized members of the order, shell external, its spire evident; aperture usually with columellar folds; an operculum present in most.

Family ACTEONIDAE

Small shells with several whorls; sculpture of fine spiral ribs with pitted interspaces; columella with one or more folds. Two Panamic genera.

Genus ACTEON MONTFORT, 1810
(ACTAEON of authors, unjustified emendation)

Outer lip simple, not thickened; a single fold on the columella.

2228. Acteon castus (Hinds, 1844). Because Hinds described this as a *Daphnella* in the Turridae, it has not been recognized by later collectors. The shell is slender, white, with a few somewhat rounded spiral ribs, the interspaces pitted. The columellar fold is low and weak. In the holotype the outer lip is broken, perhaps a major reason for Hinds's misinterpretation of form. Length, 6 mm. Type locality, Gulf of Nicoya, Costa Rica, in 42 m.

2229. Acteon panamensis Dall, 1908. A small shell from very deep water, Gulf of Panama, 2,320 m. Length, 7 mm.

2230. Acteon traskii Stearns, 1897. A flesh-colored to deep rosy shell with a central lighter band on the body whorl. Length, 18 mm; diameter, 9 mm. Southern California to the Gulf of California and possibly to Panama, in depths to 31 m.

2231. Acteon venustus (Orbigny, 1840). Resembling *A. castus* but with a lower spire and more coloration—white with one or two broad pinkish bands. Length, 10 mm. Paita, Peru.

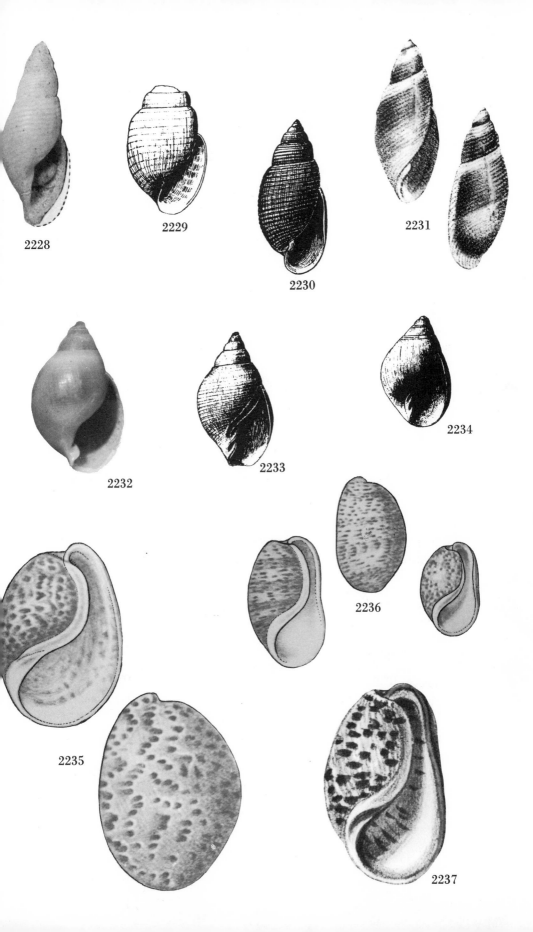

2228

2229

2230

2231

2232

2233

2234

2235

2236

2237

Genus MICROGLYPHIS DALL, 1902

Shell short and inflated, columella with two distinct folds and a siphonal furrow anteriorly; operculum apparently wanting.

2232. Microglyphis estuarinus (Dall, 1908). Small, thin, white, last whorl inflated; sculpture of fine punctate spiral lirae. Resembling the Patagonian *M. curtulus*, which is type of the genus, though the spire is more pointed, the nucleus more sunken. Length, 5.5 mm; diameter, 3.7 mm. Southern California (type locality) to the Gulf of California in depths of about 170 m. The species has not been illustrated.

2233. Microglyphis mazatlanicus (Dall, 1908). Small, white, with the spire shorter than the aperture, its nucleus immersed. Length, 5.5 mm; diameter, 3 mm. Off Mazatlán, Mexico, in 1,820 m.

2234. Microglyphis perconicus (Dall, 1890). Galápagos Islands, 1,485 m.

Superfamily BULLACEA

Shell bubble-shaped, with a sunken spire, aperture as long as the shell, rising above the spire.

Family BULLIDAE

Columella with a callus but no folds. Animal unable to withdraw completely into the shell; habit generally as burrowers living slightly below the surface of the sea floor, able, however, to move about freely.

Genus BULLA LINNAEUS, 1758
(BULLARIA RAFINESQUE, 1815; VESICA SWAINSON, 1840)

Shell generally dark-colored; ends of aperture evenly arched. Two Panamic subgenera.

Subgenus BULLA, *s. s.*

Brightly colored shallow-water forms with relatively large shells.

2235. Bulla (Bulla) gouldiana Pilsbry, 1895 (Synonym: **B. nebulosa** A. Adams, 1850, *ex* Gould, MS [not Schroeter, 1804]). The shell is pinkish, dappled with gray-black spots that are bordered with white on the left. Length, 55 mm; diameter, 37 mm. Southern California through the Gulf of California to Ecuador, intertidally, in bays.

2236. Bulla (Bulla) punctulata A. Adams in Sowerby, 1850 (Synonyms: **? B. panamensis** Philippi, 1849; **B. adamsi** Menke, 1850; **? B. aspera** A. Adams, 1850; **B. punctata** A. Adams, 1850 [not Schroeter, 1804]; **B. quoyii** A. Adams, 1850 [not Gray, 1843]; **B. quoyana** Dall, 1919). This is a smaller and sturdier shell than *B. gouldiana*. Length, 25 mm; diameter, 16 mm. Magdalena Bay, Baja California, through the Gulf and south to Peru; offshore beyond the low-tide limit.

See Color Plate XVIII.

2237. Bulla (Bulla) rufolabris A. Adams in Sowerby, 1850. Resembling *B. punctulata*; ground color pinkish buff, spotted and mottled with dark gray to black; edge of the outer lip red or russet-brown. Length, 32 mm; diameter, 20 mm. Galápagos Islands, Ecuador.

Subgenus LEUCOPHYSEMA DALL, 1908

Shells small, white; from deep water.

2238. Bulla (Leucophysema) morgana Dall, 1908. Gulf of Panama, 2,320 m.

Family ATYIDAE

Resembling the Bullidae, but animal able to withdraw completely into shell. Two Panamic genera.

Genus ATYS MONTFORT, 1810

Shell generally light-colored. Outer lip twisted into a low fold above the spire and with the columella also slightly twisted.

2239. Atys casta Carpenter, 1864. The thin white shell has the entire surface finely striate spirally. Length, 10 mm; diameter, 4.6 mm. Cape San Lucas, Baja California, to Cape Tepoca, Sonora, Mexico.

2240. Atys chimera Baker & Hanna, 1927. The shiny white shell has about nine incised spiral lines on the upper part and about sixteen on the lower part, separated by a smooth area. Length, 6.8 mm; diameter, 3.3 mm. Gulf of California, mainly on the east coast of Baja California, to Acapulco, Mexico.

2241. Atys exarata (Carpenter, 1857). The type specimen in the British Museum is brown, possibly from a stain. The columellar lip has no umbilical chink and no fold. There are four thin, brown, punctate, spiral lines near the apex, a smooth band, then ten more close-set lines toward the base. Length, 3.8 mm. Mazatlán, Mexico.

2242. Atys liriope Hertlein & Strong, 1951. White and translucent, the ovoid shell has fine spiral lines cut by equally fine axials. Length, about 10 mm; diameter, about 4 mm. Gulf of California, southern part, in 91 m depth.

Genus HAMINOEA TURTON & KINGSTON in CARRINGTON, 1830

Shell thin and fragile, oval, green or yellow, the spire concealed.

2243. Haminoea angelensis Baker & Hanna, 1927. The grayish-yellow shell is sculptured with microscopic incised lines that are stronger toward the base. Except for striate sculpture and, according to Baker and Hanna, the color and smaller size, this is very similar to *H. vesicula*. Length, 7.2 mm; diameter, 5.6 mm. Bahía de los Angeles, Gulf of California (type locality), to Puerto Peñasco, Sonora, Mexico.

2244. Haminoea vesicula (Gould, 1855). The shell is grayish yellow, smooth, with a broadly oval to truncate outline above, not constricted near the apex. Length, to 20 mm (mostly about 15 mm); diameter, to 13 mm. Southeastern Alaska to the southern end of the Gulf of California, where it is often identified as *H. angelensis*. The two may prove to be synonymous.

2245. Haminoea virescens (Sowerby, 1833) (Synonyms: **H. cymbiformis** Carpenter, 1857; **H. strongi** Baker & Hanna, 1927). The shell is white to pale green, with a constriction of the upper part of the whorl near the apex and a widely flaring aperture anteriorly. There may be some fine spiral striations or the shell may be smooth. Length, to 15 mm (very large northern specimen, 20 mm); diameter, to 11 mm. Southern Alaska to Panama. Sowerby's species was described

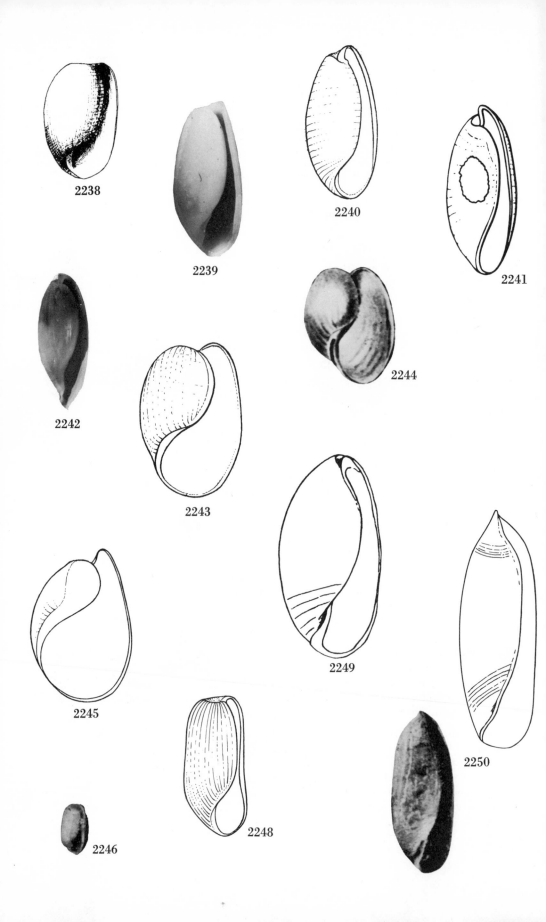

2238

2239

2240

2241

2242

2243

2244

2245

2246

2248

2249

2250

without locality; later some specimens in the British Museum were erroneously labeled as from Pitcairn Island, which led to confusion. However, it now seems clear that authors who identified Sowerby's figure with West American material were justified, and the species may be accepted as eastern Pacific in distribution, with a wide range. Marcus (1967) could detect no difference between Californian specimens of *H. virescens* and topotypes of *H. strongi* (type locality, Gulf of California). One specimen in the Stanford collection was collected in Panama Bay.

Family RETUSIDAE

Cylindrical small shells with some fine spiral sculpture; nucleus hyperstrophic (ultradextral), often concealed; columella smooth or with only weak folds; animal able to withdraw into the shell. Two Panamic genera.

Genus SULCORETUSA BURCH, 1945
(SULCULARIA DALL, 1921 [not RAFINESQUE, 1831])

Small, subcylindrical, with weak spiral lirae and a few well-marked vertical striae on the upper part of the whorls, fading out below. Zilch (1959) considers that this is a synonym of *Cylichnina* Monterosato, 1884, the type of which is European in distribution.

2246. Sulcoretusa carpenteri (Hanley, 1859). Length, 1.5 mm. Mazatlán, Mexico.

2247. S. galapagana (Dall, 1919). Length, 3 mm. Galápagos Islands.

2248. S. paziana (Dall, 1919). Length, 3 mm. San Luis Gonzaga Bay to La Paz, Gulf of California (type locality) in depths of 22 to 34 m.

Genus VOLVULELLA NEWTON, 1891
(RHIZORUS of authors, not of MONTFORT, 1810;
VOLVULA A. ADAMS, 1850 [not GISTL, 1848])

Small, subcylindrical, colorless, with a long, narrow aperture, the outer lip forming a spine above the mostly concealed nucleus at apex; columella smooth; sculpture weak to microscopic. Two Panamic subgenera.

Subgenus VOLVULELLA, s. s.

Outer lip smoothly arched to its junction with the inner lip, not sinuate.

2249. Volvulella (Volvulella) catharia Dall, 1919. Thick-shelled, apical spine low, not completely concealing the nuclear whorls, which are in a narrow chink; sculpture of a few spiral grooves. Length, 2.8 mm; diameter, 1.8 mm. Panama Bay, 113 m. Figured by Harry (1967), who points out that a similar species in the Atlantic is *V. paupercula* (Watson, 1883).

2250. Volvulella (Volvulella) cylindrica (Carpenter, 1864) (Synonyms: **V. callicera** and **cooperi** Dall, 1919; **V. lowei** Hertlein & Strong, 1937). The shell is slender, its apical end tapering into a spine; sculpture of fine spiral lirae at both ends, with finer lines in the middle on fresh material. Length, 6 mm; diameter, 2 mm. Southern California through the Gulf of California and south to Panama and the Galápagos Islands, in depths to 75 m. Similar species in the Atlantic are *V. acuminata* (Bruguière, 1792) and *V. persimilis* (Mörch, 1875).

Subgenus PARAVOLVULELLA HARRY, 1967

Outer lip sinuate to truncate just below its junction with the inner lip, spine formed by inner lip.

2251. Volvulella (Paravolvulella) panamica Dall, 1919 (Synyonm: **V. tenuissima** Willett, 1944). Cylindrical, with the posterior end truncate except for the central spine; ends of shell with some fine spiral grooving. Length, 4.25 mm; diameter, 1.5 mm. Southern California through the Gulf of California to Panama. A similar species in the Atlantic is *V. texasiana* Harry, 1967.

Superfamily DIAPHANACEA

Small, few-whorled, thin, spire low to sunken, aperture somewhat flaring.

Family DIAPHANIDAE

Columella smooth; base umbilicate; operculum wanting.

Genus WOODBRIDGEA BERRY, 1953
(? CLISTAXIS of DALL, 1908 [not COSSMANN, 1895]; BROCKTONIA of DALL, 1921 [not IREDALE, 1915])

Resembling *Bulla* or *Haminoea* in outline but with a narrow umbilical chink along the edge of the inner lip.

2252. Woodbridgea williamsi Berry, 1953. The affinities of this possibly immature form are yet uncertain, and it may prove not to belong in the Diaphanidae. The shell is nearly transparent, greenish in color, with faint, diverging spiral sculpture. The aperture is widely flaring. Length, 1.3 mm; diameter, 1 mm. Cedros Island, Baja California, depth 45 m, to the Gulf of California.

Superfamily PHILINACEA

Shells thin, cylindrical, and somewhat involutely coiled when external; disjunct when internal; body somewhat wedge-shaped. The typical family, Philinidae, has not yet been recorded in the Panamic province (it is mainly a cold-water group, but specimens might occur on the outer coast of Baja California, where it should be watched for); shells are very thin, like a tiny roll of tissue paper.

Family AGLAJIDAE

Shell internal, forming a flattened spiral coil. The type genus, *Aglaja*, is customarily cited as of Renier, 1807; however, Renier's work has been rejected by the International Commission on Zoological Nomenclature (ICZN Opinion 427) as not validly published, but *Aglaja* was provisionally retained, pending further decisions. No species of *Aglaja* occur in the Panamic province.

Genus NAVANAX PILSBRY, 1895
(STRATEGUS COOPER, 1863 [not KIRBY & SPENCE, 1828]; NAVARCHUS COOPER, 1863 [not FILIPPI & VERANY, 1859])

Shell in a loose or disconnected spiral coil, as in *Aglaja*. Body with an elongate head shield produced into earlike folds at the anterior corners. Principal differences from *Aglaja* are anatomical. Another closely related genus is *Chelidonura* A. Adams, 1850, the type species of which is Indo-Pacific. Marcus (1967) implied, and Marcus and Marcus (1970) even recommend, that *Navanax* should be synony-

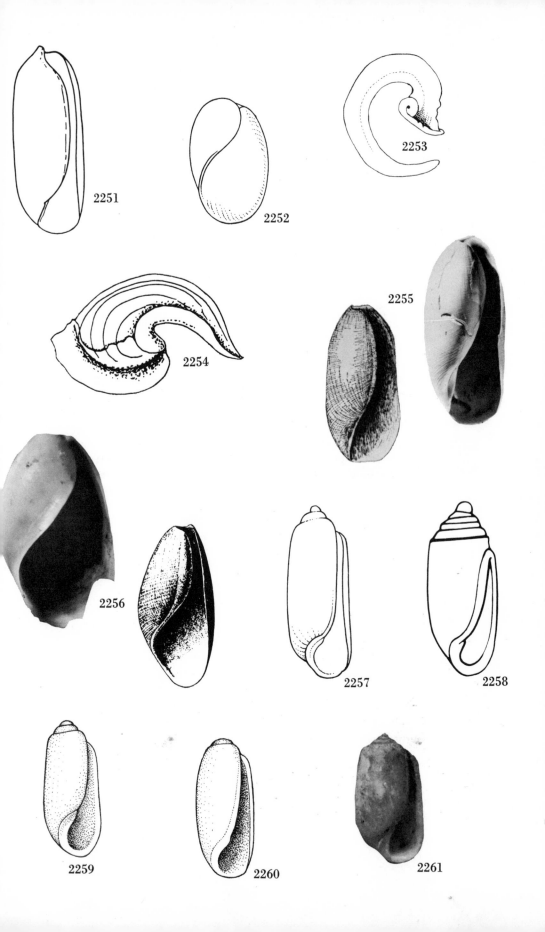

2251

2252

2253

2254

2255

2256

2257

2258

2259

2260

2261

mized under *Chelidonura*. However, the genus is retained here because its type species is West American. Obviously more work is needed on the group.

2253. Navanax aenigmaticus (Bergh, 1894). Animal yellowish white, mottled with black and gray, the folds under the head shield conspicuous. Shell thin-edged, somewhat chalky. Length, 25 mm (preserved specimen). Panama.

2254. Navanax inermis (Cooper, 1863) (Synonym: **Aglaja bakeri** MacFarland, 1924). Animal dark brown to black, variously spotted and striped. Marcus (1967) has given some anatomical figures. Length of the animal, 75 to 225 mm. Diameter of shell, about 20 mm, nearly the length of the head shield. Monterey Bay, California, to the head of the Gulf of California. The animal feeds voraciously on *Haminoea* and *Bulla*. This is the type of the genus.

See Color Plate XXI.

Family SCAPHANDRIDAE

Cylindrical, spire evident to sunken, nuclear whorls hyperstrophic; sculpture mostly of spiral punctations; columella with or without folds; animal mostly able to withdraw into shell. Five Panamic genera.

Genus SCAPHANDER MONTFORT, 1810

Medium-sized white shells, some with a yellow periostracum; sculpture of fine spiral grooves and punctations; aperture inflated, especially anteriorly; columellar edge overhanging as a false umbilicus in some. Only deepwater forms are as yet reported in the Panamic province.

2255. Scaphander cylindrellus Dall, 1908. The outer lip is truncate above, so that the outline appears more cylindrical than in most other species of the genus. Length, 33 mm; diameter, 16 mm. Off Magdalena Bay, Baja California, 3,380 to 3,600 m (Stanford University collection, from California Institute of Technology), to Callao, Peru (type locality), at 5,200 m.

2256. Scaphander interruptus Dall, 1889. More conical than *S. cylindrellus*, the aperture wider below; sculpture relatively coarse. Length, 33 mm; diameter, 19 mm. Off Magdalena Bay, Baja California, 3,380 to 3,600 m (Stanford University collection, from California Institute of Technology), to Panama and the Galápagos Islands, in depths to 2,320 m.

Genus ACTEOCINA GRAY, 1847

Spire visible, its apex hyperstrophic, somewhat immersed, at an angle to the later whorls. Shell cylindrical, nearly smooth; columella with one fold.

2257. Acteocina angustior Baker & Hanna, 1927. A small, solid, white shell with irregular incised spiral lines. Length, 5.4 mm; diameter, 2 mm. Gulf of California, mostly offshore in depths of 6 to 30 m.

2258. Acteocina carinata (Carpenter, 1857). With a double carina at the shoulder, especially in immature specimens. Length, 3 mm. Throughout the Gulf of California in depths of 2 to 45 m; to Mazatlán, Mexico (type locality).

2259. Acteocina inculta (Gould & Carpenter, 1857). A Californian form that ranges throughout the Gulf of California in shallow bays. Length, 5.5 mm; diameter, 2.5 mm.

2260. Acteocina infrequens (C. B. Adams, 1852). Shell more slender than in other species and more smoothly tapering. Length, 7.2 mm; diameter, 2.8 mm. Scammon's Lagoon, Baja California, through the Gulf of California, in depths of 2 to 30 m, to Panama.

2261. Acteocina smirna Dall, 1919. Length, 4 mm. The range is from southern California to El Salvador.

Genus CYLICHNA LOVÉN, 1846
(BULLINELLA NEWTON, 1891; CYLICHNIUM DALL, 1908)

Cylindrical, spire sunken; aperture long; columella simple or with one low fold.

2262. Cylichna atahualpa (Dall, 1908). White under a yellow periostracum, with a few punctate incised lines at ends of shell. Length, 9 mm; diameter, 4 mm. Gulf of Panama, 590 m.

2263. Cylichna fantasma (Baker & Hanna, 1927). The white shell is rather more tapering at the ends than in most species of the genus. Length, 9 mm; diameter, 4 mm. Gulf of California, in shallow bays.

2264. Cylichna inca (Dall, 1908). White, spire deeply immersed. Gulf of Panama, depth 2,320 m.

2265. Cylichna luticola (C. B. Adams, 1852). Length, 5 mm. Panama.

2266. Cylichna pizarro (Dall, 1908). White under a yellow periostracum. Length, 10 mm; diameter, 5 mm. Gulf of Panama, depth 2,320 m.

2267. Cylichna stephensae Strong & Hertlein, 1939. Length, 5 mm. Panama, 5–15 m.

2268. Cylichna veleronis Strong & Hertlein, 1939. Length, 5 mm. Panama, 5–15 m.

Genus CYLICHNELLA GABB, 1873

Cylindrical, spire sunken, aperture long; columella with two folds; spiral sculpture stronger toward ends of shell.

2269. Cylichnella defuncta Baker & Hanna, 1927. Length, 2.6 mm. San José Island, Gulf of California.

2270. Cylichnella gonzagensis (Baker & Hanna, 1927). Length, 2.9 mm. San Luis Gonzaga Bay (type locality) to Bahía de los Angeles, depths 20 to 30 m, Gulf of California.

2271. Cylichnella tabogaensis Strong & Hertlein, 1939. Length, 3 mm. Panama.

2272. Cylichnella zeteki Bartsch, 1918. Length, 2.5 mm. Panama.

Genus MICRAENIGMA BERRY, 1953

Small, bubble-shaped, thin, nearly transparent, smooth; aperture long, flaring below; columella with one heavy laminar fold.

2273. Micraenigma oxystoma Berry, 1953. This may prove to be the juvenile stage of one of the larger forms. Length, 1.8 mm; diameter, 1.1 mm. Cedros Island, Baja California (type locality).

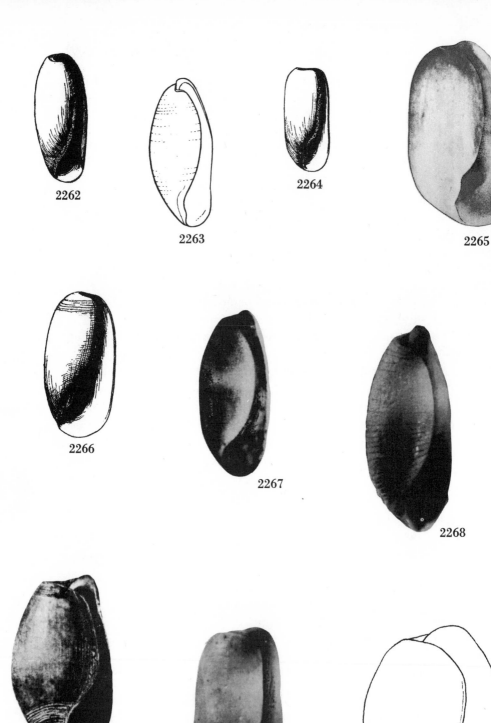

2262

2263

2264

2265

2266

2267

2268

2269

2271

2273

Order THECOSOMATA
(PTEROPODA of authors, in part)

Pelagic mollusks with the epipodia modified into swimming wings. Shells variously shaped, mostly calcareous. Foot reduced; mouth with jaws and a small radula; generally with a pallial cavity (some with a ctenidium). Because most of these are obtainable only in plankton tows well offshore, the species will be listed with only a sample generic figure.

The assistance of Dr. Eugene V. Coan and Dr. John McGowan in the organization of data is here gratefully acknowledged. Dr. McGowan also supplied notes on species distribution hitherto unpublished. Useful references: Van der Spoel (1967), McGowan (1968).

Suborder EUTHECOSOMATA

An external shell present in the adult.

Family CAVOLINIIDAE

Shells variable in shape, never coiled, bilaterally symmetrical, with a median axis; adult lacking operculum. Fins notched on the free edge, middle lobe of the foot broad and semicircular; columellar muscle dorsal, intestinal loop ventral. Six Panamic genera.

Genus CAVOLINIA ABILDGAARD, 1791 (ICZN pending)
(CAVOLINA ABILDGAARD, 1791, original spelling; emended to *Cavolinia* by Philippi, 1853; decision on acceptance of the emendation pending, ICZN)

Shell dorso-ventrally inflated, aperture with a hoodlike projection of the dorsal shell, with spinous elongations at the side.

2274. Cavolinia globulosa (Gray, 1850, *ex* Rang, MS). Temperate to tropical seas; recorded off the Galápagos Islands. Length, 6 mm.

2275. Cavolinia inflexa (Lesueur, 1813). Epipelagic, well offshore, Baja California and Galápagos Islands. Length, 5 mm.

2276. Cavolinia longirostris (Blainville, 1821, *ex* Lesueur, MS). Epipelagic, offshore. Southern Mexico to Peru. Length to 9 mm.

2277. Cavolinia tridentata (Niebuhr, 1775, *ex* Forskäl, MS). Off Galápagos Islands. Length, 10 to 20 mm.

2278. Cavolinia uncinata (Rang, 1829). Mexico to Peru. Length to 11 mm.

Genus CLIO LINNAEUS, 1767

Pyramidal; shell with a pronounced central midrib dorsally. Aperture unobstructed; posterior end with a sharply pointed larval tip.

2279. Clio chaptalii Gray, 1850. Panama to Galápagos Islands. Length to 19 mm.

2280. Clio pyramidata Linnaeus, 1767. Worldwide in warm seas; Baja California and southern Mexico to Galápagos Islands. Length about 20 mm.

Genus CRESEIS RANG, 1828

Conical; shell with smooth exterior; straight or flexed, circular in cross section.

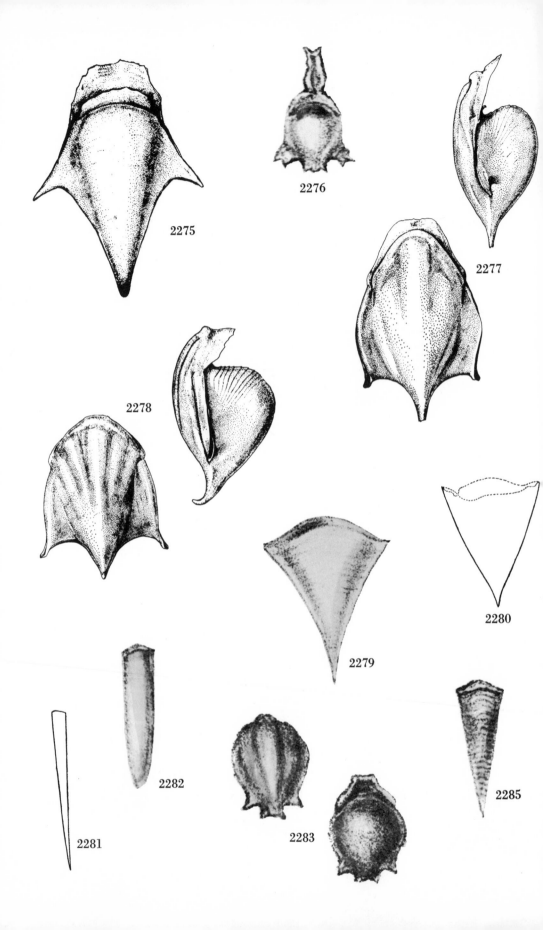

2275

2276

2277

2278

2279

2280

2281

2282

2283

2285

2281. Creseis acicula (Rang, 1828). Widely distributed in warm seas; Baja California to Peru. Length, 6 mm.

2282. Creseis virgula (Rang, 1828). Widely distributed in warm seas; southern California to Peru. Length about 6 mm.

Genus DIACRIA GRAY, 1840

Resembling *Cavolinia* but edges of aperture rolled, looking like fat lips (McGowan, 1968).

2283. Diacria quadridentata (Blainville, 1821, *ex* Lesueur, MS). Epipelagic. Southern Mexico to Peru. Length, 3 mm.

2284. Diacria trispinosa (Blainville, 1821, *ex* Lesueur, MS). Epipelagic, well offshore. Length, 13 mm.

Genus HYALOCYLIS FOL, 1875
(HYALOCYLIX of authors, an unjustified emendation)

Shell conical, without protoconch, the tip open and truncate, ornamented with raised annulations through the entire length.

2285. Hyalocylis striata (Rang, 1828). Widely distributed in warm seas; southern Mexico to Peru. Length about 8 mm.

Genus STYLIOLA GRAY, 1850

Conical, circular in section, with a triangular tooth on dorsal margin of aperture; visceral mass rose-red.

2286. Styliola subula (Quoy & Gaimard, 1827). Southern California; Ecuador and Peru. Length to 13 mm.

Family LIMACINIDAE
(SPIRATELLIDAE)

Shell trochoid, apparently sinistrally coiled to involute; eccentrically coiled operculum present. Paired fins large, not subdivided at edges; mantle cavity dorsal, no ctenidium.

Genus LIMACINA BOSC, 1817
(SPIRATELLA BLAINVILLE, 1817)

Small, thin, transparent, smooth; shells hyperstrophic in coiling, appearing to be sinistral; rarely found in beach drift.

The name *Limacina* is usually attributed to Lamarck, 1819, over which *Spiratella* Blainville, 1817, has priority. Van der Spoel (1967), showing that there is an earlier validation by Bosc in the same year as Blainville's name, has chosen *Limacina* as preferred. This he may do, acting as first reviser, under the ICZN Code. To establish exact priority between Bosc's and Blainville's publications might be difficult; thus the "first reviser" rule has a useful function.

2287. Limacina bulimoides (Orbigny, 1836). Panama to Peru.

2288. L. inflata (Orbigny, 1836). Off Baja California and El Salvador to Peru.

2289. L. trochiformis (Orbigny, 1836). Off Baja California to Peru.

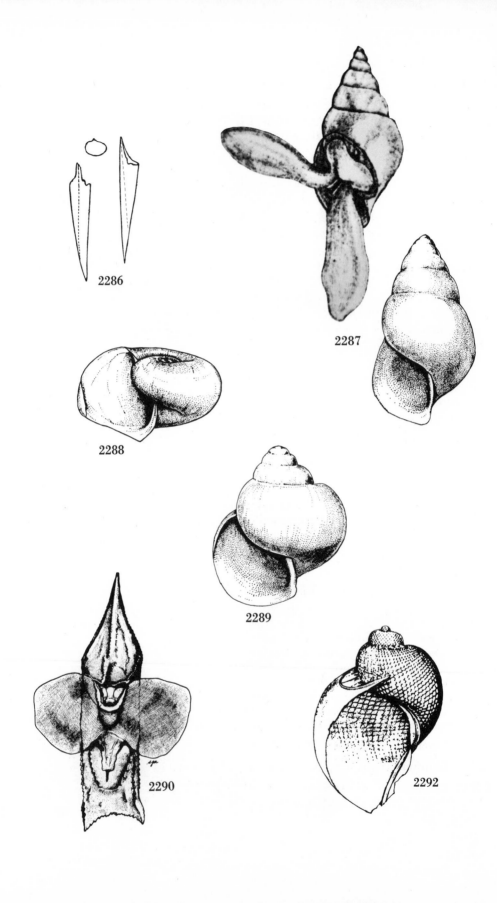

2286

2287

2288

2289

2290

2292

Suborder PSEUDOTHECOSOMATA

Shell present only in larval stages; adults with a gelatinous false shell or pseudo-conch that is relatively large, shaped like a boat, partially internal.

Family CYMBULIIDAE

Fins united into a continuous plate; tentacles symmetrical; no ctenidium.

Genus CYMBULIA PÉRON & LESUEUR, 1810

Pseudoconch attenuated anteriorly, truncate posteriorly.

2290. Cymbulia peroni Blainville, 1818. Offshore in warm seas.

Family PERACLIDAE

With sinistrally coiled shells covered with a delicate meshwork; operculum nearly circular, attached below the fins; proboscis short; mantle cavity with a ctenidium.

Genus PERACLE FORBES, 1844
(PERACLIS of authors, unjustified emendation)

Shell with marked ornamentation, often light brown in color.

2291. Peracle apicifulva Meisenheimer, 1906. Offshore, pelagic.

2292. P. bispinosa Pelseneer, 1888. Offshore, pelagic.

2293. P. reticulata (Orbigny, 1836). Offshore, pelagic.

Order ANASPIDEA

Animals lacking a cephalic shield.

Superfamily APLYSIACEA

With an internal membranous to slightly calcified shell.

Family APLYSIIDAE

Popularly called "sea hares," perhaps because the two rhinophores look a little like ears and the animal resembles a crouching rabbit in shape. The shell is a thin membranous to calcareous asymmetrical plate, entirely internal. The animals are variously colored or mottled, large (one in California measures up to 30 inches in length and weighs up to 35 pounds), of herbivorous habits, grazing on seaweeds. Useful references: Eales (1960), Marcus (1967), Beeman (1968). Four Panamic subfamilies.

Subfamily APLYSIINAE

Shell relatively large, flattened, membranous.

Genus APLYSIA LINNAEUS, 1767
(TETHYS of some authors, not of LINNAEUS, 1758; ICZN Opinion 200,1954)

Shell ovate, thin, with an anal sinus on the right posterior side; enclosed by the mantle. Skin smooth, without filaments; radula with a broad denticulate rachidian tooth and many lateral teeth; mantle glands present. There has been some confusion attending the generic name; for a time, on the basis of supposed priority, *Tethys* was substituted, but in 1954 the ICZN ruled that this name should be applied to a nudibranch and *Aplysia* conserved for the sea hares. Four Panamic subgenera.

Subgenus APLYSIA, s. s.

Mantle glands not secreting purple color; radula simple but with a maximum number of lateral teeth.

2294. Aplysia (Aplysia) juliana Quoy & Gaimard, 1832 (Synonym: **A. rangiana** Orbigny, 1835). Medium-sized to large, color variable. Length to 300 mm. Sonora, Mexico, to Paita, Peru; worldwide in warm seas.

2295. Aplysia (Aplysia) vaccaria Winkler, 1955. Large and bulky; deep purplish black, sometimes with fine gray to white spots. Length, 255 mm. Southern California to Bahía de los Angeles, Gulf of California. Perhaps the world's largest aplysiid; Lance (1967) reports subtidal specimens 75 cm (30 inches) long.

Subgenus NEAPLYSIA COOPER, 1863

Shell with a flattened calcareous apex instead of a reduced spiral or hook.

2296. Aplysia (Neaplysia) californica Cooper, 1863. Greenish gray with dark streaks and spots and white skin glands. Length, about 130 mm. Humboldt Bay, California, to the northern end of the Gulf of California. Type of the subgenus.
See Color Plate XIX.

Subgenus PRUVOTAPLYSIA ENGEL, 1936

Foot narrow, mantle glands secreting purple; radula small, with fewer than 25 laterals on each side in a row.

2297. Aplysia (Pruvotaplysia) parvula Mörch, 1863. Type of the subgenus. Of wide distribution in warm seas; reported from the Gulf of California by MacFarland in 1924 and by Marcus in 1967. The color is greenish gray with white dots. Length, about 60 mm.
See Color Plate XIX.

Subgenus VARRIA EALES, 1960

Mantle glands secreting purple; radula well developed, with many rows of teeth and numerous laterals.

2298. Aplysia (Varria) dactylomela Rang, 1828 (Synonym: **Tethys panamensis** Pilsbry, 1895). Large and bulky, dark brown to green with white flecks. Length, to 400 mm. Panama; also worldwide in warm seas.

2299. Aplysia (Varria) robertsi (Pilsbry, 1895). Medium-sized; mantle olive-green bordered with black. Length, about 110 mm. Mexico to Central America (*fide* Eales, 1960).

Subfamily DOLABELLINAE

Shell well developed, strongly calcified, flat, apex curved and covered by heavily thickened axial edge, right side concave. Skin of animal sometimes rough or with appendages; radula with all teeth narrow and unicuspid, rachidian teeth much smaller than the numerous simple laterals.

Genus DOLABELLA LAMARCK, 1801

Shell surface smooth, thinner at the edges.

2300. Dolabella auricularia [Lightfoot, 1786] (Synonym: "**D. scapula** Martyn" of authors [non-binomial]). Marcus and Marcus (1970) report a speci-

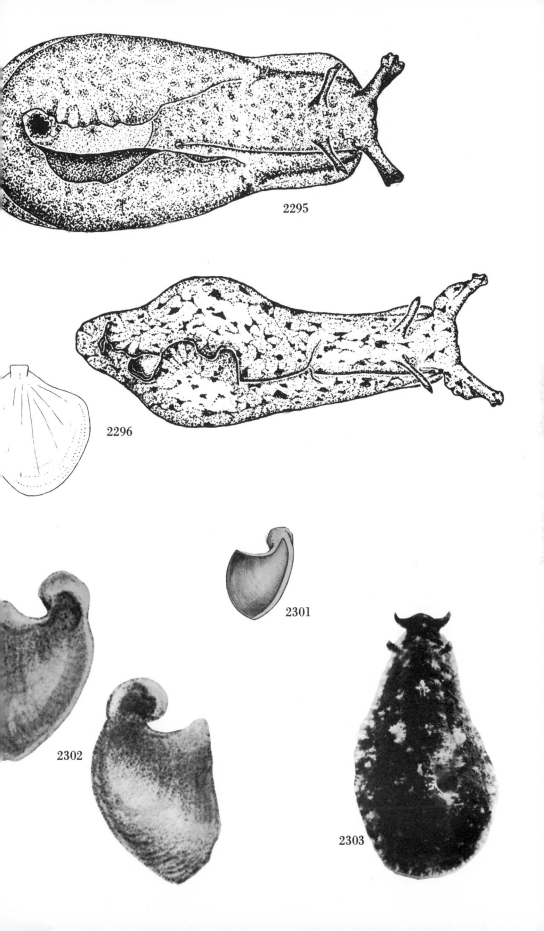

2295

2296

2301

2302

2303

men of *D. auricularia* ("Solander") from Cabo Pulmo, Baja California. It is not clear from their discussion whether this is regarded as a second species for the Gulf of California or whether they would synonymize *D. californica* with the wide ranging Indo-Pacific species.

2301. Dolabella californica Stearns, 1877. The distinctively shaped internal shell is white, with a horny brown layer on one side. Length of shell, about 45 mm. Gulf of California.

2302. Dolabella guayaquilensis Sowerby, 1868, *ex* Petit, MS. The shell is small, with a gray periostracum. Length, 21 mm. Guayaquil, Ecuador.

Subfamily DOLABRIFERINAE

Shell small, flat, calcareous; wanting in some. Skin often rough. Radula with broad rachidian tooth, larger than the serrated laterals. Represented in the Panamic province by a single genus, although two others—*Petalifera* Gray, 1847, and *Phyllaplysia* Fischer, 1872—should eventually be detected, because of their wide dispersal in warm seas.

Genus DOLABRIFERA GRAY, 1847

Posterior third of the body broadly expanded; shell narrow, with a knobbed spire, enclosed by mantle.

2303. Dolabrifera dolabrifera (Rang, 1828). Worldwide in tropics; eastern Pacific, at Las Cruces, near La Paz, Baja California (Bertsch, 1970).

2304. Dolabrifera nicaraguana Pilsbry, 1896. Length of animal, 40 mm; of shell, 9.5 mm. San Juan del Sur, Nicaragua.

Subfamily NOTARCHINAE

Soft parts similar to those of the Dolabriferinae; shell minute and orbicular but usually not present. Skin with filaments in some and often with brightly colored eye spots. The type genus *Notarchus* Cuvier, 1817, is widely dispersed in warm seas but has not been reported in the Panamic province.

Genus STYLOCHEILUS GOULD, 1852

Body sluglike, head distinct, with four tentacles. Living on floating seaweed, offshore.

2305. Stylocheilus longicauda (Quoy & Gaimard, 1824) (Synonym: **Aclesia rickettsi** MacFarland, 1966). The animal is marked with brown lines like pencil strokes and blue ocellar spots circled with orange, and the skin filaments are conspicuous. Length, to 23 mm. Cabo Pulmo, Baja California, to Sonora, Mexico; Hawaii; warm seas generally, especially in the western Pacific (type locality, New Guinea).

Order GYMNOSOMATA
(PTEROPODA of authors, in part)

Though often grouped with the "sea butterflies" or pteropods, these are not closely related; they are planktonic opisthobranchs with no mantle or mantle cavity, no shell at any stage. The head is well developed, with two tentacles and a mouth, jaw, and radula, the lateral teeth of which are long, with large basal plates. There is a sheathed proboscis and a foot distinct from the fins. Most animals are less than 30 mm in length and are vigorous swimmers; all are probably carnivorous. Little is known about the group. McGowan (1968) has summarized the scanty information available on the Californian gymnosomes; even less is known about those of the Panamic province. Two or three families, each with a few genera, seem to be represented, but as yet no species lists have been assembled (McGowan, *in litt.*).

From the published literature, Dr. Eugene V. Coan has compiled the following list of families and genera that presumably would be represented:

Family: Clionidae
  Genera: *Clione* Pallas, 1774; *Clionina* Pruvot-Fol, 1924; *Cliopsis* Troschel, 1854
Family: Pneumodermatidae
  Genera: *Pneumoderma* Roissy, 1805; *Pneumodermopsis* BRONN, 1862; *Spongiobranchea* Orbigny, 1836
Family: Thliptodontidae
  Genus: *Thliptodon* Boas, 1886

McGowan (1968) has published diagrammatic figures for some of the families in the Gymnosomata.

Order NOTASPIDEA

External shell flat to conical, calcareous, mostly smooth (lacking in a few forms).

Superfamily PLEUROBRANCHACEA

Body flat and sluglike; mantle covering dorsal surface and projecting at the margin like a skirt, with grooved rhinophores extending beyond it in front; although there is no mantle cavity, there are a ctenidium and an osphradium on the right.

Family PLEUROBRANCHIDAE

Shell partially covered by the mantle, ear-shaped, thin; rarely absent. Two Panamic subfamilies; a third that might be expected is the Pleurobranchaeinae, based on *Pleurobranchaea* Leue, 1813, which occurs in the Californian area and in the Caribbean, a group entirely lacking a shell.

Subfamily PLEUROBRANCHINAE

With a pedal gland on the posterior part of the foot.

Genus PLEUROBRANCHUS CUVIER, 1804

Shell flat, like a small abalone without holes.

2306. Pleurobranchus areolatus (Mörch, 1863). A brownish-orange to brick-red animal, mottled with white, with some wartlike tubercles. Length of animal to 75 mm; of shell, 3.6 mm. Puerto Peñasco, Sonora, Mexico, to Panama; widely distributed in the Caribbean (type locality, Virgin Islands).
  See Color Plate XIX.

2307. Pleurobranchus digueti Rochebrune, 1895. This was reported as being orange-red above, white below. The type specimen has not since been studied. Material from the type area has been figured by Pilsbry (1896), and his interpretation of the species is generally accepted. Dall's (1921) allocation to *Lamellaria* was, according to Marcus and Marcus (1967), untenable. Length of animal about 25 mm. Gulf of California (type locality near La Paz).

Subfamily BERTHELLINAE

Without a pedal gland on the posterior part of the foot.

Genus BERTHELLINA GARDINER, 1936

Shell with shallow cavity, not angulate at right or projecting above the low spire, growth lines weak, with finer radial striae; radula with bladelike teeth, distal teeth more serrate; gill plane bipinnate.

2308. Berthellina ilisima Marcus & Marcus, 1967 (Synonym: **Pleurobranchus plumula** of authors [not *Bulla plumula* Montagu, 1803]). Apricot-to orange-colored animals; shell flat, light brown, with an iridescent periostracum, the larval coil a white knob; sculpture of coarse growth lines and radiating rows of dots. Length of animal, 60 mm; of shell, 4.5 mm. Throughout the Gulf of California (type locality, Puerto Peñasco, Sonora). A similar form in the eastern Atlantic is *B. engeli* Gardiner, 1936, a species so close, in fact, that Marcus and Marcus named *B. ilisima* as a subspecies of it.

2309. Berthellina quadridens (Mörch, 1863). Animals orange to vermilion, with white dots. Length, about 30 mm. Panama Bay; otherwise Caribbean in distribution.

See Color Plate XIX.

Superfamily UMBRACULACEA

Shell conic, with a central apex, a circular muscle scar within.

Family UMBRACULIDAE

Shell platelike to conical, with an external periostracum; radula broad, with numerous teeth (up to 150,000). Feeding is by grazing on microscopic organisms in the surface film on sponges or other objects attached to the sea floor. Two Panamic genera.

Genus UMBRACULUM SCHUMACHER, 1817
(UMBRELLA LAMARCK, 1819)

Shells limpet-shaped, embedded in the dorsal surface of the much larger animal. Faint radiating ribs and a central pointed apex enhance the resemblance of the shell to a parasol. Animal unable to retract within the shell.

2310. Umbraculum ovale (Carpenter, 1856). The shell, covered by a thin, dull periostracum, is oval, smooth-margined, the apex a little off center. Length, about 50 mm; width, 40 mm. Cape San Lucas, Baja California, and Gulf of California to Panama.

Genus TYLODINA RAFINESQUE, 1819

Shell smoothly conic; embryonal whorls preserved as two apparently sinistral coils; animal nearly able to retract within shell.

2311. Tylodina fungina Gabb, 1865. The shell is horn-colored, with an adherent periostracum that may lap over the rim of the aperture. The bright yellow animal is often found on a yellow sponge of the same hue. Length of shell, 20 mm; width, 17 mm. Southern California to Guaymas, Gulf of California, Costa Rica, and Galápagos Islands (Los Angeles County Museum), intertidally and offshore to 25 m.

See Color Plate XIX.

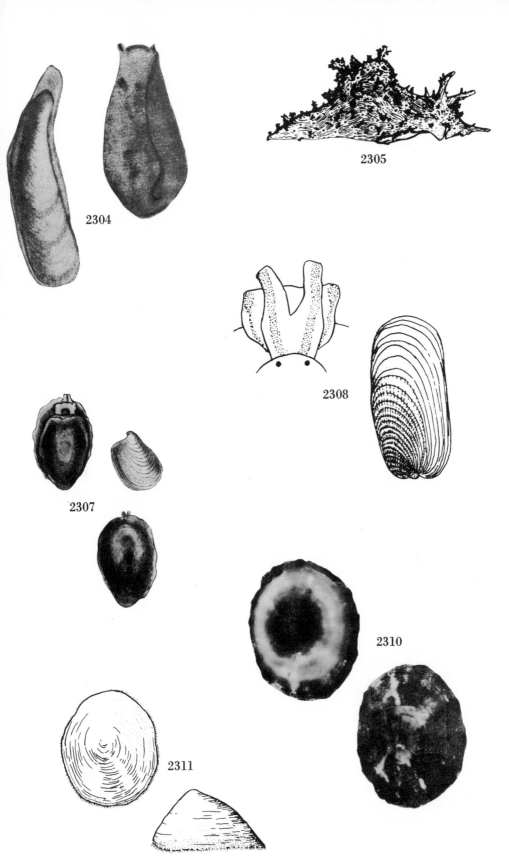

2304

2305

2308

2307

2310

2311

Order **SACOGLOSSA**
(**ASCOGLOSSA**)

Shells thin, subspiral to spiral, or wanting. Gill small. Nervous system concentrated, approaching a euthyneurous condition in some. Animals adapted to suctorial feeding on green algae, with consequent diminishing of the radula, which is uniserial, with platelike detached teeth contained in a blind pouch.

Superfamily **OXYNOACEA**

Shells present, cap-shaped to involute.

Family **OXYNOIDAE**

With bulloid shells in which the margins meet midventrally, the anterior end of the aperture flaring, the posterior end with a sutural slit. Lobes of the body cover a part of the shell on either side; foot long, posteriorly tail-like; head with two large tentacles that cannot be completely retracted.

Genus **OXYNOE** RAFINESQUE, 1819
(**ICARUS** FORBES, 1844; **LOPHOCERCUS** KROHN, 1847)

Ovate, inrolled, last whorl large, the columella concave, the outer lip sharp-edged.

2312. ? Oxynoe panamensis Pilsbry & Olsson, 1943. The shell is transparent to whitish, fragile, with a constriction below the apex, a few undulating wrinkles near the suture, otherwise smooth; microscopically fine striae can be detected over a part of the surface. Length, 5 to 7 mm; width, 4 to 5 mm. The status of the name is open to some question; the type locality was cited as "Bocas del Toro" (a Caribbean location), yet it was stated to be in the Panamic province. The only *Oxynoe* in later lists from Bocas del Toro is *O. antillarum* Mörch, 1863, a similar Atlantic form. Material that fits the description of *O. panamensis* with respect to size, shape, and distinctive features of the shell has been collected in the Gulf of California at Candelero Bay, Espíritu Santo Island (Stanford University and California Academy of Sciences collections). The species is fairly common on *Caulerpa* there and one large specimen was found on another alga, *Halimeda*. The animal is olive-green, with white spots. Presumably, the geographic range is from the southern end of the Gulf of California to Panama Bay.

See Color Plate XIX.

Family **LOBIGERIDAE**

Shell somewhat cap-shaped and asymmetrical, not forming a complete spiral turn.

Genus **LOBIGER** KROHN, 1847
(**PTERYGOPHYSIS** FISCHER, 1883; **DIPTEROPHYSIS** PILSBRY, 1896)

Shell small, ovate, thin, the aperture expanded, its outer lip arched, inner lip sharp. Animal with two pairs of pleuropodial lobes projecting at the sides.

2313. Lobiger souverbii Fischer, 1857 (Synonyms: **L. viridis** Pease, 1863; **L. pictus** Pease, 1868). The shell is dark green, flaring anteriorly. Soft parts are green also, with black and white markings. Length of shell about 10 mm. The first specimens to be detected in the eastern Pacific were taken by Faye Howard and Gale Sphon in 1962; additional material and observations on habitat were obtained in 1970, at Santa Cruz, Nayarit, Mexico. Here, as elsewhere in the tropics where this widely distributed species occurs, the food plant is *Caulerpa*. The species is known in the Caribbean (type locality, Guadeloupe, West Indies), the Hawaiian Islands, and the western Pacific. One wonders why it remained so long unnoticed by West American collectors and whether its distribution is limited to a narrow segment of the Mexican coast. Supposed differences in the form and number of the pleuropodial lobes (signalized in the subgeneric names *Pterygophysis* and *Dipterophysis*) prove to be illusory, for when roughly handled or disturbed, the animal may autotomize or discard one or more of the lobes.

See Color Plate XIX.

Family VOLVATELLIDAE

Bulloid shells in which the margins meet midventrally. The family comprises two genera, only one of which has yet been reported in the Panamic province. Collectors should watch for the other—*Cylindrobulla* Fischer, 1857, in which the coil overlaps ventrally.

Genus VOLVATELLA PEASE, 1860
(ARTHESSA EVANS, 1950, in part)

Shell thin and elastic; coiled portion sunken along the left posterior margin; the posterior end produced as a spout.

2314. Volvatella cumingii (A. Adams in Sowerby, 1850). The shell is horn-colored and fragile, its spire concealed. Length, 15 mm. Santa Elena, Ecuador, depth 11 m.

Superfamily JULIACEA

Shell present, in two hinged parts.

Family JULIIDAE

"Bivalved gastropods" sounds like a contradition in terms, but the snails of this group are definitely bivalved. One valve has the normal spiral coiling, with an embryonal shell well preserved in most; the other valve develops from a lobe at the left side of the aperture and comes to be hinged along the columellar margin with an articulation of crude denticles and ligamental material derived from the periostracum. For a long while they have been classed as pelecypods, especially when both valves were found intact. A few fossil forms were described as gastropods when single valves, especially the valve with the embryonic whorls, were found. The single central muscle scar on the inside made them anomalous as pelecypods. The discovery of living material by Dr. Siro Kawaguti in Japan in 1959 made correct placement at last possible. References: Keen & Smith (1961), Kay (1968). Two Panamic subfamilies.

Subfamily JULIINAE

Heart-shaped shells with a heavy hinge, a prominent toothlike ridge in one valve and a fossette-like fold in the other; apical nucleus present in young stages, obsolete in adults; adductor scar with a central constriction, sometimes dividing it horizontally into two parts.

Genus JULIA GOULD, 1862
(PRASINA DESHAYES, 1863)

Minute shells, greenish in color, posterior end of aperture ending in a sharp point, anterior end rounded.

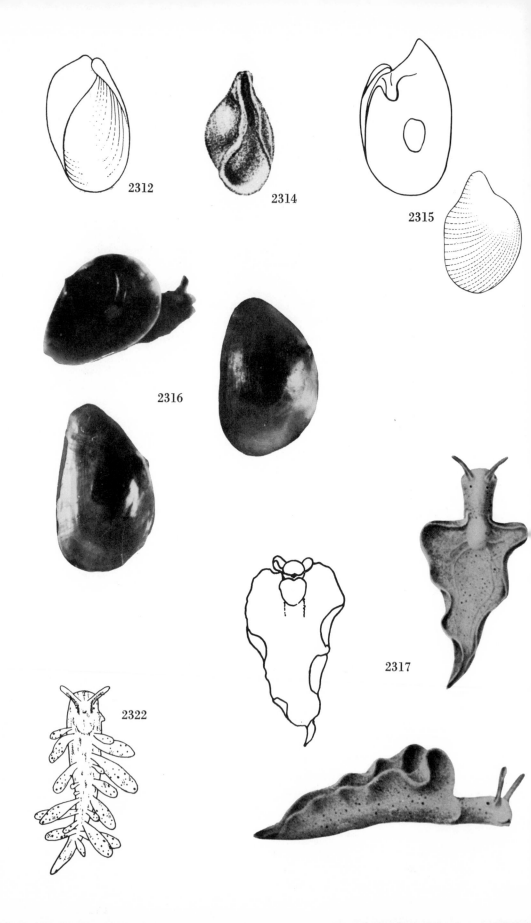

2312

2314

2315

2316

2317

2322

2315. Julia thecaphora (Carpenter, 1857) (Synonyms: **J. equatorialis** Pilsbry & Olsson, 1944; "**J. exquisita** Gould" of authors, not of Gould, 1862; a
Hawaiian form). The shell is green to greenish yellow, weathering to white;
smooth, obscurely marked with about 20 shallow radial lines. Length, 3 to 4 mm.
La Paz, Gulf of California, and Socorro Island, off west Mexico, to Tumbes, Peru,
in beach drift and offshore to 18 m, probably on green algae.

Subfamily BERTHELINIINAE

Shells lenticular in shape, ovate to quadrate, with a weak hinge; spiral nucleus
in one valve, retained in the adult; adductor muscle scar central, undivided,
circular.

Genus BERTHELINIA CROSSE, 1875

Left valve with spiral nucleus, right valve slightly smaller, without spiral apex;
outline of shell somewhat quadrate; color in living specimens greenish to yellowish. Subgenus *Berthelinia, s. s.,* was based on a minute (2 to 3 mm long) Eocene
fossil species from France, in which the hinge plate was widened at the ends.
No living species have so far been recognized.

Subgenus EDENTTELLINA GATLIFF & GABRIEL, 1911

Hinge plate of more uniform width than in *Berthelinia, s. s.*; shell nearly trigonal
in adult, with only a slight ventral sinuosity; size larger (to 9.5 mm).

2316. Berthelinia (Edenttellina) chloris (Dall, 1918) (Synonym: **B. (E.)
c. belvederica** Keen & Smith, 1961). Small, translucent green; periostracum
nearly transparent; sculpture of growth lines and a few radial striae; nucleus on
left valve of one and a quarter turns, sharply set off from the rest of the shell;
hinge weak, faint pits on the shell rim. Length, 6 to 9 mm, width, 4 to 6 mm. Bahía
Ballenas, Punta Abreojos, Baja California (Boettger, 1962) to the southern end
of the Gulf of California at La Paz and possibly at Guaymas, Mexico (type locality, Magdalena Bay, Baja California), on *Caulerpa.*

Superfamily ELYSIACEA

Shell wanting; animals feed on seaweed, mostly green algae.

Family ELYSIIDAE

Body flattened, elongate, with lateral leaflike expansions; foot narrow; tentacles
large, sometimes short.

Genus ELYSIA RISSO, 1818

Body smooth, without cerata or papillae; radular plates relatively large, compressed, finely serrate below.

2317. Elysia hedgpethi Marcus, 1961 (Synonym: **E. bedeckta** MacFarland,
1966). Body bright green with blue flecks, a white line on rhinophores and center of
foot; rhinophores inrolled, as broad as long; animal resembling a scrap of sea
lettuce. Length, about 25 mm. Puget Sound, Washington, to Bahía de los Angeles,
Gulf of California.

2318. Elysia oerstedii Mörch, 1859. The body is described as white, spotted
with green. Length, about 8 mm. Puntarenas, Costa Rica. The species remains
unillustrated, and the name has been overlooked by later authors.

2319. Elysia vreelandae Marcus & Marcus, 1970. Body dark olive-green, with lighter borders on the parapodia and with blue dots. Length, about 10 mm. San Agustin, Sonora, Mexico. Only anatomical details have been figured.

Genus TRIDACHIELLA MacFARLAND, 1924

Head rounded above, with large, inrolled tentacles; lateral expansions wide, their margins sinuous.

2320. Tridachiella diomedea (Bergh, 1894). Pale yellow to translucent gray, with black velvety triangular spots along the parapodia, which extend the length of the animal and are thrown into six folds. Length, to 35 mm. Throughout the Gulf of California (type locality, Cerralvo Island) and south to Panama.

See Color Plate XX.

Family HERMAEIDAE
(STILIGERIDAE)

Body elongate, pointed posteriorly; without distinct oral tentacles, but with rhinophores and two rows of cerata, the inner row larger; radular teeth serrate to smooth-edged. Two Panamic genera.

Genus HERMAEA LOVÉN, 1844

Rhinophore inrolled, sometimes bifurcated; no cephalic tentacles; dorsal cerata small; radular plates smooth-edged.

2321. Hermaea hillae Marcus & Marcus, 1967. The sluglike animals are translucent, with black eyes and mahogany digestive annexes in the body. Length of animal, 12 mm. Type locality, Puerto Peñasco, Sonora, Mexico. Unfigured except for internal anatomy.

Genus STILIGER EHRENBERG, 1831

With cephalic tentacles and rhinophores; dorsal cerata large, carrying branches of the midgut gland; radular plates smooth-edged.

2322. Stiliger fuscovittatus Lance, 1962. Creamy white with brown streaks. Length, 10 mm; width, 1.1 mm. San Juan Island, Washington, to Bahía de los Angeles, Gulf of California (type locality, San Diego, California).

Order NUDIBRANCHIA
(NUDIBRANCHIATA of authors)

Nudibranchs are snails in which all vestiges of shells are lost in the adult stage, and in which there is thus a return to bilateral symmetry. The gills (branchiae) are mostly on the dorsal surface toward the posterior end of the sluglike body, looking like a bundle of plumes. The body may be variously colored and elaborated into festooned appendages or cerata.

Food habits range from simple grazing on sessile animals, such as hydroids and sponges, to active swimming and capture of small planktonic prey. All nudibranchs are carnivorous, and the radulas and jaws may be modified for managing the food. In those that consume soft-bodied coelenterates, a sucking mechanism or pharyngeal bulb is developed. A remarkable adaptation in some of these grazers on coelenterates is their ability to ingest and store in the cerata of the dorsal surface the stinging cells or nematocysts of the coelenterate without rupturing the deli-

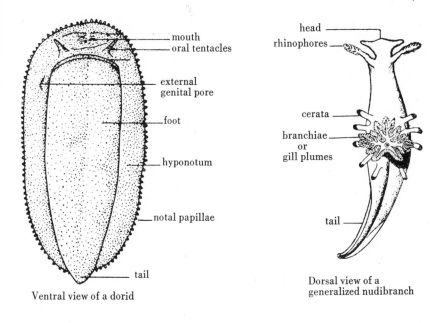

mouth
oral tentacles

external
genital pore

foot

hyponotum

notal papillae

tail

Ventral view of a dorid

head
rhinophores

cerata

branchiae
or
gill plumes

tail

Dorsal view of a
generalized nudibranch

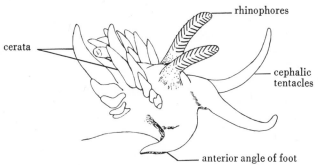

rhinophores

cerata

cephalic
tentacles

anterior angle of foot

Side view of head of an aeolid

cate capsule; thus the nematocyst can serve as a protective device for the mollusk, discharging its barbed thread into any predator that detaches or damages one of the cerata. How the nudibranch remains unaffected or can swallow the nematocyst without triggering it is a feat not easy to explain.

Classification of the nudibranchs is far from satisfactory as yet, for many of the distinctions between groups are based on anatomical features that can be observed only upon dissection. Thiele's work (1929–35) remains the standard reference on the group, although his twofold division was rather generally superseded by the classification of a Swedish zoologist, Dr. Nils Odhner, in 1939. This has in recent years been modified by other workers, notably Eveline and Ernst Marcus in their studies of American opisthobranchs. Because in this group of animals there are several organ systems that can be used to show differentiation—gills, liver, digestive tract, buccal apparatus, reproductive organs, etc.— a complex hierarchy of groupings has developed. Taylor and Sohl (1962) attempted to reconcile this complex with standard categories, and their scheme has been in the main accepted here, for it sets aside such unconventional terms as division, subdivision, and group, and disregards the tribe. However, until specialists supply guidance on the grouping of certain families into superfamilies, some makeshift rank between family and suborder is necessary; the term infraorder, used by Taylor and Sohl, is a temporary expedient that is retained here.

The present review of Panamic nudibranchs began as a list of known Panamic species, compiled by Hans Bertsch, Wesley Farmer, and Stephen Long, in support of the plea of Dr. Dwight Taylor that even though, strictly speaking, nudibranchs are not "sea shells," a census of the named species should be included. Diagnoses of the generic and higher taxa were added from available literature. The resulting systematic list was then fleshed out by James Lance, who undertook the task of supplying descriptions for all of the species, with notes on their habitats and geographic distribution. He also furnished color photographs of many of them. All of this collaboration is gratefully acknowledged.

Some of the terms used in describing nudibranchs are illustrated in the accompanying text figure. Useful references that have appeared in recent years on Panamic nudibranchs include Collier & Farmer (1964), Farmer & Collier (1963), Farmer (1963, 1967), Lance (1966, 1968), MacFarland (1966), Ernst Marcus (1961), Eveline Marcus & Ernst Marcus (1967).

Suborder DORIDOIDA
(HOLOHEPATICA)

Ovate to rounded, dorsoventrally flattened forms with smooth or papillose surface, the edge of the back (notum) projecting beyond the foot; midgut gland compact, unbranched.

Infraorder CRYPTOBRANCHIA

Anus posterior to the middle of the dorsal area, more or less encircled by the feathery gills, which are retractile into a pocket.

Superfamily DORIDACEA

Foot wide; dorsum without cerata.

Family DORIDIDAE

Rhinophores retractable into pits; dorsal surface warty, stiffened by spicules in some. Three Panamic subfamilies.

Subfamily DORIDINAE

Lips smooth or unarmed; teeth of radula hook-shaped, not denticulate.

Genus DORIS LINNAEUS, 1758

Ovate, anterior end broadly rounded; tentacles short; rhinophore sheaths and border of dorsum with warty papillae.

2323. Doris pickensi Marcus & Marcus, 1967. White, with a translucent notal border and numerous low, opaque white papillae on the back, the rhinophores, gills, and frontal foot margins yellowish. Length, 25 mm. Puerto Lobos to Puerto Peñasco, Sonora, Mexico, under boulders intertidally. Other specimens attributed to this species in the original description were said to have light orange and brick-red rhinophores and gills—probably a composite diagnosis involving two species. The range given here is of the original lot of nine specimens.

Subfamily ALDISINAE
(THORUNNINAE of some authors)
Lips armed with flattened rodlets.

Genus ROSTANGA BERGH, 1879
Dorsal surface covered with papillae of various sizes; lips armed with hooks; gills generally unipinnate.

2324. Rostanga pulchra MacFarland, 1905. Body color varies from pale yellowish to bright red, sometimes with faint to dark brownish blotches. A hallmark is the unusual rhinophores, which bear nearly vertical lamellae and a finger-shaped apical projection. Length, 31 mm; width, 12 mm. Vancouver Island, British Columbia, to southern South America (Chiloé Island, Chile, and off Camarones Bay, Argentina, in 100 m), on red siliceous sponges; rare in the northern end of the Gulf of California.

Subfamily CONUALEVINAE
Rhinophores smooth, retractile; body soft; notum depressed; anterior edge of foot bilabiate; oral tentacles short and stout.

Genus CONUALEVIA COLLIER & FARMER, 1964
Body soft, lacking spicules; notum smooth to papillose; rachidian tooth of radula absent, the others simply hooked.

2325. Conualevia alba Collier & Farmer, 1964. White, with two irregular rows of opaque white glands in the notal margins, the back smooth or densely covered with minute tubercles. Rhinophores are smooth, the eight gills tripinnate. Flatter than *C. marcusi*. Newport Bay, California, to Bahía Tortuga, outer coast of Baja California; rare, except for local and seasonal aggregations.
See Color Plate XIX.

2326. Conualevia marcusi Collier & Farmer, 1964. Ground color pale orange to white, with slightly darker rhinophores. Numerous low tubercles are close-set over the back, giving it a textured appearance. Gills are unipinnate, sixteen in number. Length, 31 mm. Puertecitos, Baja California, to Isla Angel de la Guarda, Gulf of California; rare.

2327. Conualevia mizuna Marcus & Marcus, 1967. White, with light yellow rhinophores and gills, the latter pluripinnate, eight in number. Length, 27 mm; width, 8.5 mm. Named from a single specimen collected at Puerto Peñasco, Sonora, Mexico. It will probably prove to be a synonym of *C. alba*.

Family CHROMODORIDIDAE
(? GLOSSODORIDIDAE)
Radula often with a rudimentary central plate, the lateral teeth short, cuspate. Four Panamic subfamilies.

Subfamily CHROMODORIDINAE
Lips armed with hooks; gills unipectinate. Two Panamic genera.

Genus CHROMODORIS ALDER & HANCOCK, 1855
(? GLOSSODORIS EHRENBERG, 1831)

Back smooth, brightly colored, notal edge narrowly projecting. Radula with hooked teeth and accessory denticles. If the type species of *Glossodoris* proves to have hooked teeth, this name will take precedence over *Chromodoris*.

2328. Chromodoris banksi banksi Farmer, 1963. Body white, edged by a thin translucent line and by an inner uninterrupted bright orange line. The back is ornamented with numerous dark brown to black dots of varying size, like polka dots, interspersed with larger areas of cream color. Length, 38 mm; width, 15 mm. Puerto Peñasco, Sonora, Mexico, to Concepcion Bay, Gulf of California.

2329. Chromodoris banksi sonora Marcus & Marcus, 1967. Described on the basis of differences in the radula, the first lateral tooth being denticulate, whereas it is smooth in *C. banksi banksi*. Since the forms are not geographically separable, this should be regarded as a species, if the difference in radula proves consistent. Length, 25 mm; width, 11 mm. Puerto Peñasco, Sonora, Mexico.

2330. Chromodoris macfarlandi Cockerell, 1901. A spectacular form, bright red-violet, with an orange stripe edging the notum and another behind each rhinophore, passing back to unite at the posterior margin of the gills; another color band marks the mid-dorsal region. Length, 46 mm; width, 11 mm. Monterey Bay, California, to the Cedros Island area, Baja California.

2331. Chromodoris norrisi Farmer, 1963. Body white, with dashes of bright orange around the back margins, the rhinophores and gills orange-tipped. Over the back are intermingling bright yellow and deep violet dots in alternating patterns. Length, 61 mm; width, 32 mm. Pacific coast of Baja California, through the Gulf of California to Puerto Peñasco, Sonora, Mexico.
See Color Plate XX.

2332. Chromodoris sedna (Marcus & Marcus, 1967) (Synonym: **C. fayeae** Lance, 1968). Body pure white, with three distinct color bands bordering the notum and posterior half of the foot, the inner one opaque white, the middle one vivid orange-red, and the outer one bright yellow. Upper portions of rhinophores and gills deep red. As in several chromodorids of the Indo-Pacific, the gills are often kept in what appears to be frenzied motion. Length, 47 mm; width, 11 mm. Throughout the Gulf of California, south to Santa Cruz, Nayarit, Mexico; one of the commonest nudibranchs during spring and summer months on the mainland shore, less so along the peninsular coast, intertidally and offshore to 16 m. The species was originally placed in the genus *Casella* Adams & Adams, 1854, but is here transferred (Lance, *in litt.*) to *Chromodoris*.
See Color Plate XX.

2333. Chromodoris tura Marcus & Marcus, 1967. Dark violet to almost black medially, with a broad bluish-white margin on which there is one orange line. Red spots and dashes are scattered on the dark center, with yellowish streaks along its border. The rhinophores are bluish black, the gills white with black tips. The gills are unipinnate, seven in number. Fort Kobbe Beach, Panama Canal Zone (Pacific side).

Genus HYPSELODORIS STIMPSON, 1855
(? PTERODORIS EHRENBERG, 1831)

Similar to *Chromodoris* but with the teeth of the radula bifid, the innermost with one or two denticles on the inner margin.

2334. ? Hypselodoris aegialia (Bergh, 1905). Described from the Gulf of California, this animal has not been taken since. The description of the radula would imply allocation to *Hypselodoris*, though it was placed in *Chromodoris*.

2335. Hypselodoris californiensis (Bergh, 1879) (Synonym: **Chromodoris agassizii** Bergh, 1894). Elongate, the body dark ultramarine-blue, with spots or streaks of golden yellow, the back and foot margins pale blue to white. Length, 71 mm; width, 11 mm. Monterey Bay, California, southward along the Baja California coast and throughout the Gulf of California, south to Jalisco, Mexico, commoner in the southern part of its range, intertidally and subtidally. The type locality of *H. agassizii* is Panama; as yet there are no intermediate records. Similarly colored slugs from the Gulf of California include *Navanax inermis* and *Polycera alabe*, neither of which have retractile rhinophores and gills. The species has been cited also in *Chromodoris* and *Glossodoris*, the synonymy of which has not yet been clearly established.
See Color Plate XX.

Subfamily CADLININAE
Gills bipinnate to tripinnate.

Genus CADLINA BERGH, 1878 (ICZN Opinion 812, 1967)
(ACTINOCHILA MÖRCH, 1868; ECTINOCHILA MÖRCH, 1869; both suppressed, ICZN, 1967)
Smooth to slightly warty, with five to twelve bipinnate to tripinnate gills. The generic name has been conserved because of long usage, although there were two earlier that were overlooked until recent years; these have now been suppressed by the ICZN.

2336. Cadlina evelinae Ernst Marcus, 1958. Body white, with numerous large or small bright red to orange spots. Length, 30 mm; width, 11 mm. Pacific coast of Baja California to the Gulf of California, commoner in the northern part of the range. Type locality, São Paulo, Brazil (Atlantic).
See Color Plate XX.

2337. Cadlina flavomaculata MacFarland, 1905. Body small, white to yellowish, with contrasting dark brown or black rhinophores and a row of yellow spots along each side of the back. Length, 24 mm; width, 10 mm. Vancouver Island, British Columbia, along the outer coast of Baja California; also, more rarely, in the northern end of the Gulf of California.

2338. ? Cadlina sparsa (Odhner, 1922). Reported in southern California and on the islands Juan Fernandez and Chiloé, off Chile; not as yet from intermediate localities, in depths to 40 m.

Subfamily DISCODORIDINAE
Lip smooth or with pointed prickles; teeth of radula hook-shaped. Four Panamic genera.

Genus DISCODORIS BERGH, 1877
Body rounded-ovate, soft, notum finely papillose; lips two, armed with rows of fine rodlike prickles.

2339. Discodoris aurila Marcus & Marcus, 1967. Color light brown, freckled with white, a darker band in the midline and some vague dusky spots between mid-

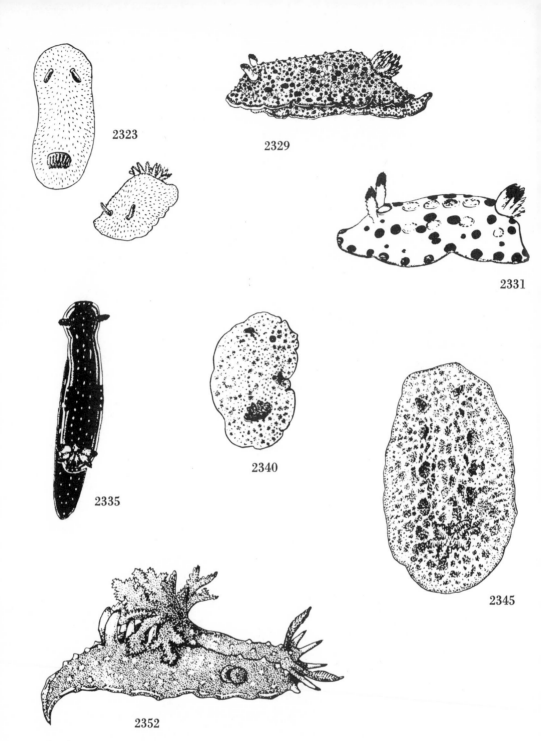

2323

2329

2331

2335

2340

2345

2352

2354

line and margin, some black dots near the midline. Rhinophore and gills uniformly brown. The underside of the body is unpigmented. Notum with tubercles and spicules, rhinophores and gills smooth. Tentacles with triangular lobes, the rhinophores with eighteen foliations; gills multipinnate, six in number. The anterior border of the foot has a transverse furrow and the upper lip a central notch. Length of a preserved specimen, 20 mm. Fort Kobbe Beach, Panama Canal Zone.

2340. Discodoris mavis Marcus & Marcus, 1967. Body orange-pink, with scattered compound brownish spots, more numerous toward the sides; rhinophores club-shaped, dark brown. Underside of the body unpigmented. Length, 24 mm; width, 17 mm. Puerto Peñasco, Sonora, Mexico; known only from a specimen found under a granite boulder. The species may be a color phase of the Californian *D. heathi* MacFarland, 1905, which is not anatomically distinguishable.

Genus DIAULULA BERGH, 1879

Body soft, foot rounded in front; lips two, the upper notched medially; gills tripinnate; radula without central teeth, laterals multicuspate, hooked.

2341. Diaulula sandiegensis (Cooper, 1863). On the white to brown ground color are a few brown or black rings, especially along the sides of the notum. Minute spiculose papillae give the back a gritty texture. Length, 80 mm; width, 45 mm. Northern Japan and the Aleutian Islands, Alaska, south to the Gulf of California, less common in its northern end; intertidally and subtidally.

? Genus GEITODORIS BERGH, 1892

Anterior border notched, lobelike; tentacles short, digitate; lips armed with prickles.

2342. ? Geitodoris immunda Bergh, 1894. Known only from some preserved specimens that were described as brown-gray, with dark rhinophores and gills. Gills eight, tripinnate. Length, 25 mm; width, 17 mm. Gulf of Panama; dredged in 6 m. Additional modern material not reported; the status of the species thus remains doubtful.

Genus TARINGA MARCUS, 1955

Radula with denticles on the outer side of the lateral teeth; male with armed penial papillae.

2343. Taringa aivica aivica Marcus & Marcus, 1967. Notum yellowish, mottled with gray and covered with minute tubercles; rhinophores and gills brownish. A few elongate papillae occur on the dorsum. Gills six, tripinnate. Length, 22 mm; width, 15 mm. Known only by a single specimen taken at Fort Kobbe Beach, Panama Canal Zone.

2344. Taringa aivica timia Marcus & Marcus, 1967. Purple-brown to brownish pink, with lighter-colored markings or brown dots. Gills dark yellow to light violet. A few elongate papillae are on the back, as in the typical subspecies. Length, 70 mm; width, 43 mm (preserved material). Puerto Peñasco, Sonora, Mexico. The radula has a large spine on the marginal teeth (which are smooth in *T. a. aivica*) and weak denticles on the lateral teeth.

Genus TAYUVA MARCUS & MARCUS, 1967

Tentacles pointed, labial plates with rodlets, teeth of radula hook-shaped; reproductive apparatus with stout penial papilla, a large vestibule stiffened by spicules lodging the penial papilla and the vaginal aperture.

2345. Tayuva ketos Marcus & Marcus, 1967. Body whitish, with irregular brown blotches on the back, which is covered with small, pointed papillae. Gills six, pluripinnate; rhinophores brown. Length, 50 mm. Puerto Peñasco to Puerto Lobos, Sonora, Mexico.

Subfamily INUDINAE
With cleft hooks on the labial cuticle.

Genus INUDA MARCUS & MARCUS, 1967
Radula multidentate, with bicuspid rachidian tooth and denticulate lateral teeth.

2346. Inuda luarna Marcus & Marcus, 1967. Described from preserved material, which shows the body as whitish, mottled with brown, the rhinophores and gills dark. The back is smooth and lacks projecting bundles of spicules. Recognition of the species probably will require dissection of the animal to obtain the radula with its distinctive features. Length, 27 mm; width, 20 mm. Puerto Peñasco, Sonora, Mexico; two specimens collected intertidally under boulders.

Doridacea of Uncertain Status

2347. "Doris" phyllophora Mörch, 1859. West coast of Central America. Type material possibly preserved in the collection of the University of Copenhagen; not studied in recent years.

2348. "Doris" punctatissima Mörch, 1859. Realejo (Corinto), Nicaragua. Type material possibly in the University of Copenhagen collection; not studied in recent years.

2349. "Doris" umbrella Rochebrune, 1895. Type material possibly in the collection of the Muséum d'Histoire Naturelle de Paris; not studied in recent years. The brief description indicates that the body is minutely tuberculate, grayish yellow above, spotted with regular patches of violet dots, white below, similarly dotted. Length, 37 mm; width, 24 mm. Lack of data on internal details prevents assignment of the species to a genus or even to a family. Type locality, Mogote, La Paz Bay, Gulf of California.

Infraorder PHANEROBRANCHIA
Branchial plumes pinnate, arranged in a circle or arc, united or separate at their bases, never retractile into a cavity.

Superfamily POLYCERATACEA
(NONSUCTORIA of authors)
Mantle brim narrowed, replaced by tentacular processes; radula adapted for grazing on sponges and bryozoa.

Family POLYCERATIDAE
(POLYCERIDAE of authors)
Rhinophores with several leaflets. Head having the more or less prominent frontal veil prolonged into horns at the margin.

Genus POLYCERA CUVIER, 1817
Body sluglike, nearly smooth, frontal margin digitate; branchial plumes few, simply pinnate, a single larger ceras or dorsal appendage on each side; tentacles short, lobiform.

2350. Polycera alabe Collier & Farmer, 1964. Body blue-black, marked by orange spots arranged loosely in rows, distinguished from the similarly colored *Hypselodoris californiensis* by being shorter, with nonretractile gills and rhinophores. Length, 25 mm; width, 5 mm. Cedros Island, Baja California (type locality) to the northern end of the Gulf of California, where it is rare.

2351. Polycera atra MacFarland, 1905. Whitish or translucent limaciform slug with numerous black longitudinal stripes, yellow or orange spots being interspersed between the bands on low tubercles. The anterior head veil is developed into prominent finger-like appendages. Length, 43 mm; width, 10 mm. Tomales Bay, California, to Los Coronados Islands, Baja California, intertidally on boat landings in bays and offshore kelp canopies, where the animals feed on compound bryozoans. Reported from Puerto Peñasco, Gulf of California, probably as a misidentification of an as-yet-undescribed striped form.

2352. Polycera hedgpethi Marcus, 1964 (Synonym: **P. gnupa** Marcus & Marcus, 1967). Body light to dark brown, without longitudinal stripes but with pigment-free borders on the back and foot; yellow-orange markings on velar appendages, foot corners, rhinophores, gills, extrabranchial appendages, caudal crest, and tip of the foot; and yellow caps on the widely scattered low tubercles. Length, 34 mm. San Francisco Bay, California, to the northern end of the Gulf of California, intertidally and on boat landings; only seasonally common, feeding on colonies of the bryozoan *Bugula*.

Genus LAILA MacFarland, 1905

Frontal and lateral margins narrow, with club-shaped papillae; a flattened submarginal ridge on each side of the anterior end of the body just behind and above the tentacles.

2353. Laila cockerelli MacFarland, 1905. The overhanging margin of the elongate body develops long, club-shaped papillae in closely packed rows that give the animal a fringed appearance. Color pure white, with red rhinophores and orange-capped papillae. Length, 20 mm; width, 7 mm. Two distinct morphological varieties are discernible: populations with low white tubercles scattered randomly over the back, ranging from Vancouver Island, British Columbia, to Point Conception, California; southern populations, with a single irregular row of large red tubercles down the midline of the dorsum, ranging from Point Conception, California, south along the Baja California coast and into the Gulf of California to Bahía de los Angeles. The two forms could qualify as geographic subspecies.
  See Color Plate XXI.

Family GYMNODORIDIDAE

Body smooth, rhinophores perfoliate, retractile; tentacles short, without ridges; radula broad, the innermost plate hook-shaped, the next awl-shaped.

Genus NEMBROTHA Bergh, 1877

Gills in middle of back, branchial plumes few, bipinnate to tripinnate; tentacles lobiform; labial armature weak; reproductive system lacking a differentiated prostate gland.

2354. Nembrotha eliora Marcus & Marcus, 1967 (Synonym: **N. hubbsi** Lance, 1968). Deep yellow-ochre to greenish black, with several longitudinal blue stripes bordered on each side by thin black bands. Rhinophores and gills deep blue-black, the gills with blue or green ribs. Length, 44 mm. Apparently at least three similarly colored species occur in the Gulf of California. The description of Marcus and Marcus is a composite of two of these. The radular teeth figured by Marcus and Marcus are incompatible with those of the figure by Lance (1968). The exact range of the species in the northern end of the Gulf of California remains to be worked out.

See Color Plate XX.

Family NOTODORIDIDAE

Body firm to the touch because of spicules in the integument; mantle edge projecting, narrow; rhinophores retractile, mostly not laminated; radula without central plates, laterals hook-shaped or arched.

Genus AEGIRES LOVÉN, 1844
(AEGIRUS, unjustified emendation)

Body arched, not depressed, highest and broadest in front of the branchial plumes, rounded in front, pointed behind.

2355. Aegires albopunctatus MacFarland, 1905. The salt-and-pepper nudibranch has a white body with black dots. Irregular rows of robust tubercles develop on the notum, some of which are as high as the extended gills. Extreme rigidity of the body results from an extraordinary development of spicules. The rhinophores are yellowish and smooth. Length, 24 mm; width, 8 mm. Vancouver Island, British Columbia, to Baja California and the northern part of the Gulf of California; rare in the latter part of its range.

See Color Plate XXI.

Superfamily ONCHIDORIDACEA
(SUCTORIA of authors)

Buccal area with pharyngeal bulb connecting to a muscular crop.

Family ONCHIDORIDIDAE

Body somewhat depressed, warty; tentacles scarcely evident; radula small. Two Panamic genera.

Genus ONCHIDORIS BLAINVILLE, 1816
(as ONCHIDORUS, emended by later authors)

Head broad, semicircular, mouth with a ring of papillae; radula with a large hooked plate and one or two smaller lateral teeth.

Onchidoris hystricina (Bergh, 1878). This species, which ranges from the Aleutian Islands, Alaska, to northern Baja California, has been reported, probably in error, from the Gulf of California.

Genus ACANTHODORIS GRAY, 1850

Body soft, notum thickly covered with short hairlike processes; margin of rhinophore aperture lobed; branchial plumes few, tripinnate, arranged in a circle; head wide, tentacles short, lobe-shaped; mouth armed with minute hooks; radula narrow.

2356. Acanthodoris pina Marcus & Marcus, 1967 (Synonym: **A. stohleri** Lance, 1968). The black body with tall cherry-red dorsal papillae and notal border is distinctive. Between the papillae are a number of opaque yellow-white flecks, and the papillae have translucent rings around their bases. Length, 21 mm; width, 11 mm. Not uncommon under stones at low tide in the northern part of the Gulf of California. The genus shows great diversity in colder waters; this is the first species recorded in the Panamic region.

Family GONIODORIDIDAE

Body oval, mantle edge reduced; rhinophores mostly sheathed, not retractile; branchial plumes numerous, usually simply pinnate; tentacles small. Two Panamic genera.

Genus ANCULA LOVÉN, 1846

Smooth, with little frontal veil; rhinophores with two anterior linear appendages; branchial plumes three, bipinnate to tripinnate; lip armature of rows of hooks; radula narrow, first pleural tooth large and broad, inner margin denticulate, outer tooth much smaller, subtriangular.

2357. Ancula lentiginosa Farmer & Sloan, 1964. The compressed body is translucent white, with varying patterns of brownish freckles. Fingerlike appendages occur singly on each side of the gills, with a pair at the base of each rhinophore. Length, 21 mm; width, 4 mm. Monterey Bay to La Jolla, California, intertidally; a single adult taken at Bahía de los Angeles, Gulf of California.

Genus OKENIA MENKE, 1830, *ex* LEUCKART in BRONN, MS [ICZN, pending] (IDALIA LEUCKART, 1828 [not HÜBNER, 1819]; CARGOA VOGEL & SCHULTZ, 1970)

Notum small, mantle edge with a row of tentacle-like projections; foot broad, pointed behind; head small, without evident tentacles; rhinophores large; radula with hooked plates.

2358. Okenia angelensis Lance, 1966. A small, drab form, difficult to detect. Body elongate, with lateral pallial ridges projecting as a few finger-like processes, five to seven others on the back. Rhinophores mostly smooth or with one or two weakly developed lamellae. Color markings of brownish, yellow-white, and green give the dorsum a mottled appearance. Length, 10 mm; width, 2.5 mm. On boat landings in southern California; in *Sargassum* beds at Bahía de los Angeles, Gulf of California, during the spring; uncommon.

See Color Plate XXI.

Infraorder POROSTOMATA

Mouth opening very small; head undeveloped, oral tentacles reduced in size.

Superfamily DENDRODORIDACEA

Branchial plumes retractile, in a pocket far back on the notum.

Family DENDRODORIDIDAE

Mantle margin undulating; jaws and radula absent; pharyngeal bulb a muscular tube. Two Panamic genera.

Genus DENDRODORIS EHRENBERG, 1831
(DORIDOPSIS ALDER & HANCOCK, 1864; DORIOPSIS PEASE, 1860, of authors)

Body soft; notum smooth to warty.

2359. Dendrodoris krebsii (Mörch, 1863) (Synonym: **Doriopsis atropos** Bergh, 1879). The velvety black body is rimmed around the notum and posterior part of the foot by a dark red band sometimes fading to yellowish. The white rhinophores and gill tips are in striking contrast. Length, 44 mm. Rare on the south shore of Sebastian Vizcaino Bay, Baja California; common throughout the Gulf of California, south to Tenacatita, Jalisco, Mexico.

See Color Plate XXI.

Genus **DORIOPSILLA** BERGH, 1880

Notum warty; buccal ganglion well forward.

2360. Doriopsilla albopunctata (Cooper, 1863). Body deep chestnut-brown, with pale yellow to orange notal margin. Back, except at edges, densely set with opaque white dots on low elevations. Head, as in other porostomes, totally undeveloped; radula lacking. Length, 60 mm. A Californian species reported in the northern end of the Gulf of California. The identification of Gulf porostomes with Cooper's species, the type locality of which is southern California, is questionable. Several morphologically similar species with very different biological patterns having come to light recently, it is probable that with more study, the need for description of additional species will become obvious. The range of Cooper's species on the outer coast is from Sonoma County, California, to Punta Eugenia, Baja California, intertidally, common from Point Conception to San Diego, California, during the spring and summer; rare in the Gulf of California.

2361. Doriopsilla janaina Marcus & Marcus, 1967. Small, light red to bright pink, with two or three ill-defined lines of black along the middorsal region. Varying patterns of red and white color the notal tubercles and their bases. The rhinophores are red, the gills orange. Length, 25 mm; width, 14 mm. Known only from a few specimens collected at Puerto Lobos, Sonora, Mexico, and in the Canal Zone, Panama, intertidally on rocks.

2362. Doriopsilla rowena Marcus & Marcus, 1967. Smallest of the Panamic porostomes, distinguished by its pale yellow to pink ground color with scattered brownish to black dots on the back. Two rows of larger white glandular spots occur between the rhinophores and gills. Length, 10 mm; width, 5 mm. Northern end of the Gulf of California, from San Felipe to Guaymas, Sonora, Mexico. The species may prove to be a synonym of the similar *D. nigromaculata* Cockerell & Eliot, 1905, recently retaken at its type locality, La Jolla, California.

Suborder **DENDRONOTOIDA**

Rhinophores with a basal sleeve or sheath developed from the mantle.

Superfamily **DENDRONOTACEA**

Dorsal surface elaborated in various ways, as by branching gill tufts or club-shaped to spoon-shaped processes.

Family **DENDRONOTIDAE**

With two rows of much-branched or arborescent cerata, mostly entered by extensions of the midgut gland.

Genus **DENDRONOTUS** ALDER & HANCOCK, 1845

Notal edge indicated by a row of cerata; frontal veil present; central plate of the radula strong, cuspate.

2356

2361

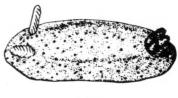

2362

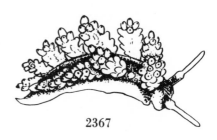

2367

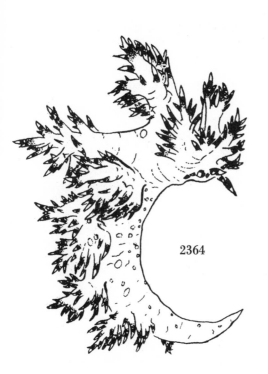

2364

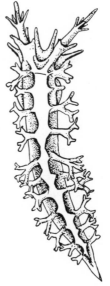

2372

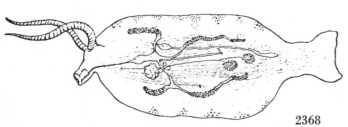

2368

2363. Dendronotus frondosus (Ascanius, 1774). Although cited as a cosmopolitan form in the northern hemisphere, this has not been definitely recorded in the Panamic province.

2364. Dendronotus nanus Marcus & Marcus, 1967. The body is translucent gray, with an opaque white line edging the foot. The cerata are distinctively marked with a white tip, an inner blackish-brown ring, and an orange base. Five simple or compound papillae border the rhinophore sheath. Length, 13 mm; width, 3 mm. Puerto Peñasco, Sonora, Mexico, known as yet by only two specimens, both found floating at the surface in 0.5 m (18 inches) of water.

Family DOTIDAE (ICZN Opinion 697, 1964)

Cerata slightly to strongly tuberculate, with or without basal gills.

Genus DOTO OKEN, 1815 (*nom. conserv.*, ICZN Opinion 697, 1964)

Small, head somewhat broadened, foot small, radula with one series of teeth; gullet short.

2365. Doto amyra Marcus, 1961 (Synonyms: **D. ganda** and **wara** Ernst Marcus, 1961). An inconspicuous white animal, with pink, orange, or orange-brown digestive gland diverticula contained within the tuberculate cerata. Length, 10.5 mm. Seasonally common among hydroids on boat docks, Monterey Bay to San Diego, California; a single specimen has been recorded at Puerto Peñasco, Sonora, Mexico (Marcus and Marcus, 1967).

2366. Doto ensifer Mörch, 1859. Unfigured. Type locality, Realejo (Corinto), Nicaragua. Holotype possibly preserved in the museum of the University of Copenhagen; until it can be studied, the species remains of doubtful status.

2367. Doto lancei Marcus & Marcus, 1967. Differing from *D. amyra* by having a light brown body, with brownish-black pigment on the head veil and rhinophore sheaths. The more distal ceratal tubercles are ringed with black near the tip and often have an apical black spot. Length, 9 mm. Bahía de los Angeles and Puerto Peñasco, Sonora, in the northern part of the Gulf of California. This may prove to be a synonym of *D. columbiana* O'Donoghue, 1921, the only other species in the eastern Pacific with subapical rings around the tubercles.

Family PHYLLIROIDAE

Pelagic, translucent, without cerata or other appendages, except for a pair of tentacle-like rhinophores. Two Panamic genera.

Genus PHYLLIROE PÉRON & LESUEUR, 1810

Body ribbon-like, laterally compressed, with elongate, simple rhinophores and a distinct tail.

2368. Phylliroe bucephala "Péron & Lesueur, 1810." Body nearly transparent, colored only by the brownish or orange digestive gland branches and by white spots that form irregular lines along the dorsal and ventral body margins. One of the few parasitic nudibranchs, it attacks pelagic hydromedusae in its larval stages, and upon penetration of the bell, it devours its host from the inside out. Length (preserved), 22 mm. Numerous free-swimming adults have been collected during plankton tows by Scripps Institution of Oceanography off the west coast of Mexico, between latitudes 14° and 24° N. The host coelenterate has not been determined for the eastern Pacific, although known elsewhere, as in the Mediterranean and Australia (Martin & Brinckmann, 1963). Although credited to Péron & Lesueur, 1810, by authors, the specific name actually was only a vernacular or common name as of their publication. Lamarck in 1822 latinized it, but several of the authors actively working on pelagic mollusks between the years 1810 and 1822 may have done so. Sherborn cites the name as of Oken, 1815, but unfortunately Oken's work has been rejected for nomenclatural purposes by the ICZN. Pending further research, the name is cited in quotation marks here to indicate its unsatisfactory status.

Genus CEPHALOPYGE HANEL, 1905

Foot small; a gland pit on the ventral surface of the head; midgut gland three-branched.

2369. Cephalopyge trematoides (Chun, 1889). Similar to *Phylliroe bucephala* in general form, coloration, and habitat, but with a more lanceolate body form. Length, 30 mm; height, 3.5 mm. Reported to prey upon hydromedusae during its developmental stages. Known only from a few specimens netted off southern California and Baja California; otherwise cosmopolitan.

Family TETHYIDAE
(FIMBRIIDAE of authors)

Anterior end greatly expanded, more or less fringed. The name of the type genus, *Tethys* Linnaeus, 1767, was for a time mistakenly used for the cephalaspidean group *Aplysia*, and the nudibranch then became "*Fimbria* Bohadsch" of authors. The name *Tethys* has been restored for use in the Nudibranchia by the International Commission on Zoological Nomenclature (ICZN Opinion 200, 1954); no species of *Tethys*, however, have as yet been reported in the Panamic province.

Genus MELIBE RANG, 1829
(CHIORAERA GOULD, 1852)

Foot relatively small, head hoodlike, fringed; cerata spatulate, without basal gills. Animals able to swim by lateral motion as well as to crawl on seaweed, feeding by entrapping small crustacea (mostly copepods) in the frontal expansion of the head, used like a net. Large specimens up to 140 mm in length.

2370. Melibe leonina (Gould, 1852). The immense globular oral hood distinguishes this form. The body is translucent greenish or yellowish white, with numerous white or yellow flecks (glands), and a network of digestive gland diverticula that appear as brown threads. Length, about 90 mm. Alaska to Punta San Hipolito, Baja California, on offshore kelp canopies and inshore *Zostera* beds, common locally and seasonally; small breeding individuals occur throughout the Gulf of California on *Sargassum* but are rare.

See Color Plate XXI.

Family TRITONIIDAE

Anterior veil not fringed, projecting beyond foot; rhinophores not leaflike, retractable into a sheath; radula with central tooth tricuspate, laterals small, slender, slightly hooked.

Genus TRITONIA CUVIER, 1797 (ICZN Opinion 668, 1963)

Body somewhat stubby, back surface smooth or pustulose; gills numerous, of uniform or alternating size; velum rounded to bilobed.

2371. Tritonia exsulans Bergh, 1894. Body deep rose-pink, nearly square in cross section, edged with contrasting white lines along the foot, dorsum, and rhinophore sheaths. An oral veil is well developed anteriorly. The red-brown gills are arranged along the back margins in an undulating line. Length, 50 mm; width, 15 mm. Vancouver Island, Canada, to Punta Santo Domingo, Baja California, intertidally and to a depth of 130 m; also in Panama Bay, on the Atlantic coast of Panama, in Florida, and in Japan.

2372. Tritonia pickensi Marcus & Marcus, 1967. Body elongate and translucent, with the pinkish internal organs visible; a median band of opaque white runs down the middle of the dorsum and widens where it meets the base of each of the ten to eleven lateral gills. The sides of the body are blue in some specimens. Length, 7 mm; width, 2 mm. Puerto Peñasco and Guaymas, Sonora, Mexico, on gorgonians; rare. Marcus and Marcus considered that this species should be assigned to the subgenus *Candiella* Gray, 1850, in which the velum is rounded, with at most eight to ten papillae.

Suborder ARMINOIDA

Body thin, flattened; dorsal surface smooth or with cerata but not warty; with a tendency to longitudinal color stripes.

Superfamily ARMINACEA

Rhinophores lacking sheaths; frontal veil present but cephalic tentacles wanting; gills on the undersurface of lateral notal margins.

Family ARMINIDAE

Without appendages on the notum; frontal veil well developed, lobed at the sides. Two Panamic genera.

Genus ARMINA RAFINESQUE, 1814
(PLEUROPHYLLIDIA MECKEL, 1816; DIPHYLLIDIA CUVIER, 1817)

Rhinophore leaflets at the border of the notum; dorsum striped.

2373. Armina californica (Cooper, 1862) (Synonym: **A. digueti** Pruvot-Fol, 1955). The ground color varies from pale gray to reddish, brown, or black, with numerous white wavy ridges running longitudinally along the dorsum, terminating along the posterolateral margins. Length, 110 mm; width, 41 mm. Vancouver Island, Canada, to Panama. Seasonally and locally abundant on muddy sand bottoms from extreme low water to 80 m; rare in the Gulf of California, although reported from San Felipe to Mazatlán.
   See Color Plate XXII.

Genus HISTIOMENA MÖRCH, 1859
(CAMARGA BERGH, 1866)

Frontal veil broad, crescent-shaped, angulate.

2374. Histiomena convolvula (Lance, 1962). Ground color deep chocolate-brown, with opaque white spots that form irregular lines on the summits of the notal ridges; ridges much more convoluted than those of *Armina californica*. Three distinct color bands border the bright pink foot: a thin white line on the outer edge, a central broad band of bright orange, and an inner wide stripe of opalescent blue-white. Length, 75 mm; width, 32 mm. Punta Diggs, Gulf of California; rare in low, sandy pools. The species may be more common than records would suggest, for the animals rarely emerge from below the surface of the sand.

2375. Histiomena marginata Mörch, 1859. This is the type of the genus. It differs from *H. convolvula* in details of the radula and in the notal sculpture. Length, 140 mm; width, 50 mm. Realejo (Corinto), Nicaragua, to Panama (type locality) in depths to 33 m.

Family ANTIOPELLIDAE

With numerous elongate spindle-shaped dorsal papillae around the margin of the notum, the longer papillae nearer the center of the body, the outermost shorter; midline smooth.

Genus ANTIOPELLA HOYLE, 1902
(ANTIOPA ALDER & HANCOCK, 1848, not MEIGEN, 1800)

2376. Antiopella barbarensis (Cooper, 1863) (Synonym: **A. aureocincta** Johnson & Snook, 1927, *ex* MacFarland, MS). Body beautifully translucent, with the many cerata ringing the dorsal margin and meeting both in front and behind, the cerata with a metallic gold band below the tip, the tip usually bright blue. A raised crest, like a cock's comb, lies between the rhinophore bases. Length, 71 mm. Vancouver Island, Canada, to Bahía San Quintín, Baja California, intertidally and on bay boat landings during late spring and summer. One specimen has been collected at Bahía de los Angeles, Gulf of California.
See Color Plate XXI.

Family DIRONIDAE

With a conspicuous frontal veil; cerata present on the notum; midgut gland not branching into cerata.

Genus DIRONA ELIOT, 1905, *ex* MACFARLAND, MS

Body broad, depressed, head with a wide, thin veil, smooth and undulating at its margin; rhinophores of several leaves, without a sheath; dorsal appendages large, in a close-set series, the papillae spindle-shaped.

2377. Dirona picta Eliot in Cockerell & Eliot, 1905, *ex* MacFarland, MS. Variable, with a translucent yellowish, greenish, or pinkish cast, the dorsal and lateral surfaces of the body, including the tuberculate papillae, dotted and splotched with white, yellow, brick-red, and dull green. The inflated papillae range along the sides of the body, nearly meeting in front of the rhinophores, the inner surfaces with an irregular longitudinal ridge that gives off side branches. Length, 54 mm; width, 9 mm. Dillon Beach, California, to Puerto Rompiente, Baja California, common during summer low tides; rare in the northern part of the Gulf of California.
See Color Plate XXII.

Suborder AEOLIDIIDA
(CLADOHEPATICA, in part)

Brightly colored, animals usually slender bodied; feeding on coelenterates and storing the nematocysts in cnidosacs in the cerata.

Infraorder PLEUROPROCTA

With the anus in a lateral position at the right.

Family FLABELLINIDAE

Body elongate, compressed; cerata numerous, on low elevations of dorsal margin; radula with large central teeth, smaller laterals.

Genus FLABELLINA VOIGHT, 1834, *ex* CUVIER, vernacular
(ICZN Opinion 781, 1966)

Tentacles and rhinophores long, with ringlike lamellae; front edge of foot pointed; radula with lateral teeth finely denticulate.

2378. Flabellina telja Marcus & Marcus, 1967. Ground color bluish, varying from individual to individual in intensity. The tips of the rhinophores, cerata, foot corners, and long head tentacles for their distal two-thirds are, by contrast, white, as is also a crest on the tail. Orange-brown to red digestive gland diverticula extend into the cerata. Length, 11 mm; width, 3 mm (preserved specimens). Puerto Peñasco, Sonora, Mexico, in rocky intertidal areas; rare.

Genus CORYPHELLA M. E. GRAY, 1850 (ICZN Opinion 781, 1966)

Cerata long, grouped into bundles; radula with small lateral teeth.

2379. Coryphella californica Bergh, 1904. The tentaculiform foot corners are longer than the rhinophores or head tentacles; the radula of the type specimen had fifteen rows of teeth. Little is known about this species, the type lot of which was dredged in the southern part of the Gulf of California.

2380. Coryphella cynara Marcus & Marcus, 1967. The animal is more or less translucent, with a pale to bright blue cast, the foot edged by a blue line. A white line between the rhinophores arches outward to the head tentacle bases, and the tail crest and tips of the foot angles also are white. The orange-brown cerata are ringed subapically with red, the tip being white. Violet patches occur near the rhinophore tips and on the foot angles and head tentacles. This is an accomplished swimmer, achieving a kind of "breast stroke" by rhythmically extending, pulling, and collapsing the cerata. Length, 42 mm. Northern part of the Gulf of California, intertidally on sand (type lot) and by dredging (off Puertecitos, with an otter trawl).

2381. Coryphella trilineata O'Donoghue, 1921 (Synonyms: **C. piunca** Ernst Marcus, 1961; **C. fisheri** MacFarland, 1966). The translucent white body has pale orange to deep red cerata cores and orange markings on the distal half of the rhinophores and head tentacles. Three opaque white lines (rarely interrupted) run from end to end, one medially and one on either side of the body; they unite in the tail region. The rhinophores are distinctly annulate, not smooth, knobbed, wrinkled, or perfoliate. Length, 41 mm. Vancouver Island, Canada, to Ensenada, Baja California, especially common on offshore kelp canopies that have stands of campanularian hydroids on which the animal feeds; rare in the northern part of the Gulf of California.

Genus FLABELLINOPSIS MACFARLAND, 1966

Tentacles long, rhinophores somewhat foliate; foot angles pointed, resembling tentacles; jaws armed with many small nodes; radula triseriate, the laterals wide, denticulate on inner margin.

2382. Flabellinopsis iodinea (Cooper, 1863). The most strikingly colored of the eastern Pacific aeolids, the body intensely violet and the cerata red-orange, with lighter tips. The perfoliate rhinophores are deep red. The body is compressed, and the animal can propel itself through the water, upside down, by bringing head and tail together in an undulating motion. Length, 70 mm; width, 12 mm. Vancouver Island, Canada, to Los Coronados Islands, Baja California, not uncommon subtidally to 30 m, and individuals may wash in at a low tide; one specimen reported at Puerto Peñasco, Sonora, Mexico.

See Color Plate XXII.

Infraorder ACLEIOPROCTA

Anus subdorsal in position.

Family EUBRANCHIDAE

Body elongate, arched above, with the anterior end of the foot rounded; tentacles long, slender, smooth; dorsum with rows of processes.

Genus CAPELLINIA TRINCHESE, 1874

2383. Capellinia rustya Marcus, 1961 (Synonym: **Eubranchus occidentalis** MacFarland, 1966). A small, inconspicuous form with a translucent body, often with a yellowish or greenish cast, lightly spotted and flecked with olive-green or brown and with chrome-yellow dots on the foot and body sides. The cerata cores are varying shades of green or brown. The head tentacles and smooth rhinophores usually show a subapical ring of metallic copper-green. Distinguished from other eastern Pacific aeolids by having two or three circular swellings on each ceras. Length, 12 mm. San Francisco Bay, California, to Punta Abreojos, Baja California; locally common on offshore kelp canopies that have campanularian hydroid colonies; common in spring on *Sargassum* beds, Bahía de los Angeles, Gulf of California.

Family FIONIDAE

Margins of foot broad, extending beyond body sides; tentacles and rhinophores long, smooth, tapering, divergent, erect.

Genus FIONA FORBES & HANLEY, 1851

Cerata numerous, with lateral expansions and cnidosacs for storage of nematocysts.

2384. Fiona pinnata (Eschscholtz in Rathke, 1831). The yellowish, reddish, or purple body results from the color of the prey. Cerata are densely situated along the margins of the notum, leaving a large part of the back clear. Each of the larger cerata has a conspicuous longitudinal membrane along its posterolateral axis, containing an undulating blood vessel. Length, 39 mm. Cosmopolitan, pelagic, on floating seaweed, reported from about 500 miles west of San Francisco, California, south at least to latitude 11° N; seasonally common, from mid-March to late May, large numbers occasionally washing ashore on blades of the algae.

Infraorder CLEIOPROCTA

Anus dorsal, on right side, behind or among a group of cerata.

Superfamily AEOLIDIACEA
(EOLIDIACEA, EOLIDACEA, unjustified emendations; CLADOHEPATICA, in part)

Rhinophores simple, without a sheath; head with cephalic tentacles, no frontal veil; cerata with a branch of the midgut and a terminal sac for storage of nematocysts from coelenterates.

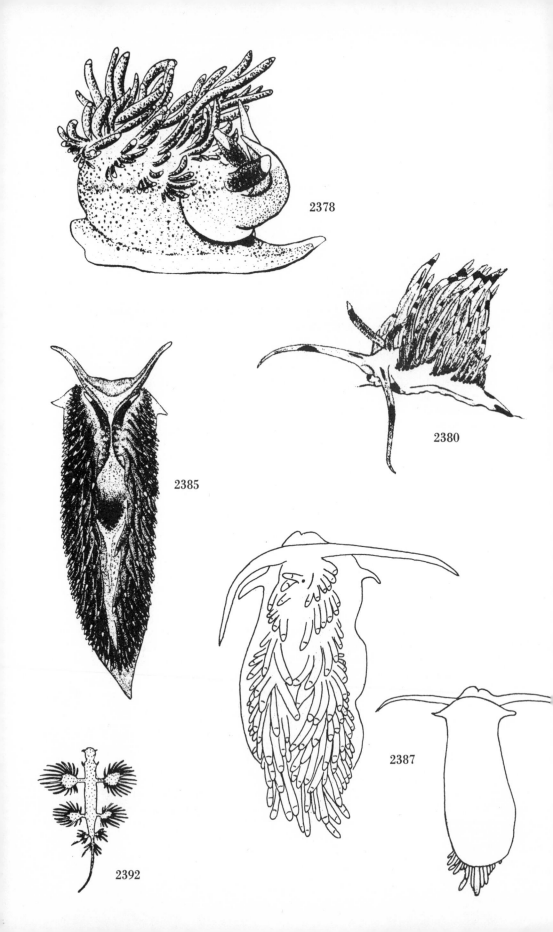

2378

2380

2385

2387

2392

Family AEOLIDIIDAE
(EOLIDIIDAE, unjustified emendation)

Cerata present in two or more rows; radula uniseriate, teeth comblike. Four Panamic genera.

Genus AEOLIDIA CUVIER, 1797 (ICZN Opinion 779, 1966)

Body somewhat flattened, cerata small, numerous; radula with arched and finely serrate teeth.

2385. Aeolidia papillosa (Linnaeus, 1761). A cosmopolitan species, recorded to depths of 750 m. Not yet positively identified in the Panamic area.

Genus BERGHIA TRINCHESE, 1877

Rhinophores finely pustulose.

2386. Berghia amakusana (Baba, 1937). A distinctively marked form, with a pale yellowish-brown body and cerata, the latter with distal rings of blue and yellow. A conspicuous dark spot surrounded by a yellowish halo occurs on the head near the cephalic tentacles. Length, 20 mm. A single specimen of this otherwise Japanese species was collected near Puertecitos, northern end of the Gulf of California by Farmer in 1966.
See Color Plate XXII.

Genus CERBERILLA BERGH, 1873

Rhinophores leafy; cerata large, numerous; radular teeth with secondary cusps.

2387. Cerberilla pungoarena Collier & Farmer, 1964. A broad-footed aeolid, this is translucent, with pale brown on the dorsum and opaque white tips on the cerata, which are compressed, elongate, translucent, carrying a slender, speckled thread of digestive gland diverticula through their whole length. The rather short perfoliate rhinophores have the upper half white, the lower part or base tan. Length, 20 mm; width, 7 mm. Gulf of California, from Puerto Refugio, north end of Isla Angel de la Guarda, where a single specimen of this burrowing aeolid was collected on the surface of sandy mud.

Genus SPURILLA BERGH, 1864

Rhinophores thick, with several leaves.

Spurilla alba (Risbec, 1928). This species, which otherwise is Indo-Pacific in distribution, ranging from East Africa to northeastern Australia, has been reported at Punta Mita, Nayarit, Mexico, by Sphon (1971).

2388. Spurilla chromosoma Cockerell & Eliot, 1905. The body is pale yellowish orange, with dark brownish cerata cores and irregular opaque white spots along the midline of the back. Patches of similar white occur on the cerata and are sprinkled over the head tentacles and rhinophores, which are long and slender and have a few shallow perfoliations running nearly parallel to the axis of the rhinophore. Length, 29 mm. Santa Barbara County, California, to Ensenada, Baja California, uncommon intertidally and on boat landings; in the Gulf of California this is the commonest aeolid, from San Felipe south to Tenacatita, Jalisco, Mexico.
See Color Plate XXII.

Family FACELINIDAE

Body long, slender, tentacles and rhinophores long, anterior end of foot angular, cerata simple, arranged in transverse rows; anus more anterior than in the Aeolidiidae, in the second right group of cerata; radula uniserial, with a single row of teeth. Two Panamic genera.

Genus HERMISSENDA BERGH, 1879

Tentacles long, rhinophores leaflike; foot angles elongate; edges of jaw plates finely toothed; central radular tooth large, finely cusped at its point.

2389. Hermissenda crassicornis (Eschscholtz in Rathke, 1831). The translucent white animal has a color pattern variable in detail, with orange, red, brown, black, or green cerata cores; invariably there is an orange stripe that begins anterior to the rhinophores, passes between them, widens immediately behind to form an elongate diamond between the first cerata groups. Length, 61 mm. Sitka, Alaska, to Baja California, the commonest aeolid in the northeastern Pacific; it is not uncommon in the northern part of the Gulf of California, rarer to the south but breeding throughout the year, at least in the southern part of its range.

See Color Plate XXII.

Genus PHIDIANA GRAY, 1850

Foot rounded in front; edges of jaw plates only slightly denticulate; radular plate with a few denticles on sides.

2390. Phidiana lynceus Bergh, 1867. The translucent gray ground color is relieved by orange on the head and a vermilion line that joins the two oral tentacles, terminating about halfway up their length as a spot. Rhinophores and oral tentacles are tipped with yellow-brown, a color pattern that distinguishes this from *P. pugnax*, as does also the smaller size. Deale (Fort Kobbe) Beach, Pacific side, Panama Canal Zone; also in the Caribbean and south to Brazil.

2391. Phidiana pugnax Lance, 1961 (Synonym: **P. nigra** MacFarland, 1966). One of the largest of the eastern Pacific aeolids, this animal is a translucent or yellowish white, with cerata cores brownish to black, and often with pink coloration on the distal one-fourth of the larger inner cerata. The robust and markedly perfoliate rhinophores are red at the base, pale yellow on the upper half. Length, 64 mm; width, 10 mm. Monterey Bay, California, to Puerto Rompiente, Baja California; uncommon both intertidally and subtidally.

Family GLAUCIDAE

Body with three pairs of lateral lobes bearing the tegumentary papillae.

Genus GLAUCUS FORSTER, 1777

Foot very narrow; animals pelagic in habitat.

2392. Glaucus atlanticus Forster, 1777 (Synonym: **Glaucilla marginata** Bergh, 1868). Bizarre in shape, intensely blue and silver in color, the animal inhabits the high seas, where it feeds on such siphonophores as *Velella* and *Porpita*. Length, 30 mm. Circumtropical, in all the warmer seas; in the eastern Pacific, numerous dip net and tow net collections have been made off West Central America, from latitude 24° N, south to latitude 8° S, by the Scripps Institution of Oceanography.

Order GYMNOPHILA
(SOLEOLIFERA of authors)

The oval to elongate body is covered by the mantle on the back and sides (the notum) and ventrally on either side of the foot (the hyponotum). A shell is present only in the embryonic stages, and there is no vestigial internal shell as there is in pulmonate slugs. There is, however, a lung at the posterior end of the body, opening in the midline, evidently not homologous with the lung of the pulmonates. The Gymnophila include not only the intertidal marine Onchidiidae but also two families of terrestrial slugs that occur in the humid tropics. The group was placed in Pulmonata by earlier workers because of the presence of the lung; modern research, however, has indicated that the closer relationships are with the Opisthobranchia.

Much of the text for this review of Panamic Gymnophila has been prepared by Dr. Dwight Taylor, whose assistance is gratefully acknowledged.

Superfamily ONCHIDIACEA

Animals oval in outline, with a dorsally arched notum. At the anterior end two contractile tentacles bearing eyes at their tips protrude from beneath the notum when the animal is active. The notum is characteristically covered with warts and papillae, and at the side (perinotum) are glands secreting a noxious fluid.

Family ONCHIDIIDAE

With the characteristics of the superfamily. Intertidal mollusks that do not necessarily retreat beneath rocks or other cover when exposed to the sun at low tide. Two Panamic genera.

Genus HOFFMANNOLA STRAND, 1932
(WATSONIELLA HOFFMANN, 1928 [not BERG, 1898])

Notum nearly smooth, without conspicuous papillae or large warts, the animals much larger than in the related genus, *Onchidella*—nearly two inches in length. The male reproductive opening is in the midline, hidden beneath the sensory veil, on the back of the head.

2393. Hoffmannola hansi Marcus & Marcus, 1967. This dark brown to black slug has lighter blotches around the low flat warts of the dorsal surface and lighter color below. Length, about 50 mm—the largest member of the family in the Gulf of California area. Kino Bay, Sonora, Mexico (type locality) to Angel de la Guarda Island, Gulf of California.

2394. Hoffmannola lesliei (Stearns, 1892). Distinguishable from *H. hansi* only by dissection; the notal glands are much smaller and more numerous, the radula has fewer teeth and fewer rows, and the kidney-lung complex is smaller. Length, averaging 40 mm. Galápagos Islands.

Genus ONCHIDELLA J. E. GRAY in M. E. GRAY, 1850
(ONCHIDIELLA, ONCIDELLA, ONCIDIELLA of authors,
unjustified emendations)

Notum with numerous projecting warts and papillae. The male reproductive opening is behind the right tentacle. Animals attain a length of about an inch and resemble small chitons, lacking, of course, any shell valves under the mantle.

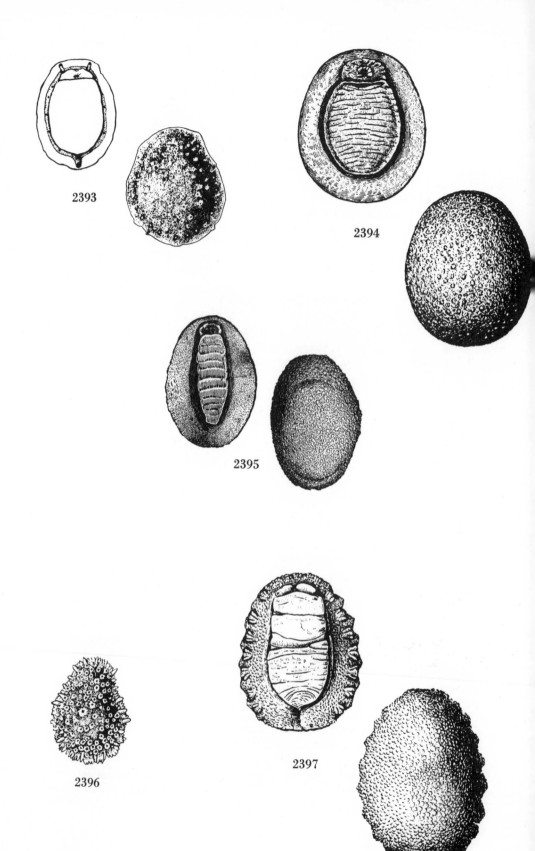

2393

2394

2395

2396

2397

2395. Onchidella binneyi Stearns, 1893. This dark gray slug is the common species of the Gulf of California, occurring in the intertidal zone, usually beneath rocks or in crevices by day but actively browsing on algae at night. In the ventral view one sees the sole of the foot as being about one-third the width of the animal, relatively wider than in *O. hildae*. Other differences between the two require dissection and study under a microscope. Length, to 30 mm. Las Cruces Bay, Gulf of California, to Puerto Peñasco, Sonora, Mexico.

2396. Onchidella hildae (Hoffmann, 1928). One obvious external difference between this and *O. binneyi* is in the width of the sole, which is narrower, only about one-fourth of the body width. The color of the warty surface may be more brownish, with a greenish-white ground color, the underside light. Length, about 25 mm. Head of the Gulf of California to Ecuador.

2397. Onchidella steindachneri (Semper, 1882). The color is dark gray to smoky black above, whitish below. The edge of the hyponotum is fringed, with fairly regular three-leafed processes. Length, about 20 mm. Galápagos Islands. This form is more like the Caribbean *O. floridana* (Dall, 1885) than any other of the eastern Pacific species.

Subclass PULMONATA

The Pulmonata, showing some relationship to both the Prosobranchia and the Opisthobranchia, have diverged by loss of the ctenidium and by the development of a functional lung in a part of the mantle cavity. Detorsion has taken place to some extent, though not completely so; the nerves, however, are never crossed. Shells are simple in construction, the aperture being entire in those that are helically coiled. The operculum is absent in most. Some forms are limpet-like. In land slugs, the shell is reduced to a vestigial plate buried in the mantle. Most of the pulmonates have moved up out of the sea, into rivers, marshes, or lakes—even onto land. Marine pulmonates (the group to be reviewed here) have remained in or near the sea; some retain the primitive pelagic stages in their life histories. Two orders of pulmonates are distinguishable by the eye position. Marine and freshwater forms have retained the basal prosobranch location of the eyes. Most land-dwelling forms—the Stylommatophora—have the eyes elevated, inside the tips of inversible cephalic tentacles.

Dr. J. P. E. Morrison of the United States National Museum, long a specialist on marine pulmonates, has supplied data on the nomenclature, classification, synonymies, and distribution of the eastern Pacific species; his help is here gratefully acknowledged.

Order BASOMMATOPHORA

Eyes at the bases of the cephalic tentacles.

Superfamily MELAMPACEA
(ELLOBIACEA)

Shell coiled; aperture entire, with plaits or other constrictions of the inner and outer lips; no operculum.

Family MELAMPIDAE
(ELLOBIIDAE of authors)

Shells ovate to conical, the spire with several whorls; not markedly thin; aperture with folds or denticles on one or both lips. The family name Ellobiidae has been commonly used, but, as Morrison (1964) has shown, Melampidae was used by Stimpson in 1851, four years before the Adams brothers introduced Ellobiidae. The three Panamic subfamilies are given here in systematic rather than alphabetical sequence, and the key is to genera and subgenera.

1. Aperture long and narrow... 2
   Aperture ovate to rounded, flaring below............................ 6
2. Outer lip smooth within.. 3
   Outer lip with teeth or heavy spiral ridges......................... 4
3. Outer lip thickened in the middle...........................*Sarnia*
   Outer lip thin or even-edged................................*Tralia*
4. Outer lip with a few teeth................................*Detracia*
   Outer lip with interrupted ribs or lirae.......................... 5
5. Periostracum evenly distributed over shell.............*Melampus, s. s.*
   Periostracum with one or more rows of setae on spire, leaving a scar pattern of small pits..*Pira*
6. Inner lip with only one to two folds.......................*Ellobium*
   Inner lip with more than two folds............................. 7
7. Surface of shell spirally sculptured..........................*Pedipes*
   Surface of shell smooth or nearly so........................... 8
8. Spire subovate; inner partitions of spire resorbed..............*Marinula*
   Spire pointed; inner whorls not resorbed......................*Phytia*

Subfamily MELAMPINAE

Anterior and posterior portions of the foot marked off by a transverse groove; radula modified for an herbivorous diet. Three Panamic genera.

Genus MELAMPUS MONTFORT, 1810

Obconic to ovate, with a bluntly tapering spire, the shell surface nearly smooth. Columellar margin with one or more folds, the outer lip with strong lirations or ribs ending abruptly inside the edge. The habitat preferred by species of this genus is in the supralittoral zone, at a level reached by the highest tides only a few times a year. Spawning occurs during these brief intervals. When seawater reaches the eggs, which are in pillow-shaped masses of soft, translucent jelly, they hatch, and the veliger larvae are released, to begin the pelagic phase of their life history in salt or brackish water. Two Panamic subgenera.

Subgenus MELAMPUS, s. s.

Periostracum evenly distributed and smooth over entire shell.

2398. Melampus (Melampus) carolianus (Lesson, 1842) (Synonyms: **Auricula ambigua** Lesson, 1842; **A. piriformis** Petit, 1843; **A. maura** Küster, 1844, *ex* Mühlfeld, MS; **A. trilineata** C. B. Adams, 1852; **A. acromelas** Troschel, 1852). The shell is dark reddish brown, with three nearly white spiral bands and a spot of rusty brown on the columella. Length, 18 mm; diameter, 11 mm. Costa Rica to Ecuador and the Galápagos Islands, in mangrove swamps. A similar species in the Atlantic is *M. (M.) coffea* (Linnaeus, 1758), type of the genus.

2399. Melampus (Melampus) mousleyi Berry, 1964. Thin, narrowly pear-shaped, white or light brown, the apex darker, spirally marked with up to five brown bands, shell surface with shallow punctate striations. Length, 10 mm; diameter, 5.4 mm. Bahía San Luis Gonzaga to Bahía de Adair (type locality),

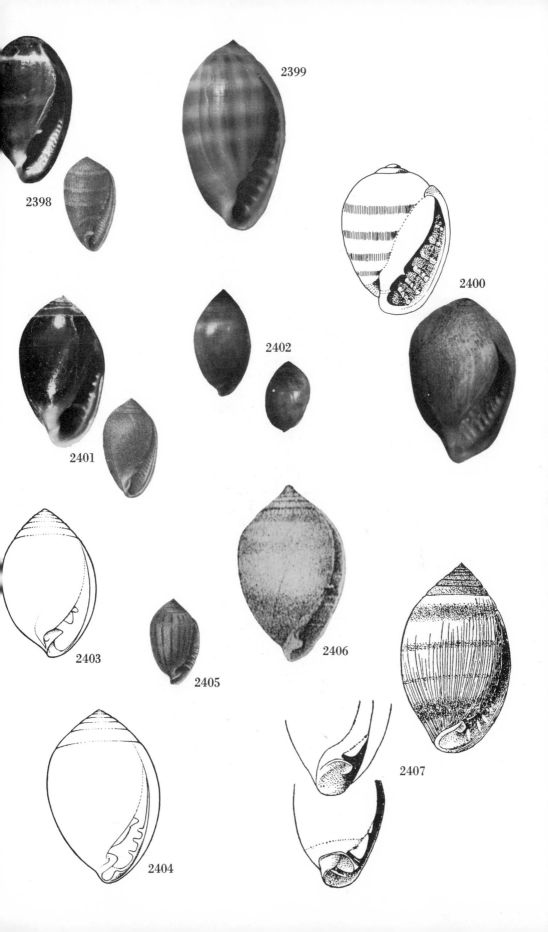

2398

2399

2400

2401

2402

2403

2404

2405

2406

2407

near the head of the Gulf of California. Morrison (*in litt.*) considers this a situs or ecologic form of *M. olivaceus*, dwarfed by life in exceptionally arid conditions. Other students regard it as a distinct species. More study is needed.

2400. Melampus (Melampus) olivaceus Carpenter, 1857 (Synonym: **? M. o. californianus** Berry, 1964). The shell is purplish brown, banded with spiral lines of lighter color under an olive-green periostracum. The outer lip has numerous lirae within that end in denticles. Columellar folds well developed. Length, 16.5 mm; diameter, 9.7 mm. Southern California southward along the Baja California coast and to the head of the Gulf of California and Mazatlán (type locality), Mexico, in bays at high-tide line, among matted saltgrass and drift. On the basis of smaller size and sharper color bands, Berry separated a northern subspecies, *M. o. californianus*, with range from southern California to the Cedros Island area, Baja California. The form from southern Baja California and the Gulf of California would thus be *M. o. olivaceus*. However, Morrison (*in litt.*) doubts that the two forms are geographically separable. More work, obviously, is needed on the *Melampus* variants.

Subgenus PIRA H. & A. ADAMS, 1855

Periostracum of the spire with a central row of bristles or setae, which leave scar pits on the midline of the spire whorls.

2401. Melampus (Pira) tabogensis C. B. Adams, 1852 (Synonym: **M. bocoronicus** Mörch, 1860). The shell is dark brownish red, with rusty brown on the columella, white on the folds and interior callus, the surface smooth and shiny except for the scar pits on the spire. Barra de Navidad, Jalisco, Mexico, to Panama (type locality) and Cocos and Galápagos Islands, on cobble beaches, under cobbles, drift, and logs at high-tide line.

Genus DETRACIA GRAY in TURTON, 1840

Aperture narrowed, not flaring anteriorly; inside of outer lip with a few teeth. The type of the genus is a West Indian species, *D. bullaeoides* (Montagu, 1808). Resembling *Melampus* except for the anterior narrowing of the aperture, *Detracia* shares also the same spawning habits and pelagic larval stages.

2402. Detracia globulus (Orbigny, 1837). The specific name means "little ball." There are, however, other species that are more spherical. In *D. globulus* the parietal callus may be weak to nearly wanting; the more anterior of the two inner lip lamellae comes nearer to bridging the anterior end of the aperture than in any other of the species. The smooth buff shell has brown color bands. Length, 9 mm; diameter, 4.5 mm. Panama and the Perlas Islands, southward to Guayaquil, Ecuador (type locality).

2403. Detracia graminea Morrison, 1946. Small, ovate, light horn-colored, banded with dark reddish brown, the outer lip with a few minute threads inside, the columella with one strong fold. Height, 5.3 mm; diameter, 3.3 mm. Panama and Perlas Islands, Panama Bay.

2404. Detracia joseana Morrison, 1946. The color is reddish brown, with spiral bands of lighter color. A few strong lirae mark the outer lip. There is a strong columellar fold and one or two low folds on the inner lip. Length, 6.2 mm; diameter, 4 mm. Panama and Perlas Islands, Panama Bay (type locality, San José Island, in a river channel above a mangrove swamp).

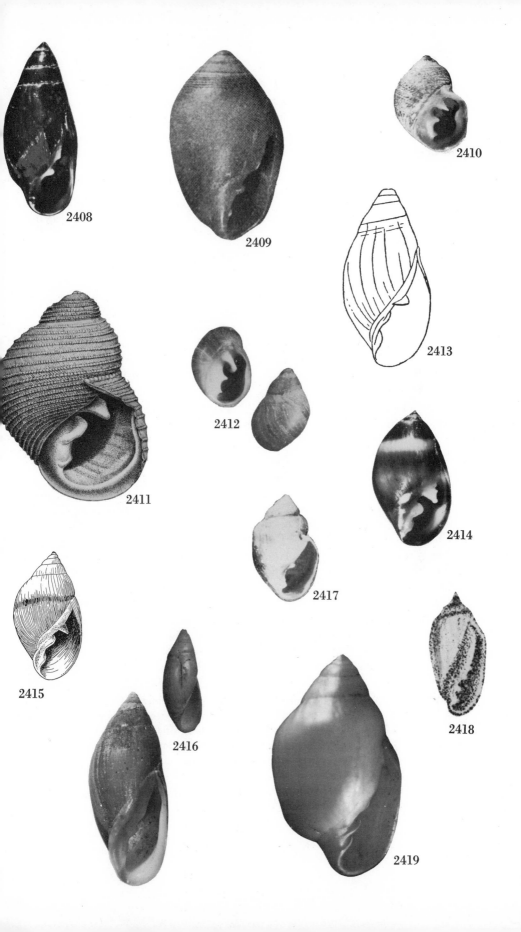

2408

2409

2410

2411

2412

2413

2414

2415

2416

2417

2418

2419

2405. Detracia strigosa (Von Martens, 1900). Adults are nearly globular, light in color, with striped markings. Length, 7.5 mm; diameter, 5 mm. Type locality, Rio Coto, Gulf of Dulce, Costa Rica.

2406. Detracia wolfii (Miller, 1879). Grayish olive, with two obscure spiral bands; columella with one fold; outer lip showing four white foldlike denticles. Length, 5 mm; diameter, 3 mm. Guayaquil, Ecuador, under drift.

2407. Detracia zeteki Pilsbry, 1920. This form varies in outline from elliptical to subglobular and in the development of the columellar fold. There is a heavy parietal callosity. Length, 8.6 mm; diameter, 5 mm. Under drift and near mangrove swamps, Panama City area, Panama, and San José Island, Perlas Islands, Panama Bay.

Genus TRALIA GRAY in TURTON, 1840

Slender, spire pointed, outer lip smooth within; columella with three folds. The type of the genus is *T. ovula* (Bruguière, 1792), a Caribbean species.

2408. Tralia panamensis (C. B. Adams, 1852) (Synonym: **Melampus bridgesii** Carpenter, 1856). Blackish red, with white columellar folds, the shell is smooth and shiny. Length, 11 mm; diameter, 6 mm. Costa Rica and Panama City (type locality) and the Perlas Islands, under stones at the high-tide line or crawling about on wet stones.

2409. Tralia vanderbilti Schwengel, 1938. Shorter and broader than *T. panamensis*, this is cinnamon-buff, darker near the suture, the spire with some engraved spiral lines. Length, 8.5 mm; diameter, 5.7 mm. Galápagos Islands, Ecuador.

Subfamily PEDIPEDINAE

Foot divided, as in Melampinae, by a transverse groove; radula adapted for an omnivorous diet. Two Panamic genera.

Genus PEDIPES FÉRUSSAC, 1821

Ovate, spire pointed, aperture large, with three heavy columellar folds, the uppermost the largest; surface of shell spirally striate. The animal lifts the front and rear halves of the foot alternately, progressing with a looping motion somewhat like that of geometrid caterpillars. The generic name *"Pedipes"* refers to this process, rather literally meaning "foot by foot."

2410. Pedipes angulatus C. B. Adams, 1852. The brownish-red shell is white within, its crowded spiral sculpture much finer on the upper part of the whorls. Length, 6.5 mm; diameter, 4.9 mm. Panama to the Galápagos Islands, under stones at the high-tide line.

2411. Pedipes liratus Binney, 1860. The straw-colored shell is more evenly spirally ribbed than the others, with a more rounded outline. Length, 3.5 mm; diameter, 2.5 mm. Scammon's Lagoon to the Gulf of California. Clench (1967) regarded this as a synonym of *P. angulatus*, but the morphological as well as distributional differences seem distinctive. It seems not to range south of Sonora on the mainland of Mexico.

2412. Pedipes unisulcatus Cooper, 1866 (Synonym: **P. biangulatus** Jaeckel, 1927). Amber-brown in color, the shell has a white columellar area and a pur-

2

4

ATE XI. 1 & 2. *Harpa crenata* Swainson (1357). 3 & 4. *Strombus peruvianus* Swainson (610).

PLATE XII. 1. *Nuculana impar* (Pilsbry & Lowe) (26) ; the serrate organ is the very flexible foot.
2. *Adrana penascoensis* (Lowe) (40) ; the soft parts are visible through the transparent shell.
3. *Laevicardium elatum* (Sowerby) (378). 4. *Trachycardium panamense* (Sowerby) (363),
with a lone *Megapitaria squalida* (Sowerby) (425). 5. *Lima pacifica* Orbigny (219) ; the elegant
fringing is on the mantle margin. 6. *Lima* sp.; possibly *L. hemphilli* Hertlein & Strong (217).
7. *Lima orbignyi* Lamy (220). 8. *Diodora inaequalis* (Sowerby) (18) ; a gastropod.

PLATE XIII. 1. *Tellina inaequistriata* Donovan (529); the siphons are quite long. 2. *Donax gracilis* Hanley (591). 3. *Tellina simulans* C. B. Adams (535); the foot and one small siphon protrude. 4. *Pitar helenae* Olsson (401); note the finely fringed mantle margin, fused siphons, and triangular foot. 5. *Semele jovis* (Reeve) (633). 6. *Semele pacifica* Dall (637). 7. *Ancistromesus mexicanus* (Broderip & Sowerby) (44); the anterior end is at the right. 8. *Calliostoma marshalli* Lowe (85); the head is emerging at the right of the upturned animal.

PLATE XIV. 1. *Niso splendidula* (Sowerby) (1752); the epithet is well chosen. 2. *Calyptraea spirata* (Forbes) (803). 3. *Natica chemnitzii* Pfeiffer (861). 4. *Polinices bifasciatus* (Griffith & Pidgeon) (873); note the broad foot, and that the soft parts partially envelop the shell; the head is withdrawn, only the tentacles showing. 5. *Epitonium billeeanum* (DuShane & Bratcher) (612); the golden-yellow proboscis is extended. 6. *Natica chemnitzii* Pfeiffer (861); the most color-variable of the Pacific Moon Snails (compare Fig. 3, above).

ATE XV. 1. *Cypraea annettae* Dall (933); the fringed mantle is characteristic of the Cypraeas.
itiscania limacina (Bergh) (172); a shell-less prosobranch related to the Nerites. 3. *Ficus
tricosa* Sowerby (952); the foot is expanded. 4. *Carinaria japonica* Okutani (851); this is
elagic species; note the swim fin at the middle left and the small shell at the right.
Cypraea albuginosa Gray (919); the White-Spotted Cowry, with freshly deposited egg mass.
Cypraea arabicula (Lamarck) (927); the Little Arabian Cowry, with egg mass.

PLATE XVI. 1. *Cassis centiquadrata* (Valenciennes) (948) ; the siphon extends upward at the right. 2. *Lamellaria sharonae* Willett (898) ; the animal's soft parts cover and obscure the almost transparent shell; tentacle and siphon emerge at the left; on eel grass. 3. *Trivia pacifica* (Sowerby) (903) ; the extended mantle partially envelops the heavily ribbed shell; lower tidal, not common. 4. *Cassis coarctata* Sowerby (947) ; the tentacles are partially extended. 5 & 6. *Casmaria vibexmexicana* (Stearns) (949) ; an elegant and very rare mollusk.

PLATE XVII. 1. *Oliva polpasta* Duclos (1363). 2. *Oliva julieta* Duclos (1361); an uncommon Olive.
Olivella dama (Wood) (1377); note the mantle filament at the anterior end of the aperture,
rmally carried in a groove or channel along the suture line. 4. *Olivella zanoeta* (Duclos) (1396).
& 6. *Oliva incrassata* [Lightfoot] (1360); these yellowish and grayish forms show the range of
lor variation of the species. 7. *Agaronia testacea* (Lamarck) (1370); an upper-beach animal.
Oliva porphyria (Linnaeus) (1364); see also Plate VIII.

PLATE XVIII. 1. *Jenneria pustulata* [Lightfoot] (940) ; a growth series, comprising two adults
with the mantle extended, partially concealing the pustular-surfaced shell, two juveniles, and
a young adult. 2. *Cyphoma emarginatum* (Sowerby) (935) ; the shell is largely enveloped by the
black-laced mantle. 3. *Nassarius tiarula* (Kiener) (1321) ; the siphons and tentacles are extended.
4. *Cancellaria tessellata* Sowerby (1458) ; rarely seen. 5. *Bulla punctulata* A. Adams (2236) ;
an opisthobranch. 6. *Simnia aequalis* (Sowerby) (936) ; feeding on a white Gorgonian.

PLATE XIX. 1. *Oxynoe panamensis* Pilsbry & Olsson (2312) ; a sacoglossan opisthobranch (all
the animals on this plate are opisthobranchs). 2. *Tylodina fungina* Gabb (2311) ; a notaspidean.
3. *Aplysia parvula* (Mörch) (2297) ; a small anaspidean. 4. *Aplysia californica* Cooper (2296).
5. *Berthellina quadridens* (Mörch) (2309) ; a notaspidean. 6. *Pleurobranchus areolatus*
(Mörch) (2306) ; a notaspidean. 7. *Lobiger souverbii* Fischer (2313) ; a sacoglossan; note the
neuropodial lobes. 8. *Conualevia alba* Collier & Farmer (2325) ; a nudibranch.

PLATE XX. 1. *Hypselodoris californiensis* (Bergh) (2335) ; a nudibranch, as are most of the animals on Plates XX–XXII. 2. *Cadlina evelinae* Ernst Marcus (2336). 3. *Nembrotha eliora* Marcus & Marcus (2354). 4. *Chromodoris norrisi* Farmer (2331). 5. *Chromodoris sedna* (Marcus & Marcus) (2332). 6. *Tridachiella diomedea* Bergh (2320) ; a sacoglossan opisthobranch.

PLATE XXI. 1. *Aegires albopunctatus* MacFarland (2355). 2. *Okenia angelensis* Lance (2358).
Navanax inermis (Cooper) (2254), a cephalaspidean opisthobranch, about to devour the smaller
ila cockerelli MacFarland (2353), a nudibranch. 4. *Antiopella barbarensis* (Cooper) (2376).
Dendrodoris krebsii (Mörch) (2359). 6. *Melibe leonina* (Gould) (2370); on seaweed.

PLATE XXII. 1. *Flabellinopsis iodinea* (Cooper) (2382) ; like *Hermissenda*, one of the most resplendent of the nudibranchs. 2. *Spurilla chromosoma* Cockerell & Eliot (2388). 3. *Armina californica* (Cooper) (2373). 4. *Dirona picta* Eliot *in* Cockerell & Eliot (2377). 5. *Berghia amakusana* (Baba) (2386) ; on coral. 6. *Hermissenda crassicornis* Eschscholtz *in* Rathke (2389).

plish outer lip. On the spire there are four strong grooves, but these die out on the body whorl, the uppermost one sometimes remaining. For the most part, the body whorl is sculptured only with growth striae. The outer lip has a weak tooth, hardly more than a thickening. Length, 8 mm; diameter, 5 mm. Southern California (type locality) to the Gulf of California and the Galápagos Islands.

Genus MARINULA KING, 1832

Columella with three folds, the posteriormost fold strongest. Externally the shell may resemble that of *Pedipes*, but the inner partitions of the spire are resorbed, as in the Melampinae. Three species are recognized here; there are other, as-yet-undescribed, species of *Marinula* on the Pacific coast of Central and South America (Morrison, *in litt.*).

2413. Marinula acuta (Orbigny, 1835) (Synonyms: **Auricula recluziana** Petit, 1843; **Phytia brevispira** Pilsbry, 1920). Ovate, elongate, brown with some darker streaks; upper two columellar folds strong, the lower fold more toothlike; outer lip straight. Length, 12.5 mm; diameter, 5.5 mm. Costa Rica to Guayaquil, Ecuador (type locality).

2414. Marinula concinna (C. B. Adams, 1852). The heavy shell is blackish brown, lighter on the periphery of the last whorl, with a dingy white spiral band below the suture. Length, 8.2 mm; diameter, 5 mm. Panama, under masses of dead leaves in mangrove swamps; only a few have been taken.

Marinula pepita King, 1832 (Synonyms: **Auricula triplicata** Anton, 1839; **A. nigra** Küster, 1841; **A. marinella** Küster, 1843). The type species of the genus, this form is more chunky than most others, with stronger axial striae. The color is brown, with white columellar teeth. Length, 15 mm; diameter, 8 mm. Coquimbo to Chiloé Island (type locality), Chile. Reports of the species at Guayaquil, Ecuador, are in error (Morrison, *in litt.*).

2415. Marinula rhoadsi Pilsbry, 1910. Pale yellowish, the thin shell is dark-banded at the shoulder and suffused with brown below the band. Length, 10 mm; diameter, 5.7 mm. The head of the Gulf of California, near the mouth of the Colorado River.

Subfamily ELLOBIINAE

Foot undivided; shells fairly sturdy, ovate to conical, with several whorls, the aperture with folds or denticles on one or both lips. Three Panamic genera.

Genus ELLOBIUM RÖDING, 1798
(AURICULA LAMARCK, 1799)

Shell elongate, thin, aperture about equal to spire in length, apertural plaits weak.

2416. Ellobium stagnalis (Orbigny, 1835). The shell of this salt-marsh snail is creamy white under a horn-colored periostracum. Sculpture and height of spire vary considerably. Length, about 25 mm. The favored habitat is rotting wood in mangrove swamps, near high-tide line, from El Salvador to Ecuador.

Genus PHYTIA GRAY, 1821

Shell ovate, with pointed spire, aperture about half the length of the shell; columellar margin with a few strong folds. The generic name is an anagram of *Pythia* Röding, 1798; even Gray later confused the two names, and many authors assumed that *Phytia* was a spelling error.

2417. Phytia (?) infrequens (C. B. Adams, 1852). The rather stubby brown shell has a paler but vague band just below the suture. There are two plaits on the columella. Length, 6 mm; width, 4 mm. Panama.

Genus **SARNIA** H. & A. ADAMS, 1855
(**SIONA** H. & A. ADAMS, 1855 [not DUPONCHEL, 1829];
PSEUDOMELAMPUS PALLARY, 1900)

Ovate-cylindrical, the spire obtuse; whorls spirally striate, aperture linear; inner lip with two anterior plaits, the outer lip thickened internally, posteriorly sinuate. Two species are distinguished here; there are one or more undescribed species from the Gulf of California and Galápagos Islands areas.

2418. Sarnia frumentum (Petit, 1842) (Synonym: **Auricula avena** Petit, 1842). For more than a century this species was known only from the original description and an early figure. Length, about 7 mm. Recent collecting proves it to be abundant (Morrison, *in litt.*) near Lima, Peru, and as far south as latitude 26° S in Chile, where it lives alongside *Marinula* under rocks and drift of cobble beaches, contrasting with *Marinula* by its slender undivided foot. The type species of the genus *Sarnia*, very similar in form to the type species of *Pseudomelampus* from the eastern Atlantic.

2419. Sarnia mexicana (Berry, 1964). The small heavy shell has a short spire and only faint traces of spiral striation below the suture. The outer lip is thickened in the middle, and the inner lip has three folds. Length, 4.2 mm; diameter, 2.7 mm; apertural length, 2.8 mm. Type locality, near Puertecitos, northern end of the Gulf of California.

Superfamily **SIPHONARIACEA**

Shell limpet-like, never coiled.

Family **SIPHONARIIDAE**

Cap-shaped, rather irregularly sculptured shells, mostly dark in color, with a siphonal groove on the right side. Muscle scar continuous at the front but open at the right. Air-breathing inhabitants of the intertidal zone, mainly tropical in distribution. Two Panamic genera.

Genus **SIPHONARIA** SOWERBY, 1824

Shells sturdy, asymmetrical, dark-colored, conspicuously sculptured, with a well-marked siphonal groove. Two Panamic subgenera; *Siphonaria, s. s.*, is not represented in the Panamic province.

Subgenus **HETEROSIPHONARIA** HUBENDICK, 1945

Sturdy, strongly sculptured shells, the apex subcentral, not hooked; relatively more symmetrical than *Siphonaria, s. s.*, and other subgenera; siphonal groove ending in a small digitation.

2420. Siphonaria (Heterosiphonaria) aequilorata Reeve, 1856 (Synonym: **S. aequilirata** Carpenter, 1857). The shell is thin, nearly white, with even ribbing, the ribs somewhat nodosely sculptured. Length, 18 mm; width, 12 mm. Southern end of the Gulf of California to Tres Marias Islands.

2421. Siphonaria (Heterosiphonaria) gigas Sowerby, 1825 (Synonyms: **S. angulata** Gray, 1825; **S. costata** Sowerby, 1835; **S. characteristica** Reeve,

1856). Probably the largest-shelled species of the genus anywhere. In its typical development the shell is dark gray exteriorly, with widely spaced, even, radial ribs, dark brown interiorly, the margin scalloped by the ribs. The form *S. characteristica* has lower ribs with inter-ribs of almost the same size, so that it has a much more subdued appearance. Length, 56 mm; width, 46 mm; height, 38 mm. Acapulco, Mexico, to Peru, on rocks. The type of the subgenus.

2422. Siphonaria (Heterosiphonaria) maura Sowerby, 1835 (Synonyms: **S. pica** Sowerby, 1835; **S. lecanium** Philippi, 1846). Apex conic, subcentral, ribs coarse to fine, outline variable, as are height of shell and color markings, which vary from light to dark brown, mottled or spotted or rayed with white. Length, about 22 mm; width, 18 mm; height, 7 to 9 mm. Guaymas, Mexico, to Peru, intertidally.

2423. Siphonaria (Heterosiphonaria) palmata Carpenter, 1857. Not only is the shell low but the apex is peculiarly flattened, not at the summit of a cone. Ribbing is coarse and variable. The shell is somewhat larger than that of *S. maura* and differs from it by being narrowed anteriorly. The ends of the ribs project to make a digitate margin. Length, 29 mm; width, 24 mm; height, 5 mm. Gulf of California to Panama.

Subgenus LIRIOLA DALL, 1870

Shell low to moderately elevated, thin; apex hooked, behind the midline to near the posterior margin; sculpture weak.

2424. Siphonaria (Liriola) brannani Stearns, 1873. The upper side is dark brown with lighter radial ribs, the interior lustrous dark brown. Major diameter, 9 mm. This is characteristically a southern Californian form, reported at Cape San Lucas, Baja California, but not found there in recent years.

Genus WILLIAMIA MONTEROSATO, 1884

Shell small, thin, semitransparent, and mostly horn-colored, symmetrically conical, with the tip of the apex twisted.

2425. Williamia peltoides (Carpenter, 1864) (Synonyms: **Nacella subspiralis** Carpenter, 1864 [first reviser, Dall, 1870]; **N. vernalis** Dall, 1878; **W. galapagana** Dall, 1917 [*nomen nudum* as of Dall, 1909]). The shell is a waxy orange-brown, with translucent rays but no ribs. Length, 10 mm; width, 8 mm; height, 6.5 mm. Intertidally, near the high-tide line, on rocks, southern California through the Gulf of California and south to Panama and the Galápagos Islands. The form *W. galapagana* is reported as occurring on floating seaweed. Whether it can be demonstrated to be distinct remains a question for further research.

Family TRIMUSCULIDAE
(GADINIIDAE)

The button shells have a limpet-like white shell with radiating ribs and a vague siphonal groove on the right side. They are to be found in the upper part of the littoral zone—air breathers living in the sea.

Genus TRIMUSCULUS SCHMIDT, 1818
(GADINIA GRAY, 1824)

The siphon, projecting from the right side, causes a break in the horseshoe-shaped muscle scar that gives the clue these are not true limpets.

2426. Trimusculus peruvianus (Sowerby, 1835). A relatively large, thin, high shell with a recurved apex; edge of shell sharp and outline round or oval. Sculpture is faint, but the siphonal groove is prominent. Length, 28 mm; diameter, 27 mm. Central America to Chile (Gulf of California specimens so labeled are probably of the next species).

2427. Trimusculus reticulatus (Sowerby, 1835). A medium-sized shell, mostly circular in shape; rarely oval. There are 40 to 50 radial ribs and some concentric sculpture. The edge of the shell is thick and faintly crenulate, the siphonal groove faint and only slightly to the right. Length, 25 mm; diameter, 23 mm. Central California to the southern part of the Gulf and south to Acapulco, Mexico, intertidally, on rocks.

2428. Trimusculus stellatus (Sowerby, 1835) (Synonym: **Gadinia pentegoniostoma** Carpenter, 1857). The thin, white shell is rather irregular in shape, varying from round to pentagonal, sculptured with numerous radial ribs but no concentric threads. Length, 15 mm; diameter, 13 mm. The Gulf to Nicaragua.

GASTROPOD DATA RECEIVED TOO LATE FOR INCLUSION IN THE TEXT

A paper by Olsson (1971), noted in the Preface, was in press almost simultaneously with this book. Although Dr. Olsson and Dr. Frederick Bayer, editor, graciously made available what they could of the illustrations and proof, the paper itself came out too late for more than a token inclusion of data. Two new pelecypods, a *Chama* and a *Tellina*, are cited in the pelecypod text. Following are the new gastropod species, notes on range extensions, and taxonomic recommendations.

New generic record and new species:

Genus **PECTINODONTA** DALL, 1882

Shell conic, with a sunken scar at apex and indistinct muscle scars within, surface with irregular concentric ribs and weak radials; radula lacking central teeth.

61a. Pectinodonta gilbertvossi Olsson, 1971. Length, 25 mm. Gulf of Panama, 3,200 m.

Range extension:

74. Turcica admirabilis Berry, 1969 (Synonym: **T. panamensis** Olsson, 1971). Gulf of Panama, 55 to 84 m.

New species:

80a. Calliostoma insignis Olsson, 1971. Height, 15 mm; diameter, 15 mm. Gulf of Panama, 59 to 77 m.

82a. Calliostoma joanneae Olsson, 1971. Height, 8.7 mm; diameter, 7.7 mm. Gulf of Panama, 57 m (type locality); also to Ecuador (Los Angeles County Museum, *teste* J. McLean). Perhaps a synonym of the more northern *C. marshalli.*

88a. Calliostoma pillsburyae Olsson, 1971. Height, 17 mm; diameter, 19 mm. Perlas Islands, Panama Bay, 57 to 64 m.

Range extensions:

90. Calliostoma sanjaimense McLean, 1970. Southward to Gulf of Panama.

92. Calliostoma veleroae McLean, 1970 (Synonym: **C. decipiens** Olsson, 1971). Bathymetric range extended to 68 to 81 m.

New species:

864a. Natica inexpectans Olsson, 1971. Height, 14 to 23 mm; diameter, 16 to 23 mm. Gulf of Panama, 117 to 119 m. See p. 475.

1044a. Trophon (Austrotrophon) panamensis (Olsson, 1971). Height, 32 mm. Gulf of Panama, 84 m. Described as *Austrotrophon panamensis*. See p. 537.

New subgenus of the genus *Cantharus*:

Subgenus **MURICANTHARUS** OLSSON, 1971

Although lacking a differential diagnosis, this is evidently proposed for the group classed on p. 558 as "*Cantharus, s. l.*" The type is no. 1104, *C. (M.) panamicus.*

New species:

1125a. Engina macleani Olsson, 1971. Height, 22 mm. Off Panama, 275 m.

New specific record:

1151. Truncaria filosa (Adams & Reeve, 1848) (Synonym: **Cominella brunneocincta** Dall, 1896). Olsson makes this a prior name for the Panamic species, concluding that the Indo-Pacific locality for the British Museum type lot is incorrect.

New genus in the family Buccinidae:

Genus **BAYERIUS** OLSSON, 1971

Slender, fusiform, with a buccinoid radula. Although described without a differentiating diagnosis, this comprises a deep-water group with only one species, *Fusinus fragilissimus* Dall, 1908.

1098a. Bayerius fragilissimus (Dall, 1908). Gulf of Panama, 3,200 m, to Ecuador (type locality). Erroneously retained under *Fusinus*, p. 617.

Range extension:

1335. Latirus tumens Carpenter, 1856. Southward to Isla la Plata, Ecuador.

New species:

1521a. Terebra argosyia Olsson, 1971. Length, 71 mm; diameter, 15 mm. Isla la Plata, Ecuador.

1722a. Carinodrillia dariena Olsson, 1971. Length, 55 mm; diameter, 10 mm. Panama Bay, 35 m. Synonym of *Compsodrillia alcestis, teste* J. McLean.

Suggested revision of synonymy:

1787. Strombinoturris crockeri Hertlein & Strong, 1951. Olsson considers that *Brachytoma stromboides* (Sowerby, 1832), with a synonym *Pleurotoma sumatrensis* Petit de la Saussaye, 1852, may prove to be an earlier name for this species. Powell (1966) has discussed the evidence, concluding that Sowerby's name applies rather to an Indo-Pacific species.

Range extension:

1857. Mangelia enora Dall, 1908. Northward to Panama Bay, in 3,200 m.

New species:

1864. Phymorhynchus speciosus Olsson, 1971. Height, 40 mm. Gulf of Panama, 3,200 m. Described as a new species in the Buccinidae, although the genus customarily has been considered a member of the Turridae. Olsson argues that some Japanese species found to have buccinoid radulas cast doubt on the allocation to the Turridae. Radular studies of American material are needed.

 # OTHER CLASSES

Class Monoplacophora

The Monoplacophora are gastropod-like mollusks with a single cap-shaped shell (the class name means "bearing one plate"), bilaterally symmetrical or only slightly curved. The soft parts do not show torsion; the ctenidia and muscles are paired. Long considered to be a Paleozoic group, the Monoplacophora were shown in 1957 to have living representatives. Three of the species now known are tropical American; others have been reported in the Red Sea, in the central Pacific, off the Hawaiian Islands (Filatova *et al.*, 1968), and in the Peru-Chile trench. All occur in very deep water, where the animal apparently grazes on microscopic sea-bottom organisms.

Order TRYBLIDIOIDEA
Scars within the shell revealing five to eight pairs of dorsal muscles.

Superfamily TRYBLIDIACEA
With the characters of the order.

Family TRYBLIDIIDAE
Aperture elongate-oval, apex mostly anterior to center.

Subfamily NEOPILININAE
Shell low; with five to six pairs of muscle scars; sculpture of fine radial striae and growth lines, rarely a few undulating concentric ribs.

Genus NEOPILINA LEMCHE, 1957
Apex marginal, slightly projecting beyond anterior margin in most. Two Panamic subgenera.

Subgenus NEOPILINA, *s. s.*
With five pairs of gills.

1. Neopilina (Neopilina) galatheae Lemche, 1957. The apex does not project beyond the anterior margin. Post-oral tentacles are branched. The shell is white and only faintly sculptured. Length, 35 mm; diameter, 30 mm. Cape San Lucas, Baja California, to off Costa Rica (type locality), in 3,590 to 3,718 m. This is the type of the genus and a very significant discovery, because it proves that a class long supposed to be extinct survives in the offshore fauna. A somewhat similar species has since been found in the Arabian Gulf.

2. Neopilina (Neopilina) veleronis Menzies & Layton, 1963. Sculpture of relatively coarse intersecting striae. Apex projecting slightly beyond the anterior margin of the shell. Post-oral tentacles simple. Length, 2.6 mm. Cedros Island, Baja California, 2,730 to 2,770 m.

Subgenus VEMA CLARKE & MENZIES, 1959

Animal with six pairs of gills.

3. Neopilina (Vema) ewingi Clarke & Menzies, 1959. Sculpture of fine intersecting striae. The apex does not project beyond the anterior margin of the shell. Length, 8 mm; width, 5 mm. Off Peru, 5,607 to 6,329 m.

Class Aplacophora

(**AMPHINEURA,** in part)

The name Aplacophora literally means "without plates." The group has been combined with the chitons as Amphineura, but the relationship is not close. Apla-
cophorans, or solenogastres as they are more commonly called, are wormlike mollusks that lack shells. They have a covering integument that often is spicular, and
in common with both the chitons and the monoplacophoran genus *Neopilina*, they
have a ladder-like nervous system. However, in other ways the solenogastres are
highly specialized and unlike other mollusks. They are mainly inhabitants of deep
water, where they work over the bottom sediment or feed on coelenterates. A
single species has been recorded from the Panamic province.

Order **NEOMENIIDA**

With a distinct longitudinal ventral groove; integument spiculose; genital glands
paired; radula, if present, with one to several sharp-cusped teeth.

Family **NEOMENIIDAE**

Body short, truncate at both ends; cloacal orifice transverse; gills present in
cloaca; integument thin; radula often wanting; intestine a straight tube; mouth
with a suctorial buccal bulb.

Genus **ALEXANDROMENIA** HEATH, 1911

Head not clearly differentiated; mouth inconspicuous; cloaca large.

1. **Alexandromenia agassizi** Heath, 1911. From 841 m depth, off Revillagigedo Islands, west of Mexico.

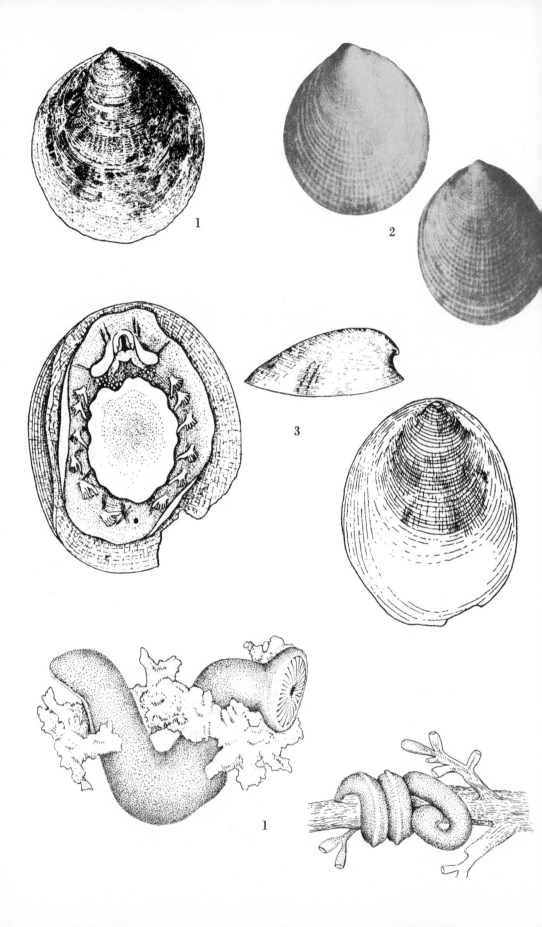

Class Polyplacophora

(**LORICATA**; **AMPHINEURA**, in part)

Popularly called chitons or sea cradles, the Polyplacophora ("having many plates") long were grouped with the Aplacophora ("without plates") as orders in the Class Amphineura ("nerves on both sides"). The modern view is that the two supposed orders are actually distinct enough to be considered separate classes. The Aplacophora occur mainly on muddy sea bottoms, in deep and cold water, whereas the Polyplacophora prefer relatively shallow water, occurring both intertidally and subtidally. They require a firm support for the unarmored ventral side, usually rock.

Except for brief notes on a few endemic Galapagan chitons from a list supplied by Allyn G. Smith, the entire chapter on Polyplacophora is the work of Spencer Thorpe, of El Cerrito, California, long a specialist on chitons. Mr. Thorpe's contribution amounts to a complete revision of the eastern Pacific chitons, which is gratefully acknowledged here. He not only has written the section on morphology and methods of study and the descriptions of species but also has supplied most of the chiton illustrations. In his introductory remarks, which follow, he acknowledges help from other colleagues.

Extensive collections of chitons from throughout the Panamic province by Dr. James McLean were made available for study and dissection through the courtesy of the Los Angeles County Museum, enabling the solution of a number of problems. More than half the photographs presented here are the joint work of Dr. McLean and the Los Angeles County Museum photographic staff. The remainder were made by Perfecto Mary, technician at Stanford University. Other individuals and institutions providing assistance are Dr. Gerald Bakus and Dr. John Garth, Allan Hancock Foundation, University of Southern California; Allyn G. Smith, California Academy of Sciences, San Francisco; Eugene Bergeron of Panama City, Panama, Canal Zone; Colonel George Hanselman of San Diego, California; and Dr. Donald R. Shasky of Redlands, California.

A chiton's shell, unlike that in other mollusks, is made up of eight separate but overlapping plates called valves. The valves are attached on their undersides and outer edges to a tough, fleshy, often armored part of the mantle termed the girdle. The jointed construction of the shell and the pliability of the girdle give chitons a flexibility unusual for so well protected a mollusk. Chitons can mold themselves almost perfectly to irregular hard surfaces, for better attachment, and when dislodged they can curl up in a ball, like a pill bug, to protect their soft parts.

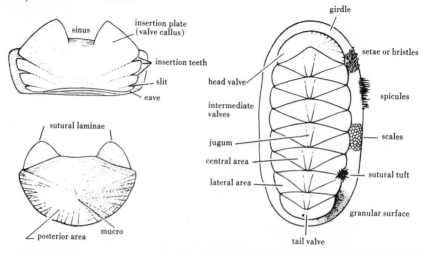

Although chitons have complex shells, the body morphology is simple and re-markably constant throughout the class. They are also a conservative group, for the fossil record shows that the chiton body plan has remained nearly constant for over 400 million years. Not surprisingly, a similar body plan (except for the shell) also occurs in the Monoplacophora, the most ancient molluscan class. When detailed comparisons between the two groups are made, striking similarities in the general morphology of the head, the nervous system, and the radula appear. The Polyplacophora and Monoplacophora are the only mollusks with more than two pairs of gills and with many paired muscle attachments to the shell.

Much of the continued success of chitons may be due to the seemingly trivial, yet unique, occurrence of magnetite as the capping material on the cutting teeth of their radulae. Magnetite is many times harder than the siliceous capping ma-terial that occurs in other molluscan radulae. It would seem to be more than coincidental that chitons normally feed on much stonier material than other mol-lusks do. The diet of many tropical West American chiton species consists of coralline algae or bryozoans or worms with calcareous tubes.

Though chitons occur commonly throughout the Panamic region wherever there is loose rock (but never where there is only sand or mud), they are seldom seen by the casual observer. In part this is because the species that live in exposed places have outstanding protective coloration, but mainly because the majority of species spend their lives on the undersides of rocks. Some chitons can be found either in the splash zone or under rocks in the mid-intertidal zone, whereas a few others can be dredged in depths exceeding 2,000 m. Most, however, will be found in the low intertidal zone—generally in tide pools or drainage channels—or in shallow water to depths of a few dozen meters.

The identification of chitons involves a study of the sculptured valve surfaces, the internal articulations of the valves, and the girdle. The accompanying text figure illustrates the important details of these structures, and gives the special terminology used for them. Although the chiton radula is essential for a general classification of the group and is sometimes a key feature for species identification, it is also very troublesome to prepare and study. Accordingly, its features will be mentioned only once in the following accounts. Dall (1879) gives a detailed account of the chiton radula.

As noted above, chitons will curl up when removed from rocks, and unless the collector quickly straps them onto flat strips of wood or plastic, he is likely to end up with permanently curled specimens. Twill tape is an excellent strapping material. Ethyl alcohol (either pure or denatured) is by far the best killing and preserving agent for chitons. Killing should be done in a 75 per cent solution; it takes eight to twenty-four hours. Final preservation should be in a 95 per cent solution. Individual valves are obtained for study as follows: first remove the foot and gut from a preserved specimen and then soak the carcass in concentrated sodium hydroxide. To prevent unnecessary corrosion of the valves, the specimen should be left in just long enough to soften the girdle, so that the valves can be removed easily. Small animals, like *Leptochiton rugatus*, require only about ten to fifteen minutes, whereas a large *Stenoplax* may require more than an hour of soaking. Sodium hydroxide is a very powerful corrosive substance, and its use requires extreme caution, especially to protect the eyes. Tweezers should be employed for handling valves until they have been thoroughly soaked in several changes of fresh water.

Two terms are used here in a special sense: the *terminal areas* are taken to be the head valve and the radially sculptured posterior part of the tail valve; the *radial areas*, to be the terminal areas plus the lateral areas of the intermediate valves. These terms have been used inconsistently by earlier workers; adoption in the sense recommended here would facilitate reference to the sculptural elements.

Order LEPIDOPLEURIDA

Valves with either unslit insertion plates or no insertion plates at all.

Family LEPIDOPLEURIDAE

Valves without insertion plates.

Genus LEPTOCHITON GRAY, 1847

Radial areas with rows of tiny, close-set, rounded beads, the central areas with similar longitudinal rows. The girdle is minutely scaly.

1. Leptochiton rugatus (Pilsbry, 1892). Small, elongate, roundly arched animals. Color uniformly cream all over, except for the fuchsia-colored foot. The gills, ten or fewer on each side, are restricted to the posterior quarter of the mantle cavity. Adult length, 6 to 9 mm. Puertecitos to Bahía de los Angeles, Baja California, and in the Guaymas region of Sonora, Mexico, under rocks, associated with *Stenoplax*.

Order CHITONIDA

Valves with slit insertion plates.

Family CHITONIDAE

Insertion teeth deeply cut or fluted on the outer and bottom surfaces. Two Panamic genera.

Genus CHITON LINNAEUS, 1767

The word *chiton* (pronounced kite'on) is from the Greek, meaning cloak or mantle, a fitting name for animals whose exposed surfaces are entirely armored. For nearly a century after Linnaeus, the taxon *Chiton* was used as the generic name for all Polyplacophora, but current usage restricts the genus to a few species found only in the tropical and subtropical Western Hemisphere. Its features are large size, very spongy shell substance, large smooth girdle scales, numerous slits on the end valves, and the absence of large eyes on the shell surface.

2. Chiton albolineatus Broderip & Sowerby, 1829. Moderately elongate, medium-sized, the radial areas jet black, with slightly undulating, radiating, white lines. Central areas violet-gray, and the smooth girdle scales uniformly gray-green. Although appearing to be strongly sculptured because of the striking color pattern, the valves are smooth except for a microscopic granulation. Adult length, 25 to 40 mm. Mazatlán to southern Mexico, under rocks and stones.

3. Chiton articulatus Sowerby, 1832 (Synonyms: **C. laevigatus** Sowerby, 1832 [not Fleming, 1813]; **Lophyrus striatosquamosus** Carpenter, 1857). Large, oval; radial areas grayish tan to slate-brown, often with fine, radiating, darker lines. Central areas tan to dull orange, with broad, black longitudinal stripes. The dull shell surface is devoid of sculpturing. Adults are most commonly eroded or overgrown with coralline algae. Adult length, 50 to 100 mm. Southern part of the Gulf of California to Acapulco, on rock surfaces and in crevices.

4. Chiton goodallii Broderip, 1832. Large, quadrate, olive-black. Length, over 100 mm. Galápagos Islands.

5. Chiton stokesii Broderip, 1832 (Synonym: **C. patulus** Sowerby, 1840). Large, oval; color dark gray to dull black-brown, the more intensely colored jugum generally with one or more longitudinal stripes of soiled white on each side. Radial areas with strong, anastomosing ribs, 30 to 40 on the end valves, and six to nine on each lateral area. Central areas with 40 to 50 longitudinal ribs on each side. Girdle with large, somewhat irregularly arranged scales. Adult length, 50 to 100 mm. Southern Mexico to Ecuador, on the tops of exposed rocks.

6. Chiton sulcatus Wood, 1815. Large, valves strongly sculptured. Length, 50 to 100 mm. Galápagos Islands.

7. Chiton virgulatus Sowerby, 1840. Large, relatively elongate; color dark grayish or brownish green. Sculpture similar to that of *C. stokesii*, but with fewer than 35 ribs on the central areas and more on the radial areas. Terminal areas with about 50, each lateral area with eight to twelve ribs. The undulating radial ribs are smooth, not beaded, and anastomose only occasionally. Girdle scales about half as large as in *C. stokesii*. Adult length, 40 to 65 mm. Throughout the Gulf of California to Guaymas and La Paz, under rocks at about midtide level.

Genus TONICIA GRAY, 1847

Head valve with seven to ten slits. Terminal and lateral areas with radiating bands or rows of minute black sensory organs. The fleshy girdle lacks scales.

8. Tonicia arnheimi Dall, 1903. Length, 15 mm; width, 7 mm. Narborough (Fernandina) Island, Galápagos Islands. Unfigured.

9. Tonicia forbesii Carpenter, 1857 (Synonym: **Chiton crenulatus** Broderip, 1832 [not Risso, 1826]). Ground color of shell buff or pinkish tan, painted with fine wrinkly lines or suffusions of blue, pink, white, olive-brown, and deep rust. Central areas with a smooth, narrow, raised longitudinal ridge along the jugum, and a deep furrow on either side. Remainder of the central areas with low, wavy, oblique or longitudinal ridges. Radial areas smoothly knobby, with radiating rows of black or glassy sensory organs covering most of the areas. Adult length, 30 to 50 mm. Mazatlán to Panama, under rocks at lowest tides, and to about 10 m.

1

2

3

7

5

9

4

6

Family ACANTHOCHITONIDAE

Insertion plates of the end valves with few slits, usually five in the head valve. Girdle with a row of about twenty tufts of spines encircling the valves.

Genus ACANTHOCHITONA GRAY, 1821

Head valve with five slits, the tail valve with a median posterior sinus and a slit on either side. Valves two to eight with narrow longitudinal plates, called the jugal tracts, along the valve ridges. The remainder of the shell surface—termed, for valves two through seven, the lateropleural areas—is uniformly sculptured with flat-topped, evenly arranged pustules.

10. Acanthochitona arragonites (Carpenter, 1857). Small, very narrow and elongate; color and color pattern variable, often strikingly beautiful. Although appearing longitudinally granular, the jugal tracts are smooth. Pustules of the lateropleural and terminal areas small, round, or oval, and flat-topped. Girdle covered all over with a downy fine spiculation and, intermixed with the spicules, medium length spines also evenly distributed over the entire girdle. Adult length, 6 to 10 mm. Puerto Peñasco to the Mazatlán region, on the sides of rocks or stones, intertidally to 20 m.

11. Acanthochitona avicula (Carpenter, 1864) (Synonym: ? A. angelica Dall, 1919). A somewhat larger and decidedly broader animal than *A. arragonites*, which it closely resembles. Color usually greenish gray with spots and flecks of black and dark green, rarely entirely black or red-orange. Jugal tracts longitudinally striate at least at the edges. Pustules of the lateropleural areas crowded, relatively large, and teardrop-shaped. The girdle spines, which are decidedly larger than those of *A. arragonites*, occur only on the outer three-fourths of the girdle, with the inner space between the valves and the spine tufts lacking large spines, and hence appearing nude. The disarticulated valves of *A. avicula* are decidedly broader than those of *A. arragonites*. Adult length (in the Gulf of California), 15 to 20 mm. Southern California to southern Mexico, intertidally on stones partly buried in sand, and subtidally to about 20 m.

12. Acanthochitona exquisita (Pilsbry, 1893). A medium-sized, rather thick, rounded form. The girdle is greatly developed, being at least twice as wide on each side as the exposed part of the valves. Color rather uniformly grayish or greenish brown. The jugal tracts weakly longitudinally striate, the pustules of the lateropleural and terminal areas low and oval. Girdle covered with a thick coat of spines, the exuberant and unusually large spine tufts having a striking bronzy cast. Adult length, 25 to 35 mm. Puertecitos to La Paz and the Bahía Kino region of Sonora, Mexico. During daylight hours *A. exquisita* can be found under rocks at the mid-tide level, often associated with *C. virgulatus*, but at night it is found moving around on the tops of rocks.

13. Acanthochitona hirudiniformis (Sowerby, 1832) (Synonyms: **A. hirundiniformis**, unjustified emendation of Gray, 1847; ? **Acanthochiton coquimboensis** and **peruvianus** Leloup, 1941; **Acanthochitona tabogensis** A. G. Smith, 1961). Medium-sized, moderately elongate, rather thick, with the girdle about as wide on each side as the exposed parts of the valves. Valve color, when not eroded or overgrown, dull greenish black. The jugal tracts smooth, the crowded pustules of the lateropleural and terminal areas small, oval, and concave on top. The girdle is tan, greenish brown, or blackish green, with rather inconspicuous

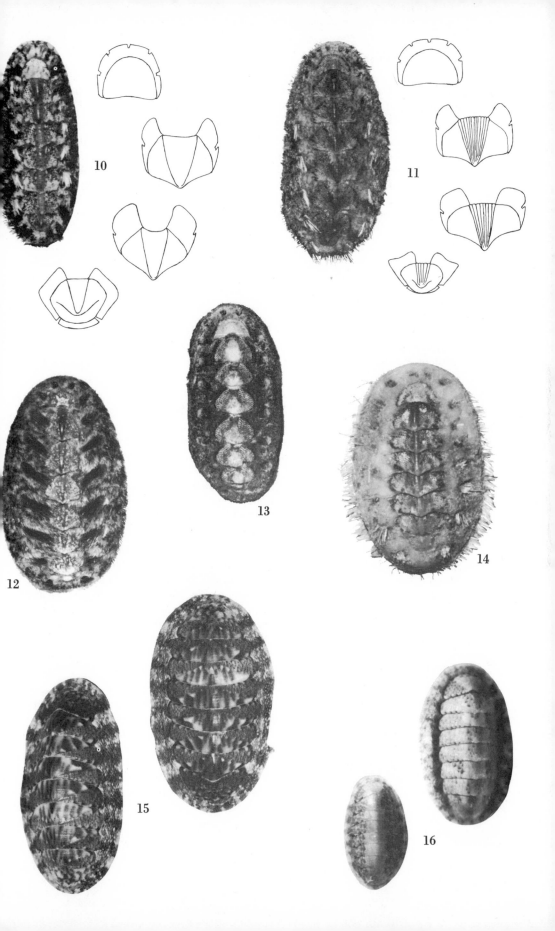

blue-green or black-brown spine tufts. The average specimen of *A. hirudiniformis* is badly worn and singularly unattractive. Adult length, 25 to 35 mm. Panama to Peru (? and Chile), on rock surfaces.

14. Acanthochitona rhodea (Pilsbry, 1893). Small to medium-sized, the white jugal tracts smooth and exceptionally narrow. Lateropleural and terminal areas colored with varying shades of pink or brown and sculptured with clearly separated yet closely spaced small pustules. Individual pustules round or roundly drop-shaped, the top surface of each concave. Girdle downy, with large white spines fringing the outer margin, and relatively small white or light brown spine tufts. Adult length, about 18 to 28 mm. Acapulco, Mexico, to Peru, subtidally at 10 to 15 m.

Family ISCHNOCHITONIDAE

Insertion teeth either smooth or with only shallow grooving. Girdle usually scaly. Three Panamic subfamilies.

Subfamily ISCHNOCHITONINAE

Girdle finely or coarsely scaly; insertion teeth smooth, sinus between the apophyses without a bridging serrate plate. Two Panamic genera.

Genus RADSIELLA PILSBRY, 1892
(RHODOPLAX THIELE in TROSCHEL, 1893)

Small; sculpture usually granular; without radial ribbing.

Most of the species here assigned to *Radsiella* have been cited as *Ischnochiton* Gray, 1847, by American malacologists. However, Thiele (1929) has shown that although there are similarities of appearance between New World and Old World species, the number of subtle but consistent differences makes possible a separation on a geographic basis. His recommendation is here followed, that *Ischnochiton* be reserved for Old World species and that the American species be allocated to *Radsiella*.

15. Radsiella dispar (Sowerby, 1832) (Synonyms: **Chiton picus** and **proprius** Reeve, 1847; **Lophyrus adamsii** Carpenter, 1863; **Ischnochiton aspidaulax** Pilsbry, 1896; **I. ophioderma** Dall, 1908). Small, oval, rather flattened; color variable, usually with gray or gray-brown background and patches and flecks of dirty white and blackish brown. Central areas granular, radial areas with large flattened warts, either arranged in a rough quincunx or amalgamated into zigzag rows. The inner parts of the warts are the most raised. They are tipped with black, giving the radial areas a peculiar scaly aspect. Girdle scales large and round, their fine striae usually worn away. Adult length, 12 to 22 mm. Nicaragua to Ecuador, on stones.

16. Radsiella guatemalensis (Thiele, 1910) (Synonym: **Ischnochiton eucosmius** Dall, 1919). Small, rather elevated, and oval to somewhat elongate and tapered. Color variable, with pink, white, and green predominating, never entirely brown. Sculpture of granules often becoming larger and somewhat irregularly arranged on the radial areas. Girdle scales finely striate and small for the size of the animal (about one-fourth the size of those of *R. tenuisculptus*). Adult length, 7 to 14 mm. Bahía Sebastian Vizcaino and Puerto Peñasco, on stones or shells in sand.

17. Radsiella muscaria (Reeve, 1847) (Synonym: **? Lepidopleurus macandrei** Carpenter, 1857). Medium-sized, elongate-oval, ground color of valves slate-gray or tawny, delicately speckled with lines and spots of brown, slate-gray, and white. Central areas granular; radial areas with rows of beads, often irregular. With about 50 ribs on the head valve, six to ten on each lateral area, and 40 on the tail valve. Girdle with finely striate, white-tipped scales. Adult length, 18 to 27 mm. Mazatlán to southern Mexico, on stones.

18. Radsiella . . . . Thorpe, MS. Minute, stubby, roundly arched; pale cream with brown flecks. Shell surface with a fine, even granulation. Bulbous tail valve the largest of the series, its mucro elevated and in the back half of the valve. Interior of the tail valve with a thick, crescentic, posterior callus, out of which the stumpy insertion teeth arise. Above the callus the valve is cavernously hollowed out. Apophyses triangular, short, and narrow, greatly separated by a sinus at least half the width of the valve. Girdle minutely scaly. Adult length, 4.0 to 4.5 mm. Bahía Kino and Guaymas, Mexico, on stones in sand at lowest tides.

19. Radsiella petaloides (Gould, 1846) (Synonyms: **Ischnochiton mariposa** Dall, 1919; **Stenoplax histrio** Berry, 1945). Small, roundly arched, elongate, color extremely variable in pattern and hue, but usually at least some valves or parts of valves have cyan blue, Indian red, and black spots on a gray, pink, or white background. Central areas polished and longitudinally grooved. Radial areas knobby and either concentrically or irregularly radially furrowed. Adult length, 9 to 21 mm. Bahía Sebastian Vizcaino, Baja California, through the Gulf of California to San Felipe and south to southern Mexico, on the sides of stones and rocks.

20. Radsiella rugulata (Sowerby, 1832) (Synonyms: **Chiton roseus** Sowerby, 1832 [not Blainville, 1825]; **C. pallidulus** Reeve, 1847; **Ischnochiton boogii** Haddon, 1886; **I. aethonus** Dall, 1919; **Stenoplax isoglypta** Berry, 1956). Small, elongate, roundly arched, often elevated; shell color usually rose-red or pink, less often brown. The minutely scaly girdle is tan to brown. All valves are sculptured with concentric ridges, sharply raised at their inner margins and sloping away gradually toward the outer margins. The effect is that of a shingle roof with the shingles mounted the wrong way. The ridges may be closely spaced, minutely plicate, and equally developed on the lateral and central areas, or wider-spaced and undulating or evenly curved. When widely spaced, the ridging is often either weak or completely lacking on the central areas. Adult length, 12 to 20 mm. Bahía de los Angeles to Peru, intertidally (very rare) and to about 20 m. The species also occurs in the western Atlantic.

21. Radsiella tenuisculpta (Carpenter, 1864) (Synonym: **Ischnochiton panamensis** Thiele, 1910). Small, rather flat, oval; color variable, with orange, brown, olive-green, or slate-gray predominating. Pattern either unicolored or variegated, with the background white. Valves granular, with at least a few granules on each radial area slightly larger and raised above the rest, giving these areas a finely warty aspect. Girdle scales rounded, with five to six coarse striae. Adult length, 10 to 15 mm. Panama, on stones.

22. Radsiella tridentata Pilsbry, 1893 (Synonym: **Ischnochiton aethalotus** Dall, 1919). Small to medium-sized, of highly variable color and shape. Green, black, white, and brown predominate in color patterns, which may be unicolored, variegated, or speckled. Central areas finely granular, the granules sometimes

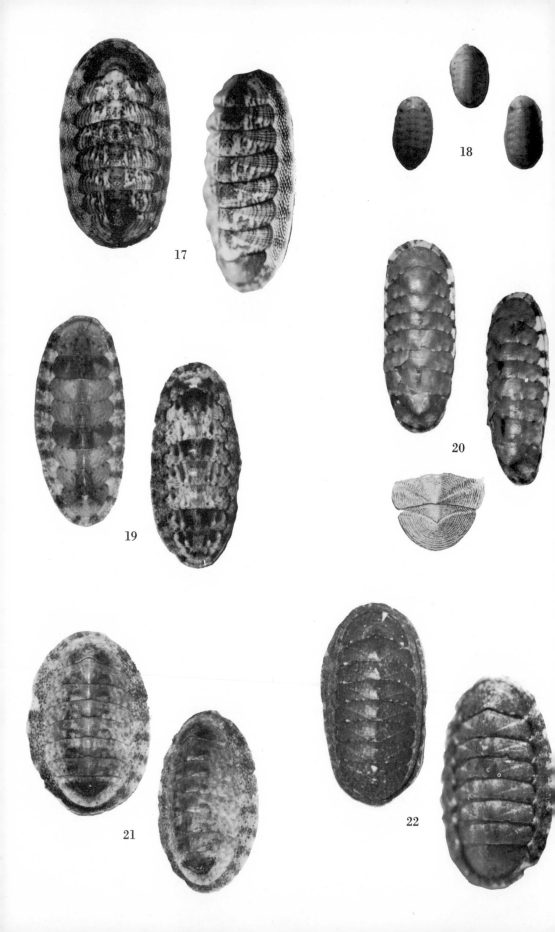

arranged into fine oblique or longitudinal striae. Radial areas with crowded oblique and V-shaped striae. Girdle scales small, about one-third the size of those of *R. muscaria*, and finely striate. Intermediate valves usually with two slits per side. Adult length, 12 to 26 mm. La Paz to Bahía Gonzaga, Gulf of California, and La Libertad to Guaymas in Sonora, Mexico, abundant on stones. The species is easily confused with either *R. muscaria* or *Lepidozona subtilis*.

Genus STENOPLAX DALL, 1879

Elongate chitons of medium to large size, with thick, scaly girdles. The tail valves are largest of the series and generally flattened. Radial sculpture usually present.

Stenoplax conspicua conspicua (Pilsbry, 1892). Larger, more brightly colored, with weaker valve sculpture than the Panamic relative. Santa Barbara County, California, to Bahía Sebastian Vizcaino, Baja California.

23. Stenoplax conspicua sonorana Berry, 1956. Similar to *S. magdalenensis* in general but narrower and more tapered, the tail valve noticeably narrower than the intermediates, radial sculpture weaker, and the central areas either coarsely granular or with only a fine weak network. Girdle scales narrow and greatly elongate, giving the girdle a bushy or furry aspect. Adult length, 30 to 60 mm. Northern end of the Gulf of California, from Bahía de los Angeles, Baja California, to Guaymas, Sonora, Mexico, under rocks.

24. Stenoplax limaciformis (Sowerby, 1832) (Synonyms: **Chiton productus** and **sanguineus** Reeve, 1847). Medium-sized, elongate, and rather arched, the color variable, with pink, green, and brown predominating, the pattern most frequently variegated or spotted. Radial areas with irregular, often beaded, concentric ridges, each ridge connecting anteriorly to one of the longitudinal ribs on the central areas. Girdle with very small, crowded, elongate scales. Adult length, 25 to 45 mm. Puertecitos and La Libertad, Mexico, to Peru, under stones or rocks intertidally, and to depths of 20 m. This species also occurs in the Caribbean.

25. Stenoplax magdalenensis (Hinds, 1845) (Synonym: **Ischnochiton acrior** Pilsbry, 1892, *ex* Carpenter, MS). Large, long, flattened, color generally brown or gray-green. Radial areas coarsely sculptured with radiating knobby ridges. Central areas crudely sculptured with longitudinal ribs, which often anastomose to form a network. Girdle with flat, curved, overlapping scales. Adult length, 50 to over 100 mm for specimens on the Pacific side of Baja California, 25 to 60 mm for those in the Gulf of California. Bahía Sebastian Vizcaino, and Cedros Island area of Baja California to San Felipe, Gulf of California, under rocks.
See Color Plate X.

Subfamily CALLISTOCHITONINAE

Girdle scaly. Valves with well-developed radial ribbing and with both longitudinal and transverse ribs in the central areas. The sinus between the apophyses bridged by a serrate plate. Two Panamic genera.

Genus CALLISTOCHITON DALL, 1879

Head valve with eight to eleven very strong radial ribs, the tail valve with slightly fewer strong ribs. Insertion teeth very short and bowed inward in a festoon-like arrangement. Girdle with very small imbricating scales.

23

24

25

26

28

26. Callistochiton colimensis (A. G. Smith, 1961). Small to medium-sized, oval, rather flat, and unusually broad for the genus. Color light to medium cinnamon-brown, clouded with dark olive-gray and tan. Each raised lateral area is divided into two broad, flat, primary ribs by a radial sulcus running from the apex to the valve margin. Each primary rib in turn divided into two secondary ribs by shallower sulci that do not extend to the apex. Each secondary rib bears a row of round, well-separated, intense orange beads. Terminal areas with eleven to twelve primary ribs, rounder, more separated, and less often twinned than those of the lateral areas. Adult length, 15 to 30 mm. Manzanillo, Mexico, to Nicaragua, intertidally and to 25 m.

27. Callistochiton duncanus Dall, 1919. Small, with a velvety girdle. Length, 10 mm; width, 4.5 mm. Galápagos Islands. This may prove not to be a *Callistochiton*; it is unfigured.

28. Callistochiton gabbi Pilsbry, 1893. A small narrow chiton with straight side slopes. Color tan or gray-green, clouded with dark olive-green and cream. The knobby radial ribs, weaker than those in the next two species, number nine or ten on the head valve, two on each lateral area, and eight to ten on the tail valve. Mucro of the tail valve slightly in front of its midpoint, the slope behind planar or slightly concave. Adult length, 15 to 20 mm. Baja California peninsula to Puerto Peñasco and Guaymas, Mexico, under rocks.

29. Callistochiton infortunatus Pilsbry, 1893. A small, roundly arched form of tan to rusty brown color. Head valve with eight or nine knobby radial ribs, the tail valve with six to eight similar ribs. Mucro in the back half of the tail valve, the slope behind at least moderately convex. Adult length, 14 to 24 mm, Gulf of California to southern Mexico, under stones and rocks. Individuals from Guaymas north to the head of the Gulf have nine ribs on the head valve, and very swollen end valves. Individuals from Cabo San Lucas and Mazatlán have only eight ribs on the head valve and only slightly swollen end valves.

30. Callistochiton pulchellus (Gray, 1828) (Synonyms: **Chiton bicostatus** Orbigny, 1841; **Callistochiton fisheri** and **periconis** Dall, 1908). Small, low-arched, with yellowish or grayish-brown color. The strong radial ribs are concentrically tiered by growth furrows, ten to eleven such ribs on the head valve, two on each lateral area, and six to nine on the tail valve. The mucro lies in the back quarter of the triangular tail valve. Middle third of the central areas sculptured with crisscrossing oblique ribs, forming a distinctive netted pattern, the outer third with wavy longitudinal ribbing only. Adult length, 10 to 15 mm. Panama to Chile, under rocks.

Genus LEPIDOZONA PILSBRY, 1892

Girdle scaly. Shell sculpture similar in pattern to *Callistochiton*, but end valves not swollen and tail valve without massive ribbing. Head valve with twelve to sixty radial ribs. Insertion teeth straight and often weakly grooved on the outer surface.

31. Lepidozona clathrata (Reeve, 1847). Medium-sized; shell color slate-brown, with intense red-brown along the ridge. Girdle alternately banded with gray and brown. Valve sculpture is sharply defined and arranged with unusual evenness. Radial ribbing of rows of oval beads or rounded bars, about twenty ribs on the head valve, four to six on each lateral area, and twelve to nineteen on

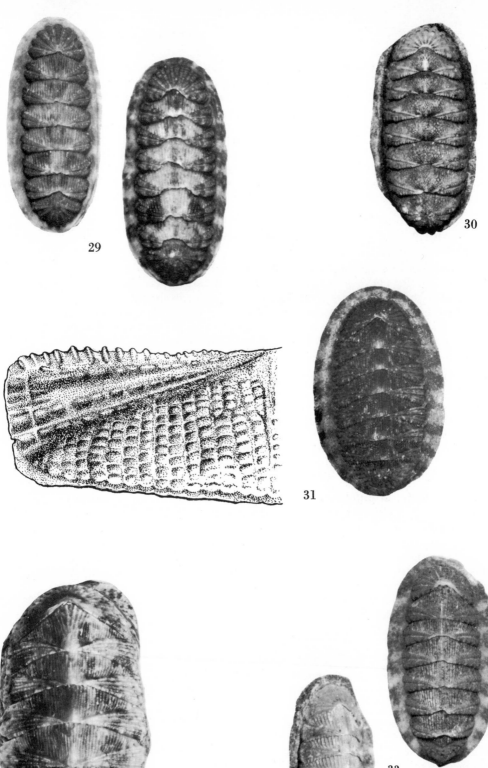

29   30   31   32   33

the tail valve. Central areas with evenly spaced longitudinal and transverse ribbing, meeting at right angles to form a neat gratelike effect. Adult length, 30 to 50 mm. Puerto Peñasco, and Baja California, under rocks.

32. Lepidozona crockeri (Willett in Hertlein & Strong, 1951). Small, similar in general appearance to *L. subtilis*, differing in the transverse ribbing of the central areas and with fewer, more widely spaced ribs on the radial areas. Southern part of the Gulf of California, 20 to 120 m. Known only from two specimens.

33. Lepidozona elenensis (Sowerby, 1832) (Synonyms: **Lepidopleurus clathratus** Carpenter, 1857 [not Reeve, 1847]; **Ischnochiton expressus** Carpenter, 1865; **I. subclathratus** Pilsbry, 1892). Small, high-arched, rather elongate. Color gray-brown, less commonly orange- or red-brown. Radial sculpture well developed, except on the tail valve, which has only eight to ten rows of indistinct beads. Head valve with eleven to seventeen beaded ribs and the strongly raised lateral areas divided into two beaded ribs by a median radial furrow. Central areas with widely spaced longitudinal ribbing and weak, crowded, transverse ribbing. Apices of valves four to seven upswept, having a peaked appearance when viewed from the side. Adult length, 10 to 15 mm. Mazatlán to Panama, under rocks and to about 20 m.

34. Lepidozona flavida (Thiele, 1910). Small, oval, high-arched; background color yellowish, with patches of black or brick-red. Valve surfaces uniformly and finely granular. Central areas with longitudinal ribbing and very weak transverse ribbing. End valves having fourteen to twenty low, indistinct, somewhat beaded radial ribs. Lateral areas with two to four similar ribs. Adult length, 5 to 8 mm. Manzanillo, Mexico, to Panama, subtidally, at 10 to 20 m depth.

35. Lepidozona .... Thorpe, MS. Small, high-arched; shell color brick-red, mottled with dark olive-green and cream. The girdle alternately banded with the lighter and darker shades. Radial sculpture of low, rounded ribs, very evenly and closely set with round beads. Head valve with 26 to 30 ribs, each lateral area with four to five, the tail valve with 18 to 19 such ribs. Central areas with rather wide-spaced longitudinal ribs and weaker, somewhat crowded transverse ribbing. Girdle with rounded overlapping scales, those nearest the valves bearing white papillae on their crowns. Adult length, 14 mm. Cabo San Lucas, Clarion Island, and Manzanillo, Mexico, 10 to 30 m.

36. Lepidozona .... Thorpe, MS. Small, oval, roundly arched, the color either uniformly tawny or buff, liberally painted with pale orange-brown. Valve surfaces conspicuously granular. Head valve with about 30 low, rounded, radial ribs, each lateral area with two to three stronger ribs, and the tail valve with 25 to 30 indistinct ribs. Middle part of the central areas coarsely and irregularly netted, the effect created by the crisscrossing or anastomosing of vermiform oblique ribs. Outer portions of the central areas with longitudinal ribbing only. Adult length, 11 to 13 mm. Cabo San Lucas, and Manzanillo, Mexico, intertidally and to about 10 m.

37. Lepidozona serrata (Carpenter, 1864). Small, arched, rather variable, the background color most often pale greenish cream, pale red-brown, or dark gray-brown, dappled with subtly lighter and darker tones and speckled with black. Central areas with the longitudinal and transverse ribbing spaced about equally far apart and of about equal strength. Middle portions of these areas worn or polished, making them appear neatly pitted rather than ribbed. Radial sculpture variable. Individuals with sharply defined beaded ribs may have as many as 35 on

the head valve, six per lateral area, and 20 on the tail valve; those with smooth low ribs may have only about half as many. Adult length, 10 to 15 mm. Cabo San Lucas, Baja California, to Puertecitos and eastward to Bahía Kino, Sonora, Mexico, under stones and rocks.

38. Lepidozona subtilis Berry, 1956 (Synonym: **L. pella** Berry, 1963). Small, arched, tapered; color variable, generally variegated with white, black, brown, orange, or green, less often uniformly black or red. Radial areas with beaded ribs, 40 to 60 on the head valve, six to nine on each lateral area, and 30 to 45 on the tail valve. Central areas with longitudinal ribbing only. Adult length, 12 to 20 mm. San Felipe to Bahía de los Angeles and Puerto Peñasco to Guaymas, Mexico, on stones.

Subfamily CHAETOPLEURINAE

Girdle spicular, interspersed with either calcareous spines or fibrous hairs. Central areas with longitudinal rows of pustules or beaded ribs. Two Panamic genera.

Genus CHAETOPLEURA SHUTTLEWORTH, 1853
(PALLOCHITON DALL, 1879)

The girdle spiculose, interspersed with scattered hairs or spines. Valve sculpture of pustules, generally arranged in rows. Serrate sinus plane present, and outer surfaces of insertion teeth grooved.

39. Chaetopleura . . . . Thorpe, MS. Medium-sized, somewhat elongate, and moderately elevated. Color variable, with gray-green, gray-brown, tan, and orange the predominant tones. Pattern either boldly striped or nearly unicolored. Shell sculpture of rounded pustules, arranged in neat longitudinal rows on the central areas, and in quincunx or very irregular radiating rows on the radial area. Tail valve with its mucro slightly behind the center, the posterior margin swollen. Girdle interspersed with thin, curved, calcareous spines. Head valve with eight or nine slits, tail valves with six to nine slits. Adult length, 25 to 35 mm. Panama, under rocks.

40. Chaetopleura euryplax Berry, 1945. Medium-sized; valve color highly variable, with lilac-brown or olive-green the most common background tones. Girdle brown or orange, flecked with black and tawny. Mucro of the tail valve near its back edge, the terminal area behind a swollen ridge. Radial areas with scattered, irregularly arranged pustules. Central areas with wide-spaced, curving, longitudinal rows of pustules. Girdle velvety, the fine spicules interspersed with short spinelets. Adult length, 27 to 43 mm. San Felipe to Bahía de los Angeles, and Puerto Peñasco to Guaymas, Mexico, under muddy rocks at lowest tides.

41. Chaetopleura lurida (Sowerby, 1832) (Synonyms: **Chiton catenulatus, columbiensis**, and **scabriculus** Sowerby, 1832; **Chiton jaspideus** Gould, 1846; **Lepidopleurus bullatus** and **bullatus** var. **calciferus** Carpenter, 1857; **Ischnochiton parallelus** and **prasinatus** Carpenter, 1864). Medium-sized; distinctly broader, rounder, and flatter than species 39. Color dusky, with gray, brown-black, and dark wine-red predominating. Pattern generally a suffusion of darker tones, the lateral areas often finely painted with wavy concentric alternating lines of black and white. Shell sculpture of round, sometimes coalescing pustules, arranged neatly in longitudinal rows on the central areas, and in radial rows on the radial areas. Tail valve with its mucro in front of the center, the slope

behind smoothly concave, not swollen. The girdle appears decidedly untidy to the naked eye, because of the irregular placement and orientation of its thin, brown hairs. Head valve with ten to twelve slits, tail valve with one to eleven slits. Adult length, 25 to 35 mm. Cabo San Lucas and Mazatlán, Mexico, to Peru, under rocks.

42. Chaetopleura mixta (Dall, 1919). Small, oval, elevated; color variable, with pink and orange tones predominating. Mucro of the tail valve peaked and slightly in front of the middle, the slope behind concave, not swollen. Radial areas with moderately numerous, regularly spaced pustules, central areas with longitudinal rows of pustules. Girdle spicular, interspersed with scattered, curved, white spines. Adult length, 10 to 18 mm. Puerto Peñasco to Guaymas and Cabo San Lucas, Mexico. Lowest intertidal zone to about 15 m, often on shells.

Genus CALLISTOPLAX DALL, 1882

Shell as in *Callistochiton*, with massive radial ribbing. The girdle having rather regularly arranged groups of fibrous bristles.

43. Callistoplax retusa (Sowerby, 1832). Elongate, arched; tan or light brown in color. Tail valve with its mucro at or near the posterior edge, the slope in front level, the surface behind convex and sloping back underneath the mucro in adults. Central areas with beaded longitudinal ribs, the lateral areas massively raised. The downy girdle shows about 40 groups of bristles, arranged more or less in a row around the valves. Adult length, 17 to 31 mm. Tres Marias Islands, Mexico, to Central America, under rocks at extreme low tide and to about 10 m.

Family LEPIDOCHITONIDAE

Shell substance spongy, its surface either smooth or granular; the central areas without strong sculpture. Girdle spiculose or spiny. Two Panamic genera.

Genus MOPALIELLA THIELE, 1909

Shell thin and exceptionally spongy. Valve surfaces granular and covered with a thin layer of tissue. Girdle with groups of curved spines at the sutures and around the end valves.

44. Mopaliella beani (Carpenter, 1857) (Synonyms: **Chiton flavescens** Carpenter, 1857; **Basiliochiton lobium** Berry, 1925). Confusing small animals, the color, shape, and ornamentation dependent largely on where the individual happens to have lived. Animals from the subtidal zone are more arched, narrower, and generally brick-red in color; the valve surface appears to be smooth, owing to a fully developed tissue covering; each spine group contains only one to three spines, and often groups in the series are missing. Individuals from the intertidal zone are lower and broader, and generally yellow or pink, with darker markings; the valves are clearly granular because the tissue layer is worn away; the spine groups are more prominent, with up to ten spines per group; and a secondary series of groups is often present. Adult length is 5 to 10 mm. Southern California to southern Mexico, intertidally on stones in sand, and to about 20 m.

Genus NUTTALLINA DALL, 1871

Valve surface with coarse granulation and rude low radial ribbing. Head valve with eight to eleven slits. Girdle irregularly spiny.

45. Nuttallina crossota Berry, 1956. Small, elongate; ground color of the valves tan or pink, mottled and streaked with black, white, and dark brown. The spicular girdle is brown, banded with white, the spines light brown. Mucro of the upswept tail valve at its back edge and in adults actually overhanging the base of the valve. Head valve with eleven slits on its insertion plate. Adult length, 10 to 12 mm for individuals from rock surfaces, up to 18 mm for solitary individuals under rocks in tide pools. Northern end of the Gulf of California: San Felipe to Bahía de los Angeles, and Puerto Peñasco to Guaymas, Mexico.

Family MOPALIIDAE

Insertion plates of the head valve with eight slits. Girdle with fibrous hairs (setae) or bristles. Three Panamic genera.

Genus CERATOZONA DALL, 1882

Head valve with seven to ten radial ribs, the slits of its insertion plate corresponding in position and number with the ribs. Girdle with fleshy, yet straplike, iridescent bristles.

46. Ceratozona angusta Thiele, 1909 (Synonym: **Chiton setosus** Sowerby, 1832 [not Tilesius, 1824]). Medium-sized, flattened; ground color of valves pink, brown, or slate-black, painted with straight or zigzag lines of white, black, or pink. Head valve with about eight to ten low, rude ribs and eight to ten corresponding slits. Tail valve with about six slits and an equal number of indistinct ribs. The pale girdle is set with pink iridescent bristles. Adult length, 20 to 32 mm. El Salvador to Panama, on exposed rock surfaces.

Genus DENDROCHITON BERRY, 1911

Head valves with seven to nine slits. Central areas showing deeply cut longitudinal grooving. Girdle with branched or spine-tipped setae.

47. Dendrochiton laurae Berry, 1963. Central areas with a few longitudinal grooves, radial areas nearly smooth. Girdle with irregularly placed, spinose setae. Length, about 10 mm. The three specimens known came from depths of 10 to 90 m in the southern part of the Gulf of California.

48. Dendrochiton lirulatus Berry, 1963. Small and rather narrow; variable and often with striking coloration, in which black, white, and green predominate. Central areas with twelve to eighteen close-set, nearly straight, longitudinal grooves. Radial areas impressed with crowded round or V-shaped pits. Girdle with numerous irregularly placed spinose setae. Adult length, 9 to 13 mm. Northern part of the Gulf of California, on stones.

Genus PLACIPHORELLA DALL, 1879

Large, flat chitons with a circular shell, the eight valves all unusually short and broad. Valves embedded in a thick, fleshy girdle, which is always widest anteriorly, often produced in front like a fan; girdle studded with scaly, unbranched setae. McLean (1962) has described the bizarre feeding behavior in this genus. The chiton raises the front part of the girdle and remains motionless until a piece of floating algae or an unsuspecting small crustacean comes close. The *Placiphorella* then closes the girdle down against the substrate, trapping the food, which is worked to the mouth by the girdle and the finger-like tentacles surrounding the head, finally being rasped to pieces by the radula.

49. Placiphorella blainvillii (Broderip, 1832). Shell color red, with darker brown markings. Girdle with bunches of white spines interspersed among the scaly setae. Adult length, 35 to 50 mm. Panama to Peru, subtidally.

50. Placiphorella velata Dall, 1879. Background color of valves tan or pink, mottled and streaked with blue, brown, and red. The girdle has scaly setae only, these most numerous at the anterior margin of the girdle. Adult length (Gulf of California specimens) 30 to 40 mm. Alaska to Bahía de los Angeles and Guaymas, Gulf of California, rare, slightly more common offshore to about 100 m. When the spine groups are worn off the girdle of *P. blainvillii*, these two very similar-appearing species cannot be separated on external features alone. The radula of *P. blainvillii* has 50 to 60 transverse rows of teeth, and the black-tipped major laterals appear decidedly small compared with the width of the radular ribbon; in *P. velata*, there are only 30 to 40 transverse rows of teeth, and the major laterals are relatively large.

Polyplacophora of Uncertain Status

The following species, all from deep water, are unfigured:

51. Lepidopleurus abbreviatus Dall, 1908. Off Acapulco, Mexico, in 900 to 1,200 m. (Described as a subspecies of *L. halistreptus*.)

52. L. farallonis Dall, 1902. Panama Bay, in 1,820 m. (Reported by Dall, 1908.)

53. L. halistreptus Dall, 1902. Off Acapulco, Mexico, in 3,440 m.

54. L. incongruus Dall, 1908. Gulf of Panama, in 590 m.

55. L. luridus Dall, 1902. Panama Bay, in 1,870 m.

56. L. opacus Dall, 1908. Gulf of Panama to Galápagos Islands and Peru, in 2,320 to 3,665 m.

Class Scaphopoda

The term scaphopod literally means "boat-footed," but the foot is more the shape of a conical plug than a boat. The popular name is tusk shell or tooth shell. The tapering, tubular, usually white shell is open at both ends, and the soft parts are modified for active life just beneath the surface of the sea floor, where the animal sifts the sand with slender club-shaped extensions of the proboscis called captacula and pulls the prey—Foraminifera and other minute invertebrates, including larvae of other mollusks—into its mouth, which is on the end of the proboscis, surrounded by a rosette of lobes. There is a radula with five teeth in each row, strong teeth that can break up the shells of the prey. The scaphopod shell increases in size by growth at the anterior end, and as it does so, a part of the smaller or posterior end is dissolved away. This part of the shell may become differently shaped at various stages of the life cycle. Rarely a specimen might be taken in the lowest intertidal zone, but mostly the scaphopods occur offshore in shallow to deep water away from areas of vigorous wave action, and some occur in very deep water; there are records of over 3,500 meters, more than two and a half miles below the surface. The Indians of the American northwest coast valued the shells and used them as a sort of currency, collecting them in some quantity with special tools, despite the offshore habitat.

A monograph by Pilsbry and Sharp (1897–98) has been the standard work on this group, and was so well done that little was added in the next half-century. In recent years, however, Dr. William K. Emerson of the American Museum of Natural History, New York, has supplemented Pilsbry & Sharp with useful papers (Emerson, 1951, 1956, 1962). The last of these presents a new classification, with keys to the taxonomic units of the Scaphopoda and detailed summaries of morphology, which are condensed here in characterizing the genera and subgenera represented in the eastern Pacific. Also, from a monographic work he has in preparation, Dr. Emerson has extracted for use here a list of Panamic species. His generosity in making this list available and his assistance in providing distributional data are gratefully acknowledged.

Family DENTALIIDAE

Shell with the greatest diameter at the oral aperture; surface smooth or sculptured; radula with central tooth twice as wide as high; foot of animal conical, not terminating in an expandable disk.

Genus DENTALIUM LINNAEUS, 1758

Shell curved, tapering, sculptured with raised ribs in juveniles, smooth or striate in adults; apical orifice modified by a notch or slit. The key is to the five Panamic subgenera, modified from one by Emerson (1962).

1. Surface with longitudinal ribs.................................... 2
   Surface smooth.. 6
2. Apex polygonal 3
   Apex not polygonal.. 4
3. Ribs fewer than twenty...............................*Dentalium, s. s.*
   Ribs more than twenty...............................*Fissidentalium*
4. Apex quadrate ....*Tesseracme*
   Apex circular 5
5. Riblets fine, almost microscopic, uniform...............*Graptacme*, in part
   Riblets not extremely fine, on apical portion only...........*Antalis*, in part
6. Apex with a V-shaped notch.........................*Antalis*, in part
   Apex with a slit on the convex side...................*Graptacme*, in part

Subgenus DENTALIUM, *s. s.*

Apex weakly to strongly polygonal; number of ribs less than twenty; ribs rounded.

1. Dentalium (Dentalium) agassizi Pilsbry & Sharp, 1897. Length, 65 mm. Off Santa Barbara Island, California, to Panama and the Galápagos Islands, in 589 to 2,320 m.

2. Dantalium (Dentalium) neohexagonum "Sharp and Pilsbry" in Pilsbry & Sharp, 1897 (Synonym: **D. pseudohexagonum** Arnold, 1903, *ex* Dall, MS). Length, 31 mm. Monterey Bay, California, to Isla Tiburon, Gulf of California, in 7 to 256 m.

3. Dentalium (Dentalium) oerstedii Mörch, 1860 (Synonym: **D. numerosum** Pilsbry & Sharp, 1897, *ex* Dall, MS). Length, 40 mm; diameter, at aperture, 3.5 mm. Punta Peñasco, Sonora, Mexico, to Santa Elena Bay, Ecuador, in 4 to 145 m.

4. Dentalium (Dentalium) vallicolens Raymond, 1904. Length, 64 mm. Straits of Juan de Fuca, Washington, to Gulf of California, in 5 to 366 m. According to Emerson (*in litt.*), this may prove to be a *Fissidentalium*.

Subgenus ANTALIS H. & A. ADAMS, 1854

Less strongly ribbed than *Dentalium, s. s.*, with a V-shaped notch on the concave side of the apical orifice, the apex partially filled by a plug.

5. Dentalium (Antalis) pretiosum berryi Smith & Gordon, 1948. Length, 47 mm. Monterey Bay, California, to the Gulf of California, in 37 to 298 m. The nominate subspecies, *D. p. pretiosum* Sowerby, 1860, is more northern in distribution, not occurring in the Panamic province.

Subgenus FISSIDENTALIUM FISCHER, 1885

Shell large, solid, circular in outline; sculpture of numerous longitudinal striae; apex mostly with a long slit on the convex side, rarely simple.

6. Dentalium (Fissidentalium) megathyris Dall, 1890. Length, 99 mm. Off California to the Gulf of California and south to Chiloé Island, Chile, in depths of 1,467 to 2,080 m.

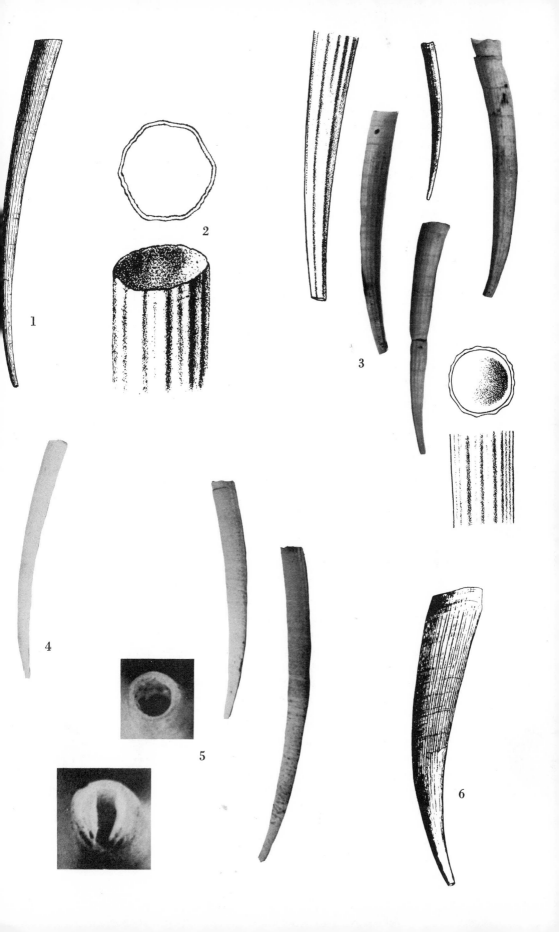

1

2

3

4

5

6

Subgenus GRAPTACME PILSBRY & SHARP, 1897

Small, slender; with very fine deeply engraved riblets on the early or posterior part of the shell, the anterior end of shell smooth; apical orifice never angular, but the slit may vary in size and position.

7. Dentalium (Graptacme) sectum Deshayes, 1826. Length, 24 mm. ? Gulf of California. According to Emerson (*in litt.*), this is a form of doubtful validity; it may be a variant of *D. semipolitum*.

8. Dentalium (Graptacme) semipolitum Broderip & Sowerby, 1829 (Synonyms: **? D. hyalinum** Philippi, 1846; **D. liratum** Carpenter, 1857; **? D. corrugatum** Carpenter, 1857 [not Hupé, in Gay, 1854]; **D. lirulatum** Mörch, 1860; **D. hannai** Baker, 1925). Length, 30 mm; diameter at aperture, 3 mm. Off southern California and through the Gulf of California south to Cocos Bay, Costa Rica, in 2 to 45 m.

Subgenus TESSERACME PILSBRY & SHARP, 1898

Cross section of shell quadrangular in youth, becoming subcircular with age; ribs at the angles rounded, the four faces smooth to sculptured with riblets that increase in number toward the aperture; apex commonly unnotched but with a short terminal pipe.

9. Dentalium (Tesseracme) hancocki Emerson, 1956. There are numerous pits in the grooves between the major ribs, sixteen to twenty riblets on the middle part of the shell. Length, 14 mm; major diameter, 1.9 mm. Santa Maria Bay, Baja California, through the Gulf of California and south to Manzanillo, Colima, Mexico, in 14 to 42 m.

10. Dentalium (Tesseracme) quadrangulare Sowerby, 1832 (Synonym: **D. fisheri** Pilsbry & Sharp, 1897, *ex* Stearns, MS). Longitudinal riblets number twenty-four to thirty in the middle part of the shell, cross-striae not pitted. Length, 20 mm; diameter at aperture, 3.8 mm. Head of the Gulf of California to La Libertad, Ecuador, in 5 to 37 m.

11. Dentalium (Tesseracme) tesseragonum Sowerby, 1832 (Synonym: **D. tetragonum** Sowerby, 1860 [not Brocchi, 1814]). The four faces of the tube are sculptured with four to eight weak longitudinal threads in the middle part of the shell; the anterior third is smooth. Length, 31 mm; diameter at aperture, 3 mm. Acapulco, Mexico, to Jipijapa, Ecuador, in 2 to 146 m.

Genus FUSTIARIA STOLICZKA, 1868

Shell slender, slightly arched, smooth, with growth lines only or with occasional annulations, no longitudinal ribs; apical orifice simple, with a short notch on the concave side. The key, to Panamic subgenera, is adapted from Emerson (1962). The subgenus *Fustiaria, s. s.*, is not represented in the Panamic province.

1. With growth lines only.. 2
   With folds, wrinkles, or annulations................................. 6
2. Shell compressed between concave and convex sides..........*Compressidens*
   Shell circular or nearly so in section............................... 3
3. Apex without any slit.....................................*Rhabdus*, in part
   Apex with a slit.. 4
4. Slit narrow to linear....................................*Fustiaria, s. s.*, in part
   Slit moderately wide.. 5

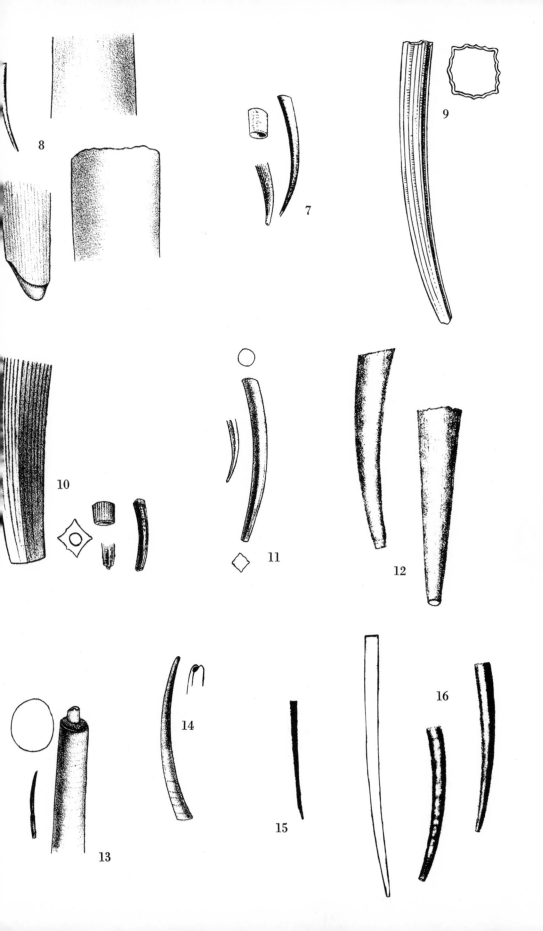

8

7

9

10

11

12

13

14

15

16

5. Tube posteriorly truncate, with a terminal pipe in adults.... *Episiphon*, in part
   Tube with a simple termination........................*Laevidentalium*
6. Tube with a long, narrow slit......................*Fustiaria, s. s.*, in part
   Tube with a moderate or short slit............................... 7
7. End of tube truncate in mature adults................*Episiphon*, in part
   End of tube attenuated..............................*Rhabdus*, in part

Subgenus COMPRESSIDENS PILSBRY & SHARP, 1897

Small, fragile, tapering, laterally compressed, elliptical in section.

12. Fustiaria (Compressidens) brevicornu (Pilsbry & Sharp, 1897).
Length, 9.5 mm. Off Mazatlán, Sinaloa, Mexico, to the Galápagos Islands, in 1,159
to 1,820 m.

Subgenus EPISIPHON PILSBRY & SHARP, 1897

Small to minute, fragile, glossy; truncate posteriorly in late maturity, with a
central pipe projecting from an apical plug.

13. Fustiaria (Episiphon) innumerabilis (Pilsbry & Sharp, 1897) (Synonym: **Dentalium (Rhabdus) cedrosense** Hertlein & Strong, 1951). Length,
17 mm; diameter at aperture, 0.8 mm. Cedros Island, Baja California, through
the Gulf of California and south to Guayaquil, Ecuador, in 18 to 165 m.

Subgenus LAEVIDENTALIUM COSSMANN, 1888

Medium-sized, slightly curved, smooth except for faint growth lines, section circular to slightly oval; apex mostly simple, with a short notch.

14. Fustiaria (Laevidentalium) splendida (Sowerby, 1832) (Synonym:
? Dentalium fissura Sowerby, 1860 [not Lamarck, 1818]). Length, 45 mm;
diameter at aperture, 3.8 mm. Puerto Peñasco, Sonora, Mexico, to La Libertad,
Ecuador, in 2 to 110 m (taken alive at extreme low tide on a sandbar at La Paz
by H. N. Lowe).

Subgenus RHABDUS PILSBRY & SHARP, 1897

Nearly straight, extremely thin, fragile, unworn surface smooth, worn surface
sometimes appearing chalky, sometimes with faint annulations. Apex simple, with
inner layer extending as a thin tube in some specimens; section circular.

15. Fustiaria (Rhabdus) aequatoria (Pilsbry & Sharp, 1897). Length, 31.5
mm. Off Manta, Ecuador, in 703 m (known only from a single specimen and possibly not a valid species).

16. Fustiaria (Rhabdus) dalli (Pilsbry & Sharp, 1897). Length, 45 to 69 mm.
Bering Sea to Aguja Point, Peru, in 139 to 1,900 m.

Family SIPHONODENTALIIDAE

Smooth shells, apertural area of the tube constricted; central tooth of radula
almost square; animal with a foot that expands into a symmetrical disk with a
crenate edge, and, in some genera, a central finger-like projection.

Genus SIPHONODENTALIUM SARS, 1859

Moderately to strongly curved, slightly tapering, largest at aperture, circular in
section, smooth; apex large, cut into lobes; foot of animal with pedal disk that
lacks a central filament or process.

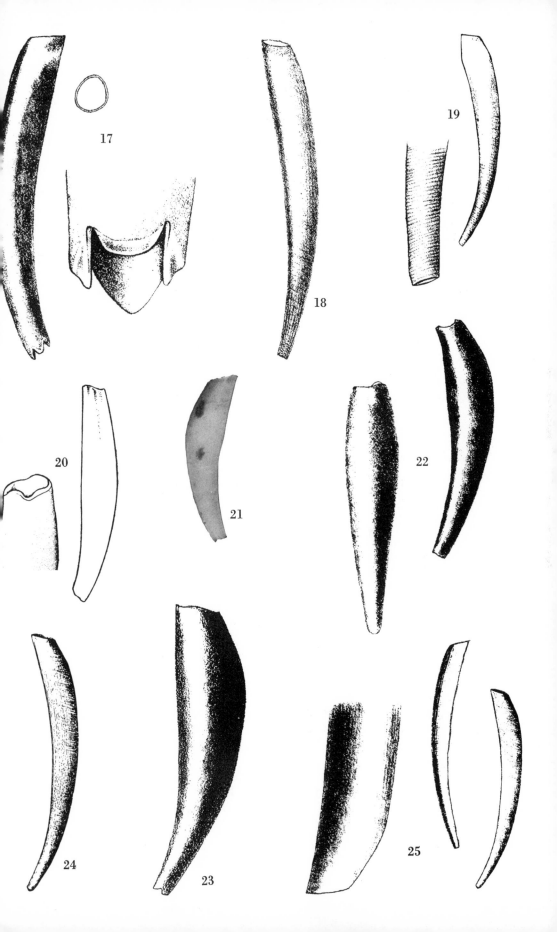

17. Siphonodentalium quadrifissatum (Pilsbry & Sharp, 1898). Length, 8.6 mm; diameter at aperture, 1 mm. Monterey Bay, California, to Los Frailes Bay, Baja California, Mexico, in 4 to 365 m.

Genus CADULUS PHILIPPI, 1844

Mostly small, somewhat arched; greatest diameter between the middle of the shell and the apertural end; aperture contracted; tube surface smooth; foot of animal with a median disk lacking a central projection. The key, modified from Emerson (1962), is to the four Panamic subgenera. The subgenus *Cadulus, s. s.,* is not represented in the Panamic province.

1. Apex lacking slits or notches.. 2
   Apex with slits or notches.. 5
2. Shell surface smooth... 3
   Shell surface somewhat sculptured.................................. 4
3. Shell cask shaped...*Cadulus, s. s.*
   Shell slender in outline.................................*Gadila,* in part
4. Shell relatively large, up to 25 mm in length.................*Striocadulus*
   Shell relatively small.....................................*Gadila,* in part
5. Slits shallow, forming wide lobes..........................*Platyschides*
   Slits moderately deep, conspicuous................................. 6
6. With two slits ....*Dischides*
   With four slits...*Polyschides*

Subgenus GADILA GRAY, 1847

Strongly curved, one side concave, the opposite convex, medially swollen; smooth (some with faint annulations).

18. Cadulus (Gadila) fusiformis Pilsbry & Sharp, 1898 (Synonym: **C. nitentior** Arnold, 1903, *ex* Carpenter, MS). Length, 10 mm; diameter at aperture, 1 mm. Monterey, California, to Cape San Lucas, Baja California, and the Gulf of California, in 7 to 365 m.

19. Cadulus (Gadila) perpusillus (Sowerby, 1832) (Synonym: **C. panamensis** "Sharp & Pilsbry" in Pilsbry & Sharp, 1898). Length, 7 mm; diameter at aperture, 0.5 mm. Cedros Island, Baja California, through the Gulf of California and south to Panama Bay and the Galápagos Islands.

Subgenus PLATYSCHIDES HENDERSON, 1920

Moderately curved, greatest swelling between middle of shell and the oral aperture; posterior portion slightly flattened; apex with four broad, shallow notches.

20. Cadulus (Platyschides) austinclarki Emerson, 1951. Length, 4.4 mm; diameter of aperture, 0.55 mm. Santa Maria Bay, Baja California, and through the Gulf of California and south to Panama and the Galápagos Islands, in 2 to 73 m.

21. Cadulus (Platyschides) peruvianus Dall, 1908. Length, 12 mm. Off Aguja Point, Peru, in 1,900 m; near Galápagos Islands in 1,485 m.

22. Cadulus (Platyschides) platystoma Pilsbry & Sharp, 1898. Length, 13 mm. Off Manta, Ecuador, in 733 m.

Subgenus POLYSCHIDES PILSBRY & SHARP, 1898

Greatest swelling near the middle; oral aperture slightly contracted; apex cut into lobes, generally four, by slits or notches.

23. Cadulus (Polyschides) californicus Pilsbry & Sharp, 1898. Length, 14 mm. Clarence Strait, Alaska, to Manta, Ecuador, in 48 to 2,320 m.

Subgenus STRIOCADULUS EMERSON, 1962

Relatively large (to 25 mm), moderately curved, surface sculptured with many longitudinal striae; aperture as in *Gadila*; orifice simple.

24. Cadulus (Striocadulus) albicomatus Dall, 1890. Length, 24 mm. Off California to Manta, Ecuador, in 733 to 3,050 m.

25. Cadulus (Striocadulus) striatus Pilsbry & Sharp, 1898, *ex* Dall, MS. Length, 25 mm. Off Acapulco, Mexico, to the Gulf of Panama, in 589 to 1,210 m.

Class Cephalopoda

Cephalopods (the "head-footed ones") are the most highly developed and agile of all mollusks. Active carnivores, they can move with astonishing rapidity, either through water by a kind of jet propulsion or over rocks by creeping. They see well, having complex eyes that are structurally similar to those of the vertebrates. Few cephalopods have any hard parts that could be called "shells," the principal exception being the Subclass Nautiloidea of the western Pacific. Their pearl-lined, chambered shells may be free-floating but never drift so far as the eastern Pacific. The Panamic cephalopods belong mostly in the Subclass Coleoidea, which is divided into several orders, especially the Sepiida (the cuttlefishes) and the Teuthidida (squids, the Teuthoidea of some authors), both of which have one pair of long tentacle-like arms and four pairs of shorter arms, all with stalked suckers; they also have an internal stiffening rod or "pen" in the body cavity. The Order Octopoda is distinguished by having four pairs of arms, all similar; nonstalked suckers; and no shell. All the Coleoidea have horny and very tough mouth parts, the "beaks." Squids and their allies (called Decapoda by some authors) are free-swimming and mostly occur offshore. The octopuses prefer rocky substrates and hide in crevices, but they also can swim actively.

Literature on this group is scanty and not readily available in most libraries. Because several specialists are at work on monographic reviews, no attempt at a complete coverage of this class in Panamic waters is made here. However, a list of the common species of octopuses that may be encountered by collectors is given, with some notes on habits and distribution.

Subclass COLEOIDEA
(DIBRANCHIA)

Without any chambered external shell; an internal membranous to calcareous support present in some; gills two.

Order OCTOPODA
(POLYPOIDEA)

Without any internal shell; with four pairs of arms, all similar in length.

Superfamily OCTOPODACEA

Sexes not markedly distinct in appearance.

Family OCTOPODIDAE

An ink sac present; bottom-dwelling to bathypelagic forms.

Genus OCTOPUS CUVIER [1797] (*Nom. conserv.*, ICZN, Opinion 233, 1954)

Body short and rounded, without fins between tentacles. Swimming is by ejection of water through the modified siphon, the funnel. Food mostly of crabs or other crustaceans. The following list of common species has been compiled by Dr. Eugene V. Coan.

1. Octopus alecto Berry, 1953. San Felipe to Puerto Peñasco, near the head of the Gulf of California.

2. O. bimaculatus Verrill, 1883. Santa Monica, California, to Panama.

3. O. chierchiae Jatta, 1889. Panama.

4. O. digueti Perrier & Rochebrune, 1894. Northern end of the Gulf of California to Mazatlán, Mexico, living inside empty shells.

5. O. fitchi Berry, 1953. San Felipe to Guaymas, Mexico. This species has been recorded by Berry and Halstead (1954) as one that will bite readily on being handled, and the saliva may be dangerously toxic.

6. O. hubbsorum Berry, 1953. Gulf of California.

7. O. occidentalis Hoyle, 1896. Galápagos Islands.

8. O. oculifer (Hoyle, 1904). Galápagos Islands.

9. O. penicillifer Berry, 1954. Punta Arena, Baja California, 31 m.

10. O. roosevelti Stuart, 1941. James Island, Galápagos Islands.

Superfamily ARGONAUTACEA

Sexes markedly unlike.

Family ARGONAUTIDAE

Pelagic; body form resembling the Octopodidae but female secreting a calcareous egg chamber that simulates a shell.

These are the "paper nautilus" of the layman. The animals are pelagic, floating near the surface of the sea, and the species may be widely distributed. Dall (1908) and Robson (1932) have reviewed the living species.

Genus ARGONAUTA LINNAEUS, 1758

Shells thin, brittle, one-chambered, involutely coiled. Although the generic name ends in *-a*, it is not of feminine gender; "*nauta*," sailor, is a masculine noun in Latin. The rules of grammar make no concession to the zoological fact that this sailor is the female of the species. Males are small, octopus-like in appearance, without a shell.

11. Argonauta cornutus Conrad, 1854 (Synonym: **A. expansus** Dall, 1872). The surface of the yellowish-white "shell" is finely granular, the spines and part of the spire dark brown, the keel relatively broad, and the two long axial expansions suffused with purplish brown. Length, 94 mm; height, 56 mm; width, 75 mm. Gulf of California to Panama.

12. Argonauta nouryi Lorois, 1852 (Synonym: **A. gruneri** Dunker, 1852). Dunker's name for this species, which is better known, apparently was published two months later than that of Lorois. The "shell" is more elliptical than that of *A. cornutus*, with only the early part of the coil moderately well tinged with brown along the wide and weakly tuberculate keel. The surface is delicately ribbed and has a finely granular texture. Length, 58 mm; height, 32 mm; width, 25 mm. The species is widespread in the equatorial Pacific, ranging on the west coast from southern California to Peru.

13. Argonauta pacificus Dall, 1869. This animal so closely resembles the Mediterranean *A. argo* Linnaeus, 1758, that it may prove to be identical, though such a distribution pattern would be strange. The whitish "shell" is rather compressed at the basal keel, which is narrow, finely tuberculate, and suffused with black. Length, 120 mm; height, 70 mm; width, 35 mm. Southern California through the Gulf of California and southward to the Galápagos Islands and Peru.

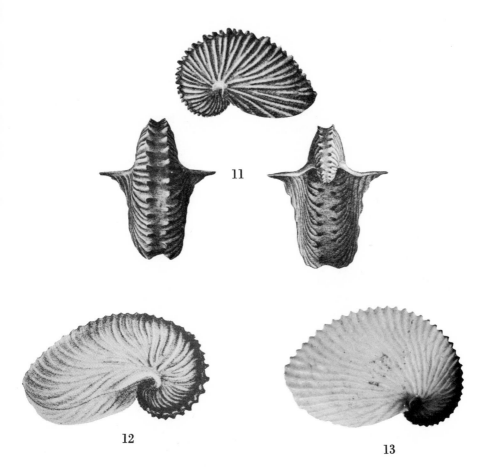

APPENDIXES

Rejected and Indeterminate Species

Most of the species in the following list have been reported from the Panamic province but have been excluded from the text proper either because the localities given appear to be in error or because there is insufficient data for establishment of the record. A few that are genuinely Panamic are rejected as indeterminate because the type material either is lost or is inadequate for recognition of the form. Taxa are cited in systematic order, that is, in the sequence they would follow were they incorporated in the text; classes and superfamilies are indicated for orientation with the text.

Class PELECYPODA
Superfamily NUCULANACEA

Nuculana rhytida (Dall, 1908) [*Leda*]. Originally cited as from off Acapulco, Mexico. The type lot in the U.S. National Museum is from Patagonia, eastern South America.

Superfamily ARCACEA

Barbatia solidula (Dunker, 1868). Described as from the Gulf of California, this seems more probably to be an Indo-Pacific form. It resembles *B. fusca* (Bruguière, 1789). No material matching Dunker's figure has since been reported in the Panamic province.

Superfamily CORBICULACEA

Cyrena convexa Deshayes, 1855. Described as from Mazatlán, and as having a shell with striate lateral teeth. The locality label was probably incorrect, for no such form has since been reported in that area.

Superfamily CHAMACEA

Chama producta Broderip, 1835. Described from the Gulf of Tehuantepec, from 18 m. Reeve, who studied the type material in 1846, concluded that it is, rather, the Indo-Pacific *C. iostoma* Conrad, 1837 (over which, one must point out, the name has priority). If the figure given by Broderip was of the sole specimen, the species might prove to be genuinely West American, because the figure resembles a worn *C. mexicana*. Broderip's specific name would then have priority. However, there are some discrepancies in the description that support Reeve's interpretation.

Superfamily VENERACEA

Ventricolaria lepidoglypta (Dall, 1902) [*Ventricola*]. Described from beach shells purchased at Acapulco, Mexico. Apparently an Oriental form, perhaps *V. foveolata* (Sowerby, 1853).

Transennella pannosa (Sowerby, 1835) [*Cytherea*] (Synonym: **Cytherea lutea** Philippi, 1845). Panamic records are open to question. The range seems to be Callao, Peru, to Chile.

Chione (Lirophora) schottii Dall, 1902. Although reported as from Panama, this was probably an import; perhaps from Europe, according to Olsson (1961).

Cytherea lubrica Broderip, 1835. Described as from Puerto Potrero [Costa Rica], but according to Sowerby (1853), it is an *Amiantis purpurata* (Lamarck, 1818), now known to be a Brazilian species.

Venus distans Philippi, 1852. Described as from Panama. The description would suggest a *Chione (Lirophora)*, although there are some problems with this interpretation. The type has not been illustrated or later studied. The name may prove to be a prior one for *C. (L.) obliterata* (Dall, 1902), which Olsson regards as a Caribbean form.

Superfamily MACTRACEA

Mulinia gabbi (Tryon, 1870) [*Mactra*]. Said to be from Baja California, this is apparently a southern South American form, with an erroneous locality.

Superfamily TELLINACEA

Tellina brevirostrata Sowerby, 1867. Said to be from San Blas, Mexico, but probably not West American.

Tellina silicula Deshayes, 1855. Described as from west Colombia, this seems to be from the East Indies. The type lot in the British Museum collection has been relabeled as a common Indo-Pacific species.

Tellina tersa Gould, 1852. Described as from Panama, this seems, however, to be a juvenile *Macoma nasuta* (Conrad, 1837) and may be dismissed from the Panamic fauna—although one valve of the Californian *M. nasuta* has been reported from off the tip of Baja California by Hertlein and Strong (1949).

Macoma gubernaculum (Hanley, 1844) [*Tellina*]. Recorded by authors as occurring on the west Central American coast. However, the type lot in the British Museum is labeled as being from India, and the species seems close to if not identical with *Psammotreta (Pseudometis) praerupta* (Salisbury, 1934), a widely distributed Indo-Pacific form.

"Macoma plebeia (Hanley, 1844)" of authors. Another of the species wrongly attributed to a West American locality, "Real Llejos." The type, in the British Museum, is now labeled as from west Africa, and it seems indeed to be *Psammotreta cumana* (Costa, 1829), from Senegal.

Donax bellus Deshayes, 1855. Described as from Acapulco, Mexico. The type, which is in the British Museum, seems rather to be an Atlantic form, for it matches well *D. variabilis* Say, 1832.

Donax semistriatus Carpenter, 1856 (not Poli, 1795). Described as from the Gulf of California. The type is lost. The Atlantic species to which Carpenter made comparison resembles *D. culter*. The name being preoccupied, we may write the species off as of doubtful status.

Tellina petalum Valenciennes in Humboldt & Bonpland, 1832. Described as from Acapulco, Mexico, and considered by later authors to be a synonym of *Heterodonax bimaculatus* (Linnaeus, 1758). Photographs of the type show that it is a non-Panamic *Macoma* (Coan, in press).

Semele mediamericana Pilsbry & Lowe, 1932. Although described as from west Central America, this is, according to Olsson (1961), actually a Caribbean form, *S. proficua* (Pulteney, 1799)—the type species of *Semele*—a specimen of which evidently became mixed with west coast material.

Superfamily PHOLADACEA

Pholas cornea Sowerby, 1834. Described as having a thin shell with a periostracum and being about 23 mm in length. It was said to have been found on the beach in western Colombia and in the trunk of a tree at low water in Chiriqui, Panama. No later author has cited or illustrated the species, which is therefore a *nomen dubium*, and its placement in the family is problematical. The unfigured type specimen has apparently been lost.

Class GASTROPODA
Superfamily FISSURELLACEA

Fissurella clypeus Sowerby, 1835. Type locality, Santa Elena, Ecuador. The species has not been recognized in recent years, nor has the holotype been located. Sowerby described the interior as tessellated, not a characteristic of the tropical *Fissurella (Cremides)* group. The locality, therefore, is suspect.

Fissurella mexicana Sowerby, 1835. Described as from "Real Llejos, Mexico," an evidently incorrect locality. If West American, this is probably a Peruvian form.

Fissurella rugosa Sowerby, 1835. Stated to have come from the Galápagos Islands. No specimen matching the original illustration and description has been collected there since. The species may well be from the Caribbean and a synonym of *F. rosea* (Gmelin, 1791). The *F. rugosa* of authors is here considered to be *F. microtrema*.

Superfamily PATELLACEA

Patella calcilus Li, 1930. Thought to have been dredged off western Panama. However, according to Pilsbry (1931), it is a common Mediterranean limpet that was accidentally intermixed with Panamic material in the laboratory.

Patella calyx Li, 1930. Another Mediterranean shell erroneously described as being Panamic (see *P. calcilus*).

Patella mazatlandica Sowerby, 1839. Said to have been from Mazatlán, Mexico; apparently a south Pacific species with an erroneous locality.

Superfamily TROCHACEA

Trochus byronianus Wood, 1828. The type lot of this supposed Pacific species, in the British Museum, proves to be an Atlantic form, and the name will fall as a synonym of *Tegula viridula* (Gmelin, 1791).

Phorcus californicus A. Adams, 1853. Specimens labeled as the type lot in the British Museum do not conform to the original description, which cites a green umbilical area. These specimens are referable to *Tegula ligulata mariamadre* Pilsbry & Lowe, 1932, a form lacking such coloration.

Phorcus liratus A. Adams, 1853. Described without locality. This was suggested as West American by Carpenter but is unrecognizable from the description alone. The type material was not found upon search at the British Museum.

Liotia carinata Carpenter, 1857. Described from Mazatlán, Mexico. The type specimen, probably immature, is in the British Museum; it is unidentifiable from available illustrations, but may be a vitrinellid. Diameter, less than 1.5 mm. No similar material has been found in recent years.

Liotia c-b-adamsii Carpenter, 1857. Unrecognizable from the original drawing. The type specimen in the British Museum has completely disintegrated from chemical corrosion; probably immature, less than 1 mm in diameter. The name must be regarded as a *nomen dubium*.

Liotia striulata Carpenter, 1857. Based on immature specimens not now readily identifiable. The type is in the British Museum. Camera lucida drawings suggest possible affinity to the Vitrinellidae. Type probably immature, less than 1 mm in diameter. Mazatlán, Mexico.

"Lunatia tenuilirata" Carpenter, 1857. Apparently a *Tricolia*, family Phasianellidae, now regarded as indeterminate. The holotype in the British Museum is worn and broken, lacking part of the spire and much of the body whorl. It measures 1.4 mm in height and came from Mazatlán, Mexico.

Superfamily NERITACEA

Neritina californica Reeve, 1855. Described from the Gulf of California, this seems instead to be a western Pacific form with an erroneous locality.

Superfamily LITTORINACEA

Lacuna porrecta Carpenter, 1864. A species of the Californian province ranging from Washington to California; reported in the Galápagos Islands by Wimmer (1880), probably in error.

Lacuna unifasciata Carpenter in Gould & Carpenter, 1856. Characteristic of southern California; reported as far south as Magdalena Bay, Baja California, and on Clarion Island.

Superfamily RISSOACEA

Syncera panamensis Bartsch, 1920. Not an assimineid but instead a freshwater species of the family Hydrobiidae now known as *Zetekina panamensis* (Bartsch), *fide* Dr. Dwight Taylor (*in litt.*).

Rissoa infrequens C. B. Adams, 1852. Panama; length, 6 mm. This was allocated to the genus *Pliciscala* De Boury, 1887 (family Epitoniidae) by Bartsch (1915), without supporting evidence. The type was figured by Turner (1956); shell form suggests *Rissoina, s. l.*, an allocation rejected by Bartsch.

Ganesa (?) atomus Pilsbry & Lowe, 1952. Named in the Vitrinellidae, although the genus *Ganesa* is generally regarded as rhipidoglossate. The minute shell may be juvenile; diameter, 1 mm. Caleta Sal, Tumbes, Peru. Indeterminate until supplemented by further collecting.

Superfamily TURRITELLACEA

Turritella sanguinea Reeve, 1849. The type lot in the British Museum shows a shell with rather flat-sided whorls, the early whorls with three carinae; intercalary ribs soon develop, and by the next to last whorl there are eight to nine spiral ribs. Markings are of somewhat rectangular reddish spots on a buff ground. Length, 85 mm. Described from "California," this has not yet been successfully recognized by West American collectors.

Superfamily CERITHIACEA

Cerithium fragaria Valenciennes, 1832. Described as from Acapulco, Mexico, but unfigured, and the type specimen has not been studied since. Not recognized subsequently and probably not Panamic.

Cerithium musica Valenciennes, 1832. Also described from Acapulco, unfigured, and not since recognized. Probably not Panamic.

Cerithium parcum Reeve, 1865, *ex* Gould, MS. Figured by Reeve as from Mazatlán; a small indeterminate form 4 mm high. The only species of this name described by Gould is *Bittium parcum* Gould, 1861, from the Ryukyu Islands of the western Pacific. Probably not Panamic.

Cerithium quadrifilatum "Carpenter" of Reeve, 1865. Figured as from Mazatlán. Indeterminate and evidently misidentified; not the *Bittium quadrifilatum* Carpenter, 1864, from California.

"Cerithium zebrum Kiener" of Reeve, 1865. Figured as from the Galapagos Islands; apparently an erroneous locality record for an Indo-Pacific form.

Superfamily EPITONIACEA

Scala replicata Sowerby, 1844. This was described from "Lord Hood's Island." There are two islands to which this name might apply, one being in the Galápagos Islands, the other in the Marquesas Group. On the basis of the sculpture of specimens in the syntype lot in the British Museum, the species is here interpreted as being from the eastern Pacific. However, other authors, probably on the basis of Sowerby's figures, are interpreting the species as southwestern Pacific. If further study proves that the species is to be rejected from the American fauna, the synonym *E. bialatum* Dall, 1917, will be available for use.

Rissoa infrequens C. B. Adams, 1852. Considered a *Pliciscala* by Bartsch. See under Rissoacea.

Recluzia rollandiana Petit, 1853. Described from "Atlantique, environs de Mazatlán"; later identified as actually from New Caledonia and as Indo-Pacific rather than West American.

Superfamily NATICACEA

Natica glabella Reeve, 1855. Although reported from Nicaragua, this is apparently not a West American species. The original figure shows a sinistral shell, but the type lot in the British Museum is dextral; the original figure was therefore reversed in printing.

Superfamily LAMELLARIACEA

Trivia pulla (Reeve, 1846, *ex* Gaskoin, MS [not Gmelin, 1791]). This was described without locality but has been considered to be West American by some authors. Schilder in 1922 renamed the homonym *T. occidentalis*, accepting a Pacific provenance. Later (1932) he synonymized it with a Caribbean species, *T. antillarum* (Schilder, 1922). Presumably, therefore, it is not a Panamic form.

"**Lamellaria digueti.**" This name has crept into lists through an unfortunate double citation by Dall (1921*b*), who listed the species name both under *Pleurobranchus*, where Rochebrune in 1895 had correctly assigned it, and under *Lamellaria*. Marcus and Marcus (1967) have shown that it is to be rejected from the Prosobranchia and retained in the Opisthobranchia, where originally placed.

Superfamily MURICACEA

Murex nitidus Broderip, 1833. Apparently the young of one of the species of *Muricanthus*, but authors are not in agreement on which one. We are saved the necessity of guessing, since the name is preoccupied by *M. nitidus* Pilkington, 1804.

Murex norrisii Reeve, 1846. Described without locality, and the type is apparently lost. It somewhat resembles a *Murexiella* but has been identified by some authors as the young of a *Muricanthus*.

Murex nucleus Broderip, 1833. Taken from the Galápagos Islands, and transferred by Sowerby in 1841 to *Fusus*. His figure is indeterminate, and the species has not since been recognized.

Murex palmarosae mexicanus Stearns, 1893. Described from the Gulf of California, this has been rejected for two reasons: first, the type specimen seems to be an Indo-Pacific form erroneously recorded as from the American coast; second, the name *Murex mexicanus* Petit, 1852, was used for a Caribbean form.

Murex pumilus Broderip, 1833. Also from the Galápagos, this was pronounced unrecognizable by Sowerby in 1841 and has remained unfigured. The name subsequently was used by A. Adams for a muricid from the Orient.

Murex taeniatus Sowerby, 1859 (? 1860). This has been considered a close relative or synonym of *M. vittatus*, but according to Vokes (1970) it is probably unrecognizable. It may even be an Atlantic rather than a Pacific form.

Murex varicosus Sowerby, 1841, of authors [not of Brocchi, 1814]. A *Homalocantha*, probably from the Indo-Pacific, which early authors erroneously attributed to the eastern Pacific, probably through a mixing of labels.

Coralliophila stearnsiana Dall, 1919. Illustrated by M. Smith (1939). This is probably not from the Panamic province, although the original locality was given as "Lower California." The specimen was from the Stearns collection, many other labels of which have proved incorrect. The species has not since been recognized. Length, 23 mm.

Purpura diadema Reeve, 1846. Reported from Paita, Peru, this probably represents an erroneous locality for *Cymia carinifera* (Lamarck, 1822), an Indo-Pacific species.

Ricinula alveolata (Kiener) of Reeve, 1846. Attributed by Reeve to Panama, this is an incorrect locality record for *Morula alveolata* (Kiener, 1835–36), an Indo-Pacific species. An unnecessary new name for it is *Engina reevei* Tryon, 1883.

Coralliophila stearnsiana; *Ricinula contracta*; *Bulla nonscripta*

"**Ricinula contracta** Reeve, 1846." Described from "Panama and Santa Elena." The shell is yellowish orange, with a white aperture. The sculpture is nodose, and the base is contracted into a narrow canal. Length, 12 mm; diameter, 6 mm. Judging from the original illustration and a photograph of the type lot in the British Museum, this may be an *Engina*. However, Tryon (1883) considered it to be an Indo-Pacific *Peristernia* (another warm-water buccinoid). Dall (1909) cited the species as an *Engina*, with a range from Panama to Ecuador. If it is genuinely Panamic, it may account for the record by Hertlein and Allison (1968) of the Hawaiian *Peristernia thaanumi* Pilsbry & Bryan, 1918, on Clipperton Island. According to R. Bullock (*in litt.*) there is a small buccinid with a characteristic radula on Clipperton Island. Obviously, more research is needed.

Ricinula heptagonalis Reeve, 1846. Said to be from Panama, this proves to be an Indo-Pacific species of *Morula*.

Ricinula zonata Reeve, 1846. This was described as from the Galápagos Islands. It, too, seems to have an erroneous locality, for the type lot in the British Museum is an *Engina* relabeled *E. melanozona* Tomlin, from the Solomon Islands.

Superfamily BUCCINACEA

Buccinum floridanum Lesson, 1842. Described as from Acapulco, Mexico. Now regarded as indeterminate.

Buccinum leiocheilus Valenciennes, 1832. Described as from Acapulco, Mexico. The type specimen is lost; the species is unidentifiable, and probably it is not from the eastern Pacific.

Buccinum phalaena Lesson, 1842. Described as from Acapulco, Mexico. The type is apparently lost; probably not from the eastern Pacific.

Buccinum tulipa Lesson, 1842. Described as from Acapulco, Mexico. The type, in the Paris Museum, proves to be an Indo-Pacific *Peristernia*.

Chrysodomus (Sipho) testudinis Dall, 1890. From deep water off the Galápagos Islands, in 1,485 m. The species has not been illustrated, and the type material was too incomplete for positive placement. Dall even implied that it might prove to belong in the family Turridae. Nomenclaturally, both the generic and the subgeneric names are unacceptable, being synonyms of earlier names, and both pertain to groups that are Arctic or boreal in distribution.

Columbella babbi Tryon, 1883 (Synonym: **Columbella lactea** Reeve, 1858, preoccupied). Originally stated to be from the Gulf of California, this seems not to be a West American species.

Columbella elata Reeve, 1859. Described without locality; later said to be from the Gulf of California. This is evidently a Caribbean form. The type is relabeled *C. terpsichore* Sowerby, 1822, in the British Museum collection, a species now allocated to *Anachis*.

Columbella livida Sowerby, 1832. This name was based on specimens from Panama that have since been lost. They were not illustrated, and the name was not again mentioned in the literature. In the British Museum collections there are three specimens from the Galápagos Islands that are labeled *C. livida*. They are the form named as *Buccinum pulchrum* Reeve, 1846, a good species of *Engina*. Because there is no proof that Sowerby wrote the label or even saw the material from the Galápagos, it seems best to regard his species as a *nomen oblitum*, a forgotten name not to be brought into the formal literature without action by the International Commission on Zoological Nomenclature.

Columbella varians Sowerby, 1832. Described as from the Galápagos Islands, although later cited as from the central Pacific. It is not a West American species.

Nassarius dentifer (Powys, 1835) (Synonyms: **? Nassa angulifera** A. Adams, 1852; **Nassa tschudii** Troschel, 1852). A member of the Peruvian province fauna, erroneously reported from Panama.

Nassarius festivus (Powys, 1835). A Japanese species sometimes cited as being from Panama.

Nassarius nodifer (Powys, 1835). Apparently an Indo-Pacific form stated to have come from the Galápagos Islands and Panama. The species has not been recognized there since.

Nassarius pallidus (Powys, 1835). Apparently an Indo-Pacific *Phos*, cited as from Panama.

Nassarius panamensis (Philippi, 1851). This species, which was proposed as *Buccinum panamense*, from "Panama and Payta, Peru," has not been figured or subsequently recognized. The description suggests a form somewhat like *N. corpulentus*. Should Philippi's type specimen come to light and represent this species, the name *panamensis* would take priority.

Fasciolaria bistriata Gould & Carpenter, 1857. Presumably from Panama. The type seems to have been lost, and the species is unrecognizable from the original description.

Fasciolaria canaliculata Valenciennes, 1832. Described as from Acapulco, Mexico. This is not recognized by later authors and is probably not from the eastern Pacific.

Fasciolaria rugosa Valenciennes, 1832. Also said to have come from Acapulco; not recognized by later authors.

Fasciolaria sulcata Lesson, 1842 [not Anton, 1839, or Lamarck, 1816]. Stated locality, Acapulco, Mexico; not subsequently recognized.

"**Turbinella nodata** Martyn." Cited by Reeve in 1847 as occurring in Panama. Martyn's nonbinomial name had been validated as *Murex nodatus* Gmelin, 1791, for an Indo-Pacific species that is not likely to be found living on the American mainland.

"**Fusinus sulcatus** (Lamarck, 1816)." Cited by Dall (1915) as occurring in Panama. Probably not a West American species.

Superfamily VOLUTACEA

Oliva petiolita Duclos, 1835. An *Olivella* that has been reported from the Gulf of California. Olsson (1956) considers it to be a Caribbean form.

Marginella albuminosa Dall, 1919. Described from "West Mexico"—apparently an erroneous locality, for the type specimen seems to be an immature shell of the Caribbean species *Prunum* (*Microspira*) *labiatum* (Kiener, 1834, *ex* Valenciennes, MS).

Marginella frumentum Sowerby, 1832. A Caribbean *Persicula* that has been erroneously reported from the Panamic province.

Marginella maculosa Kiener, 1834. Another Caribbean *Persicula* that has erroneously been reported as from the Panamic province.

Mitra orientalis Griffith & Pidgeon, 1834 (Synonym: **M. maura** Swainson, 1836). A member of the subgenus *Atrimitra* that has been confused with *M.* (*A.*) *fultoni* Smith, 1892, and *M.* (*A.*) *idae* Melvill, 1893, from California. *M. orientalis* is southern in distribution, from Peru to Chile.

Superfamily CONACEA

Crassispira bridgesi Dall, 1919. "Panama, Stearns collection" (the Stearns collection is a source of many doubtful records). This is a *Buchema*, resembling those described by Corea in 1934 from the Caribbean. The locality should probably be corrected to the Caribbean coast of Panama.

Crassispira candace Dall, 1919. "Gulf of California, Stearns collection." This has not been recognized in the eastern Pacific, but specimens from Granada, British West Indies (ANSP 296,633), are a good match for the type. It should probably be regarded as a Caribbean species.

Crassispira martinensis Dall, 1919. Attributed to Cape San Martin, Gulf of California, evidently an error for San Martin Island, outer coast of Baja California. A synonym of the California species *Pseudomelatoma grippi* (Dall, 1919).

Mangelia coniformis Reeve, 1846, *ex* Gray, MS. Described without locality. Suggested as Panamic by some authors, but the figure shows a shell unlike any now known in the Panamic province.

Mangelia rhyssa Dall, 1919. Originally attributed to the Gulf of California, this was cited by Dall (1921) only near San Diego. It has since been recognized in southern California.

Pleurotoma dysoni Reeve, 1845. "Honduras." Recognized as a Caribbean *Crassispira* (*Gibbaspira*).

Philbertia hilaira Dall, 1919. "Gulf of California, Stearns collection." Not recognized; probably not Panamic.

Clathrodrillia limans Dall, 1919. "Gulf of California, Stearns collection." Unlike any eastern Pacific group.

Pilsbryspira pilsbryi Bartsch, 1950. Supposedly Taboga Island, Panama, an evident error, for it is a common Caribbean species, *Pilsbryspira albomaculata* (Orbigny, 1842). Bartsch's species is, however, the type of *Pilsbryspira*.

Pleurotomella thiarella Kiener, 1839–40. Described without locality, it has been cited by some authors as a synonym of *Crassispira nigerrima* (Sowerby) but is not closely matched by any Panamic species.

Superfamily PHILINACEA

Bulla nonscripta A. Adams, 1850. Described without locality, this has been suggested as West American by some authors. The nearest similarity to a Panamic species seems to be to *Atys casta* Carpenter, 1864, which, however, is larger and more slender. Length, 6 mm; diameter, 3.5 mm. The type specimen is in the British Museum. It seems not to be identifiable as a Panamic form.

Superfamily DORIDACEA

Archidoris britannica (Johnson, 1838). An Atlantic species that has been cited from Panama and from Baja California in 18 m. No modern records confirm it as Panamic.

Class POLYPLACOPHORA

Nuttallina allantophora Dall, 1919. Described as from the Gulf of California; the type specimen proves to belong in the genus *Liolophura* Pilsbry, 1893, a group that occurs commonly in the western Pacific but that is absent in the Panamic province; the type locality is therefore probably in error.

Nuttallina fluxa Carpenter, 1864 (Synonym: **Chiton scaber** Reeve, 1847, not Blainville, 1825). Although cited by some authors as from the Gulf of California, this is instead a Californian form that ranges no farther south than the outer coast of Baja California.

Glossary

This Glossary is in three parts: following the list of terms used in describing morphologic features are a list of terms used in scientific nomenclature and a list of symbols used in scientific nomenclature.

TERMS USED IN MORPHOLOGICAL DESCRIPTION

Accessory plate. A secondary calcareous structure formed in some bivalves to protect the soft parts.

Adductor muscle. Commonly, one of two muscles connecting bivalve shells, tending to draw them together.

Adductor scar. An impression on the interior of a shell where an adductor muscle has been attached.

Amphidetic. Extending on both anterior and posterior sides of the beak; said of the ligamentary area in certain bivalve shells.

Anal fasciole. A band on the outer lip generated by a sinus, notch, or slit, close to the suture and anal opening of a gastropod aperture. See also **selenizone** and **slit band.**

Angulation. The edge along which two surfaces meet at an angle.

Annulation. The pattern of more or less regularly spaced growth rings or striae.

Anterior. Toward the end at or from which the head, in gastropods, or the foot, in pelecypods, tends to emerge; opposite of **posterior.**

Anterior canal. The siphonal canal, a tubular or troughlike extension of the anterior end of the gastropod aperture, enclosing the inhalant siphon.

Aperture. The principal opening in a gastropod shell.

Apex. The first-formed end of a shell, generally pointed.

Apical. Pertaining to the apex.

Apophysis. A projecting peglike or fingerlike structure, functioning as a support for a muscle; a myophore (see p. 274).

Appressed. With the whorls overlapping so that their outer surfaces converge gradually; modern authors prefer the spelling "adpressed."

Aragonitic. Composed of one of two crystalline forms of calcium carbonate—calcite or aragonite.

Arcuate. Arched or curved.

Axial. In gastropods, more or less parallel to the axis of coiling.

Banding. A color marking in continuous stripes.

Basal. Pertaining to the base.

Base. In coiled gastropods, the anterior part, excluding the aperture; in uncoiled or limpet-like shells, the apertural rim; also used, less precisely, for the flattened apertural side of cowries.

Beading. A form of sculpture resembling beads.

Beak. The small tip of a bivalve shell, near the hinge.

Biconic. With a diamond-shaped outline, the spire of about the same shape and size as the body whorl.

Bifid. Divided by a groove into two parts; applied especially to the hinge teeth in bivalves.

Bifurcation. Forking or division into two elements, especially in sculpture.

Buccal. Pertaining to the organs of the

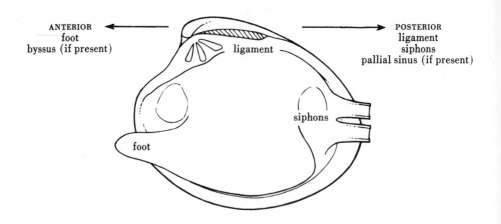

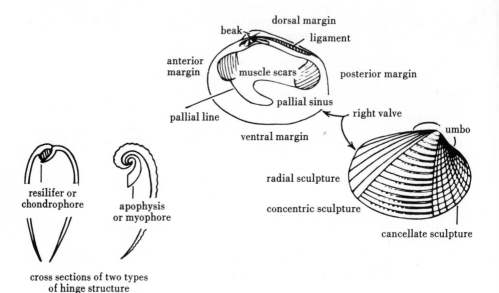

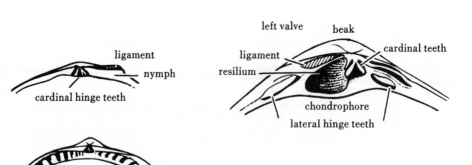

mouth area in gastropods, especially the bulging flexible mass that supports the radula.

Buccinid, buccinoid. Having the shape of shells of the gastropod family Buccinidae.

Bulloid. Bubble-shaped; in the shape of a *Bulla* shell.

Byssal foramen, gape, notch, or **sinus.** An embayment for the passage of the byssus.

Byssiferous. Having a byssus.

Byssus. A bundle of tough fibers secreted by the pelecypod foot, used for attachment.

Caecum. A blind pouch or cavity open at one end.

Calcareous. Having the shell material composed of calcium carbonate.

Callous. Coated with a smooth enamel-like layer (inductura).

Callum. Shell material filling a gape between valves in certain bivalves (see p. 274).

Callus. A shelly substance composing a thickened layer (inductura of some authors), especially around the aperture in gastropods.

Canal. In gastropods, a narrow, semitubular extension of the aperture.

Cancellate. Having sculpture lines intersecting at right angles; reticulate. See also **decussate.**

Cardinal. In bivalves, situated more or less in the central part of the hinge area; immediately below the beaks.

Carina. A keel or prominent knife-edged ridge.

Carinate. With a carina or keel.

Cartilage. An old term for the internal ligament in bivalves.

Cartilaginous. Having a flexible or horny texture, as contrasted with shelly or calcareous.

Cementation. Fixation to the substrate in sessile mollusks, especially bivalves.

Ceras, pl. **cerata.** One of the horn-shaped dorsal appendages in the Nudibranchia.

Chondrophore. In pelecypods, a large, spoon-shaped resilifer.

Clathrate. Having intersecting sculptural elements forming a broad lattice.

Collar. A raised rim bordering a suture.

Columella. The axis of coiling of a tightly spiraled gastropod.

Columellar. Pertaining to the columella.

Columellar callus. A smooth shelly layer extending over the columellar area, secreted by the mantle.

Columellar lip. The part of the inner lip nearest to the axis of coiling, comprising the visible part of the columella.

Concentric. With direction coinciding with that of the growth lines; "commarginal" is preferred by some modern authors.

Conchiolin. The proteinaceous material of which the periostracum and the organic matrix of calcareous parts of the shell are composed.

Condyle. An enlarged and prominent end of a ridge, serving as a pivot (see p. 274).

Cord. A round-topped, moderately coarse, spiral or axial sculptural element.

Coronate. Encircled by a row of spines or prominent nodes, especially at the shoulder of the last whorl in gastropods.

Costa, pl. **costae.** A round-topped sculptural element, stronger than a cord, usually formed by periodic thickening of the outer lip in gastropods.

Crenate. Having notches along the edge or along the crest of ribs.

Crenulate. Having a regularly notched edge, as on the gastropod aperture or the ventral margin of pelecypod valves.

Crural. Pertaining to crura.

Crus, pl. **crura.** Pairs of diverging ridges on the hinge of some bivalves, resembling teeth.

Ctenidium, pl. **ctenidia.** The respiratory organ in the Mollusca, modified for food-gathering in the Pelecypoda.

Cusp. A prominence or point, especially on a tooth of the radula or a denticle on the shell.

Deck. A small sheet of shelly substance in the umbonal region of a valve; also used to describe the diaphragm of slipper shells (*Crepidula*).

Decussate. Having a latticed surface formed by the intersection of fine ribs, not necessarily at right angles. See also **cancellate.**

Dentate. Having teeth; with a toothed margin.

Denticles. Small projections resembling teeth, around the margin of the gastropod aperture or the pelecypod valve.

Denticulate. With denticles.

Denticulation. An edge or pattern of fine teeth, having raised points.

Dentition. Tooth structure; referring in bivalves to the hinge teeth, in gastropods usually to the structure of the elements of the radula.

Depressed. Low in proportion to diameter.

Dextral. In gastropods, the direction of coiling in which, with the shell held upright (apex at the top), the aperture opening is at the right; and in which, with the shell viewed from above the apex, the coiling proceeds from the apex in clockwise direction; opposite of sinistral.

Distal. Relatively remote from the center of the body or point of attachment.

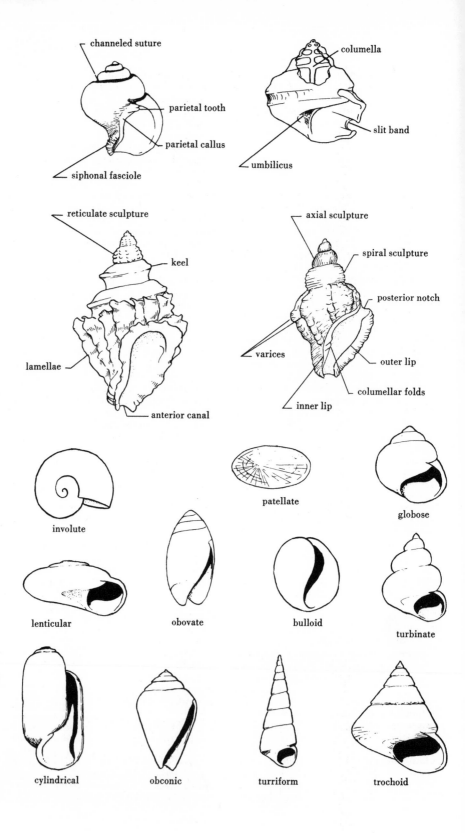

channeled suture

columella

parietal tooth

slit band

parietal callus

umbilicus

siphonal fasciole

reticulate sculpture

axial sculpture

spiral sculpture

keel

posterior notch

varices

outer lip

lamellae

columellar folds

inner lip

anterior canal

involute

patellate

globose

lenticular

obovate

bulloid

turbinate

cylindrical

obconic

turriform

trochoid

Divaricate. Having the sculpture composed of pairs of rather widely divergent costae or threads.

Dorsal. In bivalves, at or toward the hinge.

Dorsum. The back of the gastropod shell, opposite the apertural side.

Edentulous. Lacking teeth.

Emarginate. Notched.

Entire. Smoothly arched, without a reentrant curve, sinus, or crenulation.

Escutcheon. A depressed, smooth, or otherwise set-off dorsal area behind the ligament in pelecypods.

Euthyneurous. Having straight visceral nerve loops; said of gastropods.

External ligament. That part of the ligament visible from outside the bivalve shell, consisting of a lamellar layer under tensional stress.

Fasciole. A spiral band generated by a notch, bordered by successive lamellar growth striae of a canal (either anterior or posterior) in gastropods.

Fimbriate. Fringed.

Fimbriation. A fringe.

Flexure. A bending or angulation.

Fluviatile. Living in rivers or streams.

Foliaceous. Of sculpture, having a leaflike appearance.

Fold. A spirally wound ridge on the columellar wall of the gastropod shell.

Foliose, foliaceous. Furnished with or made up of foliations or leaflike small plates.

Foot. A muscular structure projecting anteriorly, used for locomotion.

Fossa. A ditchlike or trenchlike depression.

Fossette. A pitlike resilifer for the attachment of the internal ligament.

Fossula. A shallow linear depression on the inner lip of some cypraeid gastropods.

Funicle. A ridge of callus spiraling into the umbilicus in the naticid gastropods.

Fusiform, fusoid. Spindle-shaped, elongate, with spire and canal or nearly equal length and similar outline.

Fusoid. See **fusiform.**

Gape. A narrow opening remaining between the valves of a bivalve when the adductor muscles have closed the shell.

Gaping. Not closing tightly.

Gill. The respiratory organ in the mollusk. Ctenidium.

Girdle. A flexible, leathery, muscular integument holding the chiton valves in place, often ornamented with scales, spicules, or hairy processes (see p. 862).

Globose. Rounded; subspherical.

Granule. A pustular surface sculpture.

Growth line. One of several lines on the surface of a shell marking the position of the margin of the shell at a previous stage of growth.

Helical. Spirally coiled.

Heterodont. In certain bivalves, having the hinge differentiated into distinct cardinal and lateral areas.

Heterostrophic. Having apical whorls coiled in a direction apparently opposite to that of succeeding whorls.

Hinge. The interlocking, often toothed, structure in the dorsal region of the two valves of the bivalve shell that functions in the opening and closing of the shell.

Hinge plate. In pelecypods, the dorsal margin carrying the hinge teeth.

Hyperstrophic. See **ultradextral.**

Hypoplax. An accessory, ventral shell piece between the valves in some burrowing clams (see p. 274).

Immersed. Sunken; said especially of apical whorls.

Incised. Sculptured with one or more sharply cut grooves.

Incremental axials. In gastropods, the growth lines between lines of axial ribbing.

Incremental line. See **growth line.**

Inductura. A smooth shelly layer, commonly extending from the inner side of the aperture in a gastropod shell over the columellar lip and base and possibly over the entire shell exterior.

Inequilateral. In pelecypods, having the parts of the shell anterior and posterior to the beaks differing appreciably in length.

Inequivalve. Having one valve larger than the other.

Inflated. Swollen.

Inoperculate. Lacking an operculum.

Integument. An outer covering layer of the soft parts.

Intercostal. Of or in the spaces between ribs or costae.

Interspaces. Channels between ribs.

Internal ligament. See **resilium.**

Involute. Having the last whorl enveloping earlier ones, so that the height of the aperture is the height of the shell.

Keel. A **carina**, or outstanding rib, usually marking an abrupt change of slope in the shell outline.

Labium. The term "inner lip" is preferred by modern authors; see **lip.**

Labrum. The term "outer lip" is preferred by modern authors; see **lip.**

Lamella, -ae. A thin plate or scale.

Lamina, -ae. Lamella.

Lamellar. Platelike.

Lateral. In bivalves, at or of the hinge area

on either side of the cardinal area, well removed from the beaks; (n.) in gastropods, one of the series of radular teeth.

Lenticular. Lens-shaped.

Ligament. The horny elastic structure or structures joining the two valves of a bivalve shell dorsally and causing them to open when the adductor muscles relax.

Lip. The margin of the aperture; the inner lip (labium) extends from the foot of the columella to the suture and consists of the columellar lip and the parietal lip; the outer lip (labrum) is that part of the aperture farthest from the axis of coiling.

Lirate. Having threadlike sculpture.

Liration. An element or pattern of threadlike sculpture.

Lithodesma. A calcareous reinforcement of an internal ligament; same as **ossicle**.

Lobiform. Lobe-shaped.

Lunule. A heart-shaped area, set off by a difference of sculpture, in front of the beaks in the bivalve shell.

Mantle. The fleshy outer layer of the mollusk body that secretes the shell from glands along its margin and also provides the periostracum; the mantle may also form tubular folds (the siphons) for circulation of water currents.

Marginal. One of the series of radular teeth in gastropods.

Mesoplax. One of the accessory plates in the Pholadacea (see p. 274).

Metaplax. One of the accessory plates in the Pholadacea (see p. 274).

Monomyarian. Having only one adductor muscle, the posterior.

Monopectinate. Having one set of comblike filaments.

Muscle scar. An impression on the interior of a shell marking the former place of attachment of a muscle.

Myophore. See **apophysis**.

Nacre. A form of shell structure consisting of thin leaves of aragonite lying parallel to the inner surface of the shell, and exhibiting a characteristic luster.

Nacreous. Consisting of or of the character of nacre.

Nepionic. In the earliest postlarval stage.

Nestler. One of various small mollusks, mostly bivalves, that enter crevices and cavities as larvae and remain there throughout adult life.

Nodose, nodular. Bearing tubercles or knobs.

Notum. The dorsal surface in certain opisthobranchs.

Nucleus. The earliest formed part of the shell or the operculum.

Nymph. A projection along the bivalve hinge margin to support the external ligament or to reinforce the normal hinge structure.

Obconic. Approximately cone-shaped.

Obovate. Reversed ovate; having the greatest width above the aperture, toward the apex.

Obsolete. Said of a structure or sculpture that tends to disappear or remain undeveloped.

Opercular. Pertaining to the operculum.

Opercular peg. An extension of the operculum forming an apophysis.

Operculate. Having an operculum.

Operculum. The horny or calcareous structure formed by the integument of the posterior part of the foot, serving as a closure for the aperture when the animal retracts.

Opisthogyrate. In bivalves, having the beaks pointing backward or posteriorly; opposite of **prosogyrate**.

Orbicular. Circular.

Orthogyrate. In bivalves, having the beaks pointing toward each other.

Osphradium. A sensory organ in gastropods in the pallial cavity, apparently for the reception of chemical stimuli.

Ossicle, ossiculum. A small calcareous plate reinforcing the internal ligament; same as lithodesma.

Pallet. One of the two simple or compound calcareous structures at the siphonal end of some woodboring mollusks, closing the burrow when the siphons are withdrawn (see p. 281).

Pallial. Pertaining to the mantle.

Pallial sinus. In bivalves, an embayment of the pallial line marking the attachment of the marginal muscles of the mantle.

Parietal. Of or situated along the basal surface of a helically coiled shell, along the columellar border of the aperture.

Patellate. Saucer- or limpet-shaped.

Paucispiral. Having only a few whorls or turns.

Periostracum. The outermost layer of the molluscan shell, composed of an amorphous and horny organic substance called conchiolin; erroneously called "epidermis" by authors.

Periphery. The part of a shell or whorls farthest from the axis of coiling.

Peristome, peritreme. The margin of the aperture.

Pillar. The part of the body whorl of the gastropod shell adjacent to the columellar lip; also, sometimes, the columella.

Plait. One of the folds on the columella or pillar of gastropods; a plication.

Planispiral. Coiled in a single plane; loosely used also with discoidal shells having asymmetrical sides.

Planorboid. With flattened or planispiral coiling, in the manner of the gastropod family Planorbidae.

Plate. A flattened calcareous structure of the bivalve shell, e.g. the hinge plate or the accessory plates.

Plica, plication. A fold, especially on the columella of gastropods.

Plicate. Folded or twisted.

Plication. A plait.

Porcelaneous. Having a translucent, porcelain-like appearance.

Posterior. Toward the apical end of, in coiled gastropod shells, or the siphonal end of, in pelecypods; opposite of anterior.

Postnuclear. In gastropods, of the whorls other than those of the nucleus.

Prismatic. Of shell structure, consisting of prisms of calcite or aragonite.

Prodissoconch. A shell secreted by the larva or embryo and preserved at the beak of some adult bivalve shells.

Prosogyrate. In bivalve shells, curved so that the beaks point anteriorly; opposite of **opisthogyrate**.

Protoconch. The apical whorls of a shell, especially where clearly demarcated from later whorls.

Protoplax. One of the accessory plates in the Pholadacea (see p. 274).

Punctate. With pinprick-like depressions.

Pustule. A unit of knobby sculpture, generally smaller than a tubercle.

Quadrate. Rectangular in general outline.

Radial. In bivalves, in the direction of growth outward from the beak at any point on the surface of a shell, usually represented by sculptural elements, such as ribs or furrows.

Radial areas. In chitons (the Polyplacophora), the terminal areas plus the lateral areas of the intermediate valves (see p. 862).

Radula. A rasping organ in the mouth area of gastropods, comprised of serial rows of flexible teeth.

Recurved. Bent back; said especially of the anterior canal.

Reflected. Turned outward or backward.

Resilifer (or **resiliifer**). The socket-like structure (fossette or chondophore, depending upon shape) that supports the internal ligament in certain bivalves.

Resilium. An internal ligament, irrespective of composition ("cartilage" of earlier authors), in a resilifer, under compressional stress.

Reticulate. See **cancellate**.

Rhipidoglossate. Having a radular dentition in which the marginal teeth are numerous, resembling the ribs of a fan.

Rib. A moderately broad and prominent, generally elongate elevation of the surface of a shell, directed radially or otherwise; a costa.

Rostrate. Having a pointed, beaklike end.

Rostrum. An elongate or beaklike structure.

Scar. A marking on the interior of a shell indicating the former place of attachment of a muscle.

Sculpture. A more or less regular relief pattern present on the surface of many shells.

Selenizone. The spiral band of crescentic growth lines generated by a notch or narrow slit in the Archaeogastropoda; a slit band.

Septum. A decklike shelly process in the anterior end of some bivalves.

Shelly. Composed of calcium carbonate rather than conchiolin; that is, with a porcelaneous rather than a horny texture.

Shoulder. An angulation of the whorl in gastropods, forming one edge of a sutural ramp or shelf.

Sinistral. In gastropods, the direction of coiling in which, with the shell held upright (apex at the top), the aperture opening is at the left; and in which, with the shell viewed from above the apex, the coiling proceeds from the apex in counterclockwise direction; opposite of dextral.

Sinuate. Curved.

Sinus. A bend or embayment, either in growth lines or in the attachment scar of the mantle.

Siphon. A tubelike extension of the mantle for the passage of water currents, whether inhalant or exhalant.

Siphonal fasciole. A spiral roughened tract in gastropods, near the anterior end of the columella, formed by successive growth stages of the anterior canal.

Siphonoplax. One of the accessory plates in the Pholadacea (see p. 274).

Slit band. See **selenizone**.

Socket. In bivalves, a recess for the reception of a hinge tooth from the opposite valve.

Spicule. One of the small, slender, pointed, hard bodies in the integument, serving as stiffening.

Spinose. Having spines or thornlike protuberances.

Spire. The visible part of all whorls of the gastropod shell except the last or body whorl.

Stria, pl. **striae.** A line or ring on a shell indicating a growth stage.

Striate. With fine sculpture having the appearance of microscopic scratches or grooves.

Striation. An element or pattern of striate sculpturing.

Stromboid notch. A sinuation in the outer lip of shells of the family Strombidae, or a similar sinuation in other gastropod shells.

Sub-. A prefix indicating "somewhat" or "almost"; as, subglobular—almost spherical. Also, "below"; as, subsutural—below the suture.

Substrate. The sea floor or other base of attachment or living site of a mollusk.

Subdued. Weak; not evident.

Suctorial. Having the mouth parts modified for ingesting juices of plants or soft-bodied animals.

Sulcus. A slit or fissure.

Suture. The spiral line marking the junction of the whorls in the gastropod shell.

Tabulate. Shouldered; with a shelflike area between the periphery of the whorl and the suture.

Taenioglossate. Having a ribbon-like radular dentition, generally with each serial row comprised of central, lateral, and marginal teeth.

Taxodont. Having numerous short hinge teeth more or less similar in shape and mostly set at an angle to the hinge margin.

Teleoconch. The entire shell exclusive of the protoconch.

Terminal areas. In chitons (the Polyplacophora), the head valve and the radially sculptured posterior part of the tail valve (see p. 862).

Torsion. The displacement of the mantle-and-shell from the enclosed visceral mass, rotating by growth stages in a flat or extended helix, and accompanied by the cessation of development of one-half of many of the paired organs in the soft parts.

Toxoglossate. With the radular teeth having an attached poison gland, the teeth generally harpoon-shaped.

Trochoid. Having the form of a top; conical, with the base of the cone at the aperture.

Truncate. Sharply or squarely cut off.

Tubercle. A moderately prominent small rounded elevation on the surface of a shell.

Turbinate. Having a broadly conical spire and a convex base, as in the gastropod family Turbinidae.

Turreted. Tower-shaped, with a long spire and somewhat shouldered whorls.

Turriform. Tower-shaped.

Ultradextral. Having seemingly sinistral shell coiling; having the soft parts dextral, but with the shell whorls being added above instead of below the periphery; hyperstrophic.

Umbilicate. Having an umbilicus.

Umbilicus. The open axis of coiling in a loosely spiral gastropod shell.

Umbo, pl. **umbones.** The upper or early part of the bivalve shell as seen from the outside, opposite the hinge.

Umbonal. Pertaining to the umbones.

Umbonal ridge. A radial ridge that sets off the posterior from the central and anterior slopes of a bivalve shell.

Unguiculate. Claw-shaped.

Varix, pl. **varices.** A periodic growth-resting stage in certain gastropods, marked by a thickening of the outer lip.

Ventral. At or toward that part of the bivalve shell opposite the hinge.

Vermiculation. Surface sculpture of irregular wavy lines or grooves.

Volution. One of the whorls or turns of a spirally coiled shell or operculum.

Whorl. Any complete coil of a helically coiled shell.

Wing. A more or less elongate triangular terminal part of the hinge area in shells of such pelecypod groups as the Pectinacea and Pteriacea.

TERMS USED IN SCIENTIFIC NOMENCLATURE

Binomial. Conforming to the principles of binary nomenclature (with generic and specific names in combination); (n.) the full taxonomic name of a species (e.g. *Hermissenda crassicornis*).

Cotype. A term abandoned by modern taxonomists because used in two entirely different senses: for *paratype*, and for *syntype*.

Emend. An emendation or deliberate alteration in the spelling of a name.

Ex. From; as *ex* Doe, MS.—from Doe's manuscript or unpublished work.

Fide. On the faith of.

Holotype. A single specimen serving as "type" of a species; the specimen designated by the author in describing the species.

Homonym. The later of two identical names that have been given to two different species or genera.

Hypotype. A figured specimen collected subsequent to an original lot or a specimen not from the original type locality (not primary type material).

In litt. Abbreviation of "*in litteris*"—information conveyed in a letter or by other correspondence.

Lectotype. A syntype selected by a later author to serve in the place of a holotype; other specimens in the syntype lot then become lectoparatypes.

MS. Abbreviation for *manuscript*, an unpublished work.

Nonbinomial. Not conforming to the accepted principles of binary nomenclature, as set forth in the International Code of Zoological Nomenclature.

Nomen dubium. A name not applicable to any known taxon.

Nomen nudum. A name published without sufficient documentation to validate it under the International Code; such a name may be used later if given with a proper description.

Nomen oblitum. A "forgotten name" or senior synonym that has remained unnoticed in the literature for more than fifty years. This category was defined by the ICZN in 1958. Opposed by many systematists, it was, in December 1970, repealed by the ICZN, in Declaration 43.

Nominate. Said of a subordinate taxon that bears the same name and therefore contains the type species of a subdivided higher taxon; e.g., *Conus*, *s. s.*, is the nominate subgenus of the genus *Conus*; Buccinidae is the nominate family of the superfamily Buccinacea.

Paratype. A specimen from the same lot as the holotype, studied by an author but not selected as the "type" of the species.

Preoccupied. A name invalidated because some previous author had used the same term or combination for a different kind of animal.

S. l. *Sensu lato*—in the broad sense.

S. s. *Sensu stricto*—in the strict sense.

Synonym. Customarily, the later of two different names that have been bestowed upon a single species or genus; technically, both are synonyms, one senior, the other junior.

Syntype. Any one of several specimens in an original lot on which a species is based, no one specimen having been designated by the author as "type."

Taxon, pl. taxa. Any unit or level in classification, such as a class, superfamily, genus, subspecies, etc. Taxa may be grouped for some purposes: a generic taxon may be either generic or subgeneric in rank, and a family-group taxon may be a family, subfamily, or superfamily.

Teste. On the testimony of.

Topotype. A specimen from the type locality of a species, but collected subsequent to the original collection (not primary type material).

Symbols Used in Scientific Nomenclature

Brackets [ ] for inserted information; for indication of the prior use of preoccupied names; or to indicate that an author's name is not given in a particular description but can confidently be deduced from other evidence, as "[Lightfoot, 1786]."

Equality sign (=) to show that a name is equivalent to a prior one.

Italic type, by international convention, for scientific names of genera and species (and of subgeneric and subspecific taxa, but not of families or higher units).

Parentheses ( ), enclosing the name of the author and date of publication, to indicate that the species name was originally associated with some other generic name and has since been transferred to the present combination.

Periods, in abbreviating scientific names, to avoid needless repetition: *Conus brunneus* may be shortened to *C. brunneus* on second mention and *Conus brunneus pictus* to *C. b. pictus.*

Question mark (?), standing ahead of a scientific name to indicate that there is uncertainty over the relationship of the species or genus concerned (for example, whether or not it is a synonym); standing between a generic or subgeneric and a specific name, to indicate that the uncertainty is in whether the species properly belongs in this generic or subgeneric unit.

Geographic Aids

This appendix is in four parts: A. Maps and Charts; B. Spanish-English Equivalents; C. Index of Place Names; and D. Conversion Tables.

A. MAPS AND CHARTS

Maps are indispensable tools in record-keeping. For field use and for approximate spotting of localities, the maps supplied by oil companies and automobile clubs may be adequate. However, for more detailed, more precise work, especially in describing localities for publication or for interpreting those from museum labels, larger-scaled maps are needed. The following are especially useful:

American Geographical Society, Map of Hispanic America, 1 : 1,000,000. (Published by the American Geographical Society, Broadway at 156th St., New York, N.Y., 10032; cost, $2.50 per sheet, the entire area of the Panamic province comprising 15 of these sheets.)

U.S. Naval Oceanographic Office, charts and publications, especially "Sailing Directions for South America, vol. 3: West Coast between Gulf of Panama and Cabo Tres Montes," H.O. Publ. 25; "Sailing Directions for the West Coasts of Mexico and Central America," H.O. Publ. 26; and "Catalog of Nautical Charts and Publications, Region 2 (Central and South America and Antarctica)." (Available from the United States Naval Oceanographic Office, Washington, D.C. 20390.)

Simplified maps of the coastline and islands of the Panamic province are given on pp. 929–32; explanations of place-name treatment and of the map-location numbers are given in sections B and C, which follow.

B. SPANISH-ENGLISH EQUIVALENTS

Because the early explorers and colonizers of Central and South America were in the main Spanish-speaking, most of the current place names are Spanish. A list of some of the common Spanish-English equivalents for geographic features is supplied here to aid the reader who is not familiar with the Spanish language.

agua	water	azul	blue
ancón	open bay	bahía	bay, cove
archipiélago	islands, archipelago	bajo	shoal
arena	sand, sandy	barra	bar

blanco	white	laguna	lagoon
boca	mouth, passage	los	the (plural [masculine])
cabeza	head	mar	sea
cabo	cape	morro	headland, bluff
caleta	bay, cove	negro	black
canal	canal	norte	north
charca	marsh, pool, pond	nuevo	new
chico	small	oeste	west
ciudad	city	paredón	mountain ridge
colorado	reddish	península	peninsula
costa	coast	piedra	rock
cruz	cross	playa	beach, shore, strand
de	of	puerto	port
del	of the	punta	point, cape
el, los	the (singular, plural [masculine])	quebrada	ravine, gorge, stream
ensenada	bay, bight	rada	roadstead, offshore anchorage
escollo	rock, reef	redondo	round
este	east	río	river
estero	estuary	roca	rock
falso	false	saladar	salt marsh
farallón	headland, cliff, rocky islet	salado	salty
golfo	gulf	salina	saltpan
gordo	wide, big	san	[contraction of santo]
isla	island	santa, santo	saint (feminine, masculine)
isleta	islet	sud, sur	south
islote	barren islet, key	tierra	land
la, las	the (singular, plural [feminine])	verde	green
lago	lake, lagoon	villa	town

C. INDEX OF PLACE NAMES

Modern map makers have adopted the convention of citing place names in the language of the country being mapped, and that convention is followed in this Index. But because the geographic data in the text were compiled from published records and museum labels, anglicized versions of names predominate there. Thus in the Index, "Magdalena Bay" becomes "Magdalena, Bahía de." English names are cited only if markedly different from their Spanish-language equivalents (for example, in the dual sets of names for the islands of the Galápagos archipelago). Also, the indexing is, in general, to the principal word in the name—"Abreojos, Punta," rather than "Point Abreojos." There are some exceptions, especially with names compounded with the definite article—"Los Angeles," rather than "Angeles, Los." The nearshore islands are indexed with mainland numbers, but the far off-shore islands form another north-to-south sequence. The term "the Gulf" in the text is to be understood to refer to the Gulf of California ("Golfo de California"), which only locally is known as "Mar de Cortez." Parenthetically one might comment that were strict priority to govern in place names, as it does in taxonomy, we might be calling this the Vermilion Sea, since "Mar Vermiglio" was the name used by Juan Martines in 1578 on the first published map that showed this geographic feature.

Because the scale of the maps on pp. 929–32 is necessarily too small for all the cited place names to be given without undue crowding, their positions are shown by index numbers. These are assigned to the principal localities in a continuous sequence from north to south, except for a south-to-north reversal on the west sides of such large embayments as the Gulf of California and the Gulf of Panama. Stations intermediate between numbered localities are indexed under the lower number; for example, all localities between Cabo San Lucas (7) and Punta Gorda

(8) are indexed as 7, and all between Guaymas (41) and Bahía Topalobampo (42) are indexed as 41. The reference points themselves are shown in boldface type in the Index. Larger-scaled maps will, of course, show many more place names than are included here, but the present list comprises most of the names commonly cited in the literature.

This seems a proper place to emphasize that in describing species and reporting upon collecting stations, authors should cite localities in terms of latitude and longitude, whether or not place names are given. Many present-day place names, after all, may prove as ephemeral as have some famous earlier localities, such as the "Guacomayo" and "Real Llejos" of Cuming, which do not appear on modern maps. Another argument for use of latitude and longitude coordinates is that names like "San Carlos" or "Punta Arena" or "Punta Coyote" may recur even within short distances of each other along some stretches of the eastern Pacific coast; there are, for example, two islands named "Isla Partida" in the Gulf of California, almost at opposite ends of the Gulf, and there are two named "Isla Gardner" in the Galápagos. Confusion arising out of such pairs might well affect precision in range records. On the other hand, it must be acknowledged that latitude and longitude indicators are more vulnerable to catastrophic error in typesetting or transcribing.

The place names are listed twice: first, in alphabetic order; second, in coastline order. The corresponding map location numbers are given in both lists. In the first list, which follows, places with two (or more) names are listed under both (or all) names, and the corresponding names are given in each case in parentheses; and place names duplicated at separate locations in the province are listed with both map-location numbers. Where a place name is applied to a complex of geographic features, the complex is listed under the feature falling first in alphabetic sequence, with the others following semicolons—e.g. "Asunción, Bahía; Isla; Punta, 2."

In the list that follows, place names are given in coastline order (see maps on pp. 929–32). As explained above, places falling between numbered localities are given under the lower number. And again, where a place name is applied to a complex of geographic features, the complex is listed under the feature falling first in alphabetic sequence, with the others given in parentheses—e.g. "Bahía (Isla, Punta) Asunción."

23 Bahía de los Angeles
    Isla Coronado (Smith)
24 Isla Angel de la Guarda
    Puerto Refugio
    Bahía (Punta) Remedios
25 Bahía Calamajue
    Punta Final
    Ensenada San Francisquito
26 Bahía San Luis Gonzaga
    Bahía (Isla; Punta) Willard
27 Isla San Luis
    Isla Encantada (Salvatierra)
    Isla Coloradito (Lobos)
28 Puertecitos
    Punta San Fermin
    Agua de Chale (Nuevo Mazatlán)
    Punta Diggs
29 Bahía (Punta) San Felipe
    Punta Estrella
    Roca Consag
    Ensenada Blanca
    Punta Sargento
    Isla Montagu
    Isla Gore
    Santa Clara (El Golfo)
30 Bahía de Adair
    Caleta (Punta) Cholla
    Puerto (Punta) Peñasco
31 Bahía (Isla) San Jorge
32 El Desemboque
33 Bahía (Cabo) Tepoca
    Cabo (Puerto) Lobos
    Puerto Libertad
34 Cabo Tepopa
    Punta Sargento
    Bahía Agua Dulce
    Isla Patos
35 Isla Tiburón
    Isla San Esteban
36 Bahía (Punta) Kino
    Isla Pelícano (Tassne)
    Ensenada Tastiota
37 Isla San Pedro Martír
38 Isla San Pedro Nolasco
    Punta Doble
    Isla Santa Catalina
    Puerto San Carlos
    Estero Soldado
39 Ensenada Bacochibampo
    Punta Colorado
    Cabo Arco
40 Cabo Haro
41 Guaymas
    Estero de Cochore
    Empalme
    Estero (Isla; Punta) Lobos
42 Bahía de Yavaros
    Estero de Agiabampo
    Bahía (Punta) San Ignacio
43 Bahía de Topolobampo
44 Altata

45 La Cruz
46 Mazatlán
    Boca de Teacapán
    Playa Novillero
    Playa los Corchos
47 Islas las Tres Marias
    Isla María Cleofas
    Isla María Magdalena
    Isla María Madre
    Isla San Juanito
48 San Blas
    Ensenada Matenchén
    Punta de (Santa Cruzita) Sayulita
49 Punta Mita
    Las Tres Marietas
    Bahía de las Banderas
    Puerto Vallarta
    Yelapa
    Cabo Corrientes
50 Bahía Chamela
51 Bahía Tenacatita (La Manzanilla)
    Bahía (Punta de) Navidad
    Punta de Juluapan
    Cuastecomate
52 Bahía de Manzanillo
    Cuyatlán
53 Bahía de Petacalco
    Zihuatanejo (Sihuatanejo)
54 Acapulco
    Laguna Papagallo
55 Punta Maldonado (Escondida)
56 Punta Galera
57–61 Golfo de Tehuantepec
57 Puerto Angel
58 Puerto Guatulco
    Bahía (Isla) Tangola-Tangola
59 Salina Cruz
    Bahía Ventosa
60 Puerto Arista
61 Puerto Madero
    Guacomayo [not on modern maps]
62 Puerto Champerico
63 Puerto San José de Guatemala
    Puerto de Iztapa
64 Acajutla (Sonsonate)
    Punta Remedios
65 La Libertad
    Puerto el Triunfo
    Playa Tamarindo
66–69 Golfo de Fonseca
67 La Unión
    Isla Perico
    Isla Tigre
68 Puerto Amapala
    Cadeño
    Punta Condega
69 Punta Monypenny
    Punta Conseguina
70 Corinto
    Río Realejo
    Poneloya

Puerto Somoza
Puerto Maschapa
71 Puerto San Juan del Sur
Bahía de Salinas
Bahía Santa Elena
72 Golfo de Papagayo
Cabo Santa Elena
Puerto Parker
Bahía Potrero Grande
73 Bahía de Culebra
Punta Guiones
74 Cabo Blanco
Golfo (Peninsula) Nicoya
Puntarenas
Islas Bocorones
75 Punta Quepos
Punta Uvita
76 Bahía de Coronado
Puerto Dominical
Isla de Caño
Punta San Pedro
Península de Osa
Cabeza Matapalo
77 Golfo de Dulce
Puerto Jiménez
78 Punta (Isla) Burica
78–81 Golfo de Chiriquí
79 Bahía Charco Azul (David)
Puerto Armuelles
80 Horconcitos
Isla Parida
Pedregal
81 Isla Coiba (Quibo)
Bahía Honda
Isla Jicarón
82 Golfo de Montijo
Isla Cébaco
Punta Mariato
Guánico
Búcaro
83 Punta Mala
83–87 Golfo de Panamá
84 Golfo de Parita
Pedasi
Punta Bruja
85 Punta Chame
Isla Venado
Punta Farfan
Fort Kobbe
Isla Changamé
Isla Culebra
Balboa
Panamá
Isla Taboga
86 Chimán
Bella Vista
Punta Patilla
87 Golfo de San Miguel
Ensenada Garachine
Bahía Jarqué
88 Archipiélago de Perlas (Pearl Is.)

Punta Changamé
Isla Pedro Gonzales
Isla del Rey (San Miguel)
Saboga
Isla San José
San Miguel
89 Jurado
Bahía Ardita
Cabo Marzo
90 Golfo de Cupica
Bahía de Octavia
Punta Cruces
Bahía Chiri Chiri
Punta San Francisco Solano
Golfo de Tibugá
91 Cabo Corrientes
92 Bahía Cuevita (Cabita)
Ensenada Utría
93 Ensenada Docampadó
Puerto Pizarro
Punta Charambirá
Ensenada Catripe
Ensenada de Juan Chaco
94 Bahía de Buenaventura
95 Punta Guascama
Isla Gorgona
96 Tumaco
Isla del Gallo
97 Cabo Manglares (Mangles)
Bahía de Ancón de Sardinas
Bahía de San Lorenzo
98 Esmeraldas
Ensenada de Atacames
99 Punta Galera
Punta Sua
100 Bajos de Cojimenes (Cojimíes)
Cabo de San Francisco
Punta Mompiche
Jama
101 Punta Ballena
102 Cabo Pasado
Bahía de Caráquez
Morro Jaramijo
103 Bahía de Manta
Cabo San Lorenzo
Isla la Plata
Punta Canoa
Jipijapa (Xipixapa; Xipixapi; Puerto
de Cayo; Callo)
104 Colonche (Puerto Palmar)
Puerto de Machalilla
Punta Illote
Manglaralto
Salango
105 Bahía de Santa Elena
Salinas (Santa Elena)
La Libertad
106 Punta Ancón
107 Estero del Muerto (El Muerte)
108 Guayaquil
Río Guayas

109 Isla Puna
110 Bolivar
111 Bahía de Tumbes
    Tumbes
    Caleta la Cruz
    Puerto Pizarro
112 Puerto Zorritos
    Caleta Boca de Pan
    Punta de Sal
    Caleta Sal
113 Ensenada de Máncora
    Caleta los Organos
    Pena Negra
114 Cabo Blanco
    Caleta Lobitos
    Negritos
    Punta Pariñas
    Puerto Talara
115 Puerto (Punta) Paita
116 Punta de Foca
    Cabo Verde
117 Ensenada de Sechura
    Puerto Bayovar
118 Punta Aguja
    Punta Falsa (Aguja)
119 Pacasmayo
120 Escollos (Rocas) Alijos [Mexico]
121–24 Islas Revillagigedo [Mexico]
121 Isla Clarión (Santa Rosa)
122 Roca Partída
123 Isla San Benedicto
124 Isla Socorro
125 L'île Clipperton [France]
126 Isla del Coco (Cocos I.) [Costa Rica]

127 Isla del Malpelo [Colombia]
128–46 Galápagos Islands (Archipiélago de Colon) [Ecuador]
128 Isla Darwin (Culpepper I.)
129 Isla Wolf (Wenman I.)
130 Isla Pinta (Abingdon I.)
131 Isla Marchena (Bindloe I.)
132 Isla Genovesa (Tower I.)
    Bahía de Darwin
133 Roca Redonda
134 Isla Isabela (Albemarle I.)
    Bahía Tagus (Tagus Cove)
    Bahía de Banks
135 Isla Fernandina (Narborough I.)
136 Isla Tortuga (Brattle I.)
137 Isla San Salvador (James I.; Isla Santiago)
138 Isla Rábida (Jervis I.)
139 Isla Pinzón (Duncan I.)
140 Isla Seymour
141 Isla Baltra (South Seymour I.)
142 Isla Santa Cruz (Isla Chaves; Indefatigable I.)
    Bahía de la Academía (Academy Bay)
143 Isla Santa Fé (Barrington I.)
144 Isla San Cristóbal (Chatham I.)
    Bahía Naufragio (Wreck Bay)
    Bahía de Stephens
145 Isla Santa María (Charles I.; Floreana I.)
    Bahía del Correo (Post Office Bay)
146 Isla Española (Hood I.)
    Bahía de Gardner

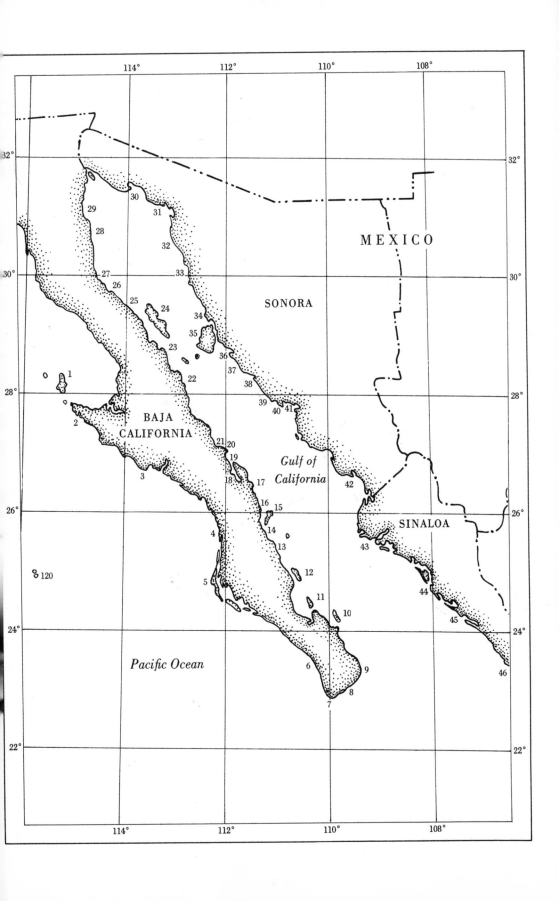

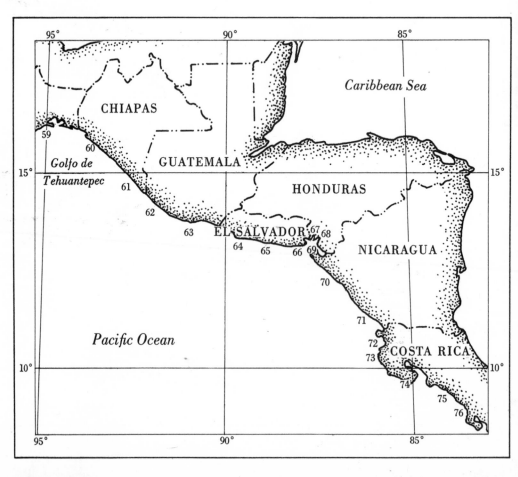

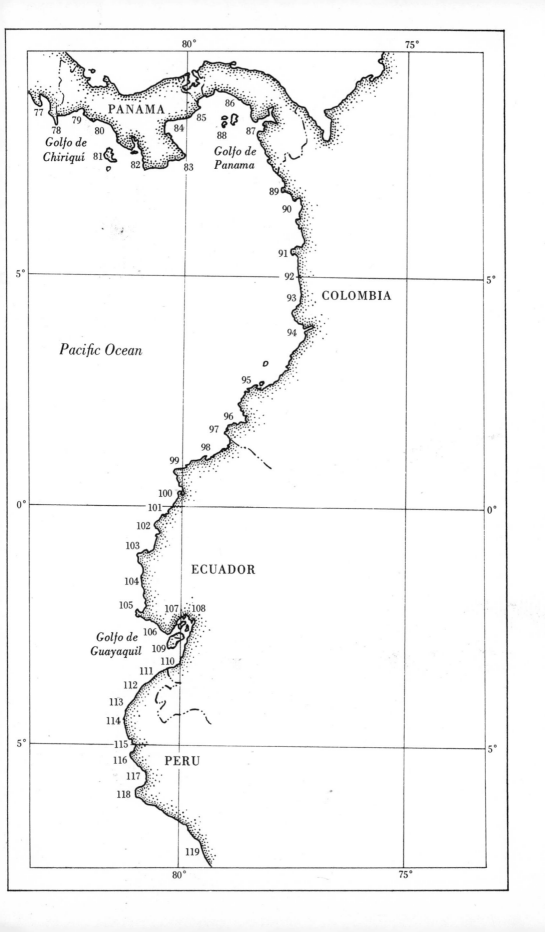

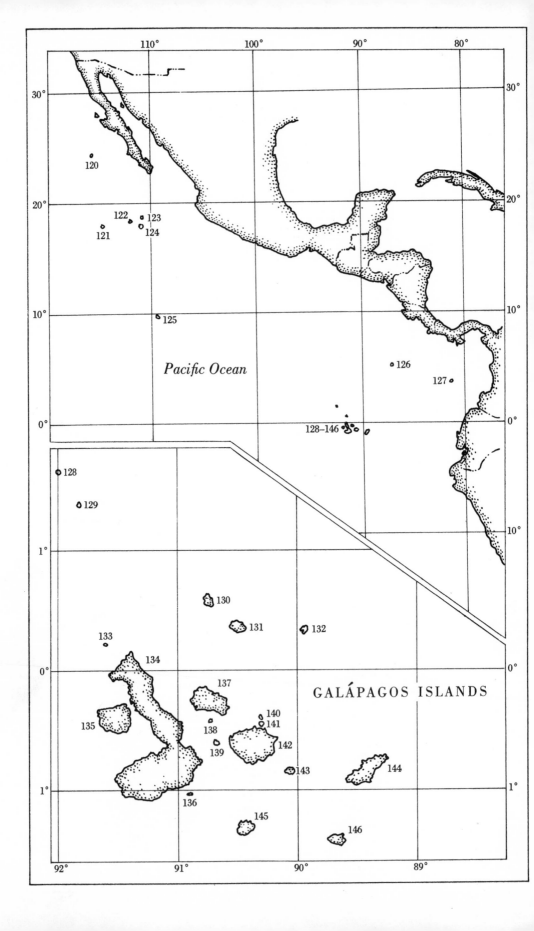

D. Conversion Tables

The metric system has been adopted throughout the text of this book, for both shell measurements and ocean depths. Conversion of millimeters to inches and vice versa is simple if one remembers that there are 25 (actually, 25.4) millimeters per inch. Dual-scale rulers are readily available. Those who are accustomed to thinking in terms of feet or fathoms rather than meters for depths may find the following conversion tables convenient.

FEET TO METERS

Feet	0	1	2	3	4	5	6	7	8	9
0	0.00	0.30	0.61	0.91	1.22	1.52	1.83	2.13	2.44	2.74
10	3.05	3.35	3.66	3.96	4.27	4.57	4.88	5.18	5.49	5.79
20	6.10	6.40	6.71	7.01	7.32	7.62	7.92	8.23	8.53	8.84
30	9.14	9.45	9.75	10.06	10.36	10.67	10.97	11.28	11.58	11.89
40	12.19	12.50	12.80	13.11	13.41	13.72	14.02	14.33	14.63	14.93
50	15.24	15.54	15.85	16.15	16.46	16.76	17.07	17.37	17.68	17.98
60	18.29	18.59	18.90	19.20	19.51	19.81	20.12	20.42	20.73	21.03
70	21.34	21.64	21.95	22.25	22.55	22.86	23.16	23.47	23.77	24.08
80	24.38	24.69	24.99	25.30	25.60	25.91	26.21	26.52	26.82	27.13
90	27.43	27.74	28.04	28.35	28.65	28.96	29.26	29.57	29.87	30.17

FATHOMS TO METERS

Fathoms	0	1	2	3	4	5	6	7	8	9
0	0.00	1.83	3.66	5.49	7.32	9.14	10.97	12.80	14.63	16.46
10	18.29	20.12	21.95	23.77	25.60	27.43	29.26	31.09	32.92	34.75
20	36.58	38.40	40.23	42.06	43.89	45.72	47.55	49.38	51.21	53.03
30	54.86	56.69	58.52	60.35	62.18	64.01	65.84	67.67	69.49	71.32
40	73.15	74.98	76.81	78.64	80.47	82.30	84.12	85.95	87.78	89.61
50	91.44	93.27	95.10	96.93	98.75	100.58	102.41	104.24	106.07	107.90
60	109.73	111.56	113.39	115.21	117.04	118.87	120.70	122.53	124.36	126.19
70	128.02	129.85	131.67	133.50	135.33	137.16	138.99	140.82	142.65	144.47
80	146.30	148.13	149.96	151.79	153.62	155.45	157.28	159.11	160.93	162.76
90	164.59	166.42	168.25	170.08	171.91	173.74	175.56	177.39	179.22	181.05

METERS TO FEET

Meters	0	1	2	3	4	5	6	7	8	9
0	0.00	3.28	6.56	9.84	13.12	16.40	19.68	22.97	26.25	29.53
10	32.81	36.09	39.37	42.65	45.93	49.21	52.49	55.77	59.06	62.34
20	65.62	68.90	72.18	75.46	78.74	82.02	85.30	88.58	91.86	95.14
30	98.42	101.71	104.99	108.27	111.55	114.83	118.11	121.39	124.67	127.95
40	131.23	134.51	137.80	141.08	144.36	147.64	150.92	154.20	157.48	160.76
50	164.04	167.32	170.60	173.88	177.16	180.45	183.73	187.01	190.29	193.57
60	196.85	200.13	203.41	206.69	209.97	213.25	216.54	219.82	223.10	226.38
70	229.66	232.94	236.22	239.50	242.78	246.06	249.34	252.62	255.90	259.19
80	262.47	265.75	269.03	272.31	275.59	278.87	282.15	285.43	288.71	291.99
90	295.28	298.56	301.84	305.12	308.40	311.68	314.96	318.24	321.52	324.80

METERS TO FATHOMS

Meters	0	1	2	3	4	5	6	7	8	9
0	0.00	0.55	1.09	1.64	2.19	2.73	3.28	3.83	4.37	4.92
10	5.47	6.01	6.56	7.11	7.66	8.20	8.75	9.30	9.84	10.39
20	10.94	11.48	12.03	12.58	13.12	13.67	14.22	14.76	15.31	15.86
30	16.40	16.95	17.50	18.04	18.59	19.14	19.68	20.23	20.78	21.33
40	21.87	22.42	22.97	23.51	24.06	24.61	25.15	25.70	26.25	26.79
50	27.34	27.89	28.43	28.98	29.53	30.07	30.62	31.17	31.71	32.26
60	32.81	33.36	33.90	34.45	35.00	35.54	36.09	36.64	37.18	37.73
70	38.28	38.82	39.37	39.92	40.46	41.01	41.56	42.10	42.65	43.20
80	43.74	44.29	44.84	45.38	45.93	46.48	47.03	47.57	48.12	48.67
90	49.21	49.76	50.31	50.85	51.40	51.95	52.49	53.04	53.59	54.13

(From U.S. Naval Oceanographic Office, Publication 26.)

Sources of Illustrations

Original photographs—especially photographs of type material—have been used as illustrations insofar as these could be obtained. The repository or collection housing type material is indicated by abbreviated terms, as shown in the list below. Notes on locality, size of specimen, or magnification of the figure are also included in some cases, where such information supplements the text.

AHF   Allan Hancock Foundation, University of Southern California, Los Angeles

AMNH   American Museum of Natural History, New York

ANSP   Academy of Natural Sciences of Philadelphia

BM   British Museum (Natural History), London

CAS   California Academy of Sciences, San Francisco

LACM   Los Angeles County Museum of Natural History

MCZ   Museum of Comparative Zoology, Harvard University, Cambridge

MHNP   Muséum d'Histoire naturelle de Paris

SBMNH   Santa Barbara Museum of Natural History, Santa Barbara, California

SDSNH   San Diego Society of Natural History

SSB   The collection of Dr. S. Stillman Berry, Redlands, California

SU   Stanford University conchological collection, Stanford, California (cited numbers refer to the Stanford University Paleontological Type Collection)

USNM   United States National Museum (Smithsonian Institution), Washington, D.C.

Where photographs have not been available, or as a supplement to the photographs, reproductions of original line figures have been given. Sources for these are given by author and date (the full references must be traced through the Bibliography) or by abbreviated title, such as "Sowerby, Thes. Conch." for the *Thesaurus Conchyliorum* or "Man. Conch." for the Tryon and Pilsbry *Manual of Conchology*. Some of the line drawings have been made either from original photographic illustrations or from other previously published illustrations; these are distinguished by the word "after." The two figures in the Glossary are from Keen (1963).

PELECYPODA

1. Above, Dall (1895); below, Dall (1908)
2. USNM, holotype
3. USNM, holotype
4. CAS, Gulf of California
5. Left, Schenck (1939), holotype, Ecuador; right, Olsson (1961)

6. Dall (1908); length, 35 mm
7. Olsson (1961)
8. CAS, west Mexico
9. Olsson (1961)
10. USNM, holotype, no. 107,649; x 8
11. Dall (1908); length, 11 mm
12. Dall (1908); length, 5 mm
13. Dall (1908); length, 22 mm
14. Dall (1908); length, 14 mm
15. Dall (1908); length, 17 mm
16. CAS, Gulf of California
17. CAS, Gulf of California
18. CAS, Guatemala
19. Hertlein & Strong, 1951
20. Below, Olsson, 1961; above, Sowerby, Conch. Icon. (1871)
21. CAS, Costa Rica
22. CAS, El Salvador
23. CAS, west Mexico
24. CAS, holotype, Costa Rica
25. Sowerby, Conch. Icon. (1871)
26. Pilsbry & Lowe (1932); see also color plate XII, fig. 1
27. Olsson (1961)
28. USNM, holotype, no. 214,448; x 5
29. Dall (1897)
30. CAS, Baja California
31. Dall (1908); length, 21 mm
32. Dall (1908); length, 8.5 mm
33. USNM, holotype, no. 122,918
34. Dall (1908); length, 16 mm
35. After Strong & Hertlein, 1937
36. No figure
37. Olsson (1961)
38. Above, Olsson (1961), ex Sowerby (1833); below, CAS, holotype, Guatemala
39. After Pilsbry & Lowe (1932)
40. Lowe (1935); see also color plate XII, fig. 2
41. Sowerby, Conch. Icon. (1871)
42. Pilsbry & Olsson (1935)
43. Sowerby, Conch. Icon. (1871)
44. After Pilsbry & Lowe (1932)
45. Dall (1908); length, 28 mm
46. Dall (1908); length, 22 mm
47. Dall (1908); length, 11 mm
48. Dall (1908); length, 9.5 mm
49. Dall (1908); length, 13 mm
50. No figure
51. Dall (1908); length, 15 mm
52. Dall (1908); length, 3.5 mm
53. Dall (1908); length, 8 mm
54. Dall (1908); length, 5.2 mm
55. No figure
56. Dall (1908); length, 5.5 mm
57. Dall (1908); length, 14 mm
58. Dall (1897)
59. USNM, holotype, no. 122,900
60. USNM, no. 211,424, Gulf of California; x 5

61. USNM, type lot, no. 122,917
62. No figure
63. USNM, holotype, no. 122,756
64. After Strong & Hertlein (1939)
65. USNM, holotype, no. 212,892, off southern California; x 15
66–67. Reinhart (1943)
68. BM, syntype
69. Reeve, Conch. Icon., vol. 1
70. After Bartsch (1931)
71. SU, Oaxaca
72. SU, holotype, no. 7,856; Mazatlán
73–74. Reinhart (1943)
75. SU, Concepción Bay
76. Frizzell, 1946
77. Olsson (1961)
78. Above, SU, Panama; below, Olsson (1961)
79. CAS, west Mexico
80. Right, CAS, off Cape San Lucas; left, above and below, Olsson (1961)
81. Left, Olsson (1961); right, CAS, Costa Rica
82. Dall (1909)
83. Olsson (1961)
84. SU, west Mexico
85. SU, Guaymas
86. Above, CAS, El Salvador; below, Olsson (1961)
87. Reinhart (1943)
88. Left, Olsson (1961); right, SDSNH, holotype, no. 16,810
89–90. Reinhart (1943)
91. Olsson (1961)
92. Above and below, Reinhart (1943); center, left and right, Rost (1955)
93. Reinhart (1943), holotype of *Arca gordita* Lowe
94. Reinhart (1943)
95. Campbell (1962)
96. SU, Peru
97. Keen (1963), from Dall (1908); length, 6 mm
98. Above, CAS, Zihuatanejo; below, Olsson (1961)
99. SU, Guaymas
100. MacNeil (1938)
101. Reinhart (1943)
102. Above, after Maury (1922); below, Olsson (1961)
103. Left, below, Reinhart (1943); right, Olsson (1961)
104. Dall (1908); length, 45 mm
105. Dall (1908); length, 4.5 mm
106. Dall (1908); length, 6.2 mm
107. No figure
108. USNM, holotype; length, 6 mm
109. No figure
110. SU, Carmen Island; see also color plate I
111. Olsson (1961)

112. SU, Gulf of California
113. SU, Tres Marias Islands
114. SU, Panama Bay
115. Pilsbry & Olsson (1941)
116. SU, Gulf of California
117. Left, SU, Manzanillo; right, CAS, Manzanillo, offshore
118. Left and below, Keen (1963); above and right, Olsson (1961)
119. Center, SU, Ecuador; others, Olsson (1961)
120. Olsson (1961)
121. Below, left, SU, paratype, Puntarenas; hinge detail from Soot-Ryen (1955); center, Olsson (1961)
122. Above, right, SU, Mazatlán; hinge detail from Soot-Ryen (1955); center, Keen (1968)
123. CAS, Tenacatita Bay; hinge detail from Soot-Ryen (1955)
124. Center, Dall (1909); above and below, Olsson (1961)
125. Below, right, Dall (1909); above, Olsson (1961); below, left, BM, holotype of *Modiola mutabilis* Carpenter
126. Olsson (1961)
127. BM, syntypes; hinge detail from Soot-Ryen (1955)
128. Above, CAS, El Salvador; center, Olsson (1961); below, Soot-Ryen (1955)
129. Below, Olsson (1961); above, Hertlein & Strong (1946); hinge detail from Soot-Ryen (1955)
130. Olsson (1961)
131. Left, Olsson (1961); right, CAS, Gulf of California
132. SU, southern California; magnified detail of a hair, Soot-Ryen (1955)
133. Olsson (1961); center, after Soot-Ryen (1955)
134. USNM, holotype, no. 210,651; x 3
135. Hertlein & Strong (1946); below, Soot-Ryen (1955)
136. BM, syntypes
137. Below, left, and above, right, Dall (1897); second from left, Soot-Ryen (1955); right, Dall (1903), holotype of *C. megas*
138. Olsson (1961)
139. Keen (1963)
140. SU, holotype, no. 8,506
141. SU, paratype, no. 8,516
142. Left, BM, holotypes; 142a, Lowe (1935), holotype of *L. abbotti*
143. Above, right, Keen (1963); below, Keen (1968), after holotypes in BM
144. Olsson (1961)
145. BM, holotype
146. Dall (1921)
147. Left, BM, holotype; right, Olsson (1961)
148. Olsson (1961)

149. Above, after Soot-Ryen (1955), showing also magnified periostracal hairs; below, left, Olsson (1961)
150. Above, CAS, Mazatlán; below, Olsson (1961)
151. Left, Keen (1963); center, Olsson (1961); detail of a periostracal hair, after Soot-Ryen (1955)
152–54. After Soot-Ryen (1955)
155. Above, Olsson (1961); below, SU, holotype, no. 8,584
156. SU, Port Parker, Costa Rica
157. CAS, Gulf of Fonseca
158. Dall (1921)
159. Hertlein & Strong (1943)
160. SU, Kino Bay, Sonora
161. Left, above and below, SU, Gulf of California; above, young specimen with long wing; below, mature individual with short wing; right, Dunker (1872), probable type of *Avicula fimbriata*, courtesy Dr. W. Kilias
162. SU, Gulf of California
163. Left, SU, Gulf of California; right, BM, syntype, Mazatlán
164. SU, San Ignacio Lagoon
165. CAS, Clarion Island; x 3
166. USNM, holotype, no. 101,935; x 3
167. Above, SU, Gulf of California; below, after holotype in MHNP
168. SU, Magdalena Bay
169. SU, Panama
170. Hertlein (1951)
171–72. SU, Oaxaca, Mexico
173. SU, Scammon's Lagoon
174. Right, SU, Magdalena Bay; left, BM, syntype
175. Olsson (1961)
176. SU, holotype, no. 9,711
177. CAS, west Mexico
178–79. LACM, holotypes
180. Olsson (1961)
181. Arnold (1906)
182. Arnold (1906); 182a, LACM, southern California
183. SSB, Angel de la Guarda Island
184–85. LACM
186. Dall (1908)
187. LACM
188. No figure
189. USNM, holotype; length, 7 mm
190. CAS, off west Mexico
191–93. No figure
194. Right, Hertlein (1935); left, Dall (1908)
195. Hertlein & Strong (1946)
196. SSB, holotype
197. SU, holotype, no. 8,505
198. SU, Guaymas
199. SSB, Acapulco
200. SU, Gulf of California

201. SU, hypotype, no. 440, Galápagos
202. Arnold (1906)
203. Dall (1908)
204. Above, Hinds (1844) ; below, Olsson (1961)
205. CAS, Gulf of California
206. SU, holotype, no. 8,500, Guaymas; right valve and exterior shown in place on a rock
207. Below, Durham (1950) ; above, CAS, Gulf of California
208. Above, SU, Gulf of California; below, BM, syntypes
209. MHNP, holotype, Gulf of California; from a photograph, x 3/4
210. Left, SU, Guaymas, adult shell; right, SU, Puerto Escondido, Gulf of California, juvenile shells
211. Left and above, BM, holotype; right and below, Olsson (1961)
212. Olsson (1961) ; see also color plate II
213. SU, holotype, no. 8,585
214. Gould (1853)
215. No figure
216. Dall (1908) ; height, 35 mm
217. Hertlein & Strong (1946)
218. USNM, holotype
219. SU, Manzanillo; see also color plate XII, fig. 5
220. SU, Panama; see also color plate XII, fig. 7
221. Reeve, Conch. Icon.
222. Olsson (1961)
223. Left, Keen (1963) ; right, Olsson (1961)
224. Olsson (1961)
225. Olsson (1942)
226. Olsson (1961)
227. Keen (1963)
228. Larger, SU, San Pedro, California; smaller, BM, syntype
229. SU, Gulf of California
230. Above, SU, Pearl Islands; below, Olsson (1961)
231–32. Olsson (1961)
233. After Pilsbry & Lowe (1932)
234. Turner (1956)
235. Left and right, Keen (1968) ; center, Olsson (1961)
236. SU, holotype, no. 9,712
237. BM, syntypes
237a. SU, Kino Bay, Sonora
238. SU, Panama
239. Above and below, SU, Tres Marias Islands; center, Dall (1908)
240. SU, Manzanillo
241. SU, Panama Bay
242. CAS, Carmen Island
243. Olsson (1961)
244. SU, Gulf of California
245. Lamy (1916)

246. Above, Univ. Copenhagen, syntype; below, Olsson (1961)
247. Left, Olsson (1961) ; right, Prime (1865)
248. Left, Olsson (1961) ; right, Pilsbry (1931)
249. Keen (1968), from syntype in BM
250. Above, CAS, Perlas Islands; below, Prime (1865)
251. Left, BM, type of *C. inflata* Deshayes; center, above, Turner (1956) ; line drawings after Morrison (1946)
252. MCZ, type of *Cyrena exquisita*, no. 176,955
253. BM, holotype
254–55. Olsson (1961)
256. Prime (1865)
257. After Morrison (1946)
258. Olsson (1961)
259. Hertlein & Hanna (1949)
260. Dall (1908) ; length, 14.5 mm
260a. Dall (1908) ; length, 75 mm
261. USNM, holotype, x 0.7; length, 37 mm
262. Dall (1908)
263. Dall (1908) ; length, 58 mm
264. Keen (1968)
265. Olsson (1961)
266. After Morrison (1946)
267. Left, after Pilsbry & Zetek (1931) ; right, Olsson (1961)
268–69. Dall (1901)
270. Center, SSB, Sonora; left and right, Keen (1968), from syntypes in BM
271. Keen (1968), from holotype in BM
272. Below, Dall (1901) ; above, Pilsbry (1931)
273. Above, Hinds (1845) ; below, CAS, Santa Inez Bay
274. Dall (1901)
275. Left, SSB, Sonora; right, Keen (1968)
276. Dall (1901)
277. Lowe (1935)
278. Dall (1901)
279–80. Dall (1901)
281. Hertlein & Strong (1946)
282. After Bartsch & Rehder (1939)
283. Dall (1901)
284. Above, Dall (1901) ; below, Keen (1968), from type of *L. pectinata* Carpenter in BM
285. Keen & Frizzell (1939)
286. Dall (1901)
287. CAS, Arena Bank
288. CAS, Gulf of California
289. Keen & Frizzell (1939)
290. Dall (1901)
291. CAS, Gulf of California
292. Below, Keen (1968) ; above, CAS, Cape San Lucas
293. Olsson (1961)

294. BM, syntypes
295. Above, Olsson (1961) ; below, CAS, Corinto
296. Above, left and center, Olsson (1961) ; right and below, Keen (1968)
297–98. Olsson (1961)
299. CAS, Zihuatanejo
300. Left, Brann (1966) ; right, SU, no. 9,515, holotype of *P. phoebe*
301. Dall (1890)
302. Dall (1901)
303. BM, holotype
304. SU, holotype, no. 9,713
305. Olsson (1961) ; Sowerby, Thes. Conch., vol. 3
306–7. Olsson (1961)
308. SU, holotype, no. 8,586
309. USNM, holotype, no. 267,403, Gulf of California
310. SU, holotype, no. 10,037
311. Keen & Frizzell (1939)
312. Above, Olsson (1961) ; below, after Arnold (1903)
313–14. Olsson (1961)
315. Left, Brann (1966) ; right, Keen (1968)
316. BM, holotype
317–18. Olsson (1961)
319. SU, hypotype, no. 10,049, Salinas, Ecuador
320. Above, Brann (1966) ; below, Keen (1968)
321. Olsson (1961)
322. Olsson (1961)
323. Sowerby (1862), Thes. Conch.
324. Dall (1909)
325. Dall (1925)
326. Brann (1966)
327. USNM, holotype, no. 214,445; also in Dall (1921)
328. Left and below, Brann (1966) ; right and above, Keen (1968)
329. Olsson (1961)
330. Brann (1966)
331. J. Q. Burch collection, off southern California
332–33. Olsson (1961)
334. Harry (1969)
335. Above, BM, holotype; below, drawing by Dr. R. Dell from holotype in BM
336–37. Olsson (1961)
338. Right, BM, holotype; left, drawing by Dr. R. Dell from holotype in BM
339. Olsson (1961)
340. Above and left, Brann (1966) ; right, Olsson (1961) ; below, Keen (1968)
341. Dall (1899)
342. Durham (1950)
343. Durham (1950)
344. After Emerson & Puffer (1957)
345. Olsson (1961)

346. Center, CAS, Galápagos Islands; left and right, Olsson (1961)
346a. Olsson (1971)
347. Reeve (1847)
348. Left, above, Reeve (1847) ; right, Olsson (1961)
349. BM, syntypes of *Chama fornicata*
350. Left, MHNP, holotype of *C. digueti*; right, SSB, west Mexico
351. Pilsbry & Lowe (1943)
352. Reeve (1847)
353. Dall (1903)
354. LACM, holotype and paratypes
355. Reeve (1847)
356. Lowe (1934)
357. BM, holotype
358. Reeve (1847)
359. CAS, Manzanillo
360. SU, Pearl Islands, Panama; see also color plate III
361. SU, Gulf of California
362. SU, Ecuador
363. SU, Kino Bay, Sonora; see also color plate XII, fig. 3
364. SU, Corinto, Nicaragua
365. SU, Gulf of California
366. SU, La Paz
367. SSB, Sonora, Mexico, in 75 m
368. Olsson (1961)
369. SU, Gulf of California
370. SU, Angeles Bay, Gulf of California
371. SU, Ecuador
372. SU, Gulf of California
373. Dall (1890)
374. CAS, Octavia Bay, Colombia
375. Left, Olsson (1961) ; right, Dall (1908)
376. CAS, Santa Inez Bay
377. CAS, Clarion Island
378. SU, Guaymas; see also color plate XII, fig. 3
379. SU, Salina Cruz, Mexico
380. SU, Oaxaca, Mexico
381. Right, CAS, Arena Bank; left, Olsson (1961)
382. Dall (1902)
383. Right, CAS, Taboga Island, Panama; upper left, Olsson (1961)
384. SU, Puntarenas
385. Above, SU, Oaxaca; left, below, BM, holotype of *Cytherea gracilior*
386. Left, CAS, Santa Inez Bay; right, Römer (1864)
387. BM, syntypes
388. Sowerby, Thes. Conch., vol. 2
389. SU, Negritos, Peru
390. SU, Acapulco
391. SU, holotype, no. 9,714
392. Hertlein & Strong (1939)
393. Keen (1968)
394. Pilsbry & Lowe (1932)

395. Above, ANSP, holotype; below, after Pilsbry & Lowe (1932)
396. SU, Acapulco
397. SU, holotype, no. 10,038
398. SU, Ecuador
399. Olsson (1961)
400. CAS, Guayabo Chiquita, Panama
401. Olsson (1961); see also color plate XIII, fig. 4
402. Fischer (1969)
403. Left, after Pilsbry & Lowe (1932); right, ANSP, syntype
404. CAS, Arena Bank
405. Olsson (1961)
406. Left, Olsson (1961); right, Reeve, Conch. Icon., vol. 14 (1864)
407. Olsson (1961)
408. SU, Gulf of California
409. Olsson (1961)
410. SU, off Mazatlán
411. BM, holotype
412. Dall (1902)
413. Upper left, SU, Ecuador; right, Olsson (1961)
414. CAS, Gulf of California
415. SU, holotype, no. 9,516
416. Left, BM, syntypes; upper right, Olsson (1961)
417. SU, Puntarenas
418. SU, Acapulco
419. Olsson (1961)
420. Fischer (1969)
421. Below, Dall (1902); above, after CAS, Colombia
422. CAS, Maldonado Point, Mexico; right, SDSNH, paratype of *P. lenis*
423. SU, Gulf of California
424. SU, Santa Inez, Gulf of California
425. SU, Gulf of California; see also color plate XII, fig. 4
426–27. SU, Gulf of California
428. Left, BM, holotype of *D. annae*; right, SU, Gulf of California
429. Olsson (1961)
430. BM, holotype
431. Olsson (1961)
432. Center, Dall (1902); above and below, Olsson (1961)
433. Above, Olsson (1961); below, left, SU, Gulf of California
434. CAS, Santa Inez Bay, Gulf of California
435. Above, Woodring, U.S. Geol. Survey Prof. Paper 147-C (1926); below, Olsson (1961)
436. No figure
437. Below, Keen (1963); above, Oldroyd (1925)
438. Above, Lamy, Jour. Conchyl., vol. 67 (1922–23); below, Sowerby, Thes. Conch., vol. 2

439. Dall (1921); USNM, syntype, no. 266,158; x 5
440. SU, Espíritu Santo Island, Gulf of California
441. Univ. Calif., no. 3,125, holotype of *C. meridionalis*
442. CAS, Port Guatulco, Mexico
443. CAS, Braxalito Bay, Costa Rica
444. SU, west Mexico
445. SU, Gulf of California; 445a, SSB
446. Upper right, BM, holotype; left and below, SU, Gulf of California
447. SU, southern California
448. SU, Gulf of California
449. Sowerby, Thes. Conch., vol. 2
450. SU, Gulf of California
451. Hertlein & Strong (1955)
452. Above, Pilsbry & Lowe (1932); below, Olsson (1961)
453. Fischer (1969)
454. CAS, Piñas Bay, Panama
455. SU, Kino Bay, Sonora
456. SU, Gulf of California
457. SU, Panama
458. BM, syntypes
459. Hinds (1845)
460. Lowe (1935)
461. Dall (1902)
462. Stearns (1890)
463. Center, Univ. Copenhagen, syntype of *Venus troglodytes*; above, left, and center, BM, syntype; below, left, Olsson (1961); right, SU, Punta Peñasco
464. Dall (1902)
465. Olsson (1961)
466. SU, Colombia
467. SU, Guaymas
468. Olsson (1961)
469. Pilsbry & Lowe (1932)
470. Fischer (1969)
471. Olsson (1961)
472. SU, Ecuador
473. SU, west Mexico
474. Dall (1902)
475. Olsson (1961)
476. Left and below, Keen (1968); above, right, Brann (1966)
477. CAS, Cape San Lucas
478. Turner (1956)
479. Olsson (1961)
480. SU, Kino Bay, Sonora
481. Left, Keen (1963); right, Pilsbry & Lowe (1932)
482. Olsson (1961)
483. Above, after Reeve, Conch. Icon.; below, Gould (1853)
484. Olsson (1961)
485. Left, Keen (1963); right, Lamy, Jour. Conchyl., vol. 67 (1922–23); below, SU, no. 6,983, California

486. SU, holotype, no. 9,517
487. Pilsbry & Lowe (1932)
488. SU, Panama
489. Dall (1894)
490. CAS, San Lucas Bay
491. CAS, La Libertad, El Salvador
492. Dall (1925)
493. CAS, Monypenny Point, Nicaragua
494. Pilsbry & Lowe (1932)
495. Pilsbry & Lowe (1932), *M. v. acymata* at left
496. Below, Pilsbry & Olsson (1935); above, Olsson (1961)
497. Below, Reeve, Conch. Icon., vol. 17; above, Olsson (1961)
498. H. & A. Adams (1858)
499. Olsson (1961)
500. CAS, Ardita Bay, Colombia
501. SU, west coast of Central America
502. Keen (1966), from holotype in Univ. Copenhagen
503. Dall (1894)
504. After Grant & Gale (1931)
505. Above, left, Grant & Gale (1931); right, Keen (1963)
506. Dall (1894)
507. Olsson (1961)
508. Dall (1900)
509. CAS, Arena Bank, holotype of *T. arenica*
510. Dall (1900)
511. Dall (1908)
512. SU, holotype, no. 10,039
513. Above, SU, Panama; below, left, CAS, Monypenny Point, Nicaragua
514. Left, ANSP, holotype; right, Pilsbry & Lowe (1932)
515. BM, syntypes, courtesy Dr. W. Cernohorsky; line drawing, *T. donacilla* Carpenter, after holotype in BM
516. Dall (1900)
517. Above, Dall (1900); below, USNM, holotype
518. Below, CAS, San Luis Gonzaga Bay, Gulf of California; above, Dall (1900)
519. BM, syntypes
520. Above, Turner (1956); below, Olsson (1961); center, Pilsbry & Lowe (1932), holotype of *T. erythronotus*
521. Below, Dall (1900); above, Olsson (1961)
522. Olsson (1961)
523. Dall (1900)
524. USNM, no. 16,101, holotype
525. Left and center, Olsson (1961); right and second from left, Pilsbry (1931), type of *T. panamanensis*
526. CAS, Champerico, Guatemala
527. Above, CAS, Manta, Ecuador; below, Olsson (1961)
528. Olsson (1961)
529. Above, Olsson (1961); below, CAS,

Santa Cruz Bay, west Mexico; see also color plate XIII, fig. 1
530. Sowerby, Conch. Icon., vol. 17 (1867); hinge, SU, Puntarenas
531. Pilsbry & Olsson (1943)
532. Below, Sowerby, Conch. Icon., vol. 17 (1867); above, BM, holotype
533. After Sowerby, Conch. Icon., vol. 17 (1867)
534. Right, Sowerby, Conch. Icon., vol. 17 (1867), with diagram showing interior
535. Turner (1956), from holotype, with diagram to show pallial sinus of maximum extent; see also color plate XIII, fig. 3
536. CAS, Ballena Bay, Costa Rica
537. Below, SU, west Mexico; above, CAS, Gulf of California
538. CAS, Champerico, Guatemala
539. After Pilsbry & Lowe (1932)
540. BM, syntypes
541. USNM, holotype
542. Olsson (1961)
543. Hertlein & Strong (1949)
544. CAS, La Libertad, El Salvador
545. USNM, holotype, no. 122,935
546. Dall (1900)
547. No figure
548. Olsson (1961)
549–50. After Pilsbry & Olsson (1932)
551. Sowerby, Conch. Icon., vol. 17 (1867)
551a. Olsson (1971)
552. Hertlein & Strong (1949)
553–54. Olsson (1961)
555. Right, Gray (1839); left, Olsson (1961)
556. After Pilsbry & Lowe (1932)
557. Left, SU, syntype, no. 6,221; right, SU, Gulf of California; hinge, above, Durham (1950)
558. Left, BM, holotype; right, Reeve, Conch. Icon., vol. 17 (1867)
559. Dall (1890)
560. After Olsson (1942)
561. Lower right, Dall (1897); above, photograph of holotype of *M. quadrana* Dall
562. CAS, Maldonado Point, Mexico
563. Left center, Sowerby, Conch. Icon., vol. 17 (1867); right, above, BM, probable syntypes, Tumbes; below, SSB, Nayarit
564. USNM, holotype
565. Pilsbry & Lowe (1932)
566. Below, Dall (1900); above, Turner (1967)
567. Hertlein & Strong (1949)
568. Palmer (1958), from holotype in USNM
569. Sowerby, Conch. Icon., vol. 17 (1867)
570. Above, after Sowerby, Thes. Conch.; below, BM, holotype
571. BM, holotype

572. After Pilsbry & Lowe (1932)
573. Sowerby, Conch. Icon., vol. 17 (1867)
574. Left, CAS, Baja California, x 1.5;
right, Olsson (1961)
575. Olsson (1961)
576. Left, Olsson (1961) ; right, Keen
(1966)
577. CAS, Panama
578–79. Olsson (1961)
580. Keen (1966), holotype, Univ. Copenhagen
581. After Sowerby, Conch. Icon., vol. 17
(1867)
582. CAS, Monypenny Point, Nicaragua
583. Dall (1909)
584. CAS, southern California
585. CAS, Corinto
586. CAS, Petatlan Bay, Mexico
587. Above, BM, syntype lot; below, right,
Sowerby, Conch. Icon., vol. 17 (1867)
588. CAS, Corinto
589. Olsson (1961)
590. No figure
591. CAS, Nicaragua; see also color plate
XIII, fig. 2
592. Olsson (1961)
593. BM, syntypes; upper right, CAS,
Nicaragua
594. Above, BM, syntype lot of *D. granifera*;
below, Reeve, Conch. Icon., vol. 8
(1855)
595. Above and left, after Orbigny (1845) ;
below, BM, holotype
596. Left, SU, Ecuador; right, Olsson (1961)
597. Olsson (1961) ; upper right center, BM,
probable holotype
598. CAS, Magdalena Bay
599. CAS, Corinto
600. After Palmer & Hertlein (1936)
601. Below, Dall (1909) ; above, Olsson
(1961)
602. Left, BM, holotype
603. Left, ANSP, holotype; right, above and
below, CAS, Punta Arena, Gulf of
California
604. SU, Costa Rica
605–6. Sowerby, Conch. Icon., vol. 10
607. Olsson (1961)
608. USNM, holotype, no. 19,409
609. Keen & Frizzell (1939)
610. BM, holotype
611. SU, Acapulco; small figure, left of
center, BM, syntype of *S. purpurea*
612. After Pilsbry & Lowe (1932)
613. After Pilsbry & Olsson (1941)
614. Lowe (1935)
615. Left, Turner (1956) ; upper right, CAS,
Panama
616. Above, CAS, San Diego, California;
below, Keen & Frizzell (1939)
617–18. Olsson (1961)

619. ANSP, holotype
620. Olsson (1961)
621. BM, holotype
622. CAS, Salina Cruz, Oaxaca
623. Olsson (1961)
624. CAS, Kino Bay, Sonora; smaller figures
at right, BM, holotype
625. Turner (1956)
626. Reeve, Conch. Icon., vol. 8
627. Olsson (1961)
628. CAS, Gulf of California
629. SU, Panama
630. SU, San Felipe
631. Below, Amherst College collection,
La Paz; above, Olsson (1961)
632. Pilsbry & Lowe (1932)
633. SU, Kino Bay, Sonora; see also color
plate XIII, fig. 5
634. Boerstler collection, Gulf of California;
see also color plate IV
635. Reeve, Conch. Icon., vol. 8
636. Olsson (1961)
637. Left, CAS, Gulf of Dulce; right, CAS,
Santa Inez Bay; see also color plate
XIII, fig. 6
638. Olsson (1961)
639. USNM, syntype, no. 96,433; x 2
640. Olsson (1961)
641. Left, CAS, Monypenny Point, Nicaragua;
right, SU, Panama
642. Olsson (1961)
643. CAS, Gulf of Fonseca
644. Olsson (1961)
645. After Pilsbry & Lowe (1932) ; detail
of sculpture below
646. Sowerby, Conch. Illus. (1833)
647. After Pilsbry & Lowe (1932) ; detail
of sculpture at left
648. Olsson (1961)
649. CAS, Gulf of Chiriquí, Panama
650. Turner (1956)
651. CAS, Tangola-Tangola Bay, Mexico
652. Univ. Copenhagen, syntype
653. Left, Olsson (1961) ; right, CAS,
Hannibal Bank, Panama
654. CAS, Gulf of Fonseca
655. USNM, holotype, no. 23,700; x 5
656. Olsson (1961)
657. Below, SU, Panama; above, BM,
holotype
658. BM, syntypes
659. USNM, holotype, no. 108,552; x 5
660. SU, Panama
661. Above, Keen (1966) ; below, Olsson
(1961)
662. CAS, Gulf of Fonseca
663. USNM, holotype, no. 120,634; x 1
664. Univ. Copenhagen, holotype
665. Lowe (1935)
666. After Sowerby, Conch. Icon., vol. 14
(1874)

667. After Weymouth, Calif. Fish. Bull. no. 4 (1921)
668. After Turner (1956)
669. After Berry (1953)
670. CAS, Pearl Islands
671. Keen & Frizzell (1939)
672. Olsson (1961)
673. Lower left, Keen & Frizzell (1939); above, center, Keen (1968); right, Olsson (1961); right, above, *Tyleria fragilis* Adams
674. Olsson (1961)
675. Durham (1950)
676. Olsson (1961)
677. Olsson (1961); right, CAS, southern end, Gulf of California
678. Olsson (1961)
679. Hinds (1844)
680. Olsson (1961)
681. Left, after Hertlein & Strong (1950); right, Olsson (1961)
682. CAS, south of Acapulco
683. Olsson (1961)
684. Below, after Reeve, Conch. Icon., vol. 2; above, Olsson (1961)
685. Left, Olsson (1961); upper right, after Reeve, Conch. Icon., vol. 2
686. Olsson (1961)
687. USNM, holotype; upper left, Olsson (1961)
688. After Morrison (1946); x 2
689. Turner (1956)
690. Olsson (1961)
691. BM, holotype
692. Right, Hinds (1844); left, Olsson (1961)
693. SU, holotype, no. 10,040, Carmen Island, Gulf of California
694. After Sowerby, Conch. Icon., vol. 20 (1878)
695. BM, holotype
696. After Sowerby, Conch. Icon., vol. 20 (1878)
697. Keen & Frizzell (1939)
698. Olsson (1961)
699. Dall (1902)
700. Sowerby, Thes. Conch., vol. 2
701. Proc. CAS, vol. 5 (1873) (*B. pacificus* Stearns)
702. Sowerby, Thes. Conch., vol. 2
703. Olsson (1961)
704. Sowerby, Thes. Conch., vol. 2
705. CAS, Baja California; center (mesoplax), Turner (1955)
706–7. Turner (1955), showing mesoplax
708. Sowerby, Thes. Conch., vol. 2
709–10. Sowerby in Conch. Icon. (1872)
711. Olsson (1961)
712. Sowerby in Conch. Icon. (1872)
713. Olsson (1961)
714. Sowerby, Thes. Conch., vol. 2
715. Turner (1955)

716–21. No figure
722. Turner (1966); pallets at lower left
723–24. After Bartsch (1922)
725. Turner (1966); pallets at right
726–31. Turner (1966); pallets below
732–33. Olsson (1961)
734. BM, syntypes
735. Olsson (1961)
736. USNM, holotype
737. After Oldroyd (1925)
738. Olsson (1961)
739. Above, USNM, holotype; below, SU, paratype, no. 6,940, off La Paz
740. Left, above, CAS, Cape San Lucas; right, BM, holotype; below, USNM, holotype of *P. convexa* Dall
741. After Pilsbry & Lowe (1932)
742. Gould (1853)
743. Dall (1908)
744. SU, Baja California
745. BM, syntypes
746. Olsson (1961)
747. SU, Punta Peñasco, Sonora
748. BM, after photograph of type, Ecuador
749. Above, Dall (1908); below, Olsson (1961)
750. Stearns (1890)
751. Olsson (1961)
752. Left, BM, syntypes; right, after Turner (1956)
753. After Grant & Gale (1931)
754. Above, Dall (1908); below, Olsson (1961)
755. After Hertlein & Strong (1946)
756. After Pilsbry & Olsson (1935)
757. Rogers (1966)
758. Olsson (1961)
759. USNM, holotype, no. 73,639, x 2
760. Above, BM, holotype; below, Oldroyd (1925)
761. Left, Keen (1968), from holotype in BM; right, Sowerby in Conch. Icon. (1869)
762. USNM, holotype, no. 16,292, Cape San Lucas
763. USNM, holotype, no. 87,583
764. Oldroyd (1925), from type in USNM, off La Paz
765. CAS, Port Guatulco, Mexico
766. SU, Kino Bay, Sonora
767–68. Dall (1908)
769. No figure
770. Dall (1908); length, 6.5 mm
771–72. No figure
773. Dall (1908)
774. Bernard (1969)
775. No figure
776. SU, San Pedro, California
777. Left, below, BM, holotype; right, Pilsbry & Lowe (1932)
778. After Hinds (1843)

779. Olsson (1961)
780. Bernard (1969)
781. After Strong & Hertlein (1937)
782. Grant & Gale (1931)
783–84. USNM, holotype
785. Dall (1908) ; length, 13.5 mm
786. Keen (1963)

787. After Howard (1950)
788. Bernard (1969)
789. After Keen & Frizzell (1939)
790. Bernard (1969)
791. Dall (1890)
792. Dall (1925)

GASTROPODA

1. LACM, Galápagos Islands
2. LACM, holotype, no. 1,368
3. LACM, holotype, no. 1,370
4. LACM, holotype, no. 1,372
5. LACM, southern California
6. LACM, Galápagos Islands
7. SU, holotype, no. 8,619
8. LACM, Cuastecomate, Mexico
9. SDSNH, holotype
10. LACM, holotype, no. 1,307
11. SDSNH, holotype, no. 13,372
12. USNM, holotype
13. USNM, no. 267,019, holotype of R. astricta McLean
14. USNM, holotype, no. 122,577
15. LACM, Puerto Peñasco, Sonora
16. AHF, Guaymas
17. Keen (1966), from holotype in BM
18. AHF, Galápagos Islands; see also color plate XII, fig. 8
19. CAS, Corinto, Nicaragua
20. AHF, Costa Rica
21. LACM, holotype
22. SU, holotype, no. 8,587
23. LACM, Acapulco
24. AHF, Cupica Bay, Colombia
25. LACM, southern California
26. USNM, Paita, Peru
27. LACM, holotype, no. 1,153
28. LACM, Cuastecomate
29. CAS, holotype, no. 9,800
30. LACM, holotype, no. 1,374
31. LACM, Bayovar, Peru
32. LACM, holotype, no. 1,313
33. LACM, holotype, no. 1,375
34. LACM, Mazatlán
35. LACM, Playa del Coco, Costa Rica
36. AHF, Jervis I., Galápagos
37. LACM, Santiago, Colima, Mexico
38. LACM, Playa Caleta, Acapulco
39. LACM, Barrington I., Galápagos
40. LACM, Guaymas
41. LACM, Salina Cruz, Oaxaca
42. AHF, Port Utria, Colombia
43. LACM, holotype, no. 1,315
44. SU, west Mexico; see also color plate XIII, fig. 7
45a. SU, holotype, no. 9,507; 45b. LACM, San Felipe, specimen showing fine sculpture
46. USNM, holotype, no. 4,019

47. LACM, Puertecitos, Baja California
48. LACM, Tres Marias Islands
49. LACM, Salina Cruz, Oaxaca
50. LACM, Acapulco
51. SU, holotype, no. 8,507
52. USNM, holotype, no. 12,594, Cape San Lucas
53. CAS, holotype, no. 10,516
54a. LACM, Guaymas; 54b. CAS, holotype, no. 10,532
55. ANSP, holotype, no. 156,854
56a. LACM, San José del Cabo; 56b. SU, holotype, no. 8,589
57. USNM, holotype, no. 15,923
58. SU, Panama
59. USNM, lectotype, no. 131,113
60. LACM, Cerralvo I., Gulf of California
61. LACM, Panama
62. USNM, holotype, no. 94,865
63. USNM, holotype, no. 96,559
64. SSB, off Cape Colnett, Baja California
65. USNM, holotype, no. 125,964
66. USNM, holotype, no. 122,959
67. LACM, holotype, no. 1,377
68. USNM, no. 194,971, Galápagos Islands
69. AHF, Bahía Honda, Panama
70. USNM, holotype, no. 122,580
71. USNM, lectotype, no. 16,281
72. USNM, holotype, no. 207,625
73. USNM, holotype, no. 97,001
74. SSB, holotype, no. 30,686
75. LACM, Cuastecomate, Jalisco
76. LACM, Panama
77. CAS, holotype, no. 6,044
78. LACM, near Puertecitos, Baja California
79. AHF, Charles I., Galápagos
80. CAS, holotype, no. 13,271
81. USNM, holotype, no. 122,957
82. SDSNH, holotype, no. 51,299, Galápagos Islands
83. LACM, holotype, no. 1,272
84. LACM, Cuastecomate, Jalisco
85. SDSNH, holotype, no. 659; see also color plate XIII, fig. 8
86. SU, holotype, no. 9,742
87. SDSNH, no. 39,799
88. USNM, holotype, no. 206,136
89. LACM, Chacahua Bay, Mexico
90. LACM, holotype, no. 1,269
91. SDSNH, holotype, no. 51,301
92. LACM, holotype, no. 1,271

93. AHF, Port Utria, Colombia
94. LACM, Puertecitos, Baja California
95. LACM, holotype, no. 1,379, Cabita Bay, Colombia
96. LACM, Galápagos Islands
97. LACM, holotype, no. 1,381
98. AHF, Lobos I., Peru
99. LACM, Magdalena Bay
100. LACM, holotype, no. 1,383
101. LACM, Cuastecomate, Jalisco
102. LACM, Mazatlán
103. ANSP, holotype, no. 156,353
104. LACM, Cocos Island
105. ANSP, no. 40,820, type of *T. turbinatus* (Pease)
106. AHF, La Libertad, Ecuador
107. SU, Panama
108. LACM, holotype, no. 1,385
109. LACM, Panama Bay
110. LACM, Fernandina I., Galápagos
111. LACM, holotype, no. 1,387
112. LACM, holotype, no. 1,389
113. USNM, no. 122,955, Galápagos
114. CAS, Nicaragua
115. USNM, no. 207,607
116. USNM, holotype, no. 109,029
117. USNM, lectotype (largest of five), no. 207,602
118. USNM, lectotype (largest of three), no. 207,624
119. USNM, holotype, no. 18,112
120. LACM, Cape San Lucas
121. LACM, Coronado Is., west coast of Baja California
122. LACM, Puertecitos, Baja California
123. CAS, holotype, Bahía Honda, Panama
124. LACM, Guaymas
125. LACM, holotype, no. 1,391
126. AMNH, holotype, no. 154,626
127. LACM, holotype, no. 1,393
128. LACM, holotype, no. 1,275
129. LACM, Cuastecomate, Jalisco
130. LACM, Espíritu Santo Island
131. USNM, no. 111,335, holotype of *Liotia litharia* Dall
132. CAS, holotype, no. 5,478
133. LACM, holotype, no. 1,394
134. LACM, Isla Otoque, Panama
135. LACM, Cuastecomate, Jalisco
136. USNM, holotype of *Liotia pacis* Dall
137. AHF, Cape San Lucas
138. LACM, holotype, no. 1,396
139. LACM, holotype, no. 1,409
140. USNM, holotype, no. 123,055
141. AHF, Isla San Pedro Nolasco, 200 m
142. AHF, San Benito Island
143. USNM, holotype, no. 122,953
144. SU, Gulf of California
145. LACM, Socorro Island
146. SU, Montijo Bay
147. LACM, Tres Marias Islands

148. SSB, Santa Cruz I., Galápagos
149. AHF, Clarion Island
150. Shasky collection, east side, Punta Ancón, Ecuador
151. USNM, no. 610,331, holotype of *A. rupicolina* Stohler
152. AHF, Natividad Island, Baja California
153. USNM, holotype, no. 222,318
154. LACM, Isla Santa Clara, Ecuador
155. AHF, Piñas Bay, Panama
156. LACM, Santa Cruz, Nayarit
157. LACM, Acapulco
158. LACM, off Cape San Lucas
159. LACM, holotype, no. 1,397
160. LACM, Mazatlán
161. LACM, Salinas, Ecuador
162. LACM, near Puertecitos, Baja California
163. AHF, Lobos Island, Peru
164. LACM, San Carlos Bay, Guaymas
165. SU, La Paz
166. SU, Montijo Bay
167. SU, Panama City
168. SU, Gulf of California
169. Vanatta (1912)
170. After Turner (1956)
171. Keen (1966), from holotype in Univ. Copenhagen
172. Marcus & Marcus (1967) ; see also color plate XV, fig. 2
173. USNM, holotype
174. Dall (1908)
175. Dall (1908)
176. Dall (1908)
177. USNM, holotype, no. 123,033; x 5
178. USNM, holotype, x 5
179. LACM, Puerto Peñasco, Sonora
180. LACM, holotype, no. 1,399
181. Three figures at lower right, BM, type lot of *L. aspera*; others, Von Martens (1901)
182. SU, Panama
183. Above, SU, Gulf of California; below, BM, type lot of *L. modesta*
184-85. Tryon, Manual Conch.
186. SSB, Cape San Lucas
187. Bartsch & Rehder (1939)
188. SU, Panama
189. SU, Puntarenas
190. Stearns (1893)
191. Bartsch & Rehder (1939) ; lower left, photograph by J. McLean, from holotype in USNM
192. Keen (1966), from holotype in Univ. Copenhagen
193-95. No figure
196. After Bartsch (1911) ; x 15
197-208. No figure
209. Univ. Copenhagen, syntype
210-14. No figure
215. Right, Brann (1966), x 30; left, Keen

(1968), x 15; center, Shasky collection, Mazatlán, x 10
216. LACM, from Shasky collection, x 10
217. Bartsch (1911) ; x 35
218–19. No figure
220. Bartsch (1911) ; x 22
221. Keen (1968) ; x 21
222. No figure
223. Bartsch (1917) ; x 14
224. Bartsch (1920) ; x 14
225. Bartsch (1920) ; x 25
226. No figure
227. Pilsbry & Olsson (1941) ; x 16
228. Keen (1968) ; x 20
229. Keen & Pearson (1942) ; x 15
230–31. No figure
232. Bartsch (1920) ; x 20
233. Keen (1968) ; x 17
234. No figure
235. Bartsch (1920) ; x 12
236–37. No figure
238. Bartsch (1920) ; x 35
239–49. No figure
250. Bartsch (1915) ; x 12
251–55. No figure
256. Bartsch (1915) ; x 9
257–59. No figure
260. BM, syntypes; x 1.7
261–70. No figure
271. Bartsch (1915), x 9; length, 9 mm
272–75. No figure
276. Bartsch (1915) ; x 27
277–80. No figure
281. SSB Collection, holotype, no. 17,383; x 16
282. No figure
283. Keen (1958), after Turner (1956) ; x 8
284. No figure
285. Keen (1968) ; x 32
286–88. No figure
289. Keen (1968) ; x 35
290–304. No figure
305. After Hertlein & Strong (1951) ; x 17
306–28. No figure
329. Pilsbry & Olsson (1945) ; x 30
330–46. No figure
347. Keen (1968), Cape San Lucas; x 8
348–49. No figure
350. After Strong (1938) ; x 21
351. Pilsbry & Olsson (1952) ; x 13
352. SU, holotype, no. 10,041; base and spire, below, x 10; apertural view, above, x 18
353. No figure
354. Keen (1968), from holotype in BM; x 28
355–56. No figure
357. After Pilsbry & Olsson (1952) ; x 20
358–61. No figure
362. After Pilsbry & Olsson (1952) ; x 21
363–74. No figure
375. After Pilsbry & Olsson (1945) ; x 35
376–84. No figure

385. Bartsch (1911) ; x 20
386–87. No figure
388. Keen (1968), from holotype in BM; x 25
389. No figure
390. Pilsbry & Olsson (1945) ; x 33
391–92. No figure
393. Keen & Pearson (1952) ; x 5
394–99. No figure
400. Keen (1968), from holotype in BM; x 33
401–10. No figure
411. Keen (1968), from holotype in BM; x 25
412–18. No figure
419. After Hertlein & Strong (1951) ; x 15
420. No figure
421. No figure
422. Pilsbry & Olsson (1945) ; x 24
423. After Pilsbry & Olsson (1951) ; x 45
424. Dall (1908) ; x 1.5
425. Below, SU, La Paz; above, Univ. Copenhagen, holotype of *A. valenciennesi*
426. Above and below, Keen (1966), from syntype in BM; center, left and right, Hinds (1844–45)
427. Bartsch (1918)
428. SU, Guaymas
429. Pilsbry & Lowe (1932)
430. Pilsbry & Lowe (1932)
431. LACM, Secas Island, Panama; diameter, 15 mm
432. USNM, holotype
433. SU, holotype, no. 8,509
434. SU, Panama
435. BM, holotype
436. Reeve, Conch. Icon.
437. CAS, holotype
438. SU, Guaymas
439. SU, La Paz
440. SU, Panama
441. Left, SU, Baja California; right, Dall (1908)
442. SU, Baja California, offshore
443. LACM, holotype, no. 1,401
444. Kiener (1843–44)
445. Below, BM, syntype; right, LACM, Panama; left, Reeve, Conch. Icon.
446. LACM, holotype
447. Right, LACM, holotype, immature; left, D. Shasky Collection, mature specimen
448. SU, Ecuador
449. SU, Puerto Peñasco
450. Left, above, juvenile, Panama, courtesy Dr. A. Olsson; below, Mörch (1860)
451. After Strong & Hertlein (1939) ; x 27
452–53. Keen (1968), from holotypes in BM; x 15
454–69. No figure
470. Keen (1968), from type lot in BM; x 17
471. No figure
472. Center, Keen (1968), from type in BM, x 20; left, after Strong & Hertlein (1939), x 35; right, Keen (1963)

473–77. No figure
478. Brann (1966) ; x 18
479–81. No figure
482. Keen (1968), from syntypes in BM;
lengths, 2.4 and 2.0 mm
483. Brann (1966) ; x 32
484. After Turner (1956) ; x 30
485–88. No figure
489. Strong & Hertlein (1939), x 20; right,
Keen (1963), x 15
490. SU, Gulf of California
491–92. SU, La Paz
493. BM, holotype, photograph and drawing
494. BM, holotype, photograph and drawing;
above, SU, Guaymas
495. USNM, holotype, no. 123,025; x 1
496. SU, Mazatlán, showing internal lirae in
enlarged detail
497. SU, Mazatlán
498 Left, BM, holotype cluster; right, Keen
(1961) ; small specimen, SU, Oaxaca
499. After Pilsbry & Lowe (1932)
500. SU, Kino Bay, Gulf of California
501. Tryon, Manual Conch., vol. 8
502. BM, holotype
503. Below, SU, Oaxaca; above, Tryon,
Manual Conch.
504. SSB, off Angel de la Guarda Island, Gulf
of California
505. Right, BM, lectotype; left, below, SU,
Oaxaca; left, above, Keen (1961),
operculum
506. SU, Acapulco
507. SU, Panama
508. USNM, holotype
509. Sowerby, Thes. Conch.
510. SU, La Paz
511. Keen (1968), from holotype in BM
512. SU, Punta Abreojos, Baja California
513. No figure
514. ANSP, holotype
515. SU, San Felipe
516. SU, Perlas Islands, Panama Bay
517. SU, holotype, no. 10,042, type locality,
Angel de la Guarda Island, Gulf of
California; length, 23 mm; diameter,
9.5 mm
518–19. No figure
520. Bartsch (1911) ; x 6
521–30. No figure
531. Bartsch (1911) ; x 8
532–34. No figure
535. Strong & Hertlein (1939) ; x 12
536–40. No figure
541. Keen (1968), from holotype in BM; x 14
542–44. No figure
545. Bartsch (1911) ; x 8
546–48. No figure
549. Bartsch (1911) ; x 7
550. USNM, holotype
551. BM, syntype lot

552–53. No figure
554. Bartsch (1911) ; x 12
555. After Baker, Hanna & Strong (1938) ;
x 20
556. No figure
557. After Turner (1956) ; x 10
558. Bartsch (1911) ; x 8
559–61. No figure
562. Bartsch (1910) ; x 10
563. No figure
564. Bartsch (1911), as *A. diomedeae*;
x 14
565–69. No figure
570. Keen & Pearson (1952) ; right, BM,
syntype of *L. divisa*
571–81. No figure
582. Baker (1926) ; x 7
583–97. No figure
598. Forbes (1851)
599. SU, Panama
600. Pilsbry & Vanatta (1902)
601. SU, Guaymas
602. Right, BM, 4 syntypes; left, SU,
Manuela Lagoon, Baja California
603. Turner (1956), type of *C. reevianum*
604–5. Turner (1956), from holotypes
606. SU, Panama
607–9. SU, Gulf of California
610. SU, Guayaquil; see also color plate XI,
figs. 3, 4
611. Left, after Baker, Hanna & Strong
(1930) ; center, after Baker, Hanna &
Strong (1930), *E. xantusi*; right,
USNM, holotype, no. 59,337
612. CAS, holotype, no. 12,729; see also color
plate XIV, fig. 5
613. Right, after Lowe (1932) ; left, USNM,
holotype
614. USNM, holotype
615. USNM, holotype, no. 211,019
616. Right, USNM, holotype, no. 194,995;
left, CAS, holotype of *E. manzanillense*
617. Right, USNM, holotype; left, Hertlein &
Strong (1951), type of *E. vivesi*
618. After Lowe (1932)
619. USNM, holotype, no. 56,055; right, after
Baker, Hanna & Strong (1930)
620. LACM, holotype, no. 1,159
621. Left, BM, syntypes; center, Sowerby,
Thes. Conch., vol. 1 (1847) ; right,
after Olsson & Smith (1951), type of
E. chalceum
622. LACM, holotype
623. USNM, holotype
624. LACM, holotype
625. USNM, holotype of *E. pacis*
626. BM, syntypes; height, 12 mm
627–28. USNM, holotypes
629. After Lowe (1932)
630. After Olsson & Smith (1951)
631. Hertlein & Strong (1951)

632. Right, USNM, holotype; left, SU, Panama
633. Left, Hertlein & Strong (1951); right, Panama, courtesy, Dr. A. Olsson
634. Left, SSB collection, holotype of *E. pentedesmium*; right, Hinds (1844)
635. Left, BM, holotype; right, Sowerby, Thes. Conch. (1847)
636. BM, holotype
637. Above, BM, syntypes; left below, Hertlein & Strong (1915), type of *E. oerstedianum*; right below, *ibid.*, type of *E. wurtsbaughi*
638. Sowerby, Thes. Conch. (1847)
639. After Baker, Hanna & Strong (1930)
640. USNM, holotype, no. 46,229
641. USNM, holotype, no. 218,096
642. After Baker, Hanna & Strong (1930)
643–44. USNM, holotypes
645. Left, after Hertlein & Strong (1951), type of *E. gissleri*; right, BM, holotype
646. Sowerby, Thes. Conch. (1847)
647. Hertlein & Strong (1951)
648. Below, BM, holotype; left, after Lowe (1936), type of *E. phanium*; right, BM, holotype of *Scala raricosta*
649. BM, syntype
650. LACM, holotype
651. Left and right, BM, syntypes; center, after Baker, Hanna & Strong (1930)
652. BM, syntypes
653. Left, BM, syntype; center, Sowerby, Thes. Conch.; right, SU, Central America
654. Left, USNM, holotype; right, USNM, holotype of *E. musidora*
655. USNM, syntypes, *E. appressicostatum*
656. USNM, holotype
657. LACM, holotype
658. Above, BM, syntypes; below, after Lowe (1932), type of *E. strongianum*
659. No figure
660. USNM, holotype
661. CAS, holotype; right, after Hertlein & Strong (1951)
662. USNM, holotype, no. 56,056
663. USNM, holotype
664. Left, BM, holotype; center, Hinds (1844–45); right, Hertlein & Strong (1951), type of *E. paradisi*
665. USNM, holotype
666. LACM, holotype
667. BM, paratype, Cape San Lucas
668. LACM, holotype
669. Above, right, Shy collection, Manzanillo; left, Pilsbry & Lowe (1932)
670. USNM, holotype; smaller figure, USNM, type of *Eglisia nebulosa*
671. LACM, holotype
672. Above, Dall (1908); below, LACM
673. USNM, holotype

674. LACM, holotype and paratype
675. LACM, holotype
676. USNM, holotype
677. Hertlein & Strong (1951)
678. Right, after Baker, Hanna & Strong (1930); left, USNM, holotype
679. BM, syntypes, Galápagos Islands
680. BM, syntypes; upper left, after Baker, Hanna & Strong (1930)
681. After Lowe (1932)
682. Baker, Hanna & Strong (1930)
683. USNM, holotype
684. USNM, holotype
685. ANSP, holotype; right, after Olsson & Smith (1951), type of *O. clarki*
686. After Palmer (1958); right, after Baker, Hanna & Strong (1930)
687. SU, Baja California
688. After Tryon, Manual Conch.; left, above and below, BM, syntypes of *J. striulata* and *J. contorta*
689. After Tryon, Manual Conch.; right, BM, syntypes of *J. decollata*
690. Left, USNM, holotype; right, Dall (1925)
691–702. No figure
703. Bartsch (1917)
704–7. No figure
708. BM, syntype
709–10. No figure
711. Bartsch (1917); x 12
712. Keen (1966), from syntypes in Univ. Copenhagen; x 6
713–15. No figure
716. USNM, holotype; length, 20 mm
717–22. No figure
723. Bartsch (1917); length, 20 mm
724–26. No figure
727. Bartsch (1917); x 25
728. Bartsch (1917); x 12
729. Bartsch (1917); x 5
730–33. No figure
734. Bartsch (1917); length, 10 mm
735. No figure
736. Bartsch (1917); x 30
737. Bartsch (1917); x 10
738. No figure
739. Bartsch (1917); x 12
740. Bartsch (1917); x 20
741. Keen & Pearson (1952), after holotype in USNM; x 15
742. No figure
743. Bartsch (1917); x 14
744. Bartsch (1917); x 13
745. Bartsch (1917); magnification not given
746. AMNH, hypotype, no. 156,819; x 2.3
747. AMNH, holotype; x 9
748. LACM, holotype, no. 1,403
749. Bartsch (1917)
750. Left, Sowerby, Thes. Conch.; right, BM, syntypes

751. LACM, off Angel de la Guarda I., Gulf of California, x 11
752. Bartsch (1917); see also color plate XIV, fig. 1
753. Bartsch (1917)
754. Left, BM, holotype; right, after Bartsch (1917)
755. Bartsch (1917); x 13
756. Bartsch (1917); x 13
757. Bartsch (1917); x 18
758. Bartsch (1917); x 13
759. Bartsch (1917); x 30
760. Bartsch (1917); x 3
761. Pilsbry (1956); x 12
762. SU, holotype, no. 7,857; x 13
763. BM, holotype; x 20
764. USNM, holotype, no. 123,041
765. SSB, La Libertad, Sonora
766. Left, SU, Mazatlán; right, above, BM, syntype of *H. serratus*
767. SU, La Paz
768. BM, syntype
769–71. No figure
772. After Turner (1956); x 11
773–75. No figure
776. USNM, holotype; x 8
777. No figure
778. After Turner (1956); x 20
779. No figure
780. BM, syntypes; x 4
781–84. No figure
785. Keen (1968), from syntype in BM; x 20
786–87. No figure
789. Pilsbry & Olsson (1952); x 21
790–96. No figure
797. USNM, holotype, no. 15,897
798. Courtesy CAS, Galápagos Islands
799. CAS, Piñas Bay, Panama
800. Right, BM, syntypes; left, Tryon, Manual Conch.
801. Below, SU, Guaymas; above, Grant & Gale (1931)
802. SU, Acapulco
803. SU, Manzanillo; see also color plate XIV, fig. 2
804. Dall (1909)
805. BM, holotype
806. Below, SU, Panama; above, BM, syntype of *C. alveolata*
807. Reeve, Conch. Icon., vol. 11
808. Center, SU, Montijo; right and left, BM, syntypes of *C. hystrix*
809. Reeve, Conch. Icon., vol. 11
810. SU, Gulf of California
811. SU, Montijo Bay
812. SU, Manta, Ecuador
813. Reeve, Conch. Icon., vol. 11
814. SU, San Ignacio
815. No figure
816. Above, SSB, Paita, Peru; below, Turner (1956)

817. Turner (1956), type of *C. nivea*
818. SSB, San Carlos Bay, Guaymas
819. Tryon, Manual Conch., vol. 8
820. Left, Keen & Pearson (1952); right, Shasky collection
821. SU, Gulf of California
822. Left and below, BM, syntypes; above, SSB, specimen on *Strombus*; right, CAS, west Mexico
823. SSB, type lot
824. SU, holotype, no. 8,498, with detail of sculpture
825. SU, Mulege, Gulf of California
826. Above, SU, Guaymas, with detail of sculpture; below, Keen & Pearson (1952)
827. SSB, San Blas, Mexico
828. Above, SU, no. 9,716, holotype of *C. castellum*; below, Reeve, Conch. Icon.
829. LACM, Mazatlán
830. Reeve, Conch. Icon., vol. 11
831. SU, holotype, no. 9,717
832. SU, holotype, no. 8,519
833. BM, holotype
834. Stearns (1890)
835. LACM, holotype, no. 1,120
836. USNM, syntypes; x 5
837. Above, SSB, Mazatlán; below, CAS, west Mexico
838–42. No figure
843. Keen & Pearson (1952); x 7
844–45. No figure
846. H. & A. Adams (1854–58); x 2
847. Tesch (1908); x 35
848–50. No figure
851. Courtesy CAS, off southern California; see also color plate XV, fig. 4
852–53. H. & A. Adams (1854–58); x 2
854. No figure
855. H. & A. Adams (1854–58); x 2
856–57. No figure
858. Lesueur (1817); x 2
859. LACM, holotype and paratype
860. CAS, holotype, no. 9,891
861. SU, Magdalena Bay; see also color plate XIV, figs. 3, 6
862. CAS, holotype
863. CAS, Cape San Lucas
864. Pilsbry & Lowe (1932)
864a. Olsson (1971)
865. USNM, syntype
866. Dall (1908)
867. LACM, holotype and paratype
868. LACM, Farfan Beach, Panama Canal Zone
869. SU, holotype, no. 9,501, with detail of aperture
870. CAS, Cape Pasado, Ecuador
871. CAS, Arena Bank, showing interior and exterior of operculum

872. CAS, Magdalena Bay
873. SU, Guaymas; see also color plate XIV, fig. 4
874. SU, Pearl Islands, Panama
875. BM, syntype
876. SU, Panama
877. CAS, Arena Bank
878. After Pilsbry (1931)
879. SU, Puntarenas
880–81. BM, syntypes
882. SU, Guaymas
882a. BM, syntypes of *P. cora* (Orbigny)
882b. BM, syntypes of *P. rapulum* (Reeve)
883. Reeve, Conch. Icon., vol. 9
884. Dall (1908); height, 26 mm
885. Dall (1908); height, 15 mm
886. No figure
887. USNM, holotype; height, 14 mm
888. SU, San Felipe
889. Oldroyd (1927)
890. Left, USNM, syntype of *S. pazianum*; x 5; right, SU, Panama
891. CAS, no. 12,601, holotype of *S. cortezi*
892. Left, below, USNM, holotype; right, SSB, Sinaloa
893. Pilsbry & Lowe (1932)
894. Dall (1885)
895. Left, CAS, Panama; right, Turner (1956)
896. Dall (1871)
897. Marcus & Marcus (Atlantic specimen)
898. CAS, west Mexico; see also color plate XVI, fig. 2
899. BM, syntypes, Galápagos Islands
900. Schilder & Tomlin (1931)
901. Dall (1908)
902. BM, syntypes
903. BM, syntypes; see also color plate XVI, fig. 3
904. USNM, holotype, x 10; below, Dall (1908)
905. Below, BM, syntype; above, Keen & Pearson (1952)
906. BM, syntypes
907. SU, holotype, no. 8,520
908. SU, Puerto Peñasco
909. Left and above, BM, syntypes; right, SU, Manta, Ecuador
910. SU, La Libertad, Sonora
911. Left, Schilder (1933); right, Dall (1908); x 8
912. Schilder (1933)
913. Schilder (1933)
914. BM, syntypes
915. Schilder (1933)
916–18. No figure
919. SU, Angel de la Guarda Island, Gulf of California; see also color plate XV, fig. 5
920–21. No figure
922. Stearns (1893b)

923–24. No figure
925. SU, Ecuador
926. No figure
927. SU, Taboga I., Panama Bay; see also color plate XV, fig. 6
928. SU, Galápagos Islands
929. SU, Panama
930. No figure
931. Emerson & Old (1968), Galápagos Islands
932. No figure
933. SU, Guaymas; see also color plate XV, fig. 1
934. Cate (1969); courtesy C. Cate
935. After Schilder (1932); see also color plate XVIII, fig. 2
936. Left, Turner (1956), type of *S. variabilis*; center, BM, syntype of *S. aequalis*; right, BM, holotype of *S. vidleri*; see also color plate XVIII, fig. 6
937. Left, Schilder (1932); right, BM, not type material
938. BM, syntype
939. Left, Turner (1956), type of *S. neglecta*; right, BM, syntype of *S. californica*
940. SU, Taboga Island, Panama; see also color plate XVIII, fig. 1
941. Schilder (1970), in Hawaiian Shell News, east Africa
942. Lower, SU, Mazatlán; upper, SU, Puntarenas
943. Dall (1908); length, 60 mm
944. No figure
945. Dall (1908); length, 45 mm
946. SU, La Paz
947. SU, Carmen Island; see also color plate XVI, fig. 4
948. SU, Acapulco; see also color plate XVI, fig. 1
949. McKibbin collection, La Paz, Baja California; see also color plate XVI, figs. 5, 6
950. SU, hypotype, no. 5,458, Gulf of California
951. Emerson (1968)
952. SU, Mazatlán; see also color plate XV, fig. 3
953. Left, SU, Central America; right, CAS, Central America
954. SU, holotype, no. 10,043, Panama Bay
955. SU, Mazatlán
956. SU, paratype, no. 10,050, off Mazatlán
957. SU, Acapulco
958. Courtesy CAS, Baltra I., Galápagos Islands
959. Emerson & Old (1963), Tres Marias Islands
960. SU, Manta, Ecuador
961. SU, La Libertad, Sonora
962. CAS, Gorda Bank
963. SU, off Mazatlán

964. SU, Panama
965. USNM, holotype, no. 123,027; x 1
966. SU, Mazatlán
967. SU, holotype, no. 9,510
968. After Strong (1938)
969. After Strong & Hertlein (1937)
970. Hertlein & Allison (1968); courtesy CAS
971. SU, Manzanillo, from Lawrence Thomas
972. SU, Panama
973. USNM, holotype
974. SU, holotype, no. 8,520
975. SU, Gulf of California
976. SU, Panama
977. Keen (1968), from syntype in BM
978. SU, holotype, no. 9,511
979. SU, Mazatlán
980. Philippi, Abbild. Conch. (1845)
981. SU, Panama
982. CAS, no. 7,780, Nicaragua
983. D'Attilio collection, probably Peru
984. SU, Pearl Islands, Panama Bay
985. USNM, holotype, courtesy Dr. E. Vokes; below, Dall (1908)
986. BM, syntypes
987. No figure
988. Vokes (1970), Guaymas
989. Vokes (1970), Panama
990. BM, syntypes; below, Sowerby, Conch. Ill.
991. Vokes (1970)
992. Smith (1947)
993–94. Hinds (1844)
995. Emerson & D'Attilio (1970)
996. Left, BM, holotype; right, Sowerby, Thes. Conch. (1871)
997. Above, left, SU, Pearl Islands; right, BM, syntypes
998. Emerson & D'Attilio (1969)
999. Left, SU, Central America; right, Reeve, Conch. Icon.
1000. Left, SU, holotype, 8,651; right, Reeve (1845)
1001. Philippi, Abbild. Conch. (1845)
1002. SU, Manzanillo
1003. SU, Panama
1004. BM, syntype, no. 1954-4.15.10-12
1005. Radwin & D'Attilio (1970), from holotype in SDSNH
1006. SU, San Jose Island, Gulf of California
1007. Left, BM, syntype; right, CAS, Panama
1008. Emerson & D'Attilio (1970)
1009. Strong, Hanna & Hertlein (1933)
1010. LACM, holotype, Loreto, Gulf of California
1011. Emerson & D'Attilio (1969)
1012. Courtesy Dr. E. Vokes; Galápagos Islands
1013. BM, syntypes
1014. Right, BM, holotype; left, SU, no. 8,496, holotype of *A. perplexa*, Panama

1015. Emerson & D'Attilio (1970), from holotype in AMNH, Nayarit
1016. Emerson & D'Attilio (1970), Nicaragua, Vokes collection
1017. CAS, Agua Verde Bay, Gulf of California
1018. Emerson & D'Attilio (1965), Galápagos Islands
1019. SU, holotype, no. 9,502
1020. Lowe (1935)
1021. Emerson (1968), from specimen in AMNH, Gulf of California; juvenile, AMNH, Guaymas
1022. BM, syntypes
1023. Left, SU, no. 8,654, holotype of *C. leucostephes*; right, CAS, Gulf of California, type of *C. bristolae*
1024. Dall (1891)
1025. Left, SU, Venado Island, Panama Bay; right, Chenu (1859)
1026. Hinds (1844–45)
1027. BM, syntypes
1028. Above, BM, syntypes; below, Sowerby, Conch. Ill.
1029. Left, Pease (1869); right, SSB, Guaymas, Sonora, 30–35 m
1030. Reeve, Conch. Icon., vol. 2
1031. Kiener (1843)
1032. BM, syntype
1033. Left, Sowerby, Conch. Ill.; right; BM, syntypes
1034. SU, San Ignacio Lagoon
1035. Courtesy Dr. E. Vokes, La Paz, Gulf of California
1036. SU, Puerto Peñasco, Sonora
1037. Left, CAS, Gulf of California; right, Hinds (1844–45)
1038. SSB, Galápagos Islands
1039. BM, syntype
1040. SU, Panama
1041. LACM, Mazatlán
1042. LACM, Galápagos Islands
1043. Hertlein & Strong (1951)
1044. Dall (1891)
1044a. Olsson (1971)
1045. Dall (1925)
1046. Hertlein & Strong (1951)
1047. Dall (1908); length, 18 mm
1048. After Sowerby, Conch. Ill. (1841)
1049. Below, CAS, Arena Bank; above, Shasky collection, off Guaymas
1050. SU, holotype, no. 9,724, Panama
1051. Left, Keen (1943); right, Hinds (1844–45)
1052. BM, holotype
1053. LACM, holotype, no. 1,194
1054. Keen & Campbell (1964), from holotype in AMNH
1055. Sowerby, Thes. Conch. (1874); right, LACM, Manzanillo
1056. LACM, holotype, no. 1,195

1057. SBMNH, holotype; SU, paratype, no. 9,726, Jalisco, Mexico
1058. ANSP, holotype
1059. BM, syntypes
1060. LACM, holotype
1061. Chenu (1859)
1062. Left, BM, syntypes; right, A. Adams (1855)
1063. Blainville (1832)
1064. Left, BM, syntypes; right, SU, Galápagos Islands
1065. CAS, Manzanillo
1066. Left, BM, syntypes; right, SU, Guaymas
1067. Sowerby, Conch. Ill.
1068. Hinds (1844–45)
1069. Emerson & D'Attilio (1970)
1070. LACM, Colombia
1071. Reeve, Conch. Icon., vol. 3
1072. Kiener (1835–36)
1073–75. Reeve, Conch. Icon., vol. 3
1076. SU, Bahía Bartolomé
1077. Dall (1909)
1078. BM, syntypes
1079–81. Reeve, Conch. Icon., vol. 3
1082. SU, holotype, no. 6,503, and paratype, Gulf of California
1083. SU, Puntarenas
1084. SU, Baja California
1085. SU, holotype, no. 8,510
1086. SU, Panama
1087–88. SU, Galápagos Islands
1089. Blainville (1832)
1090. No figure
1091. Chenu (1859)
1092. SU, Cape San Lucas
1093. Turner (1956)
1094. Kiener (1835–36)
1095. SU, Pearl Islands, Panama Bay
1096. SU, Guaymas
1097. Kiener (1843)
1098. Hinds (1844–45)
1099. LACM, Saladita Bay, Guaymas; depth, 10 m
1100. Right, BM, syntypes; left, Pilsbry & Lowe (1932)
1101. Left, BM, holotype; right, ANSP, holotype of C. tabogensis
1102. LACM, Isla Gorda Bay, Mexico
1103. BM, syntypes
1104. SU, Panama Bay
1105. Dall (1908)
1106. SSB, Guaymas shrimp boats
1107. LACM, holotype, Jalisco, Mexico
1108. Left, Reeve, Conch. Icon., vol. 3; right, BM, holotype of C. aequilirata
1109. Reeve, Conch. Icon., vol. 3
1110. LACM, Isabela I., Galápagos Islands
1111. LACM, Cape San Francisco, Ecuador
1112. Reeve, Conch. Icon., vol. 3
1113. Right, BM, syntypes; upper left, Reeve, Conch. Icon., vol. 3

1114–15. Reeve, Conch. Icon., vol. 3
1116. Left, BM, syntypes; right, Reeve, Conch. Icon., vol. 3
1117. BM, syntype
1118. SU, holotype, no. 8,512
1119. After Reeve, Conch. Icon., vol. 3
1120. Left, SU, holotype, no. 9,718; right, SSB, with egg capsules, Guaymas
1121. SU, holotype, no. 8,511
1122. SSB, Magdalena Bay
1123. BM, lectotype
1124. Upper left, USNM, holotype; right, Pilsbry & Lowe (1932)
1125. Turner (1956)
1126. USNM, holotype
1127. Right, Pilsbry & Lowe (1932) ; left, USNM, holotype of E. panamensis
1128. BM, Galápagos Islands (not type material)
1129. USNM, holotype of E. earlyi
1130. Left, Pilsbry & Lowe (1932) ; right, Bartsch (1931)
1131. Grant & Gale (1931)
1132. USNM, holotype
1133. After Vanatta (1913)
1134. Hinds (1844–45)
1135. ANSP, holotype, courtesy E. Vokes
1136. Griffith & Pidgeon (1834)
1137. Reeve, Conch. Icon., vol. 3
1138. Hinds (1844–45)
1139. Right, BM, syntype; left, Hinds (1844–45)
1140. LACM, La Paz; length, 26 mm
1141. Right, Turner (1956) ; left, USNM, holotype of P. porticus
1142. Hinds (1844–45)
1143. USNM, holotype
1144. Left, USNM, holotype of P. cocosensis; center, BM, syntypes; right, Hinds (1844–45)
1145. Emerson (1967), Galápagos Islands
1146. BM, syntypes, Colombia
1147. After Pilsbry & Lowe (1932)
1148. USNM, holotype
1149. SU, Panama
1150. SU, San Blas, Mexico, with operculum
1151. Dall (1908)
1152. USNM, holotype; x 3; height, 10.5 mm
1153. SBMNH, Puerto Peñasco
1154–58. Sowerby, Thes. Conch., vol. 1
1159. LACM, Paita, Peru
1160. Shasky (1970), from holotype in LACM
1161. Left, Keen (1966), from holotype in Univ. Copenhagen; right, Sowerby, Thes. Conch.
1162. Above and below, Sowerby, Thes. Conch., vol. 1; left, CAS, Santa Inez Bay
1163. CAS, Cape San Lucas
1164. USNM, holotype
1165. CAS, Cape San Lucas

1166. Pilsbry & Lowe (1932)
1167. USNM, holotype
1168. Hertlein & Strong (1951)
1169. Palmer (1958), from holotype in USNM
1170. Sowerby, Thes. Conch.
1171. Left, BM, syntypes of *A. pachyderma*; right, SU, Panama; center, Pilsbry & Lowe (1932)
1172. SU, Magdalena Bay
1173. Shasky (1970), from holotype in LACM
1174. SU, Panama
1175–76. Sowerby, Thes. Conch.
1177. ANSP, holotype (left) and paratype
1178. Left, SU, paratype, no. 10,044, Guaymas; below, right, SBMNH, holotype, no. 12,658; above, CAS, hypotype, San Marcos, Gulf of California
1179. SU, Santa Elena, Ecuador
1180. Sowerby, Thes. Conch.
1181. Left, Dall (1908), x 3; photograph at right courtesy J. McLean
1182. No figure
1183. Turner (1956)
1184. Hertlein & Strong (1951), from holotype in CAS
1185. Turner (1956)
1186. BM, syntypes; upper right, Reeve, Conch. Icon.
1187. SU, Mazatlán
1188. Von Martens (1897)
1189. CAS, holotype
1190. Reeve, Conch. Icon., vol. 11
1191. Left, ANSP, paratype; above, SDSNH, holotype; right, Lowe (1935)
1192. BM, syntypes
1193. Hertlein & Strong (1951), from holotype in CAS
1194. Right, CAS, holotype; photograph at left courtesy J. McLean
1195. Left, Reeve, Conch. Icon., type of *A. encaustica*; right, BM, syntypes of *A. varia*
1196–97. Reeve, Conch. Icon.
1198. LACM, Galápagos Islands; x 6
1199. Pilsbry & Lowe (1932)
1200. Left, BM, syntype; center, Turner (1956), holotype of *A. tessellata*; right, Hinds (1844–45)
1201. SU, Panama
1202. BM, syntypes
1203. Bartsch (1931)
1204. Turner (1956)
1205. Right, Keen (1968), from type lot in BM; left, USNM, holotype of *A. bartschii*
1206. Strong & Hertlein (1937)
1207. Left, Keen (1968), from lectotype in BM; right, Sowerby, Thes. Conch.
1208. Left, Pilsbry & Lowe (1932), type of *A. carmen*; right, BM, holotype

1209. Left, BM, syntype; center, Univ. Copenhagen, syntype of *A. elegantula*; right, Sowerby, Thes. Conch.
1210. Left, USNM, holotype; right, BM, specimens mislabeled "*parva*"
1211. After Strong & Hertlein (1937)
1212. Bartsch (1928)
1213. Stearns (1892)
1214. No figure
1215. Sowerby, Thes. Conch., vol. 1
1216. Reeve, Conch. Icon., vol. 3
1217–19. Dall (1925); photograph (1218) courtesy J. McLean
1220. Hertlein & Strong (1951), from holotype in CAS
1221. USNM, holotype
1222. SU, holotype and paratype, nos. 8,081–8,082
1223. Baker, Hanna & Strong (1938)
1224. LACM, holotype (right) and paratype, nos. 1,266–1,267; additional paratypes, SU, 10,045
1225. Keen (1968), BM, syntypes
1226. SU, Mazatlán
1227. ANSP, holotype
1228. Panama, courtesy R. Bullock
1229. Sowerby, Thes. Conch.
1230. Reeve, Conch. Icon.
1231. Below, Reeve, Conch. Icon.; above, Keen (1968), from holotype of *M. cervinetta* in BM
1232. CAS, holotype, no. 13,632, Puerto Ballandra, near La Paz, Baja California; additional paratypes, SU, no. 10,046
1233. Right, BM, syntypes; left, Reeve, Conch. Icon.
1234. USNM, syntype, no. 4,146; x 5
1235. Baker, Hanna & Strong (1938)
1236. BM, syntype, photograph by R. Robertson
1237. Left, USNM, holotype of *M. baileyi*; right, USNM, holotype
1238. SDSNH, holotype
1239. After Baker, Hanna & Strong (1930)
1240. After Strong & Hertlein (1937)
1241. Pilsbry & Lowe (1932)
1242. Baker, Hanna & Strong (1930)
1243. Turner (1956)
1244. Right, BM, syntype; left, Reeve, Conch. Icon.
1245. USNM, holotype, no. 59,370; x 2
1246. Left, CAS, holotype, no. 13,635; right, CAS, paratype, no. 13,636; Puerto Parker, Costa Rica, in 27 m
1247. SU, holotype, no. 10,047, Guaymas
1248. LACM, holotype, Puerto Videra, Chiapas, Mexico; additional paratypes, SU, no. 10,048
1249. Shasky (1970), from holotype in LACM
1250. Hertlein & Strong (1939)
1251. SDSNH, paratype; x 11

1252. Left, USNM, syntype, Cape San Lucas, x 11; right, ANSP, holotype of *N. fredbakeri*, x 8
1253. SU, holotype, no. 8,530
1254. After Pilsbry & Lowe (1932)
1255. Turner (1956)
1256. Three at left, Keen (1966), from syntypes in Univ. Copenhagen; right, Pilsbry & Lowe (1932), from holotype of *N. xeno*; upper right, USNM, no. 610,334, type of *N. humerosa*
1257. After Pilsbry & Lowe (1932)
1258. Palmer (1958), from lectotype in USNM
1259. After Pilsbry & Lowe (1932)
1260. USNM, holotype, no. 367,977
1261. Left, Kiener (1849–50); right, Reeve, Conch. Icon.
1262. Left, SU, Panama; right, BM, lectotype
1263. LACM, holotype
1264. USNM, holotype
1265. Left, BM, syntype; center, Hinds (1844); right, after Lowe (1935), *S. subangularis*
1266. After Strong & Hertlein (1937)
1267. Right, Lowe (1935); left, SDSNH, holotype
1268. SU, Panama
1269. Right, Sowerby, Thes. Conch.; left, Shasky collection, Gulf of Tehuantepec
1270–71. Sowerby, Thes. Conch., vol. 1
1272. Hinds (1844–45)
1273. Sowerby, Thes. Conch., vol. 1
1274. CAS, Octavia Bay, Colombia
1275. Sowerby, Thes. Conch., vol. 1
1276. Right, Dall (1925); left, USNM, holotype, courtesy J. McLean
1277. Sowerby, Thes. Conch., vol. 1
1278. Hertlein & Strong (1951)
1279. USNM, holotype, no. 130,616; x 1
1280. Left, SDSNH, Mazatlán; right, BM, syntype
1281. Sowerby, Thes. Conch., vol. 1
1282. Left, CAS, Octavia Bay, Colombia; right, SU, Galápagos Islands
1283. SU, Guatemala
1284. Right, Reeve, Conch. Icon., vol. 11; left, BM, syntypes
1285. Emerson & D'Attilio (1969)
1286. USNM, holotype; courtesy J. McLean
1287. Shasky (1970), from holotype in LACM
1288. Sowerby, Thes. Conch.
1289. Right, Dall (1919); left, Turner (1956)
1290. After M. Smith (1944)
1291. Pilsbry & Lowe (1932)
1292. Right, Dall (1908); left, USNM, holotype
1293. USNM, no. 106,533, Punta Abreojos, Baja California
1294–96. Turner (1956)

1297. USNM, holotype, no. 110,565
1298. BM, syntype
1299. After Strong & Hertlein (1937)
1300. Turner (1956)
1301. USNM, holotype, no. 110,630; x 6
1302. Pilsbry & Lowe (1932)
1303. SDSNH, holotype, no. 12,954
1304. Hertlein & Strong (1951)
1305. USNM, holotype, no. 274,095; x 5
1306. Dall (1908)
1307. BM, syntypes
1308. USNM, holotype, no. 96,827; x 5
1309. Reeve, Conch. Icon.
1310. USNM, Peru; courtesy J. McLean
1311. Turner (1956)
1312. McLean (1970), from holotype in LACM, no. 1,405
1313. After Philippi (1845)
1313a. Dall (1890)
1314–15. Turner (1956)
1316. Pilsbry & Lowe (1932b)
1317. SU, Panama
1318. SU, San Felipe
1319. After Kiener (1841)
1320. Left, Pilsbry & Lowe (1932b); right, after Kiener (1841)
1321. After Kiener (1841)
1322. Hinds (1844–45)
1323–25. After M. Smith (1944)
1326. Reeve (1847)
1327. After USNM holotype, no. 96,390
1328. Reeve, Conch. Icon.
1329–30. Hertlein & Strong (1951)
1331. LACM, off Guaymas
1332. LACM, Palo Seco, Canal Zone
1333. LACM, Isla Santa María, Galápagos
1334. Hertlein & Strong (1951)
1335. BM, holotype
1336. SU, Ecuador
1337. Panama Bay, R. G. Shaver collection
1338–39. Reeve, Conch. Icon.
1340–41. Grabau (1904), from MCZ specimens
1342. Left, USNM, locality unknown; right, BM, hypotypes of *Pisania elata*
1343. Right, Lowe (1935); left, SDSNH, paratype
1344. USNM, holotype, no. 111,111
1345. McLean (1970)
1346. Lowe (1935), from type of *F. hertleini*
1347. USNM, holotype, no. 123,007
1348. Left, Lowe (1935); right, SDSNH, paratype
1349. CAS, Santa Elena, Ecuador
1350. Dall (1908); length, 30 mm
1351. Strong & Hertlein (1937), San José del Cabo; courtesy CAS
1352. SU, Gulf of California
1353. SU, Bahía Concepción
1354. Dall (1908); length, 125 mm
1355. USNM, holotype; x 1

1356. USNM, holotype, no. 122,999; x 1
1357. SU, Panama; see also color plate XI, figs. 1, 2
1358. Dall (1908) ; length, 26 mm
1359. Rehder (1967) ; length, 18 mm
1360. SU, Panama; see also color plate XVII, figs. 5, 6
1361. SU, Panama; see also color plate XVII, fig. 2
1362. SU, Galápagos Islands
1363. SU, Panama; see also color plate XVII, fig. 1
1364. Olsson & Dance (1968), from lectotype in Linnean Society collection, London; see also color plates VIII and XVII, fig. 8
1365. Left, SU, San Jose Island; right, Keen (1968), from holotype of *O. intertincta* in BM
1366. SU, Panama
1367. SU, Gulf of California
1368. After Berry (1953)
1369. Conrad (1850)
1370. SU, Panama; see also color plate XVII, fig. 7
1371. Olsson (1956) ; specimens at right in each pair, here and in several following figures, are coated to reveal sculpture details
1372. CAS, no. 12,527
1373. Keen (1968), BM, syntypes
1374–76. Olsson (1956)
1377. Olsson (1956) ; see also color plate XVII, fig. 3
1378. SU, holotype, no. 8,652
1379. SSB, Guatemala
1380–81. Olsson (1956)
1382. CAS, holotype, no. 12,528
1383. CAS, holotype, no. 12,529
1384. Olsson (1956)
1385. SSB, no. 25,673
1386. Keen (1968), from holotype in BM
1387. Olsson (1956) ; specimen at right coated, as above
1388. SU, holotype, no. 9,719
1389. LACM, San Blas, Lowe collection; x 7
1390–95. Olsson (1956) ; specimens at right coated, as above
1396. Olsson (1956) ; see also color plate XVII, fig. 4
1397. Reeve, Conch. Icon., *Turbinella*
1398. Left, BM, syntypes; right, SU, Central America
1399. Left, BM, syntypes; right, Hinds (1844–45)
1400. SSB, off San Jose, Guatemala
1401. Coan & Roth (1966), from type lot in USNM
1402. USNM, holotype, no. 513,647, Panama
1403. LACM, holotype and paratype, nos. 1,141 and 1,142, Banderas Bay

1404. Roth & Coan (1968), from lectotype at University of Alabama
1405. Left, CAS, Guatemala; right, ANSP, holotype of *P. adamsiana*
1406. Left, SU, Oaxaca; right, Coan & Roth (1966), from syntype in BM
1407. CAS, manuscript holotype
1408. Coan & Roth (1966), from syntype of *V. parallela* in USNM
1409. Coan & Roth (1966)
1410. Roth & Coan (1968) ; x 9
1411. Roth & Coan (1968) ; x 9
1412. USNM, no. 568,110, manuscript type
1413. CAS, manuscript type
1414. Turner (1956)
1415. Right, Coan & Roth (1966), x 14; left, composite sketch by Barry Roth, Gulf of California
1416. Coan & Roth (1966)
1417. Left, Roth & Coan (1968) ; right, Keen (1968), from lectotype in BM
1418. No figure
1419. Hinds (1844–45)
1420. BM, holotype
1421. Strong, Hanna & Hertlein (1933), type of *M. zaca*
1422. Left, Reeve, Conch. Icon.; right, BM, holotype
1423. BM, holotype
1424. SSB, Galápagos Islands
1425. BM, holotype
1426. Center, after M. Smith (1947) ; above, BM, syntypes of *Tiara foraminata*
1427. Von Martens (1897)
1428. SU, holotype, no. 9,743
1429. Left, BM, syntypes; right, USNM, holotype of *M. dolorosa*
1430. Right, Reeve, Conch. Icon., vol. 2; left, BM, holotype
1431. Bartsch (1931)
1432. Right, Reeve, Conch. Icon.; left, BM, syntypes
1433. Below, left, Reeve, Conch. Icon.; above, right, BM, syntypes
1434. BM, syntypes
1435. SU, holotype, no. 9,836
1436. SU, holotype, no. 9,513
1437. SSB, Gulf of Nicoya
1438. BM, holotype
1439. SU, Panama
1440. SSB, Gulf of Nicoya
1441. SDSNH, holotype, no. 1,312
1442. SSB, Sonora, 15 to 30 m
1443. SSB, off Acapulco
1444. Swainson (1831–32)
1445. Left, BM, holotype; right, Stearns (1890)
1446. Sphon (1969)
1447. Turner (1956)
1448. Hinds (1844–45)
1449. Petit (1970)

1450. Sowerby, Thes. Conch.
1451. Left, Sowerby, Thes. Conch.; right, BM, syntypes
1452. Sowerby, Thes. Conch.
1453. Reeve, Conch. Icon.
1454. Above, Hinds (1844–45) ; right, BM, syntype
1455. Above, BM, syntype; right, Hinds (1844–45)
1456. Petit (1970), from holotype in AMNH, no. 154,676
1457. Right, CAS, holotype, no. 12,348; left, SU, juvenile, x 8
1458. BM, syntypes; see also color plate XVIII, fig. 4
1459. Above, BM, syntype; below, Sowerby, Conch. Ill.
1460. Left and right, BM, syntypes; center, Sowerby, Conch. Ill.
1461. SU, holotype, no. 8,502, Panama Bay
1462. Sowerby, Thes. Conch., vol. 2
1463. Below, Sowerby, Thes. Conch.; above, BM, syntypes
1464. Sowerby, Thes. Conch.
1465. Pilsbry (1931)
1466–67. Sowerby, Thes. Conch., vol. 2
1468. BM, syntypes
1469. Left, BM, syntype; right, Sowerby, Thes. Conch.
1470. USNM, holotype, photograph courtesy J. McLean
1471. BM, syntypes
1472. Reeve, Conch. Icon.
1473–75. Sowerby, Thes. Conch., vol. 2
1476. Below, Dall (1908) ; above, CAS, Gorda Bank
1477–78. Dall (1908)
1479. Olsson & Bergeron (1967)
1480. SU, paratype, no. 7,875, Costa Rica
1481. CAS, Gorda Bank
1482. Right, Panama Bay, R. G. Shaver collection; left, Hinds (1844–45)
1483. Above, Mazatlán, Shasky collection; below, Sowerby, Thes. Conch.
1484. CAS, east of Cedros Island
1485. Right, after M. Smith (1947) ; left, Panama, Shasky collection
1486. Sowerby, Thes. Conch., vol. 1
1487. Turner (1956)
1488. Sowerby, Thes. Conch., vol. 1
1489. Left, CAS, Cape San Lucas; right, CAS, Galápagos Islands
1490. No figure
1491. Left, CAS, Clarion Island; right, CAS, Cocos Island
1492. CAS, Clipperton Island
1493. CAS, Gulf of Fonseca
1494. Left (1494a), Hanna & Strong (1949) ; right, CAS, Gulf of California
1495. CAS, Galápagos Islands
1496. CAS, Manzanillo

1497. AMNH, holotype, no. 92,200
1498. CAS, Baja California
1499. ANSP, holotype of C. drangai
1500. CAS, Magdalena Bay
1501. CAS, Mazatlán
1502. CAS, Tres Marias Islands
1503. CAS, Magdalena Bay
1504. No figure
1505. LACM, Colombia
1506. CAS, Scammon's Lagoon
1507. Right, LACM, Guaymas; left, ANSP, holotype of C. helenae
1508. CAS, Arena Bank
1509. CAS, Sonora
1510. Left, CAS, Cape San Lucas; right, SU, hypotype, no. 8,036, Panama
1511. Left, CAS, off Cape San Lucas; right, SSB, holotype of C. chrysocestus
1512. Hanna & Strong (1949)
1513. SU, hypotype, no. 8,040
1514. CAS, Galápagos Islands
1515. CAS, Mazatlán
1516. CAS, Santa Margarita Island
1517. CAS, Angeles Bay, Gulf of California
1518. CAS, smaller, holotype, two views; larger, paratype, Puerto Peñasco
1519. D. Brown collection, San Luis Gonzaga
1520. BM, syntypes
1521. Left, Bratcher & Burch (1970), Maria Madre Island; right, Bratcher & Burch (1970), Santa Margarita Island
1522. Left, BM, syntypes; right, Bratcher collection, Sonora, Mexico
1523. Right, USNM, holotype, no. 110,599; left, Bratcher collection, Guaymas
1524. CAS, holotype, no. 12,352
1525. Bratcher & Burch (1970), from holotype in LACM, no. 1,252
1526. Right, USNM, holotype; left, Bratcher collection, Puertecitos
1527. Left, SDSNH, paratype; right, BM, holotype of T. varicosa
1528. Right, SBMNH, holotype; left, SDSNH, paratype
1529. ANSP, holotype
1530. Left, Bratcher collection, Panama Bay; right, SU, Panama
1531. After Pilsbry & Lowe (1932)
1532. Bratcher collection, Hawaii
1533. Bratcher collection, San Luis Gonzaga Bay, Gulf of California
1534. LACM, holotype, no. 1,250
1535. Right, Keen (1966), from syntype in BM; left, CAS, no. 17,737, Santa Inez Bay, 33 m
1536. BM, syntype
1537. Right, BM, holotype; left, Bratcher collection, Galápagos Islands
1538. BM: center above, three syntypes, T. aspera; below at left, holotype

of *T. radula*; right, syntype of
T. glauca; large figures, right and
left, syntypes, *T. petiveriana*

1539. BM, holotype
1540. LACM, holotype, no. 1,255
1541. CAS, holotype, no. 13,222
1542. Left, CAS, Sal si Puedes Island Gulf of
California; right, BM, syntypes
1543. Bratcher collection, Guaymas, Sonora
1544. SDSNH, paratypes
1545. CAS, holotype and paratype, no. 13,215
1546. USNM, holotype of *T. stylus*
1547. Below, Keen (1966), from syntypes
in BM; above, Bratcher collection,
Manzanillo
1548. Left, USNM, holotype, no. 96,567;
right, Bratcher collection, Gorda
Bank
1549. Bratcher collection, Red Sea
1550. LACM, Socorro Island
1551. SDSNH, paratype, no. 44,778
1552. SDSNH, paratype, no. 44,779
1553. CAS, no. 37,738, off Puerto Madera,
Chiapas
1554. Bratcher collection, Manzanillo
1555. Left, Dall (1908); right, photograph
courtesy J. McLean, from holotype
in USNM
1556. BM, holotype
1557. ANSP, holotype
1558. Left, Bratcher collection, Manzanillo;
right, SDSNH, holotype, no. 45,962
1559. LACM, holotype, no. 1,182
1560. Left, BM, syntype; right, Bratcher
collection, Guaymas
1561. SDSNH, paratype
1562. BM, syntype
1563. ANSP, holotype, no. 155,286
1564. LACM, holotype, no. 1,249
1565. BM, syntype
1566. Bratcher collection, Guaymas, Sonora
1567. LACM, holotype, no. 1,180
1568. BM, holotype
1569. CAS, no. 1,486, San Ignacio Lagoon
1570. BM, syntype
1571. Left, BM, syntype; right, BM, holotype
of *T. africana*
1572. LACM, Socorro Island
1573. BM, syntypes
1574. SU, hypotype, no. 5,013, Punta Abreojos, Baja California; radula from
Powell (1966), with large central,
massive marginal
1575. LACM, no. 65-16, Banderas Bay
1576. D. Shasky collection, off Mazatlán
1576a. LACM, Gulf of California
1577. USNM, holotype, no. 266,371, Agua
Verde Bay
1578. D. Shasky collection, Veracruz, Panama
1579. CAS, no. 9,564, holotype of *Kylix
turveri*; radula from Powell (1966),

with vestigial central, compressed
lateral, and solid marginal, shown
below
1580. LACM, Gulf of Tehuantepec
1581. LACM, holotype, no. 1,465, Tenacatita
Bay
1582. LACM, holotype, no. 1,467, Punta
Gorda, Baja California
1583. BM, syntype, no. 1963.469, Costa Rica
1584. AHF, Tepoca Bay, Sonora
1585. LACM, holotype, no. 1,468, Isla Grande
Bay, Guerrero
1586. LACM, holotype, no. 1,470, Jicarita
Island, Panama
1587. LACM, holotype, no. 1,472, Isla
Santiago, Galápagos
1588. CAS, holotype, no. 9,566, Arena Bank,
Gulf of California
1589. LACM, holotype, no. 1,473, Isla
Española, Galápagos
1590. SSB collection, holotype, San Juanico
Bay, Baja California
1591. LACM, *ex* AHF, Tagus Cove, Isabela
Island, Galápagos
1592. USNM, holotype, no. 223,150, Cabo
Lobos, Sonora
1593. LACM, holotype, no. 1,474, Guatulco
Bay, Oaxaca
1594. LACM, San Bartolomé Bay, Baja
California
1595. AHF, Puerto Peñasco, Sonora
1596. USNM, holotype, no. 212,367, Cape
Tepoca, Sonora
1597. AHF, Puerto Culebra, Costa Rica
1598. LACM, Banderas Bay, Jalisco
1599. AHF, Puerto Escondido, Baja
California
1600. LACM, Isla Santiago, Galápagos
1601. LACM, holotype, no. 1,476, Jicarita
Island, Panama
1602. AHF, Tiburon Island, Gulf of
California
1603. LACM, Isla Grande Bay, Mexico
1604. AHF, San José, Guatemala
1605. SDSNH, holotype, no. 520, type of
Clavus pembertoni
1606. USNM, holotype, no. 122,799a, Panama
Bay
1607. LACM, holotype, no. 1,478, Puerto
Utria, Colombia
1608. LACM, Fort Amador, Canal Zone
1609. LACM, holotype, no. 1,479, Taboga
Island, Panama
1610. LACM, holotype, no. 1,481, Isla Santa
Cruz, Galápagos
1611. LACM, holotype, no. 1,483, Gulf of
Tehuantepec
1612. LACM, holotype, no. 1,485, Gorgona
Island, Colombia
1613. LACM, Mazatlán
1614. LACM, Tenacatita Bay

1615. AHF, Bahía Honda, Panama
1616. AHF, Angel de la Guarda Island, Gulf of California
1617. LACM, Galápagos Islands
1618. AHF, Puerto Culebra, Costa Rica
1619. LACM, holotype, no. 1,487, Sonora, Mexico
1620. LACM, holotype, no. 1,489, Angel de la Guarda Island
1621. BM, syntype
1622. LACM, holotype, no. 1,491, Isla Santa Cruz, Galápagos
1623. LACM, holotype, no. 1,492, Banderas Bay
1624. LACM, holotype, no. 1,494, Dewey Channel, Cedros Island, Baja California
1625. AHF, Tenacatita Bay, Mexico
1626. LACM, holotype, no. 1,496, Tovari, Sonora
1627. CAS, holotype, no. 9,563, La Libertad, El Salvador
1628. LACM, Perlas Islands, Panama
1629. AHF, Tiburon Island, Gulf of California
1630. LACM, Puertecitos, Gulf of California
1631. LACM, La Paz, Gulf of California
1632. CAS, holotype, no. 9,571, Cape San Lucas
1633. LACM, holotype, no. 1,498, Isla Isabela, Galápagos
1634. LACM, San Carlos Bay, Guaymas
1635. LACM, holotype, no. 1,500, Academy Bay, Isla Santa Cruz, Galápagos
1636. CAS, holotype, no. 13,677, Port Guatulco, Oaxaca
1637. LACM, holotype, no. 1,502, Tiburon Island, Gulf of California
1638. LACM, Muertos Bay, Baja California
1639. ANSP, holotype, no. 155,203, San Juan del Sur, Nicaragua
1640. ANSP, holotype, no. 155,204, Manzanillo
1641. LACM, holotype, no. 1,193, Guaymas
1642. LACM, holotype, no. 1,366, Clarion Island
1643. LACM, Guaymas
1644. USNM, holotype, no. 123,132, Gulf of Panama
1645. USNM, holotype, no. 207,587, Manta, Ecuador
1646. LACM, Mazatlán
1647. AHF, Bahía Honda, Panama
1648. SU, no. 8,515, holotype of *P. artia*, Angel de la Guarda Island; radula from Powell (1966), with wishbone marginal only
1649. AHF, Bahía de los Angeles, Gulf of California
1650. USNM, holotype, no. 123,089, Gulf of Panama

1651. USNM, holotype, no. 123,120, Gulf of Panama
1652. USNM, holotype, no. 96,498, Galápagos Islands
1653. LACM, holotype, no. 1,504, Isla Santa Cruz, Galápagos
1654. LACM, Guaymas
1655. LACM, Guaymas; radula from Powell (1966), with large central wishbone marginal
1656. LACM, La Paz; radula from Powell (1966), modified wishbone with distal limb severed
1657. Poorman collection, Salina Cruz, Oaxaca
1658. CAS, holotype, no. 9,551, Judas Point, Costa Rica; with and without white coating to reveal sculpture details
1659. Poorman collection, Salina Cruz, Oaxaca
1660. SSB, holotype, Cedros Island
1661. LACM, Guaymas; from shrimp boats
1662. AHF, Puerto Escondido, Gulf of California
1663. AHF, Tortola Island, Panama Bay; radula from Powell (1966), modified wishbone with small distal limb
1664. USNM, holotype of Irenosyrinx persimilis, off Chile
1665. USNM, holotype, no. 123,099, Acapulco
1666. USNM, holotype, no. 123,125, Gulf of Panama
1667. USNM, holotype, no. 97,092, Patagonia
1668. USNM, holotype, no. 123,126, Gulf of Panama
1669. USNM, holotype, no. 96,499, Galápagos Islands
1670. USNM, holotype, no. 123,130, Gulf of Panama
1671. USNM, holotype, no. 123,130, Gulf of Panama
1672. USNM, holotype, no. 123,118, Acapulco
1673. USNM, holotype, no. 110,564, Peru
1674. USNM, holotype, no. 123,123, Gulf of Panama
1675. LACM, Venado Island, Canal Zone
1676. AHF, Santa Maria Bay, Baja California
1677. LACM, Veracruz, Panama
1678. LACM, Newport Bay, California
1679. ANSP, no. 155,379, holotype of *Crassispira erebus*, Corinto, Nicaragua
1680. LACM, Guaymas
1681. USNM, no. 606,738, Gulf of Nicoya, Costa Rica
1682. LACM, Banderas Bay, Mexico
1683. AHF, Puerto Utria, Colombia
1684. Shasky collection, Guaymas
1685. LACM, Guaymas
1686. Shasky collection, Cuastecomate, Jalisco

1687. AHF, Thurloe Head, Baja California; radula from Powell (1966), with duplex marginal
1688. LACM, Venado Island, Canal Zone
1689. AHF, Cape San Francisco, Ecuador; radula from Powell (1966), with duplex marginal
1690. AHF, Cape San Francisco, Ecuador
1691. LACM, Puertecitos, Baja California
1692. LACM, Cuastecomate, Jalisco
1693. LACM, Mazatlán
1694. LACM, Guaymas
1695. LACM, Mazatlán
1696. LACM, holotype, no. 1,506, Venado Island, Canal Zone
1697. AHF, Secas Islands, Panama
1698. LACM, holotype, no. 1,508, Venado Island, Canal Zone
1699. USNM, no. 267,916, holotype of *S. lucasensis*, Cape San Lucas
1700. LACM, Canelo Bay, near Cape San Lucas
1701. LACM, Bahía de los Angeles, Gulf of California
1702. CAS, holotype, no. 9,545, Cape San Lucas
1703. LACM, Canelo Bay, near Cape San Lucas
1704. LACM, holotype, no. 1,510, Sayulita, Nayarit
1705. LACM, Mazatlán
1706. Left, ANSP, holotype, no. 155,376, Guaymas; right, LACM, Gonzaga Bay
1707. SDSNH, paratype, Maria Madre Island, Tres Marias
1708. SSB, no. 17,500, holotype of *T. torquifer*, Angel de la Guarda Island
1709. AMNH, holotype, no. 150,514, Daphne Island, Galápagos
1710. AHF, Puerto Escondido, Baja California
1711. AHF, Puerto Utria, Colombia
1712. LACM, holotype, Judas Point, Costa Rica
1713. LACM, Guaymas
1714. AHF, Tiburon Island, Gulf of California
1715. LACM, holotype, no. 1,512, Guaymas
1716. ANSP, holotype, no. 155,377, Acapulco
1717. LACM, holotype, no. 1,513, Guaymas
1718. Shasky collection, Mazatlán
1719. ANSP, no. 155,196, holotype of *L. acapulcanum*, Acapulco
1720. LACM, holotype, no. 1,566, Gulf of Tehuantepec
1721. AHF, Gulf of Dulce, Costa Rica
1722. AHF, Cape San Francisco, Ecuador
1723. LACM, Guaymas
1724. AHF, Angel de la Guarda Island

1725. SDSNH, no. 527, holotype of *C. pilsbryi*, Puerto Peñasco
1726. LACM, holotype, no. 1,515, Cuastecomate, Jalisco
1727. AHF, Bahía de los Angeles, Gulf of California; radula drawing by J. McLean, showing a solid marginal with collar
1728. LACM, holotype, no. 1,567, Venado Island, Canal Zone
1729. CAS, holotype, no. 9,426, Chatham Island, Galápagos
1730. LACM, holotype, no. 1,569, Veracruz, Panama
1731. LACM, Guaymas
1732. LACM, Veracruz, Panama
1733. LACM, Punta Diggs, near San Felipe
1734. USNM, holotype, no. 212,354, Guaymas
1735. SU, holotype, no. 8,622, Espíritu Santo Island
1736. AHF, Santa Elena Peninsula, Ecuador
1737. Shasky collection, Venado Island, Canal Zone
1738. LACM, holotype, no. 1,517, Isla Santa Cruz, Galápagos
1739. AHF, Chacahua Bay, Mexico
1740. SDSNH, paratype, no. 523, Manzanillo
1741. LACM, holotype, no. 1,519, Santa Elena Bay, Ecuador
1742. LACM, holotype, no. 1,521, Cedros Island, Baja California
1743. LACM, Puertecitos
1744. LACM, holotype, no. 1,523, Isla San Salvador, Galápagos
1745. LACM, Venado Island, Canal Zone
1746. Shasky collection, Cuastecomate, Jalisco
1747. BM, syntype, Ecuador
1748. ANSP, no. 155,522, holotype of *C. fonseca*, La Union, Gulf of Fonseca
1749. LACM, Mazatlán
1750. LACM, Venado Island, Canal Zone; radula drawing by J. McLean, showing a hollow marginal, barbed
1751. Shasky collection, Bahía Santa Cruz, Oaxaca
1752. LACM, Cuastecomate, Jalisco
1753. LACM, Las Hadas, Santiago Bay, Colima
1754. USNM, holotype, no. 56,135, Bartolomé Bay, Baja California; radula from Powell (1966), showing a hollow marginal, unbarbed
1755–56. LACM, Guaymas
1757. USNM, holotype, no. 56,085, west Mexico
1758. LACM, holotype, no. 1,524, Isla Santa Maria, Galápagos
1759. USNM, holotype, no. 123,107, Gulf of Panama

1760. LACM, holotype, no. 1,525, Isla Santa Cruz, Galápagos
1761. USNM, holotype, no. 123,105, Gulf of Panama
1762. USNM, holotype, no. 109,030, Acapulco
1763. USNM, holotype, no. 123,106, Gulf of Panama
1764. AHF, Puerto Utria, Colombia; radula drawing by J. McLean, showing a hollow marginal, doubly barbed
1765. AHF, Secas Islands, Panama
1766. USNM, holotype, no. 194,965, Galápagos Islands
1767. USNM, holotype, no. 207,577, Manta, Ecuador
1768. AMNH, holotype, no. 152,601, Tagus Cove, Isabela Island, Galápagos
1769. SU, holotype, no. 8,621, Espíritu Santo Island
1770. LACM, San Pedro, California; radula from Powell (1966), showing a hollow marginal, constricted near tip
1771. LACM, holotype, no. 1,527, Cape San Lucas
1772. LACM, holotype, no. 1,528, Monserrate Island, Gulf of California
1773. AHF, Puerto Utria, Colombia
1774. LACM, holotype, no. 1,529, Guaymas
1775. DuShane collection, Mazatlán
1776. LACM, Canelo Bay, near Cape San Lucas
1777. AHF, Puerto Culebra, Costa Rica
1778. USNM, holotype, no. 194,857, Cacachitos, Gulf of California; radula from Powell (1966), showing hollow marginal
1779. LACM, Puerto Utria, Colombia
1780. LACM, holotype, no. 1,573, Jicarita Island, Panama
1781. AMNH, no. 73,442, holotype of *L. armstrongi*, Guayabo Chiquito, Panama
1782. AMNH, holotype, no. 157,263, Jervis Island, Galápagos
1783. LACM, holotype, no. 1,530, Isla Santa Cruz, Galápagos
1784. USNM, holotype, no. 110,612, Cabo Lobos, Sonora
1785. USNM, no. 224,413, holotype of *G. partefilosa*, Cabo Tepoca, Sonora
1786. USNM, holotype, no. 123,115, Gulf of Panama; radula from Powell (1966), showing hollow marginal, serrate edge
1787. CAS, holotype, no. 9,606, Arena Bank, Gulf of California
1788. ANSP, holotype, no. 155,190, Montijo Bay, Panama
1789. Left, BM, syntype, Magdalena Bay;

right, USNM, no. 268,813, San Bartolomé Bay
1790. LACM, holotype, no 1,574, Agua de Chale, Baja California
1791. LACM, Chamela Bay, Mexico
1792. USNM, holotype, no. 206,554, Punta Año Nuevo, California
1793. USNM, holotype, no. 211,384, La Paz; radula from Powell (1966), showing hollow marginal, spur at base
1794. LACM, Punta Abreojos, Baja California
1795. Shasky collection, Venado Island, Canal Zone
1796. ANSP, holotype, no. 156,341, Acapulco
1797. AHF, Santa Elena Bay, Ecuador
1798. Left, AHF, Acapulco; right, LACM, Punta Peñasco
1799. CAS, no. 9,576, holotype of *Cytharella burchi*, Arena Bank, Gulf of California
1800. MCZ, holotype, no. 186,426, Panama
1801. ANSP, holotype, no. 155,194, San Juan del Sur, Nicaragua
1802. LACM, holotype, no. 1,532, Tagus Cove, Isla Isabela, Galápagos
1803. ANSP, holotype, no. 155,211, Manzanillo
1804. LACM, holotype, no. 1,533, Gulf of Guayaquil, Ecuador
1805. USNM, holotype, no. 214,266, Panama Bay
1806. USNM, holotype, no. 267,706, Concepción Bay
1807. LACM, holotype, no. 1,534, Guaymas
1808. LACM, Venado Island, Canal Zone
1809. LACM, Guaymas
1810. LACM, Guatulco Bay, Oaxaca
1811. LACM, holotype, no. 1,535, Gulf of Guayaquil, Ecuador
1812. USNM, holotype, no. 274,107, Panama
1813. LACM, Guaymas
1814. LACM, holotype, no. 1,537, Tague Cove, Isla Isabela, Galápagos
1815. USNM, holotype, no. 73,994, Cape San Lucas
1816. USNM, no. 268,908, holotype of *Cytharella euryclea*, Agua Verde Bay, Gulf of California
1817. USNM, holotype, no. 122,125, Isla Santa Cruz, Galápagos
1818. ANSP, holotype, no. 155,212, San Juan del Sur, Nicaragua
1819. MCZ, holotype, no. 186,406, Panama
1820. USNM, holotype, no. 55,503, Gulf of California
1821. USNM, no. 127,536, holotype of *Cytharella hippolita*, San Hipolito Point, Baja California
1822. LACM, holotype, no. 1,576, Mazatlán

1823. LACM, holotype, no. 1,578, Mazatlán
1824. LACM, no. 1,539, Bahía Guasimas, Sonora
1825. USNM, holotype, no. 265,920, Mulege, Baja California
1826. LACM, holotype, no. 1,579, Agua de Chale
1827. USNM, Panama
1828. Redpath Museum, holotype, no. 95, Panama
1829. USNM, holotype, no. 286,910, Agua Verde Bay
1830. USNM, holotype, no. 266,426, Agua Verde Bay
1831. Shasky collection, Guaymas
1832. USNM, holotype, no. 106,488, Scammon's Lagoon, Baja California
1833. USNM, holotype, no. 16,219, Cape San Lucas
1834. LACM, San José, Guatemala
1835. USNM, holotype, no. 268,908a, Agua Verde Bay; radula from Powell (1966), showing hollow marginal, partially unrolled
1836. ANSP, no. 170,297, holotype of *D. thalia*, Isla San Cristóbal, Galápagos
1837. USNM, holotype, no. 267,341, "Lower California"; radula from Powell (1966), showing hollow marginal, constricted below tip
1838. LACM, holotype, no. 1,541, Isla Santa Cruz, Galápagos
1839. SDSNH, paratype, no. 508, Cape San Lucas
1840. LACM, holotype, no. 1,542, Loreto, Baja California
1841. LACM, holotype, no. 1,543, Isla Santa Cruz, Galápagos
1842. BM, holotype, Panama
1843. SU, no. 9,745, holotype of *C. crebriforma*, Guaymas
1844. LACM, holotype, no. 1,545, Monserrate Island, Gulf of California
1845. LACM, Duncan Island, Galápagos
1846. LACM, holotype, no. 1,546, Isla Isabela, Galápagos
1847. LACM, holotype, no. 1,548, Isla Isabela, Galápagos
1848. LACM, Cuastecomate, Jalisco
1849. USNM, holotype, no. 96,480, Galápagos Islands
1850. USNM, holotype, no. 96,552, Galápagos Islands
1851. USNM, holotype, no. 110,610, Peru
1852. USNM, holotype, no. 123,113, Cocos Island, Costa Rica
1853. USNM, holotype, no. 123,114
1854. USNM, holotype, no. 123,116, Gulf of Panama

1855. USNM, holotype, no. 92,553, Galápagos Islands
1856. USNM, holotype, no. 96,479, Galápagos Islands
1857. USNM, holotype, no, 123,121, Ecuador
1858. USNM, holotype, no. 96,488, Galápagos Islands
1859. USNM, holotype, no. 123,117, Gulf of Panama
1860. USNM, holotype, no. 123,135, Ecuador
1861. USNM, holotype, no. 123,134, Galápagos Islands
1862. USNM, holotype, no. 96,554, Galápagos Islands
1863. USNM, holotype, no. 123,097, Panama
1864. Olsson (1971), Panama Bay
1865. USNM, holotype, no. 96,486, Galápagos Islands
1866. USNM, holotype, no. 123,112, Gulf of Panama
1867. USNM, holotype, no. 123,122, Gulf of Panama
1868. USNM, holotype, no. 123,111, Galápagos Islands
1869–74. No figure
1875. Dall & Bartsch (1909)
1876. Strong & Hertlein (1939), Panama Bay
1877–79. Dall & Bartsch (1909)
1880. Bartsch (1924)
1881–82. Dall & Bartsch (1909)
1883. Bartsch (1926)
1884. Right, Dall & Bartsch (1909); left, BM, syntypes
1885. Bartsch (1917)
1886. Above, Dall & Bartsch (1909); below, BM, syntypes of *P. clavulus*
1887–88. Dall & Bartsch (1909)
1889. ANSP, holotype
1890. Dall (1921)
1891. No figure
1892. Keen (1968)
1893. Johnson (1964)
1894. Dall & Bartsch (1909); x 32
1895. Dall & Bartsch (1909); x 4
1896. Dall & Bartsch (1909); x 15
1897–1902. No figure
1903. Dall & Bartsch (1909); x 12
1904–57. No figure
1958. Dall & Bartsch (1909); x 25
1959. No figure
1960. Dall & Bartsch (1909); x 16
1961–71. No figure
1972. Dall & Bartsch (1909); x 23
1973. No figure
1974. Dall & Bartsch (1909); x 20
1975. No figure
1976. Dall & Bartsch (1909); x 30
1977. Dall & Bartsch (1909); x 13
1978. No figure
1979. Dall & Bartsch (1909); x 10

1980. No figure
1981. Dall & Bartsch (1909) ; x 11
1982. Keen (1968), type of *Rissoina contabulata*; x 20
1983–85. No figure
1986. Dall & Bartsch (1909) ; x 20
1987–88. No figure
1989. Right, Dall & Bartsch (1909), x 11; left, Keen (1968)
1990–93. No figure
1994. Dall & Bartsch (1909) ; x 10
1995–2004. No figure
2005. Right, Dall & Bartsch (1909), x 13; left, Keen (1968), x 15
2006–9. No figure
2010. Dall & Bartsch (1909) ; x 15
2011–12. No figure
2013. Dall & Bartsch (1909) ; x 12
2014–15. No figure
2016. Dall & Bartsch (1909) ; x 15
2017. No figure
2018. Dall & Bartsch (1909) ; x 13
2019. No figure
2020. Lowe (1935) ; x 6
2021. Dall & Bartsch (1909) ; x 7
2022. Bartsch (1926) ; x 14
2023. Bartsch (1926) ; x 13
2024–26. No figure
2027. Dall & Bartsch (1909) ; x 7
2028–30. No figure
2031. Dall & Bartsch (1909) ; x 8
2032. Dall & Bartsch (1909) ; x 10
2033–38. No figure
2039. Dall & Bartsch (1909) ; x 8
2040–47. No figure
2048. Dall & Bartsch (1909) ; x 6
2049–54. No figure
2055. Bartsch (1917)
2056–58. No figure
2059. Dall & Bartsch (1909) ; x 9
2060–72. No figure
2073. Dall & Bartsch (1909) ; x 15
2074–81. No figure
2082. Dall & Bartsch (1909) ; x 18
2083. Lowe (1935) ; x 3
2084–2149. No figure
2150. Hertlein & Strong (1951), Corinto, Nicaragua; x 11
2151–57. No figure
2158. Dall & Bartsch (1909) ; x 8
2159–82. No figure
2183. Hertlein & Strong (1951) ; x 11
2184–2211. No figure
2212. Dall & Bartsch (1909) ; x 6
2213–25. No figure
2226. Bartsch (1917) ; x 12
2227. No figure
2228. Keen (1966), from holotype in BM
2229. Dall (1908) ; length, 7 mm
2230. Stearns (1897)
2231. Tryon, Manual Conch., vol. 15

2232. USNM, holotype
2233. Dall (1908) ; length, 5.5 mm
2234. Dall (1890) ; length, 5 mm; x 6
2235–37. Sowerby, Thes. Conch., vol. 1; (2236), see also color plate XVIII, fig. 5
2238. Dall (1908), length, 5.5 mm
2239. USNM, syntype, no. 4,014; x 5
2240. After Baker & Hanna (1927)
2241. Keen (1968), from holotype in BM
2242. CAS, Arena Bank
2243. After Baker & Hanna (1927)
2244. Gould (1855)
2245. After Baker & Hanna (1927), type of *H. strongi*
2246. BM, holotype; x 10
2247. No figure
2248. Keen (1958) ; x 14
2249. Harry (1967) ; x 20
2250. Left, Strong & Hertlein (1939) ; right, Harry (1967), x 11
2251. Harry (1967) ; x 12
2252. Keen (1958), from holotype; x 25
2253. Keen & Pearson (1952)
2254. Marcus (1961) ; see also color plate XXI, fig. 3
2255. Right, SU, off Baja California, from Calif. Inst. Tech.; left, Dall (1908)
2256. Right, Dall (1908) ; left, SU, off Baja California, from Calif. Inst. Tech.
2257. After Baker & Hanna (1927)
2258. Keen (1968), after syntypes in BM
2259. Pilsbry, Manual Conch., vol. 16
2260. After Turner (1956)
2261. USNM, holotype; x 8
2262. Dall (1908) ; length, 9 mm
2263. After Baker & Hanna (1927)
2264. Dall (1908) ; length, 6 mm
2265. Turner (1956) ; x 10
2266. Dall (1908) ; length, 10 mm
2267–68. Strong & Hertlein (1939)
2269. Keen (1958), from Baker & Hanna (1927) ; x 12
2270. No figure
2271. Strong & Hertlein (1939)
2272. No figure
2273. Berry (1953) ; x 36
2274. No figure
2275. McGowan (1968) ; x 10
2276. Sowerby, Thes. Conch.
2277. McGowan (1968) ; x 4
2278. McGowan (1968) ; x 7
2279. Sowerby, Thes. Conch.; x 2
2280–81. Keen (1958) ; x 5
2282. Sowerby, Thes. Conch.; x 6
2283. Sowerby, Thes. Conch.; x 10
2284. No figure
2285. Sowerby, Thes. Conch.; x 1.5
2286. Keen (1958) ; x 3
2287. Left, H. & A. Adams (1858) ; right, McGowan (1968) ; x 15

2288. McGowan (1968) ; x 25
2289. McGowan (1968) ; x 36
2290. McGowan (1968) ; x 2
2291. No figure
2292. McGowan (1968) ; x 7
2293–94. No figure
2295. Beeman (1968)
2296. Right, Beeman (1968) ; left, Keen & Pearson (1952), x 2.5; see also color plate XIX, fig. 4
2297. See color plate XIX, fig. 3
2298–99. No figure
2300. No figure
2301. Pilsbry, Manual Conch.
2302. Sowerby, Thes. Conch.; x 2
2303. Bertsch (1970)
2304. Pilsbry, Manual Conch., vol. 16; shell, x 7
2305. Farmer (1967)
2306. See color plate XIX, fig. 6
2307. Pilsbry, Manual Conch.
2308. Marcus & Marcus (1967) ; shell, x 10
2309. See color plate XIX, fig. 5
2310. SSB, Bahía Salinas, Costa Rica
2311. Keen & Pearson (1952) ; see also color plate XIX, fig. 2
2312. Keen (1958), after Pilsbry & Olsson (1943), x 7; see also color plate XIX, fig. 1
2313. See color plate XIX, fig. 7
2314. Sowerby, Thes. Conch.
2315. Left, BM, holotype, Keen (1968) ; right, after Howard (1951)
2316. Above, SU, La Paz; below, USNM, no. 218,179, type lot, x 5
2317. Left, Marcus (1961) ; right, MacFarland (1966)
2318–19. No figure
2320. See color plate XX, fig. 6
2321. No figure
2322. Lance (1962)
2323. Marcus & Marcus (1967)
2324. No figure
2325. See color plate XIX, fig. 8
2326–28. No figure
2329. Marcus & Marcus (1967)
2330. No figure
2331. Marcus & Marcus (1967) ; see also color plate XX, fig. 4
2332. See color plate XX, fig. 5
2333. No figure
2334. No figure
2335. Marcus & Marcus (1967) ; see also color plate XX, fig. 1
2336. See color plate XX, fig. 2
2337–39. No figure
2340. Marcus & Marcus (1967)
2341–44. No figure
2345. Marcus & Marcus (1967)
2346–51. No figure
2352. Marcus (1964)

2353. See color plate XXI, fig. 3
2354. Marcus & Marcus (1967) ; see also color plate XX, fig. 3
2355. See color plate XXI, fig. 1
2356. Marcus & Marcus (1967)
2357. No figure
2358. See color plate XXI, fig. 2
2359. See color plate XXI, fig. 5
2360. No figure
2361. Courtesy Dr. Frederick Bayer, Panama Bay
2362. Marcus & Marcus (1967)
2363. No figure
2364. Farmer (1970)
2365–66. No figure
2367. Marcus & Marcus (1967)
2368. Péron & Lesueur (1810)
2369. No figure
2370. See color plate XXI, fig. 6
2371. No figure
2372. Marcus & Marcus (1967)
2373. See color plate XXII, fig. 3
2374–75. No figure
2376. See color plate XXI, fig. 4
2377. See color plate XXII, fig. 4
2378. Farmer (1970)
2379. No figure
2380. Farmer (1970)
2381. No figure
2382. See color plate XXII, fig. 1
2383–84. No figure
2385. Hoffmann (1929)
2386. See color plate XXII, fig. 5
2387. Collier & Farmer (1964), dorsal and ventral views
2388. See color plate XXII, fig. 2
2389. See color plate XXII, fig. 6
2390–91. No figure
2392. Hoffmann (1929)
2393. Marcus & Marcus (1967)
2394. Stearns (1893c)
2395. Stearns (1893b)
2396. Marcus & Marcus (1967)
2397. Stearns (1893c)
2398. Below, right, Von Martens (1900) ; left, above, Turner (1956)
2399. SU, holotype, no. 9,503
2400. Right, Keen (1968), from syntype of M. o. olivaceus in BM; above, Keen (1963), M. o. californicus Berry
2401. Below, Von Martens (1900) ; above, Turner (1956)
2402. BM, syntypes
2403–04. After Morrison (1946)
2405. Von Martens (1900)
2406. Miller (1879)
2407. Pilsbry (1920)
2408. Turner (1956)
2409. Schwengel (1938)
2410. Turner (1956)
2411. SDSNH, no. 53,266; Bahía de los

Angeles, Gulf of California; drawing
by A. D'Attilio, courtesy Dr. D. W.
Taylor
2412. SU, southern California
2413. Pilsbry (1920)
2414. Turner (1956)
2415. Pilsbry (1910)
2416. Right, BM, syntype; left, SU, Panama
2417. Turner (1956)
2418. Reeve, Conch. Icon.

2419. SSB, no. 29, 882
2420. BM, holotype of *S. aequilirata*
2421. Reeve, Conch. Icon., vol. 9
2422. Below, BM, syntypes; above, SU,
Mazatlán
2423. BM, syntypes
2424–25. Dall (1921)
2426. After Gray (1839)
2427. SU, Baja California
2428. SU, Oaxaca

MONOPLACOPHORA

1. Lemche (1957)
2. Menzies & Layton (1963)

3. Clark & Menzies (1959)

APLACOPHORA

1. After Heath (1911)

POLYPLACOPHORA

1. Thorpe coll., Puertecitos, Baja California
2–3. Thorpe coll., from G. Hanselman;
Mazatlán, Mexico
4. SU, Galápagos Islands
5. Thorpe coll., from E. Bergeron, Panama
6. SU, Galápagos Islands
7. Thorpe coll., Bahía Kino, Sonora
8. No figure
9. Thorpe coll., from E. Bergeron, Panama
10. Thorpe coll., Bahía Kino, Sonora, with
details of plate structure at right
11. Thorpe coll., Bahía Cholla, Sonora
12. Thorpe coll., Bahía Kino, Sonora
13. Thorpe coll., from E. Bergeron, Panama
14. LACM, from J. McLean, Banderas Bay,
Jalisco
15. Thorpe coll., from E. Bergeron, Panama
16. Thorpe coll., smaller specimen from
J. McLean, Panama; larger specimen,
Bahía Cholla, Sonora
17. Thorpe coll., from G. Hanselman,
Mazatlán
18. Thorpe coll., Bahía Kino region, Sonora
19. Thorpe coll., northern end, Gulf of
California
20. LACM, coll. J. McLean, Maria Cleofas
Island, Tres Marias
21. Thorpe coll., from E. Bergeron, Panama
22. Thorpe coll.: left, oblique view, Bahía
Kino; right, Guaymas
23. Thorpe coll., Sonora coast
24. Thorpe coll., Bahía Kino, Sonora
25. Thorpe coll., Percebu, Baja California
26. LACM, from J. McLean, Cuastecomate,
Jalisco

27. No figure
28. Thorpe coll., Puertecitos, Baja California
29. Thorpe coll., Bahía Kino, Sonora
30. Thorpe coll., from E. Bergeron, Panama
31. Right, Thorpe coll., Puertecitos, Baja
California; left, after Berry (1931)
32. Hertlein & Strong (1951)
33. Thorpe coll.: right, Maria Magdalena
Island, Tres Marias, from J. McLean;
left, Panama, from E. Bergeron
34. Left, LACM, from J. McLean, Panama;
right, after Thiele (1910)
35. LACM: left, oblique view, Cuastecomate,
Jalisco; right, Cabo San Lucas
36. LACM, from J. McLean, Cabo Pulmo,
Baja California
37. Thorpe coll., Bahía Kino, Sonora
38. Thorpe coll., San Felipe, Baja California
39. Thorpe coll., from E. Bergeron, Panama
40. Thorpe coll., San Felipe, Baja California
41. Thorpe coll., from E. Bergeron, Panama
42. LACM, from J. McLean, Tiburon Island,
Gulf of California
43. LACM, from J. McLean, Cuastecomate,
Jalisco; details of plates from Pilsbry,
Manual Conch.
44. Thorpe coll., Bahía Kino, Sonora
45. Thorpe coll., Guaymas
46. Thorpe coll., from E. Bergeron, Panama
47. No figure
48. Thorpe coll., San Felipe, Baja California
49. BM, syntype, Peru; courtesy BM
50. Thorpe coll., Puerto Santo Tomas, Baja
California
51–56. No figure

SCAPHOPODA

1. Dall (1908)
2. Pilsbry, Manual Conch.
3. Pilsbry, Manual Conch.; right and center,

(Keen, 1966), from syntypes in Univ.
Copenhagen
4. Oldroyd (1925)

5. Smith & Gordon (1948)
6. Dall (1908)
7-8. Pilsbry, Manual Conch., vol. 17
9. After Emerson (1956)
10-11. Pilsbry, Manual Conch.
12. Pilsbry, Manual Conch.; length, 11 mm
13-14. Pilsbry Manual Conch.
15. Pilsbry, Manual Conch.; length, 31 mm
16. Left, Oldroyd (1925) ; right, Pilsbry, Manual Conch.

17. Pilsbry, Manual Conch.
18. Oldroyd (1925)
19. Pilsbry, Manual Conch.
20. Emerson (1951)
21. USNM, holotype; length, 12 mm
22. Pilsbry, Manual Conch.; length, 12 mm
23. Pilsbry, Manual Conch.; length, 15 mm
24. Dall (1908) ; length, 24 mm
25. Dall (1908) ; length, 25 mm

CEPHALOPODA

1-10. No figure
11. Dall (1902)

12. Reeve, Conch. Icon.
13. SU, Catalina Island, California

REJECTED AND INDETERMINATE SPECIES
(see p. 905)

Coralliophila stearnsiana Dall. USNM, holotype; x 1.6; photograph by J. McLean
Ricinula contracta Reeve. BM, syntypes; x 2;
photograph by Myra Keen
Bulla nonscripta A. Adams. BM, holotype; x 5.5; photograph by Myra Keen

COLOR PLATES

Plates I-II. SU, Gulf of California
Plate III. SU, Espíritu Santo Island
Plate IV. SSB, off Sonora coast
Plate V. SU, Gulf of California
Plate VI. Pearl Islands, Panama
Plate VII. Above, SU, Guaymas; below, SU, Panama
Plate VIII. SU, Gulf of California
Plate IX. SU, Panama
Plate X. SU, San Benito Island, Baja California

PLATE XI (Gastropoda)

1-2. No. 1357. Karl H. Lust collection, northwest of Panama Canal Zone, offshore; photograph by Edward N. Hamilton
3-4. No. 610. Panama. Karl H. Lust collection; photograph by Edward N. Hamilton

PLATE XII (Pelecypoda, except Fig. 8)

1. No. 26. Dr. Thomas Burch, Cholla Bay, Sonora; 27 m
2. No. 40. Dr. Burch, Cholla Bay, Sonora
3. No. 378. Mrs. Faye Howard, San Luis Gonzaga Bay
4. Nos. 363, 425. Mrs. Howard, Kino Bay, Sonora
5. No. 219. Mrs. Howard, Cholla Bay, Sonora
6. [indeterminate] Mrs. Howard, Mazatlán
7. No. 220. Mrs. Howard, Mazatlán
8. No. 18. Mrs. Howard, San Felipe, Baja California

PLATE XIII (Pelecypoda, except Figs. 7 and 8)

1. No. 529. Dr. Thomas Burch, Cholla Bay, Sonora; 27 m

2. No. 591. Dr. Burch, Cholla Bay, Sonora
3. No. 535. Dr. Burch, Cholla Bay, Sonora
4. No. 401. Dr. Burch, Cholla Bay, Sonora
5. No. 633. Dr. Burch, Cholla Bay, Sonora
6. No. 637. Dr. Burch, Cholla Bay, Sonora
7. No. 44. Mrs. Faye Howard, Tenacatita Bay
8. No. 85. Dr. Burch, Cholla Bay; 45 m

PLATE XIV (Gastropoda: Mesogastropoda)

1. No. 752. Ralph Kettenring collection, off Guaymas; photograph by Myra Keen
2. No. 803. Mrs. Faye Howard, Bahía Audencia
3. No. 861. Dr. Thomas Burch, Cholla Bay, Sonora
4. No. 873. Dr. Burch, Cholla Bay, Sonora
5. No. 612. Don Wobber, off La Paz
6. No. 861. Mrs. Howard, Estero Soldado

PLATE XV (Gastropoda: all Mesogastropoda except Fig. 2)

1. No. 933. Mrs. Faye Howard, Santa Rosalía, Baja California
2. No. 172. Dr. Frederick Bayer, Panama
3. No. 952. Dr. Thomas Burch, Cholla Bay, Sonora
4. No. 851. Collected by Don Wobber, off Monterey, California; photograph by Allyn G. Smith
5. No. 919. Mrs. Howard, Rancho El Tule, Baja California
6. No. 927. Mrs. Howard, Mazatlán

PLATE XVI (Gastropoda: Mesogastropoda)

1. No. 948. Mrs. Faye Howard, San Luis Gonzaga Bay
2. No. 898. Mrs. Howard, San Luis Gonzaga Bay, Baja California

3. No. 903. Mrs. Howard, Tenacatita, Jalisco
4. No. 947. Mrs. Howard, Tenacatita, Jalisco
5–6. No. 949. Collection of Verona McKibbin, Tucson, off La Paz, Baja California; photograph by Dr. G Dallas Hanna

PLATE XVII (Gastropoda: all Neogastropoda, Family Olividae)

1. No. 1363. Dr. Thomas Burch, Cholla Bay, Sonora
2. No. 1361. Mrs. Faye Howard, Matenchen, Mexico
3. No. 1377. Dr. Burch, Cholla Bay, Sonora
4. No. 1396. Dr. Burch, Cholla Bay, Sonora
5. No. 1360. Dr. Burch, Cholla Bay, Sonora
6. No. 1360. Mrs. Howard, San Felipe, Baja California
7. No. 1370. Dr. Burch, Norse Beach, Sonora
8. No. 1364. Mrs. Howard, Guaymas, Sonora

PLATE XVIII (Gastropoda: all Prosobranchiata except Fig. 5)

1. No. 940. Mrs. Faye Howard, Tenacatita, Jalisco
2. No. 935. Mrs. Howard, Bahía San Carlos, Sonora
3. No. 1321. Dr. Thomas Burch, Cholla Bay, Sonora
4. No. 1458. Mrs. Howard, Mazatlán
5. No. 2236. Dr. Burch, Norse Beach, Sonora
6. No. 936. Dr. Burch, Cholla Bay, Sonora

PLATE XIX (Gastropoda: all tectibranch Opisthobranchia except Fig. 8)

1. No. 2312. Walter Harvey and James Lance, Isla San José, Gulf of California
2. No. 2311. Walter Harvey and James Lance, La Jolla, California
3. No. 2297. Dr. Peter Pickens, Puerto Peñasco, Sonora
4. No. 2296. Dr. Eugene Coan, San Carlos Bay, Guaymas
5. No. 2309. Dr. Frederick Bayer, Panama
6. No. 2306. James Lance and David Mulliner, Gulf of California

7. No. 2313. Collected by Gale Sphon at Santa Cruz, Nayarit; photograph by James Lance
8. No. 2325. James Lance, outer coast of Baja California

PLATE XX (Gastropoda: Opisthobranchia, all Nudibranchia except Fig. 6)

1. No. 2335. Walter Harvey and James Lance, Gulf of California
2. No. 2336. Wesley Farmer and James Lance, Baja California
3. No. 2354. Wesley Farmer, Isla Angel de la Guarda, Gulf of California
4. No. 2331. James Lance, Gulf of California
5. No. 2332. Richard Roller, Gulf of California
6. No. 2320. Walter Harvey and James Lance, Bahía San Luis Gonzaga, Gulf of California

PLATE XXI (Gastropoda: Opisthobranchia, all Nudibranchia except Fig. 3, in part)

1. No. 2355. James Lance, California (?)
2. No. 2358. Richard Roller, Bahía de los Angeles, Gulf of California
3. Nos. 2254, 2353. James Lance and David Mulliner, San Diego, California
4. No. 2376. James Lance, California (?)
5. No. 2359. Wesley Farmer and James Lance, Bahía de los Angeles, Gulf of California
6. No. 2370. Don Wobber, California

PLATE XXII (Gastropoda: Nudibranchia)

1. No. 2382. Walter Harvey and James Lance, La Jolla, California
2. No. 2388. James Lance, Gulf of California (?)
3. No. 2373. James Lance, California (?)
4. No. 2377. James Lance, California (?)
5. No. 2386. Wesley Farmer, Puertecitos, Baja California
6. No. 2389. Walter Harvey and James Lance, La Jolla, California

TEXT FIGURES

Mytilidae (p. 58). Adapted from Soot-Ryen (1955)

Pholadidae (p. 274). From Turner (1954)

Teredinidae (p. 281). Adapted from Bartsch (1922)

Cypraeacea (p. 491). Adapted from Schilder (1938–39), redrawn by Perfecto Mary; juvenile *Cypraea* drawn by H. D. Vernon

Conidae (p. 659). From Nybakken (1970); operculum drawn by H. D. Vernon

Turridae (p. 687). Adapted from Powell 1966); figure at lower right by J. McLean

Nudibranchia (p. 819). Adapted from Lance (1966); Cunningham in Encyclopaedia Britannica (1910); aeolid head drawn by Hans Bertsch from a color transparency by Wesley Farmer

Chitonidae (p. 862). Adapted from Keen (1963)

ENDSHEET PHOTOGRAPH: Christian Hansen

BIBLIOGRAPHY

Bibliography

This Bibliography draws heavily upon one compiled by Hertlein and Strong (1955) that was the first really comprehensive review of literature on mollusks of the Panamic province. It is the intention here to cite the titles of original references for all the species discussed in the text as well as other references mentioned in the systematic part of the book and some general literature on the area. Not all of the titles of original references for the synonyms are included, especially those in works not otherwise pertaining to the province. The finding list or topical summary, which precedes the Bibliography proper, emphasizes the literature of the past two decades and those papers that are not explicitly referred to in the text.

FINDING LIST OR TOPICAL SUMMARY

Bibliographies of authors
Carpenter, P. P.: Coan, 1969
Dall, W. H.: Bartsch, *et al.*, 1946; Boss, *et al.*, 1968
Gould, A. A.: Johnson, 1964
Hertlein, L. G.: Addicott, 1970
Pilsbry, H. A.: [Baker], 1940; Clench & Turner, 1962

Bibliographies of areas
Sonora, Mexico: Skoglund, 1970
Revillagigedo Islands: Richards, *et al.*, 1950
Cocos Island: Hertlein, 1963

Bibliographies of libraries
British Museum (Natural History): Woodward, 1903–40

Bionomics (functional morphology; geographic distribution; behavior)
Gonor, 1965; Hagberg & Kalb, 1968; Paine, 1966; Pohlo, 1963; Saunders, 1959; Seilacher, 1959; Stohler, 1963; Valentine, 1966; Yonge, 1958, 1967

Ecology, geology, coastal features
Bakus, 1968*a*; Coan, 1968; Dawson & Beaudette, 1959; Durham & Allison, 1960; Emerson, 1960, 1967; Emerson & Hertlein, 1969; Hertlein, 1957; Keen, 1964; Lindsay, 1964; Seilacher, 1959; Shasky, 1960; Vokes & Vokes, 1962

General works
Dance, 1966; Farmer, 1968; R. C. Moore, *et al.*, 1960, 1969; Morris, 1966

Geographic areas
  Dredging stations, "Albatross": Townsend, 1901–
  Gulf of California: Allison, 1964; DuShane, 1962; DuShane & Poorman, 1967; Du-
    Shane & Sphon, 1968; Parker, 1964; Violette, 1964; Western Society of Malacolo-
    gists, 1970
  Revillagigedo Is.: Richards & Brattstrom, 1959
  Clipperton I.: Hertlein & Allison, 1963, 1966; Hertlein & Emerson, 1953, 1957;
    Sachet, 1962; Salvat & Ehrhardt, 1970
  Cocos I.: Hertlein, 1963; Emerson & Old, 1964
  Panama: Olsson, 1961, 1971
  Galápagos Is.: Emerson & Old, 1965
  Ecuador: Hofstetter, 1952

Taxonomy (lists of type specimens of persons and institutions)
  C. B. Adams: Turner, 1956
  California Academy of Sciences (in part): Stasek, 1966
  Carpenter: Brann, 1966; Keen, 1968; Palmer, 1958, 1963
  Conrad: Keen, 1966a
  Dall: Boss et al., 1968
  Gould: Johnson, 1964
  Hinds: Keen, 1966b
  Los Angeles County Museum: Sphon, 1971
  Mörch: Keen, 1966c
  Orbigny: Keen, 1966d
  San Diego Natural History Museum: Wilson & Kennedy, 1967
  Santa Barbara Museum of Natural History: Sphon, 1966

BIBLIOGRAPHY

Abbott, R. T. 1954. American seashells. New York, xiv + 541 pp., 100 figs., 40 pls.
  1968. The helmet shells of the world (Cassidae). Pt. 1. Indo-Pacific Mollusca, vol. 2,
    no. 9, pp. 7–201, 187 figs., 13 col. pls. (Aug. 30).
Adams, Arthur. 1850a. Descriptions of new species of shells from the Cumingian col-
    lection. Proc. Zool. Soc. London, vol. 17, for 1849, pp. 169–70, pl. 6 (Jan.–June).
  1850b. Monograph of the family Bullidae. In G. B. Sowerby, Thesaurus conchylio-
    rum. London, vol. 2, pp. 553–608, pls. 119–25.
  1851a. A monograph of Phos, a genus of gasteropodous Mollusca. Proc. Zool. Soc.
    London, vol. 18, for 1850, pp. 152–55 (Feb. 28).
  1851b. A monograph of Modulus, a genus of gasteropodous Mollusca, of the family
    Littorinidae. Ibid., pp. 203–4 (Feb. 28).
  1852–53. Catalogue of the species of Nassa, a genus of gasteropodous Mollusca be-
    longing to the family Buccinidae, in the collection of Hugh Cuming, Esq., with the
    description of some new species. Ibid., for 1851, pp. 94–112 (Dec. 7, 1852); 113–14
    (Apr. 29, 1853).
  1853a. Contributions towards a monograph of the Trochidae, a family of gastero-
    podous mollusks. Ibid., for 1851, pp. 150–92 (June 28).
  1853b. Descriptions of sixteen new species of Rissoina . . . Ibid., for 1851, pp. 264–67
    (Dec. 7).
  1853c. Description of several new species of Murex, Rissoina, Planaxis, and Eulima
    from the Cumingian collection. Ibid., for 1851, pp. 267–72 (Dec. 7).
  1854a. Descriptions of new species of Semele, Rhizochilus, Plotia, and Tiara in the
    Cumingian collection. Ibid., for 1853, pp. 94–99 (July 25).
  1854b. Descriptions of new species of shells, in the collection of Hugh Cuming, Esq.
    Ibid., for 1853, pp. 173–76, pl. 20 (Dec. 15).
  1855a. Further contributions toward the natural history of the Trochidae: with the

description of a new genus and of several new species, from the Cumingian collection. *Ibid.*, for 1854, pp. 37–41, pl. 27 (Jan. 10).

1855b. Monographs of *Actaeon* and *Solidula*, two genera of gasteropodous Mollusca with descriptions of several new species from the Cumingian collection. *Ibid.*, for 1854, pp. 58–62 (Jan. 10).

1855c. A monograph of *Cerithidea*, a genus of Mollusca, with descriptions of several new species, from the collection of Hugh Cuming, Esq.: to which are added, descriptions of two new species of *Colina*, and one of *Donax*. *Ibid.*, for 1854, pp. 83–87 (Apr. 11).

1855d. Descriptions of thirty-nine new species of shells, from the collection of Hugh Cuming, Esq. *Ibid.*, for 1854, pp. 130–38, pl. 38 (Apr. 11).

1855e. Descriptions of twenty-seven new species of shells from the collection of Hugh Cuming, Esq. *Ibid.*, for 1854, pp. 311–17 (May 8).

1855f. Descriptions of new genera and species of gasteropodous Mollusca. *Ibid.*, for 1853, pt. 21, pp. 182–86 (May 16).

1856. Description of thirty-four new species of bivalve Mollusca (*Leda, Nucula,* and *Pythina*) from the Cumingian collection. *Ibid.*, for 1856, pp. 47–53 (July 30).

1857. Notice of a new species of *Trichotropis* from the Cumingian collection. *Ibid.*, for 1856, p. 369 (May 8).

1859. On the synonyms and habitats of *Cavolina, Diacria* and *Pleuropus*. Ann. Mag. Nat. Hist., ser. 3, vol. 3, pp. 44–46.

Adams, Arthur, and L. Reeve. 1848–50. Mollusca. *In* A. Adams, The zoology of the voyage of H.M.S. Samarang; under the command of Captain Sir Edward Belcher. London, pt. 1, pp. i–x, [i–ii], 1–24, pls. 1–9 (Nov. 1848) ; pt. 2, pp. 25–44, pls. 10–17 (May 1850) ; pt. 3, pp. 45–87, pls. 18–24 (Aug. 1850).

Adams, C. B. 1845. Specierum novarum conchyliorum, in Jamaica repertorum, synopsis. Proc. Boston Soc. Nat. Hist., vol. 2, pp. 1–17 (Jan.).

1852a. Catalogue of shells collected at Panama, with notes on synonymy, station, and habitat. . . . Ann. Lyceum Nat. Hist. New York, vol. 5, pp. 229–96 (June) ; 297–549 [549 unnumbered] (July).

1852b. Catalogue of shells collected at Panama, with notes on their synonymy, station, and geographical distribution. New York (R. Craighead). pp. i–viii + 1–334.

Adams, H., and A. Adams. 1853–58. The genera of Recent Mollusca, arranged according to their organization. London: vol. 1, pp. vi–xl, 1–484; vol. 2, pp. 1–661; vol. 3, pls. 1–138 [Dating of parts, vol. 2, p. 661].

1854. Description of a new genus of bivalve Mollusca. Ann. Mag. Nat. Hist., ser. 2, vol. 14, p. 418.

1864. Descriptions of new species of shells chiefly from the Cumingian collection. Proc. Zool. Soc. London, for 1863, pp. 428–35 (Apr.).

Addicott, W. O. 1970. Bibliography of Leo G. Hertlein for the period 1925 to 1970. The Nautilus, vol. 84, no. 2, pp. 52–69. (Oct.).

Afshar, Freydoun. 1969. Taxonomic revision of the superspecific groups of the Cretaceous and Cenozoic Tellinidae. Geol. Soc. America, Memoir 119: 215 pp., 45 pls.

Allison, E. C. 1959. Distribution of *Conus* on Clipperton Island. Veliger, vol. 1, no. 4, pp. 32–34 (Apr. 1).

1964. Geology of areas bordering Gulf of California. Marine geology of the Gulf of California—a Symposium, Mem. no. 3, American Assoc. Petrol. Geol., pp. 3–29.

Anton, H. E. 1839. Verzeichniss der Conchylien welche sich in der Sammlung von Hermann Eduard Anton befinden. Halle, xvi + 110 pp.

Arnold, Ralph. 1903. The paleontology and stratigraphy of the marine Pliocene and Pleistocene of San Pedro, California. Mem. California Acad. Sci., vol. 3, pp. 1–420, pls. 1–37 (June 27). [For notes on the date of issuance, *see* L. G. Hertlein, Veliger, vol. 6, no. 3, p. 172, 1964].

1906. The Tertiary and Quaternary pectens of California. Prof. Paper U.S. Geol. Surv., no. 47, pp. 1–264, 2 figs., pls. 1–53.

Ascanius, P. 1774. [*Amphitrite frondosus*]. K. Nordske Vid. Selsk. Skrift., vol. 5, p. 158.

Baba, A. 1937. Opisthobranchia of Japan (II). Jour. Dept. Agric. Kyushu Univ., vol. 5, pt. 7, pp. 289–344, pls. 1–2.

Baily, J. L. 1950. *Maxwellia*, genus novum of Muricidae. Nautilus, vol. 64, no. 1, pp. 9–14 (July 5).

Baker, Fred. 1926. Mollusca of the family Triphoridae. Proc. Calif. Acad. Sci., ser. 4, vol. 15, no. 6, pp. 223–39, pl. 24 (Apr. 26).

    1929. A new name for a California shell—*Liotia acuticosta* Carpenter, var. *radiata* Dall. Nautilus, vol. 43, p. 72.

Baker, Fred, and G. D. Hanna. 1927. Marine Mollusca of the Order Opisthobranchiata. Proc. Calif. Acad. Sci., ser. 4, vol. 16, no. 5, pp. 123–35, pl. 4 (Apr. 22).

Baker, Fred, G. D. Hanna, and A. M. Strong. 1928. Some Pyramidellidae from the Gulf of California. Proc. Calif. Acad. Sci., ser. 4, vol. 17, no. 7, pp. 205–46, pls. 11–12 (June 29).

    1930a. Some Rissoid Mollusca from the Gulf of California. *Ibid.*, vol. 19, no. 4, pp. 23–40, pl. 1, 4 figs. in text (July 15).

    1930b. Some Mollusca of the family Epitoniidae from the Gulf of California. *Ibid.*, vol. 19, no. 5, pp. 41–56, pls. 2, 3 (July 15).

    1938a. Some Mollusca of the families Cerithiopsidae, Cerithiidae and Cyclostrematidae from the Gulf of California and adjacent waters. *Ibid.*, vol. 23, no. 15, pp. 217–44, pls. 17–23 (May 24).

    1938b. Columbellidae from western Mexico. *Ibid.*, vol. 23, no. 16, pp. 245–54, pl. 24 (May 24).

Baker, Fred, and V. D. P. Spicer. 1935. New species of mollusks of the genus *Triphora*. Trans. San Diego Soc. Nat. Hist., vol. 8, no. 7, pp. 35–46, pl. 5 (Mar. 21).

[Baker, H. B.]. 1940. Scientific contributions made from 1882 to 1939 by Henry A. Pilsbry, Sc.D. American Malacol. Union, 63 pp. (June).

Bakus, G. J. 1968a. Zonation in marine gastropods of Costa Rica and species diversity. Veliger, vol. 10, no. 3, pp. 207–11 (Jan. 1).

    1968b. Quantitative studies on the cowries (Cypraeidae) of the Allan Hancock Foundation collections. Veliger, vol. 11, no. 2, pp. 93–97 (Oct. 1).

Barnes, D. H. 1824. Notice of several species of shells. Ann. Lyc. Nat. Hist. New York, vol. 1, pp. 131–40, pl. 9 [Issued at least as early as Apr. 5, 1824; *see* M. Meisel, 1926, Bibl. Amer. Nat. Hist., vol. 2, p. 254.]

Bartsch, Paul. 1902. A new *Rissoina* from California: Nautilus, vol. 16, no. 1, p. 9 (May).

    1907a. New mollusks of the family Vitrinellidae from the west coast of America. Proc. U.S. Nat. Mus., vol. 32, no. 1520, pp. 167–76, 11 figs. (Feb. 8).

    1907b. A new mollusk of the genus *Macromphalina* from the west coast of America: Proc. U.S. Nat. Mus., vol. 32, no. 1522, p. 233, 1 fig. (Mar. 12).

    1907c. New marine mollusks from the west coast of America: Proc. U.S. Nat. Mus., vol. 33. pp. 177–83 (Oct. 13).

    1907d. The West American mollusks of the genus *Triphoris*. Proc. U.S. Nat. Mus., vol. 33, no. 1569, pp. 249–62, pl. 16 (Dec. 12).

    1910. The West American mollusks of the genus *Alaba*. *Ibid.*, vol. 39, no. 1781, pp. 153–56, 3 figs. in text (Oct. 25).

    1911a. Descriptions of new mollusks of the family Vitrinellidae from the west coast of America: Proc. U.S. Nat. Mus., vol. 39, no. 1785, pp. 229–34, pls. 39–40 (Jan. 9).

    1911b. The Recent and fossil mollusks of the genus *Alabina* from the west coast of America. *Ibid.*, vol. 39, no. 1790, pp. 409–18, pls. 61–62 (Jan. 13).

    1911c. The West American mollusks of the genus *Eumeta*. *Ibid.*, vol. 39, no. 1799, pp. 565–68, 3 figs. in text (Feb. 15).

    1911d. The Recent and fossil mollusks of the genus *Diastoma* from the west coast of America. *Ibid.*, vol. 39, no. 1802, pp. 581–84, figs. 1–4 (Feb. 15).

    1911e. The Recent and fossil mollusks of the genus *Cerithiopsis* from the west coast of America. *Ibid.*, vol. 40, no. 1823, pp. 327–67, pls. 36–41 (May 8).

1911*f*. The Recent and fossil mollusks of the genus *Bittium* from the west coast of America. *Ibid.*, vol. 40, no. 1826, pp. 383–414, pls. 51–58 (May 12).

1911*g*. The West American mollusks of the genus *Amphithalamus*. *Ibid.*, vol. 41, no. 1854, pp. 263–65, 3 figs. in text (June 30).

1911*h*. The Recent and fossil mollusks of the genus *Alvania* from the west coast of America. *Ibid.*, vol. 41, no. 1863, pp. 333–62, pls. 29–32 (Nov. 15).

1912*a*. Additions to the West American pyramidellid mollusk fauna, with descriptions of new species. *Ibid.*, vol. 42, no. 1903, pp. 261–89, pls. 35–38 (May 17).

1912*b*. A zoogeographic study based on the pyramidellid mollusks of the west coast of America. *Ibid.*, no. 1906, pp. 297–349, pl. 40 (June 15).

1915. The Recent and fossil mollusks of the genus *Rissoina* from the west coast of America. *Ibid.*, vol. 49, no. 2094, pp. 33–62, pls. 28–33 (July 24).

1916. *Eulimastoma*, a new subgenus of Pyramidellidae and remarks on the genus *Scalenostoma*. The Nautilus, vol. 30, no. 7, pp. 73–74 (Nov.).

1917*a*. Descriptions of new West American marine mollusks and notes on previously described forms: Proc. U.S. Nat. Mus., vol. 52, no. 2193, pp. 637–81, pls. 42–47 (May 29).

1917*b*. A monograph of West American melanellid mollusks. *Ibid.*, vol. 53, no. 2207, pp. 295–356, pls. 34–49 (Aug. 13).

1918. New marine shells from Panama. *Ibid.*, vol. 54, no. 2250, pp. 571–75, pl. 88 (Dec. 23).

1920*a*. The West American mollusks of the families Rissoellidae and Synceratidae, and the rissoid genus *Barleeia*. *Ibid.*, vol. 58, no. 2331, pp. 159–76, pls. 12–13 (Nov. 9).

1920*b*. The Caecidae and other marine mollusks from the northwest coast of America. Jour. Washington Acad. Sci., vol. 10, no. 20, pp. 565–72 (Dec. 4).

1921. A new classification of the shipworms and descriptions of some new wood boring mollusks. Proc. Biol. Soc. Washington, vol. 34, pp. 25–32 (Mar. 31).

1922. A monograph of the American shipworms. Bull. U.S. Nat. Mus. no. 122, 51 pp., 37 pls.

1923. Additions to our knowledge of shipworms. Proc. Biol. Soc. Washington, vol. 36, pp. 95–102 (Mar. 28).

1924. New mollusks from Santa Elena Bay, Ecuador. Proc. U.S. Nat. Mus., vol. 66, art. 14, pp. 1–9, pls. 1–2 (Oct. 17).

1926. Additional new mollusks from Santa Elena Bay, Ecuador. *Ibid.*, vol. 69, art. 20, pp. 1–20, pls. 1–3 (Dec. 16).

1927. New West American marine mollusks. Proc. U.S. Nat. Mus., vol. 70, no. 2660, art. 11, pp. 1–36, pls. 1–6 (Apr. 8).

1928*a*. *Odostomia* (*Ividella*) *mariae*, new species: Nautilus, vol. 42, no. 3, pl. 2, (Jan.)

1928*b*. New marine mollusks from Ecuador. Jour. Washington Acad. Sci., vol. 18, pp. 66–75, 16 figs. (Feb. 4).

1931*a*. Descriptions of new marine mollusks from Panama, with a figure of the genotype of *Engina*. Proc. U.S. Nat. Mus., vol. 79, art. 15, pp. 1–10, 1 pl. (Aug. 1).

1931*b*. The West American mollusks of the genus *Acar*. *Ibid.*, vol. 80, art. 9, pp. 1–4, 1 pl. (Nov. 23).

1937. A new West American cone. Nautilus, vol. 51, no. 1, pp. 3–4, pl. 2, fig. 8 (July 3).

1941. A new shipworm from Panama. Smithsonian Misc. Coll., vol. 99, no. 21, pp. 1–2, pl. 1 (Mar. 31).

1944*a*. Some notes upon West American turrid mollusks. Proc. Biol. Soc. Washington, vol. 57, pp. 25–30 (June 28).

1944*b*. A new shipworm from the Panama Canal. Smithsonian Misc. Coll., vol. 104, no. 8, pp. 1–3, pl. 1 (Sept. 7).

1944*c*. Imaclava, a correction. Nautilus, vol. 58, no. 2, p. 67 (Oct.).

1950. New West American turrids. *Ibid.*, vol. 63, no. 3, pp. 87–97, pl. 6 (Feb. 13).

Bartsch, Paul, and H. A. Rehder. 1939. Mollusks collected on the Presidential Cruise of 1938. Smithsonian Misc. Coll., vol. 98, no. 10, pp. 1–18, pls. 1–5 (June 13).

Bartsch, Paul, H. A. Rehder, and Beulah Shields. 1946. A bibliography and short biographical sketch of William Healey Dall. Smithsonian Misc. Coll. vol. 104, no. 5, 96 pp.

Bayle, E. 1880. Liste rectificative de quelques noms de genres et d'espèces. Jour. Conchyl., vol. 28, p. 240–51 (Sept. 8).

Beeman, R. D. 1963. Notes on the California species of *Aplysia* (Gastropoda: Opisthobranchia). Veliger, vol. 5, no. 4, pp. 145–47 (Apr. 1).

1968. The Order Anaspidea. Veliger, vol. 3, Suppl., pt. 2, pp. 87–102, 12 figs. 1 col. pl. (May 1).

Benson, W. H. 1835. Account of *Oxygyrus*, new genus of pelagic shells allied to the genus *Atlanta* of Lesueur, with a note on some other pelagic shells lately taken on board the ship *Malcolm*. Jour. Asiatic Soc. Bengal, vol. 4, no. 39, pp. 173–76 (Apr.).

Bequaert, J. 1942. Random notes on American Potamididae. Nautilus, vol. 56, no. 1, pp. 20–30 (July 23).

Bergh, Rudolph. 1867. *Phidiana lynceus* og *Ismaila monstrosa*. Vedensk. Meddel. Naturhist. Fören. für 1866 [Transl. in Ann. Mag. Nat. Hist., ser. 4, vol. 2 (1868), pp. 133–38].

1870–95. Malacologische Untersuchungen, Nudibranchiaten. *In* C. G. Semper, Reisen im Archipel der Philippinen, Bd. 2, Theil 1, Heft 1–9, pp. 1–376, pls. 45–49, 1870–75; Theil 2, Heft 10–14, pp. 377–645, pls. 50–68, 1876–78; Theil 3, Heft 15–18, pp. 647–1165, pls. 69–89, 1884–92; Theil 4 (Suppl. 1–4), pp. 1–289, pls. A–Z, AE, 1880–95. [For a complete collation, see R. I. Johnson, Jour. Soc. Biblio. Nat. Hist., vol. 5, pt. 2, pp. 144–47, Apr. 1969].

1879–80. On the nudibranchiate gasteropod Mollusca of the north Pacific Ocean ... *In* Scientific Results of the exploration of Alaska ..., vol. 1, Art. 5, pt. 1, pp. 127–88, pls. 1–8 (May 1879); Art. 6, pt. 2, pp. 189–276, pls. 9–16 (Jan. 1880). [Also *in* Proc. Acad. Nat. Sci. Philadelphia, for 1879, pt. 1, Vol. 31, pp. 71–132, pls. 1–8 (1880); pt. 2, *ibid,* for 1880, vol. 32, pp. 40–127, pls. 1–8 [9–16] (1881).]

1894. Die Opisthobranchien. Bull. Mus. Comp. Zool., Harvard, vol. 25, no. 10, pp. 125–233, pls. 1–12 (Oct.).

1897–1902. Malacologische Untersuchungen, Theil 5. [Tectibranchia, etc.]. *In* C. G. Semper, Reisen in Archipel der Philippinen, vol. 7, 382 pp., 29 pls. [For dates of parts, *see* R. I. Johnson, Jour. Soc. Biblio. Nat. Hist., vol. 5, pt. 2, pp. 144–47, Apr. 1969].

1904–8. Malacologische Untersuchungen, Theil 6. [Nudibranchiata, etc.]. *In* C. G. Semper, Reisen im Archipel der Philippinen, vol. 9, 178 pp., 12 pls. [For exact dates of parts, *see* R. I. Johnson, Jour. Soc. Biblio. Nat. Hist., vol. 5, pt. 2, pp. 144–47, Apr. 1969].

Bernard, F. R. 1969. Preliminary diagnoses of new septibranch species from the eastern Pacific (Bivalvia: Anomalodesmata). Jour. Fisheries Research Board of Canada, vol. 26, no. 8, pp. 2230–34., ill.

Berry, S. S. 1912. A review of the cephalopods of western North America. Bull. Bur. Fisheries, vol. 30, Doc. 761, pp. 269–336, pls. 32–56, 18 text figs. (July 24).

1925. The species of *Basiliochiton*. Proc. Acad. Nat. Sci. Philadelphia, vol. 77, pp. 23–29, 1 pl., 2 text figs.

1936. A new *Dimya* from California. Proc. Malac. Soc. London, vol. 22, no. 3, pp. 126–28, 1 pl. (Nov. 14).

1945. Two new chitons from the Gulf of California. American Midland Naturalist, vol. 34, no. 2, pp. 491–95, 18 figs. in text (Sept.).

1946. A re-examination of the Chiton *Stenoplax magdalenensis* (Hinds), with description of a new species. Proc. Malac. Soc. London, vol. 26, pt. 6, pp. 161–66, figs. 1–6 (Jan.).

1950. A partial review of some West American species of *Crepidula*. Leaflets in Malacol., Redlands, California, vol. 1, pp. 35–40 (Nov. 14).

1953a. West American razor-clams of the genus *Ensis*. San Diego Soc. Nat. Hist. Trans., vol. 11, no. 15, pp. 393–404, 1 pl. (Aug. 14).

1953b. Notices of new West American Mollusca. *Ibid.*, no. 16, pp. 405–28, pls. 28–29 (Sept. 1).

1953c. Preliminary diagnoses of six West American species of *Octopus*. Leaflets in Malacology, vol. 1, no. 10, pp. 51–58 (Dec. 18).

1954a. *Octopus penicillifer*, new species. Leaflets in Malacology, vol. 1, no. 11, p. 66 (Jan. 28).

1954b. An hitherto unnamed West American Ark-shell. Leaflets in Malac., vol. 1, no. 12, pp. 67–69 (July 1).

1956a. A tidal flat on the Vermilion Sea. Jour. Conch., vol. 24, no. 3, pp. 81–84 (Feb.).

1956b. Diagnoses of new Eastern Pacific chitons. Leaflets in Malac., vol. 1, no. 13, pp. 71–74 (July 9).

1956c. A new West Mexican Prosobranch mollusk parasitic on Echinoids. American Midland Naturalist, vol. 56, no. 2, pp. 355–57, 2 figs. in text (Oct.).

1957. Notices of new eastern Pacific Mollusca.—I. Leaflets in Malacol., vol. 1, no. 14, pp. 75–82 (July 19).

1958a. Double-trouble in violet snails (Abstract). Amer. Malac. Union Ann. Rept. for 1957, Bull. 24, p. 27 (Jan. 1).

1958b. Notices of new eastern Pacific Mollusca.—II. Leaflets in Malacol., vol. 1, no. 15, pp. 83–90 (Mar. 28).

1958c. West American molluscan miscellany.—II. *Ibid.*, No. 16, pp. 91–98 (May 31).

1959a. Comments on some of the trivaricate muricines. Leaflets in Malacol., vol. 1, no. 17, p. 106; *ibid.*, no. 18, pp. 113–14 (July 29).

1959b. Notices of new eastern Pacific Mollusca—III. Leaflets in Malacol., vol. 1, no. 18, pp. 108–13 (July 19).

1960. Notices of new eastern Pacific Mollusca—IV. Leaflets in Malacol., vol. 1, no. 19, pp. 115–22 (Dec. 31).

1962. A note on *Cantharus*, with proposal of a new specific name. Leaflets in Malacol., vol. 1, no. 20, pp. 129–30 (Nov. 13).

1963. Diagnoses of new eastern Pacific Chitons—II. Leaflets in Malacol., vol. 1, no. 22, pp. 135–38 (March 29).

1963. Notices of new eastern Pacific Mollusca—V. Leaflets in Malacol., vol. 1, no. 23, pp. 139–46 (Sept. 30).

1964. Notices of new eastern Pacific Mollusca—VI. Leaflets in Malacol., vol. 1, no. 24, pp. 147–54 (July 29).

1968a. Some unusual mollusks, mainly Panamic. American Malacol. Union Ann. Rept., for 1967, pp. 71–72 (Mar.).

1968b. Notices of new eastern Pacific Mollusca—VII. Leaflets in Malacol., vol. 1, no. 25, pp. 155–58 (Sept. 26).

1969. Notices of new eastern Pacific Mollusca—VIII. Leaflets in Malacol. vol. 1, no. 26, pp. 159–66 (Dec. 17).

Berry, S. S., and Bruce W. Halstead. 1954. Octopus bites—a second report. Leaflets in Malacol., vol. 1, no. 11, pp. 59–65 (Jan. 28).

Bertin, Victor. 1878. Révision des tellinidés du Muséum d'Histoire Naturelle . . . Nouv. Arch. Natl. Mus. Hist. Nat. Paris, ser. 2, vol. 1, pp. 201–361, pls. 8–9.

1880. Révision des garidées du Muséum d'Histoire Naturelle . . . Nouv. Arch. Mus. Natl. Hist. Nat. ser. 2, vol. 3, pp. 57–129, pls. 4–5.

1881. Révision des donacidées du Muséum d'Histoire Naturelle . . . Nouv. Arch. Mus. Natl. Hist. Nat., Paris, ser. 2, vol. 4, pp. 57–121, pls. 3–4.

Bertsch, Hans. 1970a. *Dolabrifera dolabrifera* (Rang, 1828), range extension to the Eastern Pacific. The Veliger, vol. 13, no. 1, pp. 110–11, 1 text fig. (July 1).

1970b. Opisthobranchs from Isla San Francisco, Gulf of California, with the description of a new species. Santa Barbara Mus. Nat. Hist. Contributions in Science, no. 2, 16 pp., 13 figs., 1 col. plate (Dec. 1).

Bertsch, Hans, and A. A. Smith. 1970. Observations on opisthobranchs of the Gulf of California. The Veliger, vol. 13, no. 2, pp. 171–74 (Oct.).

Beu, Alan G. 1970. The Mollusca of the subgenus *Monoplex* (Family Cymatiidae). Trans. Royal Society of New Zealand, Biol. Sci., vol. 11, no. 17, pp. 225–37, 5 pls., 1 fig. (Mar. 11).

Binney, W. G. 1860. Notes on American shells: Proc. Acad. Nat. Sci. Philadelphia, vol. 12, no. 6, pp. 150–54.

1865. Land and fresh water shells of North America: Pt. II, Pulmonata, limnophila and thalassophila: Smithsonian Misc. Coll. no. 143 (Sept.).

Biolley, P. 1907. Mollusques de l'Isla del Coco. Museum Nacional de Costa Rica, San José. Pp. 1–30, 2 maps.

Blainville, H. M., Ducrotay de. 1816–30. Vers et Zoophytes, *in* Dictionnaire des sciences naturelles. Pt. 2. Règne organisé. Paris. 60 vols. and atlas of plates. [*Janthina*, vol. 24, 1822.]

1832. Disposition méthodique des espèces récentes et fossiles des genres Pourpre, Ricinule, Licorne, et Concholépas de M. de Lamarck, et description des espèces nouvelles ou peu connues, faisant partie de la collection du Muséum d'Histoire Naturelle de Paris. Nouv. Ann. Mus. d'Hist. Nat., Paris, vol. 1, pp. 189–263, pls. 9–12. [According to Sherborn, published after May 1832.]

Böse, E. 1910. Zur jungtertiaren Fauna von Tehuantepec. I. Stratigraphie, Beschreibung and Vergleich mit amerikanischen Tertiarfaunen. Jahrb. d. K.-K. Geologisch. Reichsanstalt., vol. 60, pp. 215–55, pls. 12–13.

Boettger, C. R. 1962. Gastropoden mit zwei Schalenklappen. Verh. d. Deutschen Zoologischen Gesellschaft in Wien (Zool. Anz. Suppl. 26), pp. 403–39, 6 figs.

Bonnevie, Kristine E. H. 1920. Heteropoda. Report on the scientific results of the "Michael Sars" North Atlantic Deep-Sea Expedition, 1910, vol. 3, pt. 2, pp. 1–17, pls. 1–5, 2 text figs. (Feb. 28).

Boone, L. 1928. Scientific results of the second oceanographic expedition of the "Pawnee," 1926. Mollusks from the Gulf of California and the Perlas Islands. Bull. Bingham Oceanogr. Coll., Peabody Mus. Nat. Hist., Yale Univ., vol. 2, art. 5, pp. 1–17, pls. 1–3 (Dec.).

1938. Scientific results of the world cruises of the yachts "Ara," 1928–29, and "Alva," 1931–32, "Alva" Mediterranean cruise, 1933, and "Alva" South American cruise, 1935, William K. Vanderbilt commanding. Bull. Vanderbilt Marine Mus., vol. 7, Mollusca, pp. 285–361, figs. 15–22, pls. 110–52 (Nov.).

Born, Ignatius. 1778. Index Rerum Naturalium Musei Caesarei Vindobonensis. Part I: Testacea. Vienna, xlii + 458 pp., 1 pl. [For notes on date of publication, see R. Rutsch, 1956, Nautilus, vol. 69, no. 3, pp. 78–79.]

1780. Testacea musei Caesarei Vindobonensis. Vienna, xxxvi + 442 + 17 pp., 18 pls.

Boss, K. J. 1964a. A note on the synonymy of *Tellina subtrigona* Sowerby, 1866 (Mollusca: Bivalvia). Veliger, vol. 6, no. 4, pp. 207–8, pl. 27, figs. 1, 1a (April 1).

1964b. The status of *Scrobicularia viridotincta* Carpenter (Mollusca: Bivalvia). Veliger, vol. 6, no. 4, pp. 208–10, pl. 27, figs. 2, 2a (April 1).

1966–69. The subfamily Tellininae in the western Atlantic. Johnsonia, vol. 4, nos. 45–47. (1) The genus *Tellina*, I (no. 45), pp. 217–72, pls. 129–42 (Oct. 31, 1966). (2) The genus *Tellina*, II (no. 46), pp. 272–49, pls. 143–63 (April 17, 1968). (3) The genus *Strigilla* (no. 47), pp. 345–66, pls. 164–71 (Feb. 7, 1969).

Boss, K. J., and D. R. Moore. 1967. Notes on *Malleus (Parimalleus) candeanus* (D'Orbigny) (Mollusca: Bivalvia). Bull. Marine Science, vol. 17, no. 1, pp. 85–94 (March).

Boss, K. J., Joseph Rosewater, and Florence Ruhoff. 1968. The zoological taxa of William Healey Dall. Smithsonian Inst., U.S. Nat. Mus., Bull. 287, 427 pp.

Boury, E. A. de. 1912–13. Description de Scalidae nouveaux ou peu connus. Jour. de Conchyl., vol. 60, pt. 2, pp. 87–107, pl. 7 (Dec. 15, 1912); pt. 4, pp. 269–328, pl. 8 (May 31, 1913).

Brann, Doris C. 1966. Illustrations to "Catalogue of the collection of Mazatlan shells" by Philip P. Carpenter. Paleontological Research Institute, Ithaca, New York. 111 pp., 56 pls. (April 1).

Bratcher, Twila, and R. D. Burch. 1970a. Five new species of *Terebra* from the eastern Pacific. The Veliger, vol. 12, no. 3, pp. 295–300, pl. 44 (Jan. 1).

——— 1970b. Four new terebrid gastropods from eastern Pacific islands. Los Angeles Co. Mus., Contr. in Sci., no. 188, 6 pp., 8 figs. (May 4).

Broderip, W. J. 1832–33. Characters of new species of Mollusca and Conchifera, collected by Mr. Cuming. Proc. Zool. Soc. London, for 1832, pp. 25–33 (Apr. 21, 1832), 50–61 (June 5, 1832), 104–8 (July 31, 1832), 124–26 (Aug. 14, 1832), 173–79 (Jan. 14, 1833), 194–202 (Mar. 13, 1833). [Some of the species cited in these pages were described by G. B. Sowerby, I.]

——— 1833. Characters of new species of Mollusca and Conchifera, collected by Mr. Cuming. *Ibid.*, pp. 4–8 (May 17), 52–56 (May 24), 82–85 (Sept. 8). [Some of the species cited in these pages were described by G. B. Sowerby, I.]

——— 1834a. [Notes on the] Genus *Placunanomia, Ibid.*, for 1834, pp. 2–3 (May).

——— 1834b. Descriptions of several new species of Calyptraeidae. *Ibid.*, pp. 35–40 (July 29).

——— 1835a. Description of some species of *Chama. Ibid.*, for 1834, pp. 148–51 (Apr. 3).

——— 1835b. Descriptions of some new species of Calyptraeidae. Trans. Zool. Soc. London, vol. 1, pp. 195–206, pls. 27–29 (Mar. 20).

——— 1835c. On the genus *Chama*, Brug., with descriptions of some species apparently not hitherto characterized. *Ibid.*, vol. 1, pp. 301–6, pls. 38–39 (Dec. 3).

——— 1835–36. Characters of new genera and species of Mollusca and Conchifera, collected by Mr. Cuming. Proc. Zool. Soc. London, for 1835, pp. 41–47 (June 1, 1835), 192–97 (Apr. 8, 1836).

Broderip, W. J., and G. B. Sowerby. 1829. Observations on new or interesting Mollusca contained, for the most part, in the Museum of the Zoological Society. Zool. Jour., London, vol. 4, pp. 359–79, pl. 9.

——— 1830. Observations on new or interesting Mollusca, contained, for the most part, in the Museum of the Zoological Society (continuation). *Ibid.*, vol. 5, pp. 46–51 (Feb.).

Brown, A. P., and H. A. Pilsbry. 1911, 1913. Fauna of the Gatun formation, Isthmus of Panama. Proc. Acad. Nat. Sci. Philadelphia, vol. 63, pp. 336–60 (June 22, 1911), 361–73 (July 27, 1911), pls. 22–29; pt. 2, vol. 64, pp. 500–519, pls. 22–26 (Jan. 30, 1913).

Bruguière, J. 1792. Histoire naturelle de vers, vol. I. *See* Encyclopédie méthodique.

Bullock, R. C. 1966. New northern record for *Papyridea mantaensis* Olsson, 1961. The Nautilus, vol. 79, no. 4, p. 143 (Apr.).

Burch, J. Q. 1949. A new *Trigonostoma* from Central America. Minutes, Conchological Club of Southern California, no. 94, 2–3 (Sept.–Oct.).

Burch, J. Q., ed. 1944–46. Distributional list of the West American marine Mollusca from San Diego, California, to the Polar Sea. [Extracts from the] Minutes of the Conchological Club of Southern California, part I (Pelecypoda), nos. 33–45 (Mar. 1944–Feb. 1945); part II, vols. I–II (Gastropoda), nos. 46–63 (Mar. 1945–Sept. 1946).

Burch, J. Q., and Rose L. Burch. 1961. A new *Capulus* from Gulf of California. Nautilus, vol. 75, no. 1, pp. 19–20, pl. 2 (July).

——— 1962. New species of *Oliva* from west Mexico. Nautilus, vol. 75, no. 4, pp. 165–66, pl. 17 (Apr.).

1963. Genus *Olivella* in eastern Pacific. Nautilus, vol. 77, no. 1, pp. 1–8, pls. 1–3 (July).

1964a. A new species of *Sinum* from the Gulf of California. Nautilus, vol. 77, no. 4, pp. 109–10, pl. 5, figs. 1, 3 (Apr.).

1964b. The genus *Agaronia* J. E. Gray, 1839. Nautilus, vol. 77, no. 4, pp. 110–12 (Apr.).

Burch, J. Q., and G. B. Campbell. 1963a. Four new *Olivella* from Gulf of California. Nautilus, vol. 76, no. 4, pp. 120–26, pls. 6,7 (Apr.).

1963b. *Shaskyus*, new genus of Pacific coast Muricidae (Gastropoda). Journal de Conchyliologie, vol. 103, no. 4, pp. 201–6, pl. VI (Dec. 31).

Burrow, E. I. 1815. Elements of conchology. London, xix + 245 pp., 28 pls. Ed. 2, 1825. [For further information on editions, *see* Tomlin, 1943, Proc. Malac. Soc. London, vol. 25, p. 143.]

Button, F. 1902. West American Cypraeidae. Jour. Conch., vol. 10, no. 8, pp. 254–58.

1906. Note on *Trivia acutidentata* Gask. Nautilus, vol. 19, no. 11, p. 132.

1908. Note on *Trivia galapagensis* Melvill. *Ibid.*, vol. 22, no. 1, p. 11.

Campbell, G. B. 1961a. Range extension for *Terebra ornata* Gray, 1834. Veliger, vol. 3, no. 4, p. 112 (Apr. 1).

1961b. Colubrariidae (Gastropoda) of tropical West America, with a new species. Nautilus, vol. 74, no. 4, pp. 136–42, pl. 10 (Apr. 6).

1961c. Four new Panamic gastropods. Veliger, vol. 4, no. 1, pp. 25–28, pl. 5 (July 1).

1961d. Range extension of *Anatina cyprinus* (Wood, 1828). *Ibid.*, vol. 4, no. 2, p. 115, 3 text figs. (Oct.).

1962. A new deep-water *Anadara* from the Gulf of California. *Ibid.*, vol. 4, no. 3, pp. 152–54, pl. 37, 1 text fig. (Jan. 1).

1963. Rediscovery of *Terebra formosa* Deshayes, 1857. *Ibid.*, vol. 5, no. 3, pp. 101–3, pls. 12, 13 (Jan. 1).

1964. New terebrid species from the eastern Pacific (Mollusca: Gastropoda). *Ibid.*, vol. 6, no. 3, pp. 132–38, pl. 17 (Jan. 1).

Carpenter, P. P. 1855. List of four hundred and forty species of shells from Mazatlan. Rept. 24th Meeting, Brit. Assoc. Adv. Sci., for 1854 [Trans.], pp. 107–8.

1856a. Descriptions of (supposed) new species and varieties of shells, from the Californian and west Mexican coasts, principally in the collection of Hugh Cuming, Esq. Proc. Zool. Soc. London, for 1855, pt. 23, pp. 228–35 (pp. 228–32, Feb. 5; pp. 233–35, Feb. 23).

1856b. Notes on the species of *Hipponyx* inhabiting the American coasts, with descriptions of new species. *Ibid.*, p. 24, for 1856, pp. 3–5 (June 16).

1856c. Description of new species of shells collected by Mr. T. Bridges in the Bay of Panama and its vicinity, in the collection of Hugh Cuming, Esq. *Ibid.*, pp. 159–66 (Nov. 11).

1856d. Description of new species and varieties of Calyptraeidae, Trochidae, and Pyramidellidae, principally in the collection of Hugh Cuming, Esq. *Ibid.*, pp. 166–71 (Nov. 11).

1857a. First steps toward a monograph of the Recent species of *Petaloconchus*, a genus of Vermetidae. *Ibid.*, for 1856, pp. 313–17, figs. 1–8 (Mar. 10).

1857b. Report on the present state of our knowledge with regard to the Mollusca of the west coast of North America. Rep. Brit. Assoc. Adv. Sci., for 1856, pp. 159–368, pls. 6–9 (before Apr. 22).

1857c. Catalogue of the collection of Mazatlan shells in the British Museum: collected by Frederick Reigen ... London (British Museum), pp. i–iv + ix–xvi + 552 (Aug. 1, *fide* Sherborn, 1934). Reprinted, Paleont. Res. Inst., 1967. [A second edition, issued simultaneously by Carpenter at Warrington (Oberlin press) is identical except for introductory pages, i–viii + i–xii; *fide* Carpenter, 1872, p. xi.]

1858. Note on peculiarities of growth in Caecidae. Rept. 27th Meeting, Brit. Assoc. Adv. Sci., for 1857 [Trans.], p. 102.

1858–59. First steps towards a monograph of the Caecidae, a family of the rostriferous Gastropoda. Proc. Zool. Soc. London, for 1858, [pt. 26], pp. 413–32 (Dec. 14, 1858) ; pp. 443–44 (Jan. to May 1859).

1860a. Notice of the shells collected by Mr. J. Xantus, at Cape St. Lucas. Proc. Acad. Nat. Sci. Philadelphia, for 1859, vol. 11, pp. 331–32 (Jan. 12).

1860b. Lectures on the shells of the Gulf of California. Ann. Rept., Smithsonian Inst., for 1859, pp. 195–219, 6 text figs. (after June 14).

1864a. Review of Prof. C. B. Adams's "Catalogue of the shells of Panama," from the type specimens. Proc. Zool. Soc. London, for 1863, pp. 339–69 (Apr.). [Reprinted, Carpenter, 1872, pp. 173–205.]

1864b. Diagnosis of new forms of mollusks collected at Cape St. Lucas, Lower California, by Mr. J. Xantus. Ann. Mag. Nat. Hist., ser. 3, vol. 13, pp. 311–15 (Apr.) ; pp. 474–79 (June) ; vol. 14, pp. 45–49 (July).

1864c. Supplementary report on the present state of our knowledge with regard to the Mollusca of the west coast of North America. Rep. Brit. Assoc. Adv. Sci., for 1863, pp. 517–686 (Aug. 1864).

1865a. Diagnoses of new forms of Mollusca from the west coast of North America, first collected by Colonel E. Jewett: Ann. Mag. Nat. Hist., 3d ser., vol. 15, pp. 177–82 (Mar.) and pp. 394–99 (May).

1865b. Diagnoses de mollusques nouveaux provenant de Californie et faisant partie du Musée de l'Institution Smithsonienne. Jour. de Conchyl., vol. 13 (ser. 3, vol. 5), pt. 2, pp. 129–49 (Apr. 4).

1865c. Contributions towards a monograph of the Pandoridae. Proc. Zool. Soc. London, for 1864, pt. 3, pp. 596–603 (May).

1865d. Diagnoses of new species and a new genus of mollusks from the Reigen Mazatlan collection; with an account of additional specimens presented to the British Museum. Ibid., for 1865, pt. 1, pp. 264–74 (June).

1865e. Descriptions of new species and varieties of Chitonidae and Acmaeidae from the Panama collection of the late Prof. C. B. Adams. Ibid., for 1865, pt. 1, pp. 274–77 (June).

1865f. Diagnoses of new species of mollusks, from the west tropical region of North America, principally collected by the Rev. J. Rowell, of San Francisco. Ibid., for 1865, pt. 1, pp. 278–82 (June).

1866. Descriptions of new marine shells from the coast of California, pt. III. Proc. California Acad. Sci., 1st ser, vol. 3, pp. 207–24 (Feb.).

1869. Catalogue of the family Pandoridae. Amer. Jour. Conch., vol. 4, no. 5, Appendix, pp. 69–71 (May 6).

1872. The mollusks of western North America. Embracing the second report made to the British Association on this subject, with other papera; reprinted by permission, with a general index. Smithsonian Inst. Misc. Coll., vol. 10, no. 252. xii + 325 + 13–121 pp. (Dec.).

Carter, R. M. 1967. The shell ornament of Hysteroconcha and Hecuba (Bivalvia) : a test case for inferential functional morphology. Veliger, vol. 10, no. 1, pp. 59–71, 3 pls., 2 text figs. (July 1).

Cate, C. N. 1961. Remarks on a variation in Cypraea annettae Dall, 1909. Ibid., vol. 4, no. 2, pp. 112–14, pl. 24 (Oct. 1).

1969a. A revision of the eastern Pacific Ovulidae. Ibid., vol. 12, no. 1, pp. 95–102, pls. 7–10 (July 1).

1969b. The eastern Pacific cowries. Ibid., vol. 12, no. 1, pp. 103–19, pls. 11–15, 3 maps (July 1).

Cate, Jean M. 1967. The radulae of nine species of Mitridae. Ibid., vol. 10, no. 2, pp. 192–95, pl. 19, 9 text figs. (Oct. 1).

Cate, Jean, and Crawford Cate. 1962. The type of Lamellaria sharonae Willett, 1939 (Gastropoda). Ibid., vol. 5, no. 2, p. 91 (Oct. 1).

Cernohorsky, W. O. 1969. List of type specimens of Terebridae in the British Museum (Natural History). *Ibid.*, vol. 11, no. 3, pp. 210–22 (Jan. 1).

1970. Systematics of the families Mitridae and Volutomitridae. Bull. Auckland Inst. & Mus., no. 8, iv + 190 pp., 18 pls., 222 text figs. (Oct. 1).

Chace, E. P. 1958. A new mollusk from San Felipe, Baja California. Trans. San Diego Soc. Nat. Hist., vol. 12, no 20, pp. 333–34, fig. 1 (Oct. 16).

Chavan, André. 1937–38. Essai critique de classification des Lucines. Jour. de Conchyl., vol. 81, pp. 133–53; 198–216; 237–82 (1937); 59–97; 105–30; 215–43 (1938).

1951. Essai critique de classification des *Divaricella*. Bull. Inst. r. des Sci. nat. de Belgique, vol. 27, no. 18, pp. 1–27, 27 figs. in text (May).

1952. Nomenclatural notes on carditids and lucinids. Jour. Washington Acad. Sci., vol. 42, pp. 116–22 (Apr. 24).

Chenu, J. C. 1842–53. Illustrations conchyliologiques ou description et figures de toutes les coquilles connues, vivantes et fossiles. Paris, vols. 1–4, 85 livr. [For collation of this work, *see* Sherborn, C. D., and E. A. Smith, 1911, Proc. Malac. Soc. London, vol. 9, pp. 264–67. According to one citation, a complete copy of the work contains 215 pages of text and 481 plates (342 colored and 139 in black and white).]

1859–62. Manuel de conchyliologie. Paris, vol. 1, vii + 508 pp., 3707 figs. (1859); vol. 2, 327 pp. 1236 figs. (1862).

Chun, C. 1889. Beriche über eine nach den Canarischen Inseln im Winter 1887/1888 ausgeführte Reise. Sitz. ber. Kgl. Preuss. Akad. Wiss (Berlin), for 1889, pp. 519–53, figs. 1–14, 1 pl.

Clapp, W. F. 1923. A new species of *Teredo* from Florida. Proc. Boston Soc. Nat. Hist., vol. 37, pp. 33–38, pls. 3–4.

Clarke, A. H., and R. J. Menzies. 1959. *Neopilina* (*Vema*) *ewingi*, a second living species of the Paleozoic Class Monoplacophora. Science, vol. 129, no. 3355, pp. 1026–27, 1 text fig. (Apr. 17).

Clench, W. J. 1947. The genera *Purpura* and *Thais* in the western Atlantic. Johnsonia, vol. 2, no. 23, pp. 61–91, figs. 32–40 (Mar. 10).

Clench, W. J., and Ruth D. Turner, 1946. The genus *Bankia* in the western Atlantic. *Ibid.*, vol. 2, no. 19, pp. 1–28, pls. 1–16 (Apr. 27).

1957. The family Cymatiidae in the western Atlantic. *Ibid.*, vol. 3, no. 36, pp. 189–244, pls. 110–35 (Dec. 20).

1962. New names introduced by H. A. Pilsbry in the Mollusca and Crustacea. Acad. Nat. Sci. Philadelphia. Special Pub. 4, 218 pp.

Clessin, S. 1900. Pyramidellidae *in* Martini and Chemnitz, 1899–1902. Systematisches Conchylien Cabinet, ed. 2: Eulimidae, Bd. 1, Abth. 28, 273 pp., 41 pls. (Pyramidellidae are pp. 241–73.)

Coan, E. V. 1962. Notes on some tropical West American mollusks. Veliger, vol. 5, no. 2, p. 92 (Oct. 1).

1964. A proposed revision of the Rissoacean families Rissoidae, Rissoinidae, and Cingulopsidae. *Ibid.*, vol. 6, no. 3, pp. 164–171, 1 text fig. (Jan. 1).

1965a. A proposed reclassification of the family Marginellidae. *Ibid.*, vol. 7, no. 3, pp. 184–94, 9 figs. (Jan.).

1965b. Kitchen midden mollusks of San Luis Gonzaga Bay. *Ibid.*, vol. 7, no. 4, pp. 216–18, 1 pl., 1 table (Apr. 1).

1965c. Generic units in the Heteropoda. *Ibid.*, vol. 8, no. 1, pp. 36–41 (July).

1966. Charles Russell Orcutt, pioneer Californian malacologist, and *The West American Scientist*. Trans. San Diego Soc. Nat. Hist., vol. 14, no. 8, pp. 85–96 (Apr. 29).

1968. A biological survey of Bahía de los Angeles, Gulf of California, Mexico, III. Benthic Mollusca. *Ibid.*, vol. 15, no. 8, pp. 107–32, 2 figs. (Sept. 25).

1969. A bibliography of the biological writings of Philip Pearsall Carpenter. Veliger, vol. 12, no. 2, pp. 222–25 (Oct. 1).

1970. The date of publication of Gould's "Description of shells from the Gulf of California," *Ibid.*, vol. 13, no. 1, p. 109 (July 1).

1971. The Northwest American Tellinidae. *Ibid.*, supplement to vol. 14, 63 pp., 12 pls., 30 text figures (July 15).

Coan, E. V., and B. Roth. 1965. A new species of *Persicula* from west Mexico. *Ibid.*, vol. 8, no. 2, pp. 67–69, pl. 12 (Oct. 1).

1966. The west American Marginellidae. *Ibid.*, vol. 8, no. 4, pp. 276–99, pls. 48–51, 5 text figs. (Apr. 1).

Cockerell, T. D. A. 1901. Three new nudibranchs from California. Jour. Malac., vol. 8, no. 3, pp. 85–87.

Cockerell, T. D. A., and C. N. Eliot. 1905. Notes on a collection of Californian nudibranchs. *Ibid.*, vol. 12, pp. 31–53, pls. 7–8.

Collier, C. L., and W. M. Farmer. 1964. Additions to the nudibranch fauna of the east Pacific and the Gulf of California. Trans. San Diego Soc. Nat. Hist., vol. 13, no. 19, pp. 377–96, 4 pls. in color, 3 figs. (Dec. 30).

Conrad, T. A. 1837. Descriptions of marine shells from Upper California, collected by Thomas Nuttall, Esq. Jour. Acad. Nat. Sci. Philadelphia, vol. 7, pt. 2, pp. 227–68, pls. 17–20. (Nov. 21).

1849. The following new and interesting shells are from the coasts of Lower California and Peru, and were presented to the Academy by Dr. Thomas B. Wilson, Proc. Acad. Nat. Sci. Philadelphia, vol. 4, pp. 155–56 (at least as early as June 16).

1850. Descriptions of new fresh water and marine shells. Jour. Acad. Nat. Sci. Philadelphia, ser. 2, vol. 1, pp. 275–80, pls. 37–39 (Feb. 5).

1854. Monograph of the genus *Argonauta* Linné, with descriptions of five new species. *Ibid.*, vol. 2, pp. 331–33, pl. 34 (Feb.).

1867. Descriptions of new West Coast shells. Amer. Jour. Conch., vol. 3, no. 2, pp. 192–93 (Sept. 5).

Contreras, F. 1932. Datos para el estudio de los Ostiones Mexicanos. An. Inst. Biol., Mexico, vol. 3, no. 3, pp. 193–212, figs. 1–25.

Cooke, A. H. 1916. The operculum of the genus *Bursa* (*Ranella*). Proc. Malac. Soc. London, vol. 12, pp. 5–11, fig. 1.

1918. On the radula of the genus *Acanthina*, G. Fischer. *Ibid.*, vol. 13, pp. 6–11, 7 figs.

1919. The radula in *Thais, Drupa, Morula, Concholepas, Cronia, Iopas*, and the allied genera. *Ibid.*, vol. 13, pp. 91–110, 38 figs.

Cooper, J. G. 1863*a*. Some new genera and species of California Mollusca. Proc. Calif. Acad. Sci., vol. 2, for 1862, pp. 202–5, 207 (Jan.).

1863*b*. On new or rare Mollusca inhabiting the coast of California. *Ibid.*, vol. 3, pp. 56–60, fig. 14 (Sept.).

Cooper, J. G. 1866. On a new species of *Pedipes*, inhabiting the coast of California. *Ibid.*, vol. 3, p. 294, fig. 29 (Sept.).

1895. Catalogue of marine shells collected chiefly on the eastern shore of Lower California for the California Academy of Sciences during 1891–92. *Ibid.*, ser. 2, vol. 5, pp. 34–48 (May 21).

Costa, O. G. 1844. Catalogo de testacei viventi nel piccolo e grande mare il Taranto redatto sul sistema del Lamarck. Ann. Accad. Sci., vol. 5, pt. 2, for 1843, pp. 14–66.

Cox, L. R., *see under* Moore, R. C., 1969.

Crosse, H. 1861. Étude sur le genre Cancellaire, suivie du catalogue des espèces vivantes et fossiles actuellement connues. Jour. de Conchyl., vol. 9, pp. 220–56 (July 14).

1863. Étude sur le genre Cancellaire et déscription d'espèces nouvelles. *Ibid.*, vol. 11, no. 1, pp. 58–69; figs. (Jan. 30).

1865*a*. Diagnoses molluscorum novorum. *Ibid.*, vol. 13, no. 1, pp. 55–57 (Jan. 27).

1865*b*. Description de cônes nouveaux provenant de la collection Cuming. *Ibid.*, no. 3, pp. 299–315, pls. 9–11 (June 28).

Dall, W. H. 1869. Notes on the Argonauta. Amer. Naturalist, vol. 3, no. 5, pp. 236–39 (July).

1870. Materials toward a monograph of the Gadiniidae. Amer. Jour. Conch., vol. 6, pt. 1, pp. 8–22, 2 pls. (July 7).

1871. Descriptions of sixty new forms of mollusks from the west coast of America and the north Pacific Ocean, with notes on others already described. *Ibid.*, vol. 7, pt. 2, pp. 93–160, pls. 13–16 (Nov. 2).

1879. Report on the limpets and chitons of the Alaskan and Arctic regions, with descriptions of genera and species believed to be new. Proc. U.S. Nat. Mus., vol. 1, no. 48, pp. 281–344, pls. 1–3 (Feb. 15–19). [Also published in Sci. Results Expl. Alaska, art. 4, pp. 63–126 (Feb.).]

1882. On the genera of chitons. Proc. U.S. Nat. Mus., vol. 4, pp. 279–91 (Feb. 1); pp. 289–91 (Mar. 13).

1885a. Notes on some Floridian land and fresh-water shells with a revision of the Auriculacea of the eastern United States. *Ibid.*, vol. 8, pp. 255–89 (July–Aug.), pls. 17–18 (Sept. 25).

1885b. [Comments, *in*] Notes on the mollusks of the vicinity of San Diego, Calif., and Todos Santos Bay, Lower California, by Charles R. Orcutt. *Ibid.*, vol. 8, pp. 534–52, pl. 24 (Oct.).

1889a. [Reports on the results of dredging . . . in the Gulf of Mexico . . . by the U.S. Coast Survey steamer "Blake," . . .] XXIV. Report on the Mollusca—Part II. Gastropoda and Scaphopoda. Bull. Mus. Comp. Zool., vol. 18, pp. 1–492, pls. 1–40 (Jan.–June).

1889b. Notes on *Lophocardium* Fischer, Nautilus, vol. 3, no. 2, pp. 13–14 (June).

1890. Scientific results of explorations by the U.S. Fish Commission steamer *Albatross.* No. VII. Preliminary report on the collection of Mollusca and Brachiopoda obtained in 1887–88. Proc. U.S. Nat. Mus., for 1889, vol. 12, no. 773, pp. 219–362, pls. 5–14 (Mar. 7).

1890–92. Contributions to the Tertiary fauna of Florida . . . Pt. 1. Pulmonate, opisthobranchiate and orthodont gastropods. Trans. Wagner Free Inst. Sci., vol. 3, pp. 1–200, pls. 1–12 (Aug. 1890); Pt. 2, Streptodont and other gastropods. *Ibid.*, pp. 201–473, pls. 13–22 (Dec. 1892).

1891. On some new or interesting West American shells obtained from the dredgings of the U.S. Fish Commission steamer *Albatross* in 1888, and from other sources. Proc. U.S. Nat. Mus., vol. 14, no. 849, pp. 173–91, pls. 5–7 (July 24).

1892. On some types new to the fauna of the Galapagos Islands. Nautilus, vol. 5, no. 9, pp. 97–99 (Jan. 14).

1894a. On some species of *Mulinia* from the Pacific coast. *Ibid.*, vol. 8, no. 1, pp. 5–6, pl. 1 (May).

1894b. Monograph of the genus *Gnathodon* Gray (*Rangia* Desmoulins). Proc. U.S. Nat. Mus., vol. 17, pp. 89–106, pl. 7 (July 23).

1894c. Synopsis of the Mactridae of Northwest America, south to Panama. Nautilus, vol. 8, no. 4, pp. 39–43 (Aug.).

1895a. New species of shells from the Galapagos Islands. *Ibid.*, vol. 8, no. 11, pp. 126–27 (Mar. 4).

1895b. Scientific results of explorations by the U.S. Fish Commission steamer "Albatross." XXXIV. Report on Mollusca and Brachiopoda dredged in deep water, chiefly near the Hawaiian Islands, with illustrations of hitherto unfigured species from northwest America. Proc. U.S. Nat. Mus., vol. 17, no. 1032, pp. 675–733, pls. 23–32 (July 2).

1896. Diagnoses of new species of mollusks from the west coast of America. *Ibid.*, vol. 18, no. 1034, pp. 7–20 (Apr. 23).

1897. Notice of some new or interesting species of shells from British Columbia and the adjacent region. Nat. Hist. Soc. Brit. Columbia, Bull. no. 2, art. 1, pp. 1–18, pls. 1–2 (Jan.).

1898. Synopsis of the Recent and Tertiary Psammobiidae of North America. Proc. Acad. Nat. Sci. Philadelphia, vol. 50, pp. 57–62 (Apr. 5).

1898–1903. Contributions to the Tertiary fauna of Florida with especial reference to the silex beds of Tampa and the Pliocene beds of Caloosahatchie River. Trans. Wagner Free Inst. Sci. vol. 3, pt. 4, pp. 571–947, pls. 23–35 (Apr. 1898) ; pt. 5, pp. 949–1218, pls. 36–47 (Dec. 1900) ; pt. 6, pp. 1219–1654, pls. 48–60 (Oct. 1903).

1899a. Synopsis of the recent and Tertiary Leptonacea of North America and the West Indies. Proc. U.S. Nat. Mus., vol. 21, no. 1177, pp. 873–97, pls. 87–88 (June 26).

1899b. Synopsis of the American species of the family Diplodontidae. Jour. Conch., vol. 9, no. 8, pp. 244–46 (Oct. 1).

1899c. Synopsis of the Solenidae of North America and the Antilles. Proc. U.S. Nat. Mus., vol. 22, no. 1185, pp. 107–12 (Oct. 9).

1900. Synopsis of the family Tellinidae and of the North American species. *Ibid.*, vol. 23, no. 1210, pp. 285–326, pls. 2–4 (Nov. 14).

1901a. Synopsis of the family Cardiidae and of the North American species. *Ibid.*, vol. 23, no. 1214, pp. 381–92 (Jan. 2).

1901b. A new *Pinna* from California. Nautilus, vol. 14, no. 12, pp. 142–43 (Apr. 6).

1901c. Synopsis of the Lucinacea and of the American species. Proc. U.S. Nat. Mus., vol. 23, no. 1237, pp. 779–833, pls. 39–42 (Aug. 22).

1902a. Illustrations and descriptions of new, unfigured, or imperfectly known shells, chiefly American, in the U.S. National Museum. *Ibid.*, vol. 24, no. 1264, pp. 499–566, pls. 27–40 (Mar. 31).

1902b. New species of Pacific coast shells. Nautilus, vol. 16, no. 4, pp. 43–44 (Aug.).

1902c. Synopsis of the family Veneridae and of the North American Recent species. Proc. U.S. Nat. Mus., vol. 26, no. 1312, pp. 335–412, pls. 12–16 (Dec. 29).

1903a. Synopsis of the Carditacea and of the American species. Proc. Acad. Nat. Sci. Philadelphia, vol. 54, pp. 696–715 (Jan. 20).

1903b. Synopsis of the family Astartidae, with a review of the American species. Proc. U.S. Nat. Mus., vol. 26, no. 1343, pp. 933–51, pls. 62–63 (July 10).

1903c. Two new mollusks from the west coast of America. Nautilus, vol. 17, no. 4, pp. 37–38 (Aug. 12).

1904. An historical and systematic review of the frog-shells and tritons. Smithsonian Inst. Misc. Coll., vol. 47, no. 1475, pp. 114–44 (Aug. 6).

1905. Note on *Lucina* (*Miltha*) *childreni* Gray and on a new species from the Gulf of California. Nautilus, vol. 18, no. 9, pp. 110–12 (Feb. 11).

1906. A new *Scala* from California. *Ibid.*, vol. 20, no. 4, p. 44 (Aug. 18).

1907. Three new species of *Scala* from California. Nautilus, vol. 20, no. 11, pp. 127–28 (Mar. 4).

1908a. Descriptions of new species of mollusks from the Pacific coast of the United States, with notes on other mollusks from the same region. Proc. U.S. Nat. Mus., vol. 34, no. 1610, pp. 245–57 (June 16).

1908b. Reports on the dredging operations off the west coast of Central America to the Galapagos, to the west coast of Mexico, and in the Gulf of California . . . XIV. The Mollusca and Brachiopoda. Bull. Mus. Comp. Zool., Harvard, vol. 43, no. 6, pp. 205–487, pls. 1–22 (Oct.).

1908c. A gigantic *Solemya* and a new *Vesicomya*. Nautilus, vol. 22, no. 7, pp. 61–63 (Nov. 14).

1909a. Contributions to the Tertiary paleontology of the Pacific Coast. I. The Miocene of Astoria and Coos Bay, Oregon. Prof. Paper U.S. Geol. Surv., no. 59, pp. 1–278, figs. 1–14, pls. 1–23 (Apr. 2).

1909b. Some notes on *Cypraea* of the Pacific coast. Nautilus, vol. 22, no. 12, pp. 125–26 (Apr. 14).

1909c. Report on a collection of shells from Peru, with a summary of the littoral marine Mollusca of the Peruvian zoological province. Proc. U.S. Nat. Mus., vol 37, no. 1704, pp. 147–294, pls. 20–28 (Nov. 24).

1910a. New species of West American shells. Nautilus, vol. 23, no. 11, pp. 133–36 (Apr. 15).

1910b. Summary of the shells of the genus *Conus* from the Pacific coast of America in the U.S. National Museum. Proc. U.S. Nat. Mus., vol. 38, no. 1741, pp. 217–28 (June 6).

1910c. New shells from the Gulf of California. Nautilus, vol. 24, no. 3, pp. 32–34 (July).

1911. Notes on California shells. II. *Ibid.*, vol. 24, no. 10, pp. 109–12 (Feb. 4).

1912. New species of fossil shells from Panama and Costa Rica collected by D. F. MacDonald. Smithsonian Inst. Misc. Coll., vol. 59, pt. 2, no. 2077, pp. 1–10 (Mar. 2).

1913a. Shells collected at Manzanillo, west of Mexico, Oct. 1910, by C. R. Orcutt, identified by W. H. Dall. Nautilus, vol. 26, no. 12, p. 143 (Apr. 2).

1913b. Diagnoses of new shells from the Pacific Ocean. Proc. U.S. Nat. Mus., vol. 45, no. 2002, pp. 587–97 (June 11).

1914a. Notes on West American oysters. Nautilus, vol. 28, no. 1, pp. 1–3 (May).

1914b. Notes on some Northwest Coast Acmaeas. *Ibid.*, no. 2, pp. 13–15 (June).

1915a. Notes on the Semelidae of the west coast of America, including some new species. Proc. Acad. Nat. Sci. Philadelphia, vol. 67, pp. 25–28 (Mar. 2).

1915b. Notes on the West American species of *Fusinus*. Nautilus, vol. 29, no. 5, pp. 54–57 (Sept. 4).

1915c. Notes on American species of *Mactrella*. *Ibid.*, vol. 29, no. 6, pp. 61–63 (Oct. 11).

1915d. A review of some bivalve shells of the group Anatinacea from the west coast of America. Proc. U.S. Nat. Mus., vol. 49, no. 2116, pp. 441–56 (Nov. 27).

1916a. Notes on the West American Columbellidae. Nautilus, vol. 30, no. 3, pp. 25–29 (July 14).

1916b. Diagnoses of new species of marine bivalve mollusks from the northwest coast of America in the United States National Museum. Proc. U.S. Nat. Mus., vol. 52, no. 2183, pp. 393–417 (Dec. 27).

1917a. Summary of the mollusks of the family Alectrionidae of the west coast of America. *Ibid.*, vol. 51, no. 2166, pp. 575–79 (Jan. 15).

1917b. Notes on the shells of the genus *Epitonium* and its allies of the Pacific coast of America. *Ibid.*, vol. 53, no. 2217, pp. 471–88 (Aug. 10).

1917c. Preliminary descriptions of new species of Pulmonata in the Galápagos Islands. Proc. California Acad. Sci., ser. 4, vol. 2, pt. 1, no. 11, pp. 375–82 (Dec. 31).

1918a. Descriptions of new species of shells, chiefly from Magdalena Bay, Lower California. Proc. Biol. Soc. Washington, vol. 31, pp. 5–8 (Feb. 27).

1918b. Notes on *Chrysodomus* and other mollusks from the North Pacific Ocean. Proc. U.S. Nat. Mus., vol. 54, no. 2234, pp. 207–34 (Apr. 5).

1918c. Notes on the nomenclature of the mollusks of the family Turritidae. *Ibid.*, vol. 54, no. 2238, pp. 313–33 (Apr. 5).

1918d. Pleistocene fossils of Magdalena Bay, Lower California, collected by Charles Russell Orcutt. Nautilus, vol. 32, no. 1, pp. 23–26 (July 20).

1919a. Descriptions of new species of chitons from the Pacific coast of America. Proc. U.S. Nat. Mus., vol. 55, no. 2283, pp. 499–516 (June 7).

1919b. Descriptions of new species of mollusks of the family Turritidae from the west coast of America and adjacent regions. *Ibid.*, vol. 56, no. 2288, pp. 1–86, pls. 1–24 (Aug. 8).

1919c. Descriptions of new species of Mollusca from the north Pacific Ocean in the collection of the United States National Museum. *Ibid.*, vol. 56, no. 2295, pp. 293–371 (Aug. 30).

1921a. Summary of marine shellbearing mollusks of the northwest coast of America . . . Bull. U.S. Nat. Mus., no. 112, 217 pp., 22 pls. (Feb. 24).

1921b. New fossil invertebrates from San Quentin Bay, Lower California. West Amer. Scientist, vol. 19, no. 2, pp. 17–18 (Apr. 27).

1921c. New shells from the Pliocene or early Pleistocene of San Quentin Bay, Lower California. *Ibid.*, vol. 19, no. 3, pp. 21–23 (June 15).

1923. Some unrecorded names in the Muricidae. Proc. Biol. Soc. Washington, vol. 36, pp. 75–77 (Mar. 28).

1925. Illustrations of unfigured types of shells in the collection of the United States National Museum. Proc. U.S. Nat. Mus., vol. 66, art. 17, pp. 1–41, pls. 1–36 (Sept. 22).

Dall, W. H., and Paul Bartsch. 1903. Systematic descriptions of Eulimidae (pp. 268–69) and Pyramidellidae (pp. 269–85) *in* R. Arnold, 1903, The paleontology and stratigraphy of the marine Pliocene and Pleistocene of San Pedro, California. Mem. California Acad. Sci., vol. III, 419 pp., pls. 1–37.

1906. Notes on Japanese, Indopacific, and American Pyramidellidae. Proc. U.S. Nat. Mus., vol. 30, no. 1452, pp. 321–69, pls. 17–26 (May 9).

1907. The pyramidellid mollusks of the Oregonian faunal area. *Ibid.*, vol. 33, no. 1574, pp. 491–534, pls. 14–18 (Dec. 31).

1909. A monograph of West American pyramidellid mollusks. Bull. U.S. Nat. Mus., no. 68, 258 pp., 30 pls. (Dec. 13).

1910. New species of shells collected by Mr. John Macoun . . . British Columbia. Canada Dept. Mines, Geol. Survey Branch, Mem. no. 14-N, pp. 5–22, pls. 1–2.

Dall, W. H., and W. H. Ochsner. 1928. Tertiary and Pleistocene Mollusca from the Galápagos Islands. Proc. Calif. Acad. Sci., ser. 4, vol. 17, no. 4, pp. 89–139, pls. 2–7, 5 figs. in text (June 22).

Dance, S. P. 1966. Shell collecting: an illustrated history. University of California Press: Berkeley. Pp. 1–344; 35 pls. (3 in color), 31 figs.

D'Asaro, C. N. 1969. The egg capsules of *Jenneria pustulata* (Lightfoot, 1786), with notes on spawning in the laboratory. Veliger, vol. 11, no. 3, pp. 182–84, 1 fig. in text (Jan. 1).

1970. Egg capsules of some prosobranchs from the Pacific coast of Panama. *Ibid.*, vol. 13, no. 1, pp. 37–43, 6 text figs. (July 1).

D'Attilio, Anthony. 1967. *Muricanthus melanamathos*, a west African muricid. Nautilus, vol. 80, no. 3, pp. 96–99, 4 figs. (Jan.).

Dawson, E. Y., and P. T. Beaudette. 1959. Field notes from the 1959 eastern Pacific cruise of the *Stella Polaris*. Pacific Naturalist, vol. 1, no. 13, 24 pp., 16 figs. (Nov. 10).

Delessert, B. 1841. Recueil de coquilles décrites par Lamarck dans son Histoire naturelle des animaux sans vertèbres et non encore figurées. Paris, 40 col. pls., with explanations of pls. (94 pp. unnumbered). [For a review of this work, *see* C. Récluz, 1842, Rev. Zool., Soc. Cuvierienne, vol. 5, pp. 319–22.]

Demond, Joan. 1951. Key to the Nassariidae of the west coast of North America. Nautilus, vol. 65, no. 1, pp. 15–17 (July).

1952. The Nassariidae of the west coast of North America between Cape San Lucas, Lower California, and Cape Flattery, Washington. Pacific Sci., Univ. Hawaii, vol. 6, pp. 300–317, pls. 1–2 (Oct.).

Deshayes, G. P. 1826. Anatomie et monographie du genre Dentale. Mém. Soc. Hist. Nat. Paris, vol. 2, pp. 324–78, pls. 15–18 (Apr.).

1830–32. Encyclopédie Méthodique. Histoire naturelle de vers, vol. 2. Paris. Pp. 1–144 (1830), 145–594 (1832). [For a collation and dates of the zoölogical portion of the Encyclopédie, see C. D. Sherborn and B. B. Woodward, Ann. Mag. Nat. Hist., ser. 7, vol. 17, pp. 577–82, 1906.]

1832 [*Pileopsis pilosus, Venericardia squamigera*], Magasin de zoologie, vol. 2, pls. 9–10.

1835–45. Histoire naturelle des animaux sans vertèbres. Deuxième édition. Paris, vols. 1–11.

1839. Nouvelles espèces de mollusques, provenant des côtes de la Californie, du Méxique, du Kamtschatka et de la Nouvelle-Zélande. Révue zoologique, par la Société Cuvierienne, vol. 2, pp. 356–61 [plates in Magasin de Zool.: 1840, pls. 12–20; 1841, pls. 25–30, 34–38].

1853–55. Catalogue of the Conchifera or bivalve shells in the British Museum. London, pt. 1, Veneridae, Cyprinidae and Glauconomidae, pp. 1–216 (1853); pt. 2, Petricoladae (concluded), Corbiculadae, pp. 217–92, "1854" (May 1855).

1854–55. Descriptions of new species of shells from the collection of H. Cuming, Esq. Proc. Zool. Soc. London, for 1854, pp. 13–16 (Dec. 30, 1854), 17–23 (Jan. 10, 1855), 62–64 (Jan. 10, 1855), 65–72 (Feb. 10, 1855), 317–20 (May 8, 1855); 321–71 (May 16, 1855).

1856a. Sur le genre *Scintilla*. *Ibid*., for 1855, pp. 171–81 (Jan. 5).

1856b. Descriptions de nouvelles espèces du genre *Erycina*. *Ibid*., for 1855, pp. 181–83 (Jan. 5).

1857. Description d'espèces nouvelles du genre *Terebra*. Jour. Conch., vol. 6, pp. 65–102, pls. 3–5 (July).

1859. A general review of the genus *Terebra*, and a description of new species. Proc. Zool. Soc. London, pp. 270–321 (issued between July and October).

Dillwyn, L. W. 1817. A descriptive catalogue of Recent shells, arranged according to the Linnaean method; with particular attention to the synonymy. London, vol. 1, pp. i–xii, 1–580; vol. 2, pp. 581–1092 and index.

Dodge, Henry. 1953–59. An historical review of the mollusks of Linnaeus. Bull. Amer. Mus. Nat. Hist.: pt. 1, Loricata and Pelecypoda, vol. 100, art. 1, pp. 1–263 (Dec. 1952); pt. 2, Cephalopoda, *Conus*, *Cypraea*, vol. 103, art. 1, pp. 1–134, (Nov. 1953); pt. 3, *Bulla*, *Voluta*, vol. 107, art. 1, pp. 1–158 (Nov. 1955); pt. 4, *Buccinum*, *Strombus*, vol. 111, art. 3, pp. 153–212 (Oct. 1956); pt. 5, *Murex*, vol. 113, art. 2, pp. 73–224 (Sept. 1957); pt. 6, *Trochus*, vol. 116, art. 2, pp. 153–224 (Dec. 1958); pt. 7, *Turbo* (part), vol. 118, art. 5, pp. 211–55 (Oct. 1959).

Donohue, Jerry. 1965. Concerning *Williamia peltoides* (Carpenter). Veliger, vol. 8, no. 1, pp. 19–21, 3 text figs. (July 1).

1966. The range of *Trivia myrae* Campbell. Ibid., vol. 9, no. 1, pp. 55–56, 1 fig. (July 1).

1967. Additional remarks on the range of *Trivia myrae* Campbell. *Ibid*., vol. 9, no. 3, p. 355 (Jan. 1).

Donohue, Jerry, and K. Hardcastle. 1962. X-ray diffraction examination of two forms of *Oliva spicata*. Nautilus, vol. 75, no. 4, pp. 162–65, 1 fig. (Apr).

Donovan, E. 1802. The natural history of British shells. London, vol. 4, pls. 109–44, text and index. [According to A. E. Salisbury, 1945, Jour. Conch., vol. 22, p. 144, pts. 31–42 with pls. 91–126 were published in 1802; pts. 43–54 with pls. 127–62, in 1803.]

Duclos, P. L. 1832. Description de quelques espèces de pourpres, servant de type à six sections établies dans ce genre. Ann. Sci. Nat., Paris, vol. 26, pp. 103–12, pls. 1, 2 (May).

1833. [Various genera.] Mag. Zool., Paris, yr. 3, cl. 5, *Oliva*, pl. 20 and text; *Purpura*, pl. 22 and text (May).

1835, 1840. Histoire naturelle ... de tous les genres de coquilles univalves marines à l'état vivant et fossile, publiée par monographies. Genre Olive. (Genre Colombelle-Strombe). Paris, Oliva: pts. 1–2; 2 sheets of text and pls. 1–12 (1835); pts. 3–6, pls. 13–33 (1840); *Columbella*: pts. 1–2, 1 sheet and pls. 1–13 (printed 1835, but published April 1840). [For collation of this work, *see* C. D. Sherborn and E. A. Smith, 1911, Proc. Malac. Soc. London, vol. 9, p. 267.]

1844–48. *Oliva*. *In* J. C. Chenu, illustrations conchyliologiques. Paris, pp. 1–31 (1844–48), pls. 1–36 (1845).

Ducros de St. Germain, A. M. P.

1857. Revue critique du genre *Oliva* de Bruguière. Pp. 120, 3 pls. Clermont.

Dunker, G. [Wm.] 1852. Aviculacea nova. Zeitschr. f. Malakozool., yr. 9, pp. 73–80.

1857. Mytilacea nova collectione Cumingianae, descripta a Guilermo Dunker. Proc. Zool. Soc. London, for 1856, pp. 358–66 (May 8).

1862. Solenacea nova collectionis Cumingianae. *Ibid*., for 1861, pp. 418–27 (Apr.).

1868–69. Mollusca marina; Abbildungen und Beschreibung ... neuer ... Meeres-

Conchylien, *in* Pfeiffer, Novitates Conchologicae, Abt. 2, Cassel: Pp. iv + 144, 45 col. pls. (1858–78). [For a complete collation of this work, *see* R. I. Johnson, Jour. Soc. Bibl. Nat. Hist., vol. 5, pt. 3, pp. 236–39, Oct. 1969.]

Durham, J. W. 1937. Gastropods of the Family Epitoniidae from Mesozoic and Cenozoic rocks of the west coast of North America. Jour. Paleont., vol. 11, no. 6, pp. 479–512, pls. 56–57 (Sept.)

   1942. Four new gastropods from the Gulf of California. Nautilus, vol. 55, no. 4, pp. 120–25, 1 pl. (May 7).

   1950. 1940 E. W. Scripps cruise to the Gulf of California. Part II. Megascopic paleontology and marine atratigraphy. Geol. Soc. America Mem. 43, pp. 1–216, pls. 1–48 (Aug. 10).

   1962. New name for *Strombus granulatus* subsp. *acutus* Durham, 1950, not Perry, 1811. Veliger, vol. 4, no. 4, p. 213 (Apr.).

Durham, J. W., and E. C. Allison. 1960. The geologic history of Baja California and its marine faunas. *In* Symposium: The biogeography of Baja California and adjacent seas. Pt. I. Geologic History. Systematic Zool., vol. 9, no. 2, pp. 47–91 (June).

DuShane, Helen. 1962. A checklist of mollusks for Puertecitos, Baja California, Mexico. Veliger, vol. 5, no. 1, pp. 39–50, 1 text fig. (July 1).

   1963. Range extensions for *Terebra robusta* Hinds, 1844, and for *Terebra formosa* Deshayes, 1857. *Ibid.*, vol. 5, no. 4, p. 159 (Apr.).

   1966*a*. A rare *Epitonium* from the Gulf of California. *Ibid.*, vol. 8, no. 4, pp. 311–12, pl. 52 (Apr. 1).

   1966*b*. Range extension for *Tylodina fungina* Gabb, 1865 (Gastropoda). *Ibid.*, vol. 9, no. 1, p. 86 (July 1).

   1967. *Epitonium (Asperiscala) billeeana* (DuShane & Bratcher, 1965) non *Scalina billeeana* DuShane & Bratcher, 1965. *Ibid.*, vol. 10, no. 1, pp. 87–88 (July 1).

   1969. A new genus and two new species of Typhinae from the Panamic province (Gastropoda: Muricidae). *Ibid.*, vol. 11, no. 4, pp. 343–44, pl. 54 (Apr. 1).

   1970. Two new Epitoniidae from the Galápagos Islands (Mollusca: Gastropoda). *Ibid.*, vol. 12, no. 3, pp. 330–32, pl. 51 (Jan. 1).

   1970. Five new epitoniid gastropods from the west coast of the Americas. Los Angeles Co. Mus., Contr. in Sci., no. 185, 6 pp., 5 figs. (Apr. 17).

   1971. The Baja California travels of Charles Russell Orcutt. Dawson's Book Shop, Baja California Travels Series, no. 23, 75 pp., 13 halftone photographs, 1 map.

DuShane, Helen, and Twila Bratcher. 1965. A new *Scalina* from the Gulf of California. Veliger, vol. 8, no. 2, pp. 160–61, pl. 24 (Oct. 1).

DuShane, Helen, and Ellen Brennan. 1969. A preliminary survey of mollusks for Consag Rock and adjacent areas, Gulf of California, Mexico. *Ibid.*, vol. 11, no. 4, pp. 351–63 (Apr. 1).

DuShane, Helen, and J. H. McLean. 1968. Three new epitoniid gastropods from the Panamic province. Contributions in Science, Los Angeles County Museum, no. 145, 6 pp., 6 figs. (June 14).

DuShane, Helen, and R. Poorman. 1967. A checklist of mollusks for Guaymas, Sonora, Mexico. Veliger, vol. 9, no. 4, pp. 413–41 (Apr. 1).

DuShane, Helen, and G. G. Sphon. 1968. A checklist of intertidal mollusks for Bahia Willard and the southwestern portion of Bahia San Luis Gonzaga State of Baja California, Mexico. *Ibid.*, vol. 10, no. 3, pp. 233–46, pl. 35 (Jan. 1).

Duval, —. 1840. Réflexions sur le genre Planaxe, et indication de deux espèces et d'une variété nouvelle. Rev. Zool., Soc. Cuvierienne, vol. 3, pp. 107.

Eales, N. B. [Nellie.] 1960. Revision of the world species of *Aplysia* (Gastropoda, Opisthobranchia). Bull. British Mus. (Nat. Hist.), Zool., vol. 5, no. 10, pp. 268–404, 52 figs. (Jan.).

Emerson, W. K. 1951. A new scaphopod mollusk, *Cadulus austinclarki*, from the Gulf of California. Jour. Washington Acad. Sci., vol. 41, no. 1, pp. 24–26, 2 figs. in text (Jan. 15).

1952. Generic and subgeneric names in the molluscan class Scaphopoda. *Ibid.*, vol. 42, pp. 296–303 (Sept. 15).

1956. A new scaphopod mollusk, *Dentalium (Tesseracme) hancocki*, from the eastern Pacific. Amer. Mus. Novitates, no. 1787, pp. 1–7, 1 fig. (Sept. 28).

1958. Results of the Puritan-American Museum of Natural History expedition to western Mexico. 1. General account. *Ibid.*, no. 1894, 25 pp., 9 figs. (July 22).

1959. The gastropod genus *Pterorytis*. *Ibid.*, no. 1974, pp. 1–8, 4 figs. (Nov. 19).

1960a. Results of the Puritan-American Museum of Natural History expedition to western Mexico. 11. Pleistocene invertebrates from Ceralvo Island. *Ibid.*, no. 1995, 6 pp. (Mar. 29).

1960b. Remarks on some eastern Pacific muricid gastropods. *Ibid.*, no. 2009, 15 pp., 7 figs. (Aug. 18).

1960c. Results of the Puritan-American Museum of Natural History expedition to western Mexico. 12. Shell middens of San Jose Island. *Ibid.*, no. 2013, 9 pp., 4 figs. (Aug. 18).

1962. A classification of the scaphopod mollusks. Jour. Paleont., vol. 36, no. 3, pp. 461–82, pls. 76–80, 2 text figs. (May).

1964. Results of the Puritan-American Museum of Natural History expedition to western Mexico. 20. The Recent mollusks: Gastropoda: Harpidae, Vasidae, and Volutidae. Amer. Mus. Novitates, no. 2202, 23 pp., 9 figs. (Dec. 15).

1965. The eastern Pacific species of *Niso* (Mollusca: Gastropoda). *Ibid.*, no. 2218, 12 pp., 11 figs. (May 6).

1967a. Indo-Pacific faunal elements in the tropical eastern Pacific, with special reference to the mollusks. "Venus," Japanese Jour. Malac., vol. 25, nos. 3/4, pp. 87–93 (July).

1967b. On the identity of *Phos laevigatus* A. Adams, 1851 (Mollusca: Gastropoda). Veliger, vol. 10, no. 2, pp. 99–102, pl. 13 (Oct. 1).

1968a. Taxonomic placement of *Coralliophila incompta* Berry, 1960, with the proposal of a new genus, *Attiliosa*. *Ibid.*, vol. 10, no. 4, pp. 379–81, 1 pl. (Apr. 1).

1968b. Familial placement of *Hanetia* Jousseaume, 1880 (Muricidae) and *Solenosteira* Dall, 1890 (Buccinidae). *Ibid.*, vol. 11, no. 1, pp. 1–3, pl. 1, 2 text figs. (July 1).

1968c. A record of the Indo-Pacific cone, *Conus ebraeus*, in Guatemala. *Ibid.*, vol. 11, no. 1, 33 pp. (July 1).

1968d. A new species of the gastropod genus *Morum* from the eastern Pacific. Jour. de Conchyl., vol. CVII, fasc. I, pp. 53–57, pl. I.

1969. Galapagan records for *Morum veleroae* (Gastropoda: Tonnacea). Nautilus, vol. 83, no. 1, pp. 19–22, 1 pl. (July).

1971. *Cadulus (Gadila) perpusillus* (Sowerby, 1832) an earlier name for *C. (G.) panamensis*. *Ibid.*, vol. 84, no. 3, pp. 77–81, 4 figs. (Jan.)

Emerson, W. K., and Anthony D'Attilio. 1965. *Aspella (Favartia) angermeyerae*, n.sp. *Ibid.*, vol. 79, no. 1, 4 pp., pl. 1 (July).

1966. Remarks on *Colubraria soverbii* (Reeve, 1844) and related species (Mollusca: Gastropoda). Veliger, vol. 8, no. 3, pp. 173–77, pl. 26 (Jan. 1).

1969a. A new species of *Strombina* from the Galapagos Islands. *Ibid.*, vol. 11, no. 3, pp. 195–97, pl. 39 (Jan. 1).

1969b. A new species of *Murexsul* from the Galapagos Islands. *Ibid.*, vol. 11, no. 4, pp. 324–25, pl. 50, 1 text fig. (Apr. 1).

1969c. Remarks on the taxonomic placement of *Purpurellus* Jousseaume, 1880, with the description of a new species (Gastropoda: Muricinae). *Ibid.*, vol. 12, no. 2, pp. 145–48, pls. 26–27, 2 figs. in text (Oct. 1).

1970a. Three new species of Muricacean gastropods from the eastern Pacific. *Ibid.*, vol. 12, no. 3, pp. 270–75, pls. 39–40, 4 figs. in text (Jan. 1).

1970b. *Aspella myrakeenae*, new species from western Mexico. Nautilus, vol. 83, pt. 3, pp. 88–94, 10 figs. (Jan. 21).

Emerson, W. K., and L. G. Hertlein. 1960. Pliocene and Pleistocene invertebrates from Punta Rosalía, Baja California, Mexico. Amer. Mus. Novitates, no. 2004, pp. 1–8, 3 text figs. (May 2).

— 1964. Invertebrate megafossils of the Belvedere expedition to the Gulf of California. Trans. San Diego Soc. Nat. Hist., vol. 13, no. 17, pp. 333–68, 6 figs. (Dec. 30).

Emerson, W. K., and W. E. Old, Jr. 1962. Results of the Puritan-American Museum of Natural History expedition to western Mexico. 16. The recent mollusks: Gastropoda, Conidae. Amer. Mus. Novitates, no. 2112, 44 pp., 20 figs. (Oct. 29).

— 1963a. Remarks on Cassis (Casmaria) vibexmexicana. Nautilus, vol. 76, no. 4, pp. 143–45, 3 figs. (Apr.).

— 1963b. Results of the Puritan-American Museum of Natural History expedition to western Mexico. 17. The recent mollusks: Gastropoda, Cypraeacea. Amer. Mus. Novitates, no. 2136, 32 pp., 18 figs. (May 7).

— 1963c. Results of the Puritan-American Museum of Natural History expedition to western Mexico. 19. The recent mollusks: Gastropoda, Strombacea, Tonnacea, and Cymatiacea. Ibid., no. 2153, 38 pp., 28 figs. (Sept. 24).

— 1964. Additional records from Cocos Island. Nautilus, vol. 77, no. 3, pp. 90–92 (Jan.).

— 1965. New molluscan records for the Galapagos Islands. Ibid., vol. 78, no. 4, pp. 116–20 (Apr.).

— 1968. An additional record for Cypraea teres in the Galapagos Islands. Veliger, vol. 11, no. 2, pp. 98–99, pl. 12 (Oct. 1).

Emerson, W. K., and E. L. Puffer. 1953. A catalogue of the molluscan genus Distorsio (Gastropoda, Cymatiidae). Proc. Biol. Soc. Washington, vol. 66, pp. 93–108 (Aug. 10).

— 1957. Recent mollusks of the 1940 "E. W. Scripps" cruise to the Gulf of California. Amer. Mus. Novitates, no. 1825, 57 pp., 1 fig. (Apr. 3).

Emerson, W. K., and G. E. Radwin. 1969. Two new species of Galapagan turrid Gastropoda. Veliger, vol. 12, no. 2, pp. 149–56, pls. 28–29, 5 text figs. (Oct. 1).

Encyclopédie méthodique. 1782–1832. By Bruguière, Lamarck, and others. Paris: 196 vols. in 186. [For collation and dates of the zoological portion, see C. D. Sherborn and B. B. Woodward, Proc. Zool. Soc. London, for 1893, pp. 582–84; ibid., for 1899, p. 595; Ann. Mag. Nat. Hist., ser. 7, vol. 17, pp. 577–82, 1906; E. V. Coan, Veliger, vol. 9, no. 2, pp. 132–33, 1966.]

Eschscholtz, J. F. 1829–31. Zoologischer Atlas... Berlin: Pts. 1–4. [For part 5, see under Rathke, 1833.]

Farmer, W. F. 1967. Notes on the Opisthobranchia of Baja California, Mexico, with range extensions—II. Veliger, vol. 9, no. 3, pp. 340–42, 1 text fig. (Jan. 1).

— 1968. Tidepool animals from the Gulf of California. Wesword Co., San Diego, California. Pp. 1–70, 170 line drawings, 24 col. photographs.

Farmer, W. M. 1963. Two new opisthobranch mollusks from Baja California. Trans. San Diego Soc. Nat. Hist., vol. 13, no. 6, pp. 81–84, 1 pl. (col.), 1 fig. (Sept. 27).

— 1970. Swimming gastropods (Opisthobranchia and Prosobranchia). Veliger, vol. 13, no. 1, pp. 73–89, 20 text figs. (July 1).

Farmer, W. M., and C. L. Collier. 1963. Notes on the Opisthobranchia of Baja California, Mexico, with range extensions. Ibid., vol. 6, no. 2, pp. 62–63 (Oct. 1).

Farmer, Wesley M., and A. J. Sloan. 1964. A new opisthobranch mollusk from La Jolla, California. Ibid., vol. 6, no. 3, pp. 148–50, 2 figs., 1 col. pl. (Jan. 1).

Filatova, Z. A., M. N. Sokalova, and R. Y. Levenstein. 1968. Mollusc of the Cambro-Devonian Class Monoplacophora found in the northern Pacific. Nature, vol. 220, pp. 1114–15, 2 text figs. (Dec. 14).

Fischer, Paul. 1857. Descriptions d'espèces nouvelles. Jour. de Conchyl., vol. 5, pt. 2, pp. 273–77 (Jan.).

— 1874. Diagnoses specierum novorum. Ibid., vol. 22, pt. 2, pp. 205–6 (May 11).

1880–87. Manuel de conchyliologie . . . Paris, pp. i–xxiv, 1–1369, 23 pls., 1138 figs. in text.

Fischer-Piette, E. 1969. Mollusques récoltés par M. Hoffstetter sur les côtes de l'Équateur et des Iles Galapagos. Veneridae. Bull. du Mus. national d'Hist. naturelle, ser. 2, tome 40, no. 5, pp. 998–1018. 3 pls.

Fischer-Piette, E., and D. Delmas. 1967. Révision des mollusques lamellibranches du genre *Dosinia* Scopoli. Mém. Mus. nat. d'Hist. nat., n.s., Sér. A, Zool., vol. 47, no. 1, 92 pp., 16 pls.

Fitch, J. E. 1953. Common marine bivalves of California. State Calif. Dept. Fish & Game, Marine Fish. Branch, Fish Bull., no. 90, pp. 1–102, 63 figs., 1 col. pl.

Folin, A. G. L. de. 1867a. Description d'espèces nouvelles de Caecidae. Jour. de Conchyl., vol. 15, no. 1, pp. 44–58, pls. 2–3 (Jan. 10).

1867b. Les méléagrinicoles. Recueil des Pub. de la Soc. Havraise d'Étude diverse, vol. 33, for 1866, pp. 41–112, 6 pls. Also published separately, 74 pp., 6 pls.

1867–76. Les fonds de la mer. Paris, vol. 1, 316 pp., 32 pls. and pl. 21 bis (1867–72) ; vol. 2, 365 pp., 11 pls. (1872–76). [For collation of this work, *see* R. Winckworth, 1941, Proc. Malac. Soc. London, vol. 24, pp. 149–51; H. A. Rehder, 1946, Proc. Malac. Soc. London, vol. 27, pp. 74–75.]

Forbes, E. 1852. On the marine Mollusca discovered during the voyages of the *Herald* and *Pandora*, by Capt. Kellett, R.N., and Lieut. Wood, R.N. Proc. Zool. Soc. London, for 1850, pp. 270–74, pls. 9, 11. [Pp. 270–72 (Jan. 24, 1852), 273–74 (1852), according to Sclater.]

Forbes, E., and S. Hanley. 1848. A history of British Mollusca and their shells. London, vol. 1, pp. i–lxxx, 1–477 pls. A–O, 1–34 (1853, but issued 1848). [For collation of this work, *see* N. Fisher and J. R. LeB. Tomlin, 1935, Jour. Conch., vol. 20, pp. 150–51.]

Ford, John. 1895. A new variety of *Olivella*. Nautilus, vol. 8, no. 9, pp. 103–4, 1 pl. (Jan.).

Forskål, Petter. 1775. Descriptiones animalium: avium, amphibiorum, piscium, insectorum, vermium; quae in itinere orientali observavit Petrus Forskål. [Posthumously published and edited by C. Niebuhr.] Copenhagen, 164 pp.

Forster, J. G. A. 1777. A voyage round the world in H.M.S. *Resolution*, commanded by Capt. J. Cook, during . . . 1772–75. London, 2 vols., illus.

Fritchman, H. K., II. 1965. The radulae of *Tegula* species from the west coast of North America and suggested intrageneric relationships. Veliger, vol. 8, no. 1, pp. 11–14, 10 text figs. (July 1).

Frizzell, D. L. 1946. A study of two arcid pelecypod species from western South America. Jour. Paleont., vol. 20, pp. 38–51, 13 figs., pl. 10 (Jan.).

Gabb, Wm. 1865. Descriptions of new species of marine shells from the coast of California. Proc. Calif. Acad. Sci., vol. 3, pp. 182–90 (Jan.).

Galtsoff, P.S. 1950. The pearl-oyster resources of Panama. U.S. Dept. Interior, Fish and Wildlife Serv., Special Sci. Rept., Fisheries, no. 28, pp. 1–53, 28 figs.

Gaskoin, J. S. 1836. Descriptions of new species of *Cypraea*. Proc. Zool. Soc. London, pt. 3, for 1835, pp. 198–204 (Apr. 8).

1852. Descriptions of twenty species of Columbellae, and one species of Cypraea. *Ibid.*, for 1851, pp. 2–14 (Oct. 28).

Gegenbaur, Carl. 1855. Untersuchungen über Pteropoden und Heteropoden. Ein Beitrag zur Anatomie und Entwicklungsgeschichte dieser Thiere. Leipzig (W. Engelmann) : pp. i–vi + 1–228, pls. 1–8.

Ghosh, Ekendranath. 1920. Taxonomic studies on the soft parts of the Solenidae. Rec., Indian Mus., vol. 19, no. 2, art. 11, pp. 47–78, 2 pls. (June 3).

Gifford, D. S., and E. W. Gifford. 1947. Color variation in *Olivella undatella*. Nautilus, vol. 60, no. 3, pp. 81–84 (Mar. 11).

1951. Olive shells on opposite shores of the Gulf of California. *Ibid.*, vol. 65, no. 2, pp. 55–56 (Nov. 9).

Glibert, Maxime. 1954. Pleurotomes du Miocène de la Belgique et du Bassin de la Loire. Inst. R. Sci. Nat. de Belgique, Mem. no. 129, pp. 1–75, pls. 1–7 (Aug. 31).

Gmelin, J. F. 1791. Caroli a Linné Systema naturae per regna tria naturae. Editio decima tertia. Leipzig ["Lipsiae"], Germany, vol. 1, pt. 6, cl. 6, Vermes, pp. 3021–3910.

Gonor, J. J. 1965. Predator-prey reactions between two marine prosobranch gastropods. Veliger, vol. 7, no. 4, pp. 228–32 (Apr. 1).

Gould, A. A. *For a review of his publications and types, see* R. I. Johnson, 1964.

1846. Expedition shells described for the work of the United States Exploring Expedition under the command of Charles Wilkes ... Proc. Boston Soc. Nat. Hist., vol. 2, pp. 142–52 (July).

1851. Descriptions of a number of California shells collected by Maj. William Rich and Lieut. Thomas P. Green. *Ibid.*, vol. 4, pp. 87–93 (Nov.).

1852. Mollusca and shells. United States Exploring Expedition, vol. 12. Boston, Mass.: pp. xv + 510 and Atlas.

1853. Descriptions of shells from the Gulf of California and the Pacific coasts of Mexico and California. Boston Jour. Nat. Hist., vol. 6, pp. 374–408, pls. 14–16 (Oct.). [Stated by Dall, 1909a, p. 204, to have been issued privately in April 1852; *see* Coan, 1970, for evidence that the correct date is Oct. 1853.]

1855. Catalogue of shells collected in California by W. P. Blake, with descriptions of the new species by Augustus A. Gould ... : from Appendix to the Prelim. Geol. Rept. of W. P. Blake, Geol. Surv. of California, Appendix, House Doc. 129, Washington, D.C.

1862a. Descriptions of new genera and species of shells. Proc. Boston Soc. Nat. Hist., vol. 8, pp. 280–84 (Feb.).

1862b. Otia Conchologia: Descriptions of shells and mollusks from 1839 to 1862. Boston, iv + 256 pp.

Gould, A. A., and P. P. Carpenter. 1857. Descriptions of shells from the Gulf of California, and the Pacific coasts of Mexico and California. Part II. Proc. Zool. Soc. London, part 24, for 1856, pp. 198–208 (Jan. 7).

Grabau, A. W. 1904. Phylogeny of *Fusus* and its allies. Smithsonian Inst. Misc. Coll., vol. 44, no. 1417, pp. 1–157, pls. 1–18.

Graham, Alastair. 1966. The fore-gut of some marginellid and cancellariid prosobranchs. Univ. of Miami (Inst. Mar. Sci.), Studies Trop. Oceanogr., vol. 4, no. 1, pp. 134–51, 7 text figs.

Grant, U. S., IV, and H. R. Gale, 1931. Catalogue of the marine Pliocene and Pleistocene Mollusca of California and adjacent regions. Mem. San Diego Soc. Nat. Hist., vol. 1, pp. 1–1036, 15 figs., pls. 1–32 (Nov. 3).

Grau, Gilbert. 1959. Pectinidae of the eastern Pacific. Univ. So. Calif. Publ., Allan Hancock Pacific Expeditions, vol. 23, 308 + viii pp., 57 pls. (Sept. 25).

Gray, J. E. 1825. A list and description of some species of shells not taken notice of by Lamarck. Ann. Philos., vol. 25, pp. 134–40, 407–15.

1825–28. Monograph on the Cypraeidae, a family of testaceous Mollusca. Zool. Jour., vol. 1, pp. 489–518 (Jan. 1825); vol. 3, no. 11, pp. 363–71 (Dec. 31, 1827); no. 12, pp. 567–76 (Apr. 1828).

1828a. Spicilegia Zoologica; or original figures and short descriptions of new and unfigured animals. Pt. 1, 8 pp., 6 pls. (July 1).

1828b. Additions and corrections to a monograph on *Cypraea*, a genus of testaceous Mollusca. Zool. Jour., vol. 4, pp. 81–88 (July).

1833. The new species of cowries in the collection of Hugh Cuming ... Proc. Zool. Soc. London, for 1832, pp. 184–86 (Jan.).

1834. Enumeration of the species of the genus *Terebra*, with characters of many hitherto undescribed. *Ibid.*, pp. 59–63 (Nov. 25).

1837. A synoptical catalogue of the species of certain tribes and genera of shells contained in the collection of the British Museum and the author's cabinet; with descriptions of the new species. Mag. Nat. Hist., n.s., vol. 1, pp. 370–76, 6 figs. (July).

1838. Catalogue of the species of the genus *Cytherea*, of Lamarck, with the description of some new genera and species. Analyst, vol. 8, pp. 302–9.

1839. Molluscous animals and their shells. *In* F. W. Beechey, The zoology of Capt. Beechey's voyage . . . to the Pacific and Behring's Straits in his Majesty's ship *Blossom* . . . London, pp. i–xii, 103–55, pls. 33–44. *See also* G. B. Sowerby, I. [For notes on the itinerary, *see* J. Rosewater, Veliger, vol. 10, no. 4, pp. 350–52, 1968.]

1843. Catalogue of the species of Mollusca. *In* E. Dieffenbach, Travels in New Zealand; with contributions to the geography, geology, botany, and natural history of that country. London, vol. 2, pp. 228–65.

1850*a*. On the species of Anomiadae. Proc. Zool. Soc. London, for 1849, pp. 113–24 (Jan. to June).

1850*b*. Catalogue of Placentadae and Anomiadae. *In* Catalogue of the bivalve Mollusca in the . . . British Museum. London, pp. 1–22 (July 6).

1850*c*. *See under* Gray, Maria E., 1842–57.

1853. On a new genus of Anomiadae in the collection of Mr. Cuming. Proc. Zool. Soc. London, vol. 19, for 1851, pp. 197–98 (June 29).

1855. List of the shells of South America in the British Museum. Collected and described by M. Alcide d'Orbigny, in the "Voyage dans l'Amérique Méridionale." London, pp. 1–89 (Jan. 13). [For dates of publication of catalogues of natural history issued by the British Museum, *see* C. D. Sherborn, 1926, Ann. Mag. Nat. Hist., ser. 9, vol. 17, pp. 271–72; 1934, *ibid.*, ser. 10, vol. 13, pp. 308–12.]

1856. Descriptions of the animals and teeth of *Tylodina* and other genera of gasteropodous Mollusca. Proc. Zool. Soc. London, pp. 41–46 (June 16).

1857. A revision of the genera of some of the families of Conchifera or bivalve shells. Pt. III. Arcadae. Ann. Mag. Nat. Hist., ser. 2, vol. 19, pp. 366–73 (May).

1858. An attempt to distribute the species of Olive (*Oliva*, Lamarck) into natural groups, and to define some of the species. Proc. Zool. Soc. London, pp. 38–48 (Mar. 9), 49–57 (Apr. 13).

1865. List of the Mollusca in the collection of the British Museum. London, pt. 2, Olividae, pp. 1–41 (Mar. 11).

Gray, Maria E. 1842–57. Figures of molluscous animals, selected from various authors . . . London: 5 vols. (vol. 1, iv + 40 pp., 78 pls., post-June, 1842; vol. 2, pls. 79–199, Aug. 1850; vol. 3, pls. 200–312, Aug. 1850; vol. 4, iv + 219 pp., including a reprint of J. E. Gray, 1847, "List of genera . . . ," Aug. 1850; vol. 5, pp. 1–49, pls. 313–81, 1857. Text by J. E. Gray). [For additional notes, *see* P. Townsend, Jour. Soc. Biblio. Nat. Hist., vol. 1, pt. 5, p. 129, 1938.]

Griffith, Edward, and Edward Pidgeon. 1833(?)–34. The Mollusca and Radiata, *in* E. Griffith, *The Animal Kingdom by Cuvier*. London: viii + 601 pp., 40 pls. Mollusca, pts. 38–40. [Probable dates and pagination of parts, construed from internal evidence and a collation by C. F. Cowan, Jour. Soc. Biblio. Nat. Hist., vol. 5, pt. 2, pp. 137–40, 1969: Pt. 38, pp. 1–192, pls. 1–27, 29–35, 38–39 (?Dec. 1833); pt. 39, pp. 193–400, pls. 28, 36–37, 40 (?Mar. 1834); pt. 40, pp. 401–601 (June 1834). All plates except 28 and 36–37 carry an 1833 imprint date.]

Haas, Fritz. 1943. Malacological Notes—III. Zool. Ser., Field Mus. Nat. Hist., vol. 29, no. 1, pp. 1–23, figs. 1–8 [especially, "A record of *Alaba interruptelineata*," pp. 17–18] (June 10).

1945. Malacological Notes—IV. *Ibid.*, vol. 30, no. 2, pp. 3–14, figs. 1–2 [especially, "On a paratype of *Circe subtrigona* Carpenter," pp. 4–5] (Sept. 19).

Haddon, A. C. 1886. Report on the Polyplacophora collected by H.M.S. *Challenger* during the years 1873–1876. Zool., *Challenger* Exped., vol. 15, pt. 43, pp. 1–50, 3 pls.

Hagberg, A. H., and Carl Kalb. 1968. Marine shelled mollusks of commercial importance in Central America. Bol. Technico, Proyecto regional de Desarrollo Pesquero en Centro America, vol. 2, no. 2 (El Salvador), 32 pp., 27 figs. (mostly un-numbered) (Oct.).

Hanley, S. C. T. 1842–56. An illustrated and descriptive catalogue of recent bivalve shells. London, pp. i–xviii, 1–392, suppl. pls. 9–24, pp. 1–24. [For dates of issue of this work, *see* pp. v–vi, *also* A. Reynell, 1918, Proc. Malac. Soc. London, vol. 13, pp. 26–27.]

    1843. Five new species of shells belonging to the genus *Donax*, a group of acepha-lous mollusks. Proc. Zool. Soc. London, for 1843, vol. 13, pp. 5–6 (July).

    1844. Description of new species of Mytilacea, etc. *Ibid.*, for 1843, vol. 13, pp. 14–17 (July).

    1844–45. Descriptions of new species of the genus *Tellina*, chiefly collected by H. Cuming, Esq., in the Philippine Islands and Central America. *Ibid.*, for 1844, pp. 59–64, 68–72 (Sept. 1844), 140–44, 147–49 (Dec. 1844), 164–66 (Feb. 1845).

    1845*a*. Descriptions of three new species of shells belonging to the genus *Artemis*. *Ibid.*, pt. 13, for 1845, pp. 11–12 (Apr.).

    1845*b*. Descriptions of six new species of *Donax*, in the collection of Hugh Cuming. *Ibid.*, for 1845, pp. 14–16 (Apr.).

    1846*a*. A description of new species of *Ostrea* in the collection of H. Cuming, Esq. *Ibid.*, for 1845, pp. 105–7 (Feb.).

    1846*b*. A monograph of the genus *Tellina*. *In* G. B. Sowerby, Thesaurus conchyli-orum. London, vol. 1, pp. 221–38, pls. 56–66 (Jan. 21); pp. 239–336 (Nov. 27).

    1854. The Conchological Miscellany of Sylvanus Hanley. Privately printed; 40 pls. [1854–58.]

    1856. Index testaceologicus, an illustrated catalogue of British and foreign shells . . . by W. Wood. A new and entirely revised edition, with ancient and modern appel-lations, synonyms, localities, etc. London, pp. i–xx, 1–196, pls. 1–38, suppl. pp. 197–234, pls. 1–8.

    1859. Description of a new *Cyrena* and *Bulla*. Proc. Zool. Soc. London, for 1858, pp. 543–44 (May 31).

    1860. Monograph of the family Nuculidae, forming the Lamarckian genus *Nucula*. *In* G. B. Sowerby, Thesaurus conchyliorum. London, vol. 3, pp. 105–68, pls. 226–30 (Nuculidae pls. 1–5).

    1863. Monograph of the recent species of the genus *Solarium* of Lamarck. *Ibid.*, pp. 227–48, pls. 250–54 (*Solarium* pls. 1–5).

Hanna, G. D. 1963. West American mollusks of the genus *Conus*—II. Occas. Papers Calif. Acad. Sci., no. 35, 103 pp., 4 figs., 11 pls. (Jan. 28).

Hanna, G. D., and L. G. Hertlein. 1927. Expedition of the California Academy of Sci-ences to the Gulf of California in 1921: Geology and paleontology. Proc. Cali-fornia Acad. Sci., ser. 4, vol. 16, no. 6, pp. 137–57, 1 pl. (Apr. 22).

    1938. Land and brackish water Mollusca of Cocos Island. Allan Hancock Founda-tion Publications of the University of Southern California. Los Angeles, vol. 2, no. 8, pp. 123–35, 1 fig. (Aug.).

    1961. Large Terebras (Mollusca) from the eastern Pacific. Proc. California Acad. Sci., ser. 4, vol. 30, no. 3, pp. 67–80, pls. 6–7 (Aug. 31).

Hanna, G. D., and A. M. Strong. 1949. West American mollusks of the genus *Conus*. Proc. California Acad. Sci., ser. 4, vol. 26, no. 9, pp. 247–322, 4 figs., pls. 5–10 (Jan. 28).

Harry, H. W. 1967. A review of the living Tectibranch snails of the genus *Volvulella*, with descriptions of a new subgenus and species from Texas. Veliger, vol. 10, no. 2, pp. 133–47, 21 text figs. (Oct. 1).

    1969*a*. A review of the living Leptonacean bivalves of the genus *Aligena*. *Ibid.*, vol. 11, no. 3, pp. 164–81, 40 text figs. (Jan. 1).

    1969*b*. Anatomical notes on the mactrid bivalve, *Raeta plicatella* Lamarck, 1818, with a review of the genus *Raeta* and related genera. *Ibid.*, vol. 12, no. 1, pp. 1–23, 20 text figs. (July 1).

Heath, Harold. 1911. The solenogastres. Mem. Mus. Comparative Zool., vol. 45, pp. 1–182, pls. 1–40.

Hemphill, H. 1894. A new species of *Bulimulus* (*Eulimella occidentalis*) : Zoe, vol. 4, pp. 395–96.

Henderson, J. B. 1915. Rediscovery of Pourtales' *Haliotis*. Proc. U.S. Nat. Mus., vol. 48, no. 2091, pp. 659–61, pls. 45–46 (May 22).

Hertlein, L. G. 1934. Pleistocene mollusks from the Tres Marias Islands, Cedros Island, and San Ignacio Lagoon, Mexico. Bull. Southern California Acad. Sci., vol. 33, pp. 59–73, pl. 21 (Aug. 31).

1935. The Templeton Crocker expedition of the California Academy of Sciences, 1932. No. 25. The Recent Pectinidae. Proc. California Acad. Sci., ser. 4, vol. 21, pp. 301–28, pls. 18–19 (Sept. 26).

1937. A note on some species of marine mollusks occurring in both Polynesia and the western Americas. Proc. Amer. Phil. Soc., vol. 78, pp. 303–12, 1 pl. (Dec.).

1941. A summary of the knowledge regarding the faunal area of tropical West America, with special reference to mollusks. Proceedings of the dedicatory exercises, Hancock Hall, Univ. Southern California, pp. 21–24.

1951. Descriptions of two new species of marine pelecypods from west Mexico. Bull. Southern California Acad. Sci., vol. 50, pt. 2, pp. 68–75, pls. 24–26 (May–Aug.).

1955. Marine mollusks collected at the Galapagos Islands during the voyage of the *Velero III*, 1931–32. Essays in Nat. Sci. in honor of Capt. Allan Hancock, Los Angeles; pp. 111–45, 1 pl.

1957. Pliocene and Pleistocene fossils from the southern portion of the Gulf of California. Bull. Southern California Acad. Sci., vol. 56, pt. 2, pp. 57–75, 1 pl.

1958. Descriptions of new species of marine mollusks from west Mexico. *Ibid.*, vol. 56, pt. 3, pp. 107–12, pl. 21 (Jan. 15).

1960. The subfamily Drupinae (Gastropoda) in the eastern Pacific. Veliger, vol. 3, no. 1, pp. 7–8 (July 1).

1963. Contribution to the biogeography of Cocos Island, including a bibliography. Proc. California Acad. Sci., ser. 4, vol. XXXII, no. 8, pp. 219–89, 4 figs. (May 20).

1968. *Tellina ulloana*, a new species from Magdalena Bay, Baja California, Mexico. Veliger, vol. 11, no. 1, p. 80 (July 1).

Hertlein, L. G., and E. C. Allison. 1960a. Species of the genus *Cypraea* from Clipperton Island. *Ibid.*, vol. 2, no. 4, pp. 94–95, pl. 22 (Apr. 1).

1960b. Gastropods from Clipperton Island. *Ibid.*, vol. 3, no. 1, pp. 13–16 (July 1).

1966. Additions to the molluscan fauna of Clipperton Island. *Ibid.*, vol. 9, no. 2, pp. 138–40 (Oct. 1).

1968. Descriptions of new species of gastropods from Clipperton Island. Occas. Papers California Acad. Sci., no. 66, 13 pp., 13 figs. (June 27).

Hertlein, L. G., and W. K. Emerson. 1953. Mollusks from Clipperton Island (eastern Pacific) with the description of a new species of gastropod. Trans. San Diego Soc. Nat. Hist., vol. 11, pp. 345–64, pls. 26, 27 (July 22).

1956. Marine Pleistocene invertebrates from near Puerto Penasco, Sonora, Mexico. *Ibid.*, vol. 12, no. 8, pp. 154–76, pl. 12, 2 maps (June 7).

1957. Additional notes on the invertebrate fauna of Clipperton Island. American Mus. Novitates, no. 1859, 9 pp., 1 fig. (Dec. 6).

1959. Results of the Puritan-American Museum of Natural History expedition to western Mexico. 5. Pliocene and Pleistocene megafossils from the Tres Marias Islands. *Ibid.*, no. 1940, 15 pp., 5 figs. (June 5).

Hertlein, L. G., and G. D. Hanna. 1949. Two new species of *Mytilopsis* from Panama and Fiji. Bull. Southern California Acad. Sci., vol. 48, pt. 1, pp. 13–18, 1 pl. (Jan.–Apr.).

Hertlein, L. G., and A. M. Strong. 1939. Marine Pleistocene mollusks from the Galapagos Islands. Proc. California Acad. Sci., ser. 4, vol. XXIII, no. 24, pp. 367–80, pl. 32 (July 20).

1940–51. Eastern Pacific expeditions of the New York Zoological Society. Mollusks from the west coast of Mexico and Central America. Pts. I–X. Zoologica, New York, pt. 1, vol. 25, pp. 369–430, pls. 1–2 (Dec. 31, 1940) ; pt. 2, vol. 28, pp. 149–68, pl. 1 (Dec. 6, 1943) ; pt. 3, vol. 31, pp. 53–76, pl. 1 (Aug. 20, 1946) ; pt. 4, vol. 31, pp. 93–120, pl. 1 (Dec. 5, 1946) ; pt. 5, vol. 31, pp. 129–50, pl. 1 (Feb. 21, 1947) ; pt. 6, vol. 33, pp. 163–98, pls. 1, 2 (Dec. 31, 1948) ; pt. 7, vol. 34, pp. 63–97, pl. 1 (Aug. 10, 1949) ; pt. 8, vol. 34, pp. 239–58, pl. 1 (Dec. 30, 1949) ; pt. 9, vol. 35, pp. 217–52, pls. 1, 2 (Dec. 30, 1950) ; pt. 10, vol. 36, pp. 67–120, pls. 1–11 (Aug. 20, 1951).

1945. Changes in the nomenclature of two West American marine bivalve mollusks. Nautilus, vol. 58, no. 3, p. 75 (Feb. 19).

1948*a*. Note on West American species of *Condylocardia. Ibid.*, vol. 61, no. 3, p. 106 (Jan.).

1948*b*. Descriptions of a new species of *Trophon* from the Gulf of California. Bull. Southern California Acad. Sci., vol. 46, pt. 2, for May–Aug. 1947, pp. 79–80, 1 pl. (Feb. 5, 1948).

1949. Note on the nomenclature of two marine gastropods from the Galapagos Islands. Nautilus, vol. 62, no. 3, pp. 102–3 (Mar. 18).

1951*a*. Descriptions of three new species of marine gastropods from West Mexico and Guatemala. Bull. Southern California Acad. Sci., vol. 50, pt. 2, for May–Aug., pp. 76–80 (Oct. 15, 1951).

1951*b*. Descriptions of two new species of marine gastropods from West Mexico and Costa Rica. *Ibid.*, vol. 50, pt. 3, pp. 152–55, 4 text figs. (Dec. 27).

1955*a*. Marine mollusks collected during the "Askoy" Expedition to Panama, Colombia, and Ecuador in 1941. Bull Amer. Mus. Nat. Hist., vol. 107, art. 2, pp. 159–318, pls. 1–3 (Nov. 28).

1955*b*. Marine mollusks collected at the Galapagos Islands during the voyage of the *Velero III*, 1931–1932. Essays in the natural sciences in honor of Capt. Allan Hancock: Univ. Southern California (Los Angeles), pp. 111–45, 1 pl.

Hidalgo, J. G. 1893–98. Obras malacológicas de J. G. Hidalgo. Pt. III. Descripción de los moluscos recogidos por la comisión científica enviada por el gobierno español á la América Meridional. Catálogo de los moluscos gasterópodos marinos. Mem. R. Acad. Sci. Fís. Nat., Madrid, vol. 19, pp. 332–432 (1893), 433–608 (1898). [For part 2 of this work, *see* F. de P. Martínez y Sáez.]

1906–8. Monografía de las especies vivientes del género *Cypraea*. Mem. R. Acad. de Ciencias Exactas, Físicas y Naturales de Madrid, no. 25, xv + 588 pp.

Hill, H. R. 1954. Variation in the Olive shells of tropical West America. Nautilus, vol. 68, no. 2, pp. 66–69 (Oct.).

1959. The cone shells of tropical west America. Veliger, vol. 2, no. 2, pp. 30–32 (Oct. 1).

1960. Key to the cone shells of western tropical America. *Ibid.*, vol. 2, no. 3, pp. 51–53 (Jan. 1).

Hinds, R. B. 1842. Descriptions of new shells. Ann. Mag. Nat. Hist., new ser., vol. 10, pp. 81–84, pl. 6 (Oct.).

1843*a*. Descriptions of new shells from the collection of Capt. Sir Edward Belcher, R.N., C.B. *Ibid.*, vol. 11, pp. 16–21 (Jan.) pp. 255–57 (Apr.).

1843*b*. On new species of shells collected by Sir Edward Belcher, C.B. Proc. Zool. Soc. London, pp. 17–19 (July).

1843*c*. On new species of *Pleurotoma, Clavatula,* and *Mangelia. Ibid.*, pp. 36–46 (Oct.)

1843*d*. Descriptions of ten new species of *Cancellaria*, from the collection of Sir Edward Belcher. *Ibid.*, pp. 47–49 (Oct.).

1843*e*. On new species of *Corbula* and *Potamomya. Ibid.*, pp. 55–59 (Nov.).

1843*f*. Descriptions of new species of *Neaera . . . Ibid.*, pp. 75–79 (Dec.).

1843g. Descriptions of new species of *Nucula*, from the collections of Sir Edward Belcher, C.B., and Hugh Cuming, Esq. *Ibid.*, pp. 97–101 (Dec.).

1844a. Description of some new species of *Scalaria* and *Murex*, from the collection of Sir Edward Belcher, C.B. *Ibid.*, for 1843, pp. 124–29 (Mar.).

1844b. Descriptions of ten new species of *Cancellaria*, from the collection of Sir Edward Belcher, Ann. Mag. Nat. Hist., new ser., vol. 13, pp. 220–24 (Mar.).

1844c. Descriptions of new shells, collected during the voyage of the *Sulphur*, and in Mr. Cuming's late visit to the Philippines. [On new species of *Terebra*.] Proc. Zool. Soc. London, for 1843, pp. 149–59 (June).

1844d. Synopsis of the known species of *Terebra. Ibid.*, for 1843, pp. 159–68 (June).

1844e. Description of new species of shells, by Mr. Hinds. Six species of *Triton*, from the collection of Sir Edward Belcher, C.B. [Descriptions of new species of *Triton, Solarium*, and *Corbula*.] *Ibid.*, for 1844, pp. 21–26 (July).

1844f. Descriptions of Marginellae collected during the voyage of H.M.S. *Sulphur* and from the collection of Mr. Cuming. *Ibid.*, pp. 72–77 (Sept.).

1844g. Six species of *Triton*, from the collection of Sir Edward Belcher, C.B. Ann. Mag. Nat. Hist., new ser., vol. 14, pp. 436–46 (Dec.).

1844–45. The zoology of the voyage of H.M.S. *Sulphur*, under the command of Capt. Sir Edward Belcher . . . during 1836–1842. London, Mollusca, pt. 1, pp. 1–24, pls. 1–7 (July, 1844); pt. 2, pp. 25–48, pls. 8–14 (Oct. 1844); pt. 3, pp. 49–72, pls. 15–21 (Jan. 1845).

1845. Monograph of the genus *Terebra*; Brugeuire [Bruguière]. *In* G. B. Sowerby, Thesaurus conchyliorum. London, vol. 1, pp. 147(bis)–190(bis), pls. 41–45 (Jan. 15).

Hoffmann, Hans. 1928–29. Zur Kenntnis der Oncidiiden (Gastrop. pulmon.). Ein Beitrag zur geographischen Verbreitung, Phylogenie, und Systematik dieser Familie. I. Teil. Untersuchung neuen Materials und Revision der Familie. Zool. Jahrb. Jena, Syst., vol. 55, pp. 29–118, pls. 2–4 (1928). II. Teil. Phylogenie und Verbreitung. *Ibid.*, vol. 57, pp. 253–302, 1 map (1929).

1932–39. Opisthobranchia, *in* Dr. H. G. Bronn's Klassen und Ordnungen des Thierreichs. Bd. 3, Mollusca, Abt. 2, Gastropoda, Teil 1, lief. 1–7, 1247 pp., 830 text figs.

Hoffstetter, Robert. 1952. Moluscos subfósiles de los estanques de sal de Salinas (Pen. de Santa Elena, Ecuador). Comparación con la fauna actual de Ecuador. Bol. Inst. Cienc. Naturales, (Quito), Ecuador, vol. 1, no. 1, pp. 5–79, 19 text figs. (June).

Houbrick, J. R. 1968. New record of *Conus ebraeus* in Costa Rica. Veliger, vol. 10, no. 3, p. 292 (Jan. 1).

Howard, A. D. 1950. A new *Verticordia* from the Pacific coast. Nautilus, vol. 63, no. 4, pp. 109–10, pl. 7 (Apr. 4).

1951. The family Juliidae. *Ibid.*, vol. 64, no. 3, pp. 84–87, pl. 6 (Feb. 15).

Howard, Faye B. 1963a. Notes on a *Mitrella* (Mollusca: Gastropoda) from the Gulf of California. Veliger, vol. 5, no. 4, pp. 159–60 (Apr. 1).

1963b. Description of a new *Pyrene* from Mexico. Santa Barbara Mus. Nat. Hist. Occ. Papers no. 7 and supplement, pp. 1–11, pl. 1 (May, June).

Howard, Faye B., and G. G. Sphon. 1960. A new Panamic species of *Trivia*. Veliger, vol. 3, no. 2, pp. 41–43, pl. 7 (Oct. 1).

Hoyle, W. E. 1886–97. A catalogue of recent Cephalopoda. Proc. R. Physical Soc. Edinburgh, pp. 205–67 (1886); supplement for 1886–97, *ibid.*, pp. 363–75 (1897).

1904. Reports on the Cephalopoda ["Albatross" reports]. Bull. Mus. Comparative Zool., Harvard, vol. 43, no. 1, 71 pp., 12 pls. (March).

Hubendick, B. 1946. Systematic monograph of the Patelliformia. K. Svenska Vetensk.-Akad. Handl., ser. 3, vol. 23, no. 5, pp. 1–93, figs. 1–20, pls. 1–6.

Hupé, L. H. 1854. Fauna Chilena, Moluscos. *In* C. Gay, Historia física y política de Chile. Santiago, Zoología, vol. 8,500 pp., atlas (vol. 2), 14 pls.

Ingram, W. M. 1947. Fossil and Recent Cypraeidae of the western regions of the Americas. Bull. Amer. Paleont., vol. 31, no. 120, pp. 43–124, pls. 5–7 (May 2).

1948. The cypraeid fauna of the Galápagos Islands. Proc. California Acad. Sci., ser. 4, vol. 26, pp. 135–45, pl. 2, figs. 10–11 (June 28).

1951. The living Cypraeidae of the Western Hemisphere. Bull. Amer. Paleont., vol. 33, no. 123, pp. 125–78, pls. 21–24 (Mar. 24).

International Code of Zoological Nomenclature adopted by the XV International Congress of Zoology. 1961. London. International Trust for Zoological Nomenclature, xviii + 176 pp. Reprinted, with minor additions, 1964. [In English and French. Obtainable for $3.00 from the Publications Office of the International Trust, 14 Belgrave Square, London, S.W. 1, England.]

Jaeckel, S. 1927. Zur marinen Molluskenfauna Sudkaliforniens. Zool. Anz. Leipzig, vol. 70, pp. 45–50, 1 text fig. (Feb. 5).

Jatta, G. 1889. Elenco cei Cefalopodi della "Vittor Pisani" Bol. Soc. Nat. Napoli, vol. 3, pp. 63–67.

Johnson, C. W. 1911. Some notes on the Olividae—III. Nautilus, vol. 24, pp. 121–24 (Mar.).

1915. Further notes on the Olividae. *Ibid.*, vol. 28, pp. 114–16 (Feb.).

Johnson, R. I. 1959. The types of Corbiculidae and Sphaeriidae (Mollusca: Pelecypoda) in the Museum of Comparative Zoology, and a bio-bibliographic sketch of Temple Prime, an early specialist of the group. Bull. Mus. Comparative Zool., Harvard, vol. 120, no. 4, pp. 431–79, 8 pls. (May).

1964. The Recent Mollusca of Augustus Addison Gould. Bull. U.S. Nat. Mus. 239, 182 pp., 45 pls.

Jonas, J. H. 1844. Neue Trochoideen. Zeit. für. Malak., yr. 1, pp. 167–72 (Nov.).

Jordan, E. K. 1924. Quaternary and Recent molluscan faunas of the west coast of Lower California. Bull. Southern California Acad. Sci., vol. 23, no. 5, pp. 145–56 (Oct. 25).

1932. A new species of Crassatellites from the Gulf of California. Nautilus, vol. 46, no. 1, pp. 9–10 (July).

1936. The Pleistocene fauna of Magdalena Bay, Lower California. Contrib. Dept. Geol. Stanford Univ., vol. 1, no. 4, pp. 107–73, pls. 17–19 (Nov. 13).

Jordan, E. K., and L. G. Hertlein. 1926. Expedition to the Revillagigedo Islands, Mexico, in 1925, VII. Contribution to the geology and paleontology of the Tertiary of Cedros Island and adjacent parts of Lower California. Proc. California Acad. Sci., ser. 4, vol. 15, no. 14, pp. 409–64, pls. 27–34 (July 22).

Jousseaume, F. P. 1879. Étude des Purpuridae et description d'espèces nouvelles. Rev. Mag. Zool., ser. 3, vol. 7, pp. 314–48.

1880. Division méthodique de la famille des Purpuridés. Le Naturaliste, yr. 2, pp. 335–36 (Dec. 15).

1887. Le famille des Cancellariidae. Le Naturaliste, ann. 9, 2ᵉ ser., pp. 155–57, 163, 192–94, 213–14, 221–23 (Sept.–Dec.). Reprinted, pp. 1–31, 1888.

Kay, E. Alison. 1968. A review of the bivalved gastropods and a discussion of evolution within the Sacoglossa. Symposium, Zool. Soc. London, No. 22, pp. 109–34, 7 figs.

Keen, A. Myra. 1942. Viability of a marine snail. Nautilus, vol. 56, no. 1, pp. 34–35.

1943. New mollusks from the Round Mountain Silt (Temblor) Miocene of California. Trans. San Diego Soc. Nat. Hist., vol. 10, no. 2, pp. 25–60, pls. 3–4 (Dec. 30).

1944. Catalogue and revision of the gastropod subfamily Typhinae. Jour. Paleont., vol. 18, no. 1, pp. 50–72, 20 figs. in text (Jan.).

1946. A new gastropod of the genus *Episcynia* Mörch. Nautilus, vol. 60, no. 1, pp. 8–11, pl. 1 (July).

1958a. New mollusks from tropical West America. Bull. of Amer. Paleont., vol. 38, no. 172, pp. 239–55, pls. 30–31 (May 23).

1958b. Sea shells of tropical West America. Stanford Univ. Press. xi + 624 pp., over 1700 illustrations.

1959. Some side notes on "Seashells of Tropical West America." Veliger, vol. 2, no. 1, pp. 1–3, pl. 1 (July 1).

1960a. New *Phyllonotus* from the eastern Pacific. Nautilus, vol. 73, no. 3, pp. 103–9, pl. 10 (Jan.).

1960b. A bivalve gastropod. Nature, vol. 186, no. 4722, pp. 406–7 (Apr. 30).

1960c. Vermetid gastropods and marine intertidal zonation. Veliger, vol. 3, no. 1, pp. 1–2 (July 1).

1960d. The riddle of the bivalved gastropod. *Ibid.*, vol. 3, no. 1, pp. 28–30 (July 1).

1961a. High-lights of a collecting trip. *Ibid.*, vol. 3, no. 3, p. 79 (Jan. 1).

1961b. A proposed reclassification of the gastropod family Vermetidae. Bull. Brit. Mus. (Nat. Hist.) Zoology, vol. 7, no. 3, pp. 181–213, pls. 54–55, 35 text figs. (Feb.).

1961c. What is *Anatina anatina*? Veliger, vol. 4, no. 1, pp. 9–12, 5 text figs. (July 1).

1962a. Nomenclatural notes on some west American mollusks, with proposal of a new species name. *Ibid.*, vol. 4, no. 4, pp. 178–80 (Apr. 1).

1962b. A new west Mexican subgenus and new species of Montacutidae (Mollusca: Pelecypoda), with a list of Mollusca from Bahía de San Quintin. Pacific Naturalist, vol. 3, no. 9, pp. 321–28, figs. 4, 5 (Oct. 16).

1963. Marine Molluscan genera of western North America: an illustrated key. Stanford Univ. Press: 126 pp., 832 unnumbered figs. (Feb. 14).

1964a. A quantitative analysis of molluscan collections from Isla Espiritu Santo, Baja California, Mexico. Proc. California Acad. Sci., ser. 4, vol. 30, no. 9, pp. 175–206, figs. 1–4 (July 1).

1964b. *Purpura, Ocenebra*, and *Muricanthus* (Gastropoda): request for clarification of status. Bull. Zool. Nomenclature, vol. 21, pt. 3, pp. 235–39 (Aug.).

1966a. West American mollusk types at the British Museum (Nat. Hist.), I. T. A. Conrad and the Nuttall collection. Veliger, vol. 8, no. 3, pp. 167–72, 1 fig. (Jan. 1).

1966b. West American mollusk types in the British Museum (Natural History) II. Species described by R. B. Hinds. *Ibid.*, vol. 8, no. 4, pp. 265–75, pls. 46, 47, 6 text figs. (Apr. 1).

1966c. Moerch's west Central American molluscan types with proposal of a new name for a species of *Semele*. Occas. Papers California Acad. Sci., no. 59, 33 pp., 41 figs. (June 30).

1966d. West American mollusk types at the British Museum (Natural History) III. Alcide d'Orbigny's South American collection. Veliger, vol. 9, no. 1, pp. 1–7, pl. 1 (July 1).

1966e. *Hippella* Moerch (Mollusca: Pelecypoda): request for suppression under the plenary powers. Z. N. (S.) 1755. Bull. Zool. Nomenclature, vol. 23, pt. 4, pp. 181–82 (Oct.).

1968. West American mollusk types at the British Museum (Natural History) IV. Carpenter's Mazatlan collection. Veliger, vol. 10, no. 4, pp. 389–439, pls. 55 to 59, 171 text figs. (Apr. 1).

1969a. An overlooked subgenus and species from Panama. *Ibid.*, vol. 11, no. 4, p. 439 (Apr. 1).

1969b. *In* R. C. Moore, ed., Treatise on Invertebrate Paleontology. Part N, Bivalvia, 952 pp., figs. [Keen, pp. N230–31, 241, 269, 518, 537, 583–639, 643–58, 664–702, 843–59] (Nov.).

1971. Two new supraspecific taxa in the Gastropoda. Veliger, vol. 13, no. 3, p. 296 (Jan. 1).

Keen, A. Myra, and G. B. Campbell. 1964. Ten new species of Typhinae (Gastropoda: Muricidae). *Ibid.*, vol. 7, no. 1, pp. 46–57, pls. 8 to 11, 3 text figs. (July 1).

Keen, A. Myra, and A. G. Smith. 1961. West American species of the bivalved gastropod genus *Berthelinia*. Proc. California Acad. Sci., ser. 4, vol. 30, no. 2, pp. 47–66, figs. 1–33, 1 pl. (Mar. 20).

Keep, Josiah. 1888. West coast shells. S. Carson & Co., San Francisco. 230 pp., col. front illustration.

Kiener, L. C. 1834–70. Spécies général et iconographie des coquilles vivantes ... (continué par ... P. Fischer). Paris, vols. 1–11, livr. 1–165. [For the dates of publication of these volumes, *see* C. D. Sherborn, and B. B. Woodward, 1901, Proc. Malac. Soc. London, vol. 4, pp. 216–19; also W. O. Cernohorsky, Veliger, vol. 10, no. 4, p. 349, 1968.]

King, P. P., and W. J. Broderip. 1832. Description of the Cirrhipeda, Conchifera and Mollusca, in a collection formed by the officers of H.M.S. *Adventure* and *Beagle* employed between the years 1826 and 1830 in surveying the southern coasts of South America, including the Straits of Magalhaens and the coast of Tierra del Fuego. Zool. Jour., vol. 5, pp. 332–49 (July 1830 to Sept. 1831). [July 1832 according to Sherborn.]

Knudsen, Jørgen. 1961. The bathyal and abyssal Xylophaga (Pholadidae, Bivalvia). Galathea Report, vol. 5, pp. 163–209, 41 figs. (Dec. 28).

1970. The systematics and biology of abyssal and hadal Bivalvia. Galathea Report, vol. 11, pp. 1–241, pls. 1–20 (Nov. 6).

Küster, H. C., W. Kobelt, and H. C. Weinkauff. 1837–1920. Systematisches Conchylien-Cabinet von Martini und Chemnitz, neu herausgegeben . . . Nuremberg: 11 vols. [For collations *see* E. A. Smith, and H. W. England, Jour. Soc. Bibliogr. Nat. Hist., vol. 1, pt. 4, pp. 89–99, Dec. 1937; and R. I. Johnson, *ibid.*, vol. 4, pt. 7, pp. 363–67, 1968.]

Lamarck, J. B. P. A. de M. de. 1810. Description du genre Porcelaine (*Cypraea*) et des espèces que le composent. Ann. Mus. hist. nat. Paris, vol. 15, pp. 443–54.

1811. [Suite de la] Détermination des espèces de mollusques testacés: continuation du genre Porcelaine et des genres Ovule, Tarrière, Ancillaire, et Olive. *Ibid.*, vol. 16, pp. 89–114, 300–338 (Jan.–Mar. 1811); Volute et Mitre, vol. 17, pp. 54–80, 195–222 (July).

1815–22. Histoire naturelle des animaux sans vertèbres. Paris, vols. 1–7. [For the dates of issue of this work, *see* C. D. Sherborn, 1922, Index animalium, sect. 2, p. lxxvii, and T. Iredale, 1922, Proc. Malac. Soc. London, vol. 15, p. 85.]

1816. Tableau encyclopédique et méthodique des trois règnes de la nature. Paris, pp. i-viii, 1–83, pls. 1–95 (1791); pp. 85–132, pls. 96–189 (1792); pls. 190–286, by Bruguière (1797); pls. 287–390 (an VI [1798]); pls. 391–488, published under superintendence of Lamarck (1816); pls. 391–488, livr. 84, with 16 pp. of text, "Liste des objects représentés," by Lamarck. [Pp. 83–84 (reprint), 133–80 (explanation to pls. 52–488, incorporating Lamarck's text), issued in 1827 by Bory St. Vincent. For collation of the "Tableau encyclopédique," *see* C. D. Sherborn, and B. B. Woodward, 1906, Ann. Mag. Nat. Hist., ser. 7, vol. 17, pp. 577–82. *See also under* Encyclopédie Méthodique.]

Lamy, E. 1909a. Pélécypodes recueillis par M. L. Diguet dans le Golfe de Californie (1894–1905). Jour. de Conchyl., vol. 57, pp. 207–54 (Sept. 12).

1909b. Gastéropodes recueillis par M. L. Diguet dans le Golfe de Californie. Bull. Mus. hist. nat., Paris, vol. 15, pp. 264–70.

1916. Description d'un lamellibranche nouveau du Golfe de Californie. *Ibid.*, vol. 22, no. 8, pp. 443–45, 1 fig.

1917. Révision des Crassatellidae vivants du Muséum d'histoire naturelle de Paris. Jour. de Conchyl., vol. 62, pt. 4, pp. 197–270, pl. 6 (Feb. 15).

1922. Révision des Carditacea vivants du Muséum national d'histoire naturelle de Paris. Jour. de Conchyl., vol. 66, pp. 218–368, text figs., pls. 7–8 (Oct. 20).

1925. Révision des Gastrochaenidae vivants du Muséum national d'histoire naturelle de Paris. *Ibid.*, vol. 68, pp. 284–319 (Mar. 20).

1929–30. Révision des *Ostrea* vivants du Muséum national d'histoire naturelle de Paris. *Ibid.*, vol. 73, pp. 1–46 (Apr. 30, 1929), 71–108 (July 20, 1929), 133–68 (Oct. 30, 1929), 233–75, pl. 1 (Feb. 28, 1930).

1930–31. Révision des Limidae vivants du Muséum national d'histoire naturelle de Paris. *Ibid.*, vol. 74, no. 2, pp. 89–114 (Sept. 15); no. 3, pp. 169–98, pl. 1 (Nov. 29, 1930); no. 4, pp. 245–69 (Feb. 9, 1931).

1931a. Révision des Thraciidae vivants du Muséum national d'histoire naturelle. *Ibid.*, vol. 75, pp. 213–41. (Sept. 30) ; 285–302 (Dec. 10).

1931b. Note sur *Leucozonia cingulifera* Lamarck et *L. cingulata* Lamarck. *Ibid.*, vol. 75, pp. 273–75.

1934. Coquilles marines recueillies par M. E. Aubert de la Rüe dans l'Amérique du Sud. Bull. Mus. hist. nat., Paris, sér. 2, vol. 6, pp. 432–35, 1 fig. (Oct.).

Lance, J. R. 1961. A distributional list of southern California opisthobranchs. Veliger, vol. 4, no. 2, pp. 64–69 (Oct. 1).

1962a. A new *Stiliger* and a new *Corambella* (Mollusca: Opisthobranchia) from the northwestern Pacific. *Ibid.*, vol. 5, no. 1, pp. 33–38, 10 figs., 1 col. pl. (July 1).

1962b. A new species of *Armina* (Gastropoda: Nudibranchia) from the Gulf of California. *Ibid.*, vol. 5, no. 1, pp. 51–54, 6 text figs. (July 1).

1966. New distributional records of some northeastern Pacific Opisthobranchiata (Mollusca: Gastropoda) with descriptions of two new species. *Ibid.*, vol. 9, no. 1, pp. 69–81, 12 text figs. (July 1).

1967. Northern and southern extensions of *Aplysia vaccaria* (Gastropoda: Opisthobranchia). *Ibid.*, vol. 9, no. 4, p. 412 (Apr. 1).

1968. New Panamic nudibranchs (Gastropoda: Opisthobranchia) from the Gulf of California. Trans. San Diego Soc. Nat. Hist., vol. 15, no. 2, 13 pp., 2 pls., 11 figs. (Jan.)

Leach, W. E. 1814–17. The Zoological Miscellany; being descriptions of new, or interesting animals. London: 3 vols. Vol. 1, pp. 1–144, pls. 60 (1814) ; vol. 2, pp. 1–154, pls. 61–120 (1815) ; vol. 3, pp. 1–149, pls. 120–49 (1817).

Leloup, Eugene. 1941. A propos de quelques Acanthochitons peu connus ou nouveaux, III. Region Pacifique, Côtes Américaines. Bull. Musée royal d'Histoire naturelle de Belgique, vol. 17, no. 61, pp. 1–9, 5 text figs. (Oct.).

Lemche, Henning. 1957. A new living deep-sea Mollusc of the Cambro-Devonian Class Monoplacophora. Nature, vol. 179, no. 4556, pp. 413–16, 4 text figs. (Feb. 23).

Lemche, Henning, and K. G. Wingstrand. 1959. The anatomy of *Neopilina galatheae* Lemche, 1957 (Mollusca, Tryblidiacea). Galathea report: Scientific results of the Danish deep-sea expedition round the world, 1950–52, vol. 3, pp. 9–72, pls. 1–56.

Lesson, R. P. 1827. Description d'un nouveau genre de mollusque nucléobranche, nommè *Pterosoma*. Bull. Universel des Sci. II: vol. 3, pt. 3, pp. 414–16, pl. 10 (also in Bull. Sci. Nat. et Geol., vol. 12, pt. 10, pp. 282–83) (Oct.).

1830. Voyage autour du monde . . . sur . . . la Coquille pendant . . . 1822–25 . . . par L. J. Duperry. Paris, Zoologie, vol. 2, pt. 1, pp. 1–471 (June 12, 1830), atlas, 157 pls. (pls. 1–16, Mollusques, ? 1831). [For dates of publication of this work, *see* C. D. Sherborn, and B. B. Woodward, 1901, Ann. Mag. Nat. Hist., ser. 7, vol. 7, p. 391.]

1832–35. Illustrations de zoologie, ou recueil de figures d'animaux. Paris, pls. 1–15 (1832), 16–33 (1833), 34–42 (1834), 43–60 (1835). [*See* C. D. Sherborn, 1922, Index animalium, sect. 2, p. lxxx.]

1842a. Mollusques recueillis dans la mer du Sud et l'Océan Atlantique, par M. Adolphe Lesson. Rev. Zool., Soc. Cuvierienne, vol. 5, pp. 184–87.

1842b. Mollusques recueillis dans la mer du Sud, par M. A. Lesson. [Genre *Buccinum*, Lin.] *Ibid.*, pp. 237–38.

Lesueur, C. A. 1813. Mémoire sur quelques nouvelles espèces d'animaux mollusques et radiaires, recueillis dans la Méditerranée, près de Nice (*Cestus veneris, Pyrosoma elegans, Hyalaea lanceolata,* et *H. inflexa*). Jour. de Phys., vol. 77, pp. 119–24.

1817a. Description of new species of the genus *Firola*, observed by Messrs. LeSueur and Peron in the Mediterranean Sea, in the months of March and April, 1809. Proc. Acad. Nat. Sci. Philadelphia, vol. 1, pt. 1, pp. 3–8, 1 pl.

1817b. Characters of a new genus, and description of three new species upon which it is formed; discovered in the Atlantic Ocean in the months of March and April, 1816. *Ibid.*, pt. 3, pp. 37–41, pl. 2 (July).

1817c. Mémoire sur deux nouveaux genres de mollusques, Atlante et Atlas. Jour. Phys. Chim. Hist. Nat. Arts, vol. 85, pp. 390–92, pl. 2 (Nov.).

Li, C. C. 1930. The Miocene and Recent Mollusca of Panama Bay. Bull. Geol., Soc. China, vol. 9, pp. 249–96, pls. 1–8 (Oct.). [*See also* commentary by H. A. Pilsbry 1931c.]

Libassi, Ignazio. 1859. Sopra alcune conchiglie fossili dei dintorni di Palermo. Reale Accademia di Scienze, lettere e belle arti di Palermo, Atti, n.s., vol. 3, 47 pp., 1 pl.

[Lightfoot, John.] 1786. *In* A Catalogue of the Portland Museum, lately the property of the Duchess Dowager of Portland, deceased: which will be sold by auction, etc. London. [For discussion of the authorship of this catalogue, usually attributed to D. Solander, *see* P. Dance, Jour. Soc. for the Bibliography of Natural History, vol. 4, pt. 1, pp. 30–34, 1962, and H. A. Rehder, Proc. U.S. Nat. Mus., vol. 121, no. 3579, pp. 1–51, 1967.]

Lindsay, G. E. 1964. Sea of Cortez expedition of the California Academy of Sciences June 20–July 4, 1964. Proc. California Acad. Sci., ser. 4, vol. 30, no. 11, pp. 211–42, figs. 1–23, 3 pls. (Dec. 31).

Linnaeus, Carl. 1758. Systema naturae per regna tria naturae. Editio decima, reformata. Stockholm, vol. 1, Regnum animale, 824 pp. (Jan. 1). [For a review of many Linnaean species, *see* Dodge, 1953–59.]

—— 1761. Fauna Suecica, sistens animalia sueciae regni. Stockholm, ed. 2, pp. 508.

—— 1766–67. Systema naturae per regna tria naturae. Editio duodecima, reformata. Stockholm, vol. 1, Regnum animale. Pt. 1, pp. 1–532 (1766); pt. 2, pp. 533–1327 (1767).

Lorois, E. L. 1852. Description d'une nouvelle espèce du genre Argonaute. Rev. Mag. de Zool., sér. 2, vol. 4, pp. 9–10, pl. 1.

—— 1857. Description d'espèces nouvelles. Jour. de Conchyl., vol. 6, pp. 388–89, 2 figs. Dec.).

Lowe, H. N. 1913. Shell collecting on the west coast of Baja California. Nautilus, vol. 27, no. 3, pp. 25–29 (July).

—— 1931. What is *Roperia roperi* Dall? With notes on Turridae and Columbellidae. *Ibid.*, vol. 45, no. 2, pp. 51–52 (Oct.).

—— 1932a. Shell collecting in West Central America. *Ibid.*, vol. 45, no. 3, pp. 73–82 (Jan.).

—— 1932b. On four new species of *Epitonium* from western Central America. *Ibid.*, vol. 45, no. 4, pp. 113–15, pl. 9 (Apr.).

—— 1933a. The cruise of the "Petrel." *Ibid.*, vol. 46, no. 3, pp. 73–76 (Jan.); no. 4, pp. 109–15, pl. 9 (Apr.).

—— 1933b. At the head of the Gulf of California. *Ibid.*, vol. 47, no. 2, pp. 45–47 (Oct.).

—— 1935. New marine Mollusca from west Mexico, together with a list of shells collected at Punta Peñasco, Sonora, Mexico. Trans. San Diego Soc. Nat. Hist., vol. 8, no. 6, pp. 15–34, pls. 1–4 (Mar. 21).

Mabille, J. 1895. Mollusques de la basse Californie recueillis par M. Diguet. Bull. Soc. Philom. Paris, ser. 8, vol. 7, pp. 54–76.

McBeth, J. W., and R. D. Bowlus. 1969. Range extension of *Tylodina fungina* in the Gulf of California. Veliger, vol. 12, no. 2, p. 229 (Oct. 1).

MacFarland, F. M. 1905. A preliminary account of the Dorididae of Monterey Bay, California. Proc. Biol. Soc. Washington, vol. 18, pp. 35–54 (Feb. 2).

—— 1924. Expedition of the California Academy of Sciences to the Gulf of California in 1921. Opisthobranchiate Mollusca. Proc. Calif. Acad. Sci., ser. 4, vol. 13, No. 25, pp. 389–420, pls. 10–12 (Nov. 29).

—— 1966. Studies of Opisthobranchiate mollusks of the Pacific coast of North America. Mem. California Acad. Sci., vol. 6, pp. 1–546, pls. 1–72 (16 col. pls.) (Apr. 8).

McGowan, J. A. 1968. Thecosomata and Gymnosomata. Veliger, vol. 3, Suppl., pt. 2, pp. 103–35, pls. 12–20 (May 1).

McLean, J. H. 1959. A new marine gastropod from west Mexico. Nautilus, vol. 73, no. 1, pp. 9–11, pl. 4 (July).

1961. Marine mollusks from Los Angeles Bay, Gulf of California. Trans. San Diego Soc. Nat. Hist., vol. 12, no. 28, pp. 449–76, figs. 1–3 (Aug. 15).

1962. Feeding behavior of the chiton *Placiphorella*. Proc. Malac. Soc. London, vol. 35, pt. 1, pp. 23–26, 2 text figs. (Apr.).

1965. New species of Recent and fossil West American Aspidobranch gastropods. Veliger, vol. 7, no. 2, pp. 129–33, 1 pl., 1 text fig. (Oct.).

1966*a*. A new genus of the Fissurellidae and a new name for a misidentified species of West American *Diodora*. Los Angeles County Museum Contributions to Science, no. 100, pp. 1–7, 4 figs. (May 5).

1966*b*. West American prosobranch Gastropoda: superfamilies Patellacea, Pleurotomariacea, and Fissurellacea. Stanford Univ. Ph.D. thesis. Pp. x + 255, 7 pls. (June). [Available through University Microfilms, Ann Arbor, Mich.]

1967*a*. West American species of *Lucapinella*. Veliger, vol. 9, no. 3, pp. 349–52, pl. 49, 3 text figs. (Jan. 1).

1967*b*. West American Scissurellidae. *Ibid.*, vol. 9, no. 4, pp. 404–10, pl. 56 (Apr. 1).

1967*c*. Note on the radula of *Mitromica* Berry, 1958. *Ibid.*, vol. 10, no. 1, p. 58, 1 text fig. (July 1).

1969. Marine shells of Southern California. Los Angeles County Museum of Natural History, Science Series 24, Zoology, No. 11, 104 pp., 54 figs. (Oct.).

1970*a*. New species of Panamic marine gastropods. Veliger, vol. 12, no. 3, pp. 310–15, pl. 46 (Jan. 1).

1970*b*. Descriptions of a new genus and eight new species of eastern Pacific Fissurellidae, with notes on other species. *Ibid.*, vol. 12, no. 3, pp. 362–67, pl. 54, 1 text fig. (Jan. 1).

1970*c*. Notes on the deep water Calliostomas of the Panamic Province, with descriptions of six new species. *Ibid.*, vol. 12, no. 4, pp. 421–26, pl. 62 (Apr. 1).

1970*d*. New species of tropical eastern Pacific Mollusca. Malac. Review, vol. 2, pt. 2, pp. 115–30, 41 text figs. (May).

1970*e*. New eastern Pacific subgenera of *Turbo* Linnaeus, 1758, and *Astraea* Röding, 1798. Veliger, vol. 13, no. 1, pp. 71–72 (July 1).

1970*f*. On the authorship of Part V of Eschscholtz's "Zoological Atlas," 1833. *Ibid.*, vol. 13, no. 1, p. 112 (July 1).

1971. A revised classification of the family Turridae, with the proposal of new subfamilies, genera, and subgenera from the eastern Pacific. *Ibid.*, vol. 14, no. 1, pp. 114–30, 4 pls. (July 1).

McLean, J. H., and W. K. Emerson. 1970. *Calotrophon*, a New World muricid genus (Gastropoda: Mollusca). *Ibid.*, vol. 13, no. 1, pp. 57–62, 1 pl., 10 text figs. (July 1).

McLean, J. H., and L. H. Poorman. 1970. Reinstatement of the turrid genus *Bellaspira* Conrad, 1868 (Mollusca: Gastropoda) with a review of the known species. Los Angeles Co. Mus., Contr. in Sci., no. 189, 11 pp., 16 figs. (May 4).

1971. New species of tropical eastern Pacific Turridae. Veliger, vol. 14, no. 1, pp. 89–113, 2 pls. (July 1).

MacNeil, F. S. 1938. Species and genera of Tertiary Noetinae. U.S. Geol. Survey, Prof. Paper 189-A, pp. 1–40, pls. 1–6, 2 figs.

Marcus, Ernst. 1958. On western Atlantic opisthobranch gastropods. Amer. Mus. Novitates, no. 1906, 82 pp., 111 text figs.

1961. Opisthobranch mollusks from California. Veliger, Suppl. 1, pp. 1–85, pls. 1–10 (Feb. 1).

1964. A new species of *Polycera* (Nudibranchia) from California. Nautilus, vol. 77, no. 4, pp. 128–31, figs. 1–4 (Apr.).

Marcus, Eveline, and Ernst Marcus. 1966. The R/V Pillsbury deep-sea biological expedition to the Gulf of Guinea, 1964–1965. 9. Opisthobranchs from tropical west

Africa. Studies in tropical Oceanography, Univ. Miami, vol. 4, no. 1, pp. 152–208, figs. 1–62.

1967. American Opisthobranch mollusks. Studies in Tropical Oceanography, no. 6: Institute of Marine Sciences, Univ. of Miami, Florida. Pp. viii + 256, 250 text figs., 1 col. pl. (Dec.). Pt. I: Tropical American Opisthobranchs, pp. 1–138, figs. 1–155. Pt. II: Opisthobranchs from the Gulf of California, pp. 141–248, figs. 1–95.

1968. On the prosobranchs *Ancilla dimidiata* and *Marginella fraterculus*. Proc. Malac. Soc. London, vol. 38, pt. 1, pp. 55–69, 12 text figs. (Apr.).

1970. Some gastropods from Madagascar and West Mexico. Malacologia, vol. 10, no. 1, pp. 181–223, 93 figs. ("May," ?Nov.).

Marks, J. G. 1949. Nomenclatural units and tropical American Miocene species of the gastropod family Cancellariidae. Jour. Paleont., vol. 23, no. 5, pp. 453–64, pl. 78 (Sept.).

1951. Miocene stratigraphy and paleontology of southwestern Ecuador. Bull. Amer. Paleont., vol. 33, no. 139, pp. 271–432, 12 figs., pls. 43–51 (Dec. 20).

Marrat, F. P. 1870–71. *Oliva*, Bruguière. *In* G. B. Sowerby, Thesaurus conchyliorum. London, vol. 4, pls. 328 [329]–41 (1870); pp. 1–46, pls. 342–51 (1871) (*Oliva* pls. 1–25).

Martens, E. von. 1863–79. Die Gattung *Neritina*. *In* Systematisches Conchylien-Cabinet von Martini und Chemnitz. Nuremberg, vol. 2, div. 10, pp. 1–303, pls. 1–24.

1890–1901. Land and freshwater Mollusca. *In* F. D. Godman, and O. Salvin, Biologia Centrali-Americana. London, pp. i–xxviii, 1–706, pls. 1–44.

1897. Conchologische Miscellen II. Archiv f. Naturgesch., yr. 63, vol. 1, pp. 157–80, pls. 15–17.

1899. Purpur-Färberei in Central-Amerika. Nachrichtsbl. Deutschen Malakozool. Gesell., yr. 31, pp. 113–22 (July–Aug.).

Martin, Rainer, and Anita Brinckmann. 1963. Zum brutparasitismus von *Phylliroe bucephala* Per. & Les. (Gastropoda, Nudibranchia) auf der Meduse *Zanclea costata* (Hydrozoa, Anthomedusae). Pubbl. staz. zool. Napoli, vol. 33, pp. 206–23.

Martínez y Saez, F. de P. 1869 [? 1870]. Moluscos del viaje al Pacífico verificado de 1862 a 1865 por una comisión de naturalistas enviado por el gobierno de Español. Parte segunda. Bivalvos marinos. Madrid. pp. 1–78 + 2, pls. 1–9. [Apparently issued in 1870. For part 3 of this work, *see* J. G. Hidalgo.]

Martini, F. H. W., and J. H. Chemnitz. 1769–95. Neues Systematisches Conchylien-Cabinet. Nuremberg, vols. 1–11. [Non-binomial.]

1837–1920. Neues Systematisches Conchylien-Cabinet, ed. 2. *See under* H. C. Küster.

Maury, C. J. 1922. The Recent arcas of the Panamic Province. Palaeontogr. Amer., vol. 1, no. 4, pp. 163–208, pls. 29–31.

Meisenheimer, J. 1906. Die Pteropoden der deutschen Süd-polar Expedition 1901–1903. Deutsch. Südpol. Exp. IX (Zool.), vol. 1, pt. 2, pp. 92–152, pls. 5–7.

Melvill, J. C. 1891. An historical account of the genus *Latirus* (Montfort) and its dependencies, with description of eleven new species, and a catalogue of *Latirus* and *Peristernia*. Mem. Manchester Mus., vol 34 (ser. 4, vol. 4), pp. 365–411, 1 pl.

1892. Description of a new species of *Latirus*. Mem. and Proc. Manchester Lit. and Philos. Soc., ser. 4, vol. 5, pp. 92–93.

1900. Descriptions of two species of *Cypraea*, both of the subgenus *Trivia* Gray. Ann. Mag. Nat. Hist., vol. 6, no. 7, pp. 207–10, 4 figs.

1927. Descriptions of eight new species of the family Turridae and of a new species of *Mitra*. Proc. Malac. Soc. London, vol. 17, no. 4, pp. 149–55, pl. 12 (May 27).

Menke, K. T. 1828. Synopsis Methodica molluscorum generum omnium et specierum earum quae in Museo Menkeana adservabitur ... Pyrmont [Germany]. Pp. xii + 91. (Ed. 2, Pp. xvi + 168, 1830).

1843. Molluscorum Novae Hollandiae specimen ... Hannover. 46 pp.

1847. Verzeichniss einer Sendung von Conchylien von Mazatlan, mit einigen kritischen Bemerkungen. Zeitschr. f. Malakozool, yr. 4, pp. 177–91 (Dec.).

1850–51. Conchylien von Mazatlan, mit kritischen Anmerkungen. *Ibid.*, yr. 7, pp. 161–73 (Nov. 1850), 177–90 (1850) [According to Sherborn, the new species in the foregoing appeared in Apr. 1851]; yr. 8, pp. 17–25, 33–38 (1851).

1853. Kritische Anzeige. Zeit. f. Malakozool., Jahrg. 10, no. 8, pp. 113–17.

1854. Noch eine *Bulla* und einige andere Konchylien. Malak. Blätter, vol. 1, pp. 26–30 (Feb.).

Menzies, R. J. ? 1968. New species of *Neopilina*, of the Cambro-Devonian class Monoplacophora from the Milne-Edwards Deep of the Peru-Chile Trench... Proc., Symposium on Mollusca held at Cochin from January 12 to 16, 1968. Part I, pp. 1–9, 4 pls., (Symposium Ser. 3, Marine Biol. Assoc. India) [no date on title page].

Menzies, R. J., Maurice Ewing, J. L. Worzel, and A. H. Clarke. 1959. Ecology of the Recent Monoplacophora. Oikos, vol. 10, no. 2, pp. 168–82, 10 text figs.

Menzies, R. J., and William Layton. 1963. A new species of the monoplacophoran mollusc *Neopilina* (*Neopilina*) *veleronis*, from the slope of the Cedros trench, Mexico. Ann. Mag. Nat. Hist., ser. 13, vol. 5, pp. 401–6, pls. 7–9 (Sept. 1).

Merriam, C. W. 1941. Fossil turritellas from the Pacific coast region of North America. Univ. California Pub. Bull. Dept. Geol. Sci., vol. 26, pp. 1–214, 19 figs., pls. 1–41, 1 map (Mar. 8).

Merriman, Jean A. 1967. Systematic implications of radular structures of West Coast species of *Tegula*. Veliger, vol. 9, no. 4, pp. 399–403, 2 pls., 2 text figs. (Apr. 1).

Michelotti, G. 1841. Monografia del genere *Murex* ossia enumerazione delle principali specie dei terreni sopracretacei dell'Italia. Vicenzia, Dalla tipografia tremeschin, 27 pp., 5 pls.

Milburn, W. P. 1959. *Neopilina* and the interpretation of the Mollusca. Veliger, vol. 2, no. 2, pp. 24–28, 1 pl. (Oct.).

1960. Further remarks on the interpretation of the Mollusca. *Ibid.*, vol. 3, no. 2, pp. 43–47, 1 pl. (Oct. 1).

Miller, Conrad. 1879. Die Binnenmollusken von Ecuador. (Schluss). Malakozool. Blätter, new ser., vol. 1, pp. 117–203, pls. 4–15.

Mörch, O. A. L. 1852–53. Catalogus conchyliorum quae reliquit D. Alphonso D'Aguirra et Gadea Comes de Yoldi. Copenhagen. Fasc. 1 [Gastropoda, etc.], 170 pp. (*post*-Aug. 1, 1852); fasc. 2 [Pelecypoda, etc.], 74 pp. (*post*-Apr. 1, 1853).

1857. Description de nouveau mollusques de l'Amérique Centrale. Jour. de Conchyl., vol. 6, pp. 281–82, pl. 10 (Oct.).

1859–61. Beiträge zur Molluskenfauna Central-Amerika's. Malakozool. Blätter, vol. 6, no. 4, pp. 102–26 (Oct. 1859); vol. 7, no. 2, pp. 66–96 (July 1890), no. 3, pp. 97–106 (Aug. 1860), no. 4, 170–92 (Dec. 1860), no. 5, 193–213 (Jan. 1861).

1860. Études sur la famille des vermets. Jour. de Conchyl., vol. 8, pp. 27–48 (Jan.).

1861–62. Review of the Vermetidae. Parts I–III. Proc. Zool. Soc. London, for 1861, pp. 145–81 (Sept. 1861); 326–64 (Apr. 1862); *Ibid.*, for 1862, pp. 54–83 (June).

1863. Contribution a la faune malacologique des Antilles danoises. Jour. de Conchyl., vol. 11, no. 1, pp. 21–34 (Jan. 30).

Molina, G. I. 1782. Saggio sulla storia naturale dei Chile. 367 pp., 1 map.

Moll, F., and F. Roch. 1931. The Teredinidae of the British Museum and the Jeffreys collection. Proc. Malac. Soc. London, vol. 19, pt. 4, pp. 201–18, pls. 22–25 (Mar. 14).

Montagu, George. 1803–8. Testacea Britannica: or an account of all the shells hitherto discovered in Britain. London: 2 vols., pp. xl + 606, pls. 1–16 (1803); suppl., pp. v + 183, pls. 17–30 (1808).

Moore, D. R. 1966. The Cyclostremellidae, a new family of prosobranch mollusks. Bull. Marine Sci., vol. 16, pp. 480–84, 6 figs.

Moore, R. C., ed. 1960. Treatise on Invertebrate Paleontology. Part I. Mollusca 1 (J. Brookes Knight *et al.*) 351 pp., 216 text figs. (Mollusca, General features; Scaphopoda; Amphineura; Monoplacophora; Archaeogastropoda).

1969. Treatise on Invertebrate Paleontology Part N: Bivalvia. 3 vols. Vols. 1–2, 952 pp., ill.

Morris, Percy. 1966. A field guide to shells of the Pacific Coast and Hawaii including shells of the Gulf of California. Boston. Ed. 2, rev. Pp. xxxiii + 297; 72 pls. (pls. 1–8 in color).

Morrison, J. P. E. 1946. The nonmarine mollusks of San José Island, with notes on those of Pedro González Island, Pearl Islands, Panamá. Smithsonian Inst. Misc. Coll., vol. 106, no. 6, pp. 1–49, pls. 1–3 (Sept. 12).

1955. Some zoogeographic problems among brackish water mollusks. Amer. Malac. Union Ann. Rept. for 1954, pp. 7–10.

1963. Notes on American *Siphonaria*. Ann. Rept., Amer. Malac. Union, Bull. 30, pp. 7–8 (Dec. 1).

1964. Notes on American Melampidae. Nautilus, vol. 77, no. 4, pp. 119–21 (Apr.).

1965. Notes on the genera of the Hipponicidae. Ann. Rept., Amer. Malac. Union, Bull. 32, pp. 33–34 (Dec. 1).

1966. On the families of Turridae. Ann. Rept., Amer. Malac. Union for 1965, pp. 1–2.

1968. Four American *Hastula* species. Texas Conchologist, vol. 4, no. 9, pp. 67–70, 1 pl. (May).

Newell, N. D., *see under* R. C. Moore, 1969.

Nicol, D. 1945. Genera and subgenera of the pelecypod family Glycymeridae. Jour. Paleont., vol. 19, no. 6, pp. 616–21, 2 figs. (Nov.).

1950. Origin of the pelecypod family Glycymeridae. *Ibid.*, vol. 24, no. 1, pp. 89–98, 2 figs., pls. 20–22 (Jan.).

1952a. Nomenclatural review of genera and subgenera of Chamidae. Jour. Washington Acad. Sci., vol. 42, pp. 154–56 (May).

1952b. Revision of the pelecypod genus *Echinochama*. Jour. Paleont., vol. 26, no. 5, pp. 803–17, 15 figs., pls. 118–19 (Sept.).

Niebuhr, C., *see under* Petter Forskål, 1775.

Nuttall, Z. 1909. A curious survival in Mexico of the use of the Purpura shell-fish for dyeing. Putnam anniversary volume anthropological essays. New York, G. E. Stechert Co., pp. 368–84, pls. 1–2.

Nybakken, J. W. 1967. Preliminary observations on the feeding behavior of *Conus purpurascens* Broderip, 1833. Veliger, vol. 10, no. 1, pp. 55–57, pl. 4 (July 1).

1968. Notes on the food of *Conus dalli* Stearns, 1873. *Ibid.*, vol. 11, no. 1, p. 50 (July 1).

1970a. Correlation of radula tooth structure and food habits of three vermivorous species of *Conus*. *Ibid.*, vol. 12, pp. 316–18, pl. 47 (Jan. 1).

1970b. Notes on the egg capsules and larval development of *Conus purpurascens* Broderip. *Ibid.*, vol. 12, no. 4, pp. 480–81, 10 text figs. (Apr. 1).

1970c. Radular anatomy and systematics of the West American Conidae. Amer. Museum Novitates, no. 2414, 29 pp., 40 figs. (May 12).

1971. The Conidae of the Pillsbury Expedition to the Gulf of Panama. Bull. Marine Sci., Univ. Miami, vol. 21, no. 1, pp. 93–110 [in press].

Nyst, P. H. 1848. Tableau synoptique et synonymique des espèces vivantes et fossiles de la famille des Arcacées. Pt. 1, Genre Arca. Mem. Acad. Roy. Sci. Lett. Beaux-Arts Belgique, vol. 22, pp. 1–79.

Odhner, N. H.-J. 1919. Studies on the morphology, the taxonomy and the relations of Recent Chamidae. K. Svenska Vetensk. Handl., vol. 59, no. 3, pp. 1–102, pls. 1–8.

1922. Mollusca of Juan Fernandez and Easter Island. *In* C. Skottsberg, The natural history of Juan Fernandez and Easter Island, vol. 3, pt. 2, pp. 219–54, pls. 8–9, 24 text figs.

O'Donoghue, C. H. 1921. Nudibranchiate Mollusca from the Vancouver Island region. Trans. Canadian Inst., Toronto, vol. 13, pp. 147–209, pls. 1–5.

Okutani, Takashi. 1955. On a new species of *Carinaria*, *C. japonica*. Venus (Japanese Jour. Malac.), vol. 18, no. 4, pp. 251–58, 3 figs. (Dec.).

Oldroyd, Ida S. 1918. List of shells from Angel and Tiburon Islands, Gulf of California, with description of a new species. Nautilus, vol. 32, no. 1, pp. 26–27 (July).

1921. A new Peruvian *Chione*. *Ibid.*, vol. 34, no. 3, p. 93, pl. 4 [part] (Jan.).

1925–27. The marine shells of the West coast of North America. Stanford Univ. Pub., Univ. Ser., Geol. Sci., vol. 1, Pelecypoda, 248 pp. 57 pls. [dated 1924, published Sept., 1925]; vol. 2, Gastropoda, Scaphopoda, and Amphineura, pt. 1, 298 pp., pls. 1–29; pt. 2, 304 pp., pls. 30–72; pt. 3, 340 pp., pls. 73–108 (1927).

Olsson, A. A. 1924. Notes on marine mollusks from Peru and Ecuador. Nautilus, vol. 37, pp. 120–30 (Apr.).

1935. *Mactra alata* Spengler var. *subalata* on the West Coast. *Ibid.*, vol. 48, no 3, p. 105 (Jan.).

1942. Tertiary and Quaternary fossils from the Burica Peninsula of Panama and Costa Rica. Bull. Am. Paleont., vol. 27, pp. 153–258, pls. 14–25 (Dec. 25).

1956. Studies on the genus *Olivella*. Proc. Acad. Nat. Sci. Philadelphia, vol. 108, pp. 155–225, pls. 8–16 (Oct. 3).

1961. Mollusks of the tropical eastern Pacific, particularly from the southern half of the Panamic-Pacific faunal province (Panama to Peru). Panamic-Pacific Pelecypoda. Paleontological Research Institution, Ithaca, N.Y. 574 pp., 86 pls. (Mar. 10).

1964. Neogene mollusks from northwestern Ecuador. Paleontological Research Institution, Ithaca, N.Y., 256 pp., 38 pls.

1970. The cancellarid radula and its interpretation. Palaeontographica Americana, vol. 7, no. 43, pp. 19–27, pls. 4–6 (Aug. 17).

1971. Mollusks from the Gulf of Panama collected by R/V John Elliott Pillsbury, 1967. Bull. Marine Sci., vol. 21, no. 1, pp. 35–92, 103 figs. (June 16).

Olsson, A. A., and E. Bergeron. 1967. *Perplicaria clarki* Maxwell Smith, a living fossil. Veliger, vol. 9, no. 4, p. 411, pl. 57.

Olsson, A. A., and S. P. Dance, 1966. The Linnaean olives. Bull. Amer. Paleont., vol. 50, no. 227, pp. 215–22, 1 pl. (Apr. 5).

Olsson, A. A., and M. Smith. 1951. New species of Epitoniidae and Vitrinellidae from Panama City. Nautilus, vol. 65, no. 2, pp. 44–46, pl. 3 (Nov. 9).

Orbigny, Alcide d'. 1834–47. Voyage dans l'Amérique Méridionale. Mollusques. Paris, vol. 5, pt. 3, xliii + 758 pp., atlas, 85 pls. [For collation of this work, *see* C. D. Sherborn and B. B. Woodward, 1901, Ann. Mag. Nat. Hist., ser. 7, vol. 7, p. 289; C. D. Sherborn and F. J. Griffin, 1934, *ibid.*, ser. 10, vol. 13, pp. 130–34; A. M. Keen, 1966, Veliger, vol. 9, no. 1, p. 7. For a catalogue of the species, *see* J. E. Gray, 1855.]

1835. Synopsis terrestrium et fluviatilium molluscorum, in suo per American meridionalem itinere. Mag. Zool., yr. 5, cl. 5, nos. 61, 62, pp. 1–32.

1841–46. Mollusques. *In* R. Sagra, Histoire physique, politique, et naturelle de l'Ile de Cuba. 2 vols. and atlas, issued as of 1853, but published in parts earlier, as follows: French ed., 1841–53: vol. 1, pp. 1–208, pts. 1–14 (1841); pp. 209–64, pts. 15–17 (1842); vol. 2, pp. 1–112, pts. 1–7 (1842); pp. 113–380, pts. 8–24, probably 1853 [imprinted date 1846, pp. 149 ff.]; Atlas (1842). Spanish ed., title page dated 1845 [pp. 149 ff. probably 1846, as imprinted].

Orcutt, C. R. 1913. Mexican shells [a check list for Oaxaca area]. Orcutt's Mexico, vol. 1, no. 1, p. 5 (Aug.).

1918. Magdalena Bay shells. West American Scientist, vol. 22, no. 1, p. 8 (Apr.). [*See also* bibliography by E. V. Coan, 1966.]

Pace, S. 1902. Contributions to the study of the Columbellidae: no. 1. Proc. Malac. Soc. London, vol. 5, pp. 36–154.

Paine, R. T. 1966. Function of labial spines, composition of diet, and size of certain marine gastropods. Veliger, vol. 9, no. 1, pp. 17–24, 2 text figs. (July 1).

Palmer, Katherine E. v.W. 1945. Molluscan types in the Carpenter collection in the Redpath Museum. Nautilus, vol. 58, no. 3, pp. 97–102 (Feb. 19).

1951. Catalogue of the first duplicate series of the Reigen Collection of Mazatlán shells in the State Museum at Albany, New York. New York State Mus. Bull. no. 342, 79 pp., 1 pl. (Jan.).

1958. Type specimens of marine Mollusca described by P. P. Carpenter from the west coast (San Diego to British Columbia). Geol. Soc. America Mem. 76, 376 pp., 35 pls. (Dec. 8).

1963. Type specimens of marine Mollusca described by P. P. Carpenter from the west coast of Mexico and Panama. Bull. Amer. Paleont., vol. 46, no. 211, pp. 289–408, pls. 58–70 (Oct. 22).

Palmer, R. H., and L. G. Hertlein. 1936. Marine Pleistocene mollusks from Oaxaca, Mexico. Bull. Southern California Acad. Sci., vol. 35, pp. 65–81, pls. 18, 19 (Sept. 10).

Parker, Pierre. 1949. Fossil and Recent species of the pelecypod genera *Chione* and *Securella* from the Pacific Coast. Jour. Paleont., vol. 23, no. 6, pp. 577–93, pls. 89–95 (Nov.).

Parker, R. H. 1964*a*. Zoogeography and ecology of some macro-invertebrates, particularly mollusks, in the Gulf of California and the continental slope off Mexico. Vidensk. Medd. fra. Dansk naturh. Foren. Bd. 126, 178 pp., pls. i–xv, 29 figs.

1964*b*. Zoogeography and ecology of macro-invertebrates of Gulf of California and continental slope of western Mexico. *In* Marine geology of the Gulf of California— a symposium memoir, no. 3, Amer. Assoc. Petroleum Geologists, pp. 331–76, 10 pls.

Pease, W. H. 1869. Remarks on marine gasteropodae, inhabiting the west coast of America; with descriptions of two new species. Amer. Jour. Conch., vol. 5, pp. 80–84, pl. 8 (Oct. 7).

Peck, J. I. 1893. Report on the pteropods and heteropods collected by the U.S. Fish Commission steamer *Albatross* during the voyage from Norfolk, Va., to San Francisco, Cal., 1887–1888. Proc. U.S. Nat. Mus., vol. 16, no. 943, pp. 451–66, pls. 53–55 (Sept. 30).

Pelseneer, P. 1887–88. Report on the Pteropoda collected by H.M.S. *Challenger*.... *Challenger* Reports, Zool., vols. 19, 23, pts. 58, 65, 66, 132 + 97 pp., 2 pls., 10 figs.

Peron, F., and C. A. Lesueur. 1810. Histoire de la famille des mollusques ptéropodes: caracters des dix genres que doivent la composer. Ann. mus. hist. nat., Paris, vol. 15, no. 85, pp. 57–69.

Perrier, E., and A. T. de Rochebrune. 1894. Sur un *Octopus* nouveaux [*O. digueti*] de la basse Californie, habitant les coquilles des mollusques bivalves. Comptes Rendus, Acad. Sci., Paris, vol. 118, pp. 770–73.

Perry, G. 1810–11. Arcana; or the museums of natural history.... 84 pls.

1811. Conchology, or the natural history of shells. London, pp. 1–4, pls. 1–61 and expl. [*See* A. T. Hopwood, 1946, Proc. Malac. Soc. London, vol. 26, pp. 152–53.]

Petit, R. E. 1967. Notes on Cancellariidae. Tulane Studies in Geology, vol. 5, no. 4, pp. 217–19, 1 fig. (Dec. 29).

1970. Notes on Cancellariidae—II. Tulane Studies in Geology and Paleontology, vol. 8, no. 2, pp. 83–88, 1 pl. (June 17).

Petit de la Saussaye, S. 1842. Description de deux auricules nouvelles (section des Conovules). Rev. Zool. (Soc. Cuvierienne), vol. 5, pp. 105–6 (Apr.).

1843. [... descriptions of new species of shells belonging to the genus *Auricula*.] Proc. Zool. Soc. London, vol. 10, for 1842, pp. 201–2 (Feb.).

1844. *Cancellaria cumingiana*, n.sp. Mag. de Zool., ser. 2, vol. 6, pl. 112.

1850. Observation sur la *Nerita scabricosta*, Lamar[c]k. Jour. de Conchyl., vol. 1, pp. 410–11, pl. 11, figs. 1, 2.

1852. Description de coquilles nouvelles. Jour. de Conchyl., vol. 3, no. 1, pp. 51–59, pl. 2 (Mar. 1).

Pfeiffer, L. 1854–82. Novitates Conchologicae. Abt. 1, Mollusca Extramarina. 5 vols. (1854–79). Abt. 2, Mollusca marina, by G. [Wilhelm] Dunker, *q.v.*, 2 vols. with 7 supplements (1858–82). Cassel. [For a complete collation, *see* R. I. Johnson, Jour. Soc. Biblio. Nat. Hist., vol. 5, pt. 3, pp. 236–39, Oct. 1969.]

1857. Descriptions of thirty-one new species of land shells in the collection of Hugh Cuming, Esq. Proc. Zool. Soc. London, for 1857, pp. 107–13 (Aug. 15).

Philippi, R. A. 1836. Enumeratio Molluscorum Siciliae. . . . Vol. 1, Berolini (Halis Saxorum). Pp. i–xiv, 1–268, pls. 1–12.

[1842]–1855. Die Kreiselschnecken oder Trochoideen . . . *In* Systematisches Conchylien-Cabinet von Martini und Chemnitz. Nuremberg, vol. 1, pts. 1–6 (pt. 1, *Turbo*, 98 pp., 20 pls., 1842–52; pts. 2–3, *Trochus*, 372 pp., 49 pls., 1846–55; pt. 4, *Phasianella* . . . , 52 pp., 6 pls., 1853; pt. 5, *Delphinula* . . . , 57 pp., 8 pls., 1852–53; pt. 6, *Adeorbis* . . . , 14 pp., 1 pl., 1853). [Dating and pagination from Sherborn and England, 1937, Jour. Soc. Biblio. Nat. Hist., vol. 1, p. 91.]

[1842] 1845–51. Abbildungen und Beschreibungen neuer oder wenig gekannter Conchylien. Kassel, 3 vols., 24 parts, illus. [For a collation, *see* Woodward, 1913, in British Museum Catalogue, vol. 4, p. 1565.]

1845. Diagnosen eininger Conchylien. Archiv. Naturg., yr. 11, vol. 1, pp. 50–71.

1846a. Diagnosen einiger neuen Conchylien-Arten. Zeit. Malak., yr. 3, pt. 1, pp. 19–24 (Feb.).

1846b. Descriptions of a new species of *Trochus*, and of eighteen new species of *Littorina*, in the collection of H. Cuming, Esq. Proc. Zool. Soc. London, for 1845, pp. 138–43 (Feb.).

1846c. Diagnosen neuer Conchylien-Arten. Zeits. Malak., yr. 3, pp. 49–55 (Apr.).

1847–48. Testaceorum novorum centuria. *Ibid.*, yr. 4 (for 1847), pp. 71–77 (May); pp. 84–96 (June); pp. 113–27 (Aug.); yr. 5 (for 1848), pp. 13–16 (Jan.), pp. 17–27 (Feb.)

1847. Beschreibung zweier neuer Conchylien-geschlechter (*Dibaphus* und *Amphichaena*). Archiv Naturg., vol. 13, pt. 1, pp. 61–66.

1849a. Centuria altera testaceorum novorum. Zeit. Malak., yr. 5 (for 1848), no. 7, pp. 99–112 (Jan. 30, 1849); no. 8, pp. 123–28 (Mar. 10); no. 9, pp. 129–44; no. 10, pp. 145–50 (Mar.).

1849b. Centuria tertia testaceorum novorum. *Ibid.*, yr. 5 (for 1848), no. 10, pp. 151–60, no. 11, pp. 161–76 (Mar. 1849); no. 12, pp. 186–92 (Apr.); yr. 6 (for 1849), no. 1, pp. 17–26 (May); no. 3, pp. 33–35 (July).

1849–51. Centuria quarta testaceorum novorum. *Ibid.*, yr. 6 (1849), no. 1, pp. 27–32 (May 1849); yr. 8 (1851), no. 2, pp. 29–32 (June 1851); no. 3, pp. 39–48 (July); no. 4, pp. 49–64 (July 15); no. 5, pp. 65–74 (July 1851).

1849–53. Die Gattung *Natica* und *Amaura*. *In* Systematisches Conchylien-Cabinet von Martini und Chemnitz, vol. 2, div. 1, pp. 1–18, pls. 1–6 (1849); pp. 19–26, pls. 7–12 (1850), pls. A, 13–18 (1851), pp. 27–64 (1852); pp. 65–120, pl. 19 (1851); pp. 121–64 (1853). [Dating from Sherborn and England, 1937, Jour. Soc. Biblio. Nat. Hist., vol. 1, p. 91.]

1850. Diagnosen mehrerer neuer Trochus-Arten. Zeit. Malak., yr. 6 (for 1849), no. 10, pp. 146–60; no. 11, pp. 168–72 (Mar. 1850); no. 12, pp. 187–92; yr. 7 (for 1850), no. 1, p. 16 (Apr.).

1851. Centuria quinta testaceorum novorum. *Ibid.*, yr. 8 (1851), no. 5, pp. 74–80 July); no. 6, pp. 81–96 (Aug.); no. 8, pp. 123–26 (?Oct.).

1853. Descriptiones naticarum quarundam novarum ex collectione Cumingiana. Proc. Zool. Soc. London, for 1851, pp. 233–34 (July 26).

1860. Reise durch die Wueste Atacama, 192 pp., 27 pls., 1 map. Holle.

Pilsbry, H. A. 1888–98. *In* G. W. Tryon, Jr., and H. A. Pilsbry, Manual of conchology. Philadelphia. [Pilsbry's contribution to this work began with vol. 10 (pt. 2, p. 161) and continued through vol. 17.]

1893. Notes on the Acanthochitidae with descriptions of new American species. Nautilus, vol. 7, no. 3, pp. 31–32 (July).

1896. Descriptions of new species of mollusks. Proc. Acad. Nat. Sci. Philadelphia, vol. 48, pp. 21–24, 1 fig. (Feb. 25).

1910. A new species of *Marinula* from near the head of the Gulf of California. *Ibid.*, vol. 62, p. 148, 1 fig. (May 23).

1920. Some Auriculidae and Planorbidae from Panama. Nautilus, vol. 33, no. 3, pp. 76–79, text figs. (Jan. 22).

1931a. Central American *Pachychilus* and *Polymesoda*. *Ibid.*, vol. 44, no. 3, pp. 84–85, 1 pl. (Jan. 27).

1931b. *Typhis lowei*, n.sp. *Ibid.*, vol. 45, no. 2, p. 72 (Oct. 14).

1931c. The Miocene and Recent Mollusca of Panama Bay. Proc. Acad. Nat. Sci. Philadelphia, vol. 83, pp. 427–40, 5 figs., pl. 41 (Nov. 13).

1932a. Notes on a Panamic corbulid clam. Nautilus, vol. 45, no. 3, p. 105 (Jan. 9).

1932b. A misidentified Lower California snail. *Ibid.*, vol. 45, no. 4, pp. 124–25, pl. 11, fig. 11 (Apr. 9).

1936. *Sanguinolaria bertini* Pilsbry and Lowe. *Ibid.*, vol. 49, no. 4, p. 140 (May 1).

1949. Drillia roseobasis (= *Pleurotoma testudinis* P. & V.) and Pleurotoma albicostata (Sowerby). *Ibid.*, vol. 62, no. 3, pp. 103–4 (Mar. 18).

1956. A gastropod domiciliary in sea urchin spines. *Ibid.*, vol. 69, no. 4, pp. 109–10, 3 figs. (Apr.).

Pilsbry, H. A., and H. N. Lowe. 1932a. West Mexican and Central American mollusks collected by H. N. Lowe, 1929–31. Proc. Acad. Nat. Sci. Philadelphia, vol. 84, pp. 33–144, 7 figs., pls. 1–17, 2 photographs (May 21).

1932b. New West American species of *Bulimulus* and *Nassa*. Nautilus, vol. 46, no. 2, pp. 49–52 [also vol. 47, pl. 8, figs. 4–5] (Oct. 22).

1934. West American Chamidae, *Periploma* and *Glycymeris*. *Ibid.*, vol. 47, pp. 81–86, pl. 8 (Jan. 26).

Pilsbry, H. A., and A. A. Olsson. 1935. New mollusks from the Panamic Province. *Ibid.*, vol. 48, pp. 116–21, pl. 6 (Apr. 24) ; vol. 49, pp. 16–19, pl. 1 (July 22).

1941. A Pliocene fauna from western Ecuador. Proc. Acad. Nat. Sci. Philadelphia, vol. 93, pp. 1–79, pls. 1–19, 1 photograph (Sept. 9).

1943. New marine mollusks from the west coast. Nautilus, vol. 56, pp. 78–81, pl. 8 (Feb. 15).

1944. A West American *Julia*. *Ibid.*, vol. 57, no. 3, pp. 86–87, 2 figs. (Feb. 9).

1945–52. Vitrinellidae and similar gastropods of the Panamic Province. Pts. I–II. I. Acad. Nat. Sci. Philadelphia, vol. 97, pp. 249–78, pls. 22–30 (Dec. 27, 1945) ; II. *Ibid.*, vol. 104, pp. 35–88, pls. 2–13 (Sept. 10, 1952).

1946. Another Pacific species of *Episcynia*. Nautilus, vol. 60, no. 1, pp. 11–12, pl. 1, figs. 6, 7, 8 (July).

Pilsbry, H. A., and B. Sharp. 1897–98. Class Scaphopoda. *In* G. W. Tryon, Jr., and H. A. Pilsbry, Manual of conchology. Philadelphia, pp. i–xxxii, 1–280, pls. 1–39.

Pilsbry, H. A., and E. G. Vanatta. 1902. Papers from the Hopkins Stanford Galapagos Expedition, 1898–99, no. 13, Marine Mollusca. Proc. Washington Acad. Sci., vol. 4, pp. 549–60, pl. 35 (Sept. 30).

Pilsbry, H. A., and James Zetek. 1931. A Panamic *Cyrenoida*. Nautilus, vol. 45, no. 2, p. 69, 1 fig. (Oct. 14).

Pohlo, R. H. 1963. Morphology and mode of burrowing in *Siliqua patula* and *Solen rosaceus* (Mollusca: Bivalvia). Veliger, vol. 6, no. 2, pp. 98–104, 6 text figs. (Oct. 1).

Ponder, W. F. 1970. Some aspects of the morphology of four species of the neogastropod family Marginellidae, with a discussion on the evolution of the toxoglossan poison gland. Jour. Malac. Soc. Australia, vol. 2, pt. 1, pp. 55–81, 4 figs. (Aug.).

Powell, A. W. B. 1942. The New Zealand Recent and fossil Mollusca of the family Turridae, with general notes on turrid nomenclature and systematics. Bull. Auckland Inst. and Mus., no. 2, 188 pp., 14 pls.

1966. The molluscan families Speightiidae and Turridae. *Ibid.*, no. 5, 184 pp., 23 pls., 179 text figs. (Nov. 1).

Powys, W. L. 1835. Undescribed shells contained in Mr. Cuming's collection . . . accompanied by characters by Mr. G. B. Sowerby and Mr. W. Lytellton Powys. Proc. Zool. Soc. London, pp. 93–96 (Sept. 25).

Prime, Temple. 1860. Synonymy of the Cyclades, a family of acephalous Mollusca. Pt. 1. Proc. Acad. Nat. Sci. Philadelphia, vol. 12, pp. 267–301 (June).

1861a. Note sur quelques espèces peu connus des genres *Batissa, Cyrena, Corbicula,* et *Sphaerium.* Jour. de Conchyl., vol. 9, no. 1, pp. 38–43, pl. 2 (Jan. 9).

1861b. Diagnoses d'espèces nouvelles. *Ibid.,* vol. 9, no. 4, pp. 354–56 (Oct. 18).

1862. Descriptions d'espèces nouvelles des genres *Glauconome, Cyrena, Batissa,* et *Corbicula. Ibid.,* vol. 10, no. 4, pp. 383–90, pl. 4 (Nov. 3).

1865. Monograph of American Corbiculadae (recent and fossil). Smithsonian Inst. Misc. Coll., no. 145, vol. 7, art. 5, pp. l–xi + 1–80, 86 text figs. (Dec.).

1867. Notes on the classification of the Corbiculadae, with figures. Ann. Lyceum Nat. Hist., New York, vol. 8, pp. 213–37 (Apr.).

1869. Catalogue of the Recent species of the family Corbiculadae. Amer. Jour. Conch., vol. 5, pp. 127–87 (Oct.).

1870. Notes on species of the family Corbiculadae, with figures. Ann. Lyceum Nat. Hist. New York, vol. 9, pp. 298–301, figs. 70–72 (Mar.).

Pruvot-Fol, A. 1955. Les Arminiadae (Pleurophyllidiadae ou Diphyllidiadae des anciens auteurs). Bull. Mus. hist. nat., Paris, ser. 2, vol. 27, pp. 462–68, 10 figs.

Quoy, J. R. C., and J. P. Gaimard. 1824–26. Voyage autour du monde ... pendant ... 1817–20 ... Uranie et Physicienne ... Zoologie. Paris. Pp. iv + 712, with atlas of 96 pls.

1827. Observations zoologiques faites à bord de "l'Astrolabe" en Mai 1826, dane le détroit de Gibraltar. ... Ann. Sci. Nat., vol. 10, pp. 225–39, pls. 7–8 + atlas.

1832–35. Voyage de découvertes de l'Astrolabe, exécuté par ordre du Roi, pendant les années 1826–29, sous le commandement de M. J. Dumont d'Urville. Paris, Zoologie, Mollusca, vol. 2, pp. 1–320 (1832), 321–686 (1833); vol. 3, pp. 1–366 (1834), 367–954 (1835), atlas, 107 pls. [For dates of publication, *see* C. D. Sherborn and B. B. Woodward, 1901, Ann. Mag. Nat. Hist., ser. 7, vol. 8, p. 333.]

Radwin, G. E. 1968. The systematic position of Glyptaesopus. Nautilus, vol. 82, no. 1, pp. 18–19, figs. A–E (July).

Radwin, G. E., and Anthony D'Attilio. 1970. A new species of *Muricopsis* from west Mexico. Veliger, vol. 12, no. 3, pp. 351–56, pl. 52, 4 text figs. (Jan. 1).

Rang, P. C. S. A. L. 1828. Notice sur quelques mollusques nouveaux appartenent au genre Cléodore, et établissement et monographie du sous-genre Créseis. Ann. Sci. Nat., vol. 13, pp. 302–19, pls. 17–18.

1829. Manuel de l'histoire naturelle des mollusques et leurs coquilles, ayant pour base de classification celle de M. le Baron Cuvier. Paris: vi + 390 pp., 8 pls. (May).

Rathke, M. H. 1833. Zoologischer Atlas ... (pt. 5) [continuation of the work begun by Eschscholtz, 1829–31]. Pp. i–viii, 1–28, pls. 21–25. [For discussion of this work, *see* J. H. McLean, 1970, Veliger, vol. 13, no. 1, p. 112.]

Raymond, W. J. 1904. A new *Dentalium* from California. Nautilus, vol. 17, no. 11, pp. 123–24 (Mar.).

Récluz, C. A. 1841. Description de quelques nouvelles espèces de Nérites vivantes. Rev. Zool., Soc. Cuvierienne, vol. 4, pp. 102–9.

1842. Description de deux coquilles nouvelles. *Ibid.,* vol. 5, pp. 305–37 (Oct.).

1844. Description of new species of *Navicella, Neritina, Nerita,* and *Natica,* in the cabinet of Hugh Cuming, Esq. Proc. Zool. Soc. London, for 1843, pp. 197–214 (June).

1850a. Notice sur le genre Nérita et sur le S.-G. Neritina, avec le catalogue synonymique des néritines. Jour. de Conchyl., vol. 1, pp. 131–64 [suite du mémoire sur le genre Nérite], pp. 277–88.

1850b. Description de natices nouvelles. *Ibid.,* vol. 1, pp. 379–402, pls. 12–14 (Dec. 15).

1853a. Description de coquilles nouvelles. *Ibid.,* vol. 4, pp. 49–54, pl. 2 (Feb. 15).

1853b. Description de coquilles nouvelles (G. *Pecten, Tellina* et *Natica*). *Ibid.,* vol. 4, pp. 152–56, pls. 5–6 (May 1).

Redfield, J. H. 1870. Catalogue of the known species, Recent and fossil, of the family Marginellidae: Amer. Jour. Conch., vol. 6, pt. 2, Appendix, pp. 215–70 (Oct. 6).

Reeve, Lovell. 1841–42. Conchologia systematica, or complete system of conchology. London, vol. 1, pp. i–vi, 1–195, pls. 1–129, 1 table (1841); vol. 2, pp. 1–337 + 1, pls. 130–300 (1842).

———1842. Descriptions of new species of shells, principally from the collection of Hugh Cuming, Esq. Proc. Zool. Soc. London, pp. 49–50 (Nov.).

———1843. Descriptions by Mr. Lovell Reeve of new species of shells figured in the "Conchologia systematica." *Ibid.*, for 1842, pp. 197–200 (Feb.).

———1843–78. Conchologia iconica: or, illustrations of the shells of molluscous animals. London, vols. 1–20, with suppl. to Conus. Continued by G. B. Sowerby, II, beginning with the genus *Pyramidella* in vol. 15, Oct. 1865.

———1844. Description of new species of *Ranella*. Proc. Zool. Soc. London, pp. 136–40 (Dec.).

———1846. Descriptions of fifty-four new species of *Mangelia* from the collection of Hugh Cuming, Esq. *Ibid.*, pt. 14, pp. 59–65 (Aug.).

———1846–49. Initiamenta conchologica or elements of conchology. London, pp. 1–160, pls. A–I, K–N, 1–37. [Pls. 36 and 37 erroneously numbered "26" and "27."] [For dates of publication of this and the next entry, *see* A. Reynell, 1916, Proc. Malac. Soc. London, vol. 12, pp. 44–46; T. Iredale, 1922, *ibid.*, vol. 15, pp. 90–91; and A. E. Salisbury, 1945, Jour. Conch., vol. 22, pp. 155–56.]

———1859–60. Elements of Conchology. London, vol. 1, pp. i–viii, 1–260, pls. A–H, 1–21; vol. 2, pp. i–vi, 1–203, pls. I, K–Q, 22–46. [For dates of publication, *see* references in note to Reeve, 1846–49, above.]

———1860. A commentary on M. Deshayes's Revision of the genus *Terebra*. Proc. Zool. Soc. London, pp. 448–50 (issued between Aug. 1860 and Mar. 1861).

Rehder, H. A. 1943*a*. The molluscan genus *Trochita* Schumacher, with a note on *Bicatillus* Swainson. Proc. Biol. Soc. Washington, vol. 56, pp. 41–46 (June 16).

———1943*b*. New marine mollusks from the Antillean region. Proc. U.S. Nat. Mus., vol. 93, no. 3161, pp. 187–203, pls. 19, 20.

———1967. A new genus and two new species in the families Volutidae and Turbinellidae (Mollusca: Gastropoda) from the western Pacific. Pacific Science, vol. 21, no. 2, pp. 182–87, 11 text figs. (Apr.).

Reinhart, P. W. 1935. Classification of the pelecypod family Arcidae, Bull. Mus. roy. d'hist. nat. de Belgique, vol. 11, no. 13, 68 pp., 4 pls. (Aug.).

———1939. The holotype of *Barbatia* (*Acar*) *gradata* (Broderip and Sowerby). Trans. San Diego Soc. Nat. Hist., vol. 9, pp. 39–46, pl. 3 (Aug. 31).

———1943. Mesozoic and Cenozoic Arcidae from the Pacific slope of North America. Spec. Paper Geol. Soc. Amer., no. 47, pp. i–xi, 1–117, pls. 1–15 (June 16).

Richards, A. F., and B. H. Brattstrom. 1959. Bibliography, cartography, discovery, and exploration of the Islas Revillagigedo. Proc. California Acad. Sci., ser. 4, vol. 29, no. 9, pp. 315–60, 4 figs. (Jan. 5).

Roberts, S. R. 1885. Monograph of the family Cypraeidae. *In* G. W. Tryon, Jr., Manual of conchology. Philadelphia, vol. 7, pp. 153–240, pls. 1–23.

Robertson, Robert. 1958. The family Phasianellidae in the western Atlantic. Johnsonia, vol. 3, no. 37, pp. 245–83, pls. 136–48 (May 8).

Robertson, Robert, and A. S. Merrill. 1963. Abnormal dextral hyperstrophy of post-larval *Heliacus* (Gastropoda: Architeconicidae). Veliger, vol. 6, no. 2, pp. 76–79, pls. 13, 14 (Oct. 1).

Robson, G. C. 1929–32. A monograph of the Recent Cephalopoda, based on the collections in the British Museum (Natural History). Part I. Octopodinae. London: British Museum Pub., pp. 1–236, pls. i–vii (1929); Part II: Octopoda, excluding Octopodinae. *Ibid.*, 359 pp., 6 pls., text figs. (Jan. 1932).

Rochebrune, A.-T. de. 1881. Sur un type nouveau de la famille Cyclostomaceae. Bull. Soc. Philomath. [Paris], ser. 7, vol. 5, pp. 108–15.

———1895. Diagnoses de mollusques nouveaux, provenant du voyage de M. Diguet en Basse-Californie. Bull. Mus. hist. nat., Paris, vol. 1, pp. 239–43.

Röding, Peter F. 1798. Museum Boltenianum . . . : pars secunda continens Conchylia. . . . Hamburg (J. C. Trappii), pp. i–vii + 109 (Dec.). [For an index to species, see Dall, 1915, Smithsonian Inst. Misc. Coll. no. 2360, 64 pp.; for evidence on publication date, see H. Rehder, 1944, Nautilus, vol. 59, pp. 50–52; for list of type species of genera, see Winckworth, 1945, Proc. Malac. Soc. London, vol. 26, pp. 136–48.]

Rogers, M. E. 1962. A new Gulf of California *Periploma*. Bull. Southern California Acad. Sci., vol. 61, pt. 4, pp. 229–31, figs. 1, 2 (Oct.–Dec.).

Römer, Edouard. 1862. Monographie der Molluskengattung *Dosinia*, Scopoli (*Artemis*, Poli). Kassel, vii + 87 pp., 16 pls.

——— 1864–69. Monographie der Molluskengattung *Venus*, Linné. Novitates conchologicae; Abbildung und Beschreibung neuer Conchylien, Suppl. 3. Kassel, vol. 1, subgenus *Cytherea* Lamarck, pp. 1–12 (June 1864), 13–24 (Jan. 1865), 25–32 (June 1865), 33–42 (Apr. 1866), 43–66 (Sept. 1866), 67–68 (Feb. 1867), 79–102 (Sept. 1867), 103–26 (Dec. 1867), 127–46 (Sept. 1868), 147–72 (Nov. 1868), 173–90 (Feb. 1869), 191–98 (June 1869), 199–206 (July 1869), 207–17 (Oct. 1869), pls. 1–59.

Rosewater, Joseph. 1970. The family Littorinidae in the Indo-Pacific. Part I. The subfamily Littorininae. Indo-Pacific Mollusca, vol. 2, no. 11, pp. 417–506, pls. 325–86 (1 col. pl.) (Nov. 30).

Rost, Helen. 1955. A report on the family Arcidae. Allan Hancock Pacific Expeditions, vol. 20, no. 2, pp. 177–249, pls. 11–16, text figs. 79–95 (Nov. 10).

Roth, B., and E. Coan. 1968. Further observations on the West American Marginellidae with the descriptions of two new species. Veliger, vol. 11, no. 1, pp. 62–69, pl. 7, 2 text figs. (July 1).

Rous, Sloman. 1908. *Cancellaria obtusa* Deshayes. Nautilus, vol. 21, no. 9, p. 105.

Rudman, W. B. 1969. Observations on *Pervicacia tristis* (Deshayes, 1859) and a comparison with other toxoglossan gastropods. Veliger, vol. 12, no. 1, pp. 53–64, 5 text figs. (July 1).

Sachet, M. H. 1962. Monographie physique et biologique de l'île de Clipperton. Ann. Inst. océanogr. Monaco, vol. 40, no. 1, pp. 1–107.

Salisbury, A. E. 1934. On the nomenclature of Tellinidae, with descriptions of new species and some remarks on distribution. Proc. Malac. Soc. London, vol. 21, pt. 2, pp. 74–91, pls. 9–14 (July 14).

Salis-Marschlins, C. U. von. 1793. Reise Neapel . . . Zurich, 8vo. (ed. 2, translated from the German by A. Aufrere, "Travels through various provinces of the Kingdom of Naples in 1789. . . . ." Pp. 527, 8 pls., with Appendix containing a catalogue of marine Mollusca, London, 1795).

Salvat, B., and J. P. Ehrhardt. 1970. Mollusques de l'île Clipperton. Bull. Mus. hist. nat., Paris, ser. 2, vol. 42, no. 1, pp. 223–31.

Saunders, P. R. 1959. Some observations on the feeding habits of *Conus californicus* Hinds. Veliger, vol. 1, no. 3, pp. 13–14 (Jan. 1).

Say, Thomas. 1822. An account of some of the marine shells of the United States. Jour. Acad. Nat. Sci. Philadelphia, vol. 2, pp. 221–48, 257–76.

——— 1830–34. American Conchology, or Descriptions of the Shells of North America. New Harmony, Indiana: Parts 1–7, 258 pp., 68 pls.

Schenck, H. G. 1926. Cassididae of Western America. Univ. California Pub. Bull. Dept. Geol. Sci., vol. 16, pp. 69–98, 1 fig., pls. 12–15 (May 4).

——— 1939. Revised nomenclature for some nuculid pelecypods. Jour. Paleont., vol. 13, no. 1, pp. 21–41, pls. 5–8 (Jan.).

Schilder, F. A. 1930. Remarks on type specimens of some Recent Cypraeidae. Proc. Malac. Soc. London, vol. 19, pp. 49–58 (Mar. 13).

——— 1931a. Beiträge zur Kenntnis der Cypraeacea (Moll. Gastr.)—IV. Zool. Anzeiger, vol. 96, no. 3–4, pp. 65–72.

——— 1931b. Zwei neue *Trivia* (Cypraeacea: Eratoidae). Arch. Mollusk., vol. 63, no. 1, pp. 42–44, pl. 3.

1932. The living species of Amphiperatinae. Proc. Malac. Soc. London, vol. 20, pp. 46–64, pls. 3–5 (Mar.).

1933a. Monograph of the subfamily Eratoinae. Proc. Malac. Soc. London, vol. 20, pp. 244–83, 85 figs. (July).

1933b. Beiträge zur Kenntnis der Cypraeacea (Moll. Gastr.)—VI. Zool. Anzeiger, vol. 101, pp. 180–93, 4 tables.

1965. The geographical distribution of cowries. Veliger, vol. 7, no. 3, pp. 171–83, 2 text figs. (Jan. 1).

1966. The higher taxa of cowries and their allies. Ibid., vol. 9, no. 1, pp. 31–35 (July 1).

1969. Zoogeographical studies on living cowries. Ibid., vol. 11, no. 4, pp. 367–77, 1 fig. (Apr. 1).

Schilder, F. A., and M. Schilder. 1938–39. Prodrome of a monograph on living Cypraeidae. Proc. Malac. Soc. London, vol. 23, pp. 119–80 (Nov. 15, 1938), 181–231, 9 maps (Mar. 15, 1939).

Schilder, F. A., and J. R. le B. Tomlin, 1931. Rediscovery of a rare cowry. Ibid., vol. 19, no. 6, pp. 274–75, figs.

Schmidt, W. J. 1959. Bemerkungen zur Schalenstruktur von Neopilina galatheae. Galathea Reports, vol. 3, pp. 73–78, pls. 1–2.

Schumacher, C. F. 1817. Essai d'un nouveau système des habitations des vers testacés. Copenhagen: pp. iv + 287, pls. 1–22 (post–Mar. 1).

Schuster, O. 1952. Olivella columellaris Sowerby, un caracol de la playa de Los Blancos. Comm. Inst. Tropical Investig. cient. Univ. El Salvador, vol. 4, pp. 10–13.

Schwengel, Jeanne S. 1938. Zoological results of the George Vanderbilt South Pacific Expedition, 1937. Part I. Galapagos Mollusca. Proc. Acad. Nat. Sci. Philadelphia, vol. 90, pp. 1–3, 3 text figs. (May 13).

1955. New Conus from Costa Rica. Nautilus, vol. 69, no. 1, pp. 13–15, pl. 2 (Aug. 1).

Scopoli, G. A. 1777. Introductio ad historiam naturalem sistens genera lapidum, plantarum, et animalium. Prague, x + 506 + 34 pp.

Seilacher, A. 1959. Schnecken im Brandungssand. Natur und Volk, 89, Heft 11, pp. 359–66 (Nov. 1).

Semper, C. G. 1867–1916. Reisen im Archipel der Philippinen. Zweiter Theil. Wissenschaftlichen Resultate. Wiesbaden. 10 vols. Vols. 2, 7, 9, Nudibranchiata, etc., by R. Bergh, q.v.; vols. 3, 8, 10, Landmollusken), vol. 3 by Semper (1870–95); vol. 8 by W. Kobelt (1898–1904), vol. 10 by W. Kobelt and G. Winter (1905–16). [For a complete collation, see R. I. Johnson, Jour. Soc. Biblio. Nat. Hist., vol. 5, pt. 2 pp. 144–47, Apr. 1969.]

Shasky, D. R. 1960. Deep water collecting off Guaymas, Mexico. Veliger, vol. 3, no. 1, pp. 22–23 (July 1).

1961a. Range extension for Cypraea (Luria) isabellamexicana Stearns, 1893. Ibid. vol. 3, no. 4, pp. 111–12 (Apr.).

1961b. New deep water mollusks from the Gulf of California. Ibid., vol. 4, no. 1, pp. 18–21, pl. 4, figs. 1–10 (July 1).

1961c. Notes on rare and little known Panamic mollusks. Ibid., vol. 4, no. 1, pp. 22–24, pl. 4, figs. 11–16 (July 1).

1968. Observations on Rosenia nidorum (Pilsbry) and Arene socorroensis (Strong). Amer. Malac. Union Ann. Repts. for 1967, p. 74 (abstr.) (Mar. 20).

1970. New gastropod taxa from tropical western America. Veliger, vol. 13, no. 2, pp. 188–95, 1 pl., 2 text figs. (Oct. 1).

1971. Ten new species of tropical eastern Pacific Turridae. Ibid., vol. 14, no. 1, pp. 67–72, 1 pl. (July 1).

Shasky, D. R., and G. B. Campbell. 1964. New and otherwise interesting species of mollusks from Guaymas, Sonora, Mexico. Ibid., vol. 7, no. 2, pp. 114–20, pls. 21, 22, 1 text fig., 1 map (Oct. 1).

Sheldon, P., and C. J. Maury. 1922. See Maury.

Sherborn, C. D. 1902–33. Index Animalium. London. [Section 1 (1902) lists generic and specific names, with references, from 1758 to 1800; Section 2, parts 1–33 (1922–33), lists names from 1800 to 1850.]

Skoglund, Carol. 1970. An annotated bibliography of references to marine Mollusca from the northern state of Sonora, Mexico. Veliger, vol. 12, no. 4, pp. 427–32 (Apr. 1).

Smith, A. G. 1946. Notes on living California crepidulas. Minutes Conch. Club Southern California, no. 60, pp. 1–8 (May).

— 1961a. Notes on the habitat of *Berthelinia* sp. nov. from the vicinity of La Paz, Baja California, Mexico. Veliger, vol. 3, no. 3, pp. 81–82 (Jan. 1).

— 1961b. Four species of chitons from the Panamic province. Proc. California Acad. Sci., ser. 4, vol. 30, no. 4, pp. 81–90, pls. 8–9 (Aug. 31).

— 1963. A revised list of chitons from Guadelupe Island, Mexico (Mollusca: Polyplacophora). Veliger, vol. 5, no. 4, pp. 147–49 (Apr. 1).

Smith, A. G., and Mackenzie Gordon. 1948. The marine mollusks and brachiopods of Monterey Bay, California, and vicinity. Proc. California Acad. Sci., ser. 4, vol. 26, no. 8, pp. 147–245, pls. 3–4, 4 text figs. (Dec. 15).

Smith, E. A. 1873. Remarks on a few species belonging to the family Terebridae and descriptions of several new forms in the collection of the British Museum. Ann. Mag. Nat. Hist., ser. 4, vol. 11, pp. 262–71 (Apr.).

— 1877. [Account of the zoological collections made during the visit of H.M.S. *Petrel* to the Galapagos Islands.] Mollusca. Proc. Zool. Soc. London, for 1877, pp. 69–73, 91–93 (June).

— 1880. Descriptions of twelve new species of shells. *Ibid.*, pp. 478–85, pl. 48 (Oct.).

— 1881. Account of the Mollusca and Molluscoidea collected during a survey of H.M.S. *Alert*, ... *Ibid.*, pp. 22–44, pls. 3–5 (June).

— 1882. Diagnoses of new species of Pleurotomidae in the British Museum. Ann. Mag. Nat. Hist., ser. 5, vol. 10, pp. 216–18 (Sept.).

— 1885. Report ... Scientific Results ... H.M.S. *Challenger*, Pt. V. Zoology, vol. 13, no. 35, Lamellibranchiata. Pp. 1–341, 25 pls.

— 1888a. Diagnoses of new species of Pleurotomidae in the British Museum. Ann. Mag. Nat. Hist., ser. 6, vol. 2, no. 10, pp. 300–317 (Oct.).

— 1888b. Report on the Heteropoda collected by H.M.S. *Challenger* during the years 1873–76 ... Zoology, vol. 23, pt. 5, no. 72, pp. 1–51, 5 figs.

— 1892. Descriptions of new species from Mauritius and California. Ann. Mag. Nat. Hist., ser. 6, vol. 9, pp. 255–56, 2 text figs.

— 1903. Note on *Macron trochlea*. Jour. Conch., vol. 10, no. 12, p. 351 (Oct.).

Smith, Maxwell. 1939. An illustrated catalog of the Recent species of the rock shells. Tropical Photographic Laboratory, Lantana, Florida. Pp. ix + 83, 21 pls.

— 1940. World-wide sea shells. Tropical Photographic Laboratory, Lantana, Florida. Pp. xviii + 139, 1629 figs., 5 portraits, 1 map. Pp. xi–xviii by J. L. Baily.

— 1944a. Panamic marine shells. Synonymy, nomenclature, range and illustrations. Tropical Photographic Laboratory, Winter Park, Florida. Pp. xiii + 127, 912 figs.

— 1944b. Shells dredged in Panama Bay. Nautilus, vol. 58, pp. 27–28, pl. 1 (Aug. 17).

— 1947. A Recent *Perplicaria* and other new Panamic marine shells. *Ibid.*, vol. 61, pp. 53–56, pl. 2 (Dec. 18).

— 1948. Triton, helmet and harp shells. Synonymy, nomenclature, range and illustrations. Tropical Photographic Laboratory, Winter Park, Florida. Pp. v + 57, 16 pls.

— 1950. New Mexican and Panamic shells. Nautilus, vol. 64, no. 2, pp. 60–61, pl. 4 (Oct. 27).

Solander, D. C. 1786. *See* J. Lightfoot.

Solem, A. 1963. On the identities of *Trivia buttoni* and *Trivia galapagensis* Melvill, 1900 (Mollusca: Gastropoda). Veliger, vol. 6, no. 1, pp. 20–22, pl. 4 (July 1).

Soot-Ryen, Tron. 1932. Pelecypoda from Floreana (Sancta Maria) Galapagos Islands. Mededelser fra det Zoologiske Museum, Oslo, no. 27, pp. 313–24, 1 pl.

— 1955. A report on the family Mytilidae (Pelecypoda). Allan Hancock Pacific Expeditions, vol. 20, no. 1 (Univ. Southern California Press, Los Angeles), 175 pp., 10 pls., 78 text figs. (Nov. 10).

1957. On a small collection of pelecypods from Peru to Panama. Lund Universitets Arsskrift, N.F., avd. 2, vol. 53, no. 10, 12 pp., 2 figs. (July 4).

Sorensen, A. 1942. Collecting in Mexico. Nautilus, vol. 55, no. 4, pp. 113–15.

1943. Traveling and collecting in Mexico. *Ibid.*, vol. 57, no. 1, pp. 1–5, pls. 1–4.

Souleyet, F. L. A. 1852. Mollusques. *In* Voyage autour du monde exécuté pendant ... 1836 et 1837 sur la corvette la Bonite commandée par M. Vaillant. Paris, Zoologie, vol. 2, pp. 1–664, atlas, pp. 2–8, pls. 1–45. [For dates of issue of this work, *see* C. D. Sherborn and B. B. Woodward, 1901, Ann. Mag. Nat. Hist., ser. 7, vol. 7, p. 391.]

Sowerby, G. B. (first of name).* 1821–34. The genera of Recent and fossil shells. London, vol. 1, pls. 1–126 and text (pages not numbered) (1821–25); vol. 2, pls. 127–262 and text (pages not numbered) (1825?–34). [Commenced by J. Sowerby. Plates by J. Sowerby and J. de C. Sowerby. For collation of this work, *see* C. D. Sherborn, 1894, Ann. Mag. Nat. Hist., ser. 6, vol. 13, pp. 370–71; E. R. Sykes, 1906, Proc. Malac. Soc. London, vol. 7, p. 194; and R. B. Newton, 1891. Systematic list of the F. E. Edwards collection of British Oligocene and Eocene Mollusca in the British Museum, London, pp. 321–22.]

1824. Descriptions, accompanied by figures, of several new species of shells. Zool. Jour., vol. 1, pp. 58–60, pl. 5 (Mar.).

1825. A catalogue of the shells contained in the collection of the late Earl of Tankerville. London, pp. i–vii, 1–92, app. pp. i–xxxiv, 9 col. pls.

1830. *In* W. Broderip, Species Conchyliorum, or concise original descriptions and observations ... on all of the species of Recent shells ... London: vol. 1, pt. 1. [*Cymba*: Broderip, 7 pp., 7 pls.; *Ancillaria, Ovulum, Pandora*: Sowerby, 23 pp., 6 pls.]

1832–33. Characters of new species of Mollusca and Conchifera, collected by Hugh Cuming. Proc. Zool. Soc. London, for 1832, pp. 25–33 (Apr. 21, 1832), 50–61 (June 5, 1832), 104–8 (July 31, 1832), 113–20 (Aug. 14, 1832), 173–79 (Jan. 14, 1833), 194–202 (Mar. 13, 1833). [Some of the species cited in these pages were described by W. J. Broderip.]

1833–34. Characters of new species of Mollusca and Conchifera. *Ibid.*, for 1833, pp. 6–8 (May 13), 16–22, 34–38 (May 17, 1833), 52–56 (May 24, 1833), 70–74 (Sept. 20, 1833), 82–85 (Sept. 8, 1833), 134–39 (Apr. 16, 1834). [Some of the species cited in these pages were described by W. J. Broderip.]

1834–35. Characters of new genera and species of Mollusca and Conchifera, collected by Mr. Cuming. *Ibid.*, for 1834, pp. 6–8 (May 14, 1834), 17–19, 21–22 (June 17, 1834), 46–47 (Sept. 26, 1834), 68–72 (Nov. 25, 1834), 87–89 (Oct. 25, 1834), 123–28 (Mar. 20, 1835).

1835. Characters of and observations on new genera and species of Mollusca and Conchifera collected by Mr. Cuming. *Ibid.*, pp. 4–7 (April 3), 21–23 (Apr. 16), 41–47 (June 1), 49–51 (July 17), 84–85, 93–96 (Sept. 25), 109–10 (Oct. 9). [Some of the species cited on pp. 93–96 were described by W. L. Powys.]

1839. Molluscous animals, and the shells: by John Edward Gray, continued by G. B. Sowerby. *In* F. W. Beechey, The zoology of Capt. Beechey's voyage ... to the Pacific and Behring's Straits in his Majesty's ship *Blossom* ... London, pp. x–xii, 103–55, pls. 33–34 [pp. 103–42 by J. E. Gray, pp. 143–55 by G. B. Sowerby I].

Sowerby, G. B., Jr. (second of name). 1832–41. The conchological illustrations. London, pts. 1–200, pls. 1–200. [The text of the catalogues of genera is separately paged. For notes and collation, *see* C. D. Sherborn, Proc. Malac. Soc. London, vol. 8, pt. 6, pp. 331–32; H. O. N. Shaw, *ibid.*, pp. 333–40 (1909); and A. Reynell, *ibid.*, vol. 9, pp. 212–13 (1910).]

* Concerning the publications on conchology by the Sowerbys I, II, and III, *see* A. E. Salisbury, 1945, Jour. Conch., vol. 22, pp. 153–54.

1840. Descriptions of some new chitons. Mag. Nat. Hist., n.s., vol. 4, pp. 287–94, Suppl. pl. 16 (June).

1841a. On some new species of the genus *Cardium*, chiefly from the collecion of H. Cuming, Esq. Proc. Zool. Soc. London, for 1840, pp. 105–11 (May).

1841b. Descriptions of some new species of *Murex*, principally from the collection of H. Cuming, Esq. *Ibid.*, for 1840, pp. 137–47 (July).

[1842] 1847–87. Thesaurus conchyliorum, or monographs of genera of shells, edited by G. B. Sowerby, Jr., completed by G. B. Sowerby [third of the name]. London, vols. 1–5.

1844. Descriptions of *Scalaria*. Proc. Zool. Soc. London, pp. 26–31 (July).

1846. Descriptions of new species of *Marginella*. *Ibid.*, vol. 14, pp. 95–97 (Nov.).

1859–60. Descriptions of new shells in the collection of Mr. Cuming. *Ibid.*, vol. 27, for 1859, pp. 428–29 (Oct. 1859–Feb. 1860).

1865–78. Conchologia iconica; or, illustrations of the shells of molluscous animals. London, vols. 15–20.

1870. Descriptions of forty-eight new species of shells. Proc. Zool. Soc. London, pp. 249–59, pls. 21–22 (Nov.).

1873. Descriptions of five new cones. *Ibid.*, pp. 145–46, pl. 15 (June).

1875. Descriptions of five new species of shells. *Ibid.* for 1874, pp. 598–600, 1 pl. (Apr.). [Reprinted, Jour. of Conch., vol. 1, pp. 78–80, 1875.]

Sowerby, J. 1812–46. Mineral conchology. London, vols. 1–7. [Vols. 5–7 are by J. de C. Sowerby. For collation of the work, *see* E. R. Sykes, 1906, Proc. Malac. Soc. London, vol. 7, pp. 191–94.]

Spengler, Lorenz. 1798. Over det toskallede slaegt tellinerne. Skr. Nat. Selsk., Copenhagen, vol. 4, no. 2, pp. 67–127, pl. 12.

1802. Beskrivelse over det toskallede Conchylie-slaegt *Mactra*. Skr. Nat. Selsk., Copenhagen, vol. 5, pt. 2, pp. 92–128.

Sphon, G. G. 1960. Range extensions of two Panamic gastropods. Veliger, vol. 3, no. 1, p. 31 (July 1).

1961. Notes on the Mitridae of the eastern Pacific, I. *Mitra fultoni* E. A. Smith. *Ibid.*, vol. 4, no. 1, pp. 32–36, pl. 7 (July 1).

1966. Material contained in the molluscan type collection of the Santa Barbara Museum of Natural History. *Ibid.*, vol. 9, no. 2, pp. 244–46 (Oct. 1).

1969. Notes on the Mitridae of the eastern Pacific, II. The genus *Thala*, with the description of a new species. *Ibid.*, vol. 12, no. 1, pp. 84–88, pl. 6 (July 1).

1971. New opisthobranch records for the eastern Pacific. *Ibid.*, vol. 13, no. 4, pp. 368–69 (Apr. 1).

1971. Type specimens of Recent mollusks in the Los Angeles County Museum of Natural History. Contributions in Science, Los Angeles Co. Mus., no. 213, 37 pp. (May 27).

Stasek, C. R. 1966. Type specimens in the California Academy of Sciences, Department of Invertebrate Zoology. Occ. Papers, Calif. Acad. Sci., no. 51, 38 pp. (Apr. 15). [List of material in alcohol.]

Stearns, R. E. C. 1873a. Description of new species of shells from California. Proc. California Acad. Sci., vol. 4, p. 249, pl. 1 (Jan.).

1873b. Descriptions of new marine mollusks from the west coast of North America. *Ibid.*, vol. 5, pp. 78–82 (May).

1879. Description of a new species of Dolabella, *Dolabella californica*, from the Gulf of California, with remarks on other rare or little-known species from the same region. Proc. Acad. Nat. Sci. Philadelphia, vol. 30, for 1878, pt. 3, pp. 395–401, pl. 7 (Jan. 28). [Plate caption, with additions and corrections, Mar. 1879.]

1890. Descriptions of new West American land, fresh-water and marine shells, etc. Scientific results of explorations by the U.S. Fish Commission Steamer "Albatross." Proc. U.S. Nat. Mus., vol. 13, pp. 205–25, pls. 15–17 (Sept. 16).

1891. List of shells collected on the west coast of South America, principally between latitudes 7° 30′ S. and 8° 49′ N., by Dr. W. H. Jones, surgeon, U.S. Navy. *Ibid.*, vol. 14, no. 854, pp. 307–35 (Aug. 29).

1892. Preliminary descriptions of new molluscan forms from West American regions, etc. Nautilus, vol. 6, no. 8, pp. 85–89 (Dec.).

1893*a*. Descriptions of a new species of Nassa (*Nassa brunneostoma*) from the Gulf of California. *Ibid.*, vol. 7, no. 1, pp. 10–11 (May).

1893*b*. On rare or little known mollusks from the west coast of North and South America, with descriptions of new species. Proc. U.S. Nat. Mus., vol. 16, no. 941, pp. 341–52, pl. 50 (Sept. 28).

1893*c*. Report on the mollusk fauna of the Galapagos Islands with descriptions of new species. *Ibid.*, vol. 16, no. 942, pp. 353–450, pls. 51, 52 (Sept. 29).

1894. The shells of the Tres Marias and other localities along the shores of Lower California and the Gulf of California. *Ibid.*, vol. 17, pp. 139–204 (July 19).

1897. Description of a new species of *Actaeon* from the Quaternary bluffs of Spanish Bight, San Diego, California. Nautilus, vol. 11, no. 2, pp. 14–15 (June).

Steinbeck, J., and E. F. Ricketts. 1941. Sea of Cortez. New York, Viking Press, x + 598 pp., 40 pls., 2 charts. Phylum Mollusca on pp. 478–560. [Partially reprinted by Steinbeck in 1951.]

Stohler, Rudolf. 1963. Studies on mollusk populations. V.—*Tegula rugosa* (A. Adams, 1853). Veliger, vol. 5, no. 3, pp. 117–21, 4 text figs. (Jan. 1).

Strong, A. M. 1922. Notes on *Acteocina*. Nautilus, vol. 35, no. 4, pp. 122–23 (Apr.).

1928. West American Mollusca of the genus *Phasianella*. Proc. Calif. Acad. Sci., ser. 4, vol. 17, no. 6, pp. 187–203, pl. 10 (June 22).

1930. Notes on some species of *Epitonium*, subgenus *Nitidiscala*, from the west coast of North America. Trans. San Diego Soc. Nat. Hist., vol. 6, pp. 183–96, pl. 20 (Aug. 30).

1934. West American species of the genus *Liotia*. *Ibid.*, vol. 7, no. 37, pp. 429–52. pls. 28–31 (May 31).

1938. New species of West American shells. Proc. California Acad. Sci., ser. 4, vol. 23, no. 14, pp. 203–16, pls. 15–16 (May 24).

1945*a*. Nassariidae. Minutes Conch. Club Southern California, no. 51, pp. 3–5 (Aug.).

1945*b*. [Keys to the] Columbellidae. *Ibid.*, no. 51, pp. 10–29 (Aug.)

1945*c*. [Keys to the] Epitoniidae. *Ibid.*, no. 52, pp. 15–26 (Sept.)

1949. Additional Pyramidellidae from the Gulf of California. Bull. Southern California Acad. Sci., vol. 48, pt. 2, pp. 71–93, pls. 11–12 (May–Aug.).

1954. A review of the eastern Pacific species in the molluscan family Cancellariidae. *Ibid.*, no. 135, pp. 7–14 (Jan.), no. 136, pp. 16–23 (Feb.), no. 137, pp. 28–32 (Mar.–Apr.); no. 138, pp. 44–47 (May); no. 139, pp. 56–59 (June).

Strong, A. M., and G. D. Hanna, 1930. (*a*) Marine Mollusca of the Revillagigedo Islands, Mexico; (*b*) Marine Mollusca of the Tres Marias Islands, Mexico. Proc. California Acad. Sci., ser. 4, vol. 19, nos. 2–3, pp. 7–12; 13–22 (June 4).

Strong, A. M., G. D. Hanna, and L. G. Hertlein. 1933. The Templeton Crocker expedition of the California Academy of Sciences, 1932. No. 10. Marine Mollusca from Acapulco, Mexico, with notes on other species. *Ibid.*, ser. 4, vol. 21, no. 10, pp. 117–30, pls. 5–6 (Dec. 21).

Strong, A. M., and L. G. Hertlein. 1937. The Templeton Crocker expedition of the California Academy of Sciences, 1932. No. 25. New species of Recent mollusks from the coast of western North America. *Ibid.*, ser. 4, vol. 22, no. 6, pp. 159–78, pls. 34–35 (Dec. 31).

1939. Marine mollusks from Panama collected by the Allan Hancock expedition to the Galapagos Islands, 1931–32. Allan Hancock Foundation Publications of the University of Southern California, Los Angeles, vol. 2, no. 12, pp. 177–245, pls. 18–23 (Aug. 21).

1947. A new name for a west American *Cyclostrema*. Nautilus, vol. 61, no. 1, p. 31 (July).

Strong, A. M., and H. N. Lowe. 1936. West American species of the genus *Phos*. Trans. San Diego Soc. Nat. Hist., vol. 8, no. 22, pp. 305–20, pl. 22 (Dec. 7).

Stuart, H. G. 1941. A new cephalopod mollusk collected on the Presidential cruise of 1938. Smithsonian Inst. Misc. Coll., vol. 99, no. 11, pp. 1–6, 2 text figs.

Swainson, William. 1820–33. Zoological illustrations, or original figures and descriptions of new, rare, or interesting animals, selected chiefly from the classes of ornithology, entomology, and conchology. London: ser. 1, vols. 1–3, pls. 1–182 (1820–23); ser. 2, vols. 1–3, pls. 1–136 (1829–33). [*See* C. D. Sherborn, 1922, Index animalium, sect. 2, p. cxx.]

1821–22. Exotic conchology. London, 4 parts. Reissued, with parts 5–6, 1834–45. Ed. 2, published by S. Hanley, 1841. [For collation, *see* Sherborn and Reynell, Proc. Malac. Soc. London, vol. 11, pt. 5, pp. 276–79, 1915.]

1822. A catalogue of the . . . shells, which formed the . . . collection . . . of Mrs. Bligh. With an appendix, containing . . . descriptions of many new species. London, pp. 58 [2] 20, 2 col. pls. [Reprinted in Swainson's Exotic conchology, ed. 2, 1841.]

1823a. The characters of several rare and undescribed shells. Philos. Mag., vol. 61, pp. 375–78 (May 31).

1823b. The specific characters of several rare and undescribed shells. *Ibid.*, vol. 62, pp. 401–3 (Dec.).

Taylor, D. W., and N. F. Sohl. 1962. An outline of gastropod classification. Malacologia, vol. 1, no. 1, pp. 7–32, 1 fig. (Oct.).

Tesch, J. J. 1949. Heteropoda. The Carlsberg Foundation's Oceanographical Expedition . . . [Dana Rept. 34]. Pp. 1–55, pls. 1–5, 44 text figs.

Thiele, Johannes. 1909–10. Revision des Systems der Chitonen, Zoologica, Bd. 22, Heft 56 (Teil 1, pp. 1–70, pls. 1–6, 1909; Teil 2, 1910), 132 pp., 10 pls.

1929–35. Handbuch der Systematischen Weichtierkunde. Jena: Gustav Fischer, Bd. I, Teil I, pp. 1–376, 470 text figs. (1929); Bd. I, Teil II, pp. 377–778, 313 text figs. (1931); Bd. II, Teil III, pp. 779–1023, 110 text figs. (1934); Bd. II, Teil IV, pp. 1023–1154, 4 text figs. (1935). [For a discussion of dates, *see* W. J. Clench, Nautilus, vol. 66, no. 1, p. 33, 1952.]

Tomlin, J. R. le B. 1916. Note on the *Marginella varia* of Sowerby. Nautilus, vol. 29, no. 11, pp. 138–39 (Apr.).

1917. A systematic list of the Marginellidae. Proc. Malac. Soc. London, vol. 12, no. 5, pp. 242–306 (Aug. 22).

1926. Note on *Donax conradi* Deshayes. Nautilus, vol. 20, no. 2, pp. 52–53 (Oct.).

1927–29. The Mollusca of the "St. George" expedition. Jour. Conch., vol. 18. (I), The Pacific coast of S. America: pp. 153–70 (Dec. 1927), 187–98 (May 1928); (II), The West Indies: pp. 307–10 (July 1929).

1931. On the name *Mitra lineata*. Nautilus, vol. 45, no. 2, pp. 53–55 (Oct.).

1932a. Notes from the British Museum. II. Arthur Adams's types of *Nassa*. Proc. Malac. Soc. London, vol. 20, pp. 41–44 (Mar. 15).

1932b. Notes from the British Museum. III. Reeve's "Monograph of the genus *Nassa*," *Ibid.*, vol. 20, pp. 95–98 (July 15).

1934a. Notes from the British Museum. V. Reeve's "Monograph of *Pleurotoma*." *Ibid.*, vol. 21, pp. 37–40 (Mar. 15).

1934b. Notes from the British Museum. VI. Reeve's "Monograph of *Mangelia*." *Ibid.*, vol. 21, pp. 40–41 (Mar. 15).

1937. Catalogue of Recent and fossil cones. *Ibid.*, vol. 22, pts. 4–5, pp. 205–330 (Mar.–July).

1944a. Some notes on Terebridae. Minutes Conch. Club Southern California, no. 41, p. 14 (Nov.).

1944b. Deshayes's review of *Terebra*. Jour Conch., vol. 22, pp. 104–8 (Nov. 30).

Townsend, C. H. [*et al.*] 1901– . Dredging and other records of the United States Fish Commission steamer *Albatross* . . . Rept., U.S. Comm. Fish and Fisheries, year ending June 30, 1900 [pt. 6], Appendix, Doc. no. 472, pp. 387–562 . . . (stations 2001–3786) (1901). [Later records without authorship, as follows: 4191–4302, Part 29 (1903) ; 4303–4748, Doc. 604 (1904–5) ; 4757–5095, Doc. 621 for 1906 (1907) ; 5096–5672, Doc. 741 (1910) ; 5675–86, Doc. 897 for 1911–20 (1921).]

Troschel, F. H. 1852. Verzeichniss der durch Herrn. Dr. v. Tschudi in Peru gesammelten conchylien. Arch. Naturgesichte, vol. 18, pt. 1, pp. 151–208, pls. 5–7.

Tryon, G. W., Jr. 1862. Synopsis of the Recent species of Gastrochaenidae, a family of acephalous Mollusca. Proc. Acad. Nat. Sci. Philadelphia, vol. 13, pp. 465–94, 1 fig. [Apparently issued between Jan. 1 and Mar. 31, 1862.]

—— 1865. Descriptions of new species of *Amnicola, Pomatiopsis, Somatogyrus, Gabbia, Hydrobia*, and *Rissoa*. Amer. Jour. Conch., vol. 1, pt. 3, pp. 217–22, 1 pl. (July 1).

—— 1866. Description of a new species of *Septifer*. *Ibid.*, vol. 2, no. 4, p. 302, 1 fig. (Oct.).

—— 1872. Catalogue and synonymy of the Recent species of the family Lucinidae. Proc. Acad. Nat. Sci. Philadelphia, for 1872, pp. 82–96 (July).

—— 1879–1913. Manual of conchology. Philadelphia, ser. 1, vols. 1–17. [Continued by H. A. Pilsbry, beginning with Monographs of the Turbinidae and Trochidae, vol. 10, pt. 2, p. 161, 1888. For the dates of issue of this work, *see* E. G. Vanatta, 1927, Nautilus, vol. 40, pp. 96–99.]

Turner, Ruth D. 1954–55. The family Pholadidae in the western Atlantic and the eastern Pacific. Part I. Pholadinae. Johnsonia, vol. 3, no. 33, pp. 1–65, 34 figs. (May 17, 1954) ; Part II. Martesiinae, Jouannetiinae, and Xylophaginae. *Ibid.*, no. 34, pp. 65–160, figs. 35–93 (Mar. 29, 1955).

—— 1956. The eastern Pacific marine mollusks described by C. B. Adams. Occ. Papers on Mollusks, Mus. Comp. Zool., Harvard, vol. 2, no. 20, pp. 21–135, pls. 5–21 (Sept. 22).

—— 1966. A survey and illustrated catalogue of the Teredinidae. Mus. Comp. Zoology, Harvard, 265 pp., 64 pls.

Valenciennes, Achille. [1821]–1833. I. Coquilles marines bivalves de l'Amérique Équinoxiale, recueillies pendant le voyage de MM. de Humboldt et Bonpland. *In* F. H. A. von Humboldt, and A. J. A. Bonpland, Voyage aux régions équinoxiales du Nouveau Continent. Paris, pt. 2, Recueil d'observations de zoologie et d'anatomie comparée, vol. 2, pp. 217–24, pls. 48–50 (1827). II. Coquilles univalves [etc.]. *Ibid.*, pp. 262–339, pl. 57 (1832). [Title page dated 1833, but the work was reviewed by Duclos in May 1832 (Ann. Sci. Nat., Paris, vol. 21, p. 110). For further notes on dating, *see* C. D. Sherborn, 1899, Ann. Mag. Nat. Hist., ser. 7, vol. 3, p. 428, and B. Walker., 1928, Nautilus, vol. 41, p. 131. A partial collation is given by Hertlein and Strong (1955).]

—— 1846. *In* Abel du Petit-Thouars, Voyage autour du monde sur . . . la Vénus, pendant . . . 1836–1839. Paris, Atlas de Zoologie, Mollusques, pls. 1–24 (No text). [*See also* J. Lockwood Chamberlin, 1960, Voyage of the Venus, Nautilus, vol. 74, no. 2, pp. 65–68, and E. Fischer-Piette, and J. Beigbeder, 1943, Catalogue des types de gastéropodes marins conservés au Laboratoire de Malacologie, Bull. Mus. hist. nat., Paris (ser. 2), vol. 15, no. 4, pp. 203–9; no. 6, pp. 429–56.]

Valentine, J. W. 1966. Numerical analysis of marine molluscan ranges on the extratropical northeastern Pacific shelf. Limnology and Oceanography, vol. 11, no. 2, pp. 198–211 (Apr.).

Vanatta, E. G. 1899. West American Eulimidae. Proc. Acad. Nat. Sci. Philadelphia, vol. 51, pp. 254–57, pl. 11, figs. 11–12 (July).

—— 1912. *Phenacolepas malonei*, n. sp. *Ibid.*, vol. 64, p. 151 (May 21).

—— 1913. Descriptions of new species of marine shells. *Ibid.*, vol. 65, pp. 22–27, 3 text figs., pl. 2 (Apr. 4).

—— 1915. Notes on *Oliva*. Nautilus, vol. 29, pp. 67–72 (Oct.).

1925. Four new species of shells. Proc. Acad. Nat. Sci. Philadelphia, vol. 76, for 1924, pp. 423–25, 5 text figs. (Feb. 15).

Van der Spoel, S. 1967. Euthecosomata: a group with remarkable developmental stages (Gastropoda, Pteropoda). J. Noordunn en Zoon N. V., Gorinchem, Netherlands. Pp. 375, figs. 1–366 (Nov. 20).

Verrill, A. E. 1870. Descriptions of shells from the Gulf of California. Amer. Jour. Sci., ser. 2, vol. 49, no. 146, pp. 217–27 (Mar.).

1883. Descriptions of two species of *Octopus* from California. Bull. Mus. Comp. Zool. Harvard, vol. 11, pp. 117–24.

Verrill, A. E., and K. J. Bush, 1898. Revision of the deep-water Mollusca of the Atlantic coast of North America, with descriptions of new genera and species. I. Bivalvia. Proc. U.S. Nat. Mus., vol. 20, no. 1139, pp. 775–901, pls. 71–97 (June 15).

Violette, P. E. 1964. Shelling in the Sea of Cortez. King, Tucson, Ariz. Pp. 96, ill.

Vokes, Emily H. 1968. Cenozoic Muricidae of the western Atlantic region. Part IV— *Hexaplex* and *Murexiella*. Tulane Studies in Geology, vol. 6, no. 3, pp. 85–126, pls. 1–8 (Nov. 22).

1969. The genus *Trajana* (Mollusca: Gastropoda) in the New World. Tulane Studies in Geology and Paleontology, vol. 7, nos. 1–2, pp. 75–83, 1 pl. (July 16).

1970. The West American species of *Murexiella* (Gastropoda: Muricidae), including two new species. Veliger, vol. 12, no. 3, pp. 325–329, pl. 50 (Jan. 1).

Vokes, H. E., and Emily H. Vokes. 1962. Pelecypods from Barra de Navidad, Mexico. Nautilus, vol. 76, no. 2, pp. 61–63 (Oct.).

Voss, G. L. 1971. Biological results of the University of Miami deep-sea expeditions. 76. Cephalopods collected by the R/V *John Elliott Pillsbury* in the Gulf of Panama in 1967. Bull. Marine Sci., Univ. Miami, vol. 21, no. 1, pp. 1–34. [In press.]

Waller, T. R. 1969. The evolution of the *Argopecten gibbus* stock (Mollusca: Bivalvia), with emphasis on the Tertiary and Quaternary species of eastern North America. Paleont. Society, Mem. 3, Jour. Paleont., vol. 43, no. 5, Suppl., pp. 1–125, pls. 1–8, 13 figs. (Sept.).

Weinkauff, H. C. 1879. *Marginella. In* Martini and Chemnitz, Systematisches Conchylien-Cabinet, ed. 2, Bd. III, Abth. 4, 166 pp., 26 pls.

1885. *Rissoina* and *Rissoa. In* Martini and Chemnitz, Systematisches Conchylien-Cabinet, ed. 2, Bd. I, Abth. 22, 205 pp., 29 pls. (1855–) 1885.

Wenz, Wilhelm. 1938–44. Gastropoda. Allgemeiner Teil und Prosobranchia. *In* O. H. Schindewolf, ed., Handbuch der Paläozoologie, Bd. 6, 1639 pp. Teil 1, pp. 1–240 (1938); Teil 2, pp. 241–480 (1938); Teil 3, pp. 481–720 (1939); Teil 4, pp. 721–960 (1940); Teil 5, pp. 961–1200 (1941); Teil 6, pp. 1201–1506 (1943); Teil 7, pp. 1507–1639 (1944) (Berlin). [*See also* continuation by Zilch (1959–60).]

Western Society of Malacologists. 1970. The Echo: Abstracts and Proceedings of the Second Annual Meeting (for June 1969), 84 pp., 1 pl., 1 map (Mar. 9).

Willett, George. 1931. Three new marine mollusks from Catalina Island, California. Nautilus, vol. 45, no. 2, pp. 65–67, pl. 4 (Oct.).

1938. A new *Pseudochama* from Clarion Island, Mexico. *Ibid.*, vol. 52, no. 2, pp. 48–49, 1 pl. (Oct.).

1939. Description of a new mollusk from California. *Ibid.*, vol. 52, no. 4, pp. 123–24, pl. 9, figs. 1, 1a, 1b (Apr.).

1944. Two new West American pelecypods. Bull. Southern California Acad. Sci., vol. 43, pt. 1, pp. 19–22, pl. 8 (May 31).

Wilson, E. C., and G. L. Kennedy. 1967. Type specimens of Recent invertebrates (except Arachnida and Insecta) in the San Diego Natural History Museum. Trans. San Diego Soc. Nat. Hist., vol. 14, no. 19, pp. 237–80 (Nov. 17).

Wimmer, A. 1880. Zur Conchylien-Fauna der Galapagos-Inseln. Sitzungsberichte der Kaiserlichen Akad. der Wiss., d. Wiener, Math.-Nat. Classe, Bd. 80, Abth. 1, no. 10, for 1879, pp. 465–514.

Winkler, L. R. 1955. A new species of *Aplysia* on the southern California coast. Bull. Southern California Acad. Sci., vol. 54, pp. 5–7, 2 pls.

Wolfson, Fay H. 1962. Comparison of two similar species of *Conus* (Gastropoda) from the Gulf of California. Veliger, vol. 5, no. 1, pp. 23–28, 6 text figs. (July 1).

———— 1968. Spawning notes, I. *Hexaplex erythrostomus. Ibid.*, vol. 10, no. 3, p. 292, 1 fig. (Jan. 1).

———— 1969a. Spawning notes, II. *Mitra dolorosa. Ibid.*, vol. 11, no. 3, pp. 282–83, 3 text figs. (Jan. 1).

———— 1969b. Spawning notes, III. *Strombina maculosa. Ibid.*, vol. 11, no. 4, pp. 440–41, 3 text figs. (Apr. 1).

———— 1969c. Spawning notes, IV. *Cerithium stercusmuscarum. Ibid.*, vol. 11, no. 4, pp. 441–42, 2 text figs. (Apr. 1).

———— 1970. Spawning notes, V. *Acanthina angelica* I. Oldroyd, 1918, and *Acanthina lugubris* (Sowerby, 1821). *Ibid.*, vol. 12, no. 3, pp. 375–77, 6 text figs. (Jan. 1).

Wood, W. 1815. General conchology; or a description of shells. London, pp. i–lxi, 1–246, 60 pls. [*See* G. B. Pritchard, and J. H. Gatliff, 1903, Proc. Roy. Soc. Victoria, new ser., vol. 16, p. 114, and T. Iredale, 1922, Proc. Malac. Soc. London, vol. 15, p. 91.]

———— 1825. Index testaceologicus; or a catalogue of shells, British and foreign. London, pp. i–xxxii, 1–188, pls. 1–38. [For information concerning this and the following entry, *see* A. Reynell, 1918, Proc. Malac. Soc. London, vol. 13, pp. 26–27.]

———— 1828. Supplement to the Index testaceologicus; or a catalogue of shells, British and foreign. London, pp. i–vi, 1–59, pls. 1–8.

Woodring, W. P. 1925–28. Miocene mollusks from Bowden, Jamaica. Pelecypods and scaphopods. Carnegie Inst. Washington, Pub. no. 366, pp. 1–122, pls. 1–28 (May 1925). Pt. 2. Gastropods and discussion of results. *Ibid.*, no. 385, pp. i–vii, 1–564, 3 figs., pls. 1–40 (Nov. 28, 1928).

———— 1945. Northern limit of *Dosinia ponderosa.* Nautilus, vol. 59, no. 1, p. 34 (July).

———— 1957. Geology and paleontology of Canal Zone and adjoining parts of Panama. U.S. Geol. Survey Prof. Paper 306-A, 145 pp., 23 pls.

———— 1959. *Ibid.*, 306-B (Vermetidae to Thaididae), pp. 147–240, pls. 24–38.

———— 1964. *Ibid.*, 306-C (Columbellidae to Volutidae), pp. 241–98, pls. 39–47.

———— 1966. The Panamic land bridge as a sea barrier. Amer. Philos. Soc., Proc., vol. 110, no. 6, pp. 425–33, 3 figs., 5 tables.

———— 1970. U.S. Geol. Survey Prof. Paper 306-D (Eulimidae, Marginellidae to Helminthoglyptidae), pp. 299–452, pls. 48–66.

Woodward, B. B. 1903–15. Catalogue of the books, manuscripts and drawings in the British Museum (Nat. Hist.). B.M.N.H., London. 5 vols., viii + 2403 pp. Supplement, 3 vols., 1480 pp. (1922–40).

Yonge, C. M. 1958. Observations in life on the pulmonate limpet *Trimusculus* (*Gadinia*) *reticulatus* (Sowerby). Proc. Malac. Soc. London, vol. 33, pt. 1, pp. 31–37, 5 figs. (Mar.).

———— 1967. Form, habit, and evolution in the Chamidae (Bivalvia) with reference to conditions in the rudists (Hippuritacea). Philos. Trans. Roy. Soc. London, ser. B, Biol. Sci., vol. 252, no. 775, pp. 49–104, 31 figs. (Feb. 16).

Young, D. K. 1967. New records of Nudibranchia (Gastropoda: Opisthobranchia: Nudibranchia) from the central and west-central Pacific with a description of a new species. Veliger, vol. 10, no. 2, pp. 159–73, 18 text figs. (Oct. 1).

Zeigler, R. F. 1969. Two infrasubspecific forms in *Oliva.* Nautilus, vol. 83, no. 1, pp. 14–19, 2 figs. (July).

Zeigler, R. F., and H. C. Porreca. 1969. Olive shells of the world. Privately printed, W. Henrietta, N.Y. Pp. 1–96, 13 pls. (Oct.).

Zetek, James. 1918. Los moluscos de la República de Panamá. Revista Nueva, Panama, vol. 5, nos. 2–6, pp. 509–75 (July–Aug.). [Also published separately, 69 pp.]

Zetek, James, and Richard A. McLean. 1936. *Hiata*, a new genus of the family Pholadidae from the Pacific at Panama, with a description of a new species. Nautilus, vol. 49, no. 4, pp. 110–11, pl. 8 (May 1).

Zilch, A. 1954. Moluscos de los Manglares de El Salvador. Con una lista de todas las especies marítimas conocidas de estas regiones. Com. Inst. Trop. Inves. Cien., El Salvador, yr. 3, nos. 2, 3, pp. 77–87, pls. 1–4.

———— 1959–60. Gastropoda: Euthyneura. *In* Handbuch der Paläozoologie. Lief. 1, pp. 1–200 (July 17, 1959); Lief. 2, pp. 201–400 (Nov. 25, 1959); Lief. 3, pp. 401–600 (Mar. 30, 1960); Lief. 4, pp. 601–835, pp. i–xii (Aug. 15, 1960). [Continuation of Wenz (1938–44).]

INDEX

Index

guaicurana, Turbonilla, 787
guanacastensis, Cerithiopsis, 413
guanacastensis, Trigoniocardia, 157
guanacastensis, Turbonilla, 787
guanica, Cardita, 109
guatemalensis, Columbella, 582
guatemalensis, Radsiella, 868
guatulcoensis, Cerithiopsis, 413
guatulcoensis, Chione, 185
guatulcoensis, Odostomia, 774
guatulcoensis, Turbonilla, 785
guayaquilensis, Alaba, 415
guayaquilensis, Dolabella, 810
guayaquilensis, Fossarus, 454
guayaquilensis, Neritina, 360
guayaquilensis, Olivella, 632
guayaquilensis, Terebra, 676
guaymasensis, Nassarius, 607
guaymasensis, Semele, 251
guaymasensis, Solecurtus, 244
guaymasensis, Tellina, 213
guaymasensis, Vitrinella, 377
guerreroensis, Anachis, 584
guilleni, Turbonilla, 784
gubernaculum, Macoma, 900
guttata, Arene, 346
guttata, Mitrella, 591
guttata, Neritina, 360
Gutturnium, 505
guyanensis, Mytella, 63
Gymnobela, 766
Gymnodorididae, 827
Gymnoglossa, 443
Gymnophila, 841
Gymnosomata, 810

haasi, Cuspidaria, 304
habeli, Epitonium, 426
haemastoma, Columbella, 574
haemastoma, Pollia, 561
haematura, Purpura, 549
haemostoma, Cancellaria, 651
Haldra, 776
haleyi, Turbonilla, 791
halia, Alvinia, 368
halia, Cerithiopsis, 413
Halicardia, 303
Halicardissa, 303
halidoma, Turbonilla, 787
Haliotidae, 308
Haliotis, 308
haliplexa, Compsodrillia, 733
Haliris, 302
halis, Carinodrillia, 729

Halistrepta, 293
halistreptus, Lepidopleurus, 882
halocydne, Kylix, 697
Halodakra, 118
hamata, Nuculana, 31
hamata, Pyrgocythara, 757
hamatus, Pterorytis, 536
hambachi, Ocenebra, 545
hambachi, Solariorbis, 384
hamillus, Anomia, 101
Haminoea, 795
hancocki, Dentalium, 886
hancocki, Epitonium, 432
hancocki, Lithophaga, 68
hancocki, Macromphalina, 454
hancocki, Nassarina, 595
hancocki, Pecten, 87
hancocki, Pyramidella, 768
hancocki, Terebra, 678
hancocki, Verticordia, 302
haneti, Columbella, 601
haneti, Mitra, 645
Hanetia, 561
hanleyi, Drillia, 721
hanleyi, Tellina, 244
hannai, Dentalium, 886
hannai, Solariorbis, 384
hannai, Triphora, 416
hannai, Turbonilla, 782
hannana, Anachis, 581
hansi, Hoffmannola, 841
Hapalorbis, 384
Haplocochlias, 343
harfordi, Mitrella, 593
harfordiana, Crassispira, 716
Harpa, 620
harpa, Voluta, 619
Harpidae, 620
harpiformis, Microcithara, 590
hartmanni, Rissoina, 374
Harvella, 206
hastasia, Lithophaga, 70
hastata, Balcis, 446
hastata, Pyramidella, 768
Hastula, 686
hastula, Aspella, 527
hastula, Cytharella, 751
Hatasia, 278
Haustellotyphis, 539
healdi, Teredo, 282
healeyi, Eulima, 445
hebe, Mangelia, 747
hecetae, Mangelia, 747
hecuba, Kylix, 697
hedgpethi, Elysia, 817
hedgpethi, Polycera, 827
hegewischii, Cerithium, 419
Heida, 776
heimi, Eunaticina, 477

heimi, Liotia, 345
helena, Cerithiopsis, 412, 413
helena, Pyrgocythara, 757
helena, Rissoina, 374
helenae, Conus, 665
helenae, Gari, 240
helenae, Nassarina, 594
helenae, Pitar, 170
Heliacus, 389
helicinoides, Atlanta, 468
helicoides, Polinices, 478
helvola, Cypraea, 492
hemicardium, Arca, 52
Hemitoma, 312
hemphilli, Balcis, 446
hemphilli, Eulima, 445
hemphilli, Latirus, 613
hemphilli, Lima, 98
hemphilli, Odostomia, 779
hemphilli, Oliva, 624
hemphilli, Teinostoma, 386
hemphilli, Triphora, 416
henoquei, Conus, 664
heptagonalis, Ricinula, 905
heptagonum, Elephantanellum, 398
herbertianum, Teinostoma, 385
Here, 120
herilda, Leucosyrinx, 714
Hermaea, 818
Hermaeidae, 818
hermanita, Maesiella, 727
hermione, Pleurotomella, 764
Hermissenda, 840
hermosa, Hemitoma, 312
hermosa, Peristichia, 780
heroica, Lucinoma, 126
herrerae, Alvinia, 368
herrerae, Odostomia, 779
Hertellina, 219
Hertleinella, 530
hertleini, Basterotia, 145
hertleini, Doxospira, 725
hertleini, Ensitellops, 145
hertleini, Fusinus, 617
hertleini, Hindsiclava, 725
hertleini, Odostomia, 780
hertleini, Pitar, 170
hertleini, Tellina, 217
hertleini, Terebra, 678
Hertleinia, 653
hespera, Cymatosyrinx, 700
Hespererato, 488
hesperia, Mazatlania, 588
hesperius, Pitar, 174
hesperus, Macoma, 229
Heteroclidus, 289
Heterodonax, 242
Heterodonta, 103
heterolopha, Turbonilla, 785

INDEX | 1047

NEW SPECIES DESCRIBED IN THIS BOOK